THE
INTERNATIONAL SERIES
OF
MONOGRAPHS ON PHYSICS

Quantum Field Theory and Critical Phenomena

Third Edition

JEAN ZINN-JUSTIN

Commissariat à l'Energie Atomique
Direction des Sciences de la matière
Gif sur Yvette, France

CLARENDON PRESS • OXFORD

OXFORD

UNIVERSITY PRESS

Great Clarendon Street, Oxford OX2 6DP

Oxford University Press is a department of the University of Oxford.
It furthers the University's objective of excellence in research, scholarship,
and education by publishing worldwide in

Oxford New York

Athens Auckland Bangkok Bogotá Buenos Aires Calcutta
Cape Town Chennai Dar es Salaam Delhi Florence Hong Kong Istanbul
Karachi Kuala Lumpur Madrid Melbourne Mexico City Mumbai
Nairobi Paris São Paulo Singapore Taipei Tokyo Toronto Warsaw

with associated companies in Berlin Ibadan

Oxford is a registered trade mark of Oxford University Press
in the UK and in certain other countries

Published in the United States
by Oxford University Press Inc., New York

© J. Zinn-Justin, 1996

A catalogue record for this book is available from the British Library

Library of Congress Cataloging in Publication Data
(Data available)
ISBN 0 19 851882 X

Printed in India by
Thomson Press (India) Ltd.

Preface

The last thirty years have witnessed the spectacular progress of quantum field theory (QFT). Originally introduced to describe quantum electrodynamics (QED), QFT has become the framework for the discussion of all fundamental interactions except gravity. Much more surprisingly, it has also provided the framework for the understanding of second order phase transitions in statistical mechanics. In fact, as will hopefully become clear in this work, QFT is the natural framework for the discussion of most systems in which an infinite number of degrees of freedom are coupled.

Therefore, although several good textbooks about QFT have already been published, I thought that it might not be completely worthless to present a work in which the common aspects of particle physics and the theory of critical phenomena are systematically emphasized. This option explains some of the choices made in the presentation. A formulation in terms of path and functional integrals has been adopted to study the properties of QFT. Less important, the space–time metric has been chosen euclidean, as is natural for statistical mechanics and convenient in general for perturbative calculations even in particle physics. The language of partition and correlation functions has been used even in applications of field theory to particle physics. Renormalization and renormalization group properties have been systematically discussed, whereas little space has been devoted to scattering theory. Only formal aspects of QED have been considered since excellent textbooks already cover this subject.

The idea of renormalizable quantum field theories first appeared empirically in Quantum electrodynamics. QED, as well as all more complete field theories describing particle physics, is plagued by a serious disease. In a straightforward calculation all physical quantities are infinite, due to the short distance singularities of the theory. This situation has to be contrasted with what happens in Classical or non-relativistic Quantum Mechanics; there the replacement of macroscopic by point-like objects leads in general to no mathematical inconsistencies and is often a very good approximation: the absence of this property would indeed have made progress in physics quite difficult. This can be summarized by saying that in the latter theories phenomena of very different scales approximately decouple.

A strange remedy to this disease has been found empirically: one artificially modifies the theory at short distance (in a way which in general leads to unphysical short distance properties), at a scale characterized by a short distance cut-off. Inspired by methods of condensed matter physics, one then re-expresses all physical quantities in terms of a small number of physical constants, like the physical masses and charges, instead of the original parameters of the lagrangian. After this change of parametrization the cut-off is removed, and somewhat miraculously all other physical quantities have a finite limit, when the theory is so-called renormalizable. Moreover this limit is independent of the precise form of the short distance modification. Applied to QED this strategy led to predictions of extraordinary accuracy. Therefore it was natural to try also to construct a renormalizable field theory for all other interactions. This led to another great achievement: a model for all three strong, weak and electromagnetic interactions. The so-called Standard Model has now successfully confronted all experimental data, for more than twenty years.

As the consequence of these truly remarkable results, renormalizability was then slowly promoted to a kind of additional law of nature. In particular, once the standard model of weak, electromagnetic and strong interactions was established, much effort was devoted to cast gravity in the same framework. Despite many ingenious attempts, no renormalizable form of quantum gravity has been found yet.

Note that very early it was realized, as a mathematical curiosity, that in massless renormalizable theories a renormalization group could be associated with transformation properties under space dilatations. Only later was it realized that this property could be used to discuss the short distance structure of some physical processes. The basic idea was to introduce a set of scale-dependent coupling constants. In *asymptotically free* field theories, these effective couplings become small at large euclidean momenta and therefore perturbation theory, improved by renormalization group, can be used. Only the theory of strong interactions, an $SU(3)$ gauge theory, shares this property. However, most of the field theories proposed to describe strong, electromagnetic and weak interactions are not asymptotically free.

More generally it was suggested by Weinberg that the existence of UV fixed points, i.e. the existence of limits for the effective short distance couplings, was a necessary condition for the consistency of a field theory. Of course the existence of other non-trivial fixed points cannot be established in the framework of perturbation theory. However many numerical simulations of field theories on the lattice, which allow for non-perturbative explorations, have failed to discover non-trivial fixed points. Therefore it seems that the Standard Model, which describes so accurately Particle Physics at present scale, is not consistent on all scales and has to be modified at short distance. This is a second indication that maybe the property of renormalizability has a different origin.

Somewhat surprisingly QFT has also become an essential tool for the understanding of some critical phenomena: second order phase transitions in condensed matter physics. Near the critical temperature cooperative phenomena generate a large scale, the correlation length, although the fundamental interactions are short range, and the large scale properties of the system become independent of most of the details of the microscopic structure. First attempts to explain these properties were based on classical ideas: a description in terms only of macroscopic degrees of freedom adapted to the scale of large distance physics. Such a description naturally emerges in simple approximations like mean field theory. The corresponding general ideas were summarized in Landau's theory of critical phenomena. Unfortunately it became slowly clear that the extremely universal predictions of such a theory were in conflict with numerical calculations of critical exponents, experimental data and finally exact results in two dimensions. All these data still supported the concept of *universality* in the sense that broad classes of systems have indeed the same large distance properties, but, unlike in mean field theory, these properties depend on a small number of qualitative features like dimension of space, number of components of the order parameter, symmetries... . Actually an analysis of leading corrections to mean field theory indeed reveals that, at least in low space dimensions, degrees of freedom associated with shorter distances never completely decouple.

To explain this remarkable situation, i.e. that large distance properties of second order phase are to a large extent short distance insensitive, although the degrees of freedom on all scales seem to be coupled, Wilson, partially inspired by some prior attempts of Kadanoff, introduced the renormalization group idea: starting from a microscopic hamiltonian one integrates out the degrees of freedom corresponding to short distance fluctuations and generates a scale-dependent effective hamiltonian. Universality relies then upon the existence of IR fixed points in hamiltonian space. One of the spectacular implications was that the universal properties of a large class of critical phenomena could be accurately predicted by the same field theory methods which had been invented for particle physics. The appearance of renormalizable field theories was there related to the fixed point structure, and the property that the effective theory relevant for long distance physics depended only on a small number of parameters.

Predictions obtained from a renormalization group analysis of simple field theories like the $(\phi^2)^2$ field theory have been successfully compared to experiments as well as numerical data from lattice models. The same field theory methods have been shown to describe vastly different physical systems at criticality, like ferromagnets, liquid–vapour, binary mixtures, superfluid helium and, even more surprisingly, statistical properties of polymers.

If quantum field theory has led to an understanding of the concept of universality and allowed the calculation of many universal physical quantities, conversely critical phenomena have shed a new light on the mysterious role of renormalization and renormalizable field theories in particle physics.

For many years the renormalization procedure has been considered as an *ad hoc* method introduced only in order to calculate physical quantities in perturbation theory in terms of a small number of parameters. One sometimes even tried to put the blame on the perturbative treatment of field theory, and one tried to hide the renormalization procedure as much as possible (like the BPHZ effort). After all it worked and still works.

However, directly inspired from the theory of critical phenomena, another interpretation, originally also proposed by Wilson, has gained strength over the years. New physics should be expected at very short distances (maybe at the Planck's scale). At this scale the familiar notion of renormalizable local field theories probably loses its meaning. However possible non-local effects are limited to this short scale (the equivalent of the condition of short range forces in statistical systems). In addition dynamical effects, of a nature which at present can only be guessed, generate long distance physics associated with the appearance of almost massless particles (compared to the Planck mass for example all known particles are essentially massless). Nevertheless, because degrees of freedom on all scales remain coupled, the short distance cut-off can never be eliminated from the theory and this explains the impossibility of constructing a finite hamiltonian formalism. Fortunately, some renormalization group is also at work here, in such a way that observations at present energies can be accurately described in terms of an effective long distance renormalizable field theory. Large distance physics is short distance insensitive but in a way that is thus much more subtle than in classical physics.

The classification of interactions in terms of renormalization group properties becomes relevant. The necessity of a UV fixed point disappears, because Green's functions no longer need to be well defined on all scales, and marginal or irrelevant interactions have to be considered. Weak and electromagnetic interactions are probably marginal and, the free field theory being an infrared fixed point, their strength is proportional to an inverse power of the logarithm of the cut-off: Green's functions are consistent on all scales only for vanishing coupling constants (the *triviality* problem). This would explain why they can be described by a renormalizable theory with small (although much larger than in gravity) coupling constants. For strong interactions the situation seems to be more complicated. One must imagine that the effective interaction first decreases at shorter distances and then increases again because the free field theory is an ultraviolet fixed point. Finally gravitation is presumably irrelevant in the sense of critical phenomena which means that it is non-renormalizable and therefore very weak because, for dimensional reasons, its strength is proportional to a power of the short distance cut-off.

However, in contrast to critical phenomena in which a control parameter, like the temperature, can be adjusted to make the correlation length large, in particle physics the existence of massless particles has to be explained from general properties of the unknown fundamental theory. This is the famous *hierarchy* problem. Spontaneous breaking of a continuous symmetry, gauge principle and chiral invariance are the known mechanisms which generate massless particles. Supersymmetry can be helpful to deal with scalar bosons. At present the set of general conditions to be imposed on any fundamental theory, i.e. in the language of critical phenomena the complete description of the universality class of particle physics, has not been formulated. This is one of the remaining fundamental problems of quantum field theory. An intriguing question among many is the following: if a theory contains a light vector boson, does the effective theory automatically take the form of a Higgs model?

On the other hand, since the large distance physics is short distance insensitive, the real nature of fundamental interactions may remain elusive in the foreseeable future, in the same way that a precise knowledge of the critical exponents of the liquid-vapour phase transitions gives very little information about the molecular interactions in water.

This work, which does not claim to shed any light on these difficult problems, simply tries to describe particle physics and critical phenomena in statistical mechanics in a unified framework. It can be roughly divided into four parts. Chapters 1–12 deal with general field theory, functional integrals and functional methods. An introduction to renormalization theory is provided on the simple example of the ϕ^4 field theory and renormalization group (Callan–Symanzik) equations are derived. In Chapters 13–22 renormalization properties of theories with symmetries are studied and specific applications to particle physics are emphasized. Chapters 23–36 are mainly devoted to critical phenomena. A brief introduction to lattice gauge theories is included and asymptotic freedom in four dimensions (at large momentum, a problem relevant to Particle Physics) is discussed (Chapter 33, 34). Chapters 36–43 describe the role of instantons in quantum mechanics and field theory, the application of instanton calculus to the analysis of large order behaviour of perturbation theory and the problem of summation of the perturbative expansion. Finally Chapter 44 contains solutions or hints of solutions for the exercises which are proposed at the end of several chapters.

I am perfectly aware that this work is largely incomplete. My ignorance or lack of understanding of many important topics is of course mostly responsible for this weakness. However I also believe that a complete survey of quantum field theory and its applications is now beyond the scope of a single physicist and can only be produced by a more collective effort.

This work incorporates notes of lectures delivered in many summer schools most notably Cargèse 1973, Bonn 1974 and Les Houches 1982, as well as notes prepared for graduate courses in Princeton, Louvain-la-Neuve, Berlin, Cambridge (Harvard) and Paris.

Acknowledgements

It is impossible to list all the physicists from whom I have benefited in my already long career and whose influence can therefore be felt in one form or another in this work. However I wish to specially thank my masters M. Froissart and D. Bessis who guided my first steps in physics, E. Brézin and J.C. Le Guillou who have collaborated with me for more than fifteen years and without whom obviously this work would never have been produced. I also think with a deep emotion of the late B.W. Lee. The year I spent working with him at Stony-Brook was one of the most exciting of my life as a physicist.

Wilson's renormalization group ideas are of course the main source of inspiration for this work. S. Coleman and the late K. Symanzik played a very important role in my understanding of several aspects of physics through their articles and lecture notes as well as in private discussions. T.D. Lee and C.N. Yang have consistently honoured me with their friendship and hospitality in their institutions. Their deep remarks have been precious to me. Finally all my colleagues of the Saclay theory group, with whom I have had so many discussions, have directly influenced this work. Let me specially mention here C. Itzykson whose untimely death still shocks me while I am writing these lines.

Several colleagues agreed to read part of the manuscript before publication and I have benefited from their criticisms, remarks and wisdom, in particular E. Brézin who directly collaborated in the first chapters, R. Stora, C. Bervillier, O. Napoly, A.N. Vasil'ev, P. Zinn-Justin and J.-B. Zuber.

Finally the many lectures I have attended in Les Houches during nine summers have provided me with additional inspiration.

S. Zaffanella typed a part of the manuscript with care and competence. M. Porneuf tracked with persistence the innumerable misprints of the manuscript.

All deserve my deepest gratitude.

3$^{\rm rd}$ edition, Saclay, 25 October 1995

General References

Several textbooks or reviews have been a direct source of inspiration or complement naturally the present work (or both):

N.N. Bogoliubov and D.V. Shirkov, *Introduction to the Theory of Quantized Fields* (Interscience, New York 1959);

J. Bjorken and S. Drell, *Relativistic Quantum Mechanics, and Relativistic Quantum Fields* (McGraw-Hill, New York 1964, 1965);

S. Weinberg, *Gravitation and Cosmology* (John Wiley & Sons, New York 1972);

E. S. Abers and B.W. Lee, *Phys. Rep.* 9C (1973) 141;

K.G. Wilson and J.B. Kogut, *Phys. Rep.* 12C (1974) 75;

R. Balian and J. Zinn-Justin eds. *Methods in Field Theory*, Proceedings of Les Houches Summer School 1975 (North-Holland, Amsterdam 1976);

Phase Transitions and Critical Phenomena, vol. 6, C. Domb and M.S. Green eds. (Academic Press, London 1976), in particular the review *Field Theoretical Approach to Critical Phenomena* by E. Brézin, J.C. Le Guillou and J. Zinn-Justin;

S.K. Ma, *Modern Theory of Critical Phenomena* (Benjamin, Reading, MA 1976);

A.N. Vasiliev, *Functional Methods in Quantum Field Theory and Statistical Physics*, (St Petersburg 1976), english translation (Gordon and Breach, Amsterdam 1998);

C. Itzykson and J.-B. Zuber, *Quantum Field Theory* (McGraw-Hill, New York 1980);

L.D. Faddeev and A.A. Slavnov, *Gauge Fields: Introduction to Quantum Field Theory* (Benjamin, Reading, MA 1980);

P. Ramond, *Field Theory: A Modern Primer* (Benjamin, Reading, MA 1980);

T.D. Lee, *Particle Physics and Introduction To Field Theory* (Harwood Academic, New York 1981);

R. Stora and J.-B. Zuber eds. *Recent Advances in Field Theory and Statistical Mechanics*, Proceedings of Les Houches Summer School 1982, (Elsevier, Amsterdam 1984);

S. Coleman, *Aspects of Symmetry* (Cambridge University Press, Cambridge 1985);

G. Parisi, *Statistical Field Theory* (Addison-Wesley, New York 1988);

A.M. Polyakov, *Gauge Fields and Strings* (Harwood Academic, New York 1988);

J. Drouffe and C. Itzykson, *Théorie Statistique des Champs* (InterEditions 1989), english version: *Statistical Field Theory*, 2 volumes, (Cambridge University Press, Cambridge 1989);

M. Le Bellac, *Des Phénomènes Critiques aux Champs de Jauge* (Editions du CNRS, 1988), english version: *Quantum and Statistical Field Theory* (Oxford Univ. Press, Oxford 1992);

L.S. Brown, *Quantum Field Theory* (Cambridge University Press, Cambridge 1992);

M.E. Peskin and D.V. Schroeder, *An Introduction to Quantum Field Theory* (Addison-Wesley, Reading, USA 1995);

S. Weinberg, *The Quantum Theory of Fields*, 2 volumes, (Cambridge University Press, Cambridge 1995, 1996);

Construction of field theories from a more rigorous point of view is discussed in

R.F. Streater and A.S. Wightman, *PCT, Spin & Statistics and All That* (Benjamin, New York 1964);

G. Glimm and A. Jaffe, *Quantum Physics: A Functional Integral Point of View*, Springer Verlag (Berlin 1981);

See also the *Critical Phenomena, Random Systems, Gauge Theories*, Proceedings of Les Houches Summer School 1984, K. Osterwalder and R. Stora eds. (Elsevier, Amsterdam 1986).

Contents

1 ALGEBRAIC PRELIMINARIES

It is somewhat unusual to begin a physics textbook with algebraic identities, which are in general hidden in appendices. However, our discussion of perturbative aspects of quantum mechanics and quantum field theory will entirely be based on path or functional integrals and more generally functional techniques. Therefore a reader not already familiar with these concepts may find it difficult to follow the algebraic manipulations which enter in the derivation of many results. Moreover we want to indicate by such a choice that the various technical difficulties which we shall meet, will in general be directly confronted rather than carefully hidden.

Therefore in this first chapter we recall a few algebraic identities about gaussian integrals. We also recall the concept of functional differentiation and the algebraic definition of the determinant of an operator.

We then define and discuss a few properties of differentiation and integration in a Grassmann, i.e. antisymmetric algebra. In particular we calculate gaussian "fermionic" integrals. Throughout the chapter all expressions are given for a finite but arbitrary number of variables, because the focus is mainly on algebraic properties. However the generalization to an infinite number of variables will be easy, as will be discussed in the following chapters.

Note that in this chapter, as well as in this whole work, *summation over repeated indices will always be meant* (except if explicitly stated otherwise).

1.1 The Gaussian Integral

In this section we briefly review a few algebraic properties of gaussian integrals in the case of a finite number of integration variables.

We consider a general gaussian integral of the form:

$$I(\mathbf{A}, \mathbf{b}) = \int \left(\prod_{i=1}^{n} \mathrm{d}x_i \right) \exp\left(- \sum_{i,j=1}^{n} \tfrac{1}{2} x_i A_{ij} x_j + \sum_{i=1}^{n} b_i x_i \right), \tag{1.1}$$

in which $\mathbf{A}$ is a complex symmetric matrix with a non-negative real part, and non-vanishing eigenvalues λ_i

$$\operatorname{Re}\mathbf{A} \geq 0, \quad \lambda_i \neq 0.$$

To calculate I one first looks for the minimum of the quadratic form:

$$\frac{\mathrm{d}}{\mathrm{d}x_k} \left(\sum_{i,j=1}^{n} \tfrac{1}{2} x_i A_{ij} x_j - \sum_{i=1}^{n} b_i x_i \right) = 0.$$

The solution is:

$$x_i = \left(A^{-1}\right)_{ij} b_j, \tag{1.2}$$

(summation over j being meant as stated above) and one sets:

$$x_i = \left(A^{-1}\right)_{ij} b_j + y_i. \tag{1.3}$$

The integral becomes:

$$I = \exp\left[\tfrac{1}{2}b_i \left(A^{-1}\right)_{ij} b_j\right] \int \left(\prod_i dy_i\right) \exp\left(-\tfrac{1}{2}y_i A_{ij} y_j\right). \tag{1.4}$$

The last integral can be calculated by changing variables, setting

$$\left(A^{1/2}\right)_{ij} y_j = y'_i,$$

the eigenvalues $\lambda_i^{1/2}$ of $\mathbf{A}^{1/2}$ being chosen such that $-\pi/4 \le \operatorname{Arg}\lambda_i^{1/2} \le +\pi/4$. The integral over the y'_i's is then straightforward and one obtains:

$$I\left(\mathbf{A}, \mathbf{b}\right) = (2\pi)^{n/2} \left(\det \mathbf{A}\right)^{-1/2} \exp\left[\sum_{i,j=1}^n \tfrac{1}{2}b_i \left(A^{-1}\right)_{ij} b_j\right]. \tag{1.5}$$

By differentiating this last expression with respect to the variables b_i, it is then possible to calculate the average of any polynomial with a gaussian weight:

$$\langle x_{k_1} x_{k_2} \ldots x_{k_\ell}\rangle \equiv \mathcal{N} \int \left(\prod_i dx_i\right) x_{k_1} x_{k_2} \ldots x_{k_\ell} \exp\left(-\sum_{i,j=1}^n \tfrac{1}{2}x_i A_{ij} x_j\right), \tag{1.6}$$

in which the normalization $\mathcal{N}$ is chosen in such a way that $\langle 1\rangle = 1$:

$$\mathcal{N}^{-1} = I\left(\mathbf{A}, 0\right).$$

Indeed from expression (1.1) one derives:

$$\frac{\partial}{\partial b_k} I\left(\mathbf{A}, \mathbf{b}\right) = \int \left(\prod_i dx_i\right) x_k \exp\left(-\tfrac{1}{2}x_i A_{ij} x_j + b_i x_i\right). \tag{1.7}$$

Repeated differentiation with respect to $\mathbf{b}$ then leads to the identity:

$$\langle x_{k_1} x_{k_2} \ldots x_{k_\ell}\rangle = (2\pi)^{-n/2} \left(\det \mathbf{A}\right)^{1/2} \left[\frac{\partial}{\partial b_{k_1}} \frac{\partial}{\partial b_{k_2}} \cdots \frac{\partial}{\partial b_{k_\ell}} I\left(\mathbf{A}, \mathbf{b}\right)\right]\Bigg|_{\mathbf{b}=0},$$

or replacing the integral $I\left(A, \mathbf{b}\right)$ by its explicit form (1.5):

$$\langle x_{k_1} \ldots x_{k_\ell}\rangle = \left\{\frac{\partial}{\partial b_{k_1}} \cdots \frac{\partial}{\partial b_{k_\ell}} \exp\left[\sum_{i,j=1}^n \tfrac{1}{2}b_i \left(A^{-1}\right)_{ij} b_j\right]\right\}\Bigg|_{\mathbf{b}=0}. \tag{1.8}$$

Wick's theorem. This identity leads to Wick's theorem. In the r.h.s. of equation (1.8) each time a differential operator acts on the exponential it generates a factor b. Another differential operator has to act on this factor, otherwise the corresponding contribution vanishes when we set $\mathbf{b} = 0$. We conclude that the average of the product $x_{k_1} \ldots x_{k_\ell}$ with the gaussian weight $\exp\left(-\tfrac{1}{2}x_i A_{ij} x_j\right)$ is obtained in the following way: one considers all possible pairings of the indices $k_1, \ldots, k_\ell$ (ℓ must thus be even). To each pair $k_p k_q$ one associates the matrix element $\left(A^{-1}\right)_{k_p k_q}$ of the matrix $\mathbf{A}^{-1}$. Then:

$$\langle x_{k_1} \ldots x_{k_\ell}\rangle = \sum_{\substack{\text{all possible pairings} \\ \text{of } \langle k_1 \ldots k_\ell\rangle}} A^{-1}_{k_{p_1} k_{p_2}} \cdots A^{-1}_{k_{p_{\ell-1}} k_{p_\ell}}. \tag{1.9}$$

Equation (1.9) which expresses Wick's theorem is, in the form adapted to Quantum Mechanics or Field Theory, the basis of perturbative calculations.

Remark. The gaussian integral has another remarkable property: if we integrate the exponential of a quadratic form over a subset of variables, the result is still the exponential of a quadratic form. This structural stability is related to some of the properties of the harmonic oscillator which will be discussed in Chapter 2.

1.2 Perturbation Theory

We now want to calculate the integral:

$$I = \int \prod_{i=1}^{n} \mathrm{d}x_i \, \exp\left(-\sum_{i,j=1}^{n} \tfrac{1}{2} x_i A_{ij} x_j - \lambda V(x) \right), \tag{1.10}$$

in which $V(x)$ is a polynomial in the variables x_i and λ is a parameter. We can expand the integrand in a formal power series in λ:

$$I = \int \prod_{i=1}^{n} \mathrm{d}x_i \, \exp\left(-\sum_{i,j=1}^{n} \tfrac{1}{2} x_i A_{ij} x_j \right) \left[\sum_{k=0}^{\infty} \frac{(-\lambda)^k}{k!} V^k(x) \right]. \tag{1.11}$$

Using equations (1.6,1.8) we can formally rewrite (1.10):

$$I = \left\{ \exp\left[-\lambda V\left(\frac{\partial}{\partial b} \right) \right] \exp\left[\sum_{i,j=1}^{n} \tfrac{1}{2} b_i \left(A^{-1} \right)_{ij} b_j \right] \right\}\Bigg|_{\mathbf{b}=0}. \tag{1.12}$$

We can also directly calculate each term in the expansion using Wick's theorem (1.9).

Steepest-descent. In the case of contour integrals in the complex domain, one sometimes uses a method, steepest-descent, which reduces their evaluation to gaussian integrals. Let us consider the integral:

$$I = \int \prod_{i=1}^{n} \mathrm{d}x_i \, \exp\left[-\frac{1}{\lambda} S(x_1, \dots, x_n) \right]. \tag{1.13}$$

In the limit $\lambda \to 0$, the integral is dominated by saddle points $\{x_i^c\}$:

$$\frac{\partial S}{\partial x_i}(x_1^c, x_2^c, \dots, x_n^c) = 0. \tag{1.14}$$

To calculate the contribution of the leading saddle point $\mathbf{x}^c$, we change variables, setting:

$$\mathbf{x} = \mathbf{x}^c + \mathbf{y}\sqrt{\lambda}. \tag{1.15}$$

We then expand $S(\mathbf{x})$ in powers of λ (and thus y):

$$\frac{1}{\lambda} S(x_1, \dots, x_n) = \frac{1}{\lambda} S(\mathbf{x}^c) + \frac{1}{2!} \frac{\partial^2 S}{\partial x_i x_j}(\mathbf{x}^c) y_i y_j$$
$$+ \sum_{k=3}^{\infty} \frac{\lambda^{k/2-1}}{k!} \frac{\partial^k S}{\partial x_{i_1} \dots \partial x_{i_k}}(\mathbf{x}^c) y_{i_1} \dots y_{i_k}. \tag{1.16}$$

The change of variables is such that the term quadratic in $\mathbf{y}$ is independent of λ. The integral becomes:

$$I = \mathrm{e}^{-S(\mathbf{x}^c)/\lambda} \int \prod_{i=1}^{n} \mathrm{d}x_i \, \exp\left[-\frac{1}{2!} \frac{\partial^2 S}{\partial x_i \partial x_j}(\mathbf{x}^c) y_i y_j - R(\mathbf{y}) \right] \tag{1.17}$$

$$R(\mathbf{y}) = \sum_{k=3}^{\infty} \frac{\lambda^{k/2-1}}{k!} \frac{\partial^k S}{\partial x_{i_1} \dots \partial x_{i_k}}(\mathbf{x}^c) y_{i_1} \dots y_{i_k}. \tag{1.18}$$

We then expand the integrand in powers of $\sqrt{\lambda}$: At each order we have to calculate the average of a polynomial with a gaussian weight.

1.3 Complex Structures

We shall often meet complex structures: we have $2n$ integration variables $\{x_i\}$ and $\{y_i\}$, $i = 1, \ldots, n$, and the integrand is invariant under a simultaneous identical rotation in all (x_i, y_i) planes. It is then natural to introduce formal complex variables z_i and $\bar{z}_i$ which, for normalization purposes, we define by:

$$z_i = (x_i + iy_i)/\sqrt{2}, \qquad \bar{z}_i = (x_i - iy_i)/\sqrt{2}. \tag{1.19}$$

Note however that z_i and $\bar{z}_i$ are *independent integration variables* and only formally complex conjugates since x_i and y_i could themselves be complex.

The generic gaussian integral now becomes:

$$I\left(\mathbf{A}; \mathbf{b}, \bar{\mathbf{b}}\right) = \int \left(\prod_{i=1}^n dz_i d\bar{z}_i\right) \exp\left[-\sum_{i,j=1}^n \bar{z}_i A_{ij} z_j + \sum_{i=1}^n \left(\bar{b}_i z_i + b_i \bar{z}_i\right)\right], \tag{1.20}$$

in which $\mathbf{A}$ is a complex matrix with non-vanishing determinant.

As before, to calculate this integral we first eliminate the terms linear in z_i and $\bar{z}_i$ by a shift of variables, setting:

$$z_i = v_i + \left(A^{-1}\right)_{ij} b_j, \qquad \bar{z}_i = \bar{v}_i + \bar{b}_j \left(A^{-1}\right)_{ji}. \tag{1.21}$$

The resulting gaussian integral can be calculated either by returning to the "real" variables (1.19) or by a change of variables like $A_{ij} v_j = v_i'$. We obtain

$$I\left(\mathbf{A}; \mathbf{b}, \bar{\mathbf{b}}\right) = (2i\pi)^n \left(\det \mathbf{A}\right)^{-1} \exp\left[\sum_{i,j=1}^n \bar{b}_i \left(A^{-1}\right)_{ij} b_j\right]. \tag{1.22}$$

By systematically differentiating with respect to b_i and $\bar{b}_j$, one establishes Wick's theorem for averages with the gaussian weight $\exp\left(-\bar{z}_i A_{ij} z_j\right)$. Only monomials with equal number of factors z and $\bar{z}$ have a non vanishing average:

$$\langle \bar{z}_{i_1} z_{j_1} \ldots \bar{z}_{i_n} z_{j_n} \rangle = \sum_{\substack{\text{all permutations} \\ P \text{ of } \{j_1, \ldots, j_n\}}} A_{j_{P_1} i_1}^{-1} A_{j_{P_2} i_2}^{-1} \ldots A_{j_{P_n} i_n}^{-1}. \tag{1.23}$$

1.4 Integral Representation of Constraints

We shall often use a simple identity about Dirac δ-functions. By definition:

$$\int \prod_{i=1}^n dy_i \, \delta\left(y_i\right) = 1. \tag{1.24}$$

If we change variables:

$$y_i = f_i\left(\mathbf{x}\right) \tag{1.25}$$

and assume that equation (1.25) defines a unique set of functions $x_i(\mathbf{y})$ for $|\mathbf{y}|$ small enough, then we obtain the identity:

$$\int \left\{ \prod_{i=1}^{n} \mathrm{d}x_i \, \delta\left[f_i(\mathbf{x})\right] \right\} J(\mathbf{x}) = 1 \,, \tag{1.26}$$

in which $J(\mathbf{x})$ is the jacobian of the change of variables (1.25):

$$J(\mathbf{x}) = \left| \det \frac{\partial f_i}{\partial x_j} \right| . \tag{1.27}$$

The identity (1.26) has a straightforward generalization: assume that we want to calculate a function $\sigma(\mathbf{x})$ for $\mathbf{x}$ solution of the equation $\mathbf{f}(\mathbf{x}) = 0$, i.e. for $\mathbf{x} = \mathbf{x}(\mathbf{y} = 0)$, without solving the equation explicitly. We can then use the identity:

$$\sigma(\mathbf{x})\,|_{\mathbf{f}(\mathbf{x})=0} = \int \left\{ \prod_{i=1}^{n} \mathrm{d}x_i \, \delta\left[f_i(\mathbf{x})\right] \right\} J(\mathbf{x})\sigma(\mathbf{x}) \,. \tag{1.28}$$

This identity, as most identities about gaussian integrals, has the useful property that it can easily be generalized to an infinite number of variables.

1.5 Algebraic Functional Techniques

1.5.1 Generating functional. Functional differentiation

In the discussion of algebraic properties of correlation functions the concept of generating functional will be extremely useful. Let $\{F^{(n)}(x_1,\ldots,x_n)\}$ be a sequence of symmetric functions of their arguments. We introduce a function $f(x)$ and consider the following formal series in f:

$$\mathcal{F}(f) = \sum_{n=0}^{\infty} \frac{1}{n!} \int \mathrm{d}x_1 \ldots \mathrm{d}x_n \, F^{(n)}(x_1,\ldots,x_n)\,f(x_1)\ldots f(x_n) \,. \tag{1.29}$$

One calls $\mathcal{F}(f)$ the generating functional of the sequence of functions $F^{(n)}$.

To recover the functions $F^{(n)}$ from $\mathcal{F}(f)$ we then need the concept of *functional derivative* $\delta/\delta f(x)$. A functional derivative is defined by the properties that it satisfies the usual algebraic rules of any differential operator:

$$\begin{aligned}
\frac{\delta}{\delta f(x)}\left[\mathcal{F}_1(f) + \mathcal{F}_2(f)\right] &= \frac{\delta}{\delta f(x)}\mathcal{F}_1(f) + \frac{\delta}{\delta f(x)}\mathcal{F}_2(f), \\
\frac{\delta}{\delta f(x)}\left[\mathcal{F}_1(f)\mathcal{F}_2(f)\right] &= \mathcal{F}_1(f)\frac{\delta}{\delta f(x)}\mathcal{F}_2(f) + \mathcal{F}_2(f)\frac{\delta}{\delta f(x)}\mathcal{F}_1(f),
\end{aligned} \tag{1.30}$$

and in addition:

$$\frac{\delta}{\delta f(y)}f(x) = \delta(x - y), \tag{1.31}$$

where again $\delta(x)$ is Dirac's δ-function. Differentiating $\mathcal{F}(f)$ for example, one finds

$$\frac{\delta}{\delta f(y)}\mathcal{F}(f) = \sum_{n=0}^{\infty} \frac{1}{n!} \int \mathrm{d}x_1 \ldots \mathrm{d}x_n \, F^{(n+1)}(y, x_1,\ldots,x_n)\,f(x_1)\ldots f(x_n) \,. \tag{1.32}$$

1.5.2 Determinants of operators

Often we shall have to calculate determinants of operators represented by some kernel $M(x,y)$ which, after some transformations, can be cast into the form $\delta(x-y)+K(x,y)$. Provided the traces of all powers of $\mathbf{K}$ exist, the following identity, valid for any matrix $\mathbf{M}$,

$$\ln \det \mathbf{M} \equiv \operatorname{tr} \ln \mathbf{M}, \qquad (1.33)$$

expanded in powers of the kernel $\mathbf{K}$:

$$\ln \det [\mathbf{1}+\mathbf{K}] = \int \mathrm{d}x\, K(x,x) - \tfrac{1}{2}\int \mathrm{d}x_1 \mathrm{d}x_2\, K(x_1,x_2)K(x_2,x_1) + \cdots$$
$$+ \frac{(-1)^{n+1}}{n}\int \mathrm{d}x_1 \cdots \mathrm{d}x_n\, K(x_1,x_2)\, K(x_2,x_3) \cdots K(x_n,x_1) + \cdots , \quad (1.34)$$

will often be useful.

1.6 Grassmann Algebras. Differential Forms

We shall also deal with theories containing fermions. Since fermion field correlation functions (or Green's functions) are antisymmetric with respect to the exchange of two arguments, the construction of generating functionals requires the introduction of anti-commuting classical functions, and thus Grassmann variables.

Grassmann algebra. We only consider Grassmann algebras over $\mathbb{R}$ or $\mathbb{C}$ (real or complex). A Grassmann algebra $\mathfrak{A}$ is an algebra constructed from a set of generators θ_i and their anticommuting products:

$$\theta_i\theta_j + \theta_j\theta_i = 0 \qquad \forall i,j. \qquad (1.35)$$

Note that as a consequence:

(i) all elements in a Grassmann algebra are first degree polynomials in each generator;

(ii) if the number n of generators is finite, the algebra forms a finite dimensional vector space on $\mathbb{R}$ or $\mathbb{C}$ of dimension 2^n.

$\mathfrak{A}$ is also a graded algebra in the sense that to any monomial $\theta_{i_1}\theta_{i_2}\dots\theta_{i_p}$ we can associate an integer p counting the number of generators in the product.

Finally let us note that elements of $\mathfrak{A}$ are invertible if and only if their expansion as a sum of products of generators contains a term of degree zero which is invertible. For example the element $1+\theta$ is invertible, and has $1-\theta$ as inverse; however θ is not invertible.

Grassmannian parity. On the algebra $\mathfrak{A}$ we can implement a simple automorphism which is a reflection P defined by:

$$\mathrm{P}\left(\theta_i\right) = -\theta_i. \qquad (1.36)$$

Then on a monomial of degree p, P acts like:

$$\mathrm{P}\left(\theta_{i_1}\dots\theta_{i_p}\right) = (-1)^p\theta_{i_1}\dots\theta_{i_p}. \qquad (1.37)$$

The reflection P divides the algebra $\mathfrak{A}$ in two eigenspaces $\mathfrak{A}^{\pm}$ containing the even or odd elements

$$\mathrm{P}\left(\mathfrak{A}^{\pm}\right) = \pm\mathfrak{A}^{\pm}. \qquad (1.38)$$

In particular $\mathfrak{A}^+$ is a subalgebra, the subalgebra of commuting elements.

Differential forms. An application of Grassmann algebras is the representation of differential forms. The language of differential forms will not be used often in this work. However it is interesting to here recall one concept, the exterior derivative of forms, whose generalization will appear in the context of BRS symmetry (see Chapter 16). Let us consider totally antisymmetric tensors $\Omega_{\mu_1,\ldots,\mu_l}(x)$, functions of n commuting variables x^μ. Associating n Grassmann variables θ^μ with x^μ, we can write the corresponding l-form Ω:

$$\Omega = \Omega_{\mu_1,\ldots,\mu_l}(x)\theta^{\mu_1}\ldots\theta^{\mu_l}\,, \tag{1.39}$$

where $l \leq n$ otherwise the form vanishes.

One can define a differential operator d acting on forms:

$$\mathrm{d} \equiv \theta^\mu \frac{\partial}{\partial x^\mu}\,. \tag{1.40}$$

We note that if Ω is a l-form, $\mathrm{d}\Omega$ is a $l+1$-form (see Chapter 22 for details). One immediately verifies that d is *nilpotent*:

$$\mathrm{d}^2 = \theta^\mu \frac{\partial}{\partial x^\mu}\theta^\nu\frac{\partial}{\partial x^\nu} = 0\,, \tag{1.41}$$

because the product $\theta^\mu\theta^\nu$ is antisymmetric in $\mu \leftrightarrow \nu$.

We also recall that a form Ω which satisfies $\mathrm{d}\Omega = 0$ is called *closed* and a form Ω which can be written $\Omega = \mathrm{d}\Omega'$ is called *exact*. The property (1.41) implies that any exact form is closed.

Note that it is customary to write in the case of forms the generators of the algebra $\mathrm{d}x^\mu$ instead of θ^μ and to then use the $\wedge$ notation for the product to show that it is antisymmetric.

1.7 Differentiation in Grassmann Algebras

It is then useful to define differentiation in Grassmann algebras. A naive definition would be inconsistent due to the non-commutative character of the algebra. The problem can be solved in the following way: Considered as functions of a generator θ_i, all elements A of $\mathfrak{A}$ can be written

$$A = A_1 + \theta_i A_2\,,$$

after some commutations, where A_1 and A_2 do not depend on θ_i. Then by definition

$$\frac{\partial A}{\partial \theta_i} = A_2\,. \tag{1.42}$$

Note that the differential operator $\partial/\partial\theta_i$ is *nilpotent*: $(\partial/\partial\theta_i)^2 = 0$, like the form differentiation (see equation (1.41)).

Remark. The equation (1.42) defines a left-differentiation in the sense that the action of $\partial/\partial\theta_i$ consists in bringing θ_i on the left in a monomial and suppressing it. Similarly a right-differentiation could have defined by commuting θ_i to the right.

Chain rule. It is easy to verify that chain rule applies to Grassmann differentiation. If $\sigma(\theta)$ belongs to $\mathfrak{A}^-$ and $x(\theta)$ belongs to $\mathfrak{A}^+$ we can write:

$$\frac{\partial}{\partial\theta}f(\sigma,x) = \frac{\partial\sigma}{\partial\theta}\frac{\partial f}{\partial\sigma} + \frac{\partial x}{\partial\theta}\frac{\partial f}{\partial x}\,. \tag{1.43}$$

For the second term in the r.h.s. the order between factors matters.

Formal construction. To show the consistency of the definition (1.42) and exhibit some properties, we now more generally define differentiation in a Grassmann algebra by some formal rules, similar but slightly different from commutative algebras. A Grassmann differential operator D (also called an anti-derivation) acting on $\mathfrak{A}$ is defined by the two properties:

(i) It is a linear mapping of $\mathfrak{A}$, considered as a vector space, into itself

$$\mathrm{D}\left(\lambda_1 A_1 + \lambda_2 A_2\right) = \lambda_1 \mathrm{D}\left(A_1\right) + \lambda_2 \mathrm{D}\left(A_2\right) \qquad \text{for } \lambda_1, \lambda_2 \in \mathbb{R} \text{ or } \mathbb{C}, \tag{1.44}$$

(ii) It satisfies the condition

$$\mathrm{D}\left(A_1 A_2\right) = \mathrm{P}\left(A_1\right)\mathrm{D}\left(A_2\right) + \mathrm{D}\left(A_1\right)A_2. \tag{1.45}$$

The unusual form of equation (1.45) compared to the differentiation rule for commuting variables is required if we want D to anticommute with P

$$\mathrm{DP} + \mathrm{PD} = 0, \tag{1.46}$$

which means that the image of $\mathfrak{A}^\pm$ by D belongs to $\mathfrak{A}^\mp$.

Note that if A belongs to $\mathfrak{A}^+$ and $F(x)$ is an ordinary function of real or complex variables:

$$\mathrm{D}\left[F(A)\right] = \mathrm{D}(A)F'(A) \qquad \text{for } A \in \mathfrak{A}^+. \tag{1.47}$$

Note finally that the form differentiation (1.40) shares all these properties, but acts on different variables.

Anticommutation relations. A short calculation shows that if D and D$'$ are two operators satisfying conditions (1.44,1.45), then the anticommutator Δ:

$$\Delta = \mathrm{DD}' + \mathrm{D}'\mathrm{D}, \tag{1.48}$$

is a usual differential operator:

$$\begin{cases} \Delta\left(\lambda_1 A_1 + \lambda_2 A_2\right) = \lambda_1 \Delta\left(A_1\right) + \lambda_2 \Delta\left(A_2\right), \\ \qquad \Delta\left(A_1 A_2\right) = \Delta\left(A_1\right)A_2 + A_1\Delta\left(A_2\right). \end{cases} \tag{1.49}$$

Furthermore:

$$\Delta \mathrm{P} = \mathrm{P}\Delta. \tag{1.50}$$

These properties, which are the consequence of the addition of relation (1.46) to the definitions (1.44,1.45), allow to extend the notion of Lie algebra and are directly relevant to the discussion of supersymmetries.

A basis. Since a differential operator satisfies conditions (1.44,1.45), it is completely defined by its action on the generators θ_i. In addition any differential operator left-multiplied by an element of $\mathfrak{A}^+$ still satisfies (1.44,1.45). We conclude, therefore, that any differential operator can be expanded on a basis of operators $\partial/\partial\theta_i$ defined by:

$$\frac{\partial}{\partial\theta_i}\theta_j = \delta_{ij}, \tag{1.51}$$

with left coefficients in $\mathfrak{A}^+$. One easily verifies that the differential operators $\partial/\partial\theta_i$ coincide with the operators defined by equation (1.42). The nilpotent differential operators $\partial/\partial\theta_i$, together with the generators θ_i considered as operators acting on $\mathfrak{A}$ by left-multiplication, satisfy the anticommutation relations:

$$\theta_i\theta_j + \theta_j\theta_i = 0, \quad \frac{\partial}{\partial\theta_i}\frac{\partial}{\partial\theta_j} + \frac{\partial}{\partial\theta_j}\frac{\partial}{\partial\theta_i} = 0, \quad \theta_i\frac{\partial}{\partial\theta_j} + \frac{\partial}{\partial\theta_j}\theta_i = \delta_{ij}, \tag{1.52}$$

and thus form a Clifford algebra.

1.8 Integration in Grassmann Algebras

We now also define integration over Grassmann variables which we denote by the integral symbol. We define the integration to be an operation identical to differentiation

$$\int d\theta_i\, A \equiv \frac{\partial}{\partial \theta_i} A, \qquad \forall A \in \mathfrak{A}. \tag{1.53}$$

The use of an integral or derivative symbol is thus just a matter of convenience.

General properties. Let us now show that this operation satisfies the formal properties we expect from a *definite* integral. Quite generally we associate to a given differential operator D an operator I which has the following defining properties. It is a linear operator acting on $\mathfrak{A}$:

$$I\left(\lambda_1 A_1 + \lambda_2 A_2\right) = \lambda_1 I\left(A_1\right) + \lambda_2 I\left(A_2\right), \tag{1.54}$$

which satisfies the three properties:

$$ID = 0, \tag{1.55}$$
$$DI = 0, \tag{1.56}$$

and

$$D(A) = 0 \Longrightarrow I(BA) = I(B)A. \tag{1.57}$$

In addition it changes the grading in the same way as a differential operator:

$$PI + IP = 0. \tag{1.58}$$

Let us explain the conditions (1.55–1.57): condition (1.55) expresses that in the absence of boundary terms the integral of a total derivative vanishes; condition (1.56) expresses that if we integrate over a variable, the result no longer depends on this variable; finally condition (1.57) implies that a factor whose derivative vanishes can be taken out of the integral.

In the case of Grassmann algebras, if D is nilpotent, D itself satisfies all conditions. The differential operators $\partial/\partial\theta_i$ are indeed nilpotent.

1.8.1 Change of variables in a Grassmann integral
Let us consider the integral I:

$$I = \int d\theta\, f(\theta), \tag{1.58}$$

and perform the (necessarily) affine change of variables:

$$\theta = a\theta' + b, \tag{1.59}$$

in which parity conservation implies that $a \in \mathfrak{A}^+$ and $b \in \mathfrak{A}^-$. The element a must be invertible, i.e. its term of degree zero in the Grassmann variables must be different from zero. Then using definition (1.53) we find:

$$\int d\theta\, f(\theta) = a^{-1} \int d\theta'\, f(a\theta' + b). \tag{1.60}$$

We find a very important property of Grassmann integrals: the jacobian is a^{-1}, while in the case of commuting variables we would have found a. This difference reflects the identity of differentiation and integration for Grassmann variables.

Generalization. More generally let us show that a change of variables

$$\theta_i = \theta_i(\theta'), \qquad \theta_i, \theta_i' \in \mathfrak{A}^-,$$

for which the matrix $\partial\theta_i/\partial\theta_j'$ has an invertible part of degree zero, leads to a jacobian which is the *inverse* of the determinant of $\partial\theta_i/\partial\theta_j'$:

$$d\theta_1 \ldots d\theta_n = d\theta_1' \ldots d\theta_n' J(\theta'), \tag{1.61}$$

with:

$$J^{-1} = \det \frac{\partial\theta_i}{\partial\theta_j'}. \tag{1.62}$$

Note that the determinant is well-defined because all elements of the matrix $\partial\theta_i/\partial\theta_j'$ belong to $\mathfrak{A}^+$.

We again start from the identity between differentiation and integration:

$$\int d\theta_1 \ldots d\theta_n f(\boldsymbol{\theta}) \equiv \prod_i \frac{\partial}{\partial\theta_i} f(\boldsymbol{\theta}),$$

in which the product in l.h.s. is ordered. We then assume that f is a function of variables θ_i' and therefore, using chain rule (1.43):

$$\prod_i \frac{\partial}{\partial\theta_i} f(\boldsymbol{\theta}) = \prod_i \frac{\partial\theta_{j_i}'}{\partial\theta_i} \frac{\partial}{\partial\theta_{j_i}'} f(\boldsymbol{\theta}).$$

We now factorize the elements $\partial\theta_{j_i}'/\partial\theta_i$ which commute. The differential operators $\partial/\partial\theta_{j_i}'$ anticommute (see equations (1.52)) and are thus all proportional to the product ordered from 1 to n. A sign is generated which is the signature of the permutation $j_1, j_2, \ldots, j_n$. We then recognize the determinant of the matrix $\partial\theta_j'/\partial\theta_i$:

$$\prod_i \frac{\partial}{\partial\theta_i} = \det \frac{\partial\theta_j'}{\partial\theta_i} \prod_k \frac{\partial}{\partial\theta_k'}.$$

The identity between differentiation and integration then immediately leads to equations (1.61,1.62).

Remark. A straightforward verification of equation (1.61) is provided by the following example:

$$1 = \int d\theta_1 \ldots d\theta_n \, \theta_n \ldots \theta_1.$$

If we make the linear change of variables:

$$\theta_i = a_{ij}\theta_j',$$

the result relies upon the identity:

$$\theta_n \ldots \theta_1 = \theta_n' \ldots \theta_1' \det \mathbf{a}.$$

1.8.2 Mixed change of variables

In this work we shall meet integrals involving both commuting and anticommuting variables (bosons and fermions). It is sometimes useful to perform mixed changes of variables.

Denoting by θ, θ' and x, x' the anticommuting and commuting variables respectively, we set (respecting parity):

$$x_a = x_a\,(x',\theta')\;\in\mathfrak{A}_+(\theta'), \qquad \theta_i = \theta_i\,(x',\theta')\;\in\mathfrak{A}_-(\theta'). \qquad (1.63)$$

We introduce the matrix $\mathbf{M}$ of partial derivatives

$$\mathbf{M} = \begin{pmatrix} \mathbf{A} & \mathbf{B} \\ \mathbf{C} & \mathbf{D} \end{pmatrix},$$

with

$$\mathbf{A}_{ab} = \frac{\partial x_a}{\partial x'_b}, \qquad \mathbf{B}_{ai} = \frac{\partial x_a}{\partial \theta'_i}, \qquad \mathbf{C}_{ia} = \frac{\partial \theta_i}{\partial x'_a}, \qquad \mathbf{D}_{ij} = \frac{\partial \theta_i}{\partial \theta'_j}.$$

It is convenient to change variables in two steps:

(i) one first passes from (θ, x) to (θ, x'). This step generates the jacobian J_1:

$$J_1 = \det \left.\frac{\partial x_a}{\partial x'_b}\right|_\theta = \det\left(\mathbf{A} - \mathbf{B}\mathbf{D}^{-1}\mathbf{C}\right). \qquad (1.64)$$

(ii) One then goes from (θ, x') to (θ', x'). The second step just gives, as explained above, the jacobian J_2:

$$J_2 = (\det \mathbf{D})^{-1}. \qquad (1.65)$$

The complete jacobian J, also called the *berezinian* of the matrix of derivatives, is thus:

$$J \equiv \frac{D(x,\theta)}{D(x',\theta')} = J_1 J_2 = \mathrm{Ber}\,\mathbf{M} \equiv \det\left(\mathbf{A} - \mathbf{B}\mathbf{D}^{-1}\mathbf{C}\right)(\det \mathbf{D})^{-1}. \qquad (1.66)$$

For the jacobian to be not singular, the matrices $\mathbf{A}$ and $\mathbf{D}$ have to be invertible (and therefore their contributions of degree zero in θ').

Trace of mixed matrices. In the case of the integration over ordinary commuting variables, if we perform a change of variables infinitesimally close to the identity:

$$x_a = x'_a + \varepsilon f_a(x'),$$

then from identity (1.33), we see that the jacobian has the form

$$J = \det \frac{\partial x_a}{\partial x'_b} = 1 + \varepsilon\,\mathrm{tr}\,\frac{\partial f_a}{\partial x'_b} + O\left(\varepsilon^2\right) = 1 + \varepsilon\frac{\partial f_a}{\partial x'_a} + O\left(\varepsilon^2\right).$$

Let us now consider the mixed case:

$$x_a = x'_a + \varepsilon f_a(x',\theta'), \qquad \theta_i = \theta'_i + \varepsilon\varphi_i(x',\theta'). \qquad (1.67)$$

Setting then

$$\mathbf{M} = 1 + \varepsilon\mathbf{M}_1 + O\left(\varepsilon^2\right), \qquad \mathbf{M}_1 = \begin{pmatrix} \mathbf{A}_1 & \mathbf{B}_1 \\ \mathbf{C}_1 & \mathbf{D}_1 \end{pmatrix},$$

we find, as a consequence of identity (1.66):

$$J = 1 + \varepsilon\left(\mathrm{tr}\,\mathbf{A}_1 - \mathrm{tr}\,\mathbf{D}_1\right) + O\left(\varepsilon^2\right), \qquad \mathrm{tr}\,\mathbf{A}_1 - \mathrm{tr}\,\mathbf{D}_1 = \frac{\partial f_a}{\partial x_a} - \frac{\partial \varphi_i}{\partial \theta_i}. \qquad (1.68)$$

To maintain the connection between jacobian and trace, we are therefore led to define the trace of a mixed matrix, denoted by Str, as the trace of the commuting block minus the trace of the anticommuting block:

$$\mathrm{Str}\,\mathbf{M}_1 = \mathrm{tr}\,\mathbf{A}_1 - \mathrm{tr}\,\mathbf{D}_1. \qquad (1.69)$$

1.9 Gaussian Integrals with Grassmann Variables

In this work we shall mainly deal with Grassmann algebras in which the generators can be separated into two conjugated sets. We then denote by θ_i and $\bar{\theta}_i$, $i = 1, \ldots, n$, these generators. In many cases one can define a complex conjugation in the algebra which exchanges θ_i and $\bar{\theta}_i$.

As in the case of commuting variables, we now calculate gaussian integrals, with the same motivation: we shall in general try to reduce any integral to a formal finite or infinite sum of gaussian integrals.

Let us consider first:

$$I(\mathbf{a}) = \int \mathrm{d}\theta_1 \mathrm{d}\bar{\theta}_1 \mathrm{d}\theta_2 \mathrm{d}\bar{\theta}_2 \ldots \mathrm{d}\theta_n \mathrm{d}\bar{\theta}_n \exp\left(\sum_{i,j=1}^{n} \bar{\theta}_i a_{ij} \theta_j \right). \qquad (1.70)$$

According to the rules of Grassmann integration, the result is simply the coefficient of the product $\bar{\theta}_n \theta_n \ldots \bar{\theta}_1 \theta_1$ in the expansion of the integrand. The integrand can be rewritten (no implicit summation over repeated indices):

$$\exp\left(\sum_{i,j=1}^{n} \bar{\theta}_i a_{ij} \theta_j \right) = \prod_{i,j=1}^{n} \exp\left(\bar{\theta}_i a_{ij} \theta_j \right) = \prod_{i=1}^{n} \prod_{j=1}^{n} \left(1 + \bar{\theta}_i a_{ij} \theta_j \right).$$

Expanding the product we see that the terms which give non-zero contribution to the integral are of the form

$$\sum_{\substack{\text{permutations} \\ \{j_1 \cdots j_n\}}} a_{n j_n} a_{n-1 j_{n-1}} \ldots a_{1 j_1} \bar{\theta}_n \theta_{j_n} \ldots \bar{\theta}_1 \theta_{j_1}.$$

Commuting the generators to put them in the standard order $\bar{\theta}_n \theta_n \ldots \bar{\theta}_1 \theta_1$ we find a sign which is the signature of the permutation, and recognize the coefficient as the determinant of a_{ij}:

$$I(\mathbf{a}) = \det \mathbf{a}. \qquad (1.71)$$

The result is the inverse of the one obtained with complex commuting variables. This will eventually lead in perturbation theory to a sign $(-1)^L$ in the front of Feynman diagrams with L fermions loops. The calculation above is mainly a verification since we could have changed variables

$$a_{ij} \theta_j = \theta_i', \qquad (1.72)$$

and used the form (1.61) of the jacobian (no summation over repeated indices)

$$I(\mathbf{a}) = \det \mathbf{a} \int \mathrm{d}\theta_1' \mathrm{d}\bar{\theta}_1 \ldots \mathrm{d}\theta_n' \mathrm{d}\bar{\theta}_n \exp\left(\sum_{i=1}^{n} \bar{\theta}_i \theta_i' \right)$$

$$= \det \mathbf{a} \int \prod_{i=1}^{n} \mathrm{d}\theta_i' \mathrm{d}\bar{\theta}_i \left(1 + \bar{\theta}_i \theta_i' \right) = \det \mathbf{a}.$$

Remark. We can also more generally calculate a gaussian integral of the form:

$$I(\mathbf{a}) = \int \mathrm{d}\theta_{2n} \ldots \mathrm{d}\theta_2 \mathrm{d}\theta_1 \exp\left(\tfrac{1}{2} \sum_{i,j=1}^{2n} \theta_i a_{ij} \theta_j \right). \qquad (1.73)$$

Since the product $\theta_i\theta_j$ is antisymmetric in (ij), the matrix a_{ij} can be chosen antisymmetric

$$a_{ij} + a_{ji} = 0. \tag{1.74}$$

Expanding the exponential in a power series we observe that only the term of order n which contains all products of degree $2n$ in θ gives a non-zero contribution:

$$I(\mathbf{a}) = \frac{1}{2^n n!} \int d\theta_{2n} \ldots d\theta_1 \, (\theta_i a_{ij} \theta_j)^n. \tag{1.75}$$

In the expansion of the product only the terms containing a permutation of $\theta_1 \ldots \theta_{2n}$ do not vanish. Ordering then all terms to factorize the product $\theta_1\theta_2 \ldots \theta_{2n}$ we find:

$$I(\mathbf{a}) = \frac{1}{2^n n!} \sum_{\substack{\text{permutations } P \\ \text{of } \{i_1 \ldots i_{2n}\}}} \varepsilon(P) \, a_{i_1 i_2} a_{i_3 i_4} \ldots a_{i_{2n-1} i_{2n}}, \tag{1.76}$$

in which $\varepsilon(P) = \pm 1$ is the signature of the permutation. The quantity in the r.h.s. called the *pfaffian* of the antisymmetric matrix $\mathbf{a}$

$$I(\mathbf{a}) = \mathrm{Pf}(\mathbf{a}). \tag{1.77}$$

Grassmann integral techniques allow to prove a well-known algebraic identity

$$\mathrm{Pf}^2(\mathbf{a}) = \det \mathbf{a}. \tag{1.78}$$

Let us indeed calculate $I^2(\mathbf{a})$:

$$I^2(\mathbf{a}) = \int d\theta_{2n} \ldots d\theta_1 \, d\theta'_{2n} \ldots d\theta'_1 \, \exp\left[\tfrac{1}{2}\left(\theta_i a_{ij}\theta_j + \theta'_i a_{ij}\theta'_j\right)\right]. \tag{1.79}$$

We change variables, setting:

$$\eta_k = \frac{1}{\sqrt{2}}\left(\theta_k + i\theta'_k\right), \qquad \bar\eta_k = \frac{1}{\sqrt{2}}\left(\theta_k - i\theta'_k\right). \tag{1.80}$$

The jacobian is $(-1)^n$. Also:

$$\theta_i\theta_j + \theta'_i\theta'_j = \bar\eta_i\eta_j - \bar\eta_j\eta_i, \tag{1.81}$$

$$d\eta_{2n} \ldots d\eta_1 d\bar\eta_{2n} \ldots d\bar\eta_1 = (-1)^{n^2} \prod_i d\eta_i d\bar\eta_i. \tag{1.82}$$

Using the antisymmetry of the matrix $\mathbf{a}$ we find:

$$\mathrm{Pf}^2(\mathbf{a}) = \int d\eta_1 d\bar\eta_1 \ldots d\eta_{2n} d\bar\eta_{2n} \, \exp\left(\bar\eta_i a_{ij}\eta_j\right) = \det \mathbf{a}. \tag{1.83}$$

The Dirac δ-function. In Grassmann algebras the role of the Dirac δ-function is played by the function:

$$\delta(\theta) \equiv \theta. \tag{1.84}$$

Indeed

$$\int \mathrm{d}\theta \, \theta f(\theta) = f(0),$$

in which $f(0)$ means the constant part of the affine function $f(\theta)$. In some applications it is useful to use an integral representation of $\delta(\theta)$ similar to the usual Fourier transform:

$$\delta(\eta) \equiv \eta = \int \mathrm{d}\theta \, \mathrm{e}^{\theta\eta}, \qquad (1.85)$$

where η is an additional Grassmann variable.

General gaussian integrals. We now consider another copy of the Grassmann algebra $\mathfrak{A}$ whose generators will be denoted by η_i and $\bar{\eta}_i$. Following the strategy of Section 1.1, we first calculate the integral:

$$Z_{\mathrm{G}}\left(\eta, \bar{\eta}\right) = \int \prod_i \mathrm{d}\theta_i \mathrm{d}\bar{\theta}_i \exp\left[\sum_{i,j=1}^{n} a_{ij}\bar{\theta}_i\theta_j + \sum_{i=1}^{n}\left(\bar{\eta}_i\theta_i + \bar{\theta}_i\eta_i\right)\right], \qquad (1.86)$$

in which the integrand is an element of the direct sum of the two Grassmann algebras.

The calculation as before relies on a change of variables:

$$\theta_i = \theta_i' + \left(a^{-1}\right)_{ij}\eta_j, \qquad \bar{\theta}_i = \bar{\theta}_i' + \bar{\eta}_j\left(a^{-1}\right)_{ji}, \qquad (1.87)$$

and leads to the result:

$$Z_{\mathrm{G}}\left(\eta, \bar{\eta}\right) = \det \mathbf{a} \exp\left[-\sum_{i,j=1}^{n} \bar{\eta}_i\left(a^{-1}\right)_{ij}\eta_j\right]. \qquad (1.88)$$

If we denote by $\langle \, \rangle$ averages with respect to the gaussian weight of equation (1.86), with our definition of Z_{G}:

$$\frac{\partial}{\partial\bar{\eta}_i} Z_{\mathrm{G}} = \det \mathbf{a} \langle\theta_i\rangle, \qquad (1.89)$$

$$\frac{\partial}{\partial\eta_i} Z_{\mathrm{G}} = \det \mathbf{a} \langle-\bar{\theta}_i\rangle. \qquad (1.90)$$

Note the sign in equation (1.90). Expressions (1.89,1.90) are the basis of perturbation theory. If we want to calculate the integral:

$$Z\left(\eta, \bar{\eta}\right) = \int \prod_i \mathrm{d}\bar{\theta}_i \mathrm{d}\theta_i \exp\left[\sum_{i,j=1}^{n} a_{ij}\bar{\theta}_i\theta_j + V\left(\bar{\theta}, \theta\right) + \sum_{i=1}^{n}\left(\bar{\eta}_i\theta_i + \bar{\theta}_i\eta_i\right)\right], \qquad (1.91)$$

we can formally expand in a power series of V and integrate term by term. We then find:

$$Z\left(\eta, \bar{\eta}\right) = \exp\left[V\left(-\frac{\partial}{\partial\eta}, \frac{\partial}{\partial\bar{\eta}}\right)\right] Z_{\mathrm{G}}\left(\eta, \bar{\eta}\right). \qquad (1.92)$$

Wick's theorem. Following the same lines, we can derive Wick's theorem for fermions. Defining now:

$$
\det a \langle \bar{\theta}_{i_1} \theta_{j_1} \bar{\theta}_{i_2} \theta_{j_2} \ldots \bar{\theta}_{i_n} \theta_{j_n} \rangle = \int \left(\prod_i d\theta_i d\bar{\theta}_i \right) \bar{\theta}_{i_1} \theta_{j_1} \ldots \bar{\theta}_{i_n} \theta_{j_n} \exp \left(\sum_{i,j=1}^{n} a_{ij} \bar{\theta}_i \theta_j \right),
\tag{1.93}
$$

we can write

$$
\det a \langle \bar{\theta}_{i_1} \theta_{j_1} \bar{\theta}_{i_2} \theta_{j_2} \ldots \bar{\theta}_{i_n} \theta_{j_n} \rangle = (-1)^n \left[\frac{\partial}{\partial \eta_{i_1}} \frac{\partial}{\partial \bar{\eta}_{j_1}} \cdots \frac{\partial}{\partial \eta_{i_n}} \frac{\partial}{\partial \bar{\eta}_{j_n}} Z_G (\eta, \bar{\eta}) \right]\Bigg|_{\eta = \bar{\eta} = 0}, \tag{1.94}
$$

or using equation (1.88):

$$
\langle \bar{\theta}_{i_1} \theta_{j_1} \bar{\theta}_{i_2} \theta_{j_2} \ldots \bar{\theta}_{i_n} \theta_{j_n} \rangle = (-1)^n
$$

$$
\times \left\{ \frac{\partial}{\partial \eta_{i_1}} \frac{\partial}{\partial \bar{\eta}_{j_1}} \cdots \frac{\partial}{\partial \eta_{i_n}} \frac{\partial}{\partial \bar{\eta}_{j_n}} \exp \left[-\sum_{i,j=1}^{n} \bar{\eta}_j \left(a^{-1} \right)_{ji} \eta_i \right] \right\}\Bigg|_{\eta = \bar{\eta} = 0}. \tag{1.95}
$$

After explicit differentiation (which is the same as integration) we obtain:

$$
\langle \bar{\theta}_{i_1} \theta_{j_1} \ldots \bar{\theta}_{i_n} \theta_{j_n} \rangle = \det a^{-1}_{j_l i_k}
$$

$$
= \sum_{\substack{\text{permutations} \\ P \text{ of } \{j_1 \ldots j_n\}}} \varepsilon(P) \left(a^{-1} \right)_{j_{P_1} i_1} \left(a^{-1} \right)_{j_{P_2} i_2} \cdots \left(a^{-1} \right)_{j_{P_n} i_n}, \tag{1.96}
$$

in which $\varepsilon(P)$ is the signature of the permutation P. This result is analogous to the expression (1.23) obtained in the case of the integral over complex commuting variables except for the sign.

Remark. Another property emphasizes the analogy between complex and Grassmann variables. It is possible to define a scalar product between entire functions by:

$$
(f, g) = \int \frac{dz \, d\bar{z}}{2i\pi} e^{-\bar{z}z} \overline{f(z)} \, g(z), \tag{1.97}
$$

where the integral is defined in Section 1.3. Normalizable entire functions are then vectors in a Hilbert space. From the point of view of Quantum Mechanics, the coefficients of the Taylor series expansion represent the components of the wave function $f(z)$ on different occupation number states (see Section 3.5).

If we impose that complex conjugation exchanges the generators θ and $\bar{\theta}$, we can in the same way introduce a scalar product between "analytic" Grassmann functions, i.e. functions of θ only:

$$
(f, g) = \int d\theta d\bar{\theta} \, e^{-\theta\bar{\theta}} \overline{f(\theta)} \, g(\theta). \tag{1.98}
$$

A function $f(\theta)$ is automatically an affine function of θ, thus

$$
f(\theta) = a + b\theta, \text{ with } \{a, b\} \in \mathbb{C}^2 - \{0, 0\} \quad \Rightarrow (f, f) = \bar{a}a + \bar{b}b > 0.
$$

Analytic Grassmann functions of one variable form a complex vector space of dimension two. This is in direct correspondence with the Fermi–Dirac statistics: the fermion occupation number can only be 0 or 1.

Bibliographical Notes

Integration over anticommuting variables was introduced in
  F.A. Berezin, *The Method of Second Quantization* (Academic Press, New York 1966).

Exercises

Exercise 1.1

Prove, using the method of fermion integration

$$\det\left[a_{ij} + s_i \delta_{ij}\right] = \det a_{ij} \left[\sum_n \frac{1}{n!} \sum_{i_1, i_2, \ldots, i_n} s_{i_1} s_{i_2} \ldots s_{i_n} \det a_{i_l i_k}^{-1}\right]. \tag{1.99}$$

Exercise 1.2

Show that the condition that P is a homomorphism of algebra is a necessary condition for the consistency of the definition of a differentiation operation in non-commutative algebras by calculating $D(ABC)$ in two different ways.

Exercise 1.3

Consider a graded algebra $\mathfrak{A}$ on $\mathbb{C}$ with one generator x satisfying $x^n = 0$. Construct a differential operator D linear, decreasing the degree by one unit, and consistent with this structure.

Exercise 1.4

Consider a graded algebra on $\mathbb{C}$ with generators x_i such that

$$x_i^n = 0, \tag{1.100}$$

and satisfying the commutation relation

$$x_j x_i = c x_i x_j \qquad \text{for } j > i,$$

where c is a complex number.
  Show that a necessary and sufficient condition for the property (1.100) to be independent on the basis in the vector space of generators is $c^n = 1$ with $c \neq 1$. In other words express the condition

$$\left(\sum a_i x_i\right)^n = 0 \qquad \forall a_i \neq \{0, 0, \ldots, 0\}.$$

2 EUCLIDEAN PATH INTEGRALS IN QUANTUM MECHANICS

In most of this work we shall study Quantum Field Theory in its euclidean formulation. This means that we shall in general discuss matrix elements of the quantum statistical operator $e^{-\beta H}$ (H is the hamiltonian and β the inverse temperature) rather than the quantum evolution operator $e^{-iHt/\hbar}$.

The former operator, whose trace is the quantum partition function $Z(\beta)$:

$$Z(\beta) = \operatorname{tr} e^{-\beta H}, \tag{2.1}$$

describes "evolution" in imaginary time, and in this sense most of the algebraic properties which will be derived, also apply to the real time evolution operator, explicit expressions being obtained by analytic continuation $\beta \mapsto it/\hbar$. (Therefore, to keep track of the $\hbar$ factors of real-time evolution, in this chapter we shall calculate $e^{-tH/\hbar}$.) This problem will be met in the calculation of scattering amplitudes.

Let us note, however, a specific property. The statistical operator provides a tool to determine the structure of the quantum mechanical ground state. For example if H is bounded from below, the ground state energy E_0 is given by:

$$E_0 = \lim_{\beta \to \infty} \left(-\frac{1}{\beta} \ln \operatorname{tr} e^{-\beta H} \right). \tag{2.2}$$

If the ground state is in addition unique and isolated, $e^{-\beta H}$ projects, for β large, onto the ground state vector $|0\rangle$:

$$e^{-\beta H} \underset{\beta \to \infty}{\sim} e^{-\beta E_0} |0\rangle \langle 0|. \tag{2.3}$$

Our basic tools to study first Quantum Mechanics and then Quantum Field Theory will be *path integrals* and *functional integrals*. We shall see that the path integral formulation of quantum mechanics is well suited to the study of systems with an arbitrary number of degrees of freedom. It allows therefore a smooth transition between quantum mechanics and quantum field theory.

One advantage of the euclidean formulation is that it is generally easier to define properly the path integral representing the operator $e^{-\beta H}$ (the Feynman–Kac formula) than $e^{-iHt/\hbar}$. Moreover the euclidean functional integral often leads to a simple and intuitive understanding of the structure of the ground state of systems with an infinite number of degrees of freedom. In particular it gives a natural interpretation to barrier penetration effects in the semiclassical approximation. Finally it emphasizes the deep connection between Quantum Field Theory and Statistical Mechanics of critical systems and phase transitions.

The main disadvantage of the euclidean presentation of quantum mechanics is that classical expressions have a somewhat unusual form because time is imaginary. We shall speak of euclidean action and euclidean lagrangian.

In this chapter we derive the path integral representation of the evolution operator only for hamiltonians of the simple form $p^2 + V(q)$. We then explicitly calculate the path integral corresponding to a harmonic oscillator in a time-dependent external force. This result can be used to reduce the evaluation of path integrals in the case of analytic potentials to perturbation theory. Comparing one-dimensional classical statistical mechanics and quantum statistical mechanics of the particle, we motivate the introduction of correlation functions and discuss their quantum mechanical meaning. Finally, returning to real time, we construct a path integral representation for the S-matrix. For illustration purpose we use this representation in the case of the Coulomb potential to derive the eikonal approximation.

2.1 Path Integrals: The General Idea

We consider a quantum mechanical hamiltonian $H(\hat{p}, \hat{q}; t)$ in which t is the time, and $\hat{p}$ and $\hat{q}$ are the usual momentum and position operators:

$$[\hat{q}, \hat{p}] = i\hbar \,, \tag{2.4}$$

The corresponding evolution operator in imaginary time $U(t'', t')$ satisfies Schrödinger's equation:

$$\hbar \frac{\partial U}{\partial t}(t, t') = -H(t)U(t, t') \,, \qquad U(t', t') = \mathbf{1} \,. \tag{2.5}$$

Since we have allowed time-dependent hamiltonians the operator U in general no longer has the simple form $\mathrm{e}^{-tH/\hbar}$. Because in most examples hamiltonians will be bounded from below but not from above, $U(t, t')$ will be defined only for $t \geq t'$.

The solution of equation (2.5) has the semi-group property:

$$U(t_3, t_2) U(t_2, t_1) = U(t_3, t_1) \,, \qquad t_3 \geq t_2 \geq t_1 \,. \tag{2.6}$$

The proof is simple. One considers the operator $\mathcal{U}(t, t_1)$ defined by:

$$\begin{cases} \mathcal{U}(t, t_1) = U(t, t_2) U(t_2, t_1) & \text{for } t \geq t_2 \,, \\ \mathcal{U}(t, t_1) = U(t, t_1) & \text{for } t \leq t_2 \,. \end{cases}$$

$\mathcal{U}(t, t_1)$ is continuous and satisfies both conditions (2.5) (for $t' = t_1$). It is thus identical to $U(t, t_1)$.

This property allows to write $U(t'', t')$ as a product of evolution operators on small time intervals:

$$U(t'', t') = \prod_{m=1}^{n} U\left[t' + m\varepsilon, t' + (m-1)\varepsilon\right], \qquad n\varepsilon = t'' - t' \,. \tag{2.7}$$

The product (2.7) is *time-ordered* according to the rule of equation (2.6). Note that the operator $U(t + \varepsilon, t)$ is the analogue of the transfer matrix in classical statistical mechanics (see Chapter 23).

If the *hamiltonian H is local in the basis in which the position operator $\hat{q}$ is diagonal*, which means that its matrix elements $\langle q_1 | H(t) | q_2 \rangle$ have a support restricted to $q_1 = q_2$, we can construct, using identity (2.7), a **path integral** representation for the matrix elements $\langle q'' | U(t'', t') | q' \rangle$.

First we write the identity (2.7), taking matrix elements:

$$\langle q'' | U(t'', t') | q' \rangle = \int \prod_{k=1}^{n-1} \mathrm{d}q_k \prod_{k=1}^{n} \langle q_k | U(t_k, t_{k-1}) | q_{k-1} \rangle \,, \tag{2.8}$$

with the conventions
$$t_k = t' + k\varepsilon, \qquad q_0 = q', \quad q_n = q''.$$

We then take the large n, small ε limit, and thus reduce the evaluation of expression (2.8) to the calculation of the matrix elements $\langle q| U (t + \varepsilon, t) |q' \rangle$ in the small time interval limit. In this limit, as a consequence of the locality of the hamiltonian, only the matrix elements with $|q - q'|$ small will contribute significantly to expression (2.8).

In this chapter we evaluate the expression (2.8) for a special class of hamiltonians. General hamiltonians will be discussed in Chapter 3.

2.2 Path Integral Representation of the Evolution Operator

Hamiltonians of the form

$$H = \hat{\mathbf{p}}^2/2m + V(\hat{\mathbf{q}}, t), \qquad [\hat{q}_\alpha, \hat{p}_\beta] = i\hbar\delta_{\alpha\beta}, \tag{2.9}$$

($\mathbf{p}, \mathbf{q}$ now being d-dimensional vectors) as well as all hamiltonians polynomial in $\hat{\mathbf{p}}$, are local. Let us write Schrödinger's equation for the evolution operator corresponding to the hamiltonian (2.9):

$$-\hbar\frac{\partial}{\partial t} \langle \mathbf{q}| U (t, t') |\mathbf{q}' \rangle = \left[-\frac{\hbar^2}{2m}\Delta_{\mathbf{q}} + V(\mathbf{q}, t) \right] \langle \mathbf{q}| U (t, t') |\mathbf{q}' \rangle, \tag{2.10}$$

with the boundary condition:

$$\langle \mathbf{q}| U (t', t') |\mathbf{q}' \rangle = \delta^{(d)}(\mathbf{q} - \mathbf{q}').$$

When the potential V vanishes, $\langle \mathbf{q}| U(t, t') |\mathbf{q}' \rangle$ is obtained by Fourier transformation:

$$\langle \mathbf{q}| U (t, t') |\mathbf{q}' \rangle = \int \frac{d^d p}{(2\pi\hbar)^d} \exp \left[\frac{1}{\hbar} \left(i(\mathbf{q} - \mathbf{q}') \cdot \mathbf{p} - (t - t')\frac{\mathbf{p}^2}{2m} \right) \right]$$

$$= \left(\frac{m}{2\pi\hbar(t - t')} \right)^{d/2} \exp \left[-\frac{1}{\hbar}\frac{m(\mathbf{q} - \mathbf{q}')^2}{2(t - t')} \right]. \tag{2.11}$$

To solve equation (2.10) in the small $t - t'$ limit, it is convenient to set

$$\varepsilon = t - t', \qquad \langle \mathbf{q}| U (t, t') |\mathbf{q}' \rangle = \exp \left[-\sigma (\mathbf{q}, \mathbf{q}'; t', \varepsilon) / \hbar \right].$$

The function $\sigma (\mathbf{q}, \mathbf{q}'; t', \varepsilon)$ at order ε is given by:

$$\sigma (\mathbf{q}, \mathbf{q}'; t', \varepsilon) = m\frac{(\mathbf{q} - \mathbf{q}')^2}{2\varepsilon} + \frac{d}{2}\hbar \ln \varepsilon + K + \varepsilon\sigma_1(\mathbf{q}, \mathbf{q}'; t') + O(\varepsilon^2), \tag{2.12}$$

$$\sigma_1(\mathbf{q}, \mathbf{q}'; t') = \int_0^1 d\lambda\, V(\mathbf{q}' + \lambda(\mathbf{q} - \mathbf{q}'), t'). \tag{2.13}$$

The constant K is determined by comparing the limit $\varepsilon \to 0$ with $\delta(\mathbf{q} - \mathbf{q}')$. This leads to:

$$\langle \mathbf{q}| U (t, t') |\mathbf{q}' \rangle = \left(\frac{m}{2\pi\hbar\varepsilon} \right)^{d/2} \exp \left[-\frac{1}{\hbar} \left(m\frac{(\mathbf{q} - \mathbf{q}')^2}{2\varepsilon} + \varepsilon\sigma_1(\mathbf{q}, \mathbf{q}'; t') + O(\varepsilon^2) \right) \right]. \tag{2.14}$$

Fundamental remark. We note that the most singular term in the ε-expansion of σ is $m(q - q')^2/2\varepsilon$ (independently of the potential). The support of the matrix element $\langle \mathbf{q}| U(t, t') |\mathbf{q}'\rangle$ is thus restricted to $|\mathbf{q}' - \mathbf{q}| \sim \sqrt{\varepsilon}$, which is typical of the brownian motion (for details see Chapter 4). For $|\mathbf{q}' - \mathbf{q}| \sim \sqrt{\varepsilon}$ we have

$$\sigma_1(\mathbf{q}, \mathbf{q}'; t') = V\big((\mathbf{q} + \mathbf{q}')/2, t'\big) + O(\varepsilon) = V(\mathbf{q}, t) + O\left(\varepsilon^{1/2}\right),$$

if the potential is differentiable. Hence replacing in the expansion (2.12) σ_1 by for instance $V(\mathbf{q}, t)$ modifies σ by terms of order $\varepsilon^{3/2}$ which, as we shall see below, are negligible. More generally for the three terms to be equivalent, the potential has to be at least continuous, other potentials require a special analysis.

From equation (2.8) we thus derive:

$$\langle \mathbf{q}''| U(t'', t') |\mathbf{q}'\rangle = \lim_{n \to \infty} \left(\frac{m}{2\pi\hbar\varepsilon}\right)^{dn/2} \int \prod_{k=1}^{n-1} \mathrm{d}^d q_k \, \exp\left[-S(\mathbf{q}, \varepsilon)/\hbar\right], \qquad (2.15)$$

with:

$$S(\mathbf{q}, \varepsilon) = \sum_{k=1}^{n} \left[m\frac{(\mathbf{q}_k - \mathbf{q}_{k-1})^2}{2\varepsilon} + \varepsilon V(\mathbf{q}_k, t_k) \right]. \qquad (2.16)$$

We observe that higher orders in ε in (2.12) give vanishing contributions in the small ε, $n\varepsilon$ fixed, limit. Let us now introduce a function $\mathbf{q}(t)$ which interpolates in time the variables $\mathbf{q}_k$:

$$\mathbf{q}_k \equiv \mathbf{q}(t_k).$$

The integral over the variables $\mathbf{q}_k$ is now equivalent to the integration over the points of the piecewise linear path represented in figure 2.1.

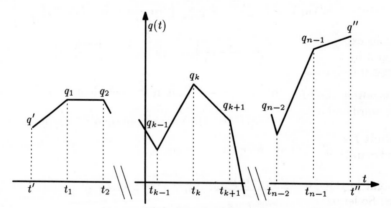

Fig. 2.1   A path contributing to integral (2.15).

In terms of the function $\mathbf{q}(t)$, the function $S(\mathbf{q}, \varepsilon)$ has the formal limit:

$$\lim_{\varepsilon \to 0} S(\mathbf{q}_k, \varepsilon) = S(\mathbf{q}) \equiv \int_{t'}^{t''} \mathrm{d}t \left[\tfrac{1}{2}m\dot{\mathbf{q}}^2(t) + V(\mathbf{q}(t), t) \right]. \qquad (2.17)$$

The functional $S(\mathbf{q})$ is the *euclidean action*, the integral of the "euclidean lagrangian" associated with the original hamiltonian.

We write the formal limit of expression (2.15) as:

$$\langle \mathbf{q}''| U (t'',t') |\mathbf{q}'\rangle = \int_{\mathbf{q}(t')=\mathbf{q}'}^{\mathbf{q}(t'')=\mathbf{q}''} [d\mathbf{q}(t)] \exp\left[-S(\mathbf{q})/\hbar\right], \tag{2.18}$$

and call it *path integral* since the r.h.s. involves a summation over all paths satisfying the prescribed boundary conditions, with a weight $\exp\left[-S/\hbar\right]$.

Note that we shall always write the integration measure $[d\mathbf{q}(t)]$ with brackets to distinguish path integrals from ordinary integrals.

The expression (2.18) immediately shows that paths which minimize the action (2.17), i.e. paths close to classical euclidean paths, which satisfy:

$$\frac{\delta S}{\delta q_i(t)} = 0, \quad \text{and} \quad \frac{\delta^2 S}{\delta q_i (t_1) \, \delta q_j (t_2)} \geq 0,$$

give the largest contributions to the path integral. This observation is at the basis of semiclassical approximations.

Discussion. Note that in action (2.17) the two terms play quite different roles. The kinetic term $\int dt \, \dot{\mathbf{q}}^2$ selects paths regular enough, i.e., as one can see from expression (2.16), those for which $[\mathbf{q}(t+\varepsilon) - \mathbf{q}(t)]^2 /\varepsilon$ remains finite when ε goes to zero (which in particular implies that the paths are continuous). The kinetic term really is part of the functional measure and determines the functional space over which to integrate. It is essential to the very existence of the path integral. The relevant paths are typical of the brownian motion (see Chapter 4). The potential weights paths according to the value of $\mathbf{q}(t)$ at each time, and determines the physical properties of the theory.

Note also that in the symbol $[d\mathbf{q}(t)]$ is buried an infinite normalization factor $\mathcal{N}$:

$$\mathcal{N} = \left(\frac{m}{2\pi\hbar\varepsilon}\right)^{dn/2}.$$

Therefore in explicit calculations we shall always normalize the result by dividing the path integral by a reference path integral for which the result is exactly known (the free motion $V \equiv 0$ for example).

Generalization. The generalization of the previous construction to several particles is straightforward and leads to a path integral involving the corresponding euclidean action.

2.3 Explicit Evaluation of a Simple Path Integral: The Harmonic Oscillator in an External Force

At this point in the discussion it may not be very clear how useful the concept of path integral is. So let us work out an example which shows that even in simple cases the path integral representation may sometimes simplify considerably algebraic manipulations. We consider the hamiltonian of a harmonic oscillator coupled linearly to a time-dependent external force:

$$H = \frac{1}{2m}p^2 + \tfrac{1}{2}m\omega^2 q^2 - qf(t). \tag{2.19}$$

The corresponding euclidean action $S_{\mathrm{G}}(q)$ is:

$$S_{\mathrm{G}} = \int_{t'}^{t''} \left[\tfrac{1}{2}m\dot{q}^2 + \tfrac{1}{2}m\omega^2 q^2 - qf(t)\right] dt, \tag{2.20}$$

which is a quadratic form in $q(t)$. When this expression is inserted into equation (2.18), it leads to a gaussian integral which can be calculated following the method explained in Section 1.1. We first eliminate the linear term in expression (2.20), changing variables:

$$q(t) = q_c(t) + r(t), \tag{2.21}$$

in which q_c is the solution of the classical equation of motion:

$$m\ddot{q}_c - m\omega^2 q_c + f(t) = 0, \qquad \text{with} \quad q_c(t') = q', \; q_c(t'') = q''. \tag{2.22}$$

The new path $r(t)$ then satisfies the boundary conditions:

$$r(t') = r(t'') = 0. \tag{2.23}$$

As a consequence of equations (2.22), the action (2.20) can be written as the sum of two terms:

$$S_G(q) = S(q_c) + S_0(r), \tag{2.24}$$

in which $S(q_c)$ is the classical action on the trajectory and $S_0(r)$ is simply:

$$S_0(r) = \int_{t'}^{t''} \left[\tfrac{1}{2}m\dot{r}^2(t) + \tfrac{1}{2}m\omega^2 r^2(t) \right] dt. \tag{2.25}$$

Thus the integral over $r(t)$ no longer depends on $f(t)$ and q', q''. We now first calculate explicitly $S(q_c)$ and then evaluate the remaining path integral.

The classical action. Owing to the linearity of equation (2.22) we can write the solution q_c as the sum of two terms, one depending on the boundary conditions q', q'', the other on $f(t)$. Setting $\beta = t'' - t'$ and

$$\sinh(\omega\beta)\, q_c(t) = x_1(t) + x_2(t),$$

we find

$$x_1(t) = q' \sinh\left[\omega(t'' - t)\right] + q'' \sinh\left[\omega(t - t')\right], \tag{2.26}$$

$$x_2(t) = \frac{1}{\omega m} \int_{t'}^{t''} \Big[\theta(t - u) \sinh\big(\omega(t'' - t)\big) \sinh\big(\omega(u - t')\big)$$
$$+ \theta(u - t) \sinh\big(\omega(t'' - u)\big) \sinh\big(\omega(t - t')\big) \Big]\, f(u)\,du, \tag{2.27}$$

$\theta(v)$ being the usual Heaviside step function.

To calculate $S(q_c)$ it is useful to note:

$$\int \dot{q}^2\, dt = q\dot{q} - \int q\ddot{q}\, dt,$$

and use the equation of motion (2.22). One then finds:

$$S(q_c) = \frac{1}{\omega m\, \sinh\omega\beta}\, (I_1 + I_2 + I_3), \tag{2.28}$$

$$
\begin{cases}
I_1 = \dfrac{(m\omega)^2}{2} \left[\left(q'^2 + q''^2 \right) \cosh \omega\beta - 2q'q'' \right], \\[2mm]
I_2 = -m\omega \displaystyle\int_{t'}^{t''} du \, \left[q'' \sinh \omega \left(u - t' \right) + q' \sinh \omega \left(t'' - u \right) \right] f\left(u \right), \\[2mm]
I_3 = \displaystyle\int_{t'}^{t''} du \, dv \, \theta \left(v - u \right) \sinh \omega \left(t'' - v \right) \sinh \omega \left(t' - u \right) f\left(u \right) f\left(v \right).
\end{cases}
\tag{2.29}
$$

The path integral. We still want to calculate the gaussian integral over $r(t)$ because it yields a ω-dependent normalization factor

$$
\mathcal{N}(\omega) = \int [dr(t)] \exp \left[-\frac{1}{\hbar} \int_{t'}^{t''} dt \left(\tfrac{1}{2}m\dot{r}^2 \left(t \right) + \tfrac{1}{2}m\omega^2 r^2 \left(t \right) \right) \right],
\tag{2.30}
$$

with $r(t') = r(t'') = 0$.

The boundary conditions imply $\int \dot{r}^2 \, dt = -\int r\ddot{r} \, dt$, and therefore the quadratic form in the exponential can be rewritten:

$$
\frac{1}{\hbar} \int_{t'}^{t''} dt \left[\tfrac{1}{2}m\dot{r}^2 \left(t \right) + \tfrac{1}{2}m\omega^2 r^2 \left(t \right) \right] = \frac{m}{2\hbar} \int dt_1 \, dt_2 \, r\left(t_1 \right) r\left(t_2 \right) \Lambda \left(t_1, t_2 \right).
$$

with:

$$
\Lambda \left(t_1, t_2 \right) = \left(-\frac{d^2}{(dt_1)^2} + \omega^2 \right) \delta \left(t_1 - t_2 \right).
$$

The gaussian integral is proportional to the determinant of the hermitian differential operator $\Lambda \left(t_1, t_2 \right)$ to the power $-1/2$ (we omit the multiplicative factor $m/\hbar$ which is ω-independent). We can calculate the determinant by diagonalizing the operator. We have to solve the equation:

$$
\left[-d^2/(dt)^2 + \omega^2 \right] r_n \left(t \right) = \lambda_n r_n(t),
\tag{2.31}
$$

in the space of functions satisfying the conditions:

$$
r_n \left(t' \right) = r_n \left(t'' \right) = 0.
$$

The normalized eigenfunctions are thus:

$$
r_n(t) = \sqrt{2/\beta} \sin \left[\pi n \left(t - t' \right) / \beta \right], \quad n > 0,
\tag{2.32}
$$

and the corresponding eigenvalues λ_n are:

$$
\lambda_n = \frac{n^2\pi^2}{\beta^2} + \omega^2.
\tag{2.33}
$$

The determinant of the operator $\left[-\left(d/dt \right)^2 + \omega^2 \right]$ is the product of eigenvalues:

$$
\det \left(-\frac{d^2}{(dt)^2} + \omega^2 \right) = \prod_{n>0} \left(\frac{n^2\pi^2}{\beta^2} + \omega^2 \right).
\tag{2.34}
$$

Evaluating explicitly the r.h.s. we get an expression for the path integral.

A difficulty remains: the infinite product in the r.h.s. of equation (2.34) is divergent. However the ratio:

$$\det \left[\left(-\frac{d^2}{dt^2} + \omega^2 \right) \Big/ \left(-\frac{d^2}{dt^2} \right) \right] = \prod_{n>0} \left(1 + \frac{\omega^2 \beta^2}{n^2 \pi^2} \right) = \frac{\sinh \omega \beta}{\omega \beta} \qquad (2.35)$$

is finite.

Thus the expression for the evolution operator is:

$$\langle q'' | U(t'', t') | q' \rangle = \mathcal{N}' \sqrt{\frac{\omega \beta}{\sinh \omega \beta}} \exp\left[-S(q_c)/\hbar \right]. \qquad (2.36)$$

The new normalization constant $\mathcal{N}'$ is independent of ω and can be fixed by comparing to the free case $\omega = 0$ (see equation (2.11)):

$$\lim_{\substack{\omega \to 0 \\ f \to 0}} \langle q'' | U(t'', t') | q' \rangle = \left(\frac{m}{2\pi \hbar \beta} \right)^{1/2} \exp\left[-\frac{m}{2\hbar \beta} (q' - q'')^2 \right],$$

obtained directly by solving Schrödinger's equation.

In this limit the classical action $S(q_c)$ becomes:

$$S(q_c) = \frac{m}{2\beta} (q' - q'')^2 .$$

Together with equation (2.36) this determines the constant $\mathcal{N}'$. The complete result is:

$$\langle q'' | U(t'', t') | q' \rangle = \langle q'' | U_0(t'', t') | q' \rangle$$

$$\times \exp\left[\frac{1}{\hbar} \int_{t'}^{t''} L(u) f(u) \, du + \frac{1}{2\hbar} \int_{t'}^{t''} K(u, v) f(u) f(v) \, du \, dv \right], (2.37)$$

with the definitions:

$$\langle q'' | U_0(t'', t') | q' \rangle = \left[\frac{m\omega}{2\pi \hbar \, \sinh \omega \beta} \right]^{1/2}$$

$$\times \exp\left\{ -\frac{m\omega}{2\hbar \sinh \omega \beta} \left[(q'^2 + q''^2) \cosh \omega \beta - 2q' q'' \right] \right\}, \qquad (2.38)$$

$$L(t) = \frac{1}{\sinh \omega \beta} \left[q'' \sinh \omega (t - t') + q' \sinh \omega (t'' - t) \right], \qquad (2.39)$$

$$K(t_1, t_2) = \frac{1}{m\omega \, \sinh \omega \beta} \left[\theta(t_2 - t_1) \sinh \omega (t'' - t_2) \sinh \omega (t_1 - t') \right.$$

$$\left. + \theta(t_1 - t_2) \sinh \omega (t'' - t_1) \sinh \omega (t_2 - t') \right]. \qquad (2.40)$$

Remark. An equivalent and useful presentation of the evaluation of the $r(t)$ path integral is the following: We expand, in the Hilbert space $\mathcal{L}^2$, the function $r(t)$ on the orthonormal basis formed by the functions $r_n(t)$:

$$r(t) = \sum_{n>0} c_n r_n(t).$$

The functional measure of integration is then

$$[dr(t)] = \prod_n dc_n , \tag{2.41}$$

and the quadratic form $S_0\left(r(t)\right)$ expressed in terms of the variables c_n reads:

$$S_0\left(r(t)\right) = \frac{m}{2\hbar} \sum_{n>0} \lambda_n c_n^2.$$

The gaussian integration over the variables c_n yields a result consistent with equation (2.34) up to an ω-independent infinite normalization factor. The functional measure can thus also be defined as the integration over the coefficients of the expansion of the function describing the path over an orthonormal basis in $\mathcal{L}^2$ as in equation (2.41). This alternative definition will be useful for the instanton calculations in Chapter 37.

The partition function. From the equations (2.37,2.38) we can calculate the partition function. Taking the trace of $U\left(\hbar\beta/2, -\hbar\beta/2\right)$ we find:

$$Z(\beta) = \operatorname{tr} U\left(\hbar\beta/2, -\hbar\beta/2\right) \equiv \int dq \,\langle q|\, U\left(\hbar\beta/2, -\hbar\beta/2\right) |q\rangle$$

$$= Z_0(\beta) \exp\left[\frac{1}{2m\hbar} \int_{-\hbar\beta/2}^{\hbar\beta/2} du\, dv\, \Delta(v-u)f(v)f(u)\right], \tag{2.42}$$

with:

$$Z_0(\beta) = \operatorname{tr} U_0(\hbar\beta/2, -\hbar\beta/2) = e^{-\beta\hbar\omega/2}\left[1 - e^{-\beta\hbar\omega}\right]^{-1}$$

and

$$\Delta(t) = \frac{1}{2\omega \,\sinh(\hbar\omega\beta/2)} \cosh\left(\omega(\hbar\beta/2 - |t|)\right). \tag{2.43}$$

The function $\Delta(t)$ is the solution of the equation:

$$-\ddot{\Delta} + \omega^2 \Delta = \delta(t),$$

with periodic boundary conditions $\Delta\left(\hbar\beta/2\right) = \Delta\left(-\hbar\beta/2\right)$.

2.4 Perturbed Harmonic Oscillator

Let us consider the hamiltonian H:

$$H = \frac{1}{2m}p^2 + \tfrac{1}{2}m\omega^2 q^2 + V_{\rm I}(q), \tag{2.44}$$

in which we impose to $V_{\rm I}(q)$ only to be expandable in powers of q:

$$V_{\rm I}(q) = \sum_n v_n q^n.$$

The corresponding evolution operator is given by (in this section we set $\hbar = 1$):

$$\langle q''|\, U\left(t'', t'\right) |q'\rangle = \int [dq] \exp\left[-\int_{t'}^{t''} \left(\tfrac{1}{2}m\dot{q}^2 + \tfrac{1}{2}m\omega^2 q^2 + V_{\rm I}(q)\right) dt\right]. \tag{2.45}$$

To evaluate expression (2.45), we apply to the expression (2.37) the generalization of identities (1.10,1.12). In Section 2.3 we have calculated the evolution operator U_G corresponding to action (2.20):

$$\langle q'' | U_G (t'',t') | q' \rangle = \int [dq] \exp \left[- \int_{t'}^{t''} \left(\tfrac{1}{2} m \dot{q}^2 + \tfrac{1}{2} m \omega^2 q^2 - q f(t) \right) dt \right]. \qquad (2.46)$$

If we differentiate with respect to $f(t)$ we obtain:

$$\frac{\delta}{\delta f(t_1)} \langle q'' | U_G (t'',t') | q' \rangle = \int [dq] \, q(t_1) \exp \left[-S_G(q,f) \right]. \qquad (2.47)$$

Therefore by differentiating p times with respect to $f(t)$ we can generate any product $q(t_1)\ldots q(t_p)$:

$$\prod_{j=1}^{p} \frac{\delta}{\delta f(t_j)} \langle q'' | U_G (t'',t') | q' \rangle = \int [dq] \prod_{j=1}^{p} q(t_j) \exp \left[-S_G(q,f) \right]. \qquad (2.48)$$

More generally a differential operator $F \left(\delta / \delta f(t) \right)$ generates in the r.h.s. the functional $F(q(t))$.

Applied to $\exp \left[- \int_{t'}^{t''} V_I(q(t)) \, dt \right]$, these identities lead to:

$$\langle q'' | U (t'',t') | q' \rangle = \left\{ \exp \left[- \int_{t'}^{t''} dt \, V_I \left(\frac{\delta}{\delta f(t)} \right) \right] \langle q'' | U_G (t'',t') | q' \rangle \right\} \bigg|_{f=0}. \qquad (2.49)$$

Using then the result of equation (2.37) we obtain:

$$\langle q'' | U (t'',t') | q' \rangle = \langle q'' | U_0 (t'',t') | q' \rangle$$
$$\times \exp \left[- \int_{t'}^{t''} dt \, V_I \left(\frac{\delta}{\delta f(t)} \right) \right] \exp \left[(L,f) + \tfrac{1}{2} (f,Kf) \right] \bigg|_{f=0}, \qquad (2.50)$$

(expression (2.37) has been rewritten in symbolic notation).

If $V_I(q)$ is a polynomial we can expand (2.50) in powers of $V_I(q)$ and calculate systematically the successive terms in the expansion using Wick's theorem (1.9). This is the basis of perturbation theory.

Example:

$$V_I(q) = \lambda q^4.$$

At first order in λ:

$$\langle q'' | U (t'',t') | q' \rangle = \langle q'' | U_0(t'',t') | q' \rangle$$
$$\times \left[1 - \lambda \int_{t'}^{t''} \left(\frac{\delta}{\delta f(t)} \right)^4 dt \right] \exp \left[(L,f) + \tfrac{1}{2} (f,Kf) \right] \bigg|_{f=0} + O(\lambda^2). \qquad (2.51)$$

Differentiating 4 times with respect to f and setting f to zero, we pick out the terms of degree 4 in f in the expansion of

$$\exp \left[(L,f) + \tfrac{1}{2} (f,Kf) \right].$$

Note that

$$\left[\frac{\delta}{\delta f(t)}\right]^4 \int v\left(t_1, t_2, t_3, t_4\right) f\left(t_1\right) f\left(t_2\right) f\left(t_3\right) f\left(t_4\right) dt_1 dt_2 dt_3 dt_4 = 4!\, v\left(t, t, t, t\right).$$

In the expansion of the exponential the terms of degree 4 in f are:

$$\frac{1}{24}\left(L, f\right)^4 + \frac{1}{4}\left(L, f\right)^2\left(f, Kf\right) + \frac{1}{8}\left(f, Kf\right)^2.$$

Combining these various expressions we find:

$$\frac{\langle q''|\, U\, |q'\rangle}{\langle q''|\, U_0\, |q'\rangle} = 1 - \lambda \int_{t'}^{t''} dt\left[L^4(t) + 6K\left(t, t\right) L^2(t) + 3K^2\left(t, t\right)\right] + O\left(\lambda^2\right). \qquad (2.52)$$

The classical limit and perturbation theory. A perturbation expansion can be generated for any decomposition of the potential into the sum of a quadratic term and a remainder, as in equation (2.44). However if we want the perturbative expansion to be associated with a formal expansion in powers of $\hbar$, then we infer from expression (2.18) that we have to expand the action, and thus the potential around a minimum. Calling q_0 a minimum of the potential we then write:

$$V\left(q\right) = V\left(q_0\right) + \tfrac{1}{2}V''\left(q_0\right)\left(q - q_0\right)^2 + V_{\mathrm{I}}(q).$$

The expansion in powers of the coefficients of V_{I} can then be organized as an expansion in powers of $\hbar$, called loopwise expansion. Problems associated with a possible degeneracy of the classical minimum will be examined in Chapter 40.

2.5 Partition Function. Correlation Functions

In the last part of the chapter we restrict ourselves, for simplicity reasons, to *time-independent* potentials.

The partition function. The quantum partition function $Z(\beta) = \operatorname{tr} e^{-\beta H}$ has a path integral representation which is immediately deduced from the representation of the evolution operator

$$Z(\beta) = \operatorname{tr} e^{-\beta H} \equiv \operatorname{tr} U(\hbar\beta, 0) = \int dq\, \langle q|\, U(\hbar\beta, 0)\, |q\rangle = \int [dq(t)] \exp\left[-S(q)/\hbar\right], \quad (2.53)$$

where the paths now satisfy periodic boundary conditions: $q(0) = q(\hbar\beta)$, and one integrates over all values of $q(0)$. For what follows it is convenient to rescale time $t \mapsto t/\hbar$. The action then reads

$$S(q)/\hbar = \int_0^\beta dt\left[\tfrac{1}{2}m\dot{\mathbf{q}}^2(t)/\hbar^2 + V\left(\mathbf{q}(t)\right)\right].$$

The large β limit. One can also set the boundary conditions at $t = \pm\beta/2$. Since the action is time translation invariant the result is the same. However in the formal large β limit (relevant for the ground state energy) in the first case one is led to integrate over paths on the positive real line with the boundary condition $q(0) = 0$, while in the second case one obtains an explicitly time translation invariant formalism on the whole real line. The latter formalism is clearly simpler.

2.5.1 Quantum and classical statistical mechanics

The time-discretized form of the path integral has an interpretation in terms of one-dimensional classical statistical mechanics on the lattice. Let us indeed consider the partition function Z

$$Z_n(T) = \int \prod_{k=1}^{n} d\rho(q_k) \exp\left[-E(q_i)/T\right], \tag{2.54}$$

where q_i characterizes the configuration on the site i of a 1D lattice, n is the size of the lattice, T the temperature, $d\rho(q)$ the distribution of the q variable and $E(q_i)$ the configuration energy which we choose of the special form

$$E(q) = \sum_{i=1}^{n} \tfrac{1}{2} J \left(q_i - q_{i-1}\right)^2, \quad J > 0. \tag{2.55}$$

Finally we assume periodic boundary conditions: $q_n = q_0$. We then recognize an expression very similar to expression (2.15). Let us show that the continuum limit $\varepsilon \to 0$ of the quantum expression corresponds here to the low temperature limit, large volume limit. It is convenient to set

$$d\rho(q) = e^{-v(q)} \, dq.$$

The simplest example corresponds to a function $v(q)$ with a unique minimum at $q = 0$ where it is regular

$$v(q) = \tfrac{1}{2} v_2 q^2 + O\left(q^3\right).$$

For T small the leading configuration is obtained by minimizing the energy (2.54) and corresponds to all q_i's equal. Then for n large only the neighbourhood of the configuration which is a maximum of the distribution, i.e. $q_i = 0$, contributes to the partition function. We find

$$Z_n(T) \sim \int \prod_{k=1}^{n} dq_k \exp\left[-\tilde{E}(q_i)\right], \tag{2.56}$$

with

$$\tilde{E}(q) = \tfrac{1}{2} \sum_{i=1}^{n} \left[J \left(q_i - q_{i-1}\right)^2 / T + v_2 q_i^2 \right]. \tag{2.57}$$

After the rescaling $q_i \mapsto q_i (Tv_2/J)^{1/4}$ we recognize that $Z_n(T)$ is also the quantum partition function $Z(\beta)$ (2.15), with $\varepsilon = \sqrt{Tv_2/J}$, $\hbar = m = 1$, of the hamiltonian H of an harmonic oscillator,

$$H = \tfrac{1}{2} \left(p^2 + q^2\right). \tag{2.58}$$

The continuum limit corresponds to the low temperature, $\varepsilon \propto \sqrt{T} \to 0$, limit.

2.5.2 Correlation functions

The analogy between classical and quantum statistical mechanics suggests to consider the integrand in path integral (2.18) as a probability distribution. In particular we can use it to calculate averages, like the $q(t)$ correlation functions:

$$\langle q(t_1) q(t_2) \ldots q(t_n) \rangle_\beta = Z^{-1}(\beta) \int [dq] \, q(t_1) \ldots q(t_n) \, e^{-S(q)/\hbar}, \tag{2.59}$$

where we have assumed periodic boundary conditions, $q(0) = q(\beta)$. The normalization $Z(\beta)$ is fixed by requiring that the average $\langle 1 \rangle$ is 1. We recognize the quantum partition function (2.53), $Z(\beta) = \text{tr}\, e^{-\beta H}$.

When the euclidean action is not real, averages of the form (2.59) no longer have a probabilistic interpretation. For reasons which we shall explain below, they still are useful quantities to consider, and we shall still call them correlation functions.

Operator interpretation. The expression (2.59) has an operator interpretation. Let us assume that we have ordered times such that:

$$0 \leq t_1 \leq t_2 \leq \ldots \leq t_n \leq \beta. \tag{2.60}$$

We decompose the interval $(0, \beta)$ into $n + 1$ subintervals $(0, t_1)$, (t_1, t_2), ...,(t_n, β). The total action is the sum of the corresponding contributions:

$$S(q)/\hbar = \sum_{i=1}^{n+1} \int_{t_{i-1}}^{t_i} \left[\tfrac{1}{2}m\dot{q}^2/\hbar^2 + V(q)\right] dt, \quad \text{with} \quad t_0 = 0, \quad t_{n+1} = \beta. \tag{2.61}$$

We then rewrite integral (2.59) with the use of the identity:

$$\prod_{i=1}^{n} q(t_i) = \int \prod_{i=1}^{n} dq_i \, \delta\left[q(t_i) - q_i\right] q_i.$$

The path integral then factorizes into a product of path integrals corresponding to the different subintervals. Returning to the very definition of the path integral (equations (2.17,2.18)), we see that the numerator in expression (2.59) is exactly (recalling the ordering (2.6))

$$\text{tr}\left[e^{-(\beta - t_n)H} \, \hat{q} \, e^{-(t_n - t_{n-1})H} \, \hat{q} \ldots e^{-(t_2 - t_1)H} \, \hat{q} \, e^{-t_1 H}\right].$$

Introducing the operator $Q(t)$, Heisenberg representation of the operator $\hat{q}$,

$$Q(t) = e^{tH} \, \hat{q} \, e^{-tH}, \tag{2.62}$$

(for t real and positive e^{tH} does not necessarily exist and then the definition is somewhat formal) we can write:

$$\langle q(t_1) \ldots q(t_n)\rangle_\beta = Z^{-1}(\beta) \, \text{tr}\left[e^{-\beta H} \, Q(t_n) \ldots Q(t_1)\right]. \tag{2.63}$$

The order of the operators in the r.h.s. reflects the time-ordering (2.60).

If we set all times equal we simply find the static statistical average of a power of the coordinate q in finite temperature quantum mechanics. More generally, after analytic continuation to real times t_i, (but not in β) we find the time-dependent quantum correlation functions.

If we introduce a time-ordering operation T, which to a set of time-dependent operators $A_1(t_1)$,...,$A_l(t_l)$ associates the time-ordered product (T-product) of these operators, for example for $l = 2$:

$$T\left[A_1(t_1) A_2(t_2)\right] = A_1(t_1) A_2(t_2)\, \theta(t_1 - t_2) + A_2(t_2) A_1(t_1)\, \theta(t_2 - t_1),$$

we can rewrite expression (2.63) irrespective now of the order between the times $t_1, \ldots, t_n$:

$$\langle q(t_1) \ldots q(t_n) \rangle = Z^{-1}(\beta) \, \mathrm{tr} \left\{ e^{-\beta H} \, \mathrm{T} \left[Q(t_1) \, Q(t_2) \ldots Q(t_n) \right] \right\}. \qquad (2.64)$$

Finally if H has a unique isolated ground state $|0\rangle$, then equation (2.2) shows that:

$$\langle q(t_1) \ldots q(t_n) \rangle \underset{\beta \to \infty}{=} \langle 0 | \mathrm{T} \left[Q(t_1) \, Q(t_2) \ldots Q(t_n) \right] | 0 \rangle. \qquad (2.65)$$

These time-ordered products are the analytic continuation in imaginary time of the time-ordered products introduced in the real time formulation of quantum field theory. After analytic continuation, they thus generate Green's functions from which one can for instance calculate scattering amplitudes (see Section 7.3). However, physically reasonable theories from the point of view of Green's functions, do not always correspond to real euclidean actions and then their analytic continuations are correlation functions only in a formal sense.

Remark. We have assumed in taking the large β limit that the ground state is unique. The existence of phase transitions, as will be discussed in Chapter 23, is related to a possible ground state degeneracy.

Perturbation theory. We have shown in Section 2.4 how to calculate perturbatively the path integral for any hamiltonian of the form $p^2/2m + V(q)$ in terms of path integral (2.46). The argument can immediately be generalized for the corresponding correlation functions.

2.6 Evolution in Real Time. The S-matrix

To calculate matrix elements of the evolution operator in real time we can proceed by analytic continuation replacing β by $\beta \, e^{i\varphi}$ in all expressions and rotating in the complex β plane from $\varphi = 0$ to $\varphi = \pi/2$ in the positive direction.

The evolution operator $U(t'', t')$ then satisfies Schrödinger's equation:

$$i\hbar \frac{\partial U}{\partial t}(t, t') = H(t) \, U(t, t'), \qquad (2.66)$$

which, for the example of a hamiltonian of the form

$$H = p^2/2m + V(q, t),$$

leads to:

$$\langle q'' | U(t'', t') | q' \rangle = \int_{q(t')=q'}^{q(t'')=q''} [dq(t)] \exp \left[i \mathcal{A}(q)/\hbar \right]. \qquad (2.67)$$

The function $\mathcal{A}(q)$ is now the usual classical action, integral of the lagrangian:

$$\mathcal{A}(q) = \int_{t'}^{t''} dt \left[\tfrac{1}{2} m \dot{q}^2 - V(q, t) \right]. \qquad (2.68)$$

Expression (2.67) relates in a beautiful way classical and quantum mechanics. In the quantum mechanical evolution all paths contribute but they are weighted with the complex weight $e^{i\mathcal{A}/\hbar}$. Therefore paths close to extrema of the action, i.e. classical paths, give the largest contributions to the path integral. In particular if the value of the classical action on classical paths is large compared to $\hbar$, regions in phase space close to the classical paths dominate the path integral.

The S-matrix. The scattering S-matrix is obtained by comparing the quantum evolution to the evolution in the absence of the potential. More precisely the S-matrix can be defined as the limit of the evolution operator in the interaction representation. For a time-independent potential:

$$S = \lim_{\substack{t' \to -\infty \\ t'' \to +\infty}} e^{iH_0 t''/\hbar} e^{-iH(t''-t')/\hbar} e^{-iH_0 t'/\hbar}, \tag{2.69}$$

in which H_0 is the free hamiltonian corresponding to H. For example in $\mathbb{R}^d$:

$$H_0 = \mathbf{p}^2/2m, \qquad H = \mathbf{p}^2/2m + V(\mathbf{q}).$$

We assume that the potential decreases fast enough for large $|\mathbf{q}|$ for the S-matrix to exist.

Note that in momentum representation the relation (2.69) between the S-matrix and the evolution operator takes the form:

$$\langle \mathbf{p}''|S|\mathbf{p}'\rangle = \lim_{\substack{t' \to -\infty \\ t'' \to +\infty}} e^{iE''t''/\hbar} \langle \mathbf{p}''|U(t'',t')|\mathbf{p}'\rangle e^{-iE't'/\hbar}, \tag{2.70}$$

where

$$E' = E(\mathbf{p}'), \quad E'' = E(\mathbf{p}''), \qquad E(\mathbf{p}) = \mathbf{p}^2/2m.$$

However the limit has to be understood mathematically in the sense of distributions (one should use test functions, i.e. wave packets).

We finally recall that the S-matrix in momentum representation is in general parametrized in terms of the scattering matrix T as:

$$\langle \mathbf{p}''|S|\mathbf{p}'\rangle = \delta^d(\mathbf{p}'' - \mathbf{p}') - 2i\pi\delta(E'' - E')\langle \mathbf{p}''|T|\mathbf{p}'\rangle. \tag{2.71}$$

2.6.1 Path integral and S-matrix in perturbation theory

Let us show first how the path integral for the evolution operator can be expanded in powers of the potential. The path integral formalism then organizes the perturbative expansion in the same way as the operator formalism recalled in Section A2.1.

We thus consider the hamiltonian H

$$H = \tfrac{1}{2}p^2 + V(x), \tag{2.72}$$

(setting for convenience $m = \hbar = 1$). The classical actions corresponding to the free hamiltonian and to H are

$$\mathcal{A}_0(x) = \tfrac{1}{2}\int_{t'}^{t''} \dot{x}^2(t)\mathrm{d}t, \qquad \mathcal{A}(x) = \int_{t'}^{t''} \left[\tfrac{1}{2}\dot{x}^2(t) - V(x(t))\right]\mathrm{d}t. \tag{2.73}$$

We want to calculate the evolution operator in a power series of the potential $V(x)$ and then obtain the S-matrix. We now assume that the potential has a Fourier representation

$$V(x) = \int \mathrm{d}k\, \tilde{V}(k)\, e^{ikx}. \tag{2.74}$$

The path integral (2.67), expanded in powers of V, then takes the form

$$\langle x''|\, U\left(t'',t'\right)|x'\rangle = \int_{x(t')=x'}^{x(t'')=x''} [\mathrm{d}x(t)] \exp\left[i\mathcal{A}(x)\right]$$

$$= \int_{x(t')=x'}^{x(t'')=x''} [\mathrm{d}x(t)]\, e^{i\mathcal{A}_0(x)} \sum_l \frac{(-i)^l}{l!} \left[\int_{t'}^{t''} V\left(x(t)\right)\mathrm{d}t\right]^l. \quad (2.75)$$

If we now introduce the Fourier representation (2.74) in the path integral we see that we have reduced perturbation theory to the calculation of gaussian integrals. The lth term becomes

$$\langle x''|\, U^{(l)}\left(t'',t'\right)|x'\rangle = \frac{(-i)^l}{l!} \int \prod_{j=1}^{l} \tilde{V}(k_j)\mathrm{d}k_j \int_{t'}^{t''} \prod_j \mathrm{d}\tau_j$$

$$\times \int_{x(t')=x'}^{x(t'')=x''} [\mathrm{d}x(t)] \exp i\left[\int_{t'}^{t''} \tfrac{1}{2}\dot{x}^2(t)\mathrm{d}t + \sum_j k_j x(\tau_j)\right]. \quad (2.76)$$

The path integral gives a physical interpretation to the terms in the perturbative expansion: The leading path contributing to the lth order is a succession of straight lines, where at times $\tau_1, \ldots, \tau_l$ the momentum changes by amounts $k_1, \ldots, k_l$. One then integrates the corresponding phase factor over all times and over all momenta weighted with $\tilde{V}(k)$.

The term of order zero in V yields $\delta(p'' - p')$, let us then calculate explicitly the first order. We can write the evolution operator as the products of two free evolutions from t' to τ and from τ to t''. We then Fourier transform with respect to x' and x''. We find

$$\langle p''|\, U^{(1)}(t'',t')|p'\rangle = -i \int \mathrm{d}k\, \tilde{V}(k) \int_{t'}^{t''} \mathrm{d}\tau$$

$$\times \int \mathrm{d}x(\tau)\, e^{-ip''^2(t''-\tau)/2} \langle p''|x(\tau)\rangle\, e^{ikx(\tau)}\, \langle x(\tau)|p'\rangle\, e^{-ip'^2(\tau-t')/2}. \quad (2.77)$$

The integration over the variable $x(\tau)$ yields $\delta(k + p' - p'')$. We then find for the first order contribution $S^{(1)}$ to the S-matrix

$$\langle p''|\, S^{(1)}\,|p'\rangle = \lim_{\substack{t'\to-\infty \\ t''\to+\infty}} -i\tilde{V}(p'' - p') \int_{t'}^{t''} \mathrm{d}\tau\, e^{i\tau(p''^2 - p'^2)/2}.$$

It is clear that this expression has no limit as a function but as a distribution we obtain a δ-function of energy conservation:

$$\langle p''|\, S^{(1)}\,|p'\rangle = -i2\pi\delta\big(E(p'') - E(p')\big)\tilde{V}(p'' - p'), \qquad E(p) \equiv p^2/2. \quad (2.78)$$

2.6.2 Path integral and the S-matrix in the semiclassical limit

The representation (2.67) of the evolution operator leads to a path integral representation for matrix elements of the scattering S-matrix which is particularly well suited to study the semiclassical limit.

Let us calculate the matrix elements of the S-matrix between two wave packets:

$$\langle \psi_2 | S | \psi_1 \rangle = \lim_{\substack{t' \to -\infty \\ t'' \to +\infty}} \int \mathrm{d}q' \mathrm{d}q'' \, \langle \psi_2 | e^{i H_0 t''/\hbar} | \mathbf{q}'' \rangle \, \langle \mathbf{q}'' | U(t'', t') | \mathbf{q}' \rangle \, \langle \mathbf{q}' | e^{-i H_0 t'/\hbar} | \psi_1 \rangle .$$

(2.79)

Introducing the two wave functions $\tilde{\psi}_1(\mathbf{p})$ and $\tilde{\psi}_2(\mathbf{p})$ in momentum representation associated with the vectors $|\psi_1\rangle$ and $|\psi_2\rangle$ we can define:

$$\psi_1(\mathbf{q}, t) = \langle q | e^{-i H_0 t/\hbar} | \psi_1 \rangle = \int \frac{\mathrm{d}^d p}{(2\pi)^d} \tilde{\psi}_1(\mathbf{p}) \exp\left[i \left(\mathbf{p} \cdot \mathbf{q} - t \frac{\mathbf{p}^2}{2m} \right) / \hbar \right],$$

(2.80)

and a similar expression for ψ_2.

When t becomes large the phase in expression (2.80) varies very rapidly, and the integral then is dominated by the stationary points of the phase:

$$\frac{\partial}{\partial \mathbf{p}} \left(\mathbf{p} \cdot \mathbf{q} - t \frac{\mathbf{p}^2}{2m} \right) = 0 \quad \Longrightarrow \quad \mathbf{q} = t \frac{\mathbf{p}}{m}.$$

(2.81)

Integral (2.80) is thus equivalent to:

$$\psi_1(\mathbf{q}, t) \underset{|t| \to \infty}{\sim} \tilde{\psi}_1(\mathbf{p}) \frac{1}{(2\pi)^{d/2}} \left(\frac{m\hbar}{|t|} \right)^{d/2} \exp\left(\frac{i\pi}{4} \operatorname{sgn} t + it \frac{\mathbf{p}^2}{2m\hbar} \right),$$

(2.82)

with

$$\mathbf{p} = \frac{m}{t} \mathbf{q}.$$

We then change variables in integral (2.79), setting:

$$\mathbf{q}' = \frac{t'}{m} \mathbf{p}', \qquad \mathbf{q}'' = \frac{t''}{m} \mathbf{p}'',$$

(2.83)

and obtain:

$$\langle \psi_2 | S | \psi_1 \rangle \propto \lim_{\substack{t' \to -\infty \\ t'' \to +\infty}} \int \mathrm{d}p' \mathrm{d}p'' \tilde{\psi}_2^*(\mathbf{p}'') \tilde{\psi}_1(\mathbf{p}') \exp\left[\frac{i}{\hbar} \left(t'' \frac{\mathbf{p}''^2}{2m} - t' \frac{\mathbf{p}'^2}{2m} \right) \right]$$

$$\times \langle t'' \mathbf{p}''/m | U(t'', t') | t' \mathbf{p}'/m \rangle .$$

(2.84)

In this equation we now introduce the path integral representation (2.67) of the evolution operator:

$$\langle t'' \mathbf{p}''/m | U(t'', t') | t' \mathbf{p}'/m \rangle = \int_{q(t') = t' \mathbf{p}'/m}^{q(t'') = t'' \mathbf{p}''/m} [\mathrm{d}q(t)] \exp\left(i \mathcal{A}(q)/\hbar \right).$$

We conclude that the S-matrix is obtained by calculating the path integral with classical scattering boundary conditions, i.e. summing over paths solutions at large time of the free equation of motion. In particular if we know how to solve the classical equations of motion with such boundary conditions we can calculate the evolution operator and thus the S-matrix for $\hbar$ small. This leads to semiclassical approximations of the S-matrix. A calculation in this spirit is presented below.

2.7 Application: The Eikonal Approximation

From the path integral representation of the S-matrix, it is easy to derive a well-known approximation for the scattering amplitude, valid in the high energy, low momentum transfer regime: the eikonal approximation.

In the absence of a potential the evolution operator is given by a gaussian path integral which we can calculate as usual by first solving the classical equation of motion. The solution which satisfies the boundary conditions implied by representation (2.67) and corresponds to the free hamiltonian:

$$H = \frac{\mathbf{p}^2}{2m}, \text{ with } \mathbf{p} \in \mathbb{R}^d,$$

is:

$$\mathbf{q}(t) = \mathbf{q}' + (\mathbf{q}'' - \mathbf{q}') \frac{t - t'}{t'' - t'}. \tag{2.85}$$

Translating the integration variables $\mathbf{q}(t)$ by the classical solution (2.85) we then have to calculate a normalization integral which we can obtain by comparing with the exact result ($\hbar = 1$):

$$\langle \mathbf{q}'' | U(t'', t') | \mathbf{q}' \rangle = \left(\frac{m}{2i\pi(t'' - t')} \right)^{d/2} \exp \left[i \frac{m}{2} \frac{(\mathbf{q}'' - \mathbf{q}')^2}{(t'' - t')} \right]. \tag{2.86}$$

The idea of the eikonal approximation is that in the momentum regime:

$$\mathbf{p}' = \mathbf{p} - \mathbf{k}/2, \quad \mathbf{p}'' = \mathbf{p} + \mathbf{k}/2, \qquad \mathbf{p}^2 \to \infty, \quad \mathbf{p}^2 \gg \mathbf{k}^2,$$

the dominant contributions come from classical trajectories which themselves can be approximated by the straight lines of the free motion (2.85). The calculation of the evolution operator is then straightforward:

$$\langle \mathbf{p} + \mathbf{k}/2 | U(t'', t') | \mathbf{p} - \mathbf{k}/2 \rangle \sim \int d\mathbf{s}\, d\mathbf{x} \exp[-i(\mathbf{p} \cdot \mathbf{s} + \mathbf{k} \cdot \mathbf{x}) + i\mathcal{A}(\mathbf{s}, \mathbf{x})], \tag{2.87}$$

in which the classical action is now:

$$\mathcal{A}(\mathbf{s}, \mathbf{x}) = \frac{im}{2} \frac{\mathbf{s}^2}{t'' - t'} - i \int_{t'}^{t''} dt\, V\left(\mathbf{x} - \frac{\mathbf{s}}{2} + \frac{t - t'}{t'' - t'} \mathbf{s} \right). \tag{2.88}$$

The normalization in equation (2.87) is determined by comparing with the result (2.86) for the free motion.

Taking the large time limit, we find that the integral over $\mathbf{s}$ is dominated by the saddle point:

$$\mathbf{s} = (t'' - t') \mathbf{p}/m. \tag{2.89}$$

We assume that the potential decreases fast enough for the integral in (2.88) to have a large time limit. Once this limit is taken we can make an arbitrary shift on the integration variable t. This has the effect on the argument of the potential V that we can translate the vector $\mathbf{x}$ by an arbitrary vector proportional to $\mathbf{p}$. We can choose this vector in such a way that V depends only on the component $\mathbf{b}$ of $\mathbf{x}$ orthogonal to $\mathbf{p}$:

$$\mathbf{b} = \mathbf{x} - \mathbf{p}(\mathbf{x} \cdot \mathbf{p}/\mathbf{p}^2). \tag{2.90}$$

The integral over the component of $\mathbf{x}$ along $\mathbf{p}$ can then be performed and implies: $\mathbf{p} \cdot \mathbf{k} = 0$ which just expresses energy conservation.

We then obtain:

$$\langle \mathbf{p} + \mathbf{k}/2 | \, U\left(t'', t'\right) | \mathbf{p} - \mathbf{k}/2 \rangle \simeq N\left(\mathbf{p}\right) \int d^{d-1} b \, e^{-i\mathbf{k} \cdot \mathbf{b}} \exp\left[-i \int_{-\infty}^{+\infty} dt \, V\left(\frac{\mathbf{p}t}{m} + \mathbf{b} \right) \right],$$

(2.91)

with:

$$N\left(\mathbf{p}\right) \sim \exp\left[i\left(t'' - t'\right) \frac{\mathbf{p}^2}{2m} \right].$$

(2.92)

Equation (2.91) yields, after Fourier transformation, the matrix elements of the transition operator T in momentum representation (defined by equation (2.71)):

$$\langle \mathbf{p} + \mathbf{k}/2 | \, T \, | \mathbf{p} - \mathbf{k}/2 \rangle \simeq \frac{i\,|\mathbf{p}|}{m} \int \frac{d^{d-1} b}{(2\pi)^d} \, e^{-i\mathbf{k} \cdot \mathbf{b}} \left\{ \exp\left[-i \int_{-\infty}^{+\infty} dt \, V\left(\frac{\mathbf{p}t}{m} + \mathbf{b} \right) \right] - 1 \right\}.$$

(2.93)

The Coulomb potential. Let us use the eikonal approximation to evaluate the scattering amplitude for a $1/q$ Coulomb-like potential. The integral over the potential has in this case no infinite time limit because the potential decreases too slowly. It is necessary to integrate over a finite time interval. We write the potential:

$$V\left(\mathbf{q}\right) = \frac{\alpha}{|\mathbf{q}|},$$

(2.94)

then:

$$\int_{t'}^{t''} dt \, V\left(\frac{\mathbf{p}t}{m} + \mathbf{b} \right) \simeq \frac{\alpha m}{p} \ln\left(-t't'' \frac{4p^2}{m^2 b^2} \right).$$

(2.95)

The appearance of this infinite phase has the following meaning: since the Coulomb potential decreases too slowly at large distance, the classical trajectory approaches too slowly the free motion which, in our definition of the S-matrix, has been taken as the reference motion. In the case of the Coulomb potential only cross-sections are well-defined, not amplitudes.

Factorizing the infinite phase, we can complete the calculation of the scattering amplitude. Integrating over the vector $\mathbf{b}$ we obtain:

$$\langle \mathbf{p} + \mathbf{k}/2 | \, T \, | \mathbf{p} - \mathbf{k}/2 \rangle \simeq \frac{i\pi^{(d-1)/2}}{(2\pi)^d} \frac{p}{m} \exp\left[-i\frac{\alpha m}{p} \ln\left(-t't'' \frac{4p^2}{m^2} \right) \right]$$
$$\times \frac{\Gamma\left[\frac{1}{2}(d-1) - \theta \right]}{\Gamma\left(\theta\right)} \left(\frac{\mathbf{k}^2}{4} \right)^{[\theta + (1-d)/2]},$$

(2.96)

with:

$$\theta = -i\alpha m/p.$$

(2.97)

In three dimensions the expression (2.96) is identical to the exact result. It contains also, for $\alpha < 0$, the correct Coulomb bound state energies E_n which are given by the poles of the scattering amplitude:

$$\theta = \frac{d-1}{2} + n \quad \Rightarrow \quad E_n = \frac{\mathbf{p}^2}{2m} = -\frac{2\alpha^2 m}{(d-1+2n)^2}.$$

(2.98)

The eikonal approximation has a relativistic generalization, and again yields remarkably interesting expressions for the energy of bound states in quantum electrodynamics. It is obtained there by an approximate summation of ladder and crossed ladder Feynman diagrams.

Bibliographical Notes

The formal analogy between time and inverse temperature has been first noticed in

F. Bloch, *Zeit. f. Phys.* 74 (1932) 295.

For the formulation of quantum mechanics in terms of path integrals, basic papers are

P.A.M. Dirac, *Physik. Z. Sowjetunion* 3 (1933) 64; R.P. Feynman, *Rev. Mod. Phys.* 367 (1948) 20; *Phys. Rev.* 80 (1950) 440.

An expanded version is found in

R.P. Feynman and A.R. Hibbs, *Quantum Mechanics and Path Integrals* (McGraw-Hill, New York 1965).

The perturbative equivalence of the operator formalism and gaussian integration is shown in

N.N. Bogoliubov and D.V. Shirkov, *Introduction to Theory of Quantized Fields* (Interscience, New York, London, 1959) ch. 7.

The approach through functional derivatives and the idea of reducing any hamiltonian problem to one in an external source is due to J. Schwinger. See for example

J. Schwinger, *Phys. Rev.* 82 (1951) 914; *Proc. Natl. Acad. Sci. USA* 37 (1951) 452.

For a discussion of the use of path integrals in statistical mechanics see for example

J. Ginibre in *Statistical Mechanics and Quantum Field Theory*, Les Houches 1970, C. De Witt and R. Stora eds. (Gordon and Breach, New York 1971).

For the Källen–Lehmann representation see

G. Källen, *Quantum Electrodynamics* (Springer-Verlag, Berlin 1972).

Several sections of this chapter have been inspired by

E.S. Abers and B.W. Lee, *Phys. Rep.* 9C (1973) 1; L.D. Faddeev in *Methods in Field Theory*, Les Houches 1975, R. Balian and J. Zinn-Justin eds. (North-Holland, Amsterdam 1976).

The argument of Subsection 2.6.2 originates from

W. Tobocman, *Nuovo Cimento* 3 (1956) 1213.

The calculation of the S-matrix in terms of functional integrals with prescribed field asymptotic behaviour was proposed in

R.P. Feynman, *Acta Phys. Polon.* 24 (1963) 697; *Magic without Magic: J.A. Wheeler*, J. Klauder ed. (Freeman, San Francisco 1972).

For a discussion of the eikonal approximation in field theory see for example

H.D.I. Abarbanel, *Cargèse Lectures in Physics* vol. 5, D. Bessis ed. (Gordon and Breach, New York 1972).

The interaction representation is discussed in:

F.J. Dyson, *Phys. Rev.* 75 (1949) 486.

Exercises

Exercise 2.1

2.1.1. Combining equations (2.42,2.43) and (2.50), calculate the energy levels E_n of the hamiltonian H:

$$H = \tfrac{1}{2}p^2 + \tfrac{1}{2}q^2 + \lambda q^4,$$

to first order in λ, from the expression of the partition function Z:

$$Z = \operatorname{tr} \mathrm{e}^{-\beta H} = \sum_{n=0}^{\infty} \mathrm{e}^{-\beta E_n}.$$

2.1.2. Evaluate the contribution of order λ^2 to E_n.

2.1.3. Calculate to order λ the energy levels for a potential of the form:

$$V(q) = \tfrac{1}{2}q^2 + \lambda q^N, \qquad \text{for } N = 6, 8.$$

Exercise 2.2

Compare the exact expression (2.38) of the evolution operator of the harmonic oscillator (set for convenience $\hbar = m = \omega = 1$) with the approximate expression obtained by using equation (2.14) and calculating explicitly the integrals in (2.15). The analysis is simplest when $V(q)$ in equation (2.14) is replaced by an equivalent symmetric function of q and q', for example $V\left[(q + q')/2\right]$. Discuss the accuracy for different equivalent functions of q and q'.

Exercise 2.3

Use expression (2.84) to obtain a semiclassical approximation for the S-matrix of the one dimensional potential $V(q) = \lambda/\cosh^2 q$. Compare with exact results (to be found for example in R.G. Newton, *Scattering Theory of Waves and Particles* (McGraw-Hill, New York 1966)) for the forward and backward S-matrix elements, S_+ and S_- at energy $E = k^2/2$:

$$S_+ = \frac{\Gamma(d + \alpha)\Gamma(\alpha + 1 - d)}{\Gamma(1 + \alpha)\Gamma(\alpha)}, \qquad S_- = \frac{\sin \pi d}{\sin \pi \alpha}\frac{\Gamma(d + \alpha)\Gamma(\alpha + 1 - d)}{\Gamma(1 + \alpha)\Gamma(\alpha)},$$

with

$$d = \tfrac{1}{2}(1 - \sqrt{1 - 8\lambda}) \qquad \alpha = -ik, \quad k > 0.$$

APPENDIX 2
PERTURBATION THEORY. THE TWO-POINT FUNCTION

A2.1 Perturbation Theory in the Operator Formalism

For completeness and to illustrate the differences with the path integral formulation, let us here recall the basis of perturbation theory in the operator formalism of quantum mechanics. To calculate the S-matrix for example, we need an expression for the operator $\Omega(t)$ (see equation (2.69)):

$$\Omega(t) = e^{iH_0 t}\, e^{-iHt}, \qquad (A2.1)$$

in which H_0 is the unperturbed hamiltonian and:

$$V = H - H_0$$

the perturbation.

The operator $\Omega(t)$ satisfies the equation:

$$\dot{\Omega}(t) = -iV_I(t)\Omega(t), \qquad (A2.2)$$

with the boundary condition:

$$\Omega(0) = \mathbf{1}\,.$$

The operator $V_I(t)$ is the perturbation in the interaction representation:

$$V_I(t) = e^{iH_0 t}\, V\, e^{-iH_0 t}\,. \qquad (A2.3)$$

It is easy to verify that the solution of equation $(A2.2)$ can be formally written:

$$\Omega(t) = \sum_{n=0}^{\infty} (-i)^n \int dt_1\, dt_2 \ldots dt_n V_I\left(t_n\right) V_I\left(t_{n-1}\right) \ldots V_I\left(t_2\right) V_I\left(t_1\right), \qquad (A2.4)$$

the domain of integration in the r.h.s. being:

$$0 \le t_1 \le t_2 \le \ldots \le t_n \le t\,.$$

If we now introduce the time-ordered product, as defined in Section 2.5, of the product of perturbation terms which appears in the l.h.s., the product becomes symmetric in the time arguments. We can therefore symmetrize the domain of integration provided we divide by a counting factor $n!$:

$$\Omega(t) = \sum_{n=0}^{\infty} \frac{(-i)^n}{n!} \int_{\substack{0 \le t_i \le t \\ 1 \le i < n}} dt_1\, dt_2 \ldots dt_n \mathrm{T}\left[V_I\left(t_1\right) V_I\left(t_2\right) \ldots V_I\left(t_n\right)\right]. \qquad (A2.5)$$

This expression can be formally rewritten:

$$\Omega(t) = \mathrm{T}\left[\exp\left(-i \int_0^t dt'\, V_I\left(t'\right)\right)\right]. \qquad (A2.6)$$

We can in particular apply these results to a hamiltonian H perturbed by a term linear in q. We can also write a path integral representation for the corresponding partition function. Comparing the expansion of the path integral in powers of the perturbation with expression $(A2.5)$, we recover the relation between correlation functions and T-products established in Section 2.5.

A2.2 The Two-Point Function: Spectral Representation

The 2-point correlation function plays a special role in quantum field theory and statistical mechanics. Let us here derive a few general properties. We assume that the hamiltonian H is hermitian, bounded from below, and, to simplify the notation, has a discrete spectrum. In the basis in which H is diagonal the 2-point correlation function:

$$Z^{(2)}(t) \equiv \langle q(0)q(t) \rangle = \langle 0| \, \hat{q} \, \mathrm{e}^{-|t|H} \, \hat{q} \, \mathrm{e}^{|t|H} \, |0\rangle \,,$$

can then be written:

$$Z^{(2)}(t) = \sum_{n \geq 0} |\langle 0| \, \hat{q} \, |n\rangle|^2 \, \mathrm{e}^{-(\varepsilon_n - \varepsilon_0)|t|} \,. \qquad (A2.7)$$

The quantities $|n\rangle$ and ε_n are respectively the eigenfunctions and eigenvalues of H. For $|t|$ large $Z^{(2)}$ has a limit, $Z^{(2)}(t) \to (\langle 0| \, \hat{q} \, |0\rangle)^2$, which can always be removed by translating $\hat{q}$: $\hat{q} \mapsto \hat{q} - \langle 0| \, \hat{q} \, |0\rangle$.

Note that, as a consequence of the hermiticity of H, the eigenvalues are real, and the sum of exponentials in the r.h.s. has positive coefficients. The Fourier transform $\widetilde{Z}^{(2)}(\omega)$ of the 2-point function has the representation:

$$\widetilde{Z}^{(2)}(\omega) = \int \mathrm{d}t \, Z^{(2)}(t) \, \mathrm{e}^{i\omega t} = 2\pi \, |\langle 0| \, \hat{q} \, |0\rangle|^2 \, \delta(\omega) + 2 \sum_{n > 0} \frac{(\varepsilon_n - \varepsilon_0) \, |\langle 0| \, \hat{q} \, |n\rangle|^2}{[\omega^2 + (\varepsilon_n - \varepsilon_0)^2]} \,. \qquad (A2.8)$$

Two properties of the Fourier transform of the 2-point function follow: except for a possible distributive part at $\omega = 0$, it is an analytic function of ω^2 with poles only on the real negative axis. Moreover the pole residues are all positive; it follows that $\widetilde{Z}^{(2)}(\omega)$ cannot decrease faster than $1/\omega^2$ for ω^2 large. More precisely let us calculate the limit when $t \to 0+$ of the derivative of $Z^{(2)}(t)$

$$\lim_{t \to 0+} \frac{\mathrm{d}}{\mathrm{d}t} Z^{(2)}(t) = \langle 0| \, \hat{q} \, [\hat{q}, H] \, |0\rangle \,.$$

Since the l.h.s. is real we can replace the operator in the r.h.s. by its hermitian part:

$$\lim_{t \to 0+} \frac{\mathrm{d}}{\mathrm{d}t} Z^{(2)}(t) = \tfrac{1}{2} \, \langle 0| \, [\hat{q} \, [\hat{q}, H]] \, |0\rangle \,.$$

For a hamiltonian quadratic in the momentum variable, $H = \frac{1}{2m}\hat{p}^2 + O(\hat{p})$, the commutators can be evaluated explicitly and one obtains

$$[\hat{q} \, [\hat{q}, H]] = -\frac{1}{m} \quad \Rightarrow \quad \lim_{t \to 0+} \frac{\mathrm{d}}{\mathrm{d}t} Z^{(2)}(t) = -\frac{1}{2m} \,.$$

Combining this result with the representation $(A2.7)$ one finds

$$\frac{1}{2m} = \sum_{n \geq 0} |\langle 0| \, \hat{q} \, |n\rangle|^2 \, (\varepsilon_n - \varepsilon_0),$$

and therefore from $(A2.8)$:

$$\widetilde{Z}^{(2)}(\omega) \underset{\omega \to \infty}{\sim} \frac{1}{m\omega^2} \,.$$

This interesting result is not surprising. The behaviour for ω large is related to short time evolution, and we have seen that the most singular part of the evolution operator is then related to the free hamiltonian.

Finally when the spectrum of H has a continuous part, the sum in $(A2.8)$ is replaced by an integral, the poles are replaced by a cut with a positive discontinuity but the other conclusions are the same. The relativistic generalization of representation $(A2.8)$ is called the Källen–Lehmann representation.

3 PATH INTEGRALS IN QUANTUM MECHANICS: GENERALIZATIONS

In Chapter 2 we have constructed a path integral representation of the operator $e^{-\beta H}$ in the case of hamiltonians of the form $p^2 + V(q)$. In this chapter we extend the construction to hamiltonians which are more general functions of phase space variables. We thus first define general path integrals on phase space. An important example is provided by hamiltonians general quadratic functions of momentum variables, a situation we analyze more thoroughly. Such hamiltonians arise in the quantization of the motion on Riemannian manifolds. We illustrate the analysis by the quantization of the free motion on the sphere S_{N-1}.

A parametrization of phase space in terms of complex variables leads to the holomorphic representation of quantum mechanics. The construction of the corresponding path integral provides a useful introduction to the construction of the path integral with fermionic degrees of freedom and to the functional representation of the S-matrix in quantum field theory which will be described in Chapter 7.

In the appendix we also discuss path integral quantization of systems for which an action generating the classical equation of motion cannot be globally defined. This problem arises when phase space, as in the quantization of spin degrees of freedom, or coordinate space, as in the example of the magnetic monopole, have non-trivial topological properties.

3.1 General Hamiltonians: Phase Space Path Integral

We now consider a general *local* quantum hamiltonian obtained by some quantization rule from a classical hamiltonian $H(p, q, t)$. If we know the quantum hamiltonian explicitly and if it is no more than quadratic in p, we can generalize the strategy of Section 2.2 and calculate the imaginary time evolution operator, i.e. the quantum statistical operator, for small time intervals, by solving the Schrödinger equation. Actually often the problem has a different formulation: we are given a classical hamiltonian which we want to quantize while preserving its symmetries. Let us in all cases establish a formal path integral representation of the evolution operator.

We again start from expression (2.8):

$$\langle q'' | U(t'', t') | q' \rangle = \int \prod_{k=1}^{n-1} \mathrm{d}q_k \prod_{k=1}^{n} \langle q_k | U(t_k, t_{k-1}) | q_{k-1} \rangle, \tag{3.1}$$

with the conventions

$$\begin{cases} t_k = t' + k\varepsilon, & t'' = t' + n\varepsilon, \\ q_0 = q', & q_n = q''. \end{cases}$$

Taking the large n, small ε limit, we again reduce the calculation of the evolution operator to the evaluation of the matrix elements $\langle q | U(t + \varepsilon, t) | q' \rangle$ in the small time interval limit. In this limit, as a consequence of the locality of the hamiltonian, only the matrix elements with $|q - q'|$ small will contribute significantly to expression (3.1). To calculate the evolution operator $U(t, t - \varepsilon)$ we solve formally the Schrödinger equation:

$$U(t, t - \varepsilon) = 1 - \frac{\varepsilon}{\hbar} \hat{H}(\hat{p}, \hat{q}, t) + O(\varepsilon^2), \tag{3.2}$$

in which $\hat{H}$ is one quantum hamiltonian which has H as the classical limit. We then take the matrix elements of this equation. The matrix elements of the identity yield a Dirac δ-function for which we use the Fourier representation:

$$\delta\left(q - q'\right) = \int \frac{\mathrm{d}p}{2\pi\hbar}\ e^{ip(q-q')/\hbar}\,.$$

Since the hamiltonian $\hat{H}$ is local, we can replace $\hat{q}$ by its average value q_{av} which tends towards the value q in the limit $q = q'$. Then $\hat{H}$ becomes a function of only one operator $\hat{p}$ and no commutation problems remain. The quantization problem is now the following: the knowledge of the classical hamiltonian H does not determine the value of q_{av} for $q' \neq q$ and instead yields only $\hat{H}\left(\hat{p}, q_{av.} = q\right)$ at $\hbar = 0$. To proceed further, a choice of quantization is thus required. We here choose, as an example, the Wigner quantization rule, which has at least the property to associate with a real classical hamiltonian a hermitian quantum hamiltonian:

$$q_{av} = \tfrac{1}{2}\left(q + q'\right) \quad \Rightarrow \quad \left\langle q \left| \hat{H} \right| q' \right\rangle = \int \frac{\mathrm{d}p}{2\pi\hbar}\ e^{ip(q-q')/\hbar}\ H\left(p, \tfrac{1}{2}(q + q'), t\right). \qquad (3.3)$$

Only in the case of hamiltonians of the form $p^2 + V(q)$ will the final result be independent of this choice of quantization, as has been discussed in Section 2.2. We shall also find out later that the situation in field theory is more favourable, in a sense, since for local hamiltonians commutators of conjugate variables are divergent (equation (5.5)) and have to be removed by renormalization. Dimensional regularization will be specially useful in this respect (see Section 9.3).

Exponentiating, we can rewrite the matrix elements of equation (3.2) as:

$$\left\langle q \left| U\left(t, t - \varepsilon\right) \right| q' \right\rangle \underset{\varepsilon \to 0}{\sim} \int \frac{\mathrm{d}p}{2\pi\hbar} \exp\left\{ \frac{1}{\hbar}\left[ip\left(q - q'\right) - \varepsilon H\left(p, \tfrac{1}{2}(q + q'), t\right) \right] \right\}. \qquad (3.4)$$

Equation (3.1) then takes the form:

$$\left\langle q'' \left| U\left(t'', t'\right) \right| q' \right\rangle = \lim_{n \to \infty} \int \prod_{k=1}^{n} \frac{\mathrm{d}p_k \mathrm{d}q_k}{2\pi\hbar} \delta\left(q_n - q''\right) \exp\left[-S_\varepsilon\left(p, q\right)/\hbar\right], \qquad (3.5)$$

in which $S_\varepsilon\left(p, q\right)$ is given by:

$$S_\varepsilon = \sum_{k=1}^{n} \left[-ip_k\left(q_k - q_{k-1}\right) + \varepsilon H\left(p_k, \tfrac{1}{2}\left(q_k + q_{k-1}\right), t_k\right) \right]. \qquad (3.6)$$

If we introduce a trajectory in phase space $\{p(t), q(t)\}$ which interpolates between the discrete set of times:

$$p\left(t_k\right) = p_k\,, \quad q\left(t_k\right) = q_k\,,$$

we can take the formal continuum time limit $\varepsilon \to 0$, $n \to \infty$ and write a path integral in phase space:

$$\left\langle q'' \left| U\left(t'', t'\right) \right| q' \right\rangle = \int_{q(t')=q'}^{q(t'')=q''} \left[\mathrm{d}p(t)\mathrm{d}q(t)\right] \exp\left[-S(p, q)/\hbar\right], \qquad (3.7)$$

in which the measure in phase space is normalized with respect to $2\pi\hbar$ and $S(p,q)$ is the imaginary time classical action written in the hamiltonian formulation:

$$S = \int_{t'}^{t''} dt\,[-ip(t)\dot{q}(t) + H(p(t), q(t), t)].\tag{3.8}$$

Note that in this formal limit all traces of the special choice of quantization (3.3) have disappeared.

Expression (3.7) is specially aesthetical since it involves only the invariant Liouville measure on phase space and the classical action. In particular it is formally invariant under canonical transformations: transformations in phase space preserving the Poisson brackets

$$(p,q) \mapsto (P(p,q), Q(p,q)),$$

with

$$\{p,q\} = \{P,Q\}.$$

The space of integration. Problems arise when one tries to characterize in general the space of trajectories in phase space which contribute to the path integral (3.7). The term which connects different time steps in expression (3.6) is now $ip_k(q_k - q_{k-1})$. It leads to oscillations in the integral (3.5) which suppress trajectories not regular enough. The typical scale of the difference $(q_k - q_{k-1})$ for contributing trajectories is given by the typical values of p_k. For example if as in Section 2.2 the hamiltonian is quadratic in p, the relevant values of p_k in integral (3.5) are of order $1/\sqrt{\varepsilon}$ and one therefore finds that $(q_k - q_{k-1})$ is of order $\sqrt{\varepsilon}$, a result which is consistent with the analysis of Section 2.2. In the same way we can rearrange expression (3.6) to transform the term $ip_k(q_k - q_{k-1})$ into $iq_k(p_k - p_{k-1})$ ("integration by parts"). To find the regularity conditions imposed to the function $p(t)$ we have to know the relevant values of q_k in integral (3.5). Considering again the example of a hamiltonian $p^2 + V(q)$ we see that if for example $V(q)$ grows for q large like q^{2N} the relevant values of q_k are such that εq^{2N} is of order 1, and therefore the difference $(p_k - p_{k-1})$ should be of order $\varepsilon^{1/2N}$. For a general hamiltonian the discussion obviously becomes rather involved.

Remark. The canonical invariance can be true only for a very restricted class of transformations. We will show elsewhere (Appendix A38.1) that in the case of a one dimensional, one degree of freedom hamiltonian H one can always find a canonical transformation which maps H onto a free hamiltonian

$$p\dot{q} - H \longmapsto P\dot{Q} - \frac{1}{2m}P^2.$$

One would then naively conclude that semiclassical approximations are always exact. It is very easy to produce counter-examples. The discrete form (3.6) shows one origin of the difficulty. A variable p_k is associated to a pair (q_k, q_{k-1}).

The conclusion is that the path integral in phase space is more difficult to handle in practice than the integral in coordinate representation defined previously and thus has found fewer applications up to now. For any non-standard example one has to return to the discrete form (3.5) and make a special analysis.

Again the extension to several degrees of freedom is straightforward. In terms of the classical action and the Liouville measure, the path integral representation of the evolution operator has the same form as in equation (3.7).

Real time formulation. In the real time formulation, which leads to the evolution operator ($e^{-itH/\hbar}$ when the hamiltonian is time independent) the representation (3.7) has to be replaced by

$$\langle q'' | U\left(t'', t'\right) | q' \rangle = \int [dp(t)dq(t)] \exp\left[\frac{i}{\hbar} \mathcal{A}\left(p, q\right)\right]. \tag{3.9}$$

The euclidean action S in the path integral has been replaced by $\mathcal{A}\left(p, q\right)$, the classical action in the hamiltonian formalism:

$$\mathcal{A}\left(p, q\right) = \int_{t'}^{t''} \left[p(t)\dot{q}(t) - H\left(p, q, t\right)\right] dt. \tag{3.10}$$

Even in this general situation the quantum mechanical evolution is obtained by summing over all paths weighted with the complex weight $e^{i\mathcal{A}/\hbar}$. Therefore paths close to extrema of the action, i.e. classical paths, still give the largest contributions to the path integral.

3.2 Hamiltonians Quadratic in Momentum Variables

We have explained the difficulties one encounters when one tries to define a path integral in phase space. To show that expression (3.8) has, however, at least some heuristic value, let us examine the case of general hamiltonians quadratic in momenta.

The consistency. Let us first verify that in the case of hamiltonians

$$H = \frac{p^2}{2m} + V(q),$$

we recover previous results after integration over $p\left(t\right)$ in expression (3.7).

The classical action is:

$$S = \int_{t'}^{t''} dt \left[-ip\dot{q} + \frac{p^2}{2m} + V(q)\right]. \tag{3.11}$$

In expression (3.7) the integral over the momentum variables $p(t)$ is then gaussian. Following the strategy explained in Chapter 1, we change variables

$$p(t) = im\dot{q}(t) + r(t). \tag{3.12}$$

The action becomes:

$$S = \int_{t''}^{t'} dt \left[\frac{1}{2m}r^2(t) + \frac{1}{2}m\dot{q}^2 + V(q)\right]. \tag{3.13}$$

The path integral thus factorizes into an integral over $r(t)$:

$$\mathcal{N}\left(t', t''\right) = \int [dr(t)] \exp\left[-\frac{1}{\hbar} \int_{t'}^{t''} \frac{r^2(t)}{2m} dt\right], \tag{3.14}$$

which does not depend on the potential $V(q)$ and only yields a normalization factor $\mathcal{N}$ function of t' and t'', and an integral over $q(t)$:

$$\langle q'' | U(t'',t') | q' \rangle = \mathcal{N} \int_{q(t')=q'}^{q(t'')=q''} [dq(t)] \exp\left[-\frac{1}{\hbar} \int_{t'}^{t''} dt\left(\tfrac{1}{2}m\dot{q}^2 + V(q)\right) \right], \quad (3.15)$$

The factor $\mathcal{N}$ can be calculated from expression (3.5):

$$\mathcal{N} = \left(\frac{m}{2\pi\hbar\varepsilon}\right)^{n/2}. \quad (3.16)$$

We have therefore verified explicitly that expression (3.7) is consistent with expression (2.18).

The electromagnetic hamiltonian. Let us consider the electromagnetic hamiltonian:

$$H = \frac{1}{2m}\left[\mathbf{p} + e\mathbf{A}(\mathbf{q})\right]^2 + V(\mathbf{q}). \quad (3.17)$$

To eliminate the terms linear in $\mathbf{p}$ in the action we now change variables

$$\mathbf{p}(t) = im\dot{\mathbf{q}}(t) - e\mathbf{A}(\mathbf{q}(t)) + \mathbf{r}(t). \quad (3.18)$$

After integration over $\mathbf{r}(t)$, we obtain an integral over the path $\mathbf{q}(t)$ with the action:

$$S(\mathbf{q}) = \int_{t'}^{t''} dt\left[\tfrac{1}{2}m\dot{\mathbf{q}}^2 + ie\mathbf{A}(\mathbf{q})\cdot\dot{\mathbf{q}} - V(\mathbf{q})\right], \quad (3.19)$$

in which we recognize the euclidean classical action. Let us verify that the path integral is consistent with gauge invariance: If we add a gradient term to $\mathbf{A}(\mathbf{q})$

$$\mathbf{A}(\mathbf{q}) \mapsto \mathbf{A}(\mathbf{q}) + \nabla\Lambda(\mathbf{q}), \quad (3.20)$$

the variation of the lagrangian is a total derivative and the variation of the action δS is:

$$\delta S = ie\left[\Lambda(\mathbf{q}'') - \Lambda(\mathbf{q}')\right]. \quad (3.21)$$

The matrix elements of the evolution operator therefore are multiplied by a phase factor $\exp\left[-ie(\Lambda(\mathbf{q}'') - \Lambda(\mathbf{q}'))/\hbar\right]$, as they should.

The quantum hamiltonian $\hat{H}$, which corresponds to the classical hamiltonian (3.17), is completely characterized by the condition of hermiticity:

$$\hat{H} = \frac{1}{2m}\left[\hat{\mathbf{p}}^2 + e\mathbf{A}(\hat{\mathbf{q}})\cdot\hat{\mathbf{p}} + e\hat{\mathbf{p}}\cdot\mathbf{A}(\hat{\mathbf{q}}) + e^2\mathbf{A}^2(\hat{\mathbf{q}})\right] + V(\hat{\mathbf{q}}), \quad (3.22)$$

A direct calculation of the evolution operator for small time intervals, starting directly from the Schrödinger equation and using hamiltonian (3.22), again yields expression (3.19). However the path integral is consistent with the (hermitian) choice of quantization (3.22) for H only if we respect the symmetry $t \to -t$, $S \to S^*$ in its calculation. This is directly related to the $\theta(0) = 1/2$ assignment ($\theta(t)$ being the step function). We shall again meet this problem in Sections 3.5, 3.6 and in Chapters 4 and 38.

The general case. A general hamiltonian quadratic in **p** can be derived from a general lagrangian quadratic in the velocities. Because, in all examples we shall meet, the quantization problem will initially be formulated in terms of a lagrangian, let us now assume that we want to quantize the real time lagrangian $\mathcal{L}(\dot{q}, q)$

$$\mathcal{L}(\dot{q}, q) = \tfrac{1}{2}\dot{q}^{\alpha} g_{\alpha\beta}(q)\dot{q}^{\beta} - \dot{q}^{\alpha} h_{\alpha}(q) - v(q), \tag{3.23}$$

where $g_{\alpha\beta}(q)$ is a positive matrix.

The corresponding classical hamiltonian is obtained by Legendre transformation. The conjugate momenta are:

$$p_{\alpha} = \frac{\partial \mathcal{L}(\dot{q}, q)}{\partial \dot{q}^{\alpha}} = g_{\alpha\beta}(q)\dot{q}^{\beta} - h_{\alpha}(q).$$

We then obtain the hamiltonian

$$\begin{aligned} H(p, q) &= p_{\alpha}\dot{q}^{\alpha} - \mathcal{L}(\dot{q}, q) \\ &= \tfrac{1}{2}\left(p_{\alpha} + h_{\alpha}(q)\right) g^{\alpha\beta}(q)\left(p_{\beta} + h_{\beta}(q)\right) + v(q), \end{aligned} \tag{3.24}$$

where we have used the traditional notation $g^{\alpha\beta}$ for the matrix inverse of $g_{\alpha\beta}$:

$$g_{\alpha\gamma}(q)g^{\gamma\beta}(q) = \delta_{\alpha}^{\beta},$$

because most interesting examples correspond to quantum mechanics on Riemannian manifolds, as we shall briefly discuss in Section 4.7 (for details see Chapter 22). The tensor $g_{\alpha\beta}(q)$ is then the metric tensor. Field theoretical generalizations will be studied in Chapters 14, 15.

Again the integration over $p(t)$ is gaussian and can be performed. However a difficulty appears with the evaluation of the determinant resulting from the gaussian integration and we therefore perform the p integration on the discretized form (3.5) or actually (3.4). Let us rewrite the expression (3.4) for hamiltonian (3.24):

$$\begin{aligned} \langle \mathbf{q} \,|U(t, t - \varepsilon)|\, \mathbf{q}' \rangle = \int \frac{dp}{2\pi\hbar} \exp\Bigg\{ & \frac{\varepsilon}{\hbar}\Bigg[\frac{i}{\varepsilon}(q - q')^{\alpha} p_{\alpha} - \tfrac{1}{2}\left(p_{\alpha} + h_{\alpha}\left(\tfrac{\mathbf{q}+\mathbf{q}'}{2}\right)\right) \\ & \times g^{\alpha\beta}\left(\tfrac{\mathbf{q}+\mathbf{q}'}{2}\right)\left(p_{\beta} + h_{\beta}\left(\tfrac{\mathbf{q}+\mathbf{q}'}{2}\right)\right) - v\left(\tfrac{\mathbf{q}+\mathbf{q}'}{2}\right)\Bigg]\Bigg\}. \end{aligned} \tag{3.25}$$

The integration yields:

$$\langle \mathbf{q} \,|U(t, t - \varepsilon)|\, \mathbf{q}' \rangle = \left[2\pi\hbar\varepsilon \det \mathbf{g}\left(\tfrac{\mathbf{q}+\mathbf{q}'}{2}\right)\right]^{1/2} \exp\left[-S\left(\mathbf{q}, \mathbf{q}'; \varepsilon\right)/\hbar\right],$$

with:

$$\begin{aligned} S/\varepsilon = \tfrac{1}{2}\left[(q^{\alpha} - q'^{\alpha})/\varepsilon\right] g_{\alpha\beta}\left(\tfrac{\mathbf{q}+\mathbf{q}'}{2}\right)\left[(q^{\beta} - q'^{\beta})/\varepsilon\right] \\ + i\left[(q^{\alpha} - q'^{\alpha})/\varepsilon\right] h_{\alpha}\left(\tfrac{\mathbf{q}+\mathbf{q}'}{2}\right) + v\left(\tfrac{\mathbf{q}+\mathbf{q}'}{2}\right), \end{aligned} \tag{3.26}$$

and we have denote by **g** the matrix with matrix elements $g_{\alpha\beta}$.

Returning to expression (3.5) and taking the formal continuum limit we verify that the exponential factor again yields the euclidean action S, integral of the classical lagrangian (3.23) in which we have made the time imaginary:

$$S\left(\mathbf{q}\right) = \int_{t'}^{t''} dt \left[\tfrac{1}{2}\dot{q}^{\alpha} g_{\alpha\beta}\left(\mathbf{q}\right) \dot{q}^{\beta} + i\dot{q}^{\alpha} h_{\alpha}\left(\mathbf{q}\right) + v\left(\mathbf{q}\right)\right]. \tag{3.27}$$

This is not surprising since, in order to integrate over p_{α} we have solved the classical equation of motion for p_{α}.

However, in contrast with the previous two cases, the integration generates a non-trivial $q(t)$ dependent normalization factor $\mathcal{N}(\mathbf{q})$:

$$\left\langle \mathbf{q}'' \left| U\left(t''t'\right)\right| \mathbf{q}' \right\rangle = \int [dq(t)]\, \mathcal{N}(\mathbf{q}) \exp\left[-S\left(\mathbf{q}\right)/\hbar\right], \tag{3.28}$$

with:

$$\mathcal{N}\left(\mathbf{q}\right) \underset{\varepsilon \to 0}{\sim} \left(2\pi\hbar\varepsilon\right)^{-n/2} \exp\left[\frac{1}{2}\sum_{i=1}^{n} \ln \det \mathbf{g}\left(\tfrac{\mathbf{q}+\mathbf{q}'}{2}\right)\right], \tag{3.29}$$

and thus yields a divergent quantum mechanical correction to the classical action (it has no factor $1/\hbar$):

$$\mathcal{N}\left(\mathbf{q}\right) \sim \exp\left[\frac{1}{2\varepsilon} \int_{t'}^{t''} \ln \det \mathbf{g}\left[\mathbf{q}(t)\right] dt\right], \tag{3.30}$$

or equivalently using the identity (1.33):

$$\ln \det \mathbf{g} = \operatorname{tr} \ln \mathbf{g},$$

$$\mathcal{N}\left(\mathbf{q}\right) \sim \exp\left[\frac{1}{2\varepsilon} \int_{t'}^{t''} \operatorname{tr} \ln \mathbf{g}\left[\mathbf{q}(t)\right] dt\right]. \tag{3.31}$$

A formal calculation, starting from expression (3.7), yields a similar result with $1/\varepsilon$ replaced by $\delta(0)$ (δ being the Dirac δ-function). This difficulty is directly related to the problem of ordering of operators. If one performs a small $\hbar$ (semiclassical) expansion of the path integral, one finds a divergent quantum correction (see Chapters 14,15). This divergence is canceled by the leading contribution coming from (3.31). However the remaining finite part is ill-defined in the formal continuous time limit. It is necessary to use the discretized form (3.5), which reflects a choice of quantization, to calculate it. Another direct way of understanding this ambiguity is to notice that, since in expression (3.30) the difference $|q - q'|$ is generically of order $\sqrt{\varepsilon}$, a replacement of $\mathbf{g}\left((q + q')/2\right)$ by any symmetric function of q and q' which has the same $q = q'$ limit changes in general this quantity at order ε. The modification of $\mathcal{N}(q)$ then generates a finite quantum mechanical correction to the classical action, typical of a commutation of momentum and position operators. Some more details about this problem can be found in Appendix A38.2.

Let us finally note that this problem worsens if the classical hamiltonian is a polynomial of higher degree in p.

Remark. In the case in which $g_{\alpha\beta}$ is the metric tensor on a Riemannian manifold, the factor (3.29) formally reconstructs the covariant measure on the manifold (see Section 22.5).

3.3 The Spectrum of the $O(2)$ Symmetric Rigid Rotator

To illustrate the discussion of general hamiltonians quadratic in momentum variables we now calculate the spectrum of the $O(N)$ rigid rotator from the path integral representation (this model is also the one-dimensional $O(N)$ non-linear σ model, see Chapter 14). We examine first the $N = 2$ case which is simpler and can be treated exactly. The $O(2)$ rotator provides actually an example of the peculiarities of the path integral when position space has non-trivial topological properties. Indeed in angular coordinates the hamiltonian of the $O(2)$ rotator is very simple:

$$H = -\frac{1}{2}\frac{\partial^2}{(\partial\theta)^2}, \tag{3.32}$$

and would be a free hamiltonian if θ were not an angular variable. The corresponding path integral for the matrix elements of the operator $\mathrm{e}^{-\beta H}$ is:

$$\langle\theta''|\,\mathrm{e}^{-\beta H}\,|\theta'\rangle = \int_{\theta(0)=\theta'}^{\theta(\beta)=\theta''}[\mathrm{d}\theta(t)]\exp\left[-\frac{1}{2}\int_0^\beta\left(\frac{\mathrm{d}\theta}{\mathrm{d}t}\right)^2\mathrm{d}t\right]. \tag{3.33}$$

Let us show how the cyclic character of the variable modifies the evaluation of the path integral. As usual we first solve the classical equation of motion. However, since θ' and θ'' are angles, we have to consider all trajectories which go from θ' to θ'' mod (2π). We then shift $\theta(t)$ by the solution:

$$\theta(t) = \theta' + \frac{t}{\beta}\left(\theta'' - \theta' + 2\pi n\right) + u(t), \quad \text{with } n \in \mathbb{Z}. \tag{3.34}$$

The path integral (3.33) then becomes a sum of contributions:

$$\langle\theta''|\,\mathrm{e}^{-\beta H}\,|\theta'\rangle = \sum_{n=-\infty}^{+\infty}\mathcal{N}(\beta)\exp\left[-\frac{1}{2\beta}\left(\theta'' - \theta' + 2\pi n\right)^2\right]. \tag{3.35}$$

The normalization $\mathcal{N}(\beta)$ is given by a path integral which is independent of θ', θ'' and n. Since the integration over $u(t)$ sums fluctuations around the classical trajectory, it is expected that the angular character of $u(t)$ is irrelevant and therefore:

$$\mathcal{N}(\beta) = \sqrt{2\pi/\beta}. \tag{3.36}$$

The expression (3.35) can be rewritten using Poisson's formula: Being a periodic function of $\theta'' - \theta'$, it has a Fourier series expansion:

$$\langle\theta''|\,\mathrm{e}^{-\beta H}\,|\theta'\rangle = \sum_{l=-\infty}^{+\infty}\mathrm{e}^{il\left(\theta''-\theta'\right)}\,\mathrm{e}^{-\beta E_l}, \tag{3.37}$$

whose coefficients are given by:

$$\exp\left[-\beta E_l\right] = \frac{1}{\sqrt{2\pi\beta}}\int_0^{2\pi}\mathrm{d}\theta\,\mathrm{e}^{-il\theta}\sum_{n=-\infty}^{+\infty}\exp\left[-\frac{1}{2\beta}\left(\theta + 2\pi n\right)^2\right]. \tag{3.38}$$

Inverting sum and integration, and changing variables, $\theta + 2n\pi \mapsto \theta$, one then finds:

$$\exp\left(-\beta E_l\right) = \frac{1}{\sqrt{2\pi\beta}} \int_{-\infty}^{+\infty} d\theta \, \exp\left[-\left(il\theta + \frac{\theta^2}{2\beta}\right)\right]. \tag{3.39}$$

The gaussian integration finally yields:

$$E_l = \tfrac{1}{2}l^2, \tag{3.40}$$

which is indeed the exact result. It has been possible to perform an exact calculation because the $O(2)$ group is abelian. The discussion of the general $O(N)$ group is more involved as we shall see now.

3.4 The Spectrum of the $O(N)$ Symmetric Rigid Rotator

The hamiltonian of the $O(N)$ rigid rotator has the form:

$$H = \tfrac{1}{2}\mathbf{L}^2, \tag{3.41}$$

in which the vector $\mathbf{L}$, the angular momentum, represents the set of generators of the Lie algebra of the $O(N)$ group. To write H explicitly we have to choose a particular parametrization of the sphere. Then the hamiltonian has the form:

$$H = \tfrac{1}{2}\left(g^{-1}\left(\mathbf{q}\right)\right)_{ij} p_i p_j, \tag{3.42}$$

in which $g_{ij}\left(\mathbf{q}\right)$ is the metric tensor on the sphere (see also Section 4.7). For example if we parametrize locally the sphere by a vector in $\mathbb{R}^N$ of components $\left(\mathbf{q}, \sqrt{1-\mathbf{q}^2}\right)$ the metric tensor reads:

$$g_{ij}\left(\mathbf{q}\right) = \delta_{ij} + q_i q_j / \left(1 - q^2\right). \tag{3.43}$$

According to the discussion of Section 3.2, the corresponding path integral representation of $e^{-\beta H}$ then is:

$$\left\langle \mathbf{q}'' \left| e^{-\beta H} \right| \mathbf{q}' \right\rangle = \int \left[\sqrt{g\left(\mathbf{q}(t)\right)} d\mathbf{q}\left(t\right)\right] \exp\left[-\frac{1}{2}\int_0^\beta dt \, g_{ij}\left(\mathbf{q}(t)\right) \dot{q}_i \dot{q}_j\right], \tag{3.44}$$

in which $g\left(\mathbf{q}\right)$ is the determinant of the matrix g_{ij}. The contribution to the measure coming from the gaussian integration over the momenta p_i has formally generated the invariant measure on the sphere. We can then rewrite the path integral (3.44) in terms of a vector $\mathbf{r}$ in $\mathbb{R}^N$ of unit length 1:

$$\left\langle \mathbf{r}'' \left| e^{-\beta H} \right| \mathbf{r}' \right\rangle = \int_{\mathbf{r}(0)=\mathbf{r}'}^{\mathbf{r}(\beta)=\mathbf{r}''} \left[d\mathbf{r}(t)\delta\left(1-\mathbf{r}^2(t)\right)\right] \exp\left[-\frac{1}{2}\int_0^\beta dt \, \dot{\mathbf{r}}^2\left(t\right)\right]. \tag{3.45}$$

Let us call θ the angle between $\mathbf{r}'$ and $\mathbf{r}''$:

$$\cos\theta = \mathbf{r}' \cdot \mathbf{r}'', \quad 0 \leq \theta \leq \pi. \tag{3.46}$$

We now introduce a matrix $\mathbf{R}(t)$ which acts on $\mathbf{r}(t)$ and rotates $\mathbf{r}'$ onto $\mathbf{r}''$ in the plane $(\mathbf{r}', \mathbf{r}'')$ in a time β. Its restriction to the two-dimensional $(\mathbf{r}', \mathbf{r}'')$ space has the form:

$$
\begin{bmatrix}
\cos(\theta t/\beta) & \sin(\theta t/\beta) \\
-\sin(\theta t/\beta) & \cos(\theta t/\beta)
\end{bmatrix}.
\tag{3.47}
$$

It is the identity in the subspace orthogonal to the $(\mathbf{r}', \mathbf{r}'')$ plane.

We then change variables, setting:

$$
\mathbf{r}(t) = \mathbf{R}(t)\boldsymbol{\rho}(t).
\tag{3.48}
$$

Let us call u and v the two components of $\boldsymbol{\rho}$ in the $(\mathbf{r}', \mathbf{r}'')$ plane, u being the component along $\mathbf{r}'$ and $\boldsymbol{\rho}_{\mathrm{T}}$ the component in the orthogonal subspace. With this notation we find:

$$
\langle \mathbf{r}'' | e^{-\beta H} | \mathbf{r}' \rangle = \int_{\boldsymbol{\rho}(0)=\mathbf{r}'}^{\boldsymbol{\rho}(\beta)=\mathbf{r}'} \left[\mathrm{d}\boldsymbol{\rho}(t) \delta \left(1 - \boldsymbol{\rho}^2(t)\right) \right] \exp\left[-S\left(\boldsymbol{\rho}\right)\right],
\tag{3.49}
$$

with:

$$
S(\boldsymbol{\rho}) = \tfrac{1}{2} \int_0^\beta \mathrm{d}t \left(\dot{\boldsymbol{\rho}}_{\mathrm{T}}^2 + \dot{u}^2 + \dot{v}^2 + \frac{\theta^2}{\beta^2}(u^2 + v^2) + 2\frac{\theta}{\beta}(\dot{v}u - \dot{u}v) \right).
\tag{3.50}
$$

We then use:

$$
u^2 + v^2 + \rho_{\mathrm{T}}^2 = 1,
\tag{3.51}
$$

to rewrite $S(\boldsymbol{\rho})$:

$$
S(\boldsymbol{\rho}) = \frac{1}{2}\frac{\theta^2}{\beta} + \frac{1}{2} \int_0^\beta \mathrm{d}t \left(\dot{\boldsymbol{\rho}}_{\mathrm{T}}^2 - \frac{\theta^2}{\beta^2}\rho_{\mathrm{T}}^2 + \dot{u}^2 + \dot{v}^2 + 2\frac{\theta}{\beta}(\dot{v}u - \dot{u}v) \right).
\tag{3.52}
$$

In contrast with the abelian case where the calculation was exact, we can here perform only a small β expansion, corresponding to the semiclassical WKB limit, and valid for large quantum numbers. We can therefore neglect contributions exponentially small in β^{-1}. At leading order only small fluctuations around the classical solution are relevant. We thus eliminate the variable u from the action (3.52) by using equation (3.51):

$$
u = \left(1 - v^2 - \rho_{\mathrm{T}}^2\right)^{1/2},
\tag{3.53}
$$

and expand the action in powers of ρ_{T} and v. The leading order result is given by the gaussian approximation and requires only the quadratic terms. In particular the term linear in θ in (3.52) is irrelevant at this order, the integral over $v(t)$ is thus independent of θ and can be absorbed into the normalization. The components of $\boldsymbol{\rho}_{\mathrm{T}}$ become independent variables, and the integral over $\boldsymbol{\rho}_{\mathrm{T}}$ is the $(N-2)$th power of the integral over one component. Since each component satisfies the boundary condition

$$
\rho_i(0) = \rho_i(\beta) = 0,
$$

we can expand the functions $\rho_i(t)$ on the basis:

$$
\rho_i(t) = \sum_{n>0} \rho_{in} \sin\left(n\pi t/\beta\right).
\tag{3.54}
$$

The gaussian integral over the variables ρ_{in} then yields:

$$\langle \mathbf{r}''| e^{-\beta H} |\mathbf{r}'\rangle \sim K(\beta)\, e^{-\theta^2/2\beta} \left[\prod_{n>0}\left(1 - \frac{\theta^2}{n^2\pi^2}\right)\right]^{-(N-2)/2} \tag{3.55}$$

The normalization constant $K(\beta) = (2\pi\beta)^{-(N-1)/2}$ is independent of θ. The infinite product can be calculated:

$$\prod_{n>0}\left(1 - \frac{\theta^2}{n^2\pi^2}\right) = \frac{\sin\theta}{\theta}, \tag{3.56}$$

and therefore:

$$\langle \mathbf{r}''| e^{-\beta H} |\mathbf{r}'\rangle \sim K(\beta)\left(\frac{\theta}{\sin\theta}\right)^{(N-2)/2} e^{-\theta^2/2\beta}. \tag{3.57}$$

To extract the eigenvalues of H, we have to project this expression over the orthogonal polynomials $P_l^N(\cos\theta)$ associated with the $O(N)$ group:

$$\int_0^\pi d\theta\,(\sin\theta)^{N-2}\,P_l^N(\cos\theta)\,P_{l'}^N(\cos\theta) = \delta_{ll'}. \tag{3.58}$$

which are proportional to the Gegenbauer polynomials $C_l^{(N-2)/2}$. For β small we need only the small θ expansion of these polynomials:

$$P_l^N(\cos\theta) = P_l^N(1)\left(1 - \frac{l\,(l+N-2)}{2\,(N-1)}\theta^2 + O\left(\theta^4\right)\right). \tag{3.59}$$

If we assume that to each value of l corresponds only one eigenvalue E_l of H then we can write:

$$e^{-\beta E_l} \propto K(\beta)\int_0^\pi d\theta\, P_l^N(\cos\theta)\,(\theta\sin\theta)^{(N-2)/2}\, e^{-\theta^2/(2\beta)}$$

$$= e^{-\beta E_0}\left(1 - \tfrac{1}{2}l(l+N-2)\beta + O\left(\beta^2\right)\right), \tag{3.60}$$

and therefore:

$$E_l = E_0 + \tfrac{1}{2}l\,(l+N-2) + O(\beta). \tag{3.61}$$

Since E_l is independent of β we can deduce from this calculation the exact result, up to an additive constant E_0. Therefore if we calculate the correction of order of β we shall find a vanishing result.

Concerning this calculation a comment is now in order: we have explained in Section 3.2 that the path integrals (3.44, 3.45) are ill-defined because the measure gives formally divergent contributions. We have stated that these divergences are cancelled by divergences in perturbation theory. However, as a consequence, the resulting expressions are ambiguous and these ambiguities reflect the problem of operator ordering in the quantization of a classical hamiltonian. Still we have here obtained some non-trivial results. The reason is that at every stage of the calculation we have assumed $O(N)$ invariance. This chooses implicitly among all possible discretizations of the path integral a subclass which corresponds to a $O(N)$ invariant quantized hamiltonian. We shall see later, when discussing the non-linear σ model, that such a hamiltonian is fully determined up to an additive constant. The ambiguities of the quantization are thus entirely contained in E_0.

3.5 Path Integral and Holomorphic Representation

We now derive another path integral representation of the evolution operator. This representation leads to interesting expressions for perturbed harmonic oscillators, provides a useful introduction to the fermion path integral, and in quantum field theory to the construction of the S-matrix and the study of the non-relativistic limit in Chapter 7. The idea is to associate classical (complex) variables $\bar{z}, z$ to creation and annihilation operators $a^\dagger, a$ rather than to position and momentum operators.

3.5.1 The holomorphic representation

To motivate the construction we first consider the harmonic oscillator

$$H_0 = \tfrac{1}{2}\hat{p}^2 + \tfrac{1}{2}\omega^2\hat{q}^2. \tag{3.62}$$

We then define the annihilation and creation operators $a, a^\dagger$ as usual by

$$\hat{p} - i\omega\hat{q} = -i\sqrt{2\omega}a, \qquad \hat{p} + i\omega\hat{q} = i\sqrt{2\omega}a^\dagger \quad \Rightarrow [a, a^\dagger] = 1. \tag{3.63}$$

In terms of $a, a^\dagger$ the hamiltonian takes the standard form

$$H_0 = \omega a^\dagger a + \tfrac{1}{2}\omega,$$

with $\omega > 0$. We omit in what follows the constant energy shift $\tfrac{1}{2}\omega$.

We introduce a complex variable $\bar{z}$ and represent $a^\dagger, a$ by the operators $\bar{z}$ and $\partial/\partial\bar{z}$ respectively, acting on functions of $\bar{z}$, which have the same commutation relations. The hamiltonian H_0 is then represented by

$$H_0 = \omega\bar{z}\frac{\partial}{\partial\bar{z}}. \tag{3.64}$$

It is clear that this complex parametrization of phase space is particularly well adapted to the description of the harmonic oscillator. In particular the eigenfunctions are simply $\bar{z}^n$:

$$\omega\bar{z}\frac{\partial}{\partial\bar{z}}\bar{z}^n = \omega n\,\bar{z}^n.$$

The action of the quantum statistical operator $U(t) = e^{-H_0 t}$ on functions of $\bar{z}$ is:

$$U(t)f(\bar{z}) = f(e^{-\omega t}\bar{z}). \tag{3.65}$$

The corresponding kernel, however, has the form of a δ-function

$$\langle\bar{z}|\,U(t)\,|\bar{z}'\rangle = \delta(\bar{z}' - e^{-\omega t}\bar{z}),$$

a representation which is not specially convenient for the study of perturbed harmonic oscillators. We therefore introduce a mixed representation of operators by performing a formal generalized Laplace transformation on the right:

$$\mathcal{O}(\bar{z}, z) \equiv \langle\bar{z}|\,\mathcal{O}\,|z\rangle = \int d\bar{z}'\,\langle\bar{z}|\,\mathcal{O}\,|\bar{z}'\rangle\,e^{\bar{z}'z}. \tag{3.66}$$

A somewhat similar representation in the case of phase space variables is a mixed coordinate momentum representation $\langle q \, | \mathcal{O}| \, p \rangle$ which is obtained by a Fourier transformation on the right argument:

$$\langle q| \, \mathcal{O} \, |p \rangle = \int \mathrm{d}q' \, \mathrm{e}^{ipq'/\hbar} \, \langle q| \, \mathcal{O} \, |q' \rangle .$$

With the definition (3.66) the identity is represented by $\langle \bar{z}|z \rangle = \mathrm{e}^{\bar{z}z}$ and the matrix elements of the hamiltonian (3.64) and the operator $U(t)$ are respectively

$$\langle \bar{z}| \, H_0 \, |z \rangle = \omega \bar{z}z \, \mathrm{e}^{\bar{z}z} \, , \qquad \langle \bar{z}| \, U(t) \, |z \rangle = \mathrm{e}^{\bar{z}z \, \mathrm{e}^{-\omega t}} \, . \tag{3.67}$$

More generally to the normal-ordered product of creation and annihilation operators $a^{\dagger m} a^n$ is associated the kernel

$$a^{\dagger m} a^n \mapsto \bar{z}^m \left(\frac{\partial}{\partial \bar{z}} \right)^n \mapsto \bar{z}^m z^n \, \mathrm{e}^{\bar{z}z} \, .$$

Let us now verify that the action of an operator $\mathcal{O}$ with kernel $\mathcal{O}(\bar{z}, z)$ on a vector $f(\bar{z})$ is given by

$$(\mathcal{O}f)(\bar{z}) = \int \frac{\mathrm{d}\bar{z}'\mathrm{d}z'}{2i\pi} \mathcal{O}(\bar{z}, z') \, \mathrm{e}^{-\bar{z}'z'} \, f(\bar{z}') .$$

By systematic commutations any operator can be written as a linear combination of normal-ordered operators. It is thus sufficient to verify the identity for normal-ordered operators:

$$\int \frac{\mathrm{d}\bar{z}'\mathrm{d}z'}{2i\pi} \, \mathrm{e}^{\bar{z}z'} \, \bar{z}^m z'^n \, \mathrm{e}^{-\bar{z}'z'} \, f(\bar{z}') = \bar{z}^m \int \frac{\mathrm{d}\bar{z}'\mathrm{d}z'}{2i\pi} \, z'^n \, \mathrm{e}^{-\bar{z}'z'} \, f(\bar{z}' + \bar{z})$$

$$= \bar{z}^m \int \frac{\mathrm{d}\bar{z}'\mathrm{d}z'}{2i\pi} \, z'^n \, \mathrm{e}^{-\bar{z}'z'} \sum_l \frac{1}{l!} \bar{z}'^l f^{(l)}(\bar{z}) .$$

The result then relies on the orthogonality conditions

$$\int \frac{\mathrm{d}\bar{z}\mathrm{d}z}{2i\pi} \, \mathrm{e}^{-\bar{z}z} \, z^n \bar{z}^m = n! \, \delta_{mn} \, ,$$

which can be easily verified, for example by changing variables, parametrizing the complex plane in terms of modulus and argument, $z = \rho \, \mathrm{e}^{i\theta}$, $\bar{z} = \rho \, \mathrm{e}^{-i\theta}$. We then find as required

$$\int \frac{\mathrm{d}\bar{z}'\mathrm{d}z'}{2i\pi} \, \mathrm{e}^{\bar{z}z'} \, \bar{z}^m z'^n \, \mathrm{e}^{-\bar{z}'z'} \, f(\bar{z}') = \bar{z}^m \left(\frac{\partial}{\partial \bar{z}} \right)^n f(\bar{z}) .$$

The orthogonality relation is a special example of the scalar product defined in Section 1.9, equation (1.97):

$$(f, g) = \int \frac{\mathrm{d}z \, \mathrm{d}\bar{z}}{2i\pi} \, \mathrm{e}^{-\bar{z}z} \, \overline{f(z)} \, g(z) .$$

We have been naturally led to the structure of Hilbert space of analytic functions.

It follows that the kernel associated with the product of the two operators is given by

$$\int \frac{\mathrm{d}z''\mathrm{d}\bar{z}''}{2i\pi} \, \langle \bar{z} \, |\mathcal{O}_2| \, z'' \rangle \, \mathrm{e}^{-\bar{z}''z''} \, \langle \bar{z}'' \, |\mathcal{O}_1| \, z' \rangle = \langle \bar{z} \, |\mathcal{O}_2\mathcal{O}_1| \, z' \rangle . \tag{3.68}$$

Finally a few remarks:

(i) To an operator $\mathcal{O}$ which has the matrix elements $\mathcal{O}_{mn}$ in the harmonic oscillator basis is associated the kernel $\sum_{m,n} \mathcal{O}_{mn}(\bar{z}^m/\sqrt{m!})(z^n/\sqrt{n!})$.

(ii) Hermitian conjugation is replaced by formal complex conjugation

$$\mathcal{O} \mapsto \mathcal{O}^{\dagger} \quad \Rightarrow \quad \mathcal{O}(\bar{z}, z) \mapsto \overline{\mathcal{O}}(\bar{z}, z) .$$

Clearly with this definition H_0 is hermitian.

3.5.2 Quantum statistical operator: path integral representation

We now use this formalism to establish a path integral representation of the quantum statistical operator first in the case of the harmonic oscillator and then for more general hamiltonians. We expect the path integral to be related to the form (3.7) by a simple change of variables of the form (3.63) (quantum operators being replaced by classical variables) but the boundary conditions and boundary terms require a more detailed analysis.

We first solve Schrödinger's equation for t small:

$$\langle \bar{z} | U(t) | z' \rangle = \left[1 - t\omega\bar{z}\frac{\partial}{\partial\bar{z}} + O\left(t^2\right) \right] e^{\bar{z}z'} = e^{\bar{z}z'(1-\omega t)} + O\left(t^2\right). \qquad (3.69)$$

We then use the group property in the form (3.68) to write the evolution operator at finite time:

$$\langle \bar{z}'' | U(t'',t') | z' \rangle = \lim_{n\to\infty} \int \prod_{k=1}^{n-1} \frac{dz_k d\bar{z}_k}{2i\pi} \exp\left[-S\left(\bar{z}_k, z_k\right) \right], \qquad (3.70)$$

with:

$$S\left(\bar{z}_k, z_k\right) = \left[\sum_{k=1}^{n-1} \bar{z}_k\left(z_k - z_{k-1}\right) - \bar{z}_n z_{n-1} + \omega\varepsilon \sum_{k=1}^{n} \bar{z}_k z_{k-1} \right], \qquad (3.71)$$

where $\varepsilon = (t'' - t')/n$ is the time step and the boundary conditions are:

$$z_0 = z', \quad \bar{z}_n = \bar{z}''. \qquad (3.72)$$

In the formal large n limit we obtain a path integral representation for the matrix elements of $U(t'',t')$:

$$U(\bar{z}'', z'; t'', t') = \int \left[\frac{dz(t)d\bar{z}(t)}{2i\pi} \right] \exp\left[-S\big(\bar{z}(t), z(t)\big) \right],$$

$$S\big(\bar{z}(t), z(t)\big) = -\bar{z}(t'')z(t'') + \int_{t'}^{t''} dt\,\bar{z}(t)\left[\dot{z}(t) + \omega z(t) \right],$$

with the boundary conditions $\bar{z}(t'') = \bar{z}''$, $z(t') = z'$. The symmetry of the action between initial and final times, which is not explicit, can be verified by an integration by parts of the term $\bar{z}\dot{z}$.

This expression generalizes immediately to a system linearly coupled to external sources $\bar{j}(t)$ and $j(t)$

$$H = \omega a^\dagger a - \bar{j}(t)a - j(t)a^\dagger. \qquad (3.73)$$

The corresponding action $S\left(\bar{z}, z\right)$ is:

$$S\big(\bar{z}(t), z(t)\big) = -\bar{z}(t'')z(t'') + \int_{t'}^{t''} dt\,\left\{ \bar{z}(t)\left[\dot{z}(t) + \omega z(t) \right] - \bar{j}(t)z(t) - \bar{z}(t)j(t) \right\}. \quad (3.74)$$

We leave the explicit calculation of the path integral as an exercise. For the hamiltonian $H + \frac{1}{2}\omega$, H being the hamiltonian (3.73), the result is

$$U(\bar{z}'', z'; t'', t') = e^{-\omega\beta/2} \exp[\bar{z}''z'\,e^{-\omega\beta} + \Sigma(\bar{j}, j)],$$

$$\Sigma(\bar{j}, j) = \int_{t'}^{t''} dt\,\left[\bar{z}''\,e^{-\omega(t''-t)}\,j(t) + z'\,e^{\omega(t'-t)}\,\bar{j}(t) \right]$$

$$+ \int_{t'}^{t''} dt d\tau\,\bar{j}(t)\theta(t-\tau)\,e^{-\omega(t-\tau)}\,j(\tau), \qquad (3.75)$$

with $\beta = t'' - t'$.

Taking the trace we also obtain the partition function

$$\operatorname{tr} U(t'', t') = \int \frac{d\bar{z}'' dz}{2i\pi} e^{-\bar{z}'' z'} U(\bar{z}'', z'; t'', t')$$

$$= \frac{1}{2\sinh(\omega\beta/2)} \exp\left[\int dt d\tau \,\bar{j}(t)\Delta(t-\tau)j(\tau)\right], \tag{3.76}$$

$$\Delta(t) = \frac{e^{-\omega t}}{1 - e^{-\omega\beta}} \left[\theta(t) + e^{-\omega\beta}\theta(-t)\right]. \tag{3.77}$$

The propagator $\Delta(t)$ is the solution of the differential equation

$$\dot{\Delta}(t) + \omega\Delta(t) = \delta(t), \tag{3.78}$$

with periodic boundary conditions on the interval $[-\beta/2, \beta/2]$.

For a general quantum hamiltonian several strategies are available, for example:

(i) We first write the phase space path integral (equation (3.7)) and then change variables

$$p - i\omega q = -i\sqrt{2\omega}z, \qquad p + i\omega q = i\sqrt{2\omega}\bar{z}. \tag{3.79}$$

Then

$$S(\bar{z}(t), z(t)) = -\bar{z}(t'')z(t'') + \int_{t'}^{t''} dt \left[\bar{z}(t)\dot{z}(t) + H\left(i\sqrt{\omega/2}(\bar{z}-z), (\bar{z}+z)/\sqrt{2\omega}\right)\right]. \tag{3.80}$$

(ii) Another strategy is to write the hamiltonian in terms of creation and annihilation operators, to normal-order it, i.e. to commute all creation operators to the left, and then replace operators by the corresponding classical variables.

The results are obviously different, a reflection again of the problem of operator ordering and quantization of classical hamiltonians. Moreover, we leave as an exercise to verify that if we use the path integral to calculate perturbatively energy levels the quantity $\Delta(0)$ will appear, as a result of self-contractions of vertices generated by the products of the non-commuting operators $a, a^\dagger$. Since $\Delta(0)$ involves $\theta(0)$ it is clearly ill-defined, and even simple hamiltonians of the form $p^2 + V(q)$ lead to difficulties. For the first path integral the solution to the problem seems to set $\theta(0) = \frac{1}{2}$ while in the second case it seems necessary to set $\theta(0_+) = 0$ (and thus $\theta(0_-) = 1 - \theta(0_+) = 1$).

Perturbed harmonic oscillator. Let us finally consider the example of an harmonic oscillator to which we add a perturbation of the form $V_I(q)$.

$$H = \tfrac{1}{2}\hat{p}^2 + \tfrac{1}{2}\omega^2\hat{q}^2 + V_I(\hat{q}).$$

The perturbative expansion in V_I then can be written in a form analogous to equation (2.50), i.e. as functional derivatives acting on the path integral representing an harmonic oscillator perturbed by a source term coupled linearly to q. We note that $q(t)$ which is proportional to $\bar{z} + z$ can be generated by a real source $j(t)$. Thus the expressions (3.75) and (3.77) can be symmetrized in time. Rescaling also $j(t) \mapsto j(t)/\sqrt{2\omega}$, one finds

$$U(\bar{z}'', z'; t'', t') = \exp\left[-V_I\left(\frac{\delta}{\delta j(t)}\right)\right] U_G(j; \bar{z}'', z'; t'', t')\Big|_{j=0}, \tag{3.81}$$

with

$$
U_G(j; \bar{z}'', z'; t'', t') = e^{-\omega\beta/2} \exp\left[\bar{z}'' z' e^{-\omega\beta} + \int_{t'}^{t''} dt \left(\bar{z}'' e^{-\omega(t''-t)} + z' e^{\omega(t'-t)} \right) \frac{j(t)}{\sqrt{2\omega}} \right.
$$

$$
\left. + \frac{1}{2} \int_{t'}^{t''} dt d\tau\, j(t) \frac{e^{-\omega|t-\tau|}}{2\omega} j(\tau) \right]. \tag{3.82}
$$

In the same way equation (3.77) can be rewritten

$$
\mathrm{tr}\, U_G(j; t'', t') = \frac{1}{2\sinh(\omega\beta/2)} \exp\left[\frac{1}{2} \int dt d\tau\, j(t)\Delta(t-\tau)j(\tau) \right], \tag{3.83}
$$

$$
\Delta(t) = \frac{\cosh\omega(\beta/2 - |t|)}{2\omega\sinh(\omega\beta/2)}, \tag{3.84}
$$

in agreement with equation (2.43).

3.5.3 Real time evolution

Let us rewrite these expressions for real time evolution. The path integral representation for the evolution operator is formally obtained by the rotation $t \mapsto it$. Then

$$
U(\bar{z}'', z'; t'', t') = \int \left[\frac{dz(t)d\bar{z}(t)}{2i\pi} \right] \exp\left[i\mathcal{A}(\bar{z}(t), z(t)) \right],
$$

$$
\mathcal{A}(\bar{z}(t), z(t)) = -i\bar{z}(t'')z(t'') + \int_{t'}^{t''} dt \left[i\bar{z}(t)\dot{z}(t) - h(\bar{z}(t), z(t)) \right],
$$

with the boundary conditions $\bar{z}(t'') = \bar{z}''$, $z(t') = z'$.

From the evolution operator we can derive the corresponding matrix S-matrix. This formalism, however, is mainly useful if the asymptotic states are eigenstates of the harmonic oscillator, a situation which we encounter in quantum field theory. Defining the S-matrix by expression (2.69), where H_0 is the hamiltonian (3.62), we find

$$
S(\bar{z}, z) = \lim_{\substack{t' \to -\infty \\ t'' \to +\infty}} \int \frac{d\bar{z}''dz''}{2i\pi} \frac{d\bar{z}'dz'}{2i\pi} e^{-\bar{z}'' z''} e^{-\bar{z}' z'} e^{i\omega t''/2} \exp\left(\bar{z}z'' e^{i\omega t''} \right)
$$

$$
\times U(\bar{z}'', z'; t'', t') e^{-i\omega t'/2} \exp\left(\bar{z}' z e^{-i\omega t'} \right).
$$

Using the equation (3.65), or integrating directly we find:

$$
S(\bar{z}, z) = \lim_{\substack{t' \to -\infty \\ t'' \to +\infty}} e^{i\omega t''/2} U(\bar{z} e^{i\omega t''}, z e^{-i\omega t'}; t'', t') e^{-i\omega t'/2}. \tag{3.85}
$$

As in the coordinate representation (Section 2.6), the configurations in the path integral which contribute to the S-matrix are for large time asymptotic to the solutions of the classical equation of motion. For the harmonic oscillator H_0 this means:

$$
z(t') \underset{t' \to -\infty}{\sim} z\, e^{-i\omega t'}, \qquad \bar{z}(t'') \underset{t'' \to +\infty}{\sim} \bar{z}\, e^{i\omega t''}.
$$

In quantum mechanics the main application is the evaluation of transitions between eigenstates of the harmonic oscillator induced by a time-dependent perturbation which vanishes for large positive and negative times. As an example let us apply the result (3.85) to the hamiltonian (3.73), where we assume that $j(t)$ vanishes at $t \to \pm\infty$. After a short calculation we find

$$
\mathcal{S}(\bar{z}, z) = \exp\left[\bar{z}z + i \int_{-\infty}^{+\infty} dt \left(\bar{z}\,e^{i\omega t}\, j(t) + z\,e^{-i\omega t}\, \bar{j}(t) \right) \right.
$$
$$
\left. - \int_{-\infty}^{+\infty} dt d\tau\, \bar{j}(t)\theta(t-\tau)\,e^{-i\omega(t-\tau)}\, j(\tau) \right]. \tag{3.86}
$$

The coefficients of the expansion of $\mathcal{S}(\bar{z}, z)$ in powers of $\bar{z}$ and z yield the matrix elements $\mathcal{S}_{mn}$ of the transition between the corresponding eigenstates of the harmonic oscillator, due to a time dependent force coupled linearly to the coordinate and momentum:

$$
\mathcal{S}(\bar{z}, z) = \sum_{m,n} \mathcal{S}_{mn} \frac{\bar{z}^m}{\sqrt{m!}} \frac{z^n}{\sqrt{n!}}.
$$

3.6 Fermionic Path Integral

The derivation of a path integral representation for a fermion hamiltonian follows closely the method of the preceding section, the main difference being that complex variables are replaced by Grassmann variables.

One-fermion hamiltonian. The one-fermion problem is a rather trivial exercise on 2×2 matrices, however the generalization of the final path integral to many fermions is straightforward.

Let us consider a hamiltonian for one fermion:

$$
H = \omega a^\dagger a\,, \tag{3.87}
$$

in which a and $a^\dagger$ are the usual annihilation and creation operators for fermions:

$$
a^2 = a^{\dagger 2} = 0\,, \qquad aa^\dagger + a^\dagger a = 1\,, \tag{3.88}
$$

whose anticommutation relations are a reflection of the Pauli principle.

We want to construct a path integral representation for the quantum statistical operator $U(t) = e^{-tH}$. We have shown in Section 1.7 how to define a differentiation operation in a Grassmann algebra $\mathfrak{A}$. Equations (1.52) exhibit a realization of the commutation relations (3.88) as multiplication and differentiation operations on $\mathfrak{A}$. We can therefore represent the hamiltonian operator H as an operator acting on functions defined on $\mathfrak{A}$, i.e. functions of Grassmann variables:

$$
H = \omega \bar{c} \frac{\partial}{\partial \bar{c}}\,, \qquad \bar{c} \in \mathfrak{A}\,. \tag{3.89}
$$

Functions of $\bar{c}$ form a vector space of dimension two: a constant has fermion number zero, the function $\bar{c}$ has fermion number one.

We here have an additional reason to introduce a mixed representation, compared to the commuting case: the matrix elements of the identity are anticommuting elements

$\delta\left(\bar{c}' - \bar{c}\right)$. We therefore introduce an additional Grassmann variable c', and perform a Grassmann Fourier transformation on the r.h.s. variable $\bar{c}'$. Then the identity is represented by $\mathrm{e}^{-\bar{c}c}$ and the hamiltonian H by $-\omega\bar{c}c\,\mathrm{e}^{-\bar{c}c}$. In the mixed representation the kernel for the product $U_2 U_1$ of two operators is given by:

$$\int \mathrm{d}c'' \mathrm{d}\bar{c}'' \left\langle \bar{c} |U_2| c'' \right\rangle \mathrm{e}^{\bar{c}''c''} \left\langle \bar{c}'' |U_1| c' \right\rangle = \left\langle \bar{c} |U_2 U_1| c' \right\rangle. \qquad (3.90)$$

This result relies on the properties of the scalar product defined in Section 1.9 (equation (1.98)), on functions of $\bar{c}$

$$(f,g) = \int \mathrm{d}c \mathrm{d}\bar{c}\,\mathrm{e}^{-c\bar{c}}\,\overline{f(c)}\,g(c).$$

Path integral. Let us first solve Schrödinger's equation for t small:

$$\left\langle \bar{c} |U(t)| c' \right\rangle = \left[1 - t\omega\bar{c}\frac{\partial}{\partial\bar{c}} + O\left(t^2\right) \right] \mathrm{e}^{-\bar{c}c'} = \mathrm{e}^{-\bar{c}c'(1-\omega t)} + O\left(t^2\right). \qquad (3.91)$$

We then use the group property in the form (3.90) to write the evolution operator at finite time:

$$\left\langle \bar{c}'' |U(t'',t')| c' \right\rangle = \lim_{n\to\infty} \int \prod_{k=1}^{n-1} \mathrm{d}c_k \mathrm{d}\bar{c}_k \exp\left[-S\left(\bar{c}_k, c_k\right)\right], \qquad (3.92)$$

with:

$$S\left(\bar{c}_k, c_k\right) = -\left[\sum_{k=1}^{n-1} \bar{c}_k \left(c_k - c_{k-1}\right) - \bar{c}_n c_{n-1} + \omega\varepsilon \sum_{k=1}^{n} \bar{c}_k c_{k-1} \right], \qquad (3.93)$$

$\varepsilon = (t'' - t')/n$, and the definitions:

$$c_0 = c', \quad \bar{c}_n = \bar{c}''. \qquad (3.94)$$

In the formal large n limit we obtain a path integral representation for the matrix elements of $U(t)$:

$$\left\langle \bar{c}'' |U\left(t'',t'\right)| c' \right\rangle = \int_{c(t')=c'}^{\bar{c}(t'')=\bar{c}''} [\mathrm{d}c(t)\mathrm{d}\bar{c}(t)] \exp\left[-S\left(\bar{c},c\right)\right], \qquad (3.95)$$

with:

$$S\left(\bar{c},c\right) = -\int_{t'}^{t''} \mathrm{d}t\,\bar{c}(t)\left[\dot{c}(t) + \omega c(t)\right] + \bar{c}\left(t''\right)c\left(t''\right). \qquad (3.96)$$

This expression generalizes immediately to a system coupled linearly to external fermionic sources $\bar{\eta}(t)$ and $\eta(t)$, i.e. sources which are generators of a Grassmann algebra and in addition anticommute with a and $a^\dagger$:

$$H = \omega a^\dagger a + \bar{\eta}(t)a - a^\dagger\eta(t). \qquad (3.97)$$

The corresponding action $S\left(\bar{c},c\right)$ is:

$$S\left(\bar{c},c\right) = -\int_{t'}^{t''} \mathrm{d}t\,\left\{\bar{c}(t)\left[\dot{c}(t) + \omega c(t)\right] + \bar{\eta}(t)c(t) + \bar{c}(t)\eta(t)\right\} + \bar{c}\left(t''\right)c\left(t''\right), \qquad (3.98)$$

where now the elements $c(t), \bar{c}(t), \eta(t)$ and $\bar{\eta}(t)$ are all independent generators of a Grassmann algebra.

Calculation of the gaussian integral. Since the integral (3.95) with action (3.98) is gaussian, it can be calculated exactly. The saddle point equation obtained by varying $\bar{c}(t)$ yields, taking into account boundary conditions:

$$\dot{c} + \omega c + \eta = 0 \;\Rightarrow\; c(t) = e^{-\omega(t-t')}\, c' - \int_{t'}^{t} e^{-\omega(t-u)}\, \eta\,(u)\, \mathrm{d}u\,. \tag{3.99}$$

Similarly the variation of $c(t)$ yields:

$$\dot{\bar{c}} - \omega\bar{c} - \bar{\eta} = 0, \;\Rightarrow\; \bar{c}(t) = e^{\omega(t-t'')}\, \bar{c}'' - \int_{t}^{t''} e^{\omega(t-u)}\, \bar{\eta}\,(u)\, \mathrm{d}u\,. \tag{3.100}$$

Translating $\bar{c}(t)$ and $c(t)$ by the solutions of the saddle point equations we then obtain:

$$\langle \bar{c}''\,|U\,(t'',t')|\,c'\rangle = \mathcal{N}\,(t',t'')\exp\left[-S_0\,(\bar{c}'',c';\bar{\eta},\eta)\right], \tag{3.101}$$

with:

$$S_0 = \bar{c}''c'\,e^{-\omega(t''-t')} - \int_{t'}^{t''}\mathrm{d}t\,\left[\bar{\eta}\,(t)\,c'\,e^{-\omega(t-t')} + \bar{c}''\eta(t)\,e^{\omega(t-t'')}\right]$$
$$+ \int_{t'}^{t''}\theta\,(t_2-t_1)\,\bar{\eta}\,(t_2)\,\eta\,(t_1)\,e^{-\omega(t_2-t_1)}\,\mathrm{d}t_1\mathrm{d}t_2\,. \tag{3.102}$$

The normalization factor $\mathcal{N}$ can be calculated by setting $\bar{\eta} = \eta = 0$ and comparing to the direct solution of the Schrödinger equation written in the mixed representation. One finds $\mathcal{N} = 1$.

As in the bosonic case, correlation functions are obtained by differentiating expressions (3.101,3.102) with respect to η and $\bar{\eta}$.

The partition function. The partition function $\operatorname{tr} U\,(\beta/2,-\beta/2)$ can be obtained from expression (3.101). With our conventions the trace is given by:

$$\operatorname{tr} U\,(\beta/2,-\beta/2) = \int \mathrm{d}\bar{c}''\,\mathrm{d}c'\,e^{-\bar{c}''c'}\,\langle \bar{c}''\,|U\,(\beta/2,-\beta/2)|\,c'\rangle\,, \tag{3.103}$$

and therefore:

$$\operatorname{tr} U\left(\frac{\beta}{2},-\frac{\beta}{2}\right) = \left(1+e^{-\omega\beta}\right)\exp\left[-\int_{-\beta/2}^{\beta/2}\mathrm{d}t_1\mathrm{d}t_2\,\bar{\eta}\,(t_2)\,\eta\,(t_1)\,\Delta\,(t_2-t_1)\right], \tag{3.104}$$

with:

$$\Delta(t) = \frac{1}{1+e^{-\omega\beta}}\left[\theta(t)\,e^{-\omega t} - \theta\,(-t)\,e^{-\omega(t+\beta)}\right]. \tag{3.105}$$

Note that, in contrast to the bosonic case (equations (2.43,3.78)) the function $\Delta(t)$ is the solution of the differential equation:

$$\dot{\Delta} + \omega\Delta = \delta(t),$$

with antiperiodic boundary conditions:

$$\Delta\left(\beta/2\right) = -\Delta\left(-\beta/2\right).$$

The solution with periodic boundary conditions appears instead in the calculation of $\mathrm{tr}\,(-1)^F\,\mathrm{e}^{-\beta H}$, in which F is the fermion number, which is obtained by integrating expression (3.101) with $\mathrm{e}^{\bar{c}''c'}$.

Generalization. The method can be immediately generalized to several species of fermions and a hamiltonian H:

$$H = h(a_i^\dagger, a_j) \quad \text{with} \quad a_i^\dagger a_j^\dagger + a_j^\dagger a_i^\dagger = a_i a_j + a_j a_i = 0 \quad \text{and} \quad a_i^\dagger a_j + a_j a_i^\dagger = \delta_{ij}\,, \quad (3.106)$$

because relations (1.52) realize the anticommutation relations of the operators $\{a_i^\dagger, a_j\}$. Let us assume that H has been normal-ordered, which means that, with the help of the commutation relations, all operators $a_i^\dagger$ have been pushed at the left of all operators a_j in all monomials contributing to $h\left(a^\dagger, a\right)$. Then equation (3.91) generalizes under the form:

$$\langle\bar{\mathbf{c}}\,|U(t)|\,\mathbf{c}'\rangle = \exp\left[-\sum_i \bar{c}_i c_i' - t\,h\left(\bar{\mathbf{c}}, -\mathbf{c}'\right) + O\left(t^2\right)\right]. \quad (3.107)$$

At finite time a path integral representation follows:

$$\langle\bar{\mathbf{c}}''\,|U\left(t'', t'\right)|\,\mathbf{c}'\rangle = \int_{\mathbf{c}(t')=\mathbf{c}'}^{\bar{\mathbf{c}}(t'')=\bar{\mathbf{c}}''} [\mathrm{d}\mathbf{c}(t)\mathrm{d}\bar{\mathbf{c}}(t)]\exp\left[-S\left(\bar{\mathbf{c}}, \mathbf{c}\right)\right], \quad (3.108)$$

with:

$$S\left(\bar{\mathbf{c}}, \mathbf{c}\right) = -\int_{t'}^{t''}\mathrm{d}t\,\{\bar{c}_i(t)\dot{c}_i(t) - h\left[\bar{\mathbf{c}}(t), -\mathbf{c}(t)\right]\}. \quad (3.109)$$

We have seen in Section 1.9 how to calculate gaussian integrals and average of polynomials. We can here use the same methods to expand the path integral (3.109) in perturbation theory. However again problems due to operator ordering arise. In particular, as in the case of the holomorphic path integral of the commuting case, perturbative calculations involve $\theta(0)$. The correct ansatz, consistent with our construction, again seems to be to set $\theta(0_+) = 0$.

Bibliographical Notes

Most of the references of Chapter 2 are still relevant here. The construction of path integrals in phase space for general hamiltonian systems can be found in

R.P. Feynman, *Phys. Rev.* 84 (1951) 108 (Appendix B); C. Garrod, *Rev. Mod. Phys.* 38 (1966) 483.

The generalization to constrained systems is given in

L.D. Faddeev, *Theor. Math. Phys.* 1 (1969) 3.

An early discussion of the problem of quantization of velocity dependent potentials in the operator formalism can be found in

T.D. Lee and C.N. Yang, *Phys. Rev.* 128 (1962) 885.

For more details about the holomorphic representation see

L. D. Faddeev in *Methods in Field Theory*, Les Houches 1975, R. Balian and J. Zinn-Justin eds. (North-Holland, Amsterdam 1976).

The fermionic path integral is discussed in

F.A. Berezin, *The Method of Second Quantization* (Academic Press, New York 1966).

Exercises

Exercise 3.1

Calculate the energy levels of the $O(\nu)$ invariant hamiltonian

$$H = \tfrac{1}{2}\mathbf{p}^2 + \tfrac{1}{2}\mathbf{q}^2 + \lambda|\mathbf{q}|^4,$$

in which $\mathbf{p}$ and $\mathbf{q}$ are vectors in $\mathbb{R}^\nu$, at order λ. Note that since levels are degenerate at leading order, it is not sufficient to calculate $\operatorname{tr} e^{-\beta H}$ as in the $\nu = 1$ case (exercise 2.1.1).

Exercise 3.2

Assume for the evolution operator a path-integral representation of the form

$$U(q'', q') = \int [dq(t)\mathcal{N}(t)] \exp\left[\frac{i}{\hbar}\int dt\, \mathcal{L}(q, \dot{q})\right].$$

The term in the exponential is then fixed to be the lagrangian by demanding the correct classical limit. This leaves the normalization factor undetermined. Show by evaluating the evolution operator for t small that $\mathcal{N}(t)$ is fixed to have the form of Section 3.2 by unitarity. Note that one has to use the property that for t small only short trajectories contribute to the path-integral.

Exercise 3.3

Discuss, in the free case, the change of variables $q = f(q')$ in the path integral (2.18). It is suggested to return to discrete forms (2.15,2.16) and then to compare with the expressions (3.25,3.26).

Exercise 3.4

3.4.1. Using expression (3.19), calculate the evolution operator for a for $\mathbf{A}(\mathbf{q}) = \tfrac{1}{2}\mathbf{B}\times\mathbf{q}$, $\mathbf{B}$ constant, i.e. a constant magnetic field, and $V(\mathbf{q}) = 0$.

3.4.2. Add to the previous action an harmonic potential term $V(\mathbf{q}) = \tfrac{1}{2}m\omega^2 q^2$ and calculate again the evolution operator.

Exercise 3.5

Following the ideas explained in Section 3.3, calculate the real time evolution operator e^{-iHT} in the case of the hamiltonian:

$$H = \tfrac{1}{2}p^2 + V(q), \quad \text{with} \quad V = 0 \text{ for } q \in [0, \pi], \text{ and } V = +\infty \text{ for } q \notin [0, \pi].$$

Note that the result obtained by adding all the classical trajectories is incorrect and that a phase factor in front of half of the contributions is required. This difficulty reflects the singular nature of the function $V(q)$.

Exercise 3.6

3.6.1. As suggested in Section 3.5 evaluate the path integral (3.74) and derive the result (3.75).

3.6.2. Verify the unitarity of the S-matrix (3.86).

3.6.3. Use the holomorphic representation to calculate the order λ correction to the energy levels of the hamiltonian

$$H = \tfrac{1}{2}p^2 + \tfrac{1}{2}q^2 + \lambda q^4.$$

Perform the calculation for the two different path integral representations, obtained either by directly replacing the classical phase space variables p, q by the complex variables $z, \bar{z}$, or by writing the hamiltonian in terms of creation and annihilation operators, commuting all creation operators to the left, and then only replacing quantum by classical variables.

Exercise 3.7

Consider, in the holomorphic representation of quantum mechanics, the states with wave function $\psi(\bar{z}) = e^{\alpha \bar{z}}$, where α is an arbitrary complex number (also called coherent states). Calculate the average of the position and momentum operators q, p and then the dispersions $\Delta q = (\langle q^2 \rangle - \langle q \rangle^2)^{1/2}$, $\Delta p = (\langle p^2 \rangle - \langle p \rangle^2)^{1/2}$. Show that the states ψ are minimal dispersion states: $\Delta q \Delta p = \tfrac{1}{2}$.

APPENDIX 3
QUANTIZATION OF SPIN DEGREES OF FREEDOM, TOPOLOGICAL ACTIONS

In Section 3.3 we have evaluated the path integral in an example where coordinate space has non-trivial topological properties. We now want to discuss two other examples where topology plays an essential role, in the sense that an action generating the classical equations of motion cannot be globally defined. We first quantize in the path integral formalism angular momentum operators in a fixed representation. One of the peculiarity of this system is that phase space itself has non-trivial topological properties. The second example is provided by the magnetic monopole which gives a non-trivial topological structure to coordinate space. In both examples one speaks about topological action. The property that a topological action cannot be globally defined leads to the quantization of its amplitude, a property specific to quantum mechanics. Indeed, in classical mechanics, a multiplication of the action by a constant does not modify the equations of motion.

Note that in this section, to simplify notations, we use a *real time* formalism.

A3.1 General Symplectic Forms and Quantization

We first discuss the general example of quantization in the case in which phase space has non-trivial topological properties. We then consider the example of spin dynamics.

In the hamiltonian formulation of classical mechanics the action has the form (3.8):

$$S = \int_{t'}^{t''} dt \left[p(t)\dot{q}(t) - H\left(p(t), q(t), t\right) \right] . \qquad (A3.1)$$

We note that the term $\int p\dot{q}\,dt$ represents the area in phase space between the classical trajectory C and the axis $p = 0$. It can be rewritten:

$$\int_{\partial D} p(t)\dot{q}(t)\,dt = \int_D dp \wedge dq , \qquad (A3.2)$$

in which ∂D, the boundary of the domain D, contains the classical trajectory C and a fixed reference curve. If we now parametrize phase space differently, introducing new coordinates u_α, the r.h.s. of the equation becomes:

$$\int_D dp \wedge dq = \int_D \omega_{\alpha\beta}(u)\,du_\alpha \wedge du_\beta .$$

In the language of forms $\omega = \omega_{\alpha\beta}\,du_\alpha \wedge du_\beta$ is a 2-form (see Section 1.6) which by construction is obtained by differentiating a 1-form, here:

$$\omega = d\omega' , \qquad (A3.3)$$

$$\omega' = p(u)\frac{dq}{u_\alpha}du_\alpha = \omega_\alpha du_\alpha , \qquad (A3.4)$$

and called the *symplectic* form.

Since the operator d acting on differential forms is nilpotent, the form ω is closed, i.e. satisfies:

$$d\omega = 0 . \qquad (A3.5)$$

Example. In Section 3.5 we have discussed the holomorphic representation of the harmonic oscillator. We have introduced a complex parametrization of phase space (equation (3.79))

$$p - iq = -i\sqrt{2}z, \qquad p + iq = i\sqrt{2}\bar{z}.$$

In terms of $z, \bar{z}$ the symplectic form becomes:

$$dp \wedge dq = \frac{1}{i}dz \wedge d\bar{z}.$$

Previous considerations immediately generalize to several degrees of freedom. Let u_α be $2n$ variables parametrizing a phase space for n degrees of freedom. The action S in the hamiltonian formulation can be written:

$$S = \int_D \omega - \int_{\partial D} dt\, H\left(u(t), t\right), \qquad (A3.6)$$

where ω, a symplectic form, is a *closed* 2-form:

$$\omega = \omega_{\alpha\beta}du_\alpha \wedge du_\beta, \qquad d\omega = 0.$$

This latter condition is a sufficient condition for the equations of motion to depend only on the boundary ∂D of the domain D but not on the interior, as was obvious for the initial action $(A3.1)$. They then take the form:

$$\omega_{\alpha\beta}\dot{u}_\beta = \frac{\partial \mathcal{H}}{\partial u_\alpha}. \qquad (A3.7)$$

Locally equation $(A3.5)$ can be integrated in the same way as $dp \wedge dq$ can be integrated into pdq. However, if phase space has non-trivial topological properties, it cannot always be integrated globally, i.e. the symplectic form is not exact. This phenomenon has peculiar consequences in quantum mechanics since the path integral involves the action explicitly in the form $e^{iS/\hbar}$. For the path integral to makes sense, this phase factor must be independent of the choice of the action S. Let us examine this problem in the example of the quantization of spin degrees of freedom.

A3.2 Spin Dynamics and Quantization

A3.2.1 Classical spin dynamics

Let us consider a vector $\mathbf{S}$ in three dimensions of fixed length s:

$$\mathbf{S}^2 = s^2.$$

The simplest dynamics we can write for $\mathbf{S}$ is:

$$\frac{d\mathbf{S}}{dt} = \mathbf{H} \times \mathbf{S}, \qquad (A3.8)$$

in which $\mathbf{H}$ is a constant vector. This equation is first order in time and involves two independent variables corresponding to a point on the sphere. These variables can be considered as a couple position and conjugate momentum. Phase space is therefore isomorphic to the sphere S_2. The hamiltonian is simply:

$$\mathcal{H} = \mathbf{H} \cdot \mathbf{S} \, . \tag{A3.9}$$

More precisely it can be verified that if we parametrize $\mathbf{S}$ as:

$$\mathbf{S} = s(\sin\theta\cos\varphi, \sin\theta\sin\varphi, \cos\theta),$$

then the action S,

$$S = \int [s\cos\theta\dot{\varphi} - \mathcal{H}] \, dt \, , \tag{A3.10}$$

generates equation (A3.8). We see in this expression that $\cos\theta$ and φ play the role of conjugate variables. However two remarks are in order: First we have integrated the symplectic 2-form but, for this purpose we have been forced to use a parametrization of the sphere which is singular at $\theta = 0$ and π. When the trajectory contains the north or the south pole of the sphere the integral is not defined. The 2-form cannot be integrated globally into a 1-form because the integral $\int d\varphi d\cos\theta$, which is the area on the sphere, is defined only (mod 4π).

In classical mechanics, of course, these properties are irrelevant since only equations of motion are physical.

Note the symplectic form has other useful representations:

$$d\cos\theta \wedge d\varphi = \tfrac{1}{2}s^{-3}\varepsilon_{ijk}S_i dS_j \wedge dS_k = 2i dz_i \wedge d\bar{z}_i \, ,$$

where z_i is a two-component complex vector of length 1, corresponding to the isomorphism between S_2 and the symmetric space CP_1 (see Chapter 15):

$$\mathbf{S} = s\bar{z}_i\boldsymbol{\sigma}z_i \, , \qquad \bar{z}_i z_i = 1 \, ,$$

where $\boldsymbol{\sigma}$ is the set of Pauli matrices. In a special *gauge* the vector $\mathbf{z}$ can also be written:

$$z_1 = e^{i\varphi/2}\cos(\theta/2),$$
$$z_2 = e^{-i\varphi/2}\sin(\theta/2).$$

A3.2.2 Quantization of spin degrees of freedom

Let us consider the path integral representation of the corresponding evolution operator in quantum mechanics. The action itself now appears explicitly and the problem discussed in the preceding section becomes relevant. The path integral exists only if the integrand $e^{iS/\hbar}$ is defined. Since the total area is defined only (mod 4π) the integrand must be invariant under such a change. This implies that $4\pi s$ must be a multiple of $2\pi\hbar$: The parameter $s/\hbar$ is quantized and can take only half-integer values. This a is a generic quantum mechanical property of *topological* contributions to the action, i.e. contributions which are not globally defined: their amplitude is *quantized*. The magnetic monopole of Section A3.3 provides another example of such a situation.

In the parametrization (A3.10) the action can be written (we now set $\hbar = 1$):

$$S = \int [(\gamma + \cos\theta)\dot{\varphi} - \mathcal{H}] \, dt \, , \tag{A3.11}$$

expression which differs from $(A3.10)$ only by a total derivative: by choosing $\gamma = 1/2, 0$ for s half-integer, integer respectively one renders e^{iS} regular near $\theta = 0, \pi$.

To relate this action to the usual operator formulation of the angular momentum relations let us first note that classically we have:

$$S_\pm = e^{\pm i\varphi} \left(s^2 - S_z^2\right)^{1/2}.$$

After quantization S_z becomes p_φ the conjugate momentum of the angular variables φ:

$$S_z \equiv p_\varphi = \frac{1}{i}\frac{d}{d\varphi}.$$

It can be verified that the quantum operator $S_\pm$ can be written:

$$S_\pm = e^{\pm i\varphi/2}\left(s^2 - p_\varphi^2\right)^{1/2}e^{\pm i\varphi/2}.$$

Using then:

$$j(j+1) = \mathbf{S}^2 = S_z^2 + S_+ S_- S_z = s^2 - 1/4,$$

we find the relation between the angular momentum j and the parameter s:

$$s = j + 1/2. \tag{A3.12}$$

In particular, since s is quantized, we recover a property of quantum mechanics, quantization of spin.

If we call m the eigenvalues of S_z we observe that in the φ configuration space the corresponding eigenvectors are of the form $e^{im\varphi}$ and the projector K on the basis is:

$$K\left(\varphi'', \varphi'\right) = \sum_{m=-j}^{m=j} e^{im\left(\varphi' - \varphi''\right)} = \frac{\sin\left[(j+1/2)\left(\varphi' - \varphi''\right)\right]}{\sin\left[\left(\varphi' - \varphi''\right)/2\right]}. \tag{A3.13}$$

We can compare this expression with the short-time path integral representation which leads to:

$$K\left(\varphi'', \varphi'\right) \propto \sum_n \int_{-s}^{s} dp_\varphi\, e^{ip_\varphi\left(\varphi' - \varphi'' + 2n\pi\right)},$$

$$\propto \sin\left[s\left(\varphi' - \varphi''\right)\right]\sum_n \frac{2(-1)^n}{\left(\varphi' - \varphi'' + 2n\pi\right)}.$$

The sum over n has to be regularized but the factor $\sin s\left(\varphi' - \varphi''\right)$ is consistent with $(A3.13)$ and the identification $(A3.12)$.

A final word of caution: Although the path integral quantization of spin variables is quite useful to study the classical limit (which is also the limit $s \to \infty$), ambiguities due to operator ordering leads to difficulties in explicit calculations.

A3.3 The Magnetic Monopole

Electromagnetism provides another example of the situation encountered in Section A3.2.1. The contribution of the vector potential to the euclidean action is of the form (see (3.19)) $ie \int \mathbf{A}(\mathbf{x}) \cdot d\mathbf{x}$, i.e. the integral of 1-form. However the only physical quantity which appears in the classical equations of motion is the electromagnetic tensor. The contribution to the action can be, as above, more generally written as the integral of a 2-form (involving the magnetic field) provided this form is closed:

$$S_{\text{elect.}} = ie \int F_{ij}\, dx_i \wedge dx_j\,, \quad F_{ij} = \partial_i A_j - \partial_j A_i = \varepsilon_{ijk} B_k\,.$$

Again if this form is not exact the vector potential cannot be globally defined. This is precisely what happens when a magnetic field is generated by a magnetic monopole.

The formal duality symmetry between magnetic and electric fields in Maxwell's equations has led Dirac to speculate about the existence of yet undiscovered isolated magnetic charges, magnetic equivalent to electric charges. An isolated magnetic charge g creates a magnetic field $\mathbf{B}$ of the form:

$$\mathbf{B} = g\frac{\mathbf{x}}{r^3}\,, \quad r = |\mathbf{x}|\,.$$

The field B also is a singular solution to the free static Maxwell's equation. It has an infinite energy and hence, in this form, it is irrelevant to physics. However in Higgs models finite energy solutions (solitons) can be found which coincide at large distance with magnetic monopoles.

The integral of the magnetic field over a closed surface containing the magnetic charge is $4\pi g$, as one immediately verifies by using polar coordinates $\{r, \theta, \varphi\}$ with the monopole as origin:

$$\int F_{ij} dx_i \wedge dx_j = g \int \frac{r}{r^3} \times r^2 d\cos\theta\, d\varphi = 4\pi g\,.$$

If the vector potential could be globally defined the integral would obviously vanish. More directly if we try to calculate the corresponding vector potential we find in a family of gauges

$$A_i(x) = g\epsilon_{ijk}n_k x_j \frac{\mathbf{n} \cdot \mathbf{x}}{r\left(r^2 - (\mathbf{n} \cdot \mathbf{x})^2\right)}$$

where $\mathbf{n}$ is a constant unit vector. We observe that the potential is singular along the line of direction $\mathbf{n}$ passing through the origin. This line of singularities is unphysical and can be displaced but not removed.

Again this property has no classical consequences. However in quantum mechanics since the action can only be defined $(\text{mod}\,4i\pi eg)$ the weight factor e^{-S} is only defined if:

$$4\pi eg = 0 \quad (\text{mod}\,2\pi) \Rightarrow 2eg = \text{integer}.$$

We recognize Dirac's quantization condition.

Note that when this condition is fulfilled, parralel transport (see Subsection 18.3.1) is globally defined in $\mathbb{R}^3$.

4 STOCHASTIC DIFFERENTIAL EQUATIONS: LANGEVIN, FOKKER–PLANCK EQUATIONS

In Chapters 2,3 we have shown how the concept of path integrals naturally arises in the calculation of the matrix elements of the quantum mechanical operator $e^{-\beta H}$ when the hamiltonian H is local. Let us note that, in the particular case of hamiltonians quadratic and even in momenta, the integrand of the path integral defines a positive measure and therefore has a probabilistic interpretation. In this chapter we discuss Langevin equations, i.e. stochastic differential equations related to diffusion processes or brownian motion. From the Langevin equation we derive the Fokker–Planck (FP) equation for probability distribution of the stochastic variables. The FP equation has the form of a Schrödinger equation in imaginary time, of the type we have studied in previous chapters (the corresponding hamiltonian is however in general non-hermitian). We then show that averaged observables can also be calculated from path integrals, whose integrands define automatically positive measures. In some cases, like brownian motion on Riemannian manifolds, difficulties appear in the proper definition of stochastic equations, quite similar to the quantization problem in quantum mechanics. Time discretization allows to study the problem.

This chapter will serve as an introduction to Chapters 17 and 35 in which *stochastic quantization* and *critical dynamics* will be discussed.

4.1 The Langevin Equation

We call the Langevin equation a first order in time stochastic differential equation of the form:

$$\dot{q}_i(t) = -\tfrac{1}{2} f_i\left(\mathbf{q}\left(t\right)\right) + \nu_i(t), \tag{4.1}$$

in which $\mathbf{q}(t)$ is a point in $\mathbb{R}^d$, $f_i\left(\mathbf{q}\right)$ a function of $\mathbf{q}$ and $\boldsymbol{\nu}(t)$ a set of stochastic functions called hereafter the "noise". The noise can be defined by a functional probability distribution $[d\rho\left(\nu\right)]$. In what follows, we shall specialize to a gaussian noise with a probability distribution of the form

$$[d\rho\left(\nu\right)] = [d\nu] \exp\left[-\frac{1}{2\Omega} \int \nu_i^2(t)\, dt\right]. \tag{4.2}$$

This particular form of the gaussian noise, called gaussian white noise, is related to Markov's processes (see Appendix A4). The constant Ω characterizes the width of the noise distribution. Note that by a rescaling of time $t \mapsto \Omega t$ (and a redefinition of the noise variable) it can be transferred in front of the driving term $f \mapsto f/\Omega$.

Alternatively the gaussian noise can be characterized by its 1- and 2-point correlation functions:

$$\langle \nu_i(t)\rangle = 0, \qquad \langle \nu_i(t)\nu_j\left(t'\right)\rangle = \Omega\, \delta_{ij}\, \delta\left(t - t'\right), \tag{4.3}$$

as can be verified immediately using the results of Chapter 2. Equation (4.1) is not the most general stochastic first order differential equation, but we want to postpone the discussion of the general form, because it involves additional problems, until Section 4.7.

Given the value of $\mathbf{q}(t)$ at initial time t_0, $\mathbf{q}(t_0) = \mathbf{q}_0$, the Langevin equation generates a time-dependent probability distribution $P(\mathbf{q}, t)$ for the stochastic vector $\mathbf{q}(t)$ which formally can be written as:

$$P(\mathbf{q}, t) = \left\langle \prod_{i=1}^{d} \delta\left[q_i(t) - q_i\right] \right\rangle. \tag{4.4}$$

In equation (4.4) as in equation (4.3) brackets indicate average over the noise. The vector $\mathbf{q}$ is the argument of $P(\mathbf{q}, t)$ and has no relation with the function $\mathbf{q}(t)$. If $\mathcal{O}(\mathbf{q})$ is an arbitrary function, definition (4.4) then implies:

$$\int P(\mathbf{q}, t) \, \mathcal{O}(\mathbf{q}) \, dq = \langle \mathcal{O}(\mathbf{q}(t)) \rangle. \tag{4.5}$$

4.2 The Fokker–Planck Equation

Let us show that equations (4.1,4.2) imply a differential equation for $P(\mathbf{q}, t)$. Differentiating with respect to time equation (4.4) and using equation (4.1) we obtain:

$$\dot{P}(\mathbf{q}, t) = \left\langle \left[-\tfrac{1}{2} f_i(\mathbf{q}(t)) + \nu_i(t) \right] \frac{\partial}{\partial q_i(t)} \delta(\mathbf{q}(t) - \mathbf{q}) \right\rangle. \tag{4.6}$$

In this equation the symmetry of the δ-function allows us to replace the differentiation $\partial/\partial q_i(t)$ by $-\partial/\partial q_i$ which can be taken out of the average. Then using the δ-function we can replace $\mathbf{q}(t)$ by $\mathbf{q}$ in $f_i(\mathbf{q}(t))$:

$$\dot{P}(\mathbf{q}, t) = \frac{\partial}{\partial q_i} \left[\tfrac{1}{2} f_i(\mathbf{q}) P(\mathbf{q}, t) - \langle \nu_i(t) \delta(\mathbf{q}(t) - \mathbf{q}) \rangle \right]. \tag{4.7}$$

Equation (4.7), which is independent of the noise distribution, implies the conservation of probability since integrating over whole $\mathbf{q}$ space we find:

$$\int dq \, \dot{P}(\mathbf{q}, t) = 0. \tag{4.8}$$

However, to be able to establish a partial differential equation for $P(\mathbf{q}, t)$, we need a noise distribution of precisely the form (4.2), i.e. local in time and gaussian.

A useful algebraic identity. Let $F(\boldsymbol{\nu})$ be an arbitrary functional of $\boldsymbol{\nu}(t)$. Then:

$$\langle F(\boldsymbol{\nu}) \, \nu_i(\tau) \rangle = \int [d\boldsymbol{\nu}] \, \nu_i(\tau) \, F(\boldsymbol{\nu}) \exp\left[-\frac{1}{2\Omega} \int_{-\infty}^{+\infty} \boldsymbol{\nu}^2(t) dt \right]. \tag{4.9}$$

Note that:

$$\nu_i(\tau) \exp\left[-\frac{1}{2\Omega} \int \boldsymbol{\nu}^2(t) dt \right] = -\Omega \frac{\delta}{\delta \nu_i(\tau)} \exp\left[-\frac{1}{2\Omega} \int \boldsymbol{\nu}^2(t) dt \right]. \tag{4.10}$$

We can therefore integrate path integral (4.9) by parts:

$$\int [d\boldsymbol{\nu}] \, F(\boldsymbol{\nu}) \, \nu_i(\tau) \exp\left[-\frac{1}{2\Omega} \int \boldsymbol{\nu}^2(t) dt \right] = \Omega \int [d\boldsymbol{\nu}] \exp\left[-\frac{1}{2\Omega} \int \boldsymbol{\nu}^2(t) dt \right] \frac{\delta}{\delta \nu_i(\tau)} F(\boldsymbol{\nu}). \tag{4.11}$$

This identity can also be written:

$$\langle F(\boldsymbol{\nu})\,\nu_i(\tau)\rangle = \Omega \left\langle \frac{\delta F(\boldsymbol{\nu})}{\delta \nu_i(\tau)} \right\rangle. \tag{4.12}$$

A similar identity can be proven for any gaussian distribution by the same argument. For example, for:

$$[d\rho(\boldsymbol{\nu})] = [d\boldsymbol{\nu}] \exp\left[-\tfrac{1}{2} \int dt\,dt'\,\nu_i(t) K_{ij}(t,t')\,\nu_j(t') \right],$$

one finds:

$$\langle F(\boldsymbol{\nu})\,\nu_i(\tau)\rangle = \int d\tau'\,\langle \nu_i(\tau)\,\nu_j(\tau')\rangle \left\langle \frac{\delta F}{\delta \nu_j(\tau')} \right\rangle. \tag{4.13}$$

Application. Using identity (4.12) and chain rule, since $\mathbf{q}(t)$ is a functional of $\boldsymbol{\nu}(t)$ through the Langevin equation, we find:

$$\langle \nu_j(t)\delta(\mathbf{q}(t)-\mathbf{q})\rangle = \Omega \left\langle \frac{\delta}{\delta \nu_j(t)}\,\delta(\mathbf{q}(t)-\mathbf{q}) \right\rangle = \Omega \left\langle \frac{\delta q_k(t)}{\delta \nu_j(t)} \frac{\partial}{\partial q_k(t)}\,\delta(\mathbf{q}(t)-\mathbf{q}) \right\rangle. \tag{4.14}$$

Again $\partial/\partial q_k(t)$ can be replaced by $-\partial/\partial q_k$ and taken out of the average. We still have to calculate the functional derivative $\delta q_k(t)/\delta \nu_j(t)$. Let us therefore formally integrate the Langevin equation (4.1):

$$q_i(t) = q_i(t_0) - \tfrac{1}{2} \int_{t_0}^{t} f_i(q(t''))\,dt'' + \int_{t_0}^{t} \nu_i(t'')\,dt''. \tag{4.15}$$

Differentiating with respect to $\nu_j(t')$ we find:

$$\frac{\delta q_i(t)}{\delta \nu_j(t')} = -\tfrac{1}{2} \int_{t'}^{t} \frac{\delta f_i(q(t''))}{\delta \nu_j(t')}\,dt'' + \theta(t-t')\,\delta_{ij}. \tag{4.16}$$

The function $\theta(t)$ is the usual step function. The Langevin equation is causal, which implies that $\mathbf{q}(t'')$ depends on $\boldsymbol{\nu}(t')$ only for $t'' > t'$. This property has been used to restrict the integration range to $t'' > t'$ in the r.h.s. of equation (4.16).

Actually we need only $\delta q_i(t)/\delta \nu_j(t')$ for $t = t'$. From equation (4.16) we see immediately that this limit is ill-defined because we find $\theta(0)$ in the r.h.s. This is a disease of our derivation, not of the equation. The difficulty is related with a well-known property of the Langevin equation (4.1): $d\langle \mathbf{q}^2(t)\rangle/dt$ is well-defined but $2\langle \mathbf{q}(t)\cdot\dot{\mathbf{q}}(t)\rangle$ is not. Due to the singular nature of the noise 2-point correlation function, time differentiation and averaging are two operations which do not commute. We shall illustrate this point later on a particular example (see equation (4.53)). To get around this difficulty it is possible either to first consider the time discretized Langevin equation (see Section 4.8) or to replace the noise 2-point function by a regularized function:

$$\langle \nu_i(t)\nu_j(t')\rangle = \delta_{ij}\,\Omega\,\eta(t-t'), \tag{4.17}$$

in which $\eta(t)$ is a function peaked around $t=0$, even by definition because the 2-point function is symmetric in the exchange $t \leftrightarrow t'$ and satisfies:

$$\int_{-\infty}^{+\infty} \eta(t)\,dt = 1. \tag{4.18}$$

Then, using identities (4.12–4.15), we find that we instead have to calculate

$$\int dt' \eta\left(t - t'\right) \frac{\delta q_k(t)}{\delta \nu_j\left(t'\right)}.$$

The quantity $\theta\left(0\right)$ is then replaced by:

$$\int_{-\infty}^{+\infty} \eta\left(t - t'\right) \theta\left(t - t'\right) dt' = \int_{-\infty}^{0} \eta\left(t'\right) dt' = \tfrac{1}{2}, \qquad (4.19)$$

because $\eta(t)$ is even.

Returning now to equation (4.16) and taking for $\eta(t)$ the δ-function limit we see that the first term of the r.h.s. goes to zero when t goes to t' and therefore:

$$\frac{\delta q_i(t)}{\delta \nu_j(t)} \longrightarrow \tfrac{1}{2}\delta_{ij}. \qquad (4.20)$$

Equation (4.7) therefore becomes:

$$\dot{P}\left(\mathbf{q}, t\right) = \frac{1}{2} \frac{\partial}{\partial q_i} \left[\Omega \frac{\partial P}{\partial q_i} + f_i\left(\mathbf{q}\right) P \right]. \qquad (4.21)$$

This is the Fokker–Planck equation associated with Langevin equation (4.1) and noise (4.2). Taking into account the boundary condition $\mathbf{q}(t_0) = \mathbf{q}_0$, we observe that the equation (4.21) is identical to the Schrödinger equation for the matrix elements of the imaginary time evolution operator of a generally non-hermitian hamiltonian H:

$$H = \tfrac{1}{2}\left(\Omega \hat{\mathbf{p}}^2 + i\hat{\mathbf{p}} \cdot \mathbf{f}\left(\hat{\mathbf{q}}\right)\right), \qquad (4.22)$$

$$P(\mathbf{q}, t) = \langle \mathbf{q}| \, e^{-(t - t_0)H} \, |\mathbf{q}_0\rangle. \qquad (4.23)$$

We have therefore established the formal relation between stochastic differential equations and euclidean quantum mechanics. All observables which can be calculated from the Langevin equation by averaging over the noise, can be recovered by methods of quantum mechanics using the Fokker–Planck (FP) hamiltonian.

4.3 Equilibrium Distribution. Correlation Functions

Methods of quantum mechanics can now be used to study several interesting questions, the first being the existence of an equilibrium distribution. We have only to remember that the FP hamiltonian is not hermitian and therefore its left and right eigenvectors do not coincide. Note that the arguments which follow can be generalized for each connected component if the system is not ergodic and the configuration space is decomposed into disconnected components, a situation we will face in Quantum Field Theory.

The equilibrium distribution $P_0\left(\mathbf{q}\right)$ is the large time limit, if it exists, of $P\left(\mathbf{q}, t\right)$:

$$P_0\left(\mathbf{q}\right) = \lim_{t \to +\infty} P\left(\mathbf{q}, t\right), \qquad (4.24)$$

This implies that $P_0\left(\mathbf{q}\right)$ is a time-independent solution of the FP equation (4.21), i.e a right eigenvector of H with eigenvalue 0, that it is positive and normalizable

$$\int P_0\left(\mathbf{q}\right) d\mathbf{q} < \infty. \qquad (4.25)$$

We now note that conservation of probability has imposed the special form of the FP hamiltonian (4.22), i.e. the factorization on the left of the differential operators $\partial/\partial q_i$. Therefore a constant is a candidate for being a left eigenvector with eigenvalue zero of the hamiltonian H. Introducing a bra and ket notation (not to be confused with the notation meaning average over the noise ν) we can write

$$\langle 0|H = 0, \quad \text{with} \quad \langle 0|\mathbf{q}\rangle = 1.$$

With the same notation we can write

$$H|0\rangle = 0, \quad \text{with} \quad \langle \mathbf{q}|0\rangle = P_0(\mathbf{q}),$$

The condition (4.25) thus is the condition that the vector $|0\rangle$ has a finite norm, a necessary condition for the eigenvalue zero to really belong to the spectrum

$$\langle 0|0\rangle = \int P_0(\mathbf{q}) \, d\mathbf{q} < \infty.$$

Since one can show that the eigenvalues of H have a non-negative real part, the vector $|0\rangle$ then is the ground state of the hamiltonian.

If no such solution exists, all eigenvalues of H have strictly positive real parts and the Langevin equation (4.1) has runaway solutions in the sense that the probability of finding $\mathbf{q}(t)$ inside a ball of arbitrary but finite radius goes to zero at large time:

(i) algebraically if the spectrum of H has a continuous part extending down to the origin (this is the case with the brownian motion, for example),

(ii) exponentially with a rate which is the inverse of the real part of the ground state eigenvalue otherwise. This rate is called the relaxation time. (For more details see Appendix A4).

Finally let us note that the sole knowledge of the equilibrium distribution:

$$P_0(q) = e^{-E(q)}, \tag{4.26}$$

does not determine the functions $f_i(q)$ uniquely. Indeed, by demanding that $P_0(q)$ is a time-independent solution of the FP equation, we only obtain the condition:

$$(\partial_i - \partial_i E)(\Omega \partial_i E - f_i) = 0, \tag{4.27}$$

which, setting:

$$f_i(q) = \Omega \partial_i E(q) + V_i(q) \, e^{E(q)}, \tag{4.28}$$

can be rewritten as a current conservation equation:

$$\partial_i V_i(q) = 0. \tag{4.29}$$

Correlation functions. Up to now we have discussed only equal time averages. However the Langevin equation can also be used to calculate correlation functions of observables at different times, like $Z^{(n)}(t_1, t_2, \ldots, t_n)$:

$$Z^{(n)}(t_1, t_2, \ldots, t_n) = \langle q(t_1) q(t_2) \cdots q(t_n) \rangle_\nu. \tag{4.30}$$

Assuming that the boundary conditions are $q(t_0) = q_0$ and ordering times $t_0 \leq t_1 \leq t_2 << \cdots \leq t_n$ we can use the definition of the FP hamiltonian to rewrite $Z^{(n)}$ as:

$$Z^{(n)}(t_1, t_2, \ldots, t_n) = \langle 0| \, \hat{Q}(t_n) \ldots \hat{Q}(t_2)\hat{Q}(t_1) \, \mathrm{e}^{Ht_0} \, |q_0\rangle \,, \tag{4.31}$$

with $\hat{Q}(t)$ being the operator $\hat{q}$ in the Heisenberg representation

$$\hat{Q}(t) = \mathrm{e}^{Ht} \, \hat{q} \, \mathrm{e}^{-Ht} \,.$$

We recognize a form similar to the representation (2.63) of correlation functions in quantum mechanics. Introducing the time-ordering operation T acting on quantum operators we can then write, irrespective now of the order between times, (see equation (2.64))

$$Z^{(n)}(t_1, t_2, \ldots, t_n) = \langle 0| \, \mathrm{T}[\hat{Q}(t_n) \ldots \hat{Q}(t_2)\hat{Q}(t_1)] \, \mathrm{e}^{Ht_0} \, |q_0\rangle \,. \tag{4.32}$$

When an equilibrium distribution exists, in the limit $t_0 \to -\infty$ (boundary conditions in far past) the correlation functions converge towards *equilibrium correlation functions*:

$$Z^{(n)}(t_1, t_2, \ldots, t_n) = \langle 0| \, \mathrm{T}[\hat{Q}(t_n) \ldots \hat{Q}(t_2)\hat{Q}(t_1)] \, |0\rangle \,. \tag{4.33}$$

We recognize the vacuum or ground state expectation value (2.65) of time-ordered products of operators introduced in Section 2.5.

Time evolution of observables. Going from the FP equation to expressions in terms of time-dependent operators, we have passed from a Schrödinger picture to a Heisenberg picture. In the Heisenberg picture one directly writes equations for the evolution of time-dependent operators. Let us give an application here. In terms of Heisenberg operators the average of an observable $\mathcal{O}(q)$ at time t can be written

$$\langle \mathcal{O}(q(t)) \rangle_\nu = \langle 0| \, \mathcal{O}[\hat{Q}(t)] \, \mathrm{e}^{Ht_0} \, |q_0\rangle \,.$$

We can write an evolution equation for the operator $\mathcal{O}[\hat{Q}(t)] \, \mathrm{e}^{Ht_0}$:

$$\frac{\mathrm{d}}{\mathrm{d}t} \mathcal{O}[\hat{Q}(t)] \, \mathrm{e}^{Ht_0} = [\, H \,, \, \mathcal{O}[\hat{Q}(t)] \, \mathrm{e}^{Ht_0} \,] \,.$$

For the matrix elements $\mathcal{O}(q, t)$ of the operator

$$\mathcal{O}(q, t) \equiv \langle 0| \mathcal{O}[\hat{Q}(t)] \, \mathrm{e}^{Ht_0} \, |q\rangle ,$$

this translates into the partial differential equation

$$\dot{\mathcal{O}}(q, t) = \frac{1}{2} \left[\Omega \frac{\partial}{\partial q} - f(q) \right] \frac{\partial}{\partial q} \mathcal{O}(q, t). \tag{4.34}$$

Then the averages $\langle \mathcal{O}(q(t)) \rangle_\nu$ are simply obtained by integrating $\mathcal{O}(q, t)$ with the initial distribution at time t_0, here

$$\langle \mathcal{O}(q(t)) \rangle_\nu = \int \mathrm{d}q \, \mathcal{O}(q, t)\delta(q - q_0). \tag{4.35}$$

4.4 A Special Case: The Purely Dissipative Langevin Equation

In one special case the hamiltonian (4.22) is equivalent to a hermitian hamiltonian, when $f_i(\mathbf{q})$ is a gradient:

$$f_i(\mathbf{q}) = \Omega \, \partial_i E(\mathbf{q}) \, . \tag{4.36}$$

This corresponds to the purely dissipative Langevin equation:

$$\dot{q}_i = -\tfrac{1}{2}\Omega \partial_i E(\mathbf{q}) + \nu_i(t) \, . \tag{4.37}$$

Then setting:

$$P(\mathbf{q},t) = e^{-\frac{1}{2}E(\mathbf{q})} \langle \mathbf{q} \,|U(t,t_0)|\, \mathbf{q}_0 \rangle \, e^{\frac{1}{2}E(\mathbf{q}_0)}, \tag{4.38}$$

we transform equation (4.21) into:

$$\frac{\partial}{\partial t} \langle \mathbf{q}\,|U(t,t_0)|\,\mathbf{q}_0 \rangle = -\frac{\Omega}{2} \left(-\frac{\partial}{\partial q_i} + \tfrac{1}{2}\partial_i E \right) \left(\frac{\partial}{\partial q_i} + \tfrac{1}{2}\partial_i E \right) \langle \mathbf{q}\,|U(t,t_0)|\,\mathbf{q}_0 \rangle \, . \tag{4.39}$$

The hamiltonian H transforms into a hamiltonian $\widetilde{H}$:

$$\widetilde{H} = \frac{\Omega}{2} A_i^\dagger A_i \equiv \frac{\Omega}{2} \left[\hat{\mathbf{p}}^2 + \tfrac{1}{4}\left(\nabla E(\hat{\mathbf{q}})\right)^2 - \tfrac{1}{2}\Delta E(\hat{\mathbf{q}}) \right], \tag{4.40}$$

in which $\mathbf{A}$ is the operator:

$$\mathbf{A}(q, \partial/\partial q) \equiv \nabla_q + \tfrac{1}{2}\nabla E \equiv i\mathbf{p} + \tfrac{1}{2}\nabla E \, , \tag{4.41}$$

and $U(t,t_0) = e^{-(t-t_0)\widetilde{H}}$ is the corresponding evolution operator in imaginary time.

We see that the hamiltonian $\widetilde{H}$ is positive. Moreover if the wave function $e^{-E(\mathbf{q})/2}$ is normalizable:

$$\langle \mathbf{q}|0\rangle = e^{-E(\mathbf{q})/2}, \qquad \langle 0|0\rangle = \int d\mathbf{q} \, e^{-E(\mathbf{q})} < \infty \, ,$$

it corresponds to the eigenvalue zero:

$$\mathbf{A}|0\rangle = 0 \quad \Rightarrow \quad \widetilde{H}|0\rangle = 0 \, ,$$

and thus is the ground state of $\widetilde{H}$. At large times the operator $e^{-(t-t_0)\widetilde{H}}$ projects onto its ground state. The interpretation of this result in terms of the probability distribution $P(\mathbf{q},t)$ is then:

$$\lim_{t\to\infty} P(\mathbf{q},t) = e^{-E(\mathbf{q})/2} \, e^{E(\mathbf{q}_0)/2} \langle \mathbf{q}|0\rangle\langle 0|\mathbf{q}_0\rangle = e^{-E(\mathbf{q})} \, .$$

The distribution $P(\mathbf{q},t)$ converges at large time towards the equilibrium distribution $e^{-E(\mathbf{q})}$.

If the wave function $e^{-E(\mathbf{q})/2}$ is not normalizable instead, there exists no equilibrium distribution, Langevin equation (4.1) has only runaway solutions.

Remark. The case of the purely dissipative Langevin equation (4.37) corresponds to detailed balance for discrete processes (see Appendix A4.2). The drift force f_i is then called conservative. In the absence of noise the Langevin equation reduces to a gradient flow:

$$\dot{q}_i(t) = -\tfrac{1}{2}\Omega \, \partial_i E(\mathbf{q}(t)) \, . \tag{4.42}$$

Taking the scalar product with the vector $\dot{\mathbf{q}}$ and integrating we obtain:

$$\int_{t_0}^{t} \dot{\mathbf{q}}^2(t')dt' = -\tfrac{1}{2}\Omega \left[E(\mathbf{q}(t)) - E(\mathbf{q}(t_0)) \right] \, . \tag{4.43}$$

Therefore, in the absence of noise, $E(\mathbf{q}(t))$ is a monotonous decreasing function of time.

4.5 A Simple Example: The Linear Langevin Equation

Let us consider the equation

$$\dot{q} = -\omega q + \nu(t),\qquad(4.44)$$

in which $\nu(t)$ is the gaussian noise of equation (4.3), with $\Omega = 1$. This equation provides an example of the special case discussed in Section 4.4:

$$\omega q = \frac{\partial}{\partial q}\left(\tfrac{1}{2}\omega q^2\right).$$

After the transformation (4.38), the associated FP hamiltonian takes the form:

$$H = \tfrac{1}{2}p^2 + \tfrac{1}{2}\omega^2 q^2 - \tfrac{1}{2}\omega.\qquad(4.45)$$

The eigenvalues ϵ_n of H are:

$$\epsilon_n = \left(n + \tfrac{1}{2}\right)|\omega| - \tfrac{1}{2}\omega,\quad n \geq 0.\qquad(4.46)$$

We see immediately that if ω is positive, ϵ_0 vanishes. At the same time $e^{-\omega q^2}$ is normalizable and is thus the equilibrium distribution.

If ω is negative:

$$\omega < 0,\qquad \epsilon_0 = -\omega,$$

the lowest eigenvalue is positive, $e^{-\omega q^2}$ is not normalizable and for large time the distribution $P(q,t)$ becomes (taking $t_0 = 0$ for convenience):

$$P(q,t) \sim e^{-|\omega|t}\, e^{-|\omega|q_0^2} + O\left(e^{-2|\omega|t}\right).\qquad(4.47)$$

This expression shows that the probability of finding q at a finite distance from the origin decreases exponentially at a rate $\tau = 1/|\omega|$.

Finally the special case $\omega = 0$ corresponds to the simple brownian motion, the spectrum of the FP hamiltonian is continuous and covers $[0, +\infty]$, and the probability of remaining at a finite distance from the origin decreases algebraically as $1/\sqrt{t}$.

Let us recover these results by solving the Langevin equation (4.44) explicitly:

$$q(t) = q_0\, e^{-\omega t} + \int_0^t e^{-\omega(t-t')}\,\nu(t')\,dt'.\qquad(4.48)$$

Averaging the equation over the noise we find:

$$\langle q(t)\rangle = q_0\, e^{-\omega t}.\qquad(4.49)$$

This expression already shows that for $\omega < 0$ no equilibrium is reached. Moreover for $q_0 \neq 0$, $\langle q(t)\rangle$ grows exponentially with time.

Since $q(t)$ is linearly related to $\nu(t)$, it has a gaussian distribution which is characterized by $\langle q(t)\rangle$ and $\langle q^2(t)\rangle$. From (4.48) we obtain:

$$\left\langle [q(t) - \langle q(t)\rangle]^2\right\rangle = \frac{1}{2\omega}\left(1 - e^{-2\omega t}\right).\qquad(4.50)$$

The gaussian distribution $P(q, t)$ of $q(t)$, determined by its two first moments, is thus given by:

$$P(q, t) = \left[\frac{\pi}{\omega} \left(1 - e^{-2\omega t} \right) \right]^{-1/2} \exp \left[-\frac{\omega}{(1 - e^{-2\omega t})} \left(q - q_0\, e^{-\omega t} \right)^2 \right]. \qquad (4.51)$$

This expression is equivalent to expression (2.37) and contains the asymptotic form (4.47).

To illustrate the difficulty encountered in the derivation of the FP equation let us calculate the $q(t)$ 2-point function for $q_0 = 0$:

$$\langle q\,(t_1)\, q\,(t_2) \rangle = \frac{1}{\omega} \left[\theta\,(t_2 - t_1)\, e^{-\omega t_2} \sinh \omega t_1 + \theta\,(t_1 - t_2)\, e^{-\omega t_1} \sinh \omega t_2 \right]. \qquad (4.52)$$

Differentiating with respect to t_1, we find:

$$\langle \dot{q}\,(t_1)\, q\,(t_2) \rangle = \theta\,(t_2 - t_1)\, e^{-\omega t_2} \cosh \omega t_1 - \theta\,(t_1 - t_2)\, e^{-\omega t_1} \sinh \omega t_2. \qquad (4.53)$$

The limit $t_1 \to t_2$ is clearly ill-defined while if we take this limit in expression (4.52) first and then differentiate we find:

$$\frac{1}{2} \frac{\mathrm{d}}{\mathrm{d}t} \langle q^2\,(t) \rangle = \frac{1}{2}\, e^{-2\omega t}.$$

The same result is obtained from equation (4.53) by setting $\theta\,(0) = \frac{1}{2}$, i.e. taking the half sum of the derivative from above and below.

Remark. It is also easy to solve a more general Langevin equation of the form:

$$\dot{q} = -\omega q + g(t) + \nu(t),$$

in which $g(t)$ is a given function of time. The same arguments as above would have led to a formula for $P\,(q, t)$ equivalent to expression (2.37) with the identification:

$$f(t) = m\,[\omega g(t) - \dot{g}(t)].$$

We leave this calculation as an exercise.

4.6 Path Integral Representation

Applying the method described in Chapter 2 to equation (4.21), we can derive path integral representations for the probability distribution $P\,(\mathbf{q}, t)$ and averaged observables. Let us however give here a more direct derivation based on Langevin equation (4.1) and the generalization of identity (1.26) to an infinite number of variables.

We calculate the generating functional $Z(j)$ of correlation functions

$$Z(\mathbf{j}) = \left\langle \exp \int \mathrm{d}t\, \mathbf{j}(t) \mathbf{q}(t) \right\rangle_{\nu}. \qquad (4.54)$$

We use the boundary condition:

$$\mathbf{q}(t_0) = \mathbf{q}_0.$$

Imposing then Langevin equation (4.1) by a product of δ-functions, we write $Z(\mathbf{j})$ as:

$$Z(\mathbf{j}) = \int [d\nu(t)] \exp\left[-\frac{1}{2\Omega}\int \nu^2(t)dt\right] \int_{\mathbf{q}(t_0)=\mathbf{q}_0} [d\mathbf{q}(t)] \exp \int dt\, \mathbf{j}(t)\mathbf{q}(t)$$
$$\times \det \mathbf{M} \prod_{i,t} \delta\left[\nu_i(t) - \dot{q}_i(t) - \tfrac{1}{2} f_i\left(\mathbf{q}(t)\right)\right]. \tag{4.55}$$

According to the identity (1.26) the operator $\mathbf{M}$ is the functional derivative of Langevin equation (4.1) with respect to $\mathbf{q}(t)$

$$M_{ij}(t,\tau) = \frac{\mathrm{d}}{\mathrm{d}t}\delta(t-\tau)\,\delta_{ij} + \tfrac{1}{2}\frac{\partial f_i\left(\mathbf{q}(t)\right)}{\partial q_j}\delta(t-\tau). \tag{4.56}$$

In expression (4.55) the integral over the noise is trivial and we obtain

$$Z(\mathbf{j}) = \int_{\mathbf{q}(t_0)=\mathbf{q}_0} [d\mathbf{q}(t)] \det \mathbf{M} \exp\left\{-\int dt\,\left[\left(\dot{\mathbf{q}} + \tfrac{1}{2}\mathbf{f}(q)\right)^2/2\Omega - \mathbf{j}(t)\mathbf{q}(t)\right]\right\}. \tag{4.57}$$

The determinant. We now evaluate the determinant of the operator $\mathbf{M}$. It can be factorized into:

$$\det \mathbf{M} = \det \frac{\mathrm{d}}{\mathrm{d}t}\delta(t-\tau)\det \widetilde{\mathbf{M}}. \tag{4.58}$$

The first factor is an irrelevant constant. The operator $\widetilde{\mathbf{M}}$ is obtained by differentiating with respect to $\mathbf{q}(t)$ the integrated form (4.15) of the Langevin equation:

$$\widetilde{M}_{ij}(t,\tau) = \delta_{ij}\delta(t-\tau) + \tfrac{1}{2}\theta(t-\tau)\frac{\partial f_i}{\partial q_j}\left(\mathbf{q}(\tau)\right), \tag{4.59}$$

in which $\theta(t)$ is again the step function.

Let us set:

$$\tfrac{1}{2}\frac{\partial f_i}{\partial q_j}\left(\mathbf{q}(\tau)\right) = v_{ij}(\tau), \tag{4.60}$$

and use identity (1.33):

$$\det = \exp \operatorname{tr} \ln,$$

to expand $\det \widetilde{\mathbf{M}}$:

$$\det \widetilde{\mathbf{M}} = \exp\left\{\theta(0)\int \operatorname{tr} v(t)dt - \tfrac{1}{2}\int \theta(t_1-t_2)\,\theta(t_2-t_1)\operatorname{tr}\left[v(t_1)\,v(t_2)\right]dt_1 dt_2 + \cdots\right.$$
$$\left. + \frac{(-1)^{n+1}}{n}\int \theta(t_1-t_2)\cdots\theta(t_n-t_1)\operatorname{tr}\left[v(t_1)\cdots v(t_n)\right]dt_1\cdots dt_n + \cdots\right\}. \tag{4.61}$$

Due to the product of θ-functions all terms vanish but the first one:

$$\det \widetilde{\mathbf{M}} = \exp\left[\theta(0)\int \operatorname{tr} v(t)\,dt\right]. \tag{4.62}$$

We again find a result involving the ill-defined quantity $\theta(0)$. Semiclassical evaluations of path integrals generate similarly ambiguous expressions (see Section 38.2). To give a meaning to expression (4.62) one can discretize time in Langevin equation (4.1) (see Section 4.8). After taking the continuum limit one finds:

$$\theta\left(0\right) = \tfrac{1}{2}\,.$$

The path integral. We thus obtain the path integral representation

$$Z(\mathbf{j}) = \int_{\mathbf{q}(t_0)=\mathbf{q}_0} [d\mathbf{q}(t)] \exp\left[-S(\mathbf{q})/\Omega + \int dt\,\mathbf{j}(t)\mathbf{q}(t)\right], \qquad (4.63)$$

with the definition:

$$S\left(\mathbf{q}\right) = \int_{t_0} \tfrac{1}{2}\left[\left(\dot{\mathbf{q}} + \tfrac{1}{2}\mathbf{f}(q)\right)^2 - \tfrac{1}{2}\Omega\partial_i f_i(q)\right] dt\,. \qquad (4.64)$$

Similarly one derives a representation for $P\left(\mathbf{q},t\right)$:

$$P\left(\mathbf{q},t\right) = \int_{\mathbf{q}(t_0)=\mathbf{q}_0}^{\mathbf{q}(t)=\mathbf{q}} [d\mathbf{q}(\tau)] \exp\left[-S\big(\mathbf{q}(\tau)\big)/\Omega\right], \qquad (4.65)$$

It is easy to verify that the same path integral representation is obtained when the method explained in Chapter 2 is applied to the Schrödinger-like equation (4.21).

It follows in particular from the path integral representation (4.65) that, as expected, a solution of equation (4.21) which is a positive distribution $P\left(\mathbf{q}_0,t_0\right)$ at initial time t_0 remains positive at any later time.

Perturbative expansion. The parameter Ω here plays the role of $\hbar$ in quantum mechanics. It orders naturally perturbation theory. We then note that, at leading order, the classical action reduces to:

$$S\left(\mathbf{q}\right) = \int \tfrac{1}{2}\left(\dot{\mathbf{q}} + \tfrac{1}{2}\mathbf{f}(q)\right)^2 dt\,. \qquad (4.66)$$

This means that in the particular example of Langevin equation (4.37), perturbation theory has to be expanded around one of the extrema of $E(q)$. However not all extrema are equivalent once one takes into account the first quantum correction due to the additional term $\Omega\partial_i f_i(q)$, as the example of Section 4.5 reveals. Expression (4.46) indeed shows that the second term $\Delta E(q)$ lifts the degeneracy between minima and maxima and that only minima are good starting points for perturbation theory.

4.7 Stochastic Processes on Riemannian Manifolds

We now consider the most general markovian Langevin equation which has the form:

$$\dot{q}^i(t) = -\tfrac{1}{2}f^i\big(\mathbf{q}(t)\big) + e_a^i\big(q(t)\big)\nu_a(t), \qquad (4.67)$$

in which $\nu_a(t)$ is again a gaussian white noise:

$$\langle\nu_a(t)\rangle = 0\,, \qquad \langle\nu_a(t)\nu_b\left(t'\right)\rangle = \delta\left(t - t'\right)\delta_{ab}\,. \qquad (4.68)$$

The linearity in the noise is a consequence of the markovian character of the noise distribution which implies that the noise correlation functions are proportional to δ-functions in time (see Section 4.8).

In contrast with the Langevin equation (4.1), equation (4.67) is somewhat ambiguous, the difficulty being of exactly the same nature as the problem of ordering of operators arising in the quantization of a hamiltonian of the form (3.24).

The problem can as usual be best understood by discretizing time (see Section 4.8 for details). Let us call ε the time step and replace equation (4.67) by:

$$q^i\left(t+\varepsilon\right) = q^i(t) - \tfrac{1}{2}\varepsilon f^i\left(q(t)\right) + e^i_a\left[\tfrac{1}{2}\left(q(t) + q\left(t+\varepsilon\right)\right)\right]\nu_a(t), \qquad (4.69)$$

with for dimensional reasons:

$$\langle \nu_a(t)\nu_b\left(t'\right)\rangle = \varepsilon\delta_{tt'}\delta_{ab}. \qquad (4.70)$$

Equations (4.69,4.70) show that, in contrast with usual differential equations, the variation $q^i\left(t+\varepsilon\right) - q^i(t)$ is of order $\sqrt{\varepsilon}$, which is typical of the brownian motion. Hence if in the discretized equation (4.69) we replace for example $e^i_a(q(t))$ by $e^i_a\left[\tfrac{1}{2}(q(t) + q(t+\varepsilon))\right]$, quantities which are formally indistinguishable in the continuum limit, we change the equation at order ε. Instead, if we make the same substitution in $f(q)$, the equation is modified by terms of order $\varepsilon^{3/2}$ which are negligible.

In what follows we adopt the symmetric Stratanovich convention (4.69), although it is unnatural from the practical point of view, because it has simpler transformation properties under a change of variables and therefore of coordinates on a manifold.

The Fokker–Planck equation. Adapting the method of Section 4.2 one derives a Fokker–Planck equation for the time-dependent distribution $P\left(\mathbf{q}, t\right)$.

Without giving all details let us just indicate the main steps. From the Langevin equation (4.67) we immediately obtain:

$$\dot{P}(q,t) = \frac{\partial}{\partial q^i}\left[\tfrac{1}{2}f^i(q)P - e^i_a(q)\left\langle \nu_a(t)\delta\left(q(t) - q\right)\right\rangle\right]. \qquad (4.71)$$

Using the same regularization for the noise 2-point function we can calculate the second term in the r.h.s.:

$$\left\langle \nu_a(t)\delta\big(q(t) - q\big)\right\rangle = -\frac{1}{2}\frac{\partial}{\partial q^j}\left(e^j_a(q)P\right). \qquad (4.72)$$

The FP equation then takes the form:

$$\dot{P}\left(q,t\right) = \frac{1}{2}\frac{\partial}{\partial q^i}\left[e^i_a(q)\frac{\partial}{\partial q^j}\left(e^j_a(q)P\right) + f^i(q)P\right]. \qquad (4.73)$$

The only examples we shall meet in this work correspond to Langevin equations on Riemannian manifolds.

Brownian motion on a Riemannian manifold. Covariance under reparametrization of the manifold implies that the FP equation for the free brownian motion on a manifold should have the form (for more details see Chapter 22):

$$\dot{D}\left(q,t\right) = \frac{1}{2\sqrt{g}}\frac{\partial}{\partial q^i}\left[\sqrt{g}\left(g^{-1}\right)^{ij}\frac{\partial}{\partial q^j}D\right], \qquad (4.74)$$

in which $g_{ij}(q)$ is the metric tensor on the manifold, g the determinant of $g_{ij}(q)$:

$$g = \det g_{ij}, \tag{4.75}$$

and $D(q,t)$ a scalar density, i.e. a density normalized with the covariant measure $\sqrt{g}\,dq$. It is therefore related to $P(q,t)$ by:

$$\sqrt{g}D(q,t) = P(q,t). \tag{4.76}$$

If the manifold is compact, a constant scalar density D is obviously a static solution to equation (4.74) and therefore corresponds to the equilibrium distribution.

Comparing equations (4.73) and (4.74) we conclude that:

$$\left(g^{-1}\right)^{ij} = e_a^i(q)e_a^j(q), \tag{4.77}$$

which means that the matrix e_a^i, if it is a square matrix, is the inverse of the vielbein (Section 22.6). We see that by construction the r.h.s. is a positive symmetric matrix.

Furthermore we observe that the free brownian motion does not correspond in general to $\mathbf{f} = 0$ since, as a short calculation shows, equations (4.73) and (4.74) are identical only if:

$$f^i(q) = -e_a^i(q)\nabla_j e_a^j(q), \tag{4.78}$$

in which ∇ is the covariant derivative on the manifold. The equation shows that $\mathbf{f}$ does not depend only on the geometry of the manifold but also on the choice of the vielbein.

Let us finally note that equation (4.73) is the most general second order stochastic differential equation: the operator $\partial/\partial q^i$ is implied by total probability conservation, the positivity of the coefficient of $\partial^2/\partial q^i\partial q^j$ by the condition that the solution remains a positive distribution at all times as will become clear once we construct the path integral representation.

Path integral representation. From now on we restrict ourselves, for simplicity, to the case in which the matrix $e_a^i(q)$ is an invertible square matrix. Equation (4.67) can then be rewritten

$$\nu_a(t) = [e^{-1}]_{ai}(q)\left(\dot{q}^i(t) + \tfrac{1}{2}f^i(q)\right). \tag{4.79}$$

We have first to calculate the determinant of $\delta\nu_a/\delta q^i$:

$$M_{ai}(t,\tau) \equiv \frac{\delta\nu_a(t)}{\delta q^i(\tau)} = [e^{-1}]_{aj}(q(t))\widetilde{M}_i^j, \tag{4.80}$$

with

$$\widetilde{M}_i^j = \left\{\delta_i^j\frac{d}{dt} + \frac{1}{2}\frac{\partial f^j}{\partial q^i} + e_b^j\frac{\partial}{\partial q^i}[e^{-1}]_{b\ell}\left[\dot{q}^\ell + \tfrac{1}{2}f^\ell(q)\right]\right\}\delta(t-\tau). \tag{4.81}$$

The determinant can thus be written:

$$\det\mathbf{M} \sim \det\widetilde{\mathbf{M}}\prod_t\left[\det\mathbf{e}\left(q(t)\right)\right]^{-1}, \tag{4.82}$$

with, as a consequence of identity (4.62)

$$\det\widetilde{\mathbf{M}} \propto \exp\left[\frac{1}{2}\int dt\left(\frac{1}{2}\frac{\partial f^i}{\partial q^i} + e_b^i\frac{\partial}{\partial q^i}[e^{-1}]_{b\ell}\left(\dot{q}^\ell + \tfrac{1}{2}f^\ell\right)\right)\right]. \tag{4.83}$$

In expression (4.82) we again find a divergent factor analogous to the normalization factor (3.29). This factor can only be defined by a limiting procedure, for example we can use the discretized Langevin equation (4.69). In both cases the appearance of this divergent factor is related to the problem of ordering of operators for the quantum mechanics or FP hamiltonian as already discussed in Section 3.2.

Collecting all factors we obtain a path integral representation for $P(\mathbf{q}, t)$:

$$P(q,t) = \int_{q(t_0)=q_0}^{q(t)=q} [dq(\tau)] \prod_\tau [\det \mathbf{e}\,(q(\tau))]^{-1} \exp\left[-S(q)\right], \tag{4.84}$$

with the definition:

$$S(q) = \frac{1}{2}\int_{t_0}^t d\tau \left[\left(\dot{q}^i + \tfrac{1}{2}f^i(q)\right)[e^{-1}]_{ai}(q)[e^{-1}]_{aj}(q)\left(\dot{q}^j + \tfrac{1}{2}f^j(q)\right) \right.$$
$$\left. - \tfrac{1}{2}\frac{\partial f^i}{\partial q^i} - e_b^i\left(\frac{\partial}{\partial q^i}[e^{-1}]_{b\ell}\right)\left(\dot{q}^\ell + \tfrac{1}{2}f^\ell(q)\right) \right]. \tag{4.85}$$

In the case of a Langevin equation on a Riemannian manifold the determinant in expression (4.84), as a consequence of relation (4.77), is given by:

$$[\det \mathbf{e}(q)]^{-1} = [\det \mathbf{g}]^{-1/2},$$

where $\mathbf{g}$ is the metric tensor. It therefore formally reconstructs the covariant measure on the manifold (see Section 22.5).

4.8 Discretized Langevin Equation

We now discuss a discretized, markovian, time translation invariant Langevin equation in the small step size limit and in the simple case of a gaussian noise. This is a useful exercise in the sense that it clarifies a few subtle points which have appeared in the discussion of the continuum equation. Let us call ε the time step. The Langevin equation, discrete analogue of equation (4.1), is then:

$$q_i(t + \varepsilon) = q_i(t) - \tfrac{1}{2}\varepsilon f_i(\mathbf{q}(t)) + \nu_i(t). \tag{4.86}$$

Note that the equation can be considered as a discretized form of the continuum equation only if the function $f(\mathbf{q})/|\mathbf{q}|$ remains bounded for $|\mathbf{q}|$ large. Otherwise it is necessary to improve it by adding higher order terms in ε.

The noise distribution is factorized and at a given time, for each degree of freedom, the distribution is:

$$d\rho(\nu) = \frac{d\nu}{\sqrt{2\pi\varepsilon\Omega}} \exp\left(-\frac{\nu^2}{2\varepsilon\Omega}\right). \tag{4.87}$$

Equivalently the gaussian distribution can be defined the 1- and 2-point correlation functions:

$$\langle \nu_i(t) \rangle = 0, \qquad \langle \nu_i(t)\nu_j(t') \rangle = \varepsilon\,\Omega\,\delta_{tt'}\delta_{ij}. \tag{4.88}$$

The dependence of the noise on ε, $\nu \propto \sqrt{\varepsilon}$, is typical of the brownian motion and it will be justified below that it yields the correct continuum limit.

From equations (4.86,4.87) it is easy to derive the discrete analogue of the FP equation. Equation (4.86) implies a relation between the time-dependent distributions $P(\mathbf{q}, t)$ at consecutive times:

$$P\left(\mathbf{q}, t+\varepsilon\right) \equiv \langle \delta\left(\mathbf{q}\left(t+\varepsilon\right)-\mathbf{q}\right)\rangle_\nu = \langle \delta\left(\mathbf{q}(t)-\tfrac{1}{2}\varepsilon\mathbf{f}(\mathbf{q}(t))+\boldsymbol{\nu}(t)-\mathbf{q}\right)\rangle_\nu . \qquad (4.89)$$

The average over the noise $\boldsymbol{\nu}$ is restricted to the noise at time t, the average over anterior times being performed by the integration over $q(t)$. Moreover the noise $\boldsymbol{\nu}(t)$ and $\mathbf{q}(t)$ at equal time are uncorrelated, as a consequence of the causality of the Langevin equation. The $\mathbf{q}(t)$ distribution by definition is $P(\mathbf{q}, t)$ and the noise distribution is given by (4.87). Thus:

$$P\left(\mathbf{q}, t+\varepsilon\right) = \int \mathrm{d}q'\, P(\mathbf{q}', t) \mathrm{d}\rho(\boldsymbol{\nu})\delta\left(\mathbf{q}'-\tfrac{1}{2}\varepsilon\mathbf{f}(\mathbf{q}')+\boldsymbol{\nu}-\mathbf{q}\right). \qquad (4.90)$$

The integral over the noise is trivial and can be performed even for a general noise distribution (we leave as an exercise to study the continuum limit in the latter case). With the gaussian noise (4.87) we obtain

$$P\left(\mathbf{q}, t+\varepsilon\right) = (2\pi\Omega)^{-d/2} \int \mathrm{d}q'\, P(\mathbf{q}', t) \exp\left[-\frac{1}{2\varepsilon\Omega}\left(\mathbf{q}'-\tfrac{1}{2}\varepsilon\mathbf{f}(\mathbf{q}')-\mathbf{q}\right)^2\right]. \qquad (4.91)$$

This equation has the general form:

$$P(q, t+\varepsilon) = \int \mathrm{d}q'\, p(q, q')P(q', t). \qquad (4.92)$$

The kernel $p(q, q')$ which acts on $P(q, t)$ is the equivalent of the evolution operator in a small time interval as given by (3.25). It plays exactly the role of a transfer matrix in a one-dimensional statistical system. A few remarks concerning equations of the form (4.92) can be found in Appendix A4.

The continuum limit. To derive the continuum FP equation it is convenient to start from equation (4.90) and to first integrate over q' using the δ-function. This leads to a jacobian $\mathcal{J}$ which we expand at first order in ε (using $\ln \det = \mathrm{tr}\ln$):

$$\mathcal{J} = \left|\det\left(\frac{\partial q_i'}{\partial \nu_j}\right)\right| = \left|\det{}^{-1}\left(\frac{\partial \nu_i}{\partial q_j'}\right)\right| = \det{}^{-1}\left(\delta_{ij}-\tfrac{1}{2}\varepsilon\frac{\partial f_i}{\partial q_j'}\right) = 1 + \tfrac{1}{2}\varepsilon\frac{\partial f_i}{\partial q_i} + O\left(\varepsilon^2\right).$$

In the same way we to expand $P(\mathbf{q}', t)$ in powers of ε, remembering that ν is of order $\sqrt{\varepsilon}$:

$$P(\mathbf{q}', t) = P(\mathbf{q}, t) + \left(\tfrac{1}{2}\varepsilon f_i(q)-\nu_i\right)\frac{\partial P}{\partial q_i} + \tfrac{1}{2}\nu_i\nu_j\frac{\partial^2 P}{\partial q_i\partial q_j} + O\left(\varepsilon^{3/2}\right).$$

Averaging then over ν and collecting all terms we find

$$P(\mathbf{q}, t+\varepsilon) = P(\mathbf{q}, t) + \tfrac{1}{2}\varepsilon\left[\Omega\Delta P + f_i(q)\frac{\partial P}{\partial q_i}+\frac{\partial f_i}{\partial q_i}P\right],$$

equation which is equivalent to equation (4.21) in the small ε limit.

Note that in equation (4.86) only the terms up to order ε are relevant in the continuum limit. In particular it is possible (although not convenient for practical use) to replace $f(q(t))$, by $f[(q(t)+q(t+\varepsilon))/2]$ since this leads to a modification of order $\varepsilon^{3/2}$ in the equation. This last form is the most convenient, however, when change of coordinates are involved because it has simpler transformation properties.

Let us finally also note that equation (4.91) iterated in time immediately leads to a discretized form of the path integral representation of $P(q,t)$. It is then possible to study the small time step limit by expanding in powers of ε up to order ε. The important point is that the typical values of $\mathbf{q} - \mathbf{q}'$ are of order $\varepsilon^{1/2}$. Then

$$\mathbf{q}' - \tfrac{1}{2}\varepsilon \mathbf{f}(\mathbf{q}') - \mathbf{q} = \mathbf{q}' - \mathbf{q} - \tfrac{1}{2}\varepsilon \mathbf{f}((\mathbf{q} + \mathbf{q}')/2) - \tfrac{1}{4}\varepsilon(q_i' - q_i)\frac{\partial \mathbf{f}}{\partial q_i} + O\left(\varepsilon^2\right)$$

$$\frac{1}{2\varepsilon}\left(\mathbf{q}' - \tfrac{1}{2}\varepsilon \mathbf{f}(\mathbf{q}') - \mathbf{q}\right)^2 = \frac{1}{2\varepsilon}\left(\mathbf{q}' - \mathbf{q} - \tfrac{1}{2}\varepsilon \mathbf{f}((\mathbf{q} + \mathbf{q}')/2)\right)^2 - \frac{1}{4}(q_i' - q_i)(q_j' - q_j)\frac{\partial f_i}{\partial q_j}.$$

We need the terms of order ε. In the last term which already if of order ε we can replace $(q_i' - q_i)(q_j' - q_j)$ by its leading order average $\Omega \varepsilon \delta_{ij}$. We then find a discretized form of the action (4.64).

Generalization. A more general form of the discretized Langevin equation is:

$$q_i(t + \varepsilon) = q_i(t) - \frac{\varepsilon}{2}f_i\left(\mathbf{q}(t)\right) + e_{ia}\left(\mathbf{q}(t)\right)\nu_a(t) + d_{iab}\left(\mathbf{q}(t)\right)\nu_a(t)\nu_b(t), \tag{4.93}$$

with the gaussian noise still defined by:

$$\langle \nu_a(t) \rangle = 0, \qquad \langle \nu_a(t)\nu_b(t') \rangle = \varepsilon \, \Omega \, \delta_{tt'}\delta_{ab}. \tag{4.94}$$

Equations (4.93) are the discrete analogues of equations (4.67,4.68).

Note that we have only expanded the r.h.s. of the Langevin equation up to second order in $\nu(t)$. The reasons are the following:

(i) the noise 2-point function must be proportional to ε for dimensional reasons; the consistency of this choice will be checked in the derivation of the FP equation;

(ii) since $\nu_a(t)$ is of order $\sqrt{\varepsilon}$, it is sufficient to consider terms at most quadratic in the noise in the Langevin equation, higher order terms give contributions of order at least $\varepsilon^{3/2}$ and therefore negligible in the small ε limit.

From the Langevin equation (4.93) we can derive by the method indicated above an equation for the probability distribution $P(\mathbf{q}, t)$. Writing equation (4.93) as:

$$q_i\left(t + \varepsilon\right) = q_i(t) - F_i\left(\mathbf{q}(t), \nu(t)\right), \tag{4.95}$$

we obtain:

$$P(\mathbf{q}, t + \varepsilon) = \int d q' \, P\left(\mathbf{q}', t\right)\langle \delta\left[\mathbf{q} - \mathbf{q}' + \mathbf{F}\left(\mathbf{q}', \nu\right)\right]\rangle_\nu. \tag{4.96}$$

We now average over the noise at time t. Since the function $\mathbf{F}(\mathbf{q}, \nu)$ is of order $\sqrt{\varepsilon}$, it is necessary to expand the r.h.s. only up to second order in $\mathbf{F}$. This leads to:

$$P(\mathbf{q}, t + \varepsilon) - P(\mathbf{q}, t)$$
$$\sim \int d q' \, P\left(\mathbf{q}', t\right)\frac{\partial}{\partial q_i}\left\langle F_i\left(\mathbf{q}, \nu\right)\delta\left(\mathbf{q} - \mathbf{q}'\right) + \tfrac{1}{2}\frac{\partial}{\partial q_j}F_i\left(\mathbf{q}, \nu\right)F_j\left(\mathbf{q}, \nu\right)\delta\left(\mathbf{q} - \mathbf{q}'\right)\right\rangle_\nu. \tag{4.97}$$

The differential operators $\partial/\partial q_i$ can be taken out of the average and factorized in front of the integral. The integral over q' is then trivial. Replacing the function $\mathbf{F}(\mathbf{q}, \nu)$ by its explicit expression taken from equation (4.93), we can perform the noise average and take the $\varepsilon = 0$ limit. We obtain the FP equation:

$$\frac{\partial}{\partial t} P\left(\mathbf{q},t\right) = \frac{1}{2} \frac{\partial}{\partial q_i} \left[\Omega \frac{\partial}{\partial q_j} \left(e_{ia} e_{ja} P\right) + \left(f_i - 2\Omega d_{iaa}\right) P \right].$$

(4.98)

This result calls for two simple observations:

(i) The term quadratic in the noise in Langevin equation (4.93) can be replaced by its average over $\nu_a(t)$. In the continuous time limit only equations linear in the noise need to be considered.

(ii) In Section 4.7 we have considered the continuum Langevin equation as a limit of a discrete equation (equation (4.69)) with a different argument in e_{ia}: we have taken the symmetric form $\frac{1}{2}\left[\mathbf{q}(t) + \mathbf{q}(t+\varepsilon)\right]$ (Stratanovich convention). If we then expand e_{ia} for ε small we get:

$$e_{ia}\left\{\tfrac{1}{2}\left[\mathbf{q}(t) + \mathbf{q}\left(t+\varepsilon\right)\right]\right\} = e_{ia}\left[\mathbf{q}(t)\right] + \tfrac{1}{2}e_{jb}\left[\mathbf{q}(t)\right]\nu_b(t)\frac{\partial}{\partial q_j}e_{ia}\left[\mathbf{q}(t)\right] + O(\varepsilon).$$

(4.99)

This corresponds to an equation (4.93) (Itô calculus) with a specific form of the term quadratic in the noise. Replacing in equation (4.93) the function d_{iab} by:

$$d_{iab}\left(\mathbf{q}\right) = \tfrac{1}{2}e_{jb}\left(\mathbf{q}\right)\frac{\partial}{\partial q_j}e_{ia}\left(\mathbf{q}\right),$$

(4.100)

we recover FP equation (4.73).

Finally from the Langevin equation which we now write without loss of generality in the small ε limit:

$$q_i\left(t+\varepsilon\right) = q_i(t) - \tfrac{1}{2}\varepsilon f_i\left(\mathbf{q}(t)\right) + e_{ia}\left(\mathbf{q}(t)\right)\nu_a(t),$$

(4.101)

we can directly derive a path integral representation in discretized form for $P\left(\mathbf{q},t\right)$. We start from equation (4.96), use the explicit form (4.93) of the Langevin equation and integrate over the noise, exactly as we have done in the continuum case, with the measure (4.87). Assuming that the matrix e_{ia} is invertible, we then obtain the equation:

$$P\left(\mathbf{q}, t+\varepsilon\right) = \int \mathrm{d}q' \, p\left(\mathbf{q},\mathbf{q}'\right) P\left(\mathbf{q}',t\right),$$

with the transition kernel $p\left(\mathbf{q},\mathbf{q}'\right)$ now given by:

$$p\left(\mathbf{q},\mathbf{q}'\right) = \frac{1}{\sqrt{2\pi\varepsilon\Omega}}\left(\det e_{ia}\left(\mathbf{q}'\right)\right)^{-1}\exp\left[-\frac{1}{2\varepsilon\Omega}g_{ij}\left(\mathbf{q}'\right)d_i d_j\right],$$

(4.102)

in which we have set:

$$\left[g^{-1}\left(\mathbf{q}\right)\right]_{ij} = e_{ia}\left(\mathbf{q}\right)e_{ja}\left(\mathbf{q}\right), \qquad \mathbf{d} = \mathbf{q} - \mathbf{q}' + \tfrac{1}{2}\varepsilon\mathbf{f}\left(\mathbf{q}'\right).$$

(4.103)

This equation directly leads to the discretized form of the path integral (4.84).

Bibliographical Notes

A historical reference is

P. Langevin, *Comptes Rendus Acad. Sci. Paris* **146** (1908) 530.

For an introduction into the subject of stochastic processes in physics see:

N.G. Van Kampen, *Stochastic Processes in Physics and Chemistry* (North-Holland, Amsterdam 1981).

A selection of early papers can be found in

Selected Papers on Noise and Stochastic Processes, N. Wax ed. (Dover, New York 1954).

Mathematical contributions are contained in

New Stochastic Methods in Physics, C. De Witt-Morette and K.D. Elworthy eds., *Phys. Rep.* **77** (1981) 121.

Examples of mathematical textbooks are

I.I. Gihman and A.V. Skorohod, *Stochastic Differential Equations* (Springer Verlag, Berlin 1972); N. Ikeda and S. Watanabe, *Stochastic Differential Equations and Diffusion Processes* (North-Holland, Amsterdam 1981);

K.D. Elworthy, *Stochastic Differential Equations on Manifolds, London Math. Soc. Lectures Notes 70* (Cambridge University Press, Cambridge 1982).

For an early reference about the application of path integrals to brownian motion and statistical mechanics see for example

S.G. Brush, *Rev. Mod. Phys.* **33** (1961) 79.

Additional comments on numerical simulations and stochastic processes as discussed in Appendix A4 can be found in

S.K. Ma, *Statistical Mechanics*, Ch. 22 (World Scientific, Singapore 1985); G. Parisi, *Statistical Field Theory*, Ch. 19,20 (Addison-Wesley, New York 1988) .

Exercises

Exercise 4.1

We want to study the Langevin equation:

$$\dot{x} = -\tfrac{1}{2}ax^n(t) + \nu(t)$$

in which $\nu(t)$ is the gaussian noise considered in equation (4.3) and a is a parameter.

4.1.1. Establish coupled differential equations for the moments $\langle x^k(t)\rangle$ using the method which leads to the Fokker–Planck equation. Deduce the possible form of the moments at equilibrium.

4.1.2. Write the Fokker–Planck equation and study the existence of the equilibrium distribution.

4.1.3. What can be said if the parameter a and $x(t)$ are complex and $\nu(t)$ is still the real noise (4.2). Begin with $n = 1$; for $n > 2$ this is a research problem.

Exercise 4.2

The goal of this exercise is to compare the properties of the ordinary Langevin equation with a second order stochastic differential equation. Consider thus the dynamical equation

$$a\ddot{x} + b\dot{x} + cx = \nu(t),$$

in which a, b and c are constants, and $\nu(t)$ still a gaussian white noise. Determine the equilibrium distribution, the relaxation time....

APPENDIX 4
MARKOV'S STOCHASTIC PROCESSES: A FEW REMARKS

We here recall a few simple properties of Markov's processes because it is useful to have them in mind when discussing Langevin or Fokker–Planck equations.

We first assume that we have a finite number of discrete states labelled by an index a, $a = 1 \ldots A$. The stochastic process is defined in terms of a matrix p_{ab} which gives the transition probability at any given time from state b to a:

$$p_{ab} \geq 0, \qquad (A4.1)$$

$$\sum_{a=1}^{A} p_{ab} = 1. \qquad (A4.2)$$

In terms of the matrix p_{ab} we can write the evolution equation for the probability $P_n(a)$ to be at time n in state a:

$$P_{n+1}(a) = \sum_{b} p_{ab} P_n(b). \qquad (A4.3)$$

Equation $(A4.2)$ expresses the conservation of probabilities. Indeed summing equation $(A4.3)$ over the index a, as a consequence of the equation $(A4.2)$, we find:

$$\sum_{a} P_{n+1}(a) = \sum_{b} P_n(b).$$

Equation $(A4.2)$ also implies that the vector v_a:

$$v_a = 1 \qquad \forall a,$$

is a left eigenvector of the matrix p_{ab} with eigenvalue 1:

$$\sum_{a} v_a p_{ab} = \sum_{a} p_{ab} = 1 = v_b.$$

We call $P(a)$ an equilibrium probability distribution if it is a right eigenvector of p_{ab} with eigenvalue 1:

$$P(a) = \sum_{b} p_{ab} P(b), \qquad (A4.4)$$

i.e. a stationary solution of equation $(A4.3)$.

A4.1 The Spectrum of the Transition Matrix

Let us show that the largest eigenvalues of the matrix p_{ab} have modulus 1. Let v_a be an eigenvector of p_{ab} and λ the corresponding eigenvalue:

$$\sum_{b} p_{ab} v_b = \lambda v_a. \qquad (A4.5)$$

Note first that summation over a using equation $(A4.2)$ yields:

$$\sum_{b} v_b = \lambda \sum_{a} v_a,$$

which implies either $\lambda = 1$, or $\sum v_a = 0$.

Comparing the modulus of both sides of equation ($A4.5$) we also find:

$$\sum_b p_{ab} |v_b| \geq |\lambda| |v_a|, \qquad (A4.6)$$

and thus using again equation ($A4.2$):

$$|\lambda| \leq 1. \qquad (A4.7)$$

Equality is possible only if inequality ($A4.6$) is an equality for all values of a:

$$\sum_b p_{ab} |v_b| = |v_a|, \qquad (A4.8)$$

which means that $|v_a|$ is an equilibrium distribution.

Let us now assume that not all components of the vector v_a can be made positive. We then decompose the set of integers $[1, A]$ into a union of subsets I_i such that within a subset the components of v_a have all the same phase and consider the vector w^i such that:

$$\begin{cases} \left(w^i\right)_a = v_a & \text{for } a \in I_i, \\ \left(w^i\right)_a = 0 & \text{otherwise.} \end{cases} \qquad (A4.9)$$

Equality in equation ($A4.7$) implies that for a given index a, p_{ab} can only be different from zero for b belonging to a unique subset I_i. The converse is also true. The matrix p_{ab} describes transitions from one subset to another.

(i) If p_{ab} has non-zero matrix elements between a subset and itself:

$$p_{ab} \neq 0 \qquad \text{for } a, b \in I_i,$$

then $\lambda = 1$ and the space of states is disconnected since starting from the states belonging to I_i one always remains in I_i.

(ii) The matrix p_{ab} is a permutation matrix between the subsets which are organized in various clusters. The corresponding eigenvalues are roots of the identity. If the space of states is connected, the matrix p_{ab} corresponds to a cyclic permutation between all the subsets I_i. This is clearly a highly non-generic case.

Connectivity assumption. From now on we assume that the space of states is connected, i.e. that starting from any state a there is a non-zero probability to reach any other state b. This means that there exists a set of states $c_1, \ldots, c_r$ such that the product:

$$\forall a, b \qquad \exists \{c_1, \ldots c_r\} : \qquad p_{bc_r} p_{c_r c_{r-1}} \cdots p_{c_1 a} \neq 0.$$

We assume in addition that we are not in the case of the cyclic permutation.

It then follows that $\lambda = 1$ is the unique largest eigenvalue and the corresponding eigenspace has dimension 1 (if the dimension were larger than 1, one could form a linear combination of eigenvectors with components with non-positive ratios).

Furthermore the corresponding eigenvector has only strictly positive components. Indeed acting repeatedly with p_{ab} on a vector with real non-negative components, we always obtain after a finite number of iterations, as a consequence of the connectivity assumption, a vector with only strictly positive components. If we act on an equilibrium distribution the result follows.

Therefore, as a consequence of the previous assumptions, any distribution converges at time infinity towards the unique equilibrium distribution $P(a)$ which can be parametrized as:

$$P(a) = e^{-E(a)} \qquad \text{with } E(a) \text{ real}. \qquad (A4.10)$$

Infinite number of states. If we let the number of states increase to infinity an important new phenomenon may occur: the components of the normalized equilibrium distribution:

$$\sum_a P(a) = 1,$$

may all go to zero.

In this case even with the previous assumptions there is no equilibrium distribution. This will be an important issue in realistic cases.

We shall also encounter another situation in the study of critical phenomena in Chapters 23–35. In the case of ferromagnetic systems the states correspond to configurations of spins on a lattice: as long as the spin system is in a finite volume, the space of states remains connected, but the space becomes disconnected in the infinite volume limit. In a finite volume the equilibrium state is unique but not in an infinite volume.

A4.2 Detailed Balance

One is sometimes confronted with the following problem: one has chosen a priori an equilibrium distribution $P(a)$ and one wants to construct a stochastic process which converges towards this distribution. This can be achieved for example by imposing on the matrix p_{ab} the condition:

$$p_{ab}P(b) = p_{ba}P(a) \quad \text{for all pairs } (a, b). \qquad (A4.11)$$

This condition is called detailed balance. It is a local condition involving only the states a and b.

Let us show that $P(a)$ is an equilibrium distribution:

$$P'(a) = \sum_b p_{ab}P(b) = \sum_b p_{ba}P(a) = P(a). \qquad (A4.12)$$

In what follows we can assume without loss of generality that the space of states is connected and that all probabilities $P(a)$ are thus strictly positive.

Let us then set in equation $(A4.3)$

$$P_n(a) = \sqrt{P(a)}\tilde{P}_n(a). \qquad (A4.13)$$

Then:

$$\tilde{P}_{n+1}(a) = \sum_b P(a)^{-1/2} p_{ab} P(b)^{1/2} \tilde{P}_n(b). \qquad (A4.14)$$

Equation $(A4.11)$ implies that the matrix p'_{ab}:

$$p'_{ab} = P(a)^{-1/2} p_{ab} P(b)^{1/2}, \qquad (A4.15)$$

is symmetric. Its spectrum is real. The distribution $P(a)$ will be an equilibrium distribution except if p_{ab} has -1 as eigenvalue. This means, according to previous discussion that the space of states is divided into two equal subsets I_+ and I_- which the matrix p_{ab} exchanges.

It is easy to see that the corresponding left eigenvector has components $+1$ for one subset and -1 for the others. This left eigenvector has to be orthogonal to the right eigenvector $P(a)$.

This implies the condition

$$\sum_{a \in I_+} P(a) = \sum_{a \in I_-} P(a).$$

This is a non-generic situation which is likely to be realized only if there exists some discrete symmetry in the space of states.

A4.3 Continuum Time Limit

In various circumstances, in particular when the matrix p_{ab} is close to the identity:

$$p_{ab} = \delta_{ab} + \varepsilon q_{ab}, \qquad \varepsilon \ll 1, \tag{A4.16}$$

equation $(A4.3)$ can be approximated by a differential equation of the form:

$$\frac{\partial P(t,a)}{\partial t} = \sum_b q_{ab} P(t,b). \tag{A4.17}$$

The matrix q_{ab} has the properties

$$q_{ab} \geq 0 \quad \text{for } a \neq b \quad \text{and} \quad \sum_a q_{ab} = 0, \tag{A4.18}$$

which ensure that if initially $P(t_0, a)$ is a probability distribution it remains one at all later times.

Results obtained in the discrete time case can be recovered by analogous methods. One first shows that the eigenvalues of q_{ab} have a positive real part. The eigenvalue equation

$$\sum_b q_{ab} v_b = \lambda v_a, \tag{A4.19}$$

implies:

$$\sum_{b \neq a} q_{ab} v_b = (\lambda - q_{aa}) v_a,$$

and thus:

$$\sum_{b \neq a} q_{ab} |v_b| \geq |\lambda - q_{aa}| |v_a|. \tag{A4.20}$$

A summation over a finally yields:

$$\sum_a (q_{aa} + |\lambda - q_{aa}|) |v_a| \leq 0. \tag{A4.21}$$

At least one term in this sum must be non-positive:

$$\exists a : q_{aa} + |\lambda - q_{aa}| \leq 0. \tag{A4.22}$$

Since as a consequence of conditions $(A4.18)$ q_{aa} is non-positive, the real part of λ must also be non-positive. Furthermore if $\mathrm{Re}\,(\lambda)$ vanishes, λ vanishes. This result shows that equation $(A4.17)$ with conditions $(A4.18)$ is slightly less general that the discrete equation $(A4.3)$. Finally if the space of states is connected by q_{ab}, the equilibrium vector is unique and has strictly positive components.

Detailed balance. Detailed balance now implies:

$$\frac{q_{ab}}{q_{ba}} = \frac{P(a)}{P(b)}. \tag{A4.23}$$

Therefore the matrix q':

$$q'_{ab} = P(a)^{-1/2} q_{ab} P(b)^{1/2}, \tag{A4.24}$$

is symmetric. It has real, negative, eigenvalues. The definition $(A4.24)$ implies that the matrices q and q' have the same eigenvalues. Thus q_{ab} also has real negative eigenvalues. This case is the discrete analogue of the case studied in Section 4.4.

A4.4 Construction of a Stochastic Process with a Given Equilibrium Distribution

There are many ways to construct stochastic processes which converge towards a given equilibrium distribution. First one has to construct the matrix p_{ab}. Let us just give the example of discrete set of states and use detailed balance.

One possibility is to connect all initial states to the same number r of final states and then take:

$$\begin{cases} p_{ab} = \dfrac{1}{r} & \text{if } P(a) \geq P(b) \text{ and } a \neq b \\[2mm] p_{ab} = \dfrac{1}{r}\dfrac{P(a)}{P(b)} & \text{if } P(a) < P(b). \end{cases} \tag{A4.25}$$

These conditions imply detailed balance. In addition as required:

$$p_{bb} = 1 - \sum_{a \neq b} p_{ab} > 0. \tag{A4.26}$$

Another idea is to take:

$$p_{ab} = pP(a)\theta_{ab}, \quad a \neq b, \tag{A4.27}$$

with

$$\theta_{ab} = \theta_{ba} \in \{0, 1\},$$

and p adjusted such that for any state b

$$p \sum_{a \neq b} P(a)\theta_{ab} < 1.$$

Depending on the structure of the space of states there are many other methods.

Once the matrix p_{ab} is given, it is easy in the discrete case to construct the corresponding stochastic process, i.e. to describe a motion in the space of states such that asymptotically at large time the probability of being at state a is just $P(a)$. At fixed initial state b, to the matrix elements p_{ab} corresponds a partition of the interval $[0, 1]$. By drawing a random number with uniform probability on $[0, 1]$, one can select the final state a with probability p_{ab}.

5 FUNCTIONAL INTEGRALS IN FIELD THEORY

In this chapter we begin our study of local quantum field theory, the relativistic generalization of quantum mechanics. We discuss the example of a neutral self-coupled scalar field $\phi(\mathbf{x}, t)$ which depends on time and a space coordinate $\mathbf{x}$ belonging to $\mathbb{R}^{d-1}$. We construct a functional integral representation of the imaginary time evolution or quantum statistical operator, natural extension of the path integral of quantum mechanics. The functional integral defines a functional measure to which correspond field correlation functions. By adding an external source to the action we obtain an integral representation of their generating functional.

As in quantum mechanics, we first calculate the gaussian integral, which corresponds to a free field theory. Then, considering a general action with an interaction expandable in powers of the field, we express the corresponding functional integral in terms of a series of gaussian integrals, and establish Wick's theorem. This provides an alternative, perturbative and algebraic definition of the functional integral. We show, because this is no longer obvious, that this definition is consistent with the usual manipulations performed on integrals like integration by parts and change of variables. We exhibit several algebraic properties of functional integrals and derive the Dyson–Schwinger field equations of motion. We also define the functional δ-function. We finally generalize these notions to spin $1/2$ fermion fields.

In the Appendix A5 we recall a few properties of spin $1/2$ fermion theories, of the spinorial representations of the $O(d)$ rotation group and the algebra of γ matrices.

5.1 Functional Integrals. Correlation Functions

We consider a classical lagrangian density $\mathcal{L}(\phi)$ for a scalar field ϕ of the form:

$$\mathcal{L}(\phi) = \tfrac{1}{2}\dot{\phi}^2(x,t) - \tfrac{1}{2}\big(\nabla\phi(x,t)\big)^2 - V\big(\phi(x,t)\big), \tag{5.1}$$

in which $V(\phi)$ is a function expandable in powers of ϕ, a polynomial in the simplest examples like:

$$V(\phi) = \tfrac{1}{2}m^2\phi^2 + \tfrac{1}{4!}g\phi^4.$$

The parameter m is called the mass because for $g = 0$ it represents the physical mass of the particle associated with the field ϕ.

The lagrangian density (5.1) is the simplest example of a lagrangian density having the following properties:

(i) It is *local* in space and time because it depends only on the field $\phi(x,t)$ and its partial derivatives (and not on product of fields at different points). This property, *locality*, plays, as we shall see, a central role in most of this work.

(ii) It is invariant under space and time translations since space and time do not appear explicitly in expression (5.1).

(iii) It is relativistic invariant, i.e. invariant under the pseudo-orthogonal group $O(d - 1, 1)$ acting linearly on $\mathbf{x}$ and t.

(iv) As we shall verify below, for a suitable class of potentials $V(\phi)$, it leads to a hermitian quantum hamiltonian bounded from below.

5.1.1 Quantization. Functional integral

To quantize the theory we first have to calculate the hamiltonian density corresponding to the lagrangian density (5.1). The lagrangian and the hamiltonian are related by a Legendre transformation involving $\dot\phi$ and $\pi(x)$ conjugate momentum of ϕ:

$$\mathcal{H}(\pi,\phi) + \mathcal{L}(\dot\phi,\phi) - \pi(x)\dot\phi(x,t) = 0. \tag{5.2}$$

The relation between π and $\dot\phi$ is obtained by expressing that the l.h.s. of the equation is stationary with respect to variations of $\dot\phi$ at π and ϕ fixed, or equivalently with respect to variations of π at $\dot\phi$ and ϕ fixed:

$$\pi(x) = \frac{\partial\mathcal{L}}{\partial\dot\phi(x,t)} \Longleftrightarrow \dot\phi(x,t) = \frac{\partial\mathcal{H}}{\partial\pi(x)}. \tag{5.3}$$

We emphasize the concept of Legendre transformation because it also plays an important role in the theory of renormalization through the introduction of the generating functional of proper vertices (see Section 6.2 and Chapter 10) and in statistical mechanics (it relates free energy and thermodynamic potential for example).

The total hamiltonian is the integral of the hamiltonian density:

$$\mathbf{H} = \int d^{d-1}x\, \mathcal{H}[\pi(x),\phi(x)]. \tag{5.4}$$

The coordinates q_i of quantum mechanics are here replaced by the field $\phi(x)$. The transition from quantum mechanics to field theory can be understood in much the same way as the transition between the discretized action (2.16) and the continuous time version (2.17).

To quantize the hamiltonian $\mathbf{H}$ we can start from the basic commutation relations between operators:

$$[\pi(x),\phi(x')] = \frac{\hbar}{i}\delta^{d-1}(x-x'), \tag{5.5}$$

and then develop a quantum theory using the standard methods of quantum mechanics. We instead work immediately with the formalism of functional integrals, which generalize the path integrals introduced in Chapter 2. We verify in Section 7.6 that this quantum field theory leads to a quantum theory of particles in the non-relativistic limit.

Let us write explicitly the hamiltonian density corresponding to the lagrangian (5.1):

$$\mathcal{H} = \tfrac{1}{2}\pi^2(x) + \tfrac{1}{2}[\nabla\phi(x)]^2 + V(\phi(x)). \tag{5.6}$$

This hamiltonian has an important property: it is quadratic in the momentum $\pi(x)$. We can therefore follow the method explained in Section 2.2 and obtain a functional integral representation for the evolution operator (in imaginary time) $U(t_2,t_1)$:

$$\langle\phi_2|U(t_2,t_1)|\phi_1\rangle = \int [d\phi(x,t)]\exp[-S(\phi)], \tag{5.7}$$

with the boundary conditions $\phi(\mathbf{x},t_1)=\phi_1(\mathbf{x})$, $\phi(\mathbf{x},t_2)=\phi_2(\mathbf{x})$.

It is important to note that states in the configuration representation are defined by the description of a complete classical field, $\phi_1(x)$ or $\phi_2(x)$ in expression (5.7), which corresponds to an infinite number of usual variables.

As we know from the analysis of Section 2.2, $S(\phi)$ is the classical euclidean action, obtained by integrating the analytic continuation to imaginary time of lagrangian density (5.1):

$$S(\phi) = \int_{t_1}^{t_2} dt \int d^{d-1}x \left\{ \tfrac{1}{2} \left[\dot{\phi}^2 + (\nabla\phi)^2 \right] + V(\phi(x,t)) \right\}. \qquad (5.8)$$

One advantage of the imaginary time formulation, obvious in expression (5.8), is that time and space now play an equivalent role (up to possible boundary conditions). In particular the original non-compact $O(d-1,1)$ pseudo-orthogonal symmetry, is replaced by the compact $O(d)$ orthogonal symmetry.

5.1.2 Regularization

We have already seen that path integrals reduced to their formal definition in continuous time are sometimes ill-defined and that in such a case it is necessary to complete the formal definition with a limiting procedure based on a time discretized version. We shall eventually find out that this problem is even more severe in field theories. This will be the subject of many chapters. Let us here note only that we may define functional integral (5.7) as the limit of an integral in which both time and space are discretized. Let us introduce a hypercubic lattice $\mathbb{Z}^d$, with lattice spacing a, and use as dynamical variables the values of the field on all lattice sites. We can replace the field derivatives $\partial_\mu \phi$ by:

$$\partial_\mu \phi \mapsto \nabla_\mu \phi = \frac{1}{a} \left[\phi(\mathbf{r} + a\mathbf{n}_\mu) - \phi(\mathbf{r}) \right], \qquad (5.9)$$

in which μ and $\mathbf{r}$ refer both to space and time and $\mathbf{n}_\mu$ is the unit vector in the μ direction. The regularized euclidean action $S_a(\phi)$ is then:

$$S_a(\phi) = a^d \sum_{\mathbf{r} \in aL[t_1,t_2]} \left\{ \tfrac{1}{2} [\nabla_\mu \phi(\mathbf{r})]^2 + V(\phi(\mathbf{r})) \right\}, \qquad (5.10)$$

in which $L(t_1, t_2)$ is the subset of $\mathbb{Z}^d$ such the time coordinate belongs to $[t_1, t_2]$.

In expression (5.10) it is clear again that the potential and the gradient squared (kinetic) term play different roles: The potential term $V(\phi)$ is ultra-local, it weights the functional integral according to the values of the field at each point independently; the gradient squared term instead suppresses the fields which are too singular in the small lattice spacing limit $a \to 0$, those for which the quantity:

$$|\phi(\mathbf{r} + a\mathbf{n}_\mu) - \phi(\mathbf{r})|^2 a^{d-2}$$

diverges.

However, this condition becomes weaker when the dimension increases. It implies continuity only for $d < 2$, i.e. quantum mechanics. This feature has deep consequences. As we shall discuss later, for $d \geq 2$ the continuum limit does not exist in general. Statistical mechanics and the theory of phase transitions provide a natural framework to discuss this problem. Indeed the discretized action can be considered as the configuration energy of a lattice model in classical statistical mechanics. The infinite $\beta = t_2 - t_1$ (zero temperature) limit corresponds to the infinite volume of the classical model. Only if the model has a continuous phase transition, can a continuum limit be defined. However, even then, this continuum limit exists only when one parameter of the theory, for example m^2 the second derivative of $V(\phi)$ at its minimum, is taken close to some special value. In the corresponding statistical model the temperature is close to the critical temperature. Various aspects of this problem will be discussed in Chapters 8–10 and in the chapters devoted to critical phenomena (Chapters 23–36).

5.1.3 Correlation functions, generating functional

The regularization proposed in Subsection 5.1.2 shows that the zero temperature quantum field theory can be considered as a formal continuum limit of a d dimensional lattice model in classical statistical mechanics. Lattice variables correlation functions have a formal limit which we therefore also call ϕ-field correlation functions. This denomination is consistent with the property that the integrand of functional integral (5.7) defines a positive measure:

$$\langle \phi\,(x_1) \ldots \phi\,(x_n) \rangle = \mathcal{Z}^{-1}(\beta = t_2 - t_1) \int [d\phi]\,\phi\,(x_1) \ldots \phi\,(x_n)\,\mathrm{e}^{-S(\phi)}, \qquad (5.11)$$

in which $\mathcal{Z}(\beta)$ is the quantum partition function $\mathrm{tr}\,\mathrm{e}^{-\beta \mathbf{H}}$. Equal-time correlation functions are also the static correlation functions of the finite temperature quantum field theory. To obtain the time-dependent quantum correlation functions one has to proceed by analytic continuation towards real time in the argument of ϕ but not in β.

In the large β-limit these finite temperature correlation functions become the ground state (usually called *vacuum* in quantum field theory) expectation values of the quantum field operator time-ordered products, as we have shown in Section 2.5. After analytic continuation towards real time, they yield the Green's functions from which S-matrix elements can be calculated.

Using the property:

$$\frac{\delta}{\delta J(y)} \exp\left[\int d^d x\, J(x)\phi(x) \right] = \phi(y) \exp\left[\int d^d x\, J(x)\phi(x) \right], \qquad (5.12)$$

where $J(x)$ is an external field or source, we verify that ϕ-field correlation functions can be obtained from a *generating functional* $Z(J)$

$$Z(J) = \int [d\phi] \exp\left[-S(\phi) + \int d^d x\, J(x)\phi(x) \right], \qquad (5.13)$$

by functional differentiation:

$$\langle \phi\,(x_1) \cdots \phi\,(x_n) \rangle = Z^{-1}(J = 0) \left[\frac{\delta}{\delta J\,(x_1)} \cdots \frac{\delta}{\delta J\,(x_n)} Z(J) \right]\Bigg|_{J=0}. \qquad (5.14)$$

5.2 Perturbative Expansion of Functional Integrals

In Section 2.4 we have shown how to calculate a path integral for a hamiltonian of the form $\frac{1}{2}p^2 + \frac{1}{2}\omega^2 q^2 + V_{\mathrm{I}}(q)$ as a series expansion in powers of $V_{\mathrm{I}}(q)$, for any function $V_{\mathrm{I}}(q)$ expandable in powers of q. The result was based on the calculation of a reference gaussian integral (in Chapter 2 the harmonic oscillator). We here use the same method for functional integrals.

In the remaining part of this chapter we work directly at zero temperature in infinite $\mathbb{R}^{d-1}$ space, and therefore in infinite $\mathbb{R}^d$ euclidean space: correlation functions therefore correspond to vacuum expectation values. As shown in Section 5.1, the euclidean action is then fully $O(d)$ invariant and therefore we no longer distinguish space and time; $\mathbf{x}$ belongs to $\mathbb{R}^d$.

Furthermore, although all results derived in this chapter will be illustrated with examples corresponding to an action of the form (5.8), these results are more general and this explains the abstract notations used below.

The gaussian integral. Let us consider a general gaussian functional integral:

$$Z_0(J) = \int [d\phi] \exp\left[-\tfrac{1}{2}\phi K\phi + J \cdot \phi\right], \tag{5.15}$$

where we have used the symbolic notations:

$$\phi K\phi \equiv \int d^d x\, d^d y\, \phi(x) K(x,y)\phi(y), \qquad J \cdot \phi \equiv \int d^d x\, J(x)\phi(x).$$

We assume that the kernel K is symmetric and positive. In expression (5.15) a normalization is implied, we have chosen $Z_0(0) = 1$.

In the example of a free massive theory with mass m, which corresponds in expression (5.8) to the choice

$$V(\phi) = \tfrac{1}{2}m^2\phi^2,$$

the kernel K is a local operator:

$$K(\mathbf{x},\mathbf{y}) = \left(-\partial^2 + m^2\right)\delta^d(\mathbf{x}-\mathbf{y}), \tag{5.16}$$

where we denote by ∂^2 the laplacian in d dimensions.

Let us introduce Δ, the inverse of K:

$$\int d^d z\, \Delta(x,z)\, K(z,y) = \delta^d(x-y). \tag{5.17}$$

In the example (5.16) Δ takes the form

$$\Delta(\mathbf{x},\mathbf{y}) = \frac{1}{(2\pi)^d} \int d^d p \frac{e^{i\mathbf{p}\cdot(\mathbf{x}-\mathbf{y})}}{p^2 + m^2}. \tag{5.18}$$

To calculate the functional integral (5.15) we simply translate ϕ by ΔJ and find after integration:

$$Z_0(J) = \exp\left(\tfrac{1}{2}J\Delta J\right), \tag{5.19}$$

with again the convention:

$$J\Delta J = \int d^d x\, d^d y\, J(x)\Delta(x,y)J(y).$$

From the identities derived in Subsection 5.1.3, we see that $Z_0(J)$ is also the generating functional of correlation functions corresponding to a general action quadratic in the field ϕ. The inverse kernel Δ is thus the 2-point function of the quadratic theory or *free field theory*

$$\Delta(x,y) = \frac{\delta^2 Z_0(J)}{\delta J(x)\delta J(y)}\bigg|_{J=0}.$$

Perturbation theory. We shall verify in the coming chapters that interacting theories with a propagator of the form (5.18) have large momentum or short distance divergences. Therefore in what follows we assume either that the field theory has been regularized by replacing continuum space by a lattice as in Subsection 5.1.2 or the propagator Δ has been replaced by a more complicated function which ensures the convergence of all terms in the perturbative expansion (see Chapter 9).

Let us now consider a more general euclidean action of the form:

$$S(\phi) = \tfrac{1}{2}\phi K\phi + V_{\mathrm{I}}(\phi). \tag{5.20}$$

Using the property (5.12)

$$\frac{\delta}{\delta J(x)}\,\mathrm{e}^{J\cdot\phi} = \phi(x)\,\mathrm{e}^{J\cdot\phi},$$

we can express the functional integral

$$Z(J) = \int [\mathrm{d}\phi]\exp\left[-S\left(\phi\right) + J\cdot\phi\right], \tag{5.21}$$

in terms of $Z_0(J)$ as:

$$Z(J) = \exp\left[-V_{\mathrm{I}}\left(\frac{\delta}{\delta J(x)}\right)\right]Z_0(J) = \exp\left[-V_{\mathrm{I}}\left(\frac{\delta}{\delta J(x)}\right)\right]\exp\left(\tfrac{1}{2}J\Delta J\right), \tag{5.22}$$

expression analogous to (2.50). Note that in this chapter we use the convention that a differential operator like $\delta/\delta J$ acts on all factors placed on its right.

We can then combine identities (5.22,5.14) to calculate all ϕ-field correlation functions as a formal series in powers of the interaction term $V_{\mathrm{I}}(\phi)$.

We see that perturbation theory involves a basic ingredient: the 2-point function Δ of the quadratic theory (equation (5.17)) which we call the *propagator*.

5.3 Wick's Theorem and Feynman Diagrams

When the interaction terms are local, for example integrals of polynomials of the field $\phi(x)$, it is possible to give a graphical representation of each term of the perturbative expansion. Indeed a contribution to the n-point correlation function is a gaussian average of the form:

$$\left\langle \phi\left(x_1\right)\cdots\phi\left(x_n\right)\int \mathrm{d}^d y_1\,\phi^{p_1}\left(y_1\right)\int \mathrm{d}^d y_2\,\phi^{p_2}\left(y_2\right)\cdots\int \mathrm{d}^d y_k\,\phi^{p_k}\left(y_k\right)\right\rangle_0$$

$$= \left[\frac{\delta}{\delta J\left(x_1\right)}\cdots\frac{\delta}{\delta J\left(x_n\right)}\int \mathrm{d}^d y_1\left(\frac{\delta}{\delta J\left(y_1\right)}\right)^{p_1}\cdots\int \mathrm{d}^d y_k\left(\frac{\delta}{\delta J\left(y_k\right)}\right)^{p_k}\exp\left(\tfrac{1}{2}J\Delta J\right)\right]\bigg|_{J=0},$$

in which the $<\ >_0$ means average with respect to the gaussian measure.

In Section 1.1 we have already discussed such expressions. A straightforward generalization of equations (1.6–1.9) yields Wick's theorem in field theory:

$$\left\langle \prod_1^{2s}\phi\left(z_i\right)\right\rangle_0 = \left[\prod_{i=1}^{2s}\frac{\delta}{\delta J\left(z_i\right)}\exp\left(\tfrac{1}{2}J\Delta J\right)\right]\bigg|_{J=0}$$

$$= \sum_{\substack{\text{all possible}\\ \text{pairings of}\\ \{1,2,\ldots,2s\}}} \Delta\left(z_{i_1}z_{i_2}\right)\ldots\Delta\left(z_{i_{2s-1}},z_{i_{2s}}\right). \tag{5.23}$$

For example, for $s = 2$ one finds:

$$\left[\prod_{i=1}^{4}\frac{\delta}{\delta J\left(z_{i}\right)}\exp\left(\tfrac{1}{2}J\Delta J\right)\right]\Bigg|_{J=0} = \Delta\left(z_{1},z_{2}\right)\Delta\left(z_{3},z_{4}\right)+\Delta\left(z_{1},z_{3}\right)\Delta\left(z_{2},z_{4}\right)$$
$$+\Delta\left(z_{1},z_{4}\right)\Delta\left(z_{2},z_{3}\right).$$

Graphically each term in the sum can be represented by a set of contractions corresponding to the particular pairing chosen:

$$\left\langle\prod_{i=1}^{4}\phi\left(z_{i}\right)\right\rangle_{0} = \overline{\phi\left(z_{1}\right)\phi\left(z_{2}\right)}\overline{\phi\left(z_{3}\right)\phi\left(z_{4}\right)} + 2\ \text{terms}.$$

Any perturbative contribution is a sum of products of propagators integrated over the points corresponding to interaction terms. We can give a graphical representation of each product: a propagator is represented by a segment joining the two points which appear as arguments; moreover any point which is common to more than one segment corresponds to an argument which has to be integrated over.

Example

$$V_{\mathrm{I}}\left(\phi\right) = \frac{g}{4!}\int\mathrm{d}^{d}x\,\phi^{4}(x).$$

The two-point function has the expansion:

$$\left\langle\phi\left(x_{1}\right)\phi\left(x_{2}\right)\right\rangle = (\mathrm{a}) - \frac{1}{2}g\,(\mathrm{b}) + \frac{g^{2}}{4}\,(\mathrm{c}) + \frac{g^{2}}{4}\,(\mathrm{d}) + \frac{g^{2}}{6}\,(\mathrm{e}) + O\left(g^{3}\right).$$

(a) is the propagator:

(a): x_1 —————————— x_2

(b), the Feynman diagram which appears at order g, is represented in figure 5.1:

(b)

Fig. 5.1   2-point function at order g.

and (c)...(e) are represented in figure 5.2:

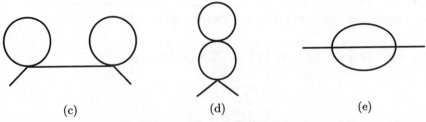

(c)   (d)   (e)

Fig. 5.2   Contributions of order g^{2} to the 2-point function.

Let us explain for example in detail the weight $1/6$ in front of diagram (e). We have to calculate the gaussian average of:

$$\left\langle \phi(x_1)\,\phi(x_2)\,\frac{g^2}{2!}\,\frac{1}{(4!)^2}\int d^d y_1 \phi^4(y_1)\int d^d y_2 \phi^4(y_2)\right\rangle_0,$$

and we apply Wick's theorem.

First $\phi(x_1)$ can be associated with any ϕ field of the interaction terms, there are 8 choices and one interaction term is distinguished. Then $\phi(x_2)$ must be contracted with a field of the remaining interaction term: 4 choices. The 3 remaining fields of the first interaction term can finally be paired with any permutation of the fields of the second one: 3! equivalent possibilities. Multiplying all factors one finds:

$$\frac{1}{2}\frac{1}{(4!)^2}\times 8\times 4\times 3! = \frac{1}{6}.$$

Let us also note that the factor $1/6$ in front of the diagram can be shown to have an interpretation as $1/3!$, the combinatorial factor in the denominator reflecting the permutation symmetry between the three lines joining the two vertices. There exist systematic expressions giving the weight factor of such terms, called Feynman diagrams, in terms of the symmetry group of the graph.

A useful practical remark is the following: the sum of all weight factors at a given order can be calculated from the "zero dimensional" field theory obtained by suppressing the arguments of the field and all derivatives and integration in the action because the propagator can then be normalized to 1. For example in the case of the ϕ^4 field considered here the action becomes:

$$S(\phi) = \tfrac{1}{2}\phi^2 + \frac{g}{4!}\phi^4,$$

and the two-point function is given by:

$$Z^{(2)} = \frac{\int d\phi\, \phi^2 \exp[-S(\phi)]}{\int d\phi \exp[-S(\phi)]} = 1 - \frac{g}{2} + \frac{2}{3}g^2 + O\left(g^3\right),$$

in which the expressions correspond to ordinary one-variable integrals.

The sum rules are satisfied. For example at order g^2:

$$\frac{2}{3} = \frac{1}{4} + \frac{1}{4} + \frac{1}{6}.$$

Let us now give the expansion of the 4-point function:

$$\langle \phi(x_1)\,\phi(x_2)\,\phi(x_3)\,\phi(x_4)\rangle = [\,(a)_{12}\,(a)_{34} + 2\,\text{terms}\,] - \frac{g}{2}[\,(a)_{12}\,(b)_{34}$$

$$+5\,\text{terms}\,] - g\,(f) + g^2\left\{(a)_{12}\left[\frac{1}{4}((c)_{34}+(d)_{34})+\frac{1}{6}(e)_{34}\right]+5\,\text{terms}\right\}$$

$$+\frac{g^2}{4}[\,(b)_{12}\,(b)_{34} + 2\,\text{terms}\,] + \frac{g^2}{2}[\,(g) + 3\,\text{terms}\,] + \frac{g^2}{2}[\,(h) + 2\,\text{terms}\,] + O\left(g^3\right).$$

The new diagrams (f), (g), (h) are represented in figure 5.3. The notation for example $(a)_{12}$ means the diagram (a), contributing to the 2-point function, with arguments x_1

Fig. 5.3   New Feynman diagrams contributing to the 4-point function.

and x_2. Finally the additional terms are obtained by exchanging the external arguments to restore the permutation symmetry of the 4-point function.

Note that the graphs which involve the 2-point functions are disconnected i.e. factorize into a product of functions depending on disjoint subsets of variables. We shall discuss this question in more details in the next chapter.

On the other hand we have omitted systematically disconnected diagrams in which one factor had no external arguments. As shall become clearer in the next chapter these diagrams are cancelled by the perturbative expansion of $Z(J = 0)$ in expression (5.14). The diagrams contributing to $Z(0)$ are called vacuum diagrams.

A final remark: local interaction terms may also involve derivatives of the field $\phi(x)$. Then in expression (5.23) derivatives of the propagator appear. The representation in terms of the Feynman diagrams given above is no longer faithful since it does not indicate where the derivatives are. A more faithful representation can be obtained by splitting points at the vertices and putting arrows on lines.

5.4 Algebraic Identities and Functional Integrals

Functional integrals: perturbative definition. To avoid the difficulties related to the precise definition of the functional integral, we can take expression (5.22) as a proper, although perturbative, definition of the functional integral. In particular we may then avoid the reference to a discretized space and time. However with this new definition, we have to prove that the usual transformations performed on standard integrals lead to identities which are also true with definition (5.22). This is one of the purposes of this section. At the same time we shall derive useful identities satisfied by correlation functions like Dyson–Schwinger equations of motion. Finally, through the formal manipulations involved in these derivations, we shall become more familiar with the algebraic properties of the perturbative expansion.

We therefore now assume that the functional integral (5.21) with action (5.20):

$$Z(J) = \int [\mathrm{d}\phi] \exp\left[-\tfrac{1}{2}\phi K\phi - V_{\mathrm{I}}(\phi) + J \cdot \phi\right], \qquad (5.24)$$

is really defined by expression (5.22):

$$Z(J) = \exp\left[-V_{\mathrm{I}}\left(\frac{\delta}{\delta J(x)}\right)\right] \exp\left(\tfrac{1}{2} J \Delta J\right).$$

We assume of course that all terms in the expansion of this expression in powers of V_{I} exist and are finite (which, as we shall see later, implies some conditions for the kernel K and the interaction term $V_{\mathrm{I}}(\phi)$). From the functional integral representation (5.21) we now heuristically derive identities satisfied by the generating functional $Z(J)$ and then prove algebraically that these identities also follow from the definition (5.22).

5.4.1 Integration by parts and equation of motion

The integral of a total derivative vanishes. Let us use this remark to derive heuristically an equation for $Z(J)$:

$$\int [\mathrm{d}\phi] \frac{\delta}{\delta\phi(x)} \exp(-S + J \cdot \phi) = 0. \tag{5.25}$$

Let us write this equation more explicitly:

$$\int [\mathrm{d}\phi] \left[J(x) - \frac{\delta S}{\delta\phi(x)} \right] \exp(-S + J \cdot \phi) = 0. \tag{5.26}$$

As a consequence of identity (5.12) this equation can be transformed into:

$$\left[\frac{\delta S}{\delta\phi(x)} \left(\frac{\delta}{\delta J} \right) - J(x) \right] Z(J) = 0, \tag{5.27}$$

which for the explicit form (5.20) of $S(\phi)$ yields:

$$\left[\int \mathrm{d}y \, K(x,y) \frac{\delta}{\delta J(y)} + \frac{\delta V_{\mathrm{I}}}{\delta\phi(x)} \left(\frac{\delta}{\delta J} \right) - J(x) \right] Z(J) = 0. \tag{5.28}$$

Equation (5.27) or (5.28) is the equation of motion or Dyson–Schwinger equation of the quantum field theory. It is equivalent to an infinite set of relations between correlation functions obtained by expanding in powers of the source $J(x)$. These equations, in turn, can be solved perturbatively, and, as we now show, determine correlation functions completely.

Solution of the equation of motion. Conversely the solution of equation (5.27) is the functional integral (5.21). To prove this statement let us write $Z(J)$, the solution of equation (5.27), as a generalized Laplace transform:

$$Z(J) = \int [\mathrm{d}\phi] \exp\left[-\tilde{S}(\phi) + J \cdot \phi \right]. \tag{5.29}$$

Equation (5.27) implies:

$$\int [\mathrm{d}\phi] \exp\left[-\tilde{S}(\phi) + J \cdot \phi \right] \left[\frac{\delta S(\phi)}{\delta\phi(x)} - J(x) \right] = 0. \tag{5.30}$$

Using equation (5.26) but with S replaced by $\tilde{S}$ we transform this equation into:

$$\int [\mathrm{d}\phi] \exp\left[-\tilde{S}(\phi) + J \cdot \phi \right] \frac{\delta}{\delta\phi(x)} \left[S(\phi) - \tilde{S}(\phi) \right] = 0. \tag{5.31}$$

Therefore:

$$S(\phi) - \tilde{S}(\phi) = \mathrm{const.}. \tag{5.32}$$

The constant is fixed by the normalization of $Z(J)$.

Example. Let us again consider the euclidean action:

$$S(\phi) = \int \mathrm{d}^d x \left\{ \tfrac{1}{2} [\partial_\mu \phi(x)]^2 + \tfrac{1}{2} m^2 \phi^2(x) + \frac{g}{4!} \phi^4(x) \right\}.$$

Equation (5.28) in this case reads:

$$\left[(-\partial_x^2 + m^2)\frac{\delta}{\delta J(x)} + \frac{g}{3!}\left(\frac{\delta}{\delta J(x)}\right)^3 - J(x)\right]Z(J) = 0\,.$$

Let us write the first equations obtained by expanding in powers of $J(x)$. Differentiating once with respect to $J(y)$ before setting J to zero we find:

$$(-\partial_x^2 + m^2)\langle\phi(x)\phi(y)\rangle + \frac{g}{3!}\langle\phi(x)^3\phi(y)\rangle = \delta^d(x-y).$$

Differentiating three times we obtain

$$(-\partial_x^2 + m^2)\langle\phi(x)\phi(y_1)\phi(y_2)\phi(y_3)\rangle + \frac{g}{3!}\langle\phi(x)^3\phi(y_1)\phi(y_2)\phi(y_3)\rangle$$
$$= \delta^d(x-y_1)\langle\phi(y_2)\phi(y_3)\rangle + 2\text{ terms}\,.$$

More generally these equations relate the $2n$, $2n+2$ and $2n+4$-point functions. They can be solved by expanding all functions in powers of g and this leads to perturbation theory. One could imagine solving these equations in a non-perturbative way, but this requires truncating the infinite set. Examples of such approximations can be found in the exercises of Chapter 9.

5.4.2 Direct algebraic proof of the equation of motion

Let us now show by purely algebraic transformations that $Z(J)$ defined by equation (5.22) indeed satisfies the equation of motion (5.28). It is here convenient to introduce a notation: when $\delta/\delta J$ appears as an argument we will denote it by D_J:

$$D_J \equiv \delta/\delta J\,.$$

We want thus to prove algebraically

$$\left[K\frac{\delta}{\delta J(y)} + \frac{\delta V_{\rm I}}{\delta\phi(x)}(D_J) - J(x)\right]\exp\left[-V_{\rm I}(D_J)\right]\exp\left(\tfrac{1}{2}J\Delta J\right) = 0\,.$$

The proof is based on the commutation relation:

$$[J(x), F(D_J)] = -\frac{\delta F}{\delta\phi(x)}(D_J), \tag{5.33}$$

which can be easily verified by expanding F in powers of $\delta/\delta J$. If we take for $F(\phi)$

$$F(\phi) = \exp\left[-V_{\rm I}(\phi)\right],$$

we obtain:

$$[J(x), \exp(-V_{\rm I}(D_J))] = \frac{\delta V_{\rm I}}{\delta\phi(x)}(D_J)\exp(-V_{\rm I}(D_J)). \tag{5.34}$$

Using this equation in expression (5.22) we obtain:

$$J(x)Z(J) = \frac{\delta V_{\rm I}}{\delta\phi(x)}(D_J)\,Z(J) + \exp\left[-V_{\rm I}(D_J)\right]J(x)\exp\left(\tfrac{1}{2}J\Delta J\right). \tag{5.35}$$

Equation (5.28) then relies on the identity:

$$\exp\left[-V_{\mathrm{I}}\left(D_J\right)\right] J(x) \exp\left(\tfrac{1}{2} J \Delta J\right) = \int \mathrm{d}y\, K\left(x,y\right) \frac{\delta}{\delta J\left(y\right)} \exp\left[-V_{\mathrm{I}}\left(D_J\right)\right] \exp\left(\tfrac{1}{2} J \Delta J\right),$$

$$(5.36)$$

which follows from:

$$\int \mathrm{d}y\, K\left(x,y\right) \frac{\delta}{\delta J\left(y\right)} \exp\left(\tfrac{1}{2} J \Delta J\right) = J(x) \exp\left(\tfrac{1}{2} J \Delta J\right), \qquad (5.37)$$

since by definition of Δ

$$\int \mathrm{d}y\, K\left(x,y\right) \Delta\left(y,z\right) = \delta\left(x-z\right).$$

5.4.3 The infinitesimal change of variables

To prove various identities about correlation functions in the case of field theories possessing symmetries, we often shall use the method of the infinitesimal change of variables. We set:

$$\phi(x) = \chi(x) + \varepsilon F\left(x; \chi\right), \qquad (5.38)$$

in which $F\left(x; \chi\right)$ is a functional of χ:

$$F\left(x; \chi\right) = \sum_1^{\infty} \frac{1}{n!} \int \mathrm{d}y_1 \ldots \mathrm{d}y_n\, \chi\left(y_1\right) \ldots \chi\left(y_n\right) F^{(n)}\left(x; y_1, \ldots, y_n\right), \qquad (5.39)$$

and ε an infinitesimal parameter.

Let us first calculate the variation of the action $S\left(\phi\right)$:

$$S\left(\phi\right) = S\left(\chi\right) + \varepsilon \int \mathrm{d}x \frac{\delta S}{\delta \chi(x)} F\left(x; \chi\right) + O\left(\varepsilon^2\right). \qquad (5.40)$$

At the same time the change of variables (5.38) in the functional integral generates a jacobian D:

$$D = \det \frac{\delta \phi(x)}{\delta \chi\left(y\right)} = \det\left[\delta\left(x-y\right) + \varepsilon \frac{\delta F\left(x; \chi\right)}{\delta \chi\left(y\right)}\right]. \qquad (5.41)$$

As a consequence of identity (1.33):

$$\det\left(1 + \varepsilon\, M\right) = 1 + \varepsilon\, \mathrm{tr}\, M + O\left(\varepsilon^2\right)$$

and thus:

$$D = 1 + \varepsilon \int \mathrm{d}x \frac{\delta F\left(x; \chi\right)}{\delta \chi(x)} + O\left(\varepsilon^2\right), \qquad (5.42)$$

it follows:

$$Z(J) = \int [\mathrm{d}\chi]\left(1 + \varepsilon \int \mathrm{d}x \frac{\delta F\left(x; \chi\right)}{\delta \chi(x)}\right)\left[1 - \varepsilon \int \mathrm{d}x \frac{\delta S}{\delta \chi(x)} F\left(x; \chi\right)\right.$$

$$\left. + \varepsilon \int \mathrm{d}x\, J(x) F\left(x; \chi\right)\right] \exp\left[-S\left(\chi\right) + J \cdot \chi\right] + O\left(\varepsilon^2\right). \qquad (5.43)$$

Collecting all terms of order ε we find the identity:

$$\int \mathrm{d}x \left[F\left(x; D_J\right) \frac{\delta S}{\delta \chi(x)}(D_J) - \frac{\delta F}{\delta \chi(x)}\left(x; D_J\right) - J(x)F\left(x; D_J\right) \right] Z(J) = 0. \qquad (5.44)$$

Algebraic proof. To prove algebraically this identity, we start from the equation of motion (5.27):

$$\left[\frac{\delta S}{\delta \phi(x)}(D_J) - J(x) \right] Z(J) = 0,$$

and act on the left with the differential operator $\int \mathrm{d}x \, F\left(x; \delta/\delta J\right)$. We then use commutation relation (5.33) under the form:

$$[F\left(y; D_J\right), J(x)] = \frac{\delta F}{\delta \chi(x)}(y; D_J).$$

Equation (5.44) follows immediately.

Note that we have dealt in this subsection with an infinitesimal change of variables at first order in ε, although the algebraic proof can be extended to all orders in ε.

5.4.4 Invariance of the functional integral with respect to the choice of the quadratic part

Let us decompose the kernel K in expression (5.20) into a sum of two terms:

$$K = K_1 + K_2 \qquad \text{with } K_1 > 0.$$

We want to prove algebraically that:

$$\begin{aligned} \exp\left[-V_{\mathrm{I}}(D_J) - \tfrac{1}{2}D_J K_2 D_J\right] &\exp\left(\tfrac{1}{2}J K_1^{-1} J\right) \\ &= \mathcal{N}\left(K_1, K_2\right) \exp\left[-V_{\mathrm{I}}(D_J)\right] \exp\left[\tfrac{1}{2}J\left(K_1 + K_2\right)^{-1} J\right], \end{aligned} \qquad (5.45)$$

where $\mathcal{N}$ is independent of the source J.

Since the operator $\exp\left[-V_{\mathrm{I}}\left(\delta/\delta J\right)\right]$ can be factorized on both sides of equation (5.45), it is sufficient to prove the identity for $V_{\mathrm{I}} = 0$. Let us calculate the quantity

$$\begin{aligned} \int \mathrm{d}y \, K_1\left(x, y\right) \frac{\delta}{\delta J(y)} &\exp\left(-\tfrac{1}{2}D_J K_2 D_J\right) \exp\left(\tfrac{1}{2}J K_1^{-1} J\right) \\ &= \exp\left(-\tfrac{1}{2}D_J K_2 D_J\right) J(x) \exp\left(\tfrac{1}{2}J K_1^{-1} J\right). \end{aligned}$$

We again commute J to bring it to the left. The r.h.s. then becomes:

$$\left[J(x) - \left(K_2 \frac{\delta}{\delta J} \right)(x) \right] \exp\left(-\tfrac{1}{2}D_J K_2 D_J\right) \exp\left(\tfrac{1}{2}J K_1^{-1} J\right).$$

Comparing the initial and final expressions, we find that the functional $Z_0(J)$:

$$Z_0(J) = \exp\left(-\tfrac{1}{2}D_J K_2 D_J\right) \exp\left(\tfrac{1}{2}J K_1^{-1} J\right),$$

satisfies the equation:

$$\left[\int \mathrm{d}y \left(K_1 + K_2\right)(x, y) \frac{\delta}{\delta J(y)} - J(x) \right] Z_0(J) = 0. \qquad (5.46)$$

Integrating the equation we thus find:

$$Z_0(J) = \mathcal{N}(K_1, K_2) \exp\left[\tfrac{1}{2} J (K_1 + K_2)^{-1} J\right],$$

proving identity (5.45). After some additional algebra one verifies that $\mathcal{N}^2 = \det(1 + K_2 K_1^{-1})$.

In the same way one can show that a part of the source term can be treated as an interaction, without changing $Z(J)$. The result follows from the identity:

$$\exp\left[\int \mathrm{d}x \, L(x) \frac{\delta}{\delta J(x)}\right] F(J) = F(J + L).$$

Of course all these identities make sense only if both sides exist.

5.4.5 The functional Dirac δ-function

Let us consider a multi-component field $\phi_i(x)$, $i = 1, \ldots, N$, satisfying a constraint:

$$F(x, \phi) = 0. \tag{5.47}$$

We assume that it is possible to solve the constraint and calculate one component, for example $\phi_N(x)$ as a formal power series in the remaining components.

We then define the functional Dirac δ-function by:

$$\delta[F] \equiv \int [\mathrm{d}\lambda(x)] \exp\left[\int \mathrm{d}x \, \lambda(x) F(x, \phi)\right]. \tag{5.48}$$

This is a generalized Fourier representation. We have on purpose omitted the factor i in the representation because the form of the contour of integration in $\lambda(x)$ is irrelevant for the algebraic considerations which follow. Also for reasons explained before we have omitted any normalization factor.

Let us now show that representation (5.48) has the properties expected from a δ-function. Let us consider a functional integral in which we wish to restrict the integration field ϕ to the manifold $F(x, \phi) = 0$:

$$Z_F(J) = \int \prod_{i=1}^{N} [\mathrm{d}\phi_i(x)] \, \delta(F) \exp\left(-S(\phi) + \sum_i J_i \phi_i\right). \tag{5.49}$$

We use the representation (5.48) to write Z_F as:

$$Z_F(J) = \int [\mathrm{d}\lambda] \prod_{i=1}^{N} [\mathrm{d}\phi_i] \exp\left[-S(\phi) + \lambda \cdot F(\phi) + J_i \phi_i\right]. \tag{5.50}$$

We now assume that the quadratic part in the fields (ϕ_i, λ) in the total action is non-singular in such a way that after adding a source term for $\lambda(x)$ we can use the equivalent of equation (5.22) to define algebraically integral (5.50) (it is sufficient in particular for the quadratic part of $S(\phi)$ to be non-singular). Then if we add a term proportional to $F(x, \phi)$ to $S(\phi)$:

$$S(\phi) \mapsto S(\phi) + \int \mu(x) F(x, \phi) \, \mathrm{d}x, \tag{5.51}$$

we can compensate this change by a change of variables:

$$\lambda(x) \mapsto \lambda(x) + \mu(x).$$ (5.52)

We have actually proven invariance by change of variables only for the infinitesimal change of variables, although for simple translations like in (5.52) it is easy to generalize the result to finite changes.

Let us finally note that with the field $\lambda(x)$ is associated a simple equation of motion. Since:

$$\int \prod_i [d\phi_i] [d\lambda] \frac{\delta}{\delta\lambda(x)} \exp[\lambda \cdot F + J_i\phi_i - S(\phi)] = 0,$$

we obtain:

$$F\left(x; \frac{\delta}{\delta J}\right) Z_F(J) = 0.$$ (5.53)

This equation expresses the constraint $F = 0$ on the generating functional of correlation functions $Z(J)$.

Examples. In Section 14.9 a functional δ-function in the form (5.48) will be used to express the non-linear σ model in terms of a N-vector field ϕ satisfying an $O(N)$ invariant constraint $\phi^2(x) = 1$.

In Chapter 18, a covariant gauge will be mentioned, which consists in imposing to the gauge field A_μ the condition $\partial_\mu A_\mu = 0$. Again the representation (5.48) is useful. The equation (5.53) then becomes $\partial_\mu^x \delta Z/\delta J_\mu(x) = 0$, where J_μ is the gauge field source.

5.5 Quantum Field Theories with Fermions

Although we shall mostly discuss boson field theories, occasionally we shall be led to consider theories with fermions, when the fermionic nature of fields plays an essential role. In Section 3.6 we already have discussed the path integral representation of the evolution operator for a system of fermions in quantum mechanics. Here we generalize the formalism to quantum field theory and present the basis of perturbation theory from the point of view of functional methods.

A deep consequence of locality, hermiticity of the hamiltonian and relativistic invariance is the spin–statistics connection: bosons must have integer spin, i.e. transform under representations of the $O(d)$ group, while fermions must have half-integer spin, i.e. transform under spinorial representations of $O(d)$. We comment in Sections 7.5, A7.2 on this relation which can be proven rigorously with a great deal of generality. We also show in Section 7.6 that the fermions we introduce below, lead in the non-relativistic limit to a theory of fermion quantum particles.

Since the spin structure leads to additional technical complications, we recall in Appendix A5 the essential properties of Dirac fermions in the euclidean representation and explain our notations and conventions.

The gaussian integral: perturbation theory. We first consider a gaussian integral with external sources:

$$Z_0(\bar\eta, \eta) = \int [d\psi(x)d\bar\psi(x)] \exp\left[\int d^dx \left(\bar\psi(x)(\slashed\partial + m)\psi(x) + \bar\eta(x)\psi(x) + \bar\psi(x)\eta(x)\right)\right],$$ (5.54)

in which ψ, $\bar{\psi}$, η and $\bar{\eta}$ are elements of a Grassmann algebra and the fermion integration rules have been defined in Section 1.8. This functional integral can be calculated by shifting variables:

$$\begin{cases} \psi(x) + (\partial\!\!\!/ + m)^{-1}\,\eta(x) = \psi'(x) \\ \bar{\psi}(x) + \bar{\eta}(x)\,(\partial\!\!\!/ + m)^{-1} = \bar{\psi}'(x). \end{cases} \tag{5.55}$$

Normalizing the functional integral (5.54) by $Z(0,0) = 1$, we obtain:

$$Z_0(\bar{\eta}, \eta) = \exp\left[-\int d^d x\, d^d y\, \bar{\eta}(y) \Delta_{\mathrm{F}}(y,x) \eta(x) \right], \tag{5.56}$$

in which the fermion propagator Δ_{F} is given by:

$$\Delta_{\mathrm{F}}(y,x) = \frac{1}{(2\pi)^d} \int d^d p\, e^{-ip(x-y)} \frac{(m - i p\!\!\!/)}{p^2 + m^2}. \tag{5.57}$$

We see that on mass-shell ($p^2 = -m^2$), $m - ip\!\!\!/$ becomes a projector on a space of dimension $2^{[d/2]-1}$. This reflects the property that physical massive fermion states can be classified according to the spinorial representation of the $O(d-1)$ subgroup of $O(d)$ which leaves the momentum p invariant.

The fermion 2-point correlation function in a free theory is then:

$$\langle \bar{\psi}_\alpha(x)\psi_\beta(y) \rangle = -\frac{\delta}{\eta_\alpha(x)} \frac{\delta}{\bar{\eta}_\beta(y)} Z_0(\bar{\eta}, \eta) = (\Delta_{\mathrm{F}})_{\beta\alpha}(y,x). \tag{5.58}$$

It is now possible to give a perturbative expansion for a functional integral of the form:

$$Z(\bar{\eta}, \eta) = \int [d\psi(x)d\bar{\psi}(x)] \exp\left[\int d^d x\, \left(\bar{\psi}(x)\,(\partial\!\!\!/ + m)\,\psi(x) + V(\psi, \bar{\psi}) \right. \right.$$
$$\left. \left. + \bar{\eta}(x)\psi(x) + \bar{\psi}(x)\eta(x) \right) \right]. \tag{5.59}$$

The functional $Z(\bar{\eta}, \eta)$ is the generating functional of ψ, $\bar{\psi}$ field correlation functions since:

$$\prod_{i=1}^n \frac{\delta}{\delta\eta(x_i)} \prod_{j=1}^n \frac{\delta}{\delta\bar{\eta}(y_j)} Z(\eta, \bar{\eta})\bigg|_{\eta=\bar{\eta}=0} = Z(0)\left[(-1)^n \langle \bar{\psi}(x_1)\ldots\bar{\psi}(x_n)\psi(y_1)\ldots\psi(y_n) \rangle \right].$$

Because the sources are Grassmann variables, the correlation functions are antisymmetric in their arguments, and this is consistent with Fermi–Dirac statistics.

Generalization of equation (1.92) then yields the identity:

$$Z(\bar{\eta}, \eta) = \exp\left[\int d^d x\, V\left(\frac{\delta}{\delta\bar{\eta}(x)}, -\frac{\delta}{\delta\eta(x)} \right) \right] Z_0(\bar{\eta}, \eta). \tag{5.60}$$

This result leads to the perturbative expansion of a field theory with self-interacting fermions. The corresponding Wick's theorem is a simple generalization of equation (1.96).

Theories with bosons and fermions. Most field theories with fermions contain both fermions and bosons. A typical example is:

$$Z(\bar{\eta}, \eta, J) = \int [d\psi\, d\bar{\psi}\, d\phi] \exp\Big\{-S(\bar{\psi}, \psi, \phi)$$
$$+ \int d^d x\, [\bar{\eta}(x)\psi(x) + \bar{\psi}(x)\eta(x) + J(x)\phi(x)]\Big\}, \qquad (5.61)$$

in which ψ, $\bar{\psi}$, η, $\bar{\eta}$ are Grassmann variables and ϕ and J usual commuting variables, and in which the action $S(\bar{\psi}, \psi, \phi)$ has the form:

$$S = \int d^d x\, \Big\{-\bar{\psi}(x)\,[\slashed{\partial} + M + g\phi(x)]\,\psi(x) + \tfrac{1}{2}(\partial_\mu\phi(x))^2 + \tfrac{1}{2}m^2\phi^2(x) + V(\phi(x))\Big\}. \qquad (5.62)$$

In equation (5.61) $Z(\bar{\eta}, \eta, J)$ is the generating functional of both ϕ field and ψ, $\bar{\psi}$ field correlation functions.

Perturbation theory is generated by:

$$Z(\bar{\eta}, \eta, J) = \exp\Big[-\int d^d x\,\Big(V\Big(\frac{\delta}{\delta J(x)}\Big) + g\frac{\delta}{\delta J(x)}\frac{\delta}{\delta\eta(x)}\frac{\delta}{\delta\bar{\eta}(x)}\Big)\Big] Z_0(\bar{\eta}, \eta, J), \quad (5.63)$$

in which $Z_0(\bar{\eta}, \eta, J)$ is the product of free fermion and free boson functionals.

$$Z_0(\bar{\eta}, \eta, J) = \exp\Big[\int d^d x\, d^d y\,\big(\tfrac{1}{2}J(x)\Delta(x,y)J(y) - \bar{\eta}(x)\Delta_F(x,y)\eta(y)\big)\Big]. \qquad (5.64)$$

All algebraic transformations we have performed on expression (5.22) can easily be generalized to the representation (5.63), in particular field equations can be derived or infinitesimal change of variables justified in perturbation theory. Moreover a functional δ-function can be defined for fermions.

In the example (5.62) the integral over fermions is gaussian. Therefore we may integrate over fermion fields to generate an additional non-local ϕ-field interaction

$$\int [d\psi\, d\bar{\psi}] \exp \int d^d x\, [\bar{\psi}(x)(\slashed{\partial} + M + g\phi(x))\psi(x) + \bar{\eta}(x)\psi(x) + \bar{\psi}(x)\eta(x)]$$
$$\propto \exp[-S_F(\phi, \eta, \bar{\eta})], \qquad (5.65)$$

with:

$$S_F = -\operatorname{tr}\ln[\slashed{\partial} + M + g\phi(x)] + \int d^d x\, d^d y\,\bar{\eta}(y)[\slashed{\partial} + M + g\phi(.)]^{-1}(y,x)\eta(x). \quad (5.66)$$

The expansion of $S_F(\phi, 0, 0)$ in powers of ϕ, generates a set of one fermion loop Feynman diagrams (see Section 5.3). A similar integral over boson fields would have generated a contribution of the form $+\operatorname{tr}\ln$. Hence, compared to boson loops, *fermion loops* are affected by an additional *minus sign*.

Bibliographical Notes

Covariant perturbation was developed in:

> S. Tomonoga, *Prog. Theor. Phys.* 1 (1946) 27; J. Schwinger, *Phys. Rev.* 74 (1948) 1439; 75 (1949) 651; 76 (1949) 790; R.P. Feynman, *Phys. Rev.* 76 (1949) 769; F.J. Dyson, *Phys. Rev.* 75 (1949) 486, 1736.

For Wick's theorem see:

> G.C. Wick, *Phys. Rev.* 80 (1950) 268.

The discussion of the perturbative definition of the functional integral is taken from

> A.A. Slavnov, *Theor. Math. Phys.* 22 (1975) 123; J. Zinn-Justin in *Trends in Elementary Particle Physics (Lectures Notes in Physics 37)* Bonn 1974, H. Rollnik and K. Dietz eds. (Springer-Verlag, Berlin 1975).

The field equations of motion appear in

> F.J. Dyson, *Phys. Rev.* 75 (1949) 1736; J. Schwinger, *Proc. Natl. Acad. Sci. USA* 37 (1951) 452, 455.

In Appendix A5 we present an elementary introduction to spin 1/2 fields. For more details see

> Y. Choquet-Bruhat and C. DeWitt-Morette, *Analysis, Manifolds and Physics: Part II* (North-Holland, Amsterdam 1989),

About spin in 4 dimensions see also

> P. Moussa and R. Stora in *Methods in Subnuclear Physics*, M. Nikolic ed. (Gordon and Breach New York 1966).

> P. Moussa in *Particle Physics*, Les Houches 1971, C. De Witt and C. Itzykson eds. (Gordon and Breach, New York 1973).

Exercises

Exercise 5.1

5.1.1. Write explicitly the field equation of motion (5.28) for the generating functional of correlation functions in the case of the action

$$S(\phi) = \int \mathrm{d}^d x \left\{ \tfrac{1}{2} \left[\partial_\mu \phi(x) \right]^2 + \tfrac{1}{2} m^2 \phi^2(x) + \frac{g}{3!} \phi^3(x) \right\}.$$

5.1.2. Deduce the equations for correlation functions obtained by differentiating zero, one, two and three times with respect to the source.

5.1.3. Use these equations to calculate the 1 and 3-point functions at order g and the 2-point function at order g^2.

Exercise 5.2

5.2.1. Establish the quantum equations of motion in the case of action (5.62) in the absence of ϕ-field self-interactions.

5.2.2. Derive from these equations a set of relations between correlation functions and use them to determine the average of $\phi(x)$ at order g and the fermion and boson 2-point functions at order g^2.

APPENDIX 5
EUCLIDEAN DIRAC FERMIONS AND γ MATRICES

In this appendix we assume a minimal knowledge of the Dirac equation and the free fermion field action. These questions are well discussed in standard textbooks of Particle Physics. We want to briefly describe instead the formalism of euclidean Dirac fermions, i.e. the continuation to imaginary time of the theory of spin $1/2$ fields. As we have noticed in Chapter 5, in this continuation the pseudo-orthogonal group $O(d-1,1)$ transforms into the orthogonal group $O(d)$, d being the euclidean space dimension. Therefore euclidean fermions transform under the spinorial representation of $O(d)$.

The appendix is organized as follows: we define the algebra of Dirac γ matrices and exhibit matrices which realize the algebra; we describe the transformations of spinors. We discuss the symmetries of the free fermion action. A section is devoted to the special example of dimension four. We finally show how to calculate traces of products of γ-matrices, quantities which appear in Feynman diagrams with fermions, and define the Fierz transformation.

A5.1 Dirac γ Matrices

Space of even dimensions d. In this appendix the dimension d of space is assumed to be even except when stated explicitly otherwise. Let γ_μ, $\mu = 1 \ldots d$, be a set of d matrices satisfying the anticommuting relations:

$$\gamma_\mu \gamma_\nu + \gamma_\nu \gamma_\mu = 2\,\delta_{\mu\nu}\mathbf{1}\,, \tag{A5.1}$$

in which $\mathbf{1}$ is the unit matrix.

These matrices are the generators of a Clifford algebra similar to the algebra of operators acting on Grassmann algebras. The operators $(\theta_i + \partial/\partial\theta_i)$, in the notations of Section 1.7, satisfy exactly relations $(A5.1)$. It follows from relations $(A5.1)$ that the γ matrices generate an algebra which, as a vector space, has a dimension 2^d. Therefore they cannot be represented as matrices of dimension smaller than $2^{d/2}$. In Section A5.2 we give an inductive construction $(d \mapsto d+2)$ of hermitian matrices satisfying $(A5.1)$. In the algebra one element plays a special role, the product of all γ matrices. The matrix γ_S:

$$\gamma_S = i^{-d/2}\gamma_1\gamma_2\cdots\gamma_d\,, \tag{A5.2}$$

anticommutes, because d is even, with all other γ matrices and

$$\gamma_S^2 = \mathbf{1}\,.$$

In calculations involving γ matrices, it is not always necessary to distinguish γ_S from other γ matrices. Identifying thus γ_S with γ_{d+1}, we have:

$$\gamma_i\gamma_j + \gamma_j\gamma_i = 2\,\delta_{ij}\mathbf{1}\,, \qquad i,j = 1,\ldots,d,d+1\,. \tag{A5.3}$$

We shall use greek letters $\mu\ \nu\ldots$ to indicate that we exclude the value $d+1$ for the index.

Space of odd dimensions. In most of what follows, we limit ourselves to spaces of even dimensions. However equation $(A5.3)$ shows that in odd dimensions we can represent the γ matrices by taking the γ matrices of dimension $d-1$, to which we add γ_S.

Note however that in this case, in contrast to the even case, the γ matrices are not all algebraically independent.

A5.2 An Explicit Construction

It is sometimes useful to have an explicit realization of the algebra of γ matrices.
For $d = 2$ the standard Pauli matrices realize the algebra:

$$\gamma_1^{(d=2)} \equiv \sigma_1 = \begin{pmatrix} 0 & 1 \\ 1 & 0 \end{pmatrix} \qquad \gamma_2^{(d=2)} \equiv \sigma_2 = \begin{pmatrix} 0 & -i \\ i & 0 \end{pmatrix} \qquad (A5.4a)$$

$$\gamma_S^{(d=2)} \equiv \gamma_3^{(d=2)} \equiv \sigma_3 = \begin{pmatrix} 1 & 0 \\ 0 & -1 \end{pmatrix} \qquad (A5.4b)$$

The three matrices are hermitian:

$$\gamma_i = \gamma_i^\dagger.$$

The matrices γ_1 and γ_3 are symmetric, and γ_2 is antisymmetric:

$$\gamma_1 = {}^T\gamma_1, \qquad \gamma_3 = {}^T\gamma_3, \qquad \gamma_2 = -{}^T\gamma_2,$$

where we have denoted by ${}^T\gamma$ the transposed of the matrix γ.

To construct the γ matrices for higher even dimensions we then proceed by induction, setting:

$$\gamma_i^{(d+2)} = \sigma_1 \otimes \gamma_i^{(d)} = \begin{pmatrix} 0 & \gamma_i^{(d)} \\ \gamma_i^{(d)} & 0 \end{pmatrix}, \qquad 1 \le i \le d+1, \qquad (A5.5)$$

$$\gamma_{d+2}^{(d+2)} = \sigma_2 \otimes \mathbf{1}_d = \begin{pmatrix} 0 & -i\mathbf{1}_d \\ i\mathbf{1}_d & 0 \end{pmatrix}, \qquad (A5.6)$$

in which $\mathbf{1}_d$ is the unit matrix in $2^{d/2}$ dimensions.

As a consequence $\gamma_S^{(d+2)}$ has the form:

$$\gamma_S^{(d+2)} \equiv \gamma_{d+3}^{(d+2)} = \sigma_3 \otimes \mathbf{1}_d = \begin{pmatrix} \mathbf{1}_d & 0 \\ 0 & -\mathbf{1}_d \end{pmatrix}. \qquad (A5.7)$$

The matrices $\gamma_i^{(d+2)}$ are the tensor products of the matrices $\gamma_i^{(d)}$ and $\mathbf{1}_d$ by the matrices σ_i. A straightforward calculation shows that if the matrices $\gamma_i^{(d)}$ satisfy relations $(A5.3)$, the $\gamma_i^{(d+2)}$ matrices satisfy the same relations.

By inspection we see that the γ matrices are all hermitian. In addition:

$$^T\gamma_i^{(d+2)} = \begin{pmatrix} 0 & {}^T\gamma_i^{(d)} \\ {}^T\gamma_i^{(d)} & 0 \end{pmatrix}, \qquad 1 \le i \le d+1.$$

Therefore if $\gamma_i^{(d)}$ is symmetric or antisymmetric, $\gamma_i^{(d+2)}$ has the same property. The matrix $\gamma_{d+2}^{(d+2)}$ is antisymmetric, and $\gamma_S^{(d+2)}$ which is also $\gamma_{d+3}^{(d+2)}$ is symmetric. It follows immediately that, in this representation, all γ matrices with odd index are symmetric, all matrices with even index are antisymmetric:

$$^T\gamma_i = (-1)^{i+1}\,\gamma_i. \qquad (A5.8)$$

A5.3 Transformation of Spinors

Let us consider the $d(d-1)/2$ hermitian traceless matrices $\sigma_{\mu\nu}$ (with for example $\mu < \nu$):

$$\sigma_{\mu\nu} = \frac{1}{2i} [\gamma_\mu, \gamma_\nu]. \tag{A5.9}$$

A short calculation shows that, as a consequence of relations $(A5.1)$, the matrices $\sigma_{\mu\nu}$ satisfy the commutation relations:

$$[\sigma_{\lambda\mu}, \sigma_{\nu\rho}] = -2i [\sigma_{\lambda\rho}\delta_{\mu\nu} + \sigma_{\mu\nu}\delta_{\lambda\rho} - \sigma_{\lambda\nu}\delta_{\mu\rho} - \sigma_{\mu\rho}\delta_{\lambda\nu}]. \tag{A5.10}$$

We recognize that the matrices $i\sigma_{\mu\nu}/2$ satisfy the commutation relations of the generators of the Lie algebra of the group $O(d)$ which in the defining representation can be chosen as $d \times d$ antisymmetric matrices $T^{\lambda\mu}$:

$$(T^{\lambda\mu})_{\alpha\beta} = \delta_{\lambda\alpha}\delta_{\mu\beta} - \delta_{\lambda\beta}\delta_{\mu\alpha}. \tag{A5.11}$$

Let us now consider the free massive fermion action $S_0\left(\bar\psi, \psi\right)$:

$$S_0\left(\bar\psi, \psi\right) = -\int \mathrm{d}^d x\, \bar\psi_\alpha(x) \left[(\gamma_\mu)_{\alpha\beta}\, \partial_\mu + m\delta_{\alpha\beta}\right] \psi_\beta(x), \tag{A5.12}$$

continuation to imaginary time of the standard action for spin $1/2$ fields.

Let us express that it is invariant under $SO\left(d\right)$ transformations which on the spinors ψ and $\bar\psi$ read:

$$\begin{cases} (\psi_R)_\alpha (x) = \Lambda^{-1}_{\alpha\beta}(\mathbf{R})\, \psi_\beta(\mathbf{R}x), \\ (\bar\psi_R)_\alpha (x) = \bar\psi_\beta(\mathbf{R}x)\, \Lambda_{\beta\alpha}(\mathbf{R}). \end{cases} \tag{A5.13}$$

The matrix $\mathbf{R}$ is orthogonal with determinant 1 and can be parametrized in terms of an antisymmetric matrix θ:

$$R_{\mu\nu} = \left(e^\theta\right)_{\mu\nu}. \tag{A5.14}$$

The invariance of the action implies that the transformation matrix Λ should satisfy:

$$\Lambda \gamma_\nu R_{\mu\nu} \Lambda^{-1} = \gamma_\mu. \tag{A5.15}$$

Expanding $R_{\mu\nu}$ at first order in θ and taking Λ close to the identity we find:

$$\Lambda = 1 + \delta\Lambda + O\left(\|\delta\Lambda\|\right)^2, \qquad [\gamma_\mu, \delta\Lambda] = \theta_{\mu\nu}\gamma_\nu.$$

A solution of this equation is:

$$\delta\Lambda = \tfrac{i}{4}\theta_{\mu\nu}\sigma_{\mu\nu},$$

and therefore:

$$\Lambda = \exp\left(\tfrac{i}{4}\theta_{\mu\nu}\sigma_{\mu\nu}\right). \tag{A5.16}$$

Since the matrix $\sigma_{\mu\nu}$ is hermitian and traceless, Λ is a unitary matrix: Λ belongs to a subgroup of $SU(2^{d/2})$ called the spin group. To illustrate the geometrical meaning of this result, let us give a simple example. From the definition $(A5.9)$ we see that $\sigma_{\mu\nu}$ $(\mu \neq \nu)$ has the property:

$$\sigma^2_{\mu\nu} = -\left(\gamma_\mu\gamma_\nu\right)^2 = 1, \tag{A5.17}$$

(no summation over μ and ν being implied). If we take an antisymmetric matrix $\theta_{\mu\nu}$ of the form

$$(\theta_{\mu\nu})_{\mu'\nu'} = \theta \left[\delta_{\mu'\mu}\delta_{\nu'\nu} - \delta_{\mu'\nu}\delta_{\nu'\mu} \right], \tag{A5.18}$$

the corresponding orthogonal matrix $\mathbf{R}$ is a rotation of angle θ in the (μ, ν) plane. The associated transformation matrix Λ is then

$$\Lambda = \cos(\theta/2) + i\sigma_{\mu\nu}\sin(\theta/2), \tag{A5.19}$$

i.e. a matrix belonging to a $U(1)$ subgroup of $SU(2)$ and corresponding to a rotation of angle $\theta/2$.

From this example we understand that the spin group defined by equation $(A5.16)$ is not isomorphic to $SO(d)$, since Λ and $-\Lambda$ correspond to the same element of $SO(d)$. Given the matrix $R_{\mu\nu}$, the equation $(A5.15)$ does not determine a unique matrix Λ, instead it determines $R_{\mu\nu}$ as a function of Λ:

$$R_{\mu\nu} = \operatorname{tr}\left(\Lambda^{-1}\gamma_\mu \Lambda \gamma_\nu \right) / \operatorname{tr} \mathbf{1}.$$

The spinorial representation is a representation only up to a sign. For example the group $SO(3)$ is associated with $SU(2)$, $SO(4)$ with $SU(2) \times SU(2)$. As an abuse of language we shall nevertheless use the expression spinorial representation.

A5.4 The Matrix γ_S

Equation $(A5.19)$ shows that $\sigma_{\mu\nu}$ is an element of the group, associated with a rotation of angle π in the $(\mu\nu)$ plane. Therefore if the dimension of space is even

$$\gamma_S = (-i)^{d/2} \prod_{\mu=1}^{d} \gamma_\mu = \sigma_{12}\sigma_{34}\ldots\sigma_{d-1,d}$$

is also a group element corresponding to the symmetry $x \mapsto -x$.

The free action $(A5.12)$ is invariant under the transformation:

$$\psi'(x) = \gamma_S \psi(-x), \qquad \bar{\psi}'(x) = \bar{\psi}(-x)\gamma_S. \tag{A5.20}$$

In addition γ_S commutes with all transformations of the spin group of $SO(d)$:

$$[\Lambda, \gamma_S] = 0. \tag{A5.21}$$

Therefore this representation of $SO(d)$ is not irreducible. Introducing the projectors $(1 \pm \gamma_S)/2$, we can reduce the representation by defining two spinors, chiral components of ψ:

$$\psi_\pm(x) = \tfrac{1}{2}\left(1 \pm \gamma_S\right)\psi(x), \tag{A5.22}$$

and correspondingly $\bar{\psi}_\pm(x)$:

$$\bar{\psi}_\pm(x) = \bar{\psi}(x)\tfrac{1}{2}\left(1 \pm \gamma_S\right), \tag{A5.23}$$

often denoted by $\psi_R(x)$, $\psi_L(x)$, $\bar{\psi}_R(x)$, $\bar{\psi}_L(x)$ for right and left components.

However, with two of these spinors it is possible to construct only either a massless theory since:

$$S\left(\bar{\psi}_-, \psi_+\right) = -\int \mathrm{d}^d x\, \bar{\psi}_-(x)\left(\slashed{\partial} + m\right)\psi_+(x) = -\int \mathrm{d}^d x\, \bar{\psi}_-(x)\slashed{\partial}\psi_+(x). \tag{A5.24}$$

or, a mass term alone if we take another pair like $\{\bar{\psi}_-, \psi_-\}$. To construct an action for a massive propagating fermion we need four spinors. In expression $(A5.24)$ we have introduced the traditional notation $\slashed{\partial}$ to represent the matrix $\partial_\mu \gamma_\mu$.

A5.5 Reflections

Let us first concentrate on spaces of even dimensions. To obtain the full orthogonal group we have still to represent reflections, which correspond to change the sign of one component of the position vector $\mathbf{x}$, and which are orthogonal matrices with determinant -1. Let us call P the matrix which changes the sign of x_1:

$$P\mathbf{x} = \tilde{\mathbf{x}} \quad \text{with} \quad \tilde{\mathbf{x}} : \begin{cases} \tilde{x}_1 = -x_1, \\ \tilde{x}_\mu = x_\mu, \quad \mu \neq 1. \end{cases} \qquad (A5.25)$$

To compensate in action $(A5.12)$ the effect of this transformation which changes ∂_1 in $-\partial_1$ we have to find a matrix which anticommutes with γ_1 and commutes with all other γ_μ matrices. We can take $\gamma_S \gamma_1$:

$$P: \qquad \psi_P(\mathbf{x}) = \gamma_S \gamma_1 \psi(\tilde{\mathbf{x}}), \qquad \bar{\psi}_P(\mathbf{x}) = \bar{\psi}(\tilde{\mathbf{x}}) \gamma_1 \gamma_S. \qquad (A5.26)$$

Note that $\gamma_S \gamma_1$ anticommutes with γ_S. In particular:

$$\gamma_S \gamma_1 \tfrac{1}{2}(1 + \gamma_S) \gamma_1 \gamma_S = \tfrac{1}{2}(1 - \gamma_S). \qquad (A5.27)$$

Therefore a reflection exchanges right and left components. The spinorial representation of $O(d)$ is irreducible. The action $(A5.24)$, in contrast with the action $(A5.12)$, is not invariant under reflection.

Odd dimensions. In odd dimensions the matrix $-\mathbf{1}$ has determinant -1. Therefore the whole group $O(d)$ can be generated by adding $-\mathbf{1}$ to $SO(d)$. Since $-\mathbf{1}$ commutes with all matrices, $O(d)$ is just the product $SO(d) \times Z_2$. For the spinors we can, for example, represent total space reflection by:

$$\psi \mapsto -\psi, \quad \bar{\psi} \mapsto \bar{\psi}.$$

We then note that a fermion mass term is not invariant under reflection and violates parity conservation in odd dimensions.

A5.6 Other Symmetries

If we start from an action in real time, it may have a set of discrete symmetries: hermiticity, parity, time reversal and charge conjugation which determine the free action as well as the coupling to other fields. After Wick's rotation, the symmetries which involve a complex conjugation are no longer directly symmetries of the euclidean action: hermiticity is lost and time reversal has another natural definition which makes it indistinguishable from the reflection as defined in Section A5.5. We therefore give below some euclidean equivalent of charge and hermitian conjugations in even dimensions.

A5.6.1 Charge conjugation

Let us set:

$$\psi_\alpha(x) = \bar{\psi}'_\beta(x) C^{-1}_{\beta\alpha}, \qquad \bar{\psi}_\alpha(x) = -C_{\alpha\beta}\psi'_\beta(x), \qquad (A5.28)$$

As a function of the new fields ψ' and $\bar{\psi}'$, the action $(A5.12)$ now reads:

$$S_0 = -\int \mathrm{d}^d x\, \bar{\psi}'(x) \left(-C^{-1}\,\overset{T}{\slashed{\partial}}C + m\right) \psi'(x). \qquad (A5.29)$$

If we want to recover the original form $(A5.12)$ we have to find a matrix C such that:

$$C^{-1\,T}\gamma_\mu C = -\gamma_\mu . \qquad (A5.30)$$

In the representation of Section A5.2, we can take for d even:

$$C = (\gamma_S)^{d/2} \prod_{\text{all } \mu \text{ odd}} \gamma_\mu .$$

Example

$$d = 2 : \quad C = \sigma_3\sigma_1 = i\sigma_2 ,$$
$$d = 4 : \quad C = \gamma_1\gamma_3 .$$

Let us also consider action $(A5.24)$. The transformation $(A5.28)$ leads to

$$\displaystyle{\not\partial} (1 + \gamma_S) \mapsto -C^{-1\,T}(1 + \gamma_S)\,{}^{T}\!\!{\not\partial} C .$$

We have chosen γ_S symmetric. If the dimension d is of the form:

$$d = 2 \quad (\text{mod } 4) ,$$

then:

$$C^{-1} \left(1 + {}^{T}\gamma_S\right) C = (1 - \gamma_S) ,$$

and therefore the action $(A5.24)$ is invariant. If the dimension is a multiple of 4 then instead:

$$\displaystyle{\not\partial} (1 + \gamma_S) \mapsto {\not\partial} (1 - \gamma_S) ,$$

and charge conjugation is not a symmetry. However reflection multiplied by charge conjugation is a symmetry. Finally equation $(A5.30)$ justifies the name charge conjugation: if ψ and $\bar\psi$ are charged fields with charges $\mp e$, and if we call $A_\mu (x)$ the electromagnetic field coupled to the charge of ψ and $\bar\psi$, the action has the form:

$$S = - \int \mathrm{d}^d x \, \bar\psi(x) \left({\not\partial} + m + ie{\not\!\!A}\right) \psi(x).$$

After charge conjugation, as a consequence of equation $(A5.30)$ the sign of the charge e has changed.

A5.6.2 Hermitian conjugation

Let us consider the effect of a complex conjugation on action $(A5.12)$:

$$S_0^* \left(\bar\psi, \psi\right) = - \int \mathrm{d}^d x \, \bar\psi(x) \left({\not\partial}^* + m\right) \psi(x).$$

We now make the transformation:

$$\psi(x) = \bar\psi'(x)\gamma_S, \qquad \bar\psi(x) = -\gamma_S\psi'(x). \qquad (A5.31)$$

Since the matrices γ_μ are hermitian

$$\gamma_\mu^* = {}^{T}\gamma_\mu ,$$

and therefore:

$$S_0^* \left(\bar{\psi}, \psi \right) = S_0 \left(\bar{\psi}', \psi' \right).$$

The determinant resulting from the integral over ψ and $\bar{\psi}$ in the functional integral is real. With each eigenvalue of the operator $\slashed{\partial} + m$ is associated an eigenvalue which is its complex conjugate.

Note that complex conjugation followed by transformation $(A5.31)$ is not a symmetry of action $(A5.24)$

$$\gamma_S[\slashed{\partial}\left(1+\gamma_S\right)]^\dagger \gamma_S = \left(1+\gamma_S\right)\slashed{\partial}.$$

However, the product of a hermitian conjugation by a reflection, is a symmetry of the action. This product of transformations is the analogue of hermitian conjugation in real time. We call the corresponding symmetry *reflection hermiticity*.

These different symmetries can be used systematically to classify the couplings of fermions to other fields.

A5.6.3 Continuous internal symmetries

If we assign a fermion number $+1$ to ψ and -1 to $\bar{\psi}$ we see that action $(A5.12)$ conserves the fermion number. To fermion number conservation corresponds a $U(1)$ invariance of the action:

$$\psi_\theta(x) = e^{i\theta}\,\psi(x), \qquad \bar{\psi}_\theta(x) = e^{-i\theta}\,\bar{\psi}(x). \tag{A5.32}$$

A massless free theory in even dimensions possesses an important additional $U(1)$ symmetry called *chiral symmetry*:

$$\psi_\theta(x) = e^{i\theta\gamma_S}\,\psi(x), \qquad \bar{\psi}_\theta(x) = \bar{\psi}(x)\,e^{i\theta\gamma_S}. \tag{A5.33}$$

This shows a deep difference between boson and fermion fields. In contrast with bosons, the property for fermions to be massless can be enforced by a symmetry of the action.

A5.7 The Example of Dimension Four

Since the dimension four plays a special role let us briefly recall what the previous results become in the case of the spinorial representation of $SO(4)$ and $O(4)$. With the conventions of Section A5.2 the 4×4 γ matrices take the form

$$\gamma_{i=1,2,3} = \begin{pmatrix} 0 & \sigma_i \\ \sigma_i & 0 \end{pmatrix}, \qquad \gamma_4 = \begin{pmatrix} 0 & -i\mathbf{1}_2 \\ i\mathbf{1}_2 & 0 \end{pmatrix},$$

(σ_i are the three Pauli matrices) and $\gamma_S \equiv \gamma_5$

$$\gamma_5 = \begin{pmatrix} \mathbf{1}_2 & 0 \\ 0 & -\mathbf{1}_2 \end{pmatrix}.$$

The matrices $\sigma_{\mu\nu}$ then become:

$$\sigma_{ij} = \epsilon_{ijk} \begin{pmatrix} \sigma_k & 0 \\ 0 & \sigma_k \end{pmatrix} \text{ for } i,j,k \leq 3,$$

$$\sigma_{i4} = \begin{pmatrix} \sigma_i & 0 \\ 0 & -\sigma_i \end{pmatrix} \text{ for } i \leq 3.$$

We recognize in the matrices

$$\sigma_i^\pm = \tfrac{1}{4}\epsilon_{ijk}\sigma_{jk} \pm \tfrac{1}{2}\sigma_{i4},$$

the generators of the group $SU(2) \times SU(2)$. The projectors $\tfrac{1}{2}(1 \pm \gamma_5)$ decompose a Dirac spinor into the sum of two vectors transforming as the $(1/2, 0)$ and $(0, 1/2)$ representations of the group (Weyl spinors). Note that a reflection exchanges the two vectors (as expected since the representation is then no longer reducible). In terms of Weyl spinors the construction of invariants with respect to the spinor group reduces to considerations about $SU(2)$. A useful remark in this context, is that the representation and its complex conjugated are equivalent since

$$U^* = \sigma_2 U \sigma_2 \ \forall \ U \in SU(2),$$

(see also Section A5.6) and thus if φ and χ are two $SU(2)$ spinors the combination

$$\varphi_\alpha \, (\sigma_2)_{\alpha\beta} \, \chi_\beta = -i\epsilon_{\alpha\beta}\varphi_\alpha\chi_\beta,$$

where $\epsilon_{\alpha\beta}$ is the antisymmetric tensor ($\epsilon_{12} = 1$), is a $SU(2)$ invariant.

A5.8 Trace of Products of γ Matrices

Perturbative calculations involving spin $1/2$ fermions often require the calculation of traces of products of γ matrices, which we therefore explain in detail.

Multiplying equation ($A5.3$) by γ_i (but not summing over the index i) and taking the trace we find:

$$\operatorname{tr}\gamma_j = \delta_{ij} \operatorname{tr}\gamma_i.$$

Taking $i \neq j$ we find a result which is obvious from the explicit realization given in Section A5.2:

$$\operatorname{tr}\gamma_i = 0. \tag{A5.34}$$

In the calculations which follow, the results always will be proportional to the trace of the unit matrix. Let us define:

$$\operatorname{tr}\mathbf{1} = N. \tag{A5.35}$$

Product of even numbers of γ matrices. To calculate the trace of the product of an even number of γ matrices, $\operatorname{tr}\gamma_{i_1} \cdots \gamma_{i_{2n}}$, we successively commute $\gamma_{i_{2n}}$ through all other γ matrices $\gamma_{i_1}, \ldots, \gamma_{i_{2n-1}}$, using anticommutation relations ($A5.3$). We then generate a set of traces of $(2n - 2)$ γ matrices. At each commutation the sign changes. After all commutations, as a consequence of the cyclic property of the trace, we get back the original expression with a minus sign. As a consequence we find:

$$\operatorname{tr}\gamma_{i_1} \cdots \gamma_{i_{2n}} = \delta_{i_1, i_{2n}} \operatorname{tr}\left(\gamma_{i_2} \cdots \gamma_{i_{2n-1}}\right) - \delta_{i_2 i_{2n}}$$
$$\times \operatorname{tr}\left(\gamma_{i_1} \gamma_{i_3} \cdots \gamma_{i_{2n-1}}\right) + \cdots + \delta_{i_{2n-1} i_{2n}} \operatorname{tr}\left(\gamma_{i_1} \cdots \gamma_{i_{2n-2}}\right). \tag{A5.36}$$

We therefore prove by induction Wick's theorem for the trace of a product of an even number of γ matrices:

$$\operatorname{tr}\gamma_{i_1} \cdots \gamma_{i_{2n}} = N \sum_{\substack{\text{all possible} \\ \text{pairings of} \\ (1,2,\ldots,2n)}} \varepsilon\left(P\right) \delta_{i_{P_1} i_{P_2}} \cdots \delta_{i_{P_{2n-1}} i_{P_{2n}}}, \tag{A5.37}$$

in which $\varepsilon\left(P\right)$ is the signature of the permutation P.

The same result can be obtained by calculating

$$\left(\theta_{i_1} + \frac{\partial}{\partial\theta_{i_1}}\right)\left(\theta_{i_2} + \frac{\partial}{\partial\theta_{i_2}}\right)\cdots\left(\theta_{i_{2n}} + \frac{\partial}{\partial\theta_{i_{2n}}}\right) N\Bigg|_{\theta_i=0}.$$

Odd number of γ matrices. Let us consider the quantity $\operatorname{tr}\gamma_{i_1}\ldots\gamma_{i_{2n-1}}$. Three cases may arise:

(i) Not all γ matrices appear in the product. For example γ_1 is absent. We can then write:

$$\operatorname{tr}\gamma_{i_1}\ldots\gamma_{i_{2n-1}} = \operatorname{tr}\gamma_1^2\gamma_{i_1}\ldots\gamma_{i_{2n-1}}.$$

We can now anticommute one matrix γ_1 with all other γ matrices. Since their number is odd we generate a minus sign and as a consequence of the cyclic property of the trace we reconstitute the original product. Therefore the trace vanishes. In particular:

$$\operatorname{tr}\gamma_{\mu_1}\ldots\gamma_{\mu_{2n-1}} = 0 \quad\text{for}\quad 2n - 1 < d. \tag{A5.38}$$

(ii) If some γ matrices appear an even number of times, we can use the anticommuting relations to cancel them pairwise and we return to the previous case.

(iii) All γ matrices appear in the product, and all an odd number of times. This corresponds in the odd dimension case to take all matrices γ_μ, and in the even case to take all matrices γ_μ and γ_S.

Anticommuting γ matrices we can relate the trace to the calculation of $\operatorname{tr}\gamma_1\ldots\gamma_d$, in the odd case, which is not determined by anticommutation relations $(A5.1)$. It is necessary to use relation $(A5.2)$ between $\gamma_S = \gamma_{d+1}$ and the other γ_μ matrices. We find:

$$\operatorname{tr}\gamma_{\mu_1}\ldots\gamma_{\mu_d} = N\, i^{(d-1)/2}\epsilon_{\mu_1\ldots\mu_d}. \tag{A5.39}$$

In the even case from relation $(A5.2)$ follows instead:

$$\operatorname{tr}\gamma_S\gamma_{\mu_1}\ldots\gamma_{\mu_d} = N\, i^{d/2}\epsilon_{\mu_1\ldots\mu_d}, \tag{A5.40}$$

in which $\epsilon_{\mu_1\ldots\mu_d}$ is the completely antisymmetric tensor normalized by:

$$\epsilon_{12\ldots d} = 1. \tag{A5.41}$$

We shall see later that relation $(A5.40)$ which depends explicitly on the number of space dimensions has deep consequences. In particular dimensional regularization does not preserve this relation and this is the source of possible chiral anomalies in field theories which are chiral invariant at the classical level.

A5.9 The Fierz Transformation

Within the algebra of γ matrices it is possible to define a basis of 2^d hermitian matrices orthogonal by the trace. Let us call these matrices Γ^A. Any fermion 2-point correlation function can then be expanded on such basis. A 4-point fermion correlation function can be expanded on a basis formed by the tensor products of these matrices. However in this case one has first to separate the 4 fermion fields in two sets of two fields and there are three ways of doing it. A connection between these different bases can be found through a Fierz transformation. Let us express that any $2^{d/2} \times 2^{d/2}$ $\mathbf{X}$ matrix can be expanded on the basis of Γ matrices:

$$X_{ab} = N^{-1} \operatorname{tr} \mathbf{X} \Gamma^A \Gamma^A_{ab}, \tag{A5.42}$$

in which N is the trace of $\mathbf{1}$. Let us now choose a matrix $\mathbf{X}$ of the form:

$$X_{ab} = \Gamma^B_{cb} \Gamma^C_{ad}. \tag{A5.43}$$

Identity $(A5.42)$ becomes:

$$\Gamma^B_{cb} \Gamma^C_{ad} = N^{-1} \left(\Gamma^B \Gamma^A \Gamma^C \right)_{cd} \Gamma^A_{ab}. \tag{A5.44}$$

By expanding the product $\Gamma^B \Gamma^A \Gamma^C$ on the basis of Γ matrices we obtain the decomposition of any element of one basis onto another.

Examples

(i) For $d = 2$ a basis is 1, γ_μ and γ_S. We leave as an exercise to verify that the set $1 \otimes 1$, $\gamma_\mu \otimes \gamma_\mu$ and $\gamma_S \otimes \gamma_S$ transforms into itself with a matrix $\mathbf{M}_2$:

$$\mathbf{M}_2 = \frac{1}{2} \begin{pmatrix} 1 & 1 & 1 \\ 2 & 0 & -2 \\ 1 & -1 & 1 \end{pmatrix}.$$

As expected the square of the matrix $\mathbf{M}_2$ is the identity.

(ii) For $d = 4$ a basis is:

$$1, \ \gamma_\mu, \ \gamma_S, \ i\gamma_S\gamma_\mu, \ \sigma_{\mu\nu}.$$

We leave as an exercise to verify that the set:

$$1 \otimes 1, \ \gamma_\mu \otimes \gamma_\mu, \ \gamma_S \otimes \gamma_S, \ i\gamma_S\gamma_\mu \otimes i\gamma_S\gamma_\mu, \ \sigma_{\mu\nu} \otimes \sigma_{\mu\nu}$$

transforms into itself with a matrix $\mathbf{M}_4$ of square $\mathbf{1}$:

$$\mathbf{M}_4 = \frac{1}{4} \begin{pmatrix} 1 & 1 & 1 & 1 & 1 \\ 4 & -2 & -4 & 2 & 0 \\ 1 & -1 & 1 & -1 & 1 \\ 4 & 2 & -4 & -2 & 0 \\ 6 & 0 & 6 & 0 & -2 \end{pmatrix}.$$

6 GENERATING FUNCTIONALS OF CORRELATION FUNCTIONS. LOOPWISE EXPANSION

In Chapter 5 we have introduced the generating functional of correlation functions $Z(J)$ which is also, from the point of view of statistical physics, the partition function in the presence of an external source (or field) $J(x)$. We have shown how to calculate it in a formal power series of the interaction. We now introduce and discuss two new generating functionals, $W(J)$ the generating functional of connected correlation functions (the free energy of statistical physics), and its Legendre transform $\Gamma(\varphi)$, the generating functional of proper vertices (also sometimes called effective potential and which is the thermodynamical potential of statistical physics). Connected correlation functions, as a consequence of locality, have cluster properties. To proper vertices contribute only one-line irreducible Feynman diagrams (diagrams which cannot be disconnected by cutting only one line), and thus in the Particle Physics sense one-particle irreducible (1PI). We therefore often call proper vertices 1PI correlation functions.

We explain how to calculate these various functionals in a reorganized perturbative expansion, called loopwise expansion.

The functional $\Gamma(\varphi)$ plays an important role in the renormalization of local quantum field theory. In Section 6.6 we give a quantum mechanical interpretation for $\Gamma(\varphi)$. We relate it to the partition function at fixed field time average. This relation explains that $\Gamma(\varphi)$ also appears in the discussion of symmetry breaking.

In the appendix we calculate the generating functional of two-loop Feynman diagrams.

6.1 Generating Functionals of Connected Correlation Functions: Cluster Properties

In Chapter 5 we have defined field correlation functions associated with an action $S(\phi)$. The n-point ϕ-field correlation function $Z^{(n)}(x_1, \ldots, x_n)$ is given by:

$$Z^{(n)}(x_1, \ldots, x_n) = \langle \phi(x_1) \ldots \phi(x_n) \rangle \equiv \mathcal{N} \int [\mathrm{d}\phi] \phi(x_1) \ldots \phi(x_n) \, \mathrm{e}^{-S(\phi)}, \qquad (6.1)$$

with

$$\mathcal{N}^{-1} = \int [\mathrm{d}\phi] \, \mathrm{e}^{-S(\phi)}.$$

We have shown in Subsection 5.1.3 (equation (5.14)) that the functional $Z(J)$:

$$Z(J) = \mathcal{N} \int [\mathrm{d}\phi] \exp\left[-S(\phi) + \int \mathrm{d}x \, J(x)\phi(x) \right], \qquad (6.2)$$

is the generating functional of correlation functions:

$$Z^{(n)}(x_1, \ldots, x_n) = \left[\frac{\delta}{\delta J(x_1)} \cdots \frac{\delta}{\delta J(x_n)} Z(J) \right]\Bigg|_{J=0}. \qquad (6.3)$$

This relation can also be written:

$$Z(J) = \sum_{n=0}^{\infty} \frac{1}{n!} \int \mathrm{d}x_1 \ldots \mathrm{d}x_n \, Z^{(n)}(x_1, \ldots, x_n) \, J(x_1) \ldots J(x_n). \qquad (6.4)$$

In Section 5.3 we have noticed that some Feynman diagrams contributing to correlation functions were disconnected in the sense of graphs. They could then be factorized into a product of the form:

$$F_1(x_1, \ldots, x_p) F_2(x_{p+1}, \ldots, x_n),$$

where the two disjoint set of arguments are not empty.

Let us define a new generating functional $W(J)$, analogous to the free energy of statistical mechanics,

$$W(J) = \ln Z(J). \tag{6.5}$$

We introduce the notation

$$W^{(n)}(x_1, \ldots, x_n) = \left[\frac{\delta}{\delta J(x_1)} \cdots \frac{\delta}{\delta J(x_n)} W(J) \right]\bigg|_{J=0}, \tag{6.6}$$

which implies

$$W(J) = \sum_{n=1}^{\infty} \frac{1}{n!} \int dx_1 \ldots dx_n \, W^{(n)}(x_1, x_2, \ldots, x_n) J(x_1) J(x_2) \ldots J(x_n). \tag{6.7}$$

We want to prove that only connected Feynman diagrams contribute to $W(J)$.

Let us first write explicitly a few relations, as implied by the definition (6.5), between correlation functions and the functions $W^{(n)}$:

$$Z^{(1)}(x) = W^{(1)}(x),$$
$$Z^{(2)}(x_1, x_2) = W^{(2)}(x_1, x_2) + W^{(1)}(x_1)W^{(1)}(x_2),$$
$$Z^{(3)}(x_1, x_2, x_3) = W^{(3)}(x_1, x_2, x_3) + W^{(1)}(x_1)W^{(2)}(x_2, x_3) + W^{(1)}(x_2)W^{(2)}(x_3, x_1)$$
$$+ W^{(1)}(x_3)W^{(2)}(x_1, x_2) + W^{(1)}(x_1)W^{(1)}(x_2)W^{(1)}(x_3),$$
$$\cdots$$

Note that the r.h.s. involves all possible products of connected functions with coefficient 1.

6.1.1 A proof of connectivity

We now give a purely algebraic proof of the connectivity property which relies on the linearity property: A linear combination of connected functions is still connected.

In Section 5.2 we have calculated the generating functional $Z(J)$ for a free field theory. We have found (equation (5.19))

$$Z(J) = \exp\left[\frac{1}{2} \int d^dx d^dy \, J(x)\Delta(x-y)J(y) \right].$$

Expanding in powers of J we immediately verify that, except the 2-point correlation function, all correlation functions are disconnected. The functional $W(J)$ instead is

$$W(J) = \ln Z(J) = \frac{1}{2} \int d^dx d^dy \, J(x)\Delta(x-y)J(y).$$

All correlations function vanish, except the 2-point function which is connected.

We now assume that for some action $S(\phi)$ we have shown that $W(J)$ generates connected correlation functions and we add a small local perturbation to the action

$$S_\varepsilon(\phi) = S(\phi) + \varepsilon \int dx \, \phi^N(x),$$

(derivative couplings leave the argument unchanged). Then

$$\exp W_\varepsilon(J) = \int [d\phi] \, e^{-S_\varepsilon(\phi) + J\phi} = \exp\left[-\varepsilon \int dx \left(\frac{\delta}{\delta J(x)} \right)^N \right] \exp W(J).$$

It follows

$$W_\varepsilon(J) = W(J) - \varepsilon \, e^{-W(J)} \int dx \left(\frac{\delta}{\delta J(x)} \right)^N e^{W(J)} + O\left(\varepsilon^2\right).$$

Examining the contribution of order ε we see that it is a linear combination of products of derivatives of $W(J)$ with respect to a source at a unique point x. For example for $N = 3$

$$W_\varepsilon(J) - W(J) = -\varepsilon \int dx \left[\frac{\delta^3 W(J)}{[\delta J(x)]^3} + 3 \frac{\delta^2 W(J)}{[\delta J(x)]^2} \frac{\delta W(J)}{\delta J(x)} + \left(\frac{\delta W(J)}{\delta J(x)} \right)^3 \right] + O\left(\varepsilon^2\right).$$

Therefore if $W(J)$ is connected, all additional terms are also connected. The corresponding diagrams are connected diagrams contributing to $W(J)$ attached to the point x.

Since $W(J)$ is connected for a general free field theory, it follows, after integration over the corresponding coupling constant, that it remains connected for any interaction.

We thus call the functions $W^{(n)}$ connected correlation functions. To insist on the connected character we sometimes use the notation

$$W^{(n)}(x_1, \ldots, x_n) = \langle \phi(x_1) \ldots \phi(x_n) \rangle_c \,, \tag{6.8}$$

where the symbol $\langle\rangle_c$ means connected part of the corresponding correlation function.

6.1.2 Connected correlation functions and cluster properties

For simplicity, we now consider only local euclidean actions $S(\phi)$ functions of a scalar field $\phi(x)$ and its derivatives, and invariant under space translations. This means that the lagrangian density does not depend on space explicitly but only through the field (as in the example (5.1)).

We have already observed in quantum mechanics that the partition function $\operatorname{tr} e^{-\beta H}$ in the large volume limit, i.e. for $\beta \to \infty$, has an exponential behaviour; more precisely (equation (2.2)):

$$\lim_{\beta \to \infty} -\frac{1}{\beta} \ln \operatorname{tr} e^{-\beta H} = E_0 \,,$$

in which E_0 is the ground state energy. Moreover the convergence towards the limit is exponential when the ground state is isolated (equation (2.3)).

Let us here generalize the quantum mechanical result to local field theories. The discussion which follows is highly intuitive and tries only to motivate results which can be proven rigorously. Peculiarities found in massless theories or due to ground state degeneracy will be ignored here and discussed only later.

The functional $Z(J)$ is a partition function in presence of an external source $J(x)$. Let us take for source $J(x)$ the sum of two terms $J_1(x)$ and $J_2(x)$,

$$J(x) = J_1(x) + J_2(x), \tag{6.9}$$

in which $J_1(x)$ and $J_2(x)$ have disjoint supports consisting of two domains Ω_1 and Ω_2 of large volumes V_1 and V_2

$$\begin{cases} J_1(x) = 0, & x \notin \Omega_1 \\ J_2(x) = 0, & x \notin \Omega_2 \end{cases} \qquad \Omega_1 \cap \Omega_2 = \emptyset.$$

We also assume that $J_1(x)$ and $J_2(x)$ fluctuate around an arbitrarily small but non-vanishing constant. Then the *locality* of $S(\phi)$ implies that we can write:

$$S(\phi) - \int dx\, J\phi = \int_{x \in \Omega_1} dx\, [\mathcal{L}(\phi(x)) - J_1(x)\phi(x)] + \int_{x \in \Omega_2} dx\, [\mathcal{L}(\phi(x)) - J_2(x)\phi(x)]$$

$$+ \int_{x \notin \Omega_1 \cup \Omega_2} dx\, \mathcal{L}(\phi(x)) + \text{contributions from boundaries}. \tag{6.10}$$

Let us then write $Z(J)$ as:

$$Z(J) = Z_1(J_1)\, Z_2(J_2)\, Z_{12}(J_1, J_2), \tag{6.11}$$

with the definitions:

$$Z_1(J_1) = \int_{x \in \Omega_1} [d\phi(x)] \exp\left[-S(\phi) + \int dx\, J_1(x)\phi(x) \right],$$

$$Z_2(J_2) = \int_{x \in \Omega_2} [d\phi(x)] \exp\left[-S(\phi) + \int dx\, J_2(x)\phi(x) \right].$$

Both functionals Z_1 and Z_2 are normalized to 1 for $J_1 = 0$ or $J_2 = 0$ respectively. The functional Z_{12} is defined by equation (6.11). Its dependence in J_1 and J_2 comes entirely from the existence of boundary terms in equation (6.10). When we scale up Ω_1 and Ω_2, these boundary terms grow like surfaces while the two first terms in equation (6.10) grow like the volumes V_1 and V_2. Therefore $\ln Z_{12}(J_1, J_2)$ becomes asymptotically negligible compared to $\ln Z_1(J_1)$ and $\ln Z_2(J_2)$ when V_1 and $V_2 \to \infty$.

To express this property it is natural to introduce the functional $W(J) = \ln Z(J)$ (equation (6.5)) which then satisfies:

$$W(J_1 + J_2) = W_1(J_1) + W_2(J_2) + \text{negligible}. \tag{6.12}$$

In particular if $S(\phi) - J\phi$ is translation invariant (which implies that J is a constant), $W(J)$ is extensive, i.e. proportional to the total volume. This property generalizes property (2.2).

Cluster properties. After having taken the infinite volume limit we can expand $W(J_1 + J_2)$ in powers of J_1 and J_2:

$$W(J_1 + J_2) = \sum_{n=1}^{\infty} \frac{1}{n!} \sum_{p=0}^{n} \binom{n}{p} \int dx_1 \ldots dx_p \, dy_{p+1} \ldots dy_n\, W^{(n)}(x_1, \ldots, x_p, y_{p+1}, \ldots, y_n)$$

$$\times J_1(x_1) \ldots J_1(x_p)\, J_2(y_{p+1}) \ldots J_2(y_n), \tag{6.13}$$

with $x_i \in \Omega_1$, $y_j \in \Omega_2$.

Property (6.12) implies that all terms with $p \neq 0$ or $p \neq n$ are negligible for V_1 and V_2 large. Considering expression (6.13) we see that this implies that the functions $W^{(n)}$ must decrease rapidly enough when the two non-empty sets of points $\{x_1 \ldots x_p\}$ and $\{y_{p+1} \ldots y_n\}$ are largely separated:

$$ W^{(n)} \left(x_1 \ldots x_p, y_{p+1} \ldots y_n \right) \to 0 \quad \text{when} \quad \min_{\substack{i=1\cdots p \\ j=p+1\cdots n}} |x_i - y_j| \to \infty. \qquad (6.14) $$

This property, which we have here described very qualitatively, is called the *cluster property*, and is a characteristic property of the connected correlation functions generated by the functional $W(J)$.

Feynman diagrams. We have seen that a Feynman diagram which is disconnected in the sense of graphs can be factorized into a product of the form:

$$ F_1 \left(x_1, \ldots, x_p \right) F_2 \left(y_1, \ldots, y_q \right). $$

It is clear that we can separate the set $\{x_i\}$ and the set $\{y_i\}$ in a translation invariant theory, in a way which leaves such a product invariant: A disconnected diagram cannot satisfy the cluster property. We recover the property that the Feynman diagrams which contribute to the perturbative expansion of $W(J)$ are all connected. Furthermore it can be verified that, in a field theory containing only massive fields, connected Feynman diagrams decrease exponentially, when points are separated, with a minimal rate which is the inverse of the smallest mass in the theory. This property is a consequence of the exponential decrease of the propagator (see Appendix A7.1 for details).

Remark. The relation between action and generating functional, which has the form of a Laplace transformation, can formally be inverted

$$ \mathrm{e}^{-S(\phi)} = \int [\mathrm{d}J] \exp \left[W(J) - \int \mathrm{d}x\, J(x)\phi(x) \right], \qquad (6.15) $$

where one integrates over imaginary sources $J(x)$. A truncated loopwise expansion (see Section 6.4) of the functional integral then yields approximate non-linear equations for correlation functions. It is actually convenient to introduce in the r.h.s. the generating functional of proper vertices defined in next section.

6.2 Proper Vertices

We now introduce a new generating functional called the generating functional of proper vertices. It is obtained by applying a Legendre transformation to the generating functional of connected correlation functions $W(J)$.

Legendre transformation. We have already defined the Legendre transformation as the transformation relating the lagrangian and the hamiltonian. We shall see later that in the context of statistical mechanics and phase transitions it is natural to consider the Legendre transform of the *free energy* $W(J)$. Here proper vertices are introduced because they have special properties from the point of view of perturbation theory as we demonstrate in Sections 6.4 and 6.5. The generating functional of proper vertices $\Gamma(\varphi)$, in which $\varphi(x)$ is a classical field argument of Γ, is obtained from $W(J)$ by:

$$ \Gamma(\varphi) + W(J) - \int \mathrm{d}x\, J(x)\, \varphi(x) = 0, \qquad (6.16) $$

$\varphi(x)$ being related to $J(x)$ by

$$\varphi(x) = \frac{\delta W}{\delta J(x)}. \tag{6.17}$$

This equation expresses that the l.h.s. of equation (6.16) is stationary with respect to variations of $J(x)$ at φ fixed.

Differentiating (6.16) with respect to $\varphi(x)$ we obtain:

$$\frac{\delta \Gamma}{\delta \varphi(x)} - J(x) + \int dy \, \frac{\delta J(y)}{\delta \varphi(x)} \frac{\delta}{\delta J(y)} \left[W(J) - \int dx \, J(x)\varphi(x) \right] = 0\,,$$

and therefore taking into account equation (6.17)

$$J(x) = \frac{\delta \Gamma}{\delta \varphi(x)}. \tag{6.18}$$

Equation (6.18) shows that the Legendre transformation is involutive. In particular equation (6.17) implies that

$$\varphi(x) = \left. \frac{\delta W}{\delta J(x)} \right|_{J=0} \equiv W^{(1)}(x) = \langle \phi(x) \rangle\,,$$

i.e. that $\varphi(x)$ in zero source becomes the expectation value of the field ϕ. Conversely the equation (6.18) then implies that the expectation value of $\phi(x)$ is an extremum of $\Gamma(\varphi)$.

Finally let us note that if $W(J)$ depends on some additional argument α then:

$$\left. \frac{\partial \Gamma}{\partial \alpha} \right|_{\varphi} + \left. \frac{\partial W}{\partial \alpha} \right|_{J} + \int dy \left. \frac{\partial J(y)}{\partial \alpha} \right|_{\varphi} \frac{\delta}{\delta J(y)} \left[W(J) - \int dx \, J(x)\varphi(x) \right] = 0\,. \tag{6.19}$$

Taking into account equation (6.17) we obtain the important result:

$$\frac{\partial W}{\partial \alpha} + \frac{\partial \Gamma}{\partial \alpha} = 0\,. \tag{6.20}$$

Note that we have derived this result for one external variable but it obviously applies also for an external field or source.

Expansion of $\Gamma(\varphi)$. If we set $J = 0$ we find from equation (6.17)

$$\varphi(x) = W^{(1)}(x) \equiv \langle \phi(x) \rangle\,.$$

Inverting the relation (6.17) we can thus expand the source $J(x)$ as a series in powers of $\chi(x)$:

$$\chi(x) = \varphi(x) - W^{(1)}(x). \tag{6.21}$$

We see that $\chi(x)$ is related to the correlation functions of the field

$$\hat{\phi}(x) = \phi(x) - \langle \phi(x) \rangle \tag{6.22}$$

which has zero expectation value.

The first terms of the expansion of equation (6.17) then are

$$\chi(x) = \int dx_1 \, W^{(2)}(x, x_1) \, J(x_1) + \frac{1}{2!} \int dx_1 dx_2 \, W^{(3)}(x, x_1, x_2) \, J(x_1) \, J(x_2) + \cdots \, .$$
(6.23)

We introduce the inverse of the connected 2-point function:

$$\int dz \, S(x, z) W^{(2)}(z, y) = \delta(x - y).$$
(6.24)

We can then write the expansion of $J(x)$:

$$J(x) = \int dx_1 S(x, x_1) \chi(x_1) - \frac{1}{2!} \int dy_1 dy_2 dy_3 dx_1 dx_2 \, S(x, y_3)$$
$$\times \, S(x_1, y_1) \, S(x_2, y_2) \, W^{(3)}(y_1, y_2, y_3) \, \chi(x_1) \, \chi(x_2) + \cdots \, .$$
(6.25)

Note that this expansion can be naturally expressed in terms of so-called amputated correlation functions $W_{\text{amp.}}^{(n)}$:

$$W_{\text{amp.}}^{(n)}(x_1, \ldots, x_n) = \int \left[\prod_{i=1}^{n} dy_i \, S(x_i, y_i) \right] W^{(n)}(y_1, \ldots, y_n).$$
(6.26)

In terms of Feynman diagrams this means that propagators and contributions to the 2-point function on external lines are omitted.

We can now use equation (6.18) to calculate $\Gamma(\varphi)$. Setting:

$$\Gamma(\varphi) = \sum_{1}^{\infty} \frac{1}{n!} \int dx_1 \ldots dx_n \, \Gamma^{(n)}(x_1, \ldots, x_n) \chi(x_1) \ldots \chi(x_n),$$
(6.27)

where the $\Gamma^{(n)}$ are associated with the field $\hat{\phi}$, we find:

$$\Gamma^{(1)}(x) = 0,$$
$$\Gamma^{(2)}(x_1, x_2) = S(x_1, x_2) = [W^{(2)}]^{-1}(x_1, x_2),$$
$$\Gamma^{(3)}(x_1, x_2, x_3) = -W_{\text{amp.}}^{(3)}(x_1, x_2, x_3),$$
$$\Gamma^{(4)}(x_1, x_2, x_3, x_4) = -W_{\text{amp.}}^{(4)}(x_1, x_2, x_3, x_4)$$
$$+ \int dy \, dz \, W_{\text{amp.}}^{(3)}(x_1, x_2, y) \, W^{(2)}(y, z) \, W_{\text{amp.}}^{(3)}(z, x_3, x_4) + 2 \text{ terms}$$
$$\cdots$$

The inverse relations are even more useful, because, as we show in Section 6.5, $\Gamma(\varphi)$ has simpler properties than $W(J)$

$$W^{(2)}(x_1, x_2) = [\Gamma^{(2)}]^{-1}(x_1, x_2),$$
$$W_{\text{amp.}}^{(3)}(x_1, x_2, x_3) = -\Gamma^{(3)}(x_1, x_2, x_3),$$
$$W_{\text{amp.}}^{(4)}(x_1, x_2, x_3, x_4) = -\Gamma^{(4)}(x_1, x_2, x_3, x_4)$$
$$+ \int dy \, dz \, \Gamma^{(3)}(x_1, x_2, y) \, W^{(2)}(y, z) \, \Gamma^{(3)}(z, x_3, x_4) + 2 \text{ terms}$$
$$\cdots$$

Let us give a graphical representation of the first equations. We define:

$$W^{(2)} = \qquad\qquad , \qquad\qquad \Gamma^{(n)} =$$

Fig. 6.1

The correlation functions $W^{(3)}$ and $W^{(4)}$ can then be represented as shown in figures 6.2 and 6.3 respectively.

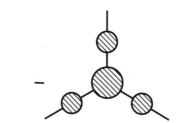

Fig. 6.2   The 3-point function $W^{(3)}$.

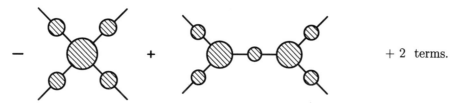

$+ \; 2 \;$ terms.

Fig. 6.3   The 4-point function $W^{(4)}$.

It follows from the set of relations between connected correlation functions and proper vertices that Feynman diagrams which contribute to proper vertices appear in the expansion of connected functions with the opposite sign except in the case of the 2-point function. Indeed let us set:

$$\Gamma^{(2)}\left(x, y\right) = K\left(x, y\right) + \Sigma\left(x, y\right),$$

in which $K\left(x, y\right)$ is just the inverse of the propagator and $\Sigma\left(x, y\right)$, called the mass operator, contains all perturbative corrections. The connected 2-point function $W^{(2)}\left(x, y\right)$ then takes the form of a geometrical series:

$$W^{(2)}\left(x, y\right) = \Delta\left(x, y\right) - \int dz_1\, dz_2\, \Delta\left(x, z_1\right) \Sigma\left(z_1, z_2\right) \Delta\left(z_2, y\right) + \cdots$$

in which $\Delta\left(x, y\right)$ is the propagator. In this case the first term in perturbation theory, the propagator, is special from the point of view of Feynman graph expansion, since it has the same sign in both functions.

Convexity. It follows from the definition of connected correlation functions that the 2-point function $W^{(2)}(x,y)$ is the kernel of a positive operator. Writing its explicit expression:

$$W^{(2)}(x,y) = \langle [\varphi(x) - \langle \varphi(x)\rangle][\varphi(y) - \langle \varphi(y)\rangle]\rangle,$$

we immediately verify:

$$\int dx\, dy\, J(x)J(y)W^{(2)}(x,y) \geq 0.$$

Since $\Gamma^{(2)}(x,y)$ is the inverse of $W^{(2)}(x,y)$ in the sense of kernels, it is also the kernel of a positive operator,

$$\int dx\, dy\, \varphi(x)\Gamma^{(2)}(x,y)\varphi(y) \geq 0. \tag{6.28}$$

For the same reasons, $\delta^2 W(J)/\delta J(x)\delta J(y)$ and $\delta^2\Gamma(\varphi)/\delta\varphi(x)\delta\phi(y)$ which are the 2-point functions in an external source, are positive operators. In particular, in the case of constant sources, $W(J)$ and $\Gamma(\varphi)$, both divided by the total space volume, are convex functions of the sources. We shall recall this property when we examine the physics of spontaneous symmetry breaking and meet functions $\Gamma(\varphi)$ which are not obviously convex (see Section 6.6).

6.3 Momentum Representation

In this work we shall mainly study theories invariant under space translations. Correlation functions then depend only on differences of space arguments. For example the connected 2-point function has the form:

$$W^{(2)}(x,y) \equiv W^{(2)}(x-y).$$

It is thus natural to introduce the Fourier transforms of correlation functions:

$$W^{(2)}(x,y) = \frac{1}{(2\pi)^d}\int d^d p\, e^{-ip(x-y)}\widetilde{W}^{(2)}(p). \tag{6.29}$$

More generally the translation invariance of $W^{(n)}(x_1,x_2,\ldots,x_n)$ is equivalent to momentum conservation of its Fourier transform $\widetilde{W}^{(n)}(p_1,\ldots,p_n)$. We set:

$$(2\pi)^d\delta\left(\sum_{i=1}^n p_i\right)\widetilde{W}^{(n)}(p_1,\ldots,p_n) = \int dx_1\ldots dx_n\, W^{(n)}(x_1,\ldots,x_n)\exp\left(i\sum_{j=1}^n x_j p_j\right). \tag{6.30}$$

Similarly we also introduce the Fourier transforms $\widetilde{\Gamma}^{(n)}(p_1,\ldots,p_n)$ of proper vertices:

$$(2\pi)^d\delta\left(\sum_{i=1}^n p_i\right)\widetilde{\Gamma}^{(n)}(p_1,\ldots,p_n) = \int dx_1\ldots dx_n\, \Gamma^{(n)}(x_1,\ldots,x_n)\exp\left(i\sum_{j=1}^n x_j p_j\right). \tag{6.31}$$

Note that the Fourier transform is defined in such a way that if we set in the generating functional $\Gamma[\varphi(x)]$:

$$\varphi(x) = \int e^{ipx}\,\tilde\varphi(p)d^d p, \tag{6.32}$$

then the l.h.s. of equation (6.31) is the coefficient of $\tilde{\varphi}(p_1) \ldots \tilde{\varphi}(p_n)$.

The explicit expressions of Section 6.2 (for example equation (6.26)) then show that amputation and Legendre transformation become in momentum representation purely algebraic operations in the sense that they involve no momentum integration because the two-point function is diagonal in this representation:

$$\widetilde{W}^{(n)}_{\text{amp.}}(p_1, \ldots, p_n) = \widetilde{W}(p_1, \ldots, p_n) \prod_{i=1}^{n} \left[\widetilde{W}^{(2)}(p_i)\right]^{-1},$$

and also:

$$\widetilde{\Gamma}^{(2)}(p) = \left[\widetilde{W}^{(2)}(p)\right]^{-1},$$

$$\widetilde{\Gamma}^{(3)}(p_1, p_2, p_3) = -\widetilde{W}^{(3)}_{\text{amp.}}(p_1, p_2, p_3),$$

$$\widetilde{\Gamma}^{(4)}(p_1, p_2, p_3, p_4) = -\widetilde{W}^{(4)}_{\text{amp.}}(p_1, p_2, p_3, p_4) + \left[\widetilde{W}^{(3)}_{\text{amp.}}(p_1, p_2, -p_1 - p_2)\widetilde{W}^{(2)}(p_1 + p_2)\right.$$

$$\left. \times \widetilde{W}^{(3)}_{\text{amp.}}(p_3, p_4, -p_3 - p_4) + \text{ cyclic permutation of } \{p_2, p_3, p_4\}\right],$$

$$\ldots .$$

This remark will be useful for the discussion of divergences in perturbation theory.

6.4 Semiclassical or Loopwise Expansion

For reasons which will become clear at the end of this section, it is useful to reorganize perturbation theory by grouping some classes of Feynman diagrams. For this purpose we shall perform a semiclassical expansion of the functional integral, i.e. a formal expansion in powers of $\hbar$ which we therefore again set in front of the classical action and the source term (the normalization $Z(0) = 1$ is implicit):

$$Z(J) = \int [\mathrm{d}\phi] \exp\left[-\frac{1}{\hbar}\left(S(\phi) - J\phi\right)\right]. \tag{6.33}$$

For $\hbar$ small, we see that we can calculate the integral by the steepest descent method.

6.4.1 Loop expansion at leading order

The saddle point equation is:

$$\frac{\delta S}{\delta \phi(x)}[\phi_c(J)] = J(x). \tag{6.34}$$

Substituting the solution $\phi_c(J)$ into the classical action we obtain $Z(J)$ at leading order

$$\ln Z(J) \sim \ln Z_0(J) \equiv \frac{1}{\hbar}\left(-S(\phi_c) + J\phi_c\right). \tag{6.35}$$

When $\hbar$ is explicit, it is convenient to normalize $W(J)$ by:

$$W(J) = \hbar \ln Z(J). \tag{6.36}$$

Then at leading order:

$$W(J) = W_0(J) \equiv -S(\phi_c) + J\phi_c. \tag{6.37}$$

Perturbation theory. Ordinary perturbation theory is recovered by expanding the solution $\phi_c(J)$ in powers of J.

Let us write $S(\phi)$ as the sum of a free part and interaction terms:

$$S(\phi) = \tfrac{1}{2}\phi K\phi + V(\phi),$$

with:

$$V(\phi) = O(\phi^3) \quad \text{for } \phi \to 0.$$

Equation (6.34) then takes the form:

$$K\phi_c + \frac{\delta V}{\delta \phi_c} = J. \tag{6.38}$$

It can be iteratively solved as:

$$\phi_c = \Delta J - \Delta \frac{\delta V(\phi_c)}{\delta \phi} = \Delta J - \Delta \frac{\delta V}{\delta \phi_c}(\Delta J) + \cdots. \tag{6.39}$$

If for example

$$V(\phi) = \frac{g}{4!}\int \mathrm{d}x\, \phi^4(x),$$

ϕ_c has the Feynman diagrams expansion:

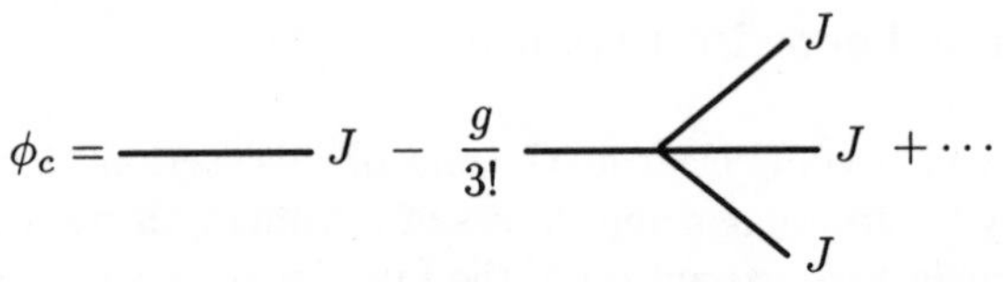

Fig. 6.4

We observe that only tree, i.e. without loops, diagrams are generated. Substituting the expansion into equation (6.37) we note that the perturbative expansion of $W_0(J)$ in powers of J also contains only connected tree Feynman diagrams. The functional $W_0(J)$ is the generating functional of connected tree diagrams.

Legendre transformation. Let us now perform the Legendre transformation:

$$\varphi(x) = \frac{\delta W_0}{\delta J(x)} = \phi_c(x) + \int \mathrm{d}y\, \frac{\delta \phi_c(y)}{\delta J(x)} \frac{\delta}{\delta \phi_c(y)}[J\phi_c - S(\phi_c)].$$

Using the saddle point equation (6.34) we find:

$$\varphi(x) = \phi_c(x). \tag{6.40}$$

Replacing ϕ_c by φ in equation (6.37) we can write:

$$W_0(J) + S(\varphi) - J\varphi = 0, \tag{6.41}$$

and therefore by definition (equation (6.16)):

$$\Gamma_0(\varphi) = S(\varphi). \tag{6.42}$$

At leading order $\Gamma(\varphi)$ is identical to the classical action. The action and the generating functional of connected tree diagrams are related by a Legendre transformation. From the point of view of Feynman diagrams, $S(\varphi)$ contains only the vertices of the theory.

6.4.2 Order $\hbar$ corrections

To calculate the order $\hbar$ corrections we have to evaluate the gaussian integral obtained by expanding around the saddle point. Setting:

$$\phi = \phi_c(J) + \sqrt{\hbar}\chi, \tag{6.43}$$

and expanding the action in powers of $\hbar$ we find:

$$S(\phi) - J\phi = S(\phi_c) - J\phi_c + \frac{\hbar}{2}\int dx_1\,dx_2\,\frac{\delta^2 S}{\delta\phi(x_1)\,\delta\phi(x_2)}\bigg|_{\phi=\phi_c}\chi(x_1)\chi(x_2) + O\left(\hbar^{3/2}\right).$$

The functional integral at this order becomes:

$$Z(J) \sim Z_0(J)\int [d\chi]\exp\left[-\frac{1}{2}\int dx_1 dx_2 \frac{\delta^2 S}{\delta\phi_c(x_1)\,\delta\phi_c(x_2)}\chi(x_1)\chi(x_2)\right],$$

and therefore:

$$Z(J) \propto Z_0(J)\left[\det\frac{\delta^2 S}{\delta\phi_c(x_1)\,\delta\phi_c(x_2)}\right]^{-1/2}. \tag{6.44}$$

Finally we normalize $Z(J)$ by the condition $Z(0) = 1$.

The connected generating functional $W(J)$ at this order is then:

$$W(J) = W_0(J) + \hbar W_1(J) + O\left(\hbar^2\right), \tag{6.45}$$

with

$$W_1(J) = -\frac{1}{2}\left[\operatorname{tr}\ln\frac{\delta^2 S}{\delta\phi_c(x_1)\,\delta\phi_c(x_2)}\bigg|_J - \operatorname{tr}\ln\frac{\delta^2 S}{\delta\phi_c(x_1)\,\delta\phi_c(x_2)}\bigg|_{J=0}\right]. \tag{6.46}$$

Let us again take the example of the ϕ^4 potential to illustrate this result:

$$\frac{\delta^2 S}{\delta\phi(x_1)\,\delta\phi(x_2)} = K(x_1, x_2) + \frac{g}{2}\phi^2(x_1)\,\delta(x_1 - x_2).$$

Therefore:

$$\operatorname{tr}\ln\frac{\delta^2 S}{\delta\phi(x_1)\,\delta\phi(x_2)}\bigg|_J - \operatorname{tr}\ln\frac{\delta^2 S}{\delta\phi(x_1)\,\delta\phi(x_2)}\bigg|_{J=0}$$
$$= \operatorname{tr}\ln\left[\delta(x_1 - x_2) + \frac{g}{2}\Delta(x_1, x_2)\,\phi_c^2(x_2)\right].$$

Let us then expand $W_1(J)$ in powers of ϕ_c:

$$W_1(J) = -\frac{1}{2}\left[\frac{g}{2}\int dx_1\,\Delta(x_1, x_1)\,\phi_c^2(x_1)\right.$$
$$\left. - \frac{g^2}{8}\int dx_1\,dx_2\,\Delta(x_1, x_2)\,\phi_c^2(x_2)\,\Delta(x_2, x_1)\,\phi_c^2(x_1) + \cdots\right].$$

It is important to note that the trace operation has generated a set of one-loop Feynman diagrams.

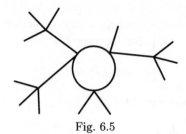

Fig. 6.5

To recover perturbation theory, we have still to expand $\phi_c(J)$ in powers of J. A typical contribution to $W_1(J)$ has the representation of figure 6.5.

Let us now perform the Legendre transformation

$$\Gamma(\varphi) = \int \mathrm{d}x\, J(x)\varphi(x) - W_0(J) - \hbar W_1(J) + O\left(\hbar^2\right). \tag{6.47}$$

By definition of the Legendre transformation the r.h.s. is stationary with respect to variations of $J(x)$ at $\varphi(x)$ fixed. Therefore a correction of order $\hbar$ to the relation between $J(x)$ and $\varphi(x)$ will produce a change of order $\hbar^2$ to the r.h.s. of equation (6.47).

Hence at order $\hbar$ we can still replace $J(x)$ by the solution of the equation

$$\varphi(x) = \phi_c(J; x).$$

We therefore conclude:

$$\Gamma(\varphi) = S(\varphi) + \hbar\Gamma_1(\varphi) + O\left(\hbar^2\right), \tag{6.48}$$

with:

$$\Gamma_1(\varphi) = \frac{1}{2}\,\mathrm{tr}\left[\ln\frac{\delta^2 S}{\delta\varphi(x_1)\,\delta\varphi(x_2)} - \ln\frac{\delta^2 S}{\delta\varphi\delta\varphi}\bigg|_{\varphi=0}\right]. \tag{6.49}$$

If $S(\phi)$ has the form:

$$S(\phi) = \tfrac{1}{2}\phi K \phi + \int V(\phi(x))\,\mathrm{d}x, \tag{6.50}$$

the expansion of $\Gamma_1(\varphi)$ in powers of V takes the form:

$$\Gamma_1(\varphi) = \frac{1}{2}\sum_{n=1}^{\infty}\frac{(-1)^{n+1}}{n}\int \mathrm{d}x_1\ldots \mathrm{d}x_n V''(\varphi(x_1))\Delta(x_1, x_2)$$
$$\times V''(\varphi(x_2))\Delta(x_2, x_3)\ldots V''(\varphi(x_n))\Delta(x_n, x_1), \tag{6.51}$$

which in terms of Feynman diagrams has the representation of a sum of one-loop diagrams (see figure 6.6).

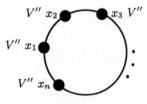

Fig. 6.6

The remarkable property of the Feynman graph expansion of the functional $\Gamma_1(\varphi)$ at this order is that all diagrams are one line irreducible (1-particle irreducible in the language of Particle Physics), i.e. they cannot be disconnected by cutting only one line. The functional $\Gamma_1(\varphi)$ is the generating functional of one line irreducible one-loop Feynman diagrams.

6.4.3 Fermions

We have discussed the Legendre transform only in the case of a boson field theory, however the extension to fermions is straightforward. Let ψ and $\bar\psi$ be two fermion fields and $S(\bar\psi, \psi)$ the corresponding local action. It is clear that $W(\eta, \bar\eta) = \ln Z(\eta, \bar\eta)$ is still the generating functional of connected correlation functions (we have called $\bar\eta$ and η the sources for ψ and $\bar\psi$). If, following the conventions of Section 5.5, we write the source terms in the functional integral $\bar\eta\psi + \bar\psi\eta$, then we define the Legendre transform of W by

$$\Gamma(\bar\psi, \psi) + W(\eta, \bar\eta) = \int \mathrm{d}x \left[\bar\eta(x)\psi(x) + \bar\psi(x)\eta(x) \right] \qquad (6.52a)$$

$$\psi(x) = \frac{\delta W}{\delta\bar\eta(x)}, \quad \bar\psi(x) = -\frac{\delta W}{\delta\eta(x)}. \qquad (6.52b)$$

The equations $(6.52b)$ are equivalent to

$$\eta(x) = \frac{\delta\Gamma}{\delta\bar\psi(x)}, \quad \bar\eta(x) = -\frac{\delta\Gamma}{\delta\psi(x)}.$$

With these conventions one easily verifies that $\Gamma(\bar\psi, \psi) = S(\bar\psi, \psi)$ at tree order. All the other algebraic properties derived for bosons generalize to the fermion case. We here recall however that a gaussian integration over fermions yields a determinant instead of the inverse of a determinant for a complex scalar. This implies that fermion loops are affected by a additional minus sign compared to boson loops. Let us, as an example, give the one-loop results corresponding to the boson-fermion action

$$S(\phi, \bar\psi, \psi) = \int \mathrm{d}^d x \left[-\bar\psi(x)\left(\slashed\partial + A(\phi)\right)\psi(x) + S_{\mathrm{B}}(\phi) \right]. \qquad (6.53)$$

We have to evaluate the functional integral:

$$Z(J, \bar\eta, \eta) = \int [\mathrm{d}\phi \mathrm{d}\bar\psi \mathrm{d}\psi] \exp\left[-S(\phi, \bar\psi, \psi) + \int \mathrm{d}^d x \left[J(x)\phi(x) + \bar\eta(x)\psi(x) + \bar\psi(x)\eta(x) \right] \right].$$
$$(6.54)$$

We first again look for the solutions $\phi_c, \bar\psi_c, \psi_c$ of the field equations. After shifting the fields we then have to calculate a gaussian functional integral over boson and fermion fields. The result can be deduced from the expressions (1.64–1.66) for the determinant of a mixed matrix involving bosons and fermions, or can be obtained directly by integrating over fermions first, and then over bosons.

After a short calculation we find for connected correlation functions

$$W(J, \bar\eta, \eta) = -S(\phi_c, \bar\psi_c, \psi_c) + \int \mathrm{d}^d x \left[J(x)\phi_c(x) + \bar\eta(x)\psi_c(x) + \bar\psi_c(x)\eta(x) \right]$$

$$- \tfrac{1}{2}\mathrm{tr}\ln\left[\frac{\delta^2 S_{\mathrm{B}}}{\delta\phi_c(x)\delta\phi_c(y)} - \bar\psi_c\frac{\delta^2 A}{\delta\phi_c(x)\delta\phi_c(y)}\psi_c + 2\bar\psi_c\frac{\delta A}{\delta\phi_c(x)}\left[\slashed\partial + A(\phi_c)\right]^{-1}\frac{\delta A}{\delta\phi_c(y)}\psi_c \right]$$

$$+ \mathrm{tr}\ln\left[\slashed\partial + A(\phi_c)\right],$$

where the fields $\phi_c, \bar{\psi}_c, \psi_c$ are solutions of the field equations

$$\frac{\delta S_{\mathrm{B}}}{\delta \phi_c(x)} - \bar{\psi}_c \frac{\delta A}{\delta \phi_c(x)} \psi_c - J(x) = 0 \,, \qquad (6.55a)$$

$$[\not{\partial} + A(\phi_c)] \, \psi_c(x) + \eta(x) = 0 \,, \qquad (6.55b)$$

$$\bar{\psi}_c(x) \, [\not{\partial} + A(\phi_c)] + \bar{\eta}(x) = 0 \,, \qquad (6.55c)$$

(by convention the operator ∂ in (6.55c) acts on the left with a minus sign).

A Legendre transformation then yields the 1PI functional at one-loop order:

$$\Gamma(\varphi, \bar{\psi}, \psi) = S(\varphi, \bar{\psi}, \psi) - \mathrm{tr}\ln\left(\not{\partial} + A(\varphi)\right)$$
$$+ \tfrac{1}{2} \mathrm{tr}\ln\left\{ \frac{\delta^2 S_{\mathrm{B}}}{\delta\varphi(x)\delta\varphi(y)} + \bar{\psi}\left[2\frac{\delta A}{\delta\varphi(x)}\left[\not{\partial} + A(\varphi)\right]^{-1}\frac{\delta A}{\delta\varphi(y)} - \frac{\delta^2 A}{\delta\varphi(x)\delta\varphi(y)} \right] \psi \right\}. \quad (6.56)$$

6.4.4 Loopwise expansion at higher orders

We now expand the functional integral

$$Z(J) = \int [\mathrm{d}\phi] \exp\left[-\frac{1}{\hbar}\left(S\left(\phi\right) - J\phi\right) \right],$$

with $S(\phi)$ of the form

$$S\left(\phi\right) = \tfrac{1}{2}\phi K \phi + V\left(\phi\right),$$

to all orders in perturbation theory. Let us determine the power of $\hbar$ in front a *connected* Feynman diagram contributing to $W(J)$. The propagator is $\hbar\Delta$. Any vertex generated by $V(\varphi)$ yields a power $\hbar^{-1}$. In the same way at the end of all external lines is attached a factor J which also yields a factor $\hbar^{-1}$. Calling I the number of internal lines of the diagram (propagators which join two vertices), E the number of external lines (propagator joining a vertex to a source J), V the number of vertices, we find the power

$$\hbar^{I+E-(V+E)+1},$$

the last factor $\hbar$ coming from our normalization of $W(J)$ (equation (6.36)).

Note that the same result is obtained for one line irreducible Feynman diagrams (i.e. as we prove in next section contributions to $\Gamma(\varphi)$) because the factor $\hbar$ coming from the source cancels the factor coming from the external propagator.

Let us now show that the power of $\hbar$ that we have found counts the number of loops of a diagram. The number of loops is defined in the following way: if by cutting a line of a connected diagram γ we obtain a new connected diagram γ' then:

$$\# \text{ loops } \gamma = \# \text{ loops } \gamma' + 1 \,.$$

From this definition follows a relation between the number of loops L, the number of internal lines I and the number of vertices V:

$$L = I - V + 1 \,. \qquad (6.57)$$

We prove this relation in the following way: We see that each time we can remove an internal line without disconnecting the diagram we decrease I by 1 and L by 1.

Eventually we get a tree diagram, i.e. a diagram in which no line can be removed without disconnecting the diagram. We then have to show:

$$I - V + 1 = 0.$$

From a tree diagram we can remove systematically a vertex at the boundary with the line connecting it to the diagram until we obtain the simplest diagram, composed of a line joining two vertices, which satisfies the equation.

We have thus shown that the expansion in powers of $\hbar$ is a reordering of perturbation theory according to the number of loops of the Feynman diagrams.

The number of loops is also the number of independent internal intensities in the corresponding electric circuit, the current being conserved at each vertex, the intensities flowing into the diagram being fixed. This can be proven by showing that equation (6.57) is again satisfied. Indeed the number of L of independent intensities is equal to the total number of intensities I minus the number of conservation equations $(V - 1)$ (because one equation gives the total conservation of the current). This remark will eventually allow us to relate the number of loops to the number of independent momentum integration variables.

6.5 Legendre Transformation and 1-Irreducibility

We have shown that the two first orders of the loop expansion of $\Gamma(\varphi)$ contain only one-line irreducible Feynman diagrams. We now suspect that at order $\hbar^L$ we shall obtain the generating functional of all L-loop diagrams. By realizing that proper vertices and connected correlation functions are affected by global combinatorial factors identical respectively to those of vertices in the action and connected tree diagrams, it is possible to give a general proof that $\Gamma(\varphi)$ is the generating functional of one line irreducible Feynman diagrams: one takes $\Gamma(\phi)$ as the action, one calculates at leading in $\hbar$, and recovers $W(J)$ as the generating functional of the corresponding connected tree diagrams. The result then follows from the considerations of the beginning of the section (equation (6.41)). However we here give a more powerful and completely algebraic proof.

To prove that proper vertices are given in perturbation theory by a sum of one line or one particle irreducible (1PI) Feynman diagrams, we directly use the definition and prove that by cutting one line in all possible ways in a diagram contributing to $\Gamma(\varphi)$, the diagram remains connected.

Let us consider the action:

$$S_\varepsilon(\phi) = \frac{1}{2} \int \mathrm{d}x \, \mathrm{d}y \, \phi(x) \, \phi(y) \left[K(x,y) + \varepsilon \right] + V(\phi). \tag{6.58}$$

The symbol ε represents a small parameter in which we expand at first order. The corresponding propagator $\Delta_\varepsilon(x,y)$ is:

$$\int \Delta_\varepsilon(x,z) \left[K(z,y) + \varepsilon \right] \mathrm{d}z = \delta(x - y),$$

$$\Delta_\varepsilon(x,y) = \Delta(x,y) - \varepsilon \eta(x)\eta(y) + O\left(\varepsilon^2\right), \tag{6.59}$$

with the definition:

$$\eta(x) = \int \Delta(x,z) \, \mathrm{d}z.$$

If we now expand a Feynman diagram with the new propagator $\Delta_\varepsilon(x, y)$ in ε, we obtain at first order a sum of terms in which in all possible ways a propagator $\Delta(x, y)$ has been replaced by the product $-\eta(x)\eta(y)$. Since in this product the dependence in x and y is factorized, this means topologically that in the Feynman diagram the corresponding line has been cut. A necessary and sufficient condition for a diagram to be 1PI is that all terms at order ε are connected.

Higher orders in ε can be used to study irreducibility with respect to cutting 2, 3,... lines.

Let us now calculate the partition function $Z_\varepsilon(J)$ at first order in ε:

$$Z_\varepsilon(J) = \int [\mathrm{d}\phi] \left(1 - \frac{\varepsilon}{2} \int \mathrm{d}x\,\mathrm{d}y\,\phi(x)\phi(y)\right) \exp\left[-S(\phi) + J\phi\right] + O\left(\varepsilon^2\right), \qquad (6.60)$$

and therefore:

$$Z_\varepsilon(J) = \left[1 - \frac{\varepsilon}{2} \int \mathrm{d}x\,\mathrm{d}y\, \frac{\delta}{\delta J(x)} \frac{\delta}{\delta J(y)} + O\left(\varepsilon^2\right)\right] Z(J). \qquad (6.61)$$

The generating functional $W_\varepsilon(J)$:

$$W_\varepsilon(J) = \ln Z_\varepsilon(J), \qquad (6.62)$$

is then given by:

$$W_\varepsilon(J) = W(J) - \frac{\varepsilon}{2} \left\{\left[\int \mathrm{d}x\, \frac{\delta W}{\delta J(x)}\right]^2 + \int \mathrm{d}x\,\mathrm{d}y\, \frac{\delta^2 W}{\delta J(y)\,\delta J(x)}\right\} + O\left(\varepsilon^2\right). \qquad (6.63)$$

The first order in ε of $W_\varepsilon(J)$ contains a contribution of the form:

$$\left[\int \mathrm{d}x\, \frac{\delta W}{\delta J(x)}\right]^2$$

which is disconnected, as expected.

Let us now perform the Legendre transformation. We use identity (6.20) under the form:

$$\frac{\partial \Gamma}{\partial \varepsilon} = -\frac{\partial W}{\partial \varepsilon},$$

and therefore:

$$\Gamma_\varepsilon(\varphi) = \Gamma(\varphi) + \frac{\varepsilon}{2} \left\{\left[\int \mathrm{d}x\,\varphi(x)\right]^2 + \int \mathrm{d}x\,\mathrm{d}y \left[\frac{\delta^2 \Gamma}{\delta\varphi(x)\delta\varphi(y)}\right]^{-1}\right\} + O\left(\varepsilon^2\right) \qquad (6.64)$$

We see that the first order in ε contains two terms, the term $\left[\int \mathrm{d}x\,\varphi(x)\right]^2$, which we have added to the action, and a second term which is the connected propagator in presence of an external field. Since the variation of $\Gamma(\varphi)$ is connected, $\Gamma(\varphi)$ is indeed one line or, in the language of particle physics, one particle irreducible. In the chapters which follow we shall use indifferently the terms of proper vertices or 1PI correlation functions. Note finally that in Section 6.2 we have shown that

$$W^{(n)}_{\text{amp.}} = -\Gamma^{(n)} + \text{reducible terms} \quad (n > 2),$$

because the difference contains only lower correlation functions related by propagators. The diagrams contributing to $\Gamma^{(n)}$ thus differ from the 1PI amputated diagrams of $W^{(n)}$ only by a sign. In the same way the diagrams contributing to $\Gamma^{(2)}$ beyond tree level are, up to a change of sign, the amputated diagrams of the mass operator.

6.6 Physical Interpretation of the 1PI Functional

In what follows we explicitly require time translation invariance: the quantum mechanical hamiltonian H is time-independent.

6.6.1 A quantum mechanical interpretation

We first give a quantum mechanical interpretation of $\Gamma(\varphi)$ when the field φ, argument of Γ, is time-independent. Let us therefore distinguish in the d dimensional space $\mathbb{R}^d$ a time direction, and now call t, x the time and space coordinates respectively. We assume that t varies in a finite interval $[0, \beta]$ and impose periodic boundary conditions on the fields in the time direction. Moreover we restrict ourselves in what follows to time-independent sources $J(x, t)$:

$$J(x, t) = J(x).$$

The functional $Z(J)$ is then the partition function $\operatorname{tr} e^{-\beta H}$ for a source dependent hamiltonian $H(\phi, J)$ (equation (5.7)):

$$H(\phi, J) = H(\phi) - \int d^{d-1}x \, \phi(x) J(x). \tag{6.65}$$

In the large β limit, the operator $e^{-\beta H}$ projects onto the ground state of the hamiltonian. Calling $E_0(J)$ the ground state energy of $H(J)$, we obtain:

$$W(J) \underset{\beta \to \infty}{\sim} -\beta E_0(J). \tag{6.66}$$

By definition $E_0(J)$ is the expectation value of the hamiltonian $H(\phi, J)$ in its ground state. Denoting by $\langle \cdot \rangle_J$ the average of an operator in the groundstate of $H(\phi, J)$ we can rewrite the relation:

$$W(J) \sim -\beta \langle H(\phi, J) \rangle_J \sim -\beta \langle H(\phi) \rangle_J - \int d^{d-1}x \, J(x) \langle \phi(x) \rangle_J. \tag{6.67}$$

Using time translation invariance it is easy to verify that $W(J)/\beta$ is the Legendre transform of $\Gamma(\varphi)/\beta$ when $\varphi(x)$ is a time-independent field:

$$\Gamma(\varphi) = -W(J) + \beta \int d^{d-1}x \, J(x) \varphi(x), \tag{6.68}$$

with

$$\varphi(x) = \frac{1}{\beta} \frac{\delta W(J)}{\delta J(x)}. \tag{6.69}$$

In the functional integral representation of the partition function $J(x)$ is a source for $\int_0^\beta dt \, \phi(x, t)$. The functional derivative of $W(J)$ with respect to $J(x)$ is the 1-point correlation function of this quantity, in a source. Due to time translation invariance $\langle \phi(x, t) \rangle$ is time-independent. In the large β limit $\varphi(x)$ thus becomes the expectation value of the field operator $\phi(x)$ in the ground state of $H(J)$ (equation (5.11)):

$$\varphi(x) = \langle \phi(x) \rangle_J. \tag{6.70}$$

Combining equations (6.67,6.70) one then finds:

$$\Gamma(\varphi) = \beta \langle H \rangle_J. \tag{6.71}$$

The 1PI functional $\Gamma(\varphi)$, restricted to time-independent sources, is proportional to the expectation value of the initial hamiltonian in the ground state of $H(\phi, J)$. As equation (6.70) shows, it is expressed in terms of the expectation value $\varphi(x)$ of the field $\phi(x)$ in the same state. This result can be used to establish variational upperbounds on the ground state energy of the hamiltonian.

6.6.2 The 1PI functional and the free energy at fixed field time average

Let us now present a related but slightly different interpretation. We calculate the partition function with the same periodic boundary condition in time but restricted to fields satisfying

$$\varphi(x) = \frac{1}{\beta} \int_0^\beta dt \, \phi(x, t). \tag{6.72}$$

Note that this implies trivially $\varphi(x) = \langle \phi(x, t) \rangle$. We call $-\beta\mathcal{G}(\varphi)$ the corresponding free energy

$$e^{-\beta\mathcal{G}(\varphi)} = \int [d\phi(x, t)] \exp[-S(\phi)]. \tag{6.73}$$

We have written the free energy in the form $-\beta\mathcal{G}(\varphi)$ because we know that in the large β limit the free energy is proportional to β.

Then the free energy corresponding to the sum over all field configurations in presence of a time-independent source $J(x)$ is given by

$$e^{W(J)} = \int [d\varphi(x)] \exp\left[-\beta\mathcal{G}(\varphi) + \beta \int dx \, J(x)\varphi(x)\right]. \tag{6.74}$$

For β large the functional integral can be calculated by steepest descent. The saddle point equation is

$$J(x) = \frac{\delta\mathcal{G}}{\delta\varphi(x)}. \tag{6.75}$$

When the equation has several solutions one has to take the stable solution which yields the largest contribution to the free energy. Then

$$W(J) \sim -\beta\mathcal{G}(\varphi) + \beta \int dx \, J(x)\varphi(x). \tag{6.76}$$

After Legendre transformation one finds

$$\beta\mathcal{G}(\varphi) = \Gamma(\varphi),$$

where again $\Gamma(\varphi)$ is the 1PI functional restricted to time-independent fields. Note however that $\mathcal{G}(\varphi)$ has in general no reasons to be convex. One may find field configurations such that the operator

$$\frac{\delta^2\mathcal{G}(\varphi)}{\delta\varphi(x)\delta\varphi(y)}$$

is not positive. On the other hand because $\Gamma(\varphi)$ is the result of a steepest descent calculation, it may coincide with $\beta\mathcal{G}(\varphi)$ only in regions of field space where the operator is positive. In general in perturbation theory one calculates a quantity which, restricted to time-independent fields, coincides with $\mathcal{G}$ rather than Γ. This explains an apparent paradox: In the several phase region one often pretends discussing the minima of $\Gamma(\varphi)$, i.e. the minima of a quantity which has convexity properties and can have only one minimum. Really one discusses the properties of $\mathcal{G}$.

Let finally note that in the framework of statistical mechanics, the Legendre transform $\Gamma(\varphi)$ is also the thermodynamical potential, a quantity which plays a central role in the discussion of critical phenomena.

Bibliographical Notes

The idea of using generating functionals with sources was originated in:

J. Schwinger, *Proc. Natl. Acad. Sci. USA* 37 (1951) 452, 455.

The Legendre transform of connected correlation functions has been discussed in the statistical mechanics context in:

C. De Dominicis, *J. Math. Phys.* 4 (1963) 255; C. De Dominicis and
P.C. Martin, *J. Math. Phys.* 5 (1964) 14, 31;

and introduced in field theory in:

G. Jona-Lasinio, *Nuovo Cimento* 34 (1964) 1790.

See also:

J. Goldstone, A. Salam and S. Weinberg, *Phys. Rev. 127* (1962) 965.

A proof of the one-line irreducibility can be found in:

S. Coleman, *Laws of Hadronic Matter* (discussion of lecture 3), Erice 1973, A. Zichichi
ed. (Academic Press, New York 1975).

The physical interpretation of the 1PI functional is taken from:

K. Symanzik, *Commun. Math. Phys.* 16 (1970) 48; see also S. Coleman, reference
above.

It has been recognized that the loop expansion is also an expansion in powers of $\hbar$ in:

Y. Nambu, *Phys. Lett.* 26B (1966) 626.

The one-loop functional is calculated by functional methods in:

B.W. Lee and J. Zinn-Justin, *Phys. Rev.* D5 (1972) 3121 Appendix B; R. Jackiw, *Phys. Rev.* D9 (1974) 1686.

Exercises

Exercise 6.1

Write explicitly the field equation of motion (5.28) for the various generating functionals in the example of the action:

$$S(\phi) = \int d^d x \left\{ \tfrac{1}{2} \left[\partial_\mu \phi(x) \right]^2 + \tfrac{1}{2} m^2 \phi^2(x) + \frac{g}{3!} \phi^3(x) \right\},$$

and deduce from the equations the one-loop correction to the one, two and three 1PI correlation functions.

Exercise 6.2

Write the field equation of motion satisfied by the generating functionals of correlation functions in the example of the action:

$$S(\phi) = \int d^d x \left(\tfrac{1}{2} \partial_\mu \phi(x) \partial_\mu \phi(x) + \tfrac{1}{2} m^2 \phi^2(x) + \frac{1}{4!} g \phi^4(x) \right).$$

Recover the form of the 1PI functional at one-loop order.

Exercise 6.3

Obtain the quantum equations of motion satisfied by the 1PI functional in the theory with bosons and fermions described by the action (5.62).

APPENDIX 6
HIGHER ORDERS IN THE LOOP EXPANSION

Let us show how successive terms in the loop expansion can be calculated, using the method of Section 6.4:

$$Z(J) = \int [d\phi] \exp\left[\frac{1}{\hbar}\left(-S\left(\phi\right) + J\phi\right)\right]. \qquad (A6.1)$$

The saddle point $\phi_c(J)$ is solution of the equation:

$$\frac{\delta S}{\delta\phi(x)}[\phi_c] = J(x). \qquad (A6.2)$$

Setting:

$$\phi(x) = \phi_c(x) + \sqrt{\hbar}\chi(x), \qquad (A6.3)$$

we can expand $S(\phi)$ in powers of χ:

$$S(\phi_c + \chi) = S(\phi_c) + \frac{\hbar}{2!}\int dx_1\, dx_2\, \chi(x_1)\,\chi(x_2)\, \frac{\delta^2 S}{\delta\phi(x_1)\,\delta\phi(x_2)}\bigg|_{\phi=\phi_c}$$

$$+ \frac{\hbar^{3/2}}{3!}\int dx_1\, dx_2\, dx_3\, \chi(x_1)\,\chi(x_2)\,\chi(x_3)\, \frac{\delta^3 S}{\delta\phi(x_1)\,\delta\phi(x_2)\,\delta\phi(x_3)}\bigg|_{\phi=\phi_c}$$

$$+ \frac{\hbar^2}{4!}\int dx_1\, dx_2\, dx_3\, dx_4\, \chi(x_1)\cdots\chi(x_4)\, \frac{\delta^4 S}{\delta\phi(x_1)\cdots\delta\phi(x_4)}\bigg|_{\phi=\phi_c} +\cdots. \qquad (A6.4)$$

The expansion in powers of $\hbar$ and the integration over χ generate vacuum Feynman diagrams with a $\phi_c(J)$ dependent propagator:

$$\int \frac{\delta^2 S}{\delta\phi_c(x)\,\delta\phi_c(y)}\Delta(y,z;\phi_c)\,dy = \delta(x-z), \qquad (A6.5)$$

and ϕ_c dependent vertices:

$$S^{(3)}(x_1,x_2,x_3;\phi_c) = \frac{\delta^3 S}{\delta\phi_c(x_1)\,\delta\phi_c(x_2)\,\delta\phi_c(x_3)}$$

$$S^{(4)}(x_1,\ldots,x_4;\phi_c) = \frac{\delta^4 S}{\delta\phi_c(x_1)\ldots\delta\phi_c(x_4)} \qquad (A6.6)$$

$$\cdots$$

Let us illustrate these remarks by an explicit two-loop calculation. Expanding the interaction terms $S^{(3)}$ and $S^{(4)}$ and integrating term by term we obtain:

$$\ln Z(J) = K + \frac{1}{\hbar}[-S(\phi_c) + J\cdot\phi_c] - \frac{1}{2}\operatorname{tr}\ln\frac{\delta^2 S}{\delta\phi_c\,\delta\phi_c}$$

$$+ \hbar\bigg\{ -\frac{1}{8}\int dx_1\ldots dx_4\, S^{(4)}(x_1,x_2,x_3,x_4)\,\Delta(x_1,x_2)\,\Delta(x_3,x_4)$$

$$+ \int dx_1\ldots dy_3\, S^{(3)}(x_1,x_2,x_3)\, S^{(3)}(y_1,y_2,y_3)\bigg[\frac{1}{8}\Delta(x_1,x_2)$$

$$\times \Delta(y_1,y_2)\,\Delta(x_3,y_3) + \frac{1}{12}\Delta(x_1,y_1)\,\Delta(x_2,y_2)\,\Delta(x_3,y_3)\bigg]\bigg\} + O(\hbar^2). \qquad (A6.7)$$

The constant K depends on the normalization of the functional integral. The expression for $W(J) = \hbar \ln Z$ follows immediately. We have now to perform the Legendre transformation:

$$\Gamma(\varphi) + W(J) = \int J(x)\varphi(x)\mathrm{d}x, \qquad \varphi(x) = \frac{\delta W}{\delta J(x)}. \tag{A6.8}$$

As we have already noticed we need $\varphi(x)$ only up to order $\hbar$ because expression $(A6.8)$ is stationary in $\varphi(x)$:

$$\varphi(x) = \phi_c(J;x) - \frac{\hbar}{2}\frac{\delta}{\delta J(x)}\left(\operatorname{tr}\ln\frac{\delta^2 S}{\delta\phi_c\,\delta\phi_c}\right) + O\left(\hbar^2\right). \tag{A6.9}$$

The order $\hbar$ correction can be rewritten:

$$\frac{\delta}{\delta J(x)}\operatorname{tr}\ln\frac{\delta^2 S}{\delta\phi_c\delta\phi_c} = \int \mathrm{d}y\,\frac{\delta\phi_c(y)}{\delta J(x)}\operatorname{tr}\left[\frac{\delta^3 S}{\delta\phi_c(y)\delta\phi_c\,\delta\phi_c}\left(\frac{\delta^2 S}{\delta\phi_c\,\delta\phi_c}\right)^{-1}\right].$$

Using then equations $(A6.2,A6.9)$ and definition $(A6.5)$, we can express ϕ_c in terms of φ:

$$\varphi(x) = \phi_c(J,x) - \frac{\hbar}{2}\int S^{(3)}(y,y_1,y_2;\varphi)\,\Delta(x,y;\varphi)\,\Delta(y_1,y_2;\varphi)\,\mathrm{d}y\,\mathrm{d}y_1\,\mathrm{d}y_2 + O\left(\hbar^2\right). \tag{A6.10}$$

We still need $J(x)$ at order $\hbar$. Using equations $(A6.2)$ and $(A6.10)$ we find:

$$J(x) = \frac{\delta S}{\delta\varphi(x)} + \frac{\hbar}{2}\int S^{(3)}(x,y_1,y_2;\varphi)\,\Delta(y_1,y_2;\varphi)\,\mathrm{d}y_1\mathrm{d}y_2 + O\left(\hbar^2\right). \tag{A6.11}$$

Equation $(A6.8)$ then yields $\Gamma(\varphi)$ at two-loop order. As expected the reducible part in expression $(A6.7)$ cancels (figure 6.7) and we obtain:

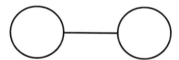

Fig. 6.7   The reducible part at two-loop order.

$$\Gamma(\varphi) = S(\varphi) + \frac{\hbar}{2}\operatorname{tr}\ln\frac{\delta^2 S}{\delta\varphi\,\delta\varphi}$$

$$+ \hbar^2\left[\frac{1}{8}\int S^{(4)}(x_1,x_2,x_3,x_4;\varphi)\,\Delta(x_1,x_2;\varphi)\,\Delta(x_3,x_4;\varphi)\,\mathrm{d}x_1\mathrm{d}x_2\mathrm{d}x_3\mathrm{d}x_4\right.$$

$$-\frac{1}{12}\int S^{(3)}(x_1,x_2,x_3;\varphi)\,S^{(3)}(y_1,y_2,y_3;\varphi)\,\Delta(x_1,y_1;\varphi)\,\Delta(x_2,y_2;\varphi)$$

$$\left. \Delta(x_3,y_3;\varphi)\,\mathrm{d}x_1\,\mathrm{d}x_2\,\mathrm{d}x_3\,\mathrm{d}y_1\,\mathrm{d}y_2\,\mathrm{d}y_3\right] + O\left(\hbar^3\right). \tag{A6.12}$$

Figure 6.8 gives a diagrammatic representation of the two-loop terms in the equation.

Fig. 6.8   The two-loop contributions to $\Gamma(\varphi)$.

7 REAL-TIME QUANTUM FIELD THEORY AND S-MATRIX

Up to now we have discussed only euclidean field theory and correlation functions. To calculate S-matrix elements, quantities relevant to Particle Physics, it is necessary to return to a real time formulation of quantum field theory. We discuss in this chapter scalar bosons and massive Dirac fermions. For scalar bosons a natural formalism is provided by the holomorphic representation. A simple generalization of the path integral of Section 3.5 then leads to a functional integral representation of the S-matrix. For Dirac fermions, we combine the considerations of Section 3.6 (after continuation to real time) with the discussion of Section 5.5. Technical details concerning the symmetries of the fermion action and the properties of γ-matrices can be found in Appendix A5. Relevant for this chapter are also the considerations of Section 2.6.

Though quantum field theory appears as a natural formal generalization of quantum mechanics, it is not quite obvious how quantum mechanics is recovered in the non-relativistic limit. Indeed one could have also thought about developing a formalism of relativistic quantum particles. The observation of physical effects due to the electro-magnetic field, which predates quantum mechanics, makes this alternative possibility less attractive since it would not have unified fields and particles. We therefore briefly indicate how quantum mechanics formally emerges in the low-energy, low-momentum of a massive quantum field theory, in the form of a quantum many-body theory.

As a warning we also have to stress that several quantities we meet in this chapter are really infinite, or at least infinite in high enough dimensions. Therefore the discussion will be rather formal. However the divergence problems will be carefully studied later, in the coming chapters. A regularization by a space lattice would have rendered most quantities meaningful but complicated the arguments.

7.1 Scalar Free Field Theory: Holomorphic Representation

The starting point of the construction is free field theory. The free field action $\mathcal{A}_0(\phi)$ for a scalar field ϕ in real time can be written

$$\mathcal{A}_0(\phi) = \int dt\, d^{d-1}x \left[\tfrac{1}{2}(\partial_t \phi)^2 - \tfrac{1}{2}(\nabla_x \phi)^2 - \tfrac{1}{2}m^2\phi^2(t,x) \right]. \tag{7.1}$$

Because the action is quadratic in $\phi(t,x)$, the field ϕ can be considered as a collection of harmonic oscillators. The different harmonic oscillators decouple in the momentum basis. We therefore perform a Fourier transformation over space variables and set:

$$\phi(x,t) = \int d^{d-1}\hat{p}\, e^{i\hat{p}x}\, \tilde{\phi}(\hat{p},t).$$

In this chapter we denote the time and space components of a d-momentum vector p as $\{p_0, \hat{p}\}$, p_0 being the energy and $\hat{p}$ the momentum.

The free action becomes

$$\mathcal{A}_0(\tilde{\phi}) = \tfrac{1}{2}(2\pi)^{d-1} \int dt\, d^{d-1}\hat{p} \left[\partial_t\tilde{\phi}(\hat{p},t)\partial_t\tilde{\phi}(-\hat{p},t) - \tilde{\phi}(\hat{p},t)\,(\hat{p}^2 + m^2)\,\tilde{\phi}(-\hat{p},t) \right].$$

The conjugate momentum $\tilde{\Pi}$ of $\tilde{\phi}$ is then

$$\tilde{\Pi}(\hat{p}) = (2\pi)^{d-1}\partial_t\tilde{\phi}(-\hat{p}).$$

The corresponding hamiltonian density $\mathcal{H}$ is

$$\mathcal{H}\left(\tilde{\Pi}, \tilde{\phi}\right) = \tfrac{1}{2}\left[(2\pi)^{1-d}\tilde{\Pi}(\hat{p})\tilde{\Pi}(-\hat{p}) + (2\pi)^{d-1}\omega^2(\hat{p})\tilde{\phi}(\hat{p})\tilde{\phi}(-\hat{p})\right],$$

with $\omega(\hat{p}) = \sqrt{\hat{p}^2 + m^2}$. The canonical commutation relations take the form

$$[\tilde{\phi}(\hat{p}_1), \tilde{\Pi}(\hat{p}_2)] = i\delta^{(d-1)}(\hat{p}_1 - \hat{p}_2).$$

We now introduce creation and annihilation operators $A^\dagger, A$ related to $\tilde{\Pi}, \tilde{\phi}$ by

$$\tilde{\phi}(\hat{p}) = \frac{1}{2\omega(\hat{p})}\left[A^\dagger(-\hat{p}) + A(\hat{p})\right], \tag{7.2a}$$

$$\tilde{\Pi}(\hat{p}) = \frac{i}{2}(2\pi)^{d-1}\left[A^\dagger(\hat{p}) - A(-\hat{p})\right], \tag{7.2b}$$

which have the commutation relations

$$[A(\hat{p}_1), A(\hat{p}_2)] = [A^\dagger(\hat{p}_1), A^\dagger(\hat{p}_2)] = 0, \quad [A(\hat{p}_1), A^\dagger(\hat{p}_2)] = (2\pi)^{1-d}2\omega(\hat{p}_1)\delta^{(d-1)}(\hat{p}_1 - \hat{p}_2).$$

Note that the transformation (7.2) is non-local in coordinate space.

The free hamiltonian $\mathbf{H}$ then reads

$$\mathbf{H}(A^\dagger, A) = \int d^{d-1}\hat{p}\,\mathcal{H} = \tfrac{1}{2}(2\pi)^{d-1}\int d^{d-1}\hat{p}\, A^\dagger(\hat{p})A(\hat{p}) + E_0,$$

where E_0 is a constant obtained by commutation, which we discuss below. We have chosen a normalization of operators such that the hamiltonian has the form

$$\mathbf{H}(A^\dagger, A) - E_0 = (2\pi)^{d-1}\int\frac{d^{d-1}\hat{p}}{2\omega(\hat{p})}\,\omega(\hat{p})A^\dagger(\hat{p})A(\hat{p}), \tag{7.3}$$

where we have introduced the $O(d-1,1)$ covariant measure $d\hat{p}/2\omega(\hat{p})$

$$\int\frac{d^{d-1}\hat{p}}{2\omega(\hat{p})}f(\hat{p}) = \int d^d p\,\delta_+(p^2 - m^2)f(\hat{p}),$$

with the notation (convenient but inconsistent with our euclidean notation)

$$p^2 = p_0^2 - \hat{p}^2, \quad \delta_+(p^2 - m^2) = \delta(p^2 - m^2)\theta(p_0).$$

We define the ground state $|0\rangle$, also called the vacuum (zero particle state), by the property that it is annihilated by all operators $A(\hat{p})$

$$\langle 0|0\rangle = 1, \quad A(\hat{p})|0\rangle = 0.$$

The vacuum energy. The quantity E_0 is thus the vacuum or ground state energy

$$\mathbf{H}|0\rangle = E_0|0\rangle.$$

It is formally proportional to $[A(\hat{p}), A^{\dagger}(\hat{p})]$ and thus to $\delta^{d-1}(0)$. To give a precise meaning to E_0, it is necessary to quantize in a large box of linear size L and to modify the theory at short distance or at large momenta so that the Fourier modes are cut-off at some momentum scale Λ (a space lattice would provide such a cut-off). The Fourier variables $\hat{p}$ are then quantized

$$\hat{p} = 2\pi \mathbf{n}/L, \qquad \mathbf{n} \in \mathbb{Z}^{d-1},$$

and the vacuum energy becomes

$$E_0 = \tfrac{1}{2} \sum_{\mathbf{n}} \omega(\hat{p}).$$

For L large sums can be replaced by integrals and $\mathbf{dn} = L^{d-1}/(2\pi)^{d-1}\mathrm{d}\hat{p}$. The space volume factorizes, showing as expected that the energy is an extensive quantity,

$$E_0/L^{d-1} = \tfrac{1}{2} \int^{\Lambda} \frac{\mathrm{d}^{d-1}\hat{p}}{(2\pi)^{d-1}} \sqrt{m^2 + \hat{p}^2}\,, \tag{7.4}$$

but the energy density is cut-off dependent. The large momentum divergence of the vacuum energy is not relevant here because in this non-gravitational theory the hamiltonian can be shifted by a constant in such a way that the vacuum has zero energy. Note, however, that if the vacuum energy itself is not a physical observable, a variation (imposed for example by a change in boundary conditions) of the vacuum energy may be (see Appendix A18.1).

We can then construct a Hilbert space by acting repeatedly with creation operators $A^{\dagger}$ on the vacuum. The vectors

$$A^{\dagger}(\hat{p}_1)A^{\dagger}(\hat{p}_2) \ldots A^{\dagger}(\hat{p}_n) \left|0\right\rangle,$$

span the space of n-particle states. The n-particle state energy is the sum of the energies of each particle: Indeed from

$$[\mathbf{H}, A^{\dagger}(\hat{p})] = \omega(\hat{p})A^{\dagger}(\hat{p}),$$

we obtain

$$(\mathbf{H} - E_0)A^{\dagger}(\hat{p}_1)A^{\dagger}(\hat{p}_2) \ldots A^{\dagger}(\hat{p}_n) \left|0\right\rangle = \sum_{i=1,n} \omega(\hat{p}_i)A^{\dagger}(\hat{p}_1)A^{\dagger}(\hat{p}_2) \ldots A^{\dagger}(\hat{p}_n) \left|0\right\rangle.$$

The direct sum of all n-particle spaces is called the Fock space.

Finally the free hamiltonian commutes with the particle number operator $\mathbf{N}$

$$\mathbf{N} = (2\pi)^{d-1} \int \frac{\mathrm{d}^{p-1}\hat{p}}{2\omega(\hat{p})} A^{\dagger}(\hat{p})A(\hat{p}) \quad \Rightarrow \quad [\mathbf{N}, \mathbf{H}] = 0,$$

a property in general no longer true in presence of interactions.

Note that if one then writes a general one-particle state $\left|\Psi\right\rangle$

$$\left|\Psi\right\rangle = (2\pi)^{d-1} \int \frac{\mathrm{d}^{d-1}\hat{p}}{2\omega(\hat{p})} \psi(\hat{p})A^{\dagger}(\hat{p}) \left|0\right\rangle,$$

the scalar product takes the relativistic covariant form

$$\langle \Psi_1 | \Psi_2 \rangle = (2\pi)^{d-1} \int \frac{\mathrm{d}^{d-1}\hat{p}}{2\omega(\hat{p})} \psi_1^*(\hat{p})\psi_2(\hat{p}) = (2\pi)^{d-1} \int \mathrm{d}^d p \, \delta_+(p^2 - m^2)\psi_1^*(\hat{p})\psi_2(\hat{p}).$$

A general n-particle vector $|\Psi\rangle$ then has the form

$$|\Psi\rangle = \frac{(2\pi)^{n(d-1)}}{\sqrt{n!}} \int \prod_i \frac{\mathrm{d}^{d-1}\hat{p}_i}{2\omega(\hat{p}_i)} \psi(\hat{p}_1, \hat{p}_2, \ldots, \hat{p}_n) \prod_i A^\dagger(\hat{p}_i) |0\rangle, \tag{7.5}$$

where $\psi(\hat{p}_1, \hat{p}_2, \ldots, \hat{p}_n)$ is a totally symmetric wave function, in momentum representation.

Two-point function. Note that the ϕ-field 2-point function, expressed as the expectation value of a time-ordered product of two fields (see for example Section 2.5), is given by

$$\langle 0| \mathrm{T}[\tilde{\phi}(t,\hat{p})\tilde{\phi}(0,\hat{p}')]|0\rangle = \langle 0|\tilde{\phi}(\hat{p})\, e^{-i\mathbf{H}|t|}\, \tilde{\phi}(\hat{p}')|0\rangle = (2\pi)^{1-d}\frac{1}{2\omega(\hat{p})}\delta^{d-1}(\hat{p}+\hat{p}')\, e^{-i\omega(\hat{p})|t|}.$$
$$\tag{7.6}$$

After Fourier transformation over time one finds

$$\frac{1}{2\pi}\int e^{ip_0 t}\,\mathrm{d}t\,\langle 0| \mathrm{T}[\tilde{\phi}(t,\hat{p})\tilde{\phi}(0,\hat{p}')]|0\rangle = \frac{1}{(2\pi)^d}\delta^{d-1}(\hat{p}+\hat{p}')\frac{i}{p_0^2 - \omega^2(\hat{p}) + i\epsilon}, \tag{7.7}$$

where the $i\epsilon$ term in the denominator indicates that we have to add a small positive imaginary part. The real time 2-point function is a distribution in the mathematical sense, boundary value of an analytic function

$$\frac{i}{p_0^2 - \omega^2(\hat{p}) + i\epsilon} \equiv 2\pi\delta(p_0^2 - \omega^2(\hat{p})) + i\mathrm{PP}\frac{1}{p_0^2 - \omega^2(\hat{p})},$$

where PP means principal part.

Holomorphic representation. We now introduce the corresponding complex classical fields $A^*(t,\hat{p}), A(t,\hat{p})$, related to the field $\tilde{\phi}$ by the classical form of the relations (7.2). The holomorphic form of the free action then is

$$\mathcal{A}_0(A^*, A) = (2\pi)^{d-1}\int \mathrm{d}t\,\frac{\mathrm{d}^{d-1}\hat{p}}{2\omega(\hat{p})}\,[iA^*(t,\hat{p})\partial_t A(t,\hat{p}) - \omega(\hat{p})A^*(t,\hat{p})A(t,\hat{p})].$$

The evolution operator is given by a functional integral of the form

$$U = \int [\mathrm{d}A^*(t,\hat{p})\mathrm{d}A(t,\hat{p})]\exp i\mathcal{A}_0(A^*, A).$$

7.2 The S-Matrix

Scattering by an external source. Let us first consider the S-matrix corresponding to the addition to the free action of a source term $\int \mathrm{d}t\mathrm{d}x\, J(t,x)\phi(t,x)$. After Fourier transformation

$$J(t,x) = \int e^{i\hat{p}x}\, \tilde{J}(t,\hat{p})\mathrm{d}\hat{p}\,,$$

this leads to the action $\mathcal{A}$

$$\mathcal{A}(A^*,A) = \mathcal{A}_0(A^*,A) + (2\pi)^{d-1}\int \mathrm{d}t\frac{\mathrm{d}\hat{p}}{2\omega(\hat{p})}\, \tilde{J}(t,-\hat{p})\left[A^*(-\hat{p},t)+A(\hat{p},t)\right].$$

The calculation closely follows the lines of Subsection 3.5.3. The result is

$$\mathcal{S}_0(J,A^*,A) = \exp\left[(2\pi)^{d-1}\int \frac{\mathrm{d}\hat{p}}{2\omega(\hat{p})}\, K(\hat{p})\right],$$

$$K(\hat{p}) = A^*(\hat{p})A(\hat{p}) + i\int \mathrm{d}t\, \tilde{J}(t,-\hat{p})\left[A^*(-\hat{p})\,e^{i\omega(\hat{p})t}+A(\hat{p})\,e^{-i\omega(\hat{p})t}\right]$$
$$-\tfrac{1}{2}\int \mathrm{d}t\mathrm{d}\tau\, \tilde{J}(t,-\hat{p})\,e^{-i|t-\tau|\omega(\hat{p})}\, \tilde{J}(\tau,\hat{p}).$$

As explained in Subsection 3.5.3, the expansion of this functional in powers of A^* and A then leads to the S-matrix elements.

It is actually more convenient to write the result in terms of the Fourier components of $\tilde{J}(t,\hat{p})$

$$\tilde{J}(t,\hat{p}) = \int \mathrm{d}p_0\, e^{-ip_0 t}\, \tilde{J}(p_0,\hat{p}).$$

Then

$$\ln \mathcal{S}_0(J,A^*,A) = (2\pi)^{d-1}\int \frac{\mathrm{d}\hat{p}}{2\omega(\hat{p})}\, A^*(\hat{p})A(\hat{p}) + i(2\pi)^d\int \mathrm{d}\hat{p}\mathrm{d}p_0\, \tilde{J}(-p_0,-\hat{p})$$
$$\times\, \delta(p_0^2-\omega^2(\hat{p}))\left[A^*(-\hat{p})\theta(-p_0)+A(\hat{p})\theta(p_0)\right]$$
$$-\tfrac{1}{2}(2\pi)^d\int \mathrm{d}\hat{p}\mathrm{d}p_0\, \tilde{J}(-p_0,-\hat{p})\frac{i}{p_0^2-\omega^2(\hat{p})+i\epsilon}\tilde{J}(p_0,\hat{p}). \tag{7.8}$$

In the coefficient of the term quadratic in J we recognize the free 2-point function (7.7).

General representation. An interaction term $V_\mathrm{I}(\phi)$ can then be added to the free action, where ϕ has to be expressed in terms of A^*,A,

$$\mathcal{A}(\phi) = \int \mathrm{d}t\, \mathrm{d}^{d-1}x\left[\tfrac{1}{2}(\partial_t\phi)^2-\tfrac{1}{2}(\nabla_x\phi)^2-\tfrac{1}{2}m^2\phi^2(t,x)\right] - V_\mathrm{I}(\phi). \tag{7.9}$$

Using the functional expressions for the perturbative expansion proven in Section 5.2 we find the form of the S-matrix for the interaction theory

$$\mathcal{S}(A^*,A) = \exp\left[-iV_\mathrm{I}\left(\frac{1}{i}\frac{\delta}{\delta J}\right)\right]\mathcal{S}_0(J,A^*,A)\bigg|_{J=0}. \tag{7.10}$$

The S-matrix thus has a Feynman diagram expansion with internal propagators Δ given by the quadratic term in J in (7.8):

$$\Delta(p_0, \hat{p}) = \frac{i}{p_0^2 - \hat{p}^2 - m^2 + i\epsilon} \equiv \frac{i}{p^2 - m^2 + i\epsilon}. \tag{7.11}$$

Unitarity. Let us note that with our conventions the unitarity of the S-matrix takes the functional form

$$\int [dA'(\hat{p})dA^{*\prime}(\hat{p})] \mathcal{S}^*(A^{*\prime}, A)\mathcal{S}(A^{*\prime}, A) \exp\left[-(2\pi)^{d-1}\int \frac{d\hat{p}}{2\omega(\hat{p})} A^{*\prime}(\hat{p})A'(\hat{p})\right]$$

$$= \exp\left[(2\pi)^{d-1}\int \frac{d\hat{p}}{2\omega(\hat{p})} A^*(\hat{p})A(\hat{p})\right]. \tag{7.12}$$

Discussion. We have constructed our basis of states from the eigenstates of the unperturbed hamiltonian. More generally we can take any similar harmonic oscillator basis, at the price of adding to the interaction quadratic terms in the field. Actually, and this will become clearer when we discuss the structure of the ground state in field theory, if we take an arbitrary basis, in general all eigenstates of the interacting hamiltonian will be orthogonal to all vectors of the basis (a property specific to systems with an infinite number of degrees of freedom, see Chapter 23). We have therefore to take as vacuum state the true ground state of the complete hamiltonian, a state we can only construct perturbatively.

Moreover the hamiltonian of a massive theory has a unique, translation invariant, lowest energy excited state (in the case of several fields this can be generalized to all superselection sectors). The physical mass m (or inverse correlation length in the statistical language) is defined as the energy of this state. We define the zero momentum creation operator as creating this state. The general creation operator $A^\dagger(\hat{p})$ is obtained by boosting the zero momentum operator, i.e. performing a $O(d-1, 1)$ transformation, and creates a one-particle state of momentum $\hat{p}$ and energy $\sqrt{m^2 + \hat{p}^2}$. Additional eigenstates have an energy at least equal to $2m$.

These conditions implicitly define a reference free theory with action $\mathcal{A}_0$, and ensures that it describes the asymptotic states at large times. Operationally, in a perturbative expansion, to calculate scattering amplitudes one has to perform a *mass renormalization*, i.e. one takes the physical mass as a parameter of the perturbative expansion, by inverting the relation between the physical mass as defined by the pole of the 2-point function, and the coefficient of ϕ^2 as it appears in the action.

Note finally that the functional integral has to be normalized by the condition $\mathcal{S}(0,0) = 1$, which means that we divide by a factor related to difference in energies between the true and the unperturbed ground state.

7.3 *S*-Matrix and Correlation Functions

Let us compare the explicit form of the S-matrix as derived from the holomorphic representation with a direct evaluation of correlation functions in the real time formalism:

$$Z(\mathcal{J}) = \int [d\phi] \exp\left[i\mathcal{A}(\phi) + i\int d^{d-1}x dt\, \mathcal{J}\phi\right].$$

Unlike the euclidean functional integral, the functional integral for the real time evolution operator has convergence problems, which we solve by considering it as obtained from the euclidean integral by analytic continuation. Such considerations lead to the $i\epsilon$ rule for real time Feynman diagrams.

The real time propagator. To the free action (7.1) formally corresponds the propagator

$$\Delta(p_0,\hat{p}) = \frac{i}{p_0^2 - \hat{p}^2 - m^2} \equiv \frac{i}{p^2 - m^2}\,,$$

which is singular on the mass-shell $p^2 = m^2$. Let us give a meaning to this form by constructing the real time theory by an analytic continuation of the euclidean theory.

The real time theory is obtained from the euclidean theory by an analytic continuation on time variables $t \mapsto t\,e^{i\theta}$ where θ varies between 0 (the euclidean theory) and $\pi/2$, (the Minkowsky theory). For Feynman diagrams in momentum representation, the corresponding rotation of the energy variable p_0 is $p_0 \mapsto p_0\,e^{-i\theta}$. However we cannot simply replace p_0^2 by $-p_0^2$ in Feynman diagrams because they then become singular on the mass-shell $p_0^2 = \hat{p}^2 + m^2$. If we follow the rotation we find that must give to p_0^2 a small positive imaginary part

$$\Delta(p_0,\hat{p}) = \frac{i}{p^2 - m^2 + i\epsilon}\,.$$

The imaginary term also ensures the convergence of the gaussian functional integral. We note that this propagator is identical to the internal propagator (7.11) which appears in the Feynman graph expansion of the S-matrix. Then in real time expression (5.22) (Section 5.2) becomes

$$Z(\mathcal{J}) = \exp\left[-iV_{\mathrm{I}}\left(\frac{1}{i}\frac{\delta}{\delta\mathcal{J}}\right)\right]\exp\left[-\tfrac{1}{2}\mathcal{J}\Delta\mathcal{J}\right]. \tag{7.13}$$

This expression can be rewritten

$$Z(\mathcal{J}) = \exp\left[-iV_{\mathrm{I}}\left(\frac{1}{i}\frac{\delta}{\delta\mathcal{J}}\right)\right]\exp\left[-\tfrac{1}{2}(\mathcal{J}+J)\Delta(\mathcal{J}+J)\right]\Big|_{J=0}.$$

We then expand

$$\tfrac{1}{2}(\mathcal{J}+J)\Delta(\mathcal{J}+J) = \tfrac{1}{2}\int\mathrm{d}p\,\tilde{\mathcal{J}}(-p)\frac{i(2\pi)^d}{p^2-m^2+i\epsilon}\tilde{\mathcal{J}}(p) + \int\mathrm{d}p\,\tilde{\mathcal{J}}(-p)\frac{i(2\pi)^d}{p^2-m^2+i\epsilon}\tilde{J}(p)$$

$$+ \tfrac{1}{2}\int\mathrm{d}p\,\tilde{J}(-p)\frac{i(2\pi)^d}{p^2-m^2+i\epsilon}\tilde{J}(p). \tag{7.14}$$

Comparing this expression with expression (7.8) we see that the main formal difference between expressions (7.13) and (7.10) comes from the term linear in J:

$$\frac{i}{p^2-m^2+i\epsilon}\tilde{\mathcal{J}}(p_0,\hat{p}) \mapsto -i\delta(p^2-m^2)\left[A^*(-\hat{p})\theta(-p_0) + A(\hat{p})\theta(p_0)\right]. \tag{7.15}$$

We therefore conclude that S-matrix elements can be obtained from real time correlation functions by first multiplying them by the product of external inverse propagators, and then restricting the external momenta to the mass-shell $p^2 = m^2$. This does not imply that the result vanishes. Indeed correlation functions have poles on the mass-shell. The final answer is proportional to the mass-shell amputated correlation functions.

Remark. The relation between correlation functions and S-matrix elements shows that the matrix elements as defined here have disconnected contributions. The new functional $\mathcal{T}$:

$$\mathcal{T}(A^*, A) = i \ln S(A^*, A), \qquad (7.16)$$

is instead the generating functional of connected scattering amplitudes.

Crossing symmetry. We see that only a linear combination of A^* and A appears, with identical coefficient, up to the sign of the energy. The sign of p_0 specifies the incoming and outgoing particles. This has deep consequences, specific to relativistic quantum field theory: in $d > 2$ dimensions, the same analytic functions lead to scattering amplitudes of different physical processes, a property known as crossing symmetry.

7.3.1 S-matrix and field asymptotic conditions

The functional representation (7.10) can be rewritten in a different way. We use the functional identity

$$\exp\left[-\tfrac{1}{2}(2\pi)^d \int \mathrm{d}p\, \tilde{J}(-p)\frac{i}{p^2 - m^2 + i\epsilon}\tilde{J}(p)\right]$$
$$= \int [\mathrm{d}\phi] \exp\left[i(2\pi)^d \int \mathrm{d}^d p \left[\tfrac{1}{2}\tilde{\phi}(-p)(p^2 - m^2)\tilde{\phi}(p) + \tilde{J}(-p)\phi(p)\right]\right].$$

Introducing this identity in (7.10) and applying the functional derivatives we find:

$$S(A^*, A) = \mathcal{I}(A^*, A) \int [\mathrm{d}\phi] \exp\left[i\mathcal{A}_0(\phi) - iV_{\mathrm{I}}(\phi + \varphi)\right],$$
$$\tilde{\varphi}(p) = \delta(p^2 - m^2)\left[A^*(-\hat{p})\theta(-p_0) + A(\hat{p})\theta(p_0)\right], \qquad (7.17)$$

with $\mathcal{I}$ representing the identity

$$\mathcal{I}(A^*, A) = \exp\left[(2\pi)^{d-1} \int \frac{\mathrm{d}\hat{p}}{2\omega(\hat{p})} A^*(\hat{p})A(\hat{p})\right].$$

The expression (7.17), up to the prefactor $\mathcal{I}$, is a functional integral in presence of a background field φ. Moreover the classical field φ is restricted to be a general solution of the free field equation

$$(p^2 - m^2)\tilde{\varphi}(p) = 0.$$

The solution is non-vanishing only on the mass hyperboloid $p^2 = m^2$. The parametrization (7.17) of the solution reflects the property that the mass hyperboloid has two disconnected components, depending on the sign of the energy p_0.

We can thus rewrite the equation (7.17)

$$S(A^*, A) = \mathcal{I}(A^*, A) \int [\mathrm{d}\phi] \exp\left[i\mathcal{A}(\phi + \varphi)\right].$$

This functional integral differs from the vacuum amplitude only by the boundary conditions which are free field boundary conditions. This result is consistent with the analysis of Subsection 2.6.2. We have shown that in quantum mechanics S-matrix elements can be calculated from the path integral representation of the evolution operator, by integrating over paths which satisfy prescribed classical scattering boundary conditions, i.e. which correspond to asymptotic free classical motion. In particular the starting point of the semiclassical expansion is a classical scattering trajectory. The arguments can be generalized to quantum field theory with massive particles (to ensure proper cluster properties and thus the existence of an S-matrix).

Let us now return to the discussion of the end of Section 7.2. We have argued that, in contrast to simple quantum mechanics with scattering states, the asymptotic states of the complete field theory are not given by leading order perturbation theory. One must therefore take for φ not a solution to the classical equation of motion, but instead a solution to

$$\frac{\delta S(\varphi)}{\delta \varphi(x)} = 0\,, \tag{7.18}$$

which contains the information about the scattering states of the complete theory. For instance the interactions modify the masses of particles. The solution of the classical equation suffices only for a the leading order calculation (the tree approximation). At higher orders, as we have already discussed, the renormalized mass has to be introduced.

Remark. Considerations based on asymptotic field boundary conditions, or the more direct considerations of Section 7.1 lead to the same perturbative S-matrix. However the preceding considerations generalize to the scattering of *solitons*, i.e. states obtained by expanding the functional integral around finite energy solutions of the classical field equations. In this case the S-matrix of soliton scattering is obtained by expanding the functional integral around classical soliton scattering solutions.

7.3.2 The ϕ^3 example

Let us illustrate the analysis by calculating a 4-point scattering amplitude in the simple ϕ^3 field theory, in the tree approximation. We consider the action

$$\mathcal{A}(\phi) = \int dt\, d^{d-1}x \left[\tfrac{1}{2}(\partial_t\phi)^2 - \tfrac{1}{2}(\nabla_x\phi)^2 - \tfrac{1}{2}m^2\phi^2(t,x) - \tfrac{1}{3!}g\phi^3(t,x)\right].$$

We introduce the Fourier transform of the field φ.

$$\varphi(t,x) = \int e^{-ip_0 t + \hat{p}x}\, \tilde{\varphi}(p) d^d p\,. \tag{7.19}$$

In the Fourier representation the field equation takes the form

$$\left(p^2 - m^2\right)\tilde{\varphi}(p) - \tfrac{1}{2}g \int d^d q\, \tilde{\varphi}(q)\tilde{\varphi}(-p-q) = 0\,. \tag{7.20}$$

This equation can be solved as a series in the coupling constant g starting from a free solution. The free solution $\tilde{\varphi}_0$ can be written

$$\tilde{\varphi}_0(p) = \delta(p^2 - m^2)\left[A^*(-\hat{p})\theta(-p_0) + A(\hat{p})\theta(p_0)\right].$$

Let us calculate the classical solution at order g

$$\tilde{\varphi}(p) = \tilde{\varphi}_0(p) + \frac{1}{2}\frac{g}{p^2 - m^2}\int d^d q\, \tilde{\varphi}_0(q)\tilde{\varphi}_0(-p-q) + O\left(g^2\right).$$

By looking for perturbative solutions of the field equation we have explicitly excluded scattering states corresponding to bound states or solitons (see the remark at the end of the preceding subsection).

Using equation (7.20) we can rewrite $\mathcal{T}$ in the tree approximation:

$$\mathcal{T}(\varphi) = i \ln \mathcal{S}(\varphi) = -\tfrac{1}{12} g (2\pi)^d \int d^d p_1 d^d p_2 d^d p_3 \, \delta(p_1 + p_2 + p_3) \tilde{\varphi}(p_1) \tilde{\varphi}(p_2) \tilde{\varphi}(p_3).$$

We then replace φ by its expansion in powers of g. The term of order g, which would describe one ϕ particle decaying into two, vanishes by energy conservation. The next term of order g^2 has the form

$$-\tfrac{1}{8} g^2 (2\pi)^d \int \delta^d(p_1 + p_2 + p_3 + p_4) \frac{1}{(p_1 + p_2)^2 - m^2} \prod_{i=1}^{4} dp_i \, \tilde{\varphi}_0(p_i).$$

The connected 4-particle scattering amplitude is then obtained by differentiating with respect to $\tilde{\varphi}_0(p)$. The result is the product of a factor which contains the momentum conservation

$$(2\pi)^d \delta^d(p_1 + p_2 + p_3 + p_4)$$

and an amplitude

$$-\frac{g^2}{(p_1 + p_2)^2 - m^2} - \frac{g^2}{(p_1 + p_3)^2 - m^2} - \frac{g^2}{(p_1 + p_4)^2 - m^2}.$$

The term we have calculated contains in principle also the decay of one particle into three but again this process vanishes by energy conservation.

Higher orders in g yields 5,6... particle scattering amplitudes.

7.3.3 S-matrix and 1PI generating functional

By comparing the general form of the S-matrix with correlation functions we have seen that S-matrix elements are related to real time analytic continuation $(t \mapsto it, p_0 \mapsto ip_0)$ of amputated correlation functions on the mass-shell

$$\mathcal{T}(\varphi) = iW\left(\Delta^{-1}J\right),$$

where J is deduced from equation (7.15), and φ is given by equation (7.17).

Let us now show that there is also a direct relation to the analytic continuation of the generating functional of proper vertices $\Gamma(\varphi)$. We start from the euclidean form

$$e^{W(J)} = \int [d\phi] \, e^{-\mathcal{A}(\phi) + J\phi}. \tag{7.21}$$

We then substitute $\Gamma(\varphi)$, the Legendre transform of $W(J)$:

$$W(J) + \Gamma(\varphi) = \int dt d^d x \, J(t,x) \varphi(t,x), \qquad \varphi(t,x) = \frac{\delta W}{\delta J(t,x)}$$

$$e^{-\Gamma(\varphi) + J\varphi} = \int [d\phi] \, e^{-\mathcal{A}(\phi) + J\phi}. \tag{7.22}$$

Using:

$$J(t,x) = \frac{\delta\Gamma}{\delta\varphi(t,x)},$$

we can write equation (7.22):

$$e^{-\Gamma(\varphi)} = \int [\mathrm{d}\phi]\exp\left[-\mathcal{A}(\phi) + \int \mathrm{d}x\mathrm{d}t(\phi - \varphi)\frac{\delta\Gamma}{\delta\varphi(t,x)}\right], \qquad (7.23)$$

or equivalently translating $\phi(t,x)$:

$$e^{-\Gamma(\varphi)} = \int [\mathrm{d}\phi]\exp\left[-\mathcal{A}(\phi + \varphi) + \int \mathrm{d}t\mathrm{d}x\,\phi(t,x)\frac{\delta\Gamma}{\delta\varphi(t,x)}\right]. \qquad (7.24)$$

We now take the limit of a vanishing source J:

$$\frac{\delta\Gamma}{\delta\varphi(t,x)} = 0. \qquad (7.25)$$

Of course this equation has propagating type solutions only after continuation to real time. We then observe that $\Gamma(\varphi)$ coincides with $i\mathcal{T}(\varphi)$, when $\varphi(x)$ is restricted to the solutions of equation (7.18).

7.4 Field Renormalization

To describe precisely how the S-matrix can be calculated in a general theory of scalar particles, we still have to discuss the field renormalization. This can more conveniently done in the euclidean form. We define the field $\phi(t,x)$ in such a way that it has zero expectation value. This can always be achieved by a constant shift $\phi(t,x) \mapsto \phi(t,x) - \langle 0|\phi|0\rangle$. We then consider the 2-point function correlation function $W^{(2)}$. It follows from the analysis of Section 2.5 that it is equal to the vacuum average value of the time-ordered product of two fields

$$\delta^{d-1}(\hat{p} + \hat{p}')W^{(2)}(t,\hat{p}) = (2\pi)^{d-1}\langle 0|\,\tilde{\phi}(\hat{p})\,e^{-|t|(\mathbf{H}-E_0)}\,\tilde{\phi}(\hat{p}')\,|0\rangle. \qquad (7.26)$$

We now adapt the arguments of Section A2.2 to the kinematical properties of a $O(d)$ invariant field theory. We introduce a complete set of eigenvectors $|\mu\rangle$ of $\mathbf{H}$, with real eigenvalues ε_μ and obtain

$$\delta^{d-1}(\hat{p} + \hat{p}')W^{(2)}(t,\hat{p}) = (2\pi)^{d-1}\int \mathrm{d}\mu\,\langle 0|\,\tilde{\phi}(\hat{p})\,|\mu\rangle\,e^{-|t|\varepsilon(\mu)}\,\langle\mu|\,\tilde{\phi}(\hat{p}')\,|0\rangle.$$

The theory being translation invariant, we know that the answer is proportional to $\delta^{d-1}(\hat{p} + \hat{p}')$. In this limit, since $\tilde{\phi}(-p) = \tilde{\phi}^\dagger(p)$, the l.h.s. is a sum of positive terms (a point we have already discussed in Section A2.2). We thus find

$$W^{(2)}(t,\hat{p}) = (2\pi)^{d-1}\int \mathrm{d}\mu\,\rho(\mu,\hat{p})\,e^{-|t|\varepsilon(\mu)},$$

where ρ is positive measure. Let us now calculate the Fourier transform over time

$$W^{(2)}(p_0,\hat{p}) = (2\pi)^{d-1}\int \mathrm{d}t\,e^{-ip_0 t}\int \mathrm{d}\mu\,\rho(\mu,\hat{p})\,e^{-|t|\varepsilon(\mu)} = (2\pi)^{d-1}\int \mathrm{d}\mu\,\frac{2\varepsilon(\mu)\rho(\mu,\hat{p})}{p_0^2 + \varepsilon^2(\mu)}.$$

Due to $O(d)$ invariance we can set $\hat{p} = 0$ and replace p_0^2 by p^2 without changing the answer. We conclude

$$W^{(2)}(p) = \int d\mu \frac{\rho(\mu)}{p^2 + \mu^2},$$

where $\rho(\mu)$, proportional to $\rho(\mu, 0)$, is a positive measure. This form is called the Källen–Lehmann (KL) representation. Since by definition the physical mass m is the lowest energy eigenstate above the ground state, the domain of integration is $\mu \geq m$. Moreover at $\hat{p} = 0$ the state is isolated. Therefore the measure has an isolated δ-function and then a continuous part starting at the threshold for scattering states (in the simple scalar field theory $2m$, or $3m$ if the parity of the number of particles is conserved)

$$\rho(\mu) = Z\delta(\mu - m) + \rho'(\mu), \quad \rho'(\mu) = 0 \text{ for } \mu < 2m. \tag{7.27}$$

Conversely let us Fourier transform with respect to the energy variable p_0

$$W^{(2)}(t, \hat{p}) = \frac{1}{2\pi} \int dp_0 \ e^{ip_0 t} \ W^{(2)}(p) = \frac{1}{2\pi} \int dp_0 \ e^{ip_0 t} \ W^{(2)}(p) \int d\mu \frac{\rho(\mu)}{p^2 + \mu^2}$$

$$= \int d\mu \frac{\rho(\mu)}{2\sqrt{\hat{p}^2 + \mu^2}} \ e^{-|t|\sqrt{\hat{p}^2 + \mu^2}}. \tag{7.28}$$

We now return to the definition (7.26), assume $t > 0$, take the derivative with respect to time, and take the limit $t = 0$. We find

$$\delta^{d-1}(\hat{p} + \hat{p}') \frac{\partial}{\partial t} W^{(2)}(t, \hat{p})\bigg|_{t \to 0_+} = -(2\pi)^{d-1} \langle 0| \ \tilde{\phi}(\hat{p})(\mathbf{H} - E_0)\tilde{\phi}(\hat{p}') \ |0\rangle. \tag{7.29}$$

The product $(\mathbf{H} - E_0)\tilde{\phi}$ in the r.h.s. can be replaced by the commutator $[\mathbf{H}, \tilde{\phi}]$. Then using the reality of the l.h.s. we can take the hermitian part of the operator. Thus

$$\delta^{d-1}(\hat{p} + \hat{p}') \frac{\partial}{\partial t} W^{(2)}(t, \hat{p})\bigg|_{t \to 0_+} = -\tfrac{1}{2}(2\pi)^{d-1} \langle 0| \ [\tilde{\phi}(\hat{p}), [\mathbf{H}, \tilde{\phi}(\hat{p}')]] \ |0\rangle.$$

If the action has the form (7.9) the commutator is proportional to the conjugated momentum. Then

$$[\tilde{\phi}(\hat{p}), [\mathbf{H}, \tilde{\phi}(\hat{p}')]] = (2\pi)^{1-d}\delta^{d-1}(\hat{p} + \hat{p}').$$

It follows

$$\frac{\partial}{\partial t} W^{(2)}(t, \hat{p})\bigg|_{t \to 0_+} = -\tfrac{1}{2}.$$

Calculating now the l.h.s. from the representation (7.28), we obtain

$$\int d\mu \, \rho(\mu) = 1.$$

We conclude that, except in a free field theory, Z, the residue of the pole at $p^2 = -m^2$, is strictly smaller than 1

$$0 < Z < 1. \tag{7.30}$$

This result has several implications, one being related to the S-matrix. Let us evaluate $W^{(2)}(t, \hat{p})$ for t large. The state of lowest energy, with momentum $\hat{p}$, gives the leading contribution. From the KL representation (7.28) and the decomposition (7.27) we then learn

$$W^{(2)}(t,\hat{p}) \underset{t\to\infty}{\sim} \frac{Z}{2\sqrt{\hat{p}^2 + m^2}} \, e^{-|t|\sqrt{\hat{p}^2+m^2}} \,.$$

If we compare this result with the contribution of a normalized one-particle eigenstate of the hamiltonian $\mathbf{H}$ (see equation (7.6)),

$$(2\pi)^{d-1} \langle 1, \hat{p}| \, e^{-|t|(\mathbf{H}-E_0)} \, |1,\hat{p}'\rangle = \delta^{d-1}(\hat{p}+\hat{p}')\frac{1}{2\sqrt{\hat{p}^2 + m^2}} \, e^{-|t|\sqrt{\hat{p}^2+m^2}},$$

we observe that the field ϕ has only a component $\sqrt{Z}$ on the one-particle states. Another way to formulate the same answer is to verify that in real time the Heisenberg field has a free field large time behaviour with an amplitude $\sqrt{Z}$ on normalized creation or annihilation operators. Let us write the real-time 2-point function:

$$\langle 0| \, \mathrm{T}[\tilde{\phi}(t,\hat{p})\tilde{\phi}(0,\hat{p}')] \, |0\rangle = \langle 0| \, \tilde{\phi}(\hat{p}) \, e^{-i\mathbf{H}|t|} \, \tilde{\phi}(\hat{p}')] \, |0\rangle$$

$$= (2\pi)^{1-d}\delta^{d-1}(\hat{p}+\hat{p}')\int \mathrm{d}\mu \, \frac{\rho(\mu)}{2\sqrt{\hat{p}^2 + \mu^2}} \, e^{-i|t|\sqrt{\hat{p}^2+\mu^2}} \,.$$

It can be verified that the large time behaviour is related to the leading singularity of the measure ρ. Since $\rho(\mu)$ is the sum of a δ-function and a continuous function (for $d \geq 2$), we obtain

$$\langle 0| \, \tilde{\phi}(\hat{p}) \, e^{-i\mathbf{H}|t|} \, \tilde{\phi}(\hat{p}')] \, |0\rangle \underset{|t|\to\infty}{=} Z(2\pi)^{1-d}\frac{1}{2\omega(\hat{p})}\delta^{d-1}(\hat{p}+\hat{p}') \, e^{-i\omega(\hat{p})|t|} + O(1/t).$$

We conclude that for large time the field $\tilde{\phi}(t,\hat{p})$ tends in weak sense (not in operator sense, but in all average values) towards

$$\tilde{\phi}(t,\hat{p}) \underset{|t|\to\infty}{\sim} \sqrt{Z}\frac{1}{2\omega(\hat{p})}\left[A^\dagger(-\hat{p}) + A(\hat{p})\right],$$

where $A^\dagger, A$ are the properly normalized creation and annihilation operators of the one-particle states. The constant $\sqrt{Z}$ is the field renormalization constant.

Normalized S-matrix elements. To calculate properly normalized S-matrix elements we can calculate with the action $\mathcal{A}(\phi\sqrt{Z})$. Alternatively if we keep the initial field we have to renormalize the matrix elements.

We express the correlation functions in terms of euclidean amputated functions

$$W^{(n)}(p_1,\ldots,p_n) = \left[\prod_{i=1}^{n} W_i^{(2)}(p_i)\right] W_{\mathrm{amp.}}^{(n)}(p_1,\ldots,p_n),$$

and call $-m_i^2$ the pole in the variable p_i^2 of $W_i(p_i)$:

$$W_i^{(2)}(p_i) \sim \frac{z_i}{p_i^2 + m_i^2} \qquad \text{for} \quad p_i^2 \to -m_i^2\,, \tag{7.31}$$

i.e. the m_i's are the physical masses of the scalar particles. Then the coefficient $\mathcal{T}^{(n)}$ of $\mathcal{T}$ in the expansion in powers of $\tilde{\varphi}$ is given by:

$$\mathcal{T}^{(n)}(p_1,\ldots,p_n) = -\prod_{i=1}^{n} \sqrt{z_i}\, W_{\mathrm{amp.}}^{(n)}(p_1,\ldots,p_n)\Big|_{p_i^2=-m_i^2}\,. \tag{7.32}$$

The factors $\sqrt{z_i}$ in this equation correspond to a finite renormalization of the field such that the residue of the 2-point function (equation (7.31)) on the physical pole $p_i^2 = -m_i^2$ is 1. They ensure that the matrix elements $\mathcal{T}^{(n)}$ satisfy the unitarity relations with the proper normalization.

7.5 Massive Dirac Fermions

First let us note that with our conventions, for a finite number of fermions, for a normal-ordered (all creation operators $\mathbf{a}^\dagger$ on the left of all annihilation operators $\mathbf{a}$) hamiltonian of the form

$$ h = h(\mathbf{a}^\dagger, \mathbf{a}), $$

the evolution operator in real time is given by the path integral

$$ U = \int [d\mathbf{c}(t) d\bar{\mathbf{c}}(t)] \exp i \mathcal{A}(\bar{\mathbf{c}}, \mathbf{c}) \tag{7.33a} $$

$$ \mathcal{A}(\bar{\mathbf{c}}, \mathbf{c}) = \int dt \left\{ \frac{1}{i} \bar{c}_\alpha(t) \dot{c}_\alpha(t) - h\left[\bar{\mathbf{c}}(t), -\mathbf{c}(t)\right] \right\}, \tag{7.33b} $$

real time continuation of equation (3.109) of Section 3.6.

We now study massive Dirac fermions, in even dimensions, since otherwise the fermion mass breaks space reflection invariance. In Appendix A5 we have verified the spin $O(d)$ and other symmetries of the fermion euclidean action. In real time the $O(d)$ symmetry becomes a $O(d-1, 1)$ symmetry. Here we will verify real time hermiticity of the hamiltonian, structure of fermion states in the free field theory, and construct the S-matrix.

7.5.1 The free massive Dirac action

Again we first discuss the free action, expressed in terms of Grassmann anticommuting spinor fields $\bar{\psi}, \psi$:

$$ \mathcal{A}\left(\bar{\psi}, \psi\right) = \int d x dt \, \bar{\psi}(t, x) \left(\frac{1}{i} \gamma_0 \partial_t + \gamma \cdot \nabla_x + m \right) \psi(t, x), $$

where we have denoted by γ_0 the matrix γ_d associated with the time variable.

To identify the action with an action resulting from a hamiltonian formalism we set

$$ \psi^\dagger = \bar{\psi} \gamma_0 . $$

Then from the construction of Section 3.6, and generalizing expression (7.33), we identify the corresponding hamiltonian density

$$ \mathcal{H} = \psi^\dagger(x) \gamma_0 \left(\gamma \cdot \nabla_x + m \right) \psi(x), $$

where $\bar{\psi}(x), \psi(x)$ are now operators with the corresponding anticommutation relations

$$ \{\psi(x), \psi(y)\} = \{\psi^\dagger(x), \psi^\dagger(y)\} = 0, \quad \{\psi_\alpha(x), \psi_\beta^\dagger(y)\} = \delta_{\alpha\beta} \delta^{d-1}(x - y). $$

To diagonalize the hamiltonian, we proceed by Fourier transformation, setting

$$ \psi(x) = \int d^{d-1}\hat{p} \, e^{i\hat{p}x} \, \tilde{\psi}(\hat{p}), \quad \psi^\dagger(x) = \int d^{d-1}\hat{p} \, e^{-i\hat{p}x} \, \tilde{\psi}^\dagger(\hat{p}). $$

We find the total hamiltonian $\mathbf{H}$

$$ \mathbf{H} = (2\pi)^{d-1} \int d^{d-1}\hat{p} \, \psi^\dagger(\hat{p}) \gamma_0 \left(i\gamma \cdot \hat{p} + m \right) \psi(\hat{p}), $$

and the anticommutation relations (from now on we write only the non-trivial part)

$$\{\tilde{\psi}_\alpha(\hat{p}), \tilde{\psi}_\beta^\dagger(\hat{p}')\} = (2\pi)^{1-d}\delta^{d-1}(\hat{p} - \hat{p}')\delta_{\alpha\beta}.$$

We see that, in contrast with the scalar case, due to the spin structure the hamiltonian is not totally diagonalized, but the diagonalization has been reduced to a simple matrix problem. Let us denote by $h(\hat{p})$ the matrix

$$h(\hat{p}) = \gamma_0\left(i\gamma \cdot \hat{p} + m\right), \qquad h(\hat{p}) = h^\dagger(\hat{p}). \tag{7.34}$$

Then we verify

$$h^2(\hat{p}) = \omega^2(\hat{p}), \quad \omega(\hat{p}) = \sqrt{\hat{p}^2 + m^2}.$$

We thus find that the two possible eigenvalues of h are $\pm\omega(\hat{p})$. The two corresponding subspaces have equal dimensions since

$$\gamma_0\gamma_S h(\hat{p})\gamma_S\gamma_0 = -h(\hat{p}).$$

Therefore if a spinor $u(\hat{p})$ is an eigenvector with eigenvalue $\omega(\hat{p})$, $\gamma_0\gamma_S u(\hat{p})$ ($\gamma_0\gamma_S$ is associated with time reflection) is an eigenvector with eigenvalue $-\omega(\hat{p})$.

Let us introduce the two orthogonal, hermitian projectors $P_\pm$ on the positive and negative energy sector:

$$P_\pm = \tfrac{1}{2}\left[1 \pm h(\hat{p})/\omega(\hat{p})\right], \quad \Rightarrow P_+ + P_- = 1, \quad P_\pm^2 = P_\pm. \tag{7.35}$$

We note that if we associate creation operators to $\psi^\dagger$ and annihilation operators to ψ, then we create states with negative energy. This simply means that we have improperly chosen the reference state which should be the ground state. As in the scalar case, and in contrast to what one could have naively guessed, we have to decompose both ψ and $\psi^\dagger$ in a sum of creation and annihilation operators, in such a way that creators always create states with positive energy.

We thus set

$$A_-^\dagger(\hat{p}) = [2\omega(\hat{p})]^{1/2}\,^T C P_-\tilde{\psi}(\hat{p}), \quad A_+(\hat{p}) = [2\omega(\hat{p})]^{1/2}P_+\tilde{\psi}(\hat{p}) \tag{7.36a}$$

$$A_+^\dagger(\hat{p}) = [2\omega(\hat{p})]^{1/2}\tilde{\psi}^\dagger(\hat{p})P_+, \quad A_-(\hat{p}) = [2\omega(\hat{p})]^{1/2}\tilde{\psi}^\dagger(\hat{p})P_- C^*, \tag{7.36b}$$

where C is a unitary matrix, $C^\dagger C = 1$. Conversely

$$\tilde{\psi}(\hat{p}) = \frac{1}{\sqrt{2\omega(\hat{p})}}\left(C^* A_-^\dagger(\hat{p}) + A_+(\hat{p})\right) \tag{7.37a}$$

$$\tilde{\psi}^\dagger(\hat{p}) = \frac{1}{\sqrt{2\omega(\hat{p})}}\left(A_-^\dagger(\hat{p})\,^T C + A_+(\hat{p})\right). \tag{7.37b}$$

Anticommutation relations follow (we write only the non-trivial part)

$$\{[A_+]_\alpha(\hat{p}), [A_+^\dagger]_\beta(\hat{p}')\} = 2\omega(\hat{p})[P_+]_{\alpha\beta}(2\pi)^{1-d}\delta^{d-1}(\hat{p} - \hat{p}'),$$

$$\{[A_-]_\alpha(\hat{p}), [A_-^\dagger]_\beta(\hat{p}')\} = 2\omega(\hat{p})[C^\dagger\,^T P_- C]_{\alpha\beta}(2\pi)^{1-d}\delta^{d-1}(\hat{p} - \hat{p}').$$

The first commutator can be rewritten

$$2\omega(\hat{p})P_+ = \frac{1}{\sqrt{2}}\gamma_0\left(p_0\gamma_0 + i\gamma \cdot \hat{p} + m\right), \qquad p_0 = \omega(\hat{p}).$$

We recognize in the parenthesis the covariant form of the projector.

The two set of anticommutation relations are different except if we can find a matrix C such that

$$C^\dagger \, {}^T\!P_- C = P_+ \, .$$

Since C is unitary this equation reduces to

$$C^\dagger \, {}^T\!h(\hat{p})C = -h(\hat{p}) \quad \Leftrightarrow \quad C^\dagger \left(i \, {}^T\!\gamma \cdot \hat{p} + m \right) {}^T\!\gamma_0 C = -\gamma_0 \left(i\gamma \cdot \hat{p} + m \right) \, .$$

In Appendix A5.6 we have constructed a charge conjugation matrix C which has the property (equation $(A5.30)$)

$$C^{-1} \, {}^T\!\gamma_\mu C = -\gamma_\mu \, .$$

This is exactly the matrix we need here. In these variables, taking into account the unitarity of the conjugation matrix C, the free hamiltonian reads

$$\mathbf{H} = \tfrac{1}{2}(2\pi)^{d-1} \int \mathrm{d}^{d-1}\mathrm{d}\hat{p} \left(A_+^\dagger(\hat{p})A_+(\hat{p}) - A_-(\hat{p})A_-^\dagger(\hat{p}) \right) \, .$$

If we now normal-order the hamiltonian by anticommuting $A_-^\dagger$ and A_-, we find

$$\mathbf{H} = (2\pi)^{d-1} \int \mathrm{d}^{d-1}\frac{\mathrm{d}\hat{p}}{2\omega(\hat{p})} \, \omega(\hat{p}) \left(A_+^\dagger(\hat{p})A_+(\hat{p}) + A_-^\dagger(\hat{p})A_-(\hat{p}) \right) + E_0(\mathrm{Dirac}) \, .$$

The ground state (vacuum) energy $E_0(\mathrm{Dirac})$ is negative, and proportional to the free scalar vacuum energy (7.4):

$$E_0(\mathrm{Dirac}) = -2^{d/2}E_0(\mathrm{scalar}) \, .$$

Note therefore that by adding $2^{d/2}$ scalar bosons of the same mass m to one Dirac fermion, one can construct a theory with zero vacuum energy. One can show that this boson-fermion free theory has then a special type of fermionic symmetry called supersymmetry.

The final form of the hamiltonian, together with the final form of the commutation relations, shows that the Dirac field carries two particles transforming under the fundamental representation of the $O(d-1)$ spin group (spin $1/2$ particles), related by charge conjugation (in the case of charged particles they will have opposite charge).

Another remark: the last step depends crucially on the anticommuting character of fermions. This is the reflection of the *connection between spin and statistics*, a specific property of local relativistic quantum field theory.

We are now in a situation quite similar to the scalar case. Starting from the ground state $|0\rangle$ which is annihilated by all operators $A_\pm(\hat{p})$

$$\langle 0|0\rangle = 1 \, , \quad A_\pm(\hat{p})\,|0\rangle = 0 \, , \quad \Rightarrow \quad \mathbf{H}\,|0\rangle = E_0\,|0\rangle \, ,$$

we can construct a basis for n-particle states by acting with creation operators on the vacuum. A general n-particle state is obtained by introducing a totally antisymmetric wave function.

The particle number operators for both particles commute with the hamiltonian

$$\mathbf{N}_\pm = (2\pi)^{d-1} \int \frac{\mathrm{d}^{d-1}\hat{p}}{2\omega(\hat{p})} \, A_\pm^\dagger(\hat{p})A_\pm(\hat{p}), \quad [\mathbf{N}_\pm, \mathbf{H}] = 0 \, .$$

S-matrix in an interacting theory. The expression of the S-matrix in an interacting theory then follows from a simple extension of the method explained in the scalar case. One has to be only careful of the signs. It is straightforward to verify that the S-matrix is given by

$$\mathcal{S}(A^\dagger, A) = \int [\mathrm{d}\bar{\psi}\mathrm{d}\psi] \exp i\mathcal{A}(\bar{\psi}\sqrt{Z} + \bar{\chi}, \psi\sqrt{Z} + \chi), \tag{7.38}$$

i.e. again a functional integral in a background field, where the classical anticommuting fields are solutions to the free field equations, which can be parametrized in the form (7.37),

$$\tilde{\chi}(\hat{p}) = P_- C^* A_-^\dagger(\hat{p}) + P_+ A_+(\hat{p}), \quad \tilde{\chi}^\dagger(\hat{p}) = A_-^\dagger(\hat{p})P_- + A_+(\hat{p})\,{}^T C P_+, \quad \bar{\chi} = \chi^\dagger \gamma_0,$$

where however the quantities $\bar{\chi}, \chi, A^\dagger, A$ are no longer field operators, but the corresponding anticommuting variables. A constant Z is here also required, to obtain S-matrix elements with the proper normalization.

In the same notation the unitarity of the S-matrix takes the form:

$$\int [\mathrm{d}A'(\hat{p})\mathrm{d}A^{\dagger\prime}(\hat{p})]\mathcal{S}^*(A^{\dagger\prime}, A)\mathcal{S}(A^{\dagger\prime}, A) \exp\left[(2\pi)^{d-1} \int \frac{\mathrm{d}\hat{p}}{2\omega(\hat{p})} \sum_{\varepsilon=\pm1} A_\varepsilon^{\dagger\prime}(\hat{p})A_\varepsilon'(\hat{p})\right]$$

$$= \exp\left[(2\pi)^{d-1} \int \frac{\mathrm{d}\hat{p}}{2\omega(\hat{p})} \sum_{\varepsilon=\pm1} A_\varepsilon^\dagger(\hat{p})A_\varepsilon(\hat{p})\right]. \tag{7.39}$$

7.6 Non-Relativistic Limit of Quantum Field Theory

Since a general discussion of the non-relativistic limit of quantum field theory would be somewhat involved, we here consider only two examples which illustrate the main point: The low-energy limit of quantum field theory is many-body quantum mechanics, and leads to a formalism naturally adapted to the statistical mechanics of quantum particles.

7.6.1 Scalar fields

We first consider a massive scalar field theory with a ϕ^4 type interaction. To discuss the non-relativistic limit it is necessary to employ the real time formalism. The real-time evolution operator is given by a functional integral of the form

$$U = \int [\mathrm{d}\phi] \exp i\mathcal{A}(\phi)$$

$$\mathcal{A}(\phi) = \int \mathrm{d}t\mathrm{d}x \left[\tfrac{1}{2}(\partial_t\phi)^2 - \tfrac{1}{2}(\nabla_x\phi)^2 - \tfrac{1}{2}m^2\phi^2 - \tfrac{1}{4!}g\phi^4\right]. \tag{7.40}$$

At least for a coupling weak enough, the integral is dominated by fields satisfying the free field equation

$$\left(\partial_t^2 - \nabla_x^2 + m^2\right)\phi(t, x) = 0.$$

In the non-relativistic limit the space variation is small compared to the time variation. If the space variations are completely neglected, the solutions to the field equation are simply $\phi(t, x) \propto e^{\pm imt}$. It is thus natural to introduce the holomorphic representation of fields, taking as the unperturbed harmonic oscillator $\mathcal{A}_0(\phi)$:

$$\mathcal{A}_0(\phi) = \int dt dx \left[\tfrac{1}{2}(\partial_t \phi)^2(t, x) - \tfrac{1}{2}m^2 \phi^2(t, x) \right].$$

Denoting by $A^*(t, x), A(t, x)$ the complex fields, in terms of which the field $\phi(t, x)$ and its conjugate momentum read

$$\phi(t, x) = (2m)^{-1/2} \big(A^*(t, x) + A(t, x) \big),$$
$$\Pi(t, x) = i\sqrt{m/2} \big(A^*(t, x) - A(t, x) \big),$$

we find the action

$$\mathcal{A}(A^*, A) = \int dt dx \left[iA^* \partial_t A - mA^* A - \frac{1}{4m} \big(\nabla_x (A^* + A) \big)^2 - \frac{1}{96m^2} (A^* + A)^4 \right].$$

To separate the fast time frequencies we then change variables $A(t, x) \mapsto e^{-imt} A(t, x)$, $A^*(t, x) \mapsto e^{imt} A^*(t, x)$, where the new A, A^* are slowly varying in time compared to the factors e^{imt}. After this change of variables the monomials of the form $A^r A^{*s}$ become multiplied by a factor $e^{im(s-r)t}$. For $r \neq s$ the corresponding time integrals give small contributions due to the rapid time oscillations. Hence at leading order the only surviving terms are those which have an equal number of A and A^* factors. The non-relativistic action is then

$$\mathcal{A}(A, A^*) = \int dt dx \left(iA^* \partial_t A - \frac{1}{2m} \nabla_x A^* \nabla_x A - \frac{g}{16m^2} A^* A^* AA \right). \tag{7.41}$$

We recognize a real time action written in terms of complex variables, of the form (3.74). Up to an infinite energy shift (the vacuum energy), it corresponds to a quantum hamiltonian of the form (Section 3.5)

$$\mathbf{H} = \int dx \left(\frac{1}{2m} \nabla_x A^\dagger \nabla_x A + \frac{g}{16m^2} A^\dagger A^\dagger AA \right), \tag{7.42}$$

where $A^\dagger, A$ now are creation and annihilation operators which satisfy the commutation relations:

$$[A^\dagger(x), A^\dagger(y)] = [A(x), A(y)] = 0, \qquad [A(x), A^\dagger(y)] = \delta(x - y).$$

The hamiltonian (7.42) has been written in normal ordered form in such a way that the corresponding vacuum energy vanishes.

We can construct a Hilbert space by acting repeatedly with creation operators on a reference state (zero particle state) $|0\rangle$ which is annihilated by all $A(x)$. The vectors

$$A^\dagger(x_1) A^\dagger(x_2) \dots A^\dagger(x_n) |0\rangle,$$

span the space of n-particle states. The important property is that the hamiltonian (7.42) then commutes with the particle number $\mathbf{N}$, a property which, in quantum field theory, in general is shared only by free field hamiltonians:

$$\mathbf{N} = \int dx\, A^\dagger(x) A(x), \qquad [\mathbf{N}, \mathbf{H}] = 0.$$

Indeed when all momenta are small compared to masses, pair creation of particles from the vacuum is no longer possible.

Acting on the n-particle vector $|\Psi\rangle$,

$$|\Psi\rangle = \int dx_1 \dots dx_n\, \psi(x_1,\dots,x_n) A^\dagger(x_1) A^\dagger(x_2) \dots A^\dagger(x_n) |0\rangle ,$$

where $\psi(x_1,\dots,x_n)$ is a totally symmetric wave function, the hamiltonian leads to the Schrödinger equation:

$$\left[-\frac{1}{2m} \sum_{i=1}^{n} \Delta_{x_i} + \frac{g}{8m^2} \sum_{i<j} \delta(x_i - x_j) \right] \psi(x_1,\dots,x_n) = E\psi(x_1,\dots,x_n).$$

This is a n-particle hamiltonian with two-body δ-function repulsive forces.

This analysis shows that the low-energy, non-relativistic limit of quantum field theory is many-body quantum mechanics.

Remark. A hamiltonian with a δ-function potential is only well defined in one dimension ($d = 2$ in the initial field theory). In higher dimensions it leads to divergences. But the initial field theory has also divergences.

7.6.2 Fermions

A similar analysis applies to the case of fermions and the result is similar, as we now show. Since a fermion mass breaks space reflection symmetry in odd dimensions we restrict ourselves below to even dimensions.

The starting point again is the non-relativistic limit of field equations. Let us take the example of a self-interacting fermion field in real time

$$\mathcal{A}\left(\bar\psi, \psi\right) = \int dx dt \left[\bar\psi \left(\frac{1}{i}\gamma_0 \partial_t + \boldsymbol{\gamma} \cdot \nabla_x + m \right) \psi + \tfrac12 G \left(\bar\psi\psi \right)^2 \right]. \tag{7.43}$$

Due to the spin structure and the linearity in ∇_x of the field equations, extracting the non-relativistic limit requires slightly more work than in the scalar case.

However, we can avoid the calculation, by extrapolating the results of the analysis of the scalar case. We transform the lagrangian into a hamiltonian written in terms of creation and annihilation operators. In the kinematic part we expand the one-particle energy

$$\sqrt{\hat p^2 + m^2} = m + \hat p^2/2m + O(m^{-3})$$

and translate the one-particle energy by the mass. In the interaction terms we take the non-relativistic limit, neglecting all momentum-dependent contributions relative to the mass. The transformation between fields $\bar\psi, \psi$ and creation, annihilation operators $A^\dagger, A$ then becomes local. The projectors $P_\pm$ defined by equation (7.35) become in the non-relativistic limit

$$P_\pm = \tfrac12(1 \pm \gamma_0).$$

The limiting fermion creation and annihilation quantum operators $A^\dagger, A$ are now defined by

$$A_-^\dagger(x) = {}^T C P_- \tilde\psi(x), \quad A_+(x) = P_+ \tilde\psi(x)$$
$$A_+^\dagger(x) = \tilde\psi^\dagger(x) P_+, \quad A_-(x) = \tilde\psi^\dagger(x) P_- C^* .$$

For γ matrices we choose a basis in which γ_0 is diagonal and restrict below the spinor indices to the non-vanishing components of $A_\pm$. The anticommutation relations then take the simple form

$$\{A^\dagger_{\varepsilon\alpha}(x), A^\dagger_{\varepsilon'\beta}(y)\} = \{A_{\varepsilon\alpha}(x), A_{\varepsilon'\beta}(y)\} = 0\,, \qquad \{A_{\varepsilon\alpha}(x), A^\dagger_{\varepsilon'\beta}(y)\} = \delta_{\varepsilon\varepsilon'}\delta_{\alpha\beta}\delta(x-y).$$

Finally we neglect all interaction terms with an unequal number of $A^\dagger$'s and A's.

Using the relation

$$\bar{\psi}(x)\psi(x) = \sum_\varepsilon A^\dagger_\varepsilon(x)A_\varepsilon(x),$$

and borrowing the result of Subsection 7.5.1, we find the non-relativistic hamiltonian, up to an infinite energy shift,

$$\mathbf{H} = \int dx \left(\frac{1}{2m} \sum_{\varepsilon=\pm 1} \nabla_x A^\dagger_\varepsilon \nabla_x A_\varepsilon - \tfrac{1}{2}G \left(\sum_\varepsilon A^\dagger_\varepsilon A_\varepsilon \right)^2 \right).$$

This action describes a many-body theory of two fermions of opposite charge and with spin, the spin playing the role of an internal quantum number decoupled from space-time.

One then proceeds in analogy with the commuting case. One again verifies that the non-relativistic theory conserves the number of particles and therefore sectors with different particle number decouple

$$\mathbf{N}_{\pm\alpha} = \int dx\, A^\dagger_{\pm\alpha}(x)A_{\pm\alpha}(x) \quad \Rightarrow \quad [\mathbf{N}_{\pm\alpha}, \mathbf{H}] = 0\,,$$

(no summation over the spinor index α and the particle index $\pm$). The vectors

$$A^\dagger_{\varepsilon_1\alpha_1}(x_1)A^\dagger_{\varepsilon_2\alpha_2}(x_2)\ldots A^\dagger_{\varepsilon_n\alpha_n}(x_n)|0\rangle\,,$$

where $|0\rangle$ is the reference state annihilated by all operators $A(x)$, span the space of the n-particle states. A general n-particle vector $|\Phi\rangle$ can be written

$$|\Phi\rangle = \int dx_1 \ldots dx_n\, \phi_{\varepsilon_1\alpha_1,\varepsilon_2\alpha_2,\ldots\varepsilon_n\alpha_n}(x_1,\ldots,x_n)A^\dagger_{\varepsilon_1\alpha_1}(x_1)A^\dagger_{\varepsilon_2\alpha_2}(x_2)\ldots A^\dagger_{\varepsilon_n\alpha_n}(x_n)|0\rangle\,,$$

where $\phi_{\varepsilon_1\alpha_1,\varepsilon_2\alpha_2,\ldots\varepsilon_n\alpha_n}(x_1,\ldots,x_n)$ is a totally antisymmetric wave function.

In the n-particle sector the Schrödinger equation reads

$$\left[-\frac{1}{2m}\sum_{i=1}^n \Delta_{x_i} - G\sum_{i<j}\delta(x_i - x_j) \right]\phi(x_1,\ldots,x_n) = E\phi(x_1,\ldots,x_n).$$

The fermions interact through a two-body $\delta(x)$ function potential which can here be repulsive or attractive. The spin and particle numbers act only through the Pauli principle which dictates the possible symmetries of the wave function ϕ.

A final remark: in two dimensions the fermion theory we have considered here, is equivalent to the well-known massive Thirring model (see Chapter 31). The non-relativistic limit, the δ-function model, can be exactly solved by the Bethe ansatz, i.e. a complete set of wave-functions is provided by a superposition of a finite number of plane waves in each of the $n!$ sectors corresponding to all possible ordering of particle positions. Its relativistic generalization is also integrable, because particle production does not arise.

Bibliographical Notes

Two important references for this chapter are

F.A. Berezin, *The Method of Second Quantization* (Academic Press, New York 1966);

L. D. Faddeev in *Methods in Field Theory*, Les Houches 1975, R. Balian and J. Zinn-Justin eds. (North-Holland, Amsterdam 1976).

Besides the work of C. Itzykson and J.B. Zuber, quoted in the introduction, another general reference is here

J.D. Bjorken and S.D. Drell, *Relativistic Quantum Mechanics*, (McGraw-Hill, New-York 1964).

The non-relativistic limit is considered in

L.L. Foldy and S.A. Wouthuysen, *Phys. Rev.* 78 (1950) 29.

See also

L.S. Brown, *Quantum Field Theory*, (Cambridge University Press, Cambridge 1992).

For a discussion of the *PCT* theorem and the spin-statistics connection, starting from first principles see

R.F. Streater and A.S. Wightman, *PCT, Spin & Statistics and All That* (Benjamin, New York 1964).

APPENDIX 7

A7.1 Cluster Properties of Correlation Functions

Let us briefly describe the cluster properties of connected Feynman diagrams in a massive field theory. We restrict ourselves to a theory with one massive scalar field, but the generalization is straightforward. Since calculations are involved, we return in this appendix to the euclidean formalism.

A7.1.1 Decay of Connected Correlation Functions

The propagator can then be written (equation (5.18)):

$$\Delta\left(x,y\right)=\frac{1}{\left(2\pi\right)^{d}}\int \mathrm{d}^{d}p\,\frac{e^{ip\cdot(x-y)}}{p^{2}+m^{2}}. \tag{A7.1}$$

To evaluate the behaviour of Δ for $|x-y|$ large, we rewrite the integral:

$$\Delta\left(x,y\right)=\frac{1}{\left(2\pi\right)^{d}}\int \mathrm{d}^{d}p\int_{0}^{+\infty}\mathrm{d}t\,e^{ip\cdot(x-y)}\,e^{-t\left(p^{2}+m^{2}\right)}. \tag{A7.2}$$

The gaussian integration over the momentum p can then be performed:

$$\Delta\left(x,y\right)=\frac{\pi^{d/2}}{\left(2\pi\right)^{d}}\int \frac{\mathrm{d}t}{t^{d/2}}\exp\left[-tm^{2}-\frac{1}{4t}\left(x-y\right)^{2}\right]. \tag{A7.3}$$

The large separation behaviour of Δ is given by the method of steepest descent. The saddle point is:

$$t=\frac{|x-y|}{2m}. \tag{A7.4}$$

The gaussian integral over fluctuations around the saddle point finally yields:

$$\Delta\left(x-y\right)\sim\frac{1}{2m}\left(\frac{m}{2\pi\left|x-y\right|}\right)^{(d-1)/2}e^{-m|x-y|}. \tag{A7.5}$$

Using this asymptotic estimate, it is not difficult to verify the following result: if in a connected diagram we separate two sets of points by a distance l, then at large l the diagram decreases as $\exp(-nml)$. In this expression n is the smallest number of lines it is necessary to cut in order to disconnect the diagram, the two sets of points being attached to different connected components (see figure 7.1).

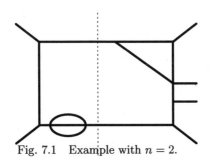

Fig. 7.1   Example with $n=2$.

In a massless theory instead ($m = 0$) the decay is algebraic when the propagator exists (this implies, in perturbation theory, $d > 2$).

A7.1.2 Threshold effects

The precise large distance behaviour of diagrams is related to the strength of the leading singularity in momentum space. For example if, in momentum space, a contribution to the 2-point function has an algebraic singularity

$$\tilde{K}^{(2)}(p) \propto (m^2 + p^2)^{-\alpha},$$

a generalization of the previous calculation yields a large distance behaviour

$$K^{(2)}(r) \propto r^{\alpha - d/2 - 1/2} \, e^{-mr}.$$

If we now consider a 1PI diagram with n internal lines, it yields for r large a contribution

$$K^{(2)}(r) \propto r^{-n(d-1)/2} \, e^{-nmr}.$$

This in turn corresponds to a singularity

$$\tilde{K}^{(2)}(p) \propto (m^2 + p^2)^{-\alpha}, \qquad \alpha = \tfrac{1}{2}[(n-1)d - n - 1].$$

In particular for $d > 1$ the singularity softens when n increases. The 2-particle threshold yields the strongest singularity $(p^2 + 4m^2)^{(d-3)/2}$. The nature of the singularity is important for the large time behaviour of correlation functions in real time: The leading large time behaviour is then related to the leading singularity in the energy variable (a property of the Fourier transformation). Therefore if we consider the 2-point function, its large time behaviour is given by the one-particle pole, then the next to leading term is related to the 2-particle threshold... .

A7.2 Connection between Spin and Statistics

We have noticed in Section 7.5.1 that spin 1/2 fields could only quantized as fermions. It can be proven more generally that as a consequence of locality, unitarity of quantum evolution and relativistic invariance bosons must have integer spin while fermions must have half integer spin. Let us just illustrate this point here.

We have shown in Section A2.2 that, as a consequence of the hermiticity of the hamiltonian the 2-point function has a spectral representation in terms of a positive measure. We have translated this result in the relativistic kinematics in Section 7.4. All possible intermediate states contribute with the same sign. This result can easily be generalized to the discontinuity in the physical domain of the diagonal scattering amplitudes. Let us show that the sign of fermion loops implies a relation between spin and statistics. The contribution to a scalar boson inverse propagator of a loop of scalar bosons of mass m has the form (see equation (6.49) and followings)

$$\Gamma^{(2)}_{1\,\mathrm{loop}}(p) = -\tfrac{1}{2} \frac{g^2}{(2\pi)^d} \int \frac{\mathrm{d}^d q}{(q^2 + m^2)\left[(p+q)^2 + m^2\right]}, \qquad (A7.6)$$

where the specific factor in front of the integral (but not the sign) corresponds to the action

$$S(\phi) = \int \mathrm{d}^d x \left[\tfrac{1}{2} \left(\partial_\mu \phi \right)^2 + \tfrac{1}{2} m^2 \phi^2 + \tfrac{1}{3!} g \phi^3 \right].$$

A convenient integral representation of the diagram is given at the end of Section 11.6 (equation (11.75)). If we take the discontinuity of the function for $s = -p^2 > 4m^2$ we find

$$\delta \Gamma^{(2)}_{1\,\mathrm{loop}}(p) = ig^2 \frac{\pi^{(3-d)/2} 2^{4-2d}}{\Gamma[\tfrac{1}{2}(d-1)]} s^{-1/2} (s - 4m^2)^{(d-3)/2}.$$

This discontinuity has a positive imaginary part, which is indeed consistent with the spectral representation of $W^{(2)}(p)$.

If we instead consider the contribution coming from scalar fermions, the tr ln generating one-loop diagrams gets a minus sign, and clearly the contribution has the opposite sign.

Let us instead consider the contribution of spin $1/2$ fermions:

$$\Gamma^{(2)}_{1\,\mathrm{loop}}(p) = \frac{g^2}{(2\pi)^d} \,\mathrm{tr} \int \frac{\mathrm{d}^d q (-i\slashed{q} + m)(-i\slashed{p} - i\slashed{q} + m)}{(q^2 + m^2) \left[(p+q)^2 + m^2 \right]}$$

$$= \frac{g^2}{(2\pi)^d} \,\mathrm{tr}\, 1 \int \frac{\mathrm{d}^d q \, (m^2 - pq - p^2)}{(q^2 + m^2) \left[(p+q)^2 + m^2 \right]}.$$

We now use the identity

$$m^2 - pq - q^2 = 2m^2 + \tfrac{1}{2} p^2 - \tfrac{1}{2} \left[(p+q)^2 + m^2 + q^2 + m^2 \right].$$

The two terms inside the brackets cancel a denominator and thus yield a constant (in general divergent, see Chapter 8) result, which has no discontinuity:

$$\Gamma^{(2)}_{1\,\mathrm{loop}}(p) = \frac{g^2 \,\mathrm{tr}\, 1}{2(2\pi)^d} \left(p^2 + 4m^2 \right) \int \frac{\mathrm{d}^d q}{(q^2 + m^2) \left[(p+q)^2 + m^2 \right]} + \mathrm{const.} . \qquad (A7.7)$$

In the region of physical emission:

$$-p^2 > (2m)^2,$$

the prefactor $\left(p^2 + 4m^2 \right)$, which reflects the spin structure, is negative and compensates the negative sign due to the fermion loop.

8 DIVERGENCES IN PERTURBATION THEORY, POWER COUNTING

A local field theory is characterized by the absence of a "small" fundamental length: the action depends only on products of fields and their derivatives at the same point. In perturbation theory the propagator has a simple power law behaviour at short distances and the interaction vertices are constants or differential operators acting on δ-functions.

We have explained in previous chapters how physical quantities can be calculated as power series in the various interactions. As we now show, perturbative calculations are affected by divergences due to severe short distance singularities. After Fourier transformation, these divergences take the form of integrals diverging at large momenta: one speaks also of UV singularities. These divergences are peculiar to Quantum Field Theory: in contrast with Classical Mechanics and Quantum Mechanics, it is impossible to define in a straightforward way a quantum field theory of point-like objects.

In Chapters 8–11 we shall explain how to deal with this problem, at least in the sense of formal perturbation theory. The study of critical phenomena by renormalization group methods will then give us a clue about the possible origin of these divergences. In this chapter we begin the discussion by analyzing the nature of divergences. We explain the problem first on the example of the ϕ^3-field theory at one-loop order. We then systematically characterize divergences in a large class of local field theories. We finally discuss the divergences of correlation functions involving also composite operators.

To analyze the divergences of Feynman diagrams it is convenient to work in the momentum representation because it is simpler to discuss divergences of integrals at large momenta rather than singularities of distributions in space representation. Moreover the explicit expressions of Section 6.3 show that amputation and Legendre transformation become in momentum representation purely algebraic operations in the sense that they involve no momentum integration because the propagator is diagonal in this representation. In particular the divergences of connected correlation functions can be simply deduced from the divergences of proper vertices. Since a short analysis reveals that the structure of divergences of proper vertices is simpler, we examine in what follows the divergences of Feynman diagrams contributing to the generating functional $\Gamma(\varphi)$, i.e. 1PI Feynman diagrams.

8.1 The Problem of Divergences and Renormalization: The ϕ^3 Field Theory at One-Loop Order

We first explain the problem of divergences of Feynman diagrams on the simple example of a scalar ϕ^3 field theory with an action:

$$S(\phi) = \int \mathrm{d}^d x \left[\tfrac{1}{2} \left(\partial_\mu \phi \right)^2 + \tfrac{1}{2} m^2 \phi^2 + \frac{1}{3!} g\, \phi^3 \right]. \tag{8.1}$$

This theory is unphysical because the potential is not bounded from below. However it has a well-defined perturbative expansion where this non-perturbative pathology does not show up (see Chapter 39). Moreover, when g is purely imaginary, it makes sense beyond perturbation theory and describes the universal properties of the Yang–Lee edge singularity of the Ising model.

8.1.1 Perturbation theory at one-loop order

Tree order. At the tree level the 1PI functional is identical to $S(\phi)$ and thus:

$$\Gamma^{(2)}\left(x,y\right) = \left(-\Delta + m^2\right)\delta\left(x-y\right).$$

The Fourier transform of the 1PI 2-point function is then:

$$\widetilde{\Gamma}^{(2)}\left(p\right) = p^2 + m^2. \tag{8.2}$$

Also:

$$\begin{cases} \widetilde{\Gamma}^{(3)}\left(p_1, p_2, -p_1 - p_2\right) = g \\ \qquad \widetilde{\Gamma}^{(n)}\left(p_1, \ldots, p_n\right) = 0 \quad \text{for } n > 3. \end{cases} \tag{8.3}$$

One-loop order. Using equation (6.49), we can now calculate the one-loop contribution $\Gamma_1\left(\varphi\right)$:

$$\Gamma_1\left(\varphi\right) = \frac{1}{2}\int d^d x \left\langle x \left| \ln\left[1 + g\left(-\Delta + m^2\right)^{-1}\varphi\right] \right| x \right\rangle. \tag{8.4}$$

If we expand this expression in powers of φ we obtain the one-loop contributions to $\Gamma^{(n)}$. After Fourier transformation we find

$$\widetilde{\Gamma}^{(n)}_{1\text{ loop}} = -\frac{(n-1)!}{2}(-g)^n \int \frac{d^d q}{(2\pi)^d}\frac{1}{q^2 + m^2}\frac{1}{\left(q+p_1\right)^2 + m^2}\cdots$$

$$\times \frac{1}{\left(q + p_1 + \cdots + p_{n-1}\right)^2 + m^2}, \tag{8.5}$$

quantity represented by the Feynman diagram of figure 8.1.

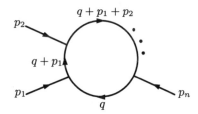

Fig. 8.1   One-loop 1PI diagram.

On this expression we see that the one-loop contribution to the n-point function, which for large momentum q behaves like $\int^\infty q^{-2n}d^d q$, thus diverges for

$$d \geq 2n.$$

Therefore it is only for $d = 1$ (quantum mechanics) that all correlation functions are finite. For $d = 2$ the 1-point function, which has no momentum dependence, diverges:

$$\widetilde{\Gamma}^{(1)}_{1\text{ loop}} = \frac{g}{2}\int \frac{d^d q}{(2\pi)^d}\frac{1}{q^2 + m^2}. \tag{8.6}$$

When d increases more correlation functions diverge. For $d = 6$ the 1, 2 and 3-point functions diverge:

$$\widetilde{\Gamma}^{(2)}_{1\,\text{loop}} = -\frac{1}{2}g^2 \int \frac{d^6q}{(2\pi)^6} \frac{1}{(q^2 + m^2)\left[(q+p)^2 + m^2\right]}, \tag{8.7}$$

$$\widetilde{\Gamma}^{(3)}_{1\,\text{loop}} = g^3 \int \frac{d^6q}{(2\pi)^6} \frac{1}{(q^2 + m^2)\left[(q+p_1)^2 + m^2\right]\left[(q+p_1+p_2)^2 + m^2\right]}. \tag{8.8}$$

If we imagine that we cut the momentum integral by integrating in a sphere $|\mathbf{q}| < \Lambda$, then the 1 and 2-point functions diverge like powers of Λ, while $\widetilde{\Gamma}^{(3)}$ diverges logarithmically.

To explicitly calculate the contributions which diverge with Λ, we can perform a Taylor series expansion of the integrand in the external momenta. It is easy to verify that the coefficients of the terms of global degree k in the momenta are given by integrals which diverge only for $d \geq k + 2n$. Therefore the divergent part of a one-loop contribution to the n-point function is a polynomial of degree $d - 2n$. For $d = 6$ we find:

$$\widetilde{\Gamma}^{(1)}_{1\,\text{loop}} = \frac{g}{2^7\pi^3}\left[\frac{\Lambda^4}{4} - \frac{m^2\Lambda^2}{2} + m^4\ln\frac{\Lambda}{m} + O(1)\right],$$

$$\widetilde{\Gamma}^{(2)}_{1\,\text{loop}} = -\frac{g^2}{2^7\pi^3}\left[\frac{\Lambda^2}{2} - \left(2m^2 + \frac{p^2}{3}\right)\ln\frac{\Lambda}{m} + O(1)\right], \tag{8.9}$$

$$\widetilde{\Gamma}^{(3)}_{1\,\text{loop}} = \frac{g^3}{2^6\pi^3}\ln\frac{\Lambda}{m} + O(1).$$

We recall that the surface of the sphere S_5 is π^3.

We see that the dimension 6 is special in the following sense: the 1PI correlation functions which diverge are all those which are already non-trivial at the tree level (a term linear in ϕ can be added to action (8.1) by translating ϕ by a constant). Moreover the divergent terms have the same momentum dependence as the tree level contributions.

On the contrary, for $d \geq 8$ the 4-point function, which vanishes at the tree level, is also divergent.

8.1.2 Empirical removal of divergences at the one-loop level

For $d \leq 6$ dimensions, the divergent parts of the one-loop correlation functions have the structure of the original action. For example in $d = 6$ dimensions the divergent part $\Gamma^{\text{div.}}_1(\varphi)$ of the one-loop 1PI functional is:

$$\Gamma^{\text{div.}}_1(\varphi) = \int d^6x \left[ga_1(\Lambda)\,\varphi(x) + \frac{1}{2}g^2a_2(\Lambda)\,(\partial_\mu\varphi)^2 \right.$$
$$\left. + \frac{1}{2}g^2a_3(\Lambda)\,\varphi^2(x) + \frac{1}{3!}g^3a_4(\Lambda)\,\varphi^3(x)\right]. \tag{8.10}$$

The functions $a_i(\Lambda)$ are derived from equations (8.9) and are therefore defined only up to additive finite parts. For later purpose (the minimal subtraction scheme) it is convenient to give a canonical definition of the divergent part of a Feynman diagram as the sum of the divergent terms in the asymptotic expansion in a dimensionless parameter. We here choose Λ/m. Then:

$$
\begin{cases}
a_1\left(\Lambda\right) = \dfrac{1}{2^7\pi^3}\left(\dfrac{\Lambda^4}{4} - \dfrac{m^2\Lambda^2}{2} + m^4\ln\dfrac{\Lambda}{m}\right), \\[2mm]
a_2\left(\Lambda\right) = \dfrac{1}{32^7\pi^3}\ln\dfrac{\Lambda}{m}, \\[2mm]
a_3\left(\Lambda\right) = \dfrac{1}{2^7\pi^3}\left(-\dfrac{\Lambda^2}{2} + 2m^2\ln\dfrac{\Lambda}{m}\right), \\[2mm]
a_4\left(\Lambda\right) = \dfrac{1}{2^6\pi^3}\ln\dfrac{\Lambda}{m}.
\end{cases}
\tag{8.11}
$$

If we now add to the original action $S(\phi)$, the quantity $\delta S_1(\phi)$:

$$
\delta S_1(\phi) = -\int \mathrm{d}^6x \left[g\,b_1\left(\Lambda\right)\phi\left(x\right) + \frac{1}{2}g^2 b_2\left(\Lambda\right)\left(\partial_\mu\phi\right)^2 \right.
$$
$$
\left. + \frac{1}{2}g^2 b_3\left(\Lambda\right)\phi^2\left(x\right) + \frac{1}{3!}g^3 b_4\left(\Lambda\right)\phi^3\left(x\right)\right],
\tag{8.12}
$$

in which the coefficients b_i are such that the differences $b_i - a_i$ have finite large cut-off limits but are otherwise arbitrary, the new action

$$
S_1(\phi) = S(\phi) + \delta S_1(\phi),
$$

differs from the original action by its parametrization, but involves the same monomials of the field and therefore depends on the same number of parameters.

$$
S_1(\phi) = \int \mathrm{d}^6x \left[\frac{1}{2}\left(1 - g^2 b_2\left(\Lambda\right)\right)\left(\partial_\mu\phi\right)^2 + \frac{1}{2}\left(m^2 - g^2 b_3\left(\Lambda\right)\right)\phi^2 \right.
$$
$$
\left. + \frac{1}{3!}\left(g - g^3 b_4\left(\Lambda\right)\right)\phi^3 - g b_1\left(\Lambda\right)\phi\right].
$$

The new 1PI functional $\Gamma\left(\varphi\right)$ at one-loop order is then:

$$
\Gamma\left(\varphi\right) = S_1\left(\varphi\right) + \Gamma_1\left(\varphi\right) + O\left(\text{two loops}\right)
$$
$$
= S\left(\varphi\right) + \Gamma_1\left(\varphi\right) + \delta S_1\left(\varphi\right) + O\left(\text{two loops}\right),
$$

where $\Gamma_1\left(\varphi\right)$ is the sum of one-loop diagrams calculated only with $S(\phi)$. Therefore $\Gamma\left(\varphi\right)$ now has a limit at one-loop order when the cut-off becomes infinite.

   This is the method we would like to generalize to arbitrary orders in the loopwise expansion and for more general field theories.

   Note that for $d \geq 8$ this method fails. Indeed to generate at the tree level a term proportional to the divergence of the 4-point function for example, we have to add a ϕ^4 term to the action. We show below that such an interaction induces in turn worse divergences which cannot be reproduced. Power counting will show that it is not possible to add terms to the action in such a way that the form of the tree level terms and the one-loop divergences become similar.

   Interpretation of perturbative divergences. To circumvent the difficulties associated with the divergences of perturbation theory, we shall try to implement the following idea which is at the basis of renormalization theory: we introduce a large momentum cut-off in the theory, or equivalently modify the field theory at short distance, as we have done here to characterize the divergences of Feynman diagrams, (the physical reality of such a cut-off is at this stage irrelevant). We then investigate the possibility of choosing the original parameters of the theory as functions of the cut-off in such a way that correlation functions have a finite large cut-off limit. When such a limit exists, we show that it is independent of the cut-off procedure (under some general conditions). We shall call the corresponding local field theories renormalizable (or super-renormalizable if some parameters in the interaction are cut-off independent). **Renormalizable theories are short distance insensitive** in the sense that even if a large mass or a microscopic scale in space provide a true physical cut-off, their long distance or low momentum properties can be described, without explicit knowledge of the short distance structure, in terms of a small number of *effective scale-dependent parameters*. We already emphasize these ideas here because they motivate all the technical analysis which follows and are at the basis of one major application of renormalization theory: the renormalization group analysis of critical phenomena (see Chapters 23–36).

8.2 Divergences in Perturbation Theory: General Analysis

Dimension of fields. We first introduce the notion of canonical, or engineering dimension of fields. The dimension $[\phi]$ of a field ϕ is related to the large momentum behaviour of the ϕ propagator. We explicitly assume that the propagator $\Delta(p)$ is $O(d)$ covariant in d dimensions, or at least has a uniform large momentum behaviour of the form:

$$C_1\lambda^{-\sigma} < |\Delta(\lambda p)| < C_2\lambda^{-\sigma} \quad \forall p \neq 0 \quad \text{for } \lambda \to \infty, \tag{8.13}$$

in which C_1 is a strictly positive constant. Other cases require a special analysis.
We then define the dimension $[\phi]$ of $\phi(x)$ by:

$$[\phi] = \tfrac{1}{2}(d - \sigma). \tag{8.14}$$

Examples. In the scalar theory taken as an example in Section 8.1:

$$\Delta(\lambda p) \sim \frac{1}{\lambda^2}, \quad \text{for } \lambda \to \infty \quad \Rightarrow \quad [\phi] = \tfrac{1}{2}(d - 2). \tag{8.15}$$

For the fermions considered in Section 5.5 the propagator in Fourier space reads:

$$\widetilde{W}^{(2)}_{\alpha\beta}(p) = \langle \bar{\psi}_\alpha(-p)\psi_\beta(p) \rangle = (m + i\not{p})^{-1}_{\beta\alpha},$$

and thus

$$\Delta(\lambda p) \sim \frac{1}{\lambda}, \quad \lambda \to \infty \quad \Rightarrow \quad [\psi] = [\bar\psi] = \tfrac{1}{2}(d - 1). \tag{8.16}$$

The definition (8.14) is such that, in the simple case of scalar and spin 1/2 fields, it coincides with the natural mass dimension of the field as deduced from the quadratic part of the action by dimensional analysis. Assigning a dimension +1 to momenta and masses:

$$[p] = [m] = 1,$$

correspondingly a dimension -1 to length and space coordinates:

$$[x] = -1 \quad \Rightarrow \quad \left[\frac{\partial}{\partial x} \right] = +1 \,,$$

and expressing that the action is dimensionless, we indeed find:

$$\left[\int \mathrm{d}^d x \, (\partial_\mu \phi \, (x))^2 \right] = 0 \quad \Rightarrow \quad -d + 2 + 2 \, [\phi] = 0 \,,$$

$$\left[\int \mathrm{d}^d x \, \bar{\psi} \, (x) \, \partial \!\!\!/ \psi \, (x) \right] = 0 \quad \Rightarrow \quad -d + 1 + 2 \, [\psi] = 0 \,.$$

This is no longer true in general for higher spin fields. For example the free action for a massive vector field A_μ is:

$$S \, (A) = \int \mathrm{d}^d x \left[\frac{1}{4} \sum_{\mu\nu} (\partial_\mu A_\nu - \partial_\nu A_\mu)^2 + \frac{1}{2} m^2 A_\mu A_\mu \right] . \tag{8.17}$$

The propagator $\Delta_{\mu\nu} \, (p)$ is then:

$$\Delta_{\mu\nu} = \frac{\delta_{\mu\nu} + p_\mu p_\nu / m^2}{p^2 + m^2} \,, \tag{8.18}$$

and therefore

$$\Delta_{\mu\nu} \, (\lambda p) \sim \lambda^0 \quad \text{for} \quad \lambda \to \infty \quad \Rightarrow \quad [A_\mu] = \frac{d}{2}, \tag{8.19}$$

while just by inspection of the action one would have concluded that A_μ has the dimension of a scalar field. The reasons can be found in the appearance of negative powers of m^2 in the propagator: the quadratic form in the action is not invertible for $m = 0$. On the pole of the propagator $p^2 = -m^2$ the numerator of $\Delta_{\mu\nu}$ is a projector orthogonal to p_μ.

The same phenomenon will also occur with higher spin fields. A spin s massive field propagator has the form:

$$\Delta \, (p) = \frac{P_{2s} \, (p/m)}{p^2 + m^2} \,, \tag{8.20}$$

in which $P_{2s} \, (p)$ is a polynomial of degree $2s$ in p, which is a projector on "mass-shell", i.e. for $p^2 = -m^2$. The dimension of the corresponding field ϕ_s is:

$$\boxed{[\phi_s] = \tfrac{1}{2} \, (d - 2 + 2s)} \tag{8.21}$$

Equation (8.21) generalizes equations (8.15, 8.16, 8.19).

Dimension of vertices. The interaction term in the action is a linear combination of vertices which are space integrals of a product of fields and their derivatives at the same points. Let us write a vertex $V(\phi)$ symbolically:

$$V(\phi) \propto \int \mathrm{d}^d x \left(\frac{\partial}{\partial x} \right)^k \phi_1^{n_1} \, (x) \phi_2^{n_2} \, (x) \ldots \phi_s^{n_s} \, (x),$$

where the k derivatives act in an unspecified way on the fields ϕ_i. We call these elementary interaction terms vertices because they are represented by vertices in Feynman diagrams.

We now define the dimension $\delta[V]$ of the vertex $V(\phi)$ by:

$$\delta[V] = -d + k + \sum_{i=1}^{s} n_i [\phi_i],\tag{8.22}$$

in which k is the total number of differential operators and n_i the number of fields ϕ_i appearing in the vertex V.

Superficial degree of divergence of a Feynman diagram. As explained before, we consider only 1PI diagrams. Let us first write the vertex $V(\phi)$ in terms of the Fourier transforms $\tilde{\phi}_i(p)$ of the fields $\phi_i(x)$ (still in symbolic notation):

$$V(\phi) \sim \int d^d p_1 \ldots d^d p_{n_1+n_2+\cdots+n_s} \delta^d (p_1 + p_2 + \cdots + p_{n_1+\cdots+n_s})$$
$$\times p^k \tilde{\phi}_1(p_1) \ldots \tilde{\phi}_s(p_{n_1+\cdots+n_s}).$$

Each vertex of this type brings a δ-function of momentum conservation. In addition this vertex adds the product of k momenta in the numerator of a Feynman diagram. If we now count the number of independent momenta over which to integrate in a diagram, taking into account momentum conservation at vertices, we just find the number of loops L. This follows directly from one of the definitions of the number of loops in a diagram given in Section 6.4. Finally a diagram contains I_i internal lines corresponding to propagators Δ_i joining the different vertices.

Therefore if all integration momenta in a diagram γ are scaled by a factor λ, for λ large the diagram is scaled by a factor $\lambda^{\delta(\gamma)}$ with:

$$\delta(\gamma) = dL - \sum_i I_i \sigma_i + \sum_\alpha v_\alpha k_\alpha \tag{8.23}$$

in which v_α is the number of vertices of type α with k_α derivatives. The number $\delta(\gamma)$ is called the superficial degree of divergence of the diagram γ. For a one-loop diagram regularized with a momentum cut-off, it characterizes the divergence of the diagram as a power of the cut-off.

More generally if $\delta(\gamma)$ is positive a regularized diagram diverges at least like $\Lambda^{\delta(\gamma)}$. If $\delta(\gamma) = 0$ it diverges at least like a power of $\ln \Lambda$. If $\delta(\gamma)$ is negative the diagram is superficially convergent, which means that divergences can come only from subdiagrams.

Let us verify equation (8.23) in the case of the ϕ^3 field theory in $d = 6$ dimensions. The three divergent one-loop diagrams are shown in figure 8.2.

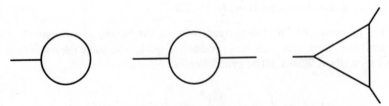

Fig. 8.2   Divergent one-loop diagrams in a ϕ^3 field theory.

Since $\sigma = 2$ and $k = 0$, expression (8.23) yields

$$\delta(\gamma) = 6 - 2I.$$

For $I = 1, 2, 3$ respectively we find the values 4, 2, 0 in agreement with equations (8.9). For $I > 3$ the diagrams are convergent.

Figure 8.3 exhibits a superficially convergent diagram with divergent subdiagrams in the same field theory:

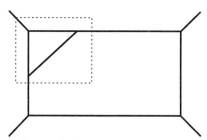

Fig. 8.3   Divergent subdiagram.

The superficial degree of divergence is -2, the diagram is superficially convergent, but the subdiagram inside the dotted box is divergent.

A different expression of the superficial degree of divergence. Various topological relations on graphs allow to write $\delta(\gamma)$ in different forms.

Consider a diagram γ contributing to a 1PI correlation function with E_i (for external line) fields ϕ_i. Then if we call n_i^α the number of fields ϕ_i at a vertex α belonging to the diagram, we have the relation:

$$\boxed{E_i + 2I_i = \sum_\alpha n_i^\alpha v_\alpha} \tag{8.24}$$

The interpretation of this formula is simple: every internal line connects two vertices while an external line is only attached to one vertex.

Figure 8.4 gives an example.

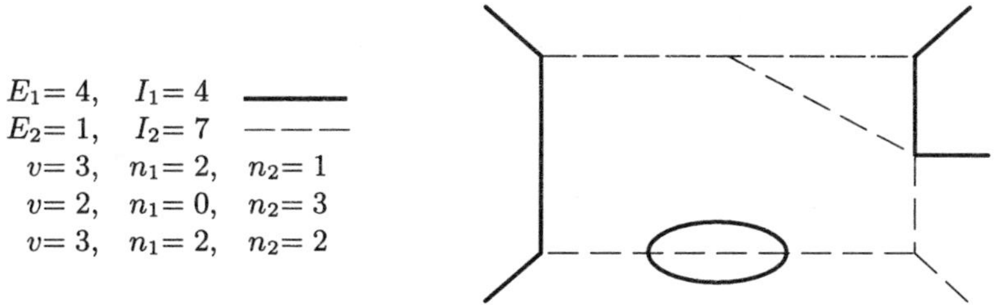

$$
\begin{aligned}
&E_1 = 4, \quad I_1 = 4 \quad \text{———}\\
&E_2 = 1, \quad I_2 = 7 \quad \text{– – –}\\
&v = 3, \quad n_1 = 2, \quad n_2 = 1\\
&v = 2, \quad n_1 = 0, \quad n_2 = 3\\
&v = 3, \quad n_1 = 2, \quad n_2 = 2
\end{aligned}
$$

Fig. 8.4

If we combine equation (8.24) with the relation (6.57) written under the form:

$$\boxed{L = \sum_i I_i - \sum_\alpha v_\alpha + 1} \tag{8.25}$$

we can eliminate L and I_i in $\delta(\gamma)$. We then obtain

$$\delta(\gamma) = d - \sum_i E_i\,[\phi_i] + \sum_\alpha v_\alpha \delta\,(V_\alpha)$$

(8.26)

in which $\delta\,(V_\alpha)$ is the dimension of the vertex α (equation (8.22))

$$\delta\,(V_\alpha) = -d + k_\alpha + \sum_i n_i^\alpha\,[\phi_i]$$

and $[\phi_i]$ the dimension of ϕ_i (equation (8.14)):

$$[\phi_i] = \tfrac{1}{2}\,(d - \sigma_i)$$

Equation (8.26) directly leads to a classification of renormalizable theories.

8.3 Classification of Renormalizable Field Theories

We now assume that the program outlined in Section 8.1 can be realized. This point of view will be justified in Chapters 9 and 10. With this assumption we can classify renormalizable field theories. In what follows we restrict ourselves to the most frequent situation: $[\phi_i] \geq 0$ for all fields. Other cases require a special analysis.

Non-renormalizable theories. If at least one vertex V has a positive dimension $\delta\,(V)$, then by considering diagrams with an increasing number v of vertices of this type we can make the degree of divergence arbitrarily large and this for any 1PI correlation function. A field theory with such a vertex is not renormalizable because in order to cancel divergences one would have to add an infinite number of new interactions to the action, and the final theory would depend on an infinite number of parameters.

Super-renormalizable theories. When only a finite number of Feynman diagrams are superficially divergent we call the field theory super-renormalizable. This happens when the dimensions of all vertices are strictly negative.

Example. In the ϕ^4 field theory in $d = 3$ dimensions

$$\delta(\gamma) = 3 - \tfrac{1}{2}E - v\,.$$

The superficially divergent diagrams are listed in figure 8.5.

Fig. 8.5   Superficially divergent diagrams in ϕ_3^4.

Renormalizable theories. These theories are characterized by the property that at least one vertex has dimension zero, and no vertex has a positive dimension. Then an infinite number of diagrams have a positive superficial degree of divergence, however the maximal degree of divergence at E_i fixed does not increase with the number of loops, and is independent of the number of insertions of the vertices of dimension zero.

In addition if all dimensions of fields $[\phi_i]$ are strictly positive, only a finite number of correlation functions are superficially divergent.

Such theories are called renormalizable by power counting.

If at least one field has dimension zero, the situation is more complicated: the degree of divergence is bounded, however an infinite number of correlation functions are superficially divergent. Generically this leads to field theories depending on an infinite number of parameters although, in contrast to the case of non-renormalizable theories, only a subclass of all possible interactions is generated by renormalization.

In addition in some cases symmetries can relate all these parameters so that only a finite number are really independent. We shall meet such a situation in Chapters 14, 15 when we discuss models defined on homogeneous spaces like the non-linear σ-model.

Let us examine conventional field theories containing fields of spin s with dimensions $[\phi_s]$ given by equation (8.21):

$$[\phi_s] = \tfrac{1}{2}\left(d - 2 + 2s\right).$$

The condition $[\phi_s] > 0$ is satisfied for $d \geq 2$ except for $s = 0$, $d = 2$, case which must be examined separately.

All vertices should satisfy:

$$-d + k + \tfrac{1}{2} \sum_s n_s \left(d - 2 + 2s\right) \leq 0. \tag{8.27}$$

This condition bounds k, the number of derivatives, n_s the number of fields of spin s at the vertex, s the spin and the dimension d.

For $k = s = 0$, the condition (8.27) implies for all vertices:

$$n \leq \frac{2d}{d - 2}.$$

The corresponding renormalizable field theories are:

$$\phi^3 \quad \text{in} \quad d = 6 \text{ dimensions},$$
$$\phi^4 \quad \text{in} \quad d = 4 \text{ dimensions},$$
$$\phi^5, \phi^6 \quad \text{in} \quad d = 3 \text{ dimensions}.$$

Note that any polynomial in ϕ is super-renormalizable in 2 dimensions.

If we allow in addition 2 derivatives, $k = 2$, the only solution is $d = 2$, but then $[\phi] = 0$.

Let us now consider vertices with one spin $1/2$ fermion pair $\bar\psi\psi$, and n scalar fields. Condition (8.27) then reads:

$$n \leq \frac{2}{d - 2}.$$

Renormalizable field theories are:

$$\psi\bar\psi\phi \quad \text{in } d = 4 \text{ dimensions},$$
$$\bar\psi\psi\phi^2 \quad \text{in } d = 3 \text{ dimensions}.$$

In addition $P(\phi)\bar{\psi}\psi$, in which $P(\phi)$ is a polynomial in ϕ, is super-renormalizable in 2 dimensions.

Finally the vertex $(\bar{\psi}\psi)^2$ is renormalizable in 2 dimensions.

The vertices $P(\phi)(\bar{\psi}\psi)^2$ and $P(\phi)\bar{\psi}\dot{\partial}\psi$ are also of dimension zero in 2 dimensions but again the dimension of ϕ vanishes.

For spin 1 vector fields, general $O(d)$ invariance leaves only dimension 2 as a possibility. The only candidate with only fermions is the vertex:

$$\bar{\psi}A_\mu\gamma_\mu\psi \equiv \bar{\psi}\slashed{A}\psi,$$

which is renormalizable in 2 dimensions. In addition the vertices $\phi\partial_\mu\phi A_\mu$ and $\phi^2 A_\mu^2$, which appear in gauge theories, are dimensionless. However, they again lead to a non-trivial renormalization problem because scalar fields are dimensionless ($[\phi] = 0$).

No higher spin field leads to renormalizable theories.

Note however that spin 1 vector fields associated with gauge symmetries do not enter in the previous classification because their propagator has in some gauges the behaviour of a scalar field propagator (for a discussion of gauge theories see Chapters 18–21).

Finally let us note that no physically acceptable, from the point of view of Particle Physics, and renormalizable theory exists above dimension 4. It is not known whether this property is logically connected with the fact that space time has just four dimensions, or is a mere coincidence.

8.4 Power Counting for Operator Insertions

Up to now we have analyzed only the divergences of the field correlation functions. However we shall also need in various places correlation functions involving local polynomials of the field, called hereafter composite fields or for historical reasons composite operators (this terminology comes from the operator formulation of quantum field theory).

Typical examples are:

$$\mathcal{O}(\phi(x)) \equiv \phi^2(x),\ \phi^4(x),\ [\partial_\mu\phi(x)]^2 \dots .$$

One insertion of operator $\mathcal{O}(\phi)$ yields the correlation functions:

$$\langle \mathcal{O}(\phi(y))\,\phi(x_1)\dots\phi(x_n)\rangle .$$

Such correlation functions can in principle be obtained from the field correlation functions by letting various points coincide. However this procedure corresponds in momentum space to additional integrations, and therefore generates new divergences.

We have to analyze the 1PI correlation functions with operator insertion, from the point of view of power counting. However we can use the results of previous section in the following way. Let us again call $S(\phi)$ the euclidean action and let us add a source (which is a space-dependent coupling constant) for the operator $\mathcal{O}(\phi(x))$:

$$S_g(\phi) = S(\phi) + \int \mathrm{d}^d x\, g(x)\, \mathcal{O}(\phi(x)). \tag{8.28}$$

To this new action S_g corresponds a generating functional $Z(J,g)$:

$$Z(J,g) = \int [\mathrm{d}\phi] \exp\left[-S_g(\phi) + \int J(x)\,\phi(x)\,\mathrm{d}^d x\right]. \tag{8.29}$$

The correlation functions with one operator $\mathcal{O}(\phi)$ insertion can be obtained from the generating functional $\delta Z/\delta g(x)$ taken at $g = 0$:

$$\langle \mathcal{O}[\phi(y)] \phi(x_1) \ldots \phi(x_n) \rangle = -\left[\frac{\delta}{\delta J(x_1)} \cdots \frac{\delta}{\delta J(x_n)} \frac{\delta}{\delta g(y)} Z(J,g)\right]\bigg|_{J=g=0}. \tag{8.30}$$

More generally successive differentiations with respect to $g(x)$ yield generating functionals of correlation functions with several operator insertions.

Performing the Legendre transformation of $W(J,g) = \ln Z(J,g)$ with respect to $J(x)$ we obtain the 1PI functional $\Gamma(\varphi, g)$. The generating functional of 1PI correlation functions with one $\mathcal{O}(\phi(y))$ insertion, $\Gamma_{\mathcal{O}}^{(n)}$, is then:

$$\frac{\partial \Gamma(\varphi, g)}{\partial g(y)}\bigg|_{g=0} = \sum_{n=1}^{\infty} \frac{1}{n!} \int dx_1 \ldots dx_n \, \varphi(x_1) \ldots \varphi(x_n) \Gamma_{\mathcal{O}}^{(n)}(y; x_1, \ldots, x_n).$$

The corresponding Feynman diagrams have the structure shown on figure 8.6.

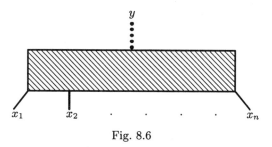

Fig. 8.6

After Fourier transformation they are just ordinary diagrams with one additional vertex $\mathcal{O}(\phi)$, except that an additional momentum enters the diagram at the vertex so that momentum is not conserved. However if we scale all integration momenta by a factor λ, in the large λ limit all external momenta are negligible. Therefore the power counting of 1PI correlation functions with one operator insertion is the same as with one vertex insertion. We assign to an operator $\mathcal{O}(\phi)$ the dimension $[\mathcal{O}]$:

$$[\mathcal{O}] = k + \sum_i n_i [\phi_i] \tag{8.31}$$

in which k is the number of derivatives in the operator, and n_i the number of fields of type i. This definition differs from the definition of the dimension of the corresponding vertex (equation (8.22)) by d.

If we then insert the operators $\mathcal{O}_1(x_1) \ldots \mathcal{O}_r(x_r)$, the superficial degree of divergence of 1PI diagram γ is:

$$\delta_\gamma[\mathcal{O}_1 \ldots \mathcal{O}_r] = \delta_\gamma + [\mathcal{O}_1] + \cdots + [\mathcal{O}_r] - rd \tag{8.32}$$

in which δ_γ is the degree of divergence of the diagram with only the "true" vertices.

For example for $d = 4$ one insertion of $\phi^m(x)$ in the ϕ^4 field theory yields

$$\begin{cases} \delta_\gamma = 4 - E + [\phi^m] - 4 \\ [\phi^m] = m \end{cases} \tag{8.33}$$

In the same field theory the 1PI n-point function with $l\,\phi^2$ insertions:

$$\Gamma^{(n,l)}(x_1, \ldots, x_n, y_1, \ldots, y_n) = \langle \phi(x_1) \ldots \phi(x_n)\, \phi^2(y_1) \ldots \phi^2(y_l) \rangle_{1PI}$$

has a degree of divergence δ:

$$\delta = 4 - n - 2l. \tag{8.34}$$

The new divergent correlation functions are shown in figure 8.7.

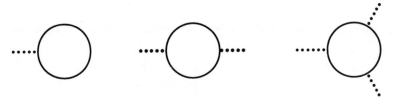

Fig. 8.7

Bibliographical Notes

For an early reference on power counting see

N.N. Bogoliubov and D.V. Shirkov, *Introduction to the Theory of Quantized Fields* (Interscience, New York 1959).

For the Yang–Lee edge singularity see

C.N. Yang and T.D. Lee, *Phys. Rev.* 87 (1952) 404; T.D. Lee and C.N. Yang, *Phys. Rev.* 87 (1952) 410; M.E. Fisher, *Phys. Rev. Lett.* 40 (1978) 1610.

9 REGULARIZATION METHODS

The divergences that we have met and analyzed in Chapter 8 show that the field theories we wanted to construct cannot be defined by a straightforward perturbative expansion. Our strategy will thus be the following: we shall modify the field theory at large momentum, short distance or otherwise in such a way that the new Feynman diagrams become well-defined finite quantities, and such that when one control parameter reaches some limit (for example the cut-off is sent to infinity), we formally recover the original perturbation theory. This procedure is called regularization. It will allow us to isolate well-defined divergent parts of diagrams and deal with them with renormalization as will be explained in Chapter 10. Many regularization methods have been introduced in the literature, but we shall only describe three of them: Pauli–Villars, lattice and dimensional regularizations, which have different advantages and shortcomings. In any particular application two criteria will guide our choice: (i) in some theories, symmetries play a crucial role and it is essential to find a regularization which preserves the symmetry; (ii) if we want to explicitly calculate Feynman diagrams, we shall look for the regularization which leads to the simplest practical calculations.

9.1 Cut-Off and Pauli–Villars's Regularization

The simplest idea is to modify the propagator in such a way that it decreases faster at large momentum. For example we can replace

$$\left(p^2 + m^2\right)^{-1} \text{ by } \left(p^2 + m^2 + \alpha_2 \frac{p^4}{\Lambda^2} + \alpha_3 \frac{p^6}{\Lambda^4} + \cdots + \alpha_n \frac{p^{2n}}{\Lambda^{2n-2}}\right)^{-1}, \qquad (9.1)$$

and choose the degree n to make all diagrams convergent. The parameter Λ is the cut-off. In the large cut-off limit the original propagator is recovered. This is the spirit of Pauli–Villars's regularization.

Note that such a propagator cannot be derived from a hermitian hamiltonian. Indeed the hermiticity of the hamiltonian leads to the representation $(A2.8)$ for the 2-point function. If the propagator is, as above, a rational fraction, it must be a sum of poles with positive residues and thus cannot decrease faster than $1/p^2$.

More general modifications are possible. Schwinger's proper time method (Appendix A9) suggests:

$$\left(\frac{1}{p^2 + m^2}\right)_{\text{reg.}} = \int_0^\infty \mathrm{d}t\, \rho\left(t\Lambda^2\right) \mathrm{e}^{-t\left(p^2 + m^2\right)}, \qquad (9.2)$$

in which $\rho(t)$ satisfies the condition

$$\lim_{t \to \infty} \rho(t) = 1, \qquad (9.3)$$

and decreases fast enough when t goes to zero.

An example is:

$$\rho(t) = \theta(t - 1), \qquad (9.4)$$

$\theta\left(t\right)$ being the step function, which yields

$$\left(\frac{1}{p^2+m^2}\right)_{\text{reg.}} = \frac{1}{p^2+m^2}\exp\left[-\frac{1}{\Lambda^2}\left(p^2+m^2\right)\right]. \tag{9.5}$$

However the regularization has to satisfy one important condition: the regularized propagator should remain a *smooth* function of the momentum **p**. Indeed singularities in the momentum representation generate, after Fourier transformation, contributions to the the large distance behaviour of the propagator, and we want to modify the theory only at short distance.

As the example (9.5) shows, it is possible to find in this more general class, propagators which have no unphysical singularities; however there is another price to pay, the theory is no longer strictly local and there exists no corresponding hamiltonian formalism.

For fermions the same method is applicable. We can replace

$$(m+i\not p)^{-1} \text{ by } \left[m+i\not p\left(1+\alpha_1\frac{p^2}{\Lambda^2}+\cdots+\alpha_n\frac{p^{2n}}{\Lambda^{2n}}\right)\right]^{-1}. \tag{9.6}$$

Regulator fields. Let us note that Pauli–Villars's regularization has another, often equivalent, formulation based on the introduction of a set of regulator fields.

To regularize the action $S\left(\phi\right)$:

$$S\left(\phi\right) = \int \mathrm{d}^d x\left[\tfrac{1}{2}\phi\left(-\Delta+m^2\right)\phi + V\left(\phi\right)\right], \tag{9.7}$$

we introduce a new set of fields ϕ_k, $k=1,\ldots,n$, over which we integrate in the functional integral, and consider the modified action $S_{\text{reg.}}\left(\phi,\phi_k\right)$:

$$S_{\text{reg.}}\left(\phi,\phi_k\right) = \int \mathrm{d}^d x\left[\frac{1}{2}\phi\left(-\Delta+m^2\right)\phi - \sum_{k=1}^{n}\frac{1}{2C_k}\phi_k\left(-\Delta+M_k^2\right)\phi_k\right.$$
$$\left. + V\left(\phi+\sum_{1}^{n}\phi_k\right)\right]. \tag{9.8}$$

With action (9.8) any internal ϕ propagator is replaced after regularization by the sum of the ϕ propagator and all the ϕ_k propagators $-C_k/\left(p^2+M_k^2\right)$. To improve the large momentum behaviour of Feynman diagrams we impose constraints to the parameters C_k. Cancellation of the $1/p^2$ term in the expansion at large momentum of the sum of propagators requires immediately:

$$\sum_{k=1}^{n}C_k = 1. \tag{9.9}$$

This implies that at least one of the C_k is positive and therefore to this field ϕ_k cannot correspond a particle in the sense of Particle Physics: the action cannot be generated by a hermitian hamiltonian, for the same reason as the regularized propagator (9.1).

On expression (9.8) the large cut-off limit is a limit in which all masses M_k become infinite at fixed relative ratios.

It is possible to use a functional argument to show that regularizations (9.1) and (9.8) are equivalent. In the integral

$$\int [\mathrm{d}\phi] \prod_1^n [\mathrm{d}\phi_k] \exp\left[-S_{\mathrm{reg.}}(\phi, \phi_k)\right], \tag{9.10}$$

we first change variables:

$$\phi = \phi' - \sum_1^n \phi_k. \tag{9.11}$$

The regularized action then becomes:

$$S_{\mathrm{reg}}(\phi', \phi_k) = \int \mathrm{d}^d x \left[\frac{1}{2} \left(\phi' - \sum_k \phi_k \right) \left(-\Delta + m^2\right) \left(\phi' - \sum_k \phi_k \right) \right.$$
$$\left. - \sum_1^n \frac{1}{C_k} \phi_k \left(-\Delta + M_k^2\right) \phi_k + V(\phi') \right].$$

The integral over the fields ϕ_k is now gaussian and can be explicitly performed. An elementary calculation leads as expected to:

$$S_{\mathrm{reg}}(\phi) = \int [\mathrm{d}\phi] \exp\left[-\int \mathrm{d}^d x \left(\tfrac{1}{2} \phi \, S^{(2)} \phi + V(\phi) \right) \right], \tag{9.12}$$

$$\left[S^{(2)}\right]^{-1} = \left(-\Delta + m^2\right)^{-1} - \sum_k C_k \left(-\Delta + M_k^2\right)^{-1}. \tag{9.13}$$

It is possible to choose the coefficients C_k in such a way that:

$$\left[\frac{1}{-\Delta + m^2} - \sum_k \frac{C_k}{-\Delta + M_k^2} \right]^{-1} = \left(-\Delta + m^2\right) \prod_k \frac{\left(-\Delta + M_k^2\right)}{\left(-m^2 + M_k^2\right)}, \tag{9.14}$$

and the ϕ propagator becomes analogous to propagator (9.1).

Pauli–Villars's inspired regularizations have one important advantage, they can be chosen in such a way that a large number of symmetries of the initial action are preserved. However, in all models which already in quantum mechanics have divergences due to operator ordering problems (see Section 3.2) a class of Feynman diagrams cannot be regularized by this method. This includes, as we shall see later, models which have non-linear (see Chapter 14) or gauge symmetries (Chapter 19).

Let us show, however, that these methods work for simple scalar field theories. We assume that the regularized propagator $\Delta_\Lambda(p)$ behaves like:

$$\Delta_\Lambda(p) \sim \frac{1}{p^{2n}}.$$

Equation (8.23) gives the degree of divergence of a regularized diagram γ:

$$\delta(\gamma) = dL - 2nI + \sum_\alpha v_\alpha k_\alpha. \tag{9.15}$$

Using topological relation (6.57):

$$L = I - \sum_\alpha v_\alpha + 1,$$

we can rewrite equation (9.15)

$$\delta\left(\gamma\right) = (d - 2n)\,L + \sum_\alpha v_\alpha\,(k_\alpha - 2n) + 2n\,.$$

It is thus necessary to choose

$$2n > d \ \text{ and } \ 2n > \sup_\alpha k_\alpha,$$

since both L and v_α can increase indefinitely.

For example the choice

$$2n > d + \left(\sup_\alpha k_\alpha\right),$$

regularizes all diagrams.

Examples. (i) In the ϕ^3 field theory in 6 dimensions, $2n = 8$ renders all diagrams finite. (ii) In the ϕ^4 field theory in 4 dimensions, $2n = 6$ suffices.

9.2 Lattice Regularization

We shall meet examples in which Pauli–Villars's regularization does not work: field theories in which the action has a definite geometrical character like models on homogeneous spaces (for example the non-linear σ-model) or gauge theories. In these theories the forms of the propagator and of the interaction terms are not independent. When the propagator is regularized, new more singular interactions have to be added to the action to preserve some symmetry and, as we shall show on examples, some one-loop diagrams cannot be regularized. Other regularization methods are needed. In most cases lattice regularization, which we have introduced in Subsection 5.1.2 to define precisely the functional integral, can be used. The advantages are the following:

(i) Lattice regularization is the only regularization which has a meaning outside perturbation theory that we know. For instance the regularized functional integral can be calculated by numerical methods, like stochastic methods (Monte Carlo calculations) or strong coupling expansions.

(ii) It preserves most global or local symmetries with the exception of the space $O(d)$ symmetry which is replaced by a hypercubic symmetry (but this is not too serious as will be argued in Chapters 24 and 25) and fermion chirality, which turns out to be a deeper problem.

The main disadvantage is that it leads to very complicated perturbative calculations.

Example of a scalar field theory. Let us examine the effect of the regularization on the action:

$$S\left(\phi\right) = \int d^d x \left[\tfrac{1}{2}\left(\partial_\mu\phi\right)^2 + \tfrac{1}{2}m^2\phi^2 + V\left(\phi\right)\right].$$

Let a be the lattice spacing. The derivative $\partial_\mu \phi$ can be replaced by the finite difference:

$$\partial_\mu \phi \mapsto \nabla_\mu \phi = \frac{1}{a}\left[\phi\left(x + an_\mu\right) - \phi(x)\right],$$

n_μ being the unit vector in the μ direction.

The action $S(\phi)$ becomes

$$S_{\text{reg}}(\phi) = a^d \sum_{x\in a\mathbb{Z}^d}\left[\tfrac{1}{2}\sum_{\mu=1}^d [\nabla_\mu \phi(x)]^2 + \tfrac{1}{2}m^2\phi^2(x) + V\left(\phi(x)\right)\right]. \qquad (9.16)$$

The arguments of the Fourier transform $\tilde{\phi}(p)$ of the field are now cyclic variables p_μ:

$$\tilde{\phi}(p) = \left(\frac{a}{2\pi}\right)^d \sum_{x\in a\mathbb{Z}^d}\phi(x)\,\mathrm{e}^{-i\mathbf{p\cdot x}} \Leftrightarrow \phi(x) = \int \mathrm{d}^d p\, \tilde{\phi}(p)\,\mathrm{e}^{i\mathbf{p\cdot x}}, \qquad (9.17)$$

$$\text{with}\quad -\frac{\pi}{a} \le p_\mu < \frac{\pi}{a};\quad \mu = 1,\dots,d.$$

The domain of variation of momenta is often called the first Brillouin zone.

In Fourier components the quadratic part of the action is:

$$a^d \sum_x \tfrac{1}{2}\left[(\nabla_\mu\phi)^2 + m^2\phi^2\right] = (2\pi)^d \int \mathrm{d}^d p\, \tfrac{1}{2}\tilde{\phi}(p)\,S^{(2)}(p)\,\tilde{\phi}(-p),$$

in which the function $S^{(2)}(p)$ is:

$$S^{(2)}(p) = m^2 + \sum_{\mu=1}^d \left|\frac{\left(\mathrm{e}^{iap\cdot n_\mu}-1\right)}{a}\right|^2.$$

The corresponding propagator $\Delta_a(p)$ is thus:

$$\Delta_a(p) = \left[m^2 + \frac{2}{a^2}\sum_{\mu=1}^d (1-\cos(ap_\mu))\right]^{-1}. \qquad (9.18)$$

In momentum representation the Feynman diagrams are now periodic functions of all momenta with period $2\pi/a$. In the small lattice spacing limit we recover the usual propagator:

$$\Delta_a(p)^{-1} = m^2 + p^2 - \tfrac{1}{12}\sum_\mu a^2 p_\mu^4 + O\left(p_\mu^6\right). \qquad (9.19)$$

Hypercubic symmetry implies $O(d)$ symmetry at order p^2.

Fermion theories. Let us briefly indicate a few problems arising when fermions are present in the action. We consider the free fermion action:

$$S(\bar{\psi},\psi) = \int \mathrm{d}^d x\, \bar{\psi}(x)\left(\slashed{\partial} + m\right)\psi(x).$$

If we want to regularize this action by a lattice and preserve parity (see Sections A5.5, A5.6) we can replace $\partial_\mu \psi(x)$ by

$$\frac{1}{2a}\left(\psi\left(x + an_\mu\right) - \psi\left(x - an_\mu\right)\right).$$

Then after Fourier transformation the action becomes:

$$S = (2\pi)^d \int d^d p\, \bar{\psi}\left(-p\right)\left(i \sum_\mu \gamma_\mu \frac{\sin ap_\mu}{a} + m\right)\psi\left(p\right), \tag{9.20}$$

and the fermion propagator is:

$$\Delta\left(p\right) = \left(m + i \sum_\mu \gamma_\mu \frac{\sin ap_\mu}{a}\right)^{-1}. \tag{9.21}$$

The problem with this propagator is that the equations relevant to the small lattice spacing limit:

$$\sin(a\,p_\mu) = 0$$

have each two solutions $p_\mu = 0$ and $p_\mu = \pi/a$ within one period, i.e. within the Brillouin zone. Therefore the propagator (9.21) propagates 2^d fermions. To remove this degeneracy it is possible to add to the regularized action an additional scalar term δS involving second derivatives:

$$\delta S\left(\bar{\psi}, \psi\right) = \frac{M}{2} \sum_{x,\mu} \left[2\bar{\psi}(x)\psi(x) - \bar{\psi}\left(x + an_\mu\right)\psi(x) - \bar{\psi}(x)\psi\left(x + an_\mu\right)\right]. \tag{9.22}$$

After Fourier transformation the new fermion propagator reads:

$$\Delta\left(p\right) = \left[m + M \sum_\mu \left(1 - \cos ap_\mu\right) - \frac{i}{a} \sum_\mu \gamma_\mu \sin ap_\mu\right] D^{-1}\left(p\right),$$

with:

$$D\left(p\right) = \left[m + M \sum_\mu \left(1 - \cos ap_\mu\right)\right]^2 + \frac{1}{a^2}\left(\sum_\mu \sin ap_\mu\right)^2.$$

Therefore we have broken the degeneracy between the different states. For each component p_μ which takes the value π/a the mass is increased by M. By choosing M of order $1/a$ we eliminate the spurious states in the continuum limit. This is the recipe of Wilson's fermions.

However a serious problem arises if one wants to construct a theory with massless fermions and chiral symmetry (equation $(A5.33)$)

$$\psi'(x) = e^{i\theta\gamma_5}\psi(x), \qquad \bar{\psi}'(x) = \bar{\psi}(x)\,e^{i\theta\gamma_5}.$$

Then both the mass term and the term (9.22) are excluded.

Of course one could think modifying the fermion propagator by adding terms connecting fermions separated by more than one lattice spacing. The propagator would then have the form:

$$\Delta\left(p\right) = \left[\sum_{\mu}\gamma_{\mu}f_{\mu}\left(p\right)\right]^{-1},$$

in which $f_{\mu}\left(p\right)$ is a smooth function (singularities introduce violation of locality), periodic with period $2\pi/a$, which vanishes linearly for $|\mathbf{p}|$ small

$$f_{\mu}\left(\mathbf{p}\right) \sim p_{\mu}, \quad \text{for } |\mathbf{p}| \to 0.$$

In one dimension it is easy to understand that this modification does not solve the problem: if $f\left(p\right)$ is periodic and continuous, it has to vanish linearly an even number of times in each period. This argument can be generalized to any dimension. This doubling of the number of fermion degrees of freedom is related to the problem of anomalies (see Chapter 20).

Since the most naive form of the propagator yields 2^d fermion states, one tries in practical calculations to reduce this number to a smaller multiple of two. The idea of staggered fermions introduced by Kogut and Susskind is often used: first by modifying the action one is able to decrease the multiplication factor from 2^d to $2^{d/2}$ with respect to form (9.20). Then the remaining degeneracy is interpreted as the reflection of an internal symmetry $SU(2^{d/2})$ of the action. The discussion is slightly involved and will not be given here.

More recently, however, a decisive advance on the problem of chiral fermions has been achieved, following the discovery of lattice Dirac operators solutions of the Ginsparg–Wilson relation.

9.3 Dimensional Regularization

In contrast with lattice regularization, dimensional regularization seems to have no meaning outside perturbation theory since it involves continuation of Feynman diagrams in the parameter d (d is the space dimension) to arbitrary complex values. However this regularization very often leads to the simplest perturbative calculations.

It is, however, not applicable when some essential property of the field theory is specific to the initial dimension. An example is provided by violations of parity symmetry involving the complete antisymmetric tensor $\epsilon_{\mu_1\cdots\mu_d}$, for example through the relation between γ_S (called γ_5 in 4 dimensions) and the other γ matrices (see Subsection 9.3.3):

$$d!\,\gamma_S = i^{-d/2}\epsilon_{\mu_1\ldots\mu_d}\gamma_{\mu_1}\cdots\gamma_{\mu_d}.$$

Before discussing the dimensional continuation of an arbitrary Feynman diagram, let us explain the idea on a particular example.

A simple example. Let us first consider the integral I

$$I = \int \mathrm{d}^d q\, f\left(q^2\right). \tag{9.23}$$

Invariance under rotations allows us to integrate over angular variables:

$$I = \frac{2\pi^{d/2}}{\Gamma\left(d/2\right)}\int_0^{\infty}\mathrm{d}q\,q^{d-1}f\left(q^2\right). \tag{9.24}$$

If $f(x)$ is an analytic function and if there exists a domain in the d complex plane for which the integral converges, then expression (9.24) gives the continuation to any dimension of integral (9.23).

However if we consider an integral involving several momenta, we need a more systematic method.

We shall therefore define an integral in d dimensions by a set of conditions which, if d is an integer, lead to the usual integral.

9.3.1 Defining properties of d dimensional integrals

Let us define the dimensional continuation of integrals by the three conditions:

$$
\begin{array}{lll}
\text{(i)} & \int d^d p\, F(p+q) = \int d^d p\, F(p) & \text{translation} \\[2mm]
\text{(ii)} & \int d^d p\, F(\lambda p) = |\lambda|^{-d} \int d^d p\, F(p) & \text{dilatation} \\[2mm]
\text{(iii)} & \int d^d p\, d^{d'} q\, f(p)\, g(q) = \int d^d p\, f(p) \int d^{d'} q\, g(q) & \text{factorization}
\end{array}
$$

Let us show that these simple rules provide a dimensional continuation to Feynman diagrams.

From property (iii) we derive:

$$
\int d^d p\, e^{-tp^2} = \left(\int dp_1\, e^{-tp_1^2} \right)^d = \left(\frac{\pi}{t} \right)^{d/2}. \tag{9.25}
$$

We now write any scalar propagator $\Delta(\mathbf{p})$ as:

$$
\Delta(\mathbf{p}) = \int_0^\infty dt\, \rho(t)\, e^{-\mathbf{p}^2 t}. \tag{9.26}
$$

For example:

$$
\rho(t) = e^{-tm^2} \Rightarrow \Delta(\mathbf{p}) = \left(\mathbf{p}^2 + m^2 \right)^{-1}.
$$

Using this representation and the property that any momentum on a line is a linear combination of loop momenta and external momenta, we can write a scalar Feynman diagram γ with constant vertices under the form:

$$
I_\gamma(\mathbf{p}) = (2\pi)^{-Ld} \int \prod_{i=1}^I dt_i\, \rho_i(t_i) \prod_{\ell=1}^L d^d q_\ell\, \exp\left[-\sum_1^L \mathbf{q}_\ell \cdot \mathbf{q}_{\ell'} M_{\ell\ell'}(t_i) \right.
$$
$$
\left. -2\sum_1^L \mathbf{q}_\ell \cdot \mathbf{k}_\ell(\mathbf{p}, t_i) - S(\mathbf{p}, t_i) \right]. \tag{9.27}
$$

Properties (i), (ii) and equation (9.25) allow us to integrate over all loop momenta q_ℓ and we find:

$$
I_\gamma(\mathbf{p}) = (4\pi)^{-Ld/2} \int \prod_{i=1}^I dt_i\, \rho_i(t_i)\, (\det \mathbf{M})^{-d/2} \exp\left[\sum_1^L \mathbf{k}_\ell \left(M^{-1} \right)_{\ell\ell'} \mathbf{k}_{\ell'} - S(\mathbf{p}, t_i) \right].
$$
$$
\tag{9.28}
$$

In this expression the dependence in d is completely explicit, and therefore continuation in d (at generic momenta if the theory is massless) can be achieved.

Example: One-loop contribution to the 2-point function in the massless ($m = 0$) ϕ^3 field theory

$$
\begin{aligned}
I_\gamma(\mathbf{p}) &= \int \frac{d^d q}{(2\pi)^d} \frac{1}{\mathbf{q}^2 (\mathbf{p}+\mathbf{q})^2} \\
&= \int dt_1 dt_2 \frac{d^d q}{(2\pi)^d} e^{-t_1 \mathbf{q}^2 - t_2 (\mathbf{p}+\mathbf{q})^2} \\
&= \int_0^\infty dt_1 \, dt_2 \frac{d^d q}{(2\pi)^d} \exp\left[-(t_1 + t_2)\left(\mathbf{q} + \frac{t_2 \mathbf{p}}{t_1 + t_2}\right)^2 - \frac{t_1 t_2}{t_1 + t_2}\mathbf{p}^2 \right] \\
&= \frac{\pi^{d/2}}{(2\pi)^d} \int_0^\infty \frac{dt_1 dt_2}{(t_1 + t_2)^{d/2}} \exp\left(-\frac{t_1 t_2}{t_1 + t_2}\mathbf{p}^2\right).
\end{aligned}
$$

For $\mathbf{p} \neq 0$, the integral converges for $2 < d < 4$.

We can complete the calculation by setting

$$
t_1 = ts, \qquad t_2 = (1 - t)\, s\,.
$$

The integral becomes:

$$
I_\gamma = \frac{\pi^{d/2}}{(2\pi)^d} \int_0^1 dt \int_0^\infty ds \, s^{1-d/2} e^{-st(1-t)\mathbf{p}^2}\,.
$$

We integrate over s:

$$
I_\gamma = \frac{\pi^{d/2}}{(2\pi)^d} (\mathbf{p}^2)^{(d/2)-2} \Gamma\left(2 - \frac{d}{2}\right) \int_0^1 dt \, [t(1-t)]^{(d/2)-2}\,,
$$

and therefore finally:

$$
I_\gamma = \frac{\pi^{d/2}}{(2\pi)^d} \Gamma(2 - d/2) \frac{\Gamma^2((d/2) - 1)}{\Gamma(d - 2)} (\mathbf{p}^2)^{(d/2)-2}\,. \tag{9.29}
$$

This expression has a pole at $d = 2$ corresponding to IR (low momentum) singularities because the theory is massless and poles at $d = 4, 6, ...$ which clearly are consequences of the UV (large momentum) divergences of the Feynman diagram.

It is interesting to explain on this example the interplay between dimensional continuation and cut-off regularization. If we regularize the propagator, for example by the method of Pauli–Villars, the Feynman diagram I_γ becomes a regular function of d for $d > 2$ up to some even integer larger than 4.

In the neighbourhood of $d = 4$, it has the form:

$$
I_\gamma \sim \frac{1}{8\pi^2 (4 - d)} \left[(p^2)^{(d/2)-2} - \Lambda^{d-4} \right] \quad \text{for } d \sim 4\,.
$$

If, at d fixed $d < 4$, we sent the cut-off to infinity we obtain the continuation of the initial diagram with a pole at $d = 4$. If at cut-off fixed we take the limit $d = 4$, we get a finite result in which $\ln \Lambda$ has replaced the pole in $(d - 4)$.

Important remark

The property (ii) implies:

$$\int \mathrm{d}^d p = 0, \quad \int \frac{\mathrm{d}^d p}{p^2} = 0 \ldots \tag{9.30}$$

for these integrals which exist for no value of d.

A first important consequence of this property is the following: if we consider models in which the hamiltonian density contains terms which are products of the field $\phi(x)$ and its conjugated momentum $\pi(x)$ taken, due to locality, at the same point x, the question of operator ordering arises. However equation (5.5) shows that the commutator of these operators is of the form:

$$[\pi(x), \phi(x)] = \frac{\hbar}{i} \delta^{d-1}(0) = \frac{\hbar}{i} (2\pi)^{1-d} \int \mathrm{d}^{d-1} p \,,$$

and therefore vanishes identically in dimensional regularization for any dimension $d > 1$, i.e. except in quantum mechanics. The precise quantization rule and the order of operators are irrelevant in dimensional regularization. This feature will be used in the quantization of models defined on homogeneous manifolds whose hamiltonians have the generic form (3.24) (see Chapters 14, 15).

Let us point out another consequence. The result:

$$\int \frac{\mathrm{d}^d p}{p^2} = 0 \,,$$

can be interpreted as a cancelation between an UV and IR divergence:

$$\int \frac{\mathrm{d}^d p}{p^2} = \frac{2\pi^{d/2}}{\Gamma(d/2)} \left[\int_1^\infty p^{d-3} \mathrm{d}p + \int_0^1 p^{d-3} \mathrm{d}p \right] \,.$$

However this property of dimensional continuation has one dangerous consequence: in a field theory involving massless fields, having for example a propagator $1/p^2$, IR divergences appear in 2 dimensions. If this theory is renormalizable in 2 dimensions, it has also UV divergences. In such a case UV and IR singularities get mixed. Therefore to be able to identify poles coming from the large momentum region, it is then necessary to introduce an IR cut-off, for example by giving a mass to the field.

Finally since the property (9.30) of dimensional regularization is at first sight somewhat strange, let us check its consistency (guaranteed *a priori* by the consistency of the defining rules) on an explicit example:

$$I = \int \frac{\mathrm{d}^d p}{p^2 (p^2 + 1)} \,,$$

which can be calculated in two ways: first one notes

$$I = \int \mathrm{d}^d p \left[\frac{1}{p^2} - \frac{1}{p^2 + 1} \right] = - \int \frac{\mathrm{d}^d p}{p^2 + 1} \,.$$

This yields:

$$I = -\pi^{d/2} \Gamma(1 - d/2) \,.$$

Second, one uses the transformation (9.26) on the initial expression. It is easy to verify that one obtains the same result.

9.3.2 Continuation of tensorial structures

Up to now we have considered diagrams corresponding to scalar fields. Any diagram which is not a scalar can be expanded on a set of tensors with scalar coefficients. For example:

$$\int d^d q \, q_\mu q_\nu f \left(\mathbf{q}^2, \mathbf{p}^2, \mathbf{p} \cdot \mathbf{q}\right) = A \left(\mathbf{p}^2\right) p_\mu p_\nu + B \left(\mathbf{p}^2\right) \delta_{\mu\nu} . \tag{9.31}$$

The scalar diagrams contributing to $A\left(\mathbf{p}^2\right)$ and $B\left(\mathbf{p}^2\right)$ can be obtained by taking the trace and the scalar product with p_μ:

$$\begin{cases} A\left(\mathbf{p}^2\right) = \dfrac{1}{d-1}\dfrac{1}{\left(p^2\right)^2} \int d^d q \left[d\left(\mathbf{p}\cdot\mathbf{q}\right)^2 - \mathbf{p}^2\mathbf{q}^2\right] f, \\[4mm] B\left(\mathbf{p}^2\right) = \dfrac{1}{d-1}\dfrac{1}{p^2} \int d^d q \left[-\left(\mathbf{p}\cdot\mathbf{q}\right)^2 + \mathbf{p}^2\mathbf{q}^2\right] f. \end{cases} \tag{9.32}$$

We have reduced the problem to the calculation of integrals of the form (9.27) with additional factors polynomial in momenta. The integration over momenta can then also be performed to yield the continuation in d.

9.3.3 Fermions

For fermions belonging to the spinorial representation of $O(d)$ the strategy is the same. The spin problem can be reduced to the calculation of traces of γ matrices. Therefore only an additional prescription for the trace of the unit matrix is needed. There is no natural continuation since odd and even dimensions behave differently. However we have shown in Appendix A5.8 that no algebraic manipulation depends on the explicit value of the trace. Thus any smooth continuation in the neighbourhood of the relevant dimension will be satisfactory. A convenient choice which we shall always adopt is to take the trace constant. In even dimension as long as only γ_μ matrices are involved no other problem arises. However if the diagrams involve γ_S and if it becomes necessary to use the identity

$$d! \gamma_S = i^{-d/2} \epsilon_{\mu_1 \cdots \mu_d} \gamma_{\mu_1} \cdots \gamma_{\mu_d} ,$$

then no dimensional continuation which preserves all properties of γ_S exists. As we have seen before, lattice regularization is equally impossible in this case. This difficulty is the source of *chiral anomalies*.

Since we have to calculate traces, one possibility is to define γ_S by its expression in terms of γ_μ matrices in the initial dimension; for example if we start from 4 dimensions we define $\gamma_5 \equiv \gamma_S$ as the product:

$$4! \gamma_5 = -\epsilon_{\mu_1 \cdots \mu_4} \gamma_{\mu_1} \cdots \gamma_{\mu_4} .$$

It is then easy to verify that, with this definition, γ_5 anticommutes with all other γ_μ matrices only in 4 dimensions . If for example we start from dimension n (n even) and evaluate the product $\gamma_\nu \gamma_S \gamma_\nu$ in d dimensions, we find:

$$\gamma_\nu \gamma_S \gamma_\nu = (d - 2n)\gamma_S .$$

Anticommuting properties of the γ_S would have led to a factor $-d$ instead of $d - 2n$ in the r.h.s..

9.3.4 Dimensional regularization and UV divergences

The principle of dimension regularization is the following: we define all diagrams by analytic continuation in the dimension parameter d as explained before. When the dimension d approaches the initial dimension, Feynman diagrams become singular as a consequence of the original UV divergences. The singular contributions can be isolated by performing a Laurent expansion of the diagram. For example expression (9.29) is the value of a Feynman diagram of the massless ϕ^4 field theory which is renormalizable in $d = 4$. The Laurent expansion is:

$$I_\gamma = N_d \left[\frac{1}{4-d} + \frac{1}{2} - \frac{1}{2} \ln p^2 + O(d-4) \right].$$

As we have implicitly done above, we shall in general include a factor

$$N_d = \frac{2\pi^{d/2}}{(2\pi)^d \, \Gamma(d/2)},$$

product of the surface of the sphere S_{d-1} by $(2\pi)^{-d}$, in the definition of the loopwise expansion parameter, because it is generated naturally by each loop integration. As we have already shown in an example, powers of $\ln \Lambda$ (Λ being the cut-off) which would appear in a cut-off regularization in the large Λ limit, are replaced by powers of $1/(d-4)$. However as a consequence of identity (9.30), no divergent contribution equivalent to a power of Λ can appear.

Bibliographical Notes

For the Pauli–Villars's regularization see
   W. Pauli and F. Villars, *Rev. Mod. Phys.* 21 (1949) 434.
For the proper time method see
   J. Schwinger, *Phys. Rev.* 82 (1951) 664.
The consistency of the lattice regularization is proven (except for theories with chiral fermions) in
   T. Reisz, *Commun. Math. Phys.* 117 (1988) 79, 639.
The doubling phenomenon for lattice fermions has been proven quite generally by
   H.B. Nielsen and M. Ninomiya, *Nucl. Phys.* B185 (1981) 20.
Wilson's solution to the fermion doubling problem is described in
   K.G. Wilson in *New Phenomena in Subnuclear Physics*, Erice 1975,
   A. Zichichi ed. (Plenum, New York 1977).
Staggered fermions have been proposed in
   T. Banks, L. Susskind and J. Kogut, *Phys. Rev.* D13 (1976) 1043.
Discussion of solutions of the Ginsparg–Wilson relation
   P.H. Ginsparg and K.G. Wilson, *Phys. Rev.* D25 (1982) 2649,
and consequences can be found in
   H. Neuberger, *Phys. Lett.* B417 (1998) 141, hep-lat/9707022, *ibidem* B427 (1998) 353, hep-lat/9801031; P. Hasenfratz, V. Laliena, F. Niedermayer, *Phys. Lett.* B427 (1998) 125, hep-lat/9801021; M. Lüscher, *Phys. Lett.* B428 (1998) 342, hep-lat/9802011, hep-lat/9811032.
Dimensional regularization has been introduced by:
   J. Ashmore, *Lett. Nuovo Cimento* 4 (1972) 289; G. 't Hooft and M. Veltman, *Nucl. Phys.* B44 (1972) 189; C.G. Bollini and J.J. Giambiagi, *Phys. Lett.* 40B (1972) 566, *Nuovo Cimento* 12B (1972).

Its use in problems with chiral anomalies has been proposed in

D.A. Akyeampong and R. Delbourgo, *Nuovo Cimento* 17A (1973) 578.

For an early review see

G. Leibbrandt, *Rev. Mod. Phys.* 47 (1975) 849.

For other schemes see also

E.R. Speer in *Renormalization Theory*, Erice 1975, G. Velo and A.S. Wightman eds. (D. Reidel, Dordrecht, Holland 1976).

Exercises

Exercise 9.1

We consider the example of the $O(N)$ invariant action:

$$S(\phi) = \int \mathrm{d}^d x \left(\tfrac{1}{2} \partial_\mu \phi \partial_\mu \phi + \tfrac{1}{2} m^2 \phi^2 + \frac{1}{4!} \frac{g}{N} (\phi^2)^2 \right),$$

where ϕ is a N-component vector.

9.1.1. Write the field equation of motion (see Chapter 5) satisfied by the generating functional of correlation functions.

9.1.2. Deduce a relation between the two and four-point functions.

9.1.3. As an approximation, assume now that the two correlation functions correspond to an unknown quadratic action (i.e. to a gaussian functional measure) and calculate the two-point function.

Exercise 9.2

Repeat the whole analysis (equation of motion and self-consistent free action approximation) for the so-called Gross–Neveu model (GN) discussed in Appendix A30.1, which is a model of N massless self-interacting Dirac fermions $\{\psi^i, \bar{\psi}^i\}$. The $U(N)$ symmetric action is:

$$S(\bar{\psi}, \psi) = - \int \mathrm{d}^d x \left[\bar{\psi} \cdot \partial\!\!\!/ \psi + \tfrac{1}{2} g \left(\bar{\psi} \cdot \psi \right)^2 \right].$$

The GN model has in even dimensions a discrete chiral symmetry:

$$\psi \mapsto \gamma_S \psi, \quad \bar{\psi} \mapsto -\bar{\psi} \gamma_S,$$

which forbids a fermion mass term.

APPENDIX 9
SCHWINGER'S PROPER TIME REGULARIZATION

In Section 9.1 we have alluded to Schwinger's proper time regularization. Let us show here that it leads to a compact representation of a regularized one-loop contribution to the generating functional of proper vertices $\Gamma(\varphi)$. In Section 6.4 we have shown that $\Gamma_{1\,\text{loop}}(\varphi)$ is given by:

$$\Gamma_{1\,\text{loop}}(\varphi) = \frac{1}{2}\,\text{tr}\left[\ln\frac{\delta^2 S}{\delta\varphi(x_1)\,\delta\varphi(x_2)} - \ln\left.\frac{\delta^2 S}{\delta\varphi(x_1)\,\delta\varphi(x_2)}\right|_{\varphi=0}\right]. \qquad (A9.1)$$

The quantity $\delta^2 S/\delta\varphi(x_1)\,\delta\varphi(x_2)$ is a local operator. For example if $S(\phi)$ is:

$$S(\phi) = \int \mathrm{d}^d x \left[\tfrac{1}{2}(\partial_\mu\phi)^2 + \tfrac{1}{2}m^2\phi^2 + g\frac{\phi^N}{N!}\right], \qquad (A9.2)$$

it takes the form of a simple quantum mechanical hamiltonian H:

$$H \equiv \frac{\delta^2 S}{\delta\varphi(x_1)\,\delta\varphi(x_2)} = \left[-\Delta + m^2 + \frac{g}{(N-2)!}\phi^{N-2}(x_1)\right]\delta(x_1 - x_2). \qquad (A9.3)$$

Let us also define H_0:

$$H_0 \equiv \left.\frac{\delta^2 S}{\delta\varphi\delta\varphi}\right|_{\varphi=0} = \left(-\Delta + m^2\right)\delta(x_1 - x_2). \qquad (A9.4)$$

Using then the general identity:

$$\text{tr}\,(\ln H - \ln H_0) = -\int_0^\infty \frac{\mathrm{d}t}{t}\,\text{tr}\left(\mathrm{e}^{-tH} - \mathrm{e}^{-tH_0}\right), \qquad (A9.5)$$

we rewrite $\Gamma_{1\,\text{loop}}(\varphi)$ as:

$$\Gamma_{1\,\text{loop}}(\varphi) = -\frac{1}{2}\int_0^\infty \frac{\mathrm{d}t}{t}\,\text{tr}\left(\mathrm{e}^{-tH} - \mathrm{e}^{-tH_0}\right). \qquad (A9.6)$$

Expression $(A9.6)$ is the representation of the one-loop functional as an integral over Schwinger's proper time. Large momentum divergences appear as divergences at $t = 0$. If we multiply the integrand by a cutting factor $\rho(t\Lambda^2)$ we recover the regularization of equation (9.2).

Expression $(A9.6)$, once regularized, provides a compact representation of the one-loop functional which often leads to a rapid evaluation of its divergent part. The general idea is to write Schrödinger's equation for the matrix elements $\langle x\,|\mathrm{e}^{-tH}|\,x'\rangle$ and use it to derive their small t expansion. This is a problem we have already encountered in Section 2.2. To determine the trace, we only need the coefficients of the expansion in the limit $x = x'$.

Let us for illustration purpose expand up to order t^3 in the case of action $(A9.2)$. We first solve an auxiliary problem. We consider the Schrödinger equation:

$$[-\Delta_x + V(x)]\langle x\,|\mathrm{e}^{-tH}|\,x'\rangle = -\frac{\partial}{\partial t}\langle x\,|\mathrm{e}^{-tH}|\,x'\rangle. \qquad (A9.7)$$

We set:

$$\langle x \left| \mathrm{e}^{-tH} \right| x' \rangle = \mathrm{e}^{-\sigma(x,x';t)} \, . \tag{A9.8}$$

The Schrödinger equation then takes the form:

$$\Delta\sigma - (\partial_\mu\sigma)^2 + V(x) = \frac{\partial\sigma}{\partial t} \, . \tag{A9.9}$$

We expand σ for t small:

$$\sigma = \frac{1}{4t}(x-x')^2 + \frac{d}{2}\ln 4\pi t + At + Bt^2 + Ct^3 + O\left(t^4\right), \tag{A9.10}$$

and obtain equations for the coefficients A, B and C

$$\begin{cases} A + (x-x')_\mu \, \partial_\mu A = V(x), \\ 2B + (x-x')_\mu \, \partial_\mu B = \Delta A \, , \\ 3C + (x-x')_\mu \, \partial_\mu C = \Delta B - (\partial_\mu A)^2 \, . \end{cases} \tag{A9.11}$$

The equation requires the calculation of A up to order $(x-x')^2$:

$$\begin{cases} A = V\left(\tfrac{1}{2}(x-x')\right) + \dfrac{1}{24}(x-x')_\mu \, (x-x')_\nu \, \partial_\mu\partial_\nu V + O\left(|x-x'|\right)^3 \, , \\ B = \dfrac{1}{6}\Delta V(x) + O\left(|x-x'|\right), \\ C = -\dfrac{1}{12}\left(\partial_\mu V(x)\right)^2 + \Delta B + O\left(|x-x'|\right) \, . \end{cases} \tag{A9.12}$$

We do not need ΔB at this order because it disappears in the trace. We calculate now the regularized expression:

$$\Gamma_{\varepsilon 1\text{ loop}}(\varphi) = -\frac{1}{2} \int_\varepsilon^\infty \frac{\mathrm{d}t}{t} \operatorname{tr}\left(\mathrm{e}^{-tH} - \mathrm{e}^{-tH_0}\right). \tag{A9.13}$$

To compare with other regularizations we can eventually set $\varepsilon = 1/\Lambda^2$.

Since we are here only interested in divergences, we only calculate contributions coming from the lower bound:

$$\begin{aligned} I_\varepsilon = {}& -\frac{1}{2}\frac{1}{(4\pi)^{d/2}} \int_\varepsilon \frac{\mathrm{d}t}{t^{1+d/2}} \int \mathrm{d}^d x \Big\{ \left[m^2 - A(x,x)\right] t + \frac{1}{2}\left[A^2(x,x) - B(x,x) - m^4\right] t^2 \\ & - \frac{1}{6}\left[A^3(x,x) - 6AB + 6C - m^6\right] t^3 \Big\} + O\left(t^4\right). \end{aligned} \tag{A9.14}$$

Replacing A, B, C by expressions $(A9.12)$ and integrating over t, we then find

$$\begin{aligned} I_\varepsilon = {}& \frac{1}{2}\frac{1}{(4\pi)^{d/2}} \Bigg\{ -\frac{\varepsilon^{1-d/2}}{1-d/2} \int \mathrm{d}^d x \left(V(x) - m^2\right) + \frac{1}{2}\frac{\varepsilon^{2-d/2}}{2-d/2} \int \mathrm{d}^d x \left(V^2(x) - m^4\right) \\ & - \frac{1}{6}\frac{\varepsilon^{3-d/2}}{3-d/2} \int \mathrm{d}^d x \left[V^3(x) - m^6 + \tfrac{1}{2}\left(\partial_\mu V(x)\right)^2\right] \Bigg\} + \cdots . \end{aligned} \tag{A9.15}$$

When d is an even integer, $\varepsilon^0/0$ has to be replaced by $\ln(1/\varepsilon)$. This expression gives all divergences for $d \leq 6$.

For example let us apply this result to the ϕ^3 field theory in 6 dimensions:

$$\Gamma^{\text{div}}_{1\,\text{loop}}(\varphi) = \frac{1}{2^7\pi^3} \int d^6x \left\{ \frac{\Lambda^4}{2} g\varphi(x) - \frac{\Lambda^2}{2} \left[g^2\varphi^2(x) + gm^2\varphi(x) \right] \right.$$
$$\left. + \frac{1}{3}\ln\frac{\Lambda}{m} \left[g^3\varphi^3(x) + 3g^2m^2\varphi^2(x) + 3gm^4\varphi(x) + \frac{g^2}{2}(\partial_\mu\varphi(x))^2 \right] \right\}. \quad (A9.16)$$

This expression should be compared with expressions (8.10,8.11). Because the regularizations are different, only the logarithmic divergent terms coincide.

We can also consider the ϕ^4 field theory in 4 dimensions:

$$\Gamma^{\text{div}}_{1\,\text{loop}} = \frac{1}{32\pi^2} \left\{ \frac{\Lambda^2}{2} g \int d^4x\, \varphi^2(x) - \ln\frac{\Lambda}{m} \int d^4x \left[\frac{g^2}{4}\varphi^4(x) + gm^2\varphi^2(x) \right] \right\}. \quad (A9.17)$$

An identical expression will be recovered in Section 10.2, equation (10.8). Both in equations $(A9.16)$ and $(A9.17)$, we have defined the divergent part of $\Gamma(\varphi)$ as the sum of the divergent terms in the asymptotic expansion for Λ/m large. Finally we can also apply equation $(A9.15)$ to $d = 2$. If $V_I(\phi(x))$ is the interaction term, we find:

$$\Gamma^{\text{div}}_{1\,\text{loop}} = \frac{1}{4\pi}\ln\frac{\Lambda}{m} \int d^2x\, V_I''(\varphi(x)). \quad (A9.18)$$

Although in those examples the results can be easily recovered from the Feynman graph expansion, in more complicated cases, in which symmetries play an essential role, this method to evaluate divergences of one-loop diagrams can be quite useful.

ζ-*function regularization.* A variant of the previous regularization method is to replace expression $(A9.6)$ by:

$$\Gamma^{\text{reg}}_{1\,\text{loop}}(\varphi) = -\frac{1}{2\Gamma(1+\sigma)} \int_0^\infty dt\, t^{\sigma-1} \operatorname{tr}\left(e^{-tH} - e^{-tH_0} \right), \quad (A9.19)$$

and to take, after analytic continuation in σ, the limit $\sigma = 0$ in the spirit of the dimensional regularization.

Let us consider for example again the ϕ^4 field theory in four dimensions and calculate $\Gamma_{1\,\text{loop}}$ per unit volume for a constant field φ:

$$\frac{1}{V}\Gamma^{\text{reg}}_{1\,\text{loop}}(\varphi) = -\frac{1}{2\Gamma(1+\sigma)} \int_0^\infty dt\, t^{\sigma-1} \int \frac{d^4p}{(2\pi)^4} \left[e^{-t(p^2+m^2+g\varphi^2/2)} \right.$$
$$\left. -(\varphi = 0) \right].$$

The integration over the momentum p is simple and yields:

$$\frac{1}{V}\Gamma^{\text{reg}}_{1\,\text{loop}}(\varphi) = -\frac{1}{32\pi^2\Gamma(1+\sigma)} \int_0^\infty dt\, t^{\sigma-3} \left[e^{-t(m^2+g\varphi^2/2)} -(\varphi = 0) \right].$$

The integration over t can then also be performed

$$\frac{1}{V}\Gamma^{\text{reg}}_{1\,\text{loop}}(\varphi) = -\frac{1}{32\pi^2\sigma(1-\sigma)(2-\sigma)} \left[(m^2 + g\varphi^2/2)^{(2-\sigma)} - (\varphi = 0) \right].$$

Expanding this expression for σ small and keeping only the divergent and finite parts, we finally obtain:

$$\frac{1}{V}\Gamma^{\text{reg}}_{1\,\text{loop}}(\varphi) = -\frac{1}{64\pi^2}\left(m^2 + \frac{g}{2}\varphi^2 \right)^2 \left[-\frac{1}{\sigma} + \ln\left(m^2 + \frac{g}{2}\varphi^2 \right) - \frac{3}{2} \right] - (\varphi = 0).$$

The coefficient of the divergent part differs by a factor 2 from the one obtained in dimensional regularization. This is due to the choice of the normalization of σ.

10 INTRODUCTION TO RENORMALIZATION THEORY. RENORMALIZATION GROUP EQUATIONS

We shall not enter into an extensive and general discussion of renormalization theory, but instead present the essential steps of the proof of the renormalizability of a simple scalar field theory: the ϕ^4 field theory in $d = 4$ dimensions. However, all the fundamental difficulties of renormalization theory are already present in this particular example and it will eventually become clear that the extension to other theories is not difficult. We have followed the elegant presentation of Callan (Les Houches 1975) which allows renormalizability and renormalization group (Callan–Symanzik) equations to be proved at once. This presentation is specially suited to our general purpose since a large part of this work is devoted to applications of renormalization group (RG). Moreover, it emphasizes already at this technical level the equivalence between renormalizability and the existence of a renormalization group.

One drawback of our proof of renormalizability is that it directly applies only to massive theories and the existence of a massless theory requires a specific discussion. Section 10.9 is devoted to this problem. A different form (homogeneous) of RG equations follows.

Finally Section 10.11 contains a few remarks about the covariance of RG functions.

In the appendix we briefly outline another method which more cleanly separates the small and large momentum region and had been employed to give another proof of renormalizability. It relies on a partial integration of large momentum modes. Finally we discuss some super-renormalizable theories and the background field method.

10.1 Power Counting. Renormalized Action

We consider the local action for a scalar field $\phi(x)$ in four dimensions:

$$S(\phi) = \int \mathrm{d}^4 x \left(\tfrac{1}{2} \partial_\mu \phi \partial_\mu \phi + \tfrac{1}{2} m^2 \phi^2 + \frac{1}{4!} g \phi^4 \right). \tag{10.1}$$

In Section 8.3 we have shown that the ϕ^4 vertex has dimension zero in four dimensions and this action is thus renormalizable from the power counting point of view: the superficial degree of divergence of correlation functions is independent of the order in perturbation theory.

Power counting. For a proper vertex with n external lines (1PI n-point correlation function) the degree of divergence δ is:

$$\delta = 4 - n.$$

We also consider insertion of the $\phi^2(x)$ operator. The degree of divergence of the function $\Gamma^{(l,n)}$:

$$\Gamma^{(l,n)}(y_1, \ldots, y_l; x_1, \ldots, x_n) = 2^{-l} \left\langle \phi^2(y_1) \ldots \phi^2(y_l) \phi(x_1) \ldots \phi(x_n) \right\rangle_{1\mathrm{PI}},$$

is (equation (8.34)):

$$\delta = 4 - n - 2l.$$

Fig. 10.1   The ϕ-field 2-point function: $l = 0$, $n = 2$, $\delta = 2$.

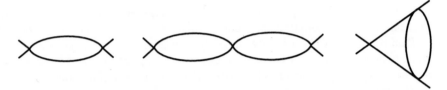

Fig. 10.2   The ϕ-field 4-point function: $l = 0$, $n = 4$, $\delta = 0$.

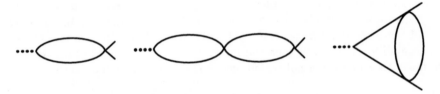

Fig. 10.3   The $\langle \phi^2 \phi \phi \rangle$ correlation function: $l = 1$, $n = 2$, $\delta = 0$.

Fig. 10.4   The $\langle \phi^2 \rangle$ average which is a constant: $l = 1$, $n = 0$, $\delta = 2$.

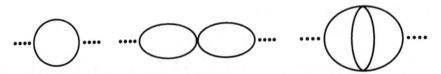

Fig. 10.5   The $\langle \phi^2 \phi^2 \rangle$ correlation function: $l = 1$, $n = 0$, $\delta = 0$.

In figures 10.1–5 are listed the superficially divergent functions and the corresponding first few diagrams.

The diagrams with $n = 0$ never arise as subdiagrams since a subdiagram has necessarily external ϕ lines. Their renormalization can be discussed separately.

To give a meaning to perturbation theory we now replace the action $S(\phi)$ by a regularized action $S_\Lambda(\phi)$, using one of the regularization schemes presented in Chapter 9. In what follows we explicitly refer to a cut-off because it is more in the spirit of this work, and some arguments would have to be slightly modified in the case of dimensional regularization.

Dimensional analysis. In what follows we systematically use mass dimensional analysis. Let us rescale in action (10.1), after regularization, mass and cut-off by a factor ρ and calculate the generating functional $W(\rho, J)$ of correlation functions of $\phi(x/\rho)$:

$$m, \Lambda \mapsto \rho m, \rho\Lambda, \qquad \int \mathrm{d}^4 x \, \phi(x) J(x) \mapsto \int \mathrm{d}^4 x \, \phi(x/\rho) J(x).$$

We then make in the functional integral the change of variables on the field:

$$\phi(x) = \rho \phi'(\rho x).$$

The source term becomes:

$$\int \mathrm{d}^4 x \, \phi(x/\rho) \, J(x) = \int \mathrm{d}^4 x \, \phi'(x) \left[\rho J(x) \right].$$

After the change of variables $x' = \rho x$, the action (10.1) resumes its initial form (the dummy variables ϕ' and x' being renamed ϕ and x). Since the source term is now multiplied by a factor ρ, the generating functional $W(\rho, J)$ satisfies:

$$W(\rho, J) = W(J\rho).$$

For the Legendre transform $\Gamma(\varphi)$ this implies:

$$\Gamma(\rho, \varphi) = \Gamma(\varphi\rho^3).$$

If we now expand $\Gamma(\varphi)$ in powers of φ we find:

$$\Gamma^{(n)}(x_1/\rho, \ldots, x_n/\rho, \rho m, \rho\Lambda) = \rho^{3n} \Gamma^{(n)}(x_1, \ldots, x_n, m, \Lambda).$$

After Fourier transformation and factorization of the δ-function of momentum conservation the relation becomes:

$$\Gamma^{(n)}(\rho p_1, \ldots, \rho p_n, \rho m, \rho\Lambda) = \rho^{4-n} \Gamma^{(n)}(p_1, \ldots, p_n, m, \Lambda).$$

We find a mass dimension $4-n$ which is also the dimension in the sense of power counting of $\Gamma^{(n)}$. The argument generalizes to $\Gamma^{(l,n)}$.

The renormalized action: counterterms. We now consider the regularized action $S_\Lambda(\phi)$. We introduce two *renormalized parameters* m_r and g_r. We then want to show that it is possible to rescale the field ϕ:

$$\phi \mapsto Z^{1/2}(\Lambda, m_\mathrm{r}, g_\mathrm{r}) \, \phi_\mathrm{r},$$

and to choose the *bare parameters* m and g as functions of m_r, g_r and Λ in such a way that all ϕ_r correlation functions have a finite limit, order by order in the loopwise expansion, when the cut-off Λ becomes infinite at m_r and g_r fixed. The field ϕ_r is called the renormalized field.

More precisely let us introduce the notion of *renormalized action* $S_r(\phi_r)$ which is the original action expressed in terms of renormalized variables:

$$S_\Lambda(\phi) \equiv S_r(\phi_r) = \int \mathrm{d}^d x \left[\tfrac{1}{2}\phi_r \left(m_r^2 - \Delta\right)_\Lambda \phi_r + \frac{1}{4!}g_r\phi_r^4 \right.$$
$$\left. + \tfrac{1}{2}(Z-1)\,\partial_\mu\phi_r\partial_\mu\phi_r + \tfrac{1}{2}\delta m^2\phi_r^2 + \frac{1}{4!}g_r(Z_g-1)\phi_r^4 \right]. \qquad (10.2)$$

The notation $\left(m_r^2 - \Delta\right)_\Lambda$ is there to remind that the action is regularized. For Pauli–Villars's regularization it means:

$$\left(m_r^2 - \Delta\right)_\Lambda = m_r^2 - \Delta + \alpha_1\frac{\Delta^2}{\Lambda^2} - \alpha_2\frac{\Delta^3}{\Lambda^4} + \cdots.$$

The identity between the renormalized action (10.2) and the regularized action (10.1), called the *bare action*, is expressed by the set of relations between renormalized and bare quantities:

$$\begin{cases} \phi = Z^{1/2}\phi_r, \\ g = g_r Z_g/Z^2, \\ m^2 = \left(m_r^2 + \delta m^2\right)/Z. \end{cases} \qquad (10.3)$$

Z and Z_g/Z^2 are respectively called the field amplitude and coupling constant renormalization constants, δm^2 characterizes the mass renormalization.

In action (10.2) we have explicitly separated a *tree level action*:

$$S_\Lambda(\phi_r) = \int \mathrm{d}^4 x \left[\tfrac{1}{2}\phi_r \left(m_r^2 - \Delta\right)_\Lambda \phi_r + \frac{1}{4!}g_r\phi_r^4 \right], \qquad (10.4)$$

from the set of *counterterms* parametrized in terms of the renormalization constants δm^2, Z_g, and Z which are formal series in g_r:

$$\begin{cases} \delta m^2 = a_1(\Lambda)\,g_r + a_2(\Lambda)\,g_r^2 + \cdots \\ Z_g = 1 + b_1(\Lambda)\,g_r + b_2(\Lambda)\,g_r^2 + \cdots \\ Z = 1 + c_1(\Lambda)\,g_r + c_2(\Lambda)\,g_r^2 + \cdots. \end{cases} \qquad (10.5)$$

We want to prove that the coefficients $a_n(\Lambda)$, $b_n(\Lambda)$ and $c_n(\Lambda)$ can be chosen in such a way that all correlation functions have a finite large Λ limit, order by order in g_r.

10.2 One-Loop Divergences

We have already shown in Section 8.1 that the ϕ_6^3 (the lower index is the space dimension) field theory could be renormalized at one-loop order. Let us here repeat the argument in the case of the ϕ_4^4 field theory.

First we note that if we rescale the field ϕ_r:

$$\phi_r \mapsto \phi_r/\sqrt{g_r},$$

the dependence in g_r is factorized in front of the tree level action:

$$S_\Lambda\left(\phi_r\right)=\frac{1}{g_r}\int d^4x\left[\tfrac{1}{2}\left(\sqrt{g_r}\phi_r\right)\left(m_r^2-\Delta\right)_\Lambda\left(\sqrt{g_r}\phi_r\right)+\frac{1}{4!}\left(\sqrt{g_r}\phi_r\right)^4\right].\qquad(10.6)$$

In the ϕ^4 field theory, the loopwise expansion is an expansion in powers of g_r at $\left(\sqrt{g_r}\phi_r\right)$ fixed.

We now write the generating functional of proper vertices $\Gamma\left(\varphi\right)$ at one-loop order:

$$\Gamma\left(\varphi_r\right)=\Gamma_0\left(\varphi_r\right)+g_r\Gamma_1\left(\varphi_r\right)+O\left(g_r^2\right).$$

At the tree level the counterterms by definition do not contribute and therefore:

$$\Gamma_0\left(\varphi_r\right)=\lim_{\Lambda\to\infty}S_\Lambda\left(\phi_r\right)=\int d^4x\left[\tfrac{1}{2}\left(\partial_\mu\varphi_r\right)^2+\tfrac{1}{2}m_r^2\varphi_r^2+\frac{1}{4!}g_r\varphi_r^4\right].\qquad(10.7)$$

At one-loop order the contributions coming from the tree level action are (equation (6.49)):

$$\tfrac{1}{2}\operatorname{tr}\ln\left[1+\left(m_r^2-\Delta\right)_\Lambda^{-1}\frac{g_r}{2}\varphi_r^2\right].$$

Expanding in powers of φ_r^2 we find two divergent diagrams which are represented in figure 10.6.

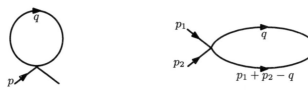

Fig. 10.6   One-loop divergent diagrams.

Using the regularized propagator $e^{-p^2/\Lambda^2}\left(p^2+m_r^2\right)^{-1}$ we can calculate the divergent contributions:

(i) $n=2$: the coefficient (a) of $g_r\varphi_r^2$:

$$\frac{1}{4}\,(a)=\frac{1}{4}\int\frac{d^4q}{(2\pi)^4}\frac{e^{-q^2/\Lambda^2}}{(q^2+m_r^2)}=\frac{1}{32\pi^2}\left(\frac{\Lambda^2}{2}-m_r^2\ln\frac{\Lambda}{m_r}\right)+O\left(\Lambda^0\right).$$

(ii) $n=4$: the coefficient (b) of $g_r^2\varphi_r^4$:

$$-\frac{1}{16}(b)=-\frac{1}{16}\int\frac{d^4q}{(2\pi)^4}\frac{e^{-[q^2+(q-p_1-p_2)^2]/\Lambda^2}}{(q^2+m_r^2)\left[(q-p_1-p_2)^2+m_r^2\right]}\sim-\frac{1}{128\pi^2}\ln\frac{\Lambda}{m_r}.$$

The divergent part $\Gamma_1^{\mathrm{div}}\left(\varphi\right)$ of $\Gamma_1\left(\varphi\right)$ in the absence of counterterms is thus:

$$\Gamma_1^{\mathrm{div}}=\frac{1}{32\pi^2}\int d^4x\left[\left(\frac{\Lambda^2}{2}-m_r^2\ln\frac{\Lambda}{m_r}\right)g\left(\varphi_r(x)\right)^2-\frac{1}{4}\ln\frac{\Lambda}{m_r}g^2\left(\varphi_r(x)\right)^4\right].\qquad(10.8)$$

Note the absence of a term proportional to $\int d^4x \, (\partial_\mu \varphi_r)^2$. This is a peculiarity of the ϕ^4 field theory at one-loop order.

Let us now take into account the counterterms at one-loop order:

$$\frac{1}{2}(Z-1)(\partial_\mu \phi_r)^2 + \frac{1}{2}\delta m^2 \phi_r^2 + \frac{1}{4!} g_r (Z_g - 1) \phi_r^4 \Big|_{\text{one loop}}$$

$$= \frac{1}{2} c_1(\Lambda) g_r (\partial_\mu \phi_r)^2 + \frac{1}{2} a_1(\Lambda) g_r \phi_r^2 + \frac{1}{4!} b_1(\Lambda) g_r^2 \phi_r^4 \,.$$

At leading order these counterterms contribute additively to $\Gamma(\varphi_r)$. We can choose them to eliminate the divergences and therefore renormalize the ϕ^4 field theory at one-loop order. Of course the condition of finiteness of correlation functions determines only the divergent part of the counterterms. It is possible to add for example to a_1, b_1 and c_1 any arbitrary finite constant. The difference between the 1PI correlation functions corresponding to two different choices is of the form of the tree level functions, i.e. a constant for the 4-point function in Fourier space and a first degree polynomial in p^2, p being the momentum, for the 2-point function. In this chapter it is convenient to determine the renormalization constants by a set of renormalization conditions imposed to the renormalized 1PI correlation functions:

$$\Gamma(\varphi_r) = \sum \frac{1}{n!} \int d^4x_1 \dots d^4x_n \, \Gamma_r^{(n)}(x_1, \dots, x_n) \, \varphi_r(x_1) \dots \varphi_r(x_n) \,,$$

which in Fourier space read:

$$\begin{cases} \tilde{\Gamma}_r^{(2)}(p=0) = m_r^2 \,, \\[2mm] \dfrac{\partial}{\partial p^2} \tilde{\Gamma}_r^{(2)}(p)\big|_{p=0} = 1 \,, \\[2mm] \tilde{\Gamma}_r^{(4)}(0,0,0,0) = g_r \,. \end{cases} \tag{10.9}$$

We have of course chosen conditions consistent with the tree approximation. Since we have three renormalization constants, by imposing three conditions to the superficially divergent correlation functions, we determine them completely. The renormalization constants are then given by:

$$\begin{cases} a_1(\Lambda) = -\dfrac{1}{2} \int \dfrac{d^4q}{(2\pi)^4} \dfrac{e^{-q^2/\Lambda^2}}{(q^2+m_r^2)} = -\dfrac{1}{16\pi^2}\left(\dfrac{\Lambda^2}{2} - m_r^2 \ln \dfrac{\Lambda}{m_r}\right) + O(\Lambda^0) \,, \\[4mm] b_1(\Lambda) = \dfrac{3}{2} \int \dfrac{d^4q}{(2\pi)^4} \dfrac{e^{-2q^2/\Lambda^2}}{(q^2+m_r^2)^2} = \dfrac{3}{16\pi^2} \ln \dfrac{\Lambda}{m_r} + O(\Lambda^0) \,, \\[4mm] c_1(\Lambda) = 0 \,. \end{cases} \tag{10.10}$$

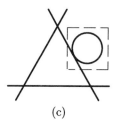
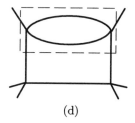

(c)   (d)

Fig. 10.7

10.3 Divergences Beyond One-Loop: Skeleton Diagrams

At any order in perturbation theory the 2-point function is quadratically divergent and the 4-point function logarithmically divergent as power counting shows. However at higher orders a new difficulty, which we have already mentioned in Section 8.2, arises: superficially convergent diagrams have divergent subdiagrams. Let us take the example of the 6-point function: at one-loop order it is given by a convergent diagram, but at two-loops the diagrams shown in figure 10.7 appear.

We recognize inside the dashed boxes divergent subdiagrams. However we identify them with one-loop divergences of the 2-point function (c) and the 4-point function (d), for which counterterms have already been provided. Indeed at this order a diagram (c') appears in which the one-loop counterterm for the 2-point function is inserted on a propagator and another one (d') in which the vertex of the tree level action is replaced by the one-loop counterterm of the 4-point function (figure 10.8).

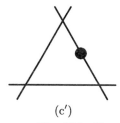
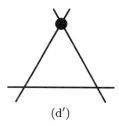

(c')   (d')

Fig. 10.8   Two-loop contributions from one-loop counterterms.

We want to show that this is generally true, i.e. that counterterms which render the divergent functions finite, at higher orders also cancel the divergence of subdiagrams of superficially convergent functions.

For this purpose we introduce the notion of *skeleton diagram*: a skeleton diagram is a really convergent diagram, i.e. it is superficially convergent and has no divergent subdiagram.

For example the one-loop diagrams of the $2n$-point functions, $n > 2$, all are skeleton diagrams (see figure 10.9).

An arbitrary superficially convergent diagram, can then be obtained from a skeleton diagram by replacing all vertices by $\Gamma^{(4)}$ and all propagators by $(\Gamma^{(2)})^{-1}$ and expanding in powers of the coupling constant g_r.

For example the diagrams (c) and (d) are generated by the expansion of the *dressed skeleton diagram* of figure 10.10.

An important property is the following: if in a dressed skeleton diagram, $\Gamma^{(4)}$ and $\Gamma^{(2)}$ are replaced by the renormalized functions $\Gamma_r^{(4)}$ and $\Gamma_r^{(2)}$ the dressed skeleton diagram

Fig. 10.9

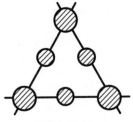

Fig. 10.10

is finite. This is a direct consequence of the following bounds on the large momentum behaviour:

$$
\left.
\begin{aligned}
&\left| \tilde{\Gamma}_{\mathrm{r}}^{(2)} (\lambda p) \right| \le \lambda^2 \times \ \text{power of} \ \ln \lambda, \\[2mm]
&\left| \tilde{\Gamma}_{\mathrm{r}}^{(4)} (\lambda p_i) \right| \le \ \text{power of} \ \ln \lambda, \\[2mm]
&\left| \tilde{\Gamma}_{\mathrm{r}}^{(1,2)} (\lambda q; \lambda p_1, \lambda p_2) \right| \le \ \text{power of} \ \ln \lambda,
\end{aligned}
\right\} \quad \text{at any finite order for } \lambda \to \infty. \qquad (10.11)
$$

We do not derive these bounds here but a few comments can be found at the end of Section 10.6. These bounds for the large momentum behaviour of the various renormalized functions, which are valid for arbitrary momenta, differ from the tree level behaviour only by powers of logarithms (at any finite order in the loop expansion). Therefore power counting arguments still apply and superficially convergent diagrams are then convergent.

Note that similar estimates exist for superficially convergent functions but are then valid only for generic momenta (see Sections 12.3–12.5).

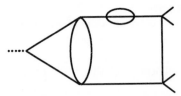

Fig. 10.11   Divergent contribution to $\Gamma^{(1,4)}$.

The bounds (10.11) together with the skeleton expansion completely reduce the problem of renormalization of superficially convergent proper vertices to the renormalization of the divergent proper vertices. The argument also applies to the proper vertices $\Gamma^{(l,n)}$ with ϕ^2 insertion. Let us for example consider the diagram of figure 10.11, which contributes to the superficially convergent function $\Gamma^{(1,4)}$: It has divergent subdiagrams and is generated from a skeleton diagram by replacing propagators by $(\Gamma^{(2)})^{-1}$, vertices by $\Gamma^{(4)}$ and the $\phi^2\phi\phi$ vertex by $\Gamma^{(1,2)}$ as shown in figure 10.12.

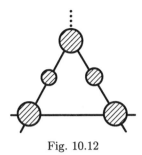

Fig. 10.12

In the next section we therefore examine the renormalization of $\Gamma^{(2)}$, $\Gamma^{(4)}$ and $\Gamma^{(1,2)}$. The diagrams contributing to these functions are superficially divergent but have also divergent subdiagrams corresponding to the divergence of the same functions at lower orders. Figure 10.13 provides an example.

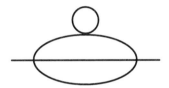

Fig. 10.13   Three-loop contribution to the 2-point function.

However another problem arises, the problem of overlapping divergences. Let us consider for example the diagram of figure 10.14.

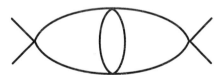

Fig. 10.14   Three-loop contribution to the 4-point function.

Figure 10.15 shows the set of divergent subdiagrams. The three subdiagrams have a common part. The concept of insertion of divergent diagrams of lower order is therefore no longer well-defined. This is the problem of so-called overlapping divergences.

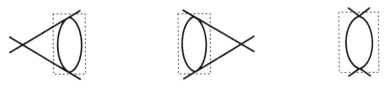

Fig. 10.15   Overlapping divergent subdiagrams.

In the next section we develop a specific technique to deal with this problem, based on a differentiation of the diagrams with respect to the mass.

10.4 Bare and Renormalized Correlation Functions, Operator ϕ^2 Insertions

Before we begin the discussion of the renormalization of the superficially divergent proper vertices a few remarks are in order.

Bare and renormalized correlation functions. By definition the bare correlation functions are those calculated with action (10.1) properly regularized. The renormalized correlation functions are calculated with action (10.2). The relation between bare and renormalized fields:

$$\phi = Z^{1/2}\phi_r \,,$$

implies immediately the relation:

$$W\left(J/\sqrt{Z}\right) = W_r\left(J\right), \tag{10.12}$$

in which $W\left(J\right)$ and $W_r\left(J\right)$ are respectively the generating functionals of connected bare and renormalized correlation functions. Since in the Legendre transformation J and φ are dual, it follows immediately:

$$\Gamma_r\left(\varphi\right) = \Gamma\left(\varphi\sqrt{Z}\right), \tag{10.13}$$

a relation which on the proper vertices translates into:

$$\Gamma_r^{(n)} = Z^{n/2}\Gamma^{(n)}. \tag{10.14}$$

Operator ϕ^2 insertions. Actually we need also the bare and renormalized ϕ^2 insertions. We therefore add to the bare action (10.1) a source term $\frac{1}{2}\int K(x)\phi^2(x)dx$:

$$S(\phi, K) = S(\phi) + \tfrac{1}{2}\int dx\, K(x)\phi^2(x). \tag{10.15}$$

We now consider the functional integral:

$$Z(J, K) = \int [d\phi]\exp\left[-S\left(\phi, K\right) + \int dx\, J(x)\phi(x)\right]. \tag{10.16}$$

Functional differentiation with respect to $K(x)$ generates insertions of the operator $-\frac{1}{2}\phi^2(x)$ (Section 8.4). In the same way if we consider the action $S_r\left(\phi_r, K\right)$:

$$S_r\left(\phi_r, K\right) = S_r\left(\phi_r\right) + \tfrac{1}{2}Z_2\int K(x)\phi_r^2(x)dx\,, \tag{10.17}$$

in which Z_2 is a new renormalization constant, the functional integral:

$$Z_r\left(J, K\right) = \int [d\phi_r]\exp\left[-S_r\left(\phi_r, K\right) + \int dx\, J(x)\phi_r(x)\right], \tag{10.18}$$

generates the renormalized correlation functions with $-\frac{1}{2}\phi^2$ insertions (we temporarily normalize $Z(J, K)$ to $Z(0, K) = 1$ to eliminate the pure $\phi^2(x)$ correlation functions).

The relation between renormalized and bare functionals is then:

$$W_{\text{r}}(J, K) = W\left(J/\sqrt{Z}, KZ_2/Z\right). \tag{10.19}$$

After Legendre transformation this relation becomes:

$$\Gamma_{\text{r}}(\varphi, K) = \Gamma\left(\varphi\sqrt{Z}, KZ_2/Z\right), \tag{10.20}$$

or in terms of proper vertices:

$$\Gamma_{\text{r}}^{(l,n)}(y_1, \ldots, y_l; x_1, \ldots, x_n) = Z^{(n/2)-l}Z_2^l\Gamma^{(l,n)}(y_1, \ldots, y_l; x_1, \ldots, x_n). \tag{10.21}$$

Since we now have a new superficially divergent function $\Gamma^{(1,2)}$ and a new renormalization constant, we impose an additional renormalization condition:

$$\tilde{\Gamma}_{\text{r}}^{(1,2)}(q = 0; p_1 = p_2 = 0) = 1, \tag{10.22}$$

which again is consistent with the tree approximation.

Let us here note that if the source K is a constant then, according to equation (10.15), it generates just a shift of m^2, the bare mass squared:

$$S(\phi, K) = \int \mathrm{d}^4x \left[\frac{1}{2}(\partial_\mu\phi)^2 + \frac{1}{2}(m^2 + K)\phi^2 + \frac{1}{4!}g\phi^4\right]. \tag{10.23}$$

Therefore in this case differentiation with respect to K is equivalent to differentiation with respect to m^2. On the other hand if $K(x)$ is a constant, its Fourier transform is proportional to $\delta(p)$, which means that it generates correlation functions with ϕ^2 insertions at zero momentum:

$$\left.\frac{\partial}{\partial m^2}\right|_{g,\Lambda} \tilde{\Gamma}^{(l,n)}(q_1, \ldots, q_l; p_1, \ldots, p_n) = \tilde{\Gamma}^{(l+1,n)}(0, q_1, \ldots, q_l; p_1, \ldots, p_n). \tag{10.24}$$

This equation has a diagrammatic interpretation: the diagrams contributing to the r.h.s. are obtained from the diagrams contributing to $\Gamma^{(l,n)}$ by doubling a propagator in all possible ways (up to a sign). In figure 10.16 we give the example of $\Gamma^{(4)}$ and $\Gamma^{(1,4)}$.

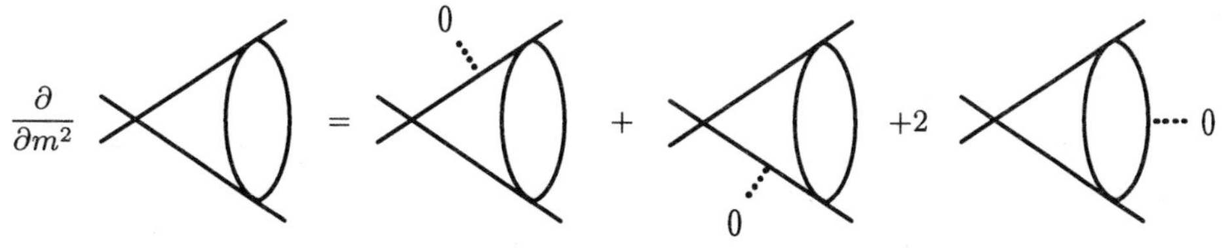

Fig. 10.16

By differentiating with respect to the bare mass, we are able to relate superficially divergent correlation functions to functions which have a skeleton expansion. Furthermore, at a given number of loops a function which has a skeleton expansion is only expressed in terms of divergent functions which have at least *one loop less*. We see there a mechanism to prove renormalizability by induction. However we want to insert into the skeleton expansion renormalized proper vertices. This introduces some additional difficulties which we will discover once we have transformed equation (10.24) into an equation for the renormalized proper vertices.

10.5 Callan–Symanzik Equations

The starting point of our analysis is equation (10.24) which shows that by differentiating with respect to the bare mass we improve the large momentum behaviour of Feynman diagrams. We first translate equation (10.24) into an equation for renormalized functions. Let us rewrite it under the form:

$$
m_{\mathrm{r}} \left. \frac{\partial}{\partial m_{\mathrm{r}}} \right|_{g,\Lambda} \tilde{\Gamma}^{(l,n)} = \left(m_{\mathrm{r}} \left. \frac{\partial}{\partial m_{\mathrm{r}}} \right|_{g,\Lambda} m^2 \right) \frac{\partial}{\partial m^2} \tilde{\Gamma}^{(l,n)} = \left(m_{\mathrm{r}} \left. \frac{\partial}{\partial m_{\mathrm{r}}} \right|_{g,\Lambda} m^2 \right) \tilde{\Gamma}^{(l+1,n)} (0,\ldots).
$$
(10.25)

We can then use relation (10.21), replace bare by renormalized functions in equation (10.25) and apply chain rule to transform differentiation with respect to m_{r} at g and Λ fixed, into differentiation at g_{r} and Λ fixed.

We now introduce a set of definitions, and immediately take into account dimensional analysis:

$$
m_{\mathrm{r}} \left. \frac{\partial}{\partial m_{\mathrm{r}}} \right|_{g,\Lambda} g_{\mathrm{r}} = \beta \left(g_{\mathrm{r}}, \frac{m_{\mathrm{r}}}{\Lambda} \right),
$$
(10.26)

$$
m_{\mathrm{r}} \left. \frac{\partial}{\partial m_{\mathrm{r}}} \right|_{g,\Lambda} \ln Z \left(g_{\mathrm{r}}, \frac{m_{\mathrm{r}}}{\Lambda} \right) = \left(m_{\mathrm{r}} \frac{\partial}{\partial m_{\mathrm{r}}} + \beta \frac{\partial}{\partial g_{\mathrm{r}}} \right) \ln Z = \eta \left(g_{\mathrm{r}}, \frac{m_{\mathrm{r}}}{\Lambda} \right),
$$
(10.27)

$$
m_{\mathrm{r}} \left. \frac{\partial}{\partial m_{\mathrm{r}}} \right|_{g,\Lambda} \ln (Z_2/Z) = \left(m_{\mathrm{r}} \frac{\partial}{\partial m_{\mathrm{r}}} + \beta \frac{\partial}{\partial g_{\mathrm{r}}} \right) \ln (Z_2/Z) = \eta_2 \left(g_{\mathrm{r}}, \frac{m_{\mathrm{r}}}{\Lambda} \right),
$$
(10.28)

$$
Z Z_2^{-1} \left(m_{\mathrm{r}} \left. \frac{\partial}{\partial m_{\mathrm{r}}} \right|_{g,\Lambda} m^2 \right) = m_{\mathrm{r}}^2 \sigma \left(g_{\mathrm{r}}, \frac{m_{\mathrm{r}}}{\Lambda} \right).
$$
(10.29)

In terms of the differential operator D_{CS}:

$$
\mathrm{D}_{\mathrm{CS}} = m_{\mathrm{r}} \frac{\partial}{\partial m_{\mathrm{r}}} + \beta \left(g_{\mathrm{r}}, \frac{m_{\mathrm{r}}}{\Lambda} \right) \frac{\partial}{\partial g_{\mathrm{r}}} - \frac{n}{2} \eta \left(g_{\mathrm{r}}, \frac{m_{\mathrm{r}}}{\Lambda} \right) - l \eta_2 \left(g_{\mathrm{r}}, \frac{m_{\mathrm{r}}}{\Lambda} \right),
$$

we find:

$$
\mathrm{D}_{\mathrm{CS}} \tilde{\Gamma}_{\mathrm{r}}^{(l,n)} (q_1, \ldots, q_l; p_1, \ldots, p_n) = m_{\mathrm{r}}^2 \sigma \left(g_{\mathrm{r}}, \frac{m_{\mathrm{r}}}{\Lambda} \right) \tilde{\Gamma}_{\mathrm{r}}^{(l+1,n)} (0, q_1, \ldots, q_l; p_1, \ldots, p_n).
$$
(10.30)

Equation (10.30), in the infinite cut-off limit, then yields an equation for the renormalized proper vertices, first derived by Callan and Symanzik, called therefore the Callan–Symanzik (CS) equation, which will play a central role in the part of this work devoted to phase transitions (Chapters 25–36).

To prove renormalizability, we shall prove inductively on the number of loops both the existence of the CS equation and the finiteness of correlation functions.

Note that the CS equation in the form (10.30) expresses only that we have rescaled the correlation functions and made an arbitrary change of parametrization. To be able to prove that the renormalized correlation functions have a finite cut-off limit it is necessary to determine the renormalization constants and therefore to impose on equation (10.30) the consequences of renormalization conditions (10.9) and (10.22):

(i) $n = 2, l = 0$

At zero momentum we obtain:

$$\left[m_r \frac{\partial}{\partial m_r} - \eta \left(g_r, \frac{m_r}{\Lambda} \right) \right] m_r^2 = m_r^2 \sigma \left(g_r, \frac{m_r}{\Lambda} \right).$$

The function σ is related to η:

$$\sigma = 2 - \eta. \tag{10.31}$$

If we then differentiate with respect to momentum squared we find:

$$-\eta = m_r^2 \left(2 - \eta \right) \frac{\partial}{\partial p^2} \tilde{\Gamma}_r^{(1,2)} \left(0; p, -p \right) \Big|_{p^2=0}. \tag{10.32}$$

(ii) $n = 4, l = 0$

At zero momentum we obtain:

$$\beta - 2g_r \eta = m_r^2 \left(2 - \eta \right) \tilde{\Gamma}_r^{(1,4)} \left(0; 0, 0, 0, 0 \right). \tag{10.33}$$

(iii) $n = 2, l = 1$

Again at zero momentum we get:

$$-\eta - \eta_2 = m_r^2 \left(2 - \eta \right) \tilde{\Gamma}_r^{(2,2)} \left(0, 0; 0, 0 \right). \tag{10.34}$$

We have related all coefficients of the partial differential equation (10.30) to values of proper vertices at zero momentum. From these relations it follows that if we can show that the renormalized proper vertices have a limit when the cut-off becomes infinite, the functions β, η and η_2 will also have a limit.

Note also that if we know the coefficients of the CS equations, we can calculate the renormalization constants from equations (10.26–10.29).

Preliminary remarks

(i) $\tilde{\Gamma}^{(1,2)} \left(0; p, -p \right)$ at order g does not depend on p:

$$\tilde{\Gamma}^{(1,2)} \left(0; p - p \right) = 1 - \frac{1}{2} \frac{g}{(2\pi)^4} \int \frac{\mathrm{d}^4 p}{\left(p^2 + m^2 \right)_\Lambda^2} + O \left(g^2 \right).$$

Equations (10.21,10.22) then imply:

$$Z^{-1} Z_2 = 1 + \frac{1}{2} \frac{g_r}{(2\pi)^4} \int \frac{\mathrm{d}^4 p}{\left(p^2 + m_r^2 \right)_\Lambda^2} + O \left(g_r^2 \right),$$

$$\tilde{\Gamma}_r^{(1,2)} \left(0; p, -p \right) = 1 + O \left(g_r^2 \right).$$

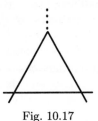

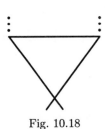

Fig. 10.17

Fig. 10.18

We conclude that the expansion of $\eta\,(g_{\mathrm{r}})$, which can be calculated from equation (10.32), begins at order g_{r}^2.

(ii) The first diagram contributing to $\tilde{\Gamma}^{(1,4)}$ is of order g_{r}^2 (see figure 10.17). It then follows from equation (10.33) and the previous remark that the expansion of the function β begins at order g_{r}^2. Thus the operator $\beta\partial/\partial g_{\mathrm{r}}$ which appears in the CS equation is of g_{r}.

(iii) The function $\tilde{\Gamma}^{(2,2)}$ has a first contribution of order g_{r} (see figure 10.18). Equation (10.34) shows that η_2 also begins at order g_{r}.

Cluster properties and analyticity at low momentum. In Section 8.1 we have used the regularity of the one-loop diagrams near zero momentum to show that the divergent contributions are polynomials in the momentum variables. This is more generally true: in a massive theory (with a mass $m \le \Lambda$), connected and 1PI correlation functions are analytic functions around $\mathbf{p} = 0$ as can be seen on the expression of regularized Feynman diagrams. This property, which will again be needed in the inductive proof, implies cluster properties: connected correlation functions decrease exponentially in space for large separations of the arguments (for details see Appendix A7.1).

10.6 Inductive Proof of Renormalizability

We have shown in Section 10.2 that we could render the theory one-loop finite. We now assume we have shown that the correlation functions defined by equations (10.21) and renormalization conditions (10.9,10.22) have a finite limit up to loop order L, at m_{r} and g_{r} fixed, when the cut-off Λ becomes infinite. This means that $\tilde{\Gamma}_{\mathrm{r}}^{(2)}$, $\tilde{\Gamma}_{\mathrm{r}}^{(4)}$, $\tilde{\Gamma}_{\mathrm{r}}^{(1,2)}$ have an infinite cut-off limit up to order g_{r}^L, g_{r}^{L+1} and g_{r}^L respectively. As we have shown in Section 10.5, from equations (10.21,10.9,10.22) follow the CS equations (10.30) and the relations (10.31–10.34).

We now use CS equation (10.30) in the form:

$$m_{\mathrm{r}}\frac{\partial}{\partial m_{\mathrm{r}}}\tilde{\Gamma}_{\mathrm{r}}^{(l,n)} = \left(-\beta\frac{\partial}{\partial g_{\mathrm{r}}} + \frac{n}{2}\eta + l\eta_2\right)\tilde{\Gamma}_{\mathrm{r}}^{(l,n)} + m_{\mathrm{r}}^2\,(2-\eta)\,\tilde{\Gamma}_{\mathrm{r}}^{(l+1,n)}, \qquad (10.35)$$

and show that the r.h.s. is finite at $(L+1)$ loop order. We note that $\Gamma^{(l,n)}$ in the r.h.s. is only needed at loop order L because its coefficient is of order g_r. For $\Gamma^{(l+1,n)}$ two cases arise: either it has a skeleton expansion and is therefore finite at loop order $L+1$, or the CS equation has to be iterated. However before discussing the correlation functions in detail let us examine the coefficient functions.

10.6.1 Coefficients of the CS equation

(i) Because $\partial\tilde{\Gamma}^{(1,2)}/\partial p^2$ is of order g_r^2, equation (10.32) then implies that η is finite up to order g_r^L.

(ii) The function $\tilde{\Gamma}^{(1,4)}$ is superficially convergent. It has therefore a skeleton expansion. The first dressed skeleton diagram contributing to $\tilde{\Gamma}^{(1,4)}$ is shown in figure 10.19.

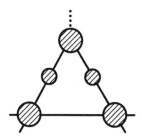

Fig. 10.19

If the functions $\tilde{\Gamma}_r^{(2)}$, $\tilde{\Gamma}_r^{(4)}$ and $\tilde{\Gamma}_r^{(1,2)}$ are finite up to L loops, $\tilde{\Gamma}_r^{(1,4)}$ is finite up to $(L+1)$ loop order, which means up to order g_r^{L+2}. Equation (10.33) then shows that the combination $\beta - 2g_r\eta$ is finite up to order g_r^{L+2}. Since η is finite up to order g_r^L, this implies that β is finite up to order g_r^{L+1}.

(iii) The function $\tilde{\Gamma}^{(2,2)}$ is also superficially convergent. It has a skeleton expansion. The first dressed skeleton is shown in figure 10.20.

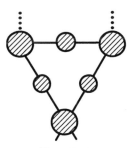

Fig. 10.20

The function $\tilde{\Gamma}_r^{(2,2)}$ is therefore also finite up to $(L+1)$ loop order which means up to order $(L+1)$ in g_r. Since $\tilde{\Gamma}^{(2,2)}$ is of order g_r, at order g_r^{L+1} the sum $\eta + \eta_2$ calculated from equation (10.34) involves only η at order g_r^L and $\tilde{\Gamma}_r^{(2,2)}$ at order g_r^{L+1} and thus is finite. The function η_2 is then finite up to order g_r^L.

We now prove that the functions $\tilde{\Gamma}_r^{(2)}$, $\tilde{\Gamma}_r^{(4)}$ and $\tilde{\Gamma}_r^{(1,2)}$ are, with the stated assumptions, finite up to $(L+1)$ loop order.

10.6.2 The $\langle\phi\phi\phi\phi\rangle$ correlation function ($l = 0$, $n = 4$)

Let us consider the coefficient of order g_r^{L+2} in equation (10.35):

$$m_r \frac{\partial}{\partial m_r} \left[\tilde{\Gamma}_r^{(4)} \right]_{L+2} = \left[\left(-\beta \frac{\partial}{\partial g_r} + 2\eta \right) \tilde{\Gamma}_r^{(4)} \right]_{L+2} + m_r^2 \left[(2 - \eta) \tilde{\Gamma}_r^{(1,4)} \right]_{L+2}.$$

Since $\beta \partial / \partial g_r$ is of order g_r and η of order g_r^2 for g_r small, in the first term of the r.h.s. we need $\tilde{\Gamma}_r^{(4)}$ only up to order g_r^{L+1} which is finite by assumption. Let us now separate in $\tilde{\Gamma}_r^{(4)}$ the leading term g_r and a remainder of order g_r^2. For the terms of order g_r^2 and higher we need η only up to order L and β up to order $L+1$ which are finite. The leading term in $\tilde{\Gamma}_r^{(4)}$ then involves the combination:

$$\left[\left(-\beta \frac{\partial}{\partial g_r} + 2\eta \right) g_r \right]_{L+2} = [-\beta + 2\eta g_r]_{L+2},$$

and we have shown above that $\beta - 2g_r\eta$ is finite up to order g_r^{L+2}.

Finally $\tilde{\Gamma}_r^{(1,4)}$ is finite up to order g_r^{L+2}. In addition its expansion in powers of g_r begins only at order g_r^2. Therefore the factor $(2 - \eta)$ is only needed up to order g_r^L. The conclusion is that $m_r \partial \tilde{\Gamma}_r^{(4)} / \partial m_r$ is finite at $(L + 1)$ loop order.

Perturbative integration of CS equations. By integrating equation (10.35) we now want to show that $\tilde{\Gamma}_r^{(4)}$ itself is finite at $(L + 1)$ loop order. The function $\tilde{\Gamma}_r^{(4)}$ is dimensionless, i.e. invariant in a dilatation

$$(\mathbf{p}, m_r, \Lambda) \mapsto (\rho\mathbf{p}, \rho m_r, \rho\Lambda),$$

(in what follows we denote by $\mathbf{p}$ all four momenta $\mathbf{p}_1...\mathbf{p}_4$).

Therefore we can write:

$$m_r \frac{\partial}{\partial m_r} \tilde{\Gamma}_r^{(4)} = f^{(4)} \left(\frac{\mathbf{p}}{m_r}, \frac{\Lambda}{m_r}, g_r \right), \tag{10.36}$$

with:

$$\begin{cases} \lim_{\Lambda\to\infty} f^{(4)} \left(\frac{\mathbf{p}}{m_r}, \frac{\Lambda}{m_r}, g_r \right) < \infty, \\ f^{(4)} \left(0, \frac{\Lambda}{m_r}, g_r \right) = m_r \frac{\partial}{\partial m_r} g_r = 0. \end{cases}$$

Let us now integrate equation (10.36) between m_r and Λ (Λ can be replaced by any scale proportional to Λ):

$$\tilde{\Gamma}_r^{(4)} \left(\frac{\mathbf{p}}{m_r}, \frac{\Lambda}{m_r}, g_r \right) = \tilde{\Gamma}_r^{(4)} \left(\frac{\mathbf{p}}{\Lambda}, 1, g_r \right) - \int_{m_r}^{\Lambda} \frac{d\rho}{\rho} f^{(4)} \left(\frac{\mathbf{p}}{\rho}, \frac{\Lambda}{\rho}, g_r \right).$$

Renormalization conditions (10.9) together with the regularity at low momentum imply:

$$\lim_{\Lambda\to\infty} \tilde{\Gamma}_r^{(4)} \left(\frac{\mathbf{p}}{\Lambda}, 1, g_r \right) = \tilde{\Gamma}_r^{(4)} (0, 1, g_r) = g_r.$$

The integral has a large cut-off limit if we restrict the domain of integration to $[m_r, \mu]$, $m_r \ll \mu \ll \Lambda$:

$$\lim_{\substack{\Lambda \to \infty \\ \mu \ll \Lambda}} \int_{m_r}^{\mu} \frac{d\rho}{\rho} f^{(4)} \left(\frac{\mathbf{p}}{\rho}, \frac{\Lambda}{\rho}, g_r \right) < \infty .$$

We have still to examine the values of ρ of order Λ. Then Λ/ρ is of order 1, while $\mathbf{p}/\rho$ is small. The integral depends on the small momentum behaviour of the 4-point correlation function. Again we use the property that in a massive theory (the mass is $\rho \sim \Lambda$) the correlation functions are analytic around $\mathbf{p} = 0$.

Since in addition $f^{(4)}$ vanishes at $\mathbf{p} = 0$:

$$\left| f^{(4)} \left(\frac{\mathbf{p}}{\rho}, \frac{\Lambda}{\rho}, g_r \right) \right| < \frac{C(\mathbf{p}, g_r)}{\rho^2} \quad \text{for} \quad \rho \gg |\mathbf{p}| ,$$

the remaining integral is then bounded by:

$$\int_{\mu}^{\Lambda} \frac{d\rho}{\rho} \left| f^{(4)} \left(\frac{\mathbf{p}}{\rho}, \frac{\Lambda}{\rho}, g_r \right) \right| < \frac{1}{2} C(\mathbf{p}, g_r) \left(\frac{1}{\mu^2} - \frac{1}{\Lambda^2} \right) .$$

We conclude that $\tilde{\Gamma}_r^{(4)}$ has a large cut-off limit at $(L+1)$ loop order given by:

$$\tilde{\Gamma}_r^{(4)} (\mathbf{p}, m_r, g_r) = g_r - \int_1^\infty \frac{d\rho}{\rho} f^{(4)} \left(\frac{\mathbf{p}}{\rho m_r}, g_r \right) , \tag{10.37}$$

both sides being expanded up to g_r^{L+2}. We now repeat the argument for $\tilde{\Gamma}_r^{(1,2)}$.

10.6.3 The $\langle \phi^2 \phi \phi \rangle$ correlation function $(l = 1, n = 2)$

We consider the term of order g_r^{L+1} in equation (10.35):

$$m_r \frac{\partial}{\partial m_r} \left[\tilde{\Gamma}_r^{(1,2)} \right]_{L+1} = - \left[\beta \frac{\partial}{\partial g_r} \tilde{\Gamma}_r^{(1,2)} \right]_{L+1} + \left[(\eta + \eta_2) \tilde{\Gamma}_r^{(1,2)} \right]_{L+1} + m_r^2 \left[(2 - \eta) \tilde{\Gamma}_r^{(2,2)} \right]_{L+1} .$$

The first term in the r.h.s. involves $\tilde{\Gamma}_r^{(1,2)}$ up to order L since $\beta \partial/\partial g_r$ is of order g_r, and β up to order $(L+1)$. Both are finite. The second term again involves $\tilde{\Gamma}_r^{(1,2)}$ up to order L since $(\eta + \eta_2)$ is of order g_r, and $(\eta + \eta_2)$ at order $(L+1)$ which is finite (although η and η_2 separately are not). The last term involves $\tilde{\Gamma}_r^{(2,2)}$ up to order $(L+1)$, and η up to order L since $\tilde{\Gamma}_r^{(2,2)}$ is of order g_r. We conclude that $m_r \partial \tilde{\Gamma}_r^{(1,2)}/\partial m_r$ is finite at $(L+1)$ loop order.

The function $\tilde{\Gamma}_r^{(1,2)}$ is also dimensionless. Its value at zero momentum is fixed by the renormalization condition (10.22), therefore:

$$m_r \frac{\partial}{\partial m_r} \tilde{\Gamma}_r^{(1,2)} (0; 00) = 0 .$$

The analysis is then the same as for $\tilde{\Gamma}_r^{(4)}$ and we conclude that $\tilde{\Gamma}_r^{(1,2)}$ has a finite limit at infinite cut-off at loop order $(L+1)$. We can now use equation (10.32) and argument (i): Since $\tilde{\Gamma}^{(1,2)}$ is finite up to order g_r^{L+1}, η is finite up to the same order g_r^{L+1}.

10.6.4 The $\langle\phi\phi\rangle$ correlation function ($l = 0$, $n = 2$)

The term of order g_r^{L+1} in equation (10.35) is:

$$m_r \frac{\partial}{\partial m_r}\left[\tilde{\Gamma}_r^{(2)}\right]_{L+1} = \left[\left(-\beta\frac{\partial}{\partial g_r} + \eta\right)\tilde{\Gamma}_r^{(2)}\right]_{L+1} + m_r^2\left[(2-\eta)\tilde{\Gamma}_r^{(1,2)}\right]_{L+1}. \qquad (10.38)$$

In the first term of the r.h.s. we need only $\tilde{\Gamma}_r^{(2)}$ up to order g_r^L since it is multiplied by terms of order g_r. We have also shown above that η is finite up to order g_r^{L+1} and β is finite at this order by argument (ii). For the second term we have just shown above that the two factors are finite up to order g_r^{L+1}. Let us consider the quantity $\tilde{\Gamma}_r^{(2)}(p) - m_r^2 - p^2$. Renormalization conditions imply that it vanishes as $(p^2)^2$ for $|\mathbf{p}|$ small. It has mass dimension 2. We set:

$$m_r \frac{\partial}{\partial m_r}\left[\tilde{\Gamma}_r^{(2)}(p) - m_r^2 - p^2\right] = m_r^2 f^{(2)}\left(\frac{p}{m_r}, \frac{\Lambda}{m_r}, g_r\right), \qquad (10.39)$$

with:

$$\begin{cases} \lim_{\Lambda\to\infty} f^{(2)}\left(\dfrac{p}{m_r}, \dfrac{\Lambda}{m_r}, g_r\right) < \infty \\[2mm] f^{(2)}\left(\dfrac{p}{m_r}, \dfrac{\Lambda}{m_r}, g_r\right) = O\left((p^2)^2\right) \text{ for } |p| \to 0. \end{cases}$$

Let us then integrate equation (10.39):

$$\tilde{\Gamma}_r^{(2)}(p, m_r, g_r, \Lambda) - m_r^2 - p^2 = \Lambda^2\left[\tilde{\Gamma}_r^{(2)}\left(\frac{p}{\Lambda}, 1, g_r, 1\right) - 1 - p^2/\Lambda^2\right]$$

$$- \int_{m_r}^{\Lambda} \rho\,\mathrm{d}\rho\, f^{(2)}\left(\frac{p}{\rho}, \frac{\Lambda}{\rho}, g_r\right). \qquad (10.40)$$

The integrated term in the r.h.s. decreases like $(p^2)^2/\Lambda^4$ for Λ large and therefore goes to zero.

For the same reason $f^{(2)}(p/\rho, \Lambda/\rho, g_r)$ is of order $1/\rho^4$, for $\rho \sim \Lambda$, and therefore the infinite cut-off limit can be taken in the integral.

This concludes the induction. The advantage of this method is that at the same time we have derived the CS equations:

$$\left[m_r\frac{\partial}{\partial m_r} + \beta(g_r)\frac{\partial}{\partial g_r} - \frac{\eta}{2}\eta(g_r) - l\eta_2(g_r)\right]\tilde{\Gamma}_r^{(l,n)}(q_j; p_i)$$

$$= [2 - \eta(g_r)]\, m_r^2 \tilde{\Gamma}_r^{(l+1,n)}(0, q_j; p_i)\,. \qquad (10.41)$$

10.6.5 The large momentum behaviour of superficially divergent correlation functions

We see from this derivation that we can also use induction to estimate the large momentum behaviour of correlation functions. If we assume the bounds (10.11) at loop order L, then $\Gamma^{(1,4)}$ and $\Gamma^{(2,2)}$ are given by convergent integrals at loop order $L+1$. It is not too difficult to bound their large momentum behaviour by powers of logarithms.

Let us now again consider the example of the 4-point functions. Once we have established representation (10.37), the induction hypothesis tells us that the function $f^{(4)}$ in the r.h.s. is bound by:

$$\left|f^{(4)}(\lambda\mathbf{p}/m_r, g_r)\right| < \text{const.}\ (\ln\lambda)^{k(L)} \quad \text{for } \lambda \to \infty.$$

We then decompose the integral over ρ into the sum of two terms:

$$\int_1^\infty \frac{d\rho}{\rho} f^{(4)}\left(\frac{\lambda \mathbf{p}}{\rho m_r}, g_r\right) = \int_\lambda^\infty \frac{d\rho}{\rho} f^{(4)}\left(\frac{\lambda \mathbf{p}}{\rho m_r}, g_r\right) + \int_1^\lambda \frac{d\rho}{\rho} f^{(4)}\left(\frac{\lambda \mathbf{p}}{\rho m_r}, g_r\right).$$

The first integral can be bound by a constant. The second integral can be bound using the large momentum behaviour of $f^{(4)}$:

$$\int_1^\lambda \frac{d\rho}{\rho}\left|f^{(4)}\left(\frac{\lambda \mathbf{p}}{\rho m_r}, g_r\right)\right| < \text{const.} \int_1^\lambda \frac{d\rho}{\rho}\left[\ln\left(\frac{\lambda}{\rho}\right)\right]^{k(L)} \sim (\ln \lambda)^{k(L)+1}.$$

The argument is the same for $\Gamma^{(1,2)}$. This last bound can then be used to bound $\Gamma^{(2)}$.

10.7 The $\langle \phi^2 \phi^2 \rangle$ Correlation Function

We have seen that the $\langle \phi^2 \phi^2 \rangle$ proper vertex $\Gamma^{(2,0)}$ has superficial degree of divergence zero. Actually even in free field theory ($g_r = 0$) it is divergent (see figure 10.21). An additional renormalization is needed. We impose the renormalization condition:

Fig. 10.21

$$\tilde{\Gamma}_r^{(2,0)}(p = 0, m_r, g_r) = 0, \tag{10.42}$$

obviously consistent with the tree level approximation. We expect the relation between bare and renormalized proper vertex to be now:

$$\tilde{\Gamma}_r^{(2,0)}(p) = (Z_2/Z)^2\left[\tilde{\Gamma}^{(2,0)}(p) - \tilde{\Gamma}^{(2,0)}(0)\right]. \tag{10.43}$$

Differentiating with respect to m_r at g and Λ fixed we then obtain:

$$\left[m_r\frac{\partial}{\partial m_r} + \beta(g_r)\frac{\partial}{\partial g_r} - 2\eta_2(g_r)\right]\tilde{\Gamma}_r^{(2,0)}(p; m_r, g_r)$$
$$= m_r^2\left[2 - \eta(g_r)\right]\left[\tilde{\Gamma}_r^{(3,0)}(p, -p, 0; m_r, g_r) - \tilde{\Gamma}_r^{(3,0)}(0, 0, 0; m_r, g_r)\right]. \tag{10.44}$$

The derivation of this equation follows the same lines as in previous cases. It uses the properties that $\Gamma^{(3,0)}$ has a skeleton expansion and that the L loop order of $\Gamma^{(2,0)}$ is of order g_r^{L-1}. Finally in the integration with respect to m_r one takes into account the renormalization condition (10.42) and notes that $\tilde{\Gamma}^{(2,0)}$ is dimensionless in mass units.

It is possible then to summarize all CS equations by:

$$\left[m_r\frac{\partial}{\partial m_r} + \beta(g_r)\frac{\partial}{\partial g_r} - \frac{\eta}{2}\eta(g_r) - l\eta_2(g_r)\right]\tilde{\Gamma}_r^{(l,n)}(q_j; p_i; m_r, g_r)$$
$$= m_r^2\left(2 - \eta(g_r)\right)\tilde{\Gamma}_r^{(l+1,n)}(0, q_j; p_i; m_r, g_r) + \delta_{n0}\delta_{l2}B(g_r), \tag{10.45}$$

with the notation:

$$-m_r^2\left(2 - \eta(g_r)\right)\tilde{\Gamma}_r^{(3,0)}(0, 0, 0; m_r, g_r) = B(g_r), \tag{10.46}$$

since the l.h.s. is dimensionless.

10.8 The Renormalized Action

The proof of the renormalizability of arbitrary massive local field theories is a rather simple generalization of the proof we have given above for the ϕ^4 field theory in 4 dimensions. We therefore only explain the results.

We now consider an arbitrary field theory renormalizable by power counting. We assume that we have added to the tree level action $S_\Lambda(\phi)$ all counterterms needed to render the theory finite up to loop order L. If we then perform a loopwise expansion of the generating functional of proper vertices $\Gamma(\varphi)$:

$$\Gamma(\varphi) = S_\Lambda(\varphi) + \sum_{l=1}^{\infty} \Gamma_l(\varphi),$$

the functionals $\Gamma_l(\varphi)$ have a finite limit for $l \leq L$ when the cut-off becomes infinite, and the diagrams contributing to $\Gamma_{L+1}(\varphi)$ have no divergent subdiagrams.

$\Gamma_{L+1}^{\text{div}}(\varphi)$, the divergent part of $\Gamma_{L+1}(\varphi)$, is therefore a general local functional linear combination of all vertices of non-positive canonical dimensions (except if symmetries forbid some terms). By adding to the renormalized action the counterterms $-\Gamma_{L+1}^{\text{div}}(\varphi)$ we render the theory finite at $(L+1)$ loop order. The counterterms are of course only defined up to an arbitrary finite part, linear combination of the same vertices which appear in $\Gamma_{L+1}^{\text{div}}$. It is sometimes convenient, in order to fix this arbitrariness, to impose specific renormalization conditions, as we have done in previous sections. *The resulting renormalized action is then also a general local functional of the fields, linear combination of all vertices of non-positive canonical dimension.*

Let us show in the case of the ϕ^4 field theory in 4 dimensions that this statement indeed summarizes the result derived in Section 10.6. The divergent part of $\Gamma(\varphi)$ at loop order L, after renormalization up to order $L-1$, should have the form:

$$\Gamma_L^{\text{div}}(\varphi) = -\int d^4x \left[\tfrac{1}{2}\delta m_L^2 \varphi^2 + \tfrac{1}{2}\delta Z_L (\partial_\mu \varphi)^2 + \frac{1}{4!} g_r \delta Z_{g,L} \varphi^4 \right], \tag{10.47}$$

because the vertex ϕ^2 have dimension -2, and the vertices $(\partial_\mu \phi)^2$ and ϕ^4 dimension 0. No odd powers of ϕ appear because the tree level action is symmetric in $\phi \mapsto -\phi$.

Let us now take another example, a field theory containing a boson field ϕ, and fermion fields ψ, $\bar\psi$:

$$S(\phi,\psi,\bar\psi) = \int d^4x \left[\tfrac{1}{2}(\partial_\mu \phi)^2 + \tfrac{1}{2}m^2\phi^2 - \bar\psi(\slashed{\partial}+M)\psi - g\bar\psi\psi\phi \right]. \tag{10.48}$$

We have shown in Section 8.2 that the dimensions of fields in this theory are:

$$[\phi] = 1, \quad [\psi] = [\bar\psi] = 3/2,$$

and that the theory is renormalizable by power counting. Using power counting we can write the most general divergent term:

$$-\Gamma^{\text{div}} = \int d^4x \Big[\tfrac{1}{2}\delta Z_\phi (\partial_\mu \varphi)^2 + \tfrac{1}{2}\delta m^2 \varphi^2 - \delta Z_\psi \bar\psi\slashed{\partial}\psi$$
$$- \delta M \bar\psi\psi - g\delta Z_g \bar\psi\psi\varphi + \frac{1}{4!}\delta\lambda_4 \varphi^4 + \frac{1}{3!}\delta\lambda_3 \varphi^3 + \delta c\varphi \Big]. \tag{10.49}$$

From these expressions we see that terms not present in the tree level action (10.48) have been generated, proportional to $\int \phi^4 \mathrm{d}^4 x$, $\int \phi^3 \mathrm{d}^4 x$ and $\int \phi \, \mathrm{d}^4 x$. We say that although action (10.48) is renormalizable, in contrast to the ϕ^4 field theory it is not multiplicatively renormalizable, in the sense that not all coefficients of the renormalized action can be obtained by a rescaling of those of the tree level action. To avoid this difficulty and to always make the theory multiplicatively renormalizable, we shall in general try to include in the tree level action all terms which we expect renormalization to generate.

10.9 The Massless Theory

We have given a derivation of the renormalizability of a field theory which applies only to massive field theories, since we have used in an essential manner the mass insertion operation to decrease the degree of divergence of Feynman diagrams.

We now want to justify the existence of renormalized massless field theories. The correlation functions of a massless theory can be obtained by rescaling, at cut-off and momenta fixed, the mass

$$m_{\mathrm{r}} \mapsto m_{\mathrm{r}}/\rho, \qquad \rho \to \infty.$$

At fixed cut-off the limit exists, as will be discussed extensively in Chapters 24–27, in dimensions larger than or equal to the dimension in which the theory is exactly renormalizable, provided this dimension is larger than 2 (because the propagator is $1/p^2$). In addition the set of arguments of the correlation function in momentum representation should be non-exceptional, i.e. all non-trivial subsets of momenta should have a non-vanishing sum.

These conditions are met in the ϕ_4^4 field theory (at non-exceptional momenta) and we now examine what happens when the cut-off is removed. We shall use the CS equations combined with Weinberg's theorem. From now on we omit the symbol $\sim$ indicating a Fourier transform when there is no ambiguity. The arguments x, y, z will indicate space variables and k, p, q momenta, arguments of the Fourier transform.

10.9.1 Large momentum behaviour and massless theory

Let us write the CS equations:

$$\left[m_{\mathrm{r}} \frac{\partial}{\partial m_{\mathrm{r}}} + \beta\left(g_{\mathrm{r}}\right) \frac{\partial}{\partial g_{\mathrm{r}}} - \frac{n}{2}\eta\left(g_{\mathrm{r}}\right) \right] \Gamma_{\mathrm{r}}^{(n)}\left(p_i\right) = m_{\mathrm{r}}^2 \left(2 - \eta\right) \Gamma_{\mathrm{r}}^{(1,n)}\left(0; p_i\right).$$

Weinberg's theorem states that if we scale all momenta $p_i \mapsto \rho p_i$, the large ρ behaviour at *non-exceptional momenta*, at any finite order of perturbation theory, is given by the canonical dimension up to powers of logarithms. Thus:

$$\Gamma_{\mathrm{r}}^{(l,n)}\left(\rho p_i\right) \underset{\to \infty}{\sim} \rho^{4-n-2l} \times \text{power of } \ln \rho,$$

$$\Gamma_{\mathrm{r}}^{(l+1,n)}\left(0; \rho p_i\right) \underset{\to \infty}{\sim} \frac{1}{\rho^2}\rho^{4-n-2l} \times \text{power of } \ln \rho.$$

If therefore in the asymptotic expansion of $\Gamma_{\mathrm{r}}^{(n)}\left(\rho p_i\right)$ for ρ large, we eliminate all terms subleading by a power ρ^{-2} up to powers of $\ln \rho$, we find a set of 1PI correlation functions $\Gamma_{\mathrm{r,as.}}^{(n)}$ which satisfy a homogeneous CS equation:

$$\left[m_{\mathrm{r}} \frac{\partial}{\partial m_{\mathrm{r}}} + \beta\left(g_{\mathrm{r}}\right) \frac{\partial}{\partial g_{\mathrm{r}}} - \frac{n}{2}\eta\left(g_{\mathrm{r}}\right) \right] \Gamma_{\mathrm{r,as.}}^{(n)}\left(p_i\right) = 0. \tag{10.50}$$

However we know from dimensional analysis

$$\Gamma_r^{(n)}(p_i, m_r) = m_r^{4-n} \Gamma_r^{(n)}(p_i/m_r, 1),$$

(10.51)

therefore scaling all momenta by a factor ρ, is equivalent, up to a global factor, to scale the mass by a factor ρ^{-1}. The solutions of equation (10.50) are thus the correlation functions of the massless ϕ^4 field theory.

Perturbative solution of the homogeneous CS equations. It is actually interesting to study the structure of correlation functions implied by equation (10.50). Let us just here consider, for illustration purpose, the 2-point function $\Gamma_{r,as.}^{(2)}(p)$. It is convenient to introduce the function $\zeta(g_r)$

$$\ln \zeta(g_r) = \int_0^{g_r} \frac{dg'\, \eta(g')}{\beta(g')}.$$

Note that since both η and β are of order g_r^2 this function has a perturbative expansion.

We then set

$$\Gamma_{r,as.}^{(2)}(p) = p^2 \zeta(g_r) A(g_r, p/m_r).$$

The function A satisfies

$$\left[m_r \frac{\partial}{\partial m_r} + \beta(g_r) \frac{\partial}{\partial g_r} \right] A(g_r, p/m_r) = 0.$$

(10.52)

We then expand A and $\beta(g_r)$ in powers of g_r, setting

$$A(g_r, p/m_r) = 1 + \sum_1^\infty g_r^n a_n(p/m_r), \qquad \beta(g_r) = \sum_2^\infty \beta_n g_r^n.$$

(10.53)

Introducing these expansions in equation (10.52) we find for $n \geq 2$:

$$-za_n'(z) + \sum_{m=1}^{n-1} ma_m(z)\beta_{n-m+1} = 0.$$

(10.54)

The equation $n = 1$ is special:

$$-za_1'(z) = 0 \quad \Rightarrow \quad a_1(z) = C_1.$$

Let us examine $n = 2$:

$$-za_2'(z) + C_1\beta_2 = 0, \Rightarrow a_2(z) = C_1\beta_2 \ln z + C_2.$$

From this example we understand the general structure: $a_n(z)$ is a polynomial of degree $(n-1)$ in $\ln z$:

$$a_n(z) = P_{n-1}(\ln z),$$

(10.55)

$$P_{n-1}'(x) = \sum_{m=1}^{n-1} mP_{m-1}(x)\beta_{n-m+1}.$$

(10.56)

Note in particular that the new information specific to the order n is characterized by two constants β_n, which enter in the coefficient of $\ln z$ and C_n which is the integration constant (to which one should add the coefficients of $\eta(g_r)$ which appear in the function $\zeta(g_r)$). Moreover the term of highest degree in P_n is entirely determined by one-loop results, the next term by one and two-loop and so on. Finally the $\Gamma^{(2)}_{r,as.}(p)$ is entirely determined by the functions $\beta(g_r)$ and $\eta(g_r)$ and for example $\Gamma^{(2)}_{r,as.}(1,g_r)/m_r^2$ which is a third function of g_r.

It also follows from these equations that $\Gamma^{(2)}_{r,as.}(p)$ has a limit for $p = 0$:

$$\Gamma^{(2)}_{r,as.}\left(p^2 = 0\right) = 0\,, \tag{10.57}$$

confirming, as expected, that the theory is massless. However its derivative $\partial \Gamma^{(2)}_{r,as.}/\partial p^2$ has no zero momentum limit. It is easy to verify that no other correlation function has a zero momentum limit either.

We have here constructed a massless theory by scaling a massive theory and shown that the corresponding 1PI functions satisfy a homogeneous CS equation, called also renormalization group (RG) equation. Let us show that such an equation can also be derived directly from the assumption of the existence of a renormalized massless field theory.

10.9.2 RG equations in a massless field theory

Renormalization conditions. We have shown that the renormalized massless ϕ^4 field theory exists. If we want to directly determine renormalization constants by renormalization conditions, we have to impose them at non-exceptional momenta, in particular we cannot use zero momentum except for the 2-point function as we have indicated above. We therefore introduce a mass scale μ and impose:

$$\Gamma^{(2)}_r\left(p^2 = 0\right) = 0\,, \tag{10.58}$$

$$\frac{\partial}{\partial p^2}\Gamma^{(2)}_r\left(p^2 = \mu^2\right) = 1\,, \tag{10.59}$$

$$\Gamma^{(4)}_r\left(p_i = \mu\theta_i\right) = g_r\,, \tag{10.60}$$

in which the θ_i form a set of arbitrary non-exceptional numerical vectors.

The bare correlation functions in a massless theory depend only on the cut-off and momenta, since the bare mass parameter is fixed by imposing that the renormalized mass m_r vanishes. The renormalized correlation functions depend on the arbitrary scale μ and momenta. They are related for Λ large by:

$$\Gamma^{(n)}_r\left(p_i; \mu, g_r\right) = Z^{n/2}\left(\Lambda/\mu, g_r\right) \Gamma^{(n)}\left(p_i; \Lambda, g\right). \tag{10.61}$$

Remarks. While in a massive theory the value of the renormalized 2-point function at zero momentum depends on the renormalization conditions, in a massless theory the bare 2-point function and the renormalized function, independently of its normalization, vanish at zero momentum as equation (10.61) shows.

Note also that when m_r vanishes, the bare mass squared m^2 begins with a term of order g_r. The first of equations (10.10) then shows that m^2, the square of the bare mass, is negative for g_r small. For traditional reasons, we shall often denote by m^2 the coefficient of ϕ^2 in the bare action but the reader should be aware that this does not imply that this coefficient is positive.

Renormalization group equations. The bare theory of course is completely independent of the parameter μ which has just been introduced to fix the normalization of the renormalized functions:

$$\mu \frac{\partial}{\partial \mu} \Gamma^{(n)}\left(p_i; \Lambda, g\right)\bigg|_{g,\Lambda} = 0.$$

Therefore if we differentiate equation (10.61) with respect to μ at Λ and g fixed, we find using chain rule:

$$\left[\mu \frac{\partial}{\partial \mu} + \tilde{\beta}\left(g_{\mathrm{r}}\right) \frac{\partial}{\partial g_{\mathrm{r}}} - \frac{n}{2} \tilde{\eta}\left(g_{\mathrm{r}}\right)\right] \Gamma_{\mathrm{r}}^{(n)} = 0, \tag{10.62}$$

with the definitions:

$$\tilde{\beta}\left(g_{\mathrm{r}}\right) = \mu \frac{\partial}{\partial \mu} g_{\mathrm{r}}\bigg|_{g,\Lambda}, \qquad \tilde{\eta}\left(g_{\mathrm{r}}\right) = \mu \frac{\partial}{\partial \mu} \ln Z\bigg|_{g,\Lambda}. \tag{10.63}$$

A priori $\tilde{\beta}$ and $\tilde{\eta}$ could also depend of the dimensionless ratio Λ/μ, but since they can be expressed in terms of the renormalized correlation functions themselves, they must have a large cut-off limit.

Equation (10.62) is analogous to equation (10.50). They differ only in the definition of g_{r} and a finite field amplitude renormalization.

Both sets of equations will be an essential tool for the analysis of the large momentum behaviour of correlation functions. In addition equation (10.62) will be used in the third part of this work to discuss the small momentum behaviour of massless theories.

In the massless theory we can also define renormalized correlation functions with ϕ^2 insertions:

$$\Gamma_{\mathrm{r}}^{(l,n)}\left(y_1, \ldots, y_l; x_1, \ldots, x_n\right) = Z^{(n/2)-l} Z_2^l \Gamma^{(l,n)}\left(y_1, \ldots, y_l; x_1, \ldots, x_n\right). \tag{10.64}$$

The ϕ^2 renormalization constant Z_2 can be fixed by a renormalization condition of the form:

$$\tilde{\Gamma}_{\mathrm{r}}^{(1,2)}\left(q; p_1, p_2\right)\big|_{p_1^2=p_2^2=q^2=\mu^2} = 1. \tag{10.65}$$

As before, the $\langle\phi^2\phi^2\rangle$ correlation function needs an additional renormalization, which can be fixed by a renormalization condition at non-zero momentum.

From equation (10.64) RG equations can be derived by differentiating with respect to μ. Introducing the RG function $\tilde{\eta}_2\left(g_{\mathrm{r}}\right)$:

$$\tilde{\eta}_2\left(g_{\mathrm{r}}\right) = \mu \frac{\partial}{\partial \mu} \ln Z_2\bigg|_{g,\Lambda}, \tag{10.66}$$

we obtain:

$$\left[\mu \frac{\partial}{\partial \mu} + \tilde{\beta}\left(g_{\mathrm{r}}\right) \frac{\partial}{\partial g_{\mathrm{r}}} - \frac{n}{2} \tilde{\eta}\left(g_{\mathrm{r}}\right) - l\tilde{\eta}_2\left(g_{\mathrm{r}}\right)\right] \Gamma_{\mathrm{r}}^{(l,n)}\left(q_1, \ldots, q_l; p_1, \ldots, p_n\right) = 0, \tag{10.67}$$

equation valid except for $n = 0, l \le 2$ where the r.h.s. is a function of g_{r} as in equation (10.45).

10.10 Homogeneous Renormalization Group (RG) Equations in a Massive Theory

Once the renormalized correlation functions $\Gamma_r^{(l,n)}$ of the massless theory have been constructed, a natural question arises: since a constant source term for ϕ^2 at zero momentum generates a mass shift (equation (10.23)), is it possible to express the correlation functions of a massive theory in terms of the correlation functions with ϕ^2 insertions of the massless theory?

An immediate difficulty arises: it is easy to verify that insertions at zero momentum in a massless theory are IR divergent. However, because the resulting theory is massive, this difficulty can be circumvented by first using a non-constant source for ϕ^2, then performing a partial summation of the 2-point function and finally taking the constant source limit. After summation the propagator becomes massive and the limit is no longer IR divergent.

Let us therefore consider the renormalized action with a source $K(x)$ for renormalized ϕ^2 insertions:

$$S\left(\phi, K\right) = \int \mathrm{d}^4 x \left[\tfrac{1}{2} Z \left(\partial_\mu \phi\right)^2 + \tfrac{1}{2} \left(\delta m^2 + Z_2 K(x)\right) \phi^2 + \frac{1}{4!} g_r Z_g \phi^4 \right]. \tag{10.68}$$

A correlation function in presence of the source $K(x)$ has the expansion:

$$\Gamma^{(n)}\left(p_1, p_2, \ldots p_n; K\right) = \sum_{l=0} \frac{1}{2^l} \frac{1}{l!} \int \mathrm{d}q_1 \mathrm{d}q_2 \ldots \mathrm{d}q_l \, \tilde{K}\left(q_1\right) \tilde{K}\left(q_2\right) \ldots \tilde{K}\left(q_l\right)$$
$$\times \Gamma_r^{(l,n)}\left(q_1, q_2, \ldots, q_l; p_1, p_2, \ldots, p_n\right), \tag{10.69}$$

in which $\tilde{K}$ is the Fourier transform of the source.

Let us apply the differential operator D:

$$\mathrm{D} \equiv \mu \frac{\partial}{\partial \mu} + \tilde{\beta}\left(g_r\right) \frac{\partial}{\partial g_r}, \tag{10.70}$$

on equation (10.69) and use the RG equations (10.67). Noting that:

$$\int \mathrm{d}q \, \tilde{K}(q) \frac{\delta}{\delta \tilde{K}(q)} = l, \tag{10.71}$$

we obtain:

$$\left[\mu \frac{\partial}{\partial \mu} + \tilde{\beta}\left(g_r\right) \frac{\partial}{\partial g_r} - \frac{n}{2} \tilde{\eta}\left(g_r\right) - \tilde{\eta}_2\left(g_r\right) \int \mathrm{d}q \, \tilde{K}\left(q\right) \frac{\delta}{\delta \tilde{K}\left(q\right)} \right] \Gamma_r^{(n)}\left(K; p_1, \ldots, p_n\right) = 0. \tag{10.72}$$

After the summation of the propagator we set:

$$K(x) = m_r^2, \tag{10.73}$$

and equation (10.72) becomes:

$$\left[\mu \frac{\partial}{\partial \mu} + \tilde{\beta}\left(g_r\right) \frac{\partial}{\partial g_r} - \frac{n}{2} \tilde{\eta}\left(g_r\right) - \tilde{\eta}_2\left(g_r\right) m_r^2 \frac{\partial}{\partial m_r^2} \right] \Gamma_r^{(n)}\left(p_1, \ldots, p_n; \mu, g_r, m_r\right) = 0. \tag{10.74}$$

We have therefore derived a new RG equation for a massive theory which differs from the original CS equations in various respects:

(i) Correlation functions depend on two mass parameters while we know that only one is necessary. However, in this parametrization the massless limit of correlation functions is directly obtained by setting $m_r = 0$ and no asymptotic expansion at large momenta is needed.

(ii) In contrast to the CS equations, these RG equations are homogeneous. We shall exploit this property later when we solve the RG equations.

10.11 Covariance of RG Functions

In the example of the massless ϕ^4 we have been naturally led to consider two different renormalization schemes. It is thus necessary to understand how renormalization group functions of different schemes are related. Both theories differ by a redefinition of the coupling constant and a finite field amplitude renormalization. Let us call g and $\tilde{g}$ the renormalized coupling constants in the two schemes. Comparing the two renormalization group equations, we obtain:

$$\beta(g)\frac{\partial}{\partial g} = \tilde{\beta}(\tilde{g})\frac{\partial}{\partial\tilde{g}},$$

and therefore using chain rule:

$$\beta(g)\frac{\partial\tilde{g}}{\partial g} = \tilde{\beta}(\tilde{g}). \tag{10.75}$$

In a ϕ^4 like field theory, the function $\beta(g)$ has the expansion:

$$\beta(g) = \beta_2 g^2 + \beta_3 g^3 + O(g^4). \tag{10.76}$$

Expanding $\tilde{g}$ in terms of g:

$$\tilde{g} = g + \gamma_2 g^2 + O(g^3), \tag{10.77}$$

and using equation (10.76), after a short calculation we find:

$$\tilde{\beta}(\tilde{g}) = \beta_2 \tilde{g}^2 + \beta_3 \tilde{g}^3 + O(\tilde{g}^4). \tag{10.78}$$

The first two terms in the expansion of the β-function are universal and all others are formally arbitrary. *One should not conclude from this result that the physical consequences derived from the form of the RG β-function are also to a large extent arbitrary.* Only change of variables $g \mapsto \tilde{g}$, which are regular mappings, are allowed. This implies in particular that $\partial\tilde{g}/\partial g$ in equation (10.75) must remain strictly positive. Therefore the sign and zeros of the β-function are properties of the theory. Equation (10.75) shows also that if the β-function vanishes with a finite slope the slope is scheme-independent.

The difficulty arises from the fact that we do not know in general which renormalization scheme leads to regular functions of the coupling constant. Our intuition is that "natural" definitions as induced by momentum or minimal subtraction have the most chance to satisfy this criterion.

In the same way if we call $\zeta(\tilde{g})$ the additional finite field amplitude renormalization we find:

$$\tilde{\eta}(\tilde{g}) = \eta(g) + \tilde{\beta}(\tilde{g})\frac{\partial}{\partial\tilde{g}}\ln\zeta(\tilde{g}). \tag{10.79}$$

Since the field amplitude renormalization appears at order g^2, $\ln \zeta\,(\tilde{g})$ is of order $\tilde{g}^2$ and therefore the modification to $\tilde{\eta}$ of order $\tilde{g}^3$. The coefficient of order g^2 is universal. The value of η at a zero of $\beta(g)$ is also universal.

Several coupling constants. We shall meet actions depending on several fields and coupling constants g_i. The transformation law under a change of parametrization of the coupling space of the RG β-functions becomes then

$$\tilde{\beta}_i\,(\tilde{g}) = \frac{\partial \tilde{g}_i}{\partial g_j} \beta_j(g), \qquad (10.80)$$

where the mapping should satisfy $\det(T_{ij} \equiv \partial \tilde{g}_i / \partial g_j) \neq 0$. Then the existence of zeros of the β-function is universal and at a zero g^*:

$$\frac{\partial \tilde{\beta}_i}{\partial \tilde{g}_j} = T_{ik} \frac{\partial \beta_k}{\partial g_l} T_{lj}^{-1},$$

and the eigenvalues of $\partial \beta_i / \partial g_j$ at a zero thus are parametrization independent.

Bibliographical Notes

The method we have adopted here to prove renormalizability of the ϕ^4 field theory follows the presentation of

C.G. Callan in *Methods in Field Theory*, Les Houches 1975, R. Balian and J. Zinn-Justin eds. (North-Holland, Amsterdam 1976).

The idea to combine proof of renormalizability and RG equations is due to

A.S. Blaer and K. Young, *Nucl. Phys.* B83 (1974) 493.

A systematic discussion of renormalization theory in a different language as well as historical references can be found in

W. Zimmermann, *Lectures on Elementary Particles and Quantum Field Theory, Brandeis 1970*, S. Deser, M. Grisaru and H. Pendleton eds. (MIT Press, Cambridge 1970); *Ann. of Phys. (NY)* 77 (1973) 536.

Among the early work see for example

F.J. Dyson, *Phys. Rev.* 75 (1949) 1736; N.N. Bogoliubov and O.S. Parasiuk, *Acta Math.* 97 (1957) 227; and the textbook N.N. Bogoliubov and D.V. Shirkov, *Introduction to the Theory of Quantized Fields*, (Interscience, New York 1959).

The asymptotic behaviour of renormalized integrals was studied in

S. Weinberg, *Phys. Rev.* 118 (1960) 838; K. Hepp, *Commun. Math. Phys.* 2 (1966) 301.

For different aspects of this problem in parametric space see also

T. Appelquist, *Ann. of Phys. (NY)* 54 (1969) 27; M.C. Bergère and J.-B. Zuber, *Commun. Math. Phys.* 35 (1974) 113; M. Bergère and Y.M.P. Lam, *Commun. Math. Phys.* 39 (1974) 1.

Early references about RG equations are

E.C.G. Stueckelberg and A. Peterman, *Helv. Phys. Acta* 26 (1953) 499; M. Gell-Mann and F.E. Low, *Phys. Rev.* 95 (1954) 1300.

See also N.N. Bogoliubov and D.V. Shirkov as quoted above and

K.G. Wilson, *Phys. Rev.* 179 (1969) 1499.

In their modern form the RG or CS equations have been derived independently by:

C.G. Callan, *Phys. Rev.* D2 (1970) 1541; K. Symanzik, *Commun. Math. Phys.* 18 (1970) 227.

A pedagogical presentation has been given in

S. Coleman, *Dilatations, Erice Lectures 1971* reprinted in *Aspects of Symmetry* (Cambridge University Press, Cambridge 1985).

Massless theories have been discussed in

K. Symanzik, *Commun. Math. Phys.* 7 (1973) 34.

Homogeneous RG equations have been proposed in

S. Weinberg, *Phys. Rev.* D8 (1973) 3497; G. 't Hooft, *Nucl. Phys.* B61 (1973) 455; J. Zinn-Justin, *Cargèse Lectures 1973* unpublished, incorporated in the review article: E. Brézin, J.C. Le Guillou and J. Zinn-Justin, *Field Theoretical Approach to Critical Phenomena* in *Phase Transitions and Critical Phenomena* vol. 6, C. Domb and M.S. Green eds. (Academic Press, New York 1976).

See also

J.C. Collins, *Nucl. Phys.* B80 (1974) 341.

In the appendix we briefly outline a different very interesting approach which combines renormalization and renormalization group see

J. Polchinski, *Nucl. Phys.* B231 (1984) 269.

For recent applications of such ideas see for example

M. Bonini, M. D'Attanasio and G. Marchesini, *Nucl. Phys.* B409 (1993) 441, *ibidem* B444 (1995) 602; J. Adams, J. Berges, S. Bornholdt, F. Freire, N. Tetradis and C. Wetterich, *Mod. Phys. Lett.* A10 (1995) 2367, hep-th/9507093; T.R. Morris, *Prog. Theor. Phys. Suppl.* 131 (1998) 395, hep-th/9802039,

and references therein.

The use of the solution of the classical field equation as argument of the generating functional or background field method (see Appendix A10.3) was suggested in:

B.S. DeWitt in *Relativity, Groups and Topology*, Les Houches 1963, C. DeWitt and B. DeWitt eds. (Gordon and Breach, New York 1964); *Phys. Rev.* 162 (1967) 1195; G. 't Hooft, *Nucl. Phys.* B62 (1973) 444.

The generating functional for the *S*-matrix was also discussed in

B. Zumino in *Theory and Phenomenology in Particle Physics*, Erice 1968, A. Zichichi ed. (Academic Press, New York 1968).

Exercises

Exercise 10.1

Calculate $\Gamma^{(2)}_{r,as.}$, as defined in Subsection 10.9.1, in an approximation in which only the term of highest degree in $\ln(p/m_r)$ is kept at each order (the leading log approximation). Discuss the result.

APPENDIX 10

A10.1 Large Momentum Mode Integration and General RG Equations

We here briefly describe a direct approach to renormalization and renormalization group, closer to Wilson's ideas, which was first proposed by Polchinski. This approach leads to another proof of renormalizability. The physical motivation for such a method will become more apparent when we discuss RG and critical phenomena starting in Chapter 25. The idea is to study the cut-off dependence of bare correlation functions by integrating systematically on the large momentum modes of the field. In such a way one constructs a renormalization group is the space of all local interactions, expressing the equivalence between a variation of the cut-off and of the coefficients of the interaction terms, equivalence valid only for long distance, low momentum (in the cut-off scale) physics.

A10.1.1 A simple equivalence

In Section 9.1 we have shown by a gaussian integration that the two actions $S(\phi)$:

$$S(\phi) = \tfrac{1}{2} \int \mathrm{d}x \, \phi \Delta^{-1}\phi + V(\phi), \tag{A10.1}$$

and $S(\phi_1, \phi_2)$:

$$S(\phi_1, \phi_2) = \tfrac{1}{2} \int \mathrm{d}x \left(\phi_1 \Delta_1^{-1}\phi_1 + \phi_2 \Delta_2^{-1}\phi_2 \right) + V(\phi_1 + \phi_2), \tag{A10.2}$$

with:

$$\Delta = \Delta_1 + \Delta_2, \tag{A10.3}$$

generate the same perturbation theory.

Let us now use this equivalence in the limit in which the propagator Δ_2 goes to zero. Then only small values of ϕ_2 contribute to the partition function. Expanding the interaction for ϕ_2 small:

$$V(\phi_1 + \phi_2) = V(\phi_1) + \int \mathrm{d}x \frac{\delta V(\phi_1)}{\delta\phi(x)} \phi_2(x) + \tfrac{1}{2} \int \mathrm{d}x \, \mathrm{d}y \frac{\delta^2 V}{\delta\phi(x)\delta\phi(y)} \phi_2(x)\phi_2(y) + \cdots,$$

we can integrate over ϕ_2 to obtain the leading order correction.

$$\int [\mathrm{d}\phi_2] \exp \left\{ -\left[\tfrac{1}{2} \int \mathrm{d}x \, \phi_2 \Delta_2^{-1}\phi_2 + V(\phi + \phi_2) - V(\phi) \right] \right\} \sim 1$$

$$+ \tfrac{1}{2} \int \mathrm{d}x \, \mathrm{d}y \, \Delta_2(x,y) \left[\frac{\delta V}{\delta\phi(x)} \frac{\delta V}{\delta\phi(y)} - \frac{\delta^2 V}{\delta\phi(x)\delta\phi(y)} \right] + \cdots . \tag{A10.4}$$

Taking the logarithm of expression $(A10.4)$ we can rewrite the partition function as a functional integral over a field ϕ (the index 1 is no longer useful) with an effective action $S'(\phi)$:

$$S'(\phi) = \tfrac{1}{2} \int \mathrm{d}x \, \phi \Delta_1^{-1}\phi + V(\phi) + \tfrac{1}{2} \int \mathrm{d}x \, \mathrm{d}y \, \Delta_2(x,y) \left[\frac{\delta^2 V}{\delta\phi(x)\delta\phi(y)} - \frac{\delta V}{\delta\phi(x)} \frac{\delta V}{\delta\phi(y)} \right] + \cdots . \tag{A10.5}$$

We have thus established that the actions $(A10.1)$ and $(A10.5)$ lead to the same partition function, up to a trivial renormalization.

We show below that this equivalence can be used to partially integrate out the large momenta in a field theory with a cut-off Λ.

A10.1.2 Large momentum mode partial integration and RG equations

Let us take for propagator Δ a massless propagator of the form:

$$\Delta(k) = \frac{C\left(k^2/\Lambda^2\right)}{k^2}, \qquad (A10.6)$$

in which the function $C(t)$ is smooth, goes to 1 for t small and decreases faster than any power for t large (see Section 9.1). We then take for Δ_1 the propagator corresponding to an infinitesimal rescaling of the cut-off Λ:

$$\Delta_1(k) = \frac{C\left(k^2/\Lambda^2(1+\sigma)^2\right)}{k^2}. \qquad (A10.7)$$

At leading order for σ small Δ_2 is then:

$$\Delta_2(k) = \frac{2\sigma}{\Lambda^2} C'\left(k^2/\Lambda^2\right) = \sigma D(k). \qquad (A10.8)$$

We see that the propagator Δ_2 has no pole at $k = 0$. Moreover if we choose a function $C(t)$ which is very close to 1 for t small:

$$|C(t) - 1|\,t^{-p} \to 0 \quad \forall p > 0\,,$$

then Δ_2 is only large for k of order Λ. The integration over ϕ_2 thus corresponds to an integration over the large momentum modes of the field ϕ.

The equivalence between actions $(A10.1)$ and $(A10.5)$, which is the starting point of a renormalization group, can be written:

$$\Lambda\frac{\mathrm{d}}{\mathrm{d}\Lambda} V(\phi, \Lambda) = \tfrac{1}{2} \int \mathrm{d}^d x\, \mathrm{d}^d y\, D(x, y) \left[\frac{\delta^2 V}{\delta\phi(x)\delta\phi(y)} - \frac{\delta V}{\delta\phi(x)}\frac{\delta V}{\delta\phi(y)}\right], \qquad (A10.9)$$

or after Fourier transformation:

$$\Lambda\frac{\mathrm{d}}{\mathrm{d}\Lambda} V(\phi, \Lambda) = \tfrac{1}{2} \int \frac{\mathrm{d}^d k}{(2\pi)^d} D(k) \left[\frac{\delta^2 V}{\delta\phi(k)\delta\phi(-k)} - \frac{\delta V}{\delta\phi(k)}\frac{\delta V}{\delta\phi(-k)}\right]. \qquad (A10.10)$$

To study the existence of fixed points we start with a given interaction $V_0(\phi)$ at a scale Λ_0 and use equation $(A10.9)$ to calculate the effective interaction $V(\phi, \Lambda)$ at a scale $\Lambda \ll \Lambda_0$. A fixed point is defined by the property that $V(\phi, \Lambda)$, after a suitable rescaling of ϕ, goes to a limit.

Let us call $\tilde{V}^{(n)}(p_1, p_2, \ldots, p_n)$ the coefficients of $V(\phi, \Lambda)$ in an expansion in powers of $\tilde{\phi}(p)$, the Fourier transform of the field. Equation $(A10.10)$ can then be written in component form (assuming translation invariance):

$$\Lambda\frac{\mathrm{d}}{\mathrm{d}\Lambda} \tilde{V}^{(n)}(p_1, p_2, \ldots, p_n) = \tfrac{1}{2} \int \frac{\mathrm{d}^d k}{(2\pi)^d} D(k)\tilde{V}^{(n+2)}(p_1, p_2, \ldots, p_n, k, -k)$$

$$- \tfrac{1}{2} D(p_0) \sum_I \tilde{V}^{(l+1)}(p_{i_1}, \ldots, p_{i_l}, p_0)\, \tilde{V}^{(n-l+1)}(p_{i_{l+1}}, \ldots, p_{i_n}, -p_0) \quad (A10.11)$$

in which the momentum p_0 is determined by momentum conservation and the set $I \equiv \{i_1, i_2, \ldots, i_l\}$ runs over all distinct subsets of $\{1, 2, \ldots, n\}$.

We see in these equations that even if we start with a pure $g\phi^4$ interaction, at scale Λ we obtain a general local interaction because all functions $\tilde{V}^{(n)}$ are coupled. However, in the spirit of the perturbative methods used so far, it is possible to solve equation $(A10.9)$ as an expansion in the coupling constant g with the ansatz that the terms of $V(\phi, \Lambda)$ quadratic and quartic in ϕ are of order g and the general term of degree $2n$ is of order g^{n-1}.

Correlation functions. To generate correlation functions we have to add a source to the interaction $V(\phi)$:

$$V(\phi) \mapsto V(\phi) - \int \mathrm{d}x\, J(x)\phi(x).$$

However, equation $(A10.5)$ then shows that $S'(\phi)$ becomes in general a complicated functional of the source $J(x)$. A solution to this problem is the following: one takes a source whose Fourier transform $\tilde{J}(k)$ vanishes for $k^2 \geq \Lambda^2$ together with a propagator Δ_2 which propagates only momenta such that $k^2 \geq \Lambda^2$. This implies that $C'(t)$ vanishes identically for $t \leq 1$ (unfortunately such cut-off functions are inconvenient for practical calculations). Then $\int \mathrm{d}x\, J(x)\phi_2(x)$ does not contribute in integral $(A10.4)$ and $S'(\phi) - S(\phi)$ does not depend on $J(x)$.

We note however that the RG transformation is then such that the correlation functions corresponding to the action $S(\phi)$ and $S'(\phi)$ are only identical when all momenta are smaller than the cut-off. The differences between correlation functions are smooth functions of momenta and thus decay at large distances in space faster than any power.

A10.2 Super-renormalizable Field Theories: The Normal-Ordered Product

In most of this work we discuss strictly renormalizable field theories. Let us however make a few simple comments about the super-renormalizable case. We first take the example of the ϕ^4 field theory in three dimensions. We then examine the special properties of two dimensional field theories.

The ϕ^4 theory in 3 dimensions. As derived in Chapter 8, the ϕ^4 field theory in 3 dimensions has only three superficially divergent diagrams, all contributing to the 2-point function, which are shown in figure 10.22.

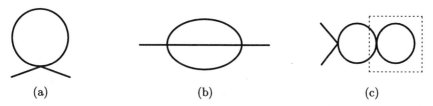

(a)   (b)   (c)

Fig. 10.22   The three divergent diagrams in the ϕ_3^4 theory.

$$(\mathrm{a}) = \frac{1}{(2\pi)^3} \int \frac{\mathrm{d}^3 q}{(q^2 + m^2)_\Lambda}$$

$$(\mathrm{b}) = \frac{1}{(2\pi)^6} \int \frac{\mathrm{d}^3 q_1 \mathrm{d}^3 q_2}{(q_1^2 + m^2)_\Lambda\, (q_2^2 + m^2)_\Lambda \left[(p - q_1 - q_2)^2 + m^2\right]_\Lambda}.$$

By adding a counterterm of the form:

$$\tfrac{1}{2}\left[a_1\left(\Lambda\right)g+a_2\left(\Lambda\right)g^2\right]\phi^2,$$

we render the first two diagrams finite.

The diagram (c) is also superficially divergent, however it is clear that the counterterm which renders diagram (a) finite, renormalizes also diagram (c). Therefore the counterterms which renormalize diagrams (a) and (b) render the whole theory finite.

Actually the divergence (a), which corresponds to a self-contraction of the vertex ϕ^4, can be eliminated *a priori* by replacing the vertex ϕ^4 by a *normal-ordered vertex* $:\phi^4:$

$$:\phi^4:(x)=\phi^4(x)-6\phi^2(x)\left\langle\phi^2(x)\right\rangle+3\left(\left\langle\phi^2(x)\right\rangle\right)^2,\qquad(A10.12)$$

in which the average $\left\langle\phi^2(x)\right\rangle$ is calculated in a free field theory with a mass μ which may or may not be equal to the renormalized mass m:

$$\left\langle\phi^2(x)\right\rangle_\mu=\frac{1}{(2\pi)^3}\int\frac{\mathrm{d}^3q}{(q^2+\mu^2)_\Lambda}.\qquad(A10.13)$$

The denomination normal ordering comes from the operator language. The quantity $:\phi^4(x):$ is such that

$$\begin{cases}\left\langle:\phi^4\left(x\right):\right\rangle=0\\\left\langle:\phi^4(x):\phi\left(y_1\right)\phi\left(y_2\right)\right\rangle=0,\end{cases}\qquad(A10.14)$$

in which again the averages are calculated with the action $S_\mu(\phi)$:

$$S_\mu(\phi)=\tfrac{1}{2}\int\mathrm{d}^3x\left[(\nabla\phi)^2+\mu^2\phi^2\right].\qquad(A10.15)$$

The averages calculated with another mass are then finite.

Of course we still have to add a counterterm for diagram (b).

Super-renormalizable scalar field theories in two dimensions. An action $S(\phi)$ of the form

$$S(\phi)=\int\mathrm{d}^2x\left[\tfrac{1}{2}\left(\nabla\phi\right)^2+\tfrac{1}{2}m^2\phi^2+V\left(\phi\right)\right],\qquad(A10.16)$$

in which $V(\phi)$ is an arbitrary function of $\phi(x)$ (but not of its derivatives) is super-renormalizable. If $V(\phi)$ is a polynomial, only a finite number of diagrams are superficially divergent. However it is a peculiarity of dimension 2 that the field is dimensionless and therefore the interaction $V(\phi)$ may have an infinite series expansion in powers of ϕ. Then, although the theory is super-renormalizable, one finds an infinite number of superficially divergent diagrams. However all divergences come from self-contractions of the vertex (see figure 10.23), therefore the operation $:V(\phi):$ removes all divergences.

Fig. 10.23

To obtain an explicit expression for $: V(\phi):$ let us first consider the interesting example of the interaction:

$$V(\phi) = e^{\lambda\phi}. \qquad (A10.17)$$

The average of $V(\phi)$ in presence of a source term can then be calculated explicitly:

$$\int [d\phi(y)] \exp\left\{-\int dy\left[\left(\tfrac{1}{2}(\nabla\phi)^2 + \mu^2\phi^2\right) - (J(y) + \lambda\delta(x-y))\phi(y)\right]\right\}$$

$$= \exp\left[\tfrac{1}{2}\int dy\,dy'\,J(y)\Delta(y,y')\,J(y') + \lambda\int dy\,\Delta(x,y)\,J(y) + \tfrac{1}{2}\lambda^2\Delta(x,x)\right] (A10.18)$$

in which $\Delta(x,y)$ is the free propagator with mass μ. The normal ordering operation has to suppress the term coming from self-contractions which is proportional to $\Delta(x,x)$. It is thus clear that the normal ordered interaction is:

$$: \exp[\lambda\phi(x)] := \exp\left[\lambda\phi(x) - \lambda^2\langle\phi^2(x)\rangle/2\right], \qquad (A10.19)$$

in which we have used:

$$\langle\phi^2(x)\rangle = \Delta(x,x). \qquad (A10.20)$$

We then write an arbitrary interaction term as a Laplace transform:

$$V(\phi(x)) = \int d\rho(\lambda)\,e^{\lambda\phi(x)}. \qquad (A10.21)$$

The normal ordering is a linear operation. Thus:

$$: V(\phi) := \int d\rho(\lambda) : e^{\lambda\phi(x)} : \ .$$

We can then use the result $(A10.19)$ for the exponential interaction and obtain:

$$: V(\phi) := \left\{\exp\left[-\tfrac{1}{2}\langle\phi^2\rangle(\partial/\partial\phi)^2\right]\right\}V(\phi), \qquad (A10.22)$$

or more explicitly:

$$: V(\phi) := V(\phi) + \left[\sum_{n=1}^{\infty}\frac{(-1)^n}{2^n n!}\langle\phi^2\rangle^n\left(\frac{\partial}{\partial\phi}\right)^{2n}\right]V(\phi).$$

The existence of the Laplace transform of a given interaction is irrelevant in this argument since the final identity is purely algebraic.

A10.3 Background Field Method and the S-Matrix

The motivation. We have examined in Chapter 10 the renormalization of a field theory, regarding the correlation functions of the field as fundamental physical objects. However in some cases all local functionals of the field which have a non-trivial linear part in ϕ are equivalent. Two important examples can be given:

(i) Some models are defined on Riemannian manifolds and the fields $\phi_i(x)$ correspond to a system of coordinates on the manifold. In some situations only quantities intrinsic to the manifold are physical.

(ii) In Particle Physics, S-matrix elements are calculated from connected correlation functions by taking the residues of the poles of the external propagator (see Section 7.3). Let us consider for simplicity the case of only one species of field ϕ, which we have defined in such a way that it has a vanishing expectation value. Then in terms of mass-shell amputated correlation functions:

$$W^{(n)}_{\text{amp.}}(p_1, \ldots, p_n)\big|_{p^2=-m^2} = \lim_{p^2 \to -m^2} W^{(n)}(p_1, \ldots, p_n) \prod_{i=1}^{n} \Gamma^{(2)}(p_i),$$

where $-m^2$ is the location of the pole in the variable p^2 of $W^{(2)}(p)$:

$$W^{(2)}(p) \sim \frac{Z}{p^2 + m^2} \qquad \text{for } p^2 \to -m^2. \tag{A10.23}$$

the matrix element $S^{(n)}$ is given by:

$$S^{(n)}(p_1, \ldots, p_n) = Z^{n/2} W^{(n)}_{\text{amp.}}(p_1, \ldots, p_n)\big|_{p^2=-m^2}. \tag{A10.24}$$

The factor $\sqrt{Z}$ in this equation corresponds to a finite renormalization of the field such that the residue of the 2-point function (equation $(A10.23)$) on the physical pole $p^2 = -m^2$ becomes 1 (see Section 7.4).

S-matrix and field representation. We now compare the S-matrix obtained from the ϕ field correlation functions to the S-matrix derived from the correlation functions of a different field $\phi'(x)$ related to $\phi(x)$ by:

$$\phi'(x) = \sum_{1}^{\infty} \frac{C_k}{k!} \phi^k(x), \qquad C_1 \neq 0. \tag{A10.25}$$

We assume, when necessary, that the theory has been regularized in such a way that the new correlation functions exist.

Using relation $(A10.25)$ we can reexpress the ϕ' correlation functions in terms of the ϕ correlation functions.

The expansion of the ϕ' propagator shows immediately that the ϕ and ϕ' propagators have poles at the same position (see figure 10.24).

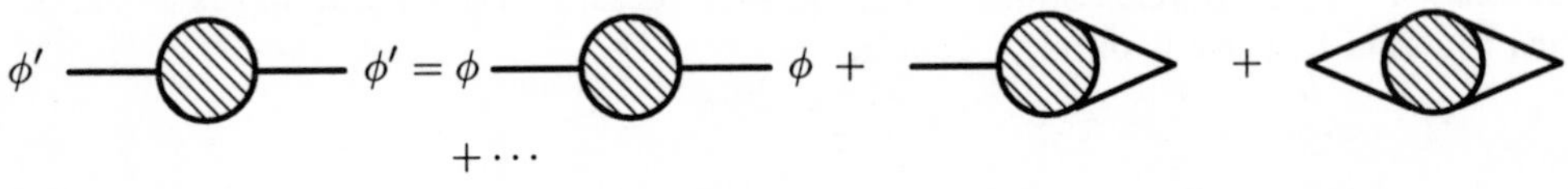

Fig. 10.24

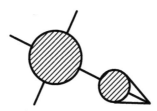

Fig. 10.25

The contributions to n-point functions which have poles on the external lines then have the form shown in figure 10.25.

In the mass shell limit $(p_i^2 = -m^2)$, the ϕ' and ϕ correlation functions become proportional. The S-matrix elements are identical. Here again all fields related by transformation $(A10.23)$ are equivalent.

In such situations not all parameters of the theory are physical. For example the field amplitude renormalization is obviously unphysical. The same physical theory may have renormalizable and non-renormalizable realizations.

Using the background field method one can avoid the calculation of unphysical quantities (however to prove renormalizability the study of correlation functions cannot be avoided).

The background field method. Let us consider a field theory with an action $S(\phi)$:

$$e^{W(J)} = \int [d\phi]\, e^{-S(\phi)+J\phi}\,. \qquad (A10.26)$$

Let us introduce immediately the Legendre transform of $W(J)$:

$$e^{-\Gamma(\varphi)+J\varphi} = \int [d\phi]\, e^{-S(\phi)+J\phi}\,. \qquad (A10.27)$$

Using:

$$J(x) = \frac{\delta\Gamma}{\delta\varphi(x)}\,,$$

we can write equation $(A10.27)$:

$$e^{-\Gamma(\varphi)} = \int [d\phi]\exp\left[-S(\phi) + \int dx\,(\phi(x) - \varphi(x))\frac{\delta\Gamma}{\delta\varphi(x)}\right]\,, \qquad (A10.28)$$

or equivalently translating $\phi(x)$:

$$e^{-\Gamma(\varphi)} = \int [d\phi]\exp\left[-S(\phi+\varphi) + \int dx\,\phi(x)\frac{\delta\Gamma}{\delta\varphi(x)}\right]\,. \qquad (A10.29)$$

Let us now assume that the equation:

$$\frac{\delta\Gamma}{\delta\varphi(x)} = 0\,, \qquad (A10.30)$$

has a non-trivial solution $\varphi_c(x)$ which at leading order in perturbation theory is a solution $\varphi_c^{(0)}(x)$ of the classical equation of motion:

$$\frac{\delta S}{\delta\varphi(x)}\left[\varphi_c^{(0)}\right] = 0\,. \qquad (A10.31)$$

Then equation ($A10.29$) becomes:

$$e^{-\Gamma(\varphi_c)} = \int [d\phi]\, e^{-S(\phi+\varphi_c)}\,. \qquad (A10.32)$$

The quantity $\Gamma(\varphi_c)$ is obviously independent of the representation of the field ϕ, and contains therefore only physical informations in the sense defined at the beginning of this section. Let us introduce renormalized quantities:

$$\Gamma_{\mathrm{r}}(\varphi_c) = -\ln \int [d\phi]\, \exp\left[-S_0\left(\phi + \varphi_c\right) + \text{counterterms}\right],$$

in which $S_0(\phi)$ is the tree level action. The solution φ_c of:

$$\frac{\delta\Gamma_{\mathrm{r}}}{\delta\varphi_c} = 0$$

is expanded around the solution of $\varphi_c^{(0)}(x)$ of

$$\frac{\delta S_0}{\delta\varphi_c(x)} = 0\,.$$

In real time, as shown in Subsection 7.3.3, the background field method yields the S-matrix. We shall provide other examples of such calculations in the coming chapters.

11 DIMENSIONAL REGULARIZATION, MINIMAL SUBTRACTION: CALCULATION OF RG FUNCTIONS

Dimensional regularization is one of the most powerful regularization techniques, and the most convenient for practical perturbative calculations, when applicable. We have discussed its implementation in Section 9.3. We want to here show how it leads naturally to the concept of renormalization by minimal subtraction. We first discuss the form of renormalization constants and renormalization group functions $\beta(g)$, $\eta(g)$, $\eta_2(g)$ when dimensional regularization is used, and then show that this form is specially simple in the minimal subtraction scheme. We perform explicit calculations at two-loop order in the ϕ^4 field theory. The calculations in this theory are simple and useful for the theory of Critical Phenomena. Moreover the results can be compared with those obtained in previous chapters. We then generalize to a several component four field interaction. Finally, to give an example of a theory involving fermions, we calculate the RG functions at one-loop order in a theory containing fermions interacting through a Yukawa-like interaction with a scalar boson: the linearized Gross–Neveu model.

11.1 Bare and Renormalized Actions: Renormalization Group (RG) Functions

We again discuss the example of the ϕ^4 field theory with the bare action:

$$S(\phi_0) = \int d^d x \left[\tfrac{1}{2} (\partial_\mu \phi_0)^2 + \tfrac{1}{2} m_0^2 \phi_0^2 + \frac{1}{4!} g_0 \phi_0^4 \right]. \tag{11.1}$$

When dimensional regularization is used, the only dimensional parameters are the bare parameters $\{m_0, g_0\}$: In $d = 4 - \varepsilon$ dimension the bare coupling constant has a dimension $[g_0]$

$$[g_0] = d - 4[\phi_0] = d - 2(d-2) = 4 - d = \varepsilon. \tag{11.2}$$

It is then convenient to introduce, besides the renormalized mass m, a dimensionless renormalized coupling constant g, in such a way that the renormalized action can be written:

$$S_r(\phi) = \int d^d x \left[\tfrac{1}{2} Z(g) (\partial_\mu \phi)^2 + \tfrac{1}{2} m^2 Z_m(g) \phi^2 + \frac{1}{4!} m^{4-d} g Z_g(g) \phi^4 \right]. \tag{11.3}$$

Indeed with this parametrization the renormalization constants $Z(g)$, $Z_m(g)$ and $Z_g(g)$, are dimensionless, and thus depend on the only dimensionless parameter available, the coupling constant g. In particular the mass is multiplicatively renormalizable.

Since the renormalized action is obtained from the bare action by the field rescaling $\phi_0 = \phi \sqrt{Z}$, the bare and renormalized parameters are related by:

$$m_0^2 = m^2 Z_m(g) / Z(g), \tag{11.4}$$

$$g_0 = g \, m^{4-d} Z_g(g) / Z^2(g). \tag{11.5}$$

Remember that in the dimensional regularization scheme renormalization constants depend on an additional hidden parameter, the dimension d.

From the relation (11.4) we can now calculate the RG β-function, the coefficient of the CS equation. Setting

$$g\,Z_g\,/Z^2 = G(g),\qquad(11.6)$$

and differentiating equation (11.5) with respect to m at g_0 fixed, we find (equation (10.26)):

$$0 = (4-d)\,G(g) + \beta(g)\frac{\partial}{\partial g}G(g),$$

or:

$$\beta(g) = -(4-d)\left(\frac{d\ln G(g)}{dg}\right)^{-1}.\qquad(11.7)$$

Using the notation of Section 10.4, we call $Z_2(g)$ the renormalization constant associated with the renormalization of ϕ^2. From equations (10.27,10.28) we then derive:

$$\eta(g) = \beta(g)\frac{d}{dg}\ln Z(g),\qquad(11.8)$$

$$\eta_2(g) = \beta(g)\frac{d}{dg}\ln\left[Z_2(g)\,/Z(g)\right],\qquad(11.9)$$

and finally from equations (11.4) and (10.29):

$$\frac{Z_m}{Z_2}\left[2 + \beta(g)\frac{d}{dg}\ln\left(Z_m\,/Z\right)\right] = \sigma(g).\qquad(11.10)$$

This equation shows that Z_m/Z_2 is a finite function of g, and therefore Z_m is not a new renormalization constant.

The renormalizability of the ϕ^4 field theory in 4 dimensions implies that the renormalized correlation functions and therefore also the RG functions $\beta(g)$, $\eta(g)$, $\eta_2(g)$ and $\sigma(g)$ have a finite limit when the deviation

$$\varepsilon = 4 - d$$

from the dimension 4 goes to 0. Since

$$G(g) = g + O\left(g^2\right),$$

the function $\beta(g)$ can thus be written:

$$\beta(g) = -\varepsilon g + \beta_2\left(\varepsilon\right)g^2 + \beta_3\left(\varepsilon\right)g^3 + \cdots,\qquad(11.11)$$

in which all the functions $\beta_n\left(\varepsilon\right)$ have regular Taylor series expansion at $\varepsilon = 0$.

$$\beta_n\left(\varepsilon\right) = \beta_n\left(0\right) + \varepsilon\beta_n'\left(0\right) + \cdots.$$

11.2 Dimensional Regularization: The Form of Renormalization Constants

Conversely let us now derive the form of $G(g)$ from the knowledge of $\beta(g)$:

$$g\frac{G'(g)}{G(g)} = -\frac{\varepsilon g}{\beta(g)} \equiv \left[1 - \frac{1}{\varepsilon}\beta_2(\varepsilon)\,g - \frac{1}{\varepsilon}\beta_3(\varepsilon)\,g^2 - \cdots\right]^{-1}.$$

If we expand the r.h.s. in powers of g, we observe that, at a fixed order in g, the most singular term in ε comes from the term of order g^2 in $\beta(g)$:

$$\frac{G'(g)}{G(g)} = \frac{1}{g} + \left(\frac{\beta_2(0)}{\varepsilon} + O(1)\right) + g\left(\frac{\beta_2^2(0)}{\varepsilon^2} + O\left(\frac{1}{\varepsilon}\right)\right) + \cdots.$$

Integrating the expansion term by term we get:

$$G(g) = g + \sum_{n=2} g^n \left[\left(\frac{\beta_2(0)}{\varepsilon}\right)^{n-1} + \text{ less singular terms}\right].$$

The coefficient $\tilde{G}_n(\varepsilon)$ of the expansion of $G(g)$ in powers of g:

$$G(g) = g + \sum_{2}^{\infty} g^n \tilde{G}_n(\varepsilon),$$

has thus a Laurent series expansion in ε, for ε small, of the form:

$$\tilde{G}_n(\varepsilon) = \frac{\beta_2^{n-1}(0)}{\varepsilon^{n-1}} + \frac{G_{n,2-n}}{\varepsilon^{n-2}} + \cdots + G_{n,0} + G_{n,1}\varepsilon + \cdots.$$

The finiteness of $\eta(g)$ and $\eta_2(g)$ leads to similar conclusions for $Z(g)$ and $Z_2(g)$ which can be written:

$$Z(g) = 1 + \sum_{1}^{\infty} \frac{\alpha^{(n)}(g)}{\varepsilon^n} + \text{ regular terms in } \varepsilon,$$

with: $\alpha^{(n)}(g) = O\left(g^{n+1}\right)$,

$$Z_2(g) = 1 + \sum_{1}^{\infty} \frac{\alpha_2^{(n)}(g)}{\varepsilon^n} + \text{ regular terms in } \varepsilon,$$

with: $\alpha_2^{(n)} = O(g^n)$.

We conclude that at order L in the loopwise expansion the divergent part of $\Gamma(\varphi)$ is a polynomial of degree L in $1/\varepsilon$.

11.3 Minimal Subtraction Scheme

Although the minimal subtraction idea can be used in any regularization scheme (see for example equation (8.11)), it is specially useful, in particular for critical phenomena, in dimensional regularization. Renormalization constants are determined in the following way: instead of imposing renormalization conditions to divergent 1PI correlation functions, one just subtracts, at each order in the loop expansion, the singular part of Laurent expansion in ε. Let us call $\Gamma_L^{\mathrm{div}}(\varphi)$ the divergent part of the generating functional of proper vertices, renormalized up to $L-1$ loops:

$$\Gamma_L^{\mathrm{div}}(\varphi) = \sum_{\ell=1}^{L} \frac{\gamma_{L,\ell}}{\varepsilon^\ell}. \tag{11.12}$$

We add as a counterterm to the action just $-\Gamma_L^{\mathrm{div}}(\varphi)$ as defined above.

Example. Let us again consider the ϕ^4 field theory. At one-loop order only two diagrams contributing to $\Gamma^{(2)}$ and $\Gamma^{(4)}$ are divergent:

$$\frac{1}{(2\pi)^d} \int \frac{\mathrm{d}^d p}{p^2 + m^2} = m^{d-2} \frac{1}{(4\pi)^{d/2}} \Gamma\left(1 - \frac{d}{2}\right) = -\frac{m^2}{8\pi^2 \varepsilon} + O(1), \tag{11.13}$$

$$\frac{1}{(2\pi)^d} \int \frac{\mathrm{d}^d p}{(p^2 + m^2)^2} = \frac{1}{8\pi^2 \varepsilon} + O(1). \tag{11.14}$$

The one-loop divergent part of $\Gamma(\varphi)$ is then:

$$\Gamma_1^{\mathrm{div}}(\varphi) = -\frac{1}{32\pi^2 \varepsilon}\left[m^2 g \int \varphi^2(x)\mathrm{d}^d x + \frac{1}{4}g^2 m^\varepsilon \int \varphi^4(x)\mathrm{d}^d x\right]. \tag{11.15}$$

This expression should be compared to equations $(A9.17)$ and (10.8). The functions Z, Z_m, Z_g at this order are then:

$$Z = 1 + O(g^2), \tag{11.16}$$

$$Z_g = 1 + \frac{3}{16\pi^2}\frac{g}{\varepsilon} + O(g^2), \tag{11.17}$$

$$Z_m = 1 + \frac{1}{16\pi^2}\frac{g}{\varepsilon} + O(g^2). \tag{11.18}$$

The calculation of the function $\langle \phi^2 \phi \phi \rangle$ at one-loop also leads to:

$$Z_2 = 1 + \frac{1}{16\pi^2}\frac{g}{\varepsilon} + O(g^2). \tag{11.19}$$

The RG functions are then:

$$\beta(g) = -\varepsilon g + \frac{3}{16\pi^2}g^2 + O(g^3), \tag{11.20}$$

$$\eta(g) = O(g^2), \tag{11.21}$$

$$\eta_2(g) = -\frac{g}{16\pi^2} + O(g^2). \tag{11.22}$$

We note that at this order $Z_m = Z_2$. Since the minimal subtraction scheme eliminates any possible finite renormalization and Z_2/Z_m is finite, this relation remains true to all orders.

Let us explore more generally the consequences of this choice for the RG functions. Writing for instance:

$$G(g) = g + \sum_1^\infty \frac{G_n(g)}{\varepsilon^n}, \quad G_n(g) = O\left(g^{n+1}\right), \tag{11.23}$$

we can calculate $\beta(g)$:

$$\beta(g) = -\varepsilon \left[g + \sum_1^\infty \frac{G_n(g)}{\varepsilon^n}\right]\left[1 + \sum_1^\infty \frac{G_n'(g)}{\varepsilon^n}\right]^{-1}.$$

Since $G_n'(g)$ is of order g^n, we can expand the denominator:

$$\beta(g) = -\varepsilon \left[g + \sum_1^\infty \frac{G_n(g)}{\varepsilon^n}\right]\left[1 - \frac{G_1'(g)}{\varepsilon} + \frac{[G_1'(g)]^2}{\varepsilon^2} - \frac{G_2'(g)}{\varepsilon^2} + \cdots\right].$$

This expression can be rewritten:

$$\beta(g) = -\varepsilon g - G_1(g) + gG_1'(g) + \sum_1^\infty \frac{b_n(g)}{\varepsilon^n}.$$

The finiteness of $\beta(g)$ then implies:

$$\beta(g) = -\varepsilon g + gG_1'(g) - G_1(g), \tag{11.24}$$
$$b_n(g) = 0 \quad \forall n. \tag{11.25}$$

The functions $\beta(g)$ and $G_n(g)$, $n \geq 2$, are uniquely determined by the function $G_1(g)$, i.e. the coefficients of $1/\varepsilon$ in the divergences.

Similar arguments apply for the other RG functions and renormalization constants.

In the expansion of equation (11.12) the whole new L-loop information about divergences is contained in $\gamma_{L,1}(\varphi)$. All other functions are determined by the counterterms of previous orders.

In addition the RG functions $\beta(g)$, $\eta(g)$ and $\eta_2(g)$ have a very simple dependence in ε. For $\beta(g)$ it is given by equation (11.24). Let us calculate η and η_2. The renormalization constants Z and Z_2 now have the form:

$$Z(g) = 1 + \sum_1^\infty \frac{\alpha^{(n)}(g)}{\varepsilon^n}, \quad \alpha^{(n)}(g) = O\left(g^{n+1}\right), \tag{11.26}$$

$$Z_2(g) = 1 + \sum_1^\infty \frac{\alpha_2^{(n)}(g)}{\varepsilon^n}, \quad \alpha_2^{(n)}(g) = O\left(g^n\right). \tag{11.27}$$

Using relation (11.8) and the form (11.24) of $\beta(g)$ we obtain:

$$\eta(g) = [-\varepsilon g + gG_1'(g) - G_1(g)]\left[\frac{1}{\varepsilon}\frac{\mathrm{d}}{\mathrm{d}g}\alpha^{(1)}(g) + O\left(\frac{1}{\varepsilon^2}\right)\right].$$

Since $\eta(g)$ has a finite limit it is given by:

$$\eta(g) = -g \frac{\mathrm{d}}{\mathrm{d}g} \alpha^{(1)}(g).$$

(11.28)

Similarly for $\eta_2(g)$ we find:

$$\eta_2(g) = -g \frac{\mathrm{d}}{\mathrm{d}g} \alpha_2^{(1)}(g).$$

(11.29)

Finally the explicit dependence of the renormalization constants in ε can be obtained by calculating them from β, η and η_2. For example:

$$G(g) = g \exp\left(-\varepsilon \int_0^g \left[\frac{1}{-\varepsilon g' + b(g')} + \frac{1}{\varepsilon g'}\right] \mathrm{d}g'\right),$$

(11.30)

in which we have set

$$\beta(g, \varepsilon) = -\varepsilon g + b(g).$$

(11.31)

11.4 The Massless Theory

Dimensional regularization can also be used to define the massless ϕ^4 field theory. Unlike, however, the massive theory, due to small momentum (IR) divergences the regularized theory only exists in an infinitesimal neighbourhood of the dimension four. This problem will be discussed at length in the Chapters 25–26 devoted to critical phenomena. Then since the massless theory is renormalizable, the minimal subtraction scheme is also applicable.

The bare action can be written

$$S(\phi_0) = \int \mathrm{d}^d x \left[\tfrac{1}{2} (\partial_\mu \phi_0(x))^2 + \frac{1}{4!} g_0 \phi_0^4(x)\right].$$

(11.32)

Note the absence of a bare mass term. Indeed if the propagator is massless no mass can be generated because there no dimensional parameter, besides the coupling constant: All diagrams contributing to the two-point function have a power-law behaviour given by simple dimensional considerations:

$$\tilde{\Gamma}^{(2)}(p) = p^2 + \sum_{n=2} C_n(\varepsilon) g_0^n p^{2-n\varepsilon},$$

(remember that ε is infinitesimal).

To define a renormalized theory it is necessary to introduce a mass scale μ which will take care of the dimension of the ϕ^4 coupling constant:

$$S_\mathrm{r}(\phi) = \int \mathrm{d}^d x \left[\frac{1}{2} Z(g) (\partial_\mu \phi)^2 + \frac{1}{4!} \mu^{4-d} g \, Z_g \phi^4\right].$$

(11.33)

To the action corresponds a set of relations between bare and renormalized correlation functions:

$$Z^{n/2}(g) \Gamma_\mathrm{r}^{(n)}(p_i; \mu, g) = \Gamma^{(n)}(p_i; g_0),$$

(11.34)

with:

$$g_0 = \mu^{4-d} g Z_g / Z^2.$$

(11.35)

Differentiation of equation (11.34) with respect to μ at g_0 fixed, yields the RG equations which express that correlation functions depend on μ and g only through the combination $g_0(\mu, g)$:

$$\left[\mu \frac{\partial}{\partial \mu} + \beta(g) \frac{\partial}{\partial g} - \frac{n}{2} \eta(g) \right] \tilde{\Gamma}_{\mathrm{r}}^{(n)}(p_i; \mu, g) = 0 \,, \tag{11.36}$$

with for β and η expressions similar to equations (11.7,11.8):

$$\beta(g) = -(4 - d) \left(\frac{\mathrm{d} \ln G(g)}{\mathrm{d} g} \right)^{-1} , \qquad \eta(g) = \beta(g) \frac{\mathrm{d}}{\mathrm{d} g} \ln Z(g). \tag{11.37}$$

As we have discussed in Section 10.10, it is then possible to define a massive theory by adding to the renormalized action a mass term of the form

$$S_{\mathrm{r}}(m) = S_{\mathrm{r}} + \tfrac{1}{2} m^2 \int \mathrm{d}^d x \, Z_2(g) \phi^2(x).$$

Note that since in the minimal subtraction scheme the renormalization constants are uniquely defined and independent on the ratio m/μ one concludes by setting $m = \mu$ that the renormalization constants in the massless theory and the massive theory of Section 11.1 are the same.

11.5 A Few Technical Remarks

Feynman parameters. Before presenting explicit calculations of Feynman diagrams at two-loop order, let us derive a simple identity which is often useful. One starts from:

$$\frac{1}{a^\alpha} = \frac{1}{\Gamma(\alpha)} \int_0^\infty \mathrm{d}t \, t^{\alpha-1} \, \mathrm{e}^{-at} \,. \tag{11.38}$$

Therefore:

$$\prod_{i=1}^n (a_i)^{-\alpha_i} = \prod_{i=1}^n (\Gamma(\alpha_i))^{-1} \int_0^\infty \left(\prod_{i=1}^n \mathrm{d}t \, t_i^{\alpha_i-1} \right) \exp\left(- \sum_{i=1}^n a_i t_i \right). \tag{11.39}$$

Setting then:

$$t_i = s u_i \,, \tag{11.40}$$

with

$$u_i \geq 0 \,, \qquad \sum_{i=1}^n u_i = 1 \,,$$

we can integrate over s to obtain:

$$\prod_{i=1}^n (a_i)^{-\alpha_i} = \Gamma(\alpha_1 + \alpha_2 + \cdots + \alpha_n) \prod_{i=1}^n (\Gamma(\alpha_i))^{-1}$$

$$\times \int_0^\infty \delta\left(\sum_i u_i - 1 \right) \prod_{i=1}^n \mathrm{d}u_i u_i^{\alpha_i-1} \left(\sum_{i=1}^n a_i u_i \right)^{-\sum_i \alpha_i}. \tag{11.41}$$

If the quantities $a_1,...,a_n$ correspond to propagators, $a_i \equiv p_i^2 + m_i^2$, the integral over momenta can then be explicitly performed. At one-loop order only one integral is needed:

$$\frac{1}{(2\pi)^d} \int \frac{\mathrm{d}^d p}{(p^2+1)^{n+1}} = \frac{1}{(2\pi)^d \, \Gamma(n+1)} \int_0^\infty t^n \mathrm{d}t \int \mathrm{d}^d p \, \mathrm{e}^{-tp^2 - t} = \frac{\Gamma(n+1-d/2)}{(4\pi)^{d/2}\Gamma(n+1)}.$$

(11.42)

Note that representation (9.28) leads to an expression similar to (11.41). When $\rho(t)$ is of the form e^{-tm^2}, the argument of the exponential is linear in the variables t_i and after the change of variables (11.40), the integral over the homogeneous variable s can be performed.

In what follows we also need the integral

$$\int_0^1 x^{\alpha-1} (1-x)^{\beta-1} \, \mathrm{d}x = \frac{\Gamma(\alpha)\Gamma(\beta)}{\Gamma(\alpha+\beta)}.$$

(11.43)

The modified minimal subtraction scheme. In the calculation of Feynman diagrams, a factor $(N_d)^L$:

$$N_d = \frac{\text{area of the sphere } S_{d-1}}{(2\pi)^d} = \frac{2}{(4\pi)^{d/2}\Gamma(d/2)},$$

(11.44)

L being the number of loops of the diagram, is generated naturally. It is therefore convenient to rescale the loopwise expansion parameter to suppress this factor. In the ϕ^4 field theory for example two choices are possible. Either one multiplies each Feynman diagram by a factor $(N_4/N_d)^L$, then the normalizations of field and coupling constant in 4 dimensions are only modified by the change in the renormalization constants. Or one completely absorbs the factor N_d in a coupling constant and field redefinition.

11.6 Two-Loop Calculation of Renormalization Constants and RG Functions in the ϕ^4 Field Theory

As an exercise, we now explicitly calculate the two-loop renormalization constants in the ϕ^4 field theory, and in the minimal subtraction scheme. We work with the massive theory though the calculations are somewhat easier in the massless theory. Note that for higher order calculations even more sophisticated methods can be used, where the masses in diagrams are chosen is the following way: Only diagrams to which all divergent subdiagrams have been subtracted are considered. The remaining global divergence is then independent of momenta and internal masses. One sets to zero as many masses and momenta as possible, as long as one does not encounter IR divergences, consistently in the diagram and the subtracted subdiagrams.

11.6.1 The diagrams

In Section 5.3 we have already given the diagrams contributing to the 2-point function at this order. A short calculation gives also $\Gamma^{(4)}$:

$$\Gamma^{(2)}(p) = p^2 + m^2 + \frac{1}{2}g\,(\mathrm{a}) - \frac{g^2}{4}\,(\mathrm{b}) - \frac{g^2}{6}\,(\mathrm{c}) + O\left(g^3\right),$$

(11.45)

$$\Gamma^{(4)}(0,0,0,0) = g - \frac{3}{2}g^2\,(\mathrm{d}) + \frac{3}{4}g^3\,(\mathrm{e}) + 3g^3\,(\mathrm{f}) - \frac{3}{2}g^3\,(\mathrm{g}) + O\left(g^4\right).$$

(11.46)

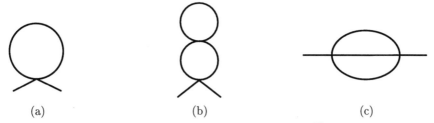

Fig. 11.1   Diagrams contributing to $\Gamma^{(2)}$.

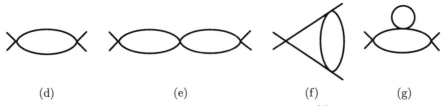

Fig. 11.2   Diagrams contributing to $\Gamma^{(4)}$.

To determine all renormalization constants we thus need the diagrams displayed in figures 11.1 and 11.2.

The diagram (g) contributing to $\Gamma^{(4)}$ corresponds to the insertion of the one-loop $\Gamma^{(2)}$ diagram into the one-loop $\Gamma^{(4)}$ diagram. When the one-loop diagrams are renormalized, it is automatically finite. At this order only the diagram (c) has a divergence which is momentum dependent:

$$\text{(c)} \equiv \Sigma^{(2)}(p,m) = \frac{1}{(2\pi)^{2d}} \int \frac{\mathrm{d}^d q_1 \, \mathrm{d}^d q_2}{(q_1^2 + m^2)\,(q_2^2 + m^2)\left[(p + q_1 + q_2)^2 + m^2\right]}. \tag{11.47}$$

To renormalize the 2-point function we need (a), (b), $\Sigma(0,m)$ and $\partial \Sigma(p,m)/\partial p^2$ at $\mathbf{p}=0$. We first set:

$$\text{(a)} \equiv \frac{1}{(2\pi)^d} \int \frac{\mathrm{d}^d q}{q^2 + m^2} = m^{d-2} I_1, \tag{11.48}$$

$$\Sigma^{(2)}(0,m) = m^{2d-6} I_2, \tag{11.49}$$

$$\left. \frac{\partial}{\partial p^2} \Sigma^{(2)}(p,m) \right|_{\mathbf{p}=0} = m^{2d-8} I_3. \tag{11.50}$$

The diagram (b) as well as the diagrams (d) ... (f) at zero momentum can be expressed in terms of the integrals I_1 and I_2:

$$\text{(d)} \equiv \frac{1}{(2\pi)^d} \int \frac{\mathrm{d}^d q}{(q^2 + m^2)^2} = -\frac{\partial}{\partial m^2}\text{(a)} = (1 - d/2)\, m^{d-4} I_1,$$

$$\text{(b)} \equiv \text{(a)(d)} = (1 - d/2)\, m^{2d-6} I_1^2,$$

$$\text{(e)} \equiv [\text{(d)}]^2 = (1 - d/2)^2\, m^{2d-8} I_1^2, \tag{11.51}$$

and finally:

$$\text{(f)} = \frac{1}{(2\pi)^{2d}} \int \frac{\mathrm{d}^d q_1 \, \mathrm{d}^d q_2}{(q_1^2 + m^2)\,(q_2^2 + m^2)\left[(q_1 + q_2)^2 + m^2\right]^2} = -\frac{1}{3}\frac{\partial}{\partial m^2}\Sigma^{(2)}(0,m)$$

$$= -\tfrac{1}{3}(d - 3)m^{2d-8} I_2. \tag{11.52}$$

The integral I_1. The integral is a special example of (11.42). As explained in Section 11.5, we extract a factor N_d (equation (11.44)) for each loop. Setting then:

$$d = 4 - \varepsilon ,$$

we find

$$I_1 = N_d \frac{\pi}{2 \sin \pi d/2} = N_d \left(-\frac{1}{\varepsilon} + O\left(\varepsilon\right) \right). \tag{11.53}$$

Calculation of I_2. To derive an expression both for I_2 and I_3 we consider $\Sigma^{(2)}(p,m)$:

$$\Sigma^{(2)}\left(p, m = 1\right) = \frac{1}{(2\pi)^{2d}} \int \frac{d^d q_1\, d^d q_2}{\left(q_1^2 + 1\right)\left(q_2^2 + 1\right)\left[\left(p + q_1 + q_2\right)^2 + 1\right]}. \tag{11.54}$$

Using the method explained in Section 9.3 we write:

$$\Sigma^{(2)}(p,1) = \frac{1}{(2\pi)^{2d}} \int_0^\infty dt_1 dt_2 dt_3 \int d^d q_1 d^d q_2 \exp\left[-A\left(q_1, q_2, p\right)\right], \tag{11.55}$$

with:

$$A\left(q_1; q_2, p\right) = \left(t_1 + t_3\right) q_1^2 + \left(t_2 + t_3\right) q_2^2 + 2t_3 q_1 q_2 + 2t_3 p \left(q_1 + q_2\right) + t_3 p^2 + \left(t_1 + t_2 + t_3\right). \tag{11.56}$$

Integrating over q_1 and q_2 we find (equation (9.28)):

$$\Sigma^{(2)}(p,1) = \frac{1}{(4\pi)^d} \int_0^\infty \frac{dt_1\, dt_2\, dt_3}{\Delta^{d/2}} \exp\left[-\left(t_1 + t_2 + t_3\right) - t_1 t_2 t_3 p^2 / \Delta\right], \tag{11.57}$$

in which Δ is the determinant associated with the quadratic form:

$$\Delta = t_1 t_2 + t_2 t_3 + t_3 t_1 . \tag{11.58}$$

We only need the two first terms of the expansion in powers of p^2. Let us first calculate I_2 which is more complicated:

$$I_2 = \Sigma^{(2)}(0,1) = \frac{1}{(4\pi)^d} \int_0^\infty \frac{dt_1\, dt_2\, dt_3}{\Delta^{d/2}} e^{-(t_1+t_2+t_3)}. \tag{11.59}$$

As explained in Section 11.5 we now change variables, setting:

$$\begin{cases} t_1 = stu , \\ t_2 = st\left(1 - u\right), \\ t_3 = s\left(1 - t\right). \end{cases} \tag{11.60}$$

The integral becomes

$$I_2 = \frac{1}{(4\pi)^d} \int_0^\infty ds\, e^{-s}\, s^{2-d} \int_0^1 du\, dt \frac{t^{1-d/2}}{\left[1 - t + tu\left(1 - u\right)\right]^{d/2}}. \tag{11.61}$$

The integral over s can be performed. The remaining integral J, from which we want to extract the divergent and the finite parts, has divergences coming from $t = 0$ and $t = 1$, $u = 0$ and $u = 1$.

$$J = \int_0^1 du\, dt\, t^{1-d/2} \left[1 - t + tu\,(1 - u)\right]^{-d/2}. \tag{11.62}$$

To separate divergences, we subtract and add 1 to each factor in the product:

$$\begin{aligned}
J = \int_0^1 du\, dt\, &\left\{1 + \left(t^{1-d/2} - 1\right) + \left[1 - t + tu\,(1 - u)\right]^{-d/2} - 1\right] \\
&+ \left(t^{1-d/2} - 1\right)\left[(1 - t + tu\,(1 - u))^{-d/2} - 1\right]\right\}.
\end{aligned} \tag{11.63}$$

The last term is finite and we can set $d = 4$. The others can be integrated immediately over t:

$$\begin{aligned}
J = \frac{2}{\varepsilon} - 1 &+ \frac{1}{(1 - \varepsilon/2)} \int_0^1 du\, \left\{\left[u\,(1 - u)\right]^{1-d/2} - 1\right\} \left[1 - u\,(1 - u)\right]^{-1} \\
&+ \int_0^1 du\, dt\, \frac{(1 - t)}{t} \left[(1 - t + tu\,(1 - u))^{-2} - 1\right].
\end{aligned} \tag{11.64}$$

The last integral in the expression yields:

$$\int_0^1 du\, dt\, \frac{(1 - t)}{t} \left[(1 - t + tu\,(1 - u))^{-2} - 1\right] = -\int_0^1 du\, \ln\left[u\,(1 - u)\right] = 2.$$

A last piece has to be worked out:

$$\begin{aligned}
\int_0^1 \frac{du}{1 - u\,(1 - u)} &\left\{\left[u\,(1 - u)\right]^{1-d/2} - 1\right\} = \int_0^1 du\, \left\{\left[u\,(1 - u)\right]^{1-d/2} - 1\right\} \\
&+ \int_0^1 \frac{du\, u\,(1 - u)}{1 - u\,(1 - u)} \left[\frac{1}{u\,(1 - u)} - 1\right] + O\,(\varepsilon).
\end{aligned}$$

Therefore:

$$\int_0^1 \frac{du}{1 - u\,(1 - u)} \left\{\left[u\,(1 - u)\right]^{1-d/2} - 1\right\} = \int_0^1 du\, \left[u\,(1 - u)\right]^{1-d/2} + O\,(\varepsilon). \tag{11.65}$$

We now use the identity (11.43) and collect all contributions:

$$J = \frac{2}{\varepsilon} - 1 + \frac{1}{(1 - \varepsilon/2)} \frac{\Gamma^2\,(\varepsilon/2)}{\Gamma\,(\varepsilon)} + 2 + O\,(\varepsilon). \tag{11.66}$$

Since:

$$\frac{\Gamma^2\,(\varepsilon/2)}{\Gamma\,(\varepsilon)} = \frac{4}{\varepsilon} + O\,(\varepsilon), \tag{11.67}$$

we obtain the final result

$$J = \frac{6}{\varepsilon} + 3 + O\,(\varepsilon). \tag{11.68}$$

This leads for I_2 to:

$$I_2 = N_d^2 \frac{1}{4} \Gamma^2 (d/2) \Gamma (3-d) \frac{6}{\varepsilon} \left(1 + \frac{\varepsilon}{2}\right) + O(1). \tag{11.69}$$

Using:

$$\Gamma^2 (d/2) \Gamma (3-d) = -\frac{(1-\varepsilon/2)^2}{\varepsilon (1-\varepsilon)} + O(\varepsilon), \tag{11.70}$$

we deduce the singular part of integral I_2:

$$I_2 = -N_d^2 \frac{3}{2\varepsilon^2} \left(1 + \frac{\varepsilon}{2}\right) + O(1). \tag{11.71}$$

The integral I_3. The calculation of I_3 relies on the same method but is simpler. I_3 is the coefficient of p^2 in expression (11.57):

$$I_3 = \frac{\partial}{\partial p^2} \Sigma^{(2)}(p,1)\bigg|_{\mathbf{p}=0} = -\frac{1}{(4\pi)^d} \int_0^\infty \frac{dt_1\, dt_2\, dt_3\, t_1 t_2 t_3}{\Delta^{1+d/2}} \, \mathrm{e}^{-(t_1+t_2+t_3)}.$$

After the change of variables (11.60) and the integration over s we find:

$$I_3 \sim -\frac{1}{(4\pi)^d} \Gamma(\varepsilon) \int_0^1 dt\, du \frac{(1-t)\, u\, (1-u)}{[1-t+ut(1-u)]^3}. \tag{11.72}$$

A short calculation then leads to:

$$I_3 \sim -N_d^2 \frac{1}{8\varepsilon} + O(1). \tag{11.73}$$

The 4-point function at one-loop order. As a final exercise let us also calculate explicitly the finite part of the one-loop 4-point function which is given by the integral $I_4(p)$:

$$I_4(p) = \frac{m^{4-d}}{(2\pi)^d} \int \frac{d^d q}{(q^2 + m^2)\left[(p+q)^2 + m^2\right]}. \tag{11.74}$$

The calculation has been already performed in Section 9.3 for the case $m = 0$. The same method leads to:

$$I_4(p) = N_d \frac{1}{2} (1-d/2) \frac{\pi}{\sin \pi d/2} \int_0^1 dt \left[\frac{p^2}{m^2} t (1-t) + 1\right]^{d/2-2}. \tag{11.75}$$

Expanding now for ε small we obtain:

$$I_4(p) = N_d \left[\frac{1}{\varepsilon} - \frac{1}{2} - \frac{1}{2} \int_0^1 dt \ln \left(\frac{p^2}{m^2} t (1-t) + 1\right) + O(\varepsilon)\right]. \tag{11.76}$$

Taking the finite part and performing explicitly the last integral we find:

$$\left(I_4(p) - \frac{N_d}{\varepsilon}\right)_{\varepsilon=0} = -\frac{1}{8\pi^2} \left[\frac{1}{2} \sqrt{\frac{p^2 + 4m^2}{p^2}} \ln \frac{\sqrt{p^2 + 4m^2} + \sqrt{p^2}}{\sqrt{p^2 + 4m^2} - \sqrt{p^2}} - \frac{1}{2}\right]. \tag{11.77}$$

11.6.2 Renormalization constants and RG functions

We now introduce the renormalization constants, i.e. replace in expressions (11.45) and (11.46) the propagator $\left(p^2 + m^2\right)^{-1}$ by:

$$\left(p^2 + m^2\right)^{-1} \mapsto \left(Zp^2 + m^2 Z_2\right)^{-1},$$

and the coupling constant g by:

$$g \mapsto g Z_g.$$

We then expand the renormalization constants in powers of the coupling constant g and adjust the coefficients to cancel all divergences.

Renormalization constants. Integral I_3 immediately yields Z:

$$Z = 1 - N_d^2 \frac{g^2}{48\varepsilon} + O\left(g^3\right). \tag{11.78}$$

Expressing that $\Gamma^{(4)}$ is finite, we obtain Z_g:

$$Z_g = 1 + N_d \frac{3g}{2\varepsilon} + N_d^2 \left(\frac{9}{4\varepsilon^2} - \frac{3}{4\varepsilon}\right) g^2 + O\left(g^3\right). \tag{11.79}$$

Finally the propagator at zero momentum determines Z_2:

$$Z_2 = 1 + N_d \frac{g}{2\varepsilon} + N_d^2 \left(\frac{1}{2\varepsilon^2} - \frac{1}{8\varepsilon}\right) g^2 + O\left(g^3\right). \tag{11.80}$$

RG functions. Equations (11.7–11.9) then yield the RG functions:

$$\tilde{\beta}\left(\tilde{g}\right) = N_d \beta\left(\tilde{g}\right) = -\varepsilon\tilde{g} + \frac{3}{2}\tilde{g}^2 - \frac{17}{12}\tilde{g}^3 + O\left(\tilde{g}^4\right), \tag{11.81}$$

$$\eta\left(\tilde{g}\right) = \frac{\tilde{g}^2}{24} + O\left(\tilde{g}^3\right), \tag{11.82}$$

$$\eta_2\left(\tilde{g}\right) = -\frac{\tilde{g}}{2} + \frac{5\tilde{g}^2}{24} + O\left(\tilde{g}^3\right), \tag{11.83}$$

with the notation:

$$\tilde{g} = N_d g. \tag{11.84}$$

Field amplitude renormalization: Positivity of the RG function. Note that for g small, i.e. in the perturbative domain, the field renormalization (11.78) satisfies $Z < 1$. This is a general property in unitary theories implied by the spectral representation of the 2-point function (see Section 7.4). It implies $\eta(g) > 0$ for g small.

11.7 Generalization to Several Component Fields

We now generalize previous calculations to a renormalized action of the form:

$$S\left(\phi\right) = \int \mathrm{d}^d x \left[\tfrac{1}{2} Z_{ij} \partial_\mu \phi_i \partial_\mu \phi_j + \frac{1}{4!} G_{ijkl} \phi_i \phi_j \phi_k \phi_l \right]. \tag{11.85}$$

We here consider only the massless theory and call μ the renormalization scale. We again use dimensional regularization and minimal subtraction. This explains the absence of mass counterterms. We call g_{ijkl} the renormalized coupling constant, which is a symmetric tensor in its four indices. The quantity G_{ijkl} which includes coupling constant renormalization has the form:

$$G_{ijkl} = \mu^\varepsilon g_{ijkl} + O\left(g^2\right). \tag{11.86}$$

To calculate the generalization of the renormalization constant Z_2 we use the renormalization of the insertion of the operator $\frac{1}{2}\phi_i(x)\phi_j(x)$.

  The calculation of the renormalization constants reduces to the calculation of weight factors, the values of the divergent parts of the diagrams being the same as in Section 11.6.

11.7.1 Diagrams and renormalization constants

The 2-point function has at two-loop order the expansion:

$$\Gamma_{ij}^{(2)} = Z_{ij} p^2 - \frac{1}{6} G_{iklm} G_{jklm} \ (\mathrm{c}') \ + O\left(G^3\right). \tag{11.87}$$

The 4-point function is at the same order:

$$\Gamma_{ijkl}^{(4)} = G_{ijkl} - \frac{1}{2}\left(G_{ijmn} G_{mnkl} \,(\mathrm{d}') + 2 \text{ terms}\right) + \frac{1}{4}\left(G_{ijmn} G_{mnpq} G_{pqkl} \,(\mathrm{e}') + 2 \text{ terms}\right)$$

$$+ \frac{1}{2}\left(G_{ijmn} G_{mpqk} G_{npql} \,(\mathrm{f}') + 5 \text{ terms}\right) + O\left(G^4\right). \tag{11.88}$$

The additional terms restore the permutation symmetry of the 4-point function in its four indices. The quantities (c'), ..., (f') are, in the massless theory, the corresponding of the diagrams of figures 11.1,2 (this accounts for the primes).

  Let us finally calculate the $\frac{1}{2}\langle \phi_i(x)\phi_j(x)\phi_k\left(y_1\right)\phi_l\left(y_2\right)\rangle$ correlation function $\Gamma_{ij,kl}^{(1,2)}$:

$$\Gamma_{ij,kl}^{(1,2)} = \frac{1}{2}\left(\delta_{ik}\delta_{jl} + \delta_{il}\delta_{jk}\right) - \frac{1}{2} G_{ijkl} \,(\mathrm{d}') + \frac{1}{4} G_{ijmn} G_{mnkl} \,(\mathrm{e}')$$

$$+ \frac{1}{4}\left(G_{ikmn} G_{jlmn} + G_{jkmn} G_{ilmn}\right) (\mathrm{f}') + O\left(G^3\right). \tag{11.89}$$

As we have argued in Section 11.4, the divergent parts of the massive diagrams of Section 11.6 and the massless theory are the same. Expanding the renormalization constants in powers of g_{ijkl}, and using the values of the divergent parts of diagrams (c), (d) and (f) of figures 11.1,2, we obtain:

$$\mu^{-\varepsilon} G_{ijkl} = g_{ijkl} + \frac{N_d}{2\varepsilon}\left(g_{ijmn} g_{mnkl} + 2 \text{ terms}\right) + \frac{N_d^2}{4\varepsilon^2}\left(g_{ijmn} g_{mnpq} g_{pqkl} + 2 \text{ terms}\right)$$

$$+ \frac{N_d^2}{4\varepsilon^2}\left(1 - \frac{\varepsilon}{2}\right)\left(g_{ijmn} g_{mpqk} g_{npql} + 5 \text{ terms}\right) + O\left(g^4\right), \tag{11.90}$$

$$Z_{ij} = \delta_{ij} - \frac{N_d^2}{48\varepsilon} g_{iklm} g_{jklm} + O\left(g^3\right). \tag{11.91}$$

From equations (11.90,11.91) we deduce the expansion of the bare coupling constant which we parametrize as $\mu^\varepsilon \gamma_{ijkl}$:

$$\gamma_{ijkl} = \mu^{-\varepsilon} G_{mnpq} \left(Z^{-1/2} \right)_{mi} \left(Z^{-1/2} \right)_{jn} \left(Z^{-1/2} \right)_{kp} \left(Z^{-1/2} \right)_{lq}. \tag{11.92}$$

We obtain:

$$\gamma_{ijkl} = g_{ijkl} + \frac{N_d}{2\varepsilon} \left(g_{ijmn} g_{mnkl} + 2 \text{ terms} \right) + \frac{N_d^2}{4\varepsilon^2} \left(g_{ijmn} g_{mnpq} g_{pqkl} + 2 \text{ terms} \right)$$

$$+ \frac{N_d^2}{4\varepsilon^2} \left(1 - \frac{\varepsilon}{2} \right) \left(g_{ijmn} g_{mpqk} g_{npql} + 5 \text{ terms} \right)$$

$$+ \frac{N_d^2}{96\varepsilon} \left(g_{ijkm} g_{mnpq} g_{npql} + 3 \text{ terms} \right) + O\left(g^4 \right). \tag{11.93}$$

Calling $Z_{ij,kl}^{(2)}$ the renormalization constant of the operator $\frac{1}{2}\phi_i(x)\phi_j(x)$, we derive from equation (11.89):

$$Z_{ij,kl}^{(2)} = \frac{1}{2} \left(\delta_{ik}\delta_{jl} + \delta_{il}\delta_{jk} \right) + \frac{N_d}{2\varepsilon} g_{ijkl} + \frac{N_d^2}{4\varepsilon^2} g_{ijmn} g_{mnkl}$$

$$+ \frac{N_d^2}{8\varepsilon^2} \left(1 - \frac{\varepsilon}{2} \right) \left(g_{ikmn} g_{jlmn} + g_{jkmn} g_{ilmn} \right) + O\left(g^3 \right). \tag{11.94}$$

Actually we need the matrix $\zeta_{ij,kl}^{(2)}$ which expresses the renormalized operator in terms of the bare fields:

$$\zeta_{ij,kl}^{(2)} = Z_{ij,mn}^{(2)} \left(Z^{-1/2} \right)_{mk} \left(Z^{-1/2} \right)_{nl}. \tag{11.95}$$

At two-loop order we find:

$$\zeta_{ij,kl}^{(2)} = Z_{ij,kl}^{(2)} + \frac{N_d^2}{192\varepsilon} \left(g_{imnp} g_{kmnp} \delta_{jl} + 3 \text{ terms} \right) + O\left(g^3 \right). \tag{11.96}$$

The additional terms restore the symmetry of exchange $(i \leftrightarrow j)$ and $(k \leftrightarrow l)$ and of the two pairs $(ij) \leftrightarrow (kl)$.

11.7.2 Renormalization group equations

Let us first derive the RG equations for a multicomponent massless field theory. The relation between bare and renormalized correlation functions now takes the form:

$$\Gamma_{i_1 i_2 \ldots i_n}^{(n)} (p_k, g, \mu) = \left(Z^{1/2} \right)_{i_1 j_1} \left(Z^{1/2} \right)_{i_2 j_2} \cdots \left(Z^{1/2} \right)_{i_n j_n} \Gamma_{B, j_1 j_2 \ldots j_n}^{(n)} (p_k, g_0, \Lambda), \tag{11.97}$$

in which B stands for bare, g for g_{ijkl} and g_0 for $\mu^\varepsilon \gamma_{ijkl}$.

If we differentiate equation (11.97) with respect to μ at g_0 and Λ fixed, we obtain the RG equation:

$$\left(\mu \frac{\partial}{\partial \mu} + \beta_{i'j'k'l'} \frac{\partial}{\partial g_{i'j'k'l'}} \right) \Gamma_{i_1 i_2 \ldots i_n}^{(n)} - \frac{1}{2} \sum_{m=1}^{n} \eta_{i_m j_m} \Gamma_{i_1 i_2 \ldots j_m \ldots i_n}^{(n)} = 0, \tag{11.98}$$

with the definitions:

$$\beta_{i'j'k'l'} \frac{\partial \gamma_{ijkl}}{\partial g_{i'j'k'l'}} = -\varepsilon \gamma_{ijkl}, \tag{11.99}$$

$$\eta_{ij} = 2\beta_{i'j'k'l'} \left(\frac{\partial Z^{1/2}}{\partial g_{i'j'k'l'}} Z^{-1/2} \right)_{ij} . \qquad (11.100)$$

Similarly to renormalize correlation functions with $\frac{1}{2}\phi_i(x)\phi_j(x)$ insertions we have to multiply each insertion by the matrix $\zeta_{ij,kl}^{(2)}$. This leads to the RG equation:

$$\left(\mu \frac{\partial}{\partial \mu} + \beta_{i'j'k'l'} \frac{\partial}{\partial g_{i'j'k'l'}} \right) \Gamma^{(l,n)}_{j_1 k_1 \dots j_l k_l, i_1 \dots i_n} - \frac{1}{2} \sum_{m=1}^{n} \eta_{i_m a_m} \Gamma^{(l,n)}_{j_1 k_1 \dots j_l k_l, i_1 \dots a_m \dots i_n}$$

$$- \sum_{m=1}^{l} \eta_{j_m k_m, b_m c_m}^{(2)} \Gamma^{(l,n)}_{j_1 k_1 \dots b_m c_m \dots j_l k_l, i_1 \dots i_n} = 0 , \quad (11.101)$$

with the definition:

$$\eta_{ij,kl}^{(2)} = \beta_{i'j'k'l'} \left[\frac{\partial \zeta^{(2)}}{\partial g_{i'j'k'l'}} \left(\zeta^{(2)} \right)^{-1} \right]_{ij,kl} . \qquad (11.102)$$

Renormalization group functions. From the expressions (11.91–11.96) we derive the expansion of the RG functions at two-loop order:

$$\beta_{ijkl} = -\varepsilon g_{ijkl} + \frac{N_d}{2} \left(g_{ijmn} g_{mnkl} + 2 \text{ terms} \right) + \frac{N_d^2}{4} \left(g_{ijmn} g_{mpqk} g_{npql} + 5 \text{ terms} \right)$$

$$+ \frac{N_d^2}{48} \left(g_{ijkm} g_{mnpq} g_{npql} + 3 \text{ terms} \right) + O \left(g^4 \right) , \qquad (11.103)$$

$$\eta_{ij} = \frac{N_d^2}{24} g_{iklm} g_{jklm} + O \left(g^3 \right) , \qquad (11.104)$$

$$\eta_{ij,kl}^{(2)} = -\frac{N_d}{2} g_{ijkl} + \frac{N_d^2}{8} \left(g_{ikmn} g_{jlmn} + g_{jkmn} g_{ilmn} \right)$$

$$- \frac{N_d^2}{96} \left(g_{imnp} g_{kmnp} \delta_{jl} + 3 \text{ terms} \right) + O \left(g^3 \right) . \qquad (11.105)$$

These expressions will be used in Chapter 26 in their general form. Let us here specialize to the $\left(\phi^2 \right)^2$ field theory with $O(N)$ symmetry. We then have to substitute:

$$g_{ijkl} = \frac{g}{3} \left(\delta_{ij} \delta_{kl} + \delta_{ik} \delta_{jl} + \delta_{il} \delta_{jk} \right) . \qquad (11.106)$$

A short calculation leads to:

$$\tilde{\beta}(\tilde{g}) = N_d \beta(\tilde{g}) = -\varepsilon \tilde{g} + \frac{1}{6}(N+8)\tilde{g}^2 - \frac{(3N+14)}{12}\tilde{g}^3 + O\left(\tilde{g}^4\right) , \qquad (11.107)$$

$$\eta(\tilde{g}) = \frac{(N+2)}{72}\tilde{g}^2 + O\left(\tilde{g}^3\right) , \qquad (11.108)$$

in the notation of equation (11.84).

The matrix $\eta_{ij,kl}^{(2)}$ becomes a linear combination of the identity matrix **I** and the projector **P**:

$$I_{ij,kl} = \frac{1}{2} \left(\delta_{ik} \delta_{jl} + \delta_{il} \delta_{jk} \right) , \qquad P_{ij,kl} = \frac{1}{N} \delta_{ij} \delta_{kl} . \qquad (11.109)$$

One finds:

$$\eta^{(2)} = -\left(N\mathbf{P} + 2\mathbf{I}\right)\frac{\tilde{g}}{6} + \left[(N+10)\,\mathbf{I} + 4N\mathbf{P}\right]\frac{\tilde{g}^2}{72} + O\left(\tilde{g}^3\right). \tag{11.110}$$

The eigenvalue corresponding to a mass insertion ϕ^2 is given by the eigenvalue $+1$ of $\mathbf{P}$:

$$\eta^{(2)}_1 = -\frac{1}{6}\left(N+2\right)\tilde{g}\left(1 - \frac{5}{12}\tilde{g}\right) + O\left(\tilde{g}^3\right). \tag{11.111}$$

The other eigenvalue related to the eigenvalue 0 of $\mathbf{P}$ corresponds to the insertion of a mass term breaking the $O\left(N\right)$ symmetry:

$$\eta^{(2)}_2 = -\frac{\tilde{g}}{3} + \frac{(N+10)}{72}\tilde{g}^2 + O\left(\tilde{g}^3\right). \tag{11.112}$$

11.8 Renormalization and RG Functions at One-Loop in a Theory with Bosons and Fermions

We now consider a simple field theory involving one scalar field $\phi(x)$ and N massless Dirac fermion fields $\psi^i(x)$ and $\bar{\psi}^i(x)$, coupled through a Yukawa type $\phi\bar{\psi}\psi$ interaction. The action is

$$S\left(\bar{\psi}, \psi, \phi\right) = \int \mathrm{d}^d x \left[-\bar{\psi} \cdot \left(\slashed{\partial} + g\phi\right)\psi + \tfrac{1}{2}\left(\partial_\mu \phi\right)^2 + \tfrac{1}{2}m^2\phi^2 + \frac{\lambda}{4!}\phi^4\right]. \tag{11.113}$$

It has an obvious $U(N)$ symmetry and in addition a discrete chiral symmetry:

$$\psi \mapsto \gamma_S \psi, \qquad \bar{\psi} \mapsto -\bar{\psi}\gamma_S, \qquad \phi(x) \mapsto -\phi(x), \tag{11.114}$$

which forbids a fermion mass term and odd powers of ϕ as ϕ self-interaction. The physics of this model, spontaneous chiral symmetry breaking and fermion mass generation, will be discussed in Appendix A30.

We have already shown that such a field theory is renormalizable in 4 dimensions by power counting. To calculate renormalization constants we use dimensional regularization and minimal subtraction. We render the renormalized coupling constants dimensionless in d dimensions by setting:

$$\lambda_{0i} = m^\varepsilon f_i\left(\lambda\right), \tag{11.115}$$

in which λ_{0i} represents symbolically the bare coupling constants $\left(g_0^2, \lambda_0\right)$ and λ_i the renormalized coupling constants $\left(g^2, \lambda\right)$.

We have given in Section 6.4 (equation (6.56)) the 1PI functional $\Gamma\left(\varphi, \psi, \bar{\psi}\right)$ at one-loop order for an action which contains the action (11.113), $N = 1$, as a special example. Here we obtain:

$$\Gamma_{1\,\mathrm{loop}} = -N\,\mathrm{tr}\ln\left[1 + g\slashed{\partial}^{-1}\varphi(x)\right] + \tfrac{1}{2}\,\mathrm{tr}\ln\left[1 + \tfrac{1}{2}\lambda\left(-\Delta + m^2\right)^{-1}\varphi^2(x)\right]$$

$$+ 2g^2\left(-\Delta + m^2\right)^{-1}\bar{\psi}(x) \cdot \left(\slashed{\partial} + g\varphi(x)\right)^{-1}\psi(x)\Big]. \tag{11.116}$$

Expanding in powers of φ, ψ and $\bar{\psi}$, we get all 1PI correlation functions at one-loop order. We consider only the divergent functions and omit all contributions involving only the ϕ^4 vertex which have already been evaluated in Section 11.3:

$$\langle \bar{\psi}^1(-p)\psi^1(p)\rangle^{-1} = i\slashed{p} + \frac{g^2}{(2\pi)^d}\int \frac{d^d q}{(p-q)^2 + m^2}\frac{i\slashed{q}}{q^2}, \tag{11.117}$$

$$\langle \bar{\psi}^1(p_1)\psi^1(p_2)\phi(k)\rangle_{1PI} = g + g^3 \int \frac{d^d q}{(q+p_2)^2 + m^2}\frac{i\slashed{q}i(\slashed{q}-\slashed{k})}{q^2(q-k)^2}, \tag{11.118}$$

$$\langle \phi(-p)\phi(p)\rangle^{-1} = \langle \phi(-p)\phi(p)\rangle^{-1}\Big|_{g=0} + \frac{Ng^2}{(2\pi)^d}\int d^d q \frac{\mathrm{tr}\, i\slashed{q}i(\slashed{q}+\slashed{p})}{q^2(p+q)^2}, \tag{11.119}$$

$$\langle \phi(p_1)\phi(p_2)\phi(p_3)\phi(p_4)\rangle_{1PI} = \langle \phi_1\phi_2\phi_3\phi_4\rangle_{1PI}\big|_{g=0}$$

$$+ N\frac{g^4}{(2\pi)^d}\int d^d q \frac{\mathrm{tr}\,[i\slashed{q}i(\slashed{q}+\slashed{p}_1)i(\slashed{q}+\slashed{p}_1+\slashed{p}_2)i(\slashed{q}-\slashed{p}_4)]}{q^2(q+p_1)^2(q+p_1+p_2)^2(q-p_4)^2}$$

$$+ 5 \text{ diagrams corresponding to permutations of } (p_2, p_3, p_4) . \tag{11.120}$$

Figure 11.3 displays the corresponding Feynman diagrams.

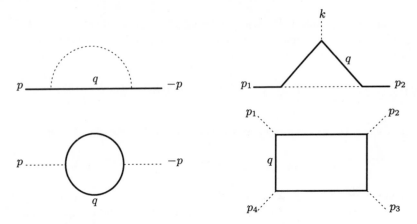

Fig. 11.3   Boson fermion diagrams (the fermions and bosons correspond to continuous and dotted lines respectively).

11.8.1 Explicit calculations of the 2-point functions

As an exercise let us first calculate completely the contributions to the two-point functions.

Fermions. We have to compute the integral $I_1(p)$:

$$I_1(p) = \frac{1}{(2\pi)^d}\int \frac{d^d q}{(p-q)^2 + m^2}\frac{i\slashed{q}}{q^2}. \tag{11.121}$$

Rotation invariance implies that $I_1(p)$ is a linear combination of the form $A(p) + iB(p)\slashed{p}$. To obtain A and B we can take the trace of the integrand multiplied successively by 1 and $\slashed{p}$. Applying the rules explained in Appendix A5.8 we immediately see that $A = 0$, and thus no fermion mass is generated.

Using the method of Feynman parameters as explained in Section 11.5 we transform (11.121) into:

$$I_1(p) = \frac{1}{(2\pi)^d} \int_0^1 dx \int \frac{d^dq \, i\slashed{q}}{\left[x \left((p-q)^2 + m^2 \right) + (1-x)q^2 \right]^2}. \tag{11.122}$$

We then shift q, $q - xp \mapsto q$, and integrate over q. We find

$$I_1(p) = i\slashed{p} \frac{\Gamma(2-d/2)}{(4\pi)^{d/2}} \int_0^1 dx \, x^{d/2-1} \left[m^2 + (1-x)p^2 \right]^{d/2-2}. \tag{11.123}$$

Expanding at first order in $\varepsilon = 4 - d$ and integrating over x we finally obtain

$$I_1(p) = \frac{i\slashed{p}}{2} m^{-\varepsilon} \frac{N_d}{\varepsilon} + i\slashed{p} \frac{1}{32\pi^2} \frac{p^2+m^2}{p^2} \left[1 - \frac{p^2+m^2}{p^2} \ln \left(\frac{p^2+m^2}{m^2} \right) \right]. \tag{11.124}$$

Bosons. We have to calculate the integral $I_2(p)$:

$$I_2(p) = \frac{N}{(2\pi)^d} \int d^dq \, \mathrm{tr} \, \frac{i\slashed{q}\,(i\slashed{q} + i\slashed{p})}{q^2 \,(p+q)^2}. \tag{11.125}$$

Evaluating the trace we get:

$$I_2(p) = -\frac{4N}{(2\pi)^d} \int d^dq \, \frac{q^2 + p \cdot q}{q^2 \,(p+q)^2}. \tag{11.126}$$

Using then:

$$q^2 + p \cdot q = \tfrac{1}{2} \left[q^2 + (p+q)^2 \right] - \tfrac{1}{2}p^2,$$

we can decompose $I_2(p)$ into a sum of simple integrals:

$$I_2(p) = -\frac{4N}{(2\pi)^d} \int \frac{d^dq}{q^2} + \frac{4N}{(2\pi)^d} \frac{p^2}{2} \int \frac{d^dq}{q^2 \,(p+q)^2}. \tag{11.127}$$

The first integral vanishes in dimensional regularization and the second is the massless limit of an integral already calculated in Section 11.6. We find:

$$I_2(p) = 2Np^2 m^{-\varepsilon} \frac{N_d}{\varepsilon} + + \frac{N}{4\pi^2} p^2 \left[\tfrac{1}{2} - \ln \left(\frac{p}{m} \right) \right]. \tag{11.128}$$

11.8.2 Other divergences. Renormalization group functions

All one-loop diagrams can be calculated by similar methods. We here evaluate only the divergent part. Taking into account the results of Section 11.3 we find:

$$\langle \bar{\psi}\psi\phi \rangle_{\mathrm{1PI, 1\ loop,\ div.}} = -g^3 \frac{N_d}{\varepsilon}, \tag{11.129}$$

$$\langle \phi\phi\phi\phi \rangle_{\mathrm{1PI, 1\ loop,\ div.}} = \left(-\frac{3}{2}\lambda^2 + 24Ng^4 \right) \frac{N_d}{\varepsilon}. \tag{11.130}$$

Renormalization constants. We first replace action (11.113) by the renormalized action:

$$S_r\left(\phi, \psi, \bar{\psi}\right) = \int \mathrm{d}^d x \left\{ -Z_\psi \left[\bar{\psi}(x) \cdot \left(\slashed{\partial} + g_0 Z_\phi^{1/2} \phi(x) \right) \psi(x) \right] \right.$$
$$\left. + \tfrac{1}{2} Z_\phi \left[(\partial_\mu \phi(x))^2 + m_0^2 \phi^2(x) \right] + \frac{1}{4!} \lambda_0 Z_\phi^2 \phi^4(x) \right\}. \tag{11.131}$$

We call m_0, g_0, λ_0 the bare parameters, m, g, λ the renormalized parameters, and use equation (11.115) to expand the bare coupling constants in terms of g and λ.

Substituting into the Feynman diagrams and identifying the divergent parts we obtain equations for the renormalization constants:

$$i\slashed{p}\left(Z_\psi + \frac{1}{2} g^2 \frac{N_d}{\varepsilon} \right) = \text{finite } O\,(2 \text{ loops})\,,$$

$$m^{-\varepsilon/2} g_0 Z_\psi Z_\phi^{1/2} - g^3 \frac{N_d}{\varepsilon} = \text{finite } O\,(2 \text{ loops})\,,$$

$$Z_\phi \left(p^2 + m_0^2 \right) + \left(-\frac{\lambda}{2} m^2 + 2N p^2 g^2 \right) \frac{N_d}{\varepsilon} = \text{finite } O\,(2 \text{ loops})\,,$$

$$m^{-\varepsilon} \lambda_0 Z_\phi^2 + \left(-\frac{3}{2} \lambda^2 + 24 N g^4 \right) \frac{N_d}{\varepsilon} = \text{finite } O\,(2 \text{ loops})\,.$$

It follows that:

$$Z_\psi = 1 - \frac{1}{2} g^2 \frac{N_d}{\varepsilon} + O\,(2 \text{ loops})\,, \tag{11.132}$$

$$Z_\phi = 1 - 2N g^2 \frac{N_d}{\varepsilon} + O(2 \text{ loops})\,, \tag{11.133}$$

$$g_0 = m^{\varepsilon/2} g \left(1 + \frac{(2N+3)}{2} g^2 \frac{N_d}{\varepsilon} \right) + O(2 \text{ loops})\,, \tag{11.134}$$

$$\lambda_0 = m^\varepsilon \left[\lambda + \left(\frac{3}{2} \lambda^2 + 4N\lambda g^2 - 24 N g^4 \right) \frac{N_d}{\varepsilon} \right] + O(2 \text{ loops})\,, \tag{11.135}$$

$$m_0^2 = m^2 \left[1 + \left(\frac{\lambda}{2} + 2N g^2 \right) \frac{N_d}{\varepsilon} \right] + O(2 \text{ loops})\,. \tag{11.136}$$

RG-functions. When bare and renormalized coupling constants are related by an equation of the form (11.115), the coupling constant RG functions $\beta_i(\lambda)$ are given by:

$$\beta_i\,(\lambda) = m \frac{\partial}{\partial m}\bigg|_{\lambda_0} \lambda_i\,, \tag{11.137}$$

and thus using equation (11.115):

$$0 = m \frac{\partial}{\partial m} \lambda_{0i}\bigg|_{\lambda_0} = m^\varepsilon \left(\varepsilon f_i\,(\lambda) + \beta_j\,(\lambda) \frac{\partial f_i}{\partial \lambda_j} \right)\,. \tag{11.138}$$

A short calculation then yields:

$$\beta_{g^2} = -\varepsilon g^2 + N_d(2N+3)g^4 + \cdots\,, \tag{11.139a}$$

$$\beta_\lambda = -\varepsilon\lambda + N_d \left(\frac{3}{2} \lambda^2 + 4N\lambda g^2 - 24 N g^4 \right) + \cdots\,. \tag{11.139b}$$

In the minimal subtraction scheme:

$$\eta(\lambda) = -\lambda_i \frac{\partial}{\partial \lambda_i} \eta^{(1)}(\lambda),$$

where $\eta^{(1)}(\lambda)$ is the coefficient of $1/\varepsilon$ in the corresponding renormalization constant. Then:

$$\eta_\phi = 2N N_d g^2, \tag{11.140}$$

$$\eta_\psi = \tfrac{1}{2} N_d g^2, \tag{11.141}$$

$$\eta_2 = -\tfrac{1}{2} N_d \lambda - 2N N_d g^2. \tag{11.142}$$

Again note that the functions η_ψ, η_ϕ are positive for small couplings, a result consistent with the general property $Z_\psi, Z_\phi < 1$ (see Section 7.4).

Bibliographical Notes

Topological properties and parametric representations of Feynman diagrams are studied in
   N. Nakanishi, *Graph Theory and Feynman Integrals* (Gordon and Breach, New York 1971).
Feynman diagrams and dimensional regularization are discussed in
   G. 't Hooft and M. Veltman, in *Particle Interactions at Very High Energies*, D. Speiser, F. Halzen and J. Weyers eds. (Plenum Press, New York 1973) vol. 4B.
The properties of RG functions in the minimal subtraction scheme is discussed in
   G. 't Hooft, *Nucl. Phys.* B61 (1973) 455.
A good presentation can be found in
   D.J. Gross in *Methods in Field Theory*, Les Houches 1975, R. Balian and J. Zinn-Justin eds. (North-Holland, Amsterdam 1976).
For more systematic techniques to calculate RG functions at higher orders see:
   A.A. Vladimirov, *Teor. Mat. Fiz.* 43 (1980) 210.

Exercises

Exercise 11.1

Characterize the form of renormalization constants in a minimal subtraction scheme with cut-off regularization, the parameter being Λ/m in which m is a mass scale (see Subsection 8.1.1).

Exercise 11.2

Pseudoscalar Yukawa interaction. Consider the action

$$S\left(\phi, \psi, \bar{\psi}\right) = \int \mathrm{d}^d x \Big\{ -\bar{\psi}(x) \left[\not{\partial} + M + ig\gamma_S \phi(x) \right] \psi(x) + \tfrac{1}{2} \left(\partial_\mu \phi(x) \right)^2$$

$$+ \tfrac{1}{2} m^2 \phi^2(x) + \frac{1}{4!} \lambda \phi^4(x) \Big\}, \tag{11.143}$$

in which ϕ is a pseudoscalar boson field and $\psi, \bar{\psi}$ Dirac fermions.

The factor i in front of γ_S is imposed by invariance under reflection hermiticity, i.e. euclidean hermitian conjugation followed by space reflection (see Appendix A5.6) which implies:

$$\gamma_1 \gamma_S^\dagger \left(ig\phi(x)\right)^* \gamma_1 = \gamma_S ig\phi(x).$$

In addition the action is reflection symmetric (see Appendix A5.5):

$$\psi(x) \mapsto \gamma_1 \psi(\tilde{x}), \qquad \bar{\psi}(x) \mapsto \bar{\psi}(\tilde{x})\, \gamma_1\,, \qquad \phi(x) \mapsto -\phi(\tilde{x})\,,$$

with $\tilde{x} = (-x_1, x_2, \ldots, x_d)$.

The transformation of $\phi(x)$ is imposed by the factor γ_S in front of the $\phi \bar{\psi}\psi$ interaction and justifies the denomination pseudoscalar for the field $\phi(x)$.

Calculations can be done using dimensional regularization and minimal subtraction because it is never necessary to evaluate any trace of products of γ_S by an even number of γ_μ matrices.

11.2.1. Calculate the $\bar{\psi}\psi$ 2-point function at one-loop order. It is suggested to expand on the basis $\{1, \rlap{/}{p}\}$ and to use identities of the form

$$p \cdot q = \tfrac{1}{2}\left[q^2 + M^2 - (p-q)^2 - m^2\right] + \tfrac{1}{2}\left(p^2 + m^2 - M^2\right).$$

11.2.2. Calculate the divergent part of the ϕ 2-point function at one-loop order.

11.2.3. Writing the renormalized action:

$$S_r\left(\phi, \psi, \bar{\psi}\right) = \int \mathrm{d}^d x \Big\{ -Z_\psi \left[\bar{\psi}(x)\left(\rlap{/}{\partial} + M_0 + i\gamma_S g_0 Z_\phi^{1/2}\,\phi(x)\right)\psi(x)\right]$$
$$+ \tfrac{1}{2}Z_\phi\left[(\partial_\mu \phi(x))^2 + m_0^2 \phi^2(x)\right] + \frac{1}{4!}\lambda_0 Z_\phi^2 \phi^4(x)\Big\}, \qquad (11.144)$$

calculate all renormalization constants and RG functions. One may introduce a renormalization scale μ and define the β-functions by

$$\beta_i(\boldsymbol{\lambda}) = \mu \left.\frac{\partial}{\partial \mu}\right|_{\lambda_0} \lambda_i\,, \qquad \lambda_1 \equiv \lambda_1\,, \quad \lambda_2 \equiv g^2\,. \qquad (11.145)$$

Compare with the results of Section 11.8 for $N = 1$.

11.2.4. Relate coupling and field renormalizations of models (11.113) and (11.143) by performing in action (11.143), restricted to $M = 0$, a chiral rotation on the fields $\psi, \bar{\psi}$ (see Appendix A5.6).

Exercise 11.3

Explain how the inductive proof of renormalizability of the ϕ^4 field theory given in Chapter 10 has to be modified in the case of dimensional regularization. Note that it is essential to take into account the additional term of order g_r (g_r is the dimensionless renormalized coupling constant) in the CS-β-function

$$\beta\left(g_r, \varepsilon = 4 - d\right) = -\varepsilon g_r + O\left(g_r^2\right).$$

12 RENORMALIZATION OF COMPOSITE OPERATORS. SHORT DISTANCE EXPANSION

In this chapter we discuss two related problems: The renormalization of composite operators, i.e. local polynomials of the field, and the short distance expansion (SDE) of the product of operators. The relation is easy to understand: we can consider the insertion of a product of operators $A(x)B(y)$ as a regularization by point splitting of the operator $A[\frac{1}{2}(x+y)]B[\frac{1}{2}(x+y)]$. Therefore in the limit $x \to y$ we expect the product to become dominated by a linear combination of the local operators which appear in the renormalization of the product AB, with singular coefficients functions of $(x-y)$ replacing the usual cut-off dependent renormalization constants.

We first discuss the renormalization of composite operators in general from the point of view of power counting. We use the relations between bare and renormalized operators to establish a set of CS equations for the insertion of operators of dimension four in the ϕ_4^4 field theory. We then show that in a given field theory there exist linear relations between operators due to the equation of motions and relations derived in Section 5.4.

In the second part of this chapter, devoted to the SDE, we first establish the existence of a SDE for the product of two basic fields, and discuss the SDE at leading order in the ϕ_4^4 field theory.

We then pass from short distance behaviour to large momentum behaviour and derive a CS equation for the coefficient of the expansion at leading order. We finally briefly discuss the generalization of this analysis to the SDE beyond leading order, to the SDE of arbitrary operators, and to the light cone expansion which appears when one studies the large momentum behaviour of real time correlation functions (in contrast to euclidean imaginary time).

12.1 Renormalization of Operator Insertions

We have shown in Section 10.6 how to renormalize insertions of the operator ϕ^2. We could have considered other vertices like ϕ^4, $(\partial_\mu \phi)^2$... They all generate new divergences which have to be eliminated by additional renormalizations.

In Section 8.4 we have explained how to calculate the superficial degree of divergence of the insertion of a local operator $\mathcal{O}(x, \phi)$ by adding a source $g(x)$ for this operator in the action:

$$S(\phi) \mapsto S(\phi) + \int \mathrm{d}x\, \mathcal{O}(x, \phi)\, g(x).$$

With this choice of sign, differentiation of $W(J, g)$ with respect to $g(x)$ generates insertions of $-\mathcal{O}(x, \phi)$ in connected correlation functions. However since $\delta\Gamma/\delta g(x) = -\delta W/\delta g(x)$, $\delta\Gamma/\delta g(x)$ corresponds to the insertion of $\mathcal{O}(x, \phi)$ in proper vertices.

As a convention we assign a canonical dimension $[g(x)]$ to the source $g(x)$, opposite to the dimension of the vertex associated to $\mathcal{O}(\phi)$ and thus related to the dimension $[\mathcal{O}(\phi)]$ of the operator $\mathcal{O}(\phi)$ by:

$$[g] = d - [\mathcal{O}(\phi)].$$

With this convention the consequences of power counting and renormalization theory have a simple formulation: after the field theory has been renormalized up to $(L-1)$

loops, the divergent part Γ_L^{div} of $\Gamma(\phi, g)$ at L loop order is the most general local functional of $\phi(x)$ and $g(x)$ allowed by power counting, i.e. *the most general linear combination of all vertices in $\phi(x)$ and $g(x)$ of non-positive dimensions*. Note that this in particular implies that an operator of a given dimension inserted one time in general mixes under renormalization with all operators of equal or lower dimension. It is thus natural to study the renormalization of operators of increasing dimension.

Let us thus first consider again the insertion of $\phi^2(x)$ in a ϕ^4 field theory in 4 dimensions which we have already discussed in Chapter 10.

12.1.1 The $\phi^2(x)$ insertion

We use the conventions of Chapter 10 for the bare and renormalized actions:

$$S(\phi) = \int \mathrm{d}^d x \left[\tfrac{1}{2} \left(\partial_\mu \phi \right)^2 + \tfrac{1}{2} m^2 \phi^2 + \frac{1}{4!} g \phi^4 \right], \tag{12.1}$$

$$S_{\mathrm{r}}(\phi_{\mathrm{r}}) = \int \mathrm{d}^d x \left[\tfrac{1}{2} Z \left(\partial_\mu \phi_{\mathrm{r}} \right)^2 + \tfrac{1}{2} \left(m_{\mathrm{r}}^2 + \delta m^2 \right) \phi_{\mathrm{r}}^2 + \frac{1}{4!} g_{\mathrm{r}} Z_g \phi_{\mathrm{r}}^4 \right]. \tag{12.2}$$

In the ϕ^4 field theory in 4 dimensions the operator $\phi^2(x)$ has canonical dimension 2:

$$[\phi^2] = 2.$$

Let us call $t(x)$ the source for $\phi^2(x)$. Its dimension $[t]$ is:

$$[t(x)] = 2.$$

Then in addition to the vertices involving only the field ϕ, the following vertices arise as counterterms

$$\int t(x) \phi^2(x) \mathrm{d}^4 x \quad \text{which has dimension zero}$$

$$\int t^2(x) \mathrm{d}^4 x \quad \text{which has also dimension zero}$$

$$\int t(x) \mathrm{d}^4 x \quad \text{which has dimension } -2.$$

The renormalized action then has the form:

$$S_{\mathrm{r}}(\phi_{\mathrm{r}}, t) = S_{\mathrm{r}}(\phi_{\mathrm{r}}) + \tfrac{1}{2} Z_2 \int \mathrm{d}^4 x \, t(x) \phi_{\mathrm{r}}^2(x) + \int \mathrm{d}^4 x \left[\tfrac{1}{2} a t^2(x) + b t(x) \right]. \tag{12.3}$$

The last two terms contribute only to the vacuum amplitude. Expression (12.3) implies a set of relations between bare and renormalized generating functionals:

$$Z_{\mathrm{r}}(J, t) = Z \left(J/\sqrt{Z}, t \, Z_2/Z \right) \exp \left[- \int \mathrm{d}^4 x \left(\tfrac{1}{2} a t^2 + b t \right) \right]. \tag{12.4}$$

For the connected functions this gives:

$$W_{\mathrm{r}}(J, t) = W \left(J/\sqrt{Z}, t \, Z_2/Z \right) - \int \mathrm{d}^4 x \left(\tfrac{1}{2} a t^2 + b t \right). \tag{12.5}$$

After Legendre transformation with respect to J one obtains:

$$\Gamma_{\mathrm{r}}(\varphi, t) = \Gamma \left(\varphi \sqrt{Z}, t Z_2/Z \right) + \int \mathrm{d}^4 x \left(\tfrac{1}{2} a t^2 + b t \right). \tag{12.6}$$

Expanding in powers of t and φ, one recovers the relations between bare and renormalized proper vertices proven in Chapter 10.

12.1.2 The ϕ^3 insertion

Let us now consider the example of the ϕ^3 insertion. The operator $\phi^3(x)$ has dimension 3, the corresponding source $t(x)$ has dimension 1. The renormalized action $S_{\mathrm{r}}(\phi, t)$ thus has the form:

$$S_{\mathrm{r}}(\phi, t) = S_{\mathrm{r}}(\phi) + \int \mathrm{d}^4 x \left[t(x) \left(\frac{1}{3!} Z_3 \phi^3(x) + a \partial^2 \phi(x) + b\phi(x) \right) \right.$$
$$\left. + c \left(\partial_\mu t(x) \right)^2 + t^2(x) \left(d\phi^2(x) + e \right) + f t^3(x) \phi(x) + g t^4(x) \right]. \quad (12.7)$$

This expression in particular shows that the operator $\phi^3(x)$ mixes under renormalization with $\phi(x)$, that the double insertion of ϕ^3 generates a counterterm proportional to $\phi^2 \dots$. To be able to write the equivalent of relations (12.4–12.6) we have therefore to explicitly introduce a source for $\phi^2(x)$. We leave it to the reader to write the renormalized action with sources for ϕ^3 and ϕ^2 and the relations between renormalized and bare proper vertices. We postpone the discussion of the CS equations of correlation functions with ϕ^3 insertion until Section 12.2 in order to be able to incorporate information coming from the field equation of motion.

12.1.3 Operators of dimension 4

Quite generally if an operator has a dimension strictly smaller than the space dimension d, the source has a strictly positive dimension and the renormalized action is a polynomial in the source. If the source is coupled to operators of dimension d, corresponding to vertices of dimension zero (here $\phi^4(x)$, $(\partial_\mu \phi(x))^2$), an infinite series in the source is generated by the renormalization procedure, together with all operators of lower or equal dimensions.

In the ϕ_4^4 field theory, if one inserts once an operator of dimension 4, one has to consider the mixing of all linearly independent operators of dimensions 4 and 2 (parity in ϕ excludes odd dimensions). For example the 4 operators:

$$\begin{cases} \mathcal{O}_1(\phi) = \frac{1}{2} m_{\mathrm{r}}^2 \phi^2(x), & \mathcal{O}_2(\phi) = -\frac{1}{2} \partial^2 \left(\phi^2(x) \right), \\ \mathcal{O}_3(\phi) = \frac{1}{2} \left[\partial_\mu \phi(x) \right]^2, & \mathcal{O}_4(\phi) = \frac{1}{4!} \phi^4(x), \end{cases} \quad (12.8)$$

form a basis of linearly independent operators mixing under renormalization. There exists another operator $\phi(x) \partial^2 \phi(x)$ of dimension 4, but it is a linear combination of $\mathcal{O}_2$ and $\mathcal{O}_3$:

$$\frac{1}{2} \partial^2 \left(\phi^2(x) \right) = \phi(x) \partial^2 \phi(x) + \left[\partial_\mu \phi(x) \right]^2.$$

Operator $\mathcal{O}_1(\phi)$, and therefore also operator $\mathcal{O}_2(\phi)$, are multiplicatively renormalizable.

We can thus write the relation between bare and renormalized correlation functions $\Gamma_{\mathcal{O}_i}^{(n)}$ with $\mathcal{O}_i$ insertion under the form:

$$\left(\Gamma_{\mathcal{O}_i}^{(n)} \right)_{\mathrm{r}} = Z^{n/2} \sum_j Z_{ij} \Gamma_{\mathcal{O}_j}^{(n)}. \quad (12.9)$$

The renormalization matrix Z_{ij} has the form:

$$\begin{pmatrix} (Z_2/Z) \, \mathbf{1}_2 & 0 \\ \mathbf{B} & \mathbf{A} \end{pmatrix},$$

in which $\mathbf{A}$ and $\mathbf{B}$ are 2×2 matrices. We have used for the renormalization of ϕ^2 the notation of Section 10.4.

CS equations. From equation (12.9) we can derive a CS equation for $\left(\Gamma_{\mathcal{O}_i}^{(n)}\right)_{\mathrm{r}}$. However, here some care is required. The CS operation involves ϕ^2 insertions and the product for example $\left(\phi^4\left(x\right)\right)_{\mathrm{r}} \left(\phi^2(y)\right)_{\mathrm{r}}$ inserted in a correlation function is not finite: since the source for ϕ^4 has dimension zero and the source for ϕ^2 dimension 2 the product of ϕ^2 by both sources has dimension 4. Figure 12.1 gives the first divergent diagram.

Fig. 12.1   Divergent contribution to $\phi^2\phi^4$ insertion.

This implies:

$$\tfrac{1}{4!}\left[\left(\phi^4(x)\right)_{\mathrm{r}} \left(\phi^2(y)\right)_{\mathrm{r}}\right]_{\mathrm{r}} = \tfrac{1}{4!}\left(\phi^4(x)\right)_{\mathrm{r}}\left(\phi^2(y)\right)_{\mathrm{r}} + C_4 \delta\left(x-y\right)\left(\phi^2(x)\right)_{\mathrm{r}}, \qquad (12.10)$$

in which C_4 is a new renormalization constant. Identity (12.10) is only true as an insertion in an n-point correlation function, $n \neq 0$.

After Fourier transformation and for an insertion of ϕ^2 at zero momentum, equation (12.10) becomes:

$$\left[\left(\tilde{\mathcal{O}}_4\left(p\right)\right)_{\mathrm{r}}\left(\tilde{\mathcal{O}}_1\left(0\right)\right)_{\mathrm{r}}\right]_{\mathrm{r}} = \left(\tilde{\mathcal{O}}_4\left(p\right)\right)_{\mathrm{r}}\left(\tilde{\mathcal{O}}_1\left(0\right)\right)_{\mathrm{r}} + C_4\left(\tilde{\mathcal{O}}_1\left(p\right)\right)_{\mathrm{r}}. \qquad (12.11)$$

A similar equation holds for the operator $\mathcal{O}_3$.

Let us now apply the CS operator, $m_{\mathrm{r}}\partial/\partial m_{\mathrm{r}}$ at g and Λ fixed, on equation (12.9). A new set of RG functions is generated involving the matrix $\tilde{\eta}_{ij}$:

$$\tilde{\eta}_{ij}\left(g_{\mathrm{r}}, \frac{\Lambda}{m_{\mathrm{r}}}\right) = \left(m_{\mathrm{r}}\frac{\partial}{\partial m_{\mathrm{r}}} Z_{ik}\right) Z_{kj}^{-1}. \qquad (12.12)$$

As a consequence of relation (12.11), two elements of the matrix $\tilde{\eta}_{ij}$ are not finite when the cut-off becomes infinite: $\tilde{\eta}_{31}$ and $\tilde{\eta}_{41}$. Their divergent part cancels the divergences coming from the insertion of ϕ^2 as represented by equation (12.11). Defining then:

$$\eta_{i1}\left(g_{\mathrm{r}}\right) = \tilde{\eta}_{i1} - m_{\mathrm{r}}^2 \sigma\left(g_{\mathrm{r}}\right) C_i, \qquad (12.13)$$

we now obtain two finite RG functions.

For the other matrix elements we just set:

$$\eta_{ij} = \tilde{\eta}_{ij}. \qquad (12.14)$$

The CS equations then read:

$$\left\{\left[m_{\mathrm{r}}\frac{\partial}{\partial m_{\mathrm{r}}} + \beta\left(g_{\mathrm{r}}\right)\frac{\partial}{\partial g_{\mathrm{r}}} - \frac{n}{2}\eta\left(g_{\mathrm{r}}\right)\right]\delta_{ij} - \eta_{ij}\left(g_{\mathrm{r}}\right)\right\}\left(\Gamma_{\mathcal{O}_j}^{(n)}\right)_{\mathrm{r}} = m_{\mathrm{r}}^2\sigma\left(g_{\mathrm{r}}\right)\left(\Gamma_{\mathcal{O}_i}^{(1,n)}\right)_{\mathrm{r}}\left(0;\ \right). \qquad (12.15)$$

The matrix η_{ij} has the form:

$$\eta_{ij} = \begin{bmatrix} \mathbf{c} & 0 \\ \mathbf{b} & \mathbf{a} \end{bmatrix},$$

in which $\mathbf{a}$, $\mathbf{b}$, and $\mathbf{c}$ are 2×2 matrices, and $\mathbf{c}$ is diagonal:

$$\mathbf{c} = \begin{bmatrix} 2 + \eta_2 & 0 \\ 0 & \eta_2 \end{bmatrix}.$$

This completes the discussion of the insertion one time of the operators of dimension 4 in ϕ_4^4 field theory. It reveals the general features of the insertion of any other operator of higher dimension.

Double insertion of operators of dimension 4. Let us now briefly discuss the double $\phi^4(x)$ or $(\partial_\mu \phi(x))^2$ insertion. It is similar to the $\phi^4 \phi^2$ insertion. The relation between product of renormalized operators and renormalized product now is:

$$\left[\left(\phi^4(x) \right)_{\mathrm{r}} \left(\phi^4(y) \right)_{\mathrm{r}} \right]_{\mathrm{r}} = \left(\phi^4(x) \right)_{\mathrm{r}} \left(\phi^4(y) \right)_{\mathrm{r}} + \delta\left(x - y\right) \sum_{i=1}^{4} D_{4i} \left[\mathcal{O}_i \left(\phi(x) \right) \right]_{\mathrm{r}}$$

$$+ \partial_\mu \delta(x - y) E_4 \partial_\mu \mathcal{O}_1 \left(\phi\left(x\right) \right)_{\mathrm{r}} + \partial^2 \delta\left(x - y\right) F_4 \mathcal{O}_1 \left(\phi\left(x\right) \right)_{\mathrm{r}} , \quad (12.16)$$

in which D_{4i}, E_4 and F_4 are new renormalization constants. A similar equation is valid for $\left[\partial_\mu \phi(x) \right]^2$. Again equation (12.16) is valid only as an insertion.

12.1.4 Operator insertion: general case

Power counting arguments, based on the dimension of operators and sources, tell us quite generally that if $\mathcal{O}\left(\phi(x) \right)$ is an operator of canonical dimension D:

$$[\mathcal{O}(\phi)] = D ,$$

then it renormalizes as:

$$[\mathcal{O}\left(\phi(x) \right)]_{\mathrm{r}} = \sum_{\alpha:\ [\mathcal{O}_\alpha] \leq D} Z_\alpha \mathcal{O}_\alpha \left(\phi(x) \right) . \qquad (12.17)$$

If we now consider the product of two operators $\mathcal{O}(\phi)$ and $\mathcal{O}'(\phi)$ of dimensions D and D' at different points x and y then

$$\{ [\mathcal{O}\left(\phi(x) \right)]_{\mathrm{r}} \, [\mathcal{O}'\left(\phi(y) \right)]_{\mathrm{r}} \}_{\mathrm{r}} = [\mathcal{O}\left(\phi(x) \right)]_{\mathrm{r}} \, [\mathcal{O}'\left(\phi(y) \right)]_{\mathrm{r}}$$

$$+ \sum_{\alpha: [\mathcal{O}_\alpha] + [P_\alpha] \leq D + D' - d} C_\alpha \mathcal{O}_\alpha \left(\phi(x) \right) P_\alpha \left(\partial_\mu \right) \delta\left(x - y\right) , \quad (12.18)$$

in which $P_\alpha \left(\partial_\mu \right)$ is a polynomial in ∂_μ. For example in the ϕ_4^4 field theory, the product $\left(\phi^6 \left(x\right) \right)_{\mathrm{r}} \left(\phi^8(y) \right)_{\mathrm{r}}$ involves all operators of dimension lower than or equal to 10.

12.2 Equations of Motion

We have discussed the renormalization of composite operators. However not all renormalizations are independent in a given field theory, because the equation of motion and other identities discussed in Section 5.4 imply relations between operators.

12.2.1 Insertion of the ϕ^3 operator

Let us again consider the example of the $\phi^3(x)$ operator in the framework of the ϕ^4 field theory.

We therefore consider the action $S(\phi, t, u)$:

$$S(\phi, t) = \int \mathrm{d}^4 x \left[\tfrac{1}{2}\left(\partial_\mu \phi\right)^2 + \tfrac{1}{2}m^2\phi^2 + \frac{1}{4!}g\phi^4 + \tfrac{1}{2}t(x)\phi^2 + \frac{1}{3!}u(x)\phi^3 \right]. \qquad (12.19)$$

Differentiation with respect to $u(x)$ and $t(x)$ generates $-\frac{1}{3!}\phi^3$ and $-\frac{1}{2}\phi^2$ insertions.

A regularization is assumed. The simplest equation of motion is obtained from the identity:

$$\int [\mathrm{d}\phi] \left[\frac{\delta S}{\delta \phi(x)} - J(x) \right] \exp\left(-S + J\phi\right) = 0.$$

Since we have introduced sources for ϕ^2 and ϕ^3, we can use them to express the terms ϕ^2 and ϕ^3 in $\delta S/\delta \phi$ as functional derivatives with respect to $t(x)$ and $u(x)$. We then obtain:

$$\left[\left(-\Delta + m^2\right)_\Lambda + t(x)\right]\frac{\delta Z}{\delta J(x)} - g\frac{\delta Z}{\delta u(x)} - u(x)\frac{\delta Z}{\delta t(x)} = J(x)Z(J, u, t). \qquad (12.20)$$

This yields an equation for $W = \ln Z$:

$$\left[\left(-\Delta + m^2\right)_\Lambda + t(x)\right]\frac{\delta W}{\delta J(x)} - g\frac{\delta W}{\delta u(x)} - u(x)\frac{\delta W}{\delta t(x)} = J(x). \qquad (12.21)$$

The Legendre transformation is straightforward. We remember that:

$$\frac{\delta W}{\delta u(x)} = -\frac{\delta \Gamma}{\delta u(x)}, \qquad \frac{\delta W}{\delta t(x)} = -\frac{\delta \Gamma}{\delta t(x)},$$

and find:

$$\left[\left(-\Delta + m^2\right)_\Lambda + t(x)\right]\varphi(x) + g\frac{\delta \Gamma}{\delta u(x)} + u(x)\frac{\delta \Gamma}{\delta t(x)} - \frac{\delta \Gamma}{\delta \varphi(x)} = 0. \qquad (12.22)$$

To explore a first consequence of this relation, let us set $t(x) = u(x) = 0$:

$$\left.\frac{\delta \Gamma}{\delta u(x)}\right|_{u=t=0} = \Gamma_{\phi^3}, \qquad (12.23)$$

Γ_{ϕ^3} is the generating functional of 1PI ϕ-field correlation functions with one $\frac{1}{3!}\phi^3$ insertion.

$$g\Gamma_{\phi^3(x)} = \frac{\delta \Gamma}{\delta \varphi(x)} - \left(-\Delta + m^2\right)_\Lambda \varphi(x). \qquad (12.24)$$

This relation shows that up to explicit subtractions affecting only the $\langle \phi^3 \phi \rangle$ correlation function, the insertion of ϕ^3 is equivalent to the insertion of ϕ itself.

The diagrammatic interpretation of equation (12.24) is simple: the insertion of ϕ^3 is indistinguishable from the addition of a ϕ^4 vertex with one of the lines attached to the vertex being an external line (see figure 12.2).

Fig. 12.2 ϕ^3 insertion.

Of course diagrams without a ϕ^4 vertex cannot be generated and this explains the subtractions.

Let us now introduce the generating functional of renormalized proper vertices $\Gamma_r(\phi)$:

$$\Gamma_r(\varphi) = \Gamma\left(\varphi\sqrt{Z}\right),$$

and thus:

$$\frac{\delta\Gamma_r(\varphi)}{\delta\varphi(x)} = \sqrt{Z}\,\frac{\delta\Gamma\left(\varphi\sqrt{Z}\right)}{\delta\varphi(x)}.$$

We use this relation in equation (12.24):

$$\left(-\Delta + m^2\right)_\Lambda Z\varphi(x) + g\sqrt{Z}\Gamma_{\phi^3(x)}\left(\varphi\sqrt{Z}\right) = \frac{\delta\Gamma_r(\varphi)}{\delta\varphi(x)}. \tag{12.25}$$

The r.h.s. is finite in the infinite cut-off limit. We now introduce the ϕ^4 field theory renormalization constants as defined by equations (12.1,12.2):

$$\left[-\Delta + Z^{-2}\left(m_r^2 + \delta m^2\right)\right] Z\varphi(x) + g_r Z_g Z^{-3/2} \Gamma_{\phi^3}\left(\varphi\sqrt{Z}\right) = \frac{\delta\Gamma_r(\varphi)}{\delta\varphi(x)}. \tag{12.26}$$

This relation shows that all ϕ-field correlation functions with one insertion of the operator $g_r Z_g \phi_r^3(x)$ are finite except the $\langle\phi^3\phi\rangle$ correlation function which needs two additional subtractions. We can determine the corresponding renormalization constants by imposing:

$$\left(\Gamma_{\phi^3}^{(1)}\right)_r (p, -p)\bigg|_{p=0} = 0,$$

$$\frac{\partial}{\partial p^2}\left(\Gamma_{\phi^3}^{(1)}\right)_r (p, -p)\bigg|_{p=0} = 0. \tag{12.27}$$

Equation (12.26) expanded in powers of φ then yields explicitly:

$$\Gamma_r^{(n+1)}(q, p_1, \ldots, p_n) = g_r\left(\Gamma_{\phi^3}^{(n)}\right)_r (q; p_1, \ldots, p_n) + \delta_{n1}\left(p^2 + m^2\right). \tag{12.28}$$

Note that with this definition $\left(\Gamma_{\phi^3}^{(3)}\right)_r$ satisfies the renormalization condition:

$$\left(\Gamma_{\phi^3}^{(3)}\right)_r (0; 0, 0, 0) = 1. \tag{12.29}$$

Equation (12.22) also contains information about multiple insertions of ϕ^3. For example after some algebraic manipulations one finds:

$$g^2\Gamma_{\phi^3(x_1)\phi^3(x_2)} + g\delta(x_1 - x_2)\Gamma_{\phi^2(x_1)} + \left(-\Delta + m^2\right)\delta(x_1 - x_2) = \frac{\delta^2\Gamma}{\delta\varphi(x_1)\delta\varphi(x_2)}. \tag{12.30}$$

The equation relates two insertions of ϕ^3 to two insertions of ϕ, again with subtraction terms which now involve the insertion of ϕ^2.

12.2.2 Other relations: renormalization of operators of dimension 4

We have shown in Section 5.4 that more general equations could be obtained by performing infinitesimal changes of variables. We can use them to establish relations between operators. For example in the change of variables

$$\phi'(x) = \phi(x) + \varepsilon(x)\phi(x),$$

the variation of the action (12.1) in the presence of a source is:

$$\delta\left[S(\phi) - J\phi\right] = \varepsilon(x)\left[\phi(x)\left(-\Delta + m^2\right)_\Lambda \phi(x) + \frac{g}{3!}\phi^4(x) - J(x)\phi(x)\right].$$

From the invariance of the functional integral follows:

$$W_{\phi(x)(-\Delta+m^2)_\Lambda\phi(x)} + \frac{g}{3!}W_{\phi^4(x)} = J(x)\frac{\delta W}{\delta J(x)}. \tag{12.31}$$

The discussion of the large cut-off limit of this equation, or the corresponding one obtained after Legendre transformation:

$$\Gamma_{\phi(x)(-\Delta+m^2)_\Lambda\phi(x)} + \frac{g}{3!}\Gamma_{\phi^4(x)} = \varphi(x)\frac{\delta\Gamma}{\delta\varphi(x)}, \tag{12.32}$$

is more delicate than in the case of equation (12.24): the operator $\phi(x)\left(-\Delta + m^2\right)_\Lambda \phi(x)$ which in Pauli–Villars regularization is:

$$\phi(x)\left(-\Delta + m^2\right)_\Lambda \phi(x) \equiv \phi(x)\left[-\Delta + m^2 + \alpha_1\frac{\Delta^2}{\Lambda^2} - \alpha_2\frac{\Delta^3}{\Lambda^3} + \cdots\right]\phi(x),$$

contains operators of canonical dimensions larger than 4 divided by powers of the cut-off. We shall discuss in Chapter 27 the problem of irrelevant operators which is directly related with the large cut-off limit of such operators. Let us here only state the result: in the large cut-off limit the operator $\phi\left(-\Delta + m^2\right)_\Lambda \phi$ is equivalent to a linear combination of all operators of dimensions 4 and 2.

Equation (12.32) implies after renormalization an identity satisfied by the operators $(\mathcal{O}_i(\phi))_r$ as defined in equations (12.8,12.9):

$$\sum_{i=1}^{4} C_i\left(g_r\right)\left(\Gamma_{\mathcal{O}_i}^{(n)}\right)_r (q; p_1, \ldots, p_n) = \sum_{m=1}^{n} \Gamma^{(n)}\left(p_1, \ldots, p_m + q, \ldots, p_n\right). \tag{12.33}$$

If we impose some renormalization conditions to define explicitly the insertions of operators $\mathcal{O}_i$, in the spirit of Chapter 10:

$$\left(\Gamma_{\mathcal{O}_1}^{(2)}\right)_r (0; 0, 0) = m_r^2,$$

$$\left(\Gamma_{\mathcal{O}_3}^{(2)}\right)_r (q; p_1, p_2) = -p_1 \cdot p_2 + O\left(p^4\right),$$

$$\left(\Gamma_{\mathcal{O}_3}^{(4)}\right)_r (0; 0, 0, 0, 0) = 0, \tag{12.34}$$

$$\left(\Gamma_{\mathcal{O}_4}^{(2)}\right)_r (q; p_1, p_2) = O\left(p^4\right),$$

$$\left(\Gamma_{\mathcal{O}_4}^{(4)}\right)_r (0; 0, 0, 0, 0) = 1,$$

then we can calculate the coefficients $C_i\left(g_r\right)$ only from $\left(\Gamma_{\mathcal{O}_1}^{(n)}\right)_r$ and its derivatives at zero momentum.

12.3 Short Distance Expansion (SDE) of Operator Products

Several chapters will be devoted to the discussion of the large or small momentum behaviour of correlation functions. Our essential tool will be the CS or RG equations. However these equations are directly useful only for non-exceptional momenta. For large momenta we shall be able to characterize the behaviour of

$$\Gamma^{(n)}\left(\rho p_1 + r_1, \rho p_2 + r_2, \dots, \rho p_n + r_n\right), \quad \rho \to \infty,$$

provided no subset of momenta p_i has a vanishing sum:

$$\sum_{i \in I \neq \emptyset} p_i \neq 0 \quad \forall I \not\equiv (1, 2, \dots n).$$

For exceptional momenta a new tool is needed: the short distance expansion (SDE) of product of operators. In this section we shall mainly discuss the SDE of the product of two fields at leading order, but we shall indicate how the method we use can be generalized to more complicated cases. It is entirely based on the theory of renormalization of composite operators we have just presented.

Definition. We consider the 1PI correlation function:

$$\Gamma^{(n+2)}\left(x + \frac{v}{2}, x - \frac{v}{2}, y_1, \dots, y_n\right) = \left\langle \phi\left(x + \frac{v}{2}\right)\phi\left(x - \frac{v}{2}\right)\phi(y_1)\cdots\phi(y_n)\right\rangle_{1\,\mathrm{PI}},$$
$$(12.35)$$

in which all arguments are fixed, except the vector **v** which goes to zero. We want to study the $\mathbf{v} \to \mathbf{0}$ limit. In a theory which is sufficiently regularized, an expansion in powers of v can be obtained by expanding the product of fields $\phi\left(x + v/2\right)\phi\left(x - v/2\right)$ in powers of v:

$$\phi\left(x + \frac{v}{2}\right)\phi\left(x - \frac{v}{2}\right) = \phi^2(x) + \frac{1}{4}v_{\mu_1}v_{\mu_2}\left[\phi(x)\partial_{\mu_1}\partial_{\mu_2}\phi(x) - \partial_{\mu_1}\phi(x)\partial_{\mu_2}\phi(x)\right]$$
$$+ \sum_{n=2}^{\infty}\frac{1}{2^{2n}}\frac{1}{2n!}v_{\mu_1}\cdots v_{\mu_{2n}}\mathcal{O}_{\mu_1\dots\mu_{2n}}\left(\phi(x)\right), \qquad (12.36)$$

in which $\mathcal{O}_{\mu_1\dots\mu_{2n}}(\phi)$ is a local operator quadratic in ϕ with $2n$ derivatives.

If we now insert expansion (12.36) into a correlation function, even after field renormalization all terms of the expansion in general diverge when the cut-off is removed. As we have extensively discussed in Section 12.1, the various composite operators appearing in expansion (12.36) have to be renormalized. Therefore we have to expand each bare operator on a basis of renormalized operators. In terms of renormalized operators $\mathcal{O}_r^\alpha(\phi)$ expansion (12.36) then takes the form:

$$\phi_r\left(x + \frac{v}{2}\right)\phi_r\left(x - \frac{v}{2}\right) = \sum_\alpha C_\alpha\left(v, \Lambda\right)\mathcal{O}_r^\alpha(x), \qquad (12.37)$$

in which, at cut-off Λ fixed, the coefficients $C_\alpha\left(v, \Lambda\right)$ are regular even functions of the vector **v**. An $\mathcal{O}_r^\alpha(x)$ receives contributions from bare operators of equal or higher dimensions. We conclude that for $|\mathbf{v}|$ small the coefficient functions $C_\alpha\left(v, \Lambda\right)$ behave like:

$$C_\alpha\left(\lambda v, \Lambda\right) \sim \lambda^{[\mathcal{O}_\alpha] - 2[\phi]}, \quad \text{for } \lambda \to 0. \qquad (12.38)$$

When the cut-off becomes infinite, the coefficients of the expansion of C_α in powers of v, being renormalization constants, diverge, and if C_α has a limit, the limiting function is singular at $v = 0$. The coefficients are functions of the cut-off which for $\mathbf{v} = \mathbf{0}$ diverge in a way predicted by power counting. Since $\mathbf{v}$ is small but non-vanishing, the coefficients will grow with the cut-off until Λ is of order $1/|\mathbf{v}|$. In this range all contributions to a given coefficient are of the same order, up to powers of logarithms because powers of v compensate the powers of Λ. Therefore, at least in perturbation theory, the ordering of operators consequence of equation (12.38) will survive, the operators of lowest dimensions will dominate expansion (12.37) for $|\mathbf{v}|$ small and the behaviour of the coefficients $C_\alpha(v, \Lambda)$ is given by equation (12.38) up to powers of $\ln v$.

The expansion (12.37) is the short distance expansion (SDE) of the product of two operators ϕ.

12.3.1 SDE at leading order

Further insight is gained into the structure of the SDE by realizing that the product $\phi_r(x + v/2)\,\phi_r(x - v/2)$ can be considered as a regularization by point splitting of the composite operator $\left(\phi^2(x)\right)_r$. Let us discuss this point in detail in the framework of the ϕ_4^4 field theory. We then expect:

$$\phi_r\left(x + \frac{v}{2}\right)\phi_r\left(x - \frac{v}{2}\right) \underset{|\mathbf{v}|\to 0}{\sim} C_1(v)\left[\phi^2(x)\right]_r . \tag{12.39}$$

The singularities of $C_1(v)$ for $|\mathbf{v}|$ small should be directly related to the divergences of the renormalization constant Z_2 which renders $\phi_r^2(x)$ finite.

In what follows it will therefore be convenient to treat the product:

$$\tfrac{1}{2}\phi_r\left(x + v/2\right)\phi_r\left(x - v/2\right)$$

as one composite operator depending on the point x and the parameter v, in particular from the point of view of connectivity and one-particle irreducibility. To make this explicit we shall introduce a notation for this operator: $\frac{1}{2}\left[\phi_r\left(x + v/2\right)\phi_r\left(x - v/2\right)\right]$.

By definition we have:

$$\tfrac{1}{2}\left\langle \left[\phi_r\left(x + \frac{v}{2}\right)\phi_r\left(x - \frac{v}{2}\right)\right]\phi_r\left(y_1\right)\ldots\phi_r\left(y_n\right)\right\rangle$$
$$\equiv \tfrac{1}{2}\left\langle \phi_r\left(x + \frac{v}{2}\right)\phi_r\left(x - \frac{v}{2}\right)\phi_r\left(y_1\right)\ldots\phi_r\left(y_n\right)\right\rangle. \tag{12.40}$$

However the relation between connected correlation functions is different:

$$\left\langle \left[\phi_r\left(x + \frac{v}{2}\right)\phi_r\left(x - \frac{v}{2}\right)\right]\phi_r\left(y_1\right)\ldots\phi_n\left(y_n\right)\right\rangle_c$$
$$= \tfrac{1}{2}\left\langle \phi_r\left(x + \frac{v}{2}\right)\phi_r\left(x - \frac{v}{2}\right)\phi_r\left(y_1\right)\ldots\phi_r\left(y_n\right)\right\rangle_c$$
$$+ \tfrac{1}{2}\sum_{I\cup J=(1,\ldots,n)}\left\langle \phi_r\left(x + \frac{v}{2}\right)\phi_r\left(y_{i_1}\right)\ldots\phi_r\left(y_{i_p}\right)\right\rangle_c$$
$$\times \left\langle \phi_r\left(x - \frac{v}{2}\right)\phi_r\left(y_{j_1}\right)\ldots\phi_r\left(y_{j_{n-p}}\right)\right\rangle_c, \tag{12.41}$$

in which I and J are all non empty partitions of $(1,\ldots,n)$.

Rather than writing explicitly the corresponding relations between proper vertices, we give the relation in terms of generating functionals. Calling $Z\left(x + v/2, x - v/2; J\right)$ the generating functional of correlation functions with the operator insertion:

$$Z\left(x + \frac{v}{2}, x - \frac{v}{2}; J\right) = \frac{1}{2}\frac{\delta^2 Z(J)}{\delta J\left(x + \frac{v}{2}\right)\delta J\left(x - \frac{v}{2}\right)},\qquad(12.42)$$

we find for connected correlation functions with obvious notation:

$$W\left(x + \frac{v}{2}, x - \frac{v}{2}; J\right) = Z^{-1}(J)\,Z\left(x + \frac{v}{2}, x - \frac{v}{2}; J\right)$$
$$= \frac{1}{2}\frac{\delta^2 W(J)}{\delta J\left(x + \frac{v}{2}\right)\delta J\left(x - \frac{v}{2}\right)} + \frac{1}{2}\frac{\delta W}{\delta J\left(x + \frac{v}{2}\right)}\frac{\delta W}{\delta J\left(x - \frac{v}{2}\right)},\qquad(12.43)$$

and finally for the generating functional of proper vertices:

$$\Gamma\left(x + \frac{v}{2}, x - \frac{v}{2}; \varphi\right) = \frac{1}{2}\varphi\left(x + \frac{v}{2}\right)\varphi\left(x - \frac{v}{2}\right) + \frac{1}{2}\left[\frac{\delta^2\Gamma}{\delta\varphi\left(x + \frac{v}{2}\right)\delta\varphi\left(x - \frac{v}{2}\right)}\right]^{-1}.$$
$$(12.44)$$

Note that this equation is similar to equation (12.64) which we have used to prove the irreducibility of $\Gamma(\varphi)$.

To now ensure the limit (12.39) we determine $C_1(v)$ by imposing that the insertion of the operator $\frac{1}{2}C_1^{-1}(v)\left[\phi\left(x + v/2\right)\phi\left(x - v/2\right)\right]$ in the 2-point function satisfies for any v the renormalization condition (10.22). Defining:

$$C_1^{-1}(v)\int \mathrm{d}x\,\mathrm{d}y_1\,\mathrm{d}y_2\; \mathrm{e}^{ipx + iq_1 y_1 + iq_2 y_2}\left.\frac{\delta^2\Gamma_{\mathrm{r}}\left(x + v/2, x - v/2; \varphi\right)}{\delta\varphi\left(y_1\right)\delta\varphi\left(y_2\right)}\right|_{\varphi = 0}$$
$$= \left(2\pi\right)^4\delta\left(p + q_1 + q_2\right)\tilde{\Gamma}_{\mathrm{r}}^{(1,2)}\left(v; p; q_1, q_2\right),\qquad(12.45)$$

we impose:

$$\tilde{\Gamma}_{\mathrm{r}}^{(1,2)}\left(v; 0; 0, 0\right) = 1\,.\qquad(12.46)$$

Setting $q_1 = q_2 = 0$ in equation (12.45), we derive:

$$C_1(v) = \left.\frac{\delta^2\Gamma_{\mathrm{r}}\left(v/2, -v/2; \varphi\right)}{\delta\tilde{\varphi}\left(0\right)\delta\tilde{\varphi}\left(0\right)}\right|_{\varphi = 0},\qquad(12.47)$$

in which $\tilde{\varphi}(q)$ is the Fourier transform of $\varphi(x)$.

By differentiating equation (12.44) twice with respect to φ, we can relate the r.h.s. of equation (12.47) to the φ-field proper vertices:

$$C_1(v) = 1 - \frac{1}{2}\int\frac{\mathrm{d}^4 k}{\left(2\pi\right)^4}\,\mathrm{e}^{-ikv}\left[W_{\mathrm{r}}^{(2)}(k)\right]^2\Gamma_{\mathrm{r}}^{(4)}\left(k, -k, 0, 0\right).\qquad(12.48)$$

The equation can also be rewritten:

$$C_1(v) = 1 + \frac{1}{2}m_{\mathrm{r}}^4\left\langle\phi_{\mathrm{r}}\left(v/2\right)\phi_{\mathrm{r}}\left(-v/2\right)\tilde{\phi}_{\mathrm{r}}\left(0\right)\tilde{\phi}_{\mathrm{r}}\left(0\right)\right\rangle_{\mathrm{c}}.\qquad(12.49)$$

We have introduced a mixed connected correlation functions, $\tilde{\phi}(p)$ being the Fourier transform of the field $\phi(x)$, and used renormalization conditions (10.9).

The coefficient $C_1(v)$ is defined in such a way that the renormalized operator $\phi_{\mathrm{r}}(x + v/2)\phi_{\mathrm{r}}(x - v/2)C_1^{-1}(v)$ then converges towards the operator $\left(\phi_{\mathrm{r}}^2(x)\right)_{\mathrm{r}}$ correctly normalized.

Let us for example write the consequence for the 4-point function:

$$C_1(v)\Gamma_{\mathrm{r}}^{(1,2)}\left(p; q_1, q_2\right) \underset{|v|\to 0}{\sim} 1 - \frac{1}{2}\int \frac{\mathrm{d}^4 k}{(2\pi)^4}\, \mathrm{e}^{-ikv}\, W_{\mathrm{r}}^{(2)}\left(\frac{p}{2} - k\right) W_{\mathrm{r}}^{(2)}\left(\frac{p}{2} + k\right)$$

$$\times\, \Gamma_{\mathrm{r}}^{(4)}\left(\frac{p}{2} + k, \frac{p}{2} - k, q_1, q_2\right). \tag{12.50}$$

The neglected contributions are of order $v^2 \left(\ln v^2\right)^p$ at any finite order in perturbation theory.

Renormalization theory tells us that equation (12.39) is valid as long as the replacement of the operator product by $\phi^2(x)$ does not generate new renormalizations. We can therefore use equation (12.39) in all $\phi(x)$ and $\phi^2(x)$ correlation functions except

(i) the 2-point function $\langle \phi\left(x + v/2\right)\phi\left(x - v/2\right)\rangle$ which leads to $\langle \phi^2(x) \rangle$;

(ii) the 4-point function $\langle \phi\left(x + v/2\right)\phi\left(x - v/2\right)\phi^2(y)\rangle$ which leads to the ϕ^2 2-point function $\langle \phi^2(x)\phi^2(y)\rangle$.

Both require an additional additive renormalization. The strategy in such cases is to first apply the CS differential operator on the correlation to generate additional $\int \phi^2(x)\mathrm{d}x$ insertions until relation (12.39) can be used. As a consequence the SDE is modified by contributions which are solutions of the homogeneous CS equations.

12.3.2 One-loop calculation of the leading coefficient of the SDE

Equation (12.48) can be used to calculate the coefficient function $C_1(v)$ in perturbation theory. At one-loop order it is sufficient to replace correlation functions by their tree level values:

$$C_1(v) = 1 - \frac{g}{2}\int \frac{\mathrm{d}^4 k}{16\pi^2}\frac{\mathrm{e}^{-ikv}}{\left(k^2 + m^2\right)^2} + O\left(g^2\right). \tag{12.51}$$

It is clear from this expression that, as expected, $C_1(v)$ has at one-loop order the form:

$$C_1(v) \sim A\ln\left(|\mathbf{v}|\, m\right) + B + O\left(v^2\right). \tag{12.52}$$

An often useful idea to extract an asymptotic expansion of this form is to calculate the Mellin transform $\mu\left(\alpha\right)$ of the function:

$$\mu\left(\alpha\right) = \int_0^\infty \mathrm{d}v\, v^{\alpha-1}C_1(v), \tag{12.53}$$

in which v is the length $|\mathbf{v}|$ of the vector $\mathbf{v}$. The expansion (12.52) then implies:

$$\mu\left(\alpha\right) = -\frac{A}{\alpha^2} + \left(A\ln m + B\right)\frac{1}{\alpha} + O\left(1\right) \quad \text{for} \quad \alpha \to 0. \tag{12.54}$$

Applying this technique one has to evaluate:

$$f\left(\alpha\right) = \int_0^\infty \mathrm{d}v\, v^{\alpha-1}\int \frac{\mathrm{d}^4 k}{16\pi^2}\frac{\mathrm{e}^{-ikv}}{\left(k^2 + m^2\right)^2}. \tag{12.55}$$

As usual we rewrite the integral as:

$$f(\alpha) = \int_0^\infty dv \, v^{\alpha-1} \int_0^\infty t \, dt \frac{d^4k}{16\pi^2} \, e^{-t(k^2+m^2)-ikv} . \tag{12.56}$$

Integrating over k, v and t, in this order, we finally obtain:

$$f(\alpha) = \frac{1}{8} \left(\frac{2}{m} \right)^\alpha \frac{\Gamma^2 \left(1 + \alpha/2\right)}{\alpha^2} . \tag{12.57}$$

An expansion for $\alpha \sim 0$ yields the coefficients A and B:

$$C_1(v) = 1 - \frac{g}{16} \left[-\ln \left(\frac{vm}{2} \right) + \psi\left(1\right) \right] + O\left(g^2\right), \tag{12.58}$$

in which $\psi\left(z\right)$ is the logarithmic derivative of the function $\Gamma\left(z\right)$.

At this order the bare parameters can be replaced by renormalized parameters.

12.4 Large Momentum Expansion of the SDE Coefficients

To the behaviour of the product of fields $\phi\left(x + v/2\right) \phi\left(x - v/2\right)$ at short distance is associated, after Fourier transformation, the behaviour at large relative momentum k of the product $\tilde{\phi}\left(p/2 - k\right) \tilde{\phi}\left(p/2 + k\right)$. However some information is lost in the transformation. The large momentum behaviour is only sensitive to the singular part of the short distance behaviour. For instance the constant terms in the asymptotic expansion of $C_1(v)$, yield, after Fourier transformation, terms proportional to $\delta^4(k)$ which do not contribute to the large momentum behaviour. At the same time the algebraic structure is for the same reason somewhat simplified.

Let us take the example of equation (12.50). We introduce the Fourier transform $C_1(k)$ of $C_1(v)$ (as stated before, we omit the tilde sign indicating Fourier transformation when, due to the change of arguments, x, y, z, v to k, p, q, there is no ambiguity):

$$C_1(k) = \int e^{ikv} C_1(v) d^4v . \tag{12.59}$$

After Fourier transformation, in the large k limit, equation (12.50) yields:

$$\Gamma_{\mathrm{r}}^{(4)} \left(\frac{p}{2} + k, \frac{p}{2} - k, q_1, q_2 \right) \sim -2C_1(k) \left[\Gamma_{\mathrm{r}}^{(2)}(k) \right]^2 \Gamma_{\mathrm{r}}^{(1,2)} \left(p; q_1, q_2\right), \tag{12.60}$$

in which the neglected terms are of order $(\ln k)^p / k^2$ at any finite order in perturbation theory.

More generally, due to momentum conservation, the disconnected contributions in equation (12.43) coming from $\delta W / \delta J \left(x + v/2\right) \delta W / \delta J \left(x - v/2\right)$ do not contribute to the large momentum behaviour.

In addition if we expand the r.h.s. of equation (12.44) in powers of φ, we obtain two types of contributions: one term which becomes 1PI after amputation of the lines corresponding to the fields $\phi\left(x + v/2\right)$ and $\phi\left(x - v/2\right)$, and the other terms which remain reducible.

Figure 12.3 gives the example of the 6-point function.

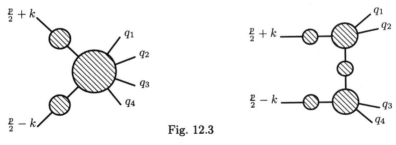

Fig. 12.3

Due to momentum conservation, in reducible terms the propagator which connects the proper vertices carries a momentum of order k for k large. The corresponding contributions are thus suppressed by a factor $1/k^2$ (up to powers of $\ln k$) and can be neglected at leading order.

At leading order we can therefore write for $n > 0$:

$$\Gamma_{\mathrm{r}}^{(n+2)}\left(\frac{p}{2} + k, \frac{p}{2} - k, q_1, \ldots, q_n\right) \underset{|k| \to \infty}{\sim} -2C_1(k)\left[\Gamma_{\mathrm{r}}^{(2)}(k)\right]^2 \Gamma_{\mathrm{r}}^{(1,n)}(p; q_1, \ldots, q_n).$$
(12.61)

As explained in Section 12.3, this equation generalizes to ϕ^2 insertions provided either n is positive or $l > 1$:

$$\Gamma_{\mathrm{r}}^{(l,n+2)}\left(p_1, \ldots, p_l; \frac{p}{2} + k, \frac{p}{2} - k, q_1, \ldots, q_n\right) \sim -2C_1(k)\left[\Gamma_{\mathrm{r}}^{(2)}(k)\right]^2$$
$$\times \Gamma_{\mathrm{r}}^{(l+1,n)}(p_1, \ldots, p_l, p; q_1, \ldots, q_n). \quad (12.62)$$

12.5 Callan–Symanzik Equation for the First Coefficient of the SDE

In this section all quantities are assumed to be renormalized and we omit the subscript r.

Let us first note by comparing equations (12.48) and (12.61) that in momentum representation the relevant function is $\Gamma^{(4)}(k, -k, 0, 0)$ which we shall call $B(k)$ in what follows

$$B(k) \equiv \Gamma^{(4)}(k, -k, 0, 0) \underset{|\mathbf{k}| \to \infty}{\sim} -2C_1(k)\left[\Gamma^{(2)}(k)\right]^2. \quad (12.63)$$

Equation (12.61) becomes:

$$\Gamma^{(n+2)}\left(\frac{p}{2} + k, \frac{p}{2} - k, q_1, \ldots, q_n\right) \sim B(k)\Gamma^{(1,n)}(p; q_1, \ldots, q_n). \quad (12.64)$$

Similarly:

$$\Gamma^{(1,n+2)}\left(0; \frac{p}{2} + k, \frac{p}{2} - k, q_1, \ldots, q_n\right) \sim B(k)\Gamma^{(2,n)}(0, p; q_1, \ldots, q_n). \quad (12.65)$$

Let us introduce a notation for the CS differential operator:

$$\mathrm{d} \equiv m\frac{\partial}{\partial m} + \beta(g)\frac{\partial}{\partial g}. \quad (12.66)$$

We write the CS equations for $\Gamma^{(n+2)}$:

$$\left[\mathrm{d} - \tfrac{1}{2}(n+2)\eta(g)\right]\Gamma^{(n+2)} = m^2\sigma(g)\Gamma^{(1,n+2)}(0; \ldots). \quad (12.67)$$

We now use equations (12.64) and (12.65) in the large $|k|$ limit:

$$\left[\mathrm{d} - \tfrac{1}{2}(n+2)\eta(g)\right]\left[B(k)\Gamma^{(1,n)}(p; q_1, \ldots, q_n)\right] = m^2\sigma(g)B(k)\Gamma^{(2,n)}(0, p; q_1, \ldots, q_n).$$
(12.68)

Using the CS equation for $\Gamma^{(1,n)}$:

$$d\Gamma^{(1,n)} = \left[\frac{n}{2}\eta(g) + \eta_2(g)\right]\Gamma^{(1,n)} + m^2\sigma(g)\Gamma^{(2,n)}(0,\dots)\,, \qquad (12.69)$$

we finally obtain an equation for $B(k)$:

$$[d - \eta(g) + \eta_2(g)]\,B(k) \sim 0\,. \qquad (12.70)$$

We can also write an equation for $C_1(k)$, using relation (12.63):

$$[d + \eta(g) + \eta_2(g)]\,C_1\left(\frac{k}{m}, g\right) \sim 0\,. \qquad (12.71)$$

We can compare this equation with equations (10.27, 10.28) which imply:

$$[d - \eta(g) - \eta_2(g)]\,Z_2\left(\frac{\Lambda}{m}, g\right) = 0\,. \qquad (12.72)$$

This indeed shows that $C_1(k/m, g)$ plays the same role as $Z_2^{-1}(\Lambda/m, g)$.

12.6 SDE beyond Leading Order

In this section also all quantities except the operators are assumed to be renormalized.

The product $\phi(x + v/2)\,\phi(x - v/2)$ is not only a regularization of $\phi^2(x)$ by point splitting but also of operators of higher dimensions which can be obtained by differentiation.

At next order, which means taking into account in expansion (12.37) all operators of dimensions 2 and 4, the SDE of $\phi(x + v/2)\,\phi(x - v/2)$ is an expansion of a regularized bare operator of dimension 4 on a basis of renormalized operators of dimensions 2 and 4 as discussed in previous sections. Indeed let us differentiate expansion (12.37) twice with respect to v:

$$\frac{1}{4}\left[\phi\left(x + \frac{v}{2}\right)\partial_\mu\partial_\nu\phi\left(x - \frac{v}{2}\right) + \phi\left(x - \frac{v}{2}\right)\partial_\mu\partial_\nu\phi\left(x + \frac{v}{2}\right) - \partial_\mu\phi\left(x + \frac{v}{2}\right)\partial_\nu\phi\left(x - \frac{v}{2}\right)\right.$$
$$\left. - \partial_\mu\phi\left(x - \frac{v}{2}\right)\partial_\nu\phi\left(x + \frac{v}{2}\right)\right] = \sum_\alpha \partial_\mu\partial_\nu C_\alpha(v)\mathcal{O}_r^\alpha(\phi(x))\,. \qquad (12.73)$$

The product in the r.h.s. can be considered as a form regularized by point splitting of $\frac{1}{2}(\phi(x)\partial_\mu\partial_\nu\phi(x) - \partial_\mu\phi(x)\partial_\nu\phi(x))$ which is a linear combination of operators of dimension 4, and spins 0 and 2. In Section 12.1 we have discussed the renormalization of operators of dimension 4 and spin zero. The operators of spin 2 introduce two new linearly independent operators which can be chosen to be the traceless part of $\partial_\mu\phi(x)\partial_\nu\phi(x)$ and $\partial_\mu\partial_\nu(\phi^2(x))$. Rotation invariance in space tells us that operators of different spin do not mix under renormalization. Therefore in addition to relations (12.9) we have:

$$\mathcal{O}_5(\phi(x)) \equiv \partial_\mu\phi(x)\partial_\nu\phi(x) - \tfrac{1}{4}\delta_{\mu\nu}(\partial\phi(x))^2$$
$$= Z_5^{-1}(\mathcal{O}_5(\phi(x)))_r - B_5\left(\partial_\mu\partial_\nu - \tfrac{1}{4}\delta_{\mu\nu}\Delta\right)(\phi^2(x))_r\,. \qquad (12.74)$$

We can impose the renormalization conditions:

$$\Gamma^{(2)}_{\mathcal{O}_5}(q; p_1, p_2) = -\left(p_1^\mu p_2^\nu + p_1^\nu p_2^\mu - \tfrac{1}{2}\delta_{\mu\nu}p_1 p_2\right) + O\left(p^4\right)\,,$$
$$\Gamma^{(4)}_{\mathcal{O}_5}(0; 0, 0, 0, 0) = 0\,. \qquad (12.75)$$

We can now use exactly the same strategy as for the leading term in the SDE. We insert expansion (12.73) truncated by omitting all operators of dimension larger than 4 which correspond to vanishing coefficients, in the 2 and 4-point correlation functions. We go over to the 1PI correlation functions in the way explained in Section 12.3, i.e. considering the product of fields in (12.73) as a composite operator. We finally use the renormalization conditions (12.34) and (12.75) to determine all the coefficients $\partial_\mu \partial_\nu C_\alpha(v)$ in the truncated SDE.

Note that we have lost no information by differentiating twice since only $C_1(v)$, which has been already determined, has a constant part when $|v|$ goes to zero. In this limit the coefficients $\partial_\mu \partial_\nu C_\alpha(v)$ have singularities in $\mathbf{v}$ which are similar to the divergences in Λ/m of the renormalization constants which appear in the expansion of operators $\mathcal{O}_5$ on a basis of dimension 4 renormalized operators. To obtain the asymptotic behaviour of the coefficients $C_\alpha(v)$ we then have to establish RG equations for them, by introducing the SDE in the CS equations for 1PI correlation functions. We have to worry about the SDE expansion in presence of a $\phi^2(x)$ insertion. We have to use the analogue of equation (12.10), the difference being that the new renormalization constant which appears in front of the contact term now is a function of $\mathbf{v}$.

We conclude that the coefficients of the SDE satisfy RG equations which are formally identical to the relations between the renormalization constants and the finite RG functions which arise in the CS equations for the operators of dimension 4 and spins 0 and 2, in complete analogy with the correspondence between equations (12.71) and (12.72).

12.7 SDE of Products of Arbitrary Local Operators

For general local operators A and B we expect:

$$A\left(x + \frac{v}{2}\right) B\left(x - \frac{v}{2}\right) = \sum_\alpha C_{AB}^\alpha(v) \mathcal{O}_r^\alpha(x), \qquad (12.76)$$

in which at any finite order in perturbation theory:

$$C_{AB}^\alpha(\lambda v) \sim \lambda^{[\mathcal{O}^\alpha] - [A] - [B]} \quad \text{for} \quad \lambda \to 0, \qquad (12.77)$$

up to powers of $\ln \lambda$ and the $\mathcal{O}_r^\alpha(x)$ form a complete basis of local operators.

Let us take the example of:

$$A \equiv B \equiv \left(\phi^2(x)\right)_r .$$

The product $\phi^2(x + v/2)\,\phi^2(x - v/2)$ is a regularized form of $\phi^4(x)$, however $\phi^4(x)$ by renormalization mixes with all operators of dimension 4 and $\phi^2(x)$. Power counting tells us that among these operators $\phi^2(x)$ has the most divergent coefficient. Therefore at leading order:

$$\left(\phi^2\left(x + \frac{v}{2}\right)\right)_r \left(\phi^2\left(x - \frac{v}{2}\right)\right)_r \underset{|\mathbf{v}| \to 0}{\sim} C_{\phi^2\phi^2}^1(v)\left(\phi^2(x)\right)_r , \qquad (12.78)$$

in which the coefficient can be determined by using the renormalization condition for $\left(\phi^2\right)_r$:

$$\Gamma^{(1,2)}(q; p_1, p_2) = 1 \quad \text{for} \quad q = p_1 = p_2 = 0,$$

and expressing that equation (12.78) should be exact when inserted in the 2-point function at the subtraction point where all momenta vanish.

It is then easy to derive RG equations for this new coefficient by inserting the relation in the CS equations for the proper vertices $\Gamma^{(l,n)}$.

We do not wish to go into further detail since the discussion is very technical. The most important idea to keep in mind is the complete parallelism between the SDE of operator products and the renormalization equations of the corresponding composite operators.

12.8 Light Cone Expansion (LCE) of Operator Products

After analytic continuation to real time, the length of the vector squared x^2 may vanish, while the vector x_μ remains finite. In such a situation the relevant expansion for a product of fields is no longer the SDE but instead the light cone expansion (LCE).

It is necessary to classify all operators not only according to their canonical dimensions, but also their spin s which characterizes their transformation properties under space rotations. The LCE takes the form:

$$\phi\left(x + \frac{v}{2}\right) \phi\left(x - \frac{v}{2}\right) = \sum_{\alpha,s} C_\alpha^s\left(v^2\right) P_{\mu_1\dots\mu_s}^s(v) \mathcal{O}_{\mu_1\dots\mu_s}^{s,\alpha}(x). \tag{12.79}$$

The polynomial $P_{\mu_1\dots\mu_s}^s(v)$ is a homogeneous, traceless for $s > 0$, polynomial of the vector v_μ and the operators $\mathcal{O}_{\mu_1\dots\mu_s}^{s,\alpha}$ form a complete basis of local operators.

For example:

$$P_{\mu_1,\mu_2}^2(v) = v_{\mu_1} v_{\mu_2} - \frac{1}{d} v^2 \delta_{\mu_1\mu_2}.$$

When v^2 goes to zero with v_μ finite, the polynomials P^s have a finite limit and therefore the coefficients $C_\alpha^s(v^2)$ contain the whole non-trivial dependence in v^2. The analysis already performed for the SDE can be extended and shows that in perturbation theory $C_\alpha^s(v^2)$ behaves like:

$$C_\alpha^s\left(\lambda^2 v^2\right) \sim \lambda^{[\mathcal{O}^{s,\alpha}]-2[\phi]-s} \text{ up to powers of } \ln\lambda \text{ for } \lambda \to 0. \tag{12.80}$$

Therefore the important parameter is no longer the dimension of the operator $\mathcal{O}^{s,\alpha}$ but a quantity called the twist τ:

$$\tau = [\mathcal{O}^{s,\alpha}] - s. \tag{12.81}$$

The operators of lowest twist dominate the LCE of product of operators.

In expansion (12.79) the lowest twist is 2 which is the twist of $\phi^2(x)$. Each new factor $\phi(x)$ increases the twist by one unit, while additional derivatives either increase the twist or leave it unchanged.

Therefore the most general operator of twist 2 has the form:

$$\phi(x)\left(\partial_{\mu_1}\dots\partial_{\mu_2}\dots\partial_{\mu_n} - \text{ traces}\right)\phi(x),$$

or is a combination of derivatives of a twist 2 operators.

Operators of twist 2 and spin s, since they are the operators of lowest dimension for a given spin, renormalize among themselves. Using previous considerations about the SDE it is easy to write RG equations for the corresponding coefficients $C_\alpha^s(v^2)$.

Bibliographical Notes

The renormalization of composite operators is discussed by:

W. Zimmermann in *Lectures in Elementary Particles and Quantum Field Theory* already quoted in Chapter 10.

The short distance expansion was postulated in:

L.P. Kadanoff, *Phys. Rev. Lett.* 23 (1969) 1430; K.G. Wilson, *Phys. Rev.* 179 (1969) 1499, D3 (1971) 1818.

RG equations for the leading coefficients of the SDE are established in:

K. Symanzik, *Commun. Math. Phys.* 23 (1971) 49; C.G. Callan, *Phys. Rev.* D5 (1972) 3202.

A thorough investigation is carried out in:

W. Zimmermann, *Ann. Phys. (NY)* 77 (1973) 536 and 570.

See also

C.G. Callan in *Methods in Field Theory,* already quoted in Chapter 9.

The extension of Wilson's work to the light cone expansion is found in

R.A. Brandt and G. Preparata, *Nucl. Phys.* B27 (1971) 541.

Exercises

Exercise 12.1

12.1.1. Study the renormalization of the ϕ^3 field theory in six dimensions,

$$S(\phi) = \int d^6x \left[\tfrac{1}{2}\left(\partial_\mu \phi\right)^2 + \tfrac{1}{2}m^2\phi^2 + \frac{1}{3!}g\phi^3 \right],$$

and derive the renormalized equation of motion.

12.1.2. Establish the corresponding Callan–Symanzik equations.

Exercise 12.2

In the ϕ^4_4 field theory, write explicitly the relation (12.33) in the case of dimensional regularization and minimal subtraction. Obtain two other equations by differentiating the functional integral with respect to the coupling constant and the mass, and determine the RG functions of CS equation (12.15).

13 SYMMETRIES AND RENORMALIZATION

Up to now we have discussed the problem of renormalization from the sole point of view of power counting. In Chapters 13–21 we shall discuss the consequences for renormalization of symmetries of the action. Indeed when the tree level action has some symmetry properties, we expect that the renormalized action will not be of the most general form allowed by pure power counting arguments but will instead keep some trace of the initial symmetry. Technically this means that, as a consequence of the symmetry, the divergences generated in perturbation theory are not of generic form, and therefore the renormalization constants are not all independent. In this chapter we deal only with *linearly realized continuous symmetries* corresponding to compact Lie groups because they lead to very interesting formal properties; consequences of discrete symmetries can also be studied with, however, somewhat different methods (see the remark at the end of Section 13.5). Also we consider below only infinitesimal group transformations and therefore topological properties of groups will play no role.

Our general strategy will be as follows:

(i) We first introduce a regularization which preserves the symmetry.

(ii) We then prove identities, generally called Ward–Takahashi (WT) identities, consequences of the symmetry of the action and satisfied by the generating functional of 1PI correlation functions.

(iii) These identities imply relations between the divergences of correlation functions and thus between the counterterms which render the theory finite. From these relations we derive the *generic* form of the counterterms. Such an analysis is based on a loopwise expansion of perturbation theory.

(iv) We finally read off the properties of the renormalized action.

More generally non-trivial identities survive when terms in the action induce a *soft breaking* of symmetry. We specifically consider the examples of linear symmetry breaking and the important limiting case of spontaneous symmetry breaking, and quadratic symmetry breaking.

Finally in Section 13.6 we apply the formalism to the physical example of chiral symmetry breaking in low-energy effective models of hadrons.

In the appendix we outline the relation between WT identities and current conservation. We derive the energy-momentum tensor and dilatation current.

13.1 Preliminary Remarks

Before entering the discussion, we describe our notations and conventions for group and Lie algebras. A few additional algebraic remarks concerning the representation of the Lie algebra in terms of differential operators are also useful.

13.1.1 Conventions and notations

We consider continuous symmetries corresponding to various Lie groups and algebras. In this context we adopt the following set of conventions except if explicitly stated otherwise: for continuous symmetries we only consider compact Lie groups. Therefore we can represent the generators of the corresponding Lie algebras as $N \times N$ real antisymmetric matrices t^α. The trace of two antisymmetric matrices defines a scalar product. We can use it to normalize the matrices by:

$$\text{tr}\, t^\alpha t^\beta = -N\delta_{\alpha\beta}\,. \tag{13.1}$$

With this convention the Lie algebra structure constants $f_{\alpha\beta\gamma}$:

$$[t^\alpha, t^\beta] = f_{\alpha\beta\gamma} t^\gamma, \tag{13.2}$$

are completely antisymmetric in the three indices. The basis of the Lie algebra is fixed up to an orthogonal transformation. In the special case of unitary groups, we also sometimes represent the generators by hermitian or anti-hermitian matrices (this will be a matter of convenience) and then normalize them by

$$\text{tr}\, t^\alpha t^\beta = N\delta_{\alpha\beta}\,,$$

(in the hermitian case).

As a consequence, as in the orthogonal case, the structure constants defined by:

$$[t^\alpha, t^\beta] = if_{\alpha\beta\gamma} t^\gamma,$$

and thus:

$$f_{\alpha\beta\gamma} = \frac{1}{2N}\,\text{Im}\,\text{tr}\left(t^\alpha t^\beta t^\gamma\right),$$

are completely antisymmetric.

13.1.2 Lie algebra and differential operators

Let us first establish a simple group theoretical property which will become increasingly useful when we discuss the renormalization of symmetries in more complicated situations. Let ϕ_i be a vector transforming linearly under a representation $\mathcal{D}(G)$ of a Lie group G:

$$\phi_i' = D_{ij}\left(g\right)\phi_j\,. \tag{13.3}$$

Let us write also the corresponding infinitesimal group transformations. In terms of the matrices t_{ij}^α representing the generators of the Lie algebra, the variation $\delta\phi$ of ϕ under an infinitesimal transformation (13.3) takes the form:

$$\delta\phi_i = t_{ij}^\alpha \phi_j \omega_\alpha\,, \tag{13.4}$$

in which ω_α are the infinitesimal parameters of the transformation.

The variation of a function $S\left(\phi\right)$ under such a transformation is then:

$$\delta S(\phi) = t_{ij}^\alpha \phi_j \frac{\partial S}{\partial \phi_i} \omega_\alpha\,.$$

In particular an invariant function $S(\phi)$ satisfies:

$$t_{ij}^\alpha \phi_j \frac{\partial S}{\partial \phi_i} = 0\,. \tag{13.5}$$

The differential operators Δ_α:

$$\Delta_\alpha = t^\alpha_{ij}\phi_j\frac{\partial}{\partial\phi_i} \tag{13.6}$$

are thus the generators of the Lie algebra of the group G realized as differential operators acting on functions of ϕ_i. It is easy to verify by direct calculation, using the commutation relations (13.2) of the generators t^α_{ij}:

$$[\Delta_\alpha, \Delta_\beta] = f_{\alpha b\gamma}\Delta_\gamma\,. \tag{13.7}$$

Note finally that equation (13.5) can be considered as a system of differential equations for $S(\phi)$.

$$\Delta_\alpha S(\phi) = 0\,. \tag{13.8}$$

Quite generally the commutators of first order differential operators are again first order differential operators. Therefore if a function is solution of a system of first order partial differential equations described in terms of operators Δ_α, it is also solution of all equations corresponding to operators belonging to the Lie algebra generated by Δ_α. The system (13.8) is said to be compatible if no new independent equation is obtained from the commutators $[\Delta_\alpha, \Delta_\beta]$. This condition is verified if all commutators are linear combinations of the operators Δ_α, i.e. if the Δ_α form a basis of the Lie algebra they generate. Therefore the Lie algebra commutation relations (13.7) are the compatibility conditions of the linear system (13.8).

We shall be concerned with cases in which ϕ is a field depending on an additional variable x, S, the action, is a functional of ϕ, and Δ_α has the typical form:

$$\Delta_\alpha = \int \mathrm{d}x\, t^\alpha_{ij}\phi_j(x)\frac{\delta}{\delta\phi_i(x)}\,,$$

but the analysis is the same.

13.2 Linearly Realized Global Symmetries

Definition. We call *global symmetry* a symmetry which corresponds to a transformation of the fields whose parameters are space-independent. More precisely let $\phi_i(x)$ be a set of fields transforming linearly under a representation $\mathcal{D}(G)$ of a compact Lie group G:

$$\phi'_i(x) = D_{ij}(g)\,\phi_j(x). \tag{13.9}$$

The transformation (13.9) is global if the group element g does not depend on the space variable x. Sometimes the expression *rigid symmetry* is also used to avoid confusions with "global" in the sense of global topological properties of the symmetry group. In what follows we explore the consequences of invariance only under infinitesimal group transformations. In the notations of Section 13.1.2, we can write the variation $\delta\phi$ of ϕ under transformation (13.9) as:

$$\delta\phi_i(x) = t^\alpha_{ij}\phi_j(x)\omega_\alpha\,, \tag{13.10}$$

in which ω_α are the space-independent parameters of the transformation.

A classical action $S(\phi)$ invariant under such a transformation then satisfies:

$$\int \mathrm{d}x\, t^\alpha_{ij}\phi_j(x)\frac{\delta S}{\delta\phi_i(x)} = 0\,. \tag{13.11}$$

Regularization. In the case of linearly realized global symmetries, it is always possible to find a regularization which preserves the symmetry of the action. For purely boson field theories, we can use dimensional, lattice or Pauli–Villars's type regularizations. In the latter case, we modify the propagator by adding to the tree action $S(\phi)$ quadratic invariant terms involving higher order derivatives (Section 9.1):

$$S_\Lambda(\phi) = S(\phi) + \frac{1}{2} \int \mathrm{d}x\, \phi_i(x) \left(\alpha_2 \frac{\Delta^2}{\Lambda^2} - \alpha_3 \frac{\Delta^3}{\Lambda^4} + \cdots \right) \phi_i(x), \qquad (13.12)$$

in which Λ is the cut-off. By introducing enough terms it is always possible to render the theory finite. The regularization terms are obviously symmetric since they are invariant under arbitrary orthogonal transformations.

In the case of massless chiral fermions, if the transformation law involves the matrix γ_S:

$$\delta \psi_i(x) = \gamma_S t_{ij}^\alpha \psi_j(x),$$

the substitution

$$\bar{\psi}(x) \slashed{\partial} \psi(x) \mapsto \bar{\psi}(x) \slashed{\partial} \left(1 - \alpha_1 \frac{\Delta}{\Lambda^2} + \cdots \right) \psi(x),$$

preserves the chiral symmetry.

Derivation of WT identities. Let us now consider the generating functional of correlation functions $Z(J)$ corresponding to the symmetric action $S(\phi)$:

$$Z(J) = \int [\mathrm{d}\phi] \exp \left[-S(\phi) + \int \mathrm{d}x\, J_i(x)\, \phi_i(x) \right]. \qquad (13.13)$$

We have shown in Section 5.4 that general identities satisfied by the generating functional $Z(J)$, like the equation of motion, can be obtained by expressing that the functional integral is invariant in an infinitesimal change of variables. We here use this observation to derive the consequences of equation (13.11) for $Z(J)$. We perform a change of variables of the form of a transformation (13.10), setting:

$$\phi_i(x) = \phi_i'(x) + t_{ij}^\alpha \phi_j'(x) \omega_\alpha\,, \qquad (13.14)$$

in the functional integral (13.13). As a consequence of the symmetry as expressed by equation (13.11), the action $S(\phi)$ and therefore the regularized action $S_\Lambda(\phi)$ are left invariant by the transformation (13.14). The measure of integration $[\mathrm{d}\phi_i]$ in the functional integral (13.13) is the flat euclidean measure which is invariant in an orthogonal transformation (more generally in any unimodular transformation, i.e. corresponding to matrices of determinant 1). The only variation comes from the source term. This implies:

$$0 = \delta Z(J) = \int [\mathrm{d}\phi'] \,\delta\,[\text{source term}] \exp \left[-S(\phi') + \int \mathrm{d}x\, J_i(x)\phi_i'(x) \right].$$

The variation of the source term is:

$$\delta\,[\text{source term}] = \int \mathrm{d}x\, J_i(x) t_{ij}^\alpha \phi_j'(x) \omega_\alpha\,.$$

This leads to the equation:

$$0 = \omega_\alpha \int [\mathrm{d}\phi] \int \mathrm{d}x \, J_i(x) t_{ij}^\alpha \phi_j(x) \exp\left[-S(\phi) + \int \mathrm{d}x \, J_k(x) \phi_k(x)\right]. \qquad (13.15)$$

We now have replaced the notation ϕ_i' by ϕ_i since ϕ is a dummy integration variable.

Equation (13.15), being valid for any set of parameters ω_α, can be rewritten for each component α. Finally the identity:

$$\int [\mathrm{d}\phi] \, \phi_i(x) \exp\left[-S(\phi) + \int \mathrm{d}y \, J_k(y) \phi_k(y)\right]$$
$$= \frac{\delta}{\delta J_i(x)} \int [\mathrm{d}\phi] \exp\left[-S(\phi) + \int \mathrm{d}y \, J_k(y) \phi_k(y)\right], \qquad (13.16)$$

allows us to rewrite equation (13.15) as an identity for the functional $Z(J)$:

$$\int \mathrm{d}x \, t_{ij}^\alpha J_i(x) \frac{\delta Z(J)}{\delta J_j(x)} = 0. \qquad (13.17)$$

From equation (13.17) we immediately derive an equation for the generating functional $W(J) = \ln Z(J)$ of connected correlation functions:

$$\int \mathrm{d}x \, t_{ij}^\alpha J_i(x) \frac{\delta W(J)}{\delta J_j(x)} = 0. \qquad (13.18)$$

Expanding equation (13.18) in a power series of the source $J(x)$, we obtain identities between the connected correlation functions which describe the physical implications of the symmetry of the action.

However for renormalization purpose it is more convenient to derive an equation for the 1PI functional $\Gamma(\varphi)$. We therefore perform a Legendre transformation:

$$\begin{cases} \Gamma(\varphi) + W(J) = \int \mathrm{d}x \, J_i(x) \varphi_i(x), \\ \varphi_i(x) = \dfrac{\delta W}{\delta J_i(x)}, \end{cases} \qquad (13.19)$$

and immediately obtain the equation:

$$\int \mathrm{d}x \, t_{ij}^\alpha \varphi_i(x) \frac{\delta\Gamma}{\delta\varphi_j(x)} = 0, \qquad (13.20)$$

which, expanded in powers of φ, yields WT identities for proper vertices. The equation implies that the regularized functional $\Gamma(\varphi)$ is invariant under the transformation (13.10).

Renormalization. We now perform a loopwise expansion of $\Gamma(\varphi)$:

$$\Gamma(\varphi) = \sum_{l=0}^{\infty} \Gamma_l(\varphi) g^l. \qquad (13.21)$$

The parameter g is any coupling constant playing the role of $\hbar$ and introduced to order the loopwise expansion. Since equation (13.20) is linear in $\Gamma(\varphi)$ and independent of g, all functionals $\Gamma_l(\varphi)$ also satisfy equation (13.20).

The functional $\Gamma_0(\varphi)$ is just the action $S(\varphi)$ and satisfies by assumption equation (13.20). The regularized one-loop functional $\Gamma_1(\varphi)$ satisfies (13.20):

$$\int \mathrm{d}x\, t_{ij}^\alpha \varphi_i(x) \frac{\delta \Gamma_1(\varphi)}{\delta \varphi_j(x)} = 0. \tag{13.22}$$

We can now perform an asymptotic expansion of $\Gamma_1(\varphi)$ in terms of the regularizing parameter (large cut-off expansion or $1/\varepsilon$ in dimensional regularization for example). Because equation (13.22) is valid for any value of the regularizing parameter, it is valid for each term in the expansion and thus for the sum of the divergent contributions $\Gamma_1^{\mathrm{div}}(\varphi)$:

$$\int \mathrm{d}x\, t_{ij}^\alpha \varphi_i(x) \frac{\delta \Gamma_1^{\mathrm{div}}(\varphi)}{\delta \varphi_j(x)} = 0. \tag{13.23}$$

General renormalization theory tells us that $\Gamma_1^{\mathrm{div}}(\varphi)$ is a general local functional of the fields restricted only by power counting; equation (13.23) tells us in addition that it is symmetric. Adding $-\Gamma_1^{\mathrm{div}}(\varphi)$ to the action renders the theory one-loop finite. The one-loop renormalized action is still symmetric and therefore the new two-loop functional $\Gamma_2(\varphi)$ still satisfies equation (13.20). After one-loop renormalization $\Gamma_2(\varphi)$ has only local divergences which also satisfy equation (13.23) and all arguments can be repeated. It is clear that the arguments extend to all orders.

The conclusion is that the renormalized action S_{r} is the most general local functional of the field $\phi_i(x)$ compatible with power counting and invariant under the transformation (13.10).

Each reader familiar with perturbative calculations will of course have realized that this is a pedantic derivation of a rather obvious result. However, since the same strategy, suitably adapted, allows us to discuss much more general situations, we believe that it has been useful to expose it first in a case in which it can be easily understood.

Note finally that we have renormalized using a minimal subtraction scheme. Additional finite renormalizations which are consistent with the symmetry can still be performed.

13.3 Linear Symmetry Breaking

For some applications (see for example Sections 13.6, 26.5) it is useful to consider the following situation: the action $S(\phi)$ is the sum of a symmetric part $S_{\mathrm{sym}}(\phi)$, i.e. invariant under the transformation (13.9), and a term breaking the symmetry linear in the fields $\phi_i(x)$:

$$S(\phi) = S_{\mathrm{sym}}(\phi) - \int c_i \phi_i(x)\mathrm{d}x, \tag{13.24}$$

in which $\mathbf{c}$ is a constant vector.

An example of such a situation is provided by the action:

$$S[\phi] = \int \mathrm{d}x \left[\tfrac{1}{2}(\partial_\mu \phi)^2 + \tfrac{1}{2}m^2\phi^2 + \frac{1}{4!}g(\phi^2)^2 - \mathbf{c}\cdot\phi \right] \tag{13.25}$$

in which $\phi(x)$ is a N-component vector. The action $S(\phi)$ is the sum of an $O(N)$ invariant part and a linear symmetry breaking term.

The perturbative expansion corresponding to action (13.25) is obtained by the following method: one first looks for a classical minimum of the action which corresponds to a constant field $\mathbf{v}_0$ satisfying:

$$\frac{\delta S_{\mathrm{sym}}(\mathbf{v}_0)}{\delta \phi_i} - c_i = 0, \tag{13.26}$$

with the condition:

$$\frac{\delta^2 S_{\text{sym}}(\mathbf{v}_0)}{\delta\phi_i\delta\phi_j} \geq 0,$$

in the matrix sense.

In the example (13.25) v_{0i} satisfies the equation:

$$\left(m^2 + \frac{g}{6}\mathbf{v}_0^2\right)v_{0i} = c_i.\tag{13.27}$$

If the action has several minima, one is in general instructed to choose the absolute minimum of the potential but this is irrelevant from the point of view of formal perturbation theory. The quantity $\mathbf{v}_0$ is, at the tree order, the expectation value (vacuum expectation value in the particle physics language) of the field ϕ.

One then translates the field ϕ setting:

$$\phi(x) = \mathbf{v}_0 + \chi(x).\tag{13.28}$$

After translation, the action no longer contains a linear term and the perturbative calculation proceeds in the standard manner. However the example (13.25) shows that after translation the mass term is no longer symmetric and a non-symmetric χ^3 interaction has been generated. Correlation functions will no longer be symmetric and the form of the UV divergences from the point of view of the symmetry is *a priori* unknown. It is thus important to understand whether the structure of the renormalized action reflects in some way the structure of the action (13.24).

The answer here follows from a very simple argument. With obvious notation we have:

$$Z(\mathbf{J}) = Z_{\text{sym}}(\mathbf{J} + \mathbf{c}),\tag{13.29}$$

and thus:

$$W(\mathbf{J}) = W_{\text{sym}}(\mathbf{J} + \mathbf{c}).\tag{13.30}$$

Equation (13.18) then in particular implies:

$$\int dx\, t_{ij}^\alpha\, [J_i(x) + c_i]\,\frac{\delta W(\mathbf{J})}{\delta J_j(x)} = 0.\tag{13.31}$$

Expanding in powers of $J_i(x)$, we obtain a set of relations (WT identities) between connected correlation functions which can be most conveniently expressed in momentum representation:

$$c_i t_{ij}^\alpha \tilde{W}_{jk_1,\ldots,k_n}^{(n+1)}(0, p_1, \ldots, p_n)$$
$$+ \sum_{r=1}^n t_{k_r j}^\alpha \tilde{W}_{k_1,\ldots,k_{r-1},j,k_{r+1},\ldots,k_n}^{(n)}(p_1, \ldots, p_n) = 0.\tag{13.32}$$

Let us now perform the Legendre transformation:

$$\begin{cases} \Gamma(\varphi) + W(\mathbf{J}) = \displaystyle\int dx\, J_i(x)\varphi_i(x), \\[2mm] \varphi_i(x) = \dfrac{\delta W}{\delta J_i(x)} = \dfrac{\delta W_{\text{sym}}(\mathbf{J} + \mathbf{c})}{\delta J_i(x)}. \end{cases}\tag{13.33}$$

Let us compare these relations with those of the symmetric case:

$$\begin{cases} \Gamma_{\text{sym}}\left(\xi\right) + W_{\text{sym}}\left(\mathbf{J}\right) = \displaystyle\int dx\, J_i(x)\xi_i(x), \\[2mm] \xi_i(x) = \dfrac{\delta W_{\text{sym}}\left(\mathbf{J}\right)}{\delta J_i(x)}\,. \end{cases} \tag{13.34}$$

Replacing $J_i(x)$ by $J_i(x) + c_i$ in relations (13.34) we obtain:

$$\begin{cases} \Gamma_{\text{sym}}\left(\varphi\right) + W_{\text{sym}}\left(\mathbf{J}+\mathbf{c}\right) = \displaystyle\int dx\, \left(J_i(x) + c_i\right)\varphi_i(x), \\[2mm] \varphi_i(x) = \dfrac{\delta W_{\text{sym}}\left(\mathbf{J}+\mathbf{c}\right)}{\delta J_i(x)}\,, \end{cases} \tag{13.35}$$

and therefore comparing (13.33) with (13.35):

$$\Gamma\left(\varphi\right) = \Gamma_{\text{sym}}\left(\varphi\right) - \int dx\, c_i\varphi_i(x). \tag{13.36}$$

This identity shows that the divergences of the functionals $\Gamma\left(\varphi\right)$ and $\Gamma_{\text{sym}}\left(\varphi\right)$ are identical. If we therefore replace the regularized symmetric action by the renormalized symmetric action, the theory is finite for any value of c_i. This casually is expressed by saying that the linear breaking term is not renormalized.

Of course to obtain the 1PI correlation functions of ϕ_i, we have to translate φ by the ϕ field expectation value setting (see Section 6.2):

$$\varphi_i(x) = v_i + \chi_i(x), \tag{13.37}$$

with:

$$\left.\frac{\delta\Gamma}{\delta\varphi_i(x)}\right|_{\varphi_i(x)=v_i} = 0 \implies \frac{\delta\Gamma_{\text{sym}}}{\delta\varphi_i(x)}\left(v_i\right) = c_i\,, \tag{13.38}$$

and $\delta^2\Gamma\left(\mathbf{v}\right)/\delta\phi_i\delta\phi_j \geq 0$.

The 1PI correlation functions are then the coefficients of the expansion of $\Gamma\left(\varphi\right)$ in powers of χ. In the tree approximation one recovers $v_i = v_{0i}$.

The WT identities for $\Gamma\left(\varphi\right)$ are immediately deduced from the identity (13.20) for Γ_{sym}:

$$\int dx\, t^{\alpha}_{ij}\left[\frac{\delta\Gamma}{\delta\varphi_i(x)} + c_i\right]\varphi_j(x) = 0\,, \tag{13.39}$$

which after the translation (13.37) becomes:

$$\int dx\, t^{\alpha}_{ij}\left[\frac{\delta\Gamma}{\delta\chi_i}\left(\chi+\mathbf{v}\right) + c_i\right]\left(\chi_j + v_j\right) = 0\,. \tag{13.40}$$

Application. Let us show that this identity leads to some non-trivial relations between the 1PI correlation functions. Setting $\chi = 0$ we obtain:

$$t^{\alpha}_{ij}c_iv_j = 0\,, \tag{13.41}$$

which shows the breaking vector $\mathbf{c}$ and the expectation value $\mathbf{v}$ are left invariant by the same subgroup of G. In the example of the $O(N)$ symmetry, equation (13.41) implies that the vector $\mathbf{v}$ is proportional to the vector $\mathbf{c}$.

Differentiating once with respect to $\chi_k(y)$ and setting then χ equal to zero, we relate the 1 and 2-point functions:

$$\int \mathrm{d}x \left[v_j t_{ij}^\alpha \Gamma_{ik}^{(2)}(x,y) + t_{ik}^\alpha c_i \delta(x-y) \right] = 0, \tag{13.42}$$

with

$$\Gamma_{ij}^{(2)}(x,y) = \left. \frac{\delta^2 \Gamma(\chi+v)}{\delta\chi_i(x)\delta\chi_j(y)} \right|_{\chi=0}.$$

Let us introduce the Fourier transform of the 2-point function:

$$\Gamma_{ij}^{(2)}(x,y) = \int \frac{\mathrm{d}^d p}{(2\pi)^d} e^{-ip(x-y)} \tilde{\Gamma}_{ij}^{(2)}(p). \tag{13.43}$$

In terms of $\tilde{\Gamma}^{(2)}(p)$ equation (13.42) becomes:

$$v_j t_{ji}^\alpha \tilde{\Gamma}_{ik}^{(2)}(0) + t_{ki}^\alpha c_i = 0. \tag{13.44}$$

This equation determines the geometrical structure of the zero momentum propagator in the presence of the linear symmetry breaking term.

In the example (13.25), the identity (13.44) yields the value of the propagator of the components of the field orthogonal to the vector $\mathbf{c}$, at zero momentum:

$$\tilde{\Gamma}_T^{(2)}(0) = \frac{c}{v}.$$

Equation (13.44) is the last equation which involves c_i explicitly. The terms of higher degree in χ are functions only of the expectation value v_i. By identifying the coefficient of degree $(n+1)$ in χ, one obtains a relation between the Fourier transform of the $(n+1)$-point function $\tilde{\Gamma}^{(n+1)}$ with one momentum set to zero and the n-point function $\tilde{\Gamma}^{(n)}$:

$$v_j t_{jk}^\alpha \tilde{\Gamma}_{ki_1\ldots i_n}^{(n+1)}(0,p_1,\ldots,p_n) + \sum_{r=1}^n t_{i_r k}^\alpha \tilde{\Gamma}_{i_1,\ldots,i_{r-1},k,i_{r+1},\ldots,i_n}^{(n)}(p_1,\ldots,p_n) = 0. \tag{13.45}$$

For example this equation for $n=2$ reads:

$$v_j t_{ji}^\alpha \tilde{\Gamma}_{ikl}^{(3)}(0,p,-p) + t_{li}^\alpha \tilde{\Gamma}_{ik}^{(2)}(p) + t_{ki}^\alpha \tilde{\Gamma}_{il}^{(2)}(p) = 0.$$

If we choose to renormalize by fixing the value of the primitively divergent correlation functions at some given point in momentum space, then the set of WT identities implies relations between the different parameters. Apart from the vector $\mathbf{v}$, the non-symmetric theory depends on as many independent parameters as the symmetric theory. In the example of the $O(N)$ symmetric $(\phi^2)^2$ field theory in four dimensions, it is possible to impose one arbitrary renormalization condition to $\Gamma_{1111}^{(4)}(p_i)$ and two conditions to $\Gamma_{11}^{(2)}(p)$. All others are given by the WT identities (13.45) used for $n=1$ to 4.

13.4 Spontaneous Symmetry Breaking

Spontaneous symmetry breaking (SSB) is a possible limit of linear symmetry breaking when the breaking parameter goes to zero. As we discuss below, in this limit the action becomes symmetric, but depending on the values of other parameters (in our examples the coefficient of the ϕ^2 operator), the physics may or may not be symmetric. Many physical models in particle physics are based on the concept of SSB. The reason is that the mechanism of SSB allows one to construct models with broken symmetries which depend on as many parameters as the symmetric models. The appearance of massless particles is in general the most characteristic feature of such models (in the absence of gauge symmetries).

In statistical mechanics also many phase transitions are related to SSB, as we shall extensively discuss in the Chapters 23–30 devoted to critical phenomena. We shall determine when, from the functional integral point of view, we should expand perturbation theory around one of the degenerate minima of the classical potential, situation of spontaneous symmetry breaking (SSB) and degeneracy of the quantum mechanical ground state, and when it is necessary to sum over the contributions of all minima, in which case quantum fluctuations restore the symmetry broken at the classical level, the ground state being unique. Below we assume that the situation of SSB is realized.

Classical analysis. Let us first consider the example of the $O(N)$ symmetric model and discuss the expectation value of the field at the tree level. The action per unit volume for a constant field ϕ is:

$$V\left(\phi\right) \equiv S\left(\phi\right)/\text{volume} = \frac{1}{2}m^2\phi^2 + \frac{1}{4!}g\left(\phi^2\right)^2 - \mathbf{c}\cdot\phi. \qquad (13.46)$$

As long as $\mathbf{c}$ does not vanish, it is possible to pass continuously from a situation in which the parameter m^2 is positive to a situation in which m^2 is negative (remember that m is no longer the physical mass) without encountering any singularity. For instance the expectation value $\mathbf{v}$ is at $\mathbf{c}$ fixed, a regular function of m^2 at $m^2 = 0$. If instead $\mathbf{c}$ vanishes, the expectation value $\mathbf{v}$ vanishes identically for $m^2 > 0$ and takes a non-trivial value for $m^2 < 0$ such that:

$$|\mathbf{v}| = \sqrt{-6m^2/g}, \qquad (13.47)$$

as can be easily understood by drawing the potential for both cases (see figure 13.1).

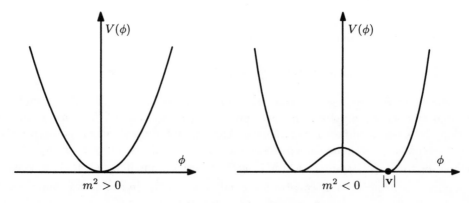

Fig. 13.1   Section of the ϕ-potential.

In the latter case the classical minimum of the potential is degenerate. Starting from a given minimum, it is possible to describe all other minima by acting on the vector $\mathbf{v}$ with the symmetry group G. In the $O(N)$ example the surface of minima is a sphere with a radius given by equation (13.47).

Assuming a situation of SSB, we construct a perturbation theory around one minimum $\mathbf{v}$ which is, at leading order, the field expectation value. We thus shift the field

$$\phi(x) = \mathbf{v} + \chi(x)\,.$$

The χ-field mass matrix is obtained by calculating the second derivatives of the potential at the minimum. Using equation (13.47) we find

$$\left.\frac{\partial V}{\partial \phi_i \partial \phi_j}\right|_{\phi=\mathbf{v}} = \tfrac{g}{3} v_i v_j\,.$$

This matrix has $N - 1$ zero eigenvalues corresponding to eigenvectors orthogonal to $\mathbf{v}$. This is not surprising since the potential is flat along a group orbit. The physical consequence is that spontaneous breaking of a continuous symmetry implies the appearance of massless (Goldstone) modes: from the point of view of particle physics massless scalar particles called Goldstone bosons.

Let us now examine a more general situation. We assume that a G-symmetric action has degenerate minima. We call $\mathbf{v}$ the minimum chosen to expand perturbation theory, and thus the field expectation value at leading order. We introduce the subgroup H of G, little group (stabilizer) of the vector $\mathbf{v}$, i.e. the subgroup of G which leaves the vector $\mathbf{v}$ invariant. By definition the p generators of the Lie algebra $\mathcal{L}(H)$ of H satisfy:

$$\mathcal{L}(H): \quad 1 \le \alpha \le p \;\Rightarrow\; t_{ij}^\alpha v_j = 0\,,$$

We denote by $\mathcal{L}(G/H)$ the vector space (it is not an algebra!) generated by the complementary set in the Lie algebra $\mathcal{L}(G)$ of G. It is characterized by:

$$\mathcal{L}(G/H): \quad \sum_{\alpha > p} t_{ij}^\alpha v_j \omega_\alpha = 0 \;\Rightarrow\; \omega_\alpha = 0 \quad \text{for all } \alpha\,.$$

For $\alpha > p$, the vectors $(v^\alpha)_i = t_{ij}^\alpha v_j$ thus are linearly independent. We then parametrize the field ϕ in the form of a group element acting on a vector:

$$\phi(x) = \exp\left(\sum_{\alpha>p} t^\alpha \xi^\alpha(x)\right)(\mathbf{v} + \rho(x)) = \mathbf{v} + \xi^\alpha(x) t^\alpha \mathbf{v} + \rho(x) + \cdots\,,$$

in which $\rho(x)$ has components only in the subspace orthogonal to all vectors $t^\alpha \mathbf{v}$. In the $O(N)$ example ρ has only one component along $\mathbf{v}$. This parametrization is such that the mapping of fields $\{\rho(x), \xi^\alpha(x)\} \mapsto \phi(x) - \mathbf{v}$ can be inverted for small fields. This property ensures that if the fluctuations of the field ϕ around its expectation value are in some sense small, perturbation theory is at least qualitatively sensible.

Inserting this parametrization into the action we note the following: the contributions to the action which are derivative-free depend only on $\rho(x)$ because they are G-invariant. The dependence in the fields $\xi^\alpha(x)$ is entirely contained in the terms with derivatives, therefore these fields are massless. We conclude that spontaneous breaking of symmetry of a group G to a subgroup H, the group which leaves the field expectation value invariant, yields a number of massless Goldstone modes (bosons) equal to the number of generators of G which do not belong to H. This result is valid in the classical approximation. Let us now generalize it to the full quantum theory.

WT identities and spontaneous symmetry breaking. To connect continuously the two phases, symmetric and with SSB, without encountering any singularity we start from the situation $m^2 > 0$, $\mathbf{c} = 0$, one gives to $\mathbf{c}$ a non-trivial value, perform the continuation from $m^2 > 0$ to $m^2 < 0$, and again take the vanishing $\mathbf{c}$ limit. We then assume the existence of non-trivial solutions to the equation

$$\left.\frac{\delta\Gamma}{\delta\varphi_i(x)}\right|_{\varphi_i(x)=v_i} = 0\,. \tag{13.48}$$

In Section 6.6 we have explained how the existence of solutions to this equation is consistent with the convexity of the function $\Gamma(\mathbf{v})$ (equation (6.28)).

Since the WT identities (13.40) hold for any value of the parameters and we have made an analytic continuation, we still have in the $m^2 < 0$, $\mathbf{c} = 0$ limit:

$$\int \mathrm{d}x\, t_{ij}^\alpha \frac{\delta\Gamma(\chi+v)}{\delta\chi_i(x)}\,(\chi_j + v_j) = 0\,, \tag{13.49}$$

the direction of v_i being fixed by equation (13.41).

Goldstone modes. Let us here discuss only one important consequence of the WT identities. Equation (13.44) becomes:

$$v_j t_{ji}^\alpha \tilde{\Gamma}_{ik}^{(2)}(0) = 0\,. \tag{13.50}$$

To explain the significance of this equation, inspired by the classical analysis, we introduce the subgroup H of G, little group (stabilizer) of the vector $\mathbf{v}$. Since for $\alpha > p$, the vectors $(v^\alpha)_i = t_{ij}^\alpha v_j$ are linearly independent, the equation (13.50) implies that the matrix $\tilde{\Gamma}_{ij}(0)$ has as many eigenvectors with eigenvalue zero as they are generators in $\mathcal{L}(G/H)$. The corresponding components of the field are Nambu–Goldstone modes associated with the spontaneous breaking of the G-symmetry, confirming the classical analysis.

13.5 Quadratic Symmetry Breaking

A symmetry may be broken by terms of higher canonical dimensions. We shall examine in detail only the case of the quadratic symmetry breaking. We shall then briefly indicate what happens with breaking terms of even higher dimensions. Let us consider the action:

$$S(\phi) = S_{\mathrm{sym}}(\phi) + \tfrac{1}{2}\mu_{ij}\int \mathrm{d}x\, \phi_i(x)\phi_j(x)\,, \tag{13.51}$$

in which μ_{ij} is a symmetric traceless constant matrix. We can assume μ_{ij} traceless without loss of generality since a term proportional to the unit matrix can always be absorbed into the symmetric action $S_{\mathrm{sym}}(\phi)$. In action (13.51) interaction terms are symmetric but the mass terms break the symmetry.

We can try to again derive WT identities by performing an infinitesimal change of variables in the functional integral:

$$\phi_i = \phi_i' + t_{ij}^\alpha \omega_\alpha \phi_j'\,.$$

The variation of the integrand now comes both from the source term and the breaking term:

$$\delta \left[S(\phi) - \int \mathrm{d}x \, J_i(x)\phi_i(x) \right] = \omega_\alpha \int \mathrm{d}x \left[\mu_{ij}\phi_i(x)t^\alpha_{jk}\phi_k(x) - J_i(x)t^\alpha_{ij}\phi_j(x) \right].$$

Expressing as usual that the result of the functional integral is not modified by a change of variables we obtain the equation:

$$\int [\mathrm{d}\phi] \, \delta \left[S(\phi) - \int \mathrm{d}x \, J_i(x)\phi_i(x) \right] \exp \left[-S(\phi) + \int \mathrm{d}x \, J_i\phi_i \right] = 0.$$

Using equation (13.51) and replacing factors of the form $\phi_i(x)$ by $\delta/\delta J_i(x)$ we derive an equation for $Z(J)$:

$$\int \mathrm{d}x \left[\mu_{ij} t^\alpha_{jk} \frac{\delta^2}{\delta J_i(x)\delta J_k(x)} - J_i(x)t^\alpha_{ij} \frac{\delta}{\delta J_j(x)} \right] Z(J) = 0. \tag{13.52}$$

Two features distinguish this equation from the equation of the symmetric case ($\mu_{ij} = 0$):

(i) It involves functional second derivatives with respect to the sources. The corresponding equation for $W(J)$ then also involves a term of the form $\delta^2 W/[\delta J(x)]^2$. If we now try to perform a Legendre transformation, we have to introduce the quantity:

$$\frac{\delta^2 W}{[\delta J(x)]^2} = \left[\frac{\delta^2 \Gamma}{\delta\varphi(x)\,\delta\varphi(y)} \right]^{(-1)} \Bigg|_{x=y}, \tag{13.53}$$

the inverse being understood in the sense of kernels. The WT identities take a very complicated form.

(ii) It is no longer a relation really between ϕ-field correlation functions because the two functional derivatives are taken at the same point; it instead also involves insertions of the composite operator $\phi_i(x)\phi_j(x)$.

The difficulties we encounter have several origins. One can be directly understood by formally expanding the correlation functions in a power series of the symmetry breaking term. This generates a sum of symmetric correlation functions with multiple insertions of the operator $\frac{1}{2}\mu_{ij} \int \mathrm{d}x \, \phi_i(x)\phi_j(x)$ (see figure 13.2).

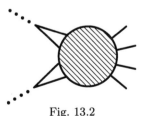

Fig. 13.2

These insertions, as we have already extensively discussed, may generate new divergences which have to be taken care of. This situation has to be contrasted with the linear case, in which only field correlation functions are generated by such an expansion.

Another difficulty stems from the fact that an infinitesimal group transformation generates a new quadratic term linearly independent of the original one, as can be seen on equation (13.51). If we want to write WT identities for renormalized quantities, we have to also renormalize the operator $\int \mathrm{d}x\, \mu_{ij} t^{\alpha}_{jk} \phi_i (x)\, \phi_k (x)$ and examine how it transforms under the group. This may generate new quadratic operators and all the operations have to be repeated. We therefore have to adapt our strategy to this new situation. We shall use a slightly different method directly inspired from the analysis of the equation of motion of Section 12.2.

The general method. The method we now introduce is fairly general and will allow us to discuss many cases of renormalization with symmetries, as the subsequent chapters devoted to different type of symmetries will show.

The idea is to perform an infinitesimal group transformation, to collect all new linearly independent composite operators generated in this way and to add source terms for them in the action. Let us call $C_\alpha (\phi)$ such operators. In our example:

$$C_\alpha (\phi) \equiv \int \mathrm{d}x\, \mu_{ij} t^{\alpha}_{jk} \phi_i(x) \phi_j (x)\,.$$

Since we have now added a source for C_α in the action, we have to worry about the effect of an infinitesimal transformation on $C_\alpha (\phi)$:

$$\delta C_\alpha (\phi) = \int \mathrm{d}x\, \frac{\delta C_\alpha}{\delta \phi_i(x)} t^{\beta}_{ij} \phi_j(x) \omega_\beta\,.$$

Two cases may arise: either $\delta C_\alpha (\phi)$ is a linear combination of $\phi_i(x)$ and $C_\alpha (\phi)$ and we proceed deriving WT identities, or new independent operators are generated and we again add sources for them in the action.

We repeat the procedure as long as necessary. It is easy to verify that under these conditions the WT identities will always be first order differential equations for the generating functional Z considered as a functional of all sources. Several examples will illustrate this point.

Application. In the example of quadratic symmetry breaking, it is clear that any infinitesimal transformation made on a linear combination of operators of the form $\phi_i(x)\phi_j(x)$ generates another linear combination of these same operators. We therefore at once consider the total action $S (\phi, K)$:

$$S (\phi, K) = S_{\text{sym}} (\phi) + \tfrac{1}{2} \int \mathrm{d}x\, K_{ij}(x)\phi_i(x)\phi_j(x), \qquad (13.54)$$

with:

$$K_{ij}(x) = K_{ji}(x), \qquad K_{ii}(x) = 0\,.$$

Actually we need only insertions at zero momentum and we could in principle restrict ourselves to constant sources K_{ij}. However it is not more difficult to use space-dependent sources. Furthermore, in the case in which the symmetric theory is massless, zero momentum insertions could lead to IR divergences which we avoid in this way. We now consider the generating functional $Z (J, K)$:

$$Z (J, K) = \int [\mathrm{d}\phi] \exp \left[-S (\phi, K) + \int \mathrm{d}x\, J_i (x)\, \phi_i(x) \right]. \qquad (13.55)$$

An infinitesimal change of variable (13.10) leads to:

$$0 = \int [\mathrm{d}\phi]\, \delta_\alpha \left[-S(\phi, K) + \int \mathrm{d}x\, J_i\phi_i \right] \exp\left[-S(\phi, K) + \int \mathrm{d}x\, J_i\phi_i \right], \qquad (13.56)$$

with:

$$\delta_\alpha \left[S(\phi, K) - \int \mathrm{d}x\, J_i\phi_i \right] = \int \mathrm{d}x\, t_{ij}^\alpha \phi_j(x)\, [K_{ik}(x)\phi_k(x) - J_i(x)]. \qquad (13.57)$$

As before the product $\phi_j(x)\phi_k(x)$ appears. However we are now able to express it in terms of $\delta/\delta K_{kj}(x)$ instead of $\delta^2/\delta J_j(x)\delta/\delta J_k(x)$.

Remark. We define derivatives with respect to complicated objects (here the symmetric traceless matrix K_{ij}) in the following way: let $F(\mathbf{K})$ be a function (or functional) of $\mathbf{K}$; we calculate the variation of $F(\mathbf{K})$ at first order when $\mathbf{K}$ varies of a quantity $\delta\mathbf{K}$,

$$F(\mathbf{K} + \delta\mathbf{K}) - F(\mathbf{K}) = \int \frac{\delta F}{\delta K_{ij}(x)}\, \delta K_{ij}(x)\mathrm{d}x + O\left(\|\delta\mathbf{K}\|^2 \right). \qquad (13.58)$$

The derivative is then defined in the sense of differential geometry: it is the linear operator acting on $\delta K_{ij}(x)$ in the r.h.s. of equation (13.58).

In the example we are considering here, differentiation with respect to K_{ij} generates the traceless part of the product $\phi_i\phi_j$. Since equation (13.57) involves only the traceless part of this same product, the definition (13.58) allows us to rewrite equation (13.56):

$$\int \mathrm{d}x \left\{ [t_{ij}^\alpha K_{ik}(x) + t_{ik}^\alpha K_{ij}(x)] \frac{\delta}{\delta K_{kj}(x)} - J_i(x) t_{ij}^\alpha \frac{\delta}{\delta J_j(x)} \right\} Z(J, K) = 0. \qquad (13.59)$$

Because equation (13.59) is a first order differential equation, an identical equation holds for $W(J, K)$.

The Legendre transformation is only performed with respect to the source $J_i(x)$, because the reducibility corresponds only to external ϕ lines.

$$\Gamma(\varphi, K) + W(J, K) = \int \mathrm{d}x\, J_i(x)\varphi_i(x), \qquad \varphi_i(x) = \frac{\delta W(J, K)}{\delta J_i(x)}.$$

The sources $K_{ij}(x)$ do not participate in the Legendre transformation and have to be considered as external parameters. This then immediately implies (equation (6.20)):

$$\left. \frac{\delta W}{\delta K_{ij}(x)} \right|_J = - \left. \frac{\delta \Gamma}{\delta K_{ij}(x)} \right|_\varphi. \qquad (13.60)$$

This relation provides an additional justification for the method we have proposed. With sources for the composite operators, the Legendre transformation becomes straightforward. The WT identity for $\Gamma(\varphi, K)$ obtained by Legendre transformation from (13.59) is then:

$$\int \mathrm{d}x \left\{ [t_{ij}^\alpha K_{ik}(x) + t_{ik}^\alpha K_{ij}(x)] \frac{\delta \Gamma(\varphi, K)}{\delta K_{kj}(x)} + \varphi_i(x) t_{ij}^\alpha \frac{\delta \Gamma(\varphi, K)}{\delta \varphi_j(x)} \right\} = 0. \qquad (13.61)$$

This equation has a very simple interpretation: $\Gamma(\varphi, K)$ is invariant under the double transformation:

$$\begin{cases} \delta\varphi_i(x) = t_{ij}^\alpha \omega_\alpha \varphi_j(x), \\ \delta K_{kj}(x) = \omega_\alpha \left[t_{ji}^\alpha K_{ik}(x) + t_{ki}^\alpha K_{ij}(x) \right]. \end{cases} \tag{13.62}$$

By performing a transformation both on ϕ and $\mathbf{K}$, we have rendered the breaking term $\frac{1}{2} \int \mathrm{d}x \, K_{ij}(x) \phi_i(x) \phi_j(x)$ group invariant. It is then almost obvious that the symmetry of $S(\phi, K)$ under the equivalent of the transformations (13.62) implies the WT identities (13.61).

Using the arguments given for the case $\mathbf{K} = 0$, we can show that if the regularized functional $\Gamma(\varphi, K)$ satisfies (13.61), the renormalized functional $\Gamma(\varphi, K)$ and the renormalized action will satisfy the same identity: the renormalized action $S_\mathrm{r}(\phi, K)$ is the most general local functional of ϕ and K compatible with power counting and is invariant under the group transformation (13.62).

For a ϕ^4-like field theory in four dimensions the action is the integral of a local function of dimension four, the field ϕ has dimension $[\phi] = 1$ and the dimension $[K]$ of the source $\mathbf{K}$ is two because it is coupled to an operator of dimension 2 (see Section 12.1)

$$[\phi] = 1, \qquad [K] = 2, \qquad [S(\phi, K)] = 4.$$

The renormalized action $S_\mathrm{r}(\phi, K)$ has thus the general form:

$$S_\mathrm{r}(\phi, K) = [S_\mathrm{sym}]_\mathrm{r}(\phi) + \frac{1}{2} \int \mathrm{d}x \, K_{ij}(x) A_{ij}(\phi(x))$$
$$+ \tfrac{1}{2} b_{ij,kl} \int \mathrm{d}x \, K_{ij}(x) K_{kl}(x), \tag{13.63}$$

in which $A_{ij}(\phi)$ is a local derivative-free polynomial of dimension 2, and $b_{ij,kl}$ is a set of constants.

Constraints on $A_{ij}(\phi)$ and $b_{ij,kl}$ are obtained by expressing the invariance of the renormalized action under transformation (13.62). In the simplest case $A_{ij}(\phi)$ has the form:

$$A_{ij}(\phi(x)) = Z_2 \left(\phi_i(x)\phi_j(x) - \frac{1}{N}\delta_{ij}\phi^2(x) \right), \tag{13.64}$$

in which N is the number of components of ϕ.

Depending on the representation content of $\phi_i(x)$, it can be a linear combination of several dimension 2 operators and also contains a contribution linear in ϕ. The term quadratic in K_{ij} does not depend on ϕ and can be factorized in front of the functional integral. It gives an additive contribution to $W(J, K)$. It therefore additively renormalizes the correlation function $\delta^2 W(J = 0)/\delta K_{ij}(x)\delta K_{kl}(y)|_{K=0}$ which has the diagrammatic form shown on figure 13.3.

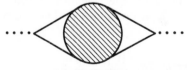

Fig. 13.3

The renormalized action for the case of quadratic symmetry breaking is then obtained by setting $K_{ij}(x)$ constant

$$K_{ij}(x) = \mu_{ij}.$$

We see from the analysis of this more complex situation that the initial form of the action is not always completely preserved. In particular a breaking term of a given dimension may generate new breaking terms of lower dimensions. This happens here when $A_{ij}(\phi)$ has a term linear in ϕ. The renormalized theory may therefore depend on more parameters than one would have naively anticipated when adding the breaking term.

Finally let us note that, as in the case of the linear symmetry breaking, one can use the WT identities (13.61) to constrain the renormalization conditions. However it is then necessary to consider together all the superficially divergent correlation functions both of the field $\phi_i(x)$ and the composite operator $\phi_i(x)\phi_j(x)$, and write all relations derived from equation (13.61) by expanding in $\varphi(x)$ and $K(x)$, in which they appear.

Breaking terms of higher canonical dimensions. The previous analysis can be easily generalized to breaking terms of higher canonical dimensions. One can verify that in a ϕ^4-like theory in four dimensions, a cubic breaking term, since it is coupled to a source of canonical dimension 1, generates in general, by renormalization, breaking terms quadratic and linear in ϕ. Breaking terms which are of canonical dimension lower than the symmetric interaction are called *soft*.

Finally it makes little sense in the context of a renormalizable ϕ_4^4 theory to speak of a breaking term of dimension 4. Indeed such a term is coupled to a source of dimension zero. The renormalized action will contain an infinite series in the source. It is easy to verify that then all traces of the initial symmetry are lost after renormalization.

Discrete symmetries. Discrete symmetries do not lead to WT identities and most of the previous analysis does not apply. However it is easy to prove that when the tree action is symmetric, the renormalized action remains symmetric. If we add to the action a linear symmetry breaking term, we can expand the new correlation functions in power series of the breaking parameter. The coefficients are symmetric correlation functions. Therefore it remains true that the counterterms which render the symmetric theory finite renormalizes the theory with linear symmetry breaking. For breaking terms of higher degree the same strategy can be applied. However we have then to renormalize symmetric correlation functions with operator insertions. This question has been examined in Section 12.1. We add sources for the operators in the action, renormalize according to power counting and use the discrete symmetry to constrain the polynomial in the sources and the field. One qualitative aspect of the previous analysis survives: symmetry breaking terms of a given dimension can only generate additional breaking terms of equal or lower dimensions.

13.6 Application to Strong Interaction Physics: Chiral Symmetry

One of the most striking feature of Strong Interactions in low energy Particle Physics is the observation of approximate spontaneously broken $SU(N) \times SU(N)$ chiral symmetries, which manifest themselves in particular in the small masses of the pseudoscalar mesons. In particular the π meson is specially light, an indication that the breaking of the $SU(2) \times SU(2)$ symmetry is quite small. With our present understanding this property is a consequence of the small masses of the **u** and **d** quarks (see for example Section A5.6) and the vector-like coupling of quarks to gluons. The mass of the **s** quark and thus the explicit breaking of the $SU(3) \times SU(3)$ symmetry are larger as can be seen from the masses of the K and η pseudoscalar mesons.

Note that, according to our previous analysis, since a fermion mass operator in a renormalizable field theory in 4 dimensions has dimension 3, the concept of a symmetry broken by fermion mass terms is indeed meaningful. However the search for analytic methods to derive low energy properties of hadrons from a fundamental theory of quarks and gluons has up to now proven elusive. Most direct results are thus obtained from computer intensive studies of discretized lattice versions (see Chapter 33). Progress is slow though quite encouraging. The most serious technical difficulties are related to the dynamics of quarks.

Therefore we here explain instead how one can construct effective low energy theories based on observed hadrons like protons, neutrons, π-mesons... In such theories the chiral symmetry is explicitly broken by linear terms in some scalar fields, which have the transformation properties of fermion mass terms, and which together with the pseudoscalars transform under representations of the chiral group. We therefore face the situation we have discussed at some length in Section 13.6.

13.6.1 The chiral symmetry: general structure

Let us first consider the action for N free massless Dirac fermions in even dimensions:

$$S\left(\psi, \bar{\psi}\right) = -\int \mathrm{d}^d x \, \bar{\psi}_i(x) \slashed{\partial} \psi_i(x). \tag{13.65}$$

It has a chiral $U(N) \times U(N)$ symmetry corresponding to the transformations:

$$\psi' = \left[\tfrac{1}{2}\left(1 + \gamma_S\right)\mathbf{U}_+ + \tfrac{1}{2}\left(1 - \gamma_S\right)\mathbf{U}_-\right]\psi\,, \tag{13.66}$$

$$\bar{\psi}' = \bar{\psi}\left[\tfrac{1}{2}\left(1 + \gamma_S\right)\mathbf{U}_-^\dagger + \tfrac{1}{2}\left(1 - \gamma_S\right)\mathbf{U}_+^\dagger\right], \tag{13.67}$$

where $\mathbf{U}_\pm$ are two $N \times N$ unitary matrices corresponding to the two $U(N)$ groups.

We now couple the fermions to scalar bosons forming a complex $N \times N$ matrix $\mathbf{M}(x)$. It is easy to verify that the interaction term:

$$-g \int \mathrm{d}^d x \, \bar{\psi}_i \left[\tfrac{1}{2}\left(1 + \gamma_S\right)M_{ij} + \tfrac{1}{2}\left(1 - \gamma_S\right)M_{ij}^\dagger\right]\psi_j\,, \tag{13.68}$$

is invariant under the transformations (13.66,13.67), if the matrix $\mathbf{M}$ transforms like:

$$\mathbf{M}' = \mathbf{U}_- \mathbf{M} \mathbf{U}_+^\dagger\,, \tag{13.69}$$

in which the two indices $\pm$ of the matrices $\mathbf{U}$ correspond to the two values of ε. The total action satisfies also reflection hermiticity as defined in Appendix A5.6. It can be made invariant under a space reflection P (Appendix A5.5) by imposing to $\mathbf{M}$ to transform like:

$$\mathbf{M}_P(x) = \mathbf{M}^\dagger\left(\tilde{x}\right), \tag{13.70}$$

in which $\tilde{x}$ is obtained from x by changing the sign of one component. Therefore the matrix $\mathbf{\Sigma} = \left(\mathbf{M} + \mathbf{M}^\dagger\right)/\sqrt{2}$ represents a set of scalar fields and $\mathbf{\Pi} = \left(\mathbf{M} - \mathbf{M}^\dagger\right)/\sqrt{2}$ a set of pseudoscalar fields. Finally under a charge conjugation C, $\mathbf{M}$ transforms like:

$$\mathbf{M}_C = \mathbf{M}^* \quad \text{if } d/2 \text{ is odd}\,, \quad \text{and} \quad \mathbf{M}_C = {}^T\mathbf{M} \quad \text{if } d/2 \text{ is even}\,. \tag{13.71}$$

It remains to construct a symmetric action for the boson fields. The action:

$$S\left(\mathbf{M}\right) = \int \mathrm{d}^d x \left(\operatorname{tr} \partial_\mu \mathbf{M} \partial_\mu \mathbf{M}^\dagger + V\left(\mathbf{M}\mathbf{M}^\dagger\right)\right),\tag{13.72}$$

in which $V\left(\varphi\right)$ is an arbitrary function of traces of powers of φ, is symmetric under $U\left(N\right) \times U\left(N\right)$ transformations. If in addition we add a term proportional to $\det \mathbf{M} + \det \mathbf{M}^\dagger$, we reduce the symmetry to $SU\left(N\right) \times SU\left(N\right) \times U\left(1\right)$ (the factor $U(1)$ corresponds to the baryonic charge).

Finally the most general symmetry breaking term linear in the boson fields, consistent with the discrete symmetries (13.70) and (13.71), is:

$$S_B\left(\mathbf{M}\right) = -\frac{1}{\sqrt{2}} \int \mathrm{d}^d x \operatorname{tr} \mathbf{C}\left(\mathbf{M} + \mathbf{M}^\dagger\right),\tag{13.73}$$

in which $\mathbf{C}$ is a hermitian matrix:

$$\mathbf{C} = \mathbf{C}^\dagger.$$

To the transformations (13.66,13.67) and (13.69) correspond two currents (for more details see Appendix A13.1). It is convenient to consider the vector current $\mathbf{V}_\mu(x)$ which is associated with the Lie algebra of the diagonal subgroup $U\left(N\right)$ of $U\left(N\right) \times U\left(N\right)$ ($\mathbf{U}_+ = \mathbf{U}_-$) which conserves parity:

$$-i V_\mu^\alpha(x) = -\bar\psi t^\alpha \gamma_\mu \psi + \operatorname{tr} t^\alpha \left\{\left[\partial_\mu \mathbf{M}^\dagger, \mathbf{M}\right] + \left[\partial_\mu \mathbf{M}, \mathbf{M}^\dagger\right]\right\},\tag{13.74}$$

and the axial current $\mathbf{A}_\mu(x)$ associated with the complementary set of generators in the Lie algebra, i.e. $\mathcal{L}\left(U\left(N\right) \times U\left(N\right) / U\left(N\right)\right)$:

$$-i A_\mu^\alpha(x) = -\bar\psi t^\alpha \gamma_S \gamma_\mu \psi + \operatorname{tr} t^\alpha \left\{\left[\partial_\mu \mathbf{M}^\dagger, \mathbf{M}\right]_+ + \left[\partial_\mu \mathbf{M}, \mathbf{M}^\dagger\right]_+\right\}.\tag{13.75}$$

The $+$ index means that the expression between brackets is an anticommutator.

If the matrix $\mathbf{C}$ is proportional to the identity, the chiral symmetry is broken, but the diagonal symmetry remains and the vector current is conserved. The axial current is conserved only if $\mathbf{C}$ vanishes:

$$\partial_\mu V_\mu^\alpha(x) = -i \operatorname{tr}\left\{\left[t^\alpha, \mathbf{C}\right] \mathbf{\Sigma}\right\},\tag{13.76}$$

$$\partial_\mu A_\mu^\alpha(x) = \operatorname{tr}\left\{\left[t^\alpha, \mathbf{C}\right]_+ \mathbf{\Pi}\right\}.\tag{13.77}$$

13.6.2 A special case: the linear σ-model

The case $N = 2$ is of special physical interest. Previous analysis leads to a theory with 8 real boson fields. However the group $SU\left(2\right)$ (but not the group $U(2)$) has the property that a representation and its complex conjugate are equivalent:

$$\mathbf{U} = \tau_2 \mathbf{U}^* \tau_2, \quad \forall\, \mathbf{U} \in SU(2),$$

in which τ_2 is the usual Pauli matrix (we denote in this section the Pauli matrices τ_i rather than σ_i, as in Appendix A5.2, to eliminate possible confusion with the traditional notation for fields). Therefore $\mathbf{M}$ and $\tau_2 \mathbf{M}^* \tau_2$ have the same transformation law. We can reduce the representation and therefore parametrize the matrix $\mathbf{M}$ in terms of two fields $\sigma(x)$ and $\boldsymbol{\pi}(x)$ under the form (see also equation $(A5.30)$):

$$\mathbf{M} = \tau_2\mathbf{M}^*\tau_2 \equiv \frac{1}{\sqrt{2}}\left(\sigma + i\boldsymbol{\tau}\cdot\boldsymbol{\pi}\right) = \frac{1}{\sqrt{2}}\begin{bmatrix} \sigma + i\pi_0 & \pi_2 + i\pi_1 \\ -\pi_2 + i\pi_1 & \sigma - i\pi_0 \end{bmatrix}. \tag{13.78}$$

The group $SU(2) \times SU(2)$ is the covering group of $O(4)$ which is also the symmetry group of the bosonic part of the action. If we break the $O(4)$ symmetry by a term linear in the boson fields, we distinguish one direction in the 4-dimensional space and therefore reduce the $O(4)$ symmetry to a residual $O(3)$ symmetry. We can assume without loss of generality that the linear breaking term is proportional to $\sigma(x)$. The action can then be written:

$$S = \int \mathrm{d}^dx \left\{ -\bar{\psi}\left[\partial\!\!\!/ + g\left(\sigma + i\gamma_S\boldsymbol{\tau}\cdot\boldsymbol{\pi}\right)\right]\psi + \tfrac{1}{2}\left((\partial_\mu\sigma)^2 + (\partial_\mu\boldsymbol{\pi})^2\right) + V\left(\sigma^2 + \boldsymbol{\pi}^2\right) - c\sigma \right\}, \tag{13.79}$$

with:

$$V(\rho) = \tfrac{1}{2}m^2\rho + \frac{1}{4!}\lambda\rho^2. \tag{13.80}$$

Action (13.79) has an exact $SU(2) \times U(1)$ symmetry to which corresponds the conservation of the vector current, and implements the idea of *partially conserved axial current* (PCAC) for $SU(2)$. In the standard normalization, which differs by a factor 2 from definition (13.75) (see equations (13.88,13.99)):

$$\partial_\mu\mathbf{A}_\mu(x) = c\boldsymbol{\pi}(x). \tag{13.81}$$

Finally from equation (13.70) it follows that $\sigma(x)$ is a scalar field and $\boldsymbol{\pi}(x)$ a pseudoscalar field. In $d = 4$ dimensions σ and π_0 correspond to neutral mesons, while the combinations:

$$\pi_\pm = (\pi_1 \pm i\pi_2)/\sqrt{2},$$

correspond to charged mesons as charge conjugation shows.

13.6.3 Tree level analysis

Bosonic sector. We discuss the pattern of symmetry breaking at the classical level. Furthermore we consider only the case $N = 2$ because it is the simplest and physically the most important. In the absence of fermions we have just the ϕ^4 field theory with $O(4)$ symmetry. Equation (13.27) gives the relation between the expectation value v of the field σ and the symmetry breaking parameter c at the tree level:

$$v\left(m^2 + \lambda v^2/6\right) = c. \tag{13.82}$$

Setting:

$$\sigma(x) = v + s(x),$$

in action (13.79), we read off the masses of the $\boldsymbol{\pi}$ and σ particles at the tree order:

$$m_\pi^2 = m^2 + \lambda v^2/6, \qquad m_\sigma^2 = m^2 + \lambda v^2/2. \tag{13.83}$$

The hypothesis which accounts for the success of PCAC phenomenology is that the explicit symmetry breaking term is small and one is close a situation of SSB. For the model (13.79) this means in particular that m_π is small compared to m_σ. With this hypothesis it is possible to predict some general features of the low energy π–π scattering. Introducing the standard variables

$$s = -\left(p_1 + p_2\right)^2 , \quad t = -\left(p_1 + p_3\right)^2 , \quad u = -\left(p_1 + p_4\right)^2 , \qquad (13.84)$$

we can write the connected amputated π-field 4-point function at this order:

$$\left[W^{(4)}_{ijkl}\right]_{\text{amp}} = \frac{s - m_\pi^2}{v^2} \frac{m_\sigma^2 - m_\pi^2}{m_\sigma^2 - s} \delta_{ij}\delta_{kl} + \frac{t - m_\pi^2}{v^2} \frac{m_\sigma^2 - m_\pi^2}{m_\sigma^2 - t} \delta_{ik}\delta_{jl}$$
$$+ \frac{u - m_\pi^2}{v^2} \frac{m_\sigma^2 - m_\pi^2}{m_\sigma^2 - u} \delta_{il}\delta_{jk} . \qquad (13.85)$$

We have used the relations (13.83) to eliminate m and λ. The physical scattering amplitude is obtained by setting all momenta on the mass shell: $p_i^2 = -m_\pi^2$.

The expectation value v is experimentally accessible from the weak π-meson decay as a consequence of relation (13.81) and is denoted traditionally f_π. Since m_σ is supposed to be large compared to m_π, the expression (13.85) makes quantitative predictions for s, t, u of order m_π^2, i.e. at low energy. Values corresponding to infinite σ-mass are often quoted. Although the π–π scattering amplitude of course cannot be measured directly, indirect methods provide an experimental confirmation of the resulting pattern.

Fermion sector. In the unbroken phase, the mass of the fermion vanishes. It is the boson field expectation value which gives a mass to the fermions. However this mass cannot be predicted because the Yukawa coupling constant g, in the model, is arbitrary. However some information about the value of g can be obtained from experiment: At this order, the parameter g can be identified with the coupling constant $g_{\pi NN}$ which governs the long range part of the N–N potential due to π exchange. We then have the relation between physical quantities:

$$m_\psi = gv \Rightarrow g_{\pi NN} = \frac{m_N}{f_\pi} . \qquad (13.86)$$

This relation is the tree level approximation of the Goldberger–Treiman relation and agrees semi-quantitatively with experiment since:

$$g_{\pi NN} = 13.6 , \quad \frac{m_N}{f_\pi} \simeq \frac{939.}{93.3} = 10. . \qquad (13.87)$$

Then all parameters but m_σ are fixed. The low energy π–N scattering amplitude for example can be calculated. A definite prediction can be made only for m_σ infinite; it agrees very well with experimental observations.

Beyond the tree approximation. Since the field theory model is renormalizable, it is possible to calculate loop corrections. Then several problems arise. First there is a question of principle. As we shall argue later, the ϕ^4 field theory, as well as the theory (13.79) with fermions, is most likely inconsistent in 4 dimensions for non-vanishing coupling (see Chapters 25–29,34). More precisely, although the theory is renormalizable in perturbation theory, it is impossible to send the cut-off Λ to infinity: The model makes sense at a mass scale μ only for renormalized couplings which are bounded by const./$\ln(\Lambda/\mu)$: this is the *triviality problem*. Therefore the addition of loop corrections is meaningful only if the momenta and the coupling constants are small enough (in a correlated way as stated above). A Landau "ghost" will typically be a manifestation of this problem. Still the loop corrections may be useful to improve the tree level amplitudes from the point of view of unitarity at low energy.

Second from the computational point of view several difficulties are encountered.

(i) Loop corrections become large at moderate energies. For example in π–π scattering one encounters the ρ resonance. Then it becomes necessary to apply a summation method to the perturbation series. Calculations have been performed using the method of Padé approximants.

(ii) Since the σ-particle is heavier than $2m_\pi$, it can decay into 2 pions and is therefore unstable. In the true π–π scattering amplitude, it leads to singularities in the second sheet of the unitarity cut in the complex s-plane. However, at any finite order in perturbation theory, the singularities associated with the σ-particle are on the real axis since the width of the particle is a non-perturbative effect. Fits of experimental data seem to impose a rather small σ mass. Therefore loop corrections are affected by unphysical singularities even at rather low energy. This problem of perturbative treatment of fields corresponding to unstable particles has never been solved in a completely satisfactory way. One possible idea is to make a systematic large m_σ expansion, but the validity of the expansion is then limited to energies smaller than $4m_\sigma^2$, i.e. very low energies.

(iii) Finally perturbative corrections to the nucleon mass are large, and this also adversely affects the position of singularities in scattering amplitudes involving fermions.

Therefore, although a lot of effort has gone into the study of the model (13.79), only limited results have been obtained, beyond the simple predictions which rely on the geometry of the model and are therefore mostly contained in WT identities as we explain below.

13.6.4 Ward–Takahashi identities

We have described the difficulties one encounters when one tries to derive consequences from a phenomenological chiral action. However some relations are valid beyond perturbation theory: the WT identities which are direct consequences of the broken symmetry. Unfortunately the equation (13.32) shows that the WT identities always involve correlation functions with one π-field at zero momentum. Therefore they would lead to relations between observables only if the π-meson were massless, i.e. if the symmetry were spontaneously broken. In reality it is necessary to extrapolate from zero momentum to pion mass-shell. This extrapolation is model-dependent, and the results are only reliable if the predictions at zero pion mass are already in qualitative agreement with experiment. Let us again first discuss the purely bosonic sector.

Boson sector. The interesting part of the WT identities corresponds to the transformations:

$$\delta\boldsymbol{\pi}\left(x\right) = -\boldsymbol{\omega}\sigma\left(x\right), \quad \delta\sigma\left(x\right) = \boldsymbol{\omega}\cdot\boldsymbol{\pi}\left(x\right). \tag{13.88}$$

Calling $\mathbf{J}\left(x\right)$ the source for the π-field and $H\left(x\right)$ the source for the σ-field, we can write the WT identities for the generating functional of connected correlation functions $W\left(\mathbf{J}, H\right)$:

$$\int \mathrm{d}x \left[J_i\left(x\right) \frac{\delta}{\delta H\left(x\right)} - \left(c + H\left(x\right)\right) \frac{\delta}{\delta J_i\left(x\right)} \right] W = 0. \tag{13.89}$$

It is convenient to introduce some additional notation to take into account the residual $O(3)$ symmetry. Let us set

$$W_{ij}^{(2)}(p) = \delta_{ij} D_\pi(p),$$

$$W^{(2)}(p) = D_\sigma(p),$$

$$\tilde{W}_{ij}^{(3)}(p_1, p_2; p_3) = \delta_{ij} D_\pi(p_1) D_\pi(p_2) D_\sigma(p_3) C(p_1, p_2; p_3),$$

$$\left[\tilde{W}_{ijkl}^{\prime(4)}(p_1, p_2, p_3, p_4)\right]_{\text{amp}} = \delta_{ij}\delta_{kl} A(p_1, p_2, p_3, p_4) + \delta_{ik}\delta_{jl} A(p_1, p_3, p_2, p_4)$$

$$+ \delta_{il}\delta_{kj} A(p_1, p_4, p_3, p_2),$$

in which our conventions are that indices correspond of course to π-fields, and in mixed π–σ correlation functions the arguments of the π-fields are placed first.

By differentiating one time with respect to J_j, and setting the sources to zero, we obtain the equivalent of equation (13.44):

$$v = \langle\sigma\rangle = cD_\pi(0) \equiv c/\mu^2, \tag{13.90}$$

where we have denoted by μ^2 the value of the inverse of the π propagator at zero momentum which is now different from the pion mass squared m_π^2.

Differentiating once with respect to J_j and H, we obtain:

$$\delta_{ij}\tilde{W}^{(2)}(p) - \tilde{W}_{ij}^{(2)}(p) = c\tilde{W}_{ij}^{(3)}(0, p; -p),$$

and thus using equation (13.90):

$$D_\pi^{-1}(p) - D_\sigma^{-1}(p) = vC(0, p; -p). \tag{13.91}$$

Setting $p = 0$, and calling now m_σ^2 the value of the inverse σ propagator at zero momentum, we get in particular:

$$\mu^2 - m_\sigma^2 = vC(0, 0; 0). \tag{13.92}$$

Finally differentiating three times with respect to J we obtain a relation between 3 and 4-point correlation functions:

$$c\tilde{W}_{ijkl}^{\prime(4)}(0, p_2, p_3, p_4) = \delta_{ij}\tilde{W}_{kl}^{(3)}(p_3, p_4; p_2) + \text{2 terms}. \tag{13.93}$$

It follows

$$vA(0, p_2, p_3, p_4) = C(p_3, p_4; p_2) D_\sigma(p_2) D_\pi^{-1}(p_2). \tag{13.94}$$

First setting $p_2^2 = -m_\pi^2$ we derive from this equation Adler's consistency condition

$$A\left(0, p_2(p_2^2 = -m_\pi^2), p_3, p_4\right) = 0. \tag{13.95}$$

Moreover setting $p_3 = 0$ and eliminating the function C between (13.91) and (13.94) we find

$$v^2 A(0, p, 0, -p) = D_\pi^{-1}(p)\left[D_\sigma(p)D_\pi^{-1}(p) - 1\right]. \tag{13.96}$$

The first term in the r.h.s. has a double zero at the pion mass. Using this property we recover Weinberg's relation

$$v^2 \frac{\partial}{\partial p^2}\left(A(0, p, 0, -p) + D_\pi^{-1}(p)\right)\Big|_{p^2 = -m_\pi^2} = 0. \tag{13.97}$$

These equations yield model- and parameter-independent constraints on the π–π scattering amplitude, which unfortunately is slightly off-shell because at least one of the π momenta vanishes. One verifies immediately that the function A in the tree approximation (13.85) satisfies both conditions (13.95,13.97).

Another condition for the π–π scattering amplitude is obtained for example by setting all momenta to zero in (13.96). Setting $m_\sigma^2 = D_\sigma^{-1}(0)$ one finds:

$$v^2 A(0,0,0,0) = \mu^2 \left(\mu^2/m_\sigma^2 - 1\right).\tag{13.98}$$

This equation, however, involves an independent free parameter m_σ. Again one verifies that expression (13.85) satisfies equation (13.98) in the tree approximation.

Fermion sector. The infinitesimal transformations of the fermion fields, corresponding to equations (13.88) are:

$$\delta\psi = \tfrac{1}{2} i\gamma_S \boldsymbol{\tau} \cdot \boldsymbol{\omega}\psi, \qquad \delta\bar\psi = \tfrac{1}{2} i\bar\psi \gamma_S \boldsymbol{\tau} \cdot \boldsymbol{\omega}.\tag{13.99}$$

We then call the sources for the fermion fields $\bar\eta$ and η . The generating functional $W(\eta, \bar\eta, J, H)$ of connected correlation functions then satisfies the following WT identity:

$$\int \mathrm{d}x \left\{ \frac{i}{2} \left[\bar\eta(x) \gamma_S \boldsymbol{\tau} \frac{\delta}{\delta\bar\eta(x)} - \eta(x) \gamma_S \boldsymbol{\tau} \frac{\delta}{\delta\eta(x)} \right] - \mathbf{J}(x) \frac{\delta}{\delta H(x)} \right.$$
$$\left. + (H(x) + c) \frac{\delta}{\delta \mathbf{J}(x)} \right\} W(\eta, \bar\eta, J, H) = 0.\tag{13.100}$$

The identities relevant for physics correspond to $H = 0$. The simplest and most famous identity is obtained by differentiating with respect to η and $\bar\eta$ and setting all sources to zero. It is actually most conveniently written in terms of 1PI functions:

$$v\tilde\Gamma_{\pi NN}^{(3)}(0; p, -p) = \frac{i\boldsymbol{\tau}}{2} \left\{ \gamma_S, \tilde\Gamma_{NN}^{(2)}(p) \right\}_+.\tag{13.101}$$

The index $+$ in the r.h.s. means anticommutator in the space of γ matrices. We have explicitly taken into account the fact that the fermion propagator is proportional to the identity in the group indices. This relation between the inverse nucleon propagator and the πNN vertex generalizes relation (13.86). It has a physical interpretation in terms of the weak current under the name of Goldberger–Treiman relation. The r.h.s. is known when the nucleons are on mass-shell. The l.h.s. can be approximately related to the nucleon weak β-decay which involves the matrix element of the axial current at zero momentum between nucleon states: since the pion has the quantum numbers of the divergence of the axial current, as can be seen on equation (13.81), one contribution to this matrix element has the pion pole. In the strict chiral limit with zero mass pions, this would be the only contribution. One assumes that since the pion mass is small, the chiral limit is a good approximation. The relation then becomes in traditional notation:

$$\frac{G_A}{G_V} \simeq g_{\pi NN} \frac{f_\pi}{m_N}.\tag{13.102}$$

Replacing by experimental numbers one finds 1.22 for the l.h.s. and 1.36 for the r.h.s., a notable improvement over the tree approximation (13.87).

More generally, one can set H to zero, differentiate once with respect to η and $\bar\eta$, and an arbitrary number of times with respect to $\mathbf{J}$ and finally set all momenta on mass-shell. One then obtains model-independent relations (generalizing equation (13.95)) involving amplitudes for the emission of one unphysical pion at zero momentum. To determine completely the low energy π–N scattering amplitude, it is however necessary to introduce also the $NN\sigma$ vertex and the result then depends on the σ-mass. The predictions for the π–N scattering lengths in the infinite σ-mass limit agree well with experimental results.

Bibliographical Notes

For early derivations of the Nambu–Goldstone theorem see

Y. Nambu, *Phys. Rev. Lett.* 4 (1960) 380; J. Goldstone, *Nuovo Cimento* 19 (1961) 155; Y. Nambu and G. Jona-Lasinio, *Phys. Rev.* 122 (1961) 345; 124 (1961) 246; J. Goldstone, A. Salam and S. Weinberg, *Phys. Rev.* 127 (1962) 965; S. Bludman and A. Klein, *Phys. Rev.* 131 (1963) 2364; G. Jona-Lasinio, *Nuovo Cimento* 34 (1964) 1790.

For a general survey

G.S. Guralnik, C.R. Hagen and T.W. Kibble, *Advances in Physics*, R.E. Marshak and R. Cool eds. (Interscience, New York 1968).

The Proceedings of the Cargèse Summer Institute 1970 contains several contributions devoted to breaking of symmetries, WT identities and renormalizations:

Cargèse Lectures in Physics vol. 5, D. Bessis ed. (Gordon and Breach, New York 1971).

Among them let us quote more specifically:

K. Symanzik, who gives a general discussion of renormalization of symmetry breaking terms; A.N. Vasil'ev (in French) who shows how WT identities can be derived from the functional integral representation of correlation functions; B.W. Lee who discusses Chiral Dynamics.

These authors have been a direct source of inspiration for the presentation we have given here, which is an update of

J. Zinn-Justin in *Trends in Elementary Particle Physics (Lectures Notes in Physics 37)*, Bonn 1974, H. Rollnik and K. Dietz eds. (Springer-Verlag, Berlin 1975).

See also the lectures of Coleman

S. Coleman in *Laws of Hadronic Matter*, Erice 1973, A. Zichichi ed. (Academic Press, New York 1975).

The linear σ-model has been proposed in

M. Gell-Mann and M. Lévy, *Nuovo Cimento* 16 (1960) 705.

Its properties have been studied in

B.W. Lee, *Nucl. Phys.* B9 (1969) 649; J.-L. Gervais and B.W. Lee, *Nucl. Phys.* B12 (1969) 627; J.L. Basdevant and B.W. Lee, *Phys. Rev.* D2 (1970) 1680.

This work is reviewed in

B.W. Lee, *Chiral Dynamics* (Gordon and Breach, New York 1972).

For a review of the application of Padé approximants to the summation of perturbation theory for phenomenological lagrangians see for example

J. Zinn-Justin, *Phys. Rep.* 1C (1970) 55.

$SU(2) \times SU(2)$ symmetry breaking has been discussed in

S. Weinberg, *Phys. Rev.* 166 (1968) 1568.

Chiral Dynamics is directly connected to Current Algebra. Among the large number of articles and text books let us just quote:

S. Adler and R. Dashen, *Current Algebras and Applications to Particle Physics* (Benjamin, New York, 1968); S. Weinberg in *Lectures on Elementary Particle Physics and Quantum Field Theory*, Brandeis 1970, S. Deser, M. Grisaru and H. Pendleton eds. (MIT Press, Cambridge 1970); S.B. Treiman, R. Jackiw and D.J. Gross, *Lectures on Current Algebra and its Applications* (Princeton University Press, Princeton 1972); C. Itzykson and J.-B. Zuber's text book as quoted in the introduction; M.L. Goldberger and S.B. Treiman, *Phys. Rev.* 110 (1958) 1178; S.L. Adler, *Phys. Rev.* 139B (1965) 1638; 140B (1965) B736; W.I. Weisberger, *Phys. Rev.* 143 (1966) 1302; S. Weinberg, *Phys. Rev. Lett.* 17 (1966) 616;

$SU(3) \times SU(3)$ symmetry breaking is discussed in

M. Gell-Mann, R.J. Oakes and B. Renner, *Phys. Rev.* 175 (1968) 2195.
For the improved energy-momentum tensor (Appendix A13.3) see:
C.G. Callan, S. Coleman and R. Jackiw, *Ann. Phys. (NY)* 59 (1970) 42.
It has been noticed that at zero of the RG β-function both dilatation and conformal invariance are restored in
B. Schroer, *Lett. Nuovo Cimento* 2 (1971) 867.
Conformal invariance and its spectacular consequences in two dimensions are thoroughly discussed in the Proceedings of Les Houches Summer Institute 1988 *Fields, Strings and Critical Phenomena*, E. Brézin and J. Zinn-Justin eds. (Elsevier, Amsterdam 1989). See in particular the lectures by
J. Cardy, *Conformal Invariance and Statistical Mechanics*; P. Ginsparg, *Applied Conformal Field Theory*.

Exercises

A useful exercise is to verify that the $SU(2) \times SU(2)$ linear σ-model with linear symmetry breaking can be renormalized at one-loop order consistently with WT identities. Some help can be found in the following references

B.W. Lee, *Chiral Dynamics*, Gordon and Breach, New-York 1972;

J.L. Basdevant and B.W. Lee, *Phys. Rev.* D2 (1970) 1680;
and the proceedings of the Cargèse summer institute 1970, D. Bessis ed. Cargèse Lectures in Physics vol. 5, Gordon and Breach, New-York 1972.

We here propose instead to study the structure of symmetry breaking, at the classical level, for an effective $SU(3) \times SU(3)$ invariant renormalizable field theory. The first reference is also relevant for this exercise. We consider as fundamental fields the scalar and pseudoscalar (π, K, η, η') bound states of the three $\mathbf{u}, \mathbf{d}, \mathbf{s}$ quarks. They can be described by a 3×3 complex matrix $\mathbf{M}$ which transforms under $SU(3) \times SU(3)$ as:

$$\mathbf{M}(x) \mapsto \mathbf{U}_1^\dagger \mathbf{M}(x) \mathbf{U}_2, \quad \mathbf{U}_1, \mathbf{U}_2 \in SU(3).$$

Exercise 13.1

Write the most general renormalizable $SU(3) \times SU(3)$ invariant scalar action, in four dimensions, and discuss the general pattern of spontaneous symmetry breaking.

Exercise 13.2

Identify the Goldstone modes in the different cases first by purely group theory considerations and then by determining the mass matrix.

Exercise 13.3

Spontaneous symmetry breaking can lead to a realistic residual symmetry but not of course to realistic masses. In particular the explicit $SU(3)$ symmetry breaking is rather large in the scale of physics we want to describe. The explicit breaking of $SU(2) \times SU(2)$ is smaller but responsible for the pion mass. Therefore to make the model more realistic we add a linear breaking term:

$$S_B = S_{\text{sym.}} - \text{tr}\,\mathbf{C}\left(\mathbf{M} + \mathbf{M}^\dagger\right),$$

where $\mathbf{C}$ is a diagonal matrix with diagonal elements c, c, c'. The model then depends on 6 parameters. Determine analytically the mass spectrum. Is it possible to adjust the parameters in such a way that the pseudoscalar decay constants and the masses agree in a reasonable way with the experimental values? This last exercise requires a particle data table and a handpocket calculator.

APPENDIX 13
CURRENTS AND NOETHER'S THEOREM

A13.1 Real Time Classical Field Theory

In this section, in order to discuss properties of the relativistic classical equations of motion, we shall adopt the covariant notation of real time field theory with the metric of signature $(- - \cdots +)$. Summation over successive upper and lower indices will be implied (see Chapter 22).

If the lagrangian density $\mathcal{L}(\phi, \partial_\mu \phi)$ depends only on the field $\phi(x)$ and its derivatives $\partial_\mu \phi(x)$, the classical equation of motion obtained by varying the action S:

$$S(\phi) = \int d^d x \, \mathcal{L}(\phi(x), \partial_\mu \phi(x)), \tag{A13.1}$$

is

$$\partial_\mu \frac{\partial \mathcal{L}}{\partial [\partial_\mu \phi(x)]} - \frac{\partial \mathcal{L}}{\partial \phi(x)} = 0, \tag{A13.2}$$

(in this notation $\phi(x)$ and $\partial_\mu \phi(x)$ are considered as independent variables).

If we perform on $\phi(x)$ a space-dependent group transformation which we parametrized by a field $\Lambda(x)$:

$$\phi \mapsto \phi_\Lambda,$$

as a consequence of the equation of motion, the action is also stationary with respect to variations of $\Lambda(x)$ at ϕ fixed:

$$\partial_\mu \frac{\partial \mathcal{L}}{\partial [\partial_\mu \Lambda(x)]} - \frac{\partial \mathcal{L}}{\partial \Lambda(x)} = 0. \tag{A13.3}$$

We define a current $J^\mu(x)$, functional of $\phi(x)$, by:

$$J^\mu(x) = \left. \frac{\partial \mathcal{L}}{\partial [\partial_\mu \Lambda(x)]} \right|_{\Lambda(x)=0}, \tag{A13.4}$$

in which we have assumed that $\Lambda(x) = 0$ corresponds in the group to the identity. By construction currents are directly associated with the generators of the Lie algebra of the symmetry group.

We can then rewrite equation $(A13.3)$ as:

$$\partial_\mu J^\mu(x) = \frac{\partial \mathcal{L}}{\partial \Lambda(x)}, \tag{A13.5}$$

which is Noether's theorem.

If in addition the lagrangian is invariant under space-independent group transformations, $\partial \mathcal{L}/\partial \Lambda$ vanishes and thus the current J_μ is conserved:

$$\partial_\mu J^\mu(x) = 0. \tag{A13.6}$$

If we take the example of a lagrangian density of the form:

$$\mathcal{L} = \tfrac{1}{2} \partial_\mu \phi_i \partial^\mu \phi_i - V[\phi(x)], \tag{A13.7}$$

(in real time covariant notation) and if the infinitesimal group transformations are:

$$\delta\phi_i(x) = t_{ij}^\alpha \Lambda^\alpha(x)\phi_j(x), \tag{A13.8}$$

the current $J_\mu^\alpha(x)$ is given by:

$$J_\mu^\alpha(x) = t_{ij}^\alpha \partial_\mu\phi_i(x)\phi_j(x). \tag{A13.9}$$

In classical mechanics the space integral of the time-component of the current is a charge $\mathcal{Q}^\alpha$ $(t \equiv x_d)$:

$$\mathcal{Q}^\alpha(t) = \int \mathrm{d}^{d-1}x\, J_d^\alpha(x). \tag{A13.10}$$

By differentiating with respect to t and using the current conservation equation $(A13.6)$ one finds:

$$\frac{\mathrm{d}}{\mathrm{d}t}\mathcal{Q}^\alpha(t) = \int \mathrm{d}^{d-1}x \sum_{\mu=1}^{d-1} \partial_\mu J_\mu(x) = 0\,.$$

The charges $\mathcal{Q}^\alpha(t)$ are constants of the classical motion.

A13.2 Euclidean Field Theory

We have already examined the consequences of symmetries for field theories and derived WT identities. These identities can also be derived in the operator formalism of quantum mechanics and in this case currents and charges, considered as quantum operators, play an important role. In our formulation, currents will appear either in the coupling at leading order of matter to gauge fields (see Chapters 18,19) or as polynomials in the fields (operators in the sense of Chapter 12) satisfying some identities which we shall derive and therefore having special renormalization properties.

Let us therefore consider the generating functional $Z(J)$:

$$Z(J) = \int [\mathrm{d}\phi] \exp\left[-S(\phi) + \int J_i(x)\phi_i(x)\,\mathrm{d}x\right], \tag{A13.11}$$

in which the action is invariant under group transformations whose infinitesimal form is given by equation $(A13.8)$ when $\Lambda(x)$ is a constant.

In what follows *dimensional* regularization is assumed.

We now perform a change of variables in integral $(A13.11)$ of the form of a transformation (13.10). We define the euclidean current $J_\mu^\alpha(x)$ by equation $(A13.4)$ in terms of the euclidean action density. If $S(\phi)$ is symmetric, the variation of the action reads:

$$\delta S(\phi) = \int \partial_\mu \Lambda^\alpha(x) J_\mu^\alpha(x)\mathrm{d}x\,. \tag{A13.12}$$

Identifying the coefficient of $\Lambda^\alpha(x)$, we obtain:

$$\int [\mathrm{d}\phi]\left[\partial_\mu J_\mu^\alpha(x) - J_i(x)t_{ij}^\alpha\phi_j(x)\right]\exp\left[-S(\phi) + \int J_i(x)\phi_i(x)\mathrm{d}x\right] = 0\,. \tag{A13.13}$$

This identity can be rewritten:

$$\partial_\mu^x Z_{J_\mu^\alpha(x)} = J_i(x)t_{ij}^\alpha \frac{\delta Z}{\delta J_j(x)}\,. \tag{A13.14}$$

$Z_{J_\mu^\alpha(x)}$ is the generating functional of correlation functions with a $J_\mu^\alpha(x)$ operator insertion.

The same equation is valid for connected correlation functions. After Legendre transformation we find:

$$\partial_\mu^x \Gamma_{J_\mu^\alpha(x)} = -\frac{\delta\Gamma}{\delta\varphi_i(x)} t_{ij}^\alpha \varphi_j(x). \qquad (A13.15)$$

Equations ($A13.14, A13.15$) are the analogues for correlation functions of the current conservation equation ($A13.6$). Integrated over all space, they yield, not surprisingly, equations (13.17–13.20), i.e. the WT identities of the symmetry.

From the point of view of renormalization, equation ($A13.15$) tells us that the insertion of $\partial_\mu J_\mu^\alpha(x)$ in a renormalized correlation function is finite.

In a simple renormalizable ϕ^4-like field theory, covariance then implies that the same must be true for the current $J_\mu^\alpha(x)$. This result is non-trivial since from expression ($A13.9$) we see that $J_\mu^\alpha(x)$ is an operator of dimension 3. A further consequence is that the insertion of a conserved current in a correlation function does not modify the form of the RG equations.

A13.3 The Energy Momentum Tensor

If the action is translation invariant, the substitution $\phi(x) \mapsto \phi(x+\varepsilon)$, in which ε is a constant, leaves the action invariant. In the spirit of Section A13.1, let us perform a space-dependent translation, which coincides of course with a general change of variables (see also Section 22.1). We thus substitute in the action $\phi(x) \mapsto \phi(x+\varepsilon(x))$. If $\phi(x)$ satisfies the equation of motion, the variation of the action ($A13.1$) at first order in ε vanishes. In the substitution the derivatives transform like

$$\partial_\mu \phi(x) \mapsto \partial_\mu \phi(x+\varepsilon) + \partial_\mu \varepsilon^\nu \partial_\nu \phi(x+\varepsilon).$$

To calculate the variation we then change variables $x + \varepsilon = y$. Translation invariance implies that the action density depends on x through the field ϕ but not explicitly. Therefore the only new effect is to change the measure of integration

$$dy^\mu = dx^\mu + \partial_\mu \varepsilon^\nu dx^\nu.$$

If we now compare the new action with the initial one ($A13.1$) we see that the modifications come only from the derivatives and the integration measure (y is a dummy integration variable). Collecting the terms of order ε and integrating by parts, we obtain the following identity:

$$\partial_\mu T_\nu^\mu(x) = 0, \qquad (A13.16)$$

in which the *energy momentum tensor* $T_\nu^\mu(x)$ is defined by:

$$T_\nu^\mu = \frac{\partial \mathcal{L}}{\partial[\partial_\mu \phi(x)]} \partial_\nu \phi(x) - \delta_\nu^\mu \mathcal{L}[\phi(x)]. \qquad (A13.17)$$

For example for the lagrangian ($A13.7$) $T_{\mu\nu}$ has the form:

$$T_{\mu\nu}(x) = \partial_\mu \phi \partial_\nu \phi - g_{\mu\nu} \left[\tfrac{1}{2} (\partial_\rho \phi)(\partial^\rho \phi) - V(\phi) \right], \qquad (A13.18)$$

in which $g_{\mu\nu}$ is the usual Minkowski metric tensor.

To the tensor $T_{\mu\nu}(x)$ correspond constants of the classical motion P_μ which are just energy and momentum, obtained by integrating the time components (with respect to one index) of $T_{\mu\nu}$ on space:

$$P_\mu = \int \mathrm{d}^{d-1}x\, T_{d\mu}(x), \qquad (A13.19)$$

with $x_d \equiv t$

$$\frac{\mathrm{d}}{\mathrm{d}t}P_\mu = 0. \qquad (A13.20)$$

We note that a space–time-dependent change of variables on x^μ is an arbitrary change of coordinates. This explains that the tensor $T_{\mu\nu}$ appears in the coupling of matter field to the metric tensor in General Relativity (for details see Chapter 22 and the corresponding references).

Also any current associated with an additional space-time symmetry of the action can be related to $T_{\mu\nu}$.

For instance the $O(d-1,1)$ pseudo-orthogonal transformations whose infinitesimal form is:

$$\delta x^\mu = \Lambda^\mu_\nu(x)x^\nu, \qquad (A13.21)$$

correspond to the choice:

$$\varepsilon^\mu = \Lambda^\mu_\nu x^\nu. \qquad (A13.22)$$

The corresponding currents $M^{\mu\nu\rho}$ are then:

$$M^{\mu\nu\rho}(x) = T^{\mu\nu}x^\rho - T^{\mu\rho}x^\nu. \qquad (A13.23)$$

Dilatation invariance. Let us, as an example, again consider the case of the ϕ^4 theory in 4 dimensions:

$$\mathcal{L} = \frac{1}{2}\partial_\mu\phi(x)\partial^\mu\phi(x) - \frac{1}{2}m^2\phi^2(x) - \frac{1}{4!}g\phi^4(x). \qquad (A13.24)$$

In the absence of the mass term, the action is scale-invariant, i.e. invariant in the substitution:

$$\phi(x) \mapsto \phi_\lambda(x) = \lambda\phi(\lambda x). \qquad (A13.25)$$

For what concerns the variation of the argument, dilatation corresponds to take ε^μ as:

$$\varepsilon^\mu = x^\mu\lambda(x). \qquad (A13.26)$$

We expect thus the dilatation current S^μ to involve $x^\nu T^\mu_\nu$. A short calculation leads to:

$$S^\mu(x) = x^\nu\left[T^\mu_\nu(x) + \frac{1}{6}\left(\partial^2\delta^\mu_\nu - \partial_\mu\partial^\nu\right)\phi^2(x)\right]. \qquad (A13.27)$$

In the presence of a mass term, the current $S^\mu(x)$ is not conserved. Instead:

$$\partial_\mu S^\mu(x) = m^2\phi^2(x). \qquad (A13.28)$$

Let us introduce the tensor $\tilde{T}_{\mu\nu}(x)$:

$$\tilde{T}_{\mu\nu}(x) = T_{\mu\nu}(x) + \frac{1}{6}\left(\partial^2 g_{\mu\nu} - \partial_\mu\partial_\nu\right)\phi^2(x). \qquad (A13.29)$$

The tensor $\tilde{T}_{\mu\nu}$ can be used as energy momentum tensor instead of $T_{\mu\nu}$: it is a polynomial in the field, symmetric as a tensor, and satisfies the conservation equation:

$$\partial_\mu \tilde{T}^\mu_\nu = 0\,. \tag{A13.30}$$

In terms of $\tilde{T}^\mu_\nu(x)$ equation (A13.27) then reads:

$$S^\mu(x) = x^\nu \tilde{T}^\mu_\nu(x)\,, \tag{A13.31}$$

and the divergence of the dilatation current is:

$$\partial_\mu S^\mu = \tilde{T}^\mu_\mu\,. \tag{A13.32}$$

In dilatation-invariant theories, the trace of the "improved" energy momentum tensor $\tilde{T}^\mu_\nu$ vanishes.

A13.4 Energy Momentum Tensor and Euclidean Field Theory

Performing the infinitesimal change of variables:

$$\phi(x) = \phi'\bigl(x + \varepsilon(x)\bigr), \tag{A13.33}$$

in the functional integral, one can easily derive WT identities for the insertion of the energy momentum tensor (also called the *stress tensor*). The variation of the action with a source is:

$$\delta\left[\int J\phi\,\mathrm{d}x - S\left(\phi\right)\right] = \varepsilon_\nu(x)\left[J(x)\partial_\nu\phi(x) + \partial_\mu T_{\mu\nu}(x)\right]. \tag{A13.34}$$

It follows that:

$$\partial^x_\mu Z_{T_{\mu\nu}(x)} + J(x)\partial^x_\nu\frac{\delta Z}{\delta J(x)} = 0\,. \tag{A13.35}$$

Integrating this identity over space yields

$$\int \mathrm{d}x\, J(x)\partial_\nu\frac{\delta Z}{\delta J(x)} = 0\,, \tag{A13.36}$$

which expresses the translation invariance of correlation functions.

After Legendre transformation, one finds:

$$\partial^x_\mu \Gamma_{T_{\mu\nu}(x)} + \frac{\delta\Gamma}{\delta\varphi(x)}\partial_\nu\varphi(x) = 0\,. \tag{A13.37}$$

Again we conclude that the insertion of the operator $\partial_\mu T_{\mu\nu}(x)$ in a renormalized correlation function is finite. However this does not imply that the insertion of $T_{\mu\nu}$ itself is finite. In the ϕ^4_4 field theory for example, $T_{\mu\nu}$ is of dimension 4. The quantity $\left(\delta_{\mu\nu}\partial^2 - \partial_\mu\partial_\nu\right)\phi^2$ is also a symmetric tensor of dimension 4 whose divergence vanishes. Therefore it can appear as an additive counterterm in the renormalization of $T_{\mu\nu}$:

$$(T_{\mu\nu})_\mathrm{r} = T_{\mu\nu} + A\left(\delta_{\mu\nu}\partial^2 - \partial_\mu\partial_\nu\right)(\phi^2)_\mathrm{r}\,. \tag{A13.38}$$

Note that the renormalized energy momentum tensor has automatically a non-vanishing trace, and it can no longer be improved since the coefficient A is divergent. The dilatation current is not conserved but this should have been expected since it is impossible to regularize the theory without breaking the classical dilatation invariance, either by introducing a cut-off, or by changing the dimension. It is nevertheless possible to derive WT identities involving the divergence of the dilatation current. By integrating them over space, one just obtains the CS equations derived in Chapter 10.

A13.5 Dilatation and Conformal Invariance

Let us now consider a general euclidean action S invariant under translation, rotation and dilatation. We perform the infinitesimal change of variables:

$$x_\mu \longmapsto x_\mu + \varepsilon_\mu(x).$$ (A13.39)

Translation invariance implies that the variation of the action involves only the partial derivatives of $\varepsilon_\mu(x)$:

$$\delta S = \int d^d x\, T_{\mu\nu}(x)\partial_\mu\varepsilon_\nu(x).$$ (A13.40)

Rotation invariance implies that δS vanishes for:

$$\varepsilon_\mu = \Lambda_{\mu\nu}x_\nu,$$ (A13.41)

in which $\Lambda_{\mu\nu}$ is an arbitrary antisymmetric matrix. Therefore the integral of the stress tensor must be symmetric:

$$\int d^d x\, (T_{\mu\nu} - T_{\nu\mu}) = 0.$$

Dilatation invariance corresponds to:

$$\varepsilon_\mu = \lambda x_\mu,$$ (A13.42)

and implies the vanishing of the integral of the trace of the stress tensor:

$$\int d^d x\, T_{\mu\mu} = 0.$$

For the simplest class of theories, like scalar field theories with an action $S(\phi)$ depending only on the field $\phi(x)$ and its *first* partial derivatives, the two integral conditions imply the existence of a symmetric, traceless stress–energy tensor:

$$T_{\mu\nu} = T_{\nu\mu}, \quad T_{\mu\mu} = 0.$$

It then follows that the variation of the action also vanishes for any function ε_μ which satisfies:

$$\partial_\mu\varepsilon_\nu + \partial_\nu\varepsilon_\mu - \tfrac{1}{2}d\delta_{\mu\nu}\partial\cdot\varepsilon = 0,$$ (A13.43)

where d is the dimension of euclidean space. The group of transformations which satisfy equation (A13.43) is larger than the product of transformations which we have considered so far: it is the whole *conformal* group. Indeed let us calculate the variation of a line element of the form:

$$(ds)^2 = g(x)dx_\mu dx_\mu,$$ (A13.44)

which corresponds to a conformally flat metric.
   We find:

$$\delta\left[(ds)^2\right] = dx_\mu dx_\mu \partial_\rho g(x)\varepsilon_\rho + dx_\mu dx_\nu\left(\partial_\mu\varepsilon_\nu + \partial_\nu\varepsilon_\mu\right)g(x).$$ (A13.45)

We now see that the equation (A13.43) is the necessary and sufficient condition for the line element to retain the form (A13.44). By definition the transformations which preserve the form of the metric (A13.44) are conformal transformations.

From equation $(A13.43)$ follows:

$$\left[\delta_{\mu\nu}\partial^2 + (d-2)\partial_\mu\partial_\nu\right]\partial\cdot\varepsilon = 0. \tag{A13.46}$$

For $d > 2$ the equation implies that all second derivatives of $\partial\cdot\varepsilon$ vanish. Returning then to equation $(A13.43)$ one shows easily that all third derivatives of ε_μ also vanish. Solutions of degree 0 correspond to translations. Solutions of degree 1 correspond to rotations and dilatations. The additional solutions of equation $(A13.43)$ are second degree polynomials of the form:

$$\varepsilon_\mu = a_\mu x^2 - 2x_\mu a \cdot x. \tag{A13.47}$$

They correspond to special conformal transformations. The integrated form of these transformations is:

$$x'_\mu = \frac{x_\mu + a_\mu x^2}{1 + 2a \cdot x + a^2 x^2}. \tag{A13.48}$$

The conformal group is isomorphic to $SO(d+1,1)$. Imposing conformal invariance to correlation functions determines in particular 2 and 3-point functions.

In dimension $d = 2$ the situation is completely different: the set of equations $(A13.43)$ are just the Cauchy conditions and express the well-known property that all analytic transformations are conformal. The conformal group has an infinite number of generators. The consequences are much more striking and lead to the classification of a whole class of conformal invariant field theories.

Of course we have seen in Chapter 10 that the scale invariance of the classical theory is broken at the quantum level. However, as we shall discuss in Chapter 25, there exist situations in which the RG β-function vanishes, at least for some values of the coupling constants. Then both the dilatation invariance and therefore the conformal invariance are restored.

Remark. The condition that the action should depend only on the field and its first derivatives can be illustrated by a simple counter-example. Consider the free action $S(\phi)$

$$S(\phi) = \int \mathrm{d}^d x \left(\partial^2 \phi(x)\right)^2.$$

The propagator in Fourier space is $1/p^4$. The theory is obviously translation, rotation and scaling invariant. However it is easy to verify that it is not conformal invariant.

We now consider models possessing global symmetries non-linearly realized on the fields. This means in particular that under an infinitesimal group transformation, the variation of the field is a non-linear function of the field. Since such models have non-trivial geometrical properties, we first extensively discuss a particular example, the non-linear σ-model, a model with an $O(N)$ symmetry, the field being a N-vector of fixed length.

A tree level analysis reveals that in the non-linear σ-model the $O(N)$ symmetry is automatically realized in the phase of spontaneous symmetry breaking, unlike a ϕ^4 like theory with the same symmetry: The action describes the interactions of on $N - 1$ massless fields, the Goldstone modes.

Power counting shows that this model is renormalizable in two dimensions. Therefore the field is dimensionless and we face a problem already mentioned in Section 8.3: the degree of divergence of Feynman diagrams is bounded, although an infinite number of counterterms are generated because all correlation functions are divergent. We prove in this chapter that, due to the special geometrical properties of the model, the coefficients of all counterterms can be calculated as a function of two of them so that the renormalized theory depends only on a finite number of parameters.

Note that since the fields are massless in two dimensions IR divergences appear in the perturbative expansion. An IR regulator is thus required.

In Section 14.8 we discuss the renormalization of composite operators. Finally in Section 14.9 we indicate how the results can be recovered from another, linear, representation of the model where the condition that the field is a vector of fixed length is enforced by a Lagrange multiplier.

In Chapter 15 we shall show how the arguments generalize to all models defined on homogeneous spaces. Actually the $O(N)$ non-linear σ model belongs to a class of models constructed on special homogeneous spaces, the symmetric spaces, which as Riemannian manifolds, admit a unique metric. We shall study them in more detail both as classical and quantum field theories.

14.1 The Non-Linear σ-Model: Definition

We consider an N-vector field $\phi(x)$ which satisfies an $O(N)$ invariant constraint:

$$\phi^2(x) = 1 \,. \tag{14.1}$$

The field ϕ can also be identified with an element of a homogeneous (symmetric) space $O(N)/O(N-1)$: let $g(x)$ be an element of $O(N)$ depending on the coordinate x, $\mathbf{D}\left(g(x)\right)$ the corresponding matrix in the fundamental representation, and $\mathbf{u}$ a fixed N-vector on which $\mathbf{D}\left(G\right)$ acts:

$$\mathbf{u} = (1, 0, \ldots, 0) \,. \tag{14.2}$$

The field ϕ can then be parametrized as:

$$\phi(x) = \mathbf{D}\left(g(x)\right)\mathbf{u} \,, \tag{14.3}$$

since the column vectors of an orthogonal matrix are by definition normalized to length 1.

The little group of $\mathbf{u}$ (or stabilizer), i.e. the subgroup of $O(N)$ which leaves $\mathbf{u}$ invariant, is $O(N-1)$. Therefore if one multiplies on the right $g(x)$ by any element of $O(N-1)$, the little group of $\mathbf{u}$, the r.h.s. of equation (14.3) is left unchanged: relation (14.3) exhibits the isomorphism between the coset (homogeneous) space $O(N)/O(N-1)$ and the sphere S_{N-1}.

To be able to generate the perturbative expansion of the corresponding field theory, it is necessary to parametrize the field in terms of independent variables. A parametrization of the sphere (14.1) convenient for renormalization purpose is:

$$\phi(x) = \{\sigma(x), \pi(x)\}, \tag{14.4}$$

in which $\pi(x)$ is a $(N-1)$-component field. The field $\sigma(x)$ is then a function of $\pi(x)$ through equation (14.1). The equation can be solved locally, for example if $\sigma(x)$ is positive:

$$\sigma(x) = \left(1 - \pi^2(x)\right)^{1/2}. \tag{14.5}$$

The consequences of the singularity of this parametrization will be discussed later.

The stabilizer group $O(N-1)$ acts linearly on $\pi(x)$. We can decompose the set of generators of the Lie algebra of $O(N)$ into the set of generators of the Lie algebra of $O(N-1)$ and the complementary set. To this complementary set correspond infinitesimal transformations of the form

$$\delta\pi_i = \omega_i \left(1 - \pi^2(x)\right)^{1/2}, \tag{14.6}$$

in which ω_i are constants, infinitesimal parameters of the transformation. The transformation law of the σ-field then is a consequence of the transformation (14.6) of the π-field:

$$\delta\sigma(x) \equiv \delta \left(1 - \pi^2(x)\right)^{1/2} = -\omega \cdot \pi(x). \tag{14.7}$$

The action and functional integral. The most general $O(N)$ symmetric action containing at most two derivatives is, up to a multiplicative constant:

$$S(\phi) = \tfrac{1}{2} \int d^d x\, \partial_\mu \phi(x) \cdot \partial_\mu \phi(x). \tag{14.8}$$

Indeed, due to the constraint (14.1), any symmetric derivative-free term reduces to a constant (and $\phi \cdot \partial_\mu \phi$ vanishes).

In terms of the field $\pi(x)$, the action (14.8) can be cast into the following geometrical form:

$$S(\pi) = \tfrac{1}{2} \int d^d x\, G_{ij}(\pi)\, \partial_\mu \pi_i \partial_\mu \pi_j, \tag{14.9}$$

in which $G_{ij}(\pi)$ is a metric tensor on the sphere:

$$G_{ij} = \delta_{ij} + \frac{\pi_i \pi_j}{1 - \pi^2}. \tag{14.10}$$

In the form (14.9) the action is covariant under a reparametrization of the sphere.

We have seen in Section 3.2 that the quantization of actions of the form (14.9) introduces an additional ill-defined (since infinite) determinant. From equations (3.27–3.30) we derive a representation of the generating functional of correlation functions:

$$Z\left(\mathbf{J}\right) = \int \left[\frac{\mathrm{d}\boldsymbol{\pi}(x)}{\left(1 - \boldsymbol{\pi}^2(x)\right)^{1/2}}\right] \exp\left[-\frac{1}{g}\left(S\left(\boldsymbol{\pi}\right) - \int \mathrm{d}^d x\, \mathbf{J}(x) \cdot \boldsymbol{\pi}\left(x\right)\right)\right], \qquad (14.11)$$

in which g is the coupling constant of the quantum field theory.

We had noticed that in the case of Riemannian manifolds the determinant is needed for geometrical reasons: here indeed it allows us to write the functional measure of integration over the field $\boldsymbol{\pi}(x)$ in an $O(N)$ invariant way.

This formal interpretation of the determinant, however, does not eliminate the difficulty. We have to find a method to deal with this infinite contribution to the action:

$$\prod_x \left(1 - \boldsymbol{\pi}^2(x)\right)^{-1/2} \sim \exp\left[-\frac{1}{2}\delta^d(0)\int \mathrm{d}^d x\, \ln\left(1 - \boldsymbol{\pi}^2(x)\right)\right]. \qquad (14.12)$$

As we have explained in Section 3.2, this difficulty is directly related to the problem of operator ordering which appears in the quantization.

14.2 Perturbation Theory. Power Counting

The minima of the classical action (14.8) are given by:

$$|\partial_\mu \boldsymbol{\phi}(x)| = 0 \quad \Rightarrow \quad \boldsymbol{\phi}(x) = \text{const.}\,.$$

Due to the $O(N)$ invariance, the action has a continuous set of degenerate and equivalent minima. To expand perturbation theory we have to choose one minimum. If we calculate only averages of $O(N)$ invariant correlation functions with action (14.8) and the $O(N)$ invariant measure of integral (14.11), all minima give exactly the same contribution. Summing over all minima yields a factor, the volume of the sphere S_{N-1}, which disappears in the normalization of the functional integral. However for non-$O(N)$ invariant correlation functions a summation over all minima is equivalent to an average over the $O(N)$ group. Therefore it would seem that all non-vanishing correlation functions are $O(N)$ invariant. This problem is directly related to the question of spontaneous symmetry breaking and breaking of ergodicity in the ordered phase. In this chapter we shall rely on rather superficial and formal arguments, postponing a deeper analysis to Chapters 23–32 in which the theory of phase transitions is discussed. In the parametrization (14.4) we choose the minimum $\boldsymbol{\pi}(x) = 0$ or $\boldsymbol{\phi}(x) = \mathbf{u}$.

14.2.1 Perturbation theory

In the functional integral (14.11) we have introduced a coupling constant g which plays the formal role of $\hbar$ and therefore orders perturbation theory. For g small, the fields $\boldsymbol{\pi}\left(x\right)$ which contribute to the functional integral then must be such that:

$$|\partial_\mu \boldsymbol{\pi}(x)| \sim \sqrt{g}\,,$$

and since we expand around $\boldsymbol{\pi}(x) = 0$, the field itself must satisfy:

$$|\boldsymbol{\pi}(x)| \sim \sqrt{g}\,. \qquad (14.13)$$

Values of $\boldsymbol{\pi}(x)$ of order 1 give exponentially small contributions to the functional integral (of order $\exp(-\text{const.}/g)$) which are negligible at any finite order of perturbation theory.

This has two consequences: the restrictions imposed by the parametrization (14.5) ($\sigma(x) > 0$) are irrelevant in perturbation theory and in addition, in the functional integral, we can freely integrate over $\boldsymbol{\pi}(x)$ from $+\infty$ to $-\infty$, disregarding the constraint:

$$|\boldsymbol{\pi}(x)| \leq 1.$$

Perturbation theory then again relies on the evaluation of simple gaussian integrals.

We can now discuss formal perturbation theory, setting temporarily aside the question of UV or IR (low momentum) divergences. We rewrite the functional integral (14.11):

$$Z(\mathbf{J}) = \int [\mathrm{d}\boldsymbol{\pi}] \exp\left[-\int \mathrm{d}^d x\, \mathcal{L}(\boldsymbol{\pi}, \mathbf{J})\right],$$

$$\mathcal{L}(\boldsymbol{\pi}, \mathbf{J}) = \frac{1}{2g}\left[(\partial_\mu \boldsymbol{\pi})^2 + \frac{(\boldsymbol{\pi} \cdot \partial_\mu \boldsymbol{\pi})^2}{1 - \boldsymbol{\pi}^2}\right] + \frac{1}{2}\delta^d(0)\ln\left(1 - \boldsymbol{\pi}^2(x)\right) - \frac{1}{g}\mathbf{J}(x) \cdot \boldsymbol{\pi}(x). \quad (14.14)$$

Note that the measure term has no $1/g$ factor and starts contributing only at one-loop order. Since $\boldsymbol{\pi}$ is of order $\sqrt{g}$, it is convenient to rescale the field:

$$\boldsymbol{\pi} \mapsto \boldsymbol{\pi}\sqrt{g}.$$

After this rescaling the action density $\mathcal{L}$ becomes

$$\mathcal{L}(\boldsymbol{\pi}, \mathbf{J}) = \frac{1}{2}\left[(\partial_\mu \boldsymbol{\pi})^2 + g\frac{(\boldsymbol{\pi} \cdot \partial_\mu \boldsymbol{\pi})^2}{1 - g\boldsymbol{\pi}^2}\right] + \frac{1}{2}\delta^d(0)\ln\left(1 - g\boldsymbol{\pi}^2\right) - \frac{1}{\sqrt{g}}\mathbf{J}(x) \cdot \boldsymbol{\pi}(x). \quad (14.15)$$

Expression (14.15) shows that the interaction term in the action, once expanded in powers of g, generates an infinite number of different vertices with arbitrary even powers of $\boldsymbol{\pi}$ and two derivatives. Still it is easy to verify that at any finite order in perturbation theory and for a given correlation function, only a finite number of vertices contribute. Formally the measure term yields additional vertices without derivatives.

The propagator $\Delta_{ij}(p)$ of the $\boldsymbol{\pi}$-field is:

$$\Delta_{ij}(p) = \frac{\delta_{ij}}{p^2}. \quad (14.16)$$

At the tree level the $\boldsymbol{\pi}$-field is massless. Returning to the analysis of Section 13.4, we understand that, at leading order in perturbation theory, the non-linear σ-model automatically realizes the $O(N)$ symmetry in the phase of spontaneous symmetry breaking, the $\boldsymbol{\pi}$-field corresponding to the Goldstone modes. The massive partner of the $\boldsymbol{\pi}$-field in the linear realization, the σ component, has been eliminated by the constraint (14.1). This constraint can be formally obtained by sending the σ mass to infinity in order to freeze the fluctuations of the σ-field.

Let us note that these properties are independent of the special choice (14.4) of parametrization of $\boldsymbol{\phi}(x)$.

14.2.2 Power counting

The general analysis has been given in Chapter 8. The form of the propagator shows that the dimension $[\pi]$ of the π-field is:

$$[\pi] = \tfrac{1}{2}(d-2).\tag{14.17}$$

Therefore the dimension of a vertex containing $2n$ π-fields is:

$$\left[\partial^2\pi^{2n}\right] = n(d-2)+2.\tag{14.18}$$

As a consequence:

(i) for $d < 2$ the theory is super-renormalizable by power counting;
(ii) for $d = 2$ it is just renormalizable;
(iii) for $d > 2$ the theory is not renormalizable.

We therefore study the model in dimension two. We have already mentioned a peculiarity of this case: although the theory is renormalizable by power counting, any local monomial in the field containing at most two derivatives and an arbitrary power of π can *a priori* appear as a counterterm. The symmetry $O(N-1)$, which is linearly realized, only restricts the counterterms to be of the general form:

$$\left(\partial_\mu\pi\cdot\pi\right)^2\left(\pi^2\right)^n,\qquad \left(\partial_\mu\pi\right)^2\left(\pi^2\right)^n,\qquad \left(\pi^2\right)^n.$$

However, at the tree level, the non-linear $O(N)$ symmetry implies that, up to a normalization constant, the action is unique. To understand the structure of the theory after renormalization, we have to investigate the implications of the non-linear $O(N)$ symmetry on the form of the divergences in perturbation theory. We have first to exhibit a regularization scheme which preserves the $O(N)$ symmetry and then derive a set of WT identities which express the consequences of the symmetry for correlation functions.

14.3 Regularization

In constructing a regularized version of the non-linear σ-model, one has to be careful to preserve the $O(N)$ symmetry. This is less trivial than in the linear case since, as a consequence of the symmetry, the interaction terms in the action (14.9) are related to the quadratic part. A simple method is to start from the description of the model in terms of the ϕ-field because the action (14.8) formally is a free field action.

14.3.1 Perturbative regularizations

Pauli–Villars's Regularization. We replace the action $S(\phi)$ by $S_\Lambda(\phi)$:

$$S_\Lambda(\phi) = \tfrac{1}{2}\int \mathrm{d}^d x\,\phi(x)\cdot\left(-\Delta + \frac{\alpha_2}{\Lambda^2}\Delta^2 - \frac{\alpha_3}{\Lambda^4}\Delta^3 + \cdots\right)\phi(x).\tag{14.19}$$

Expressing then $\phi(x)$ in terms of $\pi(x)$, we discover that the large momentum behaviour of the propagator has improved, but at the same time new more singular interactions have been generated. If the propagator behaves like:

$$\Delta(p) \propto 1/p^s,$$

then the most singular interaction has s derivatives. Using equation (8.23) which gives the superficial degree of divergence of a diagram γ:

$$\delta\left(\gamma\right) = 2L - sI + \sum_\alpha k_\alpha v_\alpha \,,$$

and eliminating the number of internal lines I through the topological relation (8.25):

$$L = I - \sum_\alpha v_\alpha + 1 \,,$$

we find:

$$\delta\left(\gamma\right) = (2-s)\,L + s + \sum_\alpha \left(k_\alpha - s\right)v_\alpha \,, \tag{14.20}$$

in which we recall that L is the number of loops and k_α the number of derivatives at vertex α.

As stated above, the worst case is $k_\alpha = s$:

$$\delta\left(\gamma\right) \le (2-s)\,L + s \,.$$

As a consequence for $L \ge 2$ it is sufficient to take $s \ge 6$ to regularize all diagrams. However the one-loop diagrams have a behaviour independent of s, and thus cannot be regularized. This property is not independent of the other limitation of Pauli–Villars's regularization, that it cannot regularize the divergent measure term. Indeed we show later that the one-loop divergences generated by the interaction are needed to cancel the divergences coming from the measure.

Pauli–Villars's regularization is mainly useful in the study of non-linear models coupled to chiral fermions.

Dimensional regularization. Dimensional regularization preserves the $O(N)$ symmetry of the action. Furthermore, as a consequence of the formal rule:

$$\int \mathrm{d}^d k = (2\pi)^d \delta^d(0) = 0 \,,$$

the measure term can be ignored, and therefore perturbation theory has no large momentum divergences for $d < 2$. Owing to his technical simplicity, this is the regularization we shall in general use for practical calculations. One theoretical drawback is that the role of the measure is hidden. Therefore for the theoretical discussion of the renormalization of the non-linear σ-model we shall consider both dimensional and lattice regularizations.

14.3.2 Lattice regularization and statistical mechanics

To construct a lattice regularized version of the non-linear σ-model, which preserves the $O(N)$ symmetry, it is again convenient to start from the description of the model in terms of the ϕ-field. We then replace derivatives by finite differences, i.e. in the notation of Section 9.2:

$$\partial_\mu \phi(x) \longmapsto \nabla_\mu \phi(x) = \frac{1}{a}\left[\phi\left(x + a n_\mu\right) - \phi(x)\right] \,,$$

in which x now belongs to a hypercubic lattice of lattice spacing a, and n_μ is the unit vector in the μ direction.

Finally, to implement condition (14.1), we integrate over $\phi(x)$ with the invariant measure on the sphere. The regularized functional integral has the form:

$$Z(\mathbf{J}) = \int \prod_{x \in a\mathbb{Z}^d} \delta\left(\phi^2(x) - 1\right) d\phi(x) \exp\left[-\frac{1}{g}S(\phi, \mathbf{J})\right], \qquad (14.21)$$

with:

$$S(\phi, \mathbf{J}) = \frac{1}{2} \sum_{x,\mu} [\nabla_\mu \phi(x)]^2 - \sum_x \mathbf{J}(x) \cdot \phi(x). \qquad (14.22)$$

Using the parametrization (14.4), we can express the lattice field $\phi(x)$ in terms of $\pi(x)$. We obtain a regularized form of functional integral (14.11). In particular the measure term now generates well-defined interactions:

$$\frac{1}{2}\delta^d(0) \int d^d x \ln\left(1 - \pi^2(x)\right) \longmapsto \frac{1}{2}\sum_x \ln\left(1 - \pi^2(x)\right).$$

In expression (14.21) we recognize the partition function of a classical spin lattice model with a nearest-neighbour ferromagnetic interaction and in presence of an external magnetic field $\mathbf{J}(x)$. The coupling constant g plays the role of the temperature. The critical properties (in the sense of phase transitions) of this model will be discussed in Chapters 23–32In particular in Chapter 30 we will use the perturbative expansion of the field theory model to study ferromagnetic order at low temperature. Expression (14.21) not only provides a regularization of perturbation theory but also allows the use of various non-perturbative methods to study the non-linear σ-model. Moreover it is the only regularization which allows a discussion of the role of the measure in perturbation theory (see Sections 14.5, 14.6).

14.4 Infrared (IR) Divergences

Since in a massless theory the propagator behaves like $1/p^2$, perturbation theory is divergent at low momentum (IR) in dimension two, the dimension in which the non-linear σ-model is renormalizable.

To generate a well-defined perturbation theory, it is necessary to introduce an IR cutoff. Since the absence of mass is a consequence of the spontaneous breaking of the $O(N)$ symmetry, it is necessary, to give a mass to the π-field, to break the symmetry explicitly. We can for example introduce an explicit mass term. However the study of symmetry breaking mechanisms in Chapter 13 suggests a more convenient method which consists in adding to the action (14.9) a constant source h for the σ-field (a magnetic field in the statistical model of interacting spins)

$$S(\pi) \longmapsto S(\pi) - h \int \sigma(x) d^d x, \qquad h > 0. \qquad (14.23)$$

A first consequence of this modification is that the minimum of the action is no longer degenerate. Instead we now have to maximize the source term and this implies $\pi = 0$ at the minimum.

Second, if we expand σ in powers of π:

$$\sigma = \left(1 - \pi^2\right)^{1/2} = 1 - \tfrac{1}{2}\pi^2 + O\left(\left(\pi^2\right)^2\right), \qquad (14.24)$$

and collect the quadratic terms in the action, we find the new π-field propagator $\Delta_{ij}(p)$

$$\Delta_{ij}(p) = \delta_{ij}\frac{g}{p^2 + h}. \tag{14.25}$$

The linear σ term has thus generated a mass $h^{1/2}$ for the π-field. At the same time it has generated new derivative-free interactions.

Let us recall that in the case of the linearly realized symmetries, the breaking term $h\sigma$ is linear in an independent field and therefore, as we have shown in Chapter 13, generates no new renormalization constant.

Finite volume. We shall discuss in Chapter 36 another IR regularization scheme, which does not break the $O(N)$ symmetry. It is based on the property that a symmetry cannot be spontaneously broken in a finite volume and therefore no Goldstone mode is generated. Technically, if we put the system into a hypercube of linear size L, the momenta after Fourier transformation are quantized. This solves the IR problem because integrals are replaced by discrete sums. In the special case of periodic boundary conditions the momenta have the form:

$$\mathbf{p} = \frac{2\pi}{L}\mathbf{n}, \quad \mathbf{n} \in \mathbb{Z}^d.$$

In this case some care is needed to handle properly the zero momentum mode $\mathbf{p} = 0$ which seems to still lead to divergences. However it has to be eliminated in favour of an integral over a constant unit vector which represents the sum over all degenerate minima. The other momenta do not lead to IR divergences and can be treated perturbatively.

IR Finiteness of $O(N)$ Invariant Correlation Functions in 2 Dimensions. Let us here mention briefly without proof an interesting result whose significance will be discussed later with the physics (in the sense of statistical mechanics) of the non-linear σ-model. We have explained that in 2 dimensions correlation functions are IR divergent and we have introduced an IR cut-off in the form of a source term giving a mass term for the π-field. At the same time, this additional term, breaking the $O(N)$ symmetry, lifts the degeneracy of the classical minimum of the action and therefore eliminates a potential difficulty with perturbation theory: is it necessary to take into account all degenerate minima of the action or can one choose one of them? However as we noted in Section 14.2 this question is irrelevant for $O(N)$ invariant correlation functions which therefore play a special role. Actually it has been conjectured by Elitzur and proven by David order by order in perturbation theory, that $O(N)$ invariant correlation functions have in 2 dimensions a finite IR limit, i.e. for example a limit when in the notation (14.23) the breaking parameter h goes to zero, or any IR cut-off is removed (see also Chapter 30).

14.5 WT Identities

Having regularized the theory both at short and long distances (large and small momenta) we can now derive a set of WT identities expressing the consequence of the $O(N)$ symmetry for correlation functions. We discuss only the part of the $O(N)$ symmetry which acts non-linearly, the consequences of the linear $O(N-1)$ symmetry have been discussed in Chapter 13 and now are obvious. We thus consider infinitesimal transformations of the form:

$$\delta\pi_i(x) = \omega_i\sigma(x) \equiv \omega_i\sqrt{1 - \boldsymbol{\pi}^2(x)}. \tag{14.26}$$

The $O(N)$ symmetric part of the action and the measure are left invariant by such a transformation. Only the source term and the breaking term are affected:

$$\delta \left[\int \mathrm{d}^d x\, \mathbf{J}(x) \cdot \boldsymbol{\pi}(x) \right] = \int \mathrm{d}^d x\, \boldsymbol{\omega} \cdot \mathbf{J}(x)\sigma(x), \qquad \delta \left[h \int \mathrm{d}^d x\, \sigma(x) \right] = -h \int \mathrm{d}^d x\, \boldsymbol{\omega} \cdot \boldsymbol{\pi}(x).$$

A new composite operator $\sigma(x) = \left(1 - \boldsymbol{\pi}^2(x) \right)^{1/2}$ is generated by the variation of the source term. Following the general strategy explained already in the case of quadratic symmetry breaking terms, we introduce a source for this operator. Then since the variation under a transformation (14.26) of this new operator as well as the variation of the breaking term are both proportional to $\boldsymbol{\pi}$ itself, no additional operator is needed. Let us note that this would not have been the case with a breaking term proportional for example to $\int \boldsymbol{\pi}^2(x)\mathrm{d}^d x$. Let us call $H(x)$ the source for σ, and $Z(\mathbf{J}, H)$ the generating functional:

$$Z(\mathbf{J}, H) = \int \left[\frac{\mathrm{d}\boldsymbol{\pi}(x)}{\sqrt{1 - \boldsymbol{\pi}^2(x)}} \right] \exp \left\{ \frac{1}{g} \left[-S(\boldsymbol{\pi}, H) + \int \mathrm{d}^d x\, \mathbf{J}(x) \cdot \boldsymbol{\pi}(x) \right] \right\}, \qquad (14.27)$$

with:

$$S(\boldsymbol{\pi}, H) = S(\boldsymbol{\pi}) - \int \mathrm{d}^d x\, H(x)\sigma(x). \qquad (14.28)$$

The generating functional of $\boldsymbol{\pi}$-field correlation functions with the linear $h\sigma$ breaking term is given by:

$$Z(\mathbf{J}) = Z(\mathbf{J}, H)|_{H(x)=h} .$$

Performing the infinitesimal transformation (14.26), we obtain the equation:

$$0 = \int \left[\frac{\mathrm{d}\boldsymbol{\pi}}{\sqrt{1 - \boldsymbol{\pi}^2}} \right] \int \mathrm{d}^d x\, [\mathbf{J} \cdot \delta\boldsymbol{\pi} + H(x)\delta\sigma] \exp \left[\frac{1}{g} \left(-S(\boldsymbol{\pi}, H) + \int \mathrm{d}^d x\, \mathbf{J}(x) \cdot \boldsymbol{\pi}(x) \right) \right],$$

or explicitly:

$$0 = \int \left[\frac{\mathrm{d}\boldsymbol{\pi}}{\sqrt{1 - \boldsymbol{\pi}^2}} \right] \int \mathrm{d}^d x\, [\sigma(x)\mathbf{J}(x) - H(x)\boldsymbol{\pi}(x)] \exp \left[\frac{1}{g} \left(-S(\boldsymbol{\pi}, H) + \int \mathrm{d}^d x\, \mathbf{J} \cdot \boldsymbol{\pi} \right) \right].$$
$$(14.29)$$

We now replace $\boldsymbol{\pi}(x)$ by $g\left(\delta/\delta\mathbf{J}(x)\right)$ and $\sigma(x)$ by $g\left(\delta/\delta H(x)\right)$. Equation (14.29) then becomes:

$$\int \mathrm{d}^d x \left[\mathbf{J}(x)\frac{\delta}{\delta H(x)} - H(x)\frac{\delta}{\delta \mathbf{J}(x)} \right] Z(\mathbf{J}, H) = 0. \qquad (14.30)$$

This first order linear differential equation makes no reference to the non-linear character of the transformation (14.26). It is identical to the equation one obtains in the case of a linearly realized $O(N)$ symmetry, when $\mathbf{J}(x)$ and $H(x)$ are the sources for the independent fields $\boldsymbol{\pi}(x)$ and $\sigma(x)$.

It is clear that $W(\mathbf{J}, H) = g \ln Z(\mathbf{J}, H)$ satisfies the same equation:

$$\int \mathrm{d}^d x \left[\mathbf{J}(x)\frac{\delta}{\delta H(x)} - H(x)\frac{\delta}{\delta \mathbf{J}(x)} \right] W(\mathbf{J}, H) = 0. \qquad (14.31)$$

We now perform a Legendre transformation. In contrast with the linear case, however, $\sigma(x)$ is here a function of the $\boldsymbol{\pi}(x)$ field, and therefore the Legendre transformation applies only to the source $\mathbf{J}(x)$ and not to $H(x)$:

$$W(\mathbf{J}, H) + \Gamma(\boldsymbol{\pi}, H) = \int \mathrm{d}^d x\, \mathbf{J}(x) \cdot \boldsymbol{\pi}(x), \qquad \boldsymbol{\pi}(x) = \frac{\delta W}{\delta \mathbf{J}(x)}. \tag{14.32}$$

We again use identity (6.20) since $H(x)$ is a set of external parameters:

$$\left. \frac{\delta W}{\delta H(x)} \right|_{\mathbf{J}} = - \left. \frac{\delta \Gamma}{\delta H(x)} \right|_{\boldsymbol{\pi}}. \tag{14.33}$$

The Legendre transform of equation (14.31) is therefore:

$$\int \mathrm{d}^d x \left[\frac{\delta \Gamma}{\delta \boldsymbol{\pi}(x)} \frac{\delta \Gamma}{\delta H(x)} + H(x) \boldsymbol{\pi}(x) \right] = 0. \tag{14.34}$$

This is the basic equation from which we shall derive the properties of the renormalized theory.

Remark. Equation (14.34) is quadratic in Γ. This is an essential difference with the case of linearly realized symmetries. Gauge theories will provide another example sharing this property. Actually it is easy to show that if one uses the strategy explained above, i.e. adds sources for all new composite operators generated in the group transformation, then the WT identities derived for the 1PI functional are at most quadratic in Γ.

The reciprocal property. It is easy to verify that the initial action satisfies equation (14.34), either directly, or by performing a loopwise expansion of equation (14.34) and remembering that:

$$\Gamma(\boldsymbol{\pi}, H) = S(\boldsymbol{\pi}, H) + O(g).$$

Let us conversely assume that the action S obeys the equation

$$\int \mathrm{d}^d x \left[\frac{\delta S}{\delta H(x)} \frac{\delta S}{\delta \boldsymbol{\pi}(x)} + H(x) \boldsymbol{\pi}(x) \right] = 0. \tag{14.35}$$

We then perform an infinitesimal change of variables in the functional integral (14.27) of the form:

$$\boldsymbol{\pi} = \boldsymbol{\pi}' + \delta \boldsymbol{\pi} \quad \text{with} \quad \delta \boldsymbol{\pi} = \frac{\delta S(\boldsymbol{\pi}', H)}{\delta H(x)} \boldsymbol{\omega}. \tag{14.36}$$

In the dimensional regularization scheme we can omit the measure term and the jacobian because S is local (for other regularizations see the remark below).

The variations of the action and the source term are (we omit the primes on the dummy variable $\boldsymbol{\pi}$):

$$\delta \left\{ \exp \left[\frac{1}{g} \left(-S(\boldsymbol{\pi}, H) + \int \mathbf{J}(x) \cdot \boldsymbol{\pi}(x) \mathrm{d}^d x \right) \right] \right\}$$
$$= \frac{\boldsymbol{\omega}}{g} \cdot \left[H(x) \frac{\delta}{\delta \mathbf{J}(x)} - \mathbf{J}(x) \frac{\delta}{\delta H(x)} \right] \exp \left[\frac{1}{g} \left(-S(\boldsymbol{\pi}, H) + \int \mathbf{J}(x) \cdot \boldsymbol{\pi}(x) \mathrm{d}^d x \right) \right]. \tag{14.37}$$

Therefore equation (14.35) by itself implies equation (14.30) for $Z(\mathbf{J}, H)$ and thus equation (14.34) for $\Gamma(\boldsymbol{\pi}, H)$.

The measure. The argument given above is valid only for dimensional regularization where the measure term vanishes identically. We now extend it to the case of lattice regularization. Indeed let us show that the invariant measure for the transformations (14.36) is

$$\prod_x \mathrm{d}\boldsymbol{\pi}(x) \left[\frac{\delta S}{\delta H(x)} \right]^{-1} . \tag{14.38}$$

First the change of variables (14.36) generates a jacobian J

$$J = 1 + \omega_i \int \mathrm{d}x \, \frac{\delta^2 S}{\delta \pi_i(x) \delta H(x)} .$$

The variation of the measure term then is

$$\left[\frac{\delta S}{\delta H(x)} \right]^{-1} \mapsto \left[\frac{\delta S}{\delta H(x)} \right]^{-1} \left\{ 1 - \left[\frac{\delta S}{\delta H(x)} \right]^{-1} \frac{\delta^2 S}{\delta \pi_i(x) \delta H(x)} \delta \pi_i(x) \right\} ,$$

which, using the explicit form (14.36), exactly cancels the jacobian. Note finally the initial measure $[\mathrm{d}\boldsymbol{\pi}(1 - \boldsymbol{\pi}^2)^{-1/2}]$ has the form (14.38).

14.6 Renormalization

Before explaining the technical details, let us describe the various steps of the proof of the renormalizability of the model.

First the stability of equation (14.35) under renormalization will be established: this means that if the action at the tree level satisfies equation (14.35), then it is possible to renormalize the theory in such a way that the renormalized action still satisfies equation (14.35). Then equation (14.35) will be solved. It is important to realize that the equation does not explicitly refer to the transformation law (14.26). This explains why the explicit form of the transformation law is modified by the renormalization although the geometrical structure does not change. Indeed solving equation (14.35) with the constraints coming from power counting, one finds that only two renormalization constants are needed, a coupling constant and a field renormalization. After renormalization the model is still $O(N)$ invariant for $h = 0$ but the field ϕ now belongs to a sphere of renormalized radius:

$$\phi^2(x) = \pi^2(x) + \sigma^2(x) = 1/Z .$$

Renormalization. We assume that the theory has been regularized. We make a loop-wise expansion, i.e. as the explicit form of the action shows, an expansion in powers of g of correlation functions:

$$\Gamma(\boldsymbol{\pi}, H) = \sum_{n=0}^{\infty} \Gamma_n g^n.$$

We then insert this expansion into equation (14.34). The functional Γ_0 is simply the original action and satisfies by itself equation (14.34). The one-loop functional Γ_1 satisfies:

$$\int \mathrm{d}^d x \left[\frac{\delta \Gamma_0}{\delta \boldsymbol{\pi}(x)} \frac{\delta \Gamma_1}{\delta H(x)} + \frac{\delta \Gamma_0}{\delta H(x)} \frac{\delta \Gamma_1}{\delta \boldsymbol{\pi}(x)} \right] = 0 . \tag{14.39}$$

This is a linear partial differential equation for Γ_1 which can be written symbolically:

$$\Delta_i \Gamma_1 (\boldsymbol{\pi}, H) = 0 \,, \tag{14.40}$$

with:

$$\Delta_i = \int d^d x \left[\frac{\delta \Gamma_0}{\delta \pi_i(x)} \frac{\delta}{\delta H(x)} + \frac{\delta \Gamma_0}{\delta H(x)} \frac{\delta}{\delta \pi_i(x)} \right] . \tag{14.41}$$

Using the property that Γ_0 satisfies the equation (14.34), it is easy to verify that Δ_i is a generator of $O(N)/O(N-1)$ acting on functionals of $\boldsymbol{\pi}$ and H. In particular the commutators $[\Delta_i, \Delta_j]$ are generators of the subgroup $O(N-1)$. Applied to an $O(N-1)$-invariant functional they vanish, which shows that the system (14.39) is integrable (see also Section 13.1.2).

We now examine the large cut-off behaviour (or the behaviour when d approaches 2). Equation (14.39) is satisfied for all values of the regularizing parameter. We conclude that the divergent part Γ_1^{div} of Γ, defined in any minimal subtraction scheme, also satisfies equation (14.39):

$$\Delta_i \Gamma_1^{\text{div}} = 0 \,.$$

By adding to the tree level action $S(\boldsymbol{\pi}, H)$, $-g\Gamma_1^{\text{div}}(\boldsymbol{\pi}, H)$ we render the theory finite at one-loop order. Actually it is necessary to also add higher order terms to construct the one-loop renormalized action $S_1(\boldsymbol{\pi}, H)$:

$$S_1(\boldsymbol{\pi}, H) = S(\boldsymbol{\pi}, H) - g\Gamma_1^{\text{div}}(\boldsymbol{\pi}, H) + \sum_{2}^{\infty} g^n \delta S_1^{(N)}(\boldsymbol{\pi}, H) \,.$$

These terms do not contribute to the one-loop order which is now finite, and are chosen in such a way that $S_1(\boldsymbol{\pi}, H)$ satisfies the non-linear equation (14.35) exactly. Indeed at order 0, equation (14.35) is verified since $S(\boldsymbol{\pi}, H)$ satisfies it. At order 1 equation (14.35) implies equation (14.39) for $\Gamma_1^{\text{div}}(\boldsymbol{\pi}, H)$ which is also satisfied. Higher order equations determine the higher order terms $\delta S_1^{(n)}$.

Let us now show that this argument can be generalized to all orders. We write equation (14.34) symbolically:

$$\Gamma * \Gamma = K \,. \tag{14.42}$$

Performing a loopwise expansion of Γ, we write the term of order n $(n > 0)$ in equation (14.42):

$$\sum_{p=0}^{n} \Gamma_p * \Gamma_{n-p} = 0 \,. \tag{14.43}$$

Our induction hypothesis is that it has been possible to construct an action $S_{n-1}(\boldsymbol{\pi}, H)$ which satisfies equation (14.35) exactly, and such that $\Gamma_1, \dots, \Gamma_{n-1}$ have been rendered finite. Then as we have shown above the generating functional $\Gamma(\boldsymbol{\pi}, H)$ renormalized up to order $(n-1)$ also satisfies equation (14.34), which implies equation (14.43). Let us rewrite this equation in the form:

$$\Gamma_0 * \Gamma_n + \Gamma_n * \Gamma_0 = -\sum_{p=1}^{n-1} \Gamma_p * \Gamma_{n-p} \,. \tag{14.44}$$

The induction hypothesis implies that the r.h.s. is finite. The divergent part of the equation satisfies therefore

$$\Gamma_0 * \Gamma_n^{\mathrm{div}} + \Gamma_n^{\mathrm{div}} * \Gamma_0 = 0 \,. \tag{14.45}$$

The form of the equation is independent of n. We then define S_n, the renormalized action at order n, by

$$S_n = S_{n-1} - g^n \Gamma_n^{\mathrm{div}} + \sum_{n+1}^{\infty} g^p \delta S_n^{(p)} \,. \tag{14.46}$$

It follows

$$
\begin{aligned}
S_n * S_n - K &= \left(S_{n-1} - g^n \Gamma_n^{\mathrm{div}}\right) * \left(S_{n-1} - g^n \Gamma_n^{\mathrm{div}}\right) - K + O\left(g^{n+1}\right) \\
&= -g^n \left(S_{n-1} * \Gamma_n^{\mathrm{div}} + \Gamma_n^{\mathrm{div}} * S_{n-1}\right) + O\left(g^{n+1}\right) \,.
\end{aligned} \tag{14.47}
$$

Since at this order in the r.h.s., S_{n-1} can be replaced by Γ_0, i.e. $S(\boldsymbol{\pi}, H)$, equation (14.45) then implies:

$$S_n * S_n - K = O\left(g^{n+1}\right) \,. \tag{14.48}$$

Hence S_n satisfies equation (14.35) at order n. As in the case $n = 1$ we then choose the higher order terms $\delta S_n^{(p)}$ in such a way that S_n satisfies equation (14.35) identically.

This completes the derivation. The renormalized 1PI functional satisfies equation (14.34) while the complete renormalized action satisfies equation (14.35).

To determine the form of the renormalized action, we now have to solve equation (14.35) taking into account locality and power counting.

The measure. In this discussion we have never mentioned the role of the measure. In the next section we shall verify that the field transformation has been renormalized, and thus the measure must have changed accordingly; our discussion was really only valid for dimensional regularization where the measure terms vanish identically. We now extend the derivation to the case of lattice regularization.

We have shown in Section 14.5 that the invariant measure for the transformations (14.36) is

$$\prod_x \mathrm{d}\boldsymbol{\pi}(x) \left[\frac{\delta S}{\delta H(x)}\right]^{-1} \,.$$

We note that the vertices generated by the measure term are not multiplied by a factor $1/g$ in contrast with those coming from the classical action. As a consequence they always contribute at the next order in comparison to the vertices coming from the action. For instance at one-loop order they contribute under the form of their tree approximation. Therefore if, once Γ_n is rendered finite, we modify the measure term by a divergent term of order g^n, to take into account the nth order renormalization of the field, this will affect $\Gamma_{n+1}, \Gamma_{n+2}, \dots$, which are not yet renormalized, and leave Γ_n unchanged. It is therefore possible to introduce the field renormalization the measure without changing the arguments given above about renormalization.

Actually the measure term cancels at next order the potential quadratic divergences which could appear according to power counting. It prevents at the same time the appearance of an induced mass term.

14.7 Solution of the WT Identities

14.7.1 Global solution

We have to solve equation (14.35):

$$\int \mathrm{d}^d x \left[\frac{\delta S}{\delta \pi(x)} \frac{\delta S}{\delta H(x)} + H(x)\pi(x) \right] = 0.$$

Power counting tells us that the dimension $[\pi]$ of π and $[H]$ of H are in two dimensions:

$$[\pi] = 0, \quad [H] = 2.$$

The action density has dimension 2, the action has therefore the general form:

$$S(\pi, H) = A(\pi) - \int \mathrm{d}^d x\, B(\pi(x)) H(x), \qquad (14.49)$$

a term quadratic in H having at least dimension 4. Moreover the coefficient $B(\pi)$ is dimensionless and thus a derivative-free function of $\pi(x)$ while $A(\pi)$ has dimension 2 and contains terms with at most two derivatives.

Let us write the coefficient of $H(x)$ in equation (14.35):

$$B(\pi)\frac{\delta B}{\delta \pi(x)} + \pi(x) = 0. \qquad (14.50)$$

Since $B(\pi)$ is derivative free and $O(N-1)$-invariant, the solution of equation (14.50) is simply:

$$B^2(\pi^2) + \pi^2(x) = Z^{-1}. \qquad (14.51)$$

This shows that $B(\pi)$ is the renormalized σ-field and Z is a first renormalization constant. Let us now write the part of the equation which is independent of $H(x)$:

$$\int \mathrm{d}^d x\, \frac{\delta A}{\delta \pi(x)} B(\pi(x)) = 0. \qquad (14.52)$$

This equation tells us that $A(\pi)$ is invariant under an infinitesimal transformation of the form:

$$\delta \pi(x) = \omega B(\pi),$$

or solving equation (14.51):

$$\delta \pi(x) = \omega \left(Z^{-1} - \pi^2(x) \right)^{1/2}. \qquad (14.53)$$

This is the renormalized form of the non-linear part of the $O(N)$ transformations. Equations (14.51) and (14.52) show that the renormalized functional $A(\pi)$ is $O(N)$-invariant, but the radius of the sphere has been renormalized. The renormalized action can therefore be written:

$$S(\pi, H) = \frac{1}{2Z_g} \int \left[(\partial_\mu \pi)^2 + (\partial_\mu \sigma)^2 \right] \mathrm{d}^d x - \int h\, \sigma(x)\, \mathrm{d}^d x, \qquad (14.54)$$

with the relation

$$\sigma(x) = \left[Z^{-1} - \boldsymbol{\pi}^2\left(x\right)\right]^{1/2}. \tag{14.55}$$

We have proven a rather remarkable result: although the interaction in the non-linear σ-model is non-polynomial, the theory can be renormalized with only two renormalization constants. Let us also stress the striking similarity with the linear $O(N)$-invariant $\left(\phi^2\right)^2$ field theory, the only differences come from the absence of the renormalization of $\left(\phi^2\right)^2$ coupling which of course has no equivalent in the non-linear theory, and the structure of the mass renormalization. In particular no spontaneous mass is generated. Finally let us note that by giving a mass to the $\boldsymbol{\pi}$-field through a term of the form $\int \sigma(x) \mathrm{d}^d x$, we have introduced no new renormalization constant. This would not have been the case for a term of the form $\int \boldsymbol{\pi}^2(x)\mathrm{d}^d x$ for example.

14.7.2 Linearized WT identities

In our discussion of the renormalization of the model we have admitted without proof that it is always possible to solve in a local way the equations determining the additional higher order terms $\delta S_n^{(p)}$ which make S_n exactly $O(N)$ invariant. This is a point which can here be easily justified.

Explicit solution of the linearized WT identities. To solve explicitly the equations (14.39) or (14.45) satisfied by the divergent part of Γ, we first discuss the general solution of an equation of the form:

$$\int \mathrm{d}^d x \left[\frac{\delta S}{\delta \pi_i\left(x\right)}\frac{\delta}{\delta H(x)} + \frac{\delta S}{\delta H(x)}\frac{\delta}{\delta \pi_i(x)}\right] \mathcal{O}\left(\boldsymbol{\pi}, H\right) = 0, \tag{14.56}$$

in which $\mathcal{O}\left(\boldsymbol{\pi}, H\right)$ is an arbitrary $O(N-1)$ symmetric local operator and S is the bare action:

$$S = \int \mathrm{d}^d x \left\{\tfrac{1}{2}\left[(\partial_\mu \boldsymbol{\pi})^2 + (\partial_\mu \sigma)^2\right] - H(x)\sigma(x)\right\}, \tag{14.57}$$

with $\sigma(x) = \left(1 - \boldsymbol{\pi}^2(x)\right)^{1/2}$. Equation (14.56) reads explicitly:

$$\int \mathrm{d}^d x \left\{\left[-\Delta \pi_i(x) + \alpha(x)\pi_i(x)\right]\frac{\delta}{\delta H(x)} - \sigma\left(x\right)\frac{\delta}{\delta \pi_i(x)}\right\}\mathcal{O}\left(\boldsymbol{\pi}, H\right) = 0, \tag{14.58}$$

in which we have defined:

$$\alpha(x) = \frac{1}{\sigma(x)}\left[H(x) + \Delta\sigma(x)\right]. \tag{14.59}$$

The operator $\alpha(x)$ is an affine function of $H(x)$. Let us choose in equation (14.58) a new set of variables and consider $\mathcal{O}\left(\boldsymbol{\pi}, H\right)$ as a functional $\tilde{\mathcal{O}}\left(\boldsymbol{\pi}, \alpha\right)$. A straightforward but careful calculation then leads to a new equation:

$$\int \mathrm{d}^d x\, \sigma(x)\frac{\delta \tilde{\mathcal{O}}\left(\boldsymbol{\pi}, \alpha\right)}{\delta \pi_i(x)} = 0, \tag{14.60}$$

which shows that the operator $\tilde{\mathcal{O}}\left(\boldsymbol{\pi}, \alpha\right)$ is $O(N)$ symmetric at $\alpha(x)$ fixed.

This result has two applications: it completes our proof and, as we discuss below, it yields the renormalized form of a general $O(N)$ invariant operator.

From power counting we know that Γ_n^{div} has dimension 2. It has therefore terms of degree 0 and 1 in α. According to the previous result it has the form:

$$\Gamma_n^{\mathrm{div}} = \tfrac{1}{2}a_n \int \left\{ [\partial_\mu \boldsymbol{\pi}(x)]^2 + [\partial_\mu \sigma(x)]^2 \right\} \mathrm{d}^d x + \tfrac{1}{2}b_n \int \alpha(x)\, \mathrm{d}^d x\,. \tag{14.61}$$

The first term can be absorbed into a coupling constant renormalization:

$$g \mapsto g\left(1 + g^n a_n\right). \tag{14.62}$$

We now calculate the variation of the action when the radius of the sphere is renormalized. The variation $\delta\sigma$ of $\sigma(x)$ is:

$$\delta\sigma(x) = \left[1 - \delta Z - \boldsymbol{\pi}^2(x)\right]^{1/2} - \left[1 - \boldsymbol{\pi}^2(x)\right]^{1/2} = -\tfrac{1}{2}\delta Z/\sigma(x).$$

It follows:

$$\delta \left\{ \int \mathrm{d}^d x \left[\tfrac{1}{2} [\partial_\mu \sigma(x)]^2 - H(x)\sigma(x) \right] \right\} = \tfrac{1}{2}\delta Z \int \alpha(x)\, \mathrm{d}^d x\,. \tag{14.63}$$

Therefore the second term can be absorbed in a field renormalization:

$$\delta Z = b_n g^n. \tag{14.64}$$

This completes altogether our proof of the renormalization of the non-linear σ-model. Nevertheless for completeness we shall outline in Section 14.9 a different derivation, whose basic idea is to return to a linear formulation of the symmetry.

14.8 Renormalization of Composite Operators

General $O(N)$ invariant operators. Let us show that by solving equation (14.56) in its most general form, we have also obtained the structure of renormalized $O(N)$ invariant operators of arbitrary dimension, inserted once in $\phi(x)$ correlation functions. Indeed to generate such insertions we can add to the action a source term of the form $\int \mathrm{d}^d x\, K(x)\mathcal{O}\left(\boldsymbol{\pi}(x)\right)$. Since $\mathcal{O}\left(\boldsymbol{\pi}\right)$ is $O(N)$ invariant, equation (14.35) holds for the complete action which includes this new term. An expansion of equation (14.35) at first order in $K(x)$ leads to equation (14.56) with $S\left(\boldsymbol{\pi}, H\right)$ replaced by the renormalized action. The renormalized operator $\mathcal{O}\left(\boldsymbol{\pi}, H\right)$ is thus the most general local functional of the renormalized fields $\boldsymbol{\pi}(x)$ and $\sigma(x)$, of dimension $[\mathcal{O}]$ and $O(N)$ invariant at $\alpha(x)$ fixed.

Renormalization of dimensionless operators and parametrization of the sphere. Dimensionless operators are derivative-free local functions of the $\boldsymbol{\pi}(x)$ field. A simple extension of previous arguments shows that they should be classified according to irreducible representations of the $O(N)$ group. To each different irreducible representation is associated a new renormalization constant. For example a mass term $\int \mathrm{d}^d x\, \boldsymbol{\pi}^2(x)$ is a component of the symmetric traceless tensor

$$\int \mathrm{d}^d x \left[\phi_i(x)\phi_j(x) - \tfrac{1}{N}\delta_{ij}\boldsymbol{\phi}^2(x) \right],$$

and introduces an additional renormalization constant. It now becomes clear why we have chosen the particular parametrization of the sphere in terms of the $\boldsymbol{\pi}$ field. Let us choose another parametrization (respecting the $O(N-1)$ symmetry) in terms of fields $\theta(x)$ related to $\pi(x)$ by:

$$\theta(x) = \pi(x) f\left(\pi^2\right).$$

We can rewrite this expansion in terms of the $(i, 0 \ldots 0)$ components of the tensors transforming under irreducible representations of $O(N)$. These components have the form $\pi(x) P_\ell^N(\sigma(x))$, in which P_ℓ^N are hyperspherical polynomials:

$$\theta(x) = \pi(x) \sum c_\ell P_\ell^N\big(\sigma(x)\big).$$

Each non-vanishing coefficient c_ℓ introduces an independent renormalization constant. In particular in the generic case in which all coefficients c_ℓ are present, the renormalization of $\theta(x)$ corresponds to an arbitrary change of parametrization:

$$\theta(x) = \theta_{\rm r}(x) Z\left[\theta_{\rm r}^2(x)\right].$$

14.9 The Linearized Representation

For all models on homogeneous spaces, it is always possible to express the relations between the components of the field in the linear representation by introducing a set of Lagrange multipliers. In the case of the non-linear σ-model this is specially easy since only one $O(N)$ invariant additional field, which we denote by $\alpha(x)$, is required to implement the constraint (14.1):

$$\phi^2(x) = 1.$$

The path integral representation (14.27) of the generating functional $Z(\mathbf{J})$ can be rewritten:

$$Z(\mathbf{J}) = \int [{\rm d}\alpha {\rm d}\phi] \exp\left[-\frac{1}{g}\left(S(\phi, \alpha) - \int {\rm d}^d x\, \mathbf{J}(x) \cdot \boldsymbol{\phi}(x)\right)\right], \tag{14.65}$$

with:

$$S(\phi, \alpha) = \frac{1}{2} \int {\rm d}^d x \left\{[\partial_\mu \phi(x)]^2 + \alpha(x)\left[\phi^2(x) - 1\right]\right\}. \tag{14.66}$$

Note that here $\mathbf{J}(x)$ is a N-component source for the N-component field $\phi(x)$.

We choose an extremum of the potential as the starting point for perturbation theory

$$\alpha(x) = 0, \qquad \sigma(x) \equiv \phi_1(x) = 1, \qquad \phi_i(x) = 0 \quad \text{for } i > 1. \tag{14.67}$$

The propagator of the $(N-1)$ remaining components $\pi(x)$ is still proportional to $1/p^2$. The component $\sigma(x)$ is coupled to $\alpha(x)$. At the tree level the connected 2-point functions are:

$$\begin{pmatrix} \langle \sigma\sigma \rangle & \langle \sigma\alpha \rangle \\ \langle \alpha\sigma \rangle & \langle \alpha\alpha \rangle \end{pmatrix}_{\rm c}(p) = \begin{pmatrix} 0 & g \\ g & gp^2 \end{pmatrix}. \tag{14.68}$$

In this framework, the model presents a spontaneously broken linearly realized $O(N)$ symmetry. The consequences of such a situation have already been analyzed in Section 13.4. Power counting tells us that the dimensions of the fields in dimension two are:

$$[\phi] = 0, \qquad [\alpha] = 2.$$

The general form of the renormalized action $S_r(\phi, \alpha)$ follows:

$$S_r(\phi, \alpha) = \tfrac{1}{2} \int d^d x \Big\{ B_1\left(\phi^2(x)\right) \left[\partial_\mu \phi(x)\right]^2 + B_2\left(\phi^2(x)\right) \left(\phi \cdot \partial_\mu \phi\right)^2$$
$$+ B_3\left(\phi^2(x)\right) \alpha(x) - B_4\left(\phi^2(x)\right) \Big\}, \qquad (14.69)$$

in which B_1, B_2, B_3 and B_4 are arbitrary functions, and therefore an infinite number of renormalization constants are needed.

However if we are only interested in the ϕ-field correlation functions (and not the α-field), we can explicitly integrate over the α-field. This fixes the value of $\phi^2(x)$ to a solution of the equation:

$$B_3\left(\phi^2(x)\right) = 0. \qquad (14.70)$$

This equation has a solution order by order in perturbation theory:

$$\phi^2(x) = Z^{-1}(g) = 1 + O(g). \qquad (14.71)$$

After integration, B_1, B_2 and B_3 become pure constants and $\phi \cdot \partial_\mu \phi$ vanishes identically.

We therefore recover the results of previous section. The main disadvantage of this formulation is that it introduces an infinite number of renormalization constants as an intermediate step, and that its generalization to other homogeneous spaces is not aesthetically appealing.

Finally the non-linear formulation emphasizes the connection with other geometric models like gauge theories. On the other hand the linear formulation clarifies the discussion of multiple insertions of general $O(N)$ invariant operators. In particular we understand that the insertions of operators of dimension larger than two will eventually generate terms of degree larger than one in $\alpha(x)$. The integration over α then no longer implies the strict constraint (14.71).

Bibliographical Notes

For early references on the σ-model see:

M. Gell-Mann and M. Lévy, *Nuovo Cimento* 16 (1960) 705; S. Weinberg, *Phys. Rev.* 166 (1968) 1568; J. Schwinger, *Phys. Rev.* 167 (1968) 1432; W.A. Bardeen and B.W. Lee, *Nuclear and Particle Physics*, Montreal 1967, B. Margolis and C.S. Lam eds. (Benjamin, New York 1969); K. Meetz, *J. Math. Phys.* 10 (1969) 589.

The quantization was discussed in:

I.S. Gerstein, R. Jackiw, B.W. Lee and S. Weinberg, *Phys. Rev.* D3 (1971) 2486; J. Honerkamp and K. Meetz, *Phys. Rev.* D3 (1971) 1996; J. Honerkamp, *Nucl. Phys.* B36 (1972) 130; A.A. Slavnov and L.D. Faddeev, *Theor. Math. Phys.* 8 (1971) 843.

Pauli–Villars's regularization for chiral models has been proposed by

A.A. Slavnov, *Nucl. Phys.* B31 (1971) 301.

The renormalization group properties were discussed in

A.M. Polyakov, *Phys. Lett.* 59B (1975) 79; E. Brézin and J. Zinn-Justin, *Phys. Rev. Lett.* 36 (1976) 691; *Phys. Rev.* B14 (1976) 3110; W.A. Bardeen, B.W. Lee and R.E. Shrock, *Phys. Rev.* D14 (1976) 985.

The renormalizability in two dimensions was proven in

E. Brézin, J.C. Le Guillou and J. Zinn-Justin, *Phys. Rev.* D14 (1976) 2615; B14 (1976) 4976.

The Elitzur conjecture has been proven in

F. David, *Commun. Math. Phys.* 81 (1981) 149.

Exercises

Exercise 14.1

Verify explicitly that when the action $S(\pi, H)$ is $O(N-1)$ invariant and satisfies the relation (14.35) then the differential operators Δ_i

$$\Delta_i = \int \mathrm{d}^d x \left[\frac{\delta S}{\delta \pi_i(x)} \frac{\delta}{\delta H(x)} + \frac{\delta S}{\delta H(x)} \frac{\delta}{\delta \pi_i(x)} \right],$$

and

$$\Delta_{ij} = \int \mathrm{d}^d x \left[\pi_i(x) \frac{\delta}{\delta \pi_j(x)} - \pi_j(x) \frac{\delta}{\delta \pi_i(x)} \right],$$

form a representation of the Lie algebra of $O(N)$.

Exercise 14.2

We consider the partition function of the $O(4) \sim SU(2) \times SU(2)$ non-linear σ-model

$$Z = \int \left[\left(f^2 - \pi^2(x) \right)^{-1/2} \mathrm{d}\pi(x) \right] \exp\left[-S(\pi, c) \right], \tag{14.72}$$

with the action

$$S(\pi, c) = \int \mathrm{d}^d x \left\{ \frac{1}{2} \left[(\partial \pi(x))^2 + \frac{(\pi \cdot \partial \pi(x))^2}{f^2 - \pi^2(x)} \right] - c\sqrt{f^2 - \pi^2(x)} \right\}. \tag{14.73}$$

It what follows it will be convenient to set $c = \mu^2 f$.

We propose to study this model at one-loop order, to verify in particular that it can be renormalized with only two renormalization constants Z, Z_f as proven in Chapter 14.

It will be convenient to set $d = 2 + \varepsilon$ and to define a dimensionless coupling constant g:

$$f^{-2} \Gamma(1 - \varepsilon/2) \mu^\varepsilon (4\pi)^{-\varepsilon/2} = g.$$

Dimensional regularization will everywhere be used.

14.2.1. Establish the Feynman rules. For the calculations we have in mind, only the propagator, the 4-point and 6-point vertices are required.

14.2.2. Write the renormalized action.

14.2.3. *One-loop order calculations.* Calculate the vacuum expectation value $\langle \sigma \rangle$ and determine the field renormalization constant. Calculate the inverse two-point function $\Gamma^{(2)}_{ij} = \left[W^{(2)} \right]^{-1}_{ij}$ and determine the coupling constant renormalization.

14.2.4. Calculate the connected amputated four-point function $W^{(4)}_{\mathrm{amp.}}$ at one-loop. In what follows we adopt the convention that all momenta enter so that $p_1 + p_2 + p_3 + p_4 = 0$. Verify that the answer is finite in two dimensions.

14.2.5. Write and verify the consequences of WT identities which involve only the 1–4-point 1PI correlation functions, using equation (14.34).

15 MODELS ON HOMOGENEOUS SPACES IN TWO DIMENSIONS

In this chapter we describe the formal properties and discuss the renormalization of a class of geometrical models: Models based on homogeneous spaces. Homogeneous spaces are associated with non-linear realizations of group representations and these models are natural generalizations of the non-linear σ-model considered in Chapter 14. They can be studied in different parametrizations corresponding to different choices of coordinates when these spaces are considered as Riemannian manifolds. However in contrast with arbitrary manifolds, there exist natural ways to embed these manifolds in flat euclidean spaces, spaces in which the symmetry group acts linearly. This is the system of coordinates that we have used in the discussion of the non-linear σ-model and again use in the first part of this chapter because the renormalization properties are simpler and the physical interpretation of correlation functions more direct. We then also examine some properties of these models in a generic parametrization. The renormalization problem is solved by the introduction of a symmetry (generally called BRS symmetry) with anticommuting (Grassmann) parameters which later will play an essential role in the renormalization of gauge theories.

In a second part of the chapter we study the more specific properties of models corresponding to a special class of homogeneous spaces: symmetric spaces. The non-linear σ-model of Chapter 14 provides the simplest example. These models are characterized by the uniqueness of the metric and thus of the classical action. The classical equations of motion generate, in two dimensions, an infinite number of non-local conservation laws. We calculate RG functions at one-loop order, and find the first examples of asymptotically free theories.

The chapter ends with a few comments about more general models based on non-compact groups and arbitrary Riemannian manifolds. The appendix contains a few additional remarks about metric and curvature in homogeneous spaces. We also briefly describe classical families of symmetric spaces.

Note finally that in the description of these models, two different formalisms and sets of notation can be employed, depending whether one wants to emphasize the group structure or the Riemannian manifold point of view.

15.1 Construction of the Model

Notation. Let us consider a field $\phi_i(x)$ transforming under a linear representation of a group G. We use the notation of Chapter 13 for the infinitesimal group transformation, denoting by t^α the generators of the Lie algebra $\mathcal{L}(G)$ in some matrix representation, and by ω_α the infinitesimal group parameters of the transformation:

$$\delta\phi_i(x) = t^\alpha_{ij}\omega_\alpha\phi_j(x). \tag{15.1}$$

Since we consider only compact groups, we can assume that the representation is orthogonal, and that the generators of $\mathcal{L}(G)$ are $N \times N$ antisymmetric matrices.

15.1.1 Spontaneous symmetry breaking: Goldstone modes

Goldstone modes. Rather than giving a purely mathematical construction, we shall try to motivate the study of these models by some physical arguments. Our starting point is, as in Chapter 13, a G-invariant action $S(\phi)$ of the form:

$$S(\phi) = \int dx \left\{ \tfrac{1}{2} [\partial_\mu \phi(x)]^2 + V(\phi(x)) \right\}. \tag{15.2}$$

We now assume that at the classical level we are in the situation of spontaneous symmetry breaking (SSB), i.e. $V(\phi)$ has non-G-invariant global minima. We distinguish one of them, ϕ^c, around which we expand perturbation theory, and call H the isotropy or little group (the stabilizer) of ϕ^c. We have shown in Section 13.4 that under these circumstances a number of field components of $\phi(x)$, which correspond to the Goldstone modes of the broken symmetry, are massless. If we are only interested in the long distance behaviour (IR limit) of correlation functions, or equivalently if the mass of the massive fields is sent to infinity (the low temperature limit of the corresponding statistical model), the fluctuations of the massive fields in the functional integral can be neglected.

In the limit the remaining massless components of the field $\phi(x)$ can be entirely parametrized in terms of a matrix $\mathbf{R}(g(x))$ of the representation of G, acting on the vector ϕ_c

$$\phi_i(x) = R_{ij}(g(x))\phi_j^c, \qquad g(x) \in G. \tag{15.3}$$

Note, however, that if we multiply $g(x)$ on the right by an element $h(x)$ of H, since ϕ_j^c is left invariant by the group H, $\phi_i(x)$ is not modified:

$$\phi_i(x) = R_{ij}(g(x)h(x))\phi_j^c = R_{ik}(g(x))R_{kj}(h(x))\phi_j^c = R_{ij}(g(x))\phi_j^c.$$

This shows that $\phi_i(x)$ is really a function of the elements of the coset space G/H. Let us then divide the set of generators of the Lie algebra $\mathcal{L}(G)$ into the set of generators belonging to the Lie algebra $\mathcal{L}(H)$, $\{t^\alpha\}$, $\alpha > l$, and the complementary set $\{t^\alpha\}$, $1 \le \alpha \le l$, which is such that:

$$\sum_{\alpha=1}^{l} c_\alpha t_{ij}^\alpha \phi_j^c = 0 \quad \Rightarrow \quad c_\alpha = 0.$$

The matrix $\mathbf{R}$ can be canonically parametrized in terms of fields $\xi_a(x)$ as:

$$\mathbf{R}(\xi_a(x)) = \exp\left(\sum_{a=1}^{l} \xi_a(x) t^a \right). \tag{15.4}$$

15.1.2 Goldstone mode effective action

Since $V(\phi)$ is derivative-free and group-invariant, it is independent of $g(x)$ and thus yields an irrelevant constant contribution to the action which can be omitted. The action $S(\phi)$ can then be written:

$$S(\phi) = \tfrac{1}{2} \int dx \, \phi_j^c \partial_\mu R_{jk}^{-1}(g(x)) \partial_\mu R_{ki}(g(x)) \phi_i^c. \tag{15.5}$$

By expanding action (15.5) for $g(x)$ close to the identity we verify indeed that all remaining fields are massless.

Notation. It will also be convenient to use a bra and ket notation to indicate vectors, denoting by $|0\rangle$ the vector ϕ^c. The equation (15.3) then takes the form:

$$|\phi(x)\rangle = \mathbf{R}\left(g(x)\right)|0\rangle ,\tag{15.6}$$

in which $|\phi(x)\rangle$ is a notation for the field $\phi_i(x)$. With this notation the classical action $S(\phi)$ can be rewritten in terms of a current $\mathbf{R}^{-1}\partial_\mu\mathbf{R}$:

$$S(\phi) = -\tfrac{1}{2}\int \mathrm{d}x \,\langle 0|\left(\mathbf{R}^{-1}\partial_\mu\mathbf{R}\right)^2|0\rangle .\tag{15.7}$$

Equation of motion. Since the field can be entirely parametrized in terms of group elements, the classical equations of motion are a consequence of the G-current conservation. A straightforward calculation yields the form of the current J_μ^α:

$$J_\mu^\alpha(x) = \langle 0|\,\mathbf{R}^{-1}t^\alpha\partial_\mu R\,|0\rangle , \qquad \partial_\mu J_\mu^\alpha(x) = 0 .$$

15.1.3 Metric and action in general coordinates

In this subsection we use a notation and conventions adapted to Riemannian geometry (see Chapter 22 for details).

We denote by φ^i an arbitrary set of coordinates on the manifold (coset space) G/H. The action (15.5) can then be written in the form:

$$S(\varphi) = \tfrac{1}{2}\int \mathrm{d}x \, g_{ij}\left(\varphi(x)\right)\partial_\mu\varphi^i(x)\partial_\mu\varphi^j(x).\tag{15.8}$$

where g_{ij} is a metric on G/H considered as a Riemannian manifold. We know from the discussion of Subsection 15.1.1 that the matrix $\mathbf{R}$ when acting on the vector $|0\rangle$ can be parametrized in the form (15.4). The variables ξ^a in equation (15.4) thus provide an example of a set of coordinates on the coset space G/H.

Comparing the expressions (15.8) and (15.7), we can relate the metric to other geometrical objects. The current $\mathbf{R}^{-1}\partial_\mu R$ can be written

$$\mathbf{R}^{-1}\partial_\mu R = \mathbf{R}^{-1}\frac{\partial\mathbf{R}}{\partial\varphi^i}\partial_\mu\varphi^i(x).\tag{15.9}$$

In what follows ∂_i with roman indices means derivative with respect to φ^i and ∂_μ with greek indices means derivative with respect to x_μ. Any matrix of the form $\mathbf{R}^{-1}\partial\mathbf{R}$ (a pure gauge) belongs to the Lie algebra of G. We can therefore expand on the basis of generators (more details can be found in Appendix A15.1):

$$\mathbf{R}^{-1}\frac{\partial\mathbf{R}}{\partial\varphi^i} \equiv \mathbf{R}^{-1}\partial_i\mathbf{R} = L_i^\alpha(\varphi)t^\alpha,\tag{15.10}$$

and therefore:

$$\mathbf{R}^{-1}\partial_\mu\mathbf{R} = L_i^\alpha\left(\varphi\right)t^\alpha\partial_\mu\varphi^i(x).\tag{15.11}$$

An expression of the metric tensor in terms of $L_i^\alpha(\varphi)$ follows:

$$g_{ij}(\varphi) = L_i^a(\varphi)L_j^b(\varphi)\mu_{ab} ,\tag{15.12}$$

with the definition

$$\mu_{ab} = -\left\langle 0 \left| t^a t^b \right| 0 \right\rangle. \tag{15.13}$$

The latin letters (a, b) indicate that the indices run only over values corresponding to generators belonging to $\mathcal{L}(G/H)$.

One verifies that the quantities L_i^α are independent of the representation (Section A15.1). Therefore the dependence in the choices of the classical vector $|0\rangle$ and the particular representation of the group G is entirely contained in the positive matrix μ_{ab}. The form of the matrix is only restricted by the symmetry under the subgroup H:

$$\left\langle 0 \left| t^a \tau^\gamma t^b \right| 0 \right\rangle = \left\langle 0 \left| t^a \left[\tau^\gamma, t^b \right] \right| 0 \right\rangle = \left\langle 0 \left| \left[t^a, \tau^\gamma \right] t^b \right| 0 \right\rangle,$$

for $t^a, t^b \in \mathcal{L}(G/H)$, $t^\gamma \in \mathcal{L}(H)$ and therefore:

$$\mu_{ac} f_{c\gamma b} - f_{a\gamma c}\mu_{cb} = 0. \tag{15.14}$$

The interpretation of this equation is simple: the vectors $t^a |0\rangle$ transform under an (in general reducible) representation of H which has the matrices $f_{a\gamma b}$ as generators.

The symmetric matrix $\boldsymbol{\mu}$ commutes with all generators $\mathbf{f}^\gamma$ ($(\mathbf{f}^\gamma)_{ab} = f_{a\gamma b}$) of $\mathcal{L}(H)$:

$$[\boldsymbol{\mu}, \mathbf{f}^\gamma] = 0.$$

As a consequence, the number of parameters on which μ_{ab} depends, is the number of different H-invariant scalar products one can form with the irreducible components of the vector $t^a |0\rangle$ (Schur's lemma).

This result exhausts all consequences of the G-symmetry.

15.1.4 Quantization and perturbation theory near dimension 2

So far we have only examined the classical theory. To quantize it, we can start from the euclidean action (15.5), derive a quantum mechanical hamiltonian and use canonical quantization. It is faster to begin with action (15.2) and freeze the massive degrees of freedom as explained above. The result is the same and yields a functional representation:

$$Z = \int [\mathrm{d}\rho(\varphi)] \exp\left[-\tfrac{1}{2} \int \mathrm{d}x \, g_{ij}\big(\varphi(x)\big) \partial_\mu \varphi^i(x) \partial_\mu \varphi^j(x)\right]. \tag{15.15}$$

The measure of the φ-integration is the G-invariant measure induced by the flat ϕ measure, and is also the restriction to the coset space G/H of the Haar group measure of G.

From the point of view of power counting the theory is clearly renormalizable in two dimensions for any parametrization since the φ-field has dimension $\tfrac{1}{2}(d-2)$ and the interaction terms contain two derivatives and arbitrary powers of the φ-field.

To prove the structural stability of action (15.5) we derive in the next sections a set of WT identities corresponding to the non-linearly realized symmetry under the group G. Dimensional or lattice regularization respecting the G-symmetry is implied in what follows. The choice of a group-invariant regularization corresponds also to a choice of quantization of the classical hamiltonian consistent with the symmetry.

15.2 WT Identities and Renormalization in Linear Coordinates

For renormalization problems it is convenient to choose a special parametrization, which consists in embedding the homogeneous space into an euclidean space, as we have explained in the preceding chapter. Let us describe this parametrization first.

Remark. Note that the vector space $\mathcal{V}$ spanned by the family of vectors of the form (15.3) may have lower dimension than the original space of representation to which ϕ belongs if this representation is reducible. The space $\mathcal{V}$ is still a space of representation for the group G. In what follows we *restrict* ϕ to its components in $\mathcal{V}$.

15.2.1 Linear coordinates

Let us consider the vector space $\mathcal{V}$ spanned by the vectors of form (15.3), as well as the subspace $\mathcal{V}'$ of dimension l spanned by the vectors $t_{ij}^{\alpha}\phi_j^c$. We choose an orthogonal basis in $\mathcal{V}$ such that the vectors of $\mathcal{V}'$ have only the first l components non-vanishing. We then distinguish the l first components of the vector ϕ of equation (15.3), calling them $\pi_a(x)$, and the others called $\sigma_i(x)$:

$$\phi(x) = \begin{cases} \pi_a(x), & 1 \le a \le l \ ; \\ \sigma_i(x), & l < i \ . \end{cases} \tag{15.16}$$

The fields $\sigma_i(x)$ and $\pi_a(x)$ are functions of the fields $\xi_a(x)$. If we expand equation (15.4) in powers of ξ:

$$\phi_i(x) = \phi_i^c + \sum_{a=1}^{l} \xi_a(x) t_{ij}^a \phi_j^c + O\left(\xi^2\right), \tag{15.17}$$

we see, comparing expressions (15.3) and (15.16), that $\pi_a(x)$ and $\xi_a(x)$ are canonically related:

$$\pi_a(x) = \xi_b(x) t_{aj}^b \phi_j^c + O\left(\xi^2\right), \tag{15.18}$$

i.e. this relation can be inverted to express the fields ξ_a as functions of the fields π_a. On the other hand the fields $\sigma_i(x)$ are of the form:

$$\sigma_i(x) = \phi_i^c + O\left(\xi^2\right),$$

and can therefore be calculated in terms of the fields π_a. The fields $\pi_a(x)$ and $\sigma_i(x)$ transform under different linear representations of the group H. However the fields $\pi_a(x)$ transform under a non-linear representation of the group G since the generators $\{t^\alpha\}$ mix the fields $\pi_a(x)$ and the $\sigma_i(x)$ which are functions of π_a:

$$\delta\pi_a = \left[t_{ab}^\alpha \pi_b + t_{aj}^\alpha \sigma_j\left(\pi\right)\right] \omega_\alpha . \tag{15.19}$$

Note that since the σ_i are functions of π_a, the transformation laws:

$$\delta\sigma_i\left(\pi\right) = \left[t_{ib}^\alpha \pi_b + t_{ij}^\alpha \sigma_j\left(\pi\right)\right] \omega_\alpha , \tag{15.20}$$

are now consequences of equation (15.19) and the functional form of the $\sigma_i\left(\pi\right)$.

15.2.2 Correlation functions, WT identities, renormalization

We now consider the generating functional of correlation functions

$$Z(J) = \int [\mathrm{d}\rho(\pi)] \exp\left[-S(\phi) + \int \mathrm{d}x\, J_i(x)\phi_i(x)\right], \qquad (15.21)$$

the integrand being expressed in terms of these special coordinates. Sources have been added for all components of ϕ in $\mathcal{V}$ for the reasons which have already been explained in the preceding chapter: when we try to derive WT identities expressing the consequences of the symmetry for correlation functions, the composite σ_a operators appears in the variation of the π-field. Moreover, since all π-fields are massless, we have to break the G-symmetry explicitly to provide an IR cut-off. This can be achieved by expanding perturbation theory at fixed constant values of the σ sources.

Let us now derive the WT identities corresponding to the non-linearly realized symmetry under the group G and show how they imply the structural stability of action (15.5).

WT identities. We perform an infinitesimal transformation (15.19) in the functional integral (15.21). Since we have introduced sources for all components of ϕ, the corresponding WT identity for $Z(J)$ and $W(J)$ is identical to the identity obtained for linearly realized symmetries:

$$\int \mathrm{d}x\, J_i(x) t^\alpha_{ij} \frac{\delta Z(J)}{\delta J_j(x)} = 0, \qquad (15.22)$$

and thus:

$$\int \mathrm{d}x\, J_i(x) t^\alpha_{ij} \frac{\delta W}{\delta J_j(x)} = 0. \qquad (15.23)$$

Again the difference appears in the Legendre transformation. Let us now call $J_a(x)$, $1 \le a \le l$, the sources for the fields π_a, and H_i, $l < i \le n$, the sources for the composite fields σ_i. The Legendre transform has to be performed only on J_a:

$$\Gamma(\pi, H) + W(J, H) = \int \mathrm{d}x\, \pi_a(x) J_a(x), \qquad (15.24)$$

$$\pi_a(x) = \frac{\delta W}{\delta J_a(x)} \quad \Leftrightarrow \quad J_a(x) = \frac{\delta \Gamma}{\delta \pi_a(x)}. \qquad (15.25)$$

Since, as explained several times,

$$\frac{\delta \Gamma}{\delta H_i(x)} = -\frac{\delta W}{\delta H_i(x)},$$

identity (15.23) implies for the 1PI functional Γ:

$$\int \mathrm{d}x \left\{ \frac{\delta \Gamma}{\delta \pi_a(x)} \left[t^\alpha_{ab}\pi_b(x) - t^\alpha_{aj}\frac{\delta \Gamma}{\delta H_j(x)} \right] + H_i(x) \left[t^\alpha_{ib}\pi_b(x) - t^\alpha_{ij}\frac{\delta \Gamma}{\delta H_j(x)} \right] \right\} = 0. \quad (15.26)$$

Renormalization. This identity has a quadratic form as in the case of the σ-model. The arguments of Section 14.6 apply and prove the stability of the identity under renormalization. The same identity is thus fulfilled by the renormalized action S_r:

$$\int \mathrm{d}x \left\{ \frac{\delta S_\mathrm{r}}{\delta \pi_a(x)} \left[t^\alpha_{ab}\pi_b(x) - t^\alpha_{aj}\frac{\delta S_\mathrm{r}}{\delta H_j(x)} \right] + H_i(x) \left[t^\alpha_{ib}\pi_b(x) - t^\alpha_{ij}\frac{\delta S_\mathrm{r}}{\delta H_j(x)} \right] \right\} = 0. \quad (15.27)$$

Power counting tells us that $S_{\rm r}(\pi, H)$ and $H_a(x)$ have dimension 2. Therefore again $S_{\rm r}$ is linear in the source $H_a(x)$. Let us write $S_{\rm r}$ as:

$$S_{\rm r}(\pi, H) = -\int \sigma_{{\rm r}i}(\pi) H_i(x){\rm d}x + \Sigma_{\rm r}(\pi), \tag{15.28}$$

in which the functions $\sigma_{{\rm r}i}(\pi)$ are derivative-free and $\Sigma_{\rm r}$ has dimension 2, i.e. has at most two derivatives. The term linear in H in equation (15.27) yields:

$$\frac{\delta\sigma_{{\rm r}i}(\pi)}{\delta\pi_a(x)}\left[t^{\alpha}_{ab}\pi_b(x) + t^{\alpha}_{aj}\sigma_{{\rm r}j}(\pi(x))\right] = t^{\alpha}_{ib}\pi_b(x) + t^{\alpha}_{ij}\sigma_{{\rm r}j}(\pi(x)). \tag{15.29}$$

These partial differential equations for the function $\sigma_{{\rm r}i}(\pi)$ imply that if the fields π_a have the transformation law:

$$\delta_{\alpha}\pi_a = t^{\alpha}_{ab}\pi_b + t^{\alpha}_{aj}\sigma_{{\rm r}j}(\pi), \tag{15.30}$$

then as a consequence:

$$\delta_{\alpha}\sigma_i(\pi) = t^{\alpha}_{ib}\pi_b + t^{\alpha}_{ij}\sigma_{{\rm r}j}(\pi). \tag{15.31}$$

Thus the field ϕ with component $(\pi_a, \sigma_{{\rm r}i})$ transforms under a linear representation of the group G. Let us now write the second equation obtained by setting H equal to zero

$$\int {\rm d}x \frac{\delta\Sigma_{\rm r}}{\delta\pi_a(x)}\left[t^{\alpha}_{ab}\pi_b(x) + t^{\alpha}_{aj}\sigma_{{\rm r}j}(x)\right] = 0. \tag{15.32}$$

This equation tells us that $\Sigma_{\rm r}(\pi)$ is the most general functional of $\pi(x)$ with two derivatives invariant under the group G.

The action can thus be constructed starting from the most general G-invariant action with two derivatives written in terms of the field $\phi(x)$, and then eliminating the fields $\sigma_{{\rm r}i}(x)$. Also we have shown in Subsection 15.1.3 that the action, expressed in terms of the metric tensor, can be parametrized in terms of the matrix μ (equation (15.13)). We here find that even if at the tree level we begin with a special choice of the matrix μ solution of equation (15.14), we obtain after renormalization the most general solution of this equation.

15.2.3 Field renormalizations

Let us here make a brief comment about the solutions of equation (15.29). For renormalization purpose, we are looking for *generic* solutions $\sigma_i(\pi)$ expandable in powers of π:

$$\sigma_i = S_i + S^{a_1}_i \pi_{a_1} + \tfrac{1}{2}S^{a_1 a_2}_i \pi_{a_1}\pi_{a_2} + \cdots, \tag{15.33}$$

in which S_i, $S^{a_1}_i$ ... are constants.

The equation of order zero in π is:

$$S^a_i t^{\alpha}_{aj}S_j = t^{\alpha}_{ij}S_j. \tag{15.34}$$

If t^{α} belongs to the Lie algebra $\mathcal{L}(H)$, t^{α}_{aj} vanishes since π and σ belong to different representations and therefore:

$$t^{\alpha}_{ij}S_j = 0 \quad \text{for all } t^{\alpha} \in \mathcal{L}(H). \tag{15.35}$$

As a consequence the generic vector S_j is the most general vector having H as an isotropy group. Note that since S_i differs from ϕ_c only at one-loop order, the isotropy group (stabilizer) of S_i cannot be larger than H.

For the same reason the generators of $\mathcal{L}(G/H)$ are such that $t_{aj}^\alpha S_j$ spans the π subspace. Therefore equation (15.34) implies:

$$S_i^a t_{aj}^\alpha S_j = t_{ij}^\alpha S_j\,, \qquad 1 \le \alpha \le l\,. \tag{15.36}$$

The scalar product of the vectors $\mathbf{S}_i$ with all vectors of a complete basis is known, therefore all $\mathbf{S}_i$ are determined. In particular if the π subspace contains no vector invariant under the action of the group H, the r.h.s. of equation (15.36) and therefore also all vectors $\mathbf{S}_i$ vanish. This property holds for symmetric spaces.

Higher order equations determine the coefficients of monomials of increasing degree in π. If we assume that we know $\sigma_i(\pi)$ at order π^k, then the equation for the coefficient of π^{k+1} takes the form:

$$S_i^{ab_1\ldots b_k}\ldots t_{aj}^\alpha S_j = \text{ known quantities}\,.$$

The coefficients $S_i^{ab_1\ldots b_k}$ are completely determined since considered as vectors $\mathbf{S}_i^{b_1\ldots b_k}$ their scalar products with all vectors of a complete basis are given.

The conclusion is that the functions $\sigma_i(\pi)$ depend on as many parameters as the number of independent vectors S_i which have H as an isotropy group. The renormalized action is then the most general "free" massless action in the linear ϕ variables. We have therefore enumerated all the renormalization constants of the model.

15.3 Renormalization in Arbitrary Coordinates, BRS Symmetry

We have shown in Section 15.2 how a special choice of parametrization leads to a rather simple discussion of the symmetry properties and renormalization of correlation functions.

We have indicated in Chapter 14, in the case of the non-linear σ-model, that with another choice we would have been forced to introduce a number, generically infinite, of additional renormalization constants, corresponding to a renormalization of the parametrization of the manifold. The field is no longer multiplicatively renormalized, the bare field becoming a function of the renormalized field.

However, as we have indicated in Appendix A10.3, in some cases the only physically relevant quantities are those which are parametrization-independent (related to geometrical properties of the manifold or to the S-matrix). It is therefore useful to also investigate models on homogeneous spaces in an arbitrary parametrization to more clearly exhibit the parametrization dependence. Moreover we shall see on the example of the calculation of the one-loop β-function that some parametrizations are more convenient for practical calculations.

In this section we derive WT identities expressing the group symmetry in an arbitrary parametrization and show that they are stable under renormalization. The general strategy, i.e. to add to the action sources for a set of composite operators which is closed under infinitesimal group transformations on the field, is only suitable if the minimum number of different operators is finite. Then the renormalization of the theory follows from a rather straightforward generalization of the arguments given in the first part of this chapter. However, for a generic parametrization an infinite number of composite fields is required, and this strategy is no longer useful. We therefore introduce a new method which, in some generalized form, will also be relevant to the renormalization of gauge theories and which is based on infinitesimal group transformations with anticommuting parameters.

15.3.1 Infinitesimal group transformations

We consider a non-linear realization of the representation of a group G acting on a field $\varphi^i(x)$. We write the infinitesimal group transformations corresponding to parameters ω^α in the form:

$$\delta_\omega \varphi^i(x) = D^i_\alpha\left(\varphi(x)\right)\omega^\alpha\,. \tag{15.37}$$

We assume that the functions $D^i_\alpha(\varphi)$ are smooth, i.e. infinitely differentiable, at $\varphi = 0$. We write the action in the form (15.8),

$$S(\varphi) = \tfrac{1}{2}\int \mathrm{d}x\, g_{ij}\left(\varphi(x)\right)\partial_\mu\varphi^i(x)\partial_\mu\varphi^j(x)\,. \tag{15.38}$$

If the action $S(\varphi)$ is invariant under group transformations, it satisfies:

$$\int \mathrm{d}x\, D^i_\alpha\left(\varphi(x)\right)\frac{\delta S(\varphi)}{\delta\varphi^i(x)} = 0\,. \tag{15.39}$$

This equation implies an identity for the metric tensor $g_{ij}(\varphi)$:

$$D^i_\alpha\frac{\partial g_{jk}}{\partial\varphi^i} + g_{ik}\frac{\partial D^i_\alpha}{\partial\varphi^j} + g_{ij}\frac{\partial D^i_\alpha}{\partial\varphi^k} = 0\,. \tag{15.40}$$

To solve this equation is equivalent to find all possible metric tensors on a given homogeneous space, compatible with the group structure. The Appendix A15.2 contains some details about the nature and the structure of this equation.

The functions $D^i_\alpha(\varphi)$ satisfy identities which can be obtained either by direct calculation or by noting that the differential operators Δ_α:

$$\Delta_\alpha = D^i_\alpha(\varphi)\frac{\partial}{\partial\varphi^i}\,, \tag{15.41}$$

acting on functions of φ, form themselves a representation of the Lie algebra (see also Section 13.1.2) and therefore have commutation relations of the form:

$$[\Delta_\alpha, \Delta_\beta] = f^\gamma_{\alpha\beta}\Delta_\gamma\,. \tag{15.42}$$

Note that these commutation relations are compatibility conditions for the equation (15.39) considered as a set of linear differential equations for $S(\varphi)$. Calculating explicitly the commutator in terms of the functions $D^i_\alpha(\varphi)$ we obtain:

$$D^j_\alpha(\varphi)\frac{\partial D^i_\beta(\varphi)}{\partial\varphi^j} - D^j_\beta(\varphi)\frac{\partial D^i_\alpha(\varphi)}{\partial\varphi^j} = f^\gamma_{\alpha\beta}D^i_\gamma(\varphi)\,. \tag{15.43}$$

This is the form of the commutation relations of the Lie algebra which is useful in the discussion of non-linear representations. Moreover in Chapter 21 we shall encounter equations which are formally identical and play an essential role in the discussion of the renormalization of gauge theories.

15.3.2 WT identities

We assume that we have introduced a group-invariant regularization and consider the generating functional $Z(J, K)$:

$$Z(J, K) = \int [\mathrm{d}\rho(\varphi)] \exp\left[-S(\varphi) + \int \mathrm{d}x \left(K(x)A\left(\varphi(x)\right) + J_i(x)\varphi^i(x)\right)\right]. \quad (15.44)$$

The measure $[\mathrm{d}\rho(\varphi)]$ is the group-invariant measure. Note that we have added to the action sources not only for the field $\varphi^i(x)$ but also for a local derivative-free function of $\varphi(x)$, $A(\varphi(x))$. The function $A(\varphi)$ has to satisfy only one condition: it has to begin with a term of order φ^2 such that, for $K(x)$ constant, masses are generated for all components of φ, which can serve as IR regulators.

To derive consequences of the non-linear symmetry (15.39), we introduce a set of anticommuting constants C^α and $\bar\varepsilon$, i.e. all belonging to an *antisymmetric algebra*, and perform a transformation:

$$\varphi^i(x) \mapsto \varphi^i(x) + \bar\varepsilon D_\alpha^i\left(\varphi(x)\right)C^\alpha. \quad (15.45)$$

The action and the measure are invariant. The variation of the source terms in expression (15.44) is:

$$\delta\left\{\int \mathrm{d}x \left[J_i(x)\varphi^i(x) + K(x)A\left(\varphi(x)\right)\right]\right\} = \bar\varepsilon \int \mathrm{d}x \left[J_i(x) + K(x)\frac{\delta A}{\delta\varphi^i(x)}\right] D_\alpha^i(\varphi)C^\alpha. \quad (15.46)$$

This variation involves two composite operators: $D_\alpha^i\left(\varphi(x)\right)C^\alpha$, because the transformation is non-linear, and $(\delta A/\delta\varphi^i) D_\alpha^i(\varphi)C^\alpha$. In accordance with our general strategy, we introduce for them two anticommuting sources $\Lambda_i(x)$ and $L(x)$. Note that if we assign a fermion number $+1$ to C^α and -1 to $\Lambda_i(x)$ and $L(x)$, this fermion number is conserved. We then calculate the variation of these operators under the transformation (15.45):

$$\delta\left[D_\alpha^i(\varphi)C^\alpha\right] = \bar\varepsilon\frac{\delta D_\alpha^i}{\delta\varphi^j}D_\beta^j C^\beta C^\alpha. \quad (15.47)$$

Since C^α and C^β anticommute, we can antisymmetrize the coefficient of $C^\alpha C^\beta$. Using then the commutation relations (15.43) we find:

$$\delta\left[D_\alpha^i(\varphi)C^\alpha\right] = \tfrac{1}{2}\bar\varepsilon f_{\alpha\beta}^\gamma C^\alpha C^\beta D_\gamma^i(\varphi). \quad (15.48)$$

It is possible to cancel this variation by performing a simultaneous transformation on the sources C^α of the form:

$$\delta C^\alpha = -\tfrac{1}{2}\bar\varepsilon f_{\beta\gamma}^\alpha C^\beta C^\gamma. \quad (15.49)$$

The geometrical origin of these transformations will be explained in Section 16.4 when we discuss *BRS symmetry*.

Let us now calculate the variation of $(\delta A/\delta\varphi^i) D_\alpha^i(\varphi)C^\alpha$ under the combined transformation (15.45, 15.49):

$$\delta\left[\frac{\delta A}{\delta\varphi^i}D_\alpha^i(\varphi)C^\alpha\right] = \bar\varepsilon\frac{\delta^2 A}{\delta\varphi^i\delta\varphi^j}D_\beta^j D_\alpha^i C^\beta C^\alpha + \frac{\delta A}{\delta\varphi_i}\delta\left[D_\alpha^i(\varphi)C_\alpha\right] = 0. \quad (15.50)$$

The first term vanishes because the coefficient of $C^\alpha C^\beta$ is symmetric in (α, β), and the second term vanishes by construction. The algebra is thus closed because the combined transformation (15.45,15.49) is nilpotent. We obtain an identity for:

$$Z(J, K, C, \Lambda, L) = \int [\mathrm{d}\rho(\varphi)] \exp\left[-S(\varphi) + \int \mathrm{d}x \left(J_i \varphi^i + KA(\varphi)\right)\right.$$
$$\left. + \int \mathrm{d}x \left(\Lambda_i + L\frac{\delta A}{\delta \varphi^i}\right) D^i_\alpha(\varphi) C^\alpha\right], \qquad (15.51)$$

which, according to the previous analysis, has the form:

$$Z(J, K, C + \delta C, \Lambda, L) = \left[1 + \bar{\varepsilon}\int \mathrm{d}x \left(J_i(x)\frac{\delta}{\delta \Lambda_i(x)} + K(x)\frac{\delta}{\delta L(x)}\right)\right] Z(J, K, C, \Lambda, L), \qquad (15.52)$$

and therefore:

$$\left[\int \mathrm{d}x \left(J_i \frac{\delta}{\delta \Lambda_i} + K\frac{\delta}{\delta L}\right) + \tfrac{1}{2}f^\alpha_{\beta\gamma}C^\beta C^\gamma \frac{\partial}{\partial C^\alpha}\right] Z = 0. \qquad (15.53)$$

A similar identity for $W = \ln Z$ follows. After Legendre transformation with respect to J_i we then obtain the WT identities for Γ:

$$\int \mathrm{d}x \left(\frac{\delta\Gamma}{\delta\varphi^i}\frac{\delta\Gamma}{\delta\Lambda_i} + K\frac{\delta\Gamma}{\delta L}\right) + \tfrac{1}{2}f^\alpha_{\beta\gamma}C^\beta C^\gamma \frac{\partial\Gamma}{\partial C^\alpha} = 0. \qquad (15.54)$$

15.3.3 The renormalized action

Arguments, which by now should be familiar to the reader, allow one to show that equation (15.54) is stable under renormalization and also satisfied by the effective renormalized action S_r:

$$\int \mathrm{d}x \left(\frac{\delta S_\mathrm{r}}{\delta\varphi^i}\frac{\delta S_\mathrm{r}}{\delta\Lambda_i} + K\frac{\delta S_\mathrm{r}}{\delta L}\right) + \tfrac{1}{2}f^\alpha_{\beta\gamma}C^\beta C^\gamma \frac{\partial S_\mathrm{r}}{\partial C^\alpha} = 0. \qquad (15.55)$$

To solve the equation, we note that, as a consequence of fermion number conservation, Λ, L and C can appear only in the combinations (ΛC) and (LC). In two dimensions the sources K, (ΛC) and (LC) have dimension 2, and therefore S_r is a linear function of these sources with derivative-free coefficients:

$$S_\mathrm{r}(\varphi, K, C, \Lambda, L) = S_\mathrm{r}(\varphi) - \int \mathrm{d}x \left[KA_\mathrm{r}(\varphi) + \Lambda_i D^i_{\mathrm{r}\alpha}(\varphi)C^\alpha + LM_\alpha(\varphi)C^\alpha\right]. \qquad (15.56)$$

Let us first write the equation obtained by identifying the coefficient of $\Lambda_i(x)$ in equation (15.55):

$$\frac{\partial D^i_{\mathrm{r}\alpha}}{\partial\varphi^j} D^j_{\mathrm{r}\beta} C^\alpha C^\beta - \tfrac{1}{2}f^\alpha_{\beta\gamma}C^\beta C^\gamma D^i_{\mathrm{r}\alpha} = 0. \qquad (15.57)$$

The antisymmetric coefficient of $C^\alpha C^\beta$ must vanish:

$$\frac{\partial D^i_{\mathrm{r}\alpha}}{\partial\varphi^j} D^j_{\mathrm{r}\beta} - \frac{\partial D^i_{\mathrm{r}\beta}}{\partial\varphi^j} D^j_{\mathrm{r}\alpha} = f^\gamma_{\alpha\beta} D^i_{\mathrm{r}\gamma}. \qquad (15.58)$$

The set of functions $D^i_{r\alpha}(\varphi)$ are thus associated with a non-linear representation of the group G.

We now identify the coefficient of K:

$$\frac{\delta A_r}{\delta \varphi^i} D^i_{r\alpha}(\varphi) C^\alpha - M_\alpha C^\alpha = 0\,, \quad \Rightarrow \quad M_\alpha(\varphi) = \frac{\delta A_r}{\delta \varphi^i} D^i_{r\alpha}(\varphi), \tag{15.59}$$

and the coefficient of L:

$$\frac{\delta M_\alpha}{\delta \varphi^i} D^i_{r\beta} - \frac{\delta M_\beta}{\delta \varphi^i} D^i_{r\alpha} = f^\gamma_{\alpha\beta} M_\gamma\,. \tag{15.60}$$

The latter equation is already implied by the two equations (15.58,15.59). We conclude that $A_r(\varphi)$ is in general an arbitrary function of φ.

Finally the last equation, independent of the different sources, is:

$$\frac{\delta S_r}{\delta \varphi^i}(\varphi) D^i_{r\alpha}(\varphi) = 0\,. \tag{15.61}$$

The renormalized action is invariant under the non-linear transformations of the group G generated by $D^i_{r\alpha}(\varphi)$.

Let us note that very often one chooses a H symmetric parametrization for homogeneous spaces G/H. This imposes additional restrictions upon the renormalized form of $D^{ri}_\alpha(\varphi)$ and implies that $S_r(\varphi)$ is H-symmetric.

Remarks. The results obtained by this method are less detailed than those obtained in the case of the linear parametrization. The method based on adding sources for composite operators should be used when applicable.

We do not need to again discuss thoroughly the solutions of the WT identities which we have reduced to equations (15.58,15.61). The latter equation implies a renormalized form of equation (15.40), which is an equation for the renormalized metric tensor.

15.4 Symmetric Spaces: Definition

Symmetric spaces are special homogeneous spaces such that the symmetry group G possesses an involutive automorphism, and the subgroup H is the subgroup of invariant elements. Considering the case in which G is compact, we show in Appendix A15.4.2 that H is then a maximal proper subgroup. Equivalently a parity can be assigned to the generators of the Lie algebra and the Lie algebra $\mathcal{L}(H)$ is the algebra of even elements. More details can be found in Appendix A15.4.

Field theory models in two dimensions, constructed on symmetric spaces, have special properties both on the classical level and after quantization, which we examine below. The non-linear σ-model provides one of the simplest examples. In particular once the parametrization of the manifold is chosen, the euclidean action is unique up to a constant multiplicative factor. This reflects the uniqueness of the metric on the manifold compatible with the group structure. The parity of generators of the Lie algebra leads to a parity assignment for the fields, $+1$ for the composite σ-field, -1 for the π-field.

The coset space G/H when it is symmetric can be constructed from a group manifold in the following way: We consider the elements g of a group $\mathfrak{G}$ of the form (see Appendix A15.4 for details)

$$g = g_2^{-1} g_0 g_1\,,$$

where g_0, g_1, g_2 are elements of $\mathfrak{G}$ which satisfy some conditions. We can then distinguish two cases:

(i) $g_0 \equiv 1$ and g_1, g_2 are two independent elements of $\mathfrak{G}$. The automorphism is $g_1 \leftrightarrow g_2$. We recognize the coset space $\mathfrak{G} \times \mathfrak{G}/\mathfrak{G}$, the automorphism exchanging the two components of $\mathfrak{G} \times \mathfrak{G}$. As manifolds they are identical to the manifold of the group $\mathfrak{G}$. The corresponding field theory models are called *principal chiral models*. They are related to the chiral models studied in Section 13.6.

(ii) The element g_0 is a fixed element different from the identity and satisfies

$$g_0^* g_0 = \epsilon \mathbf{1}, \qquad \epsilon = \pm 1,$$

in which the star operation is an involutive automorphism of the group $\mathfrak{G}$ which may be trivial (typically for unitary groups it can also be the complex conjugation). The elements g_1 and g_2 are related by

$$g_2 = \left(g_1^{-1}\right)^*,$$

which implies that the elements g have the form:

$$g = \left(f^{-1}\right)^* g_0 f, \tag{15.62}$$

where the elements f vary freely in the group $\mathfrak{G}$. As a consequence:

$$g^* g = \epsilon \mathbf{1}. \tag{15.63}$$

Then $G \equiv \mathfrak{G}$, the automorphism of G is $g \mapsto g_0^{-1} g^* g_0$ and H is thus the subgroup of elements which satisfy $h = g_0^{-1} h^* g_0$.

Note that below we extend by continuity the $*$ operation to the Lie algebra.

15.5 The Classical Action. Conservation Laws

In the case of symmetric spaces the classical action S can always be written in a simple geometrical form, since symmetric spaces can be realized in the group manifold itself:

$$S(\mathbf{g}) = \tfrac{1}{2} \int \mathrm{d}x \, \mathrm{tr} \left[\partial_\mu \mathbf{g}(x) \partial_\mu \mathbf{g}^{-1}(x)\right], \tag{15.64}$$

in which $\mathbf{g}(x)$ belongs to some matrix representation of $\mathfrak{G}$. The different symmetric spaces are characterized by the group $\mathfrak{G}$ and the constraints imposed on $\mathbf{g}(x)$.

The action can be rewritten in terms of the associated current $\mathbf{A}_\mu$ which belongs to the Lie algebra $\mathcal{L}(\mathfrak{G})$:

$$\mathbf{A}_\mu = \mathbf{g}^{-1}(x) \partial_\mu \mathbf{g}(x). \tag{15.65}$$

In the language of gauge theories (see Chapters 18,19) $\mathbf{A}_\mu$ is a pure gauge.

The action then takes the form

$$S(\mathbf{g}) = -\frac{1}{2} \int \mathrm{d}x \, \mathrm{tr} \, \mathbf{A}_\mu^2. \tag{15.66}$$

To the action (15.64) corresponds a classical equation of motion. In all cases a general variation of $\mathbf{g}(x)$ can be written:

$$\mathbf{g}(x) \mapsto (1 - \varepsilon^*(x)) \mathbf{g}(x) (1 + \varepsilon(x)), \tag{15.67}$$

in which $\varepsilon(x)$ and $\varepsilon^*(x)$ belong to $\mathcal{L}(\mathfrak{G})$, and are either independent in the case of chiral models or otherwise related by the star automorphism in order to preserve the condition (15.63):

$$\mathbf{g}^*\mathbf{g} = s\mathbf{1}\,. \tag{15.68}$$

The corresponding variation of $\mathbf{A}_\mu$ is

$$\delta\mathbf{A}_\mu = \mathbf{D}_\mu\left(\varepsilon - \mathbf{g}^{-1}\varepsilon^*\mathbf{g}\right),$$

where we have introduced the *covariant derivative* associated with the current $\mathbf{A}_\mu$, which acts on an element ω of the Lie algebra as:

$$\mathbf{D}_\mu\omega = \partial_\mu\omega + [\mathbf{A}_\mu, \omega]. \tag{15.69}$$

The variation of the action then takes the form:

$$\delta S = \operatorname{tr}\left[\partial_\mu\mathbf{A}_\mu\left(\varepsilon - \mathbf{g}^{-1}\varepsilon^*\mathbf{g}\right)\right].$$

In the case of chiral models we can restrict ourselves for example to $\varepsilon^* = 0$. In the other cases, the relation (15.68) implies

$$\partial_\mu\mathbf{A}_\mu^* = -\mathbf{g}\partial_\mu\mathbf{A}_\mu\mathbf{g}^{-1}.$$

In both cases we thus find the equation of motion

$$\partial_\mu\mathbf{A}_\mu = 0\,, \tag{15.70}$$

which expresses the conservation of the current related to the $\mathfrak{G}$-symmetry .

The non-local conserved currents. Equations (15.65,15.70) imply, in two dimensions, the existence of an infinite number of non-local conserved currents.

Let us now define a covariant derivative $\mathbf{D}_\mu$ by:

$$\mathbf{D}_\mu = \partial_\mu\mathbf{1} + \mathbf{A}_\mu\,. \tag{15.71}$$

We note that the explicit form cf the covariant derivative is different from the expression (15.69): Indeed, as also discussed in Section 22.2, this form depends on the representation.

We consider the linear partial differential equations for matrices χ:

$$\mathbf{D}_\mu\chi(x) = \kappa\varepsilon_{\mu\nu}\partial_\nu\chi(x), \tag{15.72}$$

in which χ is a function of x and the spectral parameter κ.

Let us show that the linear system (15.72) is compatible. Setting:

$$\Delta_\mu = \mathbf{D}_\mu - \kappa\varepsilon_{\mu\rho}\partial_\rho, \tag{15.73}$$

we have to calculate the commutator

$$[\Delta_\mu, \Delta_\nu] = [\mathbf{D}_\mu, \mathbf{D}_\nu] - \kappa\left(\varepsilon_{\mu\rho}[\partial_\rho, \mathbf{D}_\nu] + \varepsilon_{\nu\sigma}[\mathbf{D}_\mu, \partial_\sigma]\right). \tag{15.74}$$

In two dimensions, since $\mu \neq \nu$, in the last term of the r.h.s. only $\rho = \nu$ and $\sigma = \mu$ give a non-vanishing contribution:

$$[\Delta_\mu, \Delta_\nu] = [\mathbf{D}_\mu, \mathbf{D}_\nu] - \kappa \varepsilon_{\mu\nu} [\partial_\rho, \mathbf{D}_\rho] . \tag{15.75}$$

The equation of motion (15.70) implies:

$$[\partial_\mu, \mathbf{D}_\mu] = \partial_\mu \mathbf{A}_\mu = 0 . \tag{15.76}$$

The commutator $[\mathbf{D}_\mu, \mathbf{D}_\nu]$ is the curvature $\mathbf{F}_{\mu\nu}$ associated with the connection or gauge field $\mathbf{A}_\mu$ (these concepts will be discussed in Chapters 19,22):

$$\mathbf{F}_{\mu\nu} \equiv [\mathbf{D}_\mu, \mathbf{D}_\nu] = \partial_\mu \mathbf{A}_\nu - \partial_\nu \mathbf{A}_\mu + [\mathbf{A}_\mu, \mathbf{A}_\nu]. \tag{15.77}$$

Equation (15.65) implies that $\mathbf{A}_\mu$ is a pure gauge (for more details see Chapters 18,19). The operator $\mathbf{D}_\mu$, acting on some matrix $\mathbf{X}(x)$ can then be written:

$$\mathbf{D}_\mu \mathbf{X} = (\partial_\mu \mathbf{1} + \mathbf{A}_\mu) \mathbf{X} = \mathbf{g}^{-1} \partial_\mu (\mathbf{g}\mathbf{X}) . \tag{15.78}$$

Thus the product $\mathbf{D}_\mu \mathbf{D}_\nu$,

$$\mathbf{D}_\mu \mathbf{D}_\nu \mathbf{X} = \mathbf{g}^{-1} \partial_\mu \partial_\nu (\mathbf{g}\mathbf{X}), \tag{15.79}$$

is symmetric in the exchange (μ, ν) and the curvature vanishes

$$\mathbf{F}_{\mu\nu} \equiv [\mathbf{D}_\mu, \mathbf{D}_\nu] = 0 . \tag{15.80}$$

The conclusion is:

$$[\Delta_\mu, \Delta_\nu] = 0 , \tag{15.81}$$

and the system (15.72) is compatible.

We can now define the current $\mathbf{J}_\mu (x, \kappa)$:

$$\mathbf{J}_\mu (x, \kappa) = \mathbf{D}_\mu \chi(x, \kappa). \tag{15.82}$$

As a consequence of equations (15.72), the current is conserved:

$$\partial_\mu \mathbf{J}_\mu = 0 . \tag{15.83}$$

The solution of equation (15.72) has an expansion in powers of κ of the form:

$$\chi(x, \kappa) = 1 + \sum_{n=1}^{\infty} \chi_n(x) \kappa^{-n}. \tag{15.84}$$

The corresponding expansion of $\mathbf{J}_\mu$ then generates an infinite number of conserved currents:

$$\mathbf{J}_\mu (x, \kappa) = \mathbf{A}_\mu(x) + \sum_{n=1}^{\infty} \mathbf{J}_\mu^n(x) \kappa^{-n}. \tag{15.85}$$

The interesting question, which we will not to investigate here, then is whether these conservation laws survive quantization. Let us also mention that the corresponding quantum conservation laws lead to the factorization of the S-matrix which can then often be completely determined. We refer the interested reader to the literature for more details.

15.6 Quantum Theory: Perturbation Theory and RG Functions

To quantize the theory we consider the functional integral $Z(\mathbf{j})$:

$$Z(\mathbf{j}) = \int [\mathrm{d}\mathbf{f}(x)] \exp \left\{ -\frac{1}{\lambda} \left[S(\mathbf{g}) - \int \mathrm{d}x \, \mathrm{tr} \, \mathbf{g}(x)\mathbf{j}(x) \right] \right\}, \qquad (15.86)$$

in which $\mathbf{f}(x)$ is a group element either identical to $\mathbf{g}(x)$ for $\mathfrak{G} \times \mathfrak{G} / \mathfrak{G}$ or such that $\mathbf{g}(x)$ is related to $\mathbf{f}(x)$ by equation (15.62):

$$\mathbf{g} = \left(\mathbf{f}^{-1} \right)^* \mathbf{g}_0 \mathbf{f},$$

and $\mathrm{d}\mathbf{f}$ is the de Haar measure for the group $\mathfrak{G}$. The parameter λ is the coupling constant.

We can then parametrize $\mathbf{f}(x)$ in a form analogous to (15.4), in terms of independent field variables $\xi_a(x)$, components on the generators belonging to $\mathcal{L}G/H$. We can then expand perturbation theory around a finite value of the source $\mathbf{j}(x)$ to provide perturbation theory with an IR cut-off.

Actually for practical calculations it is convenient to return to the field representation (15.4,15.6):

$$|\phi(x)\rangle = \mathbf{R}(\xi) |0\rangle = \exp \left[\sum_a t^a \xi_a(x) \right] |0\rangle, \qquad (15.87)$$

valid for general homogeneous spaces. With these notations the generating functional $Z(J)$ reads:

$$Z(J) = \int [\mathrm{d}\rho(\xi)] \exp \left\{ -\frac{1}{\lambda} \int \mathrm{d}^d x \left[-\frac{1}{2} \langle 0| \left(\mathbf{R}^{-1} \partial_\mu \mathbf{R} \right)^2 |0\rangle - \langle J(x)| \mathbf{R} |0\rangle \right] \right\}. \qquad (15.88)$$

To calculate in perturbation theory we expand the action in powers of $\xi_a(x)$. The geometrical part of the Feynman diagram calculation then involves the evaluation of averages of the form $\langle 0| t^{a_1} t^{a_2} \dots t^{a_k} |0\rangle$.

15.6.1 RG functions at one-loop order

We shall eventually discuss the properties of these models from the point of view of the renormalization group. Therefore we here calculate the renormalization constants and the renormalization group functions at one-loop order. At the same time these calculations will illustrate some of the previous considerations.

Preliminary remarks. We normalize the vector $|0\rangle$:

$$\langle 0 | 0 \rangle = 1. \qquad (15.89)$$

To evaluate the one-loop diagrams, we need also a few tensors:

$$\langle 0| t^a t^b |0\rangle = -\delta_{ab}. \qquad (15.90)$$

This relation is a consequence (up to the normalization) of the property that in the case of symmetric spaces the vectors $t^a |0\rangle$ form an irreducible representation of the subgroup H.

In the same way because the structure constants are antisymmetric and the vector $|0\rangle$ is the unique vector having H for the stabilizer group:

$$t^a t^a |0\rangle = -\mu |0\rangle . \tag{15.91}$$

In equation (15.91) and in the equations which follow, repeated indices means summation over indices running from 1 to l, values which correspond to the generators of G/H.

Combining equations (15.89–15.91) we find the value of μ:

$$t^a t^a |0\rangle = -l |0\rangle . \tag{15.92}$$

We need two tensors with three indices. Since the generators are represented by anti-symmetric matrices:

$$\langle 0| t^a t^b t^c |0\rangle + \langle 0| t^c t^b t^a |0\rangle = 0 .$$

A few commutations and the properties of symmetric spaces lead to:

$$\langle 0| t^a t^b t^c |0\rangle = 0 . \tag{15.93}$$

A different useful quantity is obtained by replacing t^b by a generator τ^β of H:

$$\langle 0| t^a \tau^\beta t^c |0\rangle = -f_{a\beta c} . \tag{15.94}$$

We have still to evaluate three tensors with four indices. Combining commutation relations and previous relations we find:

$$\langle 0| t^a t^a t^b t^b |0\rangle = l^2, \tag{15.95}$$

$$\langle 0| t^a t^b t^a t^b |0\rangle = l^2 - \sum_{\gamma > l} f_{\gamma ab} f_{\gamma ab}, \tag{15.96}$$

$$\langle 0| t^a t^b t^b t^a |0\rangle = \langle 0| t^a t^b t^a t^b |0\rangle .$$

Using the antisymmetry of the structure constants and the special properties of symmetric spaces it is easy to verify:

$$\sum_{\alpha > l} \sum_{a=1}^{l} f_{\alpha ab} f_{\alpha ab'} = \tfrac{1}{2} \delta_{bb'} C , \tag{15.97}$$

in which C is the Casimir of the group G. Summing over $b = b'$ we find an expression for the second term in the r.h.s. of equation (15.96). Therefore:

$$\langle 0| t^a t^b t^a t^b |0\rangle = l \, (l - C/2) . \tag{15.98}$$

Renormalization constants at one-loop. To obtain the two one-loop renormalization constants, we demand that the functional $W(J)$ is one-loop finite when calculated with the renormalized action S_{r}:

$$S_{\mathrm{r}} = \frac{1}{\lambda Z_\lambda} \int \mathrm{d}^d x \left[\tfrac{1}{2} \langle \partial_\mu \phi | \partial_\mu \phi \rangle - Z_\phi^{-1/2} Z_\lambda \langle J(x) | \phi(x) \rangle \right], \tag{15.99}$$

for the special constant source:

$$|J(x)\rangle = m^2 |0\rangle . \tag{15.100}$$

Dimensional regularization will be used so that the measure term can be omitted. The action has to be expanded up to order $\xi^4(x)$. Using:

$$e^{-t\cdot\xi}\, \partial_\mu\, e^{t\cdot\xi} = t\cdot\partial_\mu\xi + \frac{1}{2}\left[t\cdot\partial_\mu\xi, t\cdot\xi\right] + \frac{1}{6}\left[\left[\partial_\mu\xi\cdot t, \xi\cdot t\right],\xi\cdot t\right] + O\left(\xi^4\right), \tag{15.101}$$

which for symmetric spaces leads to:

$$\langle\partial_\mu\phi\,|\partial_\mu\phi\rangle = (\partial_\mu\xi)^2 + \frac{1}{3}\sum_\gamma (f_{\gamma ab}\xi_a\partial_\mu\xi_b)^2 + O\left(\xi^5\right), \tag{15.102}$$

we can write the renormalized action up to order ξ^4:

$$S_r(\xi) = \frac{1}{\lambda}\int d^d x\left\{\frac{1}{Z_\lambda}\left[\frac{1}{2}(\partial_\mu\xi)^2 + \frac{1}{6}\sum_\gamma (f_{\gamma ab}\xi_a\partial_\mu\xi_b)^2\right]\right.$$
$$\left. - Z_\phi^{-1/2}m^2\left(1 - \tfrac{1}{2}\xi^2 + \frac{1}{24}\langle 0|\, t^a t^b t^c t^d\,|0\rangle\, \xi_a\xi_b\xi_c\xi_d\right)\right\} + O\left(\xi^5\right). \tag{15.103}$$

A short calculation yields $W(m^2)$ up to order λ:

$$W(m^2) = Z_\phi^{-1/2}\frac{m^2}{\lambda} + \frac{l}{2}\int d^d q \ln\left(1 + \frac{m^2}{q^2}Z_\phi^{-1/2}Z_\lambda\right) - \frac{\lambda}{8}m^2 l\,(l - C)\left(\int\frac{d^d q}{q^2 + m^2}\right)^2$$
$$+ O\left(\lambda^2\right). \tag{15.104}$$

Setting $d = 2+\varepsilon$, we make a Laurent expansion for ε small and define the renormalization constants by minimal subtraction:

$$Z_\lambda = 1 + \frac{C}{4\pi\varepsilon}\lambda + O\left(\lambda^2\right), \tag{15.105}$$

$$Z_\phi = 1 + \frac{l}{2\pi\varepsilon}\lambda + \frac{l\,(2l + C)}{(4\pi\varepsilon)^2}\lambda^2 + O\left(\lambda^3\right). \tag{15.106}$$

The coupling constant RG function $\beta(\lambda)$ and the field RG function $\eta(\lambda)$ are then:

$$\beta(\lambda) = \varepsilon\lambda\left(1 + \lambda\frac{d}{d\lambda}\ln Z_\lambda\right)^{-1} = \varepsilon\lambda - \frac{C}{4\pi}\lambda^2 + O\left(\lambda^3\right), \tag{15.107}$$

$$\eta(\lambda) = \beta(\lambda)\frac{d\ln Z_\phi}{d\lambda} = \frac{l}{2\pi}\lambda + O\left(\lambda^4\right). \tag{15.108}$$

In contrast with field theories like $\lambda\phi_4^4$, the sign of the leading term of the β-function, in the dimension in which the theory is just renormalizable ($\varepsilon = 0$), is negative. The physical significance (asymptotic freedom) of this remarkable property will be discussed in Chapter 30.

15.6.2 One-loop β-function and background field method

If we want only to calculate the coefficients of the perturbative expansion of the β-function, we can shorten the calculation by using the background field method whose principles have been explained in Appendix A10.3. As an exercise and to show some advantages of this method, we again perform the calculation at one-loop order. For this purpose we evaluate the vacuum amplitude $Z(\theta)$ and the free energy $W(\theta)$ in a finite volume with non-trivial boundary conditions (for more details see Chapter 36): in $(d-1)$ dimensions the coordinates x_μ vary in the interval:

$$0 \le x_\mu \le L_\perp, \qquad 1 \le \mu \le d-1,$$

and we impose periodic boundary conditions on the field:

$$\phi(z, x_1, \ldots, 0, \ldots, x_{d-1}) = \phi(z, x_1, \ldots, L_\perp, \ldots, x_{d-1}).$$

We have called z the last coordinate, the imaginary time coordinate, which varies in the interval:

$$0 \le z \le L,$$

and for which we impose fixed "twisted" boundary conditions:

$$|\phi_\theta(z=0, \mathbf{x})\rangle = |0\rangle, \qquad |\phi_\theta(z=L, \mathbf{x})\rangle = \mathrm{e}^{\boldsymbol{\theta}}|0\rangle, \tag{15.109}$$

in which $\boldsymbol{\theta}$ is a linear combination of generators belonging to $\mathcal{L}(G/H)$:

$$\boldsymbol{\theta} = \sum t^a \theta_a, \qquad t^a \in \mathcal{L}(G/H). \tag{15.110}$$

As a consequence, momenta in Fourier space are quantized:

$$p_\mu = \frac{2\pi}{L_\perp} n_\mu, \quad n_\mu \in \mathbb{Z}^{d-1}, \qquad p_z = \frac{\pi}{L} m, \quad m \in \mathbb{Z}.$$

The large L limit will be taken before the large $L_\perp$ limit.

To deal with the longitudinal boundary conditions it is convenient to set:

$$|\phi_\theta(z, \mathbf{x})\rangle = \mathrm{e}^{z\boldsymbol{\theta}/L} |\phi(z, \mathbf{x})\rangle. \tag{15.111}$$

The new field then satisfies:

$$|\phi(0, \mathbf{x})\rangle = |\phi(L, \mathbf{x})\rangle = |0\rangle, \tag{15.112}$$

which in the parametrization:

$$|\phi(z, \mathbf{x})\rangle = \exp[t^a \xi_a(z, \mathbf{x})] |0\rangle, \tag{15.113}$$

is equivalent to:

$$\xi_a(0, \mathbf{x}) = \xi_a(L, \mathbf{x}) = 0. \tag{15.114}$$

The renormalized action then reads $(\partial_z \equiv \partial/\partial z)$:

$$S_{\mathrm{r}}(\xi) = \frac{1}{2\lambda Z_\lambda} \int \mathrm{d}z\, \mathrm{d}^{d-1}x \left[\langle \partial_\mu \phi | \partial_\mu \phi \rangle + \frac{2}{L} \langle \partial_z \phi | \boldsymbol{\theta} | \phi \rangle - \frac{1}{L^2} \langle \phi | \boldsymbol{\theta}^2 | \phi \rangle \right]. \tag{15.115}$$

The calculation of the one-loop contribution involves only the expansion of $S_r(\xi)$ up to order ξ^2. Using the relations (15.90) and (15.93):

$$\langle 0 | t^a t^b | 0 \rangle = -\delta_{ab}, \qquad \langle 0 | t^a t^b t^c | 0 \rangle = 0,$$

we find:

$$\langle \partial_\mu \phi | \partial_\mu \phi \rangle = (\partial_\mu \boldsymbol{\xi})^2 + O\left(\xi^3\right), \tag{15.116}$$

$$\int dz \, \langle \partial_z \phi | \, \boldsymbol{\theta} \, | \phi \rangle = \int \partial_z \xi_a \theta_a dz + O\left(\xi^3\right) = O\left(\xi^3\right), \tag{15.117}$$

$$\langle \phi | \boldsymbol{\theta}^2 | \phi \rangle = -\theta_a \theta_a + \tfrac{1}{2} \xi_a \xi_b \, \langle 0 | [t^a, [t^b, \boldsymbol{\theta}^2]] | 0 \rangle + O\left(\xi^3\right). \tag{15.118}$$

Let us evaluate the last term:

$$V_{abcd} = \tfrac{1}{2} \langle 0 | [t^a, [t^b, t^c t^d]] | 0 \rangle = \tfrac{1}{2} \sum_\varepsilon \left(f_{\varepsilon ad} f_{\varepsilon bc} + f_{\varepsilon ac} f_{\varepsilon bd} \right). \tag{15.119}$$

The action in the gaussian approximation is:

$$S_r(\xi) = \frac{L_\perp^{d-1} \theta_a \theta_a}{2L\lambda Z_\lambda} + \int d^{d-1}x \, dz \left[\frac{1}{2} (\partial_\mu \boldsymbol{\xi})^2 - \frac{1}{2L^2} \xi_a \xi_b \theta_c \theta_d V_{abcd} \right] + O\left(\xi^3\right). \tag{15.120}$$

The free energy as a function of θ_a is at one-loop order:

$$W(\theta) = -\frac{L_\perp^{d-1} \theta_a \theta_a}{2L\lambda Z_\lambda} - \frac{1}{2} \operatorname{tr} \ln \left\{ -\left[\partial_z^2 + \Delta_\perp \right] \delta_{ab} - \frac{1}{L^2} V_{abcd} \theta_c \theta_d \right\}, \tag{15.121}$$

where $\Delta_\perp$ is the laplacian in $d-1$ dimensions. To compute the renormalization constant Z_λ, it is sufficient to expand up to second order in θ:

$$W(\theta) = -\frac{L_\perp^{d-1} \theta_a \theta_a}{2L\lambda Z_\lambda} + \frac{1}{2L^2} V_{aacd} \theta_c \theta_d \sum_{\substack{p_z = m\pi/L \\ \mathbf{p} = 2\pi\mathbf{n}/L_\perp}} \frac{1}{p_z^2 + \mathbf{p}^2}. \tag{15.122}$$

We have shown that in a symmetric space (equation (15.97)):

$$V_{aacd} = \sum_\varepsilon f_{\varepsilon ad} f_{\varepsilon ac} = \tfrac{1}{2} C \delta_{cd} > 0,$$

in which C is the Casimir of the group G.

Since we want to evaluate the UV divergences of the one-loop sum for $d = 2 + \varepsilon$, we can replace the sum by an integral:

$$\sum_{p_z, \mathbf{p}} \frac{1}{p_z^2 + \mathbf{p}^2} \sim \frac{L L_\perp}{(2\pi)^2} \int_{|p| > 1} \frac{d^d p}{\mathbf{p}^2} \sim -\frac{L L_\perp}{2\pi\varepsilon}. \tag{15.123}$$

It follows that Z_λ in the minimal subtraction scheme is:

$$Z_\lambda = 1 + \frac{C}{4\pi\varepsilon} \lambda + O\left(\lambda^2\right), \tag{15.124}$$

in agreement with the result (15.105) of Subsection 15.6.1.

15.7 Generalizations

Non-compact homogeneous spaces. Up to now we have restricted the discussion to homogeneous spaces based on compact groups. However most of the arguments can be generalized in a straightforward manner to a class of homogeneous spaces G/H in which G is non-compact. Obvious conditions are that the metric g_{ij} should be positive, and the group G unimodular to preserve the G invariance of the functional measure. For example we can take G semi-simple. The positivity of the metric implies that H is compact and even the maximal compact subgroup of G.

Some care has to be taken to define properly the functional integral since the volume of the group manifold is in general infinite so that appropriate boundary conditions have to be imposed.

Simple examples of models belonging to this class are provided by analytic continuations of compact symmetric spaces replacing formally in the orthogonal representation the generators t^a in G/H by it^a. For example, after this transformation, the compact Grassmannian manifold $O(N+M)/O(M) \times O(N)$ becomes the manifold $O(M,N)/O(M) \times O(N)$ in which $O(M,N)$ is the pseudo-orthogonal group leaving the metric with $M +$ signs and $N -$ signs invariant.

The perturbative expansion of the correlation functions of these models is obtained, up to global signs, by changing the sign of the coupling constant in the compact models. This changes in particular the sign of the one loop β-function in two dimensions, and has thus important consequences from the renormalization group point of view.

Arbitrary Riemannian manifolds. A much wider class of models has been studied by Friedan. One considers an arbitrary smooth (infinitely differentiable) Riemannian manifold with a smooth positive definite metric $g_{ij}(\varphi)$ and one takes the classical action $S(\varphi)$:

$$S(\varphi) = \tfrac{1}{2} \int \mathrm{d}^d x \, \partial_\mu \varphi^i(x) g_{ij}\left(\varphi(x)\right) \partial_\mu \varphi^j(x). \tag{15.125}$$

The generating functional of correlation functions $Z(J)$ is then given by:

$$Z(J) = \int [\mathrm{d}\rho(\varphi)] \exp\left[-S(\varphi) + \int \mathrm{d}x \, J_i(x)\varphi^i(x) \right], \tag{15.126}$$

in which $\mathrm{d}\rho(\varphi)$ has to be a smooth, strictly positive, covariant measure on the manifold, for example (see Chapter 22)

$$\mathrm{d}\rho(\varphi) = \sqrt{g}\, \mathrm{d}\varphi \,,$$

in which g is the determinant of the metric tensor.

As we have discussed in Sections 3.2 and 4.7, in one dimension the functional integral (15.126) is associated with a hamiltonian of the form of a Laplace operator on the manifold with metric g_{ij} and is related to brownian motion or diffusion processes on the manifold.

In two dimensions, the action (15.125) corresponds to theories renormalizable by power counting, which generalizes theories on homogeneous spaces. However since the symmetry properties are lost, and in particular the Goldstone theorem which forces fields to remain massless, important differences appear after renormalization:

(i) in general the space of all possible metrics is infinite dimensional;

(ii) with a given metric can be associated an infinite number of different covariant measure terms since one can construct an infinite number of scalars like the scalar curvature (see Chapter 22), and a covariant measure multiplied by a scalar is still covariant.

As a consequence, in contrast to homogeneous cases, derivative-free terms will be generated in the renormalization and it is only possible to maintain the form (15.125) of the action by adjusting an infinite number of parameters.

In addition the renormalized metric will be generically the most general metric on the manifold. In other words the renormalized action is the most general action allowed by power counting arguments.

Considerations based on covariance (Chapter 22) nevertheless are useful since they simplify perturbative calculations in the general situation by restricting the form of the counterterms at a given order in the loop expansion.

For example the equivalent of the coupling constant RG function is a functional of the metric $\beta\left(g_{ij}\right)$. It has the covariance of the metric tensor. At one-loop it can only involve first and second derivatives of the metric and therefore R_{ij} the Ricci tensor and Rg_{ij} in which R is the scalar curvature. By inspection it is possible to eliminate Rg_{ij} and the constant in front of R_{ij} can be obtained from a particular model. Friedan has in this way obtained the first two terms. We have expressed them in terms of g^{ij} the inverse of the metric tensor g_{ij}, because it naturally orders perturbation theory. In $2 + \varepsilon$ dimensions:

$$\beta^{ij}(g) = \varepsilon g^{ij} - \frac{1}{2\pi}R^{ij} - \frac{1}{8\pi^2}R^{iklm}R^j_{klm} + O\left(\left(g^{ij}\right)^4\right), \qquad (15.127)$$

in which R^{ij} is obtained from the Ricci tensor by raising the indices with g^{ij} and R^j_{klm} is the curvature tensor.

Bibliographical Notes

The general group theoretical problem of non-linear realizations is discussed in
    S. Coleman, J. Wess and B. Zumino, *Phys. Rev.* 177 (1969) 2239.
See also
    S. Gasiorowicz and D. Geffin, *Rev. Mod. Phys.* 41 (1969) 531.
Quantization of chiral models is described in
    L.D. Faddeev and A.A. Slavnov, *Theor. Math. Phys.* 8 (1971) 843.
Classical conservation laws and RG functions in symmetric spaces have been discussed in
    V.E. Zakharov and A.V. Mikhailov, *Zh. Eksp. Teor. Fiz.* 74 (1978) 1953, *Commun. Math. Phys.* 74 (1980) 21; V.L. Golo and A.M. Perelomov, *Phys. Lett.* 79B (1978) 112; H. Eichenherr, *Nucl. Phys.* B146 (1978) 215; A. D'Adda, P. di Vecchia and M. Lüscher, *Nucl. Phys.* B146 (1978) 63; E. Witten, *Nucl. Phys.* B149 (1979) 285; H. Eichenherr and M. Forger, *Nucl. Phys.* B155 (1979) 381; M. Lüscher, *Nucl. Phys.* B135 (1978) 1; E. Brézin, C. Itzykson, J. Zinn-Justin and J.-B. Zuber, *Phys. Lett.* 82B (1979) 442; H.J. de Vega, *Phys. Lett.* 87B (1979) 233.
For a recent discussion of quantum conservation laws and *S*-matrix in symmetric spaces see
    E. Abdalla, M.C.B. Abdalla and M. Forger, *Nucl. Phys.* B297 (1988) 374 and references therein.
Additional results concerning non-local currents can be found in
    D. Bernard, *Commun. Math. Phys.* 137 (1991) 191; D. Bernard and A. LeClair, *Commun. Math. Phys.* 142 (1991) 99.

Different properties are described and RG functions calculated in

R.D. Pisarski, *Phys. Rev.* D20 (1979) 3358; E. Brézin, S. Hikami and J. Zinn-Justin, *Nucl. Phys.* B165 (1980) 528.

For the application of these models to the metal-insulator transition of independent electrons in a random potential see

F. Wegner, *Z. Phys.* B35 (1979) 207; L. Schäfer and F. Wegner, *Z. Phys.* B38 (1980) 113; S. Hikami, *Prog. Theor. Phys. Suppl.* 84 (1985) 120.

Calculation of the RG β-functions up to three and four loops for a class of models can be found in:

S. Hikami, *Phys. Lett.* 98B (1981) 208; W. Bernreuther and F.J. Wegner, *Phys. Rev. Lett.* 57 (1986) 1383; F. Wegner, *Nucl. Phys.* B316 (1989) 663.

For a discussion of general two-dimensional models on Riemannian manifolds see

D.H. Friedan, *Ann. Phys. (NY)* 163 (1985) 318.

Exercises

Exercise 15.1

Show by explicit calculation that the differential operators

$$\Delta_\alpha \equiv D_a^\alpha \frac{\partial}{\partial \pi_a} = \left[t_{ab}^\alpha \pi_b + t_{aj}^\alpha \sigma_j(\pi) \right] \frac{\partial}{\partial \pi_a}, \qquad (15.128)$$

where the functions σ_i are defined in equation (15.19), satisfy the commutation relations (15.42).

Exercise 15.2

Verify then by direct calculation the covariance of equations (15.43) in a reparametrization of the homogeneous space.

Exercise 15.3

Prove that the tensor g^{ij} $(A15.14)$ satisfies the relation $(A15.7)$ and is therefore a symmetric inverse metric tensor.

Exercise 15.4

Verify explicitly in the case of the $O(4)$ group that the tensor $(A15.14)$ is the inverse of the metric on the sphere S_3.

APPENDIX 15
HOMOGENEOUS SPACES: A FEW ALGEBRAIC PROPERTIES

This appendix first describes some additional properties of homogeneous spaces when considered as Riemannian manifolds. It assumes some minimal familiarity with the elements of differential geometry presented in Chapter 22. The second part is devoted to some elements of classification of symmetric spaces.

A15.1 Pure Gauge. Maurer–Cartan Equations

Multiplying $\mathbf{R}(\varphi)$ by a group element $\mathbf{R}_g$ on the left we see that $\mathbf{R}^{-1}\partial_i\mathbf{R}$ transforms like

$$\mathbf{R}^{-1}\partial_i\mathbf{R} \mapsto \mathbf{R}_g^{-1}\mathbf{R}^{-1}\partial_i\mathbf{R}\mathbf{R}_g\,.$$

This implies that the matrices $\mathbf{R}^{-1}\partial_i\mathbf{R}$ transform like elements of the adjoint representation of the group G. They can therefore be expanded on the generators t^α and we set (equation (15.10)):

$$\mathbf{R}^{-1}\partial_i\mathbf{R} = L_i^\alpha(\varphi)t^\alpha\,. \tag{A15.1}$$

It is easy to verify more directly by parametrizing the l.h.s. as

$$\mathbf{R} = e^{t^\alpha\xi^\alpha}\,,$$

that the expansion of L_i^α in powers of ξ involves only commutators of generators of $\mathcal{L}(G)$ and therefore $L_i^\alpha(\varphi)$ depends on the parametrization of the group elements, but not on the representation to which $\mathbf{R}(\varphi)$ belongs.

The definition $(A15.1)$ implies that the quantities $L_i^\alpha(\varphi)$ are the components of a vector belonging to the adjoint representation and have the φ^i dependence of a pure gauge (see Chapters 19, 20 for details). We have shown in Section 15.5 that the corresponding curvature vanishes (equation (15.80)). In terms of the components L_i^α one finds

$$\partial_i\left(\mathbf{R}^{-1}\partial_j\mathbf{R}\right) - \partial_j\left(\mathbf{R}^{-1}\partial_i\mathbf{R}\right) = t^\alpha\left[\partial_i L_j^\alpha - \partial_j L_i^\alpha\right]$$
$$\left[\mathbf{R}^{-1}\partial_i\mathbf{R}, \mathbf{R}^{-1}\partial_j\mathbf{R}\right] = L_i^\alpha L_j^\beta f_{\alpha\beta}^\gamma t^\gamma\,.$$

We thus find

$$\partial_i L_j^\alpha - \partial_j L_i^\alpha + f_{\beta\gamma}^\alpha L_i^\beta L_j^\gamma = 0\,. \tag{A15.2}$$

These relations, which express that the curvature corresponding to the gauge connection L_i^α vanishes, are known as the Maurer–Cartan equations.

A15.2 Metric and Curvature in Homogeneous Spaces

A group transformation acting on the coordinates φ^i can also be considered as a reparametrization of the manifold. The infinitesimal form is given by equation (15.37):

$$\varphi^i = \varphi'^i + D_\alpha^i(\varphi')\omega^\alpha\,. \tag{A15.3}$$

The generator Δ_α of $\mathcal{L}(G)$, as defined by equation (15.41), then characterizes the corresponding infinitesimal variation of scalars

$$\Delta_\alpha S(\varphi) = D_\alpha^i(\varphi)\partial_i S(\varphi)\,. \tag{A15.4}$$

More generally equation (22.9) defines its action on all tensors on the homogeneous space. For vectors $V_i(\varphi)$ it yields

$$\Delta_\alpha V_i = D_\alpha^j \partial_j V_i + \partial_i D_\alpha^j V_j \,. \tag{A15.5}$$

This transformation law can be verified by a short calculation in the case of the gauge field $L_i^a(\varphi)$, defined by equation (15.10). As explained in Section 22.1 $\Delta_\alpha V_i$ is a vector, as expected.

For general tensors the result is

$$\Delta_\alpha V^{i_1 \ldots i_p}_{i_{p+1} \ldots i_n} = D_\alpha^j \partial_j V^{i_1 \ldots i_p}_{i_{p+1} \ldots i_n} - \sum_{\ell=1}^{p} \partial_j D_\alpha^{i_\ell} V^{i_1 \ldots j \ldots i_p}_{i_{p+1} \ldots i_n} + \sum_{\ell=p+1}^{n} \partial_{i_\ell} D_\alpha^j V^{i_1 \ldots i_p}_{i_{p+1} \ldots j \ldots i_n} \,. \tag{A15.6}$$

With this definition Δ_α obeys the usual rule of differentiation for products of tensors.

The invariance of the metric, as expressed by equation (15.40), then takes the simple form

$$\Delta_\alpha g_{jk} = 0 \quad \Leftrightarrow \quad \nabla_i D_{j\alpha} + \nabla_j D_{i\alpha} = 0 \,, \tag{A15.7}$$

where the second equation follows from equation (22.110).

Consistency with parallel transport. The metric tensor defines uniquely a torsion-free parallel transport on the manifold. If the infinitesimal change of variables $(A15.3)$ leaves the metric invariant, it leaves invariant all quantities function only of the metric. With the definition $(A15.6)$ we can write

$$\Delta_\alpha R^k_{lij} = 0 \,, \tag{A15.8}$$

$$\Delta_\alpha R_{ij} = 0 \,, \tag{A15.9}$$

$$\Delta_\alpha R = 0 \,. \tag{A15.10}$$

The Christoffel connection is also invariant, but since it is not a tensor the action of Δ_α takes the inhomogeneous form (22.33):

$$\Delta_\alpha \Gamma^i_{jk} = \partial_j D_\alpha^l \Gamma^i_{jk} - \partial_l D_\alpha^i \Gamma^l_{jk} + \partial_k D_\alpha^l \Gamma^i_{jl} + \partial_j D_\alpha^l \Gamma^i_{lk} + \partial_j \partial_k D_\alpha^i = 0 \,.$$

The compatibility between parallel transport and symmetry can then be expressed by the commutation relation

$$[\Delta_\alpha, \nabla_i] = 0 \,. \tag{A15.11}$$

To prove this relation, it is sufficient to verify it on scalars and vectors, it then follows from forming tensor products. For scalars it is an immediate consequence of the definition $(A15.6)$. For vectors one finds

$$[\Delta_\alpha, \nabla_i] V_j = V^k \Delta_\alpha \Gamma^j_{ki} = 0 \,.$$

The equation $(A15.9)$ implies that R_{ij} is an acceptable metric tensor and the equation $(A15.10)$ that the scalar curvature is a constant in homogeneous spaces. More generally all symmetric tensors with two indices constructed from the curvature tensor satisfy the equivalent of equation $(A15.9)$ and are of the form of a metric tensor. Since in the case of homogeneous spaces, as we have shown, the set of metrics form a finite dimensional vector space, only a finite number of these tensors are linearly independent.

A final remark: expression (15.12) for the metric shows that the quantities $L_i^a(\varphi)$ play essentially the role of a vielbein in the case of homogeneous spaces, the only difference being the constant internal metric μ_{ab}.

Remark. In Chapter 15 the coordinates φ^i are themselves fields depending on variables x^μ. Then $\partial_\mu \varphi^i(x)$ belongs to the space tangent to the manifold at point $\varphi^i(x)$, and thus transforms like a vector. Using the metric tensor, we have then constructed scalars like the action density of (15.8). Moreover in this situation, another covariant derivative D_μ can be defined, which involves the connection (22.20):

$$D_\mu V^i = \partial_\mu V^i + \Gamma_{kj}^i \partial_\mu \varphi^j V^k \,. \tag{$A15.12$}$$

On functions of $\varphi(x)$ only, this definition is redundant, since D_μ can be rewritten in terms of the covariant derivative, but it is useful when applied to derivatives of the field φ. Generalization to higher order tensors is straightforward: one contracts the free index with $\partial_\mu \varphi^i$.

The definition ($A15.12$) allows us to write the classical equation of motion corresponding to action (15.125) in covariant form:

$$D_\mu \partial_\mu \varphi^i(x) = 0 \,. \tag{$A15.13$}$$

A15.3 Explicit Expressions for the Metric

Let finally give several more explicit expressions for the metric.

A15.3.1 Metric tensor and transformation law

Let us show that, for a general homogeneous space, we can find an inverse metric tensor g^{ij} of the form

$$g^{ij}(\varphi) = D_\alpha^i(\varphi) m^{\alpha\beta} D_\beta^j(\varphi), \tag{$A15.14$}$$

where $m^{\alpha\beta}$ is a constant symmetric non-singular matrix. The tensor has to satisfy the equivalent of equations (15.40) or ($A15.7$):

$$D_\alpha^k \partial_k g^{ij} = g^{ik} \partial_k D_\alpha^j + g^{jk} \partial_k D_\alpha^i.$$

We replace g^{ij} by the form ($A15.14$)

$$m^{\beta\gamma} D_\alpha^k \left(\partial_k D_\beta^i D_\gamma^j + \partial_k D_\beta^j D_\gamma^i \right) = m^{\beta\gamma} D_\gamma^k \left(D_\beta^i \partial_k D_\alpha^j + D_\beta^j \partial_k D_\alpha^i \right),$$

and use the Lie algebra commutation relations (15.43). We then obtain

$$m^{\beta\gamma} f_{\alpha\beta}^\delta \left(D_\gamma^j D_\delta^i + D_\gamma^i D_\delta^j \right) = 0 \,. \tag{$A15.15$}$$

Exchanging $\gamma \leftrightarrow \delta$ in one of the terms, we see that the equation is satisfied if $m^{\alpha\beta}$ is solution to the numerical equation

$$m^{\gamma\beta} f_{\beta\alpha}^\delta = f_{\alpha\beta}^\gamma m^{\beta\delta} \,.$$

In a basis in which the generators t^α are orthogonal by the trace, the structure constants are antisymmetric and $m^{\alpha\beta} = m\delta_{\alpha\beta}$ is the solution to the last equation.

In the case of symmetric spaces expressions simplify. The vector $t^a\,|0\rangle$ transforms under an irreducible representation of H, the matrix $\boldsymbol{\mu}$ defined by equation (15.13):

$$\mu_{ab} = -\,\langle 0|\, t^a t^b\, |0\rangle \,, \tag{A15.16}$$

is diagonal

$$\mu_{ab} = \mu\delta_{ab}\,.$$

This provides another proof of the uniqueness of the metric.

The tensor $g^{ij}\,(\varphi)$ (equation $(A15.14)$) is a possible inverse metric tensor. In the case of symmetric spaces the unique metric, in a basis in which the generators t^α are orthogonal by the trace, is thus,

$$g^{ij}(\varphi) = D^i_\alpha(\varphi)D^j_\alpha(\varphi), \tag{A15.17}$$

up to the normalization.

A15.3.2 Manifolds embedded in euclidean space

If we know an embedding of a Riemannian manifold $\mathfrak{M}$ in euclidean space, we can describe it with constrained euclidean coordinates, as we have done in the case of homogeneous spaces in Section 15.2.

Let $\left(\sigma^s, \varphi^i\right)$ be such a set of euclidean coordinates. We assume that locally the σ^s can be expressed as functions of the independent coordinates φ^i:

$$\sigma^s = \sigma^s(\varphi). \tag{A15.18}$$

The metric tensor in this representation is obtained from

$$g_{ij}(\varphi)\mathrm{d}\varphi^i\mathrm{d}\varphi^j = \mathrm{d}\varphi^i\mathrm{d}\varphi^i + \mathrm{d}\sigma^s\mathrm{d}\sigma^s, \tag{A15.19}$$

and therefore:

$$g_{ij}(\varphi) = \delta_{ij} + \partial_i\sigma^s\partial_j\sigma^s. \tag{A15.20}$$

A short calculation shows that the connection has the simple form:

$$\Gamma^i_{jk} = g^{il}\partial_l\sigma^s\partial_j\partial_k\sigma^s, \tag{A15.21}$$

and the curvature tensor is given by:

$$R_{ijkl} = \left(\partial_i\partial_k\sigma^s\partial_j\partial_l\sigma^s - \partial_i\partial_l\sigma^s\partial_j\partial_k\sigma^s\right)\left(\mathrm{tr}\left(\mathbf{g}^{-1}\right) - N + 1\right), \tag{A15.22}$$

in which N is the dimension of $\mathfrak{M}$ and $\mathrm{tr}\,\mathbf{g}^{-1}$ the trace of the inverse of the metric:

$$\mathrm{tr}\,\mathbf{g}^{-1} = \mathrm{tr}_{st}\left[\delta_{st} + \partial_k\sigma^s\partial_k\sigma^t\right]^{(-1)}. \tag{A15.23}$$

A15.4 Symmetric Spaces: Classification

In this appendix we examine a few simple properties of symmetric spaces and provide some elements of classification.

A15.4.1 Definition

Let us consider a semi-simple compact Lie group G and assume that we have constructed a non-trivial involutive automorphism of G, which to an element g of G associates an element $\bar{g}$:

$$(\overline{g_1 g_2}) = \bar{g}_1 \bar{g}_2\,, \qquad \text{with } \bar{\bar{g}} = g\,. \qquad (A15.24)$$

We consider the coset space G/H obtained by taking for H the subgroup of invariant elements under the automorphism:

$$\overline{H} \equiv H\,. \qquad (A15.25)$$

The automorphism can be extended to the Lie algebra $\mathcal{L}(G)$. It then becomes a reflection and each element of $\mathcal{L}(G)$ can be decomposed into a sum of an even and an odd element. Even elements belong by definition to $\mathcal{L}(H)$ and the generators of $\mathcal{L}(H)$ are denoted by τ^α. The generators of $\mathcal{L}(G)$ not belonging to $\mathcal{L}(H)$ (we denote the corresponding vector space $\mathcal{L}(G/H)$) can be chosen odd. We denote them by t^a. We choose the Lie algebra structure constants f_{ijk} completely antisymmetric. We then have the rules:

$$\begin{cases} \bar{t}^a = -t^a\,, & t^a \in \mathcal{L}(G/H)\,; \\ \bar{\tau}^\alpha = \ \tau^\alpha\,, & \tau^\alpha \in \mathcal{L}(H)\,. \end{cases} \qquad (A15.26)$$

It follows:

$$[\tau^\alpha, \tau^\beta] = f_{\alpha\beta\gamma} \tau^\gamma\,, \qquad (A15.27)$$

$$[\tau^\alpha, t^b] = f_{\alpha b c} t^c\,, \qquad (A15.28)$$

$$[t^a, t^b] = f_{ab\gamma} \tau^\gamma\,. \qquad (A15.29)$$

Note that in the case of a compact group only the last set $(A15.29)$ of commutation relations is characteristic of a symmetric space, since $(A15.28)$ is then a consequence of $(A15.27)$ and the antisymmetry of $f_{\alpha b c}$.

Preliminary remarks

(i) We will consider symmetric spaces derived from non-simple groups G, but we want to exclude the possibility that:

$$G = G_1 \times G_2\,,$$
$$H = H_1 \times H_2 \ \text{ with } \ H_1 \subset G_1\,, H_2 \subset G_2\,, \qquad (A15.30)$$

because in this case the coset space decomposes into two independent spaces G_1/H_1 and G_2/H_2. In particular this excludes the trivial situation $G_2 \equiv H_2$ and this property will be used in what follows.

(ii) All generators of $\mathcal{L}(H)$ can be obtained as linear combinations of commutators of generators of $\mathcal{L}(G/H)$. Indeed assume a generator τ_δ cannot be obtained. We can then rearrange the generators in such a way that $f_{ab\delta} = 0$. It follows that τ_δ commutes with $\mathcal{L}(G/H)$ and thus with all generators of $\mathcal{L}(H)$ which can be obtained as commutators $(A15.29)$. We are exactly in the situation we just excluded.

A15.4.2 A Basic property

The purpose of this appendix is not to present a complete description of the mathematical properties of symmetric spaces. However a few of these properties are directly relevant to the problem of renormalization and can be derived by elementary methods. A very important property is the following:

If a homogeneous space G/H is symmetric, H is a maximal proper subgroup of G.

To prove this assertion let us assume that there exists a subgroup G' of G which contains H, and exhibit a contradiction:

$$G \supset G' \supset H.$$

Note first that G'/H is then also a symmetric space.

If t''^a belongs to $\mathcal{L}(G) - \mathcal{L}(G')$ and t'^b belongs to $\mathcal{L}(G') - \mathcal{L}(H)$, then $[t''^a, t'^b]$ belongs to $\mathcal{L}(H)$ from $(A15.29)$. However the relations $(A15.28)$ and the antisymmetry of the structure constants imply that such a commutator vanishes. Since $\mathcal{L}(H)$ is obtained from the commutators of generators in $\mathcal{L}(G'/H)$, again we find that the generators t''_a commute with $\mathcal{L}(G')$, and thus

$$G = G' \times G'',$$

in which H is a subgroup of G', situation we have excluded.

Therefore the maximality of H has been derived. Several other important properties of symmetric spaces follow.

A few consequences

(i) The maximality of H has one very important consequence: the generators $\{t^a\}$ form a real irreducible representation of the group H.

To derive this result, let us assume the converse, i.e. that the generators $\{t^a\}$ can be divided in two representations of H, $\mathcal{L}_1$ and $\mathcal{L}_2$:

$$t^a \in \mathcal{L}_1 \Rightarrow [t^a, \tau^\alpha] \in \mathcal{L}_1,$$
$$t^a \in \mathcal{L}_2 \Rightarrow [t^a, \tau^\alpha] \in \mathcal{L}_2.$$

Since as a consequence of $(A15.29)$ the commutator of two elements of $\mathcal{L}_1$ belongs to $\mathcal{L}(H)$, $\mathcal{L}(H) \oplus \mathcal{L}_1$ forms a subalgebra of $\mathcal{L}(G)$ and H is not maximal. The converse is obvious: If H is not maximal the representation is reducible.

(ii) Let us assume that we have constructed the space G/H in the manner described in Subsection 15.1.1. From the previous results we conclude that the field $\phi(x)$ of equation (15.4):

$$\phi_i(x) = \left[\exp \left(\sum_{\alpha=1}^l \xi_a(x) t^a \right) \right]_{ij} \phi_j^c,$$

belongs to an irreducible representation of G. Also simple considerations show that since the $\{t^a\}$ form an irreducible representation of H, the vector ϕ^c is unique in the following sense: given H there exists a unique vector ϕ^c in the representation which has H as little group (stabilizer).

Therefore to each symmetric space is associated a unique classical model. The quantum model will be defined in terms of a unique coupling constant, and the perturbative expansion is equally unique. The general arguments on homogeneous spaces given before tell us that two renormalization constants are sufficient to renormalize the model.

We now describe symmetric spaces corresponding to orthogonal and unitary groups G.

A15.4.3 The principal chiral models

We first examine the case in which G is not simple and factorizes into $G \equiv \mathfrak{G}_1 \times \mathfrak{G}_2$. Since we have excluded the situation $(A15.30)$, the automorphism must map elements of $\mathfrak{G}_1$ into $\mathfrak{G}_2$ and vice versa. This implies $\mathfrak{G}_1 \equiv \mathfrak{G}_2$ and the automorphism is:

$$(\overline{g_1, g_2}) = (g_2, g_1) \,, \qquad g_1 \in \mathfrak{G}_1 \,, \ \mathfrak{G}_2 \in G_2 \,.$$

The subgroup H is then given by elements of the form

$$H \equiv \{(g, g)\} \equiv \mathfrak{G} \,.$$

The symmetric space $\mathfrak{G} \times \mathfrak{G} / \mathfrak{G}$ is isomorphic to the group space $\mathfrak{G}$ itself. A canonical realization is to consider group elements of the form $g_1^{-1} g_2$.

A15.4.4 Simple groups

We now assume that the group G is simple. We decompose the automorphism into the product of an inner automorphism and a remaining irreducible involutive automorphism:

$$\bar{g} = g_0^{-1} g^* g_0 \,,$$

with:

$$g_1^* g_2^* = (g_1 g_2)^* \,,$$
$$(g^*)^* = g \,.$$

The condition $\bar{\bar{g}} = g$ then has the form:

$$g_0^{-1} \left(g_0^{-1}\right)^* g g_0^* g_0 = g \,.$$

The element $g_0^* g_0$ commutes with all elements of the group. This implies that $g_0^* g_0$ belongs to the centre of the group:

$$g_0^* g_0 = \lambda \mathbf{1} \,,$$

$\mathbf{1}$ being the unit matrix in the defining representation, and λ a phase factor which reduces to $\lambda = \pm 1$ for orthogonal groups.

The subgroup H is defined by the invariant group elements h:

$$\bar{h} = g_0^{-1} h^* g_0 = h \ \text{ or } \ h^* g_0 = h g_0 \,.$$

A realization of G/H in the group space will be given by group elements of the form

$$g = \left(f^{-1}\right)^* g_0 f \qquad \forall f \in G \,. \tag{A15.31}$$

Indeed if we multiply f on the left by an element of H, g is not modified.

Note that these group elements satisfy:

$$g^* g = \lambda \mathbf{1} \,. \tag{A15.32}$$

We can now classify symmetric spaces corresponding to orthogonal and unitary groups:

Orthogonal groups. The group G is $O(N)$, the star automorphism is the identity and $\lambda = \pm 1$.

(i) $\lambda = +1$

We take g_0 diagonal without loss of generality. It has only ± 1 as eigenvalues. If it possesses p eigenvalues $+1$ and $N - p$ eigenvalues -1, the subgroup H is clearly $O(p) \times O(N - p)$.

The symmetric space $O(N)/O(p) \times O(N - p)$ is called a real Grassmannian manifold. The $O(N)$ non-linear σ-model corresponds to $p = 1$.

(ii) $\lambda = -1$

This implies that N must be even: $N = 2N'$. Without loss of generality we can choose g_0 of the form:

$$g_0 = \begin{bmatrix} 0 & \mathbf{1}_{N'} \\ -\mathbf{1}_{N'} & 0 \end{bmatrix},$$

and the subgroup which commutes with g_0 is isomorphic to the unitary group $U(N')$.

Unitary groups. The group G is $U(N)$.

(i) The star automorphism is the identity. The phase λ is irrelevant. Taking $\lambda = 1$, one sees that g_0 has only ± 1 as eigenvalues. If it has p eigenvalues $+1$ and $N - p$ eigenvalues -1, the subgroup H is $U(p) \times U(N - p)$.

The symmetric spaces $U(N)/U(p) \times U(N - p)$ are complex Grassmannian manifolds. The case $p = 1$ corresponds to the complex projective space CP_{N-1}.

(ii) The star automorphism is the complex conjugation. The condition $g_0^* g_0 = \lambda \mathbf{1}$ then implies $\lambda = \pm 1$. If we take $\lambda = +1$, we can diagonalize g_0 by an orthogonal transformation and then set it equal to one by a diagonal unitary transformation. Since the elements h of H then satisfy:

$$h = h^*,$$

the subgroup H is the orthogonal subgroup $O(N)$ of $U(N)$.

If we take $\lambda = -1$, we again see that we must have taken N even

$$N = 2N'.$$

We can then choose g_0 of the form

$$g_0 = \begin{bmatrix} 0 & \mathbf{1}_{N'} \\ -\mathbf{1}_{N'} & 0 \end{bmatrix}.$$

The subgroup H is defined by the elements h which satisfy:

$$\begin{bmatrix} 0 & \mathbf{1}_{N'} \\ -\mathbf{1}_{N'} & 0 \end{bmatrix} h = h^* \begin{bmatrix} 0 & \mathbf{1}_{N'} \\ -\mathbf{1}_{N'} & 0 \end{bmatrix},$$

which is by definition the symplectic subgroup $Sp(N)$ of $U(N)$.

16 SLAVNOV–TAYLOR AND BRS SYMMETRY. STOCHASTIC FIELD EQUATIONS

In Section 15.3 we have naturally been led to introduce a transformation depending on anticommuting parameters, to prove the geometrical stability of homogeneous spaces under renormalization. There is a set of topics, stochastic field equations, gauge theories, in which similar transformations are met. These new problems have one common feature: They all involve a constraint equation to which, by a set of formal transformations, is associated an effective action. This action has an anticommuting type symmetry which has no geometrical origin but merely is a consequence of these transformations.

In this chapter we first discuss this mathematical structure from a rather abstract point of view. We explain the appearance of Slavnov–Taylor symmetry, which is a conventional non-linear symmetry, in the integral representation of constraint equations. We then show how it leads to a symmetry with anticommuting parameters first discovered in quantized gauge theories by Becchi, Rouet and Stora and hereafter called therefore BRS symmetry. Its generator has a vanishing square, and generalizes exterior differentiation. We show that this symmetry is remarkably stable against a number of algebraic deformations. In some cases it can be expressed in compact form by introducing Grassmann coordinates. We show how BRS symmetry can encode the compatibility conditions of a system of linear first order differential equations. We exhibit the special form BRS symmetry takes when the constraint equations apply to group manifolds. Finally we indicate how, as a generalization of BRS symmetry, a richer structure called supersymmetry emerges. Throughout most of this chapter we use a notation adapted to a finite number of degrees of freedom. However, the extension of all algebraic identities to field theories is straightforward and in the last section we discuss the role of BRS symmetry in this latter context.

16.1 Slavnov–Taylor (ST) Symmetry

Let φ^α be a set of dynamical variables satisfying a system of equations:

$$E_\alpha(\varphi) = 0 \,, \tag{16.1}$$

where the functions $E_\alpha(\varphi)$ are smooth, and $E_\alpha = E_\alpha(\varphi)$ is a one-to-one map in some neighbourhood of $E_\alpha = 0$ which can be inverted in $\varphi^\alpha = \varphi^\alpha(E)$. This implies in particular that the equation (16.1) has a unique solution φ_s^α. We then consider some function $F(\varphi)$ and we want to give a formal expression for $F(\varphi_s)$, without solving equation (16.1) explicitly. We can then use equation (1.28) and write:

$$F(\varphi_s) = \int \left\{ \prod_\alpha \mathrm{d}E^\alpha \, \delta(E_\alpha) \right\} F(\varphi(E))$$

$$= \int \left\{ \prod_\alpha \mathrm{d}\varphi^\alpha \, \delta[E_\alpha(\varphi)] \right\} \mathcal{J}(\varphi) \, F(\varphi), \tag{16.2}$$

with:

$$\mathcal{J}(\varphi) = \det \mathbf{E} \,, \quad E_{\alpha\beta} \equiv \frac{\partial E_\alpha}{\partial \varphi^\beta}. \tag{16.3}$$

We have chosen $E_\alpha(\varphi)$ such that det $\mathbf{E}$ is positive.

Notation. It will be convenient throughout this chapter to use the notation

$$\frac{\partial}{\partial\varphi^\alpha} \;\mapsto\; \partial_\alpha \,.$$

An invariant measure. The measure $\mathrm{d}\rho(\varphi)$:

$$\mathrm{d}\rho(\varphi) = \mathcal{J}(\varphi) \prod_\alpha \mathrm{d}\varphi^\alpha \,, \tag{16.4}$$

has a simple but important induced property. The measure $\prod_\alpha \mathrm{d}E_\alpha$ is the invariant measure for the group of translations $E_\alpha \mapsto E_\alpha + \nu_\alpha$. It follows that $\mathrm{d}\rho(\varphi)$ is the invariant measure for the translation group non-linearly realized on the new coordinates φ_α (provided ν_α is small enough):

$$\varphi^\alpha \mapsto \varphi'^\alpha \quad \text{with} \quad E_\alpha\left(\varphi'\right) - \nu_\alpha = E_\alpha(\varphi). \tag{16.5}$$

With the corresponding invariance is associated, in gauge theories, the Slavnov–Taylor symmetry.

The infinitesimal form of the transformation law can be written more explicitly:

$$\delta\varphi^\alpha = [E^{-1}(\varphi)]^{\alpha\beta}\nu_\beta \,. \tag{16.6}$$

We shall see later that these somewhat straightforward remarks lead, when applied to field theories, to very useful identities.

Reciprocal property. Conversely let us characterize the general form of non-linear representations of the translation group. We write an infinitesimal group transformation:

$$\delta\varphi^\alpha = [M^{-1}(\varphi)]^{\alpha\beta}\nu_\beta \,, \tag{16.7}$$

in which the matrix $M_{\alpha\beta}(\varphi)$ has to be determined. Following the strategy explained in Section 15.3, we impose to the differential operators Δ^α:

$$\Delta^\alpha = [M^{-1}(\varphi)]^{\beta\alpha}\frac{\partial}{\partial\varphi^\beta}, \tag{16.8}$$

to form a representation of the Lie algebra of the translation group, i.e. commute:

$$[\Delta^\alpha, \Delta^\beta] = 0\,. \tag{16.9}$$

This implies:

$$[M^{-1}]^{\gamma\alpha}[M^{-1}\partial_\gamma M M^{-1}]^{\delta\beta} = [M^{-1}]^{\gamma\beta}[M^{-1}\partial_\gamma M M^{-1}]^{\delta\alpha}\,.$$

Multiplying by $M_{\alpha\alpha'} M_{\beta\beta'} M_{\delta\delta'}$ and summing over α, β, δ, we find:

$$\partial_\beta M_{\delta\alpha} - \partial_\alpha M_{\delta\beta} = 0\,. \tag{16.10}$$

This implies (for a simply connected φ-manifold, an implicit assumption throughout the whole chapter) that $M_{\alpha\beta}$ has the form:

$$M_{\alpha\beta} = \partial_\beta E_\alpha \,. \tag{16.11}$$

Let us now characterize the invariant measure $\mathcal{J}(\varphi)\mathrm{d}\varphi$ for these non-linear transformations. The variation of $\mathcal{J}(\varphi)$ has to cancel the jacobian coming from the change of variables corresponding to the transformation (16.7):

$$\partial_\alpha \mathcal{J}[M^{-1}(\varphi)]^{\alpha\beta} + \mathcal{J}\partial_\alpha[M^{-1}(\varphi)]^{\alpha\beta} = 0 \,. \tag{16.12}$$

This yields a system of partial differential equations for the function $\mathcal{J}(\varphi)$:

$$\partial_\alpha \ln \mathcal{J} = [M^{-1}]^{\gamma\beta}\partial_\gamma M_{\beta\alpha} \,. \tag{16.13}$$

The equation (16.10) is, as expected, an integrability condition for this system. We can use it to rewrite equation (16.13):

$$\partial_\alpha \ln \mathcal{J} = \partial_\alpha \ln \det \mathbf{M} \,, \tag{16.14}$$

which has the solution:

$$\mathcal{J} = \text{const.} \ \det \mathbf{M} \,. \tag{16.15}$$

We shall derive similar identities later in this chapter.

16.2 Effective Action and BRS Symmetry

In Field Theory the non-linear and as we shall see later non-local character of transformations (16.5) and (16.6) leads to many technical difficulties. Remarkably enough the infinitesimal transformations (16.6) can be replaced by a linear anticommuting type transformation at the price of introducing additional variables.

Let us again start from identity (16.2) and first replace the δ-function by its Fourier representation:

$$\prod_\alpha \delta\left[E_\alpha(\varphi)\right] = \int \prod_\alpha \frac{\mathrm{d}\lambda^\alpha}{2i\pi}\, e^{-\lambda^\alpha E_\alpha(\varphi)} \,. \tag{16.16}$$

The λ-integration runs along the imaginary axis. As a consequence of the rules of fermion integration we can also write the determinant as an integral over Grassmann variables c^α and $\bar{c}^\alpha$:

$$\det \mathbf{E} = \int \prod_\alpha (\mathrm{d}c^\alpha \mathrm{d}\bar{c}^\alpha) \exp\left(\bar{c}^\alpha E_{\alpha\beta} c^\beta\right) \,. \tag{16.17}$$

Expression (16.2) then becomes:

$$F(\varphi_{\mathrm{s}}) = \mathcal{N} \int \prod_\alpha (\mathrm{d}\varphi^\alpha \mathrm{d}c^\alpha \mathrm{d}\bar{c}^\alpha \mathrm{d}\lambda^\alpha)\, F(\varphi) \exp\left[-S(\varphi, c, \bar{c}, \lambda)\right] \,, \tag{16.18}$$

in which $\mathcal{N}$ is a constant normalization factor and $S(\varphi, c, \bar{c}, \lambda)$ the quantity:

$$S(\varphi, c, \bar{c}, \lambda) = \lambda^\alpha E_\alpha(\varphi) - \bar{c}^\alpha E_{\alpha\beta}(\varphi) c^\beta \,. \tag{16.19}$$

In field theory problems we shall call S the effective action associated with the equation (16.1). Quite surprisingly the function(al) S has a new type of symmetry, which actually is a consequence of the invariance of the measure (16.4) under the group of transformations (16.6), and which we explain below.

16.2.1 BRS symmetry

The BRS symmetry, first discovered in the quantization of gauge theories by Becchi, Rouet and Stora (BRS) (see Chapters 19 and 21), is a fermionic symmetry in the sense that it transforms bosons into fermions and vice versa. The parameter of the transformation is a Grassmann variable, an anticommuting constant $\bar{\varepsilon}$. The variations of the various dynamic variables are:

$$\begin{cases} \delta\varphi^\alpha = \bar{\varepsilon}c^\alpha, & \delta c^\alpha = 0, \\ \delta\bar{c}^\alpha = \bar{\varepsilon}\lambda^\alpha, & \delta\lambda^\alpha = 0, \end{cases} \tag{16.20}$$

with:

$$\bar{\varepsilon}^2 = 0, \qquad \bar{\varepsilon}c^\alpha + c^\alpha\bar{\varepsilon} = 0, \qquad \bar{\varepsilon}\bar{c}^\alpha + \bar{\varepsilon}\bar{c}^\alpha = 0.$$

The transformation is obviously *nilpotent* of vanishing square: $\delta^2 = 0$.

The BRS transformation can be represented by a Grassmann differential operator $\mathcal{D}$, when acting on functions of $\{\varphi, c, \bar{c}, \lambda\}$:

$$\mathcal{D} = c^\alpha \frac{\partial}{\partial\varphi^\alpha} + \lambda^\alpha \frac{\partial}{\partial\bar{c}^\alpha}. \tag{16.21}$$

The nilpotency of the BRS transformation is then expressed by the identity:

$$\mathcal{D}^2 = 0. \tag{16.22}$$

The differential operator $\mathcal{D}$ is a simple generalization of the exterior differentiation of differential forms (see also Sections 1.6 and 22.1).

The equation (16.22) implies that all quantities of the form $\mathcal{D}Q(\varphi, c, c, \lambda)$, quantities we call *BRS exact*, are BRS invariant. We immediately verify that the function S defined by equation (16.19) is BRS exact:

$$S = \mathcal{D}\left[\bar{c}^\alpha E_\alpha(\varphi)\right]. \tag{16.23}$$

It follows that S is BRS invariant.

$$\mathcal{D}S = 0. \tag{16.24}$$

This property plays an important role in particular in the discussion of the renormalization of gauge theories.

The meaning and implications of the BRS symmetry will be discussed in the coming sections.

Remark. Analogous expressions can be derived for a system of Grassmann equations involving Grassmann variables. The Lagrange multiplier λ is then a Grassmann variable. The properties of change of variables in a Grassmann integral imply that the determinant det $\mathbf{E}$ is replaced by the inverse of a determinant. Therefore the auxiliary variables which we have to introduce to represent the determinant are now commuting variables. It can be verified, however, that the final integrand is still BRS symmetric and many considerations related to the BRS symmetry apply to both the commuting and anticommuting cases. We shall not discuss this question any further here because we shall encounter no equation of this type in this work.

16.2.2 Relation between ST and BRS symmetry

A simple way of understanding the relation between transformations (16.6) and (16.20) is to note that:

$$\det \mathbf{E} \left(E^{-1} \right)^{\alpha\beta} = \int dc\, d\bar{c}\, \bar{c}^{\beta} c^{\alpha} \exp \left(\bar{c}^{\gamma} E_{\gamma\delta} c^{\delta} \right).$$

(16.25)

Therefore if we factorize the integral over c and $\bar{c}$, we can rewrite inside the integral:

$$\delta\varphi^{\alpha} = \left(E^{-1} \right)^{\alpha\beta} \nu_{\beta} = -c^{\alpha} \left(\bar{c}^{\beta} \nu_{\beta} \right).$$

(16.26)

BRS transformations correspond to:

$$-\bar{c}^{\alpha} \nu_{\alpha} \longmapsto \bar{\varepsilon}.$$

(16.27)

A form of this argument can be used to prove that as long as only averages of functions of φ are concerned, the consequences of ST or BRS symmetry are the same. The BRS symmetry extends the transformations to functions of φ, c, $\bar{c}$ and λ. This extension is useful for two reasons:

(i) The transformations (16.20) are linear, while the transformation (16.6) is non-linear.

We have already seen in the case of the non-linear σ-model that non-linear transformations could be linearized at the price of introducing auxiliary fields.

(ii) More important, in field theory transformations (16.20) will be local, in contrast to transformation (16.6).

These two properties greatly simplify the discussion of renormalization of various field theories.

16.3 Grassmann Coordinates, Gradient Equations

A more compact representation of BRS transformations is obtained by introducing a Grassmann coordinate $\bar{\theta}$ and then the following two functions of $\bar{\theta}$:

$$\phi^{\alpha} \left(\bar{\theta} \right) = \varphi^{\alpha} + \bar{\theta} c^{\alpha}, \qquad \bar{C}^{\alpha} \left(\bar{\theta} \right) = \bar{c}^{\alpha} + \bar{\theta} \lambda^{\alpha}.$$

(16.28)

With this notation the transformations (16.20) simply become a translation of $\bar{\theta}$:

$$\begin{cases} \delta\phi^{\alpha} \left(\bar{\theta} \right) = \bar{\varepsilon} \dfrac{\partial \phi^{\alpha}}{\partial \bar{\theta}} = \phi^{\alpha} \left(\bar{\theta} + \bar{\varepsilon} \right) - \phi^{\alpha} \left(\bar{\theta} \right), \\[2ex] \delta\bar{C}^{\alpha} \left(\bar{\theta} \right) = \bar{\varepsilon} \dfrac{\partial \bar{C}^{\alpha}}{\partial \bar{\theta}} = \bar{C}^{\alpha} \left(\bar{\theta} + \bar{\varepsilon} \right) - \bar{C}^{\alpha} \left(\bar{\theta} \right). \end{cases}$$

(16.29)

In particular the BRS operator $\mathcal{D}$ is represented by $\partial/\partial\bar{\theta}$:

$$\mathcal{D} \longmapsto \frac{\partial}{\partial \bar{\theta}}.$$

Let us note the identity:

$$\bar{C}^{\alpha} \left(\bar{\theta} \right) E_{\alpha} \left(\phi \left(\bar{\theta} \right) \right) = \bar{c}^{\alpha} E_{\alpha}(\varphi) + \bar{\theta} \left[\lambda^{\alpha} E_{\alpha}(\varphi) - \bar{c}^{\alpha} \frac{\partial E_{\alpha}}{\partial \varphi^{\beta}} c^{\beta} \right].$$

(16.30)

We recall that the integration over $\bar{\theta}$ selects the coefficient of $\bar{\theta}$ in the integrand. Therefore $S(\varphi, c, \bar{c}, \lambda)$ can be rewritten as an integral over $\bar{\theta}$:

$$S = \int d\bar{\theta}\, \bar{C}^{\alpha}\left(\bar{\theta}\right) E_{\alpha}\left(\phi\left(\bar{\theta}\right)\right). \tag{16.31}$$

In this expression the BRS symmetry is manifest: the integrand does not depend explicitly on $\bar{\theta}$. Another consequence follows: in Chapter 1, when we have defined fermion integration, we have emphasized that integration and differentiation in the case of Grassmann variables were identical operations. Thus:

$$S(\varphi, c, \bar{c}, \lambda) = \frac{\partial}{\partial \bar{\theta}} \left[\bar{C}^{\alpha}\left(\bar{\theta}\right) E_{\alpha}\left(\phi\left(\bar{\theta}\right)\right)\right]. \tag{16.32}$$

We recover equation (16.23) in a different notation.

In addition S has the property that it involves only a fermionic combination of the form $\bar{c}c$. In a field theory language, if we assign a fermion charge $+1$ to c and -1 to $\bar{c}$, S conserves the fermionic charge. This implies that in a representation in terms of the functions (16.28), as in equation (16.31), each integration over $\bar{\theta}$ is associated with a factor $\bar{C}^{\alpha}\left(\bar{\theta}\right)$.

Gradient equations. The two Grassmann variables c^{α} and $\bar{c}^{\alpha}$ that we have introduced in the preceding section play in general totally different roles. There is however an exceptional situation in which a symmetry is established between them: it is when the matrix $E_{\alpha\beta}$ is symmetric:

$$E_{\alpha\beta} = E_{\beta\alpha} \iff \partial_{\beta}E_{\alpha} = \partial_{\alpha}E_{\beta}. \tag{16.33}$$

Hence in this case $E_{\alpha}(\varphi)$ is itself a gradient; there exists some function $A(\varphi)$ such that:

$$E_{\alpha}(\varphi) = \partial_{\alpha}A(\varphi). \tag{16.34}$$

This symmetry between c and $\bar{c}$ clearly generates an additional independent BRS symmetry. It is thus natural to introduce two Grassmann variables θ and $\bar{\theta}$, and a function $\phi^{\alpha}\left(\theta, \bar{\theta}\right)$:

$$\phi^{\alpha}\left(\theta, \bar{\theta}\right) = \varphi^{\alpha} + \bar{\theta}c^{\alpha} + \bar{c}^{\alpha}\theta + \bar{\theta}\theta\lambda^{\alpha}. \tag{16.35}$$

In terms of ϕ the expression (16.31) quite generally reads:

$$S(\phi) = \int d\theta d\bar{\theta}\, \theta \frac{\partial \phi^{\alpha}}{\partial \theta} E_{\alpha}\left[\phi\left(\theta, \bar{\theta}\right)\right]. \tag{16.36}$$

When the function $E(\varphi)$ has the particular form (16.34), it is possible to integrate by parts over θ and the function $S(\phi)$ then takes the remarkable form:

$$S(\phi) = \int d\theta d\bar{\theta}\, A\left[\phi\left(\theta, \bar{\theta}\right)\right] = \mathcal{D}\bar{\mathcal{D}}A(\varphi). \tag{16.37}$$

The two symmetries, which correspond to independent translations of $\bar{\theta}$ and θ, are here explicit and we have denoted by $\bar{\mathcal{D}}$ the generator of the second BRS symmetry.

16.4 BRS Symmetry and Compatibility Condition, Group Manifolds

The part of the BRS operator which acts on the variables φ^α is identical to the differentiation of forms. The Grassmann variables however play a different role. In the case of differential forms they are external variables, introduced for convenience to exhibit the antisymmetry of the corresponding tensors. Here instead the c^α's are additional dynamical variables. In particular it may be convenient, as we show below, to change variables and to set

$$\varphi^\alpha = F^\alpha(\varphi'), \qquad c^\alpha = U^\alpha_\beta(\varphi')c'^\beta.$$

It is of course the second equation which is characteristic of the difference. If we now rewrite the transformation (16.20) in these new variables it takes a more complicated form (omitting the primes on the new variables)

$$\delta\varphi^\alpha = \bar{\varepsilon}D^\alpha_\beta(\varphi)c^\beta \qquad \delta c^\alpha = -\tfrac{1}{2}f^\alpha_{\beta\gamma}(\varphi)\bar{\varepsilon}c^\beta c^\gamma, \tag{16.38}$$

where the functions $D^\alpha_\beta(\varphi)$ and $f^\alpha_{\beta\gamma}(\varphi)$, which is antisymmetric in $\beta \leftrightarrow \gamma$, can be expressed in terms of F and U. Note that the BRS generator $\mathcal{D}$ also now takes a more complicated form but the equation (16.23) is still true.

More generally we can consider situations in which physical quantities depend on some variables A^i, themselves functions of the φ^α. Then BRS transformations will have the form

$$\delta A^i = \bar{\varepsilon}D^i_\beta(A)c^\beta \qquad \delta c^\alpha = -\tfrac{1}{2}f^\alpha_{\beta\gamma}(A)\bar{\varepsilon}c^\beta c^\gamma. \tag{16.39}$$

Let us then express the condition that the BRS operator $\mathcal{D}$ is nilpotent, $\mathcal{D}^2 = 0$, directly in terms of D and f. We identify the terms cubic and quadratic in the Grassmann variables c^α and write that the properly antisymmetrized coefficients vanish

$$\left\{f^\delta_{\alpha\beta}f^\epsilon_{\gamma\delta} + D^i_\alpha\partial_i f^\epsilon_{\beta\gamma}\right\}_{\alpha\beta\gamma} = 0, \tag{16.40}$$

$$D^j_\alpha\partial_j D^i_\beta - D^j_\beta\partial_j D^i_\alpha = f^\delta_{\alpha\beta}D^i_\delta, \tag{16.41}$$

where the global subscript $\alpha\beta\gamma$ in (16.40) means antisymmetrized in the three indices. In equation (16.41) we immediately recognize the equation (15.43) of Section 15.3, i.e. the commutation relation between generators of a Lie algebra in a non-linear representation. Moreover if $f^\alpha_{\beta\gamma}$ is independent of A, equation (16.40) is simply the Jacobi identity for the structure constants of the Lie algebra.

Compatibility conditions. The extension we find here can be understood in the following way. Let us consider the set of first order partial differential equations

$$\Delta_\alpha S(A) = 0, \tag{16.42}$$

where the Δ_α are the differential operators

$$\Delta_\alpha = D^i_\alpha(A)\frac{\partial}{\partial A^i}.$$

As explained in Sections 13.1.2 and 15.3 the system (16.42) is called compatible if the equations $[\Delta_\alpha, \Delta_\beta]S = 0$ are linear combinations of the initial equations (16.42). Then

$$[\Delta_\alpha, \Delta_\beta] = f^\gamma_{\alpha\beta}(A)\Delta_\gamma. \tag{16.43}$$

We have encountered before only examples where the structure constants $f^\gamma_{\alpha\beta}$ were A-independent (i.e. associated with Lie algebras), but this is not the general situation.

The equation (16.43) itself has an integrability condition, the Jacobi identities since the l.h.s. is a commutator. A short calculation yields the condition (16.40). Therefore the nilpotency of the BRS operator encodes the compatibility of the linear system (16.42). Note finally that the equation (16.42) is equivalent to $\mathcal{D}S(A) = 0$. If it is has non-trivial solutions these solutions cannot be cast into the form $S(A) = \mathcal{D}S'(A, c)$, i.e. are not BRS exact.

BRS symmetry and group manifolds. Group manifolds provide a simple example of such a situation. If the variables φ^α parametrize a group element $g(\varphi)$ in some matrix representation, it is convenient to rewrite BRS transformations on $g(\varphi)$ directly. This can be most easily done by noting that with $\phi(\bar\theta)$ (defined by equation (16.28)) we can associate some group element $\mathcal{G}(\bar\theta)$, on which BRS transformations according to equation (16.29) read:

$$\delta\mathcal{G}(\bar\theta) = \bar\varepsilon \frac{\partial\mathcal{G}}{\partial\bar\theta}. \tag{16.44}$$

However, since $\mathcal{G}(\bar\theta)$ is a group element, it is natural to parametrize it under the form:

$$\mathcal{G}(\bar\theta) = \exp(\bar\theta c) g = (1 + \bar\theta c) g, \tag{16.45}$$

in which c is now a Grassmann matrix belonging to the Lie algebra of the group. In component form the transformation (16.44) then becomes

$$\delta g = \bar\varepsilon c g, \tag{16.46}$$
$$\delta c = \bar\varepsilon c^2. \tag{16.47}$$

Introducing matrices t_α generators of the Lie algebra and parametrizing c as

$$c = c^\alpha t_\alpha, \tag{16.48}$$

we can rewrite equation (16.47) as:

$$\delta c^\alpha = \tfrac{1}{2}\bar\varepsilon f^\alpha_{\beta\gamma} c^\beta c^\gamma. \tag{16.49}$$

In this form we recognize the transformation (15.49). Moreover the transformation (16.46) applied to the matrix $\mathbf{R}$ of the representation (15.6) leads to the transformation (15.45) for the field $\varphi(x)$. We have thus found a geometrical interpretation to the transformations used in the derivation of WT identities for homogeneous spaces in an arbitrary system of coordinates.

16.5 Stochastic Equations

Let us now assume that equation (16.1) depends on a set of stochastic variables ν_a with probability distribution $\mathrm{d}\rho(\nu)$:

$$E_\alpha(\varphi, \nu) = 0. \tag{16.50}$$

The solution φ^α of the equation becomes a stochastic variable. Quantities of interest are now average values of functions of φ:

$$\langle F(\varphi)\rangle = \int \mathrm{d}\rho(\nu) \prod_\alpha \mathrm{d}\varphi^\alpha \, \delta\left[E_\alpha(\varphi, \nu)\right] \det \mathbf{E}\, F(\varphi). \tag{16.51}$$

After the set of transformations described in Section 16.2, this equation becomes:

$$\langle F(\varphi) \rangle \propto \int d\rho(\nu) \prod_\alpha (d\varphi^\alpha dc^\alpha d\bar{c}^\alpha d\lambda^\alpha) \, F(\varphi) \exp\left[-S\left(\varphi, c, \bar{c}, \lambda, \nu\right)\right], \qquad (16.52)$$

with S given by equation (16.19):

$$S = \lambda^\alpha E_\alpha\left(\varphi, \nu\right) - \bar{c}^\alpha E_{\alpha\beta}\left(\varphi, \nu\right) c^\beta. \qquad (16.53)$$

We have shown that S has a BRS symmetry. The important remark is that the function $\Sigma\left(\varphi, c, \bar{c}, \lambda\right)$ obtained after noise averaging:

$$\exp\left[-\Sigma\left(\varphi, c, \bar{c}, \lambda\right)\right] = \int d\rho(\nu) \exp\left[-S\left(\varphi, c, \bar{c}, \lambda, \nu\right)\right], \qquad (16.54)$$

is still BRS symmetric, although it no longer has the simple form (16.53), i.e. a function linear in λ and $\bar{c}c$. Using the definition (16.21) of the BRS generator, we can write the BRS symmetry as:

$$\mathcal{D}\Sigma = 0. \qquad (16.55)$$

16.5.1 Stochastic equations linear in the noise: A simple example

In the coming chapters examples of stochastic equations of the simple form

$$E_\alpha\left(\nu, \varphi\right) \equiv E_\alpha\left(\varphi\right) - \nu_\alpha, \qquad (16.56)$$

will be met. Introducing the Laplace transform of the measure $d\rho(\nu)$,

$$e^{w(\lambda)} = \int d\rho(\nu) \, e^{\lambda^\alpha \nu_\alpha}, \qquad (16.57)$$

we obtain for the function Σ defined by (16.54):

$$\Sigma\left(\varphi, c, \bar{c}, \lambda\right) = \mathcal{D}\tilde{\Sigma}, \quad \tilde{\Sigma} = \bar{c}_\alpha \left[E_\alpha(\varphi) - \frac{\partial}{\partial \lambda_\alpha} \int_0^1 ds\, w(s\lambda)\right]. \qquad (16.58)$$

This is a minimal modification of expression (16.19).

Remarks.

(i) After integration over the noise, the expression of the function Σ in the notation of Grassmann coordinates is in general rather complicated. However, in the case of equation (16.56) with gaussian noise the additional term $w(\lambda) = \frac{1}{2} w_{\alpha\beta} \lambda^\alpha \lambda^\beta$ is represented in the notation (16.35) by

$$\frac{1}{2} w_{\alpha\beta} \lambda^\alpha \lambda^\beta = \int d\theta d\bar{\theta} \, \frac{1}{2} w_{\alpha\beta} \frac{\partial \phi^\alpha}{\partial \theta} \frac{\partial \phi^\beta}{\partial \bar{\theta}}.$$

(ii) In the latter case it also possible to integrate explicitly over the λ variables. The resulting integrand correspond to

$$\Sigma(\varphi, c, \bar{c}) = \frac{1}{2} E_\alpha(\varphi) [w^{-1}]^{\alpha\beta} E_\beta(\varphi) - \bar{c}^\alpha E_{\alpha\beta}(\varphi) c^\beta.$$

The BRS transformation of $\bar{c}$ is now as expected non-linear

$$\delta_{\mathrm{BRS}}\bar{c}^\alpha = \bar{\varepsilon}[w^{-1}]^{\alpha\beta}E_\beta \, .$$

We note that in this form the BRS transformation of $\bar{c}$ still is *nilpotent but no longer of vanishing square*. Therefore the latter property of the BRS transformation may not be apparent in all formulations and may require the introduction of some auxiliary variables.

A sufficient condition. We now want to characterize the most general BRS invariant function of the form

$$\Sigma\left(\varphi, c, \bar{c}, \lambda\right) = \Sigma\left(\varphi, \lambda\right) - \bar{c}^\alpha M_{\alpha\beta}(\varphi, \lambda)c^\beta \, . \tag{16.59}$$

Identifying the coefficient of $\bar{c}c^2$ and c in equation (16.55) we obtain two equations:

$$c^\alpha \bar{c}^\beta \partial_\alpha M_{\beta\gamma}c^\gamma = 0 \, , \tag{16.60a}$$

$$c^\alpha \partial_\alpha \Sigma - \lambda^\alpha M_{\alpha\beta}(\varphi, \lambda)c^\beta = 0 \, . \tag{16.60b}$$

The equation (16.60a) implies that the antisymmetric part of the coefficient of $c^\alpha c^\gamma$ vanishes

$$\partial_\alpha M_{\beta\gamma} + \partial_\gamma M_{\beta\alpha} = 0 \, , \tag{16.61}$$

and therefore the matrix **M** has the form (again assuming the φ-manifold to be simply connected):

$$M_{\alpha\beta} = \partial_\beta E_\alpha \, . \tag{16.62}$$

The equation (16.61) is the integrability condition of equation (16.60b) which now can be solved:

$$\partial_\alpha \Sigma = \lambda^\beta M_{\beta\alpha}\left(\varphi, \lambda\right) = \lambda^\beta \partial_\alpha E_\beta \, ,$$

and yields:

$$\Sigma\left(\varphi, \lambda\right) = \lambda^\alpha E_\alpha(\varphi, \lambda). \tag{16.63}$$

The complete solution thus is

$$\Sigma\left(\varphi, c, \bar{c}, \lambda\right) = \lambda^\alpha E_\alpha(\varphi, \lambda) - \bar{c}^\alpha \partial_\beta E_\alpha(\varphi, \lambda)c^\beta, \tag{16.64}$$

which is BRS exact:

$$\Sigma\left(\varphi, c, \bar{c}, \lambda\right) == \mathcal{D}\left[\bar{c}^\alpha E_\alpha(\varphi, \lambda)\right]. \tag{16.65}$$

In particular expression (16.58) is the most general BRS symmetric expression of the form (16.59) with $M_{\alpha\beta}$ independent of λ.

16.5.2 The general linear case

A slightly more general form will often be met:

$$E_\alpha(\varphi, \nu) = E_\alpha(\varphi) - e_\alpha^a(\varphi)\nu_a \, . \tag{16.66}$$

Then:

$$\partial_\beta E_\alpha(\varphi, \nu) = \partial_\beta E_\alpha(\varphi) - \partial_\beta e_\alpha^a(\varphi)\nu_a \, .$$

Let us now consider the special example of gaussian stochastic variables ν with probability distribution:

$$d\rho(\nu) = \left(\prod_a d\nu_a\right) \exp\left(-\tfrac{1}{2}\nu_a[\Omega^{-1}]^{ab}\nu_b\right). \tag{16.67}$$

After integration over ν and some algebra we find that Σ can be written:

$$\Sigma\left(\varphi, c, \bar{c}, \lambda\right) = \mathcal{D}\widetilde{\Sigma}\left(\varphi, c, \bar{c}, \lambda\right),$$
$$\widetilde{\Sigma}\left(\varphi, c, \bar{c}, \lambda\right) = \bar{c}^\alpha E_\alpha - \tfrac{1}{2}\bar{c}^\alpha\left(w_{\alpha\beta}\lambda^\beta - w_{\alpha\beta,\gamma}\bar{c}^\beta c^\gamma\right),$$

with the definitions:

$$w_{\alpha\beta} = e_\alpha^a \Omega_{ab} e_\beta^b, \qquad w_{\alpha\beta,\gamma} = e_\alpha^a \Omega_{ab} \partial_\gamma e_\beta^b.$$

Σ differs from S by the addition of a function quadratic in both λ and $\bar{c}c$.

Conversely one may ask the question: what is the most general form of a function Σ quadratic in $\{\lambda, \bar{c}c\}$ and BRS symmetric:

$$\mathcal{D}\Sigma = 0.$$

The answer, as we shall discuss in next section, is that Σ is BRS exact:

$$\Sigma = \mathcal{D}\widetilde{\Sigma}, \tag{16.68}$$

with:

$$\widetilde{\Sigma} = \bar{c}^\alpha \bar{c}^\beta c^\gamma \widetilde{\Sigma}_{\alpha\beta,\gamma}(\varphi) + \bar{c}^\alpha \lambda^\beta \widetilde{\Sigma}_{\alpha\beta}(\varphi). \tag{16.69}$$

16.6 Consequences of BRS Symmetry: General Discussion

The property that the most general BRS invariant quantity is BRS exact (in the case of simply connected manifolds), as discussed in Subsection 16.5.2, is a simple consequence of a similar property for differential forms. However because this property is important, let us discuss it in the present context. Let us again introduce an anticommuting variable $\bar{\theta}$ and the functions $\phi\left(\bar{\theta}\right)$ and $\bar{C}\left(\bar{\theta}\right)$. The equation (16.54) can then be rewritten:

$$\exp\left[\Sigma\left(\phi, \bar{C}\right)\right] = \int d\rho\left(\nu\right)\exp\left[-\int d\bar{\theta}\, \bar{C}^\alpha\left(\bar{\theta}\right) E_\alpha\left(\phi\left(\bar{\theta}\right), \nu\right)\right]. \tag{16.70}$$

Expanding the exponential in the integrand in power series and integrating term by term, we obtain for $\Sigma\left(\phi, \bar{C}\right)$ a formal expansion:

$$\Sigma\left(\phi, \bar{C}\right) = \sum_1^\infty \frac{1}{n!}\int d\bar{\theta}_1 \ldots d\bar{\theta}_n\, \bar{C}^{\alpha_1}\left(\bar{\theta}_1\right) \ldots \bar{C}_{\alpha_n}\left(\bar{\theta}_n\right) \Sigma^{(n)}_{\alpha_1 \ldots \alpha_n}\left[\phi\left(\bar{\theta}_1\right), \ldots, \phi\left(\bar{\theta}_n\right)\right]. \tag{16.71}$$

Conversely we observe that the expansion has the most general form allowed by translation invariance in $\bar{\theta}$ (BRS symmetry) and fermion number conservation.

Since the BRS operator $\mathcal{D}$ corresponds to a translation of the coordinate $\bar{\theta}$, on a function of $\bar{C}\left(\theta_i\right)$ and $\phi\left(\theta_i\right)$, as those which appear in the r.h.s. of expression (16.71), it acts as:

$$\mathcal{D} \equiv \sum_{i=1}^n \frac{\partial}{\partial\bar{\theta}_i}. \tag{16.72}$$

The identity between differentiation and integration for Grassmann variables then leads to:

$$\mathcal{D} \int d\bar{\theta}_2 \ldots d\bar{\theta}_n = \sum_i \int d\bar{\theta}_i d\bar{\theta}_2 \ldots d\bar{\theta}_n = \int d\bar{\theta}_1 \ldots d\bar{\theta}_n . \tag{16.73}$$

Hence the expression (16.71) can be rewritten:

$$\Sigma\left(\varphi, c, \bar{c}, \lambda\right) = \mathcal{D}\widetilde{\Sigma}\left(\varphi, c, \bar{c}, \lambda\right), \tag{16.74}$$

with:

$$\widetilde{\Sigma}\left(\varphi, c, \bar{c}, \lambda\right) = \sum_0^\infty \frac{\bar{c}^{\alpha_1}}{n!} \int d\bar{\theta}_2 \ldots d\bar{\theta}_n \bar{C}^{\alpha_2}\left(\bar{\theta}_2\right) \ldots \bar{C}^{\alpha_n}\left(\bar{\theta}_n\right) \Sigma_{\alpha_1 \ldots \alpha_n}^{(n)} \left[\varphi, \phi\left(\bar{\theta}_2\right) \ldots \phi\left(\bar{\theta}_n\right)\right]. \tag{16.75}$$

From this analysis we draw two conclusions:

(i) Even after the most general integration over the noise the function Σ remains BRS exact and thus, since $\mathcal{D}^2 = 0$, BRS invariant,

$$\mathcal{D}\Sigma = \mathcal{D}\left(\mathcal{D}\widetilde{\Sigma}\right) = 0 .$$

(ii) Conversely, in the case of simply connected manifolds, any BRS symmetric function is BRS exact:

$$\mathcal{D}\Sigma = 0 , \; \Rightarrow \; \Sigma = \mathcal{D}\widetilde{\Sigma} . \tag{16.76}$$

An important consequence. To appreciate the significance of this result, let us consider the following integral Z:

$$Z = \langle 1 \rangle = \int \prod_\alpha (d\varphi^\alpha dc^\alpha d\bar{c}^\alpha d\lambda^\alpha) \exp\left[-\Sigma\left(\varphi, c, \bar{c}, \lambda\right)\right], \tag{16.77}$$

and calculate the variation of δZ of Z induced by an infinitesimal variation $\delta E_\alpha(\varphi)$ of the function $E_\alpha(\varphi)$.

To this variation corresponds a variation $\delta\Sigma$ of Σ which, according to our previous result, can be written:

$$\delta\Sigma = \mathcal{D}\left[\delta\widetilde{\Sigma}\left(\varphi, c, \bar{c}, \lambda\right)\right]. \tag{16.78}$$

Therefore δZ has the form:

$$\delta Z = \int \prod_\alpha (d\varphi^\alpha d\bar{c}^\alpha dc^\alpha d\lambda^\alpha) \, \mathcal{D}[\delta\widetilde{\Sigma}] \exp\left(-\Sigma\right). \tag{16.79}$$

Since $\mathcal{D}$ is a differential operator we can integrate by parts:

$$\delta Z = \int \prod_\alpha (d\varphi^\alpha d\bar{c}^\alpha dc^\alpha d\lambda^\alpha) \, \delta\widetilde{\Sigma} \, \mathcal{D}\Sigma \exp\left(-\Sigma\right). \tag{16.80}$$

Using then the BRS symmetry condition $\mathcal{D}\Sigma = 0$ we conclude:

$$\delta Z = 0 . \tag{16.81}$$

Of course this result is not surprising since by construction Z is a constant independent of $E_\alpha(\varphi)$. However it tells us that equation (16.76) implies by itself this independence without any further assumption about the form of Σ.

This result has important implications in quantized gauge theories.

16.7 Extension of Grassmannian Symmetries

A trivial extension of BRS transformation can be constructed by considering a function of several Grassmann variables $f(\theta_1, \ldots, \theta_n)$ and performing independent translations on each variable. The generators of these translations are the differential operators $\partial/\partial\theta_i$. The equivalent of the usual commutation relations between Lie algebra generators are replaced by anticommutation relations:

$$\frac{\partial}{\partial\theta_i}\frac{\partial}{\partial\theta_j} + \frac{\partial}{\partial\theta_j}\frac{\partial}{\partial\theta_i} = 0. \tag{16.82}$$

Translations on Grassmann variables form the equivalent of usual abelian Lie groups; non-trivial extensions, "non-abelian", will correspond to non-vanishing anticommutators.

First a few general remarks are in order:

(i) If D is the unique generator of this generalized structure called "supergroup" then exponentiation implies

$$e^{\varepsilon_1 D}\, e^{\varepsilon_2 D} = e^{(\varepsilon_1 + \varepsilon_2)D}\,,$$

in which ε_1 and ε_2 are Grassmann variables. Expanding both sides we obtain the condition:

$$-\varepsilon_1\varepsilon_2 D^2 = 0 \Rightarrow D^2 = 0\,, \tag{16.83}$$

which is indeed satisfied by the generators of translations. If this condition is not satisfied the supergroup has two generators D and D^2 in which D^2 is a commuting differential operator.

(ii) More generally the anticommutators of generators are even elements of the Grassmann algebra. They form an ordinary Lie algebra. If we call D_i, L_a the anticommuting and commuting elements respectively, the general structure of a Lie superalgebra is

$$\{D_i, D_j\} = c_{ija}L_a\,, \quad [D_i, L_a] = d_{iaj}D_j\,, \quad [L_a, L_b] = f_{abc}L_c\,.$$

(iii) Triplets of generators thus satisfy a mixed Jacobi identity of the form:

$$[D_i, (D_jD_k + D_kD_j)] + \text{cyclic permutations } (ijk) = 0\,. \tag{16.84}$$

Many aspects of the theory of Lie groups and algebras can be extended to supergroups and super or graded Lie algebras. We shall not discuss them here because this goes beyond the scope of this work. We shall construct only a simple example.

Supersymmetry and space translation. We want to realize our algebra under the form of differential operators. From the previous considerations we conclude that we need at least two Grassmann variables θ and $\bar\theta$ and a usual variable which we call t.

Differential operators, odd elements of the corresponding Grassmann algebra, have the form

$$\frac{\partial}{\partial\theta}, \frac{\partial}{\partial\bar\theta}, \theta\frac{\partial}{\partial t}, \bar\theta\frac{\partial}{\partial t}.$$

If we impose the condition (16.82) we find two non-trivial combinations:

$$D = \frac{\partial}{\partial\bar\theta} + a\theta\frac{\partial}{\partial t}\,, \qquad \bar{D} = \frac{\partial}{\partial\theta} + \bar{a}\bar\theta\frac{\partial}{\partial t}\,, \tag{16.85}$$

in which a and $\bar{a}$ are two ordinary arbitrary constants. It follows:

$$\mathrm{D}\bar{\mathrm{D}} + \bar{\mathrm{D}}\mathrm{D} = (\bar{a} + a)\,\frac{\partial}{\partial t}. \tag{16.86}$$

The Lie subgroup is the group of translations on the variable t. This is a realization of quantum mechanical supersymmetry as will be discussed in next chapter.

This structure can immediately be generalized to d-dimensional space:

$$\mathrm{D}_\alpha = \frac{\partial}{\partial \bar{\theta}_\alpha} + a^\mu_{\alpha\beta}\theta_\beta\frac{\partial}{\partial x_\mu}, \qquad \bar{\mathrm{D}}_\alpha = \frac{\partial}{\partial \theta_\alpha} + \bar{\theta}_\beta\bar{a}^\mu_{\beta\alpha}\frac{\partial}{\partial x_\mu}, \tag{16.87}$$

in which x_μ corresponds to the d commuting variables, θ_α and $\bar{\theta}_\alpha$ to anticommuting variables and $a^\mu_{\alpha\beta}$, $\bar{a}^\mu_{\alpha\beta}$ are constants. Then:

$$\begin{cases} \mathrm{D}_\alpha\mathrm{D}_\beta + \mathrm{D}_\beta\mathrm{D}_\alpha = 0\,, \\[1mm] \bar{\mathrm{D}}_\alpha\bar{\mathrm{D}}_\beta + \bar{\mathrm{D}}_\beta\bar{\mathrm{D}}_\alpha = 0\,, \\[1mm] \mathrm{D}_\alpha\bar{\mathrm{D}}_\beta + \bar{\mathrm{D}}_\beta\mathrm{D}_\alpha = \left(\bar{a}^\mu_{\alpha\beta} + a^\mu_{\alpha\beta}\right)\frac{\partial}{\partial x_\mu}\,. \end{cases} \tag{16.88}$$

Again the Lie subgroup is a product of translation groups. Note that to $\bar{\mathrm{D}}, \mathrm{D}$ correspond two other supersymmetry generators $\bar{\mathrm{D}}', \mathrm{D}'$:

$$\mathrm{D}'_\alpha = \frac{\partial}{\partial \bar{\theta}_\alpha} - \bar{a}^\mu_{\alpha\beta}\theta_\beta\frac{\partial}{\partial x_\mu}, \quad \bar{\mathrm{D}}'_\alpha = \frac{\partial}{\partial \theta_\alpha} - \theta_\beta\bar{a}^\mu_{\beta\alpha}\frac{\partial}{\partial x_\mu}, \tag{16.89}$$

which anticommute with $\bar{\mathrm{D}}, \mathrm{D}$.

16.8 Application: Stochastic Field Equations

Let us now consider a stochastic field equation which we write formally as:

$$E_\alpha\left(\varphi(x)\right) = \nu_\alpha\left(x\right), \tag{16.90}$$

in which α corresponds to the various components of the field $\varphi^\alpha(x)$ and $\nu_\alpha(x)$ is a field, hereafter called the noise, for which a probability distribution is provided. This probability distribution is generally given in terms of an action and in most examples this action will be quadratic in $\nu_\alpha(x)$ and thus the noise gaussian.

We shall be concerned with local stochastic field equations and we shall find out from the example of the Langevin equation that the integration over the noise leads to large momentum divergences. It is therefore necessary to understand their properties from the point of view of renormalization and renormalization group. It is then convenient to associate with these equations a local effective action because renormalization of actions is much better understood than renormalization of equations. At the same time it is possible to explicitly perform the integration over the noise. This naturally leads to the geometrical structure we have studied up to now in this chapter, and in particular to actions with BRS symmetries.

Starting from an equation as general as equation (16.90), we shall derive results which apply to several problems which have the same formal structure, for example $E_\alpha(\varphi)$ can correspond to a classical field equation (of the form (16.34)) for a spin φ and ν_α to a random magnetic field, or to a Langevin equation, as will be discussed in Chapter 17, and x then has to be understood of the collection of space and time variables. In addition as shall see in Chapters 19,21, a similar structure emerges in the quantization of gauge theories.

16.8.1 The effective action

We write the noise probability distribution $[d\rho(\nu)]$ as:

$$[d\rho(\nu)] = [d\nu] \exp\left[-\sigma(\nu)\right]. \tag{16.91}$$

The most useful example corresponds to $\sigma(\nu)$ quadratic in $\nu_\alpha(x)$:

$$\sigma(\nu) = \frac{1}{2} \int dx\, dy\, \nu_\alpha(x)[\Omega^{-1}]^{\alpha\beta}(x,y)\,\nu_\beta(y), \tag{16.92}$$

in which the $\Omega_{\alpha\beta}(x,y)$ is a local symmetric operator.

We want to calculate the generating functional $Z(J)$ of φ-field correlation functions which is the average over the noise of:

$$Z(J) = \left\langle \exp\left[\int dx\, J_\alpha(x)\varphi^\alpha(x)\right]\right\rangle_\nu. \tag{16.93}$$

It is clear that the discussion of preceding sections can be extended to the problem of stochastic field equations. Let us rapidly again review the arguments. Using identity (16.2) we write $Z(J)$ under the form:

$$Z(J) = \int [d\rho(\nu)] \int [d\varphi]\, \delta\left[E_\alpha(\varphi) - \nu_\alpha\right] \det \mathbf{E} \exp\left[\int dx\, J_\alpha(x)\,\varphi^\alpha(x)\right], \tag{16.94}$$

with:

$$E_{\alpha\beta}(x,y) = \frac{\delta E_\alpha(\varphi(x))}{\delta\varphi^\beta(y)}. \tag{16.95}$$

In writing the identity (16.94) we have implicitly assumed that equation (16.90) has a unique solution. In field theory this condition may be a source of serious difficulties because it requires an investigation of the uniqueness of the solution beyond perturbation theory. The Langevin equation we consider in Chapter 17, is a first order differential equation in time and thus this condition is always satisfied.

We could now integrate over the noise. Instead, following the method explained in Section 16.2, to more clearly exhibit the general algebraic structure, we introduce an integral representation for equation (16.90):

$$\delta\left(E_\alpha(\varphi) - \nu_\alpha\right) = \int [d\lambda^\alpha(x)] \exp\left\{\int dx\, \lambda^\alpha(x)\left[\nu_\alpha(x) - E_\alpha(\varphi(x))\right]\right\}. \tag{16.96}$$

The integration over the noise ν involves the generating functional $w(\lambda)$ of connected ν-field correlation functions:

$$\exp\left[w(\lambda)\right] = \int [d\rho(\nu)] \exp\left[\int dx\, \lambda^\alpha(x)\,\nu_\alpha(x)\right]. \tag{16.97}$$

Finally we rewrite $\det \mathbf{E}$ as an integral over fermion fields $\bar{c}^\alpha$ and c^α:

$$\det \mathbf{E} = \int [d\bar{c}^\alpha(x)]\,[dc^\alpha(x)] \exp\left[\int dx\, dy\, \bar{c}^\alpha(x)E_{\alpha\beta}(x,y)c^\beta(y)\right]. \tag{16.98}$$

The functional representation (16.94) of $Z(J)$ then becomes:

$$Z(J) = \int [\mathrm{d}\varphi]\,[\mathrm{d}\bar{c}]\,[\mathrm{d}c]\,[\mathrm{d}\lambda]\exp\left[-S\left(\varphi,\bar{c},c,\lambda\right) + \int \mathrm{d}x\, J_\alpha(x)\varphi^\alpha(x)\right] \qquad (16.99)$$

in which $S\left(\varphi,\bar{c},c,\lambda\right)$ is the effective action:

$$S = -w(\lambda) + \int \mathrm{d}x\,\lambda^\alpha(x)E_\alpha(\varphi) - \int \mathrm{d}x\,\mathrm{d}y\,\bar{c}^\alpha(x)E_{\alpha\beta}\left(x,y\right)c^\beta(y). \qquad (16.100)$$

This expression is, in more explicit notations adapted to field theory, the expression (16.19). It therefore has a simple BRS symmetry.

The transformations of the fields can be written in terms of an anticommuting constant $\bar{\varepsilon}$ as:

$$\delta\varphi^\alpha(x) = \bar{\varepsilon}c^\alpha(x), \quad \delta c^\alpha(x) = 0\,, \qquad \delta\bar{c}^\alpha(x) = \bar{\varepsilon}\lambda^\alpha(x), \quad \delta\lambda^\alpha(x) = 0\,. \qquad (16.101)$$

The action (16.100) can also be expressed in terms of the fields:

$$\phi^\alpha\left(x,\bar{\theta}\right) = \varphi^\alpha(x) + \bar{\theta}c^\alpha(x), \qquad \bar{C}^\alpha\left(x,\bar{\theta}\right) = \bar{c}^\alpha(x) + \bar{\theta}\lambda^\alpha(x),$$

as has been done in Section 16.3, to render the BRS obvious. In some cases, as we have shown, it is also useful to combine all fields in one superfield:

$$\phi^\alpha\left(x,\bar{\theta},\theta\right) = \phi^\alpha\left(x,\bar{\theta}\right) + \bar{C}^\alpha\left(x,\bar{\theta}\right)\theta\,. \qquad (16.102)$$

In the case of gaussian noise (16.92), the effective action $S\left(\phi\right)$ then reads:

$$S\left(\phi\right) = \int \mathrm{d}\theta \mathrm{d}\bar{\theta}\left\{\int \mathrm{d}x\,\theta\frac{\partial\phi^\alpha}{\partial\theta}E_\alpha\left[\phi\left(x,\bar{\theta},\theta\right)\right] + \int \mathrm{d}x\,\mathrm{d}y\,\frac{\partial\phi^\alpha}{\partial\theta}\Omega_{\alpha\beta}\frac{\partial\phi^\beta}{\partial\bar{\theta}}\right\}. \qquad (16.103)$$

16.8.2 BRS symmetry and WT identities

To the BRS symmetry corresponds a set of WT identities which can be derived by introducing sources for all fields and performing in the functional integral a change of variables of form (16.101):

$$Z\left(J,\bar{\eta},\eta,l\right) = \int [\mathrm{d}\varphi]\,[\mathrm{d}\bar{c}]\,[\mathrm{d}c]\,[\mathrm{d}\lambda]\exp\Big[-S\left(\varphi,\bar{c},c,\lambda\right)$$
$$+ \int \mathrm{d}x\left(J_\alpha\varphi^\alpha + \bar{\eta}_\alpha c^\alpha + \bar{c}^\alpha\eta_\alpha + l_\alpha\lambda^\alpha\right)\Big]. \qquad (16.104)$$

Since the transformations are linear in the fields, the consequences are simple, the generating functional of connected correlation functions satisfies:

$$\int \mathrm{d}x\left(J_\alpha(x)\frac{\delta}{\delta\bar{\eta}_\alpha(x)} - \eta_\alpha(x)\frac{\delta}{\delta l_\alpha(x)}\right)W\left(J,\bar{\eta},\eta,l\right) = 0\,, \qquad (16.105)$$

and the 1PI generating functional $\Gamma\left(\varphi,\bar{c},c,\lambda\right)$ is BRS symmetric, i.e. is also invariant under transformation (16.101). This property is expressed by the equation:

$$\int \mathrm{d}x\left[c^\alpha(x)\frac{\delta\Gamma}{\delta\varphi^\alpha(x)} + \lambda^\alpha(x)\frac{\delta\Gamma}{\delta\bar{c}^\alpha(x)}\right] = 0\,. \qquad (16.106)$$

If the effective action has the form of a local action renormalizable by power counting, and if a BRS invariant regularization can be found, then equation (16.106) implies that the counterterms needed to render the theory finite are BRS symmetric. Equation (16.106) is stable under renormalization and the renormalized effective action S_r is BRS symmetric, i.e. also satisfies:

$$\int \mathrm{d}x\left(c^\alpha\frac{\delta S_\mathrm{r}}{\delta\varphi^\alpha} + \lambda^\alpha\frac{\delta S_\mathrm{r}}{\delta\bar{c}^\alpha}\right) = 0\,. \qquad (16.107)$$

16.8.3 Solutions of WT identities

Let us assume that we can show, using power counting arguments and fermion number conservation, that S_r has the general form:

$$S_r = -\int dx\,dy\, E^r_{\alpha\beta}(\varphi)\bar{c}^\alpha c^\beta + \Sigma(\varphi, \lambda). \tag{16.108}$$

It then follows from the analysis of Subsection 16.5.1 that $E^r_{\alpha\beta}(\varphi)$ and $\Sigma(\varphi, \lambda)$ have respectively the form:

$$E^r_{\alpha\beta} = \frac{\partial E^r_\alpha(x)}{\partial \varphi^\beta(y)}, \qquad \Sigma(\varphi, \lambda) = \int dx\, \lambda^\alpha(x) E^r_\alpha(\varphi) - w_r(\lambda). \tag{16.109}$$

Thus we have shown that the form (16.100) is preserved by the renormalization. In the parametrization (16.102) these properties have a simple interpretation, they correspond to translation symmetry on the coordinate $\bar{\theta}$.

16.8.4 An example: the random field Ising model

It can be shown that the long distance properties of an Ising model in a random magnetic field can be described by a stochastic field equation of the form:

$$\left(-\Delta + m^2\right)\varphi(x) + \frac{g}{3!}\varphi^3(x) = h(x), \tag{16.110}$$

in which the field $\varphi(x)$ represents the Ising spin and $h(x)$, the magnetic field, has a gaussian distribution:

$$\langle h(x)h(x')\rangle = \alpha\delta(x - x'), \quad \langle h(x)\rangle = 0. \tag{16.111}$$

The equation (16.110) is an example of a gradient equation of the form (16.34) and therefore leads to at least two BRS symmetries. Introducing the superfield $\phi(x, \theta, \bar{\theta})$ we can write the effective action $S(\phi)$, after averaging over the magnetic field,

$$S(\phi) = \int d\theta d\bar{\theta}\left(\frac{\alpha}{2}\int d^d x \frac{\partial \phi}{\partial \theta}\frac{\partial \phi}{\partial \bar{\theta}} + \mathcal{A}(\phi)\right), \tag{16.112}$$

with:

$$\mathcal{A}(\phi) = \int d^d x \left[\tfrac{1}{2}\left(\partial_\mu \phi\right)^2 + \tfrac{1}{2}m^2\phi^2 + \frac{g}{4!}\phi^4\right]. \tag{16.113}$$

The remarkable property of action (16.112) is that it is invariant by a large group of transformations: translations of θ and $\bar{\theta}$ as anticipated but also "rotations" which leave the line element:

$$(ds)^2 = dx_\mu dx_\mu + \alpha d\theta d\bar{\theta} \tag{16.114}$$

invariant. In addition to transformations internal to the $\{\theta, \bar{\theta}\}$ and x_μ spaces one finds the two infinitesimal transformations:

$$\theta = \theta' + \alpha a_\mu x_\mu, \quad x_\mu = x'_\mu - 2\bar{\theta}'a_\mu, \tag{16.115}$$

$$\bar{\theta} = \bar{\theta}' + \alpha \bar{a}_\mu x_\mu, \quad x_\mu = x'_\mu - 2\bar{a}_\mu\theta'. \tag{16.116}$$

The vectors a_μ and $\bar{a}_\mu$ are anticommuting elements of the Grassmann algebra. If we call D_μ and $\bar{D}_\mu$ the corresponding generators, we find the anticommutation relations:

$$D_\mu D_\nu + D_\nu D_\mu = 4\alpha \delta_{\mu\nu} \, \bar{\theta} \frac{\partial}{\partial \theta} \tag{16.117}$$

$$\bar{D}_\mu \bar{D}_\nu + \bar{D}_\nu \bar{D}_\mu = -4\alpha \delta_{\mu\nu} \, \theta \frac{\partial}{\partial \bar{\theta}} \tag{16.118}$$

$$\bar{D}_\mu D_\nu + D_\nu \bar{D}_\mu = 2\alpha \left[\delta_{\mu\nu} \left(\bar{\theta} \frac{\partial}{\partial \bar{\theta}} - \theta \frac{\partial}{\partial \theta} \right) + x_\mu \partial_\nu - x_\mu \partial_\nu \right]. \tag{16.119}$$

We recognize in the r.h.s. generators of internal transformations.

Using this symmetry it is possible to prove a property of dimensional reduction for action (16.112) by a variant of the method of Section 17.6. One shows that the correlation functions calculated from action (16.112) in d dimensions are the same as those calculated with $\mathcal{A}(\varphi(x))$ in $d-2$ dimensions. This result maps the random field Ising model in d dimensions onto the pure Ising model in $d-2$ dimensions. It has unfortunately consequences which contradict physical intuition. A reason can be found in the starting point: equation (16.110) in the region of interest, m^2 small, has certainly for some fields a large number of solutions and therefore the whole method is not applicable without modifications. A similar problem, called Gribov's ambiguity, arises in gauge theories.

Bibliographical Notes

The ST and BRS symmetries have been introduced in the context of gauge theories by
     A.A. Slavnov, *Theor. Math. Phys.* 10 (1972) 99; J.C. Taylor, *Nucl. Phys.* B33 (1971) 436; C. Becchi, A. Rouet and R. Stora, *Commun. Math. Phys.* 42 (1975) 127.
The present formulation is taken from
     J. Zinn-Justin, *Nucl. Phys.* B275 [FS18] (1986) 135.
The discussion of Section 16.1 is inspired by
     B.W. Lee and J. Zinn-Justin, *Phys. Rev.* D7 (1973) 1049 (Appendix).
Supersymmetry has been introduced in Field Theory context by
     J. Wess and B. Zumino, *Nucl. Phys.* B70 (1974) 39.
Anticommuting coordinates were used in
     A. Salam and J. Strathdee, *Phys. Lett.* 51B (1974) 353; S. Ferrara, B. Zumino and J. Wess, *Phys. Lett.* 51B (1974) 239.
Numerous textbooks have been devoted to supersymmetry see for example
     J. Wess and J. Bagger, *Supersymmetry and Supergravity* (Princeton University Press, Princeton 1983); S. Ferrara, *Supersymmetry and Supergravity (Lecture Notes in Physics 208)*, (Springer-Verlag, Berlin 1984); B.S. DeWitt, *Supermanifolds* (Cambridge University Press, Cambridge 1984).
Many articles are reprinted in
     Supersymmetry, S. Ferrara ed. (North-Holland and World Scientific 1987).
The supersymmetry and the dimensional reduction in the random-field Ising model are discussed in
     G. Parisi and N. Sourlas, *Phys. Rev. Lett.* 43 (1979) 744; J. Cardy, *Phys. Lett.* 125B (1983) 470; A. Klein and J. Fernando-Perez, *Phys. Lett.* 125B (1983) 473.
The consequences of dimensional reduction contradict physical intuition
     Y. Imry and S.K. Ma, *Phys. Rev. Lett* 35 (1975)
and are disproved in
     J. Imbrie, *Commun. Math. Phys.* 98 (1985) 145; J. Bricmont and A. Kupiainen, *Phys. Rev. Lett.* 59 (1987) 1829.

17 RENORMALIZATION AND STOCHASTIC FIELD EQUATIONS. SUPERSYMMETRY

Stochastic dynamic equations like the Langevin or Fokker–Planck equations, extensions to Field Theory of the equations introduced in Chapter 4, have been proposed to describe the dynamics of critical phenomena (see Chapter 35), or as an alternative method of quantization which could be useful in cases where ordinary methods lead to difficult problems like gauge theories. In both cases divergences appear in perturbative calculations and it is necessary to understand how these equations renormalize.

For many problems related to perturbation theory and renormalization, it is more convenient to work with an effective action (often called in this context dynamical action) associated with the Langevin equation rather than with the equation directly, because then standard methods of quantum field theory become applicable. The method has been explained in Section 16.8. We have shown quite generally that this dynamical action has a BRS symmetry. We shall use this symmetry and its consequences in the form of WT identities to prove that under some general conditions the structure of the Langevin equation is stable under renormalization.

A particular class of Langevin equations is of special interest: purely dissipative equations with gaussian noise, in which the drift force derives from a static action and is thus conservative (the corresponding stochastic processes satisfy the detailed balance condition). Then the dynamical action has a second fermionic symmetry which, combined with the first one, provides an example of quantum mechanical supersymmetry.

For completeness in Section 17.9 we discuss the peculiarities of the stochastic quantization of gauge theories.

The formalism presented here may seem rather heavy when applied to simple models, but it enables us to exhibit general structures with many different applications.

Finally in Section 17.10 we briefly review how the concept of supersymmetry, suitably generalized, can be used to construct theories potentially relevant for four-dimensional particle physics. The new technical feature is the combination of spin and supersymmetry.

The appendix is devoted to the calculation at two-loop order of some dynamical RG functions.

17.1 Langevin and Fokker–Planck Equations

To keep notations as simple as possible we expose the problem on the example of a one-component scalar field $\varphi(x,t)$. The field depends in addition to the time variable t on d euclidean variables x_μ. Its time evolution is governed by a first order in time stochastic differential equation, straightforward generalization of the Langevin equation introduced in Chapter 4:

$$\dot{\varphi}(x,t) = -\tfrac{1}{2}\Omega L\left[\varphi(x,t)\right] + \nu(x,t), \tag{17.1}$$

where $L\left[\varphi(x)\right]$ is a local functional of $\varphi(x)$ and Ω^{-1} a constant introduced for convenience, to provide a time scale. The noise field $\nu(x,t)$ is a stochastic field which we assume to have a gaussian local distribution $[\mathrm{d}\rho(\nu)]$:

$$[\mathrm{d}\rho(\nu)] = [\mathrm{d}\nu]\exp\left[-\int \mathrm{d}^d x\,\mathrm{d}t\,\nu^2(x,t)/2\Omega\right], \tag{17.2}$$

which can also be characterized by its 1 and 2-point functions:

$$\langle \nu(x,t) \rangle = 0\,, \qquad \langle \nu(x,t) \nu(x',t') \rangle = \Omega\,\delta(x-x')\,\delta(t-t')\,. \tag{17.3}$$

We discuss the problem of renormalization on this simple example though many results apply to more general equations and gaussian distributions of noise (see for example Sections 17.7,17.8).

The Fokker–Planck equation. Given the noise (17.2), and some initial conditions for the field $\varphi(x,t)$, Langevin equation (17.1) generates a time-dependent field distribution $P(t,\varphi(x))$:

$$P(t,\varphi(x)) = \langle \delta(\varphi(x,t) - \varphi(x)) \rangle_\nu\,. \tag{17.4}$$

The derivation of Section 4.2 can be immediately generalized and yields a Fokker–Planck equation, the field theory form of equation (4.21):

$$\dot{P}(\varphi,t) = -\Omega\mathcal{H}_{\mathrm{FP}}P(\varphi,t)\,, \tag{17.5}$$

where $\mathcal{H}_{\mathrm{FP}}$, the Fokker–Planck hamiltonian, is given by

$$\mathcal{H}_{\mathrm{FP}}\left(\varphi, \frac{\delta}{\delta\varphi}\right) = -\frac{1}{2}\int \mathrm{d}^d x\, \frac{\delta}{\delta\varphi(x)}\left[\frac{\delta}{\delta\varphi(x)} + L(\varphi(x))\right]\,. \tag{17.6}$$

The dissipative Langevin equation. In Section 17.5 we discuss in detail a special case, the purely dissipative Langevin equation which corresponds to the choice:

$$L[\varphi(x)] = \frac{\delta\mathcal{A}}{\delta\varphi(x)}\,, \tag{17.7}$$

and generalizes the example studied in Section 4.4. The functional $\mathcal{A}(\varphi)$ is a static (time-independent) euclidean action for the scalar field φ, for example:

$$\mathcal{A}(\varphi) = \int \mathrm{d}^d x\left[\tfrac{1}{2}\left(\partial_\mu\varphi(x)\right)^2 + V(\varphi(x))\right]\,.$$

The substitution

$$P = \mathrm{e}^{-\mathcal{A}/2}\,\widetilde{P}\,,$$

then transforms the hamiltonian (17.6) into a hermitian positive hamiltonian. This property allows to prove the convergence at time $+\infty$ of P towards the static distribution $\mathrm{e}^{-\mathcal{A}}$ (provided the latter is normalizable). It is at the basis of the idea of stochastic quantization, the Langevin equation being then only a device to generate the static distribution $\mathcal{A}$.

In Chapter 4 we have already discussed these equations for a finite number of degrees of freedom. All arguments based on purely algebraic identities can be generalized to field theories. We here discuss new problems which arise because, as we shall show, the Langevin equation in general requires renormalizations. In the case (17.7) when the action $\mathcal{A}(\varphi)$ is renormalizable, we expect that the renormalizations of the static theory together with a time scale renormalization, render the Langevin equation finite. To discuss this problem, we have to set up a formalism more directly amenable to the ordinary methods of quantum field theory. This can be done by constructing a functional integral representation of the time-dependent φ-field correlation functions in terms of an effective local action, which in this framework it is natural to also call dynamic action.

17.2 Time-Dependent Correlation Functions

Effective or dynamical action. The generating functional $Z(J)$ of dynamic correlation functions of the field $\varphi(x, t)$ solution of equation (17.1), is given by the noise average:

$$Z(J) = \left\langle \exp\left[\int d^d x\, dt\, J(x, t)\varphi(x, t)\right]\right\rangle_\nu,$$

$$= \int [d\nu] \exp\left[-\int d^d x\, dt \left(\frac{1}{2\Omega}\nu^2(x, t) - J(x, t)\varphi(x, t)\right)\right]. \tag{17.8}$$

To impose equation (17.1), following the method explained in Section 4.6, we insert the identity:

$$\int [d\varphi] \det M \delta\left[\dot\varphi + (\Omega/2)L(\varphi) - \nu\right] = 1, \tag{17.9}$$

where M is the differential operator:

$$M = \frac{\partial}{\partial t} + \frac{\Omega}{2}\frac{\delta L}{\delta \varphi(x, t)}, \tag{17.10}$$

into expression (17.8). We then find:

$$Z(J) = \int [d\nu][d\varphi] \det M \delta\left[\dot\varphi + \frac{\Omega}{2}L(\varphi) - \nu\right] \exp\left[\int d^d x\, dt\left(J\varphi - \frac{\nu^2}{2\Omega}\right)\right]. \tag{17.11}$$

The δ-function can be used to integrate over the noise ν:

$$Z(J) = \int [d\varphi] \det M \exp\left[-\int d^d x\, dt\left(\frac{1}{2\Omega}\left(\dot\varphi + \tfrac{1}{2}\Omega L(\varphi)\right)^2 - J\varphi\right)\right]. \tag{17.12}$$

For a system with a discrete set of degrees of freedom (a $d = 0$ dimensional or a lattice regularized field theory), the determinant can be calculated, using the identity:

$$\det M \propto \exp\left(\operatorname{tr}\ln\right)\left[1 + \left(\frac{\partial}{\partial t}\right)^{-1}\frac{\Omega}{2}\frac{\delta L}{\delta\varphi}\right]. \tag{17.13}$$

As a consequence of the causality of the Langevin equation the inverse of the operator $(\partial/\partial t)\,\delta(t - t')$ is the kernel $\theta(t - t')$ ($\theta(t)$ is the Heaviside step function). In an expansion in powers of Ω all terms thus vanish when one takes the trace, except the first one which yields:

$$\det M \propto \exp\left\{\theta(0)\frac{\Omega}{2}\int dt\, d^d x\left.\frac{\delta L[\varphi(x, t)]}{\delta\varphi(x', t)}\right|_{x'=x}\right\}. \tag{17.14}$$

As we have seen in Section 4.6 the symmetry $(t \leftrightarrow t')$ of the noise 2-point function imposes $\theta(0) = 1/2$. The final expression then formally reads:

$$Z(J) = \int [d\varphi] \exp\left[-S(\varphi) + \int d^d x\, dt\, J(x, t)\varphi(x, t)\right], \tag{17.15}$$

$$S(\varphi) = \frac{1}{2\Omega}\int d^d x\, dt\left(\dot\varphi + \frac{\Omega}{2}L(\varphi)\right)^2 - \frac{\Omega}{4}\int dt\, d^d x\left.\frac{\delta L[\varphi(x, t)]}{\delta\varphi(x', t)}\right|_{x'=x}. \tag{17.16}$$

When the force in the Langevin equation derives from a static action (equation (17.7)), the term linear in $\dot{\varphi}$ in the action S is a total time derivative and contributes only to boundary terms.

Note that the dynamical action (17.16) is the generalization to field theory of action (4.64), and can thus be directly obtained by writing the solution $P(\varphi, t)$ of the Fokker–Planck equation (17.5) as a functional integral.

Divergences and the problem of renormalization. In dimension $d > 0$, expression (17.16) is ill-defined when $L(\varphi)$ is a local functional because the contribution of the determinant is formally proportional to $\delta^d(0)$:

$$\int \mathrm{d}^d x \left. \frac{\delta L[\varphi(x, t)]}{\delta \varphi(x', t)} \right|_{x'=x} \propto \delta^d(0).$$

We have already encountered a similar situation in the discussion of homogeneous spaces in Chapters 14, 15 where also a divergent measure term appeared. The determinant has thus to be regularized. Again we have two choices:

(i) With dimensional regularization terms like $\delta^d(0)$ vanish and therefore the determinant can be completely omitted. With this convention expression (17.15) can be used in practical perturbative calculations.

(ii) It is, however, interesting to keep this divergent term in some regularized form, as we shall do in the discussion of the renormalization of the dynamic theory, in order to be able to understand its geometrical significance. This can be achieved by using a lattice regularization.

Let us show, however, that even with dimensional regularization a non-trivial renormalization problem arises. We take the example of a dissipative Langevin equation (equation (17.7)) with the four-dimensional action $\mathcal{A}(\varphi)$:

$$\mathcal{A}(\varphi) = \int \mathrm{d}^4 x \left[\tfrac{1}{2} (\partial_\mu \varphi)^2 + \tfrac{1}{2} m^2 \varphi^2 + (g/4!) \varphi^4 \right]. \tag{17.17}$$

From the dynamical action (17.16) we calculate the propagator Δ of the $\varphi(x, t)$ field. After Fourier transformation, as a function of $\mathbf{k}$ and ω the variables associated with space and time respectively, the propagator reads:

$$\Delta(\mathbf{k}, \omega) = \frac{\Omega}{\omega^2 + \Omega^2 (k^2 + m^2)^2 / 4}. \tag{17.18}$$

In the standard power counting analysis, as presented in Chapter 8, it is generally assumed that the propagator is, in momentum representation and when all arguments become large, a homogeneous function. This is not, at first sight, the case here. However, the propagator has a homogeneous asymptotic behaviour if we scale the frequency ω as a momentum squared, or equivalently the time t as a distance squared (the brownian motion behaviour). Then the canonical (engineering) dimensions $[\varphi]$ of the field φ in the static and dynamic theories coincide:

$$[\varphi] = 1, \tag{17.19}$$

and the interactions coming from $(\delta \mathcal{A}/\delta \varphi)^2$ have dimensions 4 and 6:

$$\frac{\delta \mathcal{A}}{\delta \varphi} = -\Delta \varphi + m^2 \varphi + g \varphi^3 / 6. \tag{17.20}$$

General renormalization theory only tells us that the renormalized action is the most general local functional of canonical dimension 6, i.e. containing all vertices of non-positive dimensions. Such a functional in general depends on more parameters than the bare action (17.16); in particular it is not a square of a local functional and can no longer be derived from a Langevin equation by the algebraic transformations (17.8–17.16). To prove that the Langevin equation can be renormalized we have thus to find identities satisfied by correlation functions, which imply relations between counterterms and ensure that the structure of action (17.16) is preserved by the renormalization.

17.3 ST Symmetry and Equilibrium Correlation Functions

Langevin equations belong to the general class of stochastic equations discussed in Section 16.8. We shall exploit more systematically this remark in next section. However, let us already note that the ST symmetry discussed in Section 16.1 implies identities for correlation functions. In the first part of this section we consider only the example (17.7). It follows from the identity (17.9) that the determinant of $\mathbf{M}$:

$$M\left(x, t; x', t'\right) = \langle x, t| \left(\frac{\partial}{\partial t} + \frac{\Omega}{2} \frac{\delta^2 \mathcal{A}}{\delta \varphi \delta \varphi}\right) |x', t'\rangle, \tag{17.21}$$

generates an invariant measure for a set of non-linear transformations which translate $\nu(x, t)$ by a function $\mu(x, t)$ independent of φ. For $\mu(x, t)$ infinitesimal the variation of φ is:

$$\delta\varphi(x, t) = \int dt' \, dx' \, M^{(-1)}\left(x, t; x', t'; \varphi\right) \mu\left(x', t'\right). \tag{17.22}$$

In an infinitesimal change of variables of the form of a transformation (17.22) the variations of the action and the source term in the functional integral (17.12) are:

$$\delta\left\{\frac{1}{2\Omega} \int dx \, dt \left(\dot\varphi + \frac{\Omega}{2} \frac{\delta\mathcal{A}}{\delta\varphi}\right)^2\right\} = \frac{1}{\Omega} \int dx \, dt \left[\dot\varphi(x, t) + \frac{\Omega}{2} \frac{\delta\mathcal{A}}{\delta\varphi(x, t)}\right] \mu(x, t)$$

$$\delta\left[\int dx \, dt \, J\varphi\right] = \int dx \, dt \, dx' \, dt' \, J(x, t) M^{-1}\left(x', t'; x, t\right) \mu\left(x', t'\right).$$

Expressing as usual the invariance of the functional integral under a change of variables we obtain the identity:

$$\left[\frac{1}{\Omega} \frac{\partial}{\partial t} \frac{\delta}{\delta J(x, t)} + \frac{1}{2} \frac{\delta\mathcal{A}}{\delta\varphi(x, t)} \left(\frac{\delta}{\delta J}\right)\right] Z\left(J\right) = \int dx' \, dt' \, J\left(x', t'\right) Z_M\left(x', t'; x, t\right), \tag{17.23}$$

with the definition:

$$Z_M\left(x, t; x', t'; J\right) = M^{(-1)}\left(x', t'; x, t; \frac{\delta}{\delta J}\right) Z\left(J\right), \tag{17.24}$$

or equivalently:

$$\left\{\frac{\partial}{\partial t} + \frac{\Omega}{2} \frac{\delta^2 \mathcal{A}}{[\delta\varphi(x, t)]^2} \left(\frac{\delta}{\delta J}\right)\right\} Z_M\left(x, t; x', t'; J\right) = \delta\left(x - x'\right) \delta\left(t - t'\right) Z\left(J\right). \tag{17.25}$$

Equation (17.23), which can of course be derived directly from the Langevin equation, implies that the large time limit of the equal-time correlation functions satisfies the field equations of the static theory.

Indeed let us take a special source:

$$J(x, t') = J(x)\,\delta(t' - \tau), \tag{17.26}$$

and set $t = \tau$ in equation (17.23). Then equation (17.23) involves only $Z_M(x', \tau; x, \tau; J)$. If we expand Z_M in powers of Ω, again due to the $\theta(t)$ function only the first term survives and Z_M becomes:

$$Z_M(x', \tau; x, \tau; J) = \tfrac{1}{2}\delta(x - x')\,Z(J)\ , \quad \left(\theta(0) = \tfrac{1}{2}\right). \tag{17.27}$$

We have again used a consequence of the causal character of the Langevin equation. Also it can be checked in perturbation theory that $Z(J)$ has a limit and that for a Langevin equation of type (17.7) the term:

$$\frac{\partial}{\partial t}\frac{\delta}{\delta J(x, t)}Z(J)\bigg|_{t=\tau}\ , \quad \left(\theta(0) = \tfrac{1}{2}\right),$$

vanishes at large time, so that the limiting functional $Z(J)$ satisfies as a consequence of equations (17.23) and (17.27):

$$\frac{\delta \mathcal{A}}{\delta \varphi(x)}\left(\frac{\delta}{\delta J}\right)Z(J) = J(x)\,Z(J), \tag{17.28}$$

which implies

$$Z(J) = \int [d\varphi(x)]\exp\left[-\mathcal{A}(\varphi) + \int dx\, J(x)\,\varphi(x)\right]. \tag{17.29}$$

From these considerations we conclude that some form of equations (17.23) and (17.24) must be used in the proof of the renormalizability of the dynamic theory, i.e. in the proof that after renormalization the structure of the Langevin equation is preserved and that the equal-time correlation functions have in the large time limit the correlation functions of the renormalized static theory corresponding to $\mathcal{A}(\varphi)$.

Remark. We can also derive a different equation directly for the generating functional $Z(J)$ of equal-time correlation functions, by using the Fokker–Planck equation (17.5):

$$Z(J, t) = \left\langle \exp\left[\int d^d x\, J(x)\,\varphi(x, t)\right]\right\rangle_\nu. \tag{17.30}$$

By definition:

$$Z(J, t) = \int [d\varphi(x)]\,P(\varphi, t)\exp\left[\int d^d x\, J(x)\,\varphi(x)\right]. \tag{17.31}$$

Therefore:

$$\frac{\partial}{\partial t}Z(J, t) = \frac{\Omega}{2}\int [d\varphi]\int d^d x\,\frac{\delta}{\delta \varphi(x)}\left[\frac{\delta P}{\delta \varphi(x)} + L(\varphi)\,P\right]\exp\left[\int d^d x\, J(x)\,\varphi(x)\right]. \tag{17.32}$$

Integrating then by parts, we obtain an equation for $Z(J, t)$:

$$\frac{\partial}{\partial t} Z(J, t) = \frac{\Omega}{2} \int \mathrm{d}^d x \left\{ J(x) \left[J(x) - L\left(\frac{\delta}{\delta J(x)} \right) \right] \right\} Z(J, t). \tag{17.33}$$

The existence of an equilibrium distribution implies that the time derivative of $Z(J, t)$ vanishes for large time. Then the limiting generating functional of correlation functions satisfies the static equation:

$$\int \mathrm{d}^d x \left\{ J(x) \left[J(x) - L\left(\frac{\delta}{\delta J(x)} \right) \right] \right\} Z(J) = 0. \tag{17.34}$$

By construction, the discussion of the solutions of this equation and of the static solutions of the Fokker–Planck equation are directly related.

17.4 Renormalization and BRS Symmetry

The Langevin equation is a special example of the stochastic equations we have discussed in Section 16.8. It is a first order differential equation in time and thus has a unique solution: The methods of Chapter 16 can be applied and all algebraic results remain valid. We now consider a particular example of Langevin equation (17.1):

$$\dot{\varphi}(x, t) + \tfrac{1}{2}\Omega \left[-\Delta\varphi + L_{\text{int.}}(\varphi) \right] = \nu(x, t), \tag{17.35}$$

in which $L_{\text{int.}}(\varphi)$ is a polynomial in φ, and $\nu(x, t)$ the gaussian noise with distribution (17.2). For notational simplicity we have written the equation for a one-component field but all arguments immediately generalize to a field φ with several components. The dynamical action corresponding to equation (17.35) is then:

$$S = \int \mathrm{d}t\, \mathrm{d}^d x \left\{ -\tfrac{1}{2}\Omega\lambda^2 + \lambda \left[\dot{\varphi} + \tfrac{1}{2}\Omega \left(-\Delta\varphi + L_{\text{int.}}(\varphi) \right) \right] - \bar{c} M c \right\}, \tag{17.36}$$

where M is the operator (17.10):

$$M = \frac{\partial}{\partial t} - \tfrac{1}{2}\Omega\Delta + \tfrac{1}{2}\Omega \frac{\partial L_{\text{int.}}[\varphi(x, t)]}{\partial \varphi}. \tag{17.37}$$

In this form the associated dynamical action (17.36) is BRS symmetric. We have seen that in the case of the Langevin equation the determinant $\det M$ can be formally calculated in such a way that the final dynamical action is in direct correspondence with the Fokker–Planck equation. However, to exhibit more clearly the general algebraic structure, it is useful to keep it in the form of a fermion integral.

Power counting and renormalization. Note that below the canonical dimension of a quantity Q is denoted by $[Q]$. As discussed in Section 17.2, from the point of view of power counting, one has to assign to frequencies ω the canonical dimension of momentum squared, i.e. 2 and thus to time the dimension -2. The dynamical action density (17.36) then has dimension $d + 2$ since the integration measure here is $\mathrm{d}t\, \mathrm{d}^d x$. The dimensions $[\lambda]$, $[\varphi]$, $[c]$ and $[\bar{c}]$ of the fields follow:

$$[\lambda] = \tfrac{1}{2}(d + 2), \quad [\varphi] = \tfrac{1}{2}(d - 2), \quad [c] + [\bar{c}] = d. \tag{17.38}$$

The theory is thus renormalizable by power counting in dimension d if the dimension of $L_{\text{int.}}(\varphi)$ is:

$$[L_{\text{int.}}(\varphi)] = \tfrac{1}{2}(d+2),$$

The dimension of M is then $[M] = [L_{\text{int.}}(\varphi)] - \tfrac{1}{2}(d-2) = 2$. If the dimension d satisfies:

$$2([c] + [\bar{c}]) = 2d > d + 2,$$

i.e. if the dimension d of space is larger than 2, the renormalized action S_{r} is at most quadratic in the fermion fields.

Moreover if the dimension d is larger than 2 the sum of dimensions $[c] + [\bar{c}] + [\lambda]$ satisfies:

$$[c] + [\bar{c}] + [\lambda] = \tfrac{3}{2}d + 1 > d + 2,$$

and therefore M_{r}, the renormalized coefficient of $\bar{c}c$, cannot depend on λ.

The results of Section 16.8 (which follow from BRS symmetry) then apply and the structure of the dynamical action (17.36) is preserved by the renormalization. It follows that it can be derived from a renormalized Langevin equation.

We have thus shown that if the canonical dimension of $L_{\text{int.}}(\varphi)$ satisfies $[L_{\text{int.}}] \leq \tfrac{1}{2}(d+2)$ and the dimension d is larger than 2, the Langevin equation (17.35) with the gaussian noise (17.2) can be renormalized. Moreover since the renormalized action remains quadratic in λ, the renormalized noise is still gaussian. Note that this result can be immediately generalized to several component fields.

The analysis can also be extended to more complicated cases, for example to models with different noise 2-point functions (see Chapter 35).

In the particular case (17.7) more precise results can be obtained which we shall describe below (for a one-component field equation the drift force in equation (17.35) can of course always be written as the variation of an action as in (17.7)).

17.5 The Purely Dissipative Langevin Equation

We have shown in Subsection 16.3 that when, in symbolic notations, the equation has the form $\delta \mathcal{A}/\delta\varphi = \nu$, the dynamical action $S(\phi)$ has two independent BRS symmetries which, in the superfield notation (16.102), correspond to translations of θ and $\bar{\theta}$. The Langevin equation in the special case (17.7):

$$\dot{\varphi}(x,t) = -\frac{\Omega}{2}\frac{\delta \mathcal{A}}{\delta\varphi(x,t)} + \nu(x,t), \tag{17.39}$$

almost shares this property. The force term is a gradient but not the time derivative $\dot{\varphi}$. Somewhat surprisingly, however, it is possible to find a slightly modified BRS transformation which leaves the dynamical action (17.36) invariant and which combines non-trivially with the first one to yield the simplest example of supersymmetry.

17.5.1 Supersymmetry

In the superfield notation (16.102), the action (17.36) corresponding to the Langevin equation (17.39) with the noise (17.2) can be rewritten:

$$S(\phi) = \int \mathrm{d}\theta\,\mathrm{d}\bar{\theta}\,\mathrm{d}t \left[\frac{2}{\Omega}\int \mathrm{d}^d x\,\bar{\mathrm{D}}\phi\mathrm{D}\phi + \mathcal{A}(\phi)\right], \tag{17.40}$$

with the definitions:

$$\bar{D} = \frac{\partial}{\partial \theta}, \quad D = \frac{\partial}{\partial \bar{\theta}} - \theta \frac{\partial}{\partial t}. \tag{17.41}$$

For convenience we have rescaled θ and $\bar{\theta}$ by a factor $(2/\Omega)^{1/2}$.

$\bar{D}$ and D satisfy the anticommutation relations:

$$D^2 = \bar{D}^2 = 0, \quad D\bar{D} + \bar{D}D = -\frac{\partial}{\partial t}. \tag{17.42}$$

Let us then introduce two other Grassmann-type differential operators $D', \bar{D}'$:

$$D' = \frac{\partial}{\partial \bar{\theta}}, \quad \bar{D}' = \frac{\partial}{\partial \theta} + \bar{\theta} \frac{\partial}{\partial t}, \tag{17.43}$$

Both *anticommute* with D and $\bar{D}$ and satisfy

$$D'^2 = \bar{D}'^2 = 0, \quad D'\bar{D}' + \bar{D}'D' = \frac{\partial}{\partial t}. \tag{17.44}$$

The two pairs $D, \bar{D}$ and $D', \bar{D}'$ provide the simplest example of generators of supersymmetry. We already know that D' (a translation of $\bar{\theta}$) is the generator of a BRS symmetry of the action (17.40). Let us verify that $\bar{D}'$ is the generator of an additional symmetry. If we perform a variation of ϕ of the form:

$$\delta\phi\left(t, \bar{\theta}, \theta\right) = \varepsilon\bar{D}'\phi, \tag{17.45}$$

which in component form reads:

$$\begin{cases} \delta\varphi = \bar{c}\varepsilon, & \delta\bar{c} = 0, \\ \delta c = (\lambda - \dot{\varphi})\varepsilon, & \delta\lambda = \dot{\bar{c}}\varepsilon, \end{cases} \tag{17.46}$$

we observe that the variation of the action density is a total derivative: This is obvious for $\mathcal{A}$ because it does not explicitly depend on t and θ. For the remaining term the additional property that $\bar{D}'$ anticommutes with D and $\bar{D}$ has to be used

$$\delta\left[\bar{D}\phi D\phi\right] = \bar{D}\left[\varepsilon\bar{D}'\phi\right]D\phi + \bar{D}\phi D\left[\varepsilon\bar{D}'\phi\right] = \varepsilon\bar{D}'\left[\bar{D}\phi D\phi\right].$$

The action is thus supersymmetric. The operators D and $\bar{D}$ can be considered as covariant derivatives from the point of view of supersymmetry.

This supersymmetry is directly connected with the property that the corresponding Fokker–Planck hamiltonian (17.5) is then equivalent to a positive hamiltonian of the form (4.40).

Remarks.

(i) It is possible to emphasize the symmetric role played by θ and $\bar{\theta}$ by performing the substitution $t \mapsto t + \frac{1}{2}\bar{\theta}\theta$. We find it more convenient to remain with the original variables but this is mainly a matter of taste.

(ii) Considering the fermions as real dynamic variables, we can write the hamiltonian associated with the supersymmetric action in boson–fermion space. We find that the functional integral describes both the Langevin equation and its time-reversed form.

17.5.2 Ward–Takahashi (WT) identities

The BRS symmetry has a simple consequence, correlation functions are invariant under a translation of the coordinate $\bar{\theta}$. The transformation (17.45) has a slightly more complicated form. It leads for the generating functional $W(J)$ to the equation:

$$\int dx\, dt\, d\theta\, d\bar{\theta} \left(\bar{\theta}\frac{\partial J}{\partial t} + \frac{\partial J}{\partial \theta} \right) \frac{\delta W}{\delta J\left(x,t,\bar{\theta},\theta\right)} = 0 \,.$$

Connected correlation functions $W^{(n)}\left(x_i, t_i, \bar{\theta}_i, \theta_i\right)$ and proper vertices $\Gamma^{(n)}\left(p_i, \omega_i, \bar{\theta}_i, \theta_i\right)$ thus satisfy the WT identities:

$$\overline{\mathcal{D}} W^{(n)}\left(x_i, t_i, \bar{\theta}_i, \theta_i\right) = 0 \,, \qquad \overline{\mathcal{D}} \Gamma^{(n)}\left(x_i, t_i, \bar{\theta}_i, \theta_i\right) = 0 \,, \qquad (17.47)$$

with

$$\overline{\mathcal{D}} \equiv \sum_{k=1}^{n} \left(\frac{\partial}{\partial \theta_k} + \bar{\theta}_k \frac{\partial}{\partial t_k} \right) \,.$$

After Fourier transformation over space and time the operator $\overline{\mathcal{D}}$ takes the form

$$\overline{\mathcal{D}} = \sum_{k=1}^{n} \left(\frac{\partial}{\partial \theta_k} - i\omega_k \bar{\theta}_k \right) \,. \qquad (17.48)$$

Example: a 2-point function. Let us explore the implications of WT identities for a 2-point function. As relations (17.43,17.44) show, supersymmetry implies translation invariance on time and $\bar{\theta}$. We can therefore write any 2-point function $W^{(2)}$:

$$W^{(2)} = A\left(t_1 - t_2\right) + \left(\bar{\theta}_1 - \bar{\theta}_2\right) \left[\left(\theta_1 + \theta_2\right) B\left(t_1 - t_2\right) + \left(\theta_1 - \theta_2\right) C\left(t_1 - t_2\right) \right]. \qquad (17.49)$$

The WT identity (17.47) then implies:

$$2B(t) = \frac{\partial A}{\partial t} \,. \qquad (17.50)$$

The WT identity does not determine the function C. An additional constraint comes from causality whose role we have already emphasized. For the 2-point function causality implies that the coefficient of $\bar{\theta}_1 \theta_2$ vanishes for $t_1 < t_2$ and the coefficient of $\bar{\theta}_2 \theta_1$ for $t_2 < t_1$. The last function is thus determined, up a possible distribution localized at $t_1 = t_2$. We find:

$$2C(t) = -\epsilon(t)\frac{\partial A}{\partial t} \,. \qquad (17.51)$$

where $\epsilon(t)$ is the sign of t, and therefore:

$$W^{(2)} = \left\{ 1 + \tfrac{1}{2}\left(\bar{\theta}_1 - \bar{\theta}_2\right) \left[\theta_1 + \theta_2 - \left(\theta_1 - \theta_2\right)\epsilon\left(t_1 - t_2\right)\right] \frac{\partial}{\partial t_1} \right\} A\left(t_1 - t_2\right). \qquad (17.52)$$

17.5.3 Renormalization of the dissipative Langevin equation

In the special case of the supersymmetric dynamical action (17.40) a comparison between the two explicit quadratic terms in ϕ of the action yields the relation between dimensions:

$$[t] - [\theta] - [\bar{\theta}] = 0 \;\Rightarrow\; [\mathrm{d}t] + [\mathrm{d}\theta] + [\mathrm{d}\bar{\theta}] = 0 \,. \tag{17.53}$$

(We recall that since integration and differentiation over anticommuting variables are equivalent operations, the dimension of $\mathrm{d}\theta$ is $-[\theta]$.)

Therefore the term proportional to $\mathcal{A}(\phi)$ in the action has the same canonical dimension as in the static case: the power counting is thus the same and the dynamic theory is always renormalizable in the same space dimension d as the static theory. Note that equation (17.53) also implies

$$2[\phi] = d + [t],$$

equation which relates the dimensions of field and time.

Let us then write in superfield notation the most general form of the renormalized action S_{r} consistent with the results derived in Section 17.4:

$$S_{\mathrm{r}} = \int \mathrm{d}^d x \, \mathrm{d}t \, \mathrm{d}\theta \, \mathrm{d}\bar{\theta} \left[\frac{2}{\Omega} \frac{\partial \phi}{\partial \theta} \left(Z' \frac{\partial \phi}{\partial \bar{\theta}} - Z\theta \frac{\partial \phi}{\partial t} \right) + \theta \frac{\partial \phi}{\partial \theta} L(\phi) \right], \tag{17.54}$$

and impose the constraints coming from the supersymmetry of the bare action. The transformation (17.45) is linearly represented on the fields ϕ and therefore the renormalized action remains symmetric. Performing the transformation (17.45) and expressing the invariance of the action we obtain two equations. First identifying the coefficient of $(\partial \phi / \partial \theta)(\partial \phi / \partial t)$ we find:

$$Z' = Z \,. \tag{17.55}$$

The second equation comes from the variation of the last term of expression (17.54):

$$\int \mathrm{d}\theta \frac{\partial \phi}{\partial \theta} L(\phi) = 0 \,. \tag{17.56}$$

Thus $L(\phi)$ is a total derivative and can be written:

$$L(\phi) = \frac{\delta(\mathcal{A}_{\mathrm{r}})}{\delta \phi} \,. \tag{17.57}$$

An integration by parts over θ of the last term in equation (17.54) finally yields the supersymmetric form of the renormalized action:

$$S_{\mathrm{r}}(\phi) = \int \mathrm{d}\theta \, \mathrm{d}\bar{\theta} \, \mathrm{d}t \left[\frac{2}{\Omega} \int \mathrm{d}^d x \frac{\partial \phi}{\partial \theta} Z \left(\frac{\partial \phi}{\partial \bar{\theta}} - \theta \frac{\partial \phi}{\partial t} \right) + \mathcal{A}_{\mathrm{r}}(\phi) \right]. \tag{17.58}$$

After renormalization the drift force in the Langevin equation is thus still of the form of the variation of an action.

Remark. Because the Fokker–Planck equation has static (time-independent) solutions which are not of the form (17.7), it is easy to construct bare Langevin equations which generate an equilibrium distribution characterized by a local static action, for which the dynamical action is not supersymmetric. Direct proofs that the renormalized equilibrium distribution still corresponds to a local static action have only be given in special cases.

17.6 Supersymmetry and Equilibrium Correlation Functions

The supersymmetry of action (17.40), leads to a direct algebraic proof that the equal-time ϕ-field correlation functions converge at large times towards the static correlation functions corresponding to the action $\mathcal{A}(\varphi)$. To simplify notations we shall consider the action (17.40), but the generalization to actions of the type examined in Sections 17.7,17.8 is straightforward. Let us assume that the initial conditions in the Langevin equation are given at time t' and let us calculate the equal-time correlation functions at time t''. The source J associated with the field ϕ then has the special form:

$$J\left(x,t,\bar{\theta},\theta\right) = J\left(x\right)\delta\left(t - t''\right)\delta\left(\bar{\theta}\right)\delta\left(\theta\right).\tag{17.59}$$

We consider the s-dependent dynamical action:

$$S\left(\phi,s\right) = (1 - s)\int_{t'}^{t''}\mathrm{d}\theta\,\mathrm{d}\bar{\theta}\,\mathrm{d}t\left[\frac{2}{\Omega}\int\mathrm{d}^d x\frac{\partial\phi}{\partial\theta}\left(\frac{\partial\phi}{\partial\bar{\theta}} - \theta\frac{\partial\phi}{\partial t}\right) + \mathcal{A}\left(\phi\right)\right]$$
$$+ s\mathcal{A}\left[\phi\left(t'',\bar{\theta} = \theta = 0\right)\right].\tag{17.60}$$

For $s = 0$ we recover the dynamical action for time-dependent correlation functions relevant when all times are in the interval $[t',t'']$ and for $s = 1$, because the source has the form (17.59), the action generates the static correlation functions. If we differentiate the connected correlation functions calculated with the action $S\left(\phi,s\right)$ with respect to s, we generate the insertion of the operator:

$$\frac{\partial}{\partial s}S\left(\phi,s\right) = -\int_{t'}^{t''}\mathrm{d}\theta\mathrm{d}\bar{\theta}\mathrm{d}t\left[\frac{2}{\Omega}\int\mathrm{d}^d x\frac{\partial\phi}{\partial\theta}\left(\frac{\partial\phi}{\partial\bar{\theta}} - \theta\frac{\partial\phi}{\partial t}\right) + \mathcal{A}\left(\phi\right)\right] + \mathcal{A}\left[\phi\left(t'',\bar{\theta} = \theta = 0\right)\right].$$
$$\tag{17.61}$$

Let us first neglect the breaking of supersymmetry due to the boundary condition at t'. Our discussion will be strictly correct only in the case in which we are already at equilibrium, i.e. $t' = -\infty$. In this case the insertion of the operator $\partial S/\partial s$ in an equal-time correlation function generates a two time correlation function. In Subsection 17.5.2 we have determined the most general two time connected correlation functions satisfying the requirements of supersymmetry and causality. Using $t'' > t$ and $\theta'' = \bar{\theta}'' = 0$, we can rewrite equation (17.52):

$$\langle\mathcal{O}(t)\mathcal{O}\left(t''\right)\rangle = \left(1 + \bar{\theta}\theta\frac{\partial}{\partial t}\right)A\left(t - t''\right) = A\left(t + \bar{\theta}\theta - t''\right).\tag{17.62}$$

We can therefore replace $\phi\left(t,\bar{\theta},\theta\right)$ by $\varphi\left(t + \bar{\theta}\theta\right)$ in the operator insertion. We then note:

$$\left(\frac{\partial}{\partial\bar{\theta}} - \theta\frac{\partial}{\partial t}\right)f\left(t + \bar{\theta}\theta\right) = 0,$$

and therefore the insertion of the first term in the r.h.s. of equation (17.61) immediately vanishes. We remain with the insertion of the operator $R\left(\phi\right)$:

$$R\left(\phi\right) = -\int_{t'}^{t''}\mathrm{d}t\,\mathrm{d}\theta\,\mathrm{d}\bar{\theta}\,\mathcal{A}\left[\varphi\left(t + \bar{\theta}\theta\right)\right] + \mathcal{A}\left[\varphi\left(t''\right)\right].\tag{17.63}$$

The integration over the variables $\bar{\theta}$ and θ can be performed and generates the time derivative of the action. The last time integration is then immediate. The contribution coming from the upper bound of the integral cancels the second term in the r.h.s. and we remain with:

$$R(\phi) = \mathcal{A}\left[\varphi(t')\right]. \tag{17.64}$$

Since $\varphi(t')$ is fixed, the insertion equals the correlation function itself multiplied by a factor independent of the correlation function. It thus corresponds to a change in the free energy or vacuum amplitude. We have therefore shown that the connected correlation functions are independent of s: the equal-time connected correlation functions are identical to the static correlation functions.

Let us now discuss the effect of breaking of supersymmetry due to the boundary condition at $t = t'$. To prove invariance of the action under transformation (17.46) one has to integrate by parts. Furthermore, since $\varphi(t')$ is fixed, one cannot perform the transformation for $t = t'$ and must therefore multiply the variations of the fields in (17.46) by some function of time which is 1 everywhere except close to t'. The result is that a supersymmetry transformation generates the insertion of an operator function of the fields taken at $t = t'$. Due to cluster properties, connected correlation functions involving the insertion of such operators vanish in the large time separation limit, $|t'' - t'| \to \infty$. This implies that the equal-time correlation functions converge at large time towards the static equilibrium correlation functions. In explicit calculations we shall always set the initial conditions in the Langevin equation at $t' = -\infty$. Then, at any time, the equal-time φ-field correlation functions are time-independent and therefore equal to the static correlation functions.

17.7 Stochastic Quantization of Two-Dimensional Chiral Models

It follows from the analysis of Section 17.4 that two-dimensional scalar field theories, which are special in the static case because the field is dimensionless, also have special dynamic properties. In particular power counting allows quartic terms in the auxiliary fermion fields.

We have shown in Chapter 15 that two-dimensional models which depend on a finite number of parameters and are strictly renormalizable are related to coset spaces G/H in which H is a subgroup of the group G. Of particular interest are the models defined on symmetric spaces discussed in Sections 15.4–15.6. We here describe Langevin equations for $G \times G/G$ chiral models, G being a simple compact group. Note that some expressions explicitly refer to unitary groups but the generalization to other groups is straightforward.

A Langevin equation for chiral fields. In chiral models the field $g(x)$ varies in some representation of a Lie group G. We have shown in Section 15.4 that a static action $\mathcal{A}$ with only two derivatives has the form:

$$\mathcal{A} = -\frac{\beta}{2} \int \mathrm{d}^d x \, \mathrm{tr} \, j_\mu^2(x), \tag{17.65}$$

where $j_\mu(x)$ is the G-current given by:

$$j_\mu(x) = g^{-1}(x) \, \partial_\mu g(x), \tag{17.66}$$

(it belongs to the Lie algebra of G) and β a coupling constant which plays the role of the inverse temperature in statistical physics. The classical equation of motion (equation (15.70)) expresses the current conservation $\partial_\mu j_\mu(x) = 0$.

It is possible to write a group covariant Langevin equation which leads to an equilibrium distribution corresponding to the static action (17.65):

$$j_0(x,t) = (\Omega\beta/2)\partial_\mu j_\mu(x,t) + g^{-1}(x,t)\nu(x,t) - \nu^\dagger(x,t)g(x,t), \tag{17.67}$$

where the current $j_0(x,t)$ is defined by:

$$j_0 = g^{-1}\partial_t g, \tag{17.68}$$

(below the index 0 will always refer to time). The noise $\nu(x,t)$ is a general complex matrix with gaussian probability distribution $[d\rho(\nu)]$:

$$[d\rho(\nu)] = [d\nu]\exp\left[-\frac{1}{2\Omega}\int d^d x\, dt\, \mathrm{tr}\big(\nu^\dagger(x,t)\nu(x,t)\big)\right]. \tag{17.69}$$

Note that $\nu(x,t)$ belongs to the linear representation of G and not only to the Lie algebra: it has more degrees of freedom than the field.

The dynamical action. To construct the dynamical action we follow the same steps as in preceding sections. However, since we have to integrate over $g(x,t)$ with the group invariant measure, the expressions are slightly modified as we expect from the analysis of Section 16.4. Similar expressions will again appear in the quantization of non-abelian gauge theories with gauge group G.

Let us introduce the covariant derivatives ∇_μ, ∇_0 associated with the currents j_μ, j_0 (for more details see Section 19.1)

$$\nabla_\mu = \partial_\mu + [j_\mu, \cdot] \qquad \nabla_0 = \partial_t + [j_0, \cdot]. \tag{17.70}$$

In an infinitesimal space–time-dependent group transformation:

$$g(x,t) \mapsto g(x,t)\,(1 + \ell(x,t)), \tag{17.71}$$

the currents transform like:

$$j_\mu \mapsto j_\mu + \nabla_\mu\ell, \qquad j_0 \mapsto j_0 + \nabla_0\ell. \tag{17.72}$$

Therefore the analogue of the operator M introduced in equation (17.37) is now defined when acting on a field ℓ as:

$$M\ell = \nabla_0\ell - \tfrac{1}{2}\beta\Omega\partial_\mu\nabla_\mu\ell + \ell g^{-1}\nu + \nu^\dagger g\ell. \tag{17.73}$$

Introducing fermion fields $\bar{c}$ and c as well as a Lagrange multiplier λ, all belonging to the Lie algebra of G, we can write in the normalization of equation (17.40) the dynamical action S resulting from the integration over the noise:

$$S = \int d^d x\, dt\, \frac{2}{\Omega}\,\mathrm{tr}\left[(\lambda + \bar{c}c)(\lambda + c\bar{c}) + \big(j_0 - \tfrac{1}{2}\Omega\beta\partial_\mu j_\mu\big)\lambda - \bar{c}\big(\nabla_0 - \tfrac{1}{2}\Omega\beta\partial_\mu\nabla_\mu\big)c\right]. \tag{17.74}$$

We have used in this expression the identity $(c\bar{c})^\dagger = -\bar{c}c.$

Note the appearance of a quartic fermion term induced by the dependence of the noise term on the field $g(x,t)$ in equation (17.67). As we expect from the analysis of Section 16.4 the set of BRS transformations which leave the action invariant is now:

$$\delta g = \bar{\varepsilon} g c , \qquad \delta c = -\bar{\varepsilon} c^2 , \tag{17.75}$$

$$\delta \bar{c} = \bar{\varepsilon} \lambda , \qquad \delta \lambda = 0 . \tag{17.76}$$

The BRS transformation of the current j_μ induced by equation (17.75) can then be written:

$$\delta j_\mu = \bar{\varepsilon} \nabla_\mu c . \tag{17.77}$$

The BRS transformations (17.75–17.77) are exactly those which will again appear in the quantization of non-abelian gauge theories in Chapter 19. The gauge field associated with the group G has the transformation law (17.77).

We now show that the Langevin equation (17.67) is the natural generalization of equation (17.39). We introduce the superfield $\mathcal{G}$, element of the group G,

$$\mathcal{G} = g \left(1 + \bar{\theta} c \right) \left(1 + \bar{c} \theta \right) \left(1 + \bar{\theta} \theta \lambda \right) . \tag{17.78}$$

The set of equations (17.75,17.76) is then equivalent to

$$\delta \mathcal{G} = \bar{\varepsilon} \frac{\partial}{\partial \bar{\theta}} \mathcal{G} . \tag{17.79}$$

From $\mathcal{G}$ we derive a set of associated currents J_μ, J_0, J_θ, $J_{\bar{\theta}}$:

$$\begin{cases} J_\mu = \mathcal{G}^{-1} \partial_\mu \mathcal{G} , & J_0 = \mathcal{G}^{-1} \partial_t \mathcal{G} , \\ J_\theta = \mathcal{G}^{-1} \dfrac{\partial}{\partial \theta} \mathcal{G} , & J_{\bar{\theta}} = \mathcal{G}^{-1} \dfrac{\partial}{\partial \bar{\theta}} \mathcal{G} . \end{cases} \tag{17.80}$$

The action (17.74) can then be rewritten in a form analogous to (17.40):

$$S = - \int d^d x \, dt \, d\theta \, d\bar{\theta} \, \mathrm{tr} \left[\frac{2}{\Omega} J_\theta \left(J_{\bar{\theta}} - \theta J_0 \right) + \frac{\beta}{2} J_\mu^2 \right] . \tag{17.81}$$

In this form it is clear that S is left invariant by the transformation (17.45) and is thus supersymmetric. This supersymmetry implies that equal-time $g(x,t)$-field correlation functions converge at time $+\infty$ towards the correlation functions of the static action (17.65), justifying the choice of the Langevin equation (17.67). We shall show in next section that equation (17.67) is a special example of a family of Langevin equations corresponding to two-dimensional models defined on Riemannian manifolds which contain all models defined on homogeneous spaces.

The form of the renormalized action is then dictated by the structure of symmetric space $G \times G/G$ and supersymmetry (equation (17.45)): two renormalization constants are needed, the usual coupling constant renormalization of the static theory and again the time scale renormalization. In addition, in general the parametrization of the group elements is also renormalized.

17.8 The Langevin Equation on Riemannian Manifolds

We have already seen in Section 4.7 that it is impossible to write a Langevin equation which only depends on the geometry on the manifold. We therefore construct a Langevin equation in the following way: we consider smooth manifolds embedded in an euclidean space and defined by a set of equations constraining the euclidean coordinates χ_α:

$$E^s(\chi_\alpha) = 0 \,. \tag{17.82}$$

To simplify calculations, we assume that we have solved locally these equations and expressed some components σ^s of χ_α as functions of a set of independent components φ^i:

$$\chi_\alpha \equiv \left\{ \sigma^s(\varphi) \quad \varphi^i \right\} \,. \tag{17.83}$$

To construct a Langevin equation on the manifold, we start from a Langevin equation of type (17.39) in the embedding euclidean space:

$$\dot{\chi}_\alpha(x,t) = -\frac{\Omega}{2} \frac{\delta \mathcal{A}}{\delta \chi_\alpha(x,t)} + \nu_\alpha(x,t) \,, \tag{17.84}$$

in which $\nu_\alpha(x,t)$ is a gaussian noise:

$$\langle \nu_\alpha(x,t)\nu_\beta(x',t') \rangle = \Omega \delta(t-t')\,\delta(x-x')\,\delta_{\alpha\beta} \,. \tag{17.85}$$

We see immediately that, as in the case of chiral fields, we are naturally led to introduce a noise with more degrees of freedom than the field.

We now project $\dot{\chi}_\alpha(x,t)$ onto the space tangent to the manifold at point $\chi_\alpha(x,t)$. For this purpose let us introduce some notations. On the manifold the variations $\delta\chi_\alpha$ of the field χ_α are constrained by:

$$\frac{\partial E^s}{\partial \chi_\alpha} \delta\chi_\alpha = 0 \,. \tag{17.86}$$

We introduce an orthogonal basis e_a^α for the variations of χ_α:

$$\frac{\partial E^s}{\partial \chi_\alpha} e_a^\alpha = 0 \,, \tag{17.87}$$

with the orthogonality conditions:

$$e_a^\alpha e_b^\alpha = \delta_{ab} \,. \tag{17.88}$$

In terms of the coordinates φ^i, equation (17.87) can be rewritten:

$$e_a^s = \partial_i \sigma^s e_a^i \,, \tag{17.89}$$

and therefore equation (17.88) becomes, after elimination of the components e_a^s,

$$e_a^i g_{ij} e_b^j = \delta_{ab} \,, \tag{17.90}$$

where we have introduced the metric tensor g_{ij} on the manifold (see Appendix A15.3.2):

$$g_{ij} = \delta_{ij} + \partial_i \sigma^s \partial_j \sigma^s \,. \tag{17.91}$$

Equation (17.90) shows that the matrix e^i_a is the inverse of the vielbein (Section 22.6).
We can now construct from equation (17.84) a Langevin equation on the manifold

$$\dot{\chi}_\alpha(x,t) = e^\alpha_a e^\beta_a \left(-\frac{\Omega}{2}\frac{\delta\mathcal{A}}{\delta\chi_\beta} + \nu_\beta \right). \tag{17.92}$$

Let us write this equation in terms of the independent components φ^i

$$\dot{\varphi}^i = -\frac{\Omega}{2}e^i_a \left(e^j_a\frac{\delta\mathcal{A}}{\delta\varphi^j} + \partial_j\sigma^s e^j_a\frac{\delta\mathcal{A}}{\delta\sigma^s} \right) + e^i_a \left(e^j_a\nu_j + e^j_a\partial_j\sigma^s\nu_s \right). \tag{17.93}$$

We then observe that:

$$\frac{\delta\mathcal{A}}{\delta\varphi^j}(\varphi,\sigma) + \partial_j\sigma^s(\varphi)\frac{\delta\mathcal{A}}{\delta\sigma^s} = \partial_j\mathcal{A}(\varphi,\sigma(\varphi)), \tag{17.94}$$

in which in the r.h.s. $\partial_j\mathcal{A}$ means total derivative of $\mathcal{A}$ with respect to φ^j.
As a consequence of equation (17.90) the inverse metric tensor g^{ij} is given by:

$$g^{ij} = e^i_a e^j_a. \tag{17.95}$$

Equation (17.93) can therefore be rewritten:

$$\dot{\varphi}^i = -\tfrac{1}{2}\Omega g^{ij}\partial_j\mathcal{A} + g^{ij}t^\alpha_j\nu_\alpha, \tag{17.96}$$

with the definition:

$$t^\alpha_i \equiv \begin{cases} t^j_i = \delta_{ij} \\ t^s_i = \partial_i\sigma^s. \end{cases} \tag{17.97}$$

Since the noise ν_α is multiplied by a function of φ, we have to specify the meaning of the product. We choose the covariant definition as in Section 4.7.
The quantities t^α_i satisfy:

$$t^\alpha_i t^\alpha_j = g_{ij}, \tag{17.98}$$

and

$$\partial_i t^\alpha_j t^\alpha_k = g_{kl}\Gamma^l_{ij}, \tag{17.99}$$

in which Γ^l_{ij} is the usual Christoffel symbol (equation (22.41)). This equation can be rewritten in covariant form:

$$\nabla_i t^\alpha_j t^\alpha_k = 0, \tag{17.100}$$

in which ∇_i is the covariant derivative on the manifold.

Dynamical action. It is convenient to rewrite equation (17.96) as:

$$g_{ij}\dot{\varphi}^j + \tfrac{1}{2}\Omega\partial_i\mathcal{A} - t^\alpha_i\nu_\alpha = 0. \tag{17.101}$$

Introducing a Lagrange multiplier λ^i and fermion fields $\bar{c}^i$ and c^i, we can write the corresponding dynamical action S before integration over the noise as the sum of two contributions S_0 and S_1:

$$S_0 = \int dx\,dt \left(\frac{1}{2\Omega}\nu_\alpha\nu_\alpha - \frac{2}{\Omega}t^\alpha_i\nu_\alpha\lambda^i + \frac{2}{\Omega}\bar{c}^i\partial_j t^\alpha_i c^j\nu_\alpha \right), \tag{17.102}$$

$$S_1 = \int dx\,dt\frac{2}{\Omega}\left[\lambda^i\left(g_{ij}\dot{\varphi}^j + \frac{\Omega}{2}\partial_i\mathcal{A} \right) - \bar{c}^i g_{ij}\dot{c}^j - \bar{c}^i\partial_k g_{ij}\dot{\varphi}^j c^k - \frac{\Omega}{2}\bar{c}^i\partial_i\partial_j\mathcal{A}\,c^j \right], \tag{17.103}$$

and:
$$S = S_0 + S_1 . \tag{17.104}$$

Integration over the noise leads after a short calculation to:

$$S = S_1 + \int \mathrm{d}x \, \mathrm{d}t \frac{2}{\Omega} \left(-\lambda^i g_{ij} \lambda^j + 2\lambda^i g_{il} \Gamma^l_{jk} \bar{c}^j c^k + \bar{c}^i c^j \bar{c}^k c^l \partial_i \partial_j g_{kl} \right) . \tag{17.105}$$

Introducing now the superfield ϕ^i:

$$\phi^i = \varphi^i + \bar{\theta} c^i + \bar{c}^i \theta + \bar{\theta} \theta \lambda^i , \tag{17.106}$$

we can rewrite the action:

$$S(\phi) = \int \mathrm{d}t \, \mathrm{d}\theta \mathrm{d}\bar{\theta} \left[\int \mathrm{d}x \frac{2}{\Omega} \frac{\partial \phi^i}{\partial \theta} g_{ij}(\phi) \left(\frac{\partial \phi^j}{\partial \bar{\theta}} - \theta \frac{\partial \phi^j}{\partial t} \right) + \mathcal{A}(\phi) \right] . \tag{17.107}$$

This supersymmetric form then implies that at equilibrium the equal-time field configurations are weighted by the measure $[\mathrm{d}\rho(\varphi)]$:

$$[\mathrm{d}\rho(\varphi)] = [\mathrm{d}\varphi^i] \sqrt{g(\varphi)} \exp\left[-\mathcal{A}(\varphi, \sigma(\varphi)) \right] . \tag{17.108}$$

This was of course the field distribution we had in mind when we wrote equation (17.96). Note that in contrast to Langevin equation (17.96), the dynamical action (17.107) and therefore the correlation functions at equilibrium depend only on the geometry of the manifold. In the case of homogeneous spaces the original action $\mathcal{A}(\chi)$ is simply:

$$\mathcal{A}(\chi) = \frac{1}{2} \int (\partial_\mu \chi_\alpha)^2 \, \mathrm{d}x , \tag{17.109}$$

which, as a functional of φ, becomes on the manifold:

$$\mathcal{A}(\varphi, \sigma(\varphi)) = \frac{1}{2} \int g_{ij}(\varphi) \, \partial_\mu \varphi^i \partial_\mu \varphi^j \mathrm{d}x . \tag{17.110}$$

It is then easy to verify that expression (17.107) contains expression (17.81) as a special case.

Note finally that the functional integral representation for the generating functional $Z(J)$ can also be rewritten introducing the superfield X in euclidean coordinates:

$$X_\alpha = \chi_\alpha + \bar{\theta} c_\alpha + \bar{c}_\alpha \theta + \lambda_\alpha \bar{\theta} \theta , \tag{17.111}$$

under the form:

$$Z(J) = \int (\mathrm{d}\phi) \exp\left[-S(\phi) + \int \mathrm{d}x \, \mathrm{d}t \, \mathrm{d}\theta \, \mathrm{d}\bar{\theta} \, J_i \phi^i \right]$$
$$= \int [\mathrm{d}X] \, \delta\left[E^s(X) \right] \exp\left(-\Sigma(X) + \int \mathrm{d}x \, \mathrm{d}t \, \mathrm{d}\theta \mathrm{d}\bar{\theta} \, J_i \phi^i \right) , \tag{17.112}$$

with for $\Sigma(X)$ the expression:

$$\Sigma(X) = \int \mathrm{d}\theta \, \mathrm{d}\bar{\theta} \mathrm{d}t \left[\int \mathrm{d}x \frac{2}{\Omega} \frac{\partial X_\alpha}{\partial \theta} \left(\frac{\partial X_\alpha}{\partial \bar{\theta}} - \theta \frac{\partial X_\alpha}{\partial t} \right) + \mathcal{A}(X) \right] . \tag{17.113}$$

These expressions emphasize the similarity between the properties of the static and the dynamic theory.

In the case of homogeneous spaces the renormalization properties of the Langevin equation then follow from the general analysis of Chapter 15 and supersymmetry.

17.9 Stochastic Quantization: the Example of Gauge Theories

We have mentioned that, as suggested by Parisi, stochastic dynamic equations can be used to quantize field theories when non-trivial quantization problems arise. The time variable in the Langevin equation is then a *fictitious* additional variable since only the equilibrium distribution is physical (it can be interpreted as the computer time of numerical simulations where stochastic methods are used to generate field configurations). As an anticipation to Chapters 18–19, where gauge theories will be extensively discussed, let us briefly review the application of this idea to gauge theories. The background of Chapter 18 will thus be assumed in this section. The problem of gauge theories is that one field degree of freedom, which corresponds to gauge transformations, is redundant and the conventional quantization method has to be adapted to this peculiar situation.

The stochastic quantization idea is to write a gauge invariant Langevin equation. Let us consider the abelian case without matter field. We use as a drift force the functional derivative of the classical gauge invariant action. The gauge field A_μ then satisfies

$$\dot{A}_\mu(t, x) = -\tfrac{1}{2}\Omega \partial_\nu F_{\mu\nu}(t, x) + N_\mu(t, x), \tag{17.114}$$

where

$$F_{\mu\nu} = \partial_\mu A_\nu - \partial_\nu A_\mu \,,$$

and $N_\mu(x, t)$ is a gaussian noise with measure

$$[\mathrm{d}\rho(N)] = [\mathrm{d}N_\mu] \exp\left[-\int \mathrm{d}^d x \, \mathrm{d}t \, N_\mu^2(x, t)/2\Omega\right]. \tag{17.115}$$

If we make a time-independent gauge transformation

$$A_\mu(x, t) \mapsto A_\mu(x, t) + \partial_\mu \Lambda(x),$$

the equation is clearly invariant. However it is not invariant under time-dependent transformations and thus provides a dynamics to all components of the gauge field. Let us solve the equation in Fourier space

$$A_\mu(t, k) = \left(\delta_{\mu\nu} - k_\mu k_\nu/k^2\right) e^{-\Omega k^2 t/2} \left[A_\nu(0, k) + \int_0^t e^{\Omega k^2 t'/2} N_\nu(t', k)\mathrm{d}t'\right]$$
$$+ \left(k_\mu k_\nu/k^2\right) \left[A_\nu(0, k) + \int_0^t N_\nu(t', k)\mathrm{d}t'\right].$$

We immediately see that the component of A_μ along k_μ follows a brownian motion and thus does not equilibrate. Let us indeed calculate the equal time 2-point function, after averaging over the noise. Calling A_L and A_T the component of A_μ respectively along and perpendicular to k_μ we find

$$\langle A_T(t, k)A_T(t, -k)\rangle = \frac{1}{k^2} + O\left(e^{-\Omega k^2 t}\right),$$
$$\langle A_L(t, k)A_L(t, -k)\rangle = A_L(0, k)A_L(0, -k) + \Omega t \,.$$

We conclude that, due to gauge invariance, the Langevin equation (17.114) does not generate an equilibrium distribution, however the gauge invariant functions have a large time limit. Since only the latter functions have a physical meaning the quantization problem has been solved.

It is clear that the same conclusion will be reached if the gauge field is interacting with matter, and in the non-abelian case.

The next problems are the formal relations with the standard quantization procedure by gauge fixing as described in Chapters 18,19 and the problem of power counting and renormalization. Let us write the gauge field propagator obtained from the effective dynamical action. In Fourier space one finds

$$\Delta_{\mu\nu}(\omega, k) = \frac{\delta_{\mu\nu} - k_\mu k_\nu / k^2}{\omega^2 + \Omega k^4/4} + \frac{k_\mu k_\nu}{k^2 \omega^2}.$$

The $1/\omega^2$ singularity reflects the absence of equilibrium distribution which now takes the form of IR divergences. It is necessary to work in a finite time interval.

We see also that the longitudinal propagator does not decrease at large momentum for fixed ω. An analogous problem will appear in the quantization with non-covariant gauges: the theory is not renormalizable by power counting. One solution to this problem is to add to the Langevin equation a non-conservative drift force of the form $D_\mu V(A)$ where D_μ is the covariant derivative, and $V(A)$ a linear function of A for example $\partial_\mu A_\mu$. It is easy to verify that such a term does not contribute to the evolution equations (equation (4.34)) for equal-time gauge invariant correlation function, and thus these functions are not modified. However with this term the Langevin equation is no longer gauge invariant, an equilibrium distribution is generated and with a suitable covariant choice of $D_\mu V(A)$ the theory is renormalizable by power counting. It can also be shown that it is equivalent to the theories obtained by canonical quantization procedure, but with a non-local gauge fixing term.

The drawback is that a lot of the aesthetic appeal of the original formulation has been lost, and the proof of renormalizability becomes as complicated as in canonical quantization. It of course remains the hope that this method will lead to the quantization of systems where no alternative method has yet been found.

17.10 Supersymmetry in Four Dimensions

In this section we present a very brief discussion of supersymmetry in four dimensions, since supersymmetry has been proposed as a principle to solve the so-called *hierarchy problem* in particle physics by relating the masses of scalar particles (like Higgs fields) to those of fermions which can be protected against "large" mass renormalization by chiral symmetry (see Subsection 20.1.4). Exact supersymmetry would also solve the problem of vanishing or at least unnatural small value of the cosmological constant (see Subsection 22.7.2). However experimentally the, as yet undiscovered, supersymmetric partners of known particles have to be much heavier. Supersymmetry is necessarily broken and the problem of the cosmological constant only slightly improved. Finally gauging of supersymmetry naturally leads to unification with gravity, because the commutators of supersymmetry currents involve the energy-momentum tensor.

The new feature of supersymmetry in higher dimensions, compared to the quantum mechanical case, is the combination of supersymmetry with spin since fermions have half-integer spins, and this leads to modifications we want to discuss here.

To introduce supersymmetry we immediately use the superfield notation. The algebra of generators has been briefly described in Section 16.7. If we combine scalar boson and spin 1/2 fermion fields into a superfield we get something like

$$\phi(x, \bar{\theta}) = A(x) + \bar{\theta}\psi(x) + \cdots ,$$

and thus Grassmann coordinates must be associated with the spinor indices. We immediately find serious restrictions coming from power counting if we want to construct a renormalizable field theory. In four dimensions these restrictions somewhat destroy the beautiful simplicity of the 1-D case. In any space dimension, the dimension $[\bar{\theta}]$ of the coordinates $\bar{\theta}$ is $[\theta] = [A] - [\psi] = -1/2$. If we give to $\bar{\theta}$ Dirac spinor indices the corresponding Grassmann algebra has 8 different generators $\theta_\alpha, \bar{\theta}_\alpha$. Therefore the volume elements $\prod_\alpha d\theta_\alpha d\bar{\theta}_\alpha$ has dimension 4 and thus only a constant action is allowed. To circumvent this difficulty one uses the special properties of the spinor group corresponding to $SO(4)$ which is $SU(2) \times SU(2)$. First the spinorial representation can be reduced into left-handed and right-handed components (Weyl spinors, see Appendix A5.7). Moreover the representations of $SU(2)$ are self-conjugated. It is thus sufficient to associate Grassmann coordinates with the two Weyl spinors. We only need four Grassmann coordinates $\theta_\alpha, \bar{\theta}_\alpha$. In this form the Grassmann volume element has only dimension 2 and an action bilinear in the scalar superfield can be written: such an action term is called a D-term. However this leads only to a free field action. As we shall explain, to construct interaction terms, one has to consider cubic polynomials of only one kind of chiral superfield (left or right-handed). Such a polynomial will only depend on two variables θ or $\bar{\theta}$, and thus can be integrated with the volume elements $d\theta_1 d\theta_2$ ($d\bar{\theta}_1 d\bar{\theta}_2$ respectively) which have dimension 1. The corresponding contribution is called a F-term.

To write the generators of supersymmetry (see Section 16.7) it is convenient to introduce a notation for the 2-spinor transforming under the conjugated representation of $SU(2)$

$$\theta^\alpha = (\sigma_2)_{\alpha\beta}\theta_\beta \equiv -i\epsilon_{\alpha\beta}\theta_\beta , \tag{17.116}$$

where $\epsilon_{\alpha\beta}$ is the antisymmetric tensor with $\epsilon_{12} = 1$. We also define the four matrices σ_μ, where for $\mu = 1, 2, 3$ σ_μ are just the usual Pauli matrices and $\sigma_4 = -i\mathbf{1}$. Two relations are useful in what follows:

$$\sigma_2\sigma_\mu\sigma_2 = -\sigma_\mu^*, \qquad \sigma_\mu^\dagger\sigma_\nu + \sigma_\nu^\dagger\sigma_\mu = 2\delta_{\mu\nu} .$$

Then under a $SU(2) \times SU(2)$ transformation

$$\theta'_\alpha = U_\alpha^\beta\theta_\beta , \quad \bar{\theta}^{\alpha\prime} = V_\beta^{*\alpha}\bar{\theta}^\beta , \quad \sigma_\mu x'_\mu = V\sigma_\mu x_\mu U^\dagger, \quad U, V \in SU(2).$$

$SU(2)$ invariant quantities are then for example $\theta^\alpha\theta_\alpha$, $\bar{\theta}^\alpha\bar{\theta}_\alpha$ or $\bar{\theta}^\alpha\sigma_{\mu\alpha}^{\ \beta}\theta_\beta$.

The supersymmetry generators

$$\bar{D}_\alpha = \frac{\partial}{\partial\bar{\theta}^\alpha} + \tfrac{1}{2}\sigma_{\mu\alpha}^{\ \beta}\theta_\beta\frac{\partial}{\partial x_\mu} , \tag{17.117a}$$

$$D^\alpha = \frac{\partial}{\partial\theta_\alpha} + \tfrac{1}{2}\bar{\theta}^\beta\sigma_{\mu\beta}^{\ \alpha}\frac{\partial}{\partial x_\mu} , \tag{17.117b}$$

belong to the representations $(0, 1/2)$ and $(1/2, 0)$ of $SU(2) \times SU(2)$ respectively. They have the anticommutation relations

$$\{\bar{D}_\alpha, D^\beta\} = \sigma_{\mu\alpha}^{\ \beta}\frac{\partial}{\partial x_\mu} \qquad \{D, D\} = \{\bar{D}, \bar{D}\} = 0 . \tag{17.118}$$

Note therefore that the supersymmetry current will have spin 3/2. This implies in particular that the spontaneous breaking of supersymmetry generates spin 1/2 Goldstone fermions. This is an unwanted feature from the point of view of Particle Physics. Although massless spin 1/2 fermions do exist (neutrinos) they do not obey the consequences of the corresponding WT identities. It is actually a problem of the supersymmetry assumption that none of the partners of known particles in supermultiplets have been found yet. Note finally that since the commutator of supersymmetries is the generator of translations any attempt to gauge supersymmetry (which would provide a solution to the unwanted Goldstone fermion) will result in a theory which contains gravitation. Therefore the problem of spontaneous supersymmetry breaking does not seem to have a solution outside of supergravity, and thus within the framework of renormalizable field theories.

As we have already indicated in Section 16.7, to $D, \bar{D}$ correspond two operators $\Delta, \bar{\Delta}$ which also satisfy the anticommutation relations of supersymmetry generators, anticommute with them and thus play the role of covariant derivatives:

$$\bar{\Delta}_\alpha = \frac{\partial}{\partial \bar{\theta}^\alpha} - \tfrac{1}{2}\sigma_{\mu\alpha}{}^\beta \theta_\beta \frac{\partial}{\partial x_\mu}, \qquad \Delta^\alpha = \frac{\partial}{\partial \theta_\alpha} - \tfrac{1}{2}\bar{\theta}^\beta \sigma_{\mu\beta}{}^\alpha \frac{\partial}{\partial x_\mu}, \qquad (17.119)$$

$$\{\bar{\Delta}_\alpha, \Delta^\beta\} = -\sigma_{\mu\beta}{}^\alpha \frac{\partial}{\partial x_\mu}, \qquad \{\Delta, \Delta\} = \{\bar{\Delta}, \bar{\Delta}\} = 0. \qquad (17.120)$$

17.10.1 Scalar chiral superfields

Because $\bar{\Delta}$ anticommutes with $\bar{D}, D$ the fields which satisfy

$$\bar{\Delta}_\alpha \phi\left(x, \theta, \bar{\theta}\right) = 0, \qquad (17.121)$$

form a space of representation for the supersymmetry generators. The general solution of equation (17.121) can be written

$$\phi\left(x, \theta, \bar{\theta}\right) = \phi\left(y, \theta\right),$$

where the new space coordinate y is

$$y_\mu = x_\mu + \tfrac{1}{2}\bar{\theta}^\alpha \sigma_{\mu\alpha}{}^\beta \theta_\beta. \qquad (17.122)$$

Note that y also satisfies the equation

$$D^\alpha y_\mu\left(x, \theta, \bar{\theta}\right) = 0. \qquad (17.123)$$

Acting on functions of y, θ, the generators $D, \bar{D}$ then take the form:

$$\bar{D}_\alpha = \frac{\partial}{\partial \bar{\theta}^\alpha} + \sigma_{\mu\alpha}{}^\beta \theta_\beta \frac{\partial}{\partial y_\mu}, \quad D^\alpha = \frac{\partial}{\partial \theta_\alpha}.$$

Note that the variable y plays a role similar to time in the 1-D case. A scalar right-handed superfield can be expanded on the θ_α basis. Since α takes only two values the most general expression has the form:

$$\phi\left(y, \theta\right) = A(y) + \psi^\alpha(y)\theta_\alpha + \tfrac{1}{2}E(y)\theta^\alpha\theta_\alpha. \qquad (17.124)$$

A and E are two complex scalar fields; A as well as ϕ itself have dimension one and thus E has dimension two. A renormalizable action can be at most quadratic in E and, as in the 1-D case, E does not propagate and can be eliminated from the action by using the corresponding equation of motion.

The action of the supersymmetry generator $\bar{\eta}^\alpha \bar{D}_\alpha - \eta_\alpha D^\alpha$ in component form then is:

$$\delta A = \eta_\alpha \psi^\alpha \,, \tag{17.125a}$$

$$\delta \psi^\beta = \bar{\eta}^\alpha \sigma_{\mu\alpha}{}^\beta \partial_\mu A - \eta^\beta E \,, \tag{17.125b}$$

$$\delta E = -\bar{\eta}^\alpha \sigma_{\mu\alpha}{}^\gamma \partial_\mu \psi_\gamma \,. \tag{17.125c}$$

For later purpose it is useful to also expand ϕ in terms of the space coordinate x:

$$\phi = A(x) + \psi^\alpha(x)\theta_\alpha + \tfrac{1}{2}\partial_\mu A(x)\bar{\theta}^\alpha \sigma_{\mu\alpha}{}^\beta \theta_\beta + \tfrac{1}{2}E(x)\theta^\alpha \theta_\alpha$$
$$- \tfrac{1}{4}\bar{\theta}^\alpha \sigma_{\mu\alpha}{}^\beta \partial_\mu \psi_\beta(x)\theta^\gamma \theta_\gamma + \tfrac{1}{16}\partial^2 A(x)\theta^\alpha \theta_\alpha \bar{\theta}^\beta \bar{\theta}_\beta \,. \tag{17.126}$$

In the same way left-handed chiral superfields can be defined which satisfy $\Delta^\alpha \bar{\phi} = 0$ and thus depend on $\bar{\theta}$ and a space variable $\bar{y}$

$$\bar{y}_\mu = x_\mu - \tfrac{1}{2}\bar{\theta}^\alpha \sigma_{\mu\alpha}{}^\beta \theta_\beta, \qquad \bar{D}_\alpha \bar{y}_\mu = 0 \,. \tag{17.127}$$

The corresponding chiral field can be written:

$$\bar{\phi}\left(\bar{y}, \bar{\theta}\right) = A^*(\bar{y}) + \bar{\psi}^\alpha(\bar{y})\bar{\theta}_\alpha + \tfrac{1}{2}E^*(\bar{y})\bar{\theta}^\alpha \bar{\theta}_\alpha \,. \tag{17.128}$$

The action of the supersymmetry generators then is:

$$\delta A^* = \bar{\eta}^\alpha \bar{\psi}_\alpha \,, \tag{17.129a}$$

$$\delta \bar{\psi}_\beta = \sigma_{\mu\beta}{}^\alpha \eta_\alpha \partial_\mu A^* + \bar{\eta}_\beta E^* \,, \tag{17.129b}$$

$$\delta E^* = -\partial_\mu \bar{\psi}^\gamma \sigma_{\mu\gamma}{}^\alpha \eta_\alpha \,. \tag{17.129c}$$

The expansion of $\bar{\phi}$ in terms of the space variable x reads:

$$\bar{\phi} = A^*(x) + \bar{\psi}^\alpha(x)\bar{\theta}_\alpha - \tfrac{1}{2}\partial_\mu A^*(x)\bar{\theta}^\alpha \sigma_{\mu\alpha}{}^\beta \theta_\beta + \tfrac{1}{2}E^*(x)\bar{\theta}^\alpha \bar{\theta}_\alpha$$
$$+ \tfrac{1}{4}\bar{\partial}_\mu \psi^\alpha(x)\sigma_{\mu\alpha}{}^\beta \theta_\beta \bar{\theta}^\gamma \bar{\theta}_\gamma + \tfrac{1}{16}\partial^2 A(x)\theta^\alpha \theta_\alpha \bar{\theta}^\beta \bar{\theta}_\beta \,. \tag{17.130}$$

The free action. We can now construct a free action

$$\mathcal{L}_D = \phi(\bar{y}, \bar{\theta})\phi(y, \theta),$$

$$\int \mathrm{d}^4 x \prod_\alpha \mathrm{d}\bar{\theta}_\alpha \mathrm{d}\theta_\alpha \, \mathcal{L}_D = \int \mathrm{d}^4 x \left(\partial_\mu A^* \partial_\mu A - \bar{\psi}^\alpha \sigma_{\mu\alpha}{}^\beta \partial_\mu \psi_\beta - E^* E \right). \tag{17.131}$$

The supersymmetry of the lagrangian $\mathcal{L}_D$ is here obvious since it does not depend explicitly on $\theta, \bar{\theta}$ and x_μ. If we have several superfields we simply add the corresponding contributions.

Interaction terms. We here write the F-term contributions for several superfields ϕ_i (we need at least two to construct a Dirac fermion). The most general renormalizable lagrangian density has the form:

$$\mathcal{L}_F(\phi) = c_i\phi_i + \tfrac{1}{2}M_{ij}\phi_i\phi_j + \tfrac{1}{3!}g_{ijk}\phi_i\phi_j\phi_k \,. \qquad (17.132)$$

Integrating over θ_1, θ_2 we obtain

$$\int i\mathrm{d}\theta_2\mathrm{d}\theta_1\,\mathcal{L}_F = c_iE_i + M_{ij}\left(A_iE_j - \tfrac{1}{2}\psi_i^\alpha\psi_{\alpha j}\right) + \tfrac{1}{2}g_{ijk}\left(A_iA_jE_k - \psi_i^\alpha\psi_{\alpha j}A_k\right). \quad (17.133)$$

Adding the kinetic term (17.131), (17.133) and its conjugated left-handed contribution

$$\bar{\mathcal{L}}_F(\bar{\phi}) = c_i^*\bar{\phi}_i + \tfrac{1}{2}M_{ij}^*\bar{\phi}_i\bar{\phi}_j + \tfrac{1}{3!}g_{ijk}^*\bar{\phi}_i\bar{\phi}_j\bar{\phi}_k \,,$$

$$\int i\mathrm{d}\bar{\theta}_2\mathrm{d}\bar{\theta}_1\,\bar{\mathcal{L}}_F = c_i^*E_i^* + M_{ij}^*\left(A_i^*E_j^* - \tfrac{1}{2}\bar{\psi}_i^\alpha\bar{\psi}_{\alpha j}\right) + \tfrac{1}{2}g_{ijk}^*\left(A_i^*A_j^*E_k^* - \bar{\psi}_i^\alpha\bar{\psi}_{\alpha j}A_k^*\right). (17.134)$$

we obtain a physically sensible supersymmetric lagrangian. A useful exercise is to verify, using the explicit expressions (17.125,17.129), that the resulting lagrangian $\mathcal{L}$ varies in a supersymmetry transformation by a total derivative and the action thus is invariant.

We can now integrate over E, E^* since the integral is gaussian. This is equivalent to use the corresponding field equations:

$$\frac{\partial\mathcal{L}}{\partial E_i} = E_i^* + c_i + M_{ij}A_j + \tfrac{1}{2}g_{ijk}A_jA_k = 0 \,, \qquad (17.135a)$$

$$\frac{\partial\mathcal{L}}{\partial E_i^*} = E_i + c_i^* + M_{ij}^*A_j^* + \tfrac{1}{2}{}^*g_{ijk}A_j^*A_k^* = 0 \,. \qquad (17.135b)$$

The lagrangian can then be cast into the form

$$\mathcal{L} = \partial_\mu A^*\partial_\mu A - \bar{\psi}^\alpha\sigma_{\mu\alpha}^\beta\partial_\mu\psi_\beta + \frac{\partial\mathcal{L}_F}{\partial\phi_i}(A)\frac{\partial\bar{\mathcal{L}}_F}{\partial\phi_i}(A^*) - \frac{1}{2}\frac{\partial^2\mathcal{L}_F}{\partial\phi_i\partial\phi_j}(A)\psi_i^\alpha\psi_{\alpha j}$$

$$- \frac{1}{2}\frac{\partial^2\bar{\mathcal{L}}_F}{\partial\bar{\phi}_i\partial\bar{\phi}_j}(\bar{A})\bar{\psi}_i^\alpha\bar{\psi}_{\alpha j}\,,$$

a form reminiscent of expression (17.16).

17.10.2 Vector superfields

Up to now we have explained how to construct a supersymmetric action containing only scalar and spin 1/2 fermion fields. For a realistic theory of particles, vector fields are also required. Therefore from now on we again assume some of the concepts which will be introduced only in Chapters 18,19.

We consider a general real superfield, which we call a vector superfield. We can parametrize it in the form:

$$V = B + \chi^\alpha\theta_\alpha + \bar{\chi}^\alpha\bar{\theta}_\alpha + \tfrac{1}{2}C\theta^\alpha\theta_\alpha + \tfrac{1}{2}C^*\bar{\theta}^\alpha\bar{\theta}_\alpha + i2^{-1/2}V_\mu\bar{\theta}^\alpha\sigma_{\mu\alpha}^\beta\theta_\beta$$

$$+ \tfrac{1}{2}\bar{\theta}^\alpha\left(\bar{\lambda}_\alpha - \tfrac{1}{2}\sigma_{\mu\alpha}^\beta\partial_\mu\chi_\beta\right)\theta^\gamma\theta_\gamma + \tfrac{1}{2}\left(\lambda^\beta + \tfrac{1}{2}\partial_\mu\bar{\chi}^\alpha(x)\sigma_{\mu\alpha}^\beta\right)\theta_\beta\bar{\theta}^\gamma\bar{\theta}_\gamma$$

$$+ \tfrac{1}{4}\left(K + \tfrac{1}{4}\partial^2 B(x)\right)\theta^\alpha\theta_\alpha\bar{\theta}^\beta\bar{\theta}_\beta\,, \qquad (17.136)$$

in which B, K are real fields. The reason for such a parametrization will become clear below. Let us first note that in this form the vector superfield contains four real scalar fields, four Weyl spinors and a vector field. However, supersymmetry naturally leads to a generalized form of gauge symmetry. We consider two scalar superfields $\bar{\phi}, \phi$ of the form (17.124,17.127). A generalized abelian supersymmetric gauge transformation then is defined by the translation:

$$V \mapsto V + \phi + \bar{\phi}.\tag{17.137}$$

In component form this leads to:

$$
\begin{aligned}
B &\mapsto B + A + A^*, & \chi &\mapsto \chi + \psi, \\
C &\mapsto C + E, & V_\mu &\mapsto V_\mu - 2^{-1/2} i \partial_\mu (A - A^*), \\
\lambda &\mapsto \lambda, & K &\mapsto K.
\end{aligned}
$$

With our parametrization λ and K are gauge invariant, while V_μ transforms like a usual abelian gauge field.

In a theory which has this kind of gauge invariance, the vector superfield can be simplified, the fields B, C, χ being eliminated by a gauge transformation. This characterizes the non-supersymmetric *Wess–Zumino (WZ) gauge*. The vector field then takes the form

$$V = i2^{-1/2} V_\mu \bar{\theta}^\alpha \sigma_{\mu\alpha}^{\ \beta} \theta_\beta + \tfrac{1}{2} \bar{\theta}^\alpha \bar{\lambda}_\alpha \theta^\beta \theta_\beta + \tfrac{1}{2} \lambda^\beta \theta_\beta \bar{\theta}^\alpha \bar{\theta}_\alpha + \tfrac{1}{4} K \theta^\alpha \theta_\alpha \bar{\theta}^\beta \bar{\theta}_\beta.\tag{17.138}$$

The physical degrees of freedom, as we shall see, reduce to a massless vector field V_μ and a massless spin $1/2$ fermion λ, since the field K does not propagate.

It is important to note that, in this gauge, powers of the vector field have the property:

$$V^2 = -\tfrac{1}{4} V_\mu V_\mu \theta^\alpha \theta_\alpha \bar{\theta}^\beta \bar{\theta}_\beta, \qquad V^n = 0 \text{ for } n \geq 3.\tag{17.139}$$

Supersymmetric curvature tensor. The vector superfield is not chiral and therefore, as in the 1-D case, the operators $\bar{\Delta}, \Delta$ can be used to construct other superfields. Note that quantities of the form $\bar{\Delta}^\beta \bar{\Delta}_\beta W$ and $\Delta^\beta \Delta_\beta W$, where W is an arbitrary superfield, are right and left-handed chiral fields respectively, because the product of three operators Δ or $\bar{\Delta}$ vanishes:

$$\bar{\Delta}_\alpha \bar{\Delta}^\beta \bar{\Delta}_\beta W = \Delta^\alpha \Delta^\beta \Delta_\beta W = 0 \quad \forall\, W.$$

In particular the quantities

$$F^\alpha = \tfrac{1}{2} \bar{\Delta}^\beta \bar{\Delta}_\beta \Delta^\alpha V, \qquad \bar{F}_\alpha = \tfrac{1}{2} \Delta^\beta \Delta_\beta \bar{\Delta}_\alpha V,\tag{17.140}$$

are chiral and gauge invariant (in the sense (17.137)) since for example

$$\bar{\Delta}^\beta \bar{\Delta}_\beta \Delta^\alpha (\phi + \bar{\phi}) = \bar{\Delta}^\beta \bar{\Delta}_\beta \Delta^\alpha \phi = -\sigma_{\mu\beta}^{\ \alpha} \partial_\mu \bar{\Delta}^\beta \phi = 0,$$

(we have used the anticommutation relation (17.120) and the chirality conditions) and thus generalize the gauge field curvature. To calculate them it is convenient to express V in terms of the variables y and $\bar{y}$:

$$
\begin{aligned}
V &= i2^{-1/2} V_\mu(y) \bar{\theta}^\alpha \sigma_{\mu\alpha}^{\ \beta} \theta_\beta + \tfrac{1}{2} \bar{\theta}^\alpha \bar{\lambda}_\alpha \theta^\beta \theta_\beta + \tfrac{1}{2} \lambda^\beta(y) \theta_\beta \bar{\theta}^\alpha \bar{\theta}_\alpha \\
&\quad + \tfrac{1}{4} \left[K(y) - i\partial_\mu V_\mu(y) \right] \theta^\alpha \theta_\alpha \bar{\theta}^\beta \bar{\theta}_\beta, \\
&= i2^{-1/2} V_\mu(\bar{y}) \bar{\theta}^\alpha \sigma_{\mu\alpha}^{\ \beta} \theta_\beta + \tfrac{1}{2} \bar{\theta}^\alpha \bar{\lambda}_\alpha \theta^\beta \theta_\beta + \tfrac{1}{2} \lambda^\beta(\bar{y}) \theta_\beta \bar{\theta}^\alpha \bar{\theta}_\alpha \\
&\quad + \tfrac{1}{4} \left[K(\bar{y}) + i\partial_\mu V_\mu(\bar{y}) \right] \theta^\alpha \theta_\alpha \bar{\theta}^\beta \bar{\theta}_\beta.
\end{aligned}
$$

We then find:

$$F^\alpha = -\lambda^\alpha(y) - i2^{-3/2}\theta^\beta \left(\sigma^\dagger_\nu \sigma_\mu\right)^\alpha_{\ \beta} F_{\mu\nu}(y) - K(y)\theta^\alpha + \tfrac{1}{2}\partial_\mu \bar\lambda^\beta(y)\sigma^\alpha_{\mu\beta}\theta^\gamma\theta_\gamma\,,$$

$$\bar F_\alpha = \bar\lambda_\alpha(\bar y) - i2^{-3/2}\left(\sigma_\mu\sigma^\dagger_\nu\right)^\beta_{\ \alpha} F_{\mu\nu}(\bar y)\bar\theta_\beta + K(\bar y)\bar\theta_\alpha + \tfrac{1}{2}\sigma^\beta_{\mu\alpha}\partial_\mu\lambda_\beta(\bar y)\bar\theta^\gamma\bar\theta_\gamma\,.$$

with the usual notation $F_{\mu\nu} = \partial_\mu A_\nu - \partial_\nu A_\mu$. We observe that indeed F and $\bar F$ depend only on the gauge invariant combinations $F_{\mu\nu}, \lambda, K$. A check of these expressions is provided by the simple relation (derived from the definitions (17.140) after a few commutations)

$$\Delta_\alpha F^\alpha + \bar\Delta^\alpha \bar F_\alpha = 0\,.$$

The field F^α is chiral and has dimension 3/2. Therefore $\mathcal{L}_F = -\tfrac{1}{4}F^\alpha F_\alpha$ is candidate to contribute to the kinetic term. We then find

$$\int \mathrm{d}\theta_1 \mathrm{d}\theta^1\, F^\alpha F_\alpha = 2\bar\lambda^\alpha \sigma^\beta_{\mu\alpha}\partial_\mu\lambda_\beta + K^2 - \tfrac{1}{2}F_{\mu\nu}F_{\mu\nu} + \tfrac{1}{4}\epsilon_{\lambda\mu\nu\rho}F_{\lambda\mu}F_{\nu\rho}\,, \qquad (17.141)$$

where we have used the identity

$$\tfrac{1}{2}\operatorname{tr}\sigma^\dagger_\lambda\sigma_\mu\sigma^\dagger_\nu\sigma_\rho = \epsilon_{\lambda\mu\nu\rho} + (\delta_{\lambda\mu}\delta_{\nu\rho} + \delta_{\mu\nu}\delta_{\rho\lambda} - \delta_{\lambda\nu}\delta_{\mu\rho})\,.$$

If we add the conjugated contribution coming from $-\tfrac{1}{4}\bar F^\alpha \bar F_\alpha$ we obtain the supersymmetric free gauge action S which can be written

$$S = \int \mathrm{d}^4x \left[\tfrac{1}{4}F^2_{\mu\nu} - \bar\lambda^\alpha \sigma^\beta_{\mu\alpha}\partial_\mu\lambda_\beta - K^2\right]\,. \qquad (17.142)$$

This action could also have been obtained by noting that the vector superfield has dimension zero. A term of the form $V\bar\Delta\bar\Delta\Delta\Delta V$ has dimension two and is thus candidate to be a D term.

Note finally that it is also possible to give a mass to vector superfield by adding the D contribution of V^2.

Gauge invariant interactions. A charged scalar superfield transforms under global $U(1)$ as $\phi \mapsto e^{ie\Lambda}\phi$. However if we want to introduce space-dependent $U(1)$ transformations consistent with supersymmetry we cannot simply replace Λ by $\Lambda(x)$. We have to introduce the constant Λ by a scalar chiral superfield:

$$\phi(y,\theta) \mapsto e^{ie\Lambda(y,\theta)}\,\phi(y,\theta) \qquad \text{with} \quad \bar\Delta_\alpha\Lambda = 0\,, \qquad (17.143a)$$

$$\bar\phi(\bar y, \bar\theta) \mapsto e^{-ie\bar\Lambda(\bar y,\bar\theta)}\,\bar\phi(\bar y, \bar\theta) \qquad \text{with} \quad \Delta^\alpha\bar\Lambda = 0\,. \qquad (17.143b)$$

We see immediately that if the charges and couplings are such that the interaction term (17.132) is invariant under global $U(1)$ transformations, it is also invariant under the transformations (17.143). However the free term (17.131) is not invariant since

$$\bar\phi\phi \mapsto \bar\phi\phi\, e^{ie(\Lambda - \bar\Lambda)}\,.$$

The render the kinetic term gauge invariant we replace it by

$$\mathcal{L}_D = \bar\phi\, e^{eV}\, \phi\,,$$

where the vector superfield V must transform like

$$V \mapsto V - i(\Lambda - \bar{\Lambda}),$$

transformation indeed consistent with (17.137).

At first sight it would seem that such a theory cannot be renormalizable, but at least in the WZ gauge the expansion of the exponential reduces to three first terms (property (17.139)) and contains no term of dimension larger than four.

We have thus presented all the ingredients necessary to construct a supersymmetric version of QED.

Non-abelian gauge theories. The extension to non-abelian gauge theories is not difficult. We denote by t_a the matrices generators of the gauge group G. The gauge transformation is now parametrized by a set of scalar superfields Λ_a and can be written

$$\phi \mapsto e^{\Lambda}\phi\,, \quad \Lambda \equiv \Lambda_a t_a\,, \qquad \bar{\phi} \mapsto \bar{\phi}\,e^{\bar{\Lambda}}\,, \quad \bar{\Lambda} \equiv \bar{\Lambda}_a t_a\,.$$

A gauge invariant lagrangian is

$$\mathcal{L}_D = \bar{\phi}\,e^V\,\phi\,, \quad \text{with } V = V_a t_a\,,$$

where the gauge field transforms like

$$e^V \mapsto e^{-\bar{\Lambda}}\,e^V\,e^{-\Lambda}\,.$$

The generalized form of the curvature F^{α} is

$$F^{\alpha} = \bar{\Delta}^{\beta}\bar{\Delta}_{\beta}\,e^{-V}\,\Delta^{\alpha}\,e^V\,, \tag{17.144}$$

which is chiral and indeed transforms like

$$F^{\alpha} \mapsto e^{\Lambda}\,F^{\alpha}\,e^{-\Lambda}\,. \tag{17.145}$$

Finally the contribution to the action takes the form $\operatorname{tr} F^{\alpha}F_{\alpha}$.

Bibliographical Notes

In the main part of this chapter we have followed

J. Zinn-Justin, *Nucl. Phys.* B275 [FS18] (1986) 135.

The relation between supersymmetry and dissipative Langevin or Fokker–Planck equations have been shown in

E. Witten, *Nucl. Phys.* B188 (1981) 513; M.V. Feigel'man and A.M. Tsvelik, *Sov. Phys.–JETP* 56 (1982) 823; H. Nakazato, M. Namiki, I. Ohba and K. Okano, *Prog. Theor. Phys.* 70 (1983) 298; E. Egorian and S. Kalitzin, *Phys. Lett.* 129B (1983) 320.

Our construction of the effective action provides an example of the so-called Nicolai map

H. Nicolai, *Phys. Lett.* 89B (1980) 341; *Nucl. Phys.* B176 (1980) 419.

The characterization of the equilibrium distribution by supersymmetry arguments is adapted from

R. Kirschner, *Phys. Lett.* 139B (1984) 180; E. Gozzi, *Phys. Lett.* 143B (1984) 183.

The non-linear σ-model has been considered in

R. Bausch, H.K. Janssen and Y. Yamazaki, *Z. Phys.* B37 (1980) 163.

Application of stochastic quantization to gauge theories is discussed in

G. Parisi and Y.S. Wu, *Acad. Sci. Sin.* 24 (1981) 483; D. Zwanziger, *Nucl. Phys.* B192 (1981) 259; B209 (1982) 336; J. Zinn-Justin and D. Zwanziger, *Nucl. Phys.* B295 [FS21] (1987) 297;

The corresponding RG β-function is given by:

K. Okano, *Nucl. Phys.* B289 (1987) 109.

See also the review

P.H. Damgaard and H. Hüffel, *Phys. Rep.* 152 (1987) 227.

In Section 17.10 we have followed

J. Wess and J. Bagger, *Supersymmetry and Supergravity* (Princeton University Press, Princeton 1983).

APPENDIX 17
RENORMALIZATION CONSTANTS AND RG FUNCTIONS AT TWO-LOOPS: SUPERSYMMETRIC PERTURBATION THEORY

From the form of the renormalized dynamical action RG equations can be derived. In the case of the purely dissipative Langevin equation (17.1,17.7) only one new RG function appears, associated with the time scale renormalization. For illustration purpose we here consider the examples of the ϕ^4 field theory and the $O(N)$ symmetric non-linear σ-model and calculate the corresponding RG functions at two-loop order.

Perturbative calculations. In perturbative calculations of dynamic quantities the formalism of Section 17.2 can be used, dimensional regularization eliminating the determinant. In the particular case of the dissipative Langevin equation, it is also possible to use the method of super-diagrams, treating the Grassmann coordinates on the same footing as the usual commuting coordinates. In this way the supersymmetry is explicit at all steps. Moreover dynamic and static perturbation theories become remarkably similar, the topology and weight factors of Feynman diagrams being the same. It is also convenient to take the boundary condition in the Langevin equation at time $-\infty$ so that the system is at equilibrium at any finite time, and therefore time translation invariance is secured.

A17.1 The $\left(\phi^2\right)^2$ Field Theory

The dynamical action. The bare dynamical action is:

$$S\left(\phi_0\right) = \int \mathrm{d}\theta\, \mathrm{d}\bar{\theta}\, \mathrm{d}t \left[\int \mathrm{d}^d x \frac{2}{\Omega_0} \frac{\partial \phi_0}{\partial \theta} \cdot \left(\frac{\partial \phi_0}{\partial \bar{\theta}} - \theta \frac{\partial \phi_0}{\partial t} \right) + \mathcal{A}\left(\phi_0\right) \right], \qquad (A17.1)$$

with:

$$\mathcal{A}\left(\phi_0\right) = \int \mathrm{d}^d x \left[\frac{1}{2} \left(\partial_\mu \phi_0 \right)^2 + \frac{1}{2} m_0^2 \phi_0^2 + \frac{1}{4!} g_0 \left(\phi_0^2\right)^2 \right]. \qquad (A17.2)$$

As we have shown in Chapter 17, the renormalized action is then:

$$S_{\mathrm{r}}\left(\phi\right) = \int \mathrm{d}\theta\, \mathrm{d}\bar{\theta}\, \mathrm{d}t \left[\int \mathrm{d}^d x \frac{2}{\Omega} Z_\omega \frac{\partial \phi}{\partial \theta} \cdot \left(\frac{\partial \phi}{\partial \bar{\theta}} - \theta \frac{\partial \phi}{\partial t} \right) + \mathcal{A}_{\mathrm{r}}\left(\phi\right) \right]. \qquad (A17.3)$$

The RG equations. To renormalize action $(A17.1)$ we have introduced, in addition to the static renormalization constants, a renormalization of the parameter Ω:

$$\Omega_0 = \Omega Z / Z_\omega \,, \qquad (A17.4)$$

where Z is the field renormalization constant. The CS operator of Section 10.5 then takes the form:

$$\mathrm{D}_{\mathrm{CS}} = m \frac{\partial}{\partial m} + \beta(g) \frac{\partial}{\partial g} + \eta_\omega(g) \Omega \frac{\partial}{\partial \Omega} - \frac{n}{2} \eta(g), \qquad (A17.5)$$

where we have defined

$$\eta_\omega(g) = m \frac{\mathrm{d}}{\mathrm{d}m} \bigg|_{g_0, \Omega_0} \ln \Omega \,. \qquad (A17.6)$$

In the case of dimensional regularization, equation $(A17.6)$ becomes:

$$\eta_\omega(g) = \beta(g)\frac{\mathrm{d}}{\mathrm{d}g}\ln\left(Z_\omega/Z\right). \qquad (A17.7)$$

As usual the solutions of RG equations have to be combined with dimensional analysis. One here finds:

$$\Gamma^{(n)}\left(\lambda p_i, \rho\omega_i, \frac{\boldsymbol{\theta}_i}{\sqrt{\rho}}, \lambda m, \frac{\rho\Omega}{\lambda^2}\right) = \lambda^{d-n(d-2)/2}\rho^{1-n}\Gamma^{(n)}\left(p_i, \omega_i, \boldsymbol{\theta}_i, m, \Omega\right), \qquad (A17.8)$$

where we have introduced the vector notation $\boldsymbol{\theta} \equiv (\theta, \bar{\theta})$.

WT identities. Let us also recall the WT identities $(17.47, 17.48)$ corresponding to transformation (17.45) satisfied by the proper vertices after Fourier transformation:

$$\sum_{i=1}^n \frac{\partial}{\partial \theta_i}\Gamma^{(n)} - i\left(\sum_{i=1}^n \omega_i \bar{\theta}_i\right)\Gamma^{(n)} = 0. \qquad (A17.9)$$

These identities provide checks in perturbative calculations.

A17.2 Perturbative Calculation of $\eta_\omega(g)$

To the action $(A17.1)$ corresponds a propagator $\Delta(\mathbf{k}, \omega, \boldsymbol{\theta}, \boldsymbol{\theta}')$:

$$\Delta(\mathbf{k}, \omega, \boldsymbol{\theta}, \boldsymbol{\theta}') = \frac{\Omega\left[1 - \frac{1}{2}i\omega\left(\bar{\theta} - \bar{\theta}'\right)\left(\theta + \theta'\right) + \frac{\Omega}{4}\left(k^2 + m^2\right)\delta^2\left(\boldsymbol{\theta} - \boldsymbol{\theta}'\right)\right]}{\omega^2 + \frac{\Omega^2}{4}\left(k^2 + m^2\right)^2}. \qquad (A17.10)$$

We have omitted the factor δ_{ij} corresponding to group indices. For practical calculations it is actually more convenient to use a mixed representation for the propagator, Fourier transformed on space but not in time:

$$\Delta(\mathbf{k}, \tau, \boldsymbol{\theta}, \boldsymbol{\theta}') = \left\{\frac{1}{k^2 + m^2} + \frac{\Omega}{4}\left(\bar{\theta} - \bar{\theta}'\right)\left[\theta - \theta' - \left(\theta + \theta'\right)\epsilon(\tau)\right]\right\}$$
$$\times \exp\left[-\frac{\Omega}{2}\left(k^2 + m^2\right)|\tau|\right], \qquad (A17.11)$$

in which $\epsilon(\tau)$ is the sign of $\tau = t - t'$. Note that, in agreement with the analysis of Subsection 17.5.2, Δ can be rewritten:

$$\Delta(\mathbf{k}, \tau, \boldsymbol{\theta}, \boldsymbol{\theta}') = \left\{1 + \tfrac{1}{2}\left(\bar{\theta} - \bar{\theta}'\right)\left[\theta + \theta' - \left(\theta - \theta'\right)\epsilon(\tau)\right]\frac{\partial}{\partial\tau}\right\}$$
$$\times \frac{1}{k^2 + m^2}\exp\left[-\frac{\Omega}{2}\left(k^2 + m^2\right)|\tau|\right].$$

This form follows directly from the supersymmetry and the causality of the Langevin equation (see also equations $(35.15-35.18)$).

From the supersymmetric form $(A17.1)$ of the dynamical action we deduce that the dynamic and the static theory have very similar perturbative expansions and differ mainly by the form of the propagator. For $N = 1$ the interaction vertex $V^{(4)}$ has the form:

$$V^{(4)} = m^\varepsilon \frac{g}{4!} \delta^2 \left(\boldsymbol{\theta}_1 - \boldsymbol{\theta}_2\right) \delta^2 \left(\boldsymbol{\theta}_1 - \boldsymbol{\theta}_3\right) \delta^2 \left(\boldsymbol{\theta}_1 - \boldsymbol{\theta}_4\right). \tag{A17.12}$$

We have already calculated all static renormalization constants to two-loop order in Sections 11.6,11.7. To calculate the new renormalization constant at leading order (i.e. two-loop order) we need only the 2-point function. Using the expressions of Section 11.7, we find:

$$\Gamma^{(2)} = -\frac{2}{\Omega}\left[2 - i\omega\left(\bar{\theta} - \bar{\theta}'\right)\left(\theta + \theta'\right)\right] + \left(k^2 + m^2\right)\delta^2\left(\boldsymbol{\theta} - \boldsymbol{\theta}'\right)$$

$$+ \frac{N+2}{6}m^\varepsilon g D_1 - \frac{(N+2)^2}{36}m^{2\varepsilon}g^2 D_2 - \frac{N+2}{18}m^{2\varepsilon}g^2 D_3, \tag{A17.13}$$

in which the three diagrams D_1, D_2 ,D_3, are given by:

$$D_1 = \delta^2 \left(\boldsymbol{\theta} - \boldsymbol{\theta}'\right) \frac{1}{(2\pi)^d} \int \frac{d^dp}{p^2 + m^2}, \tag{A17.14}$$

$$D_2 = D_1 \int \frac{d^dp}{(2\pi)^d} dt\, d^2\theta''\, e^{-\Omega(p^2 + m^2)|t|} \frac{1}{p^2 + m^2}$$

$$\times \left\{\frac{1}{p^2 + m^2} + \frac{\Omega}{2}\delta\left(\bar{\theta}'' - \bar{\theta}\right)\left[\theta'' - \theta - \varepsilon\left(t\right)\left(\theta'' + \theta\right)\right]\right\}, \tag{A17.15}$$

$$D_3 = \frac{1}{(2\pi)^{2d}} \int dt\, e^{i\omega t} \int d^dp_1 d^dp_2\, e^{-\Omega s(p_i)|t|/2} \prod_{i=1}^{3} \frac{1}{p_i^2 + m^2}$$

$$\times \left\{1 + \frac{\Omega}{4}s\left\{p_i\right\}\delta\left(\bar{\theta} - \bar{\theta}'\right)\left[\theta - \theta' - \varepsilon\left(t\right)\left(\theta + \theta'\right)\right]\right\}, \tag{A17.16}$$

with the definitions:

$$\mathbf{p}_3 = -\left(\mathbf{k} + \mathbf{p}_1 + \mathbf{p}_2\right), \tag{A17.17}$$

$$s\left(p_i\right) = \sum_{i=1}^{3}\left(p_i^2 + m^2\right). \tag{A17.18}$$

The diagrams D_1 and D_2 (after integration over t) are identical to the diagrams of the static theory. Only D_3 contains a new dynamic divergence contributing to the renormalization of Ω. Actually D_3 can be written as the sum of two terms, $D_3^{(1)}$ proportional to $\delta^2\left(\boldsymbol{\theta} - \boldsymbol{\theta}'\right)$ and which contains the static two-loop divergence and another one $D_3^{(2)}$ which after an integration by parts over t can be written:

$$D_3^{(2)} = \left[1 - \tfrac{1}{2}i\omega\left(\bar{\theta} - \bar{\theta}'\right)\left(\theta + \theta'\right)\right]\int dt\, e^{i\omega t} \int \frac{d^dp_1}{(2\pi)^d}\frac{d^dp_2}{(2\pi)^d}\, e^{-\Omega s(p_i)|t|/2} \prod_{i=1}^{3}\frac{1}{p_i^2 + m^2}. \tag{A17.19}$$

The divergent part of the integral is a constant which can be calculated at $\omega = 0$ and $\mathbf{k} = 0$. Integrating over t, p_2 and p_1 successively, one finally obtains:

$$Z_\omega = 1 - N_d^2 g^2 \ln(4/3)\frac{N+2}{24\varepsilon} + O\left(g^3\right), \tag{A17.20}$$

and therefore, using the results of Section 11.7,

$$\eta_\omega(g) = N_d^2 \frac{N+2}{72} g^2 \left(6\ln(4/3) - 1\right) + O\left(g^3\right). \tag{A17.21}$$

A17.3 The Non-Linear σ-Model

Let us now consider the case of the $O(N)$ non-linear σ-model (as defined in Chapter 14), still with the supersymmetric dynamics, to illustrate the discussion of Sections 17.7,17.8. The dynamical bare action can be written:

$$S\left(\phi_0\right) = \frac{1}{g_0} \int \mathrm{d}\theta \, \mathrm{d}\bar{\theta} \mathrm{d}t \left[\int \mathrm{d}^d x \frac{2}{\Omega_0} \frac{\partial \phi_0}{\partial \theta} \cdot \left(\frac{\partial \phi_0}{\partial \bar{\theta}} - \theta \frac{\partial \phi_0}{\partial t} \right) + \mathcal{A}\left(\phi_0\right) \right], \qquad (A17.22)$$

with:

$$\mathcal{A}\left(\phi_0\right) = \int \mathrm{d}^d x \tfrac{1}{2} \left(\partial_\mu \phi_0\right)^2 - \int \mathrm{d}^d x \, \mathbf{h}_0 \cdot \phi_0 . \qquad (A17.23)$$

Note that for practical reasons we have adopted normalizations which differ from those of Sections 17.7,17.8.

The renormalized theory is defined by the substitutions:

$$\phi_0 = \sqrt{Z}\{\sigma(x,t), \ \boldsymbol{\pi}(x,t)\}, \qquad (A17.24)$$

with then:

$$\sigma(x,t) = \sqrt{Z^{-1} - \boldsymbol{\pi}^2}, \qquad (A17.25)$$

and:

$$g_0 = \mu^{-\varepsilon} g Z_g, \quad \Omega_0 g_0 = \Omega g \mu^{-\varepsilon} Z/Z_\omega, \quad h_0 \sqrt{Z}/g_0 = h \, \mu^\varepsilon / g . \qquad (A17.26)$$

We have called μ the renormalization scale and set $\varepsilon = d - 2$.

With dimensional regularization, the dynamic RG function $\eta_\omega(g)$ is then defined by:

$$\eta_\omega(g) = \mu \frac{\mathrm{d}}{\mathrm{d}\mu}\bigg|_{g_0,\Omega_0} \ln \Omega = \beta\left(g\right) \frac{\mathrm{d}}{\mathrm{d}g} \ln \left(\frac{Z_\omega Z_g}{Z} \right) . \qquad (A17.27)$$

Dimensional analysis yields the relation:

$$\Gamma^{(n)}\left(\lambda p_i, \rho \omega_i, \frac{\boldsymbol{\theta}_i}{\sqrt{\rho}}, \lambda \mu, \frac{\rho \Omega}{\lambda^2}, \lambda^2 h \right) = \lambda^d \rho^{1-n} \Gamma^{(n)}\left(p_i, \omega_i, \boldsymbol{\theta}_i, \mu, \Omega, h \right) . \qquad (A17.28)$$

Perturbation theory. The propagator is the same as in equation $(A17.10)$ up to the normalization:

$$\Delta\left(\mathbf{k}, \omega, \theta, \theta'\right) = g \mu^{-\varepsilon} \frac{\Omega \left[1 - \tfrac{1}{2} i \omega \left(\bar{\theta} - \bar{\theta}' \right) \left(\theta + \theta' \right) + \frac{\Omega}{4} \left(k^2 + h \right) \delta^2 \left(\theta - \theta' \right) \right]}{\omega^2 + \frac{\Omega^2}{4}\left(k^2 + h \right)^2} . \qquad (A17.29)$$

To calculate the 2-point function at two-loop order we need the π^4 and the π^6 vertices:

$$V^{(4)} = \frac{\mu^\varepsilon}{8g} \delta^2\left(\boldsymbol{\theta}_1 - \boldsymbol{\theta}_2\right) \delta^2\left(\boldsymbol{\theta}_3 - \boldsymbol{\theta}_4\right) \delta_{i_1 i_2} \delta_{i_3 i_4}$$

$$\times \left\{ \frac{2}{\Omega}\left[-2 + i\left(\omega_1 + \omega_2\right)\left(\bar{\theta}_1 - \bar{\theta}_3\right)\left(\theta_1 + \theta_3\right) \right] + \left(\left(p_1 + p_2\right)^2 + h \right) \delta^2\left(\boldsymbol{\theta}_1 - \boldsymbol{\theta}_3\right) \right\}, \quad (A17.30)$$

$$V^{(6)} = \frac{\mu^\varepsilon}{16g} \delta_{i_1 i_2} \delta_{i_3 i_4} \delta_{i_5 i_6} \delta^2\left(\boldsymbol{\theta}_1 - \boldsymbol{\theta}_2\right) \delta^2\left(\boldsymbol{\theta}_3 - \boldsymbol{\theta}_4\right) \delta^2\left(\boldsymbol{\theta}_3 - \boldsymbol{\theta}_5\right) \delta^2\left(\boldsymbol{\theta}_3 - \boldsymbol{\theta}_6\right)$$

$$\times \left\{ \frac{2}{\Omega}\left[-2 + i\left(\omega_1 + \omega_2\right)\left(\bar{\theta}_1 - \bar{\theta}_3\right)\left(\theta_1 + \theta_3\right) \right] + \delta^2\left(\boldsymbol{\theta}_1 - \boldsymbol{\theta}_3\right) \left(\left(p_1 + p_2\right)^2 + h \right) \right\} . \quad (A17.31)$$

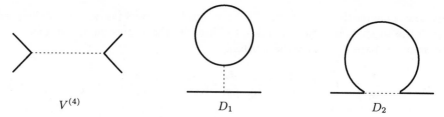

Fig. 17.1   One-loop diagrams: the dotted lines do not correspond to propagators but are used only to represent faithfully the flow of group indices.

2-point function at one-loop order. At one-loop order, the 2-point function then has the expansion:

$$\frac{g}{\mu^{\varepsilon}}\Gamma^{(2)} = -\frac{2Z_{\omega}}{\Omega}\left[2 - i\omega\left(\bar{\theta}-\bar{\theta}'\right)\left(\theta+\theta'\right)\right] + \left(\frac{Z}{Z_g}k^2 + h\sqrt{Z}\right)\delta^2\left(\boldsymbol{\theta}-\boldsymbol{\theta}'\right)$$

$$+ \frac{N-1}{2}gD_1 + gD_2 + O\left(g^2\right). \tag{A17.32}$$

The coefficients which correspond to the diagrams of figure 17.1 have the following expressions:

$$D_1 = \mu^{\varepsilon}h\delta^2\left(\boldsymbol{\theta}-\boldsymbol{\theta}'\right)I, \tag{A17.33}$$

$$D_2 = \mu^{\varepsilon}\left\{-\frac{2}{\Omega}\left[2 - i\omega\left(\bar{\theta}-\bar{\theta}'\right)\left(\theta+\theta'\right)\right] + k^2\delta^2\left(\boldsymbol{\theta}-\boldsymbol{\theta}'\right)\right\}I, \tag{A17.34}$$

$$I = \frac{1}{(2\pi)^d}\int\frac{d^dq}{q^2+h} = -\frac{N_d}{\varepsilon} + O\left(1\right). \tag{A17.35}$$

We recover the one-loop static results:

$$Z = 1 + \frac{(N-1)}{2\pi\varepsilon}g + O\left(g^2\right),$$

$$Z_g = 1 + \frac{(N-2)}{2\pi\varepsilon}g + O\left(g^2\right),$$

and in addition we find:

$$Z_{\omega} = Z/Z_g + O\left(g^2\right). \tag{A17.36}$$

This equation together with equation $(A17.27)$ implies the absence of dynamics renormalization at this order:

$$\eta_{\omega}(g) = O\left(g^2\right). \tag{A17.37}$$

Two-loop calculation. We now have to calculate the one-loop diagrams with the renormalization constants expanded at one-loop order, and the various two-loop diagrams generated by two 4-point vertices and one 6-point vertex. Since the static renormalization constants have already been calculated, to calculate $\eta_{\omega}(g)$ at two-loop order we need only $\Gamma^{(2)}$ for vanishing arguments: $\mathbf{k} = 0, \omega = 0, \theta = \bar{\theta} = 0$.

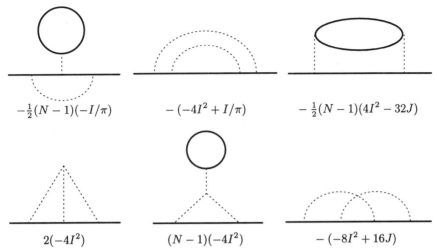

$$-\tfrac{1}{2}(N-1)(-I/\pi) \qquad -(-4I^2+I/\pi) \qquad -\tfrac{1}{2}(N-1)(4I^2-32J)$$

$$2(-4I^2) \qquad\qquad (N-1)(-4I^2) \qquad\qquad -(-8I^2+16J)$$

Fig. 17.2   Two-loop diagrams: faithful representation and contribution.

Figure 17.2 lists all contributing two-loop diagrams with their natural weight factors. The integral J is given by:

$$J = \frac{1}{(2\pi)^{2d}} \int \frac{d^d p \, d^d q}{\left(p^2+h\right)\left(p^2+q^2+(p+q)^2+3h\right)}$$
$$= \frac{N_d^2}{4\varepsilon^2}\left(1+\frac{\varepsilon}{2}\ln\frac{3}{4}+O\left(\varepsilon^2\right)\right). \tag{A17.38}$$

We obtain therefore:

$$\Gamma^{(2)} = -\frac{4Z_\omega}{\Omega}\left[1 + gZ_g\left(hZ_g/Z\right)^{\varepsilon/2}I + \left(\frac{3N-5}{2}\right)I^2g^2Jg^2 \right.$$
$$\left. - 4\left(N-2\right)Jg^2 - \frac{N-3}{4}N_d Ig^2 + O\left(g^3\right)\right]. \tag{A17.39}$$

Expanding all terms, we find finally the expression of Z_ω at two-loop order:

$$Z_\omega = 1 + \frac{g}{2\pi\varepsilon} + \left(\frac{N-1}{2\varepsilon^2}\right)\frac{g^2}{(2\pi)^2} - \frac{N-2}{2\varepsilon}\ln\frac{4}{3}\frac{g^2}{(2\pi)^2} + O\left(g^3\right). \tag{A17.40}$$

The expansion of $\eta_\omega(g)$ at order g^2 follows:

$$\eta_\omega(g) = (N-2)\left(1-\ln(4/3)\right)\frac{g^2}{(2\pi)^2} + O\left(g^3\right). \tag{A17.41}$$

18 ABELIAN GAUGE THEORIES

With this chapter we begin the study of a new class of geometrical models which have dominated the studies of fundamental interactions for more than a century: gauge theories. They are characterized by new physical properties and new technical difficulties. Gauge theories can be considered in some sense as geometrical generalizations of the principal chiral model. In particular the pure abelian gauge field leads to a free field theory, as the pure abelian chiral model.

We devote first a whole chapter to a simple and physically important example, the abelian gauge theory, whose physical realization is Quantum Electrodynamics (QED). However, since many excellent textbooks deal extensively with QED, we will mainly concentrate on the formal aspects of abelian gauge theories.

The set-up of this chapter is the following: We begin with elementary considerations about the massive vector field in perturbation theory. We show that coupling to matter field leads to field theories which are renormalizable in four dimensions only if the vector field is coupled to a conserved current. In the latter case the massless vector limit can be defined. The corresponding field theories are gauge invariant. We then discuss the specific properties of gauge invariant theories and mention the IR problem of physical observables. We quantize gauge theories starting directly from first principles. The formal equivalence between different gauges is established.

In Section 18.5 regularization methods are presented which allow to overcome the new difficulties one encounters in gauge theories. The abelian gauge symmetry, broken by gauge fixing terms, then leads to a set of WT identities which are used to prove the renormalizability of the theory. The gauge dependence of correlation functions in a set of covariant gauges is determined. Renormalization group equations follow and we calculate the RG β-function at leading order.

Finally, as an introduction to the next chapter, we analyze the abelian Higgs mechanism. The appendix contains the derivation of the Casimir effect and some more details about gauge dependence.

18.1 The Massive Vector Field

The quantization of the free massive vector field does not immediately follow from the quantization of the scalar field and thus requires a short discussion. In the first of this section we work in real time with the metric $\{+, -, -, \cdots\}$ where the first component is the time component. Space-time is denoted by $\{x_0 \equiv t, x_i\}$, $x_i \in \mathbb{R}^{d-1}$, and the vector field $A_\mu \equiv \{A_0, A_i\}$. Finally time derivative is indicated by

$$\frac{\partial A_i}{\partial t} \equiv \dot{A}_i \,.$$

The local $O(1, d-1)$ invariant classical action for a free massive vector field then can be written:

$$\mathcal{A}(A) = -\int \mathrm{d}t \, \mathrm{d}^{d-1}x \, \left[\tfrac{1}{4} F_{\mu\nu}(t,x) F^{\mu\nu}(t,x) - \tfrac{1}{2} m^2 A_\mu(t,x) A^\mu(t,x) \right] , \qquad (18.1)$$

with

$$F_{\mu\nu}(t, x) = \partial_\mu A_\nu(t, x) - \partial_\nu A_\mu(t, x). \tag{18.2}$$

One may wonder about this peculiar form of the derivative term, but is easy to verify that the additional we could think adding, $\partial_\mu A_\mu \partial^\nu A^\nu$, depending of its sign either corresponds to a A_0 field with a negative metric, or to an unbounded potential.

Quantization. The action (18.1) has one peculiar feature: the time component A_0 of the vector field has no conjugate momentum since the action does not depend on the time derivative $\dot{A}_0$. Actually the action can be rewritten

$$\mathcal{A}(A) = \int \mathrm{d}t\, \mathrm{d}^{d-1}x \left[\tfrac{1}{2}\dot{A}_i(t, x) \left(\delta_{ij} - \frac{\partial_i \partial_j}{\partial_\perp^2 - m^2} \right) \dot{A}_j(t, x) - \frac{1}{2}m^2 A_i^2(x) - \frac{1}{4}F_{ij}^2(t, x) \right.$$
$$+ \tfrac{1}{2}\left((-\partial_\perp^2 + m^2)A_0(t, x) + \partial_i \dot{A}_i(t, x) \right) (-\partial_\perp^2 + m^2)^{-1}$$
$$\left. \times \left((-\partial_\perp^2 + m^2)A_0(t, x) + \partial_i \dot{A}_i(t, x) \right) \right], \tag{18.3}$$

where roman indices denote space components, and $\partial_\perp^2$ is the space laplacian.

Therefore A_0 should not be considered as a dynamical degree of freedom and the corresponding field equation

$$\frac{\delta \mathcal{A}}{\delta A_0(t, x)} = \left(-\partial_\perp^2 + m^2 \right) A_0(t, x) + \partial_i \dot{A}_i(t, x) = 0, \tag{18.4}$$

is a constraint equation which can be used to eliminate A_0 from the action. This feature reflects the property that a massive vector field has only $d-1$ physical degrees of freedom. The reduced lagrangian density then takes the form

$$\mathcal{L}(A_i) = \frac{1}{2}\dot{A}_i(t, x) \left(\delta_{ij} - \frac{\partial_i \partial_j}{\partial_\perp^2 - m^2} \right) \dot{A}_j(t, x) - \frac{1}{2}m^2 A_i^2(t, x) - \frac{1}{4}F_{ij}^2(t, x). \tag{18.5}$$

After Legendre transformation we obtain the hamiltonian density

$$\mathcal{H}(E_i, A_i) = \frac{1}{2}E_i(x) \left(\delta_{ij} - \frac{1}{m^2}\partial_i \partial_j \right) E_j(x) + \frac{1}{2}m^2 A_i^2(x) + \frac{1}{4}F_{ij}^2(x) \tag{18.6}$$

where we have denoted by E_i (because it becomes the electric field in the massless limit) the momentum conjugated to A_i

$$E_i = \dot{A}_i - \partial_i \left(\partial_\perp^2 - m^2 \right)^{-1} \partial_j \dot{A}_j.$$

The differential operator $-\partial_i \partial_j$ being non-negative, the hamiltonian is positive. The quantization procedure from now on is standard. It leads to an euclidean functional integral in which appears the euclidean reduced lagrangian. This lagrangian has the unpleasant properties that it is non-local and not $O(d)$ space-time symmetric. We note, however, that the dependence in A_0 of the action (18.3) is quadratic. Therefore, we can proceed in the following way: We substitute in the functional integral representation of the partition function the initial euclidean lagrangian. We then perform the gaussian integral over the time component. As we know this is equivalent to solve the corresponding equation of motion, and we thus recover the reduced lagrangian. Finally the determinant resulting from the integration is field independent (see Appendix A19 for the more

More general propagators, interpretation. Let us add to the action (18.7), a term of the form of a regulator field action:

$$S\left(A_\mu, \chi\right) = S\left(A_\mu\right) - \frac{1}{2} \int \mathrm{d}^d x \left[\left(\partial_\mu \chi\right)^2 + \mu^2 \chi^2\right]. \tag{18.14}$$

In the absence of a source for χ, the generating functional

$$Z(J) = \int \left[\mathrm{d}A_\mu\right] \left[\mathrm{d}\chi\right] \exp\left[-S\left(A_\mu, \chi\right) + \int \mathrm{d}^d x \, J_\mu(x) A_\mu(x)\right], \tag{18.15}$$

is proportional to the functional (18.9).

We now change variables in the integral (18.15):

$$A_\mu(x) = A'_\mu(x) + \frac{1}{m} \partial_\mu \chi(x). \tag{18.16}$$

This change leaves $F_{\mu\nu}$ invariant. If the source satisfies the conservation equation (18.12), the source term is not modified. Only the vector field mass term is affected:

$$\frac{1}{2} m^2 A_\mu^2 = \frac{1}{2} m^2 A_\mu'^2 + m A'_\mu \partial_\mu \chi + \frac{1}{2} \left(\partial_\mu \chi\right)^2, \tag{18.17}$$

and therefore:

$$S\left(A'_\mu, \chi\right) = S\left(A'_\mu\right) + \int \mathrm{d}^d x \left(m A'_\mu \partial_\mu \chi - \frac{1}{2} \mu^2 \chi^2\right). \tag{18.18}$$

We now integrate over $\chi(x)$ and find a new action $S_\xi\left(A'_\mu\right)$:

$$S_\xi\left(A'_\mu\right) = S\left(A'_\mu\right) + \frac{1}{2\xi} \int \left(\partial_\mu A'_\mu\right)^2 \mathrm{d}^d x, \tag{18.19}$$

in which we have set:

$$\xi = \mu^2/m^2. \tag{18.20}$$

It is easy to calculate the corresponding propagator $[\Delta_\xi]_{\mu\nu}(k)$:

$$[\Delta_\xi]_{\mu\nu}(k) = \frac{\delta_{\mu\nu}}{k^2 + m^2} + \frac{(\xi - 1) k_\mu k_\nu}{\left(k^2 + m^2\right)\left(k^2 + \xi m^2\right)}. \tag{18.21}$$

For all finite values of ξ the propagator behaves at large momentum like a scalar field propagator. By varying ξ from 0 to $+\infty$ we describe a set of *gauges*: $\xi = 0$ corresponds to Landau's gauge, $\xi = 1$ is Feynman's gauge. For $\xi = \infty$ (the *unitary gauge*) we recover the original propagator of the vector field.

The massless limit. The propagator $[\Delta_\xi]_{\mu\nu}(k)$ has a zero mass limit:

$$[\Delta_\xi]_{\mu\nu}(k) = \frac{\delta_{\mu\nu}}{k^2} + (\xi - 1) \frac{k_\mu k_\nu}{\left(k^2\right)^2}. \tag{18.22}$$

However, for values of $\xi \neq 1$, the term proportional to $1/k^4$ may generate IR divergences in interacting theories for dimensions $d \leq 4$.

18.2 Action with Fermion Matter

We conclude from the previous analysis that vector fields coupled to conserved currents, and thus associated with continuous symmetries, are candidates for the construction of theories renormalizable in four dimensions. Let us now construct the simplest example.

We start from a free action for a massive fermion:

$$S_{\mathrm{F}}\left(\bar{\psi},\psi\right) = -\int \mathrm{d}^d x\, \bar{\psi}(x)\left(\slashed{\partial}+M\right)\psi(x), \tag{18.23}$$

and want to add an $O(d)$ invariant coupling to a vector field. The only interaction term which, from the power counting point of view, has a chance to be renormalizable is proportional to

$$\int \mathrm{d}^d x\, \bar{\psi}(x)\slashed{A}\psi(x)\,.$$

As we have already noted in Section A5.6, the free Dirac fermion action (18.23) has a $U(1)$ symmetry associated with the conservation of the fermionic charge:

$$\psi(x) = \mathrm{e}^{i\Lambda}\,\psi'(x), \qquad \bar{\psi}(x) = \mathrm{e}^{-i\Lambda}\,\bar{\psi}'(x). \tag{18.24}$$

This property remains true after the addition of the interaction term. To this symmetry corresponds a current whose expression is obtained by calculating the variation of the action under a space-dependent group transformation (see Appendix A13.1). If Λ is space-dependent, the variation of the action is:

$$\delta S_{\mathrm{F}} = -i\int \mathrm{d}^d x\, \bar{\psi}(x)\slashed{\partial}\Lambda(x)\psi(x), \tag{18.25}$$

and thus the corresponding conserved current $J_\mu(x)$ is:

$$J_\mu(x) = -i\bar{\psi}(x)\gamma_\mu\psi(x).$$

Therefore the interaction term has exactly the form of a vector field linearly coupled to the conserved current $J_\mu(x)$. The complete action then is

$$S_{\mathrm{F}}\left(A_\mu,\bar{\psi},\psi\right) = S\left(A_\mu\right) - \int \mathrm{d}^d x\, \bar{\psi}(x)\left(\slashed{\partial}+M+ie\slashed{A}\right)\psi(x). \tag{18.26}$$

The parameter e is the current–vector field coupling constant.

The transformations of Section 18.1, which have led to the propagator (18.21), rely on a change of variables in the functional integral of the form (18.16):

$$A_\mu(x) = -\frac{1}{e}\partial_\mu\Lambda(x) + A'_\mu(x). \tag{18.27}$$

We verify that the induced variation of $S_{\mathrm{F}}(A_\mu,\bar{\psi},\psi)$ can be cancelled by a change of the fermion variables of the form of the transformation (18.24) with a space dependent function $\Lambda(x)$. Therefore the algebraic transformations which allow to pass from a unitary non-renormalizable action to an action non-unitary but renormalizable by power counting are justified.

This result is valid for the action. Eventually we shall have to discuss the problem of correlation functions. However before we want to examine the special properties of the massless vector field theory.

Higher spins One may wonder why the strategy which has led to a renormalizable theory of vector particles does not work for higher spins. Let us take the example of the symmetric traceless rank 2 tensor. It must be coupled to a conserved current which is also a rank 2 tensor. Only the energy-momentum tensor $T_{\mu\nu}$ has the required property (see Section A13.3). But $T_{\mu\nu}$ has at least dimension four for $d = 4$ and once coupled to a field of dimension at least one generates a non-renormalizable interaction. A similar argument applies to a spin 3/2 field coupled to the supersymmetry current.

18.3 Massless Vector Field: Abelian Gauge Symmetry

The action (18.26) in the massless vector field limit $m = 0$ has a remarkable property: it has, at the classical level, a *local* (i.e. with a space-dependent parameter) $U(1)$ symmetry, called also $U(1)$ *gauge symmetry*. The vector field then is called a *gauge field*. This symmetry has a geometrical interpretation which we discuss now.

18.3.1 Gauge symmetry

The invariance of the fermion part of the action can be seen to be a consequence of the replacement of the derivative ∂_μ of the free fermion theory by the *covariant derivative* $\partial_\mu + ieA_\mu$, which allows the transformation of the gauge field to cancel the term coming from the derivative. We are thus reminded of the concepts of covariant derivative, affine connection, curvature and parallel transport introduced for Riemannian manifolds (see Chapter 22). In particular we note the similarity with the rotations of the local frame considered in Section 22.6. The difference is that the vectors which here are parallel-transported do not belong to the space tangential to the manifold (which is flat), but are vectors for Lie group transformations.

Let us now indicate the correspondence:

(i) $\psi(x)$ and $\bar\psi(x)$ are vectors for $U(1)$ transformations.

(ii) $A_\mu(x)$ is the affine connection. The connection is expected to have three indices, one, here called μ, which refers to the manifold, and two which refer to the group space. Since the group $U(1)$ corresponds to multiplication by complex numbers, they are here omitted.

(iii) The covariant derivative is D_μ:

$$D_\mu = \partial_\mu + ieA_\mu\,. \tag{18.28}$$

(iv) It follows that the curvature tensor is $ieF_{\mu\nu}$:

$$ieF_{\mu\nu} = [D_\mu, D_\nu] = ie\,(\partial_\mu A_\nu - \partial_\nu A_\mu)\,. \tag{18.29}$$

Finally since the group is abelian, it is easy to write down explicitly the parallel transporter $U(C)$, which is an element of the $U(1)$ group, in terms of a line integral:

$$U(C) = \exp\left[-ie\oint_C A_\mu(s)\mathrm{d}s_\mu\right]\,. \tag{18.30}$$

One immediately verifies that the transformation of $U(C)$ induced by a gauge transformation (18.27) of A_μ, has indeed the expected form:

$$U(C) = U'(C) \exp\left[i \oint_C \partial_\mu \Lambda\left(s\right) \mathrm{d}s_\mu\right] = U'(C) \exp\left[i\big(\Lambda(y) - \Lambda(x)\big)\right], \qquad (18.31)$$

where x and y are the end-points of C. By trying to build a renormalizable theory for a vector in 4 dimensions we have been naturally led to introduce a new geometrical structure, an abelian gauge theory, which is just the quantum version of Maxwell's electromagnetism.

Quantization of charge. If we introduce additional charged fields, we have to assign them charges which characterize their transformation properties under $U(1)$. A delicate question here arises. Since the $U(1)$ group has the same Lie algebra as the group of translations, properties depending only on infinitesimal group transformations do not require a *quantization* of charge. In particular in perturbation theory WT identities will be true even if the charges are not rationally related and the necessity of a quantization of charges will therefore never appear. The conventional wisdom is that QED does not imply charge quantization. Note however that the only known non-perturbative regularization, based on a lattice approximation, involves group elements in the form of parallel transporters and therefore requires charge quantization (see Section 18.5).

The question remains open.

Charged scalar fields. With the geometrical ideas of parallel transport in mind, it is easy to construct an gauge action for a charged scalar field. We start from a $U(1)$ invariant action $S_\mathrm{B}(\phi)$:

$$S_\mathrm{B}(\phi) = \int \mathrm{d}^d x \left[\left|\partial_\mu \phi\right|^2 + V\left(|\phi(x)|^2\right)\right], \qquad (18.32)$$

in which the field $\phi(x)$ is complex and replace the derivative ∂_μ by a covariant derivative. The explicit form of the covariant derivative depends on the charge we assign to the field $\phi(x)$. If we assume that ϕ couples to A_μ with a coupling constant e_B we find

$$S_\mathrm{B}\left(\phi, A_\mu\right) = \int \mathrm{d}^d x \left[\left|\mathrm{D}_\mu \phi\right|^2 + V\left(|\phi|^2\right)\right], \qquad (18.33)$$

where D_μ is now

$$\mathrm{D}_\mu = \partial_\mu + i e_\mathrm{B} A_\mu\,.$$

Note that power counting allows three independent interaction terms, two linear and one quadratic in the vector field. The gauge symmetry relates them.

18.3.2 The massless vector field limit

Geometrical and physical considerations single out the theory with a massless vector field. It is the only one which is exactly gauge invariant and it describes the physics of QED because the photon is, at least to a very good approximation, massless. We could therefore have restricted our discussion to the massless case, as we shall do in the non-abelian case. However, considering the massless theory as a limit of the massive theory provides us with a simple resolution to several difficulties.

First we have already seen in Section 18.1 that we could not write at once a propagator for a massless vector field. In an interacting theory this difficulty persists, and if the theory is gauge invariant it is not only a disease of perturbation theory. Indeed the gauge symmetry implies that the action does not depend on one of the dynamical variables which is related to a gauge transformation. In particular the functional integral is ill-defined because it is infinite by a factor which is the volume of the gauge group (this statement has a precise meaning only in the framework of lattice regularization).

This difficulty reflects the property that in electrodynamics only the electromagnetic tensor $F_{\mu\nu}$ is physical, and the vector field A_μ has been introduced as a mathematical device that enable us to write a local covariant action which generates the classical equations of motion. The vector field A_μ contains redundant degrees of freedom and is not uniquely determined by the equations of motion. We show in Section 18.4 how the gauge field action can be quantized starting from first principles. The procedure is less straightforward and leads to non-covariant gauges, problems whose analysis we wanted to postpone.

By giving a mass to the gauge field A_μ, we make all field components dynamical. The symmetry properties of the action allow the algebraic manipulations indicated in Section 18.1 (provided the action is used to calculate only gauge invariant observables, see Section 18.7). We can eventually take the zero mass limit. As we have shown, in the process, the gauge-dependent part of the gauge field has acquired a dynamics: we have "fixed" the gauge.

Finally the mass of the vector field provides the theory with a natural IR cut-off which somewhat simplifies the analysis. For this reason in what follows we mostly work with the massive theory, the gauge invariant theory appearing as a limiting case.

18.3.3 Physical observables and IR problems

In the limit in which the mass of the vector field goes to zero, the gauge invariant limit, correlation functions calculated in Feynman's gauge have a finite limit. In Section 18.7 we derive a relation (equation (18.94)) which exhibits the gauge dependence of the bare 2-point function. The corresponding relations for other correlation functions can be obtained by expanding expression (18.93). They imply that bare correlation functions then have a limit in all gauges. The renormalized correlation functions will then also have a limit provided one has been careful not to choose IR divergent renormalization constants (as it is usual in massless theories).

The important question is however to understand whether physical, i.e. gauge independent observables are IR finite. The averages of gauge invariant operators have certainly limits. On the contrary we do not expect scattering amplitudes to be IR finite. Actually we have shown in Section 2.7, using the eikonal approximation, that even in quantum mechanics IR divergences appear in the phase of the scattering amplitude as a consequence of the long distance behaviour of the Coulomb force. These divergences survive in the relativistic theory. However it can be shown that additional relativistic IR divergences appear which only cancel if one adds to the scattering amplitude, the amplitude for producing any number of additional low momentum gauge fields. This is physically acceptable because gauge fields which have momenta smaller than the uncertainty in the momenta measurements cannot be detected. Since this question is somewhat technical, we refer to the literature where detailed discussions of this question can be found.

18.4 Canonical Quantization and Gauge Invariance

Although we have been able to construct a gauge invariant theory as a limit of a theory of a massive vector field coupled to a conserved current it is useful to contemplate the difficulties one encounters when one tries to quantize a gauge theory starting from first principles. Moreover, as we shall show in Appendix A19, in the case of non-abelian gauge symmetries the massless limit is not continuous. We therefore show how, starting directly from the classical field equations of a gauge invariant theory, it is possible to recover the functional integral representation of the generating functional of correlation functions.

The problem can be solved by several different strategies and we shall present two of them, corresponding to so-called *Coulomb's gauge* and *temporal gauge*. We again consider the simple action

$$S\left(A_\mu, J_\mu\right) = S\left(A_\mu\right) - \int \mathrm{d}^d x \, J_\mu(x) A_\mu(x), \tag{18.34}$$

$$S\left(A_\mu\right) = \frac{1}{4} \int \mathrm{d}^d x \, F_{\mu\nu}^2(x), \tag{18.35}$$

of a gauge field coupled to a conserved current.

18.4.1 Coulomb's gauge

We first proceed as in the massive case and eliminate the field time component from the action. Taking into account current conservation (note $J_0 = iJ_d$), we then obtain the integral of a reduced lagrangian density:

$$\mathcal{L}\left(A_i\right) = \frac{1}{2}\dot{A}_i(t,x)\left(\delta_{ij} - \frac{\partial_i\partial_j}{\partial_\perp^2}\right)\dot{A}_j(t,x) - \frac{1}{4}F_{ij}^2(t,x) + J_i(t,x)\left(\delta_{ij} - \frac{\partial_i\partial_j}{\partial_\perp^2}\right)A_j(t,x).$$

In contrast with the massive case the action depends only on $\left(\delta_{ij} - \partial_i\partial_j/\partial_\perp^2\right)A_j(t,x)$. After Fourier transformation this implies that the action does not depend on the component of $\tilde{A}_i(t,\hat{k})$ along $\hat{k}$, the space component of the momentum **k**. We recover the well-known property that a massless vector field has only $d-2$ physical components. We thus expand the vector $\tilde{A}_i(t,\hat{k})$ on a transverse basis $e_i^a(\hat{k})$, calling $\tilde{B}_a$ the corresponding $d-2$ components:

$$\hat{k}\cdot e^a(\hat{k}) = 0, \quad e^a(\hat{k})\cdot e^b(\hat{k}) = \delta_{ab}, \quad \tilde{\mathbf{A}}(t,\hat{k}) = \frac{\hat{k}\cdot\tilde{\mathbf{A}}}{\hat{k}^2}\hat{k} + \sum_{a=1}^{d-2}e^a(\hat{k})\tilde{B}_a,$$

and J_a the corresponding sources. The lagrangian density in these variables becomes

$$\mathcal{L}\left(B_a\right) = \sum_{a=1}^{d-2}\left[\partial_\mu B_a(t,x)\partial^\mu B_a(t,x) + J_a(t,x)B_a(t,x)\right].$$

The quantization is now straightforward. We eventually obtain a functional integral over the fields B_a. The corresponding action, once expressed in terms of the initial current J_μ, is however non-local. We can reintroduce the components A_i of the gauge field provided we multiply the integrand by $\delta(\partial_i A_i)$. We finally return to an integral over the time component, as in the massive case. The final result is the euclidean generating functional in *Coulomb's gauge*

$$Z_{\text{Coul.}}(J) = \int [\mathrm{d}A_\mu(x)\delta\,(\partial_i A_i(x))] \exp\left[-S\,(A) + \int \mathrm{d}^d x\, J_\mu(x)A_\mu(x)\right]. \quad (18.36)$$

Coulomb's gauge, in the abelian case, has a nice physical interpretation: only physical degrees of freedom propagate, but it leads to non-covariant calculations, and this is a serious drawback. In particular let us write the gauge field propagator ($\mathbf{k} \equiv \{k_d, \mathbf{k}_\perp\}$)

$$\begin{cases} W_{dd}^{(2)}(k) = \dfrac{1}{\mathbf{k}_\perp^2}, \\[2mm] W_{id}^{(2)}(k) = 0, \\[2mm] W_{ij}^{(2)}(k) = \dfrac{1}{\mathbf{k}^2}\left(\delta_{ij} - \dfrac{k_i k_j}{\mathbf{k}_\perp^2}\right). \end{cases} \quad (18.37)$$

We see that the time component does not decrease in the large time direction. Therefore, with this propagator the theory is not explicitly renormalizable by power counting.

We still have to prove that this gauge is equivalent to the covariant gauges introduced in Section 18.1, but let us postpone this point and before discuss another quantization scheme.

18.4.2 The temporal gauge

In the non-abelian case the quantization in Coulomb's gauge is a complicated problem and therefore we now explain another, more easily generalizable, method.

Let us write the field equations in real time t which correspond to the action (18.34):

$$\partial_\mu F^{\mu\nu}(t,x) = J^\nu(t,x), \quad x \in \mathbb{R}^{d-1}. \quad (18.38)$$

The extension of the arguments which follow to a theory containing matter fields coupled in a gauge invariant manner is straightforward.

The method relies on the observation that the gauge transformed of any solution of equation (18.38) is again a solution. We thus can describe the set of all solutions by restricting the gauge field to a gauge section, considering for example only the solutions satisfying:

$$A_0(t,x) = 0, \quad (18.39)$$

in which A_0 is the time component of the field A_μ. Let us then rewrite equation (18.38), separating time and space components, and taking into account the condition (18.39):

$$\partial_j \dot{A}_j(t,x) = J_0(t,x), \quad (18.40)$$
$$\ddot{A}_i(t,x) - \partial_j F_{ji}(t,x) = J_i(t,x), \quad (18.41)$$

in which the indices i and j run only from 1 to $d-1$. The equation (18.41) is simply the field equation that can be derived from the classical lagrangian density $\mathcal{L}$:

$$\mathcal{L}(A_i) = \tfrac{1}{2}\dot{A}_i^2 - \tfrac{1}{4}F_{ij}^2 + J_i A_i. \quad (18.42)$$

The conjugated momentum $E_i(t,x)$ of the field $A_i(t,x)$ is the electric field:

$$E_i(t,x) = \dot{A}_i(t,x). \quad (18.43)$$

The expression of the hamiltonian density follows:

$$\mathcal{H}\big(E(x), A(x)\big) = \tfrac{1}{2}E_i^2(x) + \tfrac{1}{4}F_{ij}^2(x) - J_i(x)A_i(x). \tag{18.44}$$

The partition function $Z(J_i)$ then is:

$$Z(J_i) = \int [\mathrm{d}A_i] \exp\left[-S\left(A_i\right) + \int \mathrm{d}^d x\, J_i(x)A_i(x)\right], \tag{18.45}$$

in which the euclidean action $S(A_i)$ is the covariant action (18.35), in which $A_d = 0$ has been set.

We still have to implement the constraint (18.40) which is Gauss's law. After quantization, it becomes a constraint on the physically acceptable states $\Psi(\mathbf{A})$. Since the conjugated momenta E_i are represented by differential operators $-i\delta/\delta A_i$, the condition (18.40) then takes the form:

$$\frac{1}{i}\partial_i \frac{\delta}{\delta A_i(x)} \Psi(\mathbf{A}) = J_0(t,x)\Psi(\mathbf{A})\,. \tag{18.46}$$

We recognize in the l.h.s. of equation (18.46) the generator of time-independent gauge transformations of the field $\mathbf{A}(x)$ acting on Ψ. In the absence of an external source ($J_0(t,x) = 0$), the physical states must be gauge invariant. This condition is consistent with quantum mechanical evolution because in the gauge (18.39) the theory has still an invariance under time-independent gauge transformations.

For a general external source, the condition (18.46) gives us the state transformation law. Consistency with quantum mechanical evolution then requires the commutation of the operator $\partial_i E_i - J_0$ with the hamiltonian. A short calculation shows that this commutation is implied by the current conservation (see exercise 18.1).

Let us finally exhibit, for later purpose, a state satisfying the condition (18.46) in the case of two opposite static charges:

$$J_0(t,x) = e\left[\delta(x - x_2) - \delta(x - x_1)\right]\,, \quad J_i(t,x) = 0\,. \tag{18.47}$$

We consider the state:

$$\Psi(\mathbf{A}) = \exp\left[-ie \oint_C A_i(s)\mathrm{d}s_i\right], \tag{18.48}$$

in which C is an arbitrary path joining x_1 to x_2. We indeed verify:

$$\frac{1}{i}\partial_i \frac{\delta}{\delta A_i(x)}\Psi(\mathbf{A}) = \frac{1}{i}\frac{\delta}{\delta\Lambda(x)}\ \Psi(\mathbf{A} - \nabla_x\Lambda)|_{\Lambda=0}$$
$$= e\left[\delta(x - x_2) - \delta(x - x_1)\right]\Psi(\mathbf{A}),$$

a result that is consistent with equations (18.46,18.47).

Note that the representation (18.48) has the form of a parallel transporter corresponding to time-independent gauge transformations. This representation, as well as its non-abelian generalization, will be useful in Chapter 33, in the discussion of the confinement problem.

The propagator in the temporal gauge. Let us calculate the gauge field propagator in the temporal gauge (in euclidean form):

$$W_{ij}^{(2)} = \frac{1}{k^2}\left(\delta_{ij} - \frac{k_i k_j}{\mathbf{k}_\perp^2}\right) + \frac{1}{k_d^2}\frac{k_i k_j}{\mathbf{k}_\perp^2}, \tag{18.49}$$

in which $\mathbf{k}_\perp$ is the "space" part of $\mathbf{k}$, i.e. its projection on $\mathbb{R}^{d-1}$. This propagator, as in the case of Coulomb's gauge, has a large momentum behaviour which is not uniform and thus, in contrast with the covariant gauges, leads to theories which are not explicitly renormalizable in four dimensions.

18.4.3 Equivalence with covariant quantization

We have obtained different functional integral representations of the same theory. We would like to show that, at least for gauge invariant observables, they are formally equivalent to the $O(d)$ covariant representations we have discussed in Section 18.1. We here show the equivalence between the temporal gauge and a class of gauges characterized by a gauge condition of the form

$$n_\mu\left(\partial\right) A_\mu(x) = h(x). \tag{18.50}$$

The vector n_μ is a constant or a differential operator. We shall eventually either set the external field $h(x)$ to zero to enforce the strict gauge condition $n_\mu\left(\partial\right) A_\mu(x) = 0$ or integrate over it with a gaussian weight in order to generate actions of the form (18.19). This covers all the examples met so far. The arguments easily generalize to other gauges.

First let us rewrite the expression (18.45) as an integral over a d-component vector field:

$$Z(J) = \int [\mathrm{d}A_\mu] \prod_x \delta\left(A_d(x)\right) \exp\left\{ -\int \mathrm{d}^d x \left[\tfrac{1}{4} F_{\mu\nu}^2(x) - J_\mu(x) A_\mu(x) \right] \right\}. \tag{18.51}$$

We then use the identity:

$$\int [\mathrm{d}\Lambda(x)] \prod_x \delta\left[n_\mu(\partial)\left(\partial_\mu\Lambda(x) + A_\mu(x)\right) - h(x) \right] = \text{const.}, \tag{18.52}$$

inside expression (18.51):

$$Z \propto \int [\mathrm{d}A_\mu \mathrm{d}\Lambda] \prod_x \delta\left(A_d\right) \delta\left[n_\mu(\partial)\left(\partial_\mu\Lambda(x) + A_\mu(x)\right) - h(x) \right]$$
$$\times \exp\left[-S\left(A\right) + \int \mathrm{d}^d x\, J_\mu(x) A_\mu(x) \right]. \tag{18.53}$$

We then perform a change of variables of the form of a gauge transformation:

$$A_\mu(x) = A'_\mu(x) - \partial_\mu\Lambda(x).$$

Since we have assumed that the current J_μ is conserved, only the δ-functions are modified:

$$\delta\left(A_d\right) \delta\left[n_\mu\left(\partial\right)\left(\partial_\mu\Lambda(x) + A_\mu(x)\right) - h(x) \right] \mapsto \delta\left(A_d - \dot\Lambda\right) \delta\left[n_\mu\left(\partial\right) A_\mu(x) - h(x) \right].$$

The integration over Λ can again be performed:

$$\int [\mathrm{d}\Lambda] \prod_x \delta\left(A_d(x) - \dot\Lambda(x)\right) = \text{const.}, \tag{18.54}$$

and therefore:

$$Z(J_\mu) = \int [\mathrm{d}A_\mu] \,\delta\left[n_\mu\left(\partial\right) A_\mu(x) - h(x) \right] \exp\left[-S\left(A\right) + \int \mathrm{d}^d x\, J_\mu(x) A_\mu(x) \right]. \tag{18.55}$$

Since the result by construction does not depend on $h(x)$ we can either set $h(x)$ to zero or integrate over $h(x)$ with for example the gaussian measure $d\rho(h)$:

$$[d\rho(h)] = [dh]\exp\left[-\frac{1}{2}\int d^dx\, h^2(x)\right].\qquad(18.56)$$

We then obtain

$$Z(J_\mu) = \int[dA_\mu]\exp\left[-S_{\text{gauge}}(A) + \int d^dx\, J_\mu(x)A_\mu(x)\right].\qquad(18.57)$$

with

$$S_{\text{gauge}}(A) = S(A) + \frac{1}{2}\int d^dx\,[n_\mu(\partial)A_\mu(x)]^2.\qquad(18.58)$$

Specializing to $n_\mu = \xi^{-1/2}\partial_\mu$ we see that we have in particular demonstrated the equivalence between the *temporal gauge* $A_d = 0$ and the covariant gauges (18.19). If instead we choose $n_d = 0$ and $\mathbf{n}_\perp \equiv \partial_\perp$, and set $h = 0$ we find Coulomb's gauge.

The propagator. To the action (18.58) corresponds the gauge field propagator:

$$W^{(2)}_{\mu\nu}(k) = \frac{1}{k^2}\left[\delta_{\mu\nu} - \frac{(k_\mu n_\nu + k_\nu n_\mu)}{k\cdot n} + \frac{(k^2+n^2)k_\mu k_\nu}{(k\cdot n)^2}\right].\qquad(18.59)$$

Remark. The strict gauge condition is recovered in the limit $|n_\mu| \to \infty$, which exists for the propagator but not for the action. To write explicitly the limiting action one has to introduce a Lagrange multiplier $l(x)$ which imposes the gauge condition:

$$S_{\text{gauge}}(A) = S(A) + \int d^dx\, l(x)n_\mu(\partial)A_\mu(x).\qquad(18.60)$$

18.4.4 Interpretation: the Faddeev–Popov quantization

The result we have obtained has the following interpretation which can be only justified in the lattice approximation on a finite lattice (see Section 18.5). The problem of the gauge invariant theory is that locality requires an action with redundant degrees of freedom or equivalently that the local gauge invariant action does not provide a dynamics to the gauge degrees of freedom. We therefore supply them with a stochastic dynamics in the sense of Chapter 16. Let us write the gauge field A_μ in terms of a gauge field B_μ projection of A_μ on some gauge section, i.e. satisfying some gauge condition, and a gauge transformation Λ:

$$A_\mu = B_\mu + \partial_\mu\Lambda.\qquad(18.61)$$

We assume that this decomposition is unique. Gauge invariance tells us that the gauge action depends only on B_μ and specifies its dynamics.

To $\Lambda(x)$ we impose for example:

$$\partial^2\Lambda(x) + \partial_\mu B_\mu(x) = h(x),\qquad(18.62)$$

in which $h(x)$ is a stochastic field for which a probability distribution is given.

We now impose this equation in the functional integral. Normally we expect to get the determinant of the functional derivative of equation (18.62) with respect to the field $\Lambda(x)$. However here this operator is just ∂^2 and therefore the determinant is a constant which disappears in the normalization of the functional integral.

The functional integral in the presence of sources for gauge invariant operators (polynomials in the fields that are invariant in gauge transformations) only becomes:

$$Z = \int [\mathrm{d}B_\mu \mathrm{d}\Lambda] \, \delta \left[\partial^2 \Lambda(x) + \partial_\mu B_\mu(x) - h(x) \right] \exp \left[-S(B) \right]. \qquad (18.63)$$

The functional measure $[\mathrm{d}B_\mu \mathrm{d}\Lambda]$ is the decomposition of the flat measure $[\mathrm{d}A_\mu]$ into a product of measures on B_μ and Λ. The action $S(B)$ is the gauge invariant action $S(A)$ in which equation (18.61) has been used. We now recognize that we can rewrite the whole expression in terms of A_μ:

$$Z = \int [\mathrm{d}A_\mu] \, \delta \left[\partial_\mu A_\mu(x) - h(x) \right] \exp \left[-S(A) \right]. \qquad (18.64)$$

Moreover since the result of the functional integration does not depend on the dynamics of $\Lambda(x)$, the result does not depend on the field $h(x)$ either and we can integrate over $h(x)$ with for example the gaussian measure $\mathrm{d}\rho(h)$.

18.5 Perturbation Theory, Regularization

From now on we will only consider the covariant gauges of Section 18.1. With the propagator (18.21) power counting is the same as for a scalar field. We can therefore construct interacting theories renormalizable for dimensions $d \leq 4$. We shall specially consider the dimension 4. Since WT identities play an essential role in gauge theories, we have first to find gauge invariant regularizations.

Dimensional regularization. We have defined dimensional regularization in Section 9.3. This regularization is well suited to perturbative calculations in QED. Examples will be given in Section 18.9. It is inapplicable only in the case of gauge theories with chiral fermions, due to the γ_S problem.

18.5.1 Pauli–Villars's regularization

In this chapter the special problems generated by chiral fermions will not be met, and we will be perfectly happy with dimensional regularization. However for later purposes it is an instructive exercise to try to use Pauli–Villars's regularization.

Because $F_{\mu\nu}$ is gauge invariant and the action for the scalar $\chi(x)$ of equation (18.14) is arbitrary, we can arbitrarily modify the gauge field propagator and therefore to make it as convergent as we want. However the situation is quite different for charged matter fields. Since we are only allowed to use the covariant derivative, we get for example for the fermion action regularized expressions of the form:

$$S_1 \left(\bar{\psi}, \psi, A_\mu \right) = \int \mathrm{d}^d x \, \bar{\psi}(x) \left[M + \slashed{D} \left(1 + \alpha_1 \frac{\slashed{D}^2}{\Lambda^2} + \cdots + \alpha_s \frac{\slashed{D}^{2s}}{\Lambda^{2s}} \right) \right] \psi(x). \qquad (18.65)$$

Fig. 18.1   One-loop contribution to the A_μ 2-point function.

Note this method of regularization, unlike dimensional regularization, also applies to massless or chiral fermions.

In momentum space the propagator now decreases indeed as k^{-2s-1}. However, by trying to improve the behaviour of the propagator, we have at the same time introduced more singular interactions. Expanding $\slashed{D}^{2s+1}$ in powers of A_μ we obtain in particular a term proportional to $\slashed{\partial}^{2s}\slashed{A}$. Therefore the action contains a $\bar\psi\slashed{A}\psi$ vertex with $2s$ derivatives. If we now consider a one-loop diagram with only fermions in the loop and $\bar\psi\psi\slashed{A}$ vertices of this kind (figure 18.1), we see immediately that the power counting is independent of s because there are exactly as many fermion propagators as vertices and therefore the additional powers of momentum in the numerator cancel the additional powers in the denominator.

This effect is however special to one-loop matter diagrams. A several-loop diagram is formed by fermion loops joined by gauge field lines (figure 18.2).

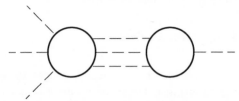

Fig. 18.2   Example of a several loop diagram.

Since the gauge field propagator can be made arbitrarily convergent, several loop diagrams can be made superficially convergent. In the case of scalar matter diagrams, scalar self-interaction vertices can be added, but then the number of matter propagators exceeds the number of gauge field vertices and again the diagrams can be made superficially convergent.

We conclude therefore that the only remaining divergent diagrams are one-loop matter field diagrams which can be generated by the determinant coming from the gaussian integration over matter field in an external gauge field. This analysis signals a difficulty in constructing in general a regularized gauge invariant expression for the determinant of operators of the form $\slashed{\partial} + m + ie\slashed{A}$ or D_μ^2 in the continuum and at fixed dimensions (see Section 20.3).

In the case of massive fermions the one-loop diagrams can finally be regularized by adding to the action a set of boson (with spin) and fermion regulator fields with suitable masses (but of the order of the cut-off):

$$\bar\psi\left(M + \slashed{D}\right)\psi \mapsto \sum_{i=1}^{r} \bar\psi_i\left(M_{\mathrm{F},i} + \slashed{D}\right)\psi_i + \varphi_i^*\left(M_{\mathrm{B},i} + \slashed{D}\right)\varphi_i\,.$$

Subsection 31.2.3 will provide an explicit example.

Let us, however, here discuss in some detail Schwinger's proper time regularization scheme (see Appendix A9):

Let us assume that the dimension of space is even. Then:

$$\det\left(\slashed{\partial} + M + ie\slashed{A}\right) = \det\gamma_S\left(\slashed{\partial} + M + ie\slashed{A}\right)\gamma_S = \det\left(-\slashed{\partial} + M - ie\slashed{A}\right).$$

It follows that we can write, at least for e small enough:

$$\det\left(\slashed{\partial} + M + ie\slashed{A}\right) = \left[\det\left(M^2 - \slashed{D}^2\right)\right]^{1/2}. \tag{18.66}$$

It is easier to deal with the operator $M^2 - \slashed{D}^2$ than $\slashed{D} + M$ because it is positive. Using the identity:

$$\gamma_\mu\gamma_\nu = \delta_{\mu\nu} + i\sigma_{\mu\nu},$$

in which the quantities $\sigma_{\mu\nu}$ have been defined by equation (A5.9), we can rewrite the operator in the r.h.s.:

$$M^2 - \slashed{D}^2 = M^2 - D_\mu^2 - \frac{i}{2}\sigma_{\mu\nu}\left[D_\mu, D_\nu\right] = M^2 - D_\mu^2 + \frac{e}{2}\sigma_{\mu\nu}F_{\mu\nu}, \tag{18.67}$$

which shows that it differs from the scalar operator only by a spin term proportional to $\sigma_{\mu\nu}$.

The regularized form of the determinant is then given by:

$$\ln\det\left[\left(\slashed{D} + M\right)\left(\slashed{\partial} + M\right)^{-1}\right]\Big|_{\text{reg.}} = \frac{1}{2}\operatorname{tr}\int_\varepsilon^\infty \frac{dt}{t}\left[V\left(t\right) - V_0\left(t\right)\right], \tag{18.68}$$

with the definitions:

$$V\left(t\right) = \exp\left[-t\left(M^2 - D_\mu^2 - \frac{e}{2}\sigma_{\mu\nu}F_{\mu\nu}\right)\right], \tag{18.69}$$

$$V_0\left(t\right) = \exp\left[-t\left(M^2 - \partial^2\right)\right]. \tag{18.70}$$

In both the boson and fermion cases, in a gauge transformation of the form:

$$A_\mu \mapsto A_\mu + \partial_\mu\Lambda, \tag{18.71}$$

the matrix elements $\langle y\left|V\left(t\right)\right|x\rangle$ become:

$$\langle y\left|V\right|x\rangle \mapsto e^{-ie\Lambda(y)}\langle y\left|V\right|x\rangle e^{ie\Lambda(x)}. \tag{18.72}$$

The trace of the integrand is therefore by itself gauge invariant.

The general strategy in the case of Pauli–Villars's regularization then is the following: we first regularize the one-loop diagrams and show that we can renormalize them by gauge invariant counterterms, we then insert the renormalized diagrams in the general diagrams regularized by Pauli–Villars's method.

In the case of the scalar determinant in which the spin term is absent, equation (3.19) leads directly to a path integral representation of $V(t)$. It then shows that the matrix elements $\langle y\left|V\left(t\right)\right|x\rangle$ are gaussian averages of the parallel transporters transporting from x to y. In the spin case, one would have to introduce additional fermionic degrees of freedom because the corresponding hamiltonian has a matrix form.

18.5.2 Lattice regularization

To understand how to construct a lattice regularization of a gauge theory, it is essential to keep in mind the geometrical interpretation of the gauge field as a connection. Since on the lattice points are split, the gauge field has to be replaced by link variables corresponding to parallel transport along links of the lattice (see Chapters 22, 33).

Gauge invariant interaction terms on the lattice then have the form:

$$\bar{\psi}\left(x + an_\mu\right)\gamma_\mu U\left(x + an_\mu, x\right)\psi(x), \tag{18.73}$$

in which U is given by equation (18.30), the curve C being the link joining the points x to $x + an_\mu$ on the lattice. We have called n_μ the unit vector in μ direction and a the lattice spacing.

The link variable $U_{i,j}$, linking site i to j is an element of the $U(1)$ group, which can therefore be parametrized in terms of an angle θ_{ij}, and is such that:

$$U_{i,j} \equiv \mathrm{e}^{i\theta_{ij}} = \left(U_{j,i}\right)^{-1}. \tag{18.74}$$

We show in Section 22.4 that the curvature tensor is associated with parallel transport around a closed curve. This suggests that we should take as a regularized form of $\int \mathrm{d}x\, F_{\mu\nu}^2$ the product of link variables on a closed curve on the lattice, the simplest being a square on a hypercubic lattice. Such a product is clearly gauge invariant. We then obtain the well-known *plaquette action*, each square forming a plaquette (for details see Chapter 33):

$$\sum_{\text{all plaquettes}} U\left(i,j\right)U\left(j,k\right)U\left(k,l\right)U\left(l,i\right). \tag{18.75}$$

We have denoted symbolically by i, j, k, l four sites forming a square on the lattice.

The typical gauge invariant lattice action corresponding to the continuum action of a gauge field coupled to fermions then has the form:

$$S\left(U, \bar{\psi}, \psi\right) = \beta \sum_{\text{plaquettes}} U_{ij}U_{jk}U_{kl}U_{li} + \kappa \sum_{\text{links}} \bar{\psi}_i \gamma_{ij} U_{ij} \psi_j + \sum_{\text{sites}} M\bar{\psi}_i\psi_i. \tag{18.76}$$

We have denoted by $i, j,\ldots$ the lattice sites, β and κ are the coupling constants. The action (18.76) is invariant under independent $U(1)$ transformations on each lattice site. These transformations are the lattice equivalents of the gauge transformations of the continuum theory. The measure of integration over the gauge variables is the group invariant measure, i.e. the flat measure $\mathrm{d}\theta_{ij}$. Note that on the lattice and in a finite volume the gauge invariant action leads to a well-defined partition function because the $U(1)$ group is compact: the volume of the gauge group is $(2\pi)^\Omega$ if Ω is the number of lattice sites. However in the continuum limit the compact character of the group is lost. It is therefore necessary to fix the gauge on the lattice in order to be able to construct a regularized perturbation theory. Since we shall devote Chapter 33 to lattice gauge theories, we postpone the discussion of this problem. Finally it is possible to add to the gauge invariant lattice action a term of the form $\sum \cos \theta_{ij}$ to give a mass to the vector field.

18.6 WT Identities, Renormalization

In this section, for convenience, we assume dimensional regularization. In gauge theories WT identities play an essential role because it is necessary to maintain the gauge symmetry in order to ensure that the theory which is not explicitly unitary, is equivalent to a unitary theory, at least for the "physical" observables, i.e. gauge invariant observables (including S-matrix elements).

Their derivation is rather straightforward. Let us take the example of the action (18.26) and call $J_\mu(x)$, $\eta(x)$ and $\bar{\eta}(x)$ the sources for the fields $A_\mu(x)$, $\bar{\psi}(x)$ and $\psi(x)$ respectively. We make infinitesimal gauge transformations (18.24,18.27) on the action in presence of sources. The terms which are not invariant are the A_μ mass term, the gauge fixing term and the sources:

$$\delta\left[S - \text{source terms}\right] = -\frac{1}{e}\int dx\, \Lambda(x)\left\{\left(\frac{1}{\xi}\partial^2 - m^2\right)\partial_\mu A_\mu(x) + \partial_\mu J_\mu(x)\right.$$

$$\left. + ie\left[\bar{\eta}(x)\psi(x) - \bar{\psi}(x)\eta(x)\right]\right\}. \tag{18.77}$$

This leads, following the usual arguments, to an equation for the generating functionals $Z\left(J, \bar{\eta}, \eta\right)$ and thus $W\left(J, \bar{\eta}, \eta\right)$ which has the form:

$$\left\{\left(m^2 - \frac{1}{\xi}\partial^2\right)\partial_\mu\frac{\delta}{\delta J_\mu(x)} - ie\left[\bar{\eta}(x)\frac{\delta}{\delta\bar{\eta}(x)} - \eta(x)\frac{\delta}{\delta\eta(x)}\right]\right\}W\left(J, \bar{\eta}, \eta\right) = \partial_\mu J_\mu(x). \tag{18.78}$$

This equation is equivalent to a set of identities for correlation functions. Let us first write the equation for the gauge field 2-point function which is special. After Fourier transformation it has the form:

$$k_\mu W^{(2)}_{\mu\nu}(k) = \xi\frac{k_\nu}{k^2 + \xi m^2}. \tag{18.79}$$

Correlation functions with only matter fields satisfy:

$$\left(\frac{k^2}{\xi} + m^2\right)k_\mu W^{(2n+1)}_\mu\left(k; p_1, \ldots, p_n; q_1, \ldots, q_n\right)$$

$$= e\sum_i\left[W^{(2n)}\left(p_1, \ldots, p_i + k, \ldots, p_n; q_1, \ldots, q_n\right)\right.$$

$$\left. - W^{(2n)}\left(p_1, \ldots, p_n; q_1, \ldots, q_i + k, \ldots, q_n\right)\right], \tag{18.80}$$

in which k is the gauge field momentum, and p_i and q_i the momenta of ψ and $\bar{\psi}$ fields respectively. The presence of additional external gauge fields does not modify the identities.

Equation (18.78) is a linear first order partial differential equation. It is therefore straightforward to transform it into an equation for the generating functional of proper vertices $\Gamma\left(A_\mu, \bar{\psi}, \psi\right)$:

$$\left(\frac{1}{\xi}\partial^2 - m^2\right)\partial_\mu A_\mu(x) + \partial_\mu\frac{\delta\Gamma}{\delta A_\mu(x)} + ie\left[\psi(x)\frac{\delta\Gamma}{\delta\psi(x)} - \bar{\psi}(x)\frac{\delta\Gamma}{\delta\bar{\psi}(x)}\right] = 0. \tag{18.81}$$

It can be verified that the equations (18.78) and (18.81) have the same content as the quantum equations of motion of the χ-field of Section 18.1.

Equation (18.81) has a simple interpretation: it tells us that Γ has the form:

$$\Gamma = \Gamma_{\text{sym.}} + \frac{1}{2} \int \left[m^2 A_\mu^2(x) + \frac{1}{\xi} \left(\partial_\mu A_\mu \right)^2 \right] \mathrm{d}x \,, \tag{18.82}$$

in which $\Gamma_{\text{sym.}}$ is gauge invariant.

Renormalization. We perform a loopwise expansion of the functional Γ. Because equation (18.81) is linear, the tree level approximation satisfies the inhomogeneous equation while all higher order terms satisfy the homogeneous equation. Calling Γ_l the l-loop contribution to Γ we have for $l > 0$:

$$\partial_\mu \frac{\delta \Gamma_l}{\delta A_\mu(x)} + ie \left[\psi(x) \frac{\delta \Gamma_l}{\delta \psi(x)} - \bar{\psi}(x) \frac{\delta \Gamma_l}{\delta \bar{\psi}(x)} \right] = 0 \,.$$

Therefore the generating functional Γ_l of l-loop proper vertices is gauge invariant. The singular part of the Laurent expansion in $\varepsilon = 4 - d$ of Γ_l is also gauge invariant, which means that the divergent part Γ_l^{div} is gauge invariant.

The conclusion is that the action can be completely renormalized by adding gauge invariant counterterms. As in the case of the linear symmetry breaking in Section 13.3, one can say that the terms which break the gauge invariance, the gauge field mass term and the gauge fixing term, are not renormalized since they are not modified by counterterms.

The full renormalized action can then be written:

$$S_{\text{r}} \left(A_\mu, \bar{\psi}, \psi \right) = \int \mathrm{d}^d x \left[\tfrac{1}{4} Z_A F_{\mu\nu}^2 + \tfrac{1}{2} m^2 A_\mu^2 + \tfrac{1}{2\xi} \left(\partial_\mu A_\mu \right)^2 \right.$$
$$\left. - Z_\psi \bar{\psi} \left(\slashed{\partial} + M + \delta M + ie\slashed{A} \right) \psi \right], \tag{18.83}$$

where Z_A is the gauge field, Z_ψ and δM are the ψ field and mass renormalization constants. We note that there is no special renormalization constant for the charge e. Indeed if we introduce the bare fields $\psi^0, \bar{\psi}^0, A_\mu^0$,

$$\psi^0 = Z_\psi^{1/2} \psi \,, \quad \bar{\psi}^0 = Z_\psi^{1/2} \bar{\psi} \,, \quad A_\mu^0 = Z_A^{1/2} A_\mu \,, \tag{18.84}$$

we see that we must set

$$e_0 = Z_e^{1/2} e \,, \quad \text{with} \quad Z_A Z_e = 1 \,. \tag{18.85}$$

In other words the transformation law has not been renormalized:

$$\partial_\mu + ie_0 A_\mu^0 = \partial_\mu + ie A_\mu \,. \tag{18.86}$$

Gauge invariance relates the renormalization of the charge and the gauge field.

18.7 Gauge Dependence

To understand the physical properties of the theory, it is necessary to characterize the gauge dependence, here the dependence on the parameter ξ which appears in the gauge fixing term, of correlation functions.

Let us therefore calculate the variation of correlation functions when we add a term proportional $\int (\partial_\mu A_\mu)^2 \, dx$ to the renormalized action.

The fermion 2-point function. We first consider the fermion 2-point function. Using twice the WT identity (18.80) we obtain a relation which, in Fourier space, reads:

$$W^{(2)}_{(\partial_\mu A_\mu)^2}(p) = 2e^2 \int \frac{d^d k}{(2\pi)^d} \frac{\xi^2}{(k^2 + \xi m^2)^2} \left[W^{(2)}(p+k) - W^{(2)}(p) \right]. \tag{18.87}$$

Two remarks are now in order: in perturbation theory only the second term in the r.h.s. is divergent. At the same time only the second term has a pole. Therefore on the fermion mass-shell $(p^2 \to -M^2)$, the only effect of the change of the gauge fixing term is a fermion field renormalization: The physical mass M is gauge independent (a gauge dependence of the mass would have been led to a double pole at $p^2 = -M^2$). From equation (18.87) we deduce the additional renormalization needed when changing ξ:

$$\frac{\partial \ln Z_\psi}{\partial \xi} = -\frac{e^2}{(2\pi)^d} \int \frac{d^d k}{(k^2 + \xi m^2)^2}, \tag{18.88}$$

and therefore integrating over ξ:

$$Z_\psi(\xi) = Z_\psi(0) \exp\left[-\frac{\xi e^2}{(2\pi)^d} \int \frac{d^d k}{k^2 (k^2 + \xi m^2)} \right]. \tag{18.89}$$

General correlation functions. We could use the WT identities in the same way to exhibit the ξ-dependence of all correlation functions. However, since the identities (18.78) legitimate at the perturbative level the set of manipulations of Section 18.1, we shall use the latter to prove a general identity. Let us add to the action $S(A, \bar{\psi}, \psi)$ a new term depending on a scalar field χ:

$$S_\xi(A, \bar{\psi}, \psi) \mapsto S_\xi(A, \bar{\psi}, \psi, \chi) = S_\xi(A, \bar{\psi}, \psi) - \tfrac{1}{2}\xi m^2 \int dx \, \chi^2(x).$$

Integrating over χ, we see that only the, irrelevant, normalization of the partition function has changed. We now shift $\chi(x)$:

$$\chi(x) \mapsto \chi(x) + \frac{1}{\xi m} \partial_\mu A_\mu(x). \tag{18.90}$$

After this change of variables the action reads:

$$S(A, \bar{\psi}, \psi) = S_\xi(A, \bar{\psi}, \psi) - \frac{\xi m^2}{2} \int dx \, \chi^2(x) + m \int A_\mu(x) \partial_\mu \chi(x) - \frac{1}{2\xi} \int dx \, (\partial_\mu A_\mu)^2. \tag{18.91}$$

The last term cancels the gauge fixing term, and the second term in the r.h.s. can be eliminated by a gauge transformation:

$$
\begin{cases}
A_\mu \mapsto A_\mu - \partial_\mu \chi / m\,, \\
\psi \mapsto \psi\, e^{ie\chi/m}, \\
\bar{\psi} \mapsto \bar{\psi}\, e^{-ie\chi/m}\,.
\end{cases}
\tag{18.92}
$$

The functional integral representation of $Z\left(J, \bar{\eta}, \eta\right)$ then takes the form:

$$
Z\left(J, \bar{\eta}, \eta\right) = \int \left[\mathrm{d}A_\mu \mathrm{d}\bar{\psi}\mathrm{d}\psi\right] \exp\left[-S\left(A, \bar{\psi}, \psi\right) - \tfrac{1}{2}m^2 \int \mathrm{d}x\, A_\mu^2\right] \int [\mathrm{d}\chi]
$$

$$
\times \exp\left\{\int \mathrm{d}x \left[\tfrac{1}{2}\left(\partial_\mu \chi\right)^2 + \tfrac{1}{2}\xi m^2 \chi^2 - \chi \partial_\mu J_\mu / m + \bar{\eta}\, e^{ie\chi/m}\, \psi + \bar{\psi}\, e^{-ie\chi/m}\, \eta\right]\right\},
\tag{18.93}
$$

in which $S\left(A, \bar{\psi}, \psi\right)$ is the gauge invariant action. Expanding in powers of η and $\bar{\eta}$, we can integrate over χ. This calculation for example in the case of the $\bar{\psi}\psi$ 2-point function yields the ratio of bare functions corresponding to two different values of ξ:

$$
W_\xi^{(2)}\left(x, y\right) = \exp\left[\frac{\xi e^2}{(2\pi)^d} \int \mathrm{d}^d k \frac{e^{ik(x-y)} - 1}{k^2 \left(k^2 + \xi m^2\right)}\right] W_{(\xi=0)}^{(2)}\left(x, y\right).
\tag{18.94}
$$

Introducing the fermion field renormalization constants, we get the relation between renormalized correlation functions:

$$
W_\xi^{(2)}\left(x, y\right) = Z_\psi^{-1}\left(\xi\right) Z_\psi\left(0\right) \exp\left[\frac{\xi e^2}{(2\pi)^d} \int \mathrm{d}^d k \frac{e^{ik(x-y)} - 1}{k^2 \left(k^2 + \xi m^2\right)}\right] W_{(\xi=0)}^{(2)}\left(x, y\right).
\tag{18.95}
$$

In particular with the choice (18.89) for $Z_\psi\left(\xi\right)$, the relation becomes finite:

$$
W_\xi^{(2)}\left(x, y\right) = \exp\left[\frac{\xi e^2}{(2\pi)^d} \int \frac{\mathrm{d}^d k\; e^{ik(x-y)}}{k^2 \left(k^2 + \xi m^2\right)}\right] W_{(\xi=0)}^{(2)}\left(x, y\right).
\tag{18.96}
$$

Equation (18.96) has additional consequences: no other ξ-dependent renormalization has been needed. It follows that the renormalization constant Z_e and thus, from equation (18.85), also Z_A are gauge independent. In the renormalized action (18.83) only Z_ψ is gauge dependent.

Other ψ-field correlation functions have a similar structure. The ξ dependence, which comes only from the χ integration, factorizes. If we now consider correlation functions with external gauge fields, the previous analysis still applies provided the gauge field is transverse, i.e. the source J_μ satisfies $\partial_\mu J_\mu = 0$: Transverse external gauge fields do not introduce any new gauge dependence.

Note that we can obtain the gauge dependence of the full gauge field correlation functions by integrating over χ after making a translation of a term proportional to $\partial_\mu J_\mu(x)$. However identities (18.78) directly relate correlation functions with external longitudinal gauge fields (in Fourier space $A_\mu(k)$ proportional to k_μ) to fermion field correlation functions.

Unitarity. Since on the mass-shell the propagator of the gauge field is a transverse projector:

$$\left(\delta_{\mu\nu} + (\xi-1)\frac{k_\mu k_\nu}{k^2+\xi m^2}\right)k_\nu = \frac{\xi\left(k^2+m^2\right)}{k^2+\xi m^2}k_\mu = 0 \quad \text{for } k^2+m^2=0\,,$$

we have obtained the gauge dependence of all matrix elements. In the mass-shell limit, the ξ-dependence of the singular part of correlation functions becomes simply a multiplicative renormalization (see Section A18.2). Therefore the properly normalized S-matrix elements are gauge independent. Being gauge independent, they cannot have the ξ-dependent singularities which we have introduced with the χ-field to make the theory renormalizable. The full S-matrix, in the subspace of physical states, is thus unitary. Note that by using WT identities one can also directly prove that the contribution of χ-field cancels in the intermediate states in the unitarity relations.

Gauge invariant operators. We have examined the gauge dependence of S-matrix elements. From the point of view of correlation functions, the only gauge independent quantities are the averages of products of gauge invariant operators, i.e. local polynomials in the field invariant under the transformations (18.24,18.27).

The simplest such operators are $F_{\mu\nu}$ which select the transverse part of the gauge field, $\bar\psi(x)\psi(x)$ or more generally $\bar\psi(x)\Gamma_A\psi(x)$ in which the matrix Γ_A is any element of the algebra of γ matrices. Equation (18.96) shows explicitly the mechanism which makes the correlation functions of ψ gauge dependent while $\bar\psi(x)\psi(x)$ is gauge independent. When in the product $\bar\psi(x)\psi(y)$, y approaches x the additional gauge dependent renormalization needed to make the product $\bar\psi(x)\psi(x)$ finite cancels the gauge dependent part of the fermion field renormalization.

To study the renormalization properties of gauge invariant operators one has to add to the action sources for them. The form of WT identities is not modified. The arguments of Section 18.6 are still valid: the counterterms are gauge invariant. This proves that gauge invariant operators mix under renormalization only with gauge invariant operators of lower or equal canonical dimensions.

Non-gauge invariant correlation functions in the unitary gauge. With the original action (18.7) for the gauge fields, all correlation functions are "physical", but the theory is not renormalizable. However we have been able to construct some correlation functions of the theory, the gauge invariant correlation functions, which have a large cut-off limit. Relation (18.94), in presence of a cut-off, leads in the large ξ limit to an explanation for this surprising property:

$$W_\infty^{(2)}(x,y) = \exp\left[-\frac{e^2}{(2\pi)^d}\int d^d k \frac{1-e^{ik(x-y)}}{m^2 k^2}\right]W_0^{(2)}(x,y)\,. \tag{18.97}$$

For $|x-y|\neq 0$, the dominant term in the large cut-off limit in the exponential is:

$$\frac{e^2}{(2\pi)^d}\int \frac{d^d k}{m^2 k^2} \sim e^2 \frac{S_d}{(2\pi)^d}\frac{\Lambda^{d-2}}{m^2}\,. \tag{18.98}$$

Therefore in the physical representation all non-gauge invariant correlation functions vanish. The explanation is the following: although the mass term breaks gauge invariance, this breaking is not sufficient to prevent fluctuations coming from the gauge degrees of freedom to suppress these correlation functions.

18.8 The RG Equations

In this section we derive RG equations in the case of the action (18.26), which corresponds to massive QED with fermions, displaying the dependence of RG functions on the gauge fixing parameter ξ. Let $\Gamma^{(l,n)}$ be the 1PI correlation functions corresponding to l gauge fields, and n fermion pairs ψ and $\bar\psi$. The relation between bare and renormalized correlation functions is:

$$\Gamma_{\mathrm{B}}^{(l,n)}\left(p_i, q_j; \alpha_0, \xi_0, m_0, M_0\right) = Z_A^{-l/2} Z_\psi^{-n} \Gamma^{(l,n)}\left(p_i, q_j; \mu, \alpha, \xi, m, M\right), \qquad (18.99)$$

in which μ is the renormalization scale, and we have called α the loop expansion parameter:

$$\alpha = e^2/4\pi. \qquad (18.100)$$

Differentiating equation (18.99) with respect to μ at bare parameters fixed, we get the RG equations:

$$\left[\mu\frac{\partial}{\partial\mu} + \beta\left(\alpha\right)\frac{\partial}{\partial\alpha} + \delta\left(\alpha\right)\xi\frac{\partial}{\partial\xi} + \eta_m\left(\alpha\right)m\frac{\partial}{\partial m} + \eta_M\left(\alpha\right)M\frac{\partial}{\partial M}\right.$$
$$\left. -\frac{l}{2}\eta_A\left(\alpha\right) - n\eta_\psi\left(\alpha\right)\right]\Gamma^{(l,n)}\left(p_i, q_j; \mu, \alpha, \xi, m, M\right) = 0. \quad (18.101)$$

Equation (18.85) relates Z_A and Z_e, the gauge field and coupling constant renormalization constants,

$$Z_A Z_e = 1.$$

Therefore:

$$\alpha = Z_A \alpha_0. \qquad (18.102)$$

Moreover we have shown in Section 18.6 that the parameters m and ξ are not renormalized. It follows that:

$$m_0^2 = m^2 Z_A^{-1}, \qquad \xi_0 = \xi Z_A. \qquad (18.103)$$

Finally in Section 18.7, we have shown that the renormalization constant Z_A can be chosen to be independent of ξ (the minimal subtraction scheme satisfies this requirement). Equations (18.102,18.103) then imply a set of relations between RG functions:

$$\beta(\alpha) = \eta_A(\alpha), \qquad (18.104)$$
$$\delta(\alpha) = -\beta(\alpha), \qquad (18.105)$$
$$\eta_m(\alpha) = \beta(\alpha)/2. \qquad (18.106)$$

In addition $\beta(\alpha)$ is independent of ξ. The function η_M can also be chosen independent of ξ, only the fermion field renormalization is necessarily gauge dependent. Actually from equation (18.89) it is even possible to determine the gauge dependence of η_ψ. A short calculation leads in the minimal subtraction scheme to:

$$\eta_\psi(\alpha, \xi) = \eta_\psi(\alpha, 0) - \alpha\xi/2\pi. \qquad (18.107)$$

18.9 The One-Loop β-Function

Fermion contribution. We now calculate the β-function at one-loop order in the case of the action (18.26) which describes the interaction of a gauge field with charged fermions. We first evaluate the gauge field renormalization constant and then deduce the coupling constant renormalization from relation (18.85). Dimensional regularization will be used in the calculation.

The one-loop contribution to the generating functional of proper vertices coming from the fermion integration is:

$$\Gamma_{1\,\text{loop}}\left(A_\mu\right) = -\operatorname{tr}\ln\left(\partial\!\!\!/ + ieA\!\!\!/ + M\right). \tag{18.108}$$

Differentiating twice with respect to A_μ, we obtain the one-loop contribution to the renormalized 1PI 2-point function:

$$\Gamma^{(2)}_{\mu\nu}\left(p\right) = Z_A\left(\delta_{\mu\nu}p^2 - p_\mu p_\nu\right) + \frac{1}{\xi}p_\mu p_\nu + e^2\int\frac{\mathrm{d}^d k}{(2\pi)^d}\frac{\operatorname{tr}\left[\gamma_\mu\left(k\!\!\!/ + iM\right)\gamma_\nu\left(k\!\!\!/ - p\!\!\!/ + iM\right)\right]}{(k^2 + M^2)\left[(p-k)^2 + M^2\right]}. \tag{18.109}$$

We verify immediately that, as expected, the one-loop contribution is transverse and therefore ξ is not renormalized. Setting:

$$\Gamma^{(2)}_{\mu\nu}\left(p\right) = \left[Z_A + e^2 N_d\operatorname{tr}\mathbf{1}\,\Sigma\left(p\right)\right]\left(p^2\delta_{\mu\nu} - p_\mu p_\nu\right) + p_\mu p_\nu/\xi, \tag{18.110}$$

in which N_d is the usual geometric factor ratio of the surface of $(d-1)$ sphere divided by $(2\pi)^d$, we find:

$$\Sigma\left(p\right) = \frac{1}{(d-1)p^2}\int\frac{\mathrm{d}^d k}{S_d}\frac{\left[(2-d)\left(k^2 - k\cdot p\right)dM^2\right]}{\left(k^2 + m^2\right)\left[(k-p)^2 + M^2\right]}. \tag{18.111}$$

Using:

$$k^2 - k\cdot p = \tfrac{1}{2}\left[\left(k^2 + M^2\right) + (k-p)^2 + M^2\right] - M^2 - \tfrac{1}{2}p^2, \tag{18.112}$$

we can rewrite $\Sigma\left(p\right)$:

$$\Sigma\left(p\right) = \frac{1}{(d-1)p^2}\int\frac{\mathrm{d}^d k}{S_d}\left\{\frac{(2-d)}{k^2 + M^2} + \frac{\tfrac{1}{2}p^2\left(d-2\right) - 2M^2}{\left(k^2 + M^2\right)\left[(k-p)^2 + M^2\right]}\right\}. \tag{18.113}$$

Using the identities:

$$(2-d)\int\frac{\mathrm{d}^d k}{k^2 + M^2} = -M\frac{\mathrm{d}}{\mathrm{d}M}\int\frac{\mathrm{d}^d k}{k^2 + M^2} = 2M^2\int\frac{\mathrm{d}^d k}{(k^2 + M^2)^2}, \tag{18.114}$$

we finally obtain:

$$\Sigma\left(p\right) = \frac{1}{2}\frac{(d-2)}{d-1}\int\frac{\mathrm{d}^d k}{S_d}\frac{1}{(k^2 + M^2)\left[(p-k)^2 + M^2\right]}$$

$$+ \frac{2M^2}{(d-1)p^2}\int\frac{\mathrm{d}^d k}{S_d}\frac{1}{(k^2 + M^2)}\left[\frac{1}{k^2 + M^2} - \frac{1}{(k-p)^2 + M^2}\right]. \tag{18.115}$$

In particular the divergent part of $\Sigma(p)$ is:

$$\Sigma(p) = \frac{1}{3\varepsilon} + O(1).$$ (18.116)

This determines the A_μ field renormalization Z_A and thus, from equation (18.85), also Z_e:

$$Z_A = 1 - N_d \operatorname{tr} \mathbf{1} \frac{e^2}{3\varepsilon} + O\left(e^4\right),$$ (18.117)

$$Z_e = 1 + N_d \operatorname{tr} \mathbf{1} \frac{e^2}{3\varepsilon} + O\left(e^4\right).$$ (18.118)

Replacing N_d and $\operatorname{tr} \mathbf{1}$ by their values for $d = 4$ ($N_4 = 1/8\pi^2$, $\operatorname{tr} \mathbf{1} = 4$), we obtain the corresponding β-function at one-loop order:

$$\beta(\alpha) = -\varepsilon \left[\frac{\mathrm{d}\ln(\alpha Z_e)}{\mathrm{d}\alpha} \right]^{-1} = -\varepsilon\alpha + \frac{4}{3}\frac{\alpha^2}{2\pi} + O\left(\alpha^3\right).$$ (18.119)

Scalar boson contribution. We now consider the action

$$S(A_\mu, \phi) = \int \mathrm{d}^d x \left(\frac{1}{4}F_{\mu\nu}^2 + |D_\mu\phi|^2 + M^2 |\phi|^2 + \frac{1}{6}g\,|\phi|^4 \right).$$ (18.120)

Again we calculate the one-loop contribution to the gauge field propagator. Since there are now two interaction terms of the form $\phi^2 A_\mu^2$ and $\phi^2 A_\mu$, two diagrams contribute at one-loop order (see figure 18.3).

Fig. 18.3   Charged bosons: the two one-loop diagrams contributing to the A_μ 2-point function.

The two terms can also be obtained from the expansion of:

$$\Gamma_{1\,\mathrm{loop}}(A_\mu) = \operatorname{tr}\ln\left[(\partial_\mu + ieA_\mu)^2 - M^2 \right] \left(\partial^2 - M^2 \right)^{-1}.$$ (18.121)

One finds:

$$\Gamma_{\mu\nu,1\,\mathrm{loop}}^{(2)} = 2e^2\delta_{\mu\nu} \int \frac{\mathrm{d}^d k}{(2\pi)^d} \frac{1}{k^2 + M^2} - e^2 \int \frac{\mathrm{d}^d k}{(2\pi)^d} \frac{(2k-p)_\mu(2k-p)_\nu}{(k^2 + M^2)\left[(k-p)^2 + M^2\right]}.$$ (18.122)

We again verify that the result is transverse. Setting:

$$\Gamma_{\mu\nu,1\,\mathrm{loop}}^{(2)} = \left(p^2\delta_{\mu\nu} - p_\mu p_\nu \right) e^2 N_d \Sigma(p),$$ (18.123)

and taking the trace of expression (18.122), we obtain:

$$p^2 (d-1) \Sigma (p) = \int \frac{d^d k}{S_d} \left\{ \frac{2d}{k^2 + M^2} - \frac{(2k-p)^2}{(k^2 + M^2)\left[(k-p)^2 + M^2\right]} \right\}, \qquad (18.124)$$

and therefore:

$$p^2 (d-1) \Sigma (p) = 2 (d-2) \int \frac{d^d k}{S_d} \frac{1}{k^2 + M^2} + \int \frac{d^d k}{S_d} \frac{(p^2 + 4M^2)}{(k^2 + M^2)\left[(k-p)^2 + M^2\right]}.$$
$$(18.125)$$

The identity already used in the fermion case leads to:

$$S_d \Sigma (p) = \frac{1}{d-1} \int \frac{d^d k}{(k^2 + M^2)\left[(k-p)^2 + M^2\right]} + \frac{4M^2}{(d-1) p^2}$$
$$\times \int \frac{d^d k}{k^2 + M^2} \left[\frac{1}{(k-p)^2 + M^2} - \frac{1}{k^2 + M^2} \right]. \qquad (18.126)$$

In particular the divergent part of $\Sigma (p)$ is:

$$\Sigma (p) = \frac{1}{3\varepsilon} + O (1). \qquad (18.127)$$

The last part of the calculation is the same as in the fermion case. The difference between fermions and bosons comes only from the trace of the identity in the space of γ matrices which yields an additional factor 4 in the contribution of fermions to the β-function. Calling $n_{\rm F}$ and $n_{\rm B}$ the number of charged fermions and scalar bosons respectively, we finally obtain:

$$\beta (\alpha) = -\varepsilon \alpha + \frac{1}{3} (4n_{\rm F} + n_{\rm B}) \frac{\alpha^2}{2\pi} + O (\alpha^3). \qquad (18.128)$$

The sign of the β-function. Note that in the domain of validity of expansion (18.117) (e^2 small), Z_A satisfies:

$$Z_A \leq 1.$$

This property also holds in the ϕ^4-field theory. The field renormalization constant Z_ϕ satisfies (equation (11.78)) $Z_\phi \leq 1$. Both are consequences of the Källen–Lehmann representation for the 2-point function (see Section 7.4). In the case of the gauge field the property is true because Z_A is related to the transverse part of the 2-point function to which unphysical states do not contribute.

Since $Z_A Z_e = 1$ we see that the *sign* of the β-function in four dimensions is *fixed* for α small enough.

Furry's theorem. For more general perturbative calculations the following observation is useful. Charge conjugation implies that correlation functions without matter field and an odd number of gauge fields vanish. The proof is based on charge conjugation. Let us consider the contribution to the effective gauge field action, $\det (\slashed{\partial} + ie\slashed{A} + M)$, which is generated by the integration over the fermion fields, and use the property of the charge conjugation matrix C introduced in Section A5.6:

$$\det\left(\partial\!\!\!/ + ie A\!\!\!/ + M\right) = \det\left[{}^T(\partial\!\!\!/ + ieA\!\!\!/ + M)\right] = \det C^{-1}\left[{}^T(\partial\!\!\!/ + ieA\!\!\!/ + M)\right] C$$
$$= \det\left(\partial\!\!\!/ - ieA\!\!\!/ + M\right).$$

Therefore the interaction between gauge fields generated by the fermions is even in A_μ. Note in particular the implication for Feynman diagrams: fermion loops with an odd number of external gauge fields can be omitted.

18.10 The Abelian Higgs Mechanism

As an introduction to the next chapter we now discuss the theory of a gauge field coupled to the charged scalar field $\phi(x)$ in an unusual phase. We start from the action (18.33). Renormalizability implies that the scalar field self-interaction is of the $|\phi|^4$ type:

$$S\left(A_\mu, \phi\right) = \int d^d x \left(\frac{1}{4}F_{\mu\nu}^2 + |D_\mu\phi|^2 + M^2|\phi|^2 + \frac{1}{6}g|\phi|^4\right). \tag{18.129}$$

The field $\phi(x)$ is complex and the covariant derivative D_μ is defined by equation (18.28).

At the classical level, if the gauge symmetry is exact, and the $U(1)$ symmetry is not spontaneously broken, the gauge field is massless and the scalar field ϕ has two real components with equal mass.

However we know that another situation is possible in which the $U(1)$ symmetry is spontaneously broken, and $\phi(x)$ has an expectation value v, which for convenience we assume real (we shall comment later on the significance of $\langle\phi\rangle$ which is not gauge invariant). Let us examine in some detail, at the classical level, the physics of such a situation. We have met an analogous problem in Section 13.4 and we have concluded that SSB of continuous symmetries leads to the appearance of a massless state, a Goldstone particle.

At the classical level this can most easily be seen by parametrizing the field ϕ:

$$\phi(x) = \frac{1}{\sqrt{2}}\left[v + \rho(x)\right]e^{i\theta(x)}, \tag{18.130}$$

and verifying that the action depends only on $\partial_\mu\theta$, the field θ being therefore massless.

However if we use the same parametrization in the gauge theory, we observe that the θ-field can be completely eliminated from the action by a gauge transformation:

$$A_\mu(x) = A'_\mu(x) + \frac{1}{e}\partial_\mu\theta(x). \tag{18.131}$$

After this transformation the action $S\left(A_\mu, \phi\right)$ indeed reads:

$$S\left(A_\mu, \rho\right) = \int d^d x \left[\frac{1}{4}F_{\mu\nu}^2 + \frac{1}{2}(\partial_\mu\rho)^2 + \frac{1}{2}e^2 A_\mu^2(\rho+v)^2 + \frac{1}{2}M^2(\rho+v)^2 + \frac{g}{24}(\rho+v)^4\right]. \tag{18.132}$$

At the tree level the spectrum of the theory is thus composed of a massive vector field of mass $e^2 v^2$ and a massive scalar field called the Higgs field:

$$m^2(A) = e^2 v^2, \qquad m^2(\rho) = \tfrac{1}{3}gv^2. \tag{18.133}$$

As a consequence of gauge invariance, no Goldstone field has been generated. This is a most remarkable property, which is also at the basis of the Meissner effect in superconductivity. It is induced by the long range forces (in the non-relativistic case the Coulomb force) generated by the massless gauge field.

Note that the total number of physical degrees of freedom has not changed between the symmetric phase and the spontaneously broken phase since one degree of freedom of the scalar field has been transferred to the vector field.

From the technical point of view, the surprising property of action (18.132) is that although it does not correspond to a massive vector field coupled to a conserved current, we nevertheless expect the physical observables to be calculable via a renormalizable field theory, while again in the so-called unitary gauge (the representation (18.132)), the theory expressed in terms of physical fields is not renormalizable. There exists however a relation between the massive vector field and the Higgs model: if we take the formal non-linear σ-model limit of the action (18.129), i.e. a limit in which the bare mass of the Higgs field becomes infinite at fixed expectation value v, we recover the action (18.7) with the identification $m = ev$.

To calculate gauge invariant observables and S-matrix elements, we have of course to return to an action in which the gauge has been fixed through a term proportional to $(\partial_\mu A_\mu)^2$. This corresponds to give a dynamics to the phase field $\theta(x)$ which plays the role of the $\Lambda(x)$ field of Section 18.1. We now examine the structure of the theory in this representation.

As a final remark let us also recall for later purposes that the mechanism of spontaneous symmetry breaking can also be used to give a mass to fermions in a theory with chiral symmetry (Section 13.6).

18.11 Quantization of the Higgs Mechanism

We now start from the action

$$S(A, \phi) = \int \mathrm{d}^d x \left[\frac{1}{4} F_{\mu\nu}^2 + \frac{1}{2\xi} (\partial_\mu A_\mu)^2 + \frac{1}{2} m^2 A_\mu^2 + \mathrm{D}_\mu \phi \mathrm{D}_\mu \phi^* + M^2 \phi \phi^* + \frac{g}{6} (\phi \phi^*)^2 \right],$$
(18.134)

in which a mass has been given to the vector field to provide an IR cut-off.

We assume that ϕ has a real expectation value v at the classical level. We introduce the real and imaginary parts of ϕ and set:

$$\phi(x) = \frac{1}{\sqrt{2}} [v + \varphi(x) + i\chi(x)].$$
(18.135)

The quadratic part S_2 of the action then is :

$$S_2(A, \phi) = \int \mathrm{d}^d x \left[\frac{1}{4} F_{\mu\nu}^2 + \frac{1}{2\xi} (\partial_\mu A_\mu)^2 + \frac{1}{2} (e^2 v^2 + m^2) A_\mu^2 \right.$$
$$\left. - ev\chi \partial_\mu A_\mu + \frac{1}{2} (\partial_\mu \varphi)^2 + \frac{1}{6} gv^2 \varphi^2 + \frac{1}{2} (\partial_\mu \chi)^2 \right].$$
(18.136)

We see that $\partial_\mu A_\mu$ is coupled to the Goldstone field χ. The corresponding propagators are:

$$W_{\mu\nu}^{(2)}(k) = \frac{\delta_{\mu\nu} - k_\mu k_\nu / k^2}{k^2 + e^2 v^2 + m^2} + \xi \frac{k_\mu k_\nu}{k^2 \left(k^2 + \xi m^2\right)},$$

$$W_{\mu\chi}^{(2)}(k) = -\xi \frac{ievk_\mu}{k^2 \left(k^2 + \xi m^2\right)}, \tag{18.137}$$

$$W_{\chi\chi}^{(2)}(k) = \frac{1}{k^2} + \frac{\xi e^2 v^2}{k^2 \left(k^2 + \xi m^2\right)},$$

and

$$W_{\varphi\varphi}^{(2)}(k) = \frac{1}{k^2 + \frac{1}{3} g v^2}. \tag{18.138}$$

The spectrum of the theory consists in 3 physical states and the usual state with negative norm coming from the regulator. We see that in the absence of a mass term for the vector field in the action (18.134), the theory is potentially IR divergent in 4 dimensions. On the other hand with the mass term the gauge symmetry is broken and the χ-field corresponds really to a Goldstone mode. Even in the physical gauge a massless field is present and coupled.

18.11.1 WT identities and renormalization

It follows from the combined analysis of Chapter 13 and Section 18.6 that after renormalization the correlation functions satisfy the equivalent of WT identities (18.78) and (18.81). As a consequence the dependence of correlation functions on the parameter ξ can be determined as in Section 18.7. In particular only correlation functions of gauge invariant operators and S-matrix elements are gauge independent.

The explicit form of the WT for correlation functions is now rather complicated. Let us here write only the identities corresponding to the (A_μ, χ) two-point proper vertices. Calling v the expectation value of the renormalized φ-field, we obtain by differentiating equation (18.81) with respect to A_μ:

$$k_\nu \Gamma_{\mu\nu}^{(2)}(k) + iev\,\Gamma_{\mu\chi}^{(2)}(k) = k_\mu \left(k^2/\xi + m^2\right). \tag{18.139}$$

Differentiating then with respect to χ we find:

$$k_\mu \Gamma_{\mu\chi}^{(2)} - iev\,\Gamma_{\chi\chi}^{(2)} = 0. \tag{18.140}$$

Let us parametrize the different functions as:

$$\Gamma_{\mu\nu}^{(2)} = a\left(k^2\right) \delta_{\mu\nu} - b\left(k^2\right) k_\mu k_\nu,$$

$$\Gamma_{\mu\chi}^{(2)} = iev\,c\left(k^2\right) k_\mu, \tag{18.141}$$

$$\Gamma_{\chi\chi}^{(2)} = d\left(k^2\right).$$

At the tree level the values of a, b, c and d are:

$$\begin{cases} a\left(k^2\right) = e^2 v^2 + m^2 + k^2, & b\left(k^2\right) = 1 - 1/\xi, \\ c\left(k^2\right) = 1, & d\left(k^2\right) = k^2. \end{cases} \tag{18.142}$$

Let us now express the identity (18.140):

$$d\left(k^2\right) = k^2 c\left(k^2\right). \tag{18.143}$$

The identity (18.139) then leads to:

$$a\left(k^2\right) - k^2 b\left(k^2\right) - e^2 v^2 c\left(k^2\right) = k^2/\xi + m^2. \qquad (18.144)$$

In the $k = 0$ limit, this equation in particular implies:

$$a(0) - e^2 v^2 c(0) = m^2. \qquad (18.145)$$

Let us calculate the corresponding connected correlation functions:

$$W^{(2)}_{\mu\nu} = \frac{1}{a}\left(\delta_{\mu\nu} - \frac{k_\mu k_\nu}{k^2}\right) + \frac{\xi k_\mu k_\nu}{k^2\left(k^2 + \xi m^2\right)},$$

$$W^{(2)}_{\mu\chi} = -\frac{\xi i e v k_\mu}{k^2\left(k^2 + \xi m^2\right)}, \qquad (18.146)$$

$$W^{(2)}_{\chi\chi} = \frac{1}{c k^2}.$$

18.11.2 Decoupling gauge

The quantization method we have used above leads to massless fields and thus IR divergences, even though the physical theory contained only massive fields. By the cleverer choice of a gauge which explicitly breaks the global $U(1)$ symmetry of the action (and therefore gets rid of the Goldstone modes), it is possible to circumvent this difficulty. In the notation of the action (18.134) let us impose the condition:

$$\partial_\mu\left(B_\mu + \partial_\mu\Lambda\right) + \lambda e v \,\mathrm{Im}\left(\phi\,\mathrm{e}^{-ie\Lambda}\right) = \xi^{1/2} h(x), \qquad (18.147)$$

in which λ is a constant which will be adjusted eventually. The important new feature is that the operator $\mathbf{M}$, functional derivative of equation (18.147) with respect to Λ, now depends on the dynamical fields:

$$\langle y\left|\mathbf{M}\right|x\rangle = \left[\partial^2 + \lambda e v\,\mathrm{Re}\left(\phi\,\mathrm{e}^{-ie\Lambda}\right)\right]\delta\left(x - y\right), \qquad (18.148)$$

and the associated determinant $\det\mathbf{M}$ is no longer a constant. This is the source of difficulties of a kind we have already encountered in Chapters 16 and 17: we have to introduce fermion fields to write $\det\mathbf{M}$ in local form, and then use the induced BRS symmetry to show that renormalization preserves the form of the action. We postpone this analysis until Chapter 21, and here give only a discussion at tree order.

As before we integrate over $h(x)$ with the distribution (18.56) and use the gauge invariance of the initial action

$$S_{\mathrm{sym}}\left(B_\mu, \phi\right) = S_{\mathrm{sym}}\left(A_\mu, \phi\,\mathrm{e}^{-ie\Lambda}\right). \qquad (18.149)$$

Setting then in the functional integral:

$$\phi\,\mathrm{e}^{-ie\Lambda} = \phi',$$

we obtain the effective action S_{eff}

$$S_{\mathrm{eff}}\left(A_\mu, \phi, \bar{C}, C\right) = \int \mathrm{d}^d x\left\{\frac{1}{4}F_{\mu\nu}^2 + \frac{1}{2\xi}\left(\partial_\mu A_\mu + \lambda e v\,\mathrm{Im}\,\phi\right)^2 + \left|D_\mu\phi\right|^2\right.$$

$$\left. + M^2\left|\phi\right|^2 + \frac{1}{6}g\left|\phi^4\right| - \bar{C}\left(\partial^2 + \lambda e v\,\mathrm{Re}\,\phi\right)C\right\}, \qquad (18.150)$$

in which C and $\bar{C}$ are two scalar fermion fields which generate the determinant $\det \mathbf{M}$. As we have explained on an example in Section A7.2, scalar fermions cannot be interpreted as physical particles but are of a nature similar to Pauli–Villars regulator fields. We now use the parametrization (18.135) and choose λ:

$$\lambda = \xi\sqrt{2} \, . \tag{18.151}$$

This is of course the value only at tree order. The propagators then are:

$$
\begin{aligned}
W^{(2)}_{\mu\nu} &= \frac{\delta_{\mu\nu}}{k^2 + e^2 v^2} + \frac{(\xi - 1)\, k_{\mu} k_{\nu}}{(k^2 + e^2 v^2)\,(k^2 + \xi e^2 v^2)}, \\[1mm]
W^{(2)}_{\chi\chi} &= \frac{1}{k^2 + \xi e^2 v^2}, \\[1mm]
W^{(2)}_{\bar{C} C} &= \frac{1}{k^2 + \xi e^2 v^2}.
\end{aligned}
\tag{18.152}
$$

The advantages of this gauge (introduced by 't Hooft) are that by construction there is no $A_{\mu}\chi$ propagator and that all unphysical fields are massive and have the same mass $\xi e^2 v^2$. It suffices to prove gauge independence of physical observables to show that the pole at $k^2 = -\xi e^2 v^2$ cancels. The price to pay here is the more complicated form of WT identities which now are mixed with BRS symmetry. We shall examine this question in Chapter 21 in detail.

18.12 Physical Observables. Unitarity of the S-Matrix

The unphysical pole at $k^2 = -\xi m^2$ can be shown to cancel in physical observables (gauge invariant operators, S-matrix) either through a gauge dependence analysis as we have done in Section 18.7, or directly by using the whole set of WT identities and showing explicitly that the pole coming from $W^{(2)}_{\mu\nu}$ cancels the contribution coming from $W^{(2)}_{\mu\chi}$ in the intermediate state in unitarity relations. As the expressions (18.146) show, the residues of the pole are related and therefore one understands that a cancellation is possible. The proof is not very difficult but tedious and we refer to the literature.

In the limit $m = 0$, we expect also the pole at $k^2 = 0$ to cancel in physical observables. According to relation (18.145), for $k^2 \to 0$ the different propagators behave like:

$$
\begin{cases}
W^{(2)}_{\mu\nu} \sim \dfrac{k_{\mu} k_{\nu}}{k^2}\left(\dfrac{1}{m^2} - \dfrac{1}{m^2 + e^2 v^2 c\,(0)}\right), \\[3mm]
W^{(2)}_{\mu\chi} \sim -iev\,\dfrac{k_{\mu}}{k^2 m^2}, \\[3mm]
W^{(2)}_{\chi\chi} \sim \dfrac{1}{c\,(0)\,k^2}.
\end{cases}
\tag{18.153}
$$

Again a direct argument based on WT identities for connected correlation functions and unitarity relations allows to prove that in the $m^2 = 0$ limit the χ-field decouples from physical observables. Here we do not have an alternative proof based upon gauge dependence. However we shall construct one in a more general context by using a different gauge.

Gauge invariant operators. When the action (18.134) is used, only averages of gauge invariant operators are physical correlation functions. The simplest examples are $F_{\mu\nu}$ and

$\sigma(x) = (v + \rho(x))^2 = \frac{1}{2}\phi(x)\phi^*(x)$. In particular the parameter v is really the square root of $\langle\sigma\rangle$. On the other hand, the action (18.132) can be rewritten in terms of this field $\sigma(x)$. One attractive feature of this representation is that the measure of integration in the functional integral is just the flat measure $[d\sigma(x)]$. Furthermore all correlation functions of the transverse part of the vector field and the scalar field correspond directly in this physical representation to gauge invariant correlation functions of the renormalizable representation. However an inspection of the action written in terms of the $A_\mu(x)$ and $\sigma(x)$ fields does not provide a direct explanation for the finiteness, after the introduction of finite number of counterterms, of the correlation functions of $F_{\mu\nu}$ and σ.

Bibliographical Notes

Many early articles about QED are reprinted in

 J. Schwinger ed., *Quantum Electrodynamics* (Dover, New York 1958).

There exists of course a vast literature devoted to QED. Among the classical textbooks let us quote

 J. Bjorken and S. Drell, *Relativistic Quantum Mechanics, and Relativistic Quantum Fields* (McGraw-Hill, New York 1964, 1965); A.I. Akhiezer and V.B. Berestetskii, *Quantum Electrodynamics* (Interscience, New York 1965); C. Itzykson and J.-B. Zuber, *Quantum Field Theory* (McGraw-Hill, New York 1980).

and references therein.

For infrared divergences see for example

 D.R. Yennie, S.C. Frautschi and H. Suura, *Ann. Phys. (NY)* 13 (1961) 379; T. Kinoshita, *J. Math. Phys.* 3 (1962) 650; T.D. Lee and M. Nauenberg, *Phys. Rev.* 133B (1964) 1549.

WT identities are derived in

 J.C. Ward, *Phys. Rev.* 78 (1950) 182; Y. Takahashi, *Nuovo Cimento* 6 (1957) 371.

Quantum Electrodynamics still provides the most accurate tests of Quantum Field Theory. For example the calculation of the eighth order contribution to the electron anomalous magnetic moment has recently been completed:

 T. Kinoshita and W.B. Lindquist, *Phys. Rev.* D42 (1990) 636.

For the muon magnetic moment, which is more sensitive to Strong and Weak Interaction corrections, see

 T. Kinoshita, B. Nizic and Y. Okamoto, *Phys. Rev.* D41 (1990) 593; T. Kinoshita, *Phys. Rev.* D47 (1993) 5013.

The abelian Higgs mechanism is discussed in

 B.W. Lee, *Phys. Rev.* D5 (1972) 823.

A general class of gauges is constructed in

 G. 't Hooft, *Nucl. Phys.* B35 (1971) 167.

The statistical physics analogue of the abelian Higgs mechanism is superconductivity. For the theory of superconductivity see

 P.G. de Gennes, *Superconductivity of Metals and Alloys* (Benjamin, New York 1966); D.R. Tilley and J. Tilley, *Superfluidity and Superconductivity* (Adam Hilger, Bristol 1986).

Effects of vacuum fluctuations are discussed in

 H.G.B. Casimir, *Proc. Kon. Ned. Akad. Wetenschap,.* B51 (1948) 793.

Exercises

Exercise 18.1

In the notation of subsection 18.4.2 verify explicitly the compatibility of Gauss's law with quantum mechanical evolution in the case of a non-vanishing external current.

Exercise 18.2

For the one-loop calculations which follow dimensional regularization will be used. For the trace of the unit matrix in the γ matrix space one can take the 4D value: $\operatorname{tr} \mathbf{1} = 4$.

Establish the Feynman rules. Calculate the fermion 2-point function at one-loop order, in the Feynman gauge, using the action (18.26). Infer the fermion mass and fermion field RG functions.

Exercise 18.3

We now consider the QED action (18.26), with a mass term for the photon and quantized in a general covariant gauge

$$S \left(A_\mu, \bar{\psi}, \psi\right) = \int \mathrm{d}^d x \left[\tfrac{1}{4} F_{\mu\nu}^2 + \tfrac{1}{2} m^2 A_\mu^2 + \frac{1}{2\xi} \left(\partial_\mu A_\mu\right)^2 - \bar{\psi} \left(\slashed{D} + M\right) \psi \right]. \qquad (18.154)$$

18.3.1. Write explicitly the WT identities involving the A 2-point function and the $\bar{\psi}\psi A$ 3-point function both for the 1PI and connected correlation functions and verify them at the tree level.

18.3.2. Calculate the inverse fermion 2-point function $\Gamma^{(2)}(p)$ at one-loop order, i.e. reduce it to a linear combination of scalar "bubble" integrals of the form

$$B \left(p; m_1, m_2\right) = \int \frac{\mathrm{d}^d k}{(2\pi)^d} \frac{1}{\left(k^2 + m_1^2\right) \left[(k - p)^2 + m_2^2\right]}.$$

It is convenient to also introduce the integral $I(\mu)$

$$I(\mu) = \int \frac{\mathrm{d}^d k}{(2\pi)^d} \frac{1}{k^2 + \mu^2} \quad \Rightarrow \quad B(0; m_1, m_2) = \frac{1}{m_1^2 - m_2^2} \left[I(m_2) - I(m_1)\right].$$

18.3.3. Calculate in the bare theory the one-loop correction to the fermion mass. What can be said of the gauge dependence of the result. Explain.

18.3.4. Characterize the divergences and calculate the corresponding renormalization constants. It is convenient to set $d = 4 - \varepsilon$ and

$$N_d = \frac{2}{\Gamma(d/2)(4\pi)^{d/2}}.$$

What can be said of the gauge dependence of Z_M and Z_ψ? Explain.

18.3.5. Write the regularized $A_\mu \bar{\psi}\psi$ vertex (1PI) at one-loop order and check the WT identity relating it to the fermion two-point function.

APPENDIX 18

A18.1 Vacuum Energy and Casimir Effect

Let us here discuss the free vector field vacuum energy. The vacuum energy is a gauge independent quantity but because the formalism depends on the gauge this is not always obvious. We then apply the result to the Casimir effect.

A18.1.1 The free electromagnetic field: Vacuum energy

The free massive vector field. Let us first consider the massive case. One way to calculate the ground state energy is to evaluate the free energy in a large euclidean volume. The result is proportional to a space-time volume factor. Dividing by the volume we obtain the vacuum energy density, the energy per unit space volume:

$$\mathcal{E}/\text{space volume} = -\ln Z/\text{space-time volume} .$$

In the initial unitary gauge one immediately finds, in Fourier variables,

$$-\ln Z = \tfrac{1}{2}\operatorname{tr}\ln\left[(k^2+m^2)\delta_{\mu\nu}-k_\mu k_\nu\right] = \tfrac{1}{2}(d-1)\operatorname{tr}\ln(k^2+m^2)+\tfrac{1}{2}\operatorname{tr}\ln m^2. \quad (A18.1)$$

Up to an irrelevant constant we find $(d-1)$ times the vacuum energy of a free scalar field of mass m, a result which is not surprising since the massive vector field has $d-1$ degrees of freedom. The result can be verified directly by diagonalizing the hamiltonian (18.6).

Let us however repeat the calculation with the action S_ξ of equation (18.19). We find instead

$$-\ln Z = \tfrac{1}{2}(d-1)\operatorname{tr}\ln(k^2+m^2)+\tfrac{1}{2}\operatorname{tr}\ln(k^2+\xi m^2)/\xi .$$

This gauge-dependent result for the vacuum energy is clearly incorrect. The reason can be simply understood. In the various algebraic manipulations which have led to the action (18.19) we have omitted field-independent normalization factors. This is justified for correlation functions but not for the vacuum energy. The additional term originates from the χ field we have added to the theory in equation (18.14). Normalizing correctly the χ integral cancels the additional unwanted factor.

The massless gauge field. We have seen that correlation functions have a smooth massless limit. This is not the case for the vacuum energy. Indeed in the massless limit one degree of freedom of the vector field, the longitudinal mode, decouples from the theory but still contributes to the vacuum energy. Let us thus evaluate the energy directly and then discuss the gauge dependence.

In the temporal gauge we can use the hamiltonian (18.44) (in a vanishing source). The hamiltonian can be partially diagonalized by a Fourier transformation (see Section 7.1)

$$\mathcal{H} = \tfrac{1}{2}\left[(2\pi)^{1-d}\tilde{E}_i(\hat{p})\tilde{E}_i(-\hat{p}) + (2\pi)^{d-1}\tilde{A}_i(\hat{p})\left(\hat{p}^2\delta_{ij}-\hat{p}_i\hat{p}_j\right)\tilde{A}_j(-\hat{p})\right], \quad (A18.2)$$

where $\hat{p}$ is a space momentum. We immediately see that we have $d-2$ harmonic oscillators with ground state energy $|\hat{p}|$ in the field direction perpendicular to the vector $\hat{p}$, and a free quantum mechanical hamiltonian in the $\hat{p}$ direction. However Gauss's law implies that $\hat{p} \cdot E(\hat{p})$ annihilates all physical states and therefore does not contribute to the energy.

We conclude that the vacuum energy is $d-2$ times the vacuum energy of a free massless scalar particle.

Covariant gauge. Let us instead evaluate the energy in a covariant gauge, using the corresponding partition function. Naively we find

$$ -\ln Z = \tfrac{1}{2}\operatorname{tr}\ln\left[k^2\delta_{\mu\nu} - k_\mu k\nu(1-1/\xi)\right] = \tfrac{1}{2}d\operatorname{tr}\ln k^2 - \tfrac{1}{2}\operatorname{tr}\ln\xi\,. $$

Up to constant we find the energy of d massless states instead of $d-2$. The discussion of the massive example give us an idea of the problem. We have omitted field independent factors in algebraic transformations. For instance in the Faddeev–Popov quantization of Section 18.4.4, if we want the multiplication by the integral over $\Lambda(x)$ to be an identity, we have to multiply the integral (18.63) by $\det(-\partial^2)$. Such a factor cancels a complex or two real massless scalar bosons. It reduces d to $d-2$. In non-abelian gauge theories "ghost fields" automatically produce the right book-keeping.

A18.1.2 Casimir effect

In Chapter 36 we discuss quantum field theory in a finite volume or more generally in restricted geometries. However the simplest example of finite size effects is the Casimir effect, i.e. the attractive force between two parallel perfectly conducting plates in the vacuum, due to the change in vacuum energy produced by a change in boundary conditions. At leading order, all charged particles can be omitted because only massless fields have a significant contribution at large plate separation. Hence the problem is reduced to a calculation of the change of vacuum energy of the free electromagnetic field due the boundary conditions. The conducting plates impose to the electric field to be perpendicular to the plates. It is easy to verify that this condition is satisfied if the vector field A_μ itself vanishes on the plates. Calling L the distance between the plates, $z=0$ and $z=L$ the plate positions, we thus integrate over fields which have the Fourier representation

$$ A_\mu(\mathbf{x}_\perp, z) = \int \mathrm{d}^{d-1}p_\perp \sum_{n\geq1} e^{i\mathbf{p}_\perp\cdot\mathbf{x}_\perp}\sin(\pi z/L)\tilde{A}(\mathbf{p}_\perp, n), \qquad (A18.3) $$

where $\mathbf{x}_\perp$ are the space-time coordinates in the remaining directions. Since all components of A_μ satisfy the same boundary conditions and since the field vacuum energy is $(d-2)$, i.e. the number of field degrees of freedom, times the free scalar vacuum energy, we now solve the problem for a free massless scalar field.

The free massless scalar theory. We consider the action for a scalar field $\phi(x)$:

$$ S(\phi) = \int \mathrm{d}^d x \big(\partial_\mu\phi(x)\big)^2, $$

where the physical situation of interest is $d=4$. We assume that in one direction the field satisfy the boundary conditions

$$ \phi(\mathbf{x}_\perp, z=0) = \phi(\mathbf{x}_\perp, z=L) = 0\,. $$

For the other directions we first assume periodic boundary conditions with finite size $L_\perp$, with $L_\perp \gg L$. Integrating over the field we know that the vacuum energy $\mathcal{E}$ is simply given by

$$ L_\perp\mathcal{E} = \sum_{n\geq1}\sum_{\mathbf{n}_\perp}\frac{1}{2}\ln\left(\frac{4\pi^2\mathbf{n}_\perp^2}{L_\perp^2} + \frac{\pi^2 n^2}{L^2}\right). \qquad (A18.4) $$

Of course this expression is only meaningful in the presence of a UV cut-off. Since

$$\int_{\varepsilon}^{\infty} \frac{dt}{t} e^{-st} \underset{\varepsilon \to 0}{=} -\ln(\varepsilon s) + \psi(1) + O(\varepsilon),$$

we replace it by the regularized form

$$L_{\perp}\mathcal{E} = -\sum_{n \geq 1} \sum_{\mathbf{n}_{\perp}} \frac{1}{2} \int_{a^2}^{\infty} \frac{dt}{t} \exp\left[-t\left(\frac{4\pi^2 \mathbf{n}_{\perp}^2}{L_{\perp}^2} + \frac{\pi^2 n^2}{L^2}\right)\right],$$

where $\Lambda = 1/a$ plays the role of a UV cut-off. We now take the large $L_{\perp}$ limit. We can replace sums by integrals:

$$L_{\perp}\mathcal{E} = -\frac{1}{2} \sum_{n \geq 1} \int_{a^2}^{\infty} \frac{dt}{t} \int d^{d-1}\mathbf{n}_{\perp} \exp\left[-t\left(\frac{4\pi^2 \mathbf{n}_{\perp}^2}{L_{\perp}^2} + \frac{\pi^2 n^2}{L^2}\right)\right].$$

The $\mathbf{n}_{\perp}$ integral is gaussian. We then obtain the energy per unit area

$$\mathcal{E}/L_{\perp}^{d-2} = -\tfrac{1}{2}(4\pi)^{(1-d)/2} \int_{a^2}^{\infty} \frac{dt}{t^{(d+1)/2}} \sum_{n \geq 1} \exp\left(-\pi^2 n^2 t/L^2\right). \qquad (A18.5)$$

We evaluate the sum, using identity (A36.26):

$$\sum_{n \geq 1} \exp\left(-\pi^2 n^2 t/L^2\right) = \frac{L}{2\sqrt{\pi t}} - \frac{1}{2} + \frac{L}{\sqrt{\pi t}} \sum_{n \geq 1} \exp\left(-L^2 n^2/t\right).$$

The first term yields a contribution proportional to $L\Lambda^d$ which is the vacuum energy $\mathcal{E}_0$ in the absence of boundaries. The second term yields a L-independent surface energy of order Λ^{d-1} due to the boundaries. Finally the remaining terms which are cut-off independent but depend on L give the interesting contribution. After integration over t it takes the form

$$(\mathcal{E} - \mathcal{E}_0)/L_{\perp}^{d-2} = \text{const.} - A(d)L^{1-d}, \qquad (A18.6)$$

with

$$A(d) = \frac{\Gamma(d/2)}{(4\pi)^{d/2}} \sum_{n \geq 1} \frac{1}{n^d}. \qquad (A18.7)$$

For $d = 4$ one finds

$$A(4) = \frac{\pi^2}{1440}.$$

Casimir effect. The resulting force between plates is therefore attractive. To pass from the scalar result to the electromagnetic result we have to take into account the $d - 2 = 2$ degrees of freedom of the gauge field. For the energy and force F per unit area we find, restoring the physical units,

$$(\mathcal{E} - \mathcal{E}_0)/L_{\perp}^2 = \text{const.} - \frac{\pi^2}{720}\frac{\hbar c}{L^3}, \quad \Rightarrow \quad F = -\frac{1}{L_{\perp}^2}\frac{d\mathcal{E}}{dL} = -\frac{\pi^2}{240}\frac{\hbar c}{L^4}. \qquad (A18.8)$$

This quantum relativistic effect is very small but measurable. It is remarkable because, though electromagnetic in nature, it is independent of the value of the electric charge.

A18.2 Gauge Dependence

To characterize the gauge dependence of a general correlation function with only matter fields (or with transverse gauge field) one has to integrate explicitly over the field χ. Setting:

$$
\begin{aligned}
W_\xi^{(2n)}(x_1,\ldots,x_n;y_1,\ldots,y_n) &= W_\infty^{(2n)}(x_1,\ldots,x_n;y_1,\ldots,y_n) \\
&\times U_\xi^{(2n)}(x_1,\ldots,x_n;y_1,\ldots,y_n),
\end{aligned}
$$

we find:

$$
U_\xi^{(2n)}(x_1,\ldots,x_n;y_1,\ldots,y_n) = \left\langle \exp\left[\frac{ie}{m} \sum_{i=1,n} \chi(x_i) - \chi(y_i) \right] \right\rangle,
$$

where $<>$ means gaussian integration over the χ-field as defined by equation (18.93). The integral can be calculated (for details see for example Section 31.1) and yields:

$$
U_\xi^{(2n)} = \exp\left\{ \frac{e^2}{m^2} \sum_{i,j} \left[\tfrac{1}{2}\Delta_\xi(x_i-x_j) + \tfrac{1}{2}\Delta_\xi(y_i-y_j) - \Delta_\xi(x_i-y_j) \right] \right\}, \qquad (A18.9)
$$

Δ_ξ being the χ-field propagator:

$$
\Delta_\xi(x) = \frac{1}{(2\pi)^d} \int \frac{d^d k\, e^{ipx}}{k^2+\xi m^2}. \qquad (A18.10)
$$

As indicated in Section 18.7, it is convenient to take the ratio of correlation functions corresponding to two finite values of ξ. The ratio between for example the generic ξ and Landau's gauge ($\xi = 0$) is then still given by an expression of the form $(A18.9)$ but with the propagator $(A18.10)$ replaced by:

$$
m^2 K_\xi(x) = \Delta_\xi(x) - \Delta_0(x) = -\frac{\xi m^2}{(2\pi)^d} \int \frac{d^d k\, e^{ipx}}{k^2(k^2+\xi m^2)}.
$$

In expression $(A18.9)$ only the diagonal terms $i=j$ with vanishing arguments yield UV divergences. One recognizes immediately the divergent factor $(Z_\psi(\xi)/Z_\psi(0))^n$ as given by equation (18.89):

$$
\frac{U_\xi^{(2n)}}{U_0^{(2n)}} = \left(\frac{Z_\psi(\xi)}{Z_\psi(0)} \right)^n \exp\left\{ e^2 \left[\sum_{i<j}(K_\xi(x_i-x_j) + K_\xi(y_i-y_j)) - \sum_{i,j} K_\xi(x_i-y_j) \right] \right\}.
$$
$$
(A18.11)
$$

If, after renormalization, we expand the ratio between correlation functions in powers of e^2 we note that all terms but the first one correspond to a matter field correlation function in which at least two external lines have been joined by a χ-field propagator. Therefore when we go on mass-shell these terms do not have the corresponding poles and thus do not contribute to the S-matrix, as indicated in Section 18.7.

19 NON-ABELIAN GAUGE THEORIES: INTRODUCTION

In Chapter 18 we have described the structure and the formal properties of abelian gauge theories which provide a framework for the discussion of Quantum Electrodynamics. However, to be able to describe other fundamental interactions, Weak and Strong Interactions, it is necessary to generalize the concept of gauge theories to non-abelian groups. In this chapter we therefore construct a field theory invariant under *local*, i.e. space-dependent, transformations of a general compact Lie group G. Inspired by the abelian example of Chapter 18, we immediately introduce the geometrical concept of parallel transport, a concept discussed more extensively in Chapter 22 in the example of Riemannian manifolds. All the required mathematical quantities then appear quite naturally. We quantize gauge theories and study some of the formal properties of the quantum theory like the BRS symmetry. We show how perturbation theory can be regularized, a somewhat non-trivial problem. We finally discuss general aspects of the Higgs mechanism.

In the appendix we quantize massive non-abelian gauge fields and briefly explain the non-renormalizability of the field theory.

19.1 Geometrical Construction

Let us consider a scalar field $\phi(x)$ transforming under a linear representation of the group G. Denoting by $\mathbf{g}$ a matrix belonging to the representation, we write the field transformation:

$$\phi'(x) = \mathbf{g}\phi(x). \tag{19.1}$$

We want to construct a model which has a *local G-symmetry*, i.e. a symmetry which is defined by the transformation (19.1) in which the matrix $\mathbf{g}$ depends on the point x. If we consider only invariant products of fields taken at the same point, global invariance implies local invariance. However if we consider invariant functions of fields and their derivatives, or more generally products of fields taken at different points this is no longer true. An analogous problem arises in the study of Riemannian manifolds. The analogue of local group transformations are there reparametrizations of Riemannian manifolds and we refer to Chapter 22 for a more detailed geometrical and algebraic discussion. To solve the problem it is necessary to introduce parallel transporters $\mathbf{U}(C)$ which are curve-dependent elements of the group G, in the same matrix representation. If C is a curve joining point y to x, and $\mathbf{g}(x)$ a space-dependent group element, we write the transformation of $\mathbf{U}(C)$:

$$\mathbf{U}'(C) = \mathbf{g}(x)\mathbf{U}(C)\mathbf{g}^{-1}(y). \tag{19.2}$$

It then follows that the vector ϕ_U defined by:

$$\phi_U = \mathbf{U}(C)\phi(y), \tag{19.3}$$

transforms by $\mathbf{g}(x)$ instead of $\mathbf{g}(y)$.

In the limit of an infinitesimal differentiable curve:

$$y_\mu = x_\mu + \mathrm{d}x_\mu, \tag{19.4}$$

we can parametrize $U(C)$ in terms of the *connection* $\mathbf{A}_\mu(x)$ which is a vector for d-space transformations, and a matrix belonging to the Lie algebra of G:

$$\mathbf{U}(C) = \mathbf{1} + \mathbf{A}_\mu(x)\mathrm{d}x_\mu + o\left(\|\mathrm{d}x_\mu\|\right). \tag{19.5}$$

The transformation law for $\mathbf{A}_\mu(x)$ is obtained by expanding equation (19.2) at first order in $\mathrm{d}x_\mu$:

$$\mathbf{A}'_\mu(x) = \mathbf{g}(x)\mathbf{A}_\mu(x)\mathbf{g}^{-1}(x) + \mathbf{g}(x)\partial_\mu\mathbf{g}^{-1}(x). \tag{19.6}$$

From this expression we learn that, from the point of view of global transformations ($\mathbf{g}(x)$ constant), the field $\mathbf{A}_\mu(x)$, which is usually called the *gauge field* or *Yang–Mills field*, transforms by the adjoint representation of the group G.

Covariant derivative. To the connection $\mathbf{A}_\mu(x)$ is associated a covariant derivative, whose explicit form depends on the tensor on which it is acting. To obtain its expression when acting on $\phi(x)$ we consider in equation (19.3) the limit (19.4) of an infinitesimal curve. Equation (19.3) becomes:

$$\begin{aligned}
\phi_U &= \left(\mathbf{1} + \mathbf{A}_\mu(x)\mathrm{d}x_\mu\right)\left(\phi(x) + \partial_\mu\phi(x)\mathrm{d}x_\mu\right) + o\left(\|\mathrm{d}x_\mu\|\right) \\
&= \left(\mathbf{1} + \mathrm{d}x_\mu\mathbf{D}_\mu\right)\phi(x),
\end{aligned} \tag{19.7}$$

where we have introduced the *covariant derivative* $\mathbf{D}_\mu$:

$$\mathbf{D}_\mu = \mathbf{1}\,\partial_\mu + \mathbf{A}_\mu. \tag{19.8}$$

$\mathbf{D}_\mu$ is both a differential operator acting on space variables and a matrix. The identity:

$$\mathbf{g}(x)\left(\mathbf{1}\,\partial_\mu + \mathbf{A}_\mu\right)\mathbf{g}^{-1}(x) = \mathbf{1}\,\partial_\mu + \mathbf{g}(x)\mathbf{A}_\mu(x)\mathbf{g}^{-1}(x) + \mathbf{g}(x)\partial_\mu\mathbf{g}^{-1}(x), \tag{19.9}$$

shows that $\mathbf{D}'_\mu$, the transform of $\mathbf{D}_\mu$ under the gauge transformation (19.6), is:

$$\mathbf{D}'_\mu = \mathbf{g}(x)\mathbf{D}_\mu\,\mathbf{g}^{-1}(x), \tag{19.10}$$

as expected. In equations (19.9,19.10) the products have to be understood as products of differential and multiplicative operators.

Infinitesimal gauge transformations. Setting:

$$\mathbf{g}(x) = \mathbf{1} + \omega(x) + o\left(\|\omega\|\right), \tag{19.11}$$

in which $\omega(x)$ belongs to the Lie algebra of G, we derive from equation (19.6) the form of the infinitesimal gauge transformation of the field $\mathbf{A}_\mu$:

$$-\delta\mathbf{A}_\mu(x) = \partial_\mu\omega + [\mathbf{A}_\mu, \omega] \equiv \mathbf{D}_\mu\omega. \tag{19.12}$$

In equation (19.8) we have given the form of the covariant derivative corresponding to the initial representation of the group, i.e. when acting on $\phi(x)$. Equation (19.12) yields the form of the covariant derivative in the adjoint representation. It is easy to verify that:

$$\partial_\mu\omega' + [\mathbf{A}'_\mu, \omega'] = \mathbf{g}(x)\{\partial_\mu\omega + [\mathbf{A}_\mu, \omega]\}\mathbf{g}^{-1}(x), \tag{19.13}$$

in which $\mathbf{A}'_\mu$ is given by equation (19.6) and ω' by:

$$\omega'(x) = \mathbf{g}(x)\omega(x)\mathbf{g}^{-1}(x). \tag{19.14}$$

Curvature tensor. From the form of the covariant derivative we can derive an expression for the curvature tensor $\mathbf{F}_{\mu\nu}$, generalization of the electromagnetic tensor of QED:

$$\mathbf{F}_{\mu\nu}(x) = [\mathbf{D}_\mu, \mathbf{D}_\nu] = \partial_\mu \mathbf{A}_\nu - \partial_\nu \mathbf{A}_\mu + [\mathbf{A}_\mu, \mathbf{A}_\nu], \tag{19.15}$$

which is again an element of the Lie algebra of G, and which transforms, as a consequence of equation (19.10), as:

$$\mathbf{F}'_{\mu\nu}(x) = \mathbf{g}(x)\mathbf{F}_{\mu\nu}(x)\mathbf{g}^{-1}(x). \tag{19.16}$$

We recognize the action of the adjoint representation of the group. As we show in Chapter 22 in a different context, the curvature tensor is associated with parallel transport along a closed curve.

Expressions in component form. In many cases it is useful to write previous expressions in component form. Let us call t^a the generators of the Lie algebra, and introduce the components $A^a_\mu(x)$ of $\mathbf{A}_\mu(x)$:

$$\mathbf{A}_\mu(x) = A^a_\mu(x)t^a. \tag{19.17}$$

The covariant derivative (19.8) then reads:

$$(D_\mu)_{ij} = \partial_\mu \delta_{ij} + A^a_\mu(x)t^a_{ij}. \tag{19.18}$$

Equation (19.12) involves the structure constants f_{abc} of the Lie algebra,

$$-\delta A^a_\mu(x) = \partial_\mu \omega_a(x) + f_{bca} A^b_\mu(x)\omega_c(x). \tag{19.19}$$

This equation yields also the representation of the covariant derivative in the case of adjoint representation.

Finally the curvature tensor can be written:

$$\mathbf{F}_{\mu\nu}(x) = F^a_{\mu\nu}(x)t^a, \tag{19.20}$$

and therefore:

$$F^a_{\mu\nu}(x) = \partial_\mu A^a_\nu(x) - \partial_\nu A^a_\mu(x) + f_{bca} A^b_\mu(x)A^c_\nu(x). \tag{19.21}$$

This last expression is independent of the group representation.

Forms and gauge fields. It is sometimes convenient to use the language of differential forms and to associate with gauge fields 1-forms, and 2-forms with the curvature tensor:

$$\mathbf{A} = \mathbf{A}_\mu \mathrm{d}x_\mu, \quad \mathbf{F} = \mathbf{F}_{\mu\nu}\mathrm{d}x_\mu \wedge \mathrm{d}x_\nu = 2\left(\mathrm{d}\mathbf{A} + \mathbf{A}^2\right), \tag{19.22}$$

where the differential operator d acting on forms has been defined in Section 1.6: $\mathrm{d} \equiv \mathrm{d}x_\mu \partial_\mu$.

19.2 The Gauge Invariant Action

The simplest gauge invariant action $S(\mathbf{A}_\mu)$ function of the gauge field $\mathbf{A}_\mu$, and generalization of the action (18.35) of the abelian case, has the form:

$$S\left(\mathbf{A}_\mu\right) = -\frac{1}{4e^2}\int \mathrm{d}^d x \,\mathrm{tr}\,\mathbf{F}_{\mu\nu}^2\left(x\right). \qquad (19.23)$$

Note that we have not added a mass term for the gauge field as in Section 18.1. We indeed show in Appendix A19 that in the non-abelian case the zero mass limit is singular. We have also changed the normalization of the gauge field, in such a way that all geometrical quantities become independent of the gauge coupling constant. The sign in front of the action takes into account that, with our definition, the matrix $\mathbf{F}_{\mu\nu}$ is anti-hermitian or antisymmetric.

Two remarks are immediately in order:

(i) As in the abelian case, the action does not provide a dynamics to all components of the gauge field. An additional dynamics has to be provided to the degrees of freedom which correspond to gauge transformations.

(ii) In contrast with the abelian case, because the gauge field transforms non-trivially under the group as equation (19.6) shows (the gauge field is "charged"), the curvature tensor $\mathbf{F}_{\mu\nu}$ is not gauge invariant, and thus not directly a physical observable. The action (19.23) is no longer a free field action; the gauge field has self-interactions and even the spectrum of the pure gauge action is non-perturbative (some analytic results can be obtained in dimension two). We indicate in Chapter 33 how lattice gauge theory provides a framework for non-perturbative investigations. The difference between the abelian and the non-abelian case is reminiscent of the non-linear chiral model.

Matter fields. It is now easy to add to the action gauge invariant terms for the matter fields. For fermions they will have the form:

$$S_{\mathrm{F}}\left(\bar{\psi},\psi\right) = -\int \mathrm{d}^d x\,\bar{\psi}\left(x\right)\left(\slashed{D} + M\right)\psi\left(x\right),$$

and for the boson fields:

$$S_{\mathrm{B}}(\phi) = \int \mathrm{d}^d x\left[\left(\mathbf{D}_\mu\phi\right)^\dagger \mathbf{D}_\mu\phi + V\left(\phi\right)\right],$$

in which $V(\phi)$ is a group invariant function of the scalar field ϕ.

19.3 Hamiltonian Formalism. Quantization

We first show that non-abelian gauge theories can be quantized, using a simple hamiltonian formalism, as we have done in the abelian case in Section 18.4. In this way we construct a theory which, at least at the formal level, is unitary because it corresponds to a hermitian hamiltonian. All expressions in this section will be written for a pure gauge theory, because the extension of all arguments to a general gauge invariant theory is straightforward.

Classical field equations. We first consider real time field theory, we denote by $t \equiv x_0 = ix_d$ time and the corresponding field component by $\mathbf{A}_0 = -i\mathbf{A}_d$. We use the notation

$\dot{Q}$ to represent the time derivative of Q. Space components will carry roman indices $(\mathbf{A}_i, x_i)$. To the continuation to real time of action (19.23) corresponds a classical field equation:

$$\mathbf{D}_\mu \mathbf{F}^{\mu\nu}(x) = 0\,, \tag{19.24}$$

in which the explicit form of $\mathbf{D}_\mu$ is given by equation (19.12). The equation (19.24) does not lead to a standard quantization because, as we have already discussed in the abelian case, the action does not depend on $\dot{\mathbf{A}}_0$, the time derivative of $\mathbf{A}_0$. Thus again $\mathbf{A}_0$ is not a dynamical variable and can be eliminated from the action by solving the corresponding field equation. The $\mathbf{A}_0$ field equation is a constraint equation. However in the absence of a mass term, as in the abelian case, the reduced action does not depend on all space components of the gauge field. Only the combination $\left[\delta_{ij} - \mathbf{D}_i(\mathbf{D}_\perp^2)^{-1}\mathbf{D}_j\right]\dot{\mathbf{A}}_j$ appears ($\mathbf{D}_\perp^2$ is the covariant space laplacian). But in contrast with the abelian case the projector acting on $\mathbf{A}_i$ depends on the field itself, and therefore the procedure which led to Coulomb's gauge here does not work, at least in its simplest form. Therefore we immediately discuss the quantization in the temporal gauge.

As in the abelian case we first note that if $\mathbf{A}_\mu(t, x)$ is a solution of the equation (19.24), any gauge transform of $\mathbf{A}_\mu(t, x)$ is also a solution. To describe all solutions we can thus make a gauge section, i.e. consider a section in the space of all gauge fields which intersects once all gauge orbits. We then represent all solutions by an element of the section and a gauge transformation. One choice of gauge condition is specially well-suited to the construction of a hamiltonian formalism,

$$\mathbf{A}_0(t, x) = 0\,, \tag{19.25}$$

which defines the *temporal gauge*.

Let us write the euclidean equation (19.24), separating the space and time components:

$$\begin{cases} \mathbf{D}_0 \mathbf{F}_{0k} - \mathbf{D}_l \mathbf{F}_{lk} = 0\,, \\ \mathbf{D}_l \mathbf{F}_{l0} = 0\,. \end{cases} \tag{19.26}$$

The indices k, l vary only from 1 to $d - 1$ and correspond to space components. In the gauge (19.25) the equations simplify and become:

$$\dot{\mathbf{E}}_k = \mathbf{D}_l \mathbf{F}_{lk}\,, \tag{19.27}$$

$$\mathbf{D}_l \mathbf{E}_l = 0\,, \tag{19.28}$$

with:

$$\mathbf{E}_k = \dot{\mathbf{A}}_k\,. \tag{19.29}$$

The equation (19.27) can be directly derived from the original lagrangian in which the condition (19.25) has been used:

$$\mathcal{L}\left(\mathbf{A}_k\right) = -\operatorname{tr}\int \mathrm{d}^{d-1}x \left[\dot{\mathbf{A}}_k^2(t, x) - \frac{1}{4e^2}\mathbf{F}_{kl}^2(t, x)\right]. \tag{19.30}$$

Expression (19.30) defines a conventional lagrangian for the space components of the gauge field: $\mathbf{E}_k$ is the conjugated momentum of $\mathbf{A}_k$ and the corresponding hamiltonian takes the form:

$$\mathcal{H}\left(\mathbf{E}, \mathbf{A}\right) = -\operatorname{tr}\int \mathrm{d}^{d-1}x \left[\mathbf{E}_k^2(x) + \frac{1}{4e^2}\mathbf{F}_{kl}^2(x)\right]. \tag{19.31}$$

However the only relevant solutions of the Hamilton equations are those which satisfy the constraint equation (19.28), non-abelian generalization of Gauss's law. It is easy to verify that equation (19.28) is compatible with the classical motion, i.e. that the Poisson brackets of the equation with the hamiltonian vanish. The reason is easy to understand: the gauge condition (19.25) is left invariant by time-independent gauge transformations. Therefore time-independent gauge transformations form a symmetry group of the lagrangian (19.30) and thus of the hamiltonian (19.31). The quantities $\mathbf{D}_l\mathbf{E}_l$ are the generators, in the sense of Poisson brackets, of the symmetry group. These considerations immediately generalize to the quantum hamiltonian, the operators $\mathbf{D}_l\mathbf{E}_l$ are the generators of a symmetry group of the hamiltonian. The space of admissible physical states $\Psi(\mathbf{A})$ is restricted by the quantum generalization of Gauss's law:

$$\mathbf{D}_l\mathbf{E}_l\Psi(\mathbf{A}) \equiv \mathbf{D}_l\frac{\delta}{\delta\mathbf{A}_l(x)}\Psi(\mathbf{A}) = 0. \tag{19.32}$$

This equation implies that physical wave functions are gauge invariant, i.e. that physical states belong to the invariant sector of the symmetry group, space which is left invariant by quantum evolution.

Quantization in the temporal gauge, as in the abelian case, then follows completely conventional lines. Returning to the *euclidean formalism*, we conclude that the partition function can be written:

$$Z = \int[\mathrm{d}\mathbf{A}_\mu]\,\delta(\mathbf{A}_d)\exp\left[\frac{1}{4e^2}\int\mathrm{d}^dx\,\mathrm{tr}\,\mathbf{F}^2_{\mu\nu}(x)\right]. \tag{19.33}$$

Remarks. The theory we have constructed is not explicitly space time covariant and this leads to serious difficulties as we have already emphasized in the abelian case (see Section 18.4). In the temporal gauge the theory it is not renormalizable by simple power counting. Indeed if we examine the propagator in this gauge

$$W^{(2)}_{ij}(\mathbf{k}_\perp, k_d) = \frac{1}{k^2}\left(\delta_{ij} - \frac{k_ik_j}{\mathbf{k}^2_\perp}\right) + \frac{1}{k_d^2}\frac{k_ik_j}{\mathbf{k}^2_\perp}, \tag{19.34}$$

in which $\mathbf{k}_\perp$ is the "space" part of $\mathbf{k}$, we observe that it does not decrease at k_d fixed for large spatial momenta.

To solve these various problems we now show that gauge invariant observables can equivalently be calculated from another quantum action which leads to a theory which is explicitly covariant and renormalizable by power counting.

Equivalence with a covariant gauge. We would like to work with a covariant gauge constraining $\partial_\mu\mathbf{A}_\mu$ rather than with the temporal gauge (19.33). We first define the functional $\Delta(\mathbf{A}_\mu, \boldsymbol{\nu})$:

$$1 = \Delta(\mathbf{A}_\mu, \boldsymbol{\nu})\int[\mathrm{dg}]\,\delta\left[\partial_\mu\mathbf{A}^g_\mu(x) - \boldsymbol{\nu}(x)\right], \tag{19.35}$$

in which $\mathbf{g}$ denotes a space-dependent group element of G and $\mathbf{A}^g_\mu$ the gauge transform of $\mathbf{A}_\mu$. The field $\boldsymbol{\nu}(x)$, which belongs to the Lie algebra of G, up to this point is arbitrary.

We then multiply the integrand in expression (19.33) by the r.h.s. of equation (19.35). Equation (19.35) implies that the partition function has not changed. We inverse the order of integrations between $\mathbf{A}_\mu$ and $\mathbf{g}$, and set:

$$\mathbf{A}^g_\mu(x) = \mathbf{A}'_\mu(x). \tag{19.36}$$

The action (19.23) is gauge invariant. The functional $\Delta\left(\mathbf{A}_{\mu}, \boldsymbol{\nu}\right)$, being an average over all gauge transformations is also gauge invariant. Changing $\mathbf{g}$ into $\mathbf{g}^{-1}$ and dropping the $'$ in $\mathbf{A}'$, we then obtain:

$$Z = \int [\mathrm{dg}]\left[\mathrm{d}\mathbf{A}_{\mu}\right] \delta\left(\mathbf{A}_d^g\right) \delta\left(\partial_\mu \mathbf{A}_\mu - \boldsymbol{\nu}\right) \Delta\left(\mathbf{A}_\mu, \boldsymbol{\nu}\right) \exp\left[\frac{1}{4e^2}\int \mathrm{d}^d x\, \mathrm{tr}\, \mathbf{F}_{\mu\nu}^2(x)\right]. \quad (19.37)$$

We now integrate over the gauges $\mathbf{g}(x)$. The result of the integral $\int [\mathrm{dg}]\, \delta\left(\mathbf{A}_d^g\right)$ is gauge invariant. We can thus calculate it only for fields satisfying the condition $\mathbf{A}_d = 0$. We then have to evaluate the determinant coming from the variation of $\mathbf{A}_d^g$ at $\mathbf{g} = \mathbf{1}$. We find

$$\delta \mathbf{A}_d = \mathbf{D}_d \delta \mathbf{g}\,.$$

However since $\mathbf{A}_d = 0$ the covariant derivative $\mathbf{D}_d$ is equal to the ordinary derivative ∂_d and the corresponding determinant is a constant independent of $\mathbf{A}_d$.

We see in expression (19.37) that we need the functional $\Delta\left(\mathbf{A}_\mu, \boldsymbol{\nu}\right)$ only for fields satisfying $\partial_\mu \mathbf{A}_\mu = \boldsymbol{\nu}$. This means in equation (19.35) that the δ-function determines $\mathbf{g} = \mathbf{1}$. The functional Δ is thus the determinant of the operator obtained as the response to an infinitesimal gauge transformation of the gauge condition:

$$\delta\left(\partial_\mu \mathbf{A}_\mu - \boldsymbol{\nu}\right) \Delta\left(\mathbf{A}_\mu, \boldsymbol{\nu}\right) = \delta\left(\partial_\mu \mathbf{A}_\mu - \boldsymbol{\nu}\right) \det \mathbf{M}\,, \quad (19.38)$$

with

$$\mathbf{M}\delta\mathbf{g} = -\partial_\mu \mathbf{D}_\mu\left(\mathbf{A}\right) \delta\mathbf{g}\,. \quad (19.39)$$

In this form the theory is no longer local. We thus rewrite $\det \mathbf{M}$ as an integral over scalar fermions $\mathbf{C}$ and $\bar{\mathbf{C}}$, the Faddeev–Popov "ghosts", transforming under the adjoint representation of G. We replace the δ-function by its Fourier representation and call $\boldsymbol{\lambda}$ the corresponding integration variable. This leads to the functional integral representation

$$Z = \int \left[\mathrm{d}\mathbf{A}_\mu \mathrm{d}\mathbf{C}\mathrm{d}\bar{\mathbf{C}}\mathrm{d}\boldsymbol{\lambda}\right] \exp\left[-S\left(\mathbf{A}_\mu, \mathbf{C}, \bar{\mathbf{C}}, \boldsymbol{\lambda}, \boldsymbol{\nu}\right)\right], \quad (19.40)$$

where the quantum action $S\left(\mathbf{A}_\mu, \mathbf{C}, \bar{\mathbf{C}}, \boldsymbol{\lambda}, \boldsymbol{\nu}\right)$ is given by

$$S\left(\mathbf{A}_\mu, \mathbf{C}, \bar{\mathbf{C}}, \boldsymbol{\lambda}, \boldsymbol{\nu}\right) = \int \mathrm{d}^d x\, \mathrm{tr}\left[-\frac{1}{4e^2}\mathbf{F}_{\mu\nu}^2 + \boldsymbol{\lambda}(x)\left[\partial_\mu \mathbf{A}_\mu(x) - \boldsymbol{\nu}(x)\right] + \bar{\mathbf{C}}(x)\partial_\mu \mathbf{D}_\mu \mathbf{C}(x)\right]. \quad (19.41)$$

Since the partition function is independent of $\boldsymbol{\nu}$ we can now average over the field $\boldsymbol{\nu}(x)$ with a gaussian distribution $[\mathrm{d}\rho(\boldsymbol{\nu})]$:

$$[\mathrm{d}\rho\left(\boldsymbol{\nu}\right)] = [\mathrm{d}\boldsymbol{\nu}] \exp\left[\frac{1}{2\xi e^2}\int \mathrm{d}^d x\, \mathrm{tr}\, \boldsymbol{\nu}^2(x)\right]. \quad (19.42)$$

We here assume for simplicity that the representation of G is orthogonal, and therefore the generators of the Lie algebra are antisymmetric matrices. The averaged partition function Z then reads:

$$Z = \int \left[\mathrm{d}\mathbf{A}_\mu \mathrm{d}\mathbf{C}\mathrm{d}\bar{\mathbf{C}}\mathrm{d}\boldsymbol{\lambda}\right] \exp\left[-S\left(\mathbf{A}_\mu, \mathbf{C}, \bar{\mathbf{C}}, \boldsymbol{\lambda}\right)\right], \quad (19.43)$$

with:

$$S\left(\mathbf{A}_\mu, \mathbf{C}, \bar{\mathbf{C}}, \boldsymbol{\lambda}\right) = \int \mathrm{d}^d x \, \mathrm{tr}\left[-\frac{1}{4e^2}\mathbf{F}^2_{\mu\nu} + \frac{\xi e^2}{2}\boldsymbol{\lambda}^2(x) + \boldsymbol{\lambda}(x)\partial_\mu \mathbf{A}_\mu(x) + \bar{\mathbf{C}}(x)\partial_\mu \mathbf{D}_\mu \mathbf{C}(x)\right].$$
(19.44)

Except in the limit in which ξ vanishes, it is also possible to integrate over $\boldsymbol{\lambda}(x)$ to find a new quantum action $S\left(\mathbf{A}_\mu, \mathbf{C}, \bar{\mathbf{C}}\right)$:

$$S\left(\mathbf{A}_\mu, \mathbf{C}, \bar{\mathbf{C}}\right) = \int \mathrm{d}^d x \, \mathrm{tr}\left\{-\frac{1}{e^2}\left[\frac{1}{4}\mathbf{F}^2_{\mu\nu} + \frac{1}{2\xi}\left(\partial_\mu \mathbf{A}_\mu\right)^2\right] + \bar{\mathbf{C}}(x)\partial_\mu \mathbf{D}_\mu \mathbf{C}(x)\right\}.$$
(19.45)

This form of the quantum action is specially suited for perturbative calculations, although geometrical properties of the action are more apparent on expression (19.44), in particular the expected BRS symmetry of the action.

We have therefore established the formal equivalence between the two expressions (19.45) and (19.33) of the partition function. The formal equivalence with other gauges can be proven by a similar method. The obvious drawback of the covariant gauge, which leads to a covariant, local and renormalizable theory, is the lack of explicit positivity and thus unitarity. In particular Faddeev–Popov fermions do not obey to the spin–statistics connection.

19.4 The Faddeev–Popov Quantization, BRS Symmetry

We now present a more heuristic form of the quantization procedure, which leads to the same results, and which can be justified in the framework of lattice regularization in a finite volume. This derivation has one advantage: it provides a simple explanation of the appearance of a BRS symmetry.

19.4.1 Faddeev–Popov quantization

As in the case of abelian gauge theories, the naive functional integral corresponding to the gauge invariant action (19.23) is not defined because the action does not depend on the degrees of freedom associated with gauge transformations. Following the method introduced in Section 18.4, we give them a stochastic dynamics. Let us assume that we can find a gauge section in the space of gauge fields which intersect once and only once all gauge orbits. This is certainly possible for small fields and therefore in perturbation theory (see the remark at the end of the section concerning this problem beyond perturbation theory). We call $\mathbf{B}_\mu$ the gauge field belonging to the section and parametrize an arbitrary gauge field $\mathbf{A}_\mu$ as:

$$\mathbf{A}_\mu(x) = [\mathbf{B}_\mu(x)]^g \equiv \mathbf{g}(x)\mathbf{B}_\mu(x)\mathbf{g}^{-1}(x) + \mathbf{g}(x)\partial_\mu \mathbf{g}^{-1}(x).$$
(19.46)

The action (19.23) depends only on $\mathbf{B}_\mu$ and takes care of its dynamics. The dynamics of $\mathbf{g}(x)$ will be given by a stochastic equation. A convenient covariant choice is:

$$\partial_\mu [\mathbf{B}_\mu(x)]^g \equiv \partial_\mu \mathbf{A}_\mu(x) = \boldsymbol{\nu}(x),$$
(19.47)

in which $\boldsymbol{\nu}(x)$ is a stochastic field.

To justify the manipulations which follow, we have to assume that the equation (19.47) has, at $\boldsymbol{\nu}(x)$ and $\mathbf{B}_\mu(x)$ fixed, a unique solution for $\mathbf{g}(x)$, condition which again is met for small fields. We can then write the partition function corresponding to the action (19.23) and gauge condition (19.47) as:

$$Z = \int [\mathrm{dg dB}_\mu \mathrm{d}\lambda]\det \mathbf{M} \exp\left\{-S(\mathbf{B}_\mu) - \mathrm{tr}\left[\lambda(x)\left(\partial_\mu \mathbf{A}_\mu(x) - \boldsymbol{\nu}(x)\right)\right]\right\}, \qquad (19.48)$$

where $[\mathrm{d}g\,\mathrm{d}B_\mu]$ is a symbolic notation for the factorized measure induced by the flat measure $[\mathrm{d}A_\mu]$. The matrix λ belongs to the Lie algebra and imposes the condition (19.47). The operator $\mathbf{M}$ comes from the variation of equation (19.47) with respect to $\mathbf{g}(x)$ (see Chapter 16 for a general discussion). From equation (19.12) it follows that $\mathbf{M}$ is the operator (19.39). Finally, since the action (19.23) is gauge invariant, the whole quantum action can be entirely expressed in terms of $\mathbf{A}$:

$$Z = \int \left[\mathrm{dA}_\mu \mathrm{dCd\bar{C}d}\lambda\right] \exp\left[-S\left(\mathbf{A}_\mu, \mathbf{C}, \bar{\mathbf{C}}, \lambda, \boldsymbol{\nu}\right)\right], \qquad (19.49)$$

with:

$$S\left(\mathbf{A}_\mu, \mathbf{C}, \bar{\mathbf{C}}, \lambda, \boldsymbol{\nu}\right) = \int \mathrm{d}^d x \, \mathrm{tr}\left[-\frac{1}{4e^2}\mathbf{F}_{\mu\nu}^2 + \lambda(x)\left[\partial_\mu \mathbf{A}_\mu(x) - \boldsymbol{\nu}(x)\right] + \bar{\mathbf{C}}(x)\partial_\mu \mathbf{D}_\mu \mathbf{C}(x)\right]. \qquad (19.50)$$

We now average over the noise $\boldsymbol{\nu}(x)$ in the special case of the gaussian distribution $[\mathrm{d}\rho(\nu)]$ of equation (19.42). The averaged partition function Z then takes the form (19.43), where the action $S\left(\mathbf{A}_\mu, \mathbf{C}, \bar{\mathbf{C}}, \lambda\right)$ is exactly the action (19.44).

Remark. As noted by Gribov, in contrast with the abelian case, depending on the value of the gauge field $\mathbf{B}_\mu(x)$, the gauge condition (19.47) does not always have a unique solution in $\mathbf{g}(x)$. Gribov's ambiguity is related to the property that there exist fields for which the operator $\partial_\mu \mathbf{D}_\mu\left(\mathbf{A}\right)$ has zero eigenvalues. This implies that the representation (19.43) is not meaningful beyond perturbation theory. The same ambiguity has been shown to arise for a large class of gauge conditions.

19.4.2 BRS symmetry

One consequence of the quantization procedure is that the action (19.45) no longer has a gauge symmetry. On the other hand we expect from the analysis of Chapter 16 applied to the formulation of the quantization procedure as explained in the preceding subsection, that the action now has a BRS symmetry, consequence of the stochastic dynamics given to the degrees of freedom of the gauge group variables. To obtain the form of the BRS transformations, it is convenient to separate the gauge group degrees of freedom which induce the BRS symmetry from the other degrees of freedom of the gauge field which play no role. We thus again use the parametrization (19.46). From the analysis of Section 16.4 we know the form of the BRS transformations in the case of group manifolds:

$$\begin{cases} \delta \mathbf{g}(x) = \bar{\varepsilon}\mathbf{C}(x)\mathbf{g}(x), & \delta\mathbf{C}(x) = \bar{\varepsilon}\mathbf{C}^2(x), \\ \delta\bar{\mathbf{C}}(x) = \bar{\varepsilon}\lambda(x), & \delta\lambda(x) = 0. \end{cases} \qquad (19.51)$$

The field $\mathbf{B}_\mu(x)$ has a dynamics provided by the gauge action and is not affected by the BRS transformation:

$$\delta \mathbf{B}_\mu(x) = 0. \qquad (19.52)$$

Let us then calculate the effect of a BRS transformation on the field $\mathbf{A}_\mu$. We find:

$$\delta \mathbf{A}_\mu(x) = \delta \mathbf{B}_\mu^g(x) = -\bar{\varepsilon} \mathbf{D}_\mu \mathbf{C}(x) . \tag{19.53}$$

This result is not surprising: the transformations (19.51) correspond to an infinitesimal change in the gauge group degrees of freedom and thus to an infinitesimal gauge transformation for $\mathbf{A}_\mu$. We recognize also in equation (19.53) the transformation law of the current associated with group elements as discussed in Section 17.7 (equations (17.75–17.77)).

We show in Chapter 21, in a more general framework, that WT identities associated with the BRS symmetry (19.51,19.53) imply the structural stability of the quantum action (19.44) under renormalization.

19.5 Perturbation Theory, Regularization

Compared with the abelian case, the new features of the action (19.45) are the presence of gauge field self-interactions and ghost terms. Let us first rewrite the different terms of the gauge action in component form to establish conventions and normalizations. The gauge action will be written:

$$S(A_\mu^a) = \frac{1}{4e^2} \int \mathrm{d}^d x \, F_{\mu\nu}^a F_{\mu\nu}^a , \tag{19.54}$$

where the curvature tensor is given by equation (19.21) and the trace of the unit matrix has been swallowed into a redefinition of the coupling constant. In the covariant gauge of Section 19.3, the vector field propagator is (equation (18.22)):

$$[\Delta_\xi]_{\mu\nu}^{ab}(k) = e^2 \delta ab \left[\frac{\delta_{\mu\nu}}{k^2} + (\xi - 1) \frac{k_\mu k_\nu}{(k^2)^2} \right] . \tag{19.55}$$

In four dimensions, as in the abelian case, the gauge field has dimension 1. The ghost field action takes the form:

$$S_{\text{ghost}} = \int \mathrm{d}^d x \, \bar{C}^a \partial_\mu \left(\partial_\mu \delta_{ac} + f_{bca} A_\mu^b \right) C^c . \tag{19.56}$$

The ghost fields thus have a simple δ_{ab}/p^2 propagator and canonical dimension 1 in four dimensions. The interaction terms have all dimension 4 and therefore the theory is renormalizable by power counting in four dimensions. The power counting for matter fields is of course the same as in the abelian case. Let us write explicitly the new vertices of the non-abelian theory. The gauge field 3-point function at leading order is:

$$[\Gamma^{(3)}]_{\mu\nu\rho}^{abc}(p,q,r) = \frac{i}{e^2} f_{abc} \left[(r-q)_\mu \, \delta_{\nu\rho} + (p-r)_\nu \, \delta_{\rho\mu} + (q-p)_\rho \, \delta_{\mu\nu} \right] . \tag{19.57}$$

The gauge 4-point function is given by:

$$\begin{aligned}
[\Gamma^{(4)}]_{\mu\nu\rho\sigma}^{abcd} = \frac{1}{e^2} \big[&f_{eab}f_{ecd} \left(\delta_{\mu\rho}\delta_{\nu\sigma} - \delta_{\mu\sigma}\delta_{\nu\rho} \right) + f_{eac}f_{ebd} \left(\delta_{\mu\nu}\delta_{\rho\sigma} - \delta_{\mu\sigma}\delta_{\nu\rho} \right) \\
&+ f_{ead}f_{ecb} \left(\delta_{\mu\rho}\delta_{\nu\sigma} - \delta_{\mu\nu}\delta_{\sigma\rho} \right) \big] .
\end{aligned} \tag{19.58}$$

All terms are obtained from the first by exchanging the indices to make the correlation function totally symmetric. Finally there is a ghost gauge field vertex:

$$\langle \bar{C}^a(p) C^b(q) A_\mu^c(r) \rangle = -i f_{abc} p_\mu \, . \tag{19.59}$$

Notice that in the generic case both ghost fields play a different role and no symmetry exchanges them. In a graphic representation of Feynman diagrams ghost propagator lines generally carry an arrow. A special case is $\xi = 0$, Landau's gauge, in which, because $\partial_\mu \mathbf{A}_\mu$ vanishes, the vertex can be antisymmetrized.

Infrared divergences. In the covariant gauge, and in the absence of a Higgs mechanism which provides a mass for the gauge fields, only the gauge $\xi = 1$ (the Feynman gauge) leads to a theory which is obviously IR finite. In contrast to the abelian case, it is impossible to give an explicit mass to the gauge field and to then construct a theory which is both unitary and renormalizable (for more details see Appendix A19). On the other hand we want eventually to prove the gauge independence of the theory and therefore we must be able to define it for more than one gauge. One way to introduce an IR regulator is to consider the theory in a finite volume. This necessitates a discussion of finite volume effects which will be given in Chapter 36.

Matter fields. The coupling to matter fields differs from the abelian case only by some geometrical factors corresponding to group indices. For example the coupling to fermions generated by the covariant derivative (19.18) is simply $\gamma_\mu t_{ij}^a$.

Regularization. The problem of regularization in non-abelian gauge theories has many features in common with the abelian case, as well as with the non-linear σ-model. Three regularization methods are of particular interest: dimensional regularization, which is the most convenient for practical calculations and works in the absence of chiral fermions; lattice regularization, which also leads to non-perturbative calculations (see Chapter 33 for details) and which again fails for chiral fermions; and finally Pauli–Villars's type regularizations which work in general except that, as we have already seen in previous geometrical models, they do not regularize all one-loop diagrams. Let us discuss briefly the latter method. The regularized gauge action takes the form:

$$S\left(\mathbf{A}_\mu\right) = \int \mathrm{d}^d x \, \mathrm{tr} \, \mathbf{F}_{\mu\nu} P\left(\mathbf{D}^2/\Lambda^2\right) \mathbf{F}_{\mu\nu} \, , \tag{19.60}$$

in which P is a polynomial of arbitrary degree. In the same way the gauge function $\partial_\mu \mathbf{A}_\mu$ is changed into:

$$\partial_\mu \mathbf{A}_\mu \longmapsto Q\left(\partial^2/\Lambda^2\right) \partial_\mu \mathbf{A}_\mu \, , \tag{19.61}$$

in which Q is a polynomial of same degree as P. As a consequence both the gauge field propagator and the ghost propagator can be made arbitrarily convergent. However, as in the abelian case, the covariant derivatives generate new interactions which are more singular. It is easy to verify that the power counting of one-loop diagrams is unchanged while higher order diagrams can be made convergent by taking the degrees of P and Q large enough. For massive fermions it is however still possible as, in the abelian case, to add a set of regulator fields, massive fermions and bosons with spin.

In the case of chiral fermions this is impossible, but the problem of the compatibility between the gauge symmetry and the quantum corrections is reduced to an explicit verification of the WT identities for the one-loop diagrams. Note that the preservation of gauge symmetry is necessary for the cancellation of unphysical states in physical amplitudes, and thus essential to the physical relevance of the quantum field theory.

WT identities and renormalization. From the BRS symmetry corresponding to the transformations (19.51–19.53) follow WT identities. Their form is somewhat complicated and we postpone the discussion to Chapter 21, where we shall derive the general form of the renormalized action. Let us here give the result only in the example of the pure gauge action in the covariant gauge. We can assume that the gauge group G is simple. Then the renormalized form of the action (19.45) is given by the substitution:

$$\begin{cases} e^2 \longmapsto Z_e e^2, & \mathbf{A}_\mu \longmapsto Z_A^{1/2} \mathbf{A}_\mu, \\ \xi \longmapsto Z_A Z_e^{-1} \xi, & \bar{\mathbf{C}} \mathbf{C} \longmapsto Z_C \bar{\mathbf{C}} \mathbf{C}. \end{cases} \tag{19.62}$$

This result has a simple interpretation: the gauge structure (19.45) is preserved and the coefficient of $(\partial_\mu \mathbf{A}_\mu)^2$ is unrenormalized exactly as in the abelian case. However, unlike the abelian case, the transformation law of the gauge field and, more generally the form of the covariant derivative, are modified by the gauge field renormalization.

19.6 The Non-Abelian Higgs Mechanism

We have already discussed the Higgs mechanism in the abelian case. The basic idea is the same in the general case: the spontaneous breaking of a global symmetry associated with a gauge invariance leads to masses for gauge fields without generating massless Goldstone modes. Simply, because the group structure is richer, more complicated situations may arise. Let us therefore consider a classical gauge invariant action for a gauge field and a scalar boson ϕ transforming under an orthogonal representation of the symmetry group:

$$S\left(\mathbf{A}_\mu, \phi\right) = \int \mathrm{d}^d x \left[-\frac{1}{4e^2} \operatorname{tr} \mathbf{F}_{\mu\nu}^2(x) + \frac{1}{2} \left(\mathbf{D}_\mu \phi\right) \cdot \left(\mathbf{D}_\mu \phi\right) + V\left(\phi\right) \right]. \tag{19.63}$$

We assume that the symmetric potential $V(\phi)$ has non-symmetric minima. In the absence of a gauge symmetry, this is the situation which we have already analyzed in Section 13.4. Since the spectrum in the classical limit depends on the group structure and the representation content of the field ϕ we shall here consider only two set of examples.

19.6.1 Simple Lie groups

Let us first assume that G, the symmetry group of the action, is simple and is thus also the gauge group. Moreover we assume for simplicity that the field ϕ belongs to an irreducible representation. We call H the stabilizer of the vector $\mathbf{v}$, the minimum of the potential. We separate the generators t^α of G in the matrix representation into two subsets $\alpha \leq p$ corresponding to the Lie algebra $\mathcal{L}(H)$ of the subgroup H, and the complementary set $\mathcal{L}(G/H)$. We parametrize the scalar field $\phi(x)$ by:

$$\phi(x) = \exp\left[\sum_{\alpha > p} t^\alpha \theta^\alpha(x) \right] \left(\mathbf{v} + \rho(x)\right), \tag{19.64}$$

in which the vectors ρ and $\{t_i^\alpha v_i\}$ span two orthogonal subspaces. The transformation:

$$\phi(x) \longmapsto \{\theta^\alpha(x), \quad \rho(x)\}, \tag{19.65}$$

is such that the new fields $\rho(x)$ and $\theta^\alpha(x)$ can be expanded in powers of $\phi(x) - \mathbf{v}$. In the absence of gauge fields, we have used the representation (19.64) to show that the fields $\theta^\alpha(x)$ correspond to massless Goldstone modes induced by the spontaneous breaking of the G-symmetry.

Equation (19.64) can also be viewed as a local group transformation relating the two fields ϕ and $\rho + \mathbf{v}$. If we perform on the field $\mathbf{A}_\mu$ a gauge transformation of the form (19.6) with:

$$\mathbf{g}(x) = \exp\left[\sum_{\alpha > p} t^\alpha \theta^\alpha(x)\right], \tag{19.66}$$

we completely eliminate the fields θ^α from the action. In fact we have fixed (at least partially) the gauge. If we now examine the scalar field contribution to the action, we see that for $\rho = 0$ it reduces to a mass term for the gauge field:

$$\tfrac{1}{2}\mathbf{D}_\mu\phi \cdot \mathbf{D}_\mu\phi|_{\rho=0} = \tfrac{1}{2}\sigma_{\alpha\beta}A_\mu^\alpha A_\mu^\beta, \tag{19.67}$$

with the mass matrix $\sigma_{\alpha\beta}$:

$$\sigma_{\alpha\beta} = t_{ij}^\alpha v_j t_{ik}^\beta v_k. \tag{19.68}$$

The matrix $\sigma_{\alpha\beta}$ is positive and has a rank equal to the number of generators of $\mathcal{L}(G/H)$ which is also the number of fields θ^α, i.e. the would-be Goldstone bosons. We conclude that the spontaneous breaking of the G-symmetry generates no Goldstone bosons but instead gives masses to all gauge fields except those which are associated with the unbroken subgroup H. In the case in which the symmetry is completely broken, all components of the gauge field acquire a mass.

If one considers directly the classical action obtained after the gauge transformation associated with group element (19.66), the set of massive vector fields can be quantized in a completely standard way. As in the abelian case, the quantized theory is however not renormalizable.

19.6.2 The $G \times G$ symmetry

Another possibility is that the symmetry group of the action is the direct product of the gauge group G by another group G'. We here consider only the simplest example where the symmetry group is $G \times G$ and G is simple. We assume that the scalar boson field ϕ is a matrix transforming under $G \times G$ by:

$$\phi' = \mathbf{g}_1 \phi \mathbf{g}_2^{-1}, \tag{19.69}$$

in which $\mathbf{g}_1$ and $\mathbf{g}_2$ are two elements of G in a matrix representation. We further assume that one minimum of the potential is proportional to the unit matrix $\phi = v\mathbf{1}$ in such a way that the subgroup H is isomorphic to G with elements of the form $(\mathbf{g}, \mathbf{g})$. We recall that the coset space G/H is then a symmetric space (see Appendix A15.4). As above the would-be Goldstone bosons correspond to gauge transformations and can thus be eliminated from the action. In this example all components of the gauge field acquire a mass since the symmetry corresponding to the gauge field is completely broken. Furthermore all components of the gauge field have the same mass m_A:

$$\mathrm{tr}^T(\mathbf{D}_\mu\phi)(\mathbf{D}_\mu\phi)|_{\phi=v} = -v^2 \mathrm{tr}\, \mathbf{A}_\mu^2 \;\Rightarrow\; m_A^2 = e^2 v^2. \tag{19.70}$$

The action obtained after the gauge transformation specified by equation (19.66) contains only physical degrees of freedom and the quantization of all vector fields is straightforward. From the point of view of the initial theory the gauge has been completely fixed, hence the name of *unitary gauge*. We have constructed an action for massive vector fields transforming under the adjoint representation of a symmetry group G. The corresponding field theory is, however, not renormalizable by power counting. The difference with the massive vector field theory we examine in Appendix A19 is that the suitable addition of some scalar fields makes this theory equivalent, at least for the physical observables, with a renormalizable theory with additional unphysical degrees of freedom.

Remark. If we formally take the non-linear model limit, i.e. send the masses of all remaining scalar fields towards infinity at v fixed, we obtain an action for a self-interacting massive vector field (see Appendix A19).

The SU(2) example. Let us discuss more specifically the important example of the $SU(2)$ group because it can be considered as a simplified version of the Standard Model of weak-electromagnetic interactions which will be described in Section 20.1. We take for scalar field ϕ a 2×2 complex matrix transforming under the $(1/2,1/2)$ representation of $SU(2) \times SU(2)$ (see also Section 13.6). The simplest action then reads:

$$S\left(\mathbf{A}_\mu, \phi\right) = \int \mathrm{d}^d x \left[\frac{1}{4e^2} \mathbf{F}_{\mu\nu}^2 + \mathrm{tr}\left(\mathbf{D}_\mu \phi\right)^\dagger \mathbf{D}_\mu \phi + a\,\mathrm{tr}\,\phi\phi^\dagger + \frac{\lambda}{6}\left(\mathrm{tr}\,\phi\phi^\dagger\right)^2 \right], \quad (19.71)$$

We have represented the gauge field as a 3-component vector $\mathbf{A}_\mu$. The curvature tensor reads:

$$\mathbf{F}_{\mu\nu} = \partial_\mu \mathbf{A}_\nu - \partial_\nu \mathbf{A}_\mu - \mathbf{A}_\mu \times \mathbf{A}_\nu, \quad (19.72)$$

and the covariant derivative acts like:

$$\mathbf{D}_\mu \phi = \left(\partial_\mu + \tfrac{1}{2} i \mathbf{A}_\mu \cdot \boldsymbol{\tau}\right)\phi. \quad (19.73)$$

The $\boldsymbol{\tau}$ matrices are identical to the σ matrices defined in Section A5.1. We choose the expectation value of ϕ proportional to the unit matrix. It is then convenient to parametrize ϕ under the form (see Section 13.6):

$$\phi = \tfrac{1}{2}\left(\sigma + i\boldsymbol{\tau} \cdot \boldsymbol{\pi}\right) \quad (19.74)$$

in which σ and $\boldsymbol{\pi}$ are real fields. Let us write the form of an infinitesimal gauge transformation in this representation:

$$\begin{cases} \delta \mathbf{A}_\mu = \partial_\mu \boldsymbol{\omega} - \mathbf{A}_\mu \times \boldsymbol{\omega}, \\ \delta \sigma = \tfrac{1}{2}\boldsymbol{\omega} \cdot \boldsymbol{\pi}, \\ \delta \boldsymbol{\pi} = -\tfrac{1}{2}\sigma\boldsymbol{\omega} + \tfrac{1}{2}\boldsymbol{\omega} \times \boldsymbol{\pi}. \end{cases} \quad (19.75)$$

The scalar field action in these variables is given by:

$$S_{\text{scalar}} = \int \mathrm{d}^d x \left[\tfrac{1}{2}\left(\partial_\mu \sigma - \tfrac{1}{2}\boldsymbol{\pi} \cdot \mathbf{A}_\mu\right)^2 + \tfrac{1}{2}\left(\partial_\mu \boldsymbol{\pi} + \tfrac{1}{2}\sigma\mathbf{A}_\mu - \tfrac{1}{2}\mathbf{A}_\mu \times \boldsymbol{\pi}\right)^2 + \tilde{V}\left(\sigma^2 + \boldsymbol{\pi}^2\right) \right]. \quad (19.76)$$

As we have already discussed in Section 13.6, if the potential $\tilde{V}$ has degenerate classical minima, the field ϕ has a non-zero expectation value. We may by convention decide that the expectation value corresponds to the component σ:

$$\langle \sigma \rangle = v.$$

Then the symmetry $SU(2) \times SU(2)$ is broken down to the diagonal $SU(2)$ group. In the absence of gauge fields, the π-field becomes a massless Goldstone boson. Here the π-field can be eliminated by a gauge transformation, in such a way that the total action written in the unitary gauge becomes:

$$S\left(\mathbf{A}_\mu, \sigma\right) = \int \mathrm{d}^d x \left[\frac{1}{4e^2}\mathbf{F}_{\mu\nu}^2 + \frac{1}{2}\left(\partial_\mu \sigma\right)^2 + \frac{1}{8}\sigma^2 \mathbf{A}_\mu^2 + \frac{a}{2}\sigma^2 + \frac{\lambda}{24}\sigma^4\right]. \tag{19.77}$$

This is a $O(3)$-invariant action in which the vector field is a $O(3)$ vector and the scalar field a $O(3)$ scalar. The gauge field mass m_A is given at tree order in terms of the σ-field expectation value v by:

$$m_A = ev/2. \tag{19.78}$$

19.6.3 Gauge fixing of the Higgs model in a covariant gauge

If we consider the contribution (19.76) to the action, we see that when the σ-field has an expectation value, a term of the form $\partial_\mu \pi \cdot \mathbf{A}_\mu$ is generated which introduces a mixing between the would-be Goldstone boson π and the longitudinal part of the vector field. This is a feature which we have already encountered in the abelian case (Section 18.10). As suggested by 't Hooft, it is possible to use the gauge function to eliminate such a term. In the $SU(2)$ example we can take as gauge function F:

$$F\left(\mathbf{A}_\mu, \pi\right) = \partial_\mu \mathbf{A}_\mu + \tfrac{1}{2}\lambda \xi \pi. \tag{19.79}$$

After a gaussian integration, the corresponding contribution to the action is:

$$S_F = S_{\text{gauge}} + S_{\text{ghost}}, \tag{19.80}$$

with:

$$S_{\text{gauge}} = \frac{1}{2\xi e^2} \int \mathrm{d}^d x \left(\partial_\mu \mathbf{A}_\mu + \tfrac{1}{2}\lambda \xi \pi\right)^2, \tag{19.81}$$

and:

$$S_{\text{ghost}} = \int \mathrm{d}^d x \left[\partial_\mu \bar{\mathbf{C}} \cdot \left(\partial_\mu \mathbf{C} - \mathbf{A}_\mu \times \mathbf{C}\right) + \frac{\lambda \xi}{4}\bar{\mathbf{C}}\left(\sigma \mathbf{C} + \pi \times \mathbf{C}\right)\right]. \tag{19.82}$$

At leading order the term $\partial_\mu \pi \cdot \mathbf{A}_\mu$ is eliminated by the choice:

$$\lambda = e^2 v. \tag{19.83}$$

This gauge has two advantages: it decouples the gauge field from the would-be Goldstone field and therefore simplifies the propagators; by explicitly breaking the global $SU(2) \times SU(2)$-symmetry, it generates a mass for the π-field which is no longer a Goldstone boson. In this gauge the propagators (equations (18.152)) have no poles at zero momentum and no IR problems are encountered:

$$W_{\mu\nu}^{(2)} = \frac{e^2 \delta_{\mu\nu}}{k^2 + m_A^2} + \frac{e^2\left(\xi - 1\right) k_\mu k_\nu}{\left(k^2 + m_A^2\right)\left(k^2 + \xi m_A^2\right)},$$

$$W_{\pi\pi}^{(2)} = \frac{1}{k^2 + \xi m_A^2}, \tag{19.84}$$

$$W_{\bar{C}C}^{(2)} = \frac{1}{k^2 + \xi m_A^2},$$

in which m_A is the mass of $\mathbf{A}_\mu$ in the tree approximation (equation (19.78)). Furthermore all unphysical states have a mass which explicitly depends on the gauge parameter ξ.

Unitarity. This property can be used to prove unitarity of the physical S-matrix: The S-matrix satisfies a generalized unitarity relation in which in the intermediate states one must include all particles both physical and unphysical. By showing that the S-matrix does not depend on the gauge, one proves simultaneously that the contributions of unphysical states cancels in the intermediate states and thus the S-matrix is unitary in the physical subspace. A general proof of this kind will be given in Chapter 21.

Bibliographical Notes

Non-abelian gauge fields were first discussed in
  C.N. Yang and R.L. Mills, *Phys. Rev.* 96 (1954) 191; R. Utiyama, *Phys. Rev.* 101 (1956) 1597; M. Gell-Mann and S. Glashow, *Ann. Phys.* (NY) 15 (1961) 437.
Original articles on the quantization of gauge theories include
  R.P. Feynman, *Acta Phys. Polon.* 24 (1963) 697; B.S. DeWitt, *Phys. Rev.* 162 (1967) 1195, 1239; L.D. Faddeev and V.N. Popov, *Phys. Lett.* 25B (1967) 29; S. Mandelstam, *Phys. Rev.* 175 (1968) 1580; M. Veltman, *Nucl. Phys.* B21 (1970) 288; G. 't Hooft, *Nucl. Phys.* B33 (1971) 173;
The axial gauge was proposed in
  R.L. Arnowitt and S.I. Fickler, *Phys. Rev.* 127 (1962) 1821.
General linear gauges were proposed in
  G. 't Hooft, *Nucl. Phys.* B35 (1971) 167;
and further discussed in
  K. Fujikawa, B.W. Lee and A.I. Sanda, *Phys. Rev.* D6 (1972) 2923; B.W. Lee and J. Zinn-Justin, *Phys. Rev.* D7 (1973) 1049.
The quantization of constrained systems is discussed in
  L.D. Faddeev, *Theor. Math. Phys.* 1 (1969) 3.
The BRS symmetry of the quantized action was discovered by
  C. Becchi, A. Rouet and R. Stora, *Comm. Math. Phys.* 42 (1975) 127.
The first discussion of the ambiguity of gauge conditions is due to
  V.N. Gribov, *Nucl. Phys.* B139 (1978) 1.
The Higgs mechanism was introduced in
  P.W. Higgs, *Phys. Lett.* 12 (1964) 132; *Phys. Rev.* 145 (1966) 1156; F. Englert and R. Brout, *Phys. Rev. Lett.* 13 (1964) 321; G.S. Guralnik, C.R. Hagen and T.W.B. Kibble, *Phys. Rev. Lett.* 13 (1964) 585; T.W.B. Kibble, *Phys. Rev.* 155 (1967) 1554.
It was cast into the framework of renormalizable field theories in
  G. 't Hooft, *Nucl. Phys.* B35 (1971) 167.
The corresponding WT identities were derived in general form in
  B.W. Lee and J. Zinn-Justin, *Phys. Rev.* D5 (1972) 3137,
an article whose Appendix C has directly inspired Appendix A19.
  A selection of original articles has been reprinted in
  Gauge Theory of Weak and Electromagnetic Interactions, C.H. Lai ed. (World Scientific, Singapore 1981).
An early review is
  E.S. Abers and B.W. Lee, *Phys. Rep.* 9C (1973) 1.
Finally among the textbooks see for example the works of Itzykson and Zuber, and Faddeev and Slavnov, quoted in the preface.

APPENDIX 19
MASSIVE YANG–MILLS FIELDS

For completeness, we briefly explain why, in contrast with the abelian case, it is not possible to construct a renormalizable field theory in which a mass is given to the gauge field by directly adding a mass term to the action. We therefore consider the real time lagrangian density:

$$\mathcal{L}\left(\mathbf{A}_\mu\right) = -\frac{1}{e^2}\,\mathrm{tr}\left[-\tfrac{1}{4}\mathbf{F}_{\mu\nu}\mathbf{F}^{\mu\nu}(t,x) + \tfrac{1}{2}m^2\mathbf{A}_\mu\mathbf{A}^\mu(t,x)\right].\qquad(A19.1)$$

The first problem we meet is quantization. We know that $\mathbf{A}_0$, the time component of the gauge field, has no conjugated momentum and thus is not a dynamical variable. It can be eliminated, using the corresponding field equation. It is actually algebraically convenient to first determine the hamiltonian, and use the equation of motion afterwards. The hamiltonian is here obtained by performing a Legendre transformation only on space components. The conjugated momenta $\mathbf{E}_i$ are (roman indices mean space components)

$$\mathbf{E}_i = \frac{\partial \mathcal{L}}{\partial \dot{\mathbf{A}}_i} = -\frac{1}{e^2}\mathbf{F}_{0i}\,.$$

After an integration by parts the hamiltonian density $\mathcal{H}$ can be written

$$\begin{aligned}
\mathcal{H} &= \mathrm{tr}\,\mathbf{E_i}\dot{\mathbf{A}}_i - \mathcal{L}\\
&= -\,\mathrm{tr}\left[\tfrac{1}{2}e^2\mathbf{E}_i^2 + \partial_i\mathbf{E}_i\mathbf{A}_0 + \frac{m^2}{2e^2}\mathbf{A}_0^2 - \frac{1}{e^2}\left(\tfrac{1}{4}\mathbf{F}_{ij}^2 + \tfrac{1}{2}m^2\mathbf{A}_i^2\right)\right].\qquad(A19.2)
\end{aligned}$$

Using a familiar property of the Legendre transformation we obtain the $\mathbf{A}_0$ equation of motion

$$0 = -\frac{\partial \mathcal{L}}{\partial \mathbf{A}_0} = \frac{\partial \mathcal{H}}{\partial \mathbf{A}_0} = \frac{m^2}{e^2}\mathbf{A}_0 + \partial_i\mathbf{E}_i\,.\qquad(A19.3)$$

This equation determines $\mathbf{A}_0$ which can be eliminated from the hamiltonian. We then write the functional integral representation of the evolution operator in terms of the reduced hamiltonian, integrating over $\{\mathbf{E}_i, \mathbf{A}_i\}$. We note however that in this expression the reduced hamiltonian can be replaced by the initial hamiltonian $(A19.2)$ provided one integrates also over $\mathbf{A}_0$. Indeed the integral over $\mathbf{A}_0$ is gaussian: The integration thus sets $\mathbf{A}_0$ to the solution of the field equation $(A19.3)$ and yields a constant determinant because the coefficient of $\mathbf{A}_0^2$ in expression $(A19.2)$ is field-independent. Finally, integrating with the initial hamiltonian over $\mathbf{E}_i$ at $\mathbf{A}_\mu$ fixed, we recover the initial action. Thus the partition function can be expressed in terms of the simple euclidean action

$$S\left(\mathbf{A}_\mu\right) = -\frac{1}{e^2}\int \mathrm{d}^d x\,\mathrm{tr}\left[\frac{1}{4}\mathbf{F}_{\mu\nu}^2(x) + \frac{1}{2}m^2\mathbf{A}_\mu^2\right],\qquad(A19.4)$$

with a flat field integration measure. In contrast with the massless gauge-invariant situation this functional integral is well-defined.

Remark. In the massless case one can follow the same strategy. The $\mathbf{A}_0$ field equation $(A19.3)$, however, no longer determines $\mathbf{A}_0$ but instead yields Gauss's law. One may then wonder why one does not impose Gauss's law by integrating over $\mathbf{A}_0$ as above. The reason is that, due to space gauge invariance, Gauss's law commutes with the hamiltonian. Therefore if it is satisfied at one time it is satisfied at all later times. An integration over $\mathbf{A}_0(t, x)$ will be infinitely redundant, as we immediately verify since this procedure leads to the initial action and thus to an undefined functional integral.

The massless limit. Provided we only consider gauge invariant correlation functions, we can introduce a gauge condition and follow all the algebraic steps of Section 19.4. The only modification, which is induced by the non-gauge invariance of the mass term, is that the field $\mathbf{g}$, associated with the gauge transformations, remains coupled. It is easy to verify that the resulting action is the sum of terms due to the gauge fixing procedure and a gauge invariant part obtained from the action (A19.4) by the substitution $\mathbf{A}_\mu \mapsto \mathbf{A}_\mu^g$ (equation (19.6)):

$$\mathbf{A}_\mu^g(x) = \mathbf{g}(x)\mathbf{A}_\mu(x)\mathbf{g}^{-1}(x) + \mathbf{g}(x)\partial_\mu\mathbf{g}^{-1}(x) = \mathbf{g}(x)\mathbf{D}_\mu\mathbf{g}^{-1}(x)\,, \qquad (A19.5)$$

with:

$$\mathbf{D}_\mu = \partial_\mu + \mathbf{A}_\mu(x)\,. \qquad (A19.6)$$

We thus find:

$$S\left(\mathbf{A}_\mu, \mathbf{g}\right) = -\frac{1}{e^2}\int \mathrm{d}^d x\, \mathrm{tr}\left[\frac{1}{4}\mathbf{F}_{\mu\nu}^2(x) - \frac{1}{2}m^2\mathbf{D}_\mu\mathbf{g}^{-1}\mathbf{D}_\mu\mathbf{g}\right]. \qquad (A19.7)$$

We recognize a $G \times G/G$ non-linear σ-model (see Section 15.4) in which one of the G components of the symmetry group has been gauged. In contrast with the abelian case the scalar field has a self-coupling and is coupled to the gauge field.

The non-linear σ-model action is not renormalizable for dimensions $d > 2$ (see Chapter 14). If we assume that the theory has a cut-off Λ, we see that the $\mathbf{g}$-field fluctuations will not be very much damped because $m \ll \Lambda$; perturbation theory is not particularly reliable. Moreover the zero gauge field mass limit appears as a strong coupling limit, and therefore we do not expect the scalar field to decouple (in perturbation theory we will get IR divergences).

We also note that the complete action can be considered as the limit of a Higgs model action in which the bare mass of the Higgs field has been sent to infinity. Calling ϕ the scalar field we can view action (A19.7) as the formal limit, when g goes to zero, of:

$$S\left(\mathbf{A}_\mu, \phi\right) = \int \mathrm{d}^d x\, \mathrm{tr}\left[-\frac{1}{4e^2}\mathbf{F}_{\mu\nu}^2(x) + \frac{1}{2}\mathbf{D}_\mu\phi^\dagger\mathbf{D}_\mu\phi + \frac{1}{g}\left(\phi^\dagger\phi - (m/e)^2\right)^2\right]. \quad (A19.8)$$

Recalling the equivalence between the non-linear σ model and the ϕ^4 field theory (see Chapter 30), we may speculate that, beyond perturbation theory, the actions (A19.7) and (A19.8) lead to the same renormalized correlation functions.

20 THE STANDARD MODEL. ANOMALIES

In Chapter 19 we have discussed the structure and the formal properties of non-abelian gauge theories. We now apply this formalism to the description of some general properties of the Standard Model of Weak, Electromagnetic and Strong Interactions. In particular we calculate the RG β-function of the strong sector (Quantum Chromodynamics or QCD) and verify the property of asymptotic freedom.

The weak–electromagnetic theory (w.e.m.), thanks to the smallness of the coupling constant at low energy, has now been tested quite systematically, in particular in e^+e^- colliders, and when radiative corrections are taken into account, provides an accurate description of all experimental results. The situation for QCD is somewhat different. High-energy so-called inclusive results can be predicted, as a consequence of the property of large momentum asymptotic freedom (see Chapter 34). However low energy properties cannot be derived from perturbation theory, the effective coupling being too large. Therefore evidence for the validity of the quark confinement scheme relies on non-perturbative, numerical, investigations of lattice gauge theories (see Chapter 33).

In some cases, when gauge fields are coupled to axial currents, the WT identities which are necessary to prove the consistency of gauge theories are not satisfied beyond the classical level. There are spoiled by *anomalies*. Therefore the second part of this chapter is devoted to the discussion of this physically important problem. Results are illustrated by some physical consequences like the π_0 decay and the solution to the $U(1)$ problem.

20.1 The Standard Model of Weak–Electromagnetic Interactions

Let us now briefly describe the Standard Model of weak and electromagnetic interactions since it provides the main physical application of the non-abelian Higgs mechanism. We shall first restrict the discussion to one generation and two flavours, and indicate at the end how it can be generalized to several generations. For detailed phenomenological applications of the model the reader is referred to the literature. The gauge group of the model is $SU(2) \times U(1)$. The form of the action for the Higgs field sector can be obtained from the action considered in the previous section by gauging a $U(1)$ subgroup of the remaining non-gauge $SU(2)$ symmetry. The scalar field is thus a $SU(2)$-doublet. The pattern of symmetry breaking is the same as before but the consequence is now that an unbroken $U(1)$ gauge symmetry remains whose generator is a linear combination of the original $U(1)$ generator and one of the $SU(2)$ generators, and which corresponds to electromagnetic interactions. All fermions are either singlets or doublets. Since the gauge group is a product of two groups, the model depends on two independent gauge couplings and therefore weak and electromagnetic interactions are combined rather than unified.

The gauge field action is a simple sum:

$$S\left(\mathbf{A}_\mu, B_\mu\right) = \frac{1}{4} \int \mathrm{d}^4 x \left(\mathbf{F}_{\mu\nu}^2 + B_{\mu\nu}^2\right), \tag{20.1}$$

in which $\mathbf{F}_{\mu\nu}$ is the curvature corresponding to the $SU(2)$ component and the gauge field $\mathbf{A}_\mu$ with the conventional normalization for this problem:

$$\mathbf{F}_{\mu\nu} = \partial_\mu \mathbf{A}_\nu - \partial_\nu \mathbf{A}_\mu - g\mathbf{A}_\mu \times \mathbf{A}_\nu , \tag{20.2}$$

and $B_{\mu\nu}$ the curvature corresponding to the $U(1)$ component and the gauge field B_μ:

$$B_{\mu\nu} = \partial_\mu B_\nu - \partial_\nu B_\mu . \tag{20.3}$$

As in Subsection 19.6.2, the scalar field ϕ can be written as a 2×2 complex matrix forming a pair of $SU(2)$ doublets (equation (19.74)), but which has in addition a charge $g'/2$ for $U(1)$ transformations. The scalar field action is then:

$$S_{\text{scal.}} = \int \mathrm{d}^4x \left[\operatorname{tr} \left(\partial_\mu \phi^\dagger - \tfrac{1}{2} i g' B_\mu \tau_3 \phi^\dagger + \tfrac{1}{2} i g \phi^\dagger \mathbf{A}_\mu \cdot \boldsymbol{\tau} \right) \right.$$
$$\left. \left(\partial_\mu \phi + \tfrac{1}{2} i g' B_\mu \tau_3 \phi - \tfrac{1}{2} i g \phi \mathbf{A}_\mu \cdot \boldsymbol{\tau} \right) + V(\phi) \right],$$

in which ϕ is a complex matrix of the form (19.74) and the potential V has the same form as in action (19.71). However in this case it is more convenient to parametrize ϕ in terms of a complex vector φ forming a $SU(2)$ doublet. The scalar action then takes the form:

$$S_{\text{scal.}} = \int \mathrm{d}^4x \left\{ \left[\partial_\mu \varphi^\dagger - \tfrac{1}{2} i \varphi^\dagger \left(g' B_\mu + g\mathbf{A}_\mu \cdot \boldsymbol{\tau} \right) \right] \left[\partial_\mu \varphi + \tfrac{1}{2} i \left(g' B_\mu + g\mathbf{A}_\mu \cdot \boldsymbol{\tau} \right) \varphi \right] \right.$$
$$\left. + V(\varphi) \right\} \tag{20.4}$$

with

$$V(\varphi) = a\varphi^\dagger \varphi + \frac{\lambda}{6} \left(\varphi^\dagger \varphi \right)^2 . \tag{20.5}$$

20.1.1 The Higgs mechanism

We now assume that the potential is such that at tree order the field φ has a non-zero expectation value proportional to the vector $(0, 1)$:

$$\langle \varphi \rangle = \frac{v}{\sqrt{2}} \begin{pmatrix} 0 \\ 1 \end{pmatrix} . \tag{20.6}$$

With this form of the expectation value, the subgroup of $SU(2) \times U(1)$ which corresponds to the transformations:

$$\varphi \longmapsto \mathrm{e}^{i\omega(1+\tau_3)} \varphi \, \mathrm{e}^{-i\omega(1+\tau_3)}, \tag{20.7}$$

leaves the expectation value of φ unchanged, and therefore, as announced, $SU(2) \times U(1)$ is broken down to $U(1)$. Replacing φ by its expectation value in action (20.4) we read off the mass terms for the gauge fields at tree order:

$$\tfrac{1}{8} v^2 \int \mathrm{d}^4x \left[\left(g' B_\mu - g A_\mu^{(3)} \right)^2 + g^2 \left| A_\mu^{(1)} + i A_\mu^{(2)} \right|^2 \right] . \tag{20.8}$$

It follows that the two components $A_\mu^{(1,2)}$ have the common mass m_W:

$$m_W = \tfrac{1}{2} g v . \tag{20.9}$$

The linear combination $g'B_\mu - gA_\mu^{(3)}$ is also massive while the orthogonal combination remains massless and thus represents the photon. Let us define the weak angle θ_W:

$$g'/g = \tan\theta_W . \tag{20.10}$$

The photon field A_μ then corresponds to:

$$A_\mu = \cos\theta_W B_\mu + \sin\theta_W A_\mu^{(3)}, \tag{20.11}$$

while the massive field Z_μ is:

$$Z_\mu = -\sin\theta_W B_\mu + \cos\theta_W A_\mu^{(3)}. \tag{20.12}$$

The Z mass is then:

$$m_Z = \tfrac{1}{2}v\sqrt{g^2 + g'^2} . \tag{20.13}$$

The components $A_\mu^{(1,2)}$ are coupled to A_μ and correspond to a charged vector field which is usually written in complex notation:

$$W_\mu^\pm = \left(A_\mu^{(1)} \pm iA_\mu^{(2)}\right)/\sqrt{2} . \tag{20.14}$$

From the coupling of the charged vector bosons with the photon, we derive the relation between the electric charge e and the coupling constants g and g':

$$e = gg'/\sqrt{g^2 + g'^2} = g\sin\theta_W = g'\cos\theta_W . \tag{20.15}$$

20.1.2 Coupling to leptons

We here consider only the electron and the corresponding neutrino since the couplings to other leptons (μ, ν_μ and τ, ν_τ) have exactly the same structure. The neutrino and the left-handed part of the electron are grouped in a left-handed doublet L of $SU(2)$:

$$L = \begin{pmatrix} \nu_e \\ e_{\mathrm{L}} \end{pmatrix}, \tag{20.16}$$

where $e_{\mathrm{L}} = \tfrac{1}{2}(1 - \gamma_5)\,e^-$, while the right-handed part of the electron $R = \tfrac{1}{2}(1 + \gamma_5)\,e^-$ forms a $SU(2)$ singlet. In addition L and R have different charges with respect to $U(1)$. The lepton contribution to the action is:

$$S_{\text{leptons}} = -\int \mathrm{d}^4x \left[\bar{R}\gamma_\mu \left(\partial_\mu - ig'B_\mu\right)R + \bar{L}\gamma_\mu \left(\partial_\mu - \tfrac{1}{2}ig'B_\mu + \tfrac{1}{2}ig\mathbf{A}_\mu \cdot \boldsymbol{\tau}\right)L\right]. \tag{20.17}$$

In this model, since the left-handed and right-handed components of the fermion fields are treated differently, the breaking of the parity symmetry is explicit. Note that the $SU(2)$ symmetry forbids a mass term for the electron field. On the other hand a coupling between the scalar field and the fermions is allowed by the symmetry:

$$S_{\text{Yukawa}} = G_e \int \mathrm{d}^4x \left(\bar{R}\varphi^\dagger L + \bar{L}\varphi R\right). \tag{20.18}$$

If we replace the φ-field by its expectation value, we see that the spontaneous breaking of the $SU(2) \times U(1)$ symmetry generates the electron mass m_e which is therefore calculable, but, in the absence of new dynamical principle, in terms of another arbitrary parameter, the Yukawa coupling constant G_e:

$$m_e = vG_e. \tag{20.19}$$

Note that this implies that the Yukawa coupling of leptons is proportional to their mass and therefore the perturbative approximation becomes worse for heavier leptons.

The coupling constant G (the Fermi constant), characteristic of the strength of weak interactions, is defined in terms of an effective low energy current–current and thus four-fermion interaction:

$$\frac{G}{\sqrt{2}} \int \mathrm{d}^4 x \, J_\mu(x) J_\mu^\dagger(x). \tag{20.20}$$

The contribution to the charged current J_μ coming from the electron and the neutrino has the form:

$$J_\mu(x) = \bar{e}\,(1 - \gamma_5)\,\gamma_\mu \nu_{\mathrm{L}} = 2\bar{e}_{\mathrm{L}} \gamma_\mu \nu_{\mathrm{L}}. \tag{20.21}$$

The relation between G and the coupling constants g and g' is obtained by taking the large W-mass limit of the tree level electron neutrino scattering amplitude. The result can be obtained by integrating over the vector fields $A_\mu^{(1,2)}$, taking only into account the mass term and neglecting the kinetic part. The corresponding part of the action is:

$$\frac{1}{8}g^2 v^2 \left[\left(A_\mu^{(1)} \right)^2 + \left(A_\mu^{(2)} \right)^2 \right] + \frac{ig}{2} \bar{L} \gamma_\mu \left(A_\mu^{(1)} \tau_1 + A_\mu^{(2)} \tau_2 \right) L. \tag{20.22}$$

Completing squares we immediately obtain the result of the integration:

$$\frac{1}{2v^2} \left[\left(\bar{L} \gamma_\mu \tau_1 L \right)^2 + \left(\bar{L} \gamma_\mu \tau_2 L \right)^2 \right] = \frac{2}{v^2} \bar{\nu}_{\mathrm{L}} \gamma_\mu e_{\mathrm{L}} \bar{e}_{\mathrm{L}} \gamma_\mu \nu_{\mathrm{L}}. \tag{20.23}$$

Comparing with definition (20.20) we conclude:

$$G/\sqrt{2} = 1/2v^2 = g^2/8m_W^2. \tag{20.24}$$

The phenomenological Fermi model of low energy charged weak interactions determines all parameters of the Standard Model but two, for example θ_W and the Higgs field self-coupling λ which have to be deduced from additional experimental results. The recent direct measurements of the W and Z masses, for example, determine the parameter θ_W. Note that at leading order the masses can be rewritten:

$$M_W^2 = \frac{e^2}{4\sqrt{2}G \sin^2 \theta_W},$$

i.e.

$$M_W = \left(\frac{\pi\alpha}{\sqrt{2}G} \right)^{1/2} = \frac{38}{\sin \theta_W} \, \mathrm{GeV}, \tag{20.25}$$

and

$$M_Z = \frac{M_W}{\cos \theta_W}. \tag{20.26}$$

The present (1998) experimental values are $m_W = 80.4 \pm 0.1$ GeV, $m_Z = 91.19 \pm 0.01$ GeV. They lead to a weak angle $\sin^2 \theta_W = 0.225 \pm 0.004$, an estimate which is in good agreement with estimates coming from the couplings to neutral currents (an even better agreement is obtained when radiative corrections are taken into account). The parameter λ mainly determines the Higgs mass, and since the Higgs particle has not yet been observed, its value is unknown.

Let us then write the coupling of the charged vector bosons $W^\pm$ to $e^- \nu_e$. We introduce the definition (20.14) into equation (20.17) and obtain:

$$\frac{g}{2} \bar{L} \gamma_\mu \left(\tau^1 A_\mu^{(1)} + \tau^2 A_\mu^{(2)} \right) L = \frac{g}{2\sqrt{2}} \left[\left(\bar{\nu}_\mathrm{L} \gamma_\mu e_\mathrm{L} + \bar{e}_\mathrm{L} \gamma_\mu \nu_\mathrm{L} \right) \left(W_\mu^+ + W_\mu^- \right) \right.$$
$$+ \left(\bar{\nu}_\mathrm{L} \gamma_\mu e_\mathrm{L} - \bar{e}_\mathrm{L} \gamma_\mu \nu_\mathrm{L} \right) \left(W_\mu^+ - W_\mu^- \right) \right]$$
$$= \frac{g}{\sqrt{2}} \left(\bar{\nu}_\mathrm{L} \gamma_\mu e_\mathrm{L} W_\mu^+ + \bar{e}_\mathrm{L} \gamma_\mu \nu_\mathrm{L} W_\mu^- \right). \qquad (20.27)$$

Using definitions (20.10–20.12) we can also write the couplings of fermions to the neutral vector fields A_μ and Z_μ:

$$\frac{e Z_\mu}{\sin 2\theta_W} \left(2 \sin^2 \theta_W \bar{e}_\mathrm{R} \gamma_\mu e_\mathrm{R} - \cos 2\theta_W \bar{e}_\mathrm{L} \gamma_\mu e_\mathrm{L} + \bar{\nu}_\mathrm{L} \gamma_\mu \nu_\mathrm{L} \right) - e A_\mu \left(\bar{e}_\mathrm{R} \gamma_\mu e_\mathrm{R} + \bar{e}_\mathrm{L} \gamma_\mu e_\mathrm{L} \right).$$
$$(20.28)$$

Note finally that recent experiments (1998) suggest that neutrinos could be massive, the mass eigenstates differing from the linear combinations of neutrinos appearing in the weak interactions (see next Subsection), thus leading to neutrino oscillations.

20.1.3 Coupling to hadrons

It is now easy to add interaction terms for quarks in the model. In what follows we first consider only one generation corresponding to two flavours. We call **u** and **d** the corresponding quarks. Each quark has a colour quantum number and forms a $SU(3)$ triplet (see next section). The left components of the quarks form a $SU(2)$ doublet $\mathbf{Q}_\mathrm{L}$.

$$\mathbf{Q}_\mathrm{L} = \{ \mathbf{u}_\mathrm{L} \ , \ \mathbf{d}_\mathrm{L} \} \ . \qquad (20.29)$$

All right-handed components form $SU(2)$ singlets. The coupling to quarks can be written:

$$S_{\text{quarks}} = - \int \mathrm{d}^4 x \left[\bar{\mathbf{Q}}_\mathrm{L} \left(\not{\partial} + \tfrac{1}{2} i g' Y_\mathrm{L} \not{B} + \tfrac{1}{2} i g \mathbf{A} \cdot \boldsymbol{\tau} \right) \mathbf{Q}_\mathrm{L} \right.$$
$$\left. + \bar{\mathbf{Q}}_{1\mathrm{R}} \left(\not{\partial} + \tfrac{1}{2} i g' Y_{1\mathrm{R}} \not{B} \right) \mathbf{Q}_{1\mathrm{R}} + \bar{\mathbf{Q}}_{2\mathrm{R}} \left(\not{\partial} + \tfrac{1}{2} i g' Y_{2\mathrm{R}} \not{B} \right) \mathbf{Q}_{2\mathrm{R}} \right]. \qquad (20.30)$$

Calling T_3 the eigenvalue of τ_3 the generator of $SU(2)$, Y the $U(1)$ charge and q the electric charge, we derive from relations (20.11,20.12,20.15) for each fermion:

$$q = \tfrac{1}{2}(T_3 + Y). \qquad (20.31)$$

The charge q_1 of the first component of the fermion being given, a set of constraints on the $U(1)$ charges Y_L, $Y_{1,2\mathrm{R}}$ follow:

$$\begin{cases} q_1 = \tfrac{1}{2} \left(Y_\mathrm{L} + 1 \right), & q_1 = \tfrac{1}{2} Y_{1\mathrm{R}}, \\ q_2 = \tfrac{1}{2} \left(Y_\mathrm{L} - 1 \right), & q_2 = \tfrac{1}{2} Y_{2\mathrm{R}}. \end{cases} \qquad (20.32)$$

We note that all $U(1)$ charges are determined by the electric charge of the first component of the fermion doublet. Let us verify these relations in the case of the leptons:

$$q_1 = 0 \Rightarrow Y_{\rm L} = -1,\ q_2 = -1,\ Y_{\rm 1R} = 0,\ Y_{\rm 2R} = -2\,. \tag{20.33}$$

The proton being a **uud** state and the neutron a **udd** state, the charges of the **u** and **d** quarks are $2/3$ and $-1/3$ respectively. Relations (20.31) then imply:

$$Y_{\rm L} = 1/3,\ Y_{\rm 1R} = 4/3,\ Y_{\rm 2R} = -2/3\,. \tag{20.34}$$

Note that the charges of the quarks are compatible with the $SU(2)$ doublet assignment. We shall verify in Section 20.5 that the $SU(3)$ triplet structure of quarks leads to the cancellation of the possible anomaly due to the chiral coupling of gauge fields to fermions in each generation and therefore secures the consistency of the gauge theory of weak electromagnetic interactions.

Couplings to the Higgs field and quark masses. As in the case of leptons direct quark mass terms are forbidden by the $SU(2)$ symmetry. The quark masses are induced by the coupling to the Higgs scalar field and the spontaneous symmetry breaking. The $SU(2) \times U(1)$ invariant Yukawa couplings are:

$$G_{q2}\bar{\bf Q}_{\rm 2R}\varphi^\dagger {\bf Q}_{\rm L} + G_{q1}\bar{\bf Q}_{\rm L} i\tau_2\varphi^* {\bf Q}_{\rm 1R} + \text{ h.c.}\,, \tag{20.35}$$

which can therefore provide masses for the two quarks. This is at least the situation for one generation. However three generations have been discovered (see table 20.1), the last quark, the "top", whose mass had been predicted from the analysis of LEP data, has recently been confirmed at Fermilab. Note moreover that from the Z decays one infers that the number of generations with "light" neutrinos (i.e. with a mass below 45 GeV) is exactly three.

In the interactions (20.35) the spinors which appear on the right and the left then need not be the same. Therefore when we replace the scalar field φ by its expectation value, we obtain a general non-diagonal mass matrix of the form:

$$\bar{\bf Q}_{\rm 1R}^\alpha M_{\alpha\beta}{\bf Q}_{\rm 1L}^\beta + \bar{\bf Q}_{\rm 1L}^\alpha M_{\alpha\beta}^\dagger {\bf Q}_{\rm 1R}^\beta\,, \tag{20.36}$$

for the quarks of charge $2/3$, and a similar one for the charge $-1/3$ quarks. Performing independent unitary transformations ${\bf U}_{\rm R,L}$ on the right and left quark components it is possible to replace the matrix ${\bf M}$ by a real diagonal matrix $\mathcal{M}$:

$$\mathbf{U}_{\rm R}^\dagger \mathbf{M} \mathbf{U}_{\rm L} = \mathcal{M}\,, \tag{20.37}$$

In this representation the quarks are mass eigenstates. However the weak interactions no longer have the simple form (20.30) because the unitary transformations on the quark components ${\bf Q}_{\rm 1L}$ and ${\bf Q}_{\rm 2L}$ are in general different. It is customary to put the blame onto the charge $-1/3$ quarks. The mismatch is expressed in terms of a 3×3 unitary matrix (because there are three generations at present), the Kobayashi–Maskawa matrix (KM) which relates the quark mass eigenstates **d**, **s** and **b** to the quarks appearing in the weak interactions:

$$\left[{\bf Q}_2^\alpha\right]_{\text{weak int.}} = U_{\alpha\beta}\left[{\bf Q}_2^\beta\right]_{\text{mass eigenst.}}\,. \tag{20.38}$$

Table 20.1

Quarks and leptons. the three generations (1998).

charge 2/3 quarks	charge -1/3 quarks	charge -1 leptons	neutrinos
u, $m = 1.5$ to 5 MeV	**d**, $m = 3$ to 9 MeV	e, $m = 0.511$ MeV	ν_e, $m < 10$ eV
c, $m = 1.1$ to 1.4 GeV	**s**, $m = 60$ to 170 MeV	μ, $m = 105.6$ MeV	ν_μ, $m < 0.17$ MeV
t, $m = 174 \pm 6$ GeV	**b**, $m = 4.1$ to 4.4 GeV	τ, $m = 1.777$ GeV	ν_τ, $m < 19$ MeV

With only two generations (**d** and **s**) it is possible to cast the matrix into the form:

$$\mathbf{U}_c = \begin{pmatrix} \cos\theta_c & \sin\theta_c \\ -\sin\theta_c & \cos\theta_c \end{pmatrix},$$

in which θ_c is the celebrated Cabibbo angle, after unobservable changes of the relative phases between the quarks. In the presence of the third **b** quark, the 3×3 KM matrix can be parametrized in terms of three rotation angles and one phase parameter. This phase may be responsible for the observed weak CP violation in kaon decay.

20.1.4 The problem of elementary scalar fields

The Standard Model has (Higgs) scalar fields as an essential ingredient. This is the source of several difficulties. The Higgs field is responsible for the masses of all particles, but in the Standard Model these masses are all given in terms of arbitrary parameters, like for example the Yukawa couplings which determine the fermions masses. This could perhaps be expected from an effective low energy theory. Yet puzzling is the diversity of these couplings. If the couplings were "natural", i.e. of order unity, all fermion masses would be in the few×100 GeV range, as the W and Z masses or the Higgs expectation value. But in this sense only the top quark (**t**) mass, which is about 174 GeV, is natural.

Another perhaps even more fundamental problem is related to the scalar field mass renormalization. Generically the scalar bare and renormalized masses are expected to be of the order of the "cut-off" which gives the scale of new physics. It is only by *fine tuning* the scalar bare mass that one can reduce the renormalized mass. As we shall extensively discuss in the chapters devoted to critical phenomena, this is quite possible in statistical physics where the bare mass represents the temperature of the system. The temperature can indeed be experimentally tuned close to the critical value where the correlation length (the inverse physical mass in the particle language) diverges. However in particle physics the parameters are given and it is somewhat unnatural for the scalar bare mass to lie accidentally close to such a critical value. One is thus led to conjecture that the scale of new physics cannot be too high (1 to 10 TeV?). Moreover at this scale the scalar field problem must be cured. Two type of schemes have been proposed so far:

(i) The Higgs boson is a bound state of a new type of fermions. This requires a specific model, hopefully not involving new scalars again. Models in which the forces are again due to gauge interactions have been proposed and fall under the name of *technicolour* (see next section). Such models have problems generating fermion masses and no widely accepted model has been constructed so far.

(ii) The Higgs boson remains associated with a fundamental field, but the mass renormalization problem is solved with the help of *supersymmetry*. Since fermions, due to chiral symmetry, can be naturally massless, the idea is to use supersymmetry (see for example Section 17.10), to relate them to scalars. In such models the scalar mass renormalization grows only logarithmically with the cut-off (it would be absent in the absence of supersymmetry breaking) and thus the problem is much less severe even if the cut-off is of the order of the Planck mass. The main difficulty with this approach is that none of the existing particles can be grouped into supermultiplets. The superpartners of all the existing particles have yet to be found. Moreover the mechanism of spontaneous supersymmetry breaking is not fully understood.

20.2 Quantum Chromodynamics (QCD)

We now concentrate on the so-called Strong Interactions which bind the quarks together to produce the observed hadrons, neglecting completely the weak and electromagnetic interactions which we have described in the preceding section. The model of Strong Interactions, as it stands at present, consists in a set of quarks characterized by a *flavour quantum number*, relevant for Weak Interactions, which are also triplets of a $SU(3)$ gauge symmetry, the *colour*, realized in the symmetric phase. Their interactions therefore are mediated by the corresponding gauge fields (called *gluons*):

$$S\left(\mathbf{A}_\mu, \bar{\mathbf{Q}}, \mathbf{Q}\right) = -\int d^4x \left[\frac{1}{4g^2} \operatorname{tr} \mathbf{F}_{\mu\nu}^2 + \sum_{\text{flavours}} \bar{\mathbf{Q}}_f \left(\slashed{D} + m_f\right) \mathbf{Q}_f\right]. \tag{20.39}$$

The most important physical arguments in favour of such a model are:

i) Quarks behave almost like free particles at short distances, as indicated by deep inelastic scattering experiments or the spectrum of bound states of heavy quarks. We shall calculate below the RG β-function and show that a pure non-abelian gauge theory is asymptotically free (AF) at large momentum in four dimensions (like the non-linear σ-model in two dimensions). In Chapter 34 we shall prove that this property survives the inclusion of a limited number of fermions and furthermore that this property is specific to non-abelian gauge theories.

ii) No free quarks have ever been observed at large distance (but they manifest themselves indirectly in the jet physics). This is consistent with the simplest picture in which the β-function (which, due to AF, is negative at small coupling) remains negative for all couplings in such a way that the effective coupling constant grows without bounds at large distances. Numerical simulations strongly support this conjecture, called the *confinement* hypothesis (see Chapter 33).

20.2.1 The RG equations in the covariant gauge

We first discuss the gauge dependence of the RG equations and functions of pure gauge theories in the covariant gauge (19.45), i.e. the dependence on the parameter ξ. A short discussion of the abelian case can be found in Section 18.8. We call Z_A the gauge field renormalization constant and Z_g the renormalization constant of the coupling constant α:

$$\alpha = g^2/4\pi.$$

α in Strong Interactions is in general denoted α_S to distinguish it from its QED analogue. In this section no confusion is possible. It will be shown in Subsection 21.6.2 that for such gauges, as in the abelian case, the gauge fixing term is not renormalized. Therefore:

$$\xi_0 = \xi Z_A/Z_g. \tag{20.40}$$

Calling μ the renormalization scale, the RG equations for the gauge field n-point function read:

$$\left[\mu\frac{\partial}{\partial\mu} + \beta(\alpha,\xi)\frac{\partial}{\partial\alpha} + \delta(\alpha,\xi)\xi\frac{\partial}{\partial\xi} - \frac{n}{2}\eta_A(\alpha,\xi)\right]\Gamma^{(n)}(\mu,\alpha,\xi) = 0, \qquad (20.41)$$

where

$$\delta(\alpha,\xi) \equiv \xi^{-1}\left.\mu\frac{\partial}{\partial\mu}\right|_{\alpha_0,\xi_0\text{ fixed}}\xi = \beta(\alpha,\xi) - \eta_A(\alpha,\xi). \qquad (20.42)$$

We prove in Section 21.7 that the bare correlation functions of gauge invariant operators are gauge independent. This in particular implies that they are independent of ξ_0. The same property applies to the renormalization constants needed to render these correlation functions finite. It is thus possible to construct renormalized correlation functions which are also ξ_0 independent. Let us call Γ such a correlation function. It satisfies:

$$\left.\frac{\partial}{\partial\xi}\right|_{\alpha_0,\text{ cut-off fixed}}\Gamma = \left(\frac{\partial}{\partial\xi} + \rho(\alpha,\xi)\frac{\partial}{\partial\alpha}\right)\Gamma(\mu,\alpha,\xi) = 0, \qquad (20.43)$$

with:

$$\rho(\alpha,\xi) = \left.\frac{\partial\alpha}{\partial\xi}\right|_{\alpha_0,\text{ cut-off fixed}}. \qquad (20.44)$$

Γ also satisfies a RG equation which we assume to be homogeneous:

$$\left[\mu\frac{\partial}{\partial\mu} + \beta(\alpha,\xi)\frac{\partial}{\partial\alpha} + \delta(\alpha,\xi)\xi\frac{\partial}{\partial\xi} - \eta_\Gamma(\alpha,\xi)\right]\Gamma(\mu,\alpha,\xi) = 0. \qquad (20.45)$$

Using equation (20.43) to eliminate $\partial/\partial\xi$, we obtain a new RG equation for Γ:

$$\left[\mu\frac{\partial}{\partial\mu} + \tilde{\beta}(\alpha,\xi)\frac{\partial}{\partial\alpha} - \eta_\Gamma(\alpha,\xi)\right]\Gamma(\mu,\alpha,\xi) = 0, \qquad (20.46)$$

with:

$$\tilde{\beta} = \beta - \xi\delta\rho. \qquad (20.47)$$

Writing then the compatibility condition between the two linear equations (20.43) and (20.46) we obtain two equations:

$$\left(\frac{\partial}{\partial\xi} + \rho(\alpha,\xi)\frac{\partial}{\partial\alpha}\right)\eta_\Gamma(\alpha,\xi) = 0, \qquad (20.48)$$

$$\left(\frac{\partial}{\partial\xi} + \rho(\alpha,\xi)\frac{\partial}{\partial\alpha}\right)\tilde{\beta} = \frac{\partial\rho}{\partial\alpha}\tilde{\beta}. \qquad (20.49)$$

The first equation expresses that, as expected, the multiplicative renormalization of Γ is independent of ξ_0. The second equation shows that the zeroes of $\tilde{\beta}$ are gauge independent. Differentiating the equation with respect to α one finds also that the slope of β at its zeroes is gauge independent. Finally let us show that in a minimal subtraction scheme the function ρ vanishes. In dimensional regularization the relation between α_0 and α takes the form:

$$\alpha_0 = \mu^\varepsilon\alpha Z_g = \mu^\varepsilon\alpha\left(1 + \frac{Z_g^1(\alpha,\xi)}{\varepsilon} + \frac{Z_g^2(\alpha,\xi)}{\varepsilon^2} + \cdots\right). \qquad (20.50)$$

The important point is that the term without pole in ε in the expansion of Z_g is ξ independent. Using the definition (20.44) of ρ we then find:

$$0 = \rho \left(1 + \frac{\partial Z_g^1}{\partial \alpha} \frac{1}{\varepsilon} + \cdots \right) + \alpha \left(\frac{\partial Z_g^1}{\partial \xi} \frac{1}{\varepsilon} + O \left(\frac{1}{\varepsilon^2} \right) \right). \qquad (20.51)$$

Therefore the expansion of ρ for ε small has only singular contributions. Since ρ is of course finite, all singular contributions must cancel and thus ρ vanishes identically. It follows that in the minimal subtraction scheme the β-function and η_Γ are independent of ξ.

20.2.2 The RG β-function at one-loop order

We shall now calculate the RG β-function at leading order in a gauge theory corresponding to a simple group G, to verify asymptotic freedom. No simple explanation has yet been proposed which allows one to understand the sign without performing the explicit calculation.

The calculation can be done by various methods, for example we could use the background field method as we have done in the case of the models on homogeneous spaces in Section 15.6.2. Here instead we calculate directly in a conventional manner the β-function from the renormalization of the gauge coupling constant as defined by the fermion-gauge field vertex. We need thus the divergent parts of the gauge field and fermion 2-point functions and the fermion gauge field 3-point function.

The gauge field 2-point function. Four diagrams contribute to the 2-point function corresponding to the gauge field loops, the Faddeev–Popov ghost loop and the fermion loops (see figure 20.1).

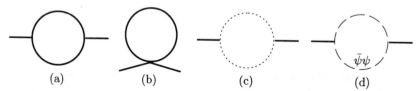

(a)   (b)   (c)   (d)

Fig. 20.1   The gauge field propagator (dotted lines represent ghosts).

The diagram (b) corresponding to the self-contraction of the gauge 4-point vertex vanishes in dimensional regularization. The fermion loop contribution (d) has already been calculated in Section 18.9 up to a simple geometrical factor. Only 2 diagrams remain. For the vertices and the propagator we use the normalizations given in Section 19.5. We use the Feynman gauge and dimensional regularization. Diagram (a) has the expression

$$(a) = \tfrac{1}{2} f_{acd} f_{bcd} \int \frac{d^d q}{(2\pi)^d} \frac{N_{\mu\nu}(k,q)}{q^2 (k+q)^2}, \qquad (20.52)$$

with

$$N_{\mu\nu}(k,q) = \delta_{\mu\nu} \left(5k^2 + 2k \cdot q + 2q^2 \right) + k_\mu k_\nu (d-6) + (q^\mu k_\nu + q^\nu k_\mu)(2d-3)$$
$$+ 2q_\mu q_\nu (2d-3). \qquad (20.53)$$

To calculate the diagrams, we project the integrand over $\delta_{\mu\nu}$ and $k_\mu k_\nu$ (see equations (9.31,9.32)), and use repeatedly the identity:

$$2k \cdot q = (k+q)^2 - k^2 - q^2.$$

We set:

$$f_{acd}f_{bcd} = C(G)\delta_{ab}. \tag{20.54}$$

A short calculation yields the divergent part:

$$(a)_{\text{div}} = \delta_{ab}\frac{C(G)}{12}\left(19k^2\delta_{\mu\nu} - 22k_\mu k_\nu\right)\frac{g^2}{8\pi^2\varepsilon}. \tag{20.55}$$

Diagram (c) is given by

$$(c) = -f_{acd}f_{bcd}\int \frac{d^d q}{(2\pi)^d}\frac{q_\mu(k+q)_\nu}{q^2\,(k+q)^2}. \tag{20.56}$$

The divergent part is

$$(c)_{\text{div}} = \delta_{ab}\left(k^2\delta_{\mu\nu} + 2k_\mu k_\nu\right)\frac{1}{12}C(G)\frac{g^2}{8\pi^2\varepsilon}. \tag{20.57}$$

Note that both divergent contributions are not separately transverse. By adding the two contributions we get the divergent part of the 2-point function in the absence of fermions, which is now transverse as expected:

$$\left[\Gamma^{(2)ab}_{\mu\nu}\right]_{\text{div}} = \delta_{ab}\left(k^2\delta_{\mu\nu} - k_\mu k_\nu\right)\frac{5}{3}C(G)\frac{g^2}{8\pi^2\varepsilon}. \tag{20.58}$$

Calling Z_A and Z_g the renormalization constants of the gauge field and the coupling constant g^2 we obtain the relation at one-loop order:

$$\frac{Z_A}{Z_g} = 1 + \frac{5}{3}C(G)\frac{g^2}{8\pi^2\varepsilon}. \tag{20.59}$$

Using result (18.117), we can add the fermion contribution and finally find:

$$\frac{Z_A}{Z_g} = 1 + \left(\frac{5}{3}C(G) - \frac{4}{3}T(R)\right)\frac{g^2}{8\pi^2\varepsilon}, \tag{20.60}$$

where the fermions belong to the representation R and $T(R)$ is the trace of the square of the generators in this representation:

$$\operatorname{tr} t^a t^b = -\delta_{ab}T(R). \tag{20.61}$$

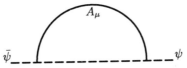

Fig. 20.2 One-loop contribution to the fermion 2-point function.

The fermion 2-point function. One diagram contributes to the fermion 2-point function $\Gamma^{(2)}$ (see figure 20.2) which differs from its QED counterpart only by a geometrical factor:

$$\Gamma^{(2)}_{1\,\text{loop}}(k) = g^2 \int \frac{d^d q}{(2\pi)^d} t^a \gamma_\mu \frac{1}{i\slashed{q} + m} t^a \gamma_\mu \frac{1}{(k-q)^2} \tag{20.62}$$

(for simplicity we have given the same mass m to all fermions since it does not affect the result). Since we need only the field renormalization we can project the integrand over $\slashed{k}$. The following identity is useful:

$$\gamma_\nu \gamma_\mu \gamma_\nu = (2-d)\gamma_\mu \,.$$

Calculating the divergent part of the integral, we obtain the fermion field renormalization Z_F at one-loop:

$$Z_F = 1 - C(R)\frac{g^2}{8\pi^2 \varepsilon}, \tag{20.63}$$

with:

$$t^a t^a = -C(R)\mathbf{1} \,. \tag{20.64}$$

The gauge field fermion vertex. Two diagrams contribute at one-loop order (see figure 20.3), the first has a QED counterpart, the second being characteristic of a non-abelian theory.

(a)   (b)

Fig. 20.3   The gauge field–fermion vertex at one-loop.

$$(a) = g^2 \int \frac{d^d q}{(2\pi)^d} t^b \gamma_\nu \frac{1}{i\slashed{q} + m} t^a \gamma_\mu \frac{1}{i\slashed{q} + i\slashed{k} + m} t^b \gamma_\nu \frac{1}{(p_1 - q)^2}. \tag{20.65}$$

To calculate the divergent part of the integral we multiply by γ_μ and take the trace. We use also the identity:

$$t^b t^a t^b = \left(\tfrac{1}{2}C(G) - C(R)\right) t^a.$$

We then find:

$$(a)_{\text{div.}} = \left(C(R) - \tfrac{1}{2}C(G)\right) t^a \gamma_\mu \frac{g^2}{8\pi^2 \varepsilon}, \tag{20.66}$$

$$(b) = i f_{abc} g^2 \int \frac{d^d q}{(2\pi)^d} t^b \gamma_\nu \frac{1}{i\slashed{p}_1 - i\slashed{q} + m} t^c \gamma_\rho \frac{V_{\mu\nu\rho}(k, q, -k-q)}{q^2(k+q)^2}, \tag{20.67}$$

with (equation (19.57)):

$$V_{\mu\nu\rho}(k, q, r) = (r-q)_\mu \delta_{\nu\rho} + (k-r)_\nu \delta_{\rho\mu} + (q-k)_\rho \delta_{\mu\nu} \,. \tag{20.68}$$

The divergent part is:

$$(a)_{\text{div.}} = \frac{3}{2}C(G)t^a\gamma_\mu\frac{g^2}{8\pi^2\varepsilon}. \tag{20.69}$$

It follows that:

$$Z_{\text{F}}Z_A^{1/2} = 1 - (C(R)+C(G))\frac{g^2}{8\pi^2\varepsilon}, \tag{20.70}$$

and therefore (equation (20.63)):

$$Z_A = 1 - 2C(G)\frac{g^2}{8\pi^2\varepsilon}. \tag{20.71}$$

Using finally the result (20.60) we obtain Z_g:

$$Z_g = 1 - \left(\frac{11}{3}C(G) - \frac{4}{3}T(R)\right)\frac{g^2}{8\pi^2\varepsilon}. \tag{20.72}$$

The β-function at one-loop order follows:

$$\beta(g^2) = -\varepsilon\left[\frac{\mathrm{d}\ln\left(g^2 Z_g\right)}{\mathrm{d}g^2}\right]^{-1} = -\left[\frac{11}{3}C(G) - \frac{4}{3}T(R)\right]\frac{g^4}{8\pi^2} + O\left(g^6\right). \tag{20.73}$$

In the case of the $SU(N)$ group with N_{F} fermions in the fundamental representation the values of $C(G)$ and $T(R)$ are:

$$C(G) = N\,, \qquad T(R) = \tfrac{1}{2}N_{\text{F}}\,,$$

and therefore:

$$\beta(g^2) = -\left(\frac{11N}{3} - \frac{2N_{\text{F}}}{3}\right)\frac{g^4}{8\pi^2} + O\left(g^6\right). \tag{20.74}$$

The theory is asymptotically free, i.e. the β-function is negative for small coupling for:

$$N_{\text{F}} < 11N/2\,, \tag{20.75}$$

which, in the case of $SU(3)$, means at most 16 flavours. If this condition is met, $g = 0$ is a UV fixed point (for details see Chapter 34).

20.3 The Abelian Anomaly

We have pointed out several times that none of the standard regularization methods can deal with one-loop diagrams in the case of chiral gauge fields. Let us now show that indeed one can find examples in which it is impossible to satisfy WT identities. These examples are physically important in two cases, the electromagnetic decay of the π_0 meson, and the theory of weak electromagnetic interactions. We first discuss the abelian case in four dimensions and then the general non-abelian case.

The only case in which anomalies may occur in gauge theories corresponds to one-loop diagrams involving fermions coupled to chiral gauge fields. This reduces the problem to the calculation of the determinant resulting from the fermion integration. In the abelian case we have only to deal with

$$D = \det\left(\slashed{\partial} + ie\slashed{A} + i\gamma_5\slashed{B}\right).$$

We can find regularizations which preserve gauge invariance associated with the vector field A_μ and, since the fermions are massless, invariance under space-independent chiral transformations. We have to examine the behaviour of the determinant under chiral gauge transformations. We have only to consider the divergent contributions. In four dimensions they correspond to the terms of degrees 2,3 and 4 in the gauge fields.

Since the axial field B_μ is a gauge field for chiral transformations, a related problem is the behaviour of the functional integral:

$$Z\left(A_\mu\right) = \int \left[\mathrm{d}\psi \mathrm{d}\bar\psi\right] \exp\left[-S\left(\psi, \bar\psi\right)\right], \tag{20.76}$$

in which $S\left(\psi, \bar\psi\right)$ is the QED action for massless fermion fields:

$$S\left(\bar\psi, \psi\right) = -\int \mathrm{d}^4 x\, \bar\psi\left(x\right)\left(\slashed{\partial} + ie\slashed{A}\right)\psi\left(x\right), \tag{20.77}$$

under a space-dependent chiral transformation:

$$\psi_\theta\left(x\right) = \mathrm{e}^{i\theta(x)\gamma_5}\,\psi\left(x\right), \qquad \bar\psi_\theta\left(x\right) = \bar\psi\left(x\right)\mathrm{e}^{i\theta(x)\gamma_5}. \tag{20.78}$$

The corresponding current $J_\mu^5(x)$ defined in terms of the variation of the action is:

$$\delta S = \int \mathrm{d}^4 x\, \partial_\mu\theta(x) J_\mu^5(x) \Rightarrow J_\mu^5(x) = i\bar\psi(x)\gamma_5\gamma_\mu\psi(x). \tag{20.79}$$

Since the determinant of $\mathrm{e}^{i\gamma_5\theta}$ is 1 we would formally conclude that $Z\left(A_\mu\right)$:

$$Z\left(A_\mu\right) = \det\left(\slashed{\partial} + ie\slashed{A}\right) \equiv \det\slashed{D}, \tag{20.80}$$

is invariant and therefore that the current $J_\mu^5(x)$ is conserved. This is a conclusion we would like to check by explicit calculation. We have to examine the expectation value of $\partial_\mu J_\mu^5(x)$ in the case of the action (20.77). Let us note that after the transformations (20.78), $Z\left(A_\mu\right)$ takes the form:

$$Z\left(A_\mu\right) = \det\left[\mathrm{e}^{i\gamma_5\theta(x)}\,\slashed{D}\,\mathrm{e}^{i\gamma_5\theta(x)}\right]. \tag{20.81}$$

For any regularization which is consistent with the hermiticity of γ_5 we find that

$$\left|Z\left(A_\mu\right)\right|^2 = \det\left(\slashed{D}\slashed{D}^\dagger\right)$$

is invariant. Therefore we expect that an anomaly can only appear in the imaginary part of $\ln Z$.

The operator $\partial_\mu J_\mu^5(x)$ has dimension 4 and since a possible anomaly is a large momentum, i.e. local effect, $\left\langle \partial_\mu J_\mu^5(x)\right\rangle$ can only be a local function of A_μ of dimension 4. In addition parity implies that it is proportional to the completely antisymmetric tensor $\epsilon_{\mu\nu\rho\sigma}$. This determines $\left\langle \partial_\mu J_\mu^5(x)\right\rangle$ up to multiplicative constant,

$$\left\langle \partial_\lambda J_\lambda^5(x)\right\rangle \propto e^2 \epsilon_{\mu\nu\rho\sigma} \partial_\mu A_\nu \partial_\rho A_\sigma.$$

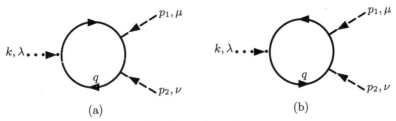

Fig. 20.4   Anomalous diagrams.

To find the constant it is thus sufficient to calculate the coefficient of degree 2 in the expansion of $\langle \partial_\lambda J^5_\lambda(x) \rangle$ in powers of A. Let us define the 3-point function:

$$\Gamma^{(3)}_{\lambda\mu\nu}(k; p_1, p_2) = \frac{\delta}{\delta A_\mu(p_1)} \frac{\delta}{\delta A_\nu(p_2)} \langle J^5_\lambda(k) \rangle \Big|_{A=0}, \tag{20.82}$$

$$= \frac{\delta}{\delta A_\mu(p_1)} \frac{\delta}{\delta A_\nu(p_2)} i \operatorname{tr} \left[\gamma_5 \gamma_\lambda \slashed{D}^{-1}(k) \right] \Big|_{A=0}.$$

$\Gamma^{(3)}$ is the sum of the two corresponding Feynman diagrams of figure 20.4.

We note that $\Gamma^{(3)}$ is formally a completely symmetric function of the three sets of external arguments. It has a linear divergence. Since the fermions are massless, the finite part of $\Gamma^{(3)}$ is ambiguous by a linear combination of the vectors $\mathbf{p}_1$ and $\mathbf{p}_2$ which, for symmetry reasons, has the form:

$$K \epsilon_{\lambda\mu\nu\rho} (p_1 - p_2)_\rho . \tag{20.83}$$

Note that such a term breaks the symmetry between the three external arguments. In addition the WT identity which follows from gauge invariance:

$$p_{1\mu} \Gamma^{(3)}_{\lambda\mu\nu}(k; p_1, p_2) = 0, \tag{20.84}$$

fixes also completely the coefficient K. Therefore, gauge invariant regularizations yield a finite result and by calculating $k_\lambda \Gamma^{(3)}_{\lambda\mu\nu}(k; p_1, p_2)$ we can check the current conservation. In particular a symmetric definition of $\Gamma^{(3)}$ in the three external arguments is either completely divergence free or breaks both gauge invariance and axial current conservation.

20.3.1 Formal verification of current conservation

To understand from the point of view of Feynman diagrams where the current conservation comes from, let us compute the contribution to the divergence coming from the diagram (a) (figure 20.4):

$$(a) \mapsto \frac{e^2}{(2\pi)^4} \operatorname{tr} \left[\gamma_5 \slashed{k} \int d^4 q \, (\slashed{q} + \slashed{k})^{-1} \gamma_\mu (\slashed{q} - \slashed{p}_2)^{-1} \gamma_\nu \slashed{q}^{-1} \right]. \tag{20.85}$$

If the integral were convergent we would proceed in the following way: We would first transform the expression, using the identity:

$$\slashed{k} (\slashed{q} + \slashed{k})^{-1} = 1 - \slashed{q} (\slashed{q} + \slashed{k})^{-1}. \tag{20.86}$$

For the contribution coming from the second term in the r.h.s. we then use the cyclic property of the trace and the *anticommutation* of $\displaystyle{\not{q}}$ and γ_5 to cancel the propagator $\displaystyle{\not{q}}^{-1}$. It follows that:

$$(a) \longmapsto \frac{e^2}{(2\pi)^4} \, \mathrm{tr} \left\{ \gamma_5 \int \mathrm{d}^4 q \, \gamma_\mu \, (\not{q} - \not{p}_2)^{-1} \, \gamma_\nu \left[\not{q}^{-1} - (\not{q} + \not{k})^{-1} \right] \right\}. \qquad (20.87)$$

The contribution coming from diagram (b) is obtained by interchanging $(\mathbf{p}_1, \mu)$ and $(\mathbf{p}_2, \nu)$. It is not too difficult to verify by shifting variables in the integrals that the contributions coming from (a) and (b) cancel (the two terms in the r.h.s. of equation (20.87) have to be shifted differently).

However since the integrals diverge all these manipulations are not necessarily legitimate. If an anomaly is generated, since it is a short distance effect, it must come from a local term, i.e. here from a linear term in momenta of the form (20.83). We could at this point do the calculation using one of the various gauge invariant regularizations, for example Pauli–Villars's regularization or dimensional regularization with γ_5 being defined as in dimension 4 and thus no longer anticommuting with other γ matrices (see Subsection 9.3.3). However, it is more convenient to use a regularization which respects the symmetry between external arguments but breaks gauge invariance, replacing in the fermion propagator:

$$(\not{q})^{-1} \longmapsto (\not{q})^{-1} \mathrm{e}^{-\varepsilon q^2}.$$

By calculating $k_\lambda \Gamma^{(3)}_{\lambda\mu\nu} (k; p_1, p_2)$ we shall check both current conservation and gauge invariance.

20.3.2 The calculation

In order to indicate how the method generalizes to arbitrary even dimensions it is convenient to consider directly the contribution $C^{(2)}(k)$ of order A^2 to $\langle k_\lambda J^5_\lambda(k) \rangle$:

$$C^{(2)}(k) = e^2 \int \mathrm{d}^4 p_1 \, \mathrm{d}^4 p_2 \, A_\mu (p_1) \, A_\nu (p_2) \int \frac{\mathrm{d}^4 q}{(2\pi)^4} \, \mathrm{e}^{-\varepsilon \left[(q+k)^2 + (q-p_2)^2 + q^2 \right]}$$
$$\times \mathrm{tr} \left[\gamma_5 \not{k} \, (\not{q} + \not{k})^{-1} \, \gamma_\mu \, (\not{q} - \not{p}_2)^{-1} \, \gamma_\nu \not{q}^{-1} \right]. \qquad (20.88)$$

We now use identity (20.86) to separate this contribution into the sum of two terms:

$$C^{(2)}(k) = e^2 \int \mathrm{d}^4 p_1 \, \mathrm{d}^4 p_2 \, A_\mu (p_1) \, A_\nu (p_2) \int \frac{\mathrm{d}^4 q}{(2\pi)^4} \, \mathrm{e}^{-\varepsilon \left[(q+k)^2 + (q-p_2)^2 + q^2 \right]}$$
$$\times \mathrm{tr} \left\{ \gamma_5 \left[(\not{q} + \not{k})^{-1} - \not{q}^{-1} \right] \gamma_\mu \, (\not{q} - \not{p}_2)^{-1} \, \gamma_\nu \right\}. \qquad (20.89)$$

In the first contribution we interchange $(\mathbf{p}_1, \mu)$ and $(\mathbf{p}_2, \nu)$. We then find:

$$C^{(2)}(k) = e^2 \int \mathrm{d}^4 p_1 \, \mathrm{d}^4 p_2 \, A_\mu (p_1) \, A_\nu (p_2) \int \frac{\mathrm{d}^4 q}{(2\pi)^4} \, \mathrm{e}^{-\varepsilon \left[(q+k)^2 + q^2 \right]}$$
$$\mathrm{tr} \left\{ \gamma_5 \left[\mathrm{e}^{-\varepsilon (q-p_1)^2} \, (\not{q} + \not{k})^{-1} \, \gamma_\nu \, (\not{q} - \not{p}_1)^{-1} \, \gamma_\mu - \mathrm{e}^{-\varepsilon (q-p_2)^2} \, \not{q}^{-1} \gamma_\mu \, (\not{q} - \not{p}_2)^{-1} \, \gamma_\nu \right] \right\}. \qquad (20.90)$$

Finally still in the first term we change $\mathbf{q}$ into $\mathbf{q} + \mathbf{p}_1$. Using the cyclic property of the trace and the anticommutation properties of γ_5 we then see that the two terms would cancel in the absence of regulators. Now instead we find a non-vanishing sum due to the exponential factors which are different:

$$C^{(2)}(k) = e^2 \int d^4p_1\, d^4p_2\, A_\mu(p_1)\, A_\nu(p_2) \int \frac{d^4q}{(2\pi)^4}\, e^{-\varepsilon[(q-p_2)^2+q^2]}$$
$$\times \operatorname{tr}\left[\gamma_5 \slashed{q}^{-1} \gamma_\mu (\slashed{q}-\slashed{p}_2)^{-1} \gamma_\nu\right]\left(e^{-\varepsilon(q+p_1)^2} - e^{-\varepsilon(q+k)^2}\right). \tag{20.91}$$

After evaluation of the trace (notice our convention ($A5.2$) for γ_5) this sum becomes:

$$C^{(2)}(k) = 4e^2 \int d^4p_1\, d^4p_2\, A_\mu(p_1)\, A_\nu(p_2) \int \frac{d^4q}{(2\pi)^4}\, e^{-\varepsilon[(q-p_2)^2+q^2]}$$
$$\times \epsilon_{\mu\nu\rho\sigma} \frac{p_{2\rho}q_\sigma}{q^2(q-p_2)^2}\left(e^{-\varepsilon(q+p_1)^2} - e^{-\varepsilon(q+k)^2}\right). \tag{20.92}$$

The relevant values of q are of order $\varepsilon^{-1/2}$. We can therefore replace the integrand in the expression by:

$$\int \frac{d^4q}{(2\pi)^4 q^4} p_{2\rho}q_\sigma\, e^{-3\varepsilon q^2}\left[2\varepsilon\mathbf{q}\cdot(\mathbf{k}-\mathbf{p}_1)\right]. \tag{20.93}$$

The identity:

$$\int d^4q\, q_\alpha q_\beta f(q^2) = \tfrac{1}{4}\delta_{\alpha\beta}\int d^4q\, q^2 f(q^2),$$

transforms the integral into:

$$-\tfrac{1}{2}p_{2\rho}(2p_1+p_2)_\sigma \int \frac{\varepsilon d^4q}{(2\pi)^4 q^2}\, e^{-3\varepsilon q^2}. \tag{20.94}$$

Since:

$$\int \frac{\varepsilon d^4q}{(2\pi)^4 q^2}\, e^{-3\varepsilon q^2} = \frac{1}{48\pi^2},$$

we finally obtain:

$$\langle k_\lambda J_\lambda^5(k)\rangle = -\frac{e^2}{12\pi^2}\epsilon_{\mu\nu\rho\sigma}\int d^4p_1\, d^4p_2\, p_{1\mu}A_\nu(p_1)\, p_{2\rho}A_\sigma(p_2). \tag{20.95}$$

From the definition (20.82) it follows that:

$$k_\lambda \Gamma^{(3)}_{\lambda\mu\nu}(k;p_1,p_2) = \frac{e^2}{6\pi^2}\epsilon_{\mu\nu\rho\sigma}p_{1\rho}p_{2\sigma}. \tag{20.96}$$

This non-vanishing result tells us that any definition of the determinant $\det\slashed{D}$ breaks at least current conservation or gauge invariance. Since gauge invariance is essential to the construction of QED we choose of course to break current conservation. Exchanging arguments, we obtain the value of $p_{1\mu}\Gamma^{(3)}_{\lambda\mu\nu}(k;p_1,p_2)$:

$$p_{1\mu}\Gamma^{(3)}_{\lambda\mu\nu}(k;p_1,p_2) = \frac{e^2}{6\pi^2}\epsilon_{\lambda\nu\rho\sigma}k_\rho p_{2\sigma}. \tag{20.97}$$

If instead we had used a gauge invariant regularization, the result for $\Gamma^{(3)}$ would have differed by a term of the form (20.83) and $p_{1\mu}\Gamma^{(3)}_{\lambda\mu\nu}(k;p_1,p_2)$ would have vanished. Imposing gauge invariance we can find the additional contribution $\delta\Gamma^{(3)}$:

$$\delta\Gamma^{(3)}_{\lambda\mu\nu}(k;p_1,p_2) = K\epsilon_{\lambda\mu\nu\rho}(p_1 - p_2)_\rho . \tag{20.98}$$

Since:

$$p_{1\mu}\delta\Gamma^{(3)}_{\lambda\mu\nu}(k;p_1,p_2) = -\frac{e^2}{6\pi^2}\epsilon_{\lambda\nu\rho\sigma}k_\rho p_{2\sigma} , \tag{20.99}$$

we find:

$$K = e^2/(6\pi^2) . \tag{20.100}$$

This gives an additional contribution to $k_\lambda\Gamma^{(3)}_{\lambda\mu\nu}(k;p_1,p_2)$:

$$k_\lambda\delta\Gamma^{(3)}_{\lambda\mu\nu}(k;p_1,p_2) = \frac{e^2}{3\pi^2}\epsilon_{\mu\lambda\rho\sigma}p_{1\rho}p_{2\sigma} . \tag{20.101}$$

Therefore in a QED-like gauge invariant field theory with massless fermions we find a non-vanishing chiral anomaly:

$$k_\lambda\Gamma^{(3)}_{\lambda\mu\nu}(k;p_1,p_2) = \left(\frac{e^2}{2\pi^2} \equiv \frac{2\alpha}{\pi}\right)\epsilon_{\mu\nu\rho\sigma}p_{1\rho}p_{2\sigma} . \tag{20.102}$$

Note that the anomaly is gauge invariant as expected.

Equation (20.102) can be rewritten after Fourier transformation as an axial current non-conservation equation:

$$\partial_\lambda J^5_\lambda(x) = -i\frac{\alpha}{4\pi}\epsilon_{\mu\nu\rho\sigma}F_{\mu\nu}F_{\rho\sigma} , \tag{20.103}$$

in which $F_{\mu\nu}$ is the electromagnetic tensor. Note that since global chiral symmetry is not broken, the integral over the whole space of the anomalous term must vanish. This condition is indeed verified since the anomaly can immediately be written as a total derivative:

$$\epsilon_{\mu\nu\rho\sigma}F_{\mu\nu}F_{\rho\sigma} = 4\,\epsilon_{\mu\nu\rho\sigma}\partial_\mu\left(\epsilon_{\mu\nu\rho\sigma}A_\nu\partial_\rho A_\sigma\right) . \tag{20.104}$$

Equation (20.103) also implies:

$$\ln\det\left[e^{i\gamma_5\theta(x)}\,\slashed{D}\,e^{i\gamma_5\theta(x)}\right] = \ln\det\slashed{D} - i\frac{\alpha}{4\pi}\int d^4x\,\theta(x)\epsilon_{\mu\nu\rho\sigma}F_{\mu\nu}(x)F_{\rho\sigma}(x). \tag{20.105}$$

Note that since the chiral group is abelian, this equation which has been derived at first order in $\theta(x)$ is actually exact.

In the form we have presented it the calculation generalizes without difficulty to higher even dimensions $2n$. Note simply that the permutation $(\mathbf{p}_1, \mu) \leftrightarrow (\mathbf{p}_2, \nu)$ in the first term of equation (20.89) is replaced by a cyclic permutation. The result for the anomaly is:

$$\partial_\lambda J^5_\lambda(x) = -2i\frac{e^n}{(4\pi)^n n!}\epsilon_{\mu_1\nu_1\ldots\mu_n\nu_n}F_{\mu_1\nu_1}\ldots F_{\mu_n\nu_n} . \tag{20.106}$$

Chiral gauge theory. Current conservation becomes a WT identity in a gauge theory. The anomaly prevents the construction of a theory which would have both an abelian gauge vector and axial symmetry. Furthermore the analysis of the *BBB* correlation function is identical and yields also an anomaly which cannot be removed since the correlation function is by definition symmetric in its three arguments. This prevents the construction of a pure axial gauge symmetry.

20.3.3 Generalization to non-abelian vector gauge theories

We still consider a unique axial current but now inside a non-abelian gauge theory. The fermion fields transform non-trivially under a gauge group G and $\mathbf{A}_\mu$ is the corresponding gauge field. The action is:

$$S\left(\bar\psi,\psi\right) = -\int \mathrm{d}^4 x\,\bar\psi\left(x\right)\slashed{D}\psi\left(x\right),\tag{20.107}$$

with:

$$\slashed{D} = \slashed\partial + \slashed{\mathbf{A}}.\tag{20.108}$$

No new calculation is needed; the result is completely fixed by the previous calculation, dimensional analysis and gauge invariance:

$$\partial_\lambda J_\lambda^5(x) = -\frac{i}{16\pi^2}\epsilon_{\mu\nu\rho\sigma}\,\mathrm{tr}\,\mathbf{F}_{\mu\nu}\mathbf{F}_{\rho\sigma},\tag{20.109}$$

in which $\mathbf{F}_{\mu\nu}$ is now the corresponding curvature. Again this expression should be a total derivative. Indeed:

$$\epsilon_{\mu\nu\rho\sigma}\,\mathrm{tr}\,\mathbf{F}_{\mu\nu}\mathbf{F}_{\rho\sigma} = 4\,\epsilon_{\mu\nu\rho\sigma}\partial_\mu\,\mathrm{tr}\left(\mathbf{A}_\nu\partial_\rho\mathbf{A}_\sigma + \tfrac{2}{3}\mathbf{A}_\nu\mathbf{A}_\rho\mathbf{A}_\sigma\right).\tag{20.110}$$

20.3.4 Anomaly and eigenvalues of the Dirac operator

Let us assume that the spectrum of $\slashed{D}$, the Dirac operator in a non-abelian gauge field (equation (20.108)), is discrete and let us call d_n and $\varphi_n(x)$ the corresponding eigenvalues and eigenvectors:

$$\slashed{D}\varphi_n = d_n\varphi_n\tag{20.111}$$

For a unitary or orthogonal group the Dirac operator is antihermitian; therefore the eigenvalues are imaginary and the eigenvectors orthogonal. In addition they satisfy:

$$\slashed{D}\gamma_5\varphi_n = -d_n\gamma_5\varphi_n.\tag{20.112}$$

Therefore either d_n is different from zero, and $\gamma_5\varphi_n$ is an eigenvector of $\slashed{D}$ with eigenvalue $-d_n$, or d_n vanishes. In the latter case the eigenvectors of $\slashed{D}$ can be chosen eigenvectors of defined chirality, i.e. with eigenvalue $\pm$ of γ_5:

$$\slashed{D}\varphi_n = 0,\quad \gamma_5\varphi_n = \pm\varphi_n.$$

Let us now calculate the variation δ_n at first order of the eigenvalues $d_n + m$ of the operator $\slashed{D} + m$ in a chiral transformation (20.78):

$$\slashed{D} + m \longmapsto \slashed{D} + m + i\gamma_5\theta(x)(\slashed{D}+m) + i(\slashed{D}+m)\gamma_5\theta(x).$$

Perturbation theory tells us:

$$\delta_n = i\int \mathrm{d}^4 x\,\varphi_n^*(x)\left[\gamma_5\theta(x)(\slashed{D}+m) + (\slashed{D}+m)\gamma_5\theta(x)\right]\varphi_n(x),$$
$$= 2i\left(d_n+m\right)\int \mathrm{d}^4 x\,\theta(x)\varphi_n^*(x)\gamma_5\varphi_n(x),\tag{20.113}$$

where we have assumed that the eigenvectors are orthonormal.

If we now take the limit of constant θ we find in the r.h.s. the scalar product of eigenvectors of $\slashed{D}$ which vanishes except when $d_n = 0$. Calling n_+ and n_- the numbers of eigenvalue zero of chirality $+$ and $-$ respectively, we can then write:

$$\ln \det \left[\left(e^{i\gamma_5\theta}(\slashed{D}+m)\,e^{i\gamma_5\theta}\right)(\slashed{D}+m)^{-1}\right] = \prod_n \left(1 + \frac{\delta_n}{d_n+m}\right)$$
$$= 2i\,(n_+ - n_-)\,\theta\,. \tag{20.114}$$

Taking the limit $m = 0$ we can relate the r.h.s. to the space integral of the chiral anomaly (20.109):

$$-\frac{1}{32\pi^2}\epsilon_{\mu\nu\rho\sigma}\int d^4x \, \mathrm{tr}\, \mathbf{F}_{\mu\nu}\mathbf{F}_{\rho\sigma} = (n_+ - n_-)\,. \tag{20.115}$$

This result has several consequences:

(i) Equation (20.110) shows that the integral of the anomaly over the whole space depends only on the behaviour at large distances of the curvature tensor $\mathbf{F}_{\mu\nu}$. The property that, in addition, the integral is quantized shows that the form of the anomaly is related to topological properties of the gauge field since the integral cannot change when the gauge field is varied continuously.

(ii) We shall see in Section 40.5 that gauge field configurations exist for which the r.h.s. of equation (20.115) does not vanish, for example instantons. We have shown above that if massless fermions are coupled to such gauge fields the determinant resulting from the fermion integration necessarily vanishes. This has some physical implications which are examined in Sections 20.5 and 40.5.

(iii) Note also that the l.h.s. of equation (20.114) is obviously 1 when $\theta = n\pi$. Also one realizes that the r.h.s. is the exact result although it has only been derived for θ small, due to the group properties of the abelian chiral group. Therefore we should have expected that the integral of the anomaly yields an even integer.

(iv) One might be surprised by the result that $\det\slashed{D}$ is not invariant under global chiral transformations. However we have just established that when the integral of the anomaly does not vanish, $\det\slashed{D}$ vanishes. This explains that to give a meaning to the r.h.s. of equation (20.114) we have been forced to introduce a mass term. In the limit $m \to 0$, we calculate the $\det\slashed{D}$ in the subspace orthogonal to eigenvectors with zero eigenvalue. When $n_+ \neq n_-$ this subspace is not chiral invariant.

20.4 Non-Abelian Anomaly

20.4.1 General axial current

Let us first consider the problem of the conservation of a general axial current in the case of a non-abelian vector gauge theory. We assume that the fermion action has a $G \times G$ chiral symmetry and discuss the conservation of the corresponding axial current. The generators of the gauge group may or may not be related to the diagonal subgroup G of $G \times G$ which correspond to vector currents. We call t^α the generators of G. The current then has the form:

$$J_\mu^{5\alpha}(x) = -\bar{\psi}\gamma_5\gamma_\mu t^\alpha \psi\,. \tag{20.116}$$

When the gauge group is connected with the chiral group, the current conservation equation involves the gauge covariant derivative:

$$\mathbf{D}_\mu J_\mu^5 = 0\,. \tag{20.117}$$

In the calculation of the contribution to the anomaly quadratic in the gauge fields the only modification in the previous arguments is the appearance of a different geometrical factor. Then the complete form of the anomaly is dictated by gauge covariance. One finds:

$$\mathbf{D}_\lambda J_\lambda^{5\alpha}(x) = -\frac{i}{16\pi^2} \epsilon_{\mu\nu\rho\sigma} \operatorname{tr} t^\alpha \mathbf{F}_{\mu\nu} \mathbf{F}_{\rho\sigma} \,. \tag{20.118}$$

In particular if the gauge group is disconnected from the chiral group the anomaly is proportional to $\operatorname{tr} t^\alpha$ and therefore only different from zero for the abelian factors of G.

20.4.2 Obstruction to gauge invariance

We now want to consider a non-abelian gauge field coupled to left or right-handed fermions:

$$S\left(\bar\psi, \psi\right) = -\int \mathrm{d}^4 x\, \bar\psi\,(x)\, \tfrac{1}{2}\left(1 + \gamma_5\right) \mathbf{\not{D}}\psi\,(x)\,, \tag{20.119}$$

$(\tfrac{1}{2}\left(1 - \gamma_5\right)$ is treated in the same way). We can construct a consistent gauge theory only if the partition function

$$Z\left(\mathbf{A}_\mu\right) = \int \left[\mathrm{d}\psi \mathrm{d}\bar\psi\right] \exp\left[-S\left(\psi, \bar\psi\right)\right] \tag{20.120}$$

is gauge invariant.

If we introduce the generators t^α of the gauge group in the fermion representation, we can write the corresponding current $\mathbf{J}_\mu$ as:

$$J_\mu^\alpha(x) = -\bar\psi \tfrac{1}{2}\left(1 + \gamma_5\right)\gamma_\mu t^\alpha \psi \,. \tag{20.121}$$

The invariance of $Z\left(\mathbf{A}_\mu\right)$ under an infinitesimal gauge transformation again leads to a covariant conservation equation for the current:

$$\langle \mathbf{D}_\mu \mathbf{J}_\mu \rangle = 0\,.$$

The calculation of the term of degree two in the gauge field of the anomaly is straightforward: the regularization adopted for the calculation in Subsection 20.3.2 is also suited to the present case since the current-gauge field 3-point function is symmetric in the external arguments. The group structure is reflected by a simple geometrical factor. The global factor can be taken from the abelian calculation. It differs from result (20.95) by a factor $1/2$ which comes from the projector $\tfrac{1}{2}\left(1 + \gamma_5\right)$. The general form of the term of degree 3 in the gauge field can also be easily found, but the calculation of the global factor is somewhat longer. We shall argue in next section that it can be obtained from consistency conditions. The complete expression reads:

$$\left(\mathbf{D}_\mu \mathbf{J}_\mu(x)\right)^\alpha = -\frac{i}{24\pi^2} \partial_\mu \epsilon_{\mu\nu\rho\sigma} \operatorname{tr}\left[t^\alpha \left(\mathbf{A}_\nu \partial_\rho \mathbf{A}_\sigma + \tfrac{1}{2}\mathbf{A}_\nu \mathbf{A}_\rho \mathbf{A}_\sigma\right)\right]\,. \tag{20.122}$$

If the projector $\tfrac{1}{2}(1 + \gamma_5)$ is replaced by $\tfrac{1}{2}(1 - \gamma_5)$ the sign of the anomaly changes.

Unless this term vanishes identically there is an obstruction to the construction of the gauge theory. It is easy to verify, taking into account the antisymmetry of the ϵ tensor, that the group factor is:

$$d_{\alpha\beta\gamma} = \tfrac{1}{2} \operatorname{tr}\left[t^\alpha \left(t^\beta t^\gamma + t^\gamma t^\beta\right)\right]\,. \tag{20.123}$$

For a unitary representation the generators t^α are, with our conventions, antihermitian. Therefore the coefficients $d_{\alpha\beta\gamma}$ are purely imaginary:

$$d_{\alpha\beta\gamma}^* = \tfrac{1}{2} \operatorname{tr}\left[t^\alpha \left(t^\beta t^\gamma + t^\gamma t^\beta\right)\right]^* = -d_{\alpha\beta\gamma}\,. \tag{20.124}$$

For all real (the t^α antisymmetric) or "pseudo-real" ($t^\alpha = -S\,{}^T t^\alpha S^{-1}$) representations these coefficients vanish. It follows that the only non-abelian groups which can lead to anomalies in four dimensions are: $SU(N)$ for $N \geq 3$, $SO(6)$ and E_6.

20.5 Physical Consequences

Electromagnetic π_0 decay. We have examined in Section 13.6 a phenomenological model for low energy Strong Interactions based on a linearly broken $SU(2) \times SU(2)$ symmetry. The current non-conservation equation for the axial current $\mathbf{J}_\mu$ was at leading order (equation (13.81)):

$$\partial_\mu \mathbf{J}_\mu = m_\pi^2 f_\pi \boldsymbol{\pi} \,. \tag{20.125}$$

We here concentrate on the third component J_μ^3 of the current which corresponds in the r.h.s. to the neutral pion π_0 field. If we introduce electromagnetic interactions in the model, this relation between the divergence of the axial current and the π_0 field allows us to calculate the electromagnetic decay rate of the neutral pion when the momentum $\mathbf{k}$ of the pion goes to zero. In the absence of anomalies, the average of relation (20.125) multiplied by two gauge fields yields immediately that the decay rate vanishes for $k = 0$ in contradiction with reasonable smoothness assumptions and experimental results. Taking instead into account anomaly equation (20.103) and a factor $1/2$ which comes from a different normalization (13.99) of chiral transformations we find:

$$\partial_\mu J_\mu^3 = m_\pi^2 f_\pi \pi_0 - i\frac{\alpha}{8\pi} \epsilon_{\mu\nu\rho\sigma} F_{\mu\nu} F_{\rho\sigma} \,. \tag{20.126}$$

Taking again the average of this equation with two photon fields and going to the limit $\mathbf{k} = 0$ to eliminate the l.h.s., one now obtains a non-vanishing decay amplitude for an unphysical π_0 at zero total momentum. In the σ-model at leading order one can extrapolate to $k^2 = -m_\pi^2$. The result is in excellent agreement with experiment. The theoretical rate Γ is given by:

$$\Gamma = \frac{\alpha^2 m_\pi^3}{64\pi^3 f_\pi^2} = 7.6 \text{ eV}$$

while $\Gamma^{\text{exp}} = (7.37 \pm 1.5)\text{eV}$. A similar estimate was first derived by Steinberger from direct Feynman graph calculation, before the relation to anomalies had been discovered.

Note that a similar theoretical estimate is obtained in the quark model with massless quarks for three colours.

The solution of the $U(1)$ problem. In a theory of Strong Interactions in which the quarks are massless and interact through a colour gauge group, the action has a chiral $U(N_{\text{F}}) \times U(N_{\text{F}})$ symmetry, in which N_{F} is the number of flavours. The spontaneous breaking of the chiral group to its diagonal subgroup $U(N_{\text{F}})$ let us expect N_{F}^2 Goldstone bosons associated with the axial currents. From the previous analysis we know that the axial current corresponding to the $U(1)$ abelian subgroup has an anomaly. Of course the WT identities which imply that the existence of Goldstone bosons correspond to constant group transformations and therefore involve only the space integral of the divergence of the current. Since the anomaly is a total derivative one might have expected the integral to vanish. However we shall see in Section 40.5 that non-abelian gauge theories admit instanton solutions which give a periodic structure to the vacuum. These instanton solutions correspond to gauge configurations which approach non-trivial pure gauges at infinity and give the set of discrete non-vanishing values one expects from equation (20.115) to the space integral of the anomaly (20.109). This indicates (but no satisfactory calculation of the instanton contribution has been performed) that for small, but non-vanishing, quark masses the $U(1)$ axial current is far from being conserved and therefore no light would-be Goldstone boson is generated. This observation resolves a long standing puzzle since experimentally no corresponding light pseudoscalar boson is found (for details see Section 40.5).

Weak–electromagnetic interactions. The cancellation of the anomaly discussed in Section 20.4.2 imposes constraints on the model of w.e.m. interactions. In the Standard Model, for example, the anomalous contribution of the leptons has to cancel the quark contribution. Let us show that within the Standard Model this cancellation occurs within each generation provided that for each flavour quarks can exist in three states. In the w.e.m. group $SU(2) \times U(1)$, $SU(2)$ alone is a safe group. Therefore the problems come from the $U(1)$ factor. We expect *a priori* two conditions coming from the vertices with one $U(1)$ and two $SU(2)$ gauge fields and with three $U(1)$ gauge fields. Actually we shall see that both conditions are equivalent. If we consider two $SU(2)$ and one $U(1)$ gauge fields, only $SU(2)$ doublets contribute and equation (20.122) leads to the condition:

$$\sum_{\text{all doublets}} Y_{\text{L}} \, \text{tr} \, \tau^\alpha \tau^\beta = 0 \,,$$

in which Y_{L} is the $U(1)$ charge (see Section 20.1). This condition reduces to:

$$\sum_{\text{all doublets}} Y_{\text{L}} = 0 \,. \qquad (20.127)$$

The three $U(1)$ gauge field vertex yields the condition:

$$\sum_{\text{left-handed parts}} Y_{\text{L}}^3 - \sum_{\text{right-handed parts}} Y_{\text{R}}^3 = 0 \,,$$

because the contributions to the anomaly of right-handed and left-handed couplings have opposite signs. However in the Standard Model the left and right charges are related (equation (20.32)). Summing the charges of one doublet and the corresponding singlets, we obtain:

$$\sum_{\text{all doublets}} (Y_{\text{L}} + 1)^3 + (Y_{\text{L}} - 1)^3 - 2Y_{\text{L}}^3 = 0 \,,$$

a condition identical to equation (20.127).

In one generation the lepton doublet has $Y_{\text{L}} = -1$ and the quark $Y_{\text{L}} = 1/3$. Therefore a cancellation requires that the quarks exist in three states. These states are provided by the colour quantum number.

20.6 Wess–Zumino Consistency Conditions

In Subsection 20.4.2 we have calculated the part of the anomaly which is quadratic in the gauge field and asserted that the remaining part could be obtained from geometrical arguments. Let us now explain the idea. The anomaly is the variation of a functional under an infinitesimal gauge transformation. As we argued in Section 15.3, this implies compatibility conditions, which here are constraints on the form of the anomaly. One convenient way to write these constraints is to express the nilpotency of BRS transformations (for details see Chapter 21). We perform on the gauge field $\mathbf{A}_\mu$ the BRS transformation (19.53):

$$\delta \mathbf{A}_\mu(x) = \mathbf{D}_\mu \mathbf{C}(x)\bar{\varepsilon} \,. \qquad (20.128)$$

The variation of $\ln Z\left(\mathbf{A}_\mu\right)$ takes the form:

$$\delta\ln Z\left(\mathbf{A}_\mu\right) = -\int \mathrm{d}^4x \,\langle\mathbf{J}_\mu(x)\rangle\,\mathbf{D}_\mu\mathbf{C}(x)\bar{\varepsilon}. \tag{20.129}$$

If we write the anomaly equation:

$$\langle\mathbf{D}_\mu\mathbf{J}_\mu(x)\rangle = \mathcal{A}\left(\mathbf{A}_\mu, x\right), \tag{20.130}$$

equation (20.129) takes the form:

$$\delta\ln Z\left(\mathbf{A}_\mu\right) = \int \mathrm{d}^4x\,\mathcal{A}\left(\mathbf{A}_\mu, x\right)\mathbf{C}(x)\bar{\varepsilon}. \tag{20.131}$$

Since $\mathcal{A}\mathbf{C}(x)$ is a BRS variation it is invariant under BRS transformations (19.53,20.128), i.e. transforming the gauge field by (20.128) and the fermion $\mathbf{C}(x)$ by:

$$\delta\mathbf{C}(x) = -\bar{\varepsilon}\,\mathbf{C}^2(x). \tag{20.132}$$

Conversely in order to find the form of possible anomalies one has to look for local functionals of $\mathbf{A}_\mu$ and $\mathbf{C}$ linear in $\mathbf{C}$ and BRS invariant. One verifies that this condition determines the term cubic in $\mathbf{A}$ in the r.h.s. of equation (20.122).

Bibliographical Notes

The Standard Model was proposed in its present form by
  S. Weinberg *Phys. Rev. Lett.* 19 (1967) 1264; A. Salam, *Elementary Particle Theory*,
  N. Svartholm ed. (Almqvist and Wiksell, Stockholm 1968).
Earlier attempts include
  S.L. Glashow, *Nucl. Phys.* 22 (1961) 579.
To explain the suppression of strangeness changing neutral currents, a fourth quark (c for charm) was conjectured in
  S.L. Glashow, J. Iliopoulos and L. Maiani, *Phys. Rev.* D2 (1970) 1285.
A selection of papers is collected in
  Gauge Theory of Weak and Electromagnetic Interactions, C.H. Lai ed. (World Scientific, Singapore 1981).
Most references given in Chapter 19 are still relevant here. See also
  E.S. Abers and B.W. Lee, *Phys. Rep.* 9C (1973) 1; J.C. Taylor, *Gauge Theories of Weak Interactions* (Cambridge University Press, Cambridge 1976).
For the KM matrix see
  M. Kobayashi and T. Maskawa, *Prog. Theor. Phys.* 49 (1973) 652.
Among the textbooks see for example
  T.D. Lee, *Particle Physics and Introduction to Field Theory* (Harwood Academic, New York 1981).
The problem of anomalies is discussed in
  J.S. Bell and R. Jackiw, *Nuovo Cimento* A60 (1969) 47; S.L. Adler, *Phys. Rev.* 177 (1969) 2426; D.J. Gross and R. Jackiw, *Phys. Rev.* D6 (1972) 477; H. Georgi and S.L. Glashow, *Phys. Rev.* D6 (1972) 429; C. Bouchiat, J. Iliopoulos and Ph. Meyer, *Phys. Lett.* 38B (1972) 519; E. Witten, *Phys. Lett.* 117B (1982) 324.
See also the lectures
  M. Peskin, *Recent Advances in Field Theory and Statistical Mechanics*, Les Houches 1982, R. Stora and J.-B. Zuber eds. (North-Holland, Amsterdam 1984),
and the volume
  S.B. Treiman, R. Jackiw, B. Zumino and E. Witten, *Current Algebra and Anomalies* (World Scientific, Singapore 1985) and references therein.

21 RENORMALIZATION OF GAUGE THEORIES: GENERAL FORMALISM

To discuss the renormalization of gauge theories in the non-abelian case in its full generality, it is necessary to use a rather abstract formalism, which allows one to understand the algebraic structure of the renormalization procedure without being overwhelmed by the notational complexity. There is however a price to pay: the translation of the general identities which then appear into usual and more concrete notations becomes a non-trivial exercise.

Our strategy is the following. We quantize the theory by providing a dynamics for the degrees of freedom corresponding to gauge transformations. We find, as expected, that the quantized action has a BRS symmetry. We then derive the corresponding set of WT identities for correlation functions. We use these identities to determine the most general form of the renormalized action.

Physical quantities in the theory should be independent of the dynamics of the gauge group degrees of freedom. Following the analysis of previous chapters, we first prove that the vacuum amplitude does not depend on this dynamics, i.e. is gauge independent. We argue that this property remains true if we add to the original action sources for gauge invariant operators of canonical dimension low enough, so that the total action remains renormalizable. As a consequence, correlation functions of these gauge invariant operators are independent of the gauge fixing procedure, and have therefore an intrinsic meaning. A similar property holds for all gauge invariant operators but the discussion is more involved and will not be given. Finally, when a S-matrix can be defined, S-matrix elements are also gauge independent.

21.1 Notations and Group Structure

Notations. In what follows, we restrict ourselves to boson fields, the generalization to fermions being straightforward. All gauge fields and other scalar boson fields are combined into one vector denoted by A^i, in which the index i stands for space variables x, Lorentz index μ, and group indices a, b:

$$A^i \equiv \left\{ A^a_\mu(x), \varphi_b(x) \right\}.$$

A gauge transformation corresponding to the compact Lie group G is written:

$$\delta A^i = D^i_\alpha(A)\omega^\alpha, \qquad (21.1)$$

in which ω^α are the infinitesimal parameters of the group transformation. As we have done in equation (21.1), we reserve greek indices for the adjoint representation of the Lie algebra. Summation over repeated indices is always meant. It includes integration over space variables. The index α also includes group indices and space variables. In more concrete form, equation (21.1) stands for:

$$\delta A^a_\mu(x) = \int \mathrm{d}y \left[\partial^x_\mu \delta(x-y)\delta^a_b + f^a_{bc} A^c_\mu(x)\delta(x-y) \right] \omega^b(y),$$

$$\delta \varphi^{a'}(x) = \int \mathrm{d}y\, \delta(x-y)[t_{b'}]^{a'}_{c'} \varphi^{c'}(x)\omega^{b'}(y), \qquad (21.2)$$

where the matrices $t_{b'}$ form a (in general reducible) representation of the Lie algebra of G and f^a_{bc} are the corresponding structure constants.

Finally $\mathcal{L}(A)$ is the gauge invariant action and thus satisfies:

$$\frac{\delta \mathcal{L}}{\delta A^i} D^i_\alpha(A) = 0 \,. \tag{21.3}$$

A group identity. We have explained in Section 15.3 that, as a consequence of the group structure, quite generally the functionals $D^i_\alpha(A)$ satisfy equations which can be regarded as compatibility conditions for the system (21.3) considered as a set of differential equations for $\mathcal{L}(A)$:

$$\frac{\delta D^i_\alpha}{\delta A^j} D^j_\beta - \frac{\delta D^i_\beta}{\delta A^j} D^j_\alpha = f^\gamma_{\alpha\beta} D^i_\gamma \,. \tag{21.4}$$

These equations are formally identical to equations (15.43) of Section 15.3 and (16.41) of Section 16.4.

Since here the functional $D^i_\alpha(A)$ is only affine, it is not difficult to verify the identities (21.4) by a direct calculation, using the commutation relations of the Lie algebra of G. Note however that because $D^i_\alpha(A)$ is a differential operator, it is necessary to carefully keep track of the δ-functions. In particular since the indices α and i include space coordinates, the structure constants $f^\gamma_{\alpha\beta}$ which are proportional to the numerical structure constants f^a_{bc} of the Lie algebra, have a non-trivial dependence on space variables:

$$f^\gamma_{\alpha\beta} \equiv f^c_{ab} \delta(x-y)\delta(y-z). \tag{21.5}$$

Due to the distribution character of the various quantities it is actually convenient to rewrite the identity (21.4) with the help of two functions $\omega^\alpha_1 \equiv \omega^a_1(x)$ and $\omega^\alpha_2 \equiv \omega^a_2(x)$ and the operator

$$\Delta(\omega) = \omega^\alpha D^i_\alpha(A) \frac{\delta}{\delta A^i} \,. \tag{21.6}$$

The identity (21.4) takes the form

$$[\Delta(\omega_1), \Delta(\omega_2)] = \Delta(\omega_{12}), \quad \omega^\gamma_{12} = f^\gamma_{\alpha\beta} \omega^\alpha_1 \omega^\beta_2 \,. \tag{21.7}$$

Let us for instance indicate here how this identity can be recovered for the gauge field part. We write $\Delta(\omega)$ in explicit notation:

$$\Delta(\omega) = \int \mathrm{d}x \left[\partial_\mu \omega^a(x) + f^a_{bc} \omega^b(x) A^c_\mu(x) \right] \frac{\delta}{\delta A^a_\mu(x)} \,.$$

We then calculate explicitly the commutator

$$[\Delta(\omega_1), \Delta(\omega_2)] = \int \mathrm{d}x \left[\partial_\mu \omega^a_1(x) + f^a_{bc} \omega^b(x) A^c_\mu(x) \right] \omega^e_2(x) f^d_{ea} \frac{\delta}{\delta A^d_\mu(x)} - \{1 \leftrightarrow 2\} \,.$$

Using the antisymmetry of f^a_{bc} and the Jacobi identity one indeed obtains equation (21.7) in explicit form, with $\omega^a_{12}(x) = f^a_{cb} \omega^b_1(x) \omega^c_2(x)$.

21.2 Quantization

We have shown in Chapters 18 and 19 that the quantization problem can be reformulated in the following way, which leads to the same results as canonical quantization: Because the action is gauge invariant, the degrees of freedom associated with gauge transformations have no dynamics. To quantize these degrees of freedom and to give a meaning to the functional integral in the continuum, it is necessary to provide one. This dynamics is somewhat arbitrary and we shall have eventually to prove that "physical results" (we shall explain later what we mean by physical results) do not depend on its choice. Gauge invariance defines classes in the field space, corresponding to fields and all their gauge transforms, i.e. orbits of the gauge group. Let us choose in some way one representative B^i for each class, i.e. a gauge section, and parametrize a general field A^i as:

$$A^i = \left(B^i\right)_g,\qquad (21.8)$$

in which B_g denotes the gauge transform of B by the gauge transformation corresponding to the group element g. Since $\mathcal{L}(A)$ is gauge invariant,

$$\mathcal{L}(A) = \mathcal{L}(B),$$

and $\mathcal{L}(B)$ provides a dynamics for the field B^i. For the group element g which is a field since it depends on space we introduce a stochastic quantization in the sense of Chapters 16 and 17:

$$F_\alpha\left(B_g\right) = \nu_\alpha.\qquad (21.9)$$

We assume that, at B fixed, the set of equations (21.9) determines g uniquely (at least for B^i small so that the equation can be solved perturbatively) as a function of ν. The field ν has a distribution $\mathrm{d}\rho(\nu)$ which we do not need specifying more precisely at this stage, but which eventually will be chosen local and gaussian. We denote the flat and thus gauge invariant measure $[\mathrm{d}A^i]$ written in terms of B and g by $[\mathrm{d}B][\mathrm{d}g]$ (gauge invariance implies that the measure can be factorized).

To avoid modifying the dynamics of B^i we must multiply the B^i integrand by a quantity independent of B^i. We thus use the identity:

$$\int [\mathrm{d}g]\,\delta\left[F_\alpha\left(B_g\right) - \nu_\alpha\right]\det M = 1,\qquad (21.10)$$

with:

$$M_{\alpha\beta}(B_g) \equiv \left(\frac{\delta F}{\delta g}\right)_{\alpha\beta} = \frac{\delta F_\alpha}{\delta A^i}\left(B_g\right) D_\beta^i\left(B_g\right).\qquad (21.11)$$

We can then write the generating functional of correlation functions $Z(J)$:

$$Z(J) = \int [\mathrm{d}\rho\left(\nu\right)]\int [\mathrm{d}B][\mathrm{d}g]\,\delta\left[F_\alpha\left(B_g\right) - \nu_\alpha\right]\det M(B_g)\exp\left[-\mathcal{L}(B) + J_i B_g^i\right].\qquad (21.12)$$

Since $\mathcal{L}(B)$ is gauge invariant we can replace B by B_g. The integrand then depends only on the argument $B_g^i = A^i$ and we can rewrite the generating functional (21.12) as:

$$Z(J) = \int [\mathrm{d}\rho\left(\nu\right)]\int [\mathrm{d}A]\det M\left(A\right)\delta\left[F_\alpha(A) - \nu_\alpha\right]\exp\left[-\mathcal{L}(A) + J_i A^i\right].\qquad (21.13)$$

Renormalization of Gauge Theories

As in Chapter 16 we impose the constraint (21.9) by integrating over an auxiliary field λ^α and introduce fermions "ghost" fields to represent the determinant:

$$Z(J) = \int [\mathrm{d}\rho(\nu)] \int [\mathrm{d}A\mathrm{d}\lambda\mathrm{d}\bar{C}\mathrm{d}C] \exp\left[-\mathcal{L}(A) + \lambda^\alpha\left(\nu_\alpha - F_\alpha(A)\right) + \bar{C}^\alpha M_{\alpha\beta}C^\beta\right.$$
$$\left. +J_iA^i\right].\tag{21.14}$$

We perform the integration over ν_α, setting

$$\int [\mathrm{d}\rho\,(\nu)] \exp \lambda^\alpha\nu_\alpha = \exp w\,(\lambda).\tag{21.15}$$

The final form of the generating functional of the A-field correlation functions is then:

$$Z(J) = \int [\mathrm{d}A]\,[\mathrm{d}\lambda]\,[\mathrm{d}C\mathrm{d}\bar{C}] \exp\left[-S\left(A,C,\bar{C},\lambda\right) + J_iA^i\right],\tag{21.16}$$

with the definitions:

$$S = \mathcal{L}(A) - w\,(\lambda) + \lambda^\alpha F_\alpha(A) - \bar{C}^\alpha M_{\alpha\beta}C^\beta\,,\qquad M_{\alpha\beta} = \frac{\partial F_\alpha}{\partial A^i}D^i_\beta\,.\tag{21.17}$$

21.3 BRS Symmetry

Let us first note that in the representation (21.13) the functional integral for $J = 0$ is invariant under gauge transformations which translate ν_α (see Section 16.1)

$$\delta A^i = D^i_\alpha[M^{-1}]^{\alpha\beta}\mu_\beta\,,\tag{21.18}$$

and thus

$$\delta\left[F_\alpha(A)\right] = \frac{\delta F_\alpha}{\delta A^i}D^i_\beta[M^{-1}]^{\beta\gamma}\mu_\gamma = \mu_\alpha\,.\tag{21.19}$$

This Slavnov–Taylor symmetry was an essential tool in the initial proof of the renormalizability of non-abelian gauge theories. However the non-local character of this transformation explains the complexity of such a derivation.

It follows from the general analysis of Chapter 16 that the final action (21.17) has a BRS symmetry. Because the field which is stochastically quantized is the gauge group, the BRS transformation for the field A^i has the form of a gauge transformation (equation (17.77)). We have verified in Chapter 19 that, since we have provided a dynamics to a group element, the BRS symmetry is similar in form to the symmetry of the dynamic action of chiral models as described in Sections 16.4, 17.7 (equations (16.46,16.47) and (17.75–17.77)). Also, in these more abstract notations, the basic equation (21.4) is formally identical to the commutation relations in the case of non-linear representation of groups of Section 15.3, and to the relations (16.41) of Section 16.4. The corresponding BRS transformations for the A and C fields thus are given by the equations (16.39). As we know from the general analysis of Chapter 16, the transformations of λ and $\bar{C}$ are independent of the dynamics. We conclude that the BRS transformations have the form

$$\delta A^i = \bar{\varepsilon}D^i_\alpha(A)C^\alpha\,,\qquad \delta C^\alpha = -\tfrac{1}{2}\bar{\varepsilon}f^\alpha_{\beta\gamma}C^\beta C^\gamma\,,\tag{21.20a}$$

$$\delta\bar{C}^\alpha = \bar{\varepsilon}\lambda^\alpha\,,\qquad\qquad \delta\lambda^\alpha = 0\,.\tag{21.20b}$$

By rewriting these transformations in more explicit notations, we immediately verify that they are identical to the BRS transformations derived in Chapter 19. For further use we set:

$$D^i = D^i_\alpha(A)C^\alpha. \tag{21.21}$$

The BRS operator. To express the BRS symmetry, it is also useful to introduce the anticommuting differential operator $\mathcal{D}_0$:

$$\mathcal{D}_0 = D^i\frac{\delta}{\delta A^i} - \tfrac{1}{2}f^\alpha_{\beta\gamma}C^\beta C^\gamma\frac{\delta}{\delta C^\alpha} + \lambda^\alpha\frac{\delta}{\bar{C}^\alpha}. \tag{21.22}$$

The essential property of the BRS operator is that it has a vanishing square:

$$[\mathcal{D}_0]^2 = 0, \tag{21.23}$$

as we know from general arguments and can again check explicitly here. Hence, apart from contributions which are gauge invariant, BRS symmetric terms include contributions which are BRS exact, i.e. of the form $\mathcal{D}_0\Phi\left(A, C, \bar{C}, \lambda\right)$. We verify that the action (21.17) has exactly such a decomposition:

$$S = \mathcal{L}(A) + \mathcal{D}_0\Phi, \tag{21.24}$$

$$\Phi = \bar{C}^\alpha\left[F_\alpha(A) - \frac{\partial}{\partial\lambda^\alpha}\int_0^1 \mathrm{d}s\, w(s\lambda)\right].$$

We are not surprised because this property has been proven quite generally in Chapter 16 (see equation (16.58)). The quantized action is therefore BRS symmetric:

$$\mathcal{D}_0 S\left(A, C, \bar{C}, \lambda\right) = 0. \tag{21.25}$$

It had been generally conjectured that any local solution of equation (21.25) can be written in the form (21.24). After several partial results, a proof now exists. Such a property plays, as we shall see later, a fundamental role in the proof of the gauge independence of physical quantities. We shall show below that this property is preserved by the renormalization procedure.

Remark. In all examples we consider here the function $w(\lambda)$ defined by equation (21.15) is quadratic in λ and thus the corresponding gaussian integral can be performed. After integration the new action is still BRS symmetric, the variation of $\bar{C}^\alpha$ taking now the form

$$\delta\bar{C}^\alpha = \bar\varepsilon\tilde{a}^{\alpha\beta}F_\beta(A),$$

$\tilde{a}$ being a constant matrix. The BRS operator is still nilpotent but no longer has a vanishing square. Therefore the property (21.23) is not shared by all realizations of BRS transformations, and may require the introduction of additional auxiliary variables. The results which follow can be proven in all formulations.

21.4 WT Identities

To study the consequence of the BRS symmetry (21.20), it is necessary to introduce sources for all fields and all operators generated by the transformation:

$$Z\left(J,\eta,\bar{\eta},l,K,L\right) = \int [dA]\,[d\lambda]\,[dCd\bar{C}]\exp\left[-S\left(A,C,\bar{C},\lambda,K,L\right)\right.$$
$$\left.+J_iA^i+\bar{\eta}_\alpha C^\alpha+\bar{C}^\alpha\eta_\alpha+l_\alpha\lambda^\alpha\right],\qquad(21.26)$$

with:

$$S\left(A,C,\bar{C},\lambda,K,L\right) = \mathcal{L}\left(A\right)-w\left(\lambda\right)+\lambda^\alpha F_\alpha(A)-\bar{C}^\alpha M_{\alpha\beta}C^\beta-K_iD^i-\tfrac{1}{2}L_\alpha f^\alpha_{\beta\gamma}C^\beta C^\gamma\,.\qquad(21.27)$$

Under a BRS transformation, only the source terms for A, C and $\bar{C}$ vary:

$$\delta\left[J_iA^i+\bar{\eta}_\alpha C^\alpha+\bar{C}^\alpha\eta_\alpha\right] = \bar{\varepsilon}\left(J_iD^i+\tfrac{1}{2}\bar{\eta}_\alpha f^\alpha_{\beta\gamma}C^\beta C^\gamma+\lambda^\alpha\eta_\alpha\right).\qquad(21.28)$$

This implies an equation for Z:

$$\left(J_i\frac{\delta}{\delta K_i}+\bar{\eta}_\alpha\frac{\delta}{\delta L_\alpha}+\eta_\alpha\frac{\delta}{\delta l_\alpha}\right)Z = 0\,.\qquad(21.29)$$

The generating functional W of connected correlation functions satisfies the same equation:

$$\left(J_i\frac{\delta}{\delta K_i}+\bar{\eta}_\alpha\frac{\delta}{\delta L_\alpha}+\eta_\alpha\frac{\delta}{\delta l_\alpha}\right)W = 0\,.\qquad(21.30)$$

We then perform a Legendre transformation:

$$\Gamma\left(A,C,\bar{C},\lambda,K,L\right)+W\left(J,\eta,\bar{\eta},l,K,L\right) = J_iA^i+\bar{\eta}_\alpha C^\alpha+\bar{C}^\alpha\eta_\alpha+l_\alpha\lambda^\alpha\,,\qquad(21.31)$$

with:

$$\begin{cases} J_i=\dfrac{\delta\Gamma}{\delta A^i}, & \bar{\eta}_\alpha=\dfrac{\delta\Gamma}{\delta C^\alpha}, \\[2mm] \eta_\alpha=\dfrac{\delta\Gamma}{\delta\bar{C}^\alpha}, & l_\alpha=\dfrac{\delta\Gamma}{\delta\lambda^\alpha}. \end{cases}\qquad(21.32)$$

Equations (21.31,21.32) imply as usual for the sources coupled to composite fields:

$$\frac{\delta\Gamma}{\delta K_i}+\frac{\delta W}{\delta K_i}=0\,,\qquad \frac{\delta\Gamma}{\delta L_\alpha}+\frac{\delta W}{\delta L_\alpha}=0\,.\qquad(21.33)$$

Therefore the generating functional of proper vertices Γ satisfies:

$$\frac{\delta\Gamma}{\delta A^i}\frac{\delta\Gamma}{\delta K_i}+\frac{\delta\Gamma}{\delta C^\alpha}\frac{\delta\Gamma}{\delta L_\alpha}-\lambda^\alpha\frac{\delta\Gamma}{\delta\bar{C}^\alpha}=0\,.\qquad(21.34)$$

As in the case of the non-linear σ-model, because the transformation laws involves composite operators, the WT identity satisfied by Γ takes the form of a quadratic equation. We thus expect that the transformation laws will in general be renormalized. As implied by equation (21.34), the original action satisfies equation (21.34), as can be easily verified by a direct calculation:

$$\frac{\delta S}{\delta A^i}\frac{\delta S}{\delta K_i} + \frac{\delta S}{\delta C^\alpha}\frac{\delta S}{\delta L_\alpha} - \lambda^\alpha\frac{\delta S}{\delta \bar{C}^\alpha} = 0. \tag{21.35}$$

Basic property. Let us show that it is sufficient to assume that the action satisfies equation (21.35) to derive equation (21.34).

Equation (21.35) implies:

$$\int [\mathrm{d}A]\,[\mathrm{d}\lambda]\,[\mathrm{d}C\mathrm{d}\bar{C}]\left[\frac{\delta S}{\delta A^i}\frac{\delta S}{\delta K_i} + \frac{\delta S}{\delta C^\alpha}\frac{\delta S}{\delta L_\alpha} - \lambda^\alpha\frac{\delta S}{\delta \bar{C}^\alpha}\right]$$
$$\times \exp\left(-S + J_i A^i + \bar{\eta}_\alpha C^\alpha + \bar{C}^\alpha\eta_\alpha + l_\alpha\lambda^\alpha\right) = 0. \tag{21.36}$$

We now integrate by parts over A^i, C^α and $\bar{C}^\alpha$ using:

$$\frac{\delta S}{\delta A^i}\,\mathrm{e}^{-S} = -\frac{\delta}{\delta A^i}\,\mathrm{e}^{-S}, \quad \frac{\delta S}{\delta C^\alpha}\,\mathrm{e}^{-S} = -\frac{\delta}{\delta C^\alpha}\,\mathrm{e}^{-S}, \quad \frac{\delta S}{\delta \bar{C}^\alpha}\,\mathrm{e}^{-S} = -\frac{\delta}{\delta \bar{C}^\alpha}\,\mathrm{e}^{-S}.$$

Equation (21.36) then becomes:

$$\int [\mathrm{d}A]\,[\mathrm{d}C\mathrm{d}\bar{C}]\,[\mathrm{d}\lambda]\left(\frac{\delta^2 S}{\delta A^i\delta K_i} + \frac{\delta^2 S}{\delta C^\alpha\delta L_\alpha} + J_i\frac{\delta S}{\delta K_i} + \bar{\eta}_\alpha\frac{\delta S}{\delta L_\alpha} - \eta_\alpha\lambda^\alpha\right)$$
$$\times \exp\left(-S + \text{sources}\right) = 0. \tag{21.37}$$

At the regularized level $\delta^2 S/\delta A^i\delta K_i$ and $\delta^2 S/\delta C^\alpha\delta L_\alpha$, which are proportional to traces of matrices of the representation of the Lie algebra, vanish because the group is compact. Then equation (21.37) can be rewritten:

$$\int [\mathrm{d}A]\,[\mathrm{d}C\mathrm{d}\bar{C}]\,[\mathrm{d}\lambda]\left(J_i\frac{\delta}{\delta K_i} + \bar{\eta}_\alpha\frac{\delta}{\delta L_\alpha} + \eta_\alpha\lambda^\alpha\right)\exp\left(-S + \text{sources}\right) = 0. \tag{21.38}$$

This equation directly leads to the WT identities (21.34).

The quantities $\delta^2 S/\delta A^i\delta K_i$ and $\delta^2 S/\delta C^\alpha\delta L_\alpha$ still vanish after renormalization if some group structure is preserved or quite generally if dimensional regularization has been used.

21.5 Renormalization: General Considerations

21.5.1 Renormalization

The method explained below is closely related to the method presented in the case of the non-linear σ-model (see Chapter 14). We assume that the local action (21.27) has been regularized in a way compatible with gauge invariance. Perturbation theory then exhibits UV divergences which have to be removed by adding counterterms to the action. The identities (21.34) imply relations among divergences. We use them to prove that equation (21.35) is stable under renormalization. In next section we then solve equation (21.35) to find the most general form of the renormalized action. As usual our analysis is based on a loopwise expansion of the regularized functional Γ, the first term being the unrenormalized action S:

$$\Gamma = S + \sum_{\ell=1}^{\infty}\Gamma_\ell. \tag{21.39}$$

To simplify notations, we introduce an auxiliary field μ_α and add to the action (21.17) and to the 1PI functional Γ the combination $-\mu_\alpha \lambda^\alpha$. The new functional then satisfies:

$$\frac{\delta\Gamma}{\delta\mu_\alpha} = -\lambda^\alpha \,, \tag{21.40}$$

and the equations (21.34) and (21.35) become homogeneous quadratic equations which can be written symbolically:

$$S * S = 0 \,, \tag{21.41}$$

$$\Gamma * \Gamma = 0 \,. \tag{21.42}$$

The general arguments presented in Section 14.6 now prove that these equations are stable under renormalization. We here recall only the main steps.

Assuming as an induction hypothesis that we have been able to construct a renormalized action $S_{\ell-1}$ which satisfies equation (21.41) and renders Γ finite at $\ell - 1$ loop order, we write the consequences of equation (21.42) at loop order ℓ:

$$S * \Gamma_\ell + \Gamma_\ell * S = -\sum_{m=1}^{\ell-1} \Gamma_m * \Gamma_{\ell-m} \,. \tag{21.43}$$

The r.h.s. of the equation is finite by induction, so that the divergent part of Γ_ℓ then satisfies:

$$S * \Gamma_\ell^{\mathrm{div}} + \Gamma_\ell^{\mathrm{div}} * S = 0 \,. \tag{21.44}$$

We recall that Γ^{div} is the sum of the divergent terms in the asymptotic expansion of Γ in terms of the regularizing parameter. Equation (21.44) shows that by defining the ℓ-loop renormalized action S_ℓ by:

$$S_\ell = S_{\ell-1} - \Gamma_\ell^{\mathrm{div}} + \text{higher orders} \,, \tag{21.45}$$

it is possible to render Γ ℓ-loop finite with a renormalized action still satisfying equation (21.41).

21.5.2 Solution of WT identities: general considerations

It now remains to solve equations (21.44) and (21.41) using power counting arguments to find the form of the counterterms and of the renormalized action. It is however useful to first exhibit several properties of equation (21.44). In explicit form equation (21.44) reads:

$$\tilde{\mathcal{D}} \Gamma_l^{\mathrm{div}} = 0 \tag{21.46}$$

where $\tilde{\mathcal{D}}$ is the differential operator:

$$\tilde{\mathcal{D}} = \frac{\delta S}{\delta A^i}\frac{\delta}{\delta K_i} + \frac{\delta S}{\delta K_i}\frac{\delta}{\delta A^i} + \frac{\delta S}{\delta C^\alpha}\frac{\delta}{\delta L_\alpha} + \frac{\delta S}{\delta L_\alpha}\frac{\delta}{\delta C^\alpha} + \frac{\delta S}{\delta \mu_\alpha}\frac{\delta}{\delta \bar{C}^\alpha} \,. \tag{21.47}$$

Quite remarkably the basic equation (21.41) implies that $\tilde{\mathcal{D}}$ is nilpotent as we now show. It is convenient to introduce some notation. We denote below by θ_i the set of all anti-commuting fields K_i, C^α, $\bar{C}^\alpha$ and x_i all commuting fields A^i, L_α, μ_α. As we see on the explicit form (21.35), the fundamental equation for the action S (and thus Γ) then takes the form:

$$\frac{\partial S}{\partial x_i}\frac{\partial S}{\partial \theta_i} = 0 ,\tag{21.48}$$

and the operator $\tilde{\mathcal{D}}$ then reads:

$$\tilde{\mathcal{D}} = \frac{\partial S}{\partial \theta_i}\frac{\partial}{\partial x_i} + \frac{\partial S}{\partial x_i}\frac{\partial}{\partial \theta_i} .\tag{21.49}$$

The nilpotency of the differential operator $\tilde{\mathcal{D}}$. Let us calculate $\tilde{\mathcal{D}}^2$. Since $\tilde{\mathcal{D}}$ is of anticommuting type only the terms generated by the non-commutation of $\partial S/\partial \theta_i$ and $\partial S/\partial x_i$ with the differential operators $\partial/\partial\theta_i$ and $\partial/\partial x_i$ survive:

$$\tilde{\mathcal{D}}^2 = \left(\frac{\partial S}{\partial \theta_i}\frac{\partial^2 S}{\partial x_i \partial \theta_j} + \frac{\partial S}{\partial x_i}\frac{\partial^2 S}{\partial \theta_i \partial \theta_j}\right)\frac{\partial}{\partial x_j} + \left(\frac{\partial S}{\partial \theta_i}\frac{\partial^2 S}{\partial x_i \partial x_j} + \frac{\partial S}{\partial x_i}\frac{\partial^2 S}{\partial \theta_i \partial x_j}\right)\frac{\partial}{\partial \theta_j} .\tag{21.50}$$

We notice that this expression can be rewritten:

$$\tilde{\mathcal{D}}^2 = \left[-\frac{\partial}{\partial \theta_j}\left(\frac{\partial S}{\partial \theta_i}\frac{\partial S}{\partial x_i}\right)\right]\frac{\partial}{\partial x_j} + \left[\frac{\partial}{\partial x_j}\left(\frac{\partial S}{\partial \theta_i}\frac{\partial S}{\partial x_i}\right)\right]\frac{\partial}{\partial \theta_j} = 0 ,\tag{21.51}$$

as a consequence of equation (21.48). The operator $\tilde{\mathcal{D}}$, which plays an essential role in solving WT identities, again is a BRS operator with vanishing square.

Canonical invariance of the fundamental equation. Equation (21.48) has properties analogous to the symplectic form $\mathrm{d}p \wedge \mathrm{d}q$ of classical mechanics, it is invariant under canonical transformations. Let us make the change of variables $(\theta, x) \mapsto (\theta', x')$:

$$x_i = \frac{\partial \varphi}{\partial \theta_i}(\theta, x') ,\tag{21.52}$$

$$\theta_i' = \frac{\partial \varphi}{\partial x_i'}(\theta, x') ,\tag{21.53}$$

in which $\varphi(x', \theta)$ is an anticommuting type function. We first eliminate x_i in equation (21.48) using equation (21.52)

$$\frac{\partial S}{\partial \theta_i}\left[\frac{\partial \varphi}{\partial \theta_i \partial x_j'}\right]^{(-1)}\frac{\partial S}{\partial x_j'} = 0 .\tag{21.54}$$

We then eliminate θ_i using (21.53). We verify that we recover equation (21.48) in the new variables

$$\frac{\partial S}{\partial \theta_i'}\frac{\partial S}{\partial x_i'} = 0 .\tag{21.55}$$

Infinitesimal transformation. Let us write these transformations in infinitesimal form:

$$\varphi = \theta_i x_i' + \varepsilon\psi(\theta, x') ,\tag{21.56}$$

calculate $S(\theta', x')$:

$$x'_i = x_i - \varepsilon \frac{\partial \psi}{\partial \theta_i}(\theta, x) + O(\varepsilon^2), \qquad \theta'_i = \theta_i + \varepsilon \frac{\partial \psi}{\partial x_i}(\theta, x) + O(\varepsilon^2), \qquad (21.57)$$

and therefore (using definition (21.49)):

$$S(\theta', x') - S(\theta, x) = -\varepsilon \frac{\partial \psi}{\partial x'_i} \frac{\partial S}{\partial \theta_i} - \varepsilon \frac{\partial \psi}{\partial \theta_i} \frac{\partial S}{\partial x_i} + O(\varepsilon^2) = -\varepsilon \tilde{D} \psi + O(\varepsilon^2). \qquad (21.58)$$

We thus find that any infinitesimal addition to S of a BRS exact term can obtained by a canonical transformation acting on S.

Let us verify that the effect of the quantization procedure can be understood as such a transformation. In our original problem the dependence on μ_α, which is an artificial variable, cannot change. This imposes the dependence on μ_α of the function φ in equations (21.52,21.53)

$$\varphi(A, C, \bar{C}, \lambda, K, L, \mu) = \bar{C}^\alpha \mu_\alpha + \tilde{\varphi}(A, C, \bar{C}, \lambda, K, L).$$

It follows that the general change of variables is equivalent to a change induced by the function $\tilde{\varphi}$ on the restricted set $\{A, C, K, L\}$ with in addition the translation

$$S \mapsto S + \lambda^\alpha \frac{\delta \tilde{\varphi}}{\delta \bar{C}^\alpha}. \qquad (21.59)$$

One verifies that the gauge invariant action with the K and L source terms, and the renormalized quantized action (21.63) are related by such a transformation with

$$\tilde{\varphi}(A, C, \bar{C}, \lambda, K, L) = \bar{C}^\alpha \left[F_\alpha(A) - \tfrac{1}{2} a_{\alpha\beta} \lambda^\beta \right] + A^i K_i + L_\alpha C^\alpha + \tfrac{1}{2} g_{\alpha\beta\gamma} \bar{C}^\alpha \bar{C}^\beta C^\gamma. \quad (21.60)$$

21.6 The Renormalized Action

We have shown that the renormalized action is the most general functional solution of equation (21.35) local in the fields and sources, and consistent with power counting. We now look for the most general solution of equation (21.35) in 4 dimensions.

21.6.1 General gauges

We have to solve equation (21.35) taking into account power counting, symmetries and locality. Also we note that in the original action only the product $\bar{C}C$ appears. This leads to ghost number conservation. If we assign a ghost number $+1$ to C and -1 to $\bar{C}$, then K_i has a ghost number -1, and L_α -2.

Power counting. In four dimensions the lagrangian density has dimension four:

$$[\mathcal{L}(A)] = 4.$$

We choose the gauge fixing term in such a way that the field A has the minimal dimension, i.e. 1 and $F(A)$ dimension 2 (we have exhibited such gauges in Chapters 18 and 19),

$$[A] = 1, \qquad [F(A)] = 2.$$

We impose $[\lambda] = 2$, by choosing $w(\lambda)$ such λ has a constant propagator:

$$w(\lambda) = \tfrac{1}{2}a_{\alpha\beta}\lambda^\alpha\lambda^\beta, \qquad a_{\alpha\beta} \text{ constants}. \tag{21.61}$$

The other dimensions follow. Since $D^i_\alpha(A)$ has a term linear in A and a constant part with one derivative it has dimension 1. The operator $M_{\alpha\beta}$ then has dimension 2. This implies that

$$[\bar{C}] + [C] = 2.$$

By convention we can choose $[C] = 0$, $[\bar{C}] = 2$. This implies:

$$[K] = 3, \quad [L] = 4.$$

Let us summarize:

$$[\mathcal{L}] = 4, \quad [A] = 1, \quad [C] = 0, \quad [\bar{C}] = 2, \quad [\lambda] = 2, \quad [K] = 3, \quad [L] = 4. \tag{21.62}$$

We immediately conclude that, since K and L have dimensions 3 and 4 respectively, the renormalized action can contain at most terms linear in K and L. Similarly, since λ has dimension 2, the renormalized action is at most a polynomial of degree 2 in λ, and the coefficient of $\lambda^\alpha\lambda^\beta$ is a constant matrix.

The solution. To solve equation (21.35) we now parametrize the solution in a way reminiscent of the initial action (21.27), but it should be kept in mind that the parameters which appear are renormalized and in general different from those parametrizing the action (21.27). The subscript "renormalized" is always implied, and is omitted only for notational simplicity.

Because the part involving $\bar{C}$ and λ is simple we expand S in powers of $\bar{C}$ and λ. Since $\bar{C}$ has dimension 2, S is a polynomial of degree 2 in $\bar{C}$

$$S = S^{(0)} + S^{(1)} + S^{(2)}.$$

Similarly, since λ has dimension 2, $S^{(0)}$ is a polynomial of degree 2 in λ and $S^{(1)}_\alpha$ of degree 1. We thus set

$$S^{(0)} = -\tfrac{1}{2}a_{\alpha\beta}\lambda^\alpha\lambda^\beta + \lambda^\alpha F_\alpha(A) + \tilde{S}(A, C, K, L), \tag{21.63a}$$

$$S^{(1)} = g_{\alpha\beta\gamma}\lambda^\alpha\bar{C}^\beta C^\gamma + \bar{C}^\alpha M_{\alpha\beta}(A)C^\beta \tag{21.63b}$$

$$S^{(2)} = \tilde{g}_{\alpha\beta,\gamma\delta}\bar{C}^\alpha\bar{C}^\beta C^\gamma C^\delta \tag{21.63c}.$$

We have immediately taken into account the dimensions of K and L: They can appear only in $S^{(0)}$.

We first deal with $S^{(0)}$ taken for $\lambda = 0$, i.e. $\tilde{S}$. The functional $\tilde{S}$ satisfies by itself a quadratic equation consequence of equation (21.35):

$$\frac{\delta\tilde{S}}{\delta A^i}\frac{\delta\tilde{S}}{\delta K_i} + \frac{\delta\tilde{S}}{\delta C^\alpha}\frac{\delta\tilde{S}}{\delta L_\alpha} = 0. \tag{21.64}$$

Power counting implies that $\tilde{S}$ is an affine function of K and L, and ghost number conservation implies that only the combinations KC and LC^2 can appear. We then set:

$$\tilde{S} = -K_i D^i_\alpha(A)C^\alpha - \tfrac{1}{2}L_\alpha f^\alpha_{\beta\gamma}C^\beta C^\gamma + \mathcal{L}(A).$$

Power counting tells us in addition that $f^\alpha_{\beta\gamma}$ has dimension 0 and is thus field independent. D^i_α has dimension 1 and can therefore only be an affine function of A.

The terms linear in L and K of equation (21.64) yield respectively:

$$f^\beta_{\alpha\gamma} f^\alpha_{\delta\varepsilon} C^\gamma C^\delta C^\varepsilon = 0\,, \tag{21.65}$$

$$\left(\frac{\delta D^i_\alpha}{\delta A^j} D^j_\beta - \tfrac{1}{2} f^\gamma_{\alpha\beta} D^i_\gamma \right) C^\alpha C^\beta = 0\,. \tag{21.66}$$

The first equation tells us that the constants $f^\alpha_{\beta\gamma}$ satisfy a Jacobi identity since the product $C^\gamma C^\delta C^\varepsilon$ is antisymmetric in $(\gamma, \delta, \varepsilon)$. They are therefore structure constants of a Lie algebra. It remains to show that the Lie algebra has not changed. This is straightforward when the gauge condition does not break the global symmetry. Otherwise when the original algebra is semi-simple this follows from a continuity argument. Finally in the general case one can still use gauge independence of physical observables as derived in Section 21.7. The structure constants are thus linearly related to the structure constants appearing in the original action (21.27).

Equation (21.65) is also the integrability condition for equation (21.66) which implies the commutation relations:

$$\frac{\delta D^i_\alpha}{\delta A^j} D^j_\beta - \frac{\delta D^i_\beta}{\delta A^j} D^j_\alpha = f^\gamma_{\alpha\beta} D^i_\gamma\,. \tag{21.67}$$

Therefore the whole group structure is recovered. The last equation reads

$$D^i_\alpha C^\alpha \frac{\delta \mathcal{L}}{\delta A^i} = 0\,. \tag{21.68}$$

This equation implies that $\mathcal{L}(A)$ is gauge invariant.

Let us return to equation (21.35) which according to the preceding analysis now reads

$$\mathcal{D}S \equiv D^i_\alpha(A) C^\alpha \frac{\delta S}{\delta A_i} - \tfrac{1}{2} f^\alpha_{\beta\gamma} C^\beta C^\gamma \frac{\delta S}{\delta C^\alpha} + \lambda^\alpha \frac{\delta S}{\delta \bar{C}^\alpha} = 0\,. \tag{21.69}$$

The next step consists in expanding the equation (21.69) in powers of $\bar{C}^\alpha$. At order 0 we find

$$\lambda^\alpha \frac{\delta S^{(1)}}{\delta \bar{C}^\alpha} = -D^i_\alpha(A) C^\alpha \frac{\delta S^{(0)}}{\delta A_i} + \tfrac{1}{2} f^\alpha_{\beta\gamma} C^\beta C^\gamma \frac{\delta S^{(0)}}{\delta C^\alpha}\,. \tag{21.70}$$

Equation (21.64) implies that the λ-independent term cancels in the r.h.s.. Only the term linear in λ contributes. We thus find

$$\lambda^\alpha \frac{\delta S^{(1)}}{\delta \bar{C}^\alpha} = -D^i_\alpha(A) C^\alpha \lambda^\beta \frac{\delta F_\beta(A)}{\delta A_i}\,. \tag{21.71}$$

The operator $\lambda^\alpha \delta/\delta \bar{C}^\alpha$ has a vanishing square. The general solution $S^{(1)}$ is the sum of a particular solution and an exact term $\lambda^\alpha \delta \Phi / \delta \bar{C}^\alpha$

$$\begin{aligned}
S^{(1)} &= \bar{C}^\alpha \frac{\delta F_\alpha(A)}{\delta A_i} D^i_\beta C^\beta + \lambda^\alpha \frac{\delta}{\delta \bar{C}^\alpha} \left[\tfrac{1}{2} g_{\beta\gamma\delta} \bar{C}^\beta \bar{C}^\gamma C^\delta \right] \\
&= \bar{C}^\alpha \frac{\delta F_\alpha(A)}{\delta A_i} D^i_\beta C^\beta + \lambda^\alpha g_{\alpha\beta\gamma} \bar{C}^\beta C^\gamma\,,
\end{aligned} \tag{21.72}$$

where the constant tensor $g_{\alpha\beta\gamma}$ is antisymmetric in the two first indices:

$$g_{\alpha\beta\gamma} = -g_{\beta\alpha\gamma} \,. \tag{21.73}$$

The second term in equation (21.72) is new in the sense that it was not present in the bare action.

We now write the equation relating $S^{(2)}$ and $S^{(1)}$:

$$\lambda^\alpha \frac{\delta S^{(2)}}{\delta \bar{C}^\alpha} = -D_\alpha^i(A)C^\alpha \frac{\delta S^{(1)}}{\delta A_i} + \tfrac{1}{2} f_{\beta\gamma}^\alpha C^\beta C^\gamma \frac{\delta S^{(1)}}{\delta C^\alpha} \,. \tag{21.74}$$

Again only the exact term, proportional to $g_{\alpha\beta\gamma}$, survives in the r.h.s.. Moreover there is a unique solution of dimension 4 because a term proportional to λ is no longer allowed. We find

$$S^{(2)} = -\tfrac{1}{4} g_{\alpha\beta\gamma} \bar{C}^\alpha \bar{C}^\beta f_{\delta\varepsilon}^\gamma C^\delta C^\varepsilon \,. \tag{21.75}$$

Again this term was not present in the bare action. The last equation for $S^{(2)}$ is automatically satisfied.

We conclude that the renormalized action has a form similar to the tree level action (21.27) except for an additional BRS invariant term $\mathcal{L}_4\left(\lambda, C, \bar{C}\right)$:

$$\mathcal{L}_4\left(\lambda, C, \bar{C}\right) = g_{\alpha\beta\gamma}\left(-\tfrac{1}{4}\bar{C}^\alpha \bar{C}^\beta f_{\delta\varepsilon}^\gamma C^\delta C^\varepsilon + \lambda^\alpha \bar{C}^\beta C^\gamma\right) \tag{21.76}$$

which is BRS exact

$$\mathcal{L}_4\left(\lambda, C, \bar{C}\right) = \mathcal{D}\Phi_4 \,, \qquad \Phi_4 = \tfrac{1}{2} g_{\alpha\beta\gamma} \bar{C}^\alpha \bar{C}^\beta C^\gamma \,. \tag{21.77}$$

This term has exactly the form given in equation (21.58) and can be associated with a shift $L_\alpha \mapsto L_\alpha + \tfrac{1}{2} g_{\beta\gamma\alpha} \bar{C}^\beta \bar{C}^\gamma$.

The quartic ghost term. A comment now is in order: since in general the renormalized action is quartic in the ghost terms, in contrast to the original action, the direct interpretation of the ghost integral as representing a determinant in local form is lost. However the following result can be proven: if one adds to the gauge function F_α a term linear in an auxiliary field transforming non-trivially under the gauge group, then the integration over this auxiliary field with an appropriate gaussian weight generates the quartic ghost terms in their most general form.

This property is expected from the general analysis of Chapter 16.

Renormalization of gauge invariant operators. To generate correlation functions with operator insertions one can add sources for them in the action. If the dimension of the gauge invariant operators is at most four the new action is still renormalizable. The general analysis is not modified; the only difference is that some coupling constants are now space dependent. In the case of operators of higher dimensions, the action with sources is no longer renormalizable. It is still possible to renormalize it at any finite order by introducing enough renormalization constants. The determination of the general form of the renormalized action, i.e. the solution of equation (21.35) is a non-trivial problem and requires the use of sophisticated cohomological techniques. In the case of compact Lie groups with semi-simple Lie algebras, the most general solution of equation (21.46) is the sum of gauge invariant terms and BRS exact contributions, i.e. of the form $\mathcal{D}\Phi$. This result first conjectured has now been rigourously proven. The part concerning $\bar{C}, \lambda$ is simple but the difficulties come from the set $\{A, C, K, L\}$. Note that the form of the renormalized operators, when inserted in field correlation functions, depends on the explicit gauge. Only the averages of products of gauge invariant operators, or the matrix elements between physical states, as we show in Section 21.7, are gauge independent.

21.6.2 Linear gauges

For a special class of gauges, the previous analysis can be simplified. This class is characterized by the property that the gauge fixing function $F_\alpha(A)$ is linear in the field A, rather than quadratic as in the most general renormalizable case:

$$F_\alpha(A) = F_{\alpha i} A^i. \tag{21.78}$$

In this case the operator $F_\alpha(A)$ is in general still of dimension two, but its correlation functions now are directly related to the correlation functions of the A^i field and therefore introduce no new independent renormalization.

To derive the consequences of equation (21.78), we use the λ-field equation of motion. Let us again explicitly parametrize $w(\lambda)$ as:

$$w(\lambda) = \tfrac{1}{2} a_{\alpha\beta} \lambda^\alpha \lambda^\beta. \tag{21.79}$$

Then the λ-field equation of motion which relies on the identity:

$$\int [\mathrm{d}\lambda] \frac{\delta}{\delta \lambda^\alpha} \left[\exp\left(-S + \text{sources}\right) \right] = 0,$$

reads:

$$\left(-a_{\alpha\beta} \frac{\delta}{\delta l_\beta} + F_{\alpha i} \frac{\delta}{\delta J_i} - l_\alpha \right) Z = 0, \tag{21.80}$$

or in terms of the functional $W = \ln Z$:

$$\left(a_{\alpha\beta} \frac{\delta}{\delta l_\beta} + F_{\alpha i} \frac{\delta}{\delta J_i} \right) W = l_\alpha. \tag{21.81}$$

Since $F(A)$ is linear in A, the λ-field equation of motion is a first order differential equation and can be easily Legendre transformed. The generating functional of proper vertices Γ thus satisfies:

$$\frac{\delta \Gamma}{\delta \lambda^\alpha} = -a_{\alpha\beta} \lambda^\beta + F_{\alpha i} A^i. \tag{21.82}$$

Equation (21.82) is satisfied by the action S and clearly is stable under renormalization. It implies that the quadratic and linear parts in λ of the action are unrenormalized. In particular no term of the form $g_{\alpha\beta\gamma} \bar{C}^\beta C^\gamma$ can be generated. The action remains quadratic in the ghost fields. The renormalized action takes the simple form

$$S(A, C, \bar{C}, \lambda, K, L) = \mathcal{L}(A) + \mathcal{D} \left[\bar{C}^\alpha F_{\alpha i} A^i - \tfrac{1}{2} a_{\alpha\beta} \bar{C}^\alpha \lambda^\beta + K_i A_i + L_\alpha C^\alpha \right], \tag{21.83}$$

($\mathcal{D}$ being defined by equation (21.69)) where in addition as stated above $a_{\alpha\beta}$ and $F_{\alpha i}$ are unrenormalized.

Remark. In the case of linear gauges another equation can be used to show that the gauge function is unrenormalized, the $\bar{C}$ ghost equation of motion. The equation

$$\frac{\delta S}{\delta \bar{C}^\alpha} = F_{\alpha i} D^i_\beta C^\beta = F_{\alpha i} \frac{\delta S}{\delta K_i}, \tag{21.84}$$

implies

$$\left(F_{\alpha i} \frac{\delta}{\delta K_i} + \eta_\alpha \right) Z = 0, \tag{21.85}$$

and thus, after Legendre transformation:

$$\left(F_{\alpha i}\frac{\delta}{\delta K_i}-\frac{\delta}{\delta\bar{C}^\alpha}\right)\Gamma=0\,.\tag{21.86}$$

Equation (21.84) is thus stable under renormalization.

We mention this property here for the following reasons: in the case of linear gauges, the introduction of the λ-field is not always useful, except for strict gauge conditions (generalized Landau gauges). If we integrate over λ^α and set l_α to zero, then equation (21.82) disappears, while equation (21.86) remains and can be used to show that the gauge fixing term $F(A)$ is not renormalized.

21.7 Gauge Independence

Physical observables should be gauge-independent. On the other hand general correlation functions are not. We show here that averages of gauge invariant operators and S-matrix elements are unaffected by infinitesimal changes of gauges. This establishes gauge independence at least for gauges which can be continuously connected.

Gauge invariant operators. To generate correlation functions of gauge invariant operators, we add source terms for them to the action. The action with these sources is still gauge invariant (equation (21.3)) and the effective action BRS symmetric. Assuming a gauge invariant regularization, let us now examine how the vacuum amplitude is affected by an infinitesimal change of gauge δF_α, before renormalization.

Since the non-gauge invariant part of the action has the general form $\mathcal{D}_0\Phi$ in which $\mathcal{D}_0$ has been defined by equation (21.22) (see equation (21.24)), the variation δS of the action takes the form $\delta S=\mathcal{D}_0(\delta\Phi)$, and thus the variation of the vacuum amplitude is:

$$\delta Z=-\int\left[\mathrm{d}A\,\mathrm{d}C\,\mathrm{d}\bar{C}\,\mathrm{d}\lambda\right]\mathcal{D}_0\left(\delta\Phi\right)\mathrm{e}^{-S}\,.\tag{21.87}$$

The operator $\mathcal{D}_0$ is a differential operator and therefore we can integrate by parts. Using again the property that the traces $f^\alpha_{\alpha\beta}$ and $\delta D^i_\alpha/\delta A^i$ vanish we obtain:

$$\delta Z=\int\left[\mathrm{d}A\,\mathrm{d}C\,\mathrm{d}\bar{C}\,\mathrm{d}\lambda\right]\delta\Phi\left(\mathcal{D}_0 S\right)\mathrm{e}^{-S},\tag{21.88}$$

and this expression vanishes as a consequence of the BRS symmetry.

We have therefore shown the bare correlation functions of gauge invariant operators are gauge independent, at least within a class of gauges which can be continuously connected. There exists therefore a renormalization procedure which produces gauge independent renormalized correlation functions of these operators.

These correlation functions contain the complete information about the physical properties of the gauge theory.

S-matrix elements. We now want to study the gauge independence of the perturbative S-matrix, when it exists. We first calculate the variation of renormalized correlation functions under an infinitesimal change of gauge. We assume that we have renormalized the theory in a given gauge, but not yet eliminated the regularizing parameter.

We then proceed as above. The variation of the action in an infinitesimal change of gauge has exactly the form $\mathcal{D}\delta\Phi$ ($\mathcal{D}$ being now the renormalized BRS operator) considered

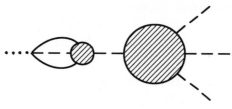

Fig. 21.1

in the preceding subsection. We can still integrate by parts, but the resulting integrand does not vanish identically because we have introduced sources for non gauge invariant fields:

$$\delta Z(J) = -\int \left[\mathrm{d}A \, \mathrm{d}C \, \mathrm{d}\bar{C} \, \mathrm{d}\lambda \right] \delta\Phi J_i D^i_\alpha(A) C^\alpha \exp\left(-S + J_i A_i\right). \qquad (21.89)$$

When we calculate correlation functions we obtain a sum of contributions in which one A_i field has been replaced by $\delta\Phi J_i D^i_\alpha(A) C^\alpha$, which is a linear combination of composite operators. When we go to the mass-shell, after amputation, we get a non-vanishing contribution only if there is a pole in each external momentum squared. For a composite operator this happens only if the line is one particle reducible. Then on the mass-shell we get a contribution proportional to the matrix element of the field itself (see figure 21.1). This argument was presented for the first time in Appendix A10.3. The final result is that an infinitesimal change of gauge renormalizes multiplicatively the S-matrix elements. This corresponds to a field amplitude renormalization. Therefore the S-matrix, properly normalized, is gauge-independent.

Bibliographical Notes

Renormalization of gauge theories is discussed in

G. 't Hooft, *Nucl. Phys.* B33 (1971) 173; B35 (1971) 167; A.A. Slavnov, *Theor. Math. Phys.* 10 (1972) 99; J.C. Taylor, *Nucl. Phys.* B33 (1971) 436; B.W. Lee and J. Zinn-Justin, *Phys. Rev.* D5 (1972) 3121, 3137, 3155; D7 (1973) 1049; G. 't Hooft and M. Veltman, *Nucl. Phys.* B50 (1972) 318; D.A. Ross and J.C. Taylor, *Nucl. Phys.* B51 (1973) 125; B.W. Lee, *Phys. Lett.* 46B (1973) 214; *Phys. Rev.* D9 (1974) 933.

The anticommuting type symmetry of the quantized action is exhibited in

C. Becchi, A. Rouet and R. Stora, *Comm. Math. Phys.* 42 (1975) 127,

where it is used to renormalize the abelian Higgs model. In

H. Kluberg-Stern and J.-B. Zuber, *Phys. Rev.* D12 (1975) 467,

quadratic WT identities generated by BRS symmetry appear and and the renormalization of a class of gauge invariant operators is discussed.

The general proof, based on BRS symmetry, of renormalizability in an arbitrary gauge, presented here, is given in the proceedings of the Bonn summer school 1974,

J. Zinn-Justin in *Trends in Elementary Particle Physics (Lectures Notes in Physics 37)*, H. Rollnik and K. Dietz eds. (Springer-Verlag, Berlin 1975). For details see also J. Zinn-Justin in *Proc. of the 12th School of Theoretical Physics, Karpacz 1975*, Acta Universitatis Wratislaviensis 368; B.W. Lee in *Methods in Field Theory*, Les Houches 1975, R. Balian and J. Zinn-Justin eds. (North-Holland, Amsterdam 1976).

An alternative proof based on BRS symmetry and the BPHZ formalism can be found in

C. Becchi, A. Rouet and R. Stora, *Ann. Phys. (NY)* 98 (1976) 287.

Renormalization of background field gauges is discussed in

    H. Kluberg-Stern and J.-B. Zuber, *Phys. Rev.* D12 (1975) 482, 3159.

A general study of gauge invariant operators can be found in

    S.D. Joglekar and B.W. Lee, *Ann. Phys. (NY)* 97 (1976) 160.

The complete proof of the conjectured result is found in

    G. Barnich, and M. Henneaux, *Phys. Rev. Lett.* 72 (1994) 1588; G. Barnich, F. Brandt and M. Henneaux, *Comm. Math. Phys.* 174 (1995) 93.

Attempts to quantize in quadratic gauges include

    G. 't Hooft and M. Veltman, *Nucl. Phys.* B50 (1972) 318; S.D. Joglekar, *Phys. Rev.* D10 (1974) 4095.

Consequences of BRS symmetry in the operator formalism are exhibited in

    T. Kugo and I. Ojima, *Phys. Lett.* B73 (1978) 459, *Nucl. Phys.* B144 (1978) 234.

The problem of non-linear gauges has been discussed in its full generality in

    J. Zinn-Justin, *Nucl. Phys.* B246 (1984) 246.

Exercises

These exercises try to illustrate the methods which have been used in the quantization of non-abelian gauge theories. However, one should be warned that the abelian example has some simplifying features which are not shared by the general non-abelian case.

We want to study the quantization and renormalization of the free abelian gauge field action,

$$S(A_\mu) = \tfrac{1}{4} \int \mathrm{d}^d x \, F_{\mu\nu}^2(x), \tag{21.90}$$

where

$$F_{\mu\nu}(x) = \partial_\mu A_\nu - \partial_\nu A_\mu \,,$$

in the covariant gauge

$$G(A) \equiv \partial A_\mu + ieA_\mu^2 = 0 \,. \tag{21.91}$$

Exercise 21.1

21.1.1. Write the quantized action. Note that after quantization the action is no longer free.

21.1.2. Verify the BRS symmetry of the quantized action.

21.1.3. Derive the corresponding WT identities.

21.1.4. Show that the additional terms in the action generated by the quantization procedure are themselves the BRS variation of a local quantity, i.e. that they can be written $\mathcal{D}\Phi(A, \lambda, \bar{C}, C)$.

Exercise 21.2

21.2.1. Calculate the different propagators.

21.2.2. Derive from the WT identities the general form of the renormalized action.

21.2.3. What can be said about the stability under renormalization of the gauge condition $G(A) = 0$? Integrate with the renormalized action over the λ-field.

21.2.4. Study the variation of the gauge field 2-point function $W_{\mu\nu}^{(2)}$ under an infinitesimal change δe of the bare parameter e in the quantized action. It is suggested to use the result of the exercise 21.1.4, i.e. express the variation of $W_{\mu\nu}^{(2)}$ in terms of the variation of Φ. An integration by parts will then lead to the result.

22 CLASSICAL AND QUANTUM GRAVITY. TENSORS AND RIEMANNIAN MANIFOLDS

This chapter has two purposes, to present the few elements of differential geometry which are required in different places in this volume and to provide, for completeness, a short introduction to the problem of quantization of gravity.

We first briefly recall a few concepts related to reparametrization (more accurately diffeomorphism) of Riemannian manifolds. We introduce the notions of parallel transport, affine connection, curvature, in analogy with gauge theories as discussed in Chapters 19–21. To define fermions on Riemannian manifolds additional mathematical objects are required, the vielbein and the spin connection. We then construct Einstein's action for classical gravity (General Relativity) and derive the equation of motion. In the last section we finally study the formal aspects of the quantization of the theory of gravity, following the lines of the quantization of non-abelian gauge theories of Chapter 19.

Because the theory of quantum gravity is not renormalizable in four dimensions (even in its extended forms like supersymmetric gravity), the general prejudice at present time is that gravity is the low energy, large distance, remnant of a more complete theory which probably no longer has the form of a quantum field theory (strings?): In the language of critical phenomena, gravity belongs to the class of irrelevant interactions (due to the presence of the massless graviton the situation can be compared with the interaction of Goldstone modes in the low temperature phase). The scale of this new physics is indicated by Planck's mass, i.e. is of the order of $\sqrt{\hbar c / G_N} \sim 10^{19}$ GeV, where G_N is Newton's gravitational constant.

The appendix is devoted to a short introduction to 2D quantum gravity, approached from the point of view of matrix models in the large N limit.

The level of mathematical rigour will be low and the notation old-fashioned. For instance we shall write most expressions in terms of *local coordinates*, ignoring, because it is not essential for our purpose, that several sets of overlapping coordinates (charts) with transition functions are in general required to fully describe a manifold. The language of fiber bundles will be avoided. The reader interested in more details is referred to the literature.

The convention of summation over repeated lower and upper indices will always be used, except when the metric is explicitly euclidean.

22.1 Change of Coordinates. Tensors

Let φ^i, $i = 1, \ldots, N$, be a set of coordinates which parametrize a manifold $\mathfrak{M}$ and let us consider the reparametrization:

$$\varphi^i = \varphi^i(\varphi'), \tag{22.1}$$

in terms of coordinates φ', the functions $\varphi^i(\varphi')$ being differentiable.

We set:

$$\mathrm{d}\varphi^i = T^i_j(\varphi')\,\mathrm{d}\varphi'^j \;\Leftrightarrow\; T^i_j(\varphi') = \frac{\partial \varphi^i}{\partial \varphi'^j} \equiv \partial_j \varphi^i. \tag{22.2}$$

As a matrix, T^i_j is an element of the defining representation of the general linear group $GL(N)$ (general invertible matrices).

We now consider a set of fields defined on $\mathfrak{M}$ and classify them according to their transformation properties in a reparametrization of $\mathfrak{M}$.

Fields $S(\varphi)$ which transform by a simple substitution:

$$S'\left(\varphi'\right) = S\big(\varphi\left(\varphi'\right)\big), \tag{22.3}$$

are called scalars.

Quantities like $d\varphi^i$, which belong to the space tangent to the manifold at point φ^i, define the transformation law of vector fields $V^i(\varphi)$ (called also contravariant vectors):

$$V^i\left(\varphi\right) = T^i_j V'^j\left(\varphi'\right). \tag{22.4}$$

Vector fields $W_i(\varphi)$ of the dual space are characterized by the property that the scalar product $V^i\left(\varphi\right) W_i\left(\varphi\right)$ is a scalar (they are also called covariant vectors). This implies the transformation law:

$$W_i\left(\varphi\right) = \left(T^{-1}\right)^j_i W'_j\left(\varphi'\right), \tag{22.5}$$

in which $\left(T^{-1}\right)^j_i$ is the inverse of the transposed of the matrix T^i_j (and therefore belongs to another representation of $GL(N)$). An example is provided by partial derivatives of a scalar. Indeed differentiating equation (22.3) we find

$$\partial_i S = \frac{\partial \varphi'^j}{\partial \varphi^i} = \left(T^{-1}\right)^j_i \partial_j S', \tag{22.6}$$

which shows that $\partial_i S$ transforms like a dual vector.

We see that the transformation of vectors is defined in terms of the substitution $\varphi \mapsto \varphi'$ and a linear transformation. More generally we can classify all quantities with respect to their transformation properties under the linear group. General n-tensors $V^{i_1 \ldots i_p}_{i_{p+1} \ldots i_n}$ transform like the tensor product of p vectors and $n - p$ dual vectors:

$$V^{i_1 \ldots i_p}_{i_{p+1} \ldots i_n}\left(\varphi\right) = T^{i_1}_{j_1} \ldots T^{i_p}_{j_p} \left(T^{-1}\right)^{j_{p+1}}_{i_{p+1}} \ldots \left(T^{-1}\right)^{j_n}_{i_n} V'^{\,j_1 \ldots j_p}_{j_{p+1} \ldots j_n}\left(\varphi'\right). \tag{22.7}$$

For what follows it is useful to introduce a notation for the element $\mathbf{T}$ of the abstract linear group $GL(N)$. Then

$$\mathbf{T} V'^{\,i_1 \ldots i_p}_{i_{p+1} \ldots i_n}\left(\varphi'\right) = T^{i_1}_{j_1} \ldots T^{i_p}_{j_p} \left(T^{-1}\right)^{j_{p+1}}_{i_{p+1}} \ldots \left(T^{-1}\right)^{j_n}_{i_n} V'^{\,j_1 \ldots j_p}_{j_{p+1} \ldots j_n}\left(\varphi'\right).$$

This representation can in general be reduced according to the representations of the permutation group acting on lower or upper indices. Moreover summing a n-tensor over a pair of upper and lower indices yields a $(n-2)$-tensor. We call this operation taking the covariant trace. Using then the property that the tensor δ^j_i, where δ is the Kronecker δ, is an invariant tensor we can render tensors traceless.

Infinitesimal change of coordinates. Often in this work we consider infinitesimal transformations. Let us then consider the change of coordinates:

$$\varphi^i = \varphi'^i + \epsilon^i(\varphi'), \tag{22.8}$$

and let us denote by $\delta_\epsilon V \sim \mathbf{T} V' - V$ the variation at first order in ϵ of all tensors. We find

$$\delta_\epsilon V^{i_1 \ldots i_p}_{i_{p+1} \ldots i_n}\left(\varphi\right) = \epsilon^j \partial_j V^{i_1 \ldots i_p}_{i_{p+1} \ldots i_n}\left(\varphi\right) - \sum_{\ell=1}^{p} \partial_j \epsilon^{i_\ell} V^{i_1 \ldots j \ldots i_p}_{i_{p+1} \ldots i_n}\left(\varphi\right) + \sum_{\ell=p+1}^{n} \partial_{i_\ell} \epsilon^j V^{i_1 \ldots i_p}_{i_{p+1} \ldots j \ldots i_n}\left(\varphi\right). \tag{22.9}$$

By construction $\delta_\epsilon V$, being the difference between two tensors, is a tensor. Applied to a tensor product δ_ϵ satisfies

$$\delta_\epsilon(V \otimes W) = (\delta_\epsilon V) \otimes W + V \otimes (\delta_\epsilon W).$$

Finally note the commutation relation

$$[\delta_{\epsilon_1}, \delta_{\epsilon_2}] = \delta_{\epsilon_3}, \quad \text{with} \quad \epsilon_3^i = \epsilon_1^j \partial_j \epsilon_2^i - \epsilon_2^j \partial_j \epsilon_1^i. \tag{22.10}$$

Differential forms. A class of tensors is often encountered which have special properties: totally antisymmetric tensors or forms. They can be conveniently represented by contracting indices with the generators θ^i of a Grassmann algebra (see also Section 1.6):

$$\Omega = \theta^{i_1}\theta^{i_2}\ldots\theta^{i_n}\Omega_{i_1 i_2 \ldots i_n}.$$

Examples are provided by gauge theories where the vector potential A_μ can be considered as a 1-form and the curvature $F_{\mu\nu}$ (the electromagnetic tensor in QED) as a 2-form (see Chapters 18,19).

Let us now define a differentiation operation denoted below by d,

$$\mathrm{d} \equiv \theta^j \frac{\partial}{\partial \varphi_j}. \tag{22.11}$$

It is clear that d is nilpotent because the θ^i's anticommute

$$\mathrm{d}^2 = \theta^j \partial_j \theta^k \partial_k = 0.$$

With this notation the transformation (22.7) can be written

$$\Omega(\varphi) = \mathrm{d}\varphi'^{i_1}\mathrm{d}\varphi'^{i_2}\ldots\mathrm{d}\varphi'^{i_n}\Omega'_{i_1 i_2 \ldots i_n}(\varphi').$$

The operator d has an important property: acting on an n-form it generates an $n+1$-form, i.e. a $n+1$ antisymmetric tensor

$$\mathrm{d}\Omega = \theta^j \partial_j\, \theta^{i_1}\theta^{i_2}\ldots\theta^{i_n}\Omega_{i_1 i_2 \ldots i_n} = \tfrac{1}{n+1}\theta^{i_1}\theta^{i_2}\ldots\theta^{i_{n+1}}\left[\partial_{i_{n+1}}\Omega_{i_1 i_2 \ldots i_n}\right]_{\text{antisym.}} \tag{22.12}$$

Let us verify that the antisymmetric quantity in the r.h.s. is indeed a tensor. Using $\mathrm{d}^2 = 0$ we find

$$\mathrm{d}\Omega(\varphi) = \mathrm{d}\left(\mathrm{d}\varphi'^{i_1}\mathrm{d}\varphi'^{i_2}\ldots\mathrm{d}\varphi'^{i_n}\Omega'_{i_1 i_2 \ldots i_n}(\varphi')\right)$$
$$= \mathrm{d}\varphi^{i_1}\mathrm{d}\varphi^{i_2}\ldots\mathrm{d}\varphi^{i_n}\mathrm{d}\varphi'^{j}\partial_j\Omega'_{i_1 i_2 \ldots i_n}(\varphi'),$$

which proves the tensor character.

In this notation the 2 form associated with the curvature tensor $F_{\mu\nu}$ can be related to the gauge field one form by

$$F = 2(\mathrm{d}A + A^2).$$

A form Ω which satisfies $\mathrm{d}\Omega = 0$ is called a *closed form*. A form Ω which can be written $\Omega = \mathrm{d}\Omega'$ is *exact*. Since d is nilpotent any exact form is closed.

Conversely a typical problem in differential forms is to know whether a closed form is exact. The corresponding partial differential equations can always be integrated locally but depending on the topology of the x-space sometimes no global solution can be found. Such a situation has been encountered in the quantization of spin variables (Appendix A3.2.1).

Finally the property that by differentiation one generates new tensors is not general as we shall argue in the next section.

22.2 Parallel Transport: Connection, Covariant Derivative

We have seen that the derivatives $\partial_i S$ of a scalar form a tensor. However the derivative of a vector is not a tensor as one easily verifies. Therefore a new concept has to be introduced, which will allow to write covariant quantities depending on products of vector fields at different points and thus also, as a limiting case, on derivatives. A similar problem has already been discussed in the context of gauge theories in Chapters 18–21.

22.2.1 Parallel transport

We consider two points, $\varphi_{(1)}, \varphi_{(2)}$ on the manifold and an oriented, continuous, almost everywhere differentiable curve C joining them. To this curve we associate a linear mapping $U(C)$ from the tangent space at $\varphi_{(1)}$ to the tangent space at $\varphi_{(2)}$. We call this operation parallel transport from $\varphi_{(1)}$ to $\varphi_{(2)}$. Let $V^i\left(\varphi_{(1)}\right)$ be a vector belonging to the space tangent at point $\varphi_{(1)}$. It is transformed into V_U^i

$$V_U^i = U_j^i(C)V_{(1)}^j\left(\varphi_{(1)}\right),\qquad(22.13)$$

which now belongs to the space tangent at the point $\varphi_{(2)}$.

We demand to $\mathbf{U}(C)$ to be a differentiable functional of the curve C, which in addition satisfy the conditions:

If a curve C_1 goes from $\varphi_{(1)}$ to $\varphi_{(2)}$ and a curve C_2 from $\varphi_{(2)}$ to $\varphi_{(3)}$ we impose:

$$\mathbf{U}\left(C_1 \cup C_2\right) = \mathbf{U}\left(C_2\right)\mathbf{U}\left(C_1\right).\qquad(22.14)$$

The definition (22.14) implies that to a curve reduced to a point corresponds the group identity:

$$\mathbf{U}\left(C \equiv 1 \text{ point}\right) = \mathbf{1} \;\Rightarrow\; U_j^i\left(C \equiv 1 \text{ point}\right) = \delta_j^i.\qquad(22.15)$$

If C is a curve going from $\varphi_{(1)}$ to $\varphi_{(2)}$, we call C^{-1} the same curve but oriented from $\varphi_{(2)}$ to $\varphi_{(1)}$. We then impose:

$$\mathbf{U}\left(C^{-1}\right) = \mathbf{U}^{-1}\left(C\right).\qquad(22.16)$$

The transformation laws for other tensors follow from the following rules: A scalar is invariant in a parallel transport. Demanding the invariance of the scalar product then determines the form of parallel transport for dual vectors

$$[V_U]_i = \left[U^{-1}\right]_i^j V_j.$$

If we denote by $\mathbf{U}(C)$ the abstract linear transformation then the tensor product $V \otimes W$ transform like

$$\mathbf{U}(C)\left(V \otimes W\right) \doteq \mathbf{U}(C)V \otimes \mathbf{U}(C)W.\qquad(22.17)$$

We can then more generally define parallel transport for any tensor, by imposing that a tensor transforms as a tensor product of vectors.

Diffeomorphisms. It is easy to verify that the transformation law (22.4) of tensors under reparametrization induces a transformation law for the parallel transporter $U(C)$ which is:

$$\mathbf{U}(C) = \mathbf{T}\left(\varphi'_{(2)}\right)\mathbf{U}'\left(C\right)\mathbf{T}^{-1}\left(\varphi'_{(1)}\right)\qquad(22.18)$$

in which we have assumed that the curve C goes from the point $\varphi'_{(1)}$ to the point $\varphi'_{(2)}$. This equation shows that $U^i_j(C)$ is a (non-local) tensor.

The affine connection. We now explain the relation between parallel transport and the notions of connection and covariant derivative. We have assumed that the matrix $\mathbf{U}(C)$ as a functional of C is differentiable. Due to the composition law (22.14) $\mathbf{U}(C)$ is then entirely determined by its value for infinitesimal curves. We thus consider an infinitesimal differentiable curve, i.e. a straight line C connecting two close points φ and $\tilde{\varphi} = \varphi + \delta\varphi$ and set:

$$\mathbf{U} = \mathbf{1} - \mathbf{\Gamma}_k \delta\varphi^k + o\,\|\delta\varphi\|\,, \qquad (22.19)$$

where $\mathbf{\Gamma}_k$ is called the affine *connection* on the manifold. The connection is entirely characterized by its action on vectors. We denote by Γ^j_{ik}, the Christoffel symbol, the matrix elements of the connection

$$U^i_j(C) = \delta^i_j - \Gamma^i_{jk}(\varphi)\,\delta\varphi^k + o\,\|\delta\varphi\|\,. \qquad (22.20)$$

Conversely the connection field completely characterizes parallel transport. Indeed once a curve C is parametrized in terms of a parameter t, the corresponding parallel-transporter $U(C)$ and $\Gamma^i_{jk}(\varphi)\mathrm{d}\varphi^k/\mathrm{d}t$ play the roles respectively of the evolution operator and the hamiltonian of quantum mechanics. Therefore, as shown in Appendix A2.1, $U(C)$ can be written as a path-ordered integral along the curve C

$$\mathbf{U} = \mathrm{P}\,\exp\left[-\oint_C \mathbf{\Gamma}_k \mathrm{d}\varphi^k\right], \qquad (22.21)$$

where P stands for path-ordered.

Using equation (22.18), it is possible to derive the transformation law of the affine connection under reparametrization. It is convenient to express it in terms of the 1-form $\mathbf{\Gamma} = \mathbf{\Gamma}_k \mathrm{d}\varphi^k$ associated with the matrix $\mathbf{\Gamma}_k$. One then finds

$$\mathbf{\Gamma} = \mathbf{T}\mathbf{\Gamma}'\mathbf{T}^{-1} - \mathrm{d}\mathbf{T}\mathbf{T}^{-1} = \mathbf{T}\mathbf{\Gamma}'\mathbf{T}^{-1} + \mathbf{T}\mathrm{d}\mathbf{T}^{-1}, \qquad (22.22)$$

where $\mathrm{d}\mathbf{T}$ is the 1-form $\mathrm{d}\varphi'^i \partial_i \mathbf{T}$. In component form, in the defining representation, the equation becomes

$$\Gamma'^{\,i}_{jk} = \left(T^{-1}\right)^i_l\,\partial_k T^l_j + \left(T^{-1}\right)^i_l\,\Gamma^l_{mn}T^n_k T^m_j\,. \qquad (22.23)$$

We note that the connection is not a tensor, because its transformation law is not linear but only affine. Let us, however, decompose the tensor Γ^i_{jk} into symmetric and antisymmetric parts, $\mathcal{G}^i_{jk}$ and $\mathcal{T}^i_{jk}$ respectively, in the exchange $j \leftrightarrow k$:

$$\Gamma^i_{jk} = \mathcal{G}^i_{jk} + \mathcal{T}^i_{jk}\,, \qquad \mathcal{G}^i_{jk} = \mathcal{G}^i_{kj} \quad \mathcal{T}^i_{jk} = -\mathcal{T}^i_{kj}\,. \qquad (22.24)$$

Because T^i_j has the form (22.2) the inhomogeneous term in (22.23) is symmetric in $j \leftrightarrow k$. Hence, both quantities transform independently under (22.23). It follows that the antisymmetric part $\mathcal{T}^i_{jk}$ is a tensor. Moreover, the restriction to connections Γ^i_{jk} such that $\mathcal{T}^i_{jk} = 0$, i.e symmetric in $j \leftrightarrow k$, is consistent with the transformation law (22.23) and characterizes the special class of parallel transports without *torsion*.

22.2.2 The covariant derivative

It is now possible to compare the vector field at point $\varphi + \delta\varphi$ with the same vector field parallel-transported from the point φ to the point $\varphi + \delta\varphi$. Using definition (22.19) and equation (22.13) we obtain:

$$\mathbf{U}(C)V^i(\varphi) - V^i(\varphi + \delta\varphi) = -\left[\partial_k V^i(\varphi) + \Gamma^i_{jk}(\varphi) V^j(\varphi)\right]\delta\varphi^k + o\left\|\delta\varphi\right\|. \quad (22.25)$$

The difference in the l.h.s. is a vector belonging to the tangent space at $\varphi + \delta\varphi$. Introducing the *covariant derivative* ∇_k whose action on vectors is

$$\nabla_k V^i = \partial_k V^i + \Gamma^i_{jk}V^j, \quad (22.26)$$

the equation can be rewritten at leading order in $\delta\varphi$:

$$\mathbf{U}(C)V^i(\varphi + \delta\varphi) - V^i(\varphi + \delta\varphi) = -\delta\varphi^k \nabla_k V^i(\varphi) + o\left\|\delta\varphi\right\|. \quad (22.27)$$

The covariant derivative characterizes infinitesimal parallel transport and allows to construct new local covariant quantities, i.e. tensors, from derivatives of tensors.

Parallel transport of general tensors yields the form of the corresponding covariant derivative

$$\nabla_i = \partial_i \mathbf{1} + \mathbf{\Gamma}_i, \quad (22.28)$$

to which we can associate the covariant form differentiation

$$\nabla = \mathrm{d} + \mathbf{\Gamma} \equiv \mathrm{d}\varphi^i\left(\mathbf{1}\partial_i + \mathbf{\Gamma}_i\right). \quad (22.29)$$

The explicit representation of the covariant derivative thus depends on the nature of the tensor it is acting on. On scalars for example $\nabla_i = \partial_i$. The form of the covariant derivative when acting on a dual vector is

$$\nabla_i V_j = \partial_i V_j - \Gamma^k_{ji}V_k. \quad (22.30)$$

From equation (22.17) one immediately that ∇_i obeys the usual rules of differential operators. It is linear and moreover, if V and W are two tensors, then for the tensor product $V \otimes W$ we find

$$\nabla(V \otimes W) = \nabla(V) \otimes W + V \otimes \nabla(W).$$

The general form of the covariant derivative of a tensor with n indices follows:

$$\nabla_i V^{j_1...j_p}_{j_{p+1}...j_n} = \partial_i V^{j_1...j_p}_{j_{p+1}...j_n} + \sum_{\ell=1}^p \Gamma^{j_\ell}_{k\ell i} V^{j_1...k_\ell...j_p}_{j_{p+1}...j_n} - \sum_{\ell=p+1}^n \Gamma^{k_\ell}_{j_\ell i} V^{j_1...j_p}_{j_{p+1}...k_\ell...j_n}, \quad (22.31)$$

which is a tensor with $n + 1$ indices.

Remarks.

(i) For the class of symmetric Christoffel connections Γ^i_{jk} (torsion-free transport) one has identity between the derivative of a form and its covariant derivative. In the notation (22.12)

$$\mathrm{d} \equiv \theta^i \partial_i = \theta^i \nabla_i,$$

as one easily verifies. In particular the ordinary curl of a vector is a 2-form and thus the covariant curl and the ordinary curl coincide

$$\partial_i V_j - \partial_j V_i = \nabla_i V_j - \nabla_j V_i. \quad (22.32)$$

(ii) The infinitesimal form of the transformation (22.23) of the Christoffel connection corresponding to (22.8), which is not homogeneous, can be written

$$\delta_\epsilon \Gamma^i_{jk} = \partial_j \partial_k \epsilon^i + \epsilon^l \partial_l \Gamma^i_{jk} - \partial_l \epsilon^i \Gamma^l_{jk} + \partial_k \epsilon^l \Gamma^i_{jl} + \partial_j \epsilon^l \Gamma^i_{lk}. \quad (22.33)$$

Although this is not obvious from this expression, $\delta_\epsilon \Gamma^i_{jk}$ is a tensor.

22.3 The Metric Tensor

We now introduce the concept of distance in $\mathfrak{M}$. It is characterized by the line element ds, distance from the point φ^i to the point $\varphi^i + d\varphi^i$

$$(ds)^2 = g_{ij}(\varphi)d\varphi^i d\varphi^j,$$

where the quadratic form is non-degenerate: $\det \mathbf{g} \neq 0$.

The invariance of distances in a reparametrization determines the transformation law of the metric tensor $g_{ij}(\varphi)$:

$$g_{ij}(\varphi) = \left(T^{-1}\right)_i^k \left(T^{-1}\right)_j^l g'_{kl}(\varphi').$$
(22.34)

It is consistent with notation (22.7) to write the inverse (in the sense of matrices) of the metric tensor as $g^{ij}(\varphi)$:

$$g_{ik}g^{kj} = \delta_i^j.$$
(22.35)

The tensors g_{ij} and g^{ij} can be used to lower or raise indices: The metric tensor establishes an isomorphism between the tangent vector space and its dual. It is thus a standard notation to use the same symbol for tensors which are deduced from one of them by lowering or raising indices with the metric tensor, for example:

$$V_i = g_{ij}V^j.$$
(22.36)

Compatibility of parallel transport with the metric. We say that parallel transport is compatible with the metric if it leaves invariant the scalar product of two vectors:

$$V^i(\varphi)\,g_{ij}(\varphi)\,W^j(\varphi) = \tilde{V}^i(\tilde{\varphi})\,g_{ij}(\tilde{\varphi})\,\widetilde{W}^j(\tilde{\varphi}),$$
(22.37)

in which $\tilde{V}^i$ and $\widetilde{W}^i$ are the parallel transported of V^i and W^i from φ to $\tilde{\varphi}$. Using the definition (22.13), we can rewrite the equation:

$$g_{ij}(\varphi) = {}^TU_i^k(C)g_{kl}(\tilde{\varphi})\,U_j^l(C),$$
(22.38)

where we denote by ${}^TU$ the transposed of the matrix U.

The interpretation of this equation is simple: the metric transported from φ to $\tilde{\varphi}$ is identical to the metric at point $\tilde{\varphi}$. Taking then the limit of an infinitesimal path we conclude:

$$\nabla_m g_{ij} = 0.$$
(22.39)

Determination of the connection in torsion-free parallel transport. We have shown that the conservation of the scalar product of vectors under parallel transport implies the compatibility condition (22.39). Using the explicit form (22.31) it can be written:

$$\partial_m g_{ij} - g_{il}\Gamma_{jm}^l - \Gamma_{im}^k g_{kj} = 0.$$
(22.40)

If, eliminating torsion in parallel transport, we assume Γ_{jk}^i symmetric we can solve the compatibility condition and express Γ_{jk}^i explicitly in terms of the metric tensor. We find:

$$\Gamma_{jk}^i = \tfrac{1}{2}g^{il}\left(\partial_j g_{lk} + \partial_k g_{lj} - \partial_l g_{jk}\right).$$
(22.41)

Compatibility with the metric *uniquely* determines torsion-free parallel transport.

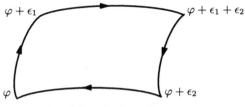

Fig. 22.1   The loop C.

22.4 The Curvature (Riemann) Tensor

Let us consider an infinitesimal parallelogram C on the manifold, joining the points: $\varphi, \varphi + \epsilon_1, \varphi + \epsilon_1 + \epsilon_2, \varphi + \epsilon_2$, and back to φ, and calculate the corresponding parallel transporter (see figure 22.1):

$$\mathbf{U}\left(C\right) = \mathbf{U}^{-1}\left(\varphi + \epsilon_2, \varphi\right) \mathbf{U}^{-1}\left(\varphi + \epsilon_1 + \epsilon_2, \varphi + \epsilon_2\right) \mathbf{U}\left(\varphi + \epsilon_1 + \epsilon_2, \varphi + \epsilon_1\right) \mathbf{U}\left(\varphi + \epsilon_1, \varphi\right). \tag{22.42}$$

Expanding $\ln \mathbf{U}\left(C\right)$ up to second order in ϵ_1 and ϵ_2 we set:

$$\ln \mathbf{U}\left(\varphi + \epsilon, \varphi\right) = -\mathbf{\Gamma}_i \epsilon^i + \tfrac{1}{2}\mathbf{\Gamma}_{ij}\epsilon^i \epsilon^j + o\left(\epsilon^2\right), \tag{22.43}$$

in which $\mathbf{\Gamma}_i$ is the connection.

To evaluate the r.h.s. of equation (22.42), we then repeatedly use Baker–Hausdorf formula (the calculation follows the lines of the derivation of the commutation relations of generators in a Lie algebra, see also Section 33.2.2):

$$e^A\, e^B = e^{(A+B+[A,B]/2+\cdots)}.$$

The first order in ϵ in the exponential of the r.h.s. vanishes as a consequence of equation (22.16). At second order we find:

$$\mathbf{U}(C) = 1 - \epsilon_1^i \epsilon_2^j \mathbf{R}_{ij} + O\left(\epsilon^3\right),$$

where the antisymmetric tensor $\mathbf{R}_{ij}$ is defined by

$$\mathbf{R}_{ij} = [\nabla_i, \nabla_j] = \partial_i \mathbf{\Gamma}_j - \partial_j \mathbf{\Gamma}_i + [\mathbf{\Gamma}_i, \mathbf{\Gamma}_j]. \tag{22.44}$$

We have obtained the expression of the *curvature tensor* $\mathbf{R}_{ij}$ (or Riemann tensor) in terms of the connection. The curvature tensor characterizes the variation of tensors in a transport along infinitesimal closed curves.

Note that we could have used the expression (22.21) to perform the calculation for an arbitrary closed curve. We fix one point φ on the curve and write a generic point $\varphi + \epsilon$. Expanding the path-ordered integral up to second order we find:

$$\mathbf{U}(C) = 1 - \oint \mathbf{\Gamma}_k(\varphi + \epsilon) d\epsilon^k + \tfrac{1}{2}\oint\oint d\epsilon^k d\epsilon^l\, \mathrm{P}[\mathbf{\Gamma}_k(\varphi + \epsilon)\mathbf{\Gamma}_l(\varphi + \epsilon)] + O\left(\epsilon^3\right).$$

We then expand $\mathbf{\Gamma}_k$ for ϵ small. In the first integral the term proportional to $\mathbf{\Gamma}_k(\varphi)$ vanishes because the curve is closed. In the second integral we can neglect the dependence in ϵ at this order. A short calculation then yields

$$\mathbf{U}(C) - 1 \sim -\tfrac{1}{2}\int_D \mathbf{R},$$

where $\mathbf{R}$ is the 2-form $\mathbf{R}_{ij}\mathrm{d}\varphi^i \wedge \mathrm{d}\varphi^j$ (see equation (22.29))

$$\mathbf{R} = 2\nabla^2 = 2\left(\mathrm{d}\boldsymbol{\Gamma} + \boldsymbol{\Gamma}^2\right),$$

and D, the domain of integration, a surface which has the curve C as boundary: $\partial D = C$.

Finally the curvature tensor $\mathbf{R}_{ij}$ is characterized by its matrix elements, when acting on vectors

$$\mathbf{R}_{ij}V^k \equiv R^k_{lij}V^l,$$

Equation (22.44) in component form then becomes

$$R^k_{lij} = \partial_i \Gamma^k_{lj} - \partial_j \Gamma^k_{li} + \Gamma^k_{mi}\Gamma^m_{lj} - \Gamma^k_{mj}\Gamma^m_{li}.$$

A general tensor transforms like

$$\mathbf{R}_{ij}V^{k_1 k_2 \ldots k_p}_{l_1 l_2 \ldots l_q} = \sum_{r=1}^{p} R^{k_r}_{mij}V^{k_1 k_2 \ldots m \ldots k_p}_{l_1 l_2 \ldots l_q} - \sum_{r=1}^{q} R^m_{l_r ij}V^{k_1 k_2 \ldots k_p}_{l_1 l_2 \ldots m \ldots l_q}. \tag{22.45}$$

Curvature tensor and metric. When parallel transport is torsion-free and compatible with the metric, the curvature tensor is determined by the metric tensor. A short calculation yields for the tensor with only lower indices, $R_{klij} = g_{km}R^m_{lij}$, the expression

$$R_{klij} = \tfrac{1}{2}\left(\partial_i \partial_l g_{kj} - \partial_i \partial_k g_{jl}\right) - \frac{1}{4}\left(\partial_i g_{mk} + \partial_k g_{mi} - \partial_m g_{ik}\right)g^{mn}\left(\partial_j g_{nl} + \partial_l g_{nj} - \partial_n g_{jl}\right)$$
$$- (i \leftrightarrow j). \tag{22.46}$$

A few properties. Being defined as a commutator, R^k_{lij} satisfies:

$$R^k_{lij} = -R^k_{lji}. \tag{22.47}$$

For the same reason the curvature tensor satisfies the consequence of a Jacobi identity:

$$[\nabla_i, [\nabla_j, \nabla_k]] + \text{cyclic permutations } (i, j, k) = 0,$$

which in terms of R^k_{lij} reads:

$$\nabla_i R^l_{mjk} + \nabla_j R^l_{mki} + \nabla_k R^l_{mij} = 0, \tag{22.48}$$

and in the framework of Riemannian geometry is called the *Bianchi identity*.

Note that the compatibility condition (22.39) implies

$$[\nabla_i, \nabla_j]g_{kl} = 0 = R^m_{kij}g_{ml} + R^m_{lij}g_{km}.$$

This equation implies that the tensor R_{klij} is antisymmetric in $k \leftrightarrow l$,

$$R_{klij} = -R_{lkij}. \tag{22.49}$$

a property can be verified on the explicit form (22.46).

We also note that R_{klij} is antisymmetric in $(k \leftrightarrow l)$ and symmetric in the exchange $(kl) \leftrightarrow (ij)$:

$$R_{klij} = R_{ijkl} \, . \tag{22.50}$$

It is easy to verify a cyclic identity:

$$R_{ijkl} + R_{kijl} + R_{jkil} = 0 \, . \tag{22.51}$$

Finally let us define the contracted curvature tensor R_{ij}:

$$R_{ij} = R^{k}_{ijk} \, , \tag{22.52}$$

which is called the *Ricci tensor*. It is symmetric:

$$R_{ij} = R_{ji} \, , \tag{22.53}$$

and has therefore properties similar to the metric. By contracting the remaining indices one obtains the *scalar curvature R*:

$$R = R^{i}_{i} \equiv R^{ij}g_{ji} \equiv R^{kl}_{kl} \, . \tag{22.54}$$

Remarks.

(i) An important problem is to classify local tensors which are functionals of the metric tensor. The compatibility condition (22.39) implies that the covariant derivative of g_{ij} is not a new tensor.

However we observe that when the connection is a functional of the metric tensor this also applies to the curvature tensor. All tensors depending only on the metric can then be obtained from the curvature tensor and its covariant derivatives.

(ii) In the case of symmetric Christoffel connections the variation (22.33) can be rewritten in a simple way to exhibit its tensor character:

$$\delta_{\epsilon}\Gamma^{i}_{jk} = \nabla_{k}\nabla_{j}\epsilon^{i} - R^{i}_{jkl}\epsilon^{l}. \tag{22.55}$$

The symmetry of the r.h.s. in the exchange $j \leftrightarrow k$ relies on the cyclic identity (22.51).

22.4.1 Holonomy variables, holonomy group

The parallel-transporters $\mathbf{U}(C)$ associated with closed curves C are often called *holonomy variables*. They in particular probe some topological properties of the manifold. This is specially obvious in the case of parallel transport associated with a vanishing curvature tensor. In the latter case the contour C can be continuously deformed without changing the corresponding holonomy variable.

In this context it can be useful to consider the group formed by holonomies $\mathbf{U}(\varphi; C)$ originating from a common fixed point φ, called the *holonomy group*.

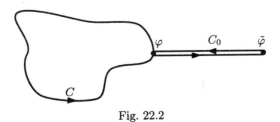

Fig. 22.2

Independence of the holonomy group of the initial point. Let us show that the holonomy groups associated with different initial points are isomorphic. If φ and $\tilde{\varphi}$ are two different initial points, we introduce a fixed curve C_0 joining them. To each curve C passing through φ we can now associate a curve $\tilde{C}$ passing through $\tilde{\varphi}$ (see figure 22.2):

$$\tilde{C} = C_0^{-1} \cup C \cup C_0 . \tag{22.56}$$

In this way we have constructed a one-to-one mapping between curves passing through φ and curves passing through $\tilde{\varphi}$. The corresponding relation between parallel transporters is:

$$\mathbf{U}\left(\tilde{C}\right) = \mathbf{U}\left(C_0\right) \mathbf{U}\left(C\right) \mathbf{U}^{-1}\left(C_0\right) . \tag{22.57}$$

It establishes an isomorphism between the holonomy groups associated with the points φ and $\tilde{\varphi}$. Therefore the abstract holonomy group is independent of the initial point and intrinsic to the set of curves on the manifold equipped with the equivalence relation (22.56).

Holonomy and metric. Let us write the compatibility condition (22.38) for a closed curve:

$$g_{ij}\left(\varphi\right) = {}^{T}U_i^k\left(C\right) g_{kl}\left(\varphi\right) U_j^l\left(C\right) . \tag{22.58}$$

If the metric is positive, equation (22.58) shows that the matrices $\mathbf{U}\left(\varphi; C\right)$ corresponding to closed curves passing through a point φ belong to a subgroup of the $O(N)$ orthogonal group (N is the dimension of the manifold $\mathfrak{M}$). Let us now consider the curve C formed by the infinitesimal parallelogram of figure 22.1. If we expand equation (22.58) up to second order in ϵ, we obtain a condition which, in terms of the curvature tensor, can be written:

$$\epsilon_1^k \epsilon_2^l \left(R_{ijkl} + R_{jikl}\right) = 0 . \tag{22.59}$$

This result provides a geometrical interpretation for the antisymmetry (22.49) of the curvature tensor: it is the antisymmetry of the generators of the orthogonal group.

22.5 Covariant Volume Element

To construct for example a classical action we need a volume element invariant under reparametrization. The euclidean measure transforms like:

$$\prod_i \mathrm{d}\varphi^i = \det T_l^k \prod_i \mathrm{d}\varphi'^i . \tag{22.60}$$

Let us denote by g the determinant of the metric g_{ij}:

$$g = \det \mathbf{g} . \tag{22.61}$$

It transforms like

$$g' = \det \mathbf{g}' = \left(\det T_j^i\right)^2 g . \tag{22.62}$$

The volume element $\mathrm{d}\rho(\varphi)$:

$$\mathrm{d}\rho\left(\varphi\right) = \sqrt{g} \prod_i \mathrm{d}\varphi^i , \tag{22.63}$$

is thus invariant. Of course if we multiply $\mathrm{d}\rho(\varphi)$ by a scalar (like a function of the scalar curvature), the volume element remains invariant.

Remark. Let us calculate the logarithmic derivative of g:

$$\ln g \equiv \ln \det \mathbf{g} = \operatorname{tr} \ln \mathbf{g}. \tag{22.64}$$

Differentiating this identity we find:

$$g^{-1}\partial_k g = g^{ij}\partial_k g_{ji}. \tag{22.65}$$

Comparing the r.h.s. with the expression (22.41) for the connection, we can verify the simple relation:

$$\Gamma^k_{ki} = \frac{1}{\sqrt{g}}\partial_i \sqrt{g}. \tag{22.66}$$

This in particular implies for the covariant divergence of a vector:

$$\nabla_i V^i = \frac{1}{\sqrt{g}}\partial_i \left(\sqrt{g}\, V^i\right), \tag{22.67}$$

and therefore:

$$\int \left(\prod_i \mathrm{d}\varphi^i\right) \sqrt{g}\, \nabla_j V^j = \int \left(\prod_i \mathrm{d}\varphi^i\right) \partial_j \left(\sqrt{g}\, V^j\right). \tag{22.68}$$

22.6 Fermions, the Vielbein

We now briefly explain how one can construct spin $1/2$ fermions living on Riemannian manifolds, because the construction is non-trivial. It is first necessary to introduce a local frame in the space tangent to the manifold (this can always be done locally, but may lead to topological obstructions). The set of vectors $e^{ai}(\varphi)$ which form the local basis is called the *vielbein*. We assume that the metric is positive. We can then choose vectors orthogonal with respect to the metric $g_{ij}(\varphi)$, of length 1:

$$e^{ai}(\varphi)\, g_{ij}(\varphi)\, e^{bj}(\varphi) = \delta_{ab}. \tag{22.69}$$

We do not distinguish upper and lower internal indices a, b because as equation (22.69) shows the internal metric is euclidean. Introducing the vectors e^a_i, obtained as usual from the e^{ai} by lowering the index with the metric tensor, we can rewrite equation (22.69) as:

$$e^{ai}e^b_i = e^a_i e^{bi} = \delta_{ab}, \tag{22.70}$$

which shows that the matrix e^{ai} is the inverse of the matrix e^b_i. Using relation (22.70), equation (22.69) can finally be written:

$$g_{ij}(\varphi) = e^a_i(\varphi)\, e^a_j(\varphi). \tag{22.71}$$

This equation yields the expression of the metric tensor in terms of the vielbein. As we have done above, in what follows we use the first letters of the alphabet $a, b, c, d, \ldots$, to represent indices corresponding to the euclidean metric and the letters $i, j, k, \ldots$ to represent tensor indices.

It is clear that the metric is invariant under orthogonal transformations acting on the local frame:

$$e^a_i(\varphi) = O_{ab}(\varphi)\, (e')^b_i(\varphi), \tag{22.72}$$

in which O_{ab} is an orthogonal matrix. Fermions with spin then transform under the spinor representation of this local $O(N)$ group (N is the dimension of $\mathfrak{M}$). As explained in Appendix A5.3, this implies that the spinors ψ and $\bar{\psi}$ transform like:

$$\begin{cases} (\psi)_\alpha (\varphi) = \Lambda_{\alpha\beta}^{-1} (\mathbf{O}) \, \psi'_\beta (\varphi), \\ (\bar{\psi})_\alpha (\varphi) = \bar{\psi}'_\beta (\varphi) \, \Lambda_{\beta\alpha} (\mathbf{O}), \end{cases} \tag{22.73}$$

in which the matrix O_{ab} is given in terms of Λ by:

$$\Lambda \gamma_b O_{ab} \Lambda^{-1} = \gamma_a. \tag{22.74}$$

With these transformation laws, the quantity $\bar{\psi}\gamma^a e^{ai}\psi$ is a tensor under reparametrization, independent of the local frame.

We recall that if Λ is written:

$$\Lambda = \exp\left(\tfrac{i}{4}\theta_{ab}\sigma_{ab}\right), \tag{22.75}$$

in which θ_{ab} is an antisymmetric real matrix, then:

$$O_{ab} = \left(e^\theta\right)_{ab}. \tag{22.76}$$

Gauge invariance. Since the choice of the local frame is arbitrary, we require invariance of all physical quantities under the transformations of the local orthogonal group (22.72). This forces us to introduce a new form of parallel transporter which is an orthogonal curve-dependent matrix $\mathbf{O}(C)$. In a change of local frame it transforms like:

$$\mathbf{O}'(C) = \mathbf{O}(\tilde{\varphi})\mathbf{O}(C)\mathbf{O}^{-1}(\varphi), \tag{22.77}$$

in which the curve C goes from φ to $\tilde{\varphi}$. Such transformations are called *gauge* transformations (for details see Chapter 19). For infinitesimal differentiable curves, $\mathbf{O}(C)$ can be expressed in terms of the spin connection ω_i^{ab}, which has the form of a gauge field. The connection is a vector on the manifold, and a matrix belonging to the Lie algebra of $O(N)$:

$$\mathbf{O}(C) = \mathbf{1} - \boldsymbol{\omega}_i(\varphi)\mathrm{d}\varphi^i + o\left(\|\mathrm{d}\varphi^i\|\right). \tag{22.78}$$

In the case of the gauge theories discussed in Chapter 19 this gauge field is a new independent mathematical quantity. However here we notice that if we consider the matrix (which is a scalar under reparametrization):

$$U^{ab}(C) = e_i^a (\tilde{\varphi}) \, U_j^i(C)e^{jb} (\varphi), \tag{22.79}$$

then, as a consequence of the compatibility condition (22.38) and equation (22.71), it satisfies:

$${}^T U^{ac}(C)U^{cb}(C) = \delta_{ab}, \tag{22.80}$$

and is therefore orthogonal. Equation (22.72) implies that it also satisfies equation (22.77). The matrix $U^{ab}(C)$ is therefore the parallel transporter we are looking for: $O^{ab}(C) \equiv U^{ab}(C)$. Taking in equation (22.79) the limit of an infinitesimal curve we obtain a relation between the connection ω_i^{ab} and the connection Γ_{ik}^j:

$$\Gamma_{ik}^j = e^{aj}\partial_k e_i^a + e^{aj}e_i^b\omega_k^{ab}. \tag{22.81}$$

Equation (22.41) expresses Γ^j_{ik} in terms of the metric tensor. It follows that the connection ω^{ab}_i can be expressed in terms of the vielbein which replaces the metric tensor as the basic geometrical quantity in theories with spin $1/2$ fermions:

$$\omega^{ab}_i = \tfrac{1}{2}e^{aj}\left(\partial_i e^b_j - \partial_j e^b_i\right) + \tfrac{1}{4}e^{aj}e^{bk}\left(\partial_k e^c_j - \partial_j e^c_k\right)e^c_i - (a \leftrightarrow b). \tag{22.82}$$

By expanding all tensors on the basis formed by the vielbein, we can completely replace the condition of covariance under reparametrization by the condition of independence of the local reference frame (gauge independence). To the connection ω^{ab}_i is associated a covariant derivative ∇_i which on a vector of components V^a:

$$V^a = V^i e^a_i, \tag{22.83}$$

acts like:

$$\nabla_i V^a = \partial_i V^a + \omega^{ab}_i V^b. \tag{22.84}$$

Finally we can define a general covariant derivative whose action on euclidean indices is given by equation (22.84) and its action on tensor indices by equation (22.31). Let us then calculate for example $\nabla_i e^a_j$:

$$\nabla_i e^a_j = \partial_i e^a_j + \omega^{ab}_i e^b_j - \Gamma^k_{ij} e^a_k. \tag{22.85}$$

With this definition, equation (22.81) takes the simple form:

$$\nabla_i e^a_j = 0, \tag{22.86}$$

and therefore:

$$\nabla_i V^a = e^a_j \nabla_i V^j. \tag{22.87}$$

This last equation directly follows from the definition of the covariant derivative and equation (22.79).

A last remark: when C is a closed curve, equation (22.79) becomes a similarity relation between matrices. In particular the curvature tensor R^{ab}_{kl}, associated with the connection ω^{ab}_i, is simply related to the Riemann tensor by the equation:

$$R^{ab}_{kl} = R_{ijkl} e^a_i e^b_j = -R^{ba}_{kl}. \tag{22.88}$$

Covariant derivative and fermions. In the case of fermions the covariant derivative $\not{\nabla}$ takes a form which can be inferred from the considerations of Chapter 19, the transformation law of the spinors and the expression of matrices of the spinor representation (22.75):

$$\not{\nabla} = e^{ai}\gamma^a\left(\partial_i + \tfrac{i}{4}\sigma_{bc}\omega^{bc}_i\right). \tag{22.89}$$

22.7 Classical Gravity. Equations of Motion

In the gravitation theory known under the name of General Relativity the metric tensor field $g_{ij}(\varphi)$ becomes a dynamical variable. Using previous considerations we can construct a classical action for a metric tensor coupled to matter which is local and purely geometrical, i.e. independent of the parametrization of the manifold.

22.7.1 The classical action

The pure gravity action $S(\mathbf{g})$ must have the form of a scalar integrated with an invariant volume element. If we look for a local action with only two derivatives we see that the scalar curvature is the only candidate. We then obtain Einstein's action

$$S\left(\mathbf{g}\right) = \int \mathrm{d}^N \varphi\, g^{1/2}(\varphi) R\big(\mathbf{g}(\varphi)\big). \tag{22.90}$$

To Einstein's gravity action we can add a derivative-free term, called a cosmological term,

$$S_{\cos.} = \int \mathrm{d}^N \varphi\, g^{1/2}(\varphi). \tag{22.91}$$

More generally covariance and locality alone allow any function of the curvature tensor and the scalars obtained by contracting covariant derivatives of the curvature tensor. Finally the simplest covariant action for a scalar matter field $\phi(\varphi)$ coupled to gravity takes the form

$$S_{\text{scalar}} = \int \mathrm{d}^N \varphi\, g^{1/2}(\varphi) \left[\tfrac{1}{2} g^{ij}(\varphi) \partial_i \phi \partial_j \phi + \mathcal{V}(\phi)\right]. \tag{22.92}$$

In the case of matter in the form of a vector field $\mathbf{A}_i$ associated to some external gauge symmetry of the type discussed in Chapters 18, 19, we note that the gauge field curvature

$$\mathbf{F}_{ij} = \partial_i \mathbf{A}_j - \partial_j \mathbf{A}_i + [\mathbf{A}_i, \mathbf{A}_j],$$

is a tensor because it is a 2-form (see (22.32)). Therefore a possible action is

$$S_{\text{gauge}} = \tfrac{1}{4} \int \mathrm{d}^N \varphi\, g^{1/2}(\varphi) \operatorname{tr} \mathbf{F}_{ij}(\varphi) \mathbf{F}^{ij}(\varphi). \tag{22.93}$$

Fermion matter coupled to gravity. The considerations of Section 22.6, in particular the definitions of the vielbein and the covariant derivative (22.89), also allow us to write an action for Dirac fermion matter coupled to gravity:

$$S_{\text{fermion}}\left(\psi, \bar\psi\right) = -\int \mathrm{d}^N \varphi\, \det\mathbf{e}\, \bar\psi(\varphi) \left(\slashed{\nabla} + M\right) \psi(\varphi). \tag{22.94}$$

22.7.2 Classical equation of motion

Let us derive the equation of motion for pure gravity, obtained from the variation of the action (22.90)

$$S_{\text{gravity}} = \int \mathrm{d}^N \varphi \sqrt{g} R.$$

Note the identity (be careful about the signs)

$$R^k_{lij} = \nabla_i \Gamma^k_{lj} - \nabla_j \Gamma^k_{li} - \Gamma^k_{mi} \Gamma^m_{lj} + \Gamma^k_{mj} \Gamma^m_{li}. \tag{22.95}$$

Using this identity and integrating by parts, we can also write the action

$$S_{\text{gravity}} = \int \mathrm{d}^N \varphi \sqrt{g} g^{il} \left(-\Gamma^j_{mi} \Gamma^m_{lj} + \Gamma^j_{mj} \Gamma^m_{li}\right).$$

If we add a cosmological term (equation (22.91)) the action becomes

$$S_{\text{cosmo.}} = S_{\text{gravity}} + \Lambda \int \mathrm{d}^N \varphi \sqrt{g} \,. \tag{22.96}$$

Equation of motion. We have therefore to calculate the variation of various quantities when the metric g_{ij} varies

$$\delta g_{ij} = h_{ij} \,. \tag{22.97}$$

Note that the variation of the connection is a tensor (the inhomogeneous part of the transformation cancels in the variation). Indeed one finds after a short calculation

$$\delta\Gamma^i_{jk} = \tfrac{1}{2} g^{il} \left(\nabla_j h_{lk} + \nabla_k h_{lj} - \nabla_l h_{jk} \right) . \tag{22.98}$$

In the same way one can calculate the variation of the curvature in terms of $\delta\Gamma$. Since $\delta\Gamma$ is a tensor one is not surprised to find

$$\delta R^k_{lij} = \nabla_i \delta\Gamma^k_{jl} - \nabla_j \delta\Gamma^k_{il}.$$

Substituting we obtain

$$\delta R^k_{lij} = \tfrac{1}{2} g^{km} \left([\nabla_i, \nabla_j] h_{ml} + \nabla_i \nabla_l h_{mj} - \nabla_j \nabla_l h_{mi} - \nabla_i \nabla_m h_{lj} + \nabla_j \nabla_m h_{il} \right) . \tag{22.99}$$

Since

$$[\nabla_i, \nabla_j] h_{ml} = -R^n_{mij} h_{nl} - R^n_{lij} h_{nm} \,,$$

we find

$$\begin{aligned} \delta R = \delta \left(g^{li} R^j_{lij} \right) &= -R^{ij} h_{ij} + g^{li} \delta R^j_{lij} \\ &= -R^{ij} h_{ij} + \tfrac{1}{2} \left(g^{il} g^{jm} - g^{im} g^{jl} \right) \nabla_i \nabla_l h_{mj} \,. \end{aligned}$$

Moreover

$$\begin{aligned} \delta \left(\sqrt{g} R \right) &= \frac{1}{2\sqrt{g}} g^{ij} h_{ij} R + \sqrt{g} \delta R \\ &= \frac{1}{2\sqrt{g}} g^{ij} h_{ij} R - \sqrt{g} R^{ij} h_{ij} + \tfrac{1}{2} \left(g^{il} g^{jm} - g^{im} g^{jl} \right) \nabla_i \nabla_l h_{mj} \,. \end{aligned}$$

Integrating by parts in the action and using $\nabla_i g_{jk} = 0$ we finally obtain

$$\delta S = \int \mathrm{d}^N \varphi \sqrt{g} \, h_{ij} \left(\tfrac{1}{2} g^{ij} R - R^{ij} \right) ,$$

and thus the equation of motion

$$R g^{ij} - 2 R^{ij} = 0 \,. \tag{22.100}$$

Note that in two dimensions the curvature tensor has only one component and thus $R^{ij} \propto g^{ij}$. Equation (22.100) reduces to only one equation. Taking then the covariant trace we find (in N dimensions)

$$(N - 2) R = 0 \,,$$

equation which is identically satisfied in two dimensions. We recover Gauss–Bonnet's theorem which implies that Einstein's action in two dimensions is topological (see Appendix A22).

The cosmological constant. The action (22.96) with a cosmological term leads to the equation of motion

$$(R + \Lambda)g^{ij} - 2R^{ij} = 0 \,. \tag{22.101}$$

Taking the covariant trace we find

$$(N - 2)R + N\Lambda = 0 \,,$$

which shows that the cosmological constant induces a non-trivial curvature of space even in the absence of matter.

Matter coupled to gravity. If we couple matter to gravity, i.e. add for example the action terms (22.92) and (22.94), the equation of motion becomes

$$(R + \Lambda)g^{ij} - 2R^{ij} + \mathcal{T}^{ij} = 0 \,, \tag{22.102}$$

where the quantity $\mathcal{T}^{ij}$

$$\mathcal{T}^{ij} = 2\frac{\delta S_{\text{matter}}}{\delta g_{ij}} \,,$$

coincides in the limit $g_{ij} \to \delta_{ij}$ with the energy-momentum tensor as defined in Appendix A13.3.

22.8 Quantization in the Temporal Gauge: Pure Gravity

It is easy to verify that, due to the covariance of the equations of motion under reparametrization, it is impossible to quantize the theory in the standard way because the time components of the metric tensor have no conjugate momenta and thus only generate constraints. This is a problem we have already encountered in gauge theories and we shall use the same strategy to solve it. We here choose to quantize the theory in the temporal gauge and we consider, for simplicity only pure gravity.

By a change of coordinates we can reduce the metric to the form $g_{00} = 1$ and $g_{0i} = 0$ for $i \neq 0$. If we then specialize the action to such metrics we obtain, as equation of motions, the space components of equations (22.100). The remaining equations have to imposed as additional constraints. As a notation we below use the letters a, b, c, d for space indices.

The action in the temporal gauge. Let us first calculate the components of the curvature tensor:

$$R_{a00d} = R_{0ad0} = \tfrac{1}{2}\partial_0^2 g_{ad} - \tfrac{1}{4}\partial_0 g_{am} g^{mn} \partial_0 g_{nd}$$
$$R_{abcd} = R^{\text{sp.}}_{abcd} - \tfrac{1}{4}\partial_0 g_{ac}\partial_0 g_{bd} + \tfrac{1}{4}\partial_0 g_{bc}\partial_0 g_{ad} \,,$$

where $R^{\text{sp.}}_{abcd}$ is the curvature tensor in the $N - 1$ space. It follows

$$R_{00} = \tfrac{1}{2}g^{ad}\partial_0^2 g_{ad} + \tfrac{1}{4}\partial_0 g^{ab}\partial_0 g_{ab}$$
$$R = R_{abcd}g^{ad}g^{bc} + 2R_{00}$$
$$= R^{\text{sp.}} + \tfrac{3}{4}\partial_0 g^{ab}\partial_0 g_{ab} + \tfrac{1}{4}\left(g^{ab}\partial_0 g_{ab}\right)^2 + g^{ab}\partial_0^2 g_{ab} \,.$$

We have to integrate by parts to eliminate the second derivatives.

$$\sqrt{g}\,g^{ab}\partial_0^2 g_{ab} = -\sqrt{g}\left[\partial_0 g_{ab}\partial_0 g^{ab} + \tfrac{1}{2}\left(g^{ab}\partial_0 g_{ab}\right)^2\right] + \text{total derivatives}.$$

We finally obtain

$$S_{\text{temp}} = \int \mathrm{d}^N\varphi\,\mathcal{L}$$
$$\mathcal{L} = \sqrt{g}\left[R^{\text{sp.}} - \tfrac{1}{4}\partial_0 g_{ab}\partial_0 g^{ab} - \tfrac{1}{4}\left(g^{ab}\partial_0 g_{ab}\right)^2\right]. \tag{22.103}$$

The corresponding conjugate momenta are:

$$\Pi^{ab} = -\tfrac{1}{2}\sqrt{g}\left(\partial_0 g^{ab} - g^{ab}g_{cd}\partial_0 g^{cd}\right) = -\frac{1}{2\sqrt{g}}\partial_0\left(g g^{ab}\right) \tag{22.104}$$

and conversely

$$\partial_0 g^{ab} = -\frac{2}{\sqrt{g}}\left(\Pi^{ab} - \frac{g^{ab}}{N-2}\Pi_c^c\right). \tag{22.105}$$

The hamiltonian density follows

$$\mathcal{H} = \Pi^{ab}\partial_0 g_{ab} - \mathcal{L}$$
$$= \frac{1}{\sqrt{g}}\left[\Pi_{ab}\Pi^{ab} - \frac{1}{N-2}\left(\Pi_c^c\right)^2\right] - \sqrt{g}R^{\text{sp.}}. \tag{22.106}$$

We note that covariance implies that the quadratic form in the momenta is a homogeneous function of the metric tensor. Therefore ordering problems cannot be avoided and this reflects in a functional measure which is not flat in metric space but multiplied by a power of g:

$$Z = \int [\mathrm{d}g_{ij}g^{N(N-5)/8}]\prod_{i=0,N-1}\delta(g_{0i})\exp[-S(\mathbf{g})].$$

It is then possible to use the same functional techniques as in the gauge theory, and introduce covariant gauges which lead to a BRS symmetric effective action.

Constraints. We must still study the constraints which are

$$R^{0c} = 0, \quad R - 2R^{00} = 0. \tag{22.107}$$

The first set is a simple generalization of Gauss's law in gauge theories

$$R^{0c} = 0 = g^{ad}g^{bc}R_{ab0d} = -\frac{1}{\sqrt{g}}\nabla_a^{\text{sp.}}\Pi^{ac}, \tag{22.108}$$

where $\nabla^{\text{sp.}}$ is the covariant derivative in $N-1$ space dimension. Note that to prove this result one can use

$$\nabla_c^{\text{sp.}}g_{ab} = 0. \tag{22.109}$$

These constraints imply that the wave functional is invariant in a change of space coordinates. Indeed in an infinitesimal coordinate transformation the variation of the metric tensor is (equation (22.9))

$$x^i \mapsto x^i + \epsilon^i(x) \Rightarrow h_{ij} = g_{ik}\partial_j\epsilon^k + g_{jk}\partial_i\epsilon^k + \partial_k g_{ij}\epsilon^k = \nabla_i\epsilon_j + \nabla_j\epsilon_i. \tag{22.110}$$

Thus the invariance of the wave functional is expressed by

$$\nabla_b^{\text{sp.}} \frac{\delta \boldsymbol{\Psi}}{\delta g_{ab}} = 0\,, \tag{22.111}$$

which is exactly the constraint. These constraints commute with the hamiltonian since time-independent reparametrization remains a symmetry of the theory in the temporal gauge.

The last constraint is related to time-independent time reparametrization and thus has no equivalent in gauge theories:

$$R - 2R^{00} = R^{\text{sp.}} + \tfrac{1}{4} \partial_0 g^{ab} \partial_0 g_{ab} + \tfrac{1}{4} \left(g^{ab} \partial_0 g_{ab}\right)^2 = 0\,. \tag{22.112}$$

A short calculation shows that

$$R - 2R^{00} = -\frac{1}{\sqrt{g}} \mathcal{H} \;\Rightarrow\; \mathcal{H}\boldsymbol{\Psi} = 0\,. \tag{22.113}$$

Therefore the constraint, known as Wheeler–DeWitt equation, implies that the physical states corresponds to wave functionals which are eigenfunctions with eigenvalue 0 of the hamiltonian. In the temporal gauge there is no time evolution in the space of physical states. Dynamics is entirely encoded in the very definition of physical states.

Remarks. The quantization method presented here has been criticized because it relies on the possibility of defining a space-like surface, a notion ill-defined in presence of strong metric fluctuations. As for gauge fields it is indeed completely justified only in perturbation theory, when the background (or classical) metric is static and typical deviations of the metric from the background (or classical) metric are small.

The definition of quantum gravity, beyond the formal level leads to a number of unsolved problems, from which we shall list only a few.

While in general the simplest way to define a quantum field theory, is first to construct the euclidean theory and then proceed by analytic continuation to real time, in quantum gravity the euclidean action is not bounded from below. Moreover the connection between euclidean and Minkowskian gravity is much less obvious than in the case of non-gravitational theories because a change in the signature of the metric is involved.

Lattice regularization of the euclidean theory by simplicial gravity, can be easily achieved in two dimensions (see Appendix A22) but remains a partially unsolved problem in four dimensions.

In the real time formulation perturbation theory is expanded around a fixed background metric η_{ij} which, in the case of asymptotic flat space, is simply the Minkowsky metric $\{+1, -1, -1, -1 \ldots\}$. How nature chooses this particular signature is a non-trivial interesting problem. In perturbation theory pure gravity then describes the self-interaction of a hypothetical spin-two massless particle, called the graviton. From the equations of motions we see that it has $\tfrac{1}{2}N(N + 1)/2 - 2N = \tfrac{1}{2}N(N - 3)$ dynamical degrees of freedom. In particular the dimension three is for gravity the analogous of dimension two for gauge theories: due to reparametrization invariance and constraints, the metric has no dynamical degrees of freedom.

However even pure quantum gravity in non-renormalizable in four dimensions. The best we can expect in a covariant gauge is that the metric has canonical dimension $\tfrac{1}{2}(N-2)$. However the action is non-polynomial. As in the case of the non-linear σ model we thus expect that pure gravity is renormalizable only in two dimensions, or more exactly in $2 + \varepsilon$ dimension since in two dimensions the action is topological. Attempts have been made to follow the methods successfully employed in the case of the non-linear σ model and to look for a non-trivial UV fixed point in $2 + \varepsilon$ dimension. However the analysis of singularities of perturbation theory when the dimension approaches two is complicated because the theory does not exist in two dimensions, and therefore remains inconclusive (but the inclusion of a scalar field called dilaton seems to improve the situation). In fact it is commonly believed that the theory, even in its supersymmetric form, remains non-renormalizable, a property which would indicate the breakdown of local quantum field theory at Planck's scale.

Note an additional outstanding problem: Vacuum energy contributions of all particles automatically generate a cosmological term in quantum gravity. However, experimentally such a term is bound by an exceedingly small number when expressed in the natural scale of the Planck mass. It has been noticed that a supersymmetric theory has an exact vanishing cosmological constant, however since experimentally supersymmetry, if it is a symmetry of nature, must be severely broken, this has not provided a solution to the problem.

Finally the peculiarities of two dimensions are discussed in the appendix.

Bibliographical Notes

For more mathematical details see for example

    E. Cartan, *Leçons sur la géométrie des espaces de Riemann* (1928), reprinted by Gauthiers-Villars (Paris 1950); A. Lichnerowicz, *Géométrie des groupes de transformations* (Dunod, Paris 1958); E. Nelson, *Tensor Analysis* (Princeton University Press, Princeton 1967); S. Helgason, *Differential Geometry, Lie Groups and Symmetric Spaces* (Academic Press, New York 1978); Y. Choquet-Bruhat, C. DeWitt-Morette and M. Dillard-Bleick, *Analysis, Manifolds and Physics* (North-Holland, Amsterdam 1977).

For a discussion of two-dimensional models on Riemannian manifolds see

    D.H. Friedan, *Ann. Phys. (NY)* 163 (1985) 318.

Finally for an introduction to the classical theory of gravitation see for example

    S. Weinberg, *Gravitation and Cosmology* (Wiley, New York 1972).

An elementary introduction to quantum gravity is found in

    M. Veltman, in *Methods in Field Theory*, Les Houches 1975, R. Balian and J. Zinn-Justin eds. (North-Holland, Amsterdam 1976).

Various issues in quantum gravity are discussed by B.S. DeWitt and S.W. Hawking in their contributions to

    Relativity, Groups and Topology, proceedings of the 1983 Les Houches school, B.S. DeWitt and R. Stora eds. (Elsevier Science, Amsterdam 1984),

and in the selected papers gathered in

    Euclidean Quantum Gravity, G.W. Gibbons and S.W. Hawking eds. (World Scientific, Singapore 1993).

For a discussion of matrix models in the large N-limit and its application to two-dimensional quantum gravity or critical models on random surfaces see for example

    G. 't Hooft, *Nucl. Phys.* B72 (1974) 461; E. Brézin, C. Itzykson, G. Parisi and J.-B. Zuber, *Comm. Math. Phys.* 59 (1978) 35; D. Bessis, C. Itzykson, and J.-B. Zuber, *Adv. Appl. Math.* 1 (1980) 109, and references therein; E. Brézin and V. Kazakov, *Phys. Lett.* B236 (1990) 144; M. Douglas and S. Shenker, *Nucl. Phys.* B235 (1990) 635; D.J. Gross

and A.A. Migdal, *Phys. Rev. Lett.* 64 (1990) 127; M. R. Douglas, *Phys. Lett.* B238 (1990) 176; P. Ginsparg, M. Goulian, M. R. Plesser, and J. Zinn-Justin, *Nucl. Phys.* B342 (1990) 539; E. Brézin, V. Kazakov and Al. A. Zamolodchikov, *Nucl. Phys.* B338 (1990) 673; P. Ginsparg and J. Zinn-Justin, *Phys. Lett.* B240 (1990) 333; D.J. Gross and N. Miljkovic, *Phys. Lett.* B238 (1990) 217; G. Parisi, *Phys. Lett.* B238 (1990) 209. See also
Random Surfaces and Quantum Gravity, proceedings of the Cargèse workshop 1990, O. Alvarez, E. Marinari and P. Windey eds. (Plenum Press B262, New York 1991), and the review
P. Di Francesco, P. Ginsparg and J. Zinn-Justin, *Phys. Rep.* 254 (1995) 1.

Exercises

Exercise 22.1

To the reader who is puzzled by the Wheeler–DeWitt equation we propose to study the quantization of the relativistic point-particle of space-time coordinates x^μ (the metric being Minkowskian, i.e. $+ - - \cdots$) in an external electromagnetic field A_μ. We consider the following lagrangian

$$-\mathcal{L}(x^\mu, \dot{x}^\mu, N) = \tfrac{1}{2}N^{-1}(\tau)\dot{x}^\mu\dot{x}_\mu + \tfrac{1}{2}N(\tau)m^2 + eA_\mu(x)\dot{x}^\mu, \qquad (22.114)$$

where τ parametrizes the particle trajectory and $N(\tau)$ is a metric factor which ensures the independence of the theory on the trajectory parametrization. Vectors have components $x^0, x^1, \ldots$.

This theory has a reparametrization invariance:

$$\tau \mapsto \tau', \qquad N'(\tau') = N(\tau)\frac{d\tau}{d\tau'}.$$

Of course it is possible to eliminate N from the lagrangian by solving the corresponding equation of motion

$$m^2N^2 = \dot{x}^\mu\dot{x}_\mu.$$

One obtains

$$-\mathcal{L}'(x^\mu, \dot{x}^\mu) = \pm m(\dot{x}^\mu\dot{x}_\mu)^{1/2} + eA_\mu(x)\dot{x}^\mu. \qquad (22.115)$$

This lagrangian has still a reparametrization invariance.

Discuss quantization from both points of view ($N(\tau) \equiv 1$ defines the temporal gauge).

Exercise 22.2

The exercises which follow try to motivate the description of discretized 2D quantum gravity in terms of triangulated surfaces as explained in the appendix A22.

22.2.1. *Random walks.* One considers brownian random walks on a hypercubic lattice (see also Appendix A28.1) with constant decay rate per unit time $1 - a$ or survival probability a, $0 < a < 1$. Calculate the probability $U(y, x)$ for a walk starting from point x to pass through the point y at any later time. Study the scaling limit in which $|x - y|$ becomes large and a approaches 1.

22.2.2. One can write directly in the continuum a path integral which weights the paths according to their length as in the preceding discretized example. One can integrate with the action

$$S(\mathbf{x}) = m \int dt \sqrt{\dot{\mathbf{x}} \cdot \dot{\mathbf{x}}}. \qquad (22.116)$$

We recognize a form analogous to expression (22.115). With this action, however, one again faces the problem of reparametrization invariance. Therefore study first the continuum limit of a natural discretization of this expression, the probability distribution $U(y, x)$ being given by

$$U(y, x) = \delta(x - y) + \sum_{n=1}^{\infty} \int d^d x_1 \, d^d x_2 \ldots d^d x_n \exp\left[-\mu \sum_{i=0}^{n} |x_i - x_{i+1}|\right],$$

with $x_0 = x$, $x_{n+1} = y$, and μ being a constant.

22.2.3. Recover now the preceding result from the continuum action (22.116).

APPENDIX 22
MATRIX MODELS IN THE LARGE N LIMIT AND 2D GRAVITY

Quantum gravity in four dimensions, beyond the formal level, is far from being well understood. It is therefore quite natural to investigate the properties of quantum gravity in lower dimensions. Two dimensions are of special interest because a speculative extension of quantum field theory, string theory, can be formulated as 2D gravity coupled to some matter. In recent years progress has been reported in the problem of 2D gravity coupled to discrete matter, in the euclidean formulation. The solution uses original techniques: space-discretized gravity is reformulated in terms of matrix integrals, and the continuum is recovered in a peculiar "double scaling limit" in which the size N of the matrices becomes large. Since most of these developments are rather technical we here briefly review only the solution of the one-matrix problem, which is relevant for two-dimensional pure quantum gravity. The interested reader is referred to the literature for details. Note that the problem of evaluating integrals over matrices of large size has also statistical applications for fluctuating surfaces, or random hamiltonians or transfer matrices.

A22.1 Quantum 2D Euclidean Gravity

A22.1.1 Classical 2D gravity

Einstein's action for pure gravity in the presence of a cosmological term reads (see equations (22.90,22.91))

$$S(g) = \int d^d\varphi\sqrt{g}(KR + \Lambda)\,,$$

in which g_{ij} is the metric tensor, $g = \det g_{ij}$, R the scalar curvature and K, Λ are two coupling constants. The parameter Λ, which multiplies the volume element, is called the cosmological constant. In $d = 2$ dimensions classical gravity is trivial because the curvature term does not contribute to the equations of motion (see Section 22.7.2). As a consequence of the Gauss–Bonnet theorem, the total curvature has a topological interpretation: the total curvature of an euclidean closed surface is proportional to the genus of the surface,

$$\int d^2\varphi\sqrt{g}R = 4\pi\chi\,,$$

where χ is the Euler–Poincaré character of the surface, related to the genus, i.e. the number of "handles" h, by $h = \frac{1}{2}(2 - \chi)$.

2D quantum gravity, however, is somewhat less trivial because large quantum fluctuations may change the genus of the surface. To calculate the partition function it is thus necessary to sum over surfaces of all genera. This problem is difficult to solve starting from the continuum field theory, and surprisingly enough it can be more easily studied in a discretized version.

A22.1.2 Matrix representation of discretized 2D gravity

Following 't Hooft's analysis of the large N limit of $SU(N)$ gauge theories it has been recognized that if one considers the integral

$$Z(\alpha_k, N) = \int dM\, e^{-N\,\mathrm{tr}\,V(M)}\,,$$

in which M is an $N \times N$ hermitian matrix and

$$V(\lambda) = \lambda^2 + \sum_{k \geq 3} \alpha_k \lambda^k,$$

then the expansion of $F \equiv \ln Z$ for N large takes the form

$$F = \ln Z = \sum N^\chi F_\chi(\alpha_k), \qquad (A22.1)$$

where F_χ is the sum of all Feynman diagrams with Euler–Poincaré character $\chi = 2 - 2h$. It follows that the dual of a Feynman diagram contributing to F_χ can be represented as an orientable surface of fixed topology, the powers of $\alpha_3, \alpha_4, \ldots, \alpha_k$ counting the number of triangles, squares,...,k-gons of the surface (taking the logarithm $F \equiv \ln Z$, yields the gravity model partition function which represents the sum only over connected surfaces). If by convention the area of each triangle, square, ..., is assumed to be 1, the power of α_k measures the total area of the surface. This gives a formulation of discrete 2D gravity in terms of a distribution of random matrices.

In what follows it will be convenient to change the normalization in the integrand and consider instead

$$Z(g, \alpha_k, N) = \int dM \, e^{-(N/g)\,\mathrm{tr}\, V(M)}, \qquad (A22.2)$$

where the coupling constant g plays the role of the cosmological constant. Note that to describe pure gravity only one α_k is needed, for instance one can use only triangles. More general models correspond to the addition of some new degrees of freedom on the surface.

The continuum limit. The continuum limit is obtained by letting the total area of the surfaces tend towards infinity. Surfaces of large area are connected with the large order behaviour of the Taylor series expansion of F_χ in powers of g and therefore to the singularity of $F_\chi(g)$ closest to the origin. At g fixed, the large N limit selects the surfaces of the topology of the sphere. It will be shown below however that it is possible to take $N \to \infty$ while simultaneously taking g to the location of the leading singularity g_c, in a coherent way so that all surfaces of infinite area and arbitrary topology contribute.

Remark. In Chapters 37–41 we shall argue that quite generally integrals of the form $(A22.2)$ are singular at $g = 0$. The result $g_c \neq 0$ reflects the property that the number of planar Feynman diagrams (selected by the large N limit) grows only geometrically instead of a factorial for generic diagrams.

A22.2 The One-Matrix Model

In contrast with vector models where the large N limit can be determined quite generally (see Chapter 29), the large N limit of matrix models has been obtained only in very simple cases, although the problem is of great potential interest (distribution of eigenvalues of random hamiltonians, QCD, critical models on random surfaces, two-dimensional quantum gravity, strings...). Problems for which solutions exist are simple integrals, already a non-trivial problem, over one matrix or several matrices with a one-dimensional nearest neighbour coupling and quantum mechanics of large hermitian matrices (only a partial solution). Most methods of solution are based on the possibility of first diagonalizing the matrices and being able to reduce the matrix integral to an integral over eigenvalues.

In contrast with vector problems even then the problem is not solved because a $N \times N$ matrix has N eigenvalues and the dependence on N is not yet explicit. At this point in the case of integrals two methods of solution have been used, steepest descent and orthogonal polynomials. In one dimension (quantum mechanics), in the sector without unitary excitations, the hamiltonian can be transformed into a hamiltonian of N independent fermions. In this work we briefly review only the solution of the one-matrix problem, which is relevant for 2D pure quantum gravity.

A22.2.1 The one-matrix model

We want to evaluate for N large the integral over a $N \times N$ hermitian matrix M

$$Z = \int \mathrm{d}^{N^2} M \; e^{-(N/g)\,\mathrm{tr}\,V(M)}, \qquad (A22.3)$$

where $\mathrm{d}^{N^2} M$ means integration with a flat measure over the N^2 independent real variables $\mathrm{Re}\,M_{ij}, \mathrm{Im}\,M_{ij}$ and $V(M)$ is a general polynomial potential

$$V(M) = M^2 + \sum_{k \geq 3} \alpha_k M^k.$$

Since the integrand depends only on the N eigenvalues λ_i of M we factorize the measure of integration $\mathrm{d}M$ into the measure of the unitary group and a measure for eigenvalues. Integrating over the unitary group one finds

$$Z = \int \prod_{i=1}^{N} \mathrm{d}\lambda_i\, \Delta^2(\lambda) \exp\left[-\sum_i (N/g) V(\lambda_i) \right], \qquad (A22.4)$$

where

$$\Delta(\lambda) = \prod_{1 \leq i < j \leq N} (\lambda_j - \lambda_i). \qquad (A22.5)$$

$\Delta(\lambda)$ can also be written as a Vandermonde determinant:

$$\Delta(\lambda) = \det \lambda_i^{j-1},$$

as one verifies using the antisymmetry of the determinant in the interchange of any two eigenvalues (the normalization is determined by comparing leading terms).

This first step is crucial for all methods used up to now. However, in contrast with the vector models, we still have to integrate over a large number of variables. Thus additional technical steps are involved in the evaluation of the large N limit.

The eigenvalues measure. Equation $(A22.4)$ may be derived via the usual Faddeev–Popov method: Let U_0 be the unitary matrix such that $M = U_0^\dagger \Lambda U_0$, where Λ is a diagonal matrix with eigenvalues λ_i. The right hand side of $(A22.4)$ follows by substituting the definition $1 = \int \mathrm{d}U\, \delta(UMU^\dagger - \Lambda)\, \Delta^2(\lambda)$ (where $\int \mathrm{d}U \equiv 1$). We first perform the integration over M, and then U decouples due to the cyclic invariance of the trace so the integration over U is trivial, leaving only the integral over the eigenvalues λ_i of Λ with flat measure. To determine $\Delta(\lambda)$, we note that only the infinitesimal neighbourhood $U = (1+T)U_0$ contributes to the U integration, so that

$$1 = \int \mathrm{d}U\, \delta(UMU^\dagger - \Lambda)\, \Delta^2(\lambda) = \int \mathrm{d}T\, \delta([T,\Lambda])\Delta^2(\lambda).$$

Now $[T,\Lambda]_{ij} = T_{ij}(\lambda_j - \lambda_i)$, and thus equation $(A22.5)$ follows (up to a sign) since we integrate over both real and imaginary parts of the off diagonal T_{ij}'s.

A22.2.2 The large N limit: steepest descent

The partition function (A22.4) can be considered as the partition function of a gas of charges of positions λ_i interacting via a repulsive 2D Coulomb force generated by the Vandermonde determinant (see Section 32.3) which prevents the charges of accumulating at the minimum of the one-body potential $V(\lambda)$. The saddle point equations, obtained by varying the λ_i's, yield the configuration of minimum energy:

$$\frac{2}{N}\sum_{j\neq i}\frac{1}{\lambda_i - \lambda_j} = \frac{1}{g}V'(\lambda_i)\,. \qquad (A22.6)$$

This equation can be solved by the following method: We introduce the trace of the resolvent $\omega(z)$ of the matrix M

$$\omega(z) = \frac{1}{N}\,\text{tr}\,\frac{1}{M-z} = \frac{1}{N}\sum_i\frac{1}{\lambda_i - z}\,. \qquad (A22.7)$$

Multiplying equation (A22.6) by $1/(\lambda_i - z)$ and summing over i, we find

$$\omega^2(z) - \frac{1}{N}\omega'(z) + \frac{1}{g}V'(z)\omega(z) = -\frac{1}{g}\sum_i\frac{V'(z) - V'(\lambda_i)}{z - \lambda_i}\,. \qquad (A22.8)$$

This equation is analogous to the Riccati form of the Schrödinger equation, the wave function ψ being related to ω by $N\omega(z) + NV'(z)/(2g) = \psi'/\psi$. The eigenvalues λ_i are the zeros of the wave function. In the large N limit, we can neglect $N^{-1}\omega'(z)$ (the WKB approximation). In this limit the distribution of eigenvalues $\rho(\lambda) = \frac{1}{N}\sum_i\delta(\lambda - \lambda_i)$ becomes continuous, and

$$\omega(z) = \int\frac{\rho(\lambda)\mathrm{d}\lambda}{\lambda - z}\,. \qquad (A22.9)$$

Note that the normalization condition $\int\rho(\lambda')\mathrm{d}\lambda' = 1$ is the analogue of the Bohr–Sommerfeld quantization condition. Conversely the eigenvalue density $\rho(\lambda)$ is extracted from $\omega(z)$ via the relation

$$\rho(\lambda) = \frac{1}{2i\pi}\big(\omega(z + i0) - \omega(z - i0)\big)\,. \qquad (A22.10)$$

In the continuum limit equation (A22.8) becomes

$$\omega^2(z) + \frac{1}{g}V'(z)\,\omega(z) + \frac{1}{4g^2}R(z) = 0\,, \qquad (A22.11)$$

where

$$R(z) = 4g\int\mathrm{d}\lambda\,\rho(\lambda)\frac{V'(z) - V'(\lambda)}{z - \lambda} \qquad (A22.12)$$

is a polynomial of degree $l - 2$ when V is of degree l. Note that the coefficient of highest degree of R is fixed by the normalization of $\rho(\lambda)$ while the remaining coefficients depend explicitly on the eigenvalue distribution.

Finally the free energy $\ln Z$ is obtained from the saddle point value of the integrand (A22.4) in the continuum limit:

$$\ln Z = N^2\left(\int\mathrm{d}\lambda\,\mathrm{d}\mu\,\rho(\lambda)\rho(\mu)\ln|\lambda - \mu| - \frac{1}{g}\int\mathrm{d}\lambda\,\rho(\lambda)V(\lambda)\right)\,. \qquad (A22.13)$$

The solution. The solution to equation $(A22.11)$ is

$$\omega(z) = \frac{1}{2g} \left(-V'(z) + \sqrt{[V'(z)]^2 - R(z)} \right). \qquad (A22.14)$$

Generically $\omega(z)$ has $2(l-1)$ branch points corresponding to the roots of the polynomial $V'^2 - R$. Therefore the support of $\rho(\lambda)$ is formed of $l-1$ disconnected pieces. In the simplest case, when the potential has only one minimum, we expect a single connected support and thus only two branch points. It follows that the polynomial $V'^2 - R$ must have $l-2$ double roots and this yields $l-2$ conditions that fully determine R.

A22.3 The Method of Orthogonal Polynomials

The steepest descent method allows a general discussion of the large N limit. It is difficult however to calculate the subleading orders in the $1/N$ expansion and therefore to discuss perturbation theory to all orders. We now present another method that allows us to recover previous results and to extend them to all orders in $1/N$.

This alternative method for solving $(A22.4)$ makes use of a set of polynomials $P_n(\lambda)$, orthogonal with respect to the measure

$$\int_{-\infty}^{\infty} d\lambda \ e^{-(N/g)V(\lambda)} \ P_n(\lambda) P_m(\lambda) = s_n \, \delta_{nm}. \qquad (A22.15)$$

The normalization of the orthogonal polynomials P_n is given by having the leading term $P_n(\lambda) = \lambda^n + \dots$, hence the constant s_n on the r.h.s. of $(A22.15)$. Due to the relation

$$\Delta(\lambda) = \det \lambda_i^{j-1} = \det P_{j-1}(\lambda_i) \qquad (A22.16)$$

(recall that arbitrary polynomials may be built up by adding linear combinations of preceding columns, a procedure that leaves the determinant unchanged), the polynomials P_n can be employed to solve $(A22.4)$. We substitute the determinant $\det P_{j-1}(\lambda_i) = \sum (-1)^\pi \prod_k P_{i_k-1}(\lambda_k)$ for each of the $\Delta(\lambda)$'s in $(A22.4)$ (where the sum is over permutations i_k and $(-1)^\pi$ is the signature of the permutation). The integrals over individual λ_i's factorize, and due to orthogonality the only contributions are from terms with all $P_i(\lambda_j)$'s paired. There are $N!$ such terms so $(A22.4)$ reduces to

$$Z = \int \prod_\ell d\lambda_\ell \ e^{-(N/g)V(\lambda_\ell)} \sum_{\pi,\pi'} (-1)^\pi (-1)^{\pi'} \prod_k P_{i_k-1}(\lambda_k) \prod_j P_{i_j-1}(\lambda_j)$$

$$= N! \prod_{i=0}^{N-1} s_i = N! \, s_0^N \prod_{k=1}^{N-1} f_k^{N-k}, \qquad (A22.17)$$

where we have defined $f_k \equiv s_k/s_{k-1}$. The solution of the original matrix integral is thus reduced to the problem of determining the normalizations s_k, or equivalently the ratios f_k.

In the naive large N limit, the rescaled index k/N becomes a continuous variable t that runs from 0 to 1, and f_k/N becomes a continuous function $f(t)$. In this limit, the free energy (up to an irrelevant additive constant) reduces to a simple one-dimensional integral:

$$\frac{1}{N^2} \ln Z = \frac{1}{N} \sum_k (1 - k/N) \ln f_k \sim \int_0^1 dt \, (1-t) \ln f(t). \qquad (A22.18)$$

To derive the functional form for $f(t)$, we assume for simplicity that the potential $V(\lambda)$ in $(A22.15)$ is even. Since the P_i's from a complete set of basis vectors in the space of polynomials, it is clear that $\lambda P_n(\lambda)$ must be expressible as a linear combination of lower P_i's, $\lambda P_n(\lambda) = \sum_{i=0}^{n+1} c_{ni} P_i(\lambda)$ (with $c_{ni} = s_i^{-1} \int e^{-V} \lambda P_n P_i$). However, terms proportional to P_i for $i < n-1$ vanish since $\int e^{-V} P_n \lambda P_i = 0$ (recall λP_i is a polynomial of order at most $i+1$ so is orthogonal to P_n for $i+1 < n$). Moreover the term proportional to P_n also vanishes due to the assumption that the potential is even, $\int e^{-V} \lambda P_n P_n = 0$. Thus the orthogonal polynomials satisfy the simple recursion relation,

$$\lambda P_n = P_{n+1} + r_n P_{n-1}, \tag{A22.19}$$

with r_n a scalar coefficient independent of λ.

By considering the quantity $P_n \lambda P_{n-1}$ with λ paired alternately with the preceding or succeeding polynomial, we derive

$$\int d\lambda \; e^{-(N/g)V} \; P_n \, \lambda \, P_{n-1} = r_n \, s_{n-1} = s_n \,.$$

This shows that the ratio $f_n = s_n/s_{n-1}$ for this simple case is identically the coefficient defined by equation $(A22.19)$, $f_n = r_n$. Similarly if we pair the λ in $P_n' \lambda P_n$ before and afterwards, an integration by parts yields

$$n s_n = \int d\lambda \; e^{-(N/g)V} \; P_n' \, \lambda \, P_n = \int e^{-(N/g)V} \; P_n' \, r_n \, P_{n-1}$$

$$= r_n \int d\lambda \; e^{-(N/g)V} \; (N/g)V' \, P_n \, P_{n-1} \,, \tag{A22.20}$$

a relation that will allow us to determine recursively r_n.

A22.3.1 The large N limit

Our intent now is to find an expression for $f_n = r_n$ and substitute into $(A22.18)$ to calculate a partition function. For definiteness, we take as an example the potential

$$V(\lambda) = \tfrac{1}{2}\lambda^2 + \tfrac{1}{2}\lambda^4, \;\; \Rightarrow \;\; V'(\lambda) = \lambda + 2\lambda^3 \,. \tag{A22.21}$$

The calculation of the r.h.s. of equation $(A22.20)$ involves then expanding $(\lambda + 2\lambda^3)P_{n-1}$ on the basis of polynomials P_i. These may be accomplished by using repeatedly equation $(A22.19)$. For the potential $(A22.21)$ this gives

$$g(n/N) = r_n + 2r_n(r_{n+1} + r_n + r_{n-1}) \,. \tag{A22.22}$$

As explained before equation $(A22.18)$, in the large N limit n/N can be treated as a continuous variable $t = n/N$, and we have $r_n \mapsto r(t)$ and $r_{n\pm1} \mapsto r(t \pm \varepsilon)$, where $\varepsilon \equiv 1/N$. To leading order in $1/N$, equation $(A22.22)$ reduces to

$$gt = r + 6r^2 = W(r) \,. \tag{A22.23}$$

It can be verified, inserting the solution of equation $(A22.23)$ into equation $(A22.18)$, that the result obtained by the method of steepest descent is recovered at this order, however the orthogonal polynomial method allows to calculate much more easily the higher order corrections.

We note that equation $(A22.23)$ has no solution when g is smaller that the minimum value g_c of $W(r)$, where $g_c = -1/24$ for $t = 1$.

A22.3.2 Large N and double scaling limit

Let us construct surfaces in terms of squares. The corresponding potential is then the potential $(A22.21)$. The large N limit, i.e. the sum of all contributions of surfaces of genus zero (planar surfaces) can be directly deduced from the results of Subsection A22.2.2. In particular we have noted the existence of a critical value $g_c = -1/24$. The limit $g \to g_c$ thus corresponds to the continuum limit. At this point the second derivative of $\ln Z$ has a square root singularity and thus the singular part of $\ln Z$ behaves like $(g - g_c)^{5/2}$.

The all genus partition function. We now search for another solution to equation $(A22.22)$ and its generalizations that describes the contribution of all genus surfaces to the partition function $(A22.18)$. We shall retain higher order terms in $1/N$ in $(A22.22)$ so that e.g. equation $(A22.23)$ instead reads

$$gt = W(r) + 2r(t)\big(r(t+\varepsilon) + r(t-\varepsilon) - 2r(t)\big)$$
$$= g_c + \tfrac{1}{2}W''|_{r=r_c}\big(r(t) - r_c\big)^2 + 2r(t)\big(r(t+\varepsilon) + r(t-\varepsilon) - 2r(t)\big) + \cdots \quad (A22.24)$$

As suggested at the end of the preceding subsection, we shall simultaneously let $N \to \infty$ and $g \to g_c$ in a particular way. Since $g - g_c$ has dimension [length]^2, it is convenient to introduce a parameter a with dimension length and let $g - g_c = \kappa^{-4/5}a^2$, with $a \to 0$. Our ansatz for a coherent large N limit will be to take $\varepsilon \equiv 1/N = a^{5/2}$ so that the quantity $\kappa^{-1} = (g - g_c)^{5/4}N$ remains finite as $g \to g_c$ and $N \to \infty$.

Moreover since the integral $(A22.18)$ is dominated by t near 1 in this limit, it is convenient to change variables from t to z, defined by $g_c - gt = a^2 z$. Our scaling ansatz in this region is $r(t) = r_c + au(z)$. If we substitute these definitions into equation $(A22.23)$, the leading terms are of order a^2 and result in the relation $u^2 \sim z$. To include the higher derivative terms, we note that

$$r(t+\varepsilon) + r(t-\varepsilon) - 2r(t) \sim \varepsilon^2 \frac{\partial^2 r}{\partial t^2} = a\frac{\partial^2}{\partial z^2}au(z) \sim a^2 u'',$$

where we have used $\varepsilon(\partial/\partial t) = -ga^{1/2}(\partial/\partial z)$ (which follows from the above change of variables from t to z). Substituting into $(A22.24)$, the vanishing of the coefficient of a^2 implies the differential equation

$$z = u^2 - \tfrac{1}{3}u'', \quad (A22.25)$$

(after a suitable rescaling of u and z). The second derivative of the partition function (the "specific heat") has a leading singular behaviour given by $f(t)$ with $t = 1$, and thus by $u(z)$ for $z = (g-g_c)/a^2 = \kappa^{-4/5}$. The solution to equation $(A22.25)$ characterizes the behaviour of the partition function of pure gravity to all orders in the genus expansion. (Notice that the leading term is $u \sim z^{1/2}$ so after two integrations the leading term in $\ln Z$ is $z^{5/2} = \kappa^{-2}$.)

Equation $(A22.25)$ is known in the mathematical literature as the Painlevé I equation. The perturbative solution in powers of $z^{-5/2} = \kappa^2$ takes the form $u = z^{1/2}(1 - \sum_{k=1} u_k z^{-5k/2})$, where the u_k are all positive. The solution of the Painlevé I equation seems to define 2D quantum gravity beyond the perturbative topological expansion. However, it can be verified that there exists no real solution to the Painlevé equation satisfying the proper boundary conditions. This property can be related to the non-Borel summability of the perturbative expansion. Its physical interpretation is that pure quantum gravity is unstable, because the number of surfaces increases too rapidly with the genus.

Chapters 23–36 are devoted to the study of second order phase transitions and the determination of their universal properties through the use of renormalization group methods.

Most of the transitions we shall consider have the following character: a system of spins on a lattice or a system of particles in the continuum interacts through short range forces. At fixed density, as long as the system is contained in a finite volume, it is *ergodic*, i.e. any finite region of available phase space has a non-vanishing probability to be explored in the course of time (in the sense of Appendix A4 the system is connected). However in the infinite volume limit, depending of the value of some control parameter which usually is the temperature, the system either remains ergodic, or experiences a breaking of ergodicity. In the latter case, phase space decomposes in disjoint sets. When the system is initially in one of the sets, it remains at later time. For example for Ising-like systems, the two sets correspond to the two possible values of the spontaneous magnetization.

In the simplest lattice models a transfer matrix can be defined. The thermodynamic limit is related to the largest eigenvalue of the matrix. In the "ergodic" phase the corresponding eigenvector is unique while in the non-ergodic phase it is degenerate.

Our goal is to analyze the behaviour of thermodynamic quantities in the neighbourhood of the phase transition, in particular their singularities as function of the control parameter.

For the simplest systems, it is possible to find local observables whose values depend on the phase in the several phase region. We call one of such observables the *order parameter*. It is for example the spin in ferromagnetic systems.

In the one phase region, for systems with short range interactions (for example decreasing exponentially or faster) the connected correlation functions decrease exponentially with distance when two non-empty sets of points are separated (this is directly related to the cluster property of Section 6.1.2). We call *correlation length* the inverse of the smallest decay rate of correlation functions (the smallest physical mass in the particle physics sense). For reasons which will become clearer later, we study only these special phase transitions for which the correlation length, measured in units of the microscopic scale (range of forces, lattice spacing), diverges at the transition point. For such systems it will be shown that, near the transition point (the critical temperature for example), some properties of thermodynamic functions are *universal* and furthermore can be described by euclidean field theories and renormalization group equations.

Several chapters will be devoted to the derivation of universal properties of correlation functions, but let us here emphasize already the deep connection between non-mean field universality in phase transitions with divergent correlation length, and renormalization in Field Theory. We have shown that renormalizable local field theories are short distance insensitive, in the sense that, although they are not finite in the limit in which the UV cut-off becomes much larger than all masses and momenta, after an infinite renormalization a unique (up to finite renormalizations) renormalized field is obtained. The translation into the phase transition language is simple: universality emerges in a regime in which the correlation length and all distances are much larger than the microscopic scale which plays the role of the inverse of the cut-off. However this universality is non-mean field like (at least in low dimensions) because the microscopic scale cannot be completely eliminated; degrees of freedom on all scales remain coupled.

Locality of field theory is the analogue, as we shall see later, of short range forces. It remains to explain the role in phase transitions of the particular dimension in which the field theory is just renormalizable.

Since renormalization group equations are a direct consequence of renormalization theory, as we have shown in Chapter 10, we shall not be too surprised to discover that they appear and play a central role in the theory of phase transitions.

Terminology. To describe critical phenomena, it has become customary to use the language of magnetic systems. Although such a presentation certainly helps our physical intuition, many systems to which the theory applies are non-magnetic. This language is therefore, in a sense, almost as abstract as the language of quantum field theory. Since we shall eventually show that all the universal problems in the theory of phase transitions can be formulated in the language of euclidean field theory, we could have stayed completely within the framework developed in the previous chapters. However, since we also wish to introduce field theory methods to readers with a background in statistical mechanics, we shall often use the statistical and magnetic language. In previous chapters we have already called correlation functions what often is called imaginary time Green's functions or Schwinger's functions. A short glossary is then:

Euclidean field theory	**Classical magnetic systems**
Fields	*Spin variables*
Source	*Magnetic field*
Euclidean action	*Configuration energy, hamiltonian*
$exp\,(-\Delta t\,H)$	*Transfer matrix in the case of finite range interactions*
Functional integral	*Sum over spin configurations*
Field vacuum expectation value	*Magnetization*
Zero momentum two-point function	*Magnetic susceptibility*
Physical mass	*Inverse correlation length*
Massless theory	*Critical theory*
Generating functional Z of correlation functions	*Partition function in a space-dependent magnetic field*
Generating functional W of connected correlation functions	*Free energy in a magnetic field*
Generating functional Γ of proper vertices	*Thermodynamic potential, function of the magnetization*

Let us now briefly explain the organization of the chapters devoted to critical phenomena:

In Chapter 24 we study ferromagnetic systems within mean field theory and show how in dimension less than or equal to four, summation of leading divergent corrections to mean field theory leads to the continuum ϕ^4 field theory. In Chapters 25–28 we then introduce renormalization group ideas. We obtain a complete description of universal properties of second order phase transitions in ferromagnetic systems near four dimensions and calculate the leading corrections to the universal behaviour in the critical domain.

In the Chapters 29 and 30 we expand the N-vector model for N large and at low temperature and verify RG predictions. The special roles of two dimensions and the abelian $O(2)$ model emerge. The Chapters 31 and 32 are thus devoted to a technique specific to two dimensions, bosonization, which gives information about the properties of several models like the sine-Gordon, Coulomb gas, Thirring or Schwinger models, and its application to the famous Kosterlitz–Thouless phase transition.

In Chapters 33 and 34 we try to gain some non-perturbative insight in the properties of gauge theories through the use of lattice regularization and RG techniques. We derive the conditions for large momentum asymptotic freedom in four dimensions. Finally in Chapter 35 we examine Critical Dynamics, i.e. the time evolution of critical systems while Chapter 36 deals with finite size effects.

However before we discuss phase transitions with more sophisticated techniques, we want to recall a few properties, from the point of view of phase transitions, of simple ferromagnetic lattice models. In systems with finite range interactions a transfer matrix can be defined. We first examine its properties in a finite volume. In the infinite volume limit, low and high temperature considerations provide convincing evidence of the existence of phase transitions in Ising-like systems. We relate the notion of order parameter to cluster properties in the low temperature broken phase. We show in a simple example that phase transitions indeed correspond to breaking of ergodicity. We finally extend the analysis to ferromagnetic systems with continuous symmetries.

The appendix contains a brief discussion of quenched disorder.

23.1 Phase Transitions and Transfer Matrix

In Chapter 2 we have constructed a path integral representation of the partition function in Quantum Mechanics by using the identity:

$$e^{-\beta H} = \left(e^{-\varepsilon H}\right)^l , \quad l\varepsilon = \beta . \tag{23.1}$$

Similarly, in the case of lattice models with finite range interactions (for example nearest neighbour interactions), distinguishing one "time" direction on the lattice, it is possible to write the partition function Z with periodic boundary conditions:

$$Z = \operatorname{tr} \mathbf{T}^l, \tag{23.2}$$

in which l is a measure of the lattice size in the time direction. The matrix $\mathbf{T}$ is the transfer matrix and plays the role of $e^{-\varepsilon H}$. For example in the case of the 1D Ising model with nearest neighbour (n.n.) interactions and in a magnetic field h the partition function is:

$$Z = \sum_{\{S_i=\pm 1\}} \exp\left[\beta \sum_{i=1}^{l} (JS_iS_{i+1} + hS_i)\right], \tag{23.3}$$

and the transfer matrix is a 2×2 matrix which can be written:

$$\langle S' |\mathbf{T}| S\rangle = \exp\left[\beta\left(JSS' + \frac{h}{2}(S + S')\right)\right]. \tag{23.4}$$

In what follows we shall restrict ourselves to isotropic interactions on a hypercubic lattice and therefore we can always choose the transfer matrix symmetric (this corresponds to hermitian hamiltonians in quantum mechanics).

Note the difference between the roles played by the inverse temperature β in Quantum and Classical Statistical Mechanics. The parameter β of quantum mechanics is the analogue of the size of a classical system in one additional dimension. In particular the large β (i.e. zero temperature) limit of quantum mechanics corresponds to the large l limit.

Since fixed temperature corresponds to fixed size in one dimension, it will become apparent later that from the point of view of phase transitions often quantum fluctuations are irrelevant. One exception is provided by zero temperature quantum statistical systems in $d - 1$ dimensions which share many properties with classical statistical systems in d dimensions. At the same time, as a consequence of the discussion of Section 2.5, we see that the infinite β quantum correlation functions are the analogues of the infinite volume statistical correlation functions.

In the thermodynamic limit, l goes to infinity and therefore the partition function is related to the largest eigenvalue of the transfer matrix. The corresponding eigenvector plays the role of the ground state of a quantum hamiltonian.

Simple properties of the transfer matrix on a finite transverse lattice. If the size of the lattice transverse to the time axis is finite, using the positivity of the coefficients of the transfer matrix and adapting the arguments of Appendix A4, we can immediately derive an important property of its spectrum. We shall give all arguments in the case in which the spin distribution is discrete and the transfer matrix is a finite matrix but the arguments generalize to continuous spin distributions.

Since $\mathbf{T}$ is real symmetric, its eigenvalues are real. Let λ_0 be the eigenvalue with largest modulus and $|0\rangle$ the corresponding eigenvector:

$$\mathbf{T}\,|0\rangle = \lambda_0\,|0\rangle\,. \tag{23.5}$$

Let us call T_{ab} the matrix elements of the transfer matrix and v_a^0 the components of the eigenvector $|0\rangle$ and rewrite equation (23.5) in component form:

$$T_{ab}v_b^0 = \lambda_0 v_a^0\,. \tag{23.6}$$

Taking the scalar product with the vector $|0\rangle$, we derive:

$$\lambda_0 = \frac{v_a^0 T_{ab} v_b^0}{v_a^0 v_a^0}\,. \tag{23.7}$$

The denominator of the r.h.s. is a sum of positive terms and the matrix elements of $\mathbf{T}$ are positive. Taking the modulus of equation (23.7) we obtain an inequality:

$$|\lambda_0| \le \frac{\left|v_a^0\right| T_{ab} \left|v_b^0\right|}{\left|v_a^0\right|\left|v_a^0\right|}\,. \tag{23.8}$$

The r.h.s. of the inequality is the average of $\mathbf{T}$ taken with the vector $\left|v_a^0\right|$. A strict inequality would imply the existence of at least one positive eigenvalue larger than $|\lambda_0|$ in contradiction with the hypothesis. Equality combined with the property that $\mathbf{T}$ has *strictly* positive coefficients implies that all components of the vector $|0\rangle$ can be chosen non-negative. Then, according to equation (23.6), λ_0 is positive and all components of $|0\rangle$ are strictly positive. It follows that the eigenvalue cannot be degenerate because two vectors with strictly positive coefficients cannot be orthogonal. Let us summarize this result:

Because the transfer matrix is real symmetric and has strictly positive coefficients, the eigenvalue with largest modulus is positive, the corresponding eigenvector is unique and has strictly positive components.

In the large l (i.e. large time) limit, the free energy W is given by:

$$W \sim l \ln \lambda_0 . \qquad (23.9)$$

Since no crossing of levels can occur, λ_0 is a regular function of β. For example for the Ising model one finds:

$$\lambda_0 = e^{\beta J} \cosh \beta h + \left(e^{2\beta J} \sinh^2 \beta h + e^{-2\beta J} \right)^{1/2} . \qquad (23.10)$$

The connected 2-point spin correlation function behaves at large time separation as:

$$\langle S_i S_j \rangle_{\mathrm{c}} = \langle S_i S_j \rangle - \langle S_i \rangle \langle S_j \rangle \sim \left(\langle 0 \left| \mathbf{S} \right| 1 \rangle \right)^2 e^{-|i-j|/\xi}, \qquad (23.11)$$

in which $|1\rangle$ is the eigenvector corresponding to the second largest eigenvalue in modulus λ_1, which we have here assumed to be positive, and ξ is given by:

$$\xi^{-1} = \ln(\lambda_0 / \lambda_1). \qquad (23.12)$$

Finally the matrix $\mathbf{S}$ is a matrix diagonal in the spin representation which corresponds to a multiplication by the spin variable. We have assumed that its matrix element $\langle 0 \left| \mathbf{S} \right| 1 \rangle$ does not vanish. Again in the example of the Ising model one finds:

$$\xi^{-1} = 2 \tanh^{-1} \left[\frac{\left(\sinh^2 \beta h + e^{-4\beta J} \right)^{1/2}}{\cosh \beta h} \right] . \qquad (23.13)$$

We conclude from this analysis that in a spin model with finite range interactions and discrete distribution no phase transition can occur, the free energy is a regular function of the inverse temperature β and the correlation length ξ remains finite except for β infinite. These results can be generalized to short range interactions and continuous spin distributions.

23.2 The Infinite Transverse Size Limit: Ising-Like Systems

When the transverse size becomes infinite new phenomena may appear, which we shall examine first on the example of the d dimensional Ising model with nearest neighbour (n.n.) interactions. The partition function can be written:

$$Z = \sum_{\{S_{\mathbf{r}} = \pm 1\}} \exp \left(\beta J \sum_{\substack{\text{n.n.} \\ \mathbf{r}, \mathbf{r}' \in \mathbb{Z}^d}} S_{\mathbf{r}} S_{\mathbf{r}'} \right). \qquad (23.14)$$

Using bracket notations to represent matrix elements, we find for the transfer matrix:

$$\left\langle \{S'_\rho\} \left| \mathbf{T} \right| \{S_\rho\} \right\rangle = \exp \left[\beta J \left(\sum_{\rho \in \mathbb{Z}^{d-1}} S_\rho S'_\rho + \sum_{\substack{\text{n.n.} \\ \rho, \rho' \in \mathbb{Z}^{d-1}}} \left(S_\rho S_{\rho'} + S'_\rho S'_{\rho'} \right) \right) \right]. \qquad (23.15)$$

The Ising model is characterized by a discrete symmetry corresponding to spin reversal. Let us call $\mathbf{P}$ the corresponding matrix. In bracket notations:

$$\mathbf{P}\,|\{S_\rho\}\rangle = |\{-S_\rho\}\rangle\,. \tag{23.16}$$

We shall study the spectrum of the transfer matrix on a finite transverse lattice of linear size L, for L large. For L finite, the general results on matrices with positive elements apply. The matrix $\mathbf{P}$ commutes with the transfer matrix, and therefore $\mathbf{P}$ and $\mathbf{T}$ can be diagonalized simultaneously. The eigenvectors of $\mathbf{T}$ are eigenvectors of $\mathbf{P}$. The eigenvalues of $\mathbf{P}$ are ± 1. The eigenvector of largest eigenvalue of $\mathbf{T}$ has positive components. Equation (23.16) shows that $\mathbf{P}$ does not change the sign of the basis vectors. Therefore the eigenvector $|0\rangle$ with the largest eigenvalue is an eigenvector of $\mathbf{P}$ with eigenvalue $+1$,

$$\mathbf{P}\,|0\rangle = |0\rangle\,. \tag{23.17}$$

Let us now examine the infinite L limit at low and high temperature, i.e. at high and low β.

High temperature. At high temperature $(\beta \to 0)$, all matrix elements of $\mathbf{T}$ become equal and $\mathbf{T}$ becomes a projector onto the eigenvector $|0\rangle$ which has equal components on all spin configurations. All eigenvalues but one vanish, thus the correlation length vanishes. This property is independent of the volume and therefore previous results apply even for L infinite.

Low temperature analysis. At low temperature the dominant configurations correspond to all spins aligned. Let us call $|+\rangle$ and $|-\rangle$ the two vectors corresponding to all spin up and down respectively. At β strictly infinite both are eigenvectors of $\mathbf{T}$ corresponding to the twice degenerate largest eigenvalue. The fact that the eigenvalue is degenerate does not contradict the general result of Section 23.1, since at zero temperature, if we normalize the largest matrix elements of $\mathbf{T}$ to 1, all matrix elements but the diagonal elements $\langle -\,|\mathbf{T}|\,-\rangle$ and $\langle +\,|\mathbf{T}|\,+\rangle$ vanish, and the general arguments of Section 23.1 no longer apply. Note that $|+\rangle$ and $|-\rangle$ are not eigenvectors of $\mathbf{P}$ since:

$$\mathbf{P}\,|+\rangle = |-\rangle\,. \tag{23.18}$$

At low but finite temperature the eigenvectors can no longer be exactly $|+\rangle$ and $|-\rangle$ but have also necessarily components on configurations in which a finite number of spins have been flipped, as low temperature perturbation theory shows. However this still allows two degenerate eigenstates close to $|+\rangle$ and $|-\rangle$ and exchanged by $\mathbf{P}$. What is relevant is the large L behaviour of the matrix elements of $\mathbf{T}$ connecting $|+\rangle$ and $|-\rangle$, more precisely the ratio δ:

$$\delta = \frac{\langle +\,|\mathbf{T}|\,-\rangle}{\langle +\,|\mathbf{T}|\,+\rangle}\,. \tag{23.19}$$

At low temperature this ratio is related to the difference of the minimal energy of the configurations corresponding to the two different boundary conditions (b.c.). The cost in energy of imposing "twisted" b.c., i.e. spins up on one side and spins down on the other side, is proportional to the area of the surface on which the spins are flipped. Therefore δ behaves like:

$$\delta \propto \mathrm{e}^{-\beta J L^{d-1}}\,. \tag{23.20}$$

The eigenvectors and eigenvalues are qualitatively given by diagonalizing a 2×2 matrix τ in the $\{|+\rangle, |-\rangle\}$ subspace:

$$\tau = \begin{pmatrix} 1 & \delta \\ \delta & 1 \end{pmatrix}.$$

(23.21)

Two different situations arise depending on the dimension d of space.

(i) $d = 1$

Then δ is finite and the eigenvector $|0\rangle$ is the linear combination

$$|0\rangle = |+\rangle + |-\rangle,$$

which is also an eigenvector of $\mathbf{P}$ with the eigenvalue $+1$.

(ii) $d > 1$

In the infinite volume limit, δ vanishes, the highest eigenvalue is twice degenerate. The finite size correlation length diverges as:

$$\xi_L \propto e^{\beta J L^{d-1}}.$$

(23.22)

Clearly in the infinite volume limit no analytic continuation is possible between the high and low temperature situations, and therefore thermodynamic quantities must have at least one singularity in β at some finite value β_c. We shall argue in Section 23.4 that at low temperature a breaking of ergodicity occurs and thus β_c also corresponds to a phase transition in the dynamical sense.

Remarks

(i) This analysis of the infinite volume limit is qualitatively correct in the whole low temperature phase. At β_c the situation is different; an infinite number of eigenvalues have the same infinite volume limit λ_0. This situation will be studied in detail in Chapter 36.

(ii) We have here analyzed a lattice model, but we will show in Section 40.1 that in the case of models defined in continuum space, instantons lift the degeneracy of the ground state in the semiclassical limit. With the correspondence:

$$\beta \longmapsto \hbar^{-1}, \quad J \longmapsto \text{instanton action},$$

the analysis is then exactly the same.

(iii) From the previous analysis we derive a criterion of spontaneous symmetry breaking (SSB). We consider the ratio r:

$$r = \lim_{l \to \infty} \frac{\operatorname{tr} \mathbf{P}\, \mathbf{T}^l}{\operatorname{tr} \mathbf{T}^l},$$

(23.23)

in which $\mathbf{P}$ in general will be an operation of the symmetry group.

In the symmetric phase, the ground state of the transfer matrix is invariant under a group operation, therefore the ratio r is 1.

On the contrary, if the symmetry is spontaneously broken, $\mathbf{P}$ exchanges the various ground states and therefore r vanishes in the infinite volume limit. In the example of the Ising model one finds at low temperature:

$$r \propto e^{-\beta J L^{d-1}}.$$

The advantage of the consideration of the ratio (23.23) is that, if we take $l = L$, the linear size of the lattice, we have just to calculate the ratio of two partition functions on a d

dimensional lattice of linear size L, with different boundary conditions: the denominator corresponds to periodic b.c. in the time direction, the numerator corresponds to twisted b.c., i.e. the configurations on the two boundaries differ by a transformation of the symmetry group. In the case of the Ising model, twisted b.c. are antiperiodic b.c.. For $l = L$ this ratio naturally incorporates the condition that the thermodynamic limit is taken by sending the sizes in all dimensions to infinity in the same way. It represents, as we shall show below, the probability in some dynamics that the system can evolve from initial conditions in which almost all spins are up to a configuration in which most of the spins are reversed.

23.3 Order Parameter and Cluster Properties

When the ground state $|0\rangle$ (vacuum state in the sense of Particle Physics) is degenerate, the determination of the infinite volume correlation functions becomes a subtle question, which depends explicitly on the way the infinite volume limit is reached; in particular it may depend on boundary conditions. This sensitivity to boundary conditions is another characteristic of the several phase region.

Let us examine in this case the problem of cluster properties of correlation functions.

We consider the two phase region of an Ising like system and call $|+\rangle$, $|-\rangle$ the two ground states which are exchanged by $\mathbf{P}$ (equation (23.18)) and orthogonal. Any vector $|\alpha\rangle$ of the form:

$$|\alpha\rangle = \cos\alpha\,|+\rangle + \sin\alpha\,|-\rangle\,, \tag{23.24}$$

is also an eigenvector of the transfer matrix with the same eigenvalue. Let us assume that the boundary conditions select the vector $|\alpha\rangle$. Correlation functions are then obtained by calculating averages of matrices in the state $|\alpha\rangle$.

It is intuitive that the spin S_σ at site σ is, according to the definition given in the introduction, an order parameter. The corresponding matrix $\mathbf{S}$, whose matrix elements between two spin configurations are

$$\langle\{S'_\rho\}|\,\mathbf{S}\,|\{S_\rho\}\rangle = S_\sigma \prod_\rho \delta_{S_\rho S'_\rho}\,,$$

is odd under reflection:

$$\mathbf{P}^{-1}\mathbf{S}\mathbf{P} = -\mathbf{S}\,. \tag{23.25}$$

In the high temperature phase, in which the vacuum state satisfies:

$$\mathbf{P}\,|0\rangle = |0\rangle\,, \tag{23.26}$$

then:

$$\begin{aligned}
\langle 0\,|\mathbf{S}|\,0\rangle &= -\,\langle 0\,|\mathbf{P}^{-1}\mathbf{S}\mathbf{P}|\,0\rangle \\
&= -\,\langle 0\,|\mathbf{S}|\,0\rangle = 0\,.
\end{aligned} \tag{23.27}$$

In the low temperature phase let us define:

$$\langle +\,|\mathbf{S}|\,+\rangle = m\,, \tag{23.28}$$

in which m approaches 1 in the Ising model in the zero temperature limit. Then:

$$\langle -\,|\mathbf{S}|\,-\rangle = \langle +\,|\mathbf{P}^{-1}\mathbf{S}\mathbf{P}|\,+\rangle = -m\,. \tag{23.29}$$

We also have:

$$\langle + |\mathbf{S}| -\rangle = \langle - \left|\mathbf{P}^{-1}\mathbf{S}\mathbf{P}\right| +\rangle = -\langle - |\mathbf{S}| +\rangle. \tag{23.30}$$

Since $\mathbf{S}$ is a symmetric matrix:

$$\langle + |\mathbf{S}| -\rangle = 0. \tag{23.31}$$

The essential property used below is that $\mathbf{S}$, restricted to the $\{|+\rangle, |-\rangle\}$ subspace, is diagonal in the $\{|+\rangle, |-\rangle\}$ basis.

It follows from equations (23.28–23.31) in particular:

$$\langle \alpha |\mathbf{S}| \alpha\rangle = m \cos 2\alpha, \tag{23.32}$$

$$\langle \alpha |\mathbf{S}| \pi/2 + \alpha\rangle = m \sin 2\alpha. \tag{23.33}$$

Except for $\alpha = \pi/4 \pmod{\pi/2}$, the average of the spin does not vanish and this characterizes the several phase region.

Let us calculate what one would naively expect to be the connected 2-point correlation function of the spin, i.e. the 2-point function of $\mathbf{S} - <\mathbf{S}>$, for two points separated by a distance l in time but at the same position in the transverse direction. It is given by:

$$W^{(2)}(l) = \frac{\langle \alpha| (\mathbf{S} - m \cos 2\alpha)\,\mathbf{T}^l\,(\mathbf{S} - m \cos 2\alpha)|\alpha\rangle}{\langle \alpha |\mathbf{T}^l| \alpha\rangle}. \tag{23.34}$$

The transfer matrix $\mathbf{T}$ projects onto the ground states for large l:

$$\begin{aligned}
\mathbf{T}^l &= \lambda_0^l \left[(|+\rangle\langle+| + |-\rangle\langle-|) + O\left(\mathrm{e}^{-l/\xi}\right)\right], \\
&= \lambda_0^l \left[|\alpha\rangle\langle\alpha| + |\pi/2 + \alpha\rangle\langle\pi/2 + \alpha| + O\left(\mathrm{e}^{-l/\xi}\right)\right].
\end{aligned} \tag{23.35}$$

It then follows:

$$W^{(2)}(l) \sim m^2 \sin^2 2\alpha + O\left(\mathrm{e}^{-l/\xi}\right). \tag{23.36}$$

It is only for $\alpha = n\pi$ that the correlation functions satisfy cluster properties. The corresponding eigenvectors are then $|+\rangle$ and $|-\rangle$ which are exchanged by $\mathbf{P}$. According to the definition of Chapter 13, the reflection symmetry is spontaneously broken.

Note that correlation functions calculated by summing over all configurations (the analogues of expressions (2.59,2.63)) correspond to average over the two ground states and do not satisfy cluster properties. This problem can be solved by adding to the configuration energy an infinitesimal interaction term coupled to the order parameter which favours one phase and lifts the degeneracy. For example in a spin system, to generate correlation functions which satisfy cluster properties in the infinite volume limit, it is possible to start with a finite volume, to add a constant magnetic field, to take first the infinite volume limit, and then the zero field limit. The ground state, $|+\rangle$ or $|-\rangle$ and low temperature, is chosen by the sign of the magnetic field.

In Subsection 6.1.2 we have added a small constant part to the sources which allowed us to define the generating functional of connected correlation functions even in the degenerate case.

One may wonder about the physical interpretation of this procedure: in next section we shall argue that a system in the spontaneously broken phase is no longer ergodic. Once prepared in one phase, it remains for ever. Therefore in the two phase region of a spin system one should only average over configurations which fluctuate around the configurations with all spins up or all spins down.

23.4 Stochastic Processes and Phase Transitions

To construct a stochastic process which converges toward the equilibrium distribution of the Ising model we can use detailed balance (see Appendix A4) and set for the transition probability p of a configuration $\{S'_\mathbf{r}\}$ towards a configuration $\{S'_r\}$, $\mathbf{r}$ belonging now to a lattice Ω of volume L^d in $\mathbb{Z}^d$:

$$
\begin{cases}
p\left(S_\mathbf{r}, S'_\mathbf{r}\right) = e^{-\beta\left[E\left(S'_\mathbf{r}\right) - E\left(S_\mathbf{r}\right)\right]} & \text{for } E\left(S_\mathbf{r}\right) < E\left(S'_\mathbf{r}\right), \\
p\left(S_\mathbf{r}, S'_r\right) = 1 \text{ otherwise},
\end{cases}
\tag{23.37}
$$

in which $E\left(S_\mathbf{r}\right)$ is the configuration energy:

$$
E\left(S\right) = \sum_{\substack{\text{n.n.} \\ \mathbf{r},\mathbf{r}' \in \Omega \subset \mathbb{Z}^d}} -J S_\mathbf{r} S'_\mathbf{r}.
\tag{23.38}
$$

For the argument which follows the precise description of which configurations are directly connected by the matrix $\mathbf{p}$ is irrelevant provided the system is globally connected. The relevant property is that the probability to go from a configuration to a configuration of higher energy is, at low temperature, of the order of $e^{-\beta\Delta E}$, in which ΔE is the energy difference.

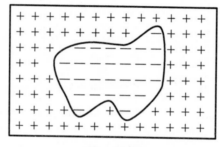

Fig. 23.1

Therefore, at low temperature, if we start from a configuration in which all spins are $+1$ for example, the probability of creating a bubble of minus spins is proportional to $e^{-\beta JA}$, in which A is the area of the bubble surface (figure 23.1). If a large fluctuation creates a bubble which fills half of the total volume, then of course there is a probability $1/2$ that afterwards all spins become equal to -1. Therefore the probability of reversing most of the spins is, at low temperature, related to the area of a surface which divides the volume Ω into two halves. Since Ω has linear size L, this probability is of the form $e^{-\sigma L^{d-1}}$.

For $d = 1$ the system is ergodic, and, as we have shown, no phase transitions can occur.

For $d > 1$ the same mechanism which allows the existence of several phases at low temperature is responsible for the breaking of ergodicity.

Note that the argument generalizes only to local dynamics, i.e. dynamics in which the probability of change of one spin on the lattice is only influenced by its neighbours (for more details see Appendix A36.1.3).

23.5 Continuous Symmetries

Let us now briefly discuss a model which has a continuous symmetry to here exhibit some difference from the case of discrete symmetries. We again consider a classical spin system, but the spins $\mathbf{S_r}$ are N-component vectors interacting through a two-body n.n. ferromagnetic interaction. The model then has a continuous $O(N)$ symmetry. The partition function is:

$$Z = \int \prod_{\mathbf{r} \in \mathbb{Z}^d} d\mathbf{S_r} \, \delta \left(\mathbf{S_r^2} - 1 \right) \exp \left[\sum_{\mathbf{r,r'} \text{n.n.} \in \mathbb{Z}^d} \beta \, \mathbf{S_r} \cdot \mathbf{S_{r'}} \right]. \tag{23.39}$$

Such a model has already been proposed as a regularization of the $O(N)$ non-linear σ-model in Chapter 14.

At high temperature ($\beta \to 0$), as in the case of the Ising model, the ground state of the transfer matrix has uniform components on all configuration vectors, and is therefore invariant under the transformations of the $O(N)$ symmetry group.

To understand the structure of low temperature phases, we now use the equivalent of ratio (23.23). The symmetry operation is here a rotation $R(\alpha)$ of angle α. Defining:

$$r_L = \frac{Z_L(\alpha)}{Z_L(0)}, \tag{23.40}$$

we examine the behaviour of r_L for β large in the large L limit. The partition function $Z_L(\alpha)$ is the partition function on a d dimensional lattice, with twisted b.c. In the time direction we impose:

$$\mathbf{S}_{t=L,\rho} \cdot \mathbf{S}_{t=0,\rho} = \cos \alpha, \qquad \rho \in \mathbb{Z}^{d-1}. \tag{23.41}$$

For convenience we take periodic b.c. in all other dimensions.

At low temperature, the configurations with minimal energy correspond to take the spins aligned in $d-1$ dimensions, and rotating by an angle α/L between two adjacent sites along the time axis. This has to be contrasted with the case of the discrete symmetry, in which the transition between the two configurations imposed by the b.c. occurs between two sites.

The cost in energy ΔE due to the rotation is thus:

$$\Delta E = L^{d-1} \times L \times [\cos(\alpha/L) - 1] \sim -\alpha^2 L^{d-2}/2, \tag{23.42}$$

and therefore:

$$r_L \propto \exp\left(-\alpha^2 \beta L^{d-2}/2 \right). \tag{23.43}$$

In the case of a continuous symmetry, it is easier to pass from one minimum of the energy to another. This property has the direct consequences that it is more difficult to break the symmetry and that Goldstone modes appear in the phase of SSB. For $d \le 2$, r_L has a finite limit and the symmetry is never broken (although, as we shall see, there is a phase transition without ordering for $d = 2$, $N = 2$). This result, for which we have given a heuristic argument, can be proven rigorously (Mermin–Wagner–Coleman theorem).

For $d > 2$ SSB occurs at low temperature. There exists some finite temperature T_c, at which a phase transition occurs.

Note that if we consider a system with a longitudinal size $l \ne L$ we obtain instead:

$$r \propto \exp\left(-\alpha^2 \beta L^{d-1}/2l \right). \tag{23.44}$$

Hence, using the results of Sections 3.3,3.4, we conclude that the finite size correlation length ξ_L behaves like:

$$\xi_L \propto \beta L^{d-1}. \tag{23.45}$$

To relate this result with the possibility of a phase transition we remember that if we assume that the correlation length is large but much smaller than the transverse size of the lattice then we have the estimate:

$$\langle S \rangle^2 \propto L^{-2d} \sum_{\mathbf{r},\mathbf{r}'} \langle \mathbf{S_r} \cdot \mathbf{S_{r'}} \rangle \propto L^{-2d} \sum_{\mathbf{r},\mathbf{r}'} \exp\left(-|\mathbf{r}-\mathbf{r}'|/\xi_L\right),$$

and therefore:

$$\langle S \rangle \propto (\xi_L/L)^d. \tag{23.46}$$

The assumption implies that $\langle S \rangle$ goes to zero in the large volume limit. A phase transition with ordering is only possible if:

$$\xi_L \underset{L\to\infty}{\geq} L \quad \Rightarrow d \geq 2.$$

The limiting case $\xi_L \propto L$, i.e. $d=2$, is quite subtle and we shall show eventually that it characterizes the critical temperature of a second order phase transition with vanishing spontaneous magnetization in the low temperature phase.

These high and low temperature analyses provide information about the existence of a phase transition, and the nature of the phases. However nothing can be inferred about the phase transition itself, or the behaviour of thermodynamic quantities in the neighbourhood of the critical temperature. These problems will be examined in the coming chapters.

Bibliographical Notes

The proof of spontaneous symmetry breaking of systems with discrete symmetries is given in
  R. Peierls, *Phys. Rev.* 54 (1938) 918.
The impossibility of spontaneous breaking of continuous symmetries in one and two dimensions has been proven by
  N.D. Mermin and H. Wagner, *Phys. Rev. Lett.* 17 (1966) 1133; S. Coleman, *Commun. Math. Phys.* 31 (1973) 259.
Two relevant textbooks are
  S.K. Ma, *Statistical Mechanics* (World Scientific, Singapore 1985);
  G. Parisi, *Statistical Field Theory* (Addison-Wesley, New York 1988).
The presentation of this chapter follows
  J. Zinn-Justin, *Basko Polje Summer School 1976*, Saclay preprint DPh-T/76-99 unpublished, and in *Recent Advances in Field Theory and Statistical Mechanics*, Les Houches 1982, Appendix C, J.-B. Zuber and R. Stora eds. (North-Holland, Amsterdam 1984).
The replica trick has been introduced in
  S.F. Edwards and P.W. Anderson, *J. Phys.* F5 (1975) 965.
For the spin glass phase transition see for instance
  M. Mézard, G. Parisi and M. Virasoro, *Spin Glass and Beyond* (World Scientific, Singapore 1987).

APPENDIX 23
QUENCHED AVERAGES

In this work we discuss only so-called annealed averages, i.e. we average over random configurations of a system which explores the whole available phase space in the course of time evolution. This leads to the concept of partition function. However another class of problems is of quite different nature, disorder is related to frozen degrees of freedom, or at least degrees of freedom (for instance impurities in crystals) which do not evolve during the time of observation. When a relaxation time can be defined (see Chapter 35), the relaxation time is large compared to the time of observation.

Then it is no longer the partition function which has to be averaged over disorder, but directly the physical observables, a procedure usually called *quenched average*. Let us here give for completeness the simple example of the gaussian model in a quenched random magnetic field with gaussian distribution. For simplicity we assume continuum space and a continuous spin distribution $\sigma(x)$ corresponding to a local free action:

$$S(\sigma) = S_0(\sigma) - \int \mathrm{d}^d x \, H(x)\sigma(x)$$

$$S_0(\sigma) = \tfrac{1}{2} \int \mathrm{d}^d x \left[(\nabla\sigma(x))^2 + \mu^2 \sigma^2(x) \right].$$

For each magnetic field distribution the partition function is given by

$$Z = \int [\mathrm{d}\sigma(x)] \exp\left[-S(\sigma)\right].$$

We assume the magnetic fields at different points uncorrelated with a gaussian distribution characterized by the 1- and 2-point functions:

$$\overline{H(x)} = h, \qquad \overline{H(x)H(y)}_{\mathrm{c}} = w^2 \delta^d(x-y),$$

h and w being two constants.

In this model we can immediately calculate the statistical average of the spin variable, at fixed field distribution, at a point x

$$\langle \sigma(x) \rangle = \int \mathrm{d}^d y \, \Delta(x-y) H(y),$$

where $\Delta(x)$ is the σ-field propagator. If we now average $\langle \sigma(x) \rangle$ over the position x in a large volume Ω, we find

$$\Omega^{-1} \int \mathrm{d}^d x \, \langle \sigma(x) \rangle = \int \mathrm{d}^d x \, \Delta(x) \Omega^{-1} \int \mathrm{d}^d y \, H(y).$$

Because the propagator in absence of disorder is invariant under translation we now find in the r.h.s. the average of the magnetic field. The result is

$$\Omega^{-1} \int \mathrm{d}^d x \, \langle \sigma(x) \rangle = \frac{1}{\mu^2} \overline{H(x)} = \frac{h}{\mu^2}.$$

We can also calculate the average over the sample of

$$\Omega^{-1}\int d^dy\,\langle\sigma(y)\rangle\,\langle\sigma(x+y)\rangle = \Omega^{-1}\int d^dy\,d^dz\,d^dt\,\Delta(y-z)\Delta(x+y-t)H(z)H(t)$$

$$= \Omega^{-1}\int d^dy\,d^dz\,\Delta(y)\Delta(x+y+z)\int d^dt\,H(z+t)H(t)$$

Again translation invariance, in the absence of field, has reduced the space average to a field average:

$$\Omega^{-1}\int d^dt\,H(z+t)H(t) = \overline{H(z+t)H(t)} = h^2 + w^2\delta^d(z).$$

It follows

$$\Omega^{-1}\int d^dy\,\langle\sigma(y)\rangle\,\langle\sigma(x+y)\rangle = \frac{h^2}{\mu^4} + \frac{w^2}{(2\pi)^d}\int\frac{e^{ipx}\,d^dp}{(p^2+\mu^2)^2}.$$

Note on the other hand that the σ-field 2-point function is independent of the magnetic field

$$\langle\sigma(y+x)\sigma(y)\rangle = \frac{1}{(2\pi)^d}\int\frac{e^{ipx}\,d^dp}{p^2+\mu^2}.$$

The replica trick. In this simple example since the observables can be calculated explicitly, it is easy to average over the quenched random variables. This is not the general situation of course, so an algebraic trick has been invented to overcome this problem. If we want to average correlation functions we have to average the free energy $W = \ln Z$ instead of the partition function. Then the following trick is used

$$\ln Z = \lim_{n\to 0}\frac{Z^n-1}{n}.$$

We do not know how to calculate Z^n for general real n, but we know how to calculate it for integer n. It is sufficient to replicate n copies of the initial variables. In our example we introduce a field σ_i with n components and find

$$Z^n(J) = \int\prod_{i=1}^{n}[d\sigma_i(x)]\exp\left[-\sum_i S(\sigma_i) + \sum_i\int d^dx\,J(x)\sigma_i(x)\right].$$

The gaussian disorder average is easy to perform

$$\overline{Z^n(J)} = \int[d\sigma_i(x)]\exp\left[-\sum_i S_0(\sigma_i)\right.$$

$$\left. +\tfrac{1}{2}w^2\int d^dx\sum_{i,j}\sigma_i(x)\sigma_j(x) + \int d^dx\sum_i(h+J(x))\sigma_i(x)\right].$$

The subtle question is then to take the $n = 0$ limit. For instance to calculate the field expectation value we minimize the potential. We find the replica symmetric solution:

$$\langle\sigma_i(x)\rangle = \frac{h}{\mu^2 - n w^2}.$$

In this simple example we set $n = 0$ and obtain the correct result

$$\overline{\langle \sigma(x) \rangle} = \lim_{n \to 0} \frac{1}{n} \sum_i \sigma_i = \frac{h}{\mu^2}.$$

Notice, however, that even here the solution does not make sense for $n > \mu^2/\omega^2$, and therefore arguments which state that under certain conditions analytic functions known for all integers are uniquely defined, do not apply. Therefore, basically the domain of validity of the replica trick is unknown.

Similarly the replica propagator can be calculated

$$\langle \tilde{\sigma}(p)\tilde{\sigma}(-p) \rangle = \frac{\delta_{ij}}{p^2 + \mu^2} + \frac{\omega^2}{(p^2 + \mu^2)(p^2 + \mu^2 - n\omega^2)}.$$

Summing over i, j, dividing by n and taking the $n = 0$ limit yields the correct result. Notice that for $m = 0$ small the replica-propagator for $n = 0$ is more singular than the free field propagator at low momentum. This has implications for critical phenomena in disordered systems.

For less trivial models the problem is very complicated. The famous spin-glass model is difficult to solve even in the mean approximation we discuss in the next chapter, and the mean field solution relies on a breaking of the symmetry between field replica, as shown by Parisi.

A final remark: to calculate correlations due to the disorder averaging, we have to evaluate instead quantities like

$$\overline{W(J_1)W(J_2)} = \overline{\ln Z(J_1) \ln Z(J_2)}.$$

The replica trick obviously generalizes with $2n$ replicas.

24 MEAN FIELD THEORY FOR FERROMAGNETIC SYSTEMS

In Chapter 23 we have discussed the existence of phase transitions in ferromagnetic systems by comparing the phase structure at low and high temperatures. However, we have used methods which are not valid near the transition itself and thus provide no information about the behaviour of thermodynamical quantities in the neighbourhood of the critical temperature. For this purpose we develop in this chapter a very simple approximation scheme, the mean field approximation. We then distinguish between first and second order phase transitions. In the latter case the correlation length diverges. This has important physical consequences: In the mean field approximation the singular behaviour of correlation functions near T_c and in small field is universal, i.e. does not depend on the details of the interactions.

The mean field approximation can be introduced by various methods, as a partial summation of the high temperature expansion (see Appendix A24), as a result of a variational method, or as the leading order in a steepest descent calculation. We shall adopt the latter point of view because it will allow us to calculate systematic corrections to the mean field approximation and therefore to discuss its validity. The role of dimension four will then emerge.

In the appendix we construct the mean field expansion for general lattice models.

24.1 Ising-like Ferromagnetic Systems

We first consider the critical behaviour of a classical spin system in a d-dimensional lattice with an attractive, translation invariant, short range, two-body interaction. The spin variable on a site i is called S_i. We assume that the system has a $S \mapsto -S$ symmetry. To discuss the effect of a magnetic field and to construct a generating functional of correlation functions it is useful to introduce in addition a lattice point dependent magnetic field.

We call $\mathcal{H}$ the energy of a spin configuration, β the inverse temperature and $\mathrm{d}\rho(S)$ the one spin distribution which weights the spin configurations at each site.

The partition function can then be written:

$$Z(H) = \int \prod_i \mathrm{d}\rho\,(S_i) \exp\left[-\beta\mathcal{H}(S)\right], \qquad (24.1)$$

with:

$$-\beta\mathcal{H}(S) = \sum_{ij} V_{ij} S_i S_j + \sum_i H_i S_i. \qquad (24.2)$$

It is technically convenient to also assume that V_{ii} vanishes (see Appendix A24). The interactions are ferromagnetic if V_{ij} is non-negative. We have already met expressions similar to (24.1) when we constructed the Feynman–Kac representation of the quantum mechanical partition function $\mathrm{tr}\,\mathrm{e}^{-\beta H}$ in Chapter 2 or introduced in field theory a lattice regularization of the generating functional of euclidean correlation functions. Note an important difference in point of view between methods inspired by high temperature considerations and field theoretical perturbation theory: at high temperature the measure $\mathrm{d}\rho(S)$ is kept fixed and the term $V_{ij} S_i S_j$, which connects different lattice sites, is the interaction term and is treated as a perturbation; in Field Theory the term $V_{ij}S_iS_j$ is rather the analogue of the kinetic term while the measure $d\rho(S)$ contains the interactions. These remarks will be clarified later.

High temperature series expansion. The simplest method to calculate $Z(H)$ is to expand expression (24.1) in powers of V (high temperature expansion) since all terms in the series can be expressed in terms of the moments of the one spin distribution in a magnetic field:

$$Z(H)/Z_0(H) = 1 + \sum_{ij} V_{ij}\langle S_i\rangle\langle S_j\rangle + \frac{1}{2!}\sum_{ijkl} V_{ij}V_{kl}\langle S_i\rangle\langle S_j\rangle\langle S_k\rangle\langle S_l\rangle + \cdots , \quad (24.3)$$

with

$$Z_0(H) = \prod_i z(H_i), \quad (24.4)$$

$$z(h) = \int d\rho(S)\,e^{Sh}, \quad (24.5)$$

$$\langle S_i^n\rangle = z^{-1}(H_i)\left(\frac{\partial}{\partial H_i}\right)^n z(H_i). \quad (24.6)$$

However this expansion is useful only if the leading term, the infinite temperature result, already is qualitatively correct: the expansion gives information about the high temperature phase and necessarily diverges at the critical temperature. It provides a full qualitative description of the phase structure only when the critical temperature vanishes. Therefore we now introduce another approximation scheme, which diagrammatically corresponds to perform a loopwise summation of the high temperature series expansion (see Appendix A24).

Remark. For what follows it is convenient to introduce the function $A(H)$:

$$A(H) = -\ln z(H). \quad (24.7)$$

For the Ising model where $S_i = \pm 1$ one finds for example:

$$A(H) = -\ln\cosh H.$$

Due to the symmetry of the distribution, $A(H)$ is even and decreasing for $H > 0$. Moreover $A''(H) < 0$ (Schwartz's inequality) and thus $-A(H)$ is convex. Finally we assume that the distribution decreases faster than a gaussian for large values of the spin S. It is then easy to verify that $z(H)$ is an entire function which implies that $A(H)$ is regular on the real axis and in some neighbourhood of the origin. Moreover:

$$\lim_{|H|\to\infty} A(H)/H^2 = 0. \quad (24.8)$$

24.2 Mean Field Expansion

Since the partition function can easily be calculated when all points are disconnected, a simple idea is to write the term which in the configuration energy connects spins on different sites as an integral over disconnected terms (Hubbard's transformation). More explicitly:

$$\int \prod_i \mathrm{d}X_i \exp\left(-\frac{1}{4}\sum_{ij} X_i V_{ij}^{-1} X_j + \sum_i S_i X_i\right) \propto \exp\left(\sum_{i,j} V_{ij} S_i S_j\right). \qquad (24.9)$$

The partition function becomes:

$$Z(H) = \int \prod_i \mathrm{d}X_i \exp\left(-\frac{1}{4}\sum_{ij} X_i V_{ij}^{-1} X_j\right) \prod_i \int \mathrm{d}\rho\left(S_i\right) \exp\left[\sum_i S_i\left(H_i + X_i\right)\right]. \qquad (24.10)$$

The sum over spin configurations can then be performed:

$$Z(H) = \int \prod_i \mathrm{d}X_i \exp\left[-\frac{1}{4}\sum_{ij} X_i V_{ij}^{-1} X_j - \sum_i A\left(X_i + H_i\right)\right], \qquad (24.11)$$

where $A(H)$ has been defined by equation (24.7).

Note that identity (24.9) is *a priori* valid only for positive matrices V_{ij}. Actually, in the case we apply it, V is not positive in general, and equation (24.9) should only be considered as a formal algebraic identity.

We now evaluate expression (24.11) by the steepest descent method. At leading order we replace X_i by a solution of the saddle point equation:

$$\frac{1}{2}\sum_j V_{ij}^{-1} X_j + A'\left(X_i + H_i\right) = 0. \qquad (24.12)$$

The meaning of this approximation is that we have replaced the interaction which connects the sites by the best (in a sense which will become clearer below) equivalent magnetic field. A more detailed analysis shows that this approximation becomes exact when the dimension d of space becomes large, so that the action of all sites on a given site can indeed be replaced by a "mean" magnetic field (this property is related to the central limit theorem).

The advantage of this algebraic formulation of mean field theory is that it allows a systematic discussion of the corrections to the mean field approximation.

Before discussing the mean field approximation it is useful to introduce a parameter which characterizes the expansion around mean field theory. Let us replace the coefficients V_{ij} in expression (24.2) by V_{ij}/l and the spin S_i by the sum of l independent spins with the same distribution $\mathrm{d}\rho(S)$:

$$S_i = \sum_{k=1}^{l} S_i^{(k)}.$$

Then:

$$\int \mathrm{d}\rho(S)\, \mathrm{e}^{HS} \longmapsto \left[\int \mathrm{d}\rho(S)\, \mathrm{e}^{HS}\right]^l.$$

The case $l = 1$ corresponds to the initial distribution. The partition function becomes:

$$Z(H) = \int \prod_i \mathrm{d}X_i \exp\left\{ -l\left[\frac{1}{4}\sum_{ij} X_i V_{ij}^{-1} X_j + \sum_i A\left(X_i + H_i\right) \right] \right\}. \qquad (24.13)$$

This expression clearly shows that a steepest descent calculation of the partition function generates an expansion in powers of $1/l$. Mean field theory is analogous to the tree approximation of quantum field theory.

Let us, for convenience, rewrite expression (24.13), setting

$$X_i + H_i = \varphi_i. \qquad (24.14)$$

This leads to:

$$Z(H) = \int \prod_i \mathrm{d}\varphi_i \exp\left\{ -l\left[\frac{1}{4}\sum_{ij} \left(\varphi_i V_{ij}^{-1}\varphi_j - 2\varphi_i V_{ij}^{-1} H_j + H_i V_{ij}^{-1} H_j\right) + \sum_i A\left(\varphi_i\right) \right] \right\}. \qquad (24.15)$$

24.3 Mean Field Approximation

In terms of the lattice variables φ_i, the saddle point equation (24.12) reads:

$$\varphi_i - H_i + 2\sum_j V_{ij} A'\left(\varphi_j\right) = 0. \qquad (24.16)$$

The Weiss free energy $W(H)$ is then:

$$W(H) \equiv \frac{1}{l}\ln Z = -\sum_{ij} \frac{1}{4}\left(\varphi_i - H_i\right) V_{ij}^{-1}\left(\varphi_j - H_j\right) - \sum_i A\left(\varphi_i\right). \qquad (24.17)$$

The magnetization M_i in the mean field approximation (24.16) is:

$$M_i = \frac{\partial W}{\partial H_i} = -A'\left(\varphi_i\right). \qquad (24.18)$$

The thermodynamic potential $\Gamma(M)$, Legendre transform of $W(H)$ is then given by:

$$\Gamma(M) = \sum_i M_i H_i - W(H) = -\sum_{ij} M_i V_{ij} M_j + \sum_i B\left(M_i\right), \qquad (24.19)$$

in which $B(M)$ is the Legendre transform of $-A\left(\varphi\right)$:

$$B(M) - A\left(\varphi\right) = \varphi M. \qquad (24.20)$$

Again in the example of the Ising model we find:

$$B(M) = \tfrac{1}{2}\left(1 + M\right)\ln\left(1 + M\right) + \tfrac{1}{2}\left(1 - M\right)\ln\left(1 - M\right). \qquad (24.21)$$

The relation between magnetic field and magnetization can also be written:

$$H_i = \frac{\partial \Gamma}{\partial M_i} = -2 \sum_j V_{ij} M_j + \frac{\partial B}{\partial M_i}. \tag{24.22}$$

In zero magnetic field, the magnetization, i.e. the spin average, is given by an extremum of the thermodynamic potential $\Gamma(M)$. Furthermore, since the partition function in zero field is $\exp[-l\Gamma(M)]$, the dominant saddle points correspond to minima of $\Gamma(M)$.

In the case of translation invariant systems, we expect the magnetization to be uniform: $M_i = M$. Let us call Ω the number of lattice sites, then:

$$\Omega^{-1}\Gamma(M) = -VM^2 + B(M), \tag{24.23}$$

in which V is a parameter defined by:

$$V = \sum_i V_{ij} > 0, \tag{24.24}$$

which is finite because the interaction is short range, and proportional to the inverse temperature β.

The Legendre transformation implies the relation:

$$B''(M) = -1/A''(\varphi). \tag{24.25}$$

Because $-A(\varphi)$ is a convex function of φ, $B(M)$ is also convex. It is analytic in the neighbourhood of the origin. Moreover the property (24.8) implies that for $|M|$ large enough, $\Gamma(M)$ in equation (24.23) is an increasing function.

Let us now find the minima of $\Gamma(M)$ when the temperature, and therefore also V, vary. For V small (high temperature) VM^2 is negligible, and the r.h.s. of equation (24.23) is convex. The minimum of $\Gamma(M)$ is $M = 0$, the magnetization vanishes. When V increases, in general at some value of V other local minima appear which eventually become the absolute minima of $\Gamma(M)$. When this occurs, the value of the magnetization M discontinuously jumps from zero to a finite value corresponding to this new absolute minima. The system undergoes a *first order phase transition*. Figure 24.1 describes the evolution. Fluctuations around the saddle point are governed by the value of the second derivative of the potential at the minimum. Generically in such a case the second derivative is strictly positive and therefore the correlation length, which, as we know from Field Theory calculations and shall again see explicitly later, is proportional to $\Omega^{1/2}[\Gamma''(M)]^{-1/2}$ at the minimum, remains finite.

Although first order phase transitions are common, they are not particularly interesting for us. When the correlation length remains finite, no universality emerges. At the same time, because the fluctuations are not critical, mean field theory gives a satisfactory qualitative description of the physics.

However if no absolute minimum appears at a finite distance from the origin, finally at a critical temperature T_c corresponding to the value V_c of V:

$$2V_c = B''(0), \tag{24.26}$$

the origin ceases to be a minimum of the potential, and below this temperature two minima move continuously away from the origin (see figure 24.2). Since the magnetization remains continuous at V_c, we have a *second order phase transition*. Because at V_c the second derivative of $\Gamma(M)$ vanishes, the correlation length diverges.

This is the situation we will analyze systematically from now on, first in the framework of the mean field approximation and then by considering corrections coming from higher order terms in the mean field expansion.

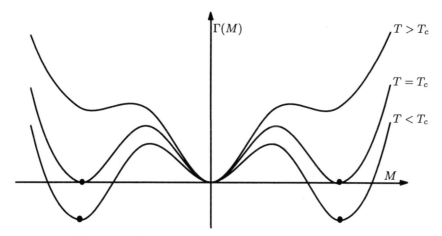

Fig. 24.1   Free energy: first order phase transition.

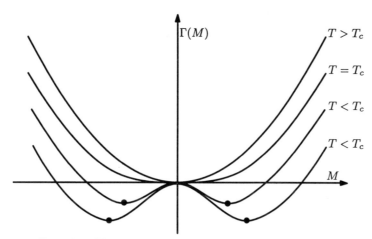

Fig. 24.2   Free energy: second order phase transition.

24.4 Universality within the Mean Field Approximation

We now examine the behaviour of various thermodynamical quantities for a temperature T close to T_c, i.e. V close to V_c. Since the transition is continuous, the magnetization is small and we can expand $\Gamma(M)$, which as $A(\varphi)$ is a smooth even function, in a Taylor series in M. We parametrize $\Gamma(M_i)$ as:

$$\Gamma(M_i) = -\sum_{ij} V_{ij} M_i M_j + \sum_i \left(\frac{a}{2!} M_i^2 + \frac{b}{4!} M_i^4 + \cdots \right). \qquad (24.27)$$

The corresponding expansion for $A(\varphi)$ then is:

$$A(\varphi) = -\frac{1}{2a} \varphi^2 + \frac{b}{24a^4} \varphi^4 + \cdots . \qquad (24.28)$$

The convexity of $B(M)$ implies that a is positive. The parameter b is also generically positive, because we have assumed that no first order transition occurs at higher temperature. In the Ising model one finds:

$$a = 1, \qquad b = 2 .$$

The critical value V_c, in the mean field approximation, is then:

$$V_c = a/2 \,. \tag{24.29}$$

For $V > V_c$, we find a spontaneous magnetization M:

$$M \sim [12\,(V - V_c)/b]^{1/2} \quad \text{for} \quad |V - V_c| \ll 1 \,. \tag{24.30}$$

Near the critical temperature T_c, the magnetization has thus a power law behaviour with a magnetic exponent β: :

$$M \sim (T_c - T)^\beta \,, \qquad \beta = 1/2 \,. \tag{24.31}$$

The inverse of the magnetic susceptibility $\chi = \partial M/\partial H$ is the second derivative of $\Gamma(M)$ with respect to M. In zero field we find:

$$\begin{aligned}
\chi_+^{-1} &= 2\,(V_c - V) & T > T_c \,, \\
\chi_-^{-1} &= 4\,(V - V_c) & T < T_c \,.
\end{aligned} \tag{24.32}$$

The magnetic susceptibility therefore diverges at T_c with susceptibility exponents γ, γ':

$$\begin{aligned}
\chi_+ &\sim C_+\,(T - T_c)^{-\gamma} \,, & \gamma &= 1 \,, \\
\chi_- &\sim C_-\,(T_c - T)^{-\gamma'} \,, & \gamma' &= 1 \,,
\end{aligned} \tag{24.33}$$

and:

$$C_+/C_- = 2 \,. \tag{24.34}$$

For V close to V_c and small constant applied magnetic field H, the equation of state, which is the relation between field, temperature and magnetization, is:

$$H = \left.\frac{\partial \Gamma}{\partial M_i}\right|_{M_i = M} = 2\left(\frac{a}{2} - V\right) M + \frac{b}{6}M^3 + O\left(M^5\right) \,. \tag{24.35}$$

For H, $T - T_c$, M small, rescaling field, temperature and magnetization, we can cast the relation into a universal scaling form:

$$H = M^\delta f\left((T - T_c)M^{-1/\beta}\right) \,, \tag{24.36}$$

where the function $f(x)$ is simply:

$$f(x) = 1 + x \,, \tag{24.37}$$

and the exponent δ has the "classical" or mean-field value:

$$\delta = 3 \,. \tag{24.38}$$

From the functional (24.27) we can also derive the 2-point correlation function $W_{ij}^{(2)}$:

$$W_{ij}^{(2)} = \left(\frac{\delta^2 \Gamma}{\partial M_i \partial M_j}\right)^{-1} \,,$$

whose Fourier transform $\widetilde{W}^{(2)}(q)$ is:

$$\widetilde{W}^{(2)}(q) = \sum_j W_{ij}^{(2)} \exp\left(i\mathbf{q}\cdot\mathbf{r}_{ij}\right) = \left[a + \frac{b}{2}M^2 - 2\tilde{V}(q)\right]^{-1}, \tag{24.39}$$

with the definition:

$$\tilde{V}(q) = \sum_j V_{ij} \exp\left(i\mathbf{q}\cdot\mathbf{r}_{ij}\right). \tag{24.40}$$

The momentum $\mathbf{q}$ has, by definition, components satisfying $|q_\mu| < \pi$. The potential V_{ij} is short ranged, positive and translation invariant. We assume in addition that it has the cubic symmetry of the lattice (this assumption is convenient but not essential since we can always make a linear transformation on the coordinates). Then the expansion of $\tilde{V}(q)$ for q small has the form:

$$\tilde{V}(q) = V\left(1 - \rho^2 q^2\right) + O\left(q^4\right), \tag{24.41}$$

in which q^2 is the length squared of the vector $\mathbf{q}$ and therefore $O(d)$ symmetric, and ρ^2 some constant. Equation (24.41) actually defines precisely what we mean by short range potential: the potential has to decrease at large distances in such a way that its Fourier transform has the expansion (24.41). Then for q small, $\widetilde{W}^{(2)}(q)$ can also be expanded:

$$\widetilde{W}^{(2)}(q) = \widetilde{W}^{(2)}(0)\left[1 + q^2\xi^2 + O\left(q^4\right)\right]^{-1}. \tag{24.42}$$

This shows that the propagator has a free field, or Ornstein and Zernike form. We have parametrized $\widetilde{W}^{(2)}(q)$ by introducing ξ^2 the second moment of W_{ij}. The length ξ characterizes, up to a numerical factor, the exponential decay of the correlation functions, and can be taken as a measure of the correlation length. From equations (24.39,24.42) we find:

$$\xi^{-2} = \frac{1}{2V\rho^2}\left(a + \frac{b}{2}M^2 - 2V\right). \tag{24.43}$$

In zero magnetic field this yields:

$$\begin{aligned} \xi_+^{-2} &= \rho^{-2}\left(1 - V/V_c\right) && \text{for } T > T_c, \\ \xi_-^{-2} &= 2\rho^{-2}\left(V/V_c - 1\right) && \text{for } T < T_c. \end{aligned} \tag{24.44}$$

Introducing general correlation length exponents ν, ν':

$$\begin{aligned} \xi_+ &= f_+\left(T - T_c\right)^{-\nu}, \\ \xi_- &= f_-\left(T_c - T\right)^{-\nu'}, \end{aligned} \tag{24.45}$$

we derive from relations (24.44) the classical values of exponents:

$$\nu = \nu' = \frac{1}{2},$$

and the ratio of amplitudes

$$f_+/f_- = \sqrt{2}. \tag{24.46}$$

Finally if we assume that the 2-point function behaves in general at T_c like:

$$W_{ij}^{(2)} \underset{|\mathbf{r}_{ij}|\to\infty}{\sim} 1 \big/ |\mathbf{r}_{ij}|^{d-2+\eta} , \tag{24.47}$$

or in Fourier transform:

$$\widetilde{W}^{(2)}(q) \sim 1 \big/ q^{2-\eta}, \tag{24.48}$$

we derive from expression (24.39) the classical or mean field value of the exponent η:

$$\eta = 0 . \tag{24.49}$$

Specific heat. Let us calculate the derivative of the free energy per unit volume with respect to V, which is a measure of the temperature,

$$\frac{1}{\Omega} \left.\frac{\partial W(H)}{\partial V}\right|_{H=0} = M^2(H=0).$$

Above T_c the average energy vanishes and below T_c it is proportional to the square of the spontaneous magnetization. Deriving again with respect to V we find the specific heat $\mathcal{C}$

$$\mathcal{C}(T \to T_{c+}) = 0, \qquad \mathcal{C}(T \to T_{c-}) = \frac{12}{b} . \tag{24.50}$$

We find that in the mean field approximation the specific heat has a non-universal jump.

Continuous symmetries. If the initial spin variable S_i is a N-component vector, and if both the interaction and the spin distribution have some continuous symmetry, most of the previous results remain clearly unchanged. The main difference comes from the appearance in zero field of Goldstone (massless) modes below T_c. Let us take the example of the $O(N)$ symmetry. We rewrite the thermodynamic potential:

$$\Gamma\left(\mathbf{M}\right) = -\sum_{ij} V_{ij}\mathbf{M}_i \cdot \mathbf{M}_j + \sum_i \left(\frac{a}{2}\mathbf{M}_i^2 + \frac{b}{24}\left(\mathbf{M}_i^2\right)^2 + \cdots\right). \tag{24.51}$$

The relation between magnetization and (constant) magnetic field still is:

$$H/M = -2V + a + \tfrac{1}{6}bM^2, \tag{24.52}$$

in which H, M are now the lengths of the vectors $\mathbf{M}, \mathbf{H}$.

From equation (24.51) we can compute the inverse 2-point correlation function $\Gamma_{ij}^{(2)\alpha\beta}$:

$$\Gamma_{ij}^{(2)\alpha\beta} = [W^{(2)}]_{ij}^{-1\alpha\beta} = \left(-2V_{ij} + a\delta_{ij}\right)\delta_{\alpha\beta} + \frac{b}{6}\left(M^2\delta_{\alpha\beta} + 2M_\alpha M_\beta\right)\delta_{ij} , \tag{24.53}$$

in which M_α $(\alpha = 1, ..., N)$ are the components of the constant magnetization vector. Let us introduce a unit vector $\mathbf{u}$ along the direction of spontaneous magnetization:

$$\mathbf{M} = M\mathbf{u}_\alpha .$$

Then the propagator (24.53) can be decomposed into a transverse and longitudinal part

$$\Gamma_{ij}^{(2)\alpha\beta} = u_\alpha u_\beta \Gamma_{\mathrm{L},ij} + \left(\delta_{\alpha\beta} - u_\alpha u_\beta\right)\Gamma_{\mathrm{T},ij} . \tag{24.54}$$

The two components are given in Fourier transform by:

$$\Gamma_L(q) = a + \tfrac{1}{2}bM^2 - 2\tilde{V}(q), \tag{24.55a}$$

$$\Gamma_T(q) = -2\tilde{V}(q) + a + \tfrac{1}{6}bM^2. \tag{24.55b}$$

The expressions (24.55a) and (24.39) are similar. Using equation (24.52) we can rewrite $\Gamma_T(q)$:

$$\Gamma_T(q) = H/M + 2\left[\tilde{V}(0) - \tilde{V}(q)\right]. \tag{24.56}$$

This equation shows that, in the low temperature phase in zero field, the inverse transverse propagator vanishes like q^2 at zero momentum for any temperature below T_c, indicating the presence of $(N-1)$ Goldstone (massless) modes. Let us recall briefly in this context the generality of this result. Going back to magnetic notations, let us assume that the thermodynamic potential $\Gamma(M)$ is a function of the square of the magnetization **M**. Then the relation between magnetic field and magnetization takes the form:

$$\Gamma(M) = \phi\left(M^2/2\right), \Rightarrow H_\alpha = M_\alpha \phi'\left(M^2/2\right). \tag{24.57}$$

The inverse 2-point function at zero momentum is given by:

$$\frac{\partial^2 \Gamma}{\partial M_\alpha \partial M_\beta} = M_\alpha M_\beta \phi''\left(M^2/2\right) + \delta_{\alpha\beta}\phi'\left(M^2/2\right). \tag{24.58}$$

Using equation (24.57), we obtain for the transverse part of the inverse propagator at zero momentum:

$$\left[W_T^{(2)}\right]^{-1}(q=0) = H/M. \tag{24.59}$$

If M does not vanish when H goes to zero (case of spontaneous magnetization) one finds that the transverse correlation length diverges below T_c, in agreement with the general results proven in Sections 13.3, 13.4 (to which we refer for an extensive discussion of this problem).

Landau's theory. All previous results have been established in the framework of a special approximation scheme, the mean field approximation but, as shown by Landau, they follow also from more general assumptions:

(i) The thermodynamical potential can be expanded in powers of M around $M = 0$.

(ii) The coefficients of the expansion are regular functions of the temperature for T close to T_c, the temperature at which the coefficient of M^2 vanishes.

(iii) The functions $\tilde{\Gamma}^{(n)}$ which appear in the expansion of $\Gamma(M)$ in powers of $\tilde{M}(q)$, the Fourier transform of the local magnetization M_i:

$$\Gamma(M) = \sum_n \frac{1}{n!} \int \mathrm{d}q_1 \ldots \mathrm{d}q_n \, \tilde{M}(q_1) \ldots \tilde{M}(q_n) \, \Gamma^{(n)}(q_1, \ldots, q_n),$$

are regular for small q_i and $\Gamma^{(4)}(0,0,0,0)$ is positive.

Hidden in these assumptions is the general idea that physics on different scales decouple and thus critical phenomena can be described at leading order in terms of a finite number of effective macroscopic variables.

24.5 Beyond the Mean Field Approximation

From the point of view of the representation (24.15) of the partition function, mean field theory is a saddle point approximation. It is therefore easy to calculate systematic corrections to the mean field approximation by expanding around the saddle point.

Since we are only interested in the behaviour of thermodynamical quantities in the critical region: T close to T_c, H, M and momenta $|q|$ small, we can somewhat simplify the expression. The term $\sum_{ij} \frac{1}{4} H_i V_{ij}^{-1} H_j$ gives an additive contribution to $W(H)$ which is negligible at leading order. We can also take the quantity J_i:

$$J_i = \tfrac{1}{2} \sum_j V_{ij}^{-1} H_j \,,$$

as a measure of the magnetic field, since the convolution of H_i with V_{ij} leads just to a short range smearing.

The simplified partition function $Z(J)$ is then given by:

$$Z(J) = \int \prod_i \mathrm{d}\varphi_i \exp\left[-l\left(\tfrac{1}{4} \sum_{ij} \varphi_i V_{ij}^{-1} \varphi_j + \sum_i A\left(\varphi_i\right) - J_i \varphi_i \right) \right]. \qquad (24.60)$$

We recognize a lattice regularized functional integral for a scalar boson field of the type studied in Chapter 5. The mean field approximation is analogous to the tree level approximation of Field Theory. Corrections to mean field are given by perturbation theory. Let us introduce the propagator Δ_{ij}:

$$\Delta_{ij}^{-1} = \tfrac{1}{2} V_{ij}^{-1} - a^{-1},$$

Its Fourier transform $\Delta(q)$ has for q small the form:

$$\Delta^{-1}(q) = \tfrac{1}{2} \tilde{V}^{-1}(q) - a^{-1} = (2V)^{-1} \left[1 - 2V/a + \rho^2 q^2 + O\left(q^4\right) \right]. \qquad (24.61)$$

In the one-loop approximation (equation (6.49)) $\Gamma(M)$ is:

$$\Gamma(M) = \frac{1}{4} \sum_{ij} M_i V_{ij}^{-1} M_j + \sum_i A\left(M_i\right) + \frac{1}{2l} \operatorname{tr} \ln \left[\frac{1}{2} V_{ij}^{-1} + A''\left(M_i\right) \delta_{ij} \right]. \qquad (24.62)$$

From this expression it is easy to derive the inverse 2-point function $\Gamma^{(2)}$ up to order $1/l$. In momentum space we find:

$$\Gamma^{(2)}\left(p\right) = \Delta^{-1}(p) + \frac{b}{2a^4 l} \int \frac{\mathrm{d}^d q}{(2\pi)^d} \Delta(q) + O\left(\frac{1}{l^2}\right). \qquad (24.63)$$

In particular at zero momentum we obtain:

$$\Gamma^{(2)}\left(0\right) = \frac{1}{2V} - \frac{1}{a} + \frac{b}{2a^4 l} \int \frac{\mathrm{d}^d q}{(2\pi)^d} \Delta(q) + O\left(\frac{1}{l^2}\right). \qquad (24.64)$$

The critical value V_c is now defined by:

$$0 = \frac{1}{2V_c} - \frac{1}{a} + \frac{b}{2a^4 l} \int \frac{\mathrm{d}^d q}{(2\pi)^d} \Delta_c(q) + O\left(\frac{1}{l^2}\right). \qquad (24.65)$$

Subtracting this equation from equation (24.64) we find:

$$\Gamma^{(2)}(0) = \frac{1}{2}\frac{(V_c - V)}{VV_c} + \frac{b}{2la^4}\int\frac{d^dq}{(2\pi)^d}\left[\Delta(q) - \Delta_c(q)\right] + O\left(\frac{1}{l^2}\right). \qquad (24.66)$$

In the $1/l$ contribution we can replace V_c by its mean field value $a/2$. Then for $V \sim V_c$ and q small:

$$\Delta(q) = (2V_c)^{-1}\left(V_c - V + \rho^2 q^2 + O\left(q^4\right)\right), \qquad (24.67)$$

and therefore:

$$\Delta(q) - \Delta_c(q) \sim \frac{(V - V_c)}{2V_c}\Delta(q)\Delta_c(q). \qquad (24.68)$$

Discussion. From these two expressions it is clear that we can safely take the limit $V = V_c$ inside the integrand provided the integral $\int d^dq/(q^2)^2$ converges at low momentum, i.e. for $d > 4$. Then $\Gamma^{(2)}(0)$ vanishes linearly in $(V - V_c)$ or $T - T_c$ as in mean field theory for $d > 4$. Since $\Gamma^{(2)}(0)$ is simply related to the inverse susceptibility in zero field, the exponent γ remains 1. However, for $d \le 4$ the limit $V = V_c$ is singular. Evaluating for V close to V_c and $2 < d < 4$ the dominant contribution to the integral, which comes entirely for the small q region, we find:

$$\Gamma^{(2)}(0) = \frac{1}{2}\frac{(V - V_c)}{V_c^2}\left[1 + \frac{b}{2la^2\rho^d}\frac{\Gamma(1 - d/2)}{(4\pi)^{d/2}}\left(\frac{V_c - V}{V_c}\right)^{d/2-2}\right] + O\left(\frac{1}{l^2}\right). \qquad (24.69)$$

For V close to V_c the correction to $\Gamma^{(2)}(0)$ always is larger than the leading term. Therefore the predictions of mean field theory can no longer be trusted:

(i) For dimension $d > 4$ the "universal" predictions of mean field theory remain unchanged.

(ii) For dimension $d \le 4$ "infrared" (IR) divergences, i.e. singularities due to the small momentum behaviour of the propagator, show that the predictions of mean field theory cannot be correct in general.

Inspection in Section 24.6 of higher order corrections will confirm this result.

24.6 Power Counting and the Role of Dimension Four

In the preceding section we have shown that the universal predictions of mean field theory cannot be trusted below four dimensions. We also note that the first divergent corrections depend on additional details of the initial microscopic model. At first sight this would suggest that large distance properties are short distance sensitive and therefore no form of universality survives, in contradiction with experimental and numerical indications. However a closer inspection of equation (24.69), for example, reveals that, at least at leading order, only one new parameter appears, related to the value of the φ^4 term in the small φ expansion of the interaction term. We now want to extend the analysis of "infrared divergences" to all orders in the mean field expansion, to understand whether the roles of dimension four and φ^4 interaction survive. Since the analysis is perturbative, we have to assume that, in some sense, deviations from mean field theory are not too large. This point will be further discussed in the coming sections.

In the spirit of what has been done in Section 24.5, we first perform, order by order in the mean field expansion, a magnetic susceptibility or mass (in field theoretical language) renormalization. We trade the parameter V for the parameter σ:

$$\sigma = \chi^{-1} = \Gamma^{(2)}(0), \qquad (24.70)$$

by inverting order by order the relation between V and σ (see Section 10.1).

The φ-propagator, for what concerns the most singular contribution at low momentum, can then be replaced by

$$\Delta(q) = \frac{1}{\sigma + q^2}. \tag{24.71}$$

We now consider the individual contributions coming from all (even due to $\varphi \mapsto -\varphi$ symmetry) powers of φ in $A(\varphi)$, allowing even for the possibility of polynomial momentum dependence in the corresponding vertices (this was not the case in the examples we have considered so far). These contributions have just the form of ordinary Feynman diagrams with the propagator (24.71), integrated up to a "cut-off" of order 1 since the momenta $\mathbf{q}$ are limited to a Brillouin zone:

$$|q_\mu| < \pi \,.$$

In Chapter 8 we have associated with each diagram a dimension given by power counting. If v_α is the number of vertex of type α, which connect n_α lines and contain k_α powers of momenta, the dimension $\delta(\gamma)$ of a diagram γ contributing to the n-point function $\Gamma^{(n)}$ is (equations (8.14,8.22,8.26)):

$$\delta(\gamma) = d - \frac{n}{2}(d-2) + \sum_\alpha v_\alpha \left[k_\alpha + n_\alpha \left(\frac{d}{2} - 1 \right) - d \right]. \tag{24.72}$$

Let us for simplicity consider the diagram γ at zero external momenta and perform the change of variables (for all integration momenta):

$$q = \sigma^{1/2} q'. \tag{24.73}$$

After this rescaling the contribution $D(\gamma)$ of the diagram γ takes the form:

$$D(\gamma) = \sigma^{\delta(\gamma)/2} D'(\gamma),$$

by definition of $\delta(\gamma)$. $D'(\gamma)$, which is calculated with a propagator $1/(q^2+1)$, is no longer IR divergent when σ goes to zero, but may have large momentum (UV) divergences since the cut-off, which is now of order $\sigma^{-1/2}$, becomes infinite in the same limit.

If the condition:

$$\sum_\alpha v_\alpha \left[k_\alpha + n_\alpha \left(\frac{d}{2} - 1 \right) - d \right] > 0 \,, \tag{24.74}$$

is satisfied for all vertices, $\delta(\gamma)$ increases with the order in perturbation theory. However, at the same time $D'(\gamma)$ has UV divergences proportional to $\sigma^{-1/2}$ to a power given by power counting. For UV divergent diagrams the two powers of σ exactly cancel and therefore all contributions in perturbation theory have the same behaviour, i.e. the behaviour of mean field theory. This in particular applies to the 2-point function for which $\delta(\gamma) > 2$. For higher correlation functions a finite number of diagrams may be UV convergent and therefore are more IR singular than predicted by the tree level action. Then the leading contribution is given by the one-loop diagram with only vertices having the smallest possible dimension.

The condition (24.74) is satisfied if all interactions are non renormalizable. For a theory with interactions even in φ this implies $d > 4$. The leading order result of Section 24.5 then extends to all orders: mean field theory is qualitatively correct order by order in perturbation theory; the effect of loop corrections is to modify the non-universal parameters of the mean field approximation.

If instead there exist interactions such that:

$$\sum_\alpha v_\alpha \left[k_\alpha + n_\alpha \left(\frac{d}{2} - 1 \right) - d \right] < 0 , \qquad (24.75)$$

then one can find diagrams in which increasingly large negative powers of σ have been factorized. These diagrams are superficially convergent at large momenta. This implies that either they have no divergent subdiagrams and their singularity is indeed given by the power $\delta(\gamma)/2$ or they are even more singular when σ goes to zero. In all cases the singularity of diagrams increases without bound with the order of perturbation theory.

Let us now try to characterize the most divergent diagrams order by order in perturbation theory. If we forget about the divergent subdiagrams we have just to find the minimum of $\delta(\gamma)/2$ at each order. This minimum is obtained for the least UV divergent interaction, i.e. for $k_\alpha = 0$ and n_α as small as possible (we assume $d > 2$). The smallest even value is $n_\alpha = 4$. This corresponds to the φ^4 interaction.

Let us then examine the problem of divergent subdiagrams: remember that we have performed a mass renormalization. According to the analysis of Chapters 8–10, the φ^4 diagrams are thus UV finite below four dimensions: The nature of the IR singularity deduced from the simple scaling (24.73) is thus the correct one.

This is no longer true for more UV divergent interactions which even below four dimensions require additional renormalizations. However, it will be shown in Chapter 27 that these interactions can be decomposed into the sum of an effective φ^4 interaction and a part which is less singular.

Therefore the most IR singular terms, order by order in perturbation theory are generated by an effective φ^4 field theory, the coefficient of the φ^4 interaction being not the coefficient read in the original hamiltonian but a coefficient renormalized by the other interactions. In Chapters 25, 26 we shall investigate the large distance behaviour of correlation functions in this effective φ^4 field theory. In Chapter 27 we shall discuss the effect of the neglected subleading divergences to show the internal consistency of the method.

Note finally that for $d = 4$, mean field behaviour is modified by logarithmic corrections.

24.7 Tricritical Points

So far we have assumed that we could vary one control parameter, the temperature, and that, therefore, the coefficient b of the term M^4 in the expansion of $\Gamma(M)$ was a number of order 1. However there are situations in which an additional physical parameter can be varied, and both the coefficients a and b of M^2 and M^4 can be made to vanish. This occurs for instance in He^3–He^4 mixtures or some metamagnetic systems. In the Ising-like models we have considered so far, this can be achieved by allowing the spin distribution to vary.

If the coefficient c of M^6 is positive:

$$\Gamma(M) = -\sum_{ij} V_{ij} M_i M_j + \sum_i \left(\frac{a}{2!} M_i^2 + \frac{b}{4!} M_i^4 + \frac{c}{6!} M_i^6 + \cdots \right), \qquad (24.76)$$

when b decreases, it is possible to follow a line of critical points until b vanishes. At this point, the tricritical point, the M^6 term becomes relevant, and a new analysis has to be performed. The various critical exponents have values different from those found for an ordinary critical point. After the tricritical point, b becomes negative and the transition becomes first order.

Corrections to the tricritical theory can be studied by the method of Section 24.6. Using power counting arguments, one now finds that the most singular corrections are reproduced by a continuum ϕ^6 field theory, and that the upper critical dimension is 3, i.e. above 3 dimensions mean field theory predicts correctly universal quantities, whereas it is definitely not valid in 3 dimensions and below.

Bibliographical Notes

In this chapter we have closely followed the review article of E. Brézin, J.C. Le Guillou and J. Zinn-Justin, published in

Phase Transitions and Critical Phenomena vol. 6, C. Domb and M.S. Green eds. (Academic Press, London 1976).

For earlier articles and reviews about mean field theory see for example

P. Weiss, *J. Phys. Radium, Paris* 6 (1907) 667; M.E. Fisher, *Rep. Prog. Phys.* 30 (1967) 615; L.P. Kadanoff, W. Götze, D. Hamblen, R. Hecht, E.A.S. Lewis, V.V. Palciauskas, M. Rayl, J. Swift, D. Aspnes and J. Kane, *Rev. Mod. Phys.* 39 (1967) 395; H.E. Stanley, *Introduction to Phase Transitions and Critical Phenomena* (Oxford University Press, New York 1971).

The scaling form of the equation of state was proposed by

B. Widom, *J. Chem. Phys.* 43 (1965) 3892, 3898.

Landau's theory is presented in

L.D. Landau, *Phys. Z. Sowjetunion* 11 (1937) 26, reprinted in *Collected Papers of L.D. Landau*, D. ter Haar ed. (Pergamon, New York 1965).

The Hubbard transformation appears in

G.A. Baker, *Phys. Rev.* 126 (1962) 2071; A.J.F. Siegert, *Brandeis Lectures* vol. 3 (Benjamin, New York 1963); J. Hubbard, *Phys. Lett.* 39A (1972) 365.

See also

R. Brout, *Phys. Rep.* 10C (1974) 1.

The role of the $\varphi^4(x)$ field theory was emphasized by

K.G. Wilson, *Phys. Rev.* B4 (1971) 3174, 3184.

For a discussion of tricritical systems see

E.H. Graf, D.M. Lee and J.D. Reppy, *Phys. Rev. Lett.* 19 (1967) 417; R.J. Birgeneau, W.B. Yelon, E. Cohen and J. Makovsky, *Phys. Rev.* B5 (1972) 2607; M. Blume, *Phys. Rev.* 141 (1966) 517; H.W. Capel, *Physica* 32 (1966) 966; E.K. Riedel and F.J. Wegner, *Phys. Rev. Lett.* 29 (1972) 349.

General spin systems have been considered by

G. Horwitz and H.B. Callen, *Phys. Rev.* 124 (1961) 1757; F. Englert, *Phys. Rev.* 129 (1963) 567.

Appendix A24 follows

J. Zinn-Justin, *Perspectives in Particles and Fields*, Cargèse 1983, vol. B126, M. Lévy *et al* eds. (Plenum Press, New York 1985).

Exercises

Exercise 24.1

The spin 1 Ising model. Consider the Ising model with a nearest neighbour (n.n.) ferromagnetic two-body interaction in which the spin S can take three values $\{0, 1, -1\}$, the common weight of the configurations $\{1, -1\}$ being p and the weight of the last configuration being therefore $1 - 2p$. Study the model in the mean field approximation as a function of temperature and p.

Exercise 24.2

The antiferromagnetic Ising model. We consider an Ising-like system on a hypercubic lattice with antiferromagnetic interactions:

$$-\beta\mathcal{H} = -\sum_{ij} V_{ij} S_i S_j, \quad V_{ij} \geq 0, \; V_{ii} = 0.$$

The lattice can be decomposed into two sublattices corresponding to odd or even sites (as defined by the parity of the sum of components of the position vector). We characterize the sublattices by a sign $\epsilon_i = (-1)^i$ assigned to each site i. We then assume

$$V_{ij}\epsilon_i\epsilon_j = -V_{ij},$$

a condition satisfied by the n.n. interaction.

Define an order parameter for the antiferromagnetic transition. Characterize then the effect of a small constant magnetic field within the mean field approximation.

Exercise 24.3

Mean field theory and effective Landau–Ginzburg action for models with Z_N symmetry. One now considers a lattice model on a cubic lattice with short range ferromagnetic interaction in which the two-body interaction has a Z_N symmetry. For example one can represent the degrees of freedom at each site i by a phase θ_i which takes N discrete values:

$$\theta = 2\pi n/N, \quad p = 0, \ldots, N-1.$$

The configuration energy then takes the form:

$$-\beta\mathcal{H} = \sum_{i,j} V_{ij} \cos(\theta_i - \theta_j),$$

with $V_{ii} = 0$, $V_{ij} \geq 0$. The N configurations on a site have the same weight.

Note that for $N = 2$ one simply recovers the Ising model. For $N = 3$ one finds a model which is equivalent to the 3-state Potts model. Define an order parameter for the Z_N model, $N > 1$. Discuss the model in the mean field approximation, in particular the order of the phase transition.

Exercise 24.4

The q-state Potts model. The model is defined in terms of spin variables which can take q values with equal weight and a configuration energy of the form:

$$-\beta\mathcal{H} = \sum_{\text{n.n.}} J\delta_{S_i S_j}.$$

For $q = 2$ the Ising model is recovered. Study the model for higher values of the integer q. A general review on the q-state Potts model and the percolation problem ($q = 1$) can be found in
F.Y. Wu, *Rev. Mod. Phys.* 54 (1982) 235.

APPENDIX 24
MEAN FIELD THEORY: GENERAL FORMALISM

We now explain how a mean field approximation can be obtained for a general spin model on the lattice. Again the mean field result will be identified with a saddle point approximation, and therefore will be the first term in a systematic expansion. A brief account of the method has already been included in the Domb and Green series review article. We shall also indicate the relation between the mean field approximation and high and low temperature expansions.

A24.1 Mean Field Theory

We consider a lattice model described in terms of some lattice energy functional $\mathcal{A}$ function of lattice variables S_i in which i is the lattice site. It is necessary for what follows to consider different powers of the same lattice variable S_i as independent variables so that we write $\mathcal{A}$ as $\mathcal{A}\left(S_i, S_i^2, ..., S_i^k\right)$. The corresponding partition function is:

$$Z = \int \prod_i d\rho\left(S_i\right) \exp\left[-\mathcal{A}\left(S_i, \ldots, S_i^k\right)\right], \qquad (A24.1)$$

in which $d\rho\left(S\right)$ is the one spin distribution.

In order to construct the mean field approximation as the leading order in a steepest descent calculation, it is necessary to first transform expression $(A24.1)$. We introduce two sets of lattice variables $\sigma_i^{(l)}$ and $H_i^{(l)}$, the quantities $H_i^{(l)}$ being Lagrange multipliers, to express the constraints:

$$\sigma_i^{(l)} = S_i^l, \qquad l = 1, \ldots, k. \qquad (A24.2)$$

Then the partition function can be rewritten:

$$Z = \int \prod_i d\rho\left(S_i\right) \prod_{i,l} d\sigma_i^{(l)} dH_i^{(l)} \exp\left[-\mathcal{A}\left(\sigma_i^{(1)}, \sigma_i^{(2)}, \ldots, \sigma_i^{(k)}\right)\right.$$
$$\left. + \sum_{i,l} H_i^{(l)}\left(\sigma_i^{(l)} - S_i^l\right)\right]. \qquad (A24.3)$$

We can now reformulate more accurately what we mean by considering all powers of the lattice variable as independent: the action $\mathcal{A}$ is a first degree polynomial in all the variables $\sigma_i^{(l)}$ and contains no product of the form $\sigma_i^{(l)}\sigma_i^{(m)}$.

In expression $(A24.3)$ the integrations over all variables S_i are decoupled. Let us introduce the function F_0:

$$\exp\left[-F_0\left(H^{(1)}, \ldots, H^{(k)}\right)\right] = \int d\rho\left(S\right) \exp\left(-\sum_{l=1}^k S^l H^{(l)}\right). \qquad (A24.4)$$

We here assume that the integration measure $d\rho\left(S\right)$ is either of compact support or decreasing fast enough so that the integral exists. The partition function can then be rewritten:

$$Z = \int \prod_{i,l} \mathrm{d}\sigma_i^{(l)} \mathrm{d}H_i^{(l)} \exp\left[-\mathcal{A}\left(\sigma_i^{(1)}, \ldots, \sigma_i^{(k)} \right) + \sum_{i,l} H_i^{(l)} \sigma_i^{(l)} \right.$$
$$\left. - \sum_i F_0\left(H_i^{(1)}, \ldots, H_i^{(k)} \right) \right]. \qquad (A24.5)$$

The mean field approximation is the leading order in the evaluation of this expression by steepest descent. The saddle point equations are:

$$\frac{\partial \mathcal{A}}{\partial \sigma_i^{(l)}} = H_i^{(l)}, \qquad (A24.6a)$$

$$\frac{\partial F_0}{\partial H_i^{(l)}} = \sigma_i^{(l)}. \qquad (A24.6b)$$

The equation $(A24.6b)$ shows that, at leading order, the original weight $\prod_i \mathrm{d}\rho(S_i)\, \mathrm{e}^{-\mathcal{A}}$ has been approximated by a product of weights for each lattice site given by the r.h.s. of equation $(A24.4)$, in which the variables $H_i^{(l)}$ have been replaced by their saddle point values, the "mean fields".

Remarks. It is clear from these equations that if $\mathcal{A}$ does not depend on a given $\sigma^{(l)}$, the corresponding field $H^{(l)}$ vanishes, and both disappear from the equations. Therefore we find the same result as if we had omitted them in the first place. Note also that if $\mathcal{A}$ is a quadratic function of the variables $\sigma^{(l)}$, the $\sigma^{(l)}$ integrations can be performed explicitly and only integrals over the corresponding $H^{(l)}$ remain. In Chapter 24 we have considered models of this form and this explains why we did not need the general formalism we are developing here.

Discussion. Let us now explain why we have introduced the k sets of variables $\sigma^{(l)}$ and $H^{(l)}$ instead of one as in the example studied in Chapter 24. In the mean field approximation the variables S_i are replaced by some average value so that the average of a product is replaced by the product of averages:

$$\langle S_{i_1} S_{i_2} \ldots S_{i_n} \rangle \mapsto \langle S_{i_1} \rangle \langle S_{i_2} \rangle \ldots \langle S_{i_n} \rangle .$$

It is plausible that in some limit this can be a good approximation if all variables in the product are different. However this cannot be true if in the product appears a power of the same variable. For example in the Ising model:

$$S_i^2 = 1 \neq \left(\langle S_i \rangle \right)^2 .$$

A variable is always correlated with itself, and the new Lagrange parameters allow to take into account these autocorrelations. To illustrate this point let us consider specific examples.

Examples. Let us assume first that the action $\mathcal{A}$ is a function only of S_i^2, then it is obvious that we should consider S_i^2 as the basic dynamical variable in the mean field approximation. The procedure explained above does it automatically since only the parameters $\sigma^{(2)}$ and $H^{(2)}$ will appear. This procedure also solves the following simple problem. Terms in the action which correspond to one-body potentials, i.e. sum of functions depending only on the variable at one site such as:

$$\sum_i \frac{a}{2} S_i^2 + \frac{b}{4!} S_i^4 + \cdots , \qquad (A24.7)$$

could be considered both as part of the measure $\rho(S_i)$ or as part of the action. Therefore one could have the feeling that the results of mean field theory depend on the formulation. The introduction of additional variables ensures that this is not the case.

Let us for example make the transformation:

$$\begin{cases} \mathcal{A}\left(S_i, S_i^2, \ldots, S_i^k\right) \mapsto \mathcal{A}\left(S_i, \cdots, S_i^k\right) + \frac{a}{2}\sum_i S_i^2, \\[2mm] \rho(S) \mapsto e^{aS^2/2}\rho(S). \end{cases} \qquad (A24.8)$$

Obviously the lattice theory is independent of a.

Consider now the modifications this transformation induces into equations ($A24.4$–$A24.6$)

$$\exp\left(-\tilde{F}_0\left(H^{(1)}, \ldots, H^{(k)}\right)\right) = \int dS \; e^{aS^2/2}\rho(S) \exp\left(-\sum_{l=1}^{k} S^l H^{(l)}\right). \qquad (A24.9)$$

Comparing the equation with equation ($A24.4$) it follows:

$$\tilde{F}_0\left(H^{(1)}, \ldots, H^{(k)}\right) = F_0\left(H^{(1)}, H^{(2)} - \frac{a}{2}, H^{(3)}, \ldots, H^{(k)}\right), \qquad (A24.10)$$

and the saddle point equations become:

$$\begin{cases} \dfrac{\partial \mathcal{A}}{\partial \sigma_i^{(l)}} + \dfrac{a}{2}\delta_{l2} = H_i^{(l)}, \\[3mm] \dfrac{\partial F_0}{\partial H_i^{(l)}}\left(H^{(1)}, H^{(2)} - \dfrac{a}{2}, H^{(3)}, \ldots, H^{(k)}\right) = \sigma_i^{(l)}. \end{cases} \qquad (A24.11)$$

We see that the equations are identical up to the change:

$$H_i^{(2)} - \frac{a}{2} \; \mapsto \; H_i^{(2)} . \qquad (A24.12)$$

Therefore mean field theory results are insensitive to such changes.

Finally we now also understand the technical assumption made in Chapter 24:

$$V_{ii} = 0 .$$

It was just to simplify the mean field formulation and avoid the introduction of variables related to S_i^2.

It is then possible to correct the mean field approximation, by systematically expanding the effective action in ($A24.3$) around the saddle points ($A24.6$) and integrating term by term as one does in usual perturbation theory. We call this expansion the mean field expansion. We have seen in Chapter 24 in a particular example that a parameter could be associated to this expansion. It is easy to show, following the same ideas, that this can also be done in this more general framework.

A24.2 Mean Field Expansion

If we set a parameter L in front of the effective action $(A24.5)$, then the mean field expansion is just a formal expansion in powers of $1/L$:

$$Z_L = \int \prod_{i,l} d\sigma_i^{(l)} dH_i^{(l)} \exp\left\{L\left[-\mathcal{A}\left(\sigma_i^{(1)}, \sigma_i^{(2)}, \ldots, \sigma_i^{(k)}\right) + \sum_{i,l} H_i^{(l)} \sigma_i^{(l)}\right.\right.$$
$$\left.\left. - \sum_i F_0\left(H_i^{(1)}, \ldots, H_i^{(k)}\right)\right]\right\}. \tag{A24.13}$$

As already explained in Chapter 24, the factor L in front of F_0 has, for integer L, the following interpretation: we have replaced each variable S by a sum of L independent variables having the same distribution $d\rho(S)$.

$$S_i = \sum_{\alpha=1}^{L} S_i^{(\alpha)}, \tag{A24.14}$$

$$\int d\rho(S_i) \mapsto \int \prod_{\alpha} d\rho\left(S_i^{(\alpha)}\right). \tag{A24.15}$$

Let us introduce the distribution $d\rho_L(S)$ of S:

$$\exp\left[-LF_0\left(H^{(1)}, \ldots, H^{(k)}\right)\right] = \int d\rho_L(S) \exp\left(-\sum_{l=1}^{k} H^{(l)} S^l\right). \tag{A24.16}$$

For L large, $d\rho_L(S)$ becomes a gaussian distribution of dispersion $\sqrt{L}$. Using the definition $(A24.16)$ we can rewrite Z_L:

$$Z_L = \int \prod_i d\rho_L(S_i) \exp\left[-L\mathcal{A}\left(\frac{S_i}{L}, \frac{S_i^2}{L}, \ldots, \frac{S_i^k}{L}\right)\right]. \tag{A24.17}$$

We see that each interaction term in $\mathcal{A}$ is multiplied by a power of L which is equal to the number of sites connected by the interaction, minus one.

The parameter L is just a formal parameter since really we want to calculate for $L = 1$, but its introduction leads to some simple considerations. It shows that mean field approximation is exact when we have a large number of identical independent variables on each site provided the coefficients in front of the interaction terms are scaled down by some appropriated factor. Such a situation is for instance approximately realized when d, the dimension of space, becomes large. Then in a cell of given linear dimension, one finds an increasing number of independent variables. By re-expanding terms in the mean field expansion, one can generate a systematic $1/d$ expansion.

A24.3 High Temperature Series and Mean Field

Expression ($A24.17$) can also be used to show some interesting connections between mean field expansion and high temperature or low temperature expansion. Let us reintroduce a factor β in front of the action $\mathcal{A}$ and perform an expansion in powers of β, which is a high temperature expansion. At order K in β we find a sum of contributions of the form

$$\beta^K L^K L^{-\sum_{l=1}^{K} C_l} L^n. \qquad (A24.18)$$

The factor L^K comes from the L factor in front of the action. The integers C_l count the number of sites linked by the corresponding interaction terms. Finally equation ($A24.16$) shows that the average of any power of the lattice variable with the weight $\mathrm{d}\rho_L\left(S\right)$ is proportional to L since it is equal to:

$$\frac{\int \mathrm{d}\rho_L\left(S\right) S^\ell}{\int \mathrm{d}\rho_L\left(S\right)} = \left[\frac{\partial}{\partial H^{(\ell)}}\left[-LF_0\left(H^{(1)},\dots,H^{(k)}\right)\right]\right]\Bigg|_{H=0}. \qquad (A24.19)$$

Therefore in equation ($A24.18$) n is the number of different lattice variables present in the high temperature diagram. For a connected tree diagram one has:

$$n = \sum_{l=1}^{K} C_l - (K-1). \qquad (A24.20)$$

Indeed each interaction term brings in C_l independent variables. But to construct a connected diagram each interaction term must have a variable in common with another interaction term. This suppresses exactly $K-1$ independent variables.

Now, each time one adds one loop to the diagram, one suppresses one additional independent variable. Calling B the number of loops of the diagram one finds:

$$n = \sum_{l=1}^{K} C_l - (K-1) - B. \qquad (A24.21)$$

The corresponding power of L for the diagram then is:

$$K - \sum_{l=1}^{K} C_l + n = 1 - B.$$

Mean field expansion is an expansion in powers of L. Therefore, at a given order, it sums all high temperature or strong coupling diagrams with the same number of loops.

A24.4 Low Temperature and Mean Field

The mean field expansion also contains the low temperature expansion. Indeed let us rewrite equations ($A24.6$) with a parameter β in front of the action:

$$\begin{cases} \beta\dfrac{\partial \mathcal{A}}{\partial \sigma_i^{(l)}} = H_i^{(l)}, \\[2mm] \dfrac{\partial F_0}{\partial H_i^{(l)}} = \sigma_i^{(l)}. \end{cases} \qquad (A24.22)$$

For large β (low temperature or weak coupling), the $H^{(l)}$ variables become large. To study this limit we have thus to evaluate F_0 for H large. This will in general select for S some classical value S_c. As a direct consequence we have:

$$\frac{\partial F}{\partial H_i^{(l)}} = (S_{i,c})^l = \sigma_i^{(l)}. \qquad (A24.23)$$

In this limit the variables σ and H play no longer role and one expands around the configuration which dominates the low temperature (large β) expansion. To show this more explicitly let us write the measure as a Fourier transform:

$$d\rho(S) = \frac{dS}{2i\pi} \int_{-i\infty}^{+i\infty} e^{\mu S} \tilde{\rho}(\mu)\, d\mu. \qquad (A24.24)$$

To be able to perform this expansion, $d\rho(S)$ has to decrease fast enough for large $|S|$ for $\tilde{\rho}(\mu)$ to be an entire function. Then for $H^{(l)}$ large, F_0 is given by the saddle point:

$$\mu = \sum_l l H^{(l)} S^{l-1}, \qquad (A24.25)$$

$$S + \frac{\partial}{\partial \mu} \ln \tilde{\rho}(\mu) = 0, \qquad (A24.26)$$

which yields:

$$F_0\left(H^{(l)}\right) = \sum_l H^{(l)} S^l - \mu S + \ln \tilde{\rho}(\mu). \qquad (A24.27)$$

The mean field saddle point equations then are:

$$\sigma_i^{(l)} = \frac{\partial F_0}{\partial H_i^{(l)}} = S_i^l, \qquad (A24.28)$$

$$H_i^{(l)} = \beta \frac{\partial \mathcal{A}}{\partial \sigma_i^{(l)}}. \qquad (A24.29)$$

In the same notations the relevant configuration at low temperature is given by the equations:

$$\beta \frac{d\mathcal{A}}{dS_i} = \sum_i l S_i^{l-1} \frac{\partial \mathcal{A}}{\partial \sigma_i^{(l)}} = \mu_i, \qquad (A24.30)$$

and

$$S_i + \frac{d}{d\mu_i} \ln \tilde{\rho}(\mu) = 0. \qquad (A24.31)$$

Now by summing equation $(A24.29)$ over l after multiplication by $l(S_i)^{l-1}$, and using equation $(A24.25)$, we reproduce equation $(A24.30)$, while equations $(A24.26)$ and $(A24.31)$ are identical.

The mean field expansion is also a partial summation of the low temperature expansion.

Two final remarks: The generalization of this formalism to several degrees of freedom per site is straightforward. Mean field theory can also be generalized to lattice models with fermions. The equivalent of the variables S_i are then products of the form $\bar{\psi}_i \Gamma_A \psi_i$, in which ψ_i and $\bar{\psi}_i$ are the fermion fields and Γ_A some element of the algebra of γ matrices. These fields correspond to "composite bosons". One also has to use a measure appropriate to Grassmann variables.

25 GENERAL RENORMALIZATION GROUP ANALYSIS.
THE CRITICAL THEORY NEAR DIMENSION FOUR

In Chapter 24 we have considered Ising-like systems (and more generally ferromagnetic systems with $O(N)$ symmetry) with short range interactions and determined the behaviour of thermodynamical functions near a second order phase transition, in the mean field approximation. We have shown that mean field theory predicts a set of *universal* properties, i.e. properties which are independent of the detailed structure of the microscopic hamiltonian. We have then systematically examined corrections to the mean field approximation. We have found that above four dimensions these corrections do not change universal quantities; on the contrary below four dimensions the corrections diverge at the critical temperature and mean field theory can certainly not be correct. Moreover we have identified in an expansion around the mean field the most singular terms at criticality and shown that they can be formally summed to give a massless ϕ^4 field theory, a result we shall use in the second part of this chapter.

Such an analysis, which reflects the property that different scales of physics cannot be decoupled, seems to indicate that below four dimensions long distance properties are sensitive to the detailed microscopic structures. Surprisingly enough, however, some universality survives, although this universality is less general than in mean field theory in the sense that long distance properties depend on the space dimension, on symmetries of the system and some other qualitative properties. To understand this universality beyond the mean field approximation it is necessary to call upon a new idea. The *renormalization group* will provide us with the essential tool we need.

The RG theory, as applied to Critical Phenomena, has been developed by Kadanoff, Wilson, Wegner and many others. We shall first describe the basic renormalization group ideas in a somewhat abstract framework. The formulation will be rather inaccurate and the arguments largely heuristic. The role of fixed points in hamiltonian space will be stressed. A simple fixed point, the gaussian fixed point, whose predictions coincide with MFT, will be described.

The abstract RG theory, although its formulation is vague, is extremely suggestive, and indeed it has been implemented in many approximate forms and induced a wealth of practical calculations. It is not our purpose to review them here. However, in the more specific context of Quantum Field Theory, assumptions at the basis of the renormalization group have been clarified and verified in many cases of physical interest, confirming in a very direct manner the deep connection, first recognized by Wilson, between Quantum Field Theory and the theory of Critical Phenomena. The methods of perturbative QFT then have allowed to efficiently calculate universal quantities for a large class of statistical models.

Therefore in Section 25.3 we shall use what we have learned both from the analysis of corrections to mean field theory in Chapter 24 and of the relevant eigenoperators at the gaussian fixed point near dimension four and construct an approximate renormalization group for an effective ϕ^4 field theory. We shall discover that the RG equations which appear as a consequence of the necessity of renormalization of local field theories are directly connected with the abstract RG equations we have introduced in Section 25.1. We shall understand that universality in the theory of critical phenomena is directly related to the property that local field theories are insensitive to the short distance structure, and physics can therefore be described by renormalized correlation functions. Conversely, in the statistical sense, quantum field theories are always close to criticality and their existence, beyond perturbation theory, relies, from the abstract RG point of view, on the presence of IR fixed points in hamiltonian space.

Finally we shall be able to solve RG equations in $d = 4 - \varepsilon$ dimensions and, following Wilson and Fisher, calculate universal quantities in an ε-expansion

25.1 The Abstract Renormalization Group: General Formulation

Let us consider a general hamiltonian $\mathcal{H}(\phi)$, function of a field $\phi(x)$ $(x \in \mathbb{R}^d)$. We assume that $\mathcal{H}$ is expandable in powers of ϕ:

$$\mathcal{H}(\phi) = \sum_{n=0}^{\infty} \frac{1}{n!} \int d^dx_1 d^dx_2 \ldots d^dx_n \, \mathcal{H}_n(x_1, x_2, \ldots, x_n) \, \phi(x_1) \ldots \phi(x_n), \qquad (25.1)$$

and that the Fourier transforms of the functions $\mathcal{H}_n$ are regular at low momenta (assumption of short range forces or locality). The hamiltonian will in general depend on an infinite number of parameters or coupling constants.

To the hamiltonian (really the configuration energy) $\mathcal{H}(\phi)$ corresponds a set of connected correlation functions $W^{(n)}(x_1, \ldots, x_n)$:

$$W^{(n)}(x_1, x_2, \ldots, x_n) = \left[\int [d\phi] \, \phi(x_1) \ldots \phi(x_n) \, e^{-\beta \mathcal{H}(\phi)} \right]_{\text{connect.}} \qquad (25.2)$$

We want to study the long distance behaviour of critical correlation functions, i.e. the behaviour of $W^{(n)}(\lambda x_1, \ldots, \lambda x_n)$ when the dilatation parameter λ becomes large.

The renormalization group idea. The RG idea is to construct a scale-dependent hamiltonian $\mathcal{H}_\lambda(\phi)$ which has correlation functions $W_\lambda^{(n)}(x_i)$ satisfying:

$$W_\lambda^{(n)}(x_1, \ldots, x_n) - Z^{-n/2}(\lambda) \, W^{(n)}(\lambda x_1, \ldots, \lambda x_n) = R_\lambda^{(n)}(x_1, \ldots, x_n), \qquad (25.3)$$

where the functions $R^{(n)}$ decrease faster than any power of λ for $\lambda \to \infty$. The mapping $\mathcal{H}(\phi) \mapsto \mathcal{H}_\lambda(\phi)$ is called a RG transformation. We define the transformation such that $\mathcal{H}_{\lambda=1}(\phi) \equiv \mathcal{H}(\phi)$.

In the case of models invariant under space translations, equation (25.3) after a Fourier transformation reads:

$$\widetilde{W}_\lambda^{(n)}(p_1, \ldots, p_n) = Z^{-n/2}(\lambda) \, \lambda^{(1-n)d} \widetilde{W}^{(n)}(p_1/\lambda, \ldots, p_n/\lambda) + \widetilde{R}_\lambda^{(n)}. \qquad (25.4)$$

Various RG transformations differ by the form of $R^{(n)}$ and the function $Z(\lambda)$. In explicit constructions the $R^{(n)}$ are generated by the integration over the large momentum modes of $\phi(x)$. When $R^{(n)} \equiv 0$ and both space and field are continuous variables, the RG transformation can be implemented by a rescaling of space and field. However, this transformation has a fixed point only in exceptional cases (gaussian models) and thus more general transformations have to be considered (see Appendix A10.1). Below we shall omit the terms $R^{(n)}$ and thus equalities between correlation functions will mean up to terms decreasing faster than any power.

The fixed point hamiltonian. The coupling constants appearing in $\mathcal{H}_\lambda$ are now all explicit functions of λ. Let us assume that we have found a RG transformation such that, when λ becomes large, the hamiltonian $\mathcal{H}_\lambda(\phi)$ has a limit $\mathcal{H}^*(\phi)$, the fixed point hamiltonian. If such a fixed point exists in hamiltonian space, then the correlation functions $W_\lambda^{(n)}$ have corresponding limits $W_*^{(n)}$ and equation (25.3) becomes:

$$W^{(n)}(\lambda x_1, \ldots, \lambda x_n) \underset{\lambda \to \infty}{\sim} Z^{n/2}(\lambda) W_*^{(n)}(x_1, \ldots, x_n). \qquad (25.5)$$

If we introduce a second scale parameter μ and calculate $W^{(n)}(\lambda \mu x_i)$ from equation (25.5) in two different ways, we obtain a relation involving only $W_*^{(n)}$:

$$W_*^{(n)}(\mu x_1, \ldots, \mu x_n) = Z_*^{n/2}(\mu) W_*^{(n)}(x_1, \ldots, x_n), \qquad (25.6)$$

with:

$$Z_*(\mu) = \lim_{\lambda \to \infty} Z(\lambda \mu)/Z(\lambda). \qquad (25.7)$$

Equation (25.6) being valid for arbitrary μ immediately implies that Z_* forms a representation of the dilatation semigroup:

$$Z_*(\lambda_1) Z_*(\lambda_2) = Z_*(\lambda_1 \lambda_2). \qquad (25.8)$$

Thus under reasonable assumptions:

$$Z_*(\lambda) = \lambda^{-2d_\phi}. \qquad (25.9)$$

The fixed point correlation functions have a power law behaviour characterized by a positive number d_ϕ which is called the dimension of the order parameter $\phi(x)$.

Returning now to equation (25.7) we conclude that $Z(\lambda)$ also has asymptotically a power law behaviour. Equation (25.5) then shows that the correlation functions $W^{(n)}$ have a scaling behaviour at large distances:

$$W^{(n)}(\lambda x_1, \ldots, \lambda x_n) \underset{\lambda \to \infty}{\sim} \lambda^{-n d_\phi} W_*^{(n)}(x_1, \ldots, x_n), \qquad (25.10)$$

with a power d_ϕ which is a property of the fixed point. The r.h.s. of the equation, which determines the critical behaviour of correlation functions, therefore depends only on the fixed point hamiltonian. In other words the correlation functions corresponding to all hamiltonians which flow after RG transformations into the same fixed point, have the same critical behaviour. This property is called *universality*. The space of hamiltonians is thus divided into *universality classes*. Universality, beyond mean field theory, relies upon the existence of IR fixed points in the space of hamiltonians.

Applied to the two-point function this result shows in particular that if $2d_\phi < d$ the correlation length ξ diverges and therefore that the corresponding hamiltonians are necessarily critical. Critical hamiltonians define in hamiltonian space the *critical surface* which is invariant under a RG flow. In the generic case where ξ is finite the correlation length ξ/λ corresponding to $\mathcal{H}_\lambda$ goes to zero. The Fourier components of correlation functions become momentum independent and thus correlation functions become δ-functions in space. This trivial fixed point corresponds to $2d_\phi = d$.

Renormalization group equations. Let us now try to write the RG transformation, consequence of equation (25.5), for the hamiltonian itself. For this purpose we assume that the dilatation parameter can be varied continuously (on the lattice this can only be implemented approximately) and perform an additional small dilatation which leads from the scale λ to the scale $\lambda \left(1 + \mathrm{d}\lambda/\lambda\right)$.

We look for RG transformations which will bring the transformed hamiltonian closer to a fixed point. This condition involves only $\mathcal{H}_\lambda$ and not the previous history which has led from $\mathcal{H}$ to $\mathcal{H}_\lambda$. Therefore we can write the RG transformation in differential form in terms of a mapping $\mathcal{T}$ of the space of hamiltonians into itself and a real function η defined on the space of hamiltonians as:

$$\lambda \frac{\mathrm{d}}{\mathrm{d}\lambda} \mathcal{H}_\lambda = \mathcal{T}\left[\mathcal{H}_\lambda\right], \tag{25.11}$$

$$\lambda \frac{\mathrm{d}}{\mathrm{d}\lambda} \ln Z\left(\lambda\right) = 2 - d - \eta\left[\mathcal{H}_\lambda\right]. \tag{25.12}$$

As a function of the "time" $\ln \lambda$ these equations define a Markov's process. Note that we have written the r.h.s. of equation (25.12) in an unnatural way for later convenience.

A fixed point hamiltonian $\mathcal{H}^*$ is necessarily a solution of the fixed point equation:

$$\mathcal{T}\left[\mathcal{H}^*\right] = 0. \tag{25.13}$$

The dimension d_ϕ of the field ϕ is then:

$$d_\phi = \tfrac{1}{2}\left(d - 2 + \eta\left[\mathcal{H}^*\right]\right). \tag{25.14}$$

To understand, at least locally, which hamiltonians flow into the fixed point it is necessary to study its stability. We therefore linearize the RG equations near the fixed point.

Linearized RG equations. The RG transformation being determined, let us apply it to a hamiltonian $\mathcal{H}$ close to the fixed point $\mathcal{H}^*$. Setting $\mathcal{H}_\lambda = \mathcal{H}^* + \Delta\mathcal{H}_\lambda$ we linearize the RG equations:

$$\lambda \frac{\mathrm{d}}{\mathrm{d}\lambda} \Delta\mathcal{H}_\lambda = L^*\left(\Delta\mathcal{H}_\lambda\right), \tag{25.15}$$

where L^* is a linear operator independent of λ, acting on the hamiltonian space. Let us assume that L^* has a discrete set of eigenvalues l_i corresponding to a set of eigenoperators $\mathcal{O}_i$. Then $\Delta\mathcal{H}_\lambda$ can be expanded on the $\mathcal{O}_i$'s:

$$\Delta\mathcal{H}_\lambda = \sum h_i\left(\lambda\right) \mathcal{O}_i, \tag{25.16}$$

and the transformation (25.15) becomes:

$$\lambda \frac{\mathrm{d}}{\mathrm{d}\lambda} h_i\left(\lambda\right) = l_i h_i\left(\lambda\right). \tag{25.17}$$

The integration then yields:

$$h_i\left(\lambda\right) = \lambda^{l_i} h_i(1). \tag{25.18}$$

Classification of eigenvectors or scaling fields. The eigenvectors $\mathcal{O}_i$ can be classified into four families depending on the values of the eigenvalues l_i:

(i) Eigenvalues which have a positive real part. The corresponding eigenoperators are called *relevant*. If $\mathcal{H}_\lambda$ has a component on one of these operators, this component will grow with λ, and $\mathcal{H}_\lambda$ will move away from the neighbourhood of $\mathcal{H}^*$. Operators associated with a deviation from criticality are clearly relevant since a dilatation decreases

570 General Renormalization Group Analysis 25.2

the effective correlation length. In Section 25.6 we calculate the corresponding eigenvalue for the ϕ^4 field theory.

(ii) Eigenvalues for which $\mathrm{Re}(l_i) = 0$. Then two situations can arise: either $\mathrm{Im}(l_i)$ does not vanish, and the corresponding component has a periodic behaviour (no example will be met in this work) or $l_i = 0$. Eigenoperators corresponding to a vanishing eigenvalue are called *marginal*. In Section 25.5 we show that in the ϕ^4 field theory the operator $\phi^4(x)$ is marginal in four dimensions. To solve the RG equation (25.11) and determine the behaviour of the corresponding component it is necessary to expand beyond the linear approximation. Generically one finds:

$$\lambda \frac{\mathrm{d}}{\mathrm{d}\lambda} h_i(\lambda) \sim B h_i^2 \,. \tag{25.19}$$

Depending on the sign of the constant B and the initial sign of h_i the fixed point then is marginally unstable or stable. In the latter case the solution takes for λ large the form:

$$h_i(\lambda) \sim -1/(B \ln \lambda) \,. \tag{25.20}$$

A marginal operator generally leads to a logarithmic approach to a fixed point.

An exceptional example is provided by the XY model in two dimensions ($O(2)$ symmetric non-linear σ-model) which instead of an isolated fixed point, has a line of fixed points. The operator which corresponds to a motion along the line is obviously marginal (see Chapter 32).

(iii) Eigenvalues which have negative real parts. The corresponding operators are called *irrelevant*. The effective components on these operators go to zero for large dilatations.

All these eigenvalues, which are characteristic of the fixed point, may appear in the asymptotic expansion at large distances of the correlation functions corresponding to some critical or near-critical hamiltonian.

(iv) Finally some operators do not affect the physics. An example is provided by the operator realizing a constant multiplicative renormalization of the dynamical variables $\phi(x)$. These operators are called *redundant*. In quantum field theory quantum equation of motions correspond to redundant operators with vanishing eigenvalue.

Classification of fixed points. Fixed points can be classified according to their local stability properties, i.e. to the number of relevant operators. This number is also the number of parameters it is necessary to fix to impose to a general hamiltonian to be on the surface which flows into the fixed point. For a non-trivial fixed point corresponding to critical hamiltonians it is the codimension of the critical surface.

25.2 The Gaussian Fixed Point

In this general framework, we have shown that universality emerges as a consequence of the existence of fixed points in hamiltonian space. To be able to quantitatively describe the critical behaviour, we therefore have to construct RG flows and to find their fixed points. A global analysis has never been performed. In general one is able only to exhibit a few fixed points and study their local stability.

However a subspace of hamiltonian space can be completely explored and the RG transformations constructed explicitly: the subspace of quadratic hamiltonians (free field theories in the field theory language) which correspond to gaussian Boltzmann weights.

In this case one can take $R^{(n)} \equiv 0$ in RG transformations.

The fixed points can be identified: critical hamiltonians converge toward the so-called *gaussian fixed point* which, as will become clear below, corresponds to mean field theory.

We thus consider a general quadratic hamiltonian invariant under space translations and rotations. In terms of the Fourier components $\tilde{\phi}(q)$ of the field $\phi(x)$ it can be written:

$$\mathcal{H}(\phi) = \tfrac{1}{2} \int \tilde{\phi}(q) u^{(2)}(q) \tilde{\phi}(-q) \mathrm{d}^d q \,. \tag{25.21}$$

For short range interactions, the class of interactions considered in Chapter 24, the Fourier transform $u^{(2)}(q)$ has a regular expansion for small momentum:

$$u^{(2)}(q) = a + bq^2 + O\left(q^4\right) \,. \tag{25.22}$$

For $a \neq 0$ the hamiltonian is non-critical. If we take for renormalization factor $Z(\lambda) = \lambda^{-d}$ we obtain the trivial fixed point $u_*^{(2)}(q) = $ constant, with a vanishing correlation length. As functions of space variables correlation functions have a δ-function limit $(d_\phi = d/2)$.

For $a = 0$ (critical hamiltonian)

$$u^{(2)}(q) = q^2 + O\left(q^4\right) \,. \tag{25.23}$$

If we take:

$$Z(\lambda) = \lambda^{-(d-2)}, \quad \Rightarrow \quad d_\phi = \tfrac{1}{2}(d-2), \tag{25.24}$$

we find instead the gaussian fixed point:

$$u_*^{(2)}(q) = q^2 \,. \tag{25.25}$$

Eigenoperators. Let us now perform the eigenoperators analysis for the gaussian fixed point (25.25) within the space of *even* hamiltonians, which span the subspace of hamiltonians having an Ising-like symmetry. We write $\Delta\mathcal{H}$ as:

$$\Delta\mathcal{H}(\phi) = \sum_2^\infty \frac{1}{n!} \int \mathrm{d}^d q_1 \dots \mathrm{d}^d q_n \, \tilde{\phi}(q_1) \dots \tilde{\phi}(q_n) \, \delta^d \left(\sum q_i\right) u^{(n)}(q_i) \,. \tag{25.26}$$

It is easy to verify that equation (25.3) leads to the relation:

$$u_\lambda^{(n)}(q_i) = \lambda^d Z^{n/2}(\lambda) u^{(n)}(q_i/\lambda) = \lambda^{d-n(d-2)/2} u^{(n)}(q_i/\lambda) \,. \tag{25.27}$$

The equivalent of equation (25.17) for the fixed point (25.24,25.25) then is:

$$\lambda \frac{\mathrm{d}}{\mathrm{d}\lambda} u_\lambda^{(n)}(q_i) = \left[-\sum_j q_j \frac{\partial}{\partial q_j} + d - \frac{n}{2}(d-2) \right] u_\lambda^{(n)}(q_i) = l u_\lambda^{(n)}(q_i) \,. \tag{25.28}$$

Equations (25.27,25.28) show that the eigenoperators are homogeneous functions of the dynamical variable ϕ and of the momenta q_i (i.e. of the number of derivatives in space). Since we have assumed that the functions $u^{(n)}(q_i)$ are regular, the eigenfunctions are homogeneous polynomials of degree $2k$ in q, and the corresponding eigenvalues $l_{n,k}$ are:

$$l_{n,k} = d - \frac{n}{2}(d-2) - 2k \,. \tag{25.29}$$

(i) The operator $n = 2$, $k = 0$ is relevant and corresponds to a deviation from the critical temperature.

(ii) The operator $n = 2$, $k = 1$ is redundant, it corresponds to a simple renormalization of the dynamical variable.

(iii) Above dimension four, all other operators are irrelevant: on the critical surface the gaussian fixed point is stable. At dimension four, $\int \phi^4(x)\mathrm{d}^d x$ is marginal and logarithmic corrections are expected.

Below dimension four $\int \phi^4(x)\mathrm{d}^d x$ is relevant and when the dimension decreases additional operators become relevant too. The gaussian fixed point is IR unstable.

It is clear that this analysis is quite similar to the analysis of corrections to mean field approximation. It is indeed straightforward to verify that the gaussian fixed point reproduces all results of mean field theory.

It is also strictly equivalent to power counting in quantum field theory as discussed in Chapter 8. The number of relevant operators is just the number of parameters on which depends the renormalized field theory.

25.3 The Critical Theory near Dimension Four: The Effective Field Theory

The main difficulty with the previous approach is that it requires an explicit construction of renormalization group transformations for hamiltonians, which has a chance to lead to a fixed point. Although the general idea is to integrate over large momentum modes of the dynamical variables, its practical implementation is far from being straightforward. We shall therefore write RG equations directly for correlation functions. The limitation of the method we use is that it is applicable only when there exists a fixed point which, in a sense which will become slowly clearer, is close to the gaussian fixed point.

In Section 24.6 we have shown that, for an Ising-like system with short range ferromagnetic interactions, in the critical domain and for $d \leq 4$, the sum of the most divergent contributions order by order in a mean field expansion can be reproduced by an effective local field theory whose action is just given by the first relevant terms of Landau–Ginzburg–Wilson hamiltonian:

$$\mathcal{H}(\phi) \equiv \beta H = \int \mathrm{d}^d x \left[\frac{1}{2} c \left(\nabla \phi \right)^2 + \frac{1}{2} a \phi^2 + b \frac{1}{4!} \phi^4 \right], \qquad (25.30)$$

with a, b and c being *regular* functions of the temperature for T close to T_c.

Consistently the analysis of the stability of the gaussian fixed point has shown that in four dimensions the ϕ^4 interaction becomes marginal while all other interactions remain irrelevant. If the dimensions of operators are *continuous* functions of the space dimension, hamiltonian (25.30) should contain all relevant operators at least in some neighbourhood of dimension four.

The hamiltonian (25.30) generates a perturbative expansion of field theory type which can be described in terms of Feynman diagrams. These have to be calculated with a cut-off of order 1, reflection of the initial microscopic structure. We shall eventually show that the precise cut-off procedure is irrelevant except that it should satisfy some general conditions. For example the propagator can be modified (as in Pauli–Villars's regularization) but the inverse propagator in momentum space must remain a regular function of momentum (the forces are short range).

In this chapter we restrict ourselves to Ising-like systems, the field ϕ has only one component. Generalizations to the N-vector model with $O(N)$ symmetry, which is straightforward, and to several-component models with more than one ϕ^4 coupling constant will be briefly analyzed in Section 26.6.

We have already seen in Chapter 24 that a convenient way to study the problem of infrared singularities is to rescale all space or momentum variables, and measure distances in units of the correlation length, or, at the critical temperature, in some arbitrary unit much larger than the lattice spacing and corresponding to the typical distances at which correlations are measured. After such a rescaling the momentum cut-off becomes a large momentum Λ analogous to the cut-off used to regularize quantum field theory.

Let us perform such a rescaling here, and rescale also the field $\phi(x)$ in such a way that the coefficient of $[\nabla\phi(x)]^2$ becomes the standard $1/2$:

$$x \mapsto \Lambda x, \tag{25.31}$$

$$\phi(x) \mapsto \zeta\phi(x). \tag{25.32}$$

After this rescaling all quantities have a dimension in units of Λ. Our choice of normalization for the gradient term implies:

$$\zeta = c^{-1/2}\Lambda^{1-d/2}, \tag{25.33}$$

which shows that ϕ now has in terms of Λ its canonical dimension $d/2 - 1$. The action $\mathcal{H}(\phi)$ then becomes:

$$\mathcal{H}(\phi) = \int d^d x \left\{ \tfrac{1}{2}[\nabla\phi(x)]^2 + \tfrac{1}{2}r\phi^2(x) + \frac{1}{4!}g\Lambda^{4-d}\phi^4(x) \right\}, \tag{25.34}$$

with:

$$r = a\Lambda^2/c, \qquad g = b/c^2. \tag{25.35}$$

Let us call r_c the parameter which corresponds, at g fixed, to the critical temperature T_c at which the correlation length ξ diverges. In terms of the scale Λ the critical domain is then defined by:

$$\text{physical mass } = \xi^{-1} \ll \Lambda \Rightarrow |r - r_c| \ll \Lambda^2$$
$$\text{distances} \gg 1/\Lambda \quad \text{or momenta} \ll \Lambda, \tag{25.36}$$
$$\text{magnetization } M \equiv \langle\phi(x)\rangle \ll \zeta^{-1} \sim \Lambda^{(d/2)-1}.$$

Note that these conditions are met if Λ is identified with the cut-off of a usual field theory. However an inspection of the action (25.34) also shows that, in contrast with conventional quantum field theory, the ϕ^4 coupling constant has a dependence in Λ given *a priori*. For $d < 4$ the ϕ^4 coupling is very large in terms of the scale relevant for the critical domain. In the usual formulation of quantum field theory instead the *bare* coupling constant is also an adjustable parameter. In the latter case the resulting quantum field theory is less generic.

25.4 Renormalization Group Equations for the Critical Theory: The ε-Expansion

The hamiltonian (25.34) can now be studied by field theoretical methods. Using a power counting argument, we have shown in Chapter 24 that the critical theory does not exist in perturbation theory for any dimension smaller than 4. If we define, by dimensional continuation, a critical theory in $d = 4 - \varepsilon$ dimensions, even for arbitrarily small ε there always exists an order in perturbation ($\sim 2/\varepsilon$) at which IR (infrared) divergences appear.

Therefore the idea, originally due to Wilson and Fisher, is to perform a double series expansion in powers of the coupling constant g and ε. Order by order in this expansion, the critical behaviour differs from the mean field behaviour only by powers of logarithm, and we can construct a perturbative critical theory by adjusting r to its critical value $r_c\,(T = T_c)$.

To study the large cut-off limit we then use methods developed for the construction of the renormalized massless ϕ^4 field theory. We introduce rescaled correlation functions, defined by renormalization conditions at a new scale $\mu \ll \Lambda$, and functions of a renormalized coupling constant g_{r}:

$$
\begin{cases}
\Gamma_{\mathrm{r}}^{(2)}\,(p; g_{\mathrm{r}}, \mu, \Lambda)|_{p^2=0} = 0\,, \\[2mm]
\dfrac{\partial}{\partial p^2}\Gamma_{\mathrm{r}}^{(2)}\,(p; g_{\mathrm{r}}, \mu, \Lambda)|_{p^2=\mu^2} = 1\,, \\[2mm]
\Gamma_{\mathrm{r}}^{(4)}\,(p_i = \mu\theta_i; g_{\mathrm{r}}, \mu, \Lambda) = \mu^{\varepsilon} g_{\mathrm{r}}\,,
\end{cases}
\tag{25.37}
$$

in which θ_i is a numerical vector. These correlation functions are related to the original ones by the equations:

$$
\Gamma_{\mathrm{r}}^{(n)}\,(p_i; g_{\mathrm{r}}, \mu, \Lambda) = Z^{n/2}\,(g, \Lambda/\mu)\,\Gamma^{(n)}\,(p_i; g, \Lambda)\,.
\tag{25.38}
$$

Renormalization theory (more precisely a slightly extended version adapted to the ε-expansion) tells us that the functions $\Gamma_{\mathrm{r}}^{(n)}\,(p_i; g_{\mathrm{r}}, \mu, \Lambda)$ of equation (25.38) have at p_i, g_{r} and μ fixed, a large cut-off limit which are the renormalized correlation functions $\Gamma_{\mathrm{r}}^{(n)}\,(p_i; g_{\mathrm{r}}, \mu)$. A detailed analysis actually shows:

$$
\Gamma_{\mathrm{r}}^{(n)}\,(p_i; g_{\mathrm{r}}, \mu, \Lambda) = \Gamma_{\mathrm{r}}^{(n)}\,(p_i; g_{\mathrm{r}}, \mu) + O\left(\Lambda^{-2}(\ln \Lambda)^L\right)\,,
\tag{25.39}
$$

in which the power of $\ln \Lambda$ increases with the order in g and ε (see equation (25.45)). Furthermore the renormalized functions $\Gamma_{\mathrm{r}}^{(n)}$ do not depend on the specific cut-off procedure and, given the normalization conditions (25.37), are therefore universal. Since the renormalized functions $\Gamma_{\mathrm{r}}^{(n)}$ and the initial ones $\Gamma^{(n)}$ are asymptotically proportional, both functions have the same small momentum or large distance behaviour. The renormalized functions thus contain the whole information about the universal critical behaviour. It follows that we could restrict ourselves from now on to the study of the renormalized field theory. And indeed it is the only one to be really useful for explicit calculations of universal quantities (see Chapter 28). However, we shall eventually be interested not only by the asymptotic critical behaviour, but also by the corrections to the asymptotic theory. Furthermore, renormalized quantities are not directly obtained in non-perturbative calculations. For these reasons it is quite useful to express the implications of equation (25.38) directly on the initial theory.

Bare RG equations. Differentiating equation (25.38) with respect to Λ at g_{r} and μ fixed, and taking into account (25.39), we can derive a new identity:

$$
\left.\Lambda\frac{\partial}{\partial\Lambda}\right|_{g_{\mathrm{r}},\mu \text{ fixed}} Z^{n/2}\,(g, \Lambda/\mu)\,\Gamma^{(n)}\,(p_i; g, \Lambda) = O\left(\Lambda^{-2}(\ln \Lambda)^L\right)\,.
\tag{25.40}
$$

In the first part of our study we neglect corrections subleading by powers of Λ. We shall return to this point in Chapter 27.

Then, using chain rule, we can rewrite equation (25.40) as:

$$\left[\Lambda\frac{\partial}{\partial\Lambda} + \beta\left(g,\Lambda/\mu\right)\frac{\partial}{\partial g} - \frac{n}{2}\eta\left(g,\Lambda/\mu\right)\right]\Gamma^{(n)}\left(p_i;g,\Lambda\right) = 0.\qquad(25.41)$$

The functions β and η, which are dimensionless and may thus depend only on the dimensionless quantities g and Λ/μ, are defined by:

$$\beta\left(g,\Lambda/\mu\right) = \Lambda\frac{\partial}{\partial\Lambda}\bigg|_{g_{\mathrm{r}},\mu}\, g\,,\qquad(25.42a)$$

$$\eta\left(g,\Lambda/\mu\right) = -\Lambda\frac{\partial}{\partial\Lambda}\bigg|_{g_{\mathrm{r}},\mu}\ln Z\left(g,\Lambda/\mu\right).\qquad(25.42b)$$

However, the functions β and η can also be directly calculated from equation (25.41) in terms of functions $\Gamma^{(n)}$ which do not depend on μ. Therefore the functions β and η cannot depend on the ratio Λ/μ. Of course, if we examine the definitions (25.42) we see that the r.h.s. has a weak dependence in Λ/μ, but consistency requires that this dependence, which goes to zero like some power of μ/Λ, should be neglected, as in equation (25.40). Then equation (25.41) can be rewritten:

$$\left[\Lambda\frac{\partial}{\partial\Lambda} + \beta\left(g\right)\frac{\partial}{\partial g} - \frac{n}{2}\eta(g)\right]\Gamma^{(n)}\left(p_i;g,\Lambda\right) = 0.\qquad(25.43)$$

Equation (25.43) is an equation satisfied when the cut-off is large by the physical correlation functions of statistical mechanics which are also the bare correlation functions of quantum field theory. It is, as we have shown, a direct consequence of the existence of a renormalized theory. It will be implicit in the solution of equation (25.43) that it also implies the existence of a renormalized theory. Both statements are thus equivalent.

Beyond leading order. Let us finally characterize more precisely the approximation involved in writing equation (25.43). In a series expansion in powers of g and ε:

$$\Gamma^{(n)}\left(p_i;g,\Lambda\right) = \sum_{r,s}\Gamma^{(n)}_{rs}g^r\varepsilon^s,\qquad(25.44)$$

the coefficients $\Gamma^{(n)}_{rs}$ have an asymptotic expansion for $|p_i|/\Lambda$ small of the form:

$$\Gamma^{(n)}_{rs} = \sum_{l=0}^{L(n,r,s)}\left[(\ln\Lambda)^l\,A^{(1)}_{lrs} + \frac{1}{\Lambda^2}\left(\ln\Lambda\right)^l A^{(2)}_{lrs} + \cdots\right],\qquad(25.45)$$

with $L(n,r,s) = 1 + r + s - n/2$ for $n = 4$ and $L(n,r,s) = r + s - n/2$ for $n \neq 4$.

The RG equations (25.43) are exact for the sum of the perturbative contributions which do not vanish for Λ large, as can be verified by expanding equations (25.43) in powers of g and ε.

25.5 Solution of the RG Equations

Equation (25.43) can be solved by the method of characteristics. One introduces a dilatation parameter λ and looks for functions $g(\lambda)$ and $Z(\lambda)$ such that:

$$\lambda \frac{\mathrm{d}}{\mathrm{d}\lambda} \left[Z^{-n/2}(\lambda) \, \Gamma^{(n)}(p_i; g(\lambda), \lambda\Lambda) \right] = 0 . \tag{25.46}$$

Differentiating explicitly with respect to λ, we find that equation (25.46) is consistent with equation (25.43) provided that:

$$\lambda \frac{\mathrm{d}}{\mathrm{d}\lambda} g(\lambda) = \beta(g(\lambda)), \qquad g(1) = g; \tag{25.47a}$$

$$\lambda \frac{\mathrm{d}}{\mathrm{d}\lambda} \ln Z(\lambda) = \eta(g(\lambda)), \qquad Z(1) = 1. \tag{25.47b}$$

The function $g(\lambda)$ is the effective coupling at the scale λ, and is governed by the flow equation (25.47a). Equations (25.46) and (25.47) imply:

$$\Gamma^{(n)}(p_i; g, \Lambda) = Z^{-n/2}(\lambda) \, \Gamma^{(n)}(p_i; g(\lambda), \lambda\Lambda) .$$

It is actually convenient to rescale Λ by a factor $1/\lambda$ and rewrite the equation:

$$\Gamma^{(n)}(p_i; g, \Lambda/\lambda) = Z^{-n/2}(\lambda) \, \Gamma^{(n)}(p_i; g(\lambda), \Lambda) . \tag{25.48}$$

Equations (25.47,25.48) implement approximately (because terms subleading by powers of Λ have been neglected) the RG ideas as presented in Section 25.1: since the coupling constant $g(\lambda)$ characterizes the hamiltonian $\mathcal{H}_\lambda$, equation (25.47a) is the equivalent of equation (25.11) (up to the change $\lambda \mapsto 1/\lambda$); equations (25.12) and (25.47b) differ only by the definition of $Z(\lambda)$.

Let us now solve equations (25.47):

$$\int_g^{g(\lambda)} \frac{\mathrm{d}g'}{\beta(g')} = \ln\lambda , \tag{25.49a}$$

$$\int_1^\lambda \frac{\mathrm{d}\sigma}{\sigma} \eta(g(\sigma)) = \ln Z(\lambda) . \tag{25.49b}$$

The equation (25.43) is the renormalization group equation in differential form. Equations (25.48) and (25.49) are the integrated RG equations. In what follows we explicitly assume that the RG functions $\beta(g)$ and $\eta(g)$ are regular functions of g for $g \geq 0$. In equation (25.48) we see that it is equivalent to increase Λ or to decrease λ. To investigate the large Λ limit we therefore study the behaviour of the effective coupling constant $g(\lambda)$ when λ goes to zero. Equation (25.49a) shows that $g(\lambda)$ increases if the function β is negative, or decreases in the opposite case. Fixed points correspond to zeros of the β-function which therefore play an essential role in the analysis of the critical behaviour. Those where the β-function has a negative slope are IR repulsive: the effective coupling moves away from such zeros, except if the initial coupling has exactly a fixed point value. Conversely those where the slope is positive are IR attractive.

The RG functions β and η can be calculated in perturbation theory. The relation between bare and renormalized coupling constant:

$$\mu^\varepsilon g_r = \Lambda^\varepsilon \left[g - \frac{3g^2}{16\pi^2} \left(\ln \frac{\Lambda}{\mu} + \text{const.} \right) \right] + O\left(g^3, g^2 \varepsilon \right), \tag{25.50}$$

($\varepsilon = 4 - d$) and the definition (25.42a) imply:

$$\beta\left(g, \varepsilon\right) = -\varepsilon g + \frac{3g^2}{16\pi^2} + O\left(g^3, g^2 \varepsilon\right). \tag{25.51}$$

We see from the explicit expression (25.51) of the β-function that above 4 dimensions, i.e. $\varepsilon < 0$, if initially g is small, $g(\lambda)$ decreases approaching the origin $g = 0$. We recover that the gaussian fixed point is IR stable.

Below 4 dimensions, if initially g is very small, expression (25.51) shows that $g(\lambda)$ first increases, a behaviour reflecting the instability of the gaussian fixed point. There are then three possibilities:

(i) The β-function remains always negative, $g(\lambda)$ increases and tends towards $+\infty$ when λ goes to zero.

(ii) The β-function remains negative for all g but the integral in equation (25.49a) converges at infinity. Then there exists a maximal value of λ for which the equation has a solution: the RG method fails, at least in this simple formulation.

(iii) The β-function has a zero g^*,

$$\beta\left(g^*\right) = 0. \tag{25.52}$$

Then equation (25.49a) shows that $g(\lambda)$ has g^* as an asymptotic limit. Let us call ω the slope at such a zero:

$$\omega = \beta'\left(g^*\right) > 0, \tag{25.53}$$

and let us linearize equation (25.49a) around the fixed point:

$$\int_g^{g(\lambda)} \frac{dg'}{\omega\left(g' - g^*\right)} \sim \ln \lambda. \tag{25.54}$$

Integrating, we get:

$$|g(\lambda) - g^*| \underset{\lambda \to 0}{=} O\left(\lambda^\omega\right). \tag{25.55}$$

The explicit expression (25.51) shows that, in the sense of an expansion in powers of ε, $\beta(g)$ has indeed such a zero:

$$g^* = \frac{16\pi^2}{3}\varepsilon + O\left(\varepsilon^2\right), \tag{25.56}$$

$$\omega = \varepsilon + O\left(\varepsilon^2\right). \tag{25.57}$$

Below dimension 4, at least for ε infinitesimal, this non-gaussian fixed point is IR stable. In dimension 4 this fixed point merges with the gaussian fixed point and the eigenvalue ω vanishes indicating the appearance of the marginal operator already identified in the analysis of the gaussian fixed point in Section 25.2.

Let us now assume that $\Gamma^{(n)}(g^*)$ and $\eta(g^*)$ are finite, conditions which are satisfied within the framework of the ε-expansion. From equation (25.49b) we derive the behaviour of $Z(\lambda)$ for λ small. The integral in the l.h.s. is dominated by small values of σ. It follows:

$$\ln Z(\lambda) \underset{\lambda \to 0}{\sim} \eta \ln \lambda, \tag{25.58}$$

where we have set:

$$\eta = \eta(g^*).$$

Equation (25.48) then determines the behaviour of $\Gamma^{(n)}(p_i; g, \Lambda)$ for Λ large:

$$\Gamma^{(n)}(p_i; g, \Lambda/\lambda) \sim \lambda^{-n\eta/2}\Gamma^{(n)}(p_i; g^*, \Lambda). \tag{25.59}$$

On the other hand from simple dimensional considerations we know that:

$$\Gamma^{(n)}(p_i; g, \Lambda/\lambda) = \lambda^{-d+(n/2)(d-2)}\Gamma^{(n)}(\lambda p_i; g, \Lambda). \tag{25.60}$$

Combining this equation with equation (25.59) we obtain:

$$\Gamma^{(n)}(\lambda p_i; g, \Lambda) \underset{\lambda \to 0}{\sim} \lambda^{d-(n/2)(d-2+\eta)}\Gamma^{(n)}(p_i; g^*, \Lambda). \tag{25.61}$$

This equation shows that the critical correlation functions have a power law behaviour for small momenta, independent of the initial value of the ϕ^4 coupling constant g.

Setting $n = 2$ in equation (25.61) we obtain in particular the small momentum behaviour of the inverse two-point function. For the correlation function $W^{(2)}(p)$ this yields:

$$W^{(2)}(p) = \left[\Gamma^{(2)}(p)\right]^{-1} \underset{|p| \to 0}{\sim} 1/p^{2-\eta}. \tag{25.62}$$

We therefore verify that the definition of equation (25.58) coincides with the usual definition of the critical exponent η. The spectral representation of the two-point function (Section 7.4) implies $\eta > 0$. Since in perturbation theory the first contribution to the field renormalization $Z(g, \Lambda/\mu)$ arises at order g^2, $\eta(g)$ is of order g^2, and η of order ε^2.

A short calculation yields:

$$\eta = \frac{\varepsilon^2}{54} + O(\varepsilon^3). \tag{25.63}$$

A semi-quantitative prediction of the theory is that η is numerically small in three dimensions.

Finally let us note that equation (25.61) can be interpreted by saying that the field $\phi(x)$, which had at the gaussian fixed point a canonical dimension $(d-2)/2$, has now acquired an anomalous dimension d_ϕ (equation (25.14)):

$$d_\phi = \tfrac{1}{2}(d - 2 + \eta).$$

All these results call for a few comments. Within the framework of the ε-expansion, we have shown that all correlation functions have, for $d < 4$, a long distance behaviour different from the one predicted by mean field theory. In addition the critical behaviour does not depend on the initial value of the ϕ^4 coupling constant g. At least for ε small we can hope that the analysis of leading IR singularities of Chapter 24 remains valid and thus it does not depend on any other coupling either (this point will be further discussed in Section 27.4). Therefore the critical behaviour is *universal*, although less universal than in MFT, in the sense that it depends only on some small number of qualitative features of the system under consideration.

Moreover the correlation functions obtained by neglecting, in perturbation theory and within the ε-expansion, power law corrections when the cut-off is large, and which satisfy exactly RG equations (25.43), define implicitly a one parameter family of critical hamiltonians which correspond to a RG trajectory which goes from the neighbourhood of the gaussian fixed point $g = 0$ which is IR unstable below 4 dimensions to a non-trivial stable fixed point g^*.

Finally the consistency of this analysis based on the ε-expansion relies on the following observation: the IR divergences found in the fixed dimension perturbation theory are generated by an expansion around the wrong, since IR repulsive, fixed point. The ε-expansion allows us to interchange limits and to follow perturbatively the attractive IR fixed point.

25.6 Critical Correlation Functions with $\phi^2(x)$ Insertions

In the next chapter we shall study the whole critical domain (25.36). In particular the following method, already explained in Section 10.10, will be used: we shall expand the correlation functions for $T \neq T_c$ in terms of *critical* correlation functions with $\int \mathrm{d}^d x\, \phi^2(x)$ insertions. The operator $\phi^2(x)$ has a direct physical interpretation. It is the most singular part (i.e. the most relevant) of the energy density (25.30).

Therefore we now discuss the long distance properties of the critical mixed correlation functions of the order parameter ϕ and the energy density $\frac{1}{2}\phi^2$. We denote by $\Gamma^{(l,n)}\left(q_1,\ldots,q_l;p_1,\ldots,p_n;g,\Lambda\right)$ the 1PI functions of $n\,\phi(x)$ fields and $l\,\frac{1}{2}\phi^2(x)$ operators, (with $(l+n) \geq 2$). Renormalization theory tells us that we can define renormalized correlation functions $\Gamma_{\mathrm{r}}^{(l,n)}\left(q_i;p_j;g_{\mathrm{r}},\mu\right)$ which in addition to conditions (25.37) satisfy:

$$
\begin{aligned}
\left.\Gamma_{\mathrm{r}}^{(1,2)}\left(q;p_1,p_2;g_{\mathrm{r}},\mu\right)\right|_{p_1^2=p_2^2=\mu^2,\ p_1\cdot p_2=-\frac{1}{3}\mu^2} &= 1\,,\\
\left.\Gamma_{\mathrm{r}}^{(2,0)}\left(q,-q;g_{\mathrm{r}},\mu\right)\right|_{q^2=\frac{4}{3}\mu^2} &= 0\,,
\end{aligned}
\tag{25.64}
$$

and are related to the original ones (see equation (10.62)) by:

$$
\lim_{\Lambda\to\infty} Z^{n/2}\left(\frac{Z_2}{Z}\right)^l\left[\Gamma^{(l,n)}\left(q_i;p_j;g,\Lambda\right)-\delta_{n0}\delta_{l2}\Lambda^{-\varepsilon}A\right]=\Gamma_{\mathrm{r}}^{(l,n)}\left(q_i;p_j;g_{\mathrm{r}},\mu\right)\,.
\tag{25.65}
$$

$Z_2\left(g,\Lambda/\mu\right)$ and $A\left(g,\Lambda/\mu\right)$ are two new renormalization constants.

Differentiating with respect to Λ at g_{r} and μ fixed, as we have done in Section 25.4, and using chain rule we obtain a set of RG equations:

$$
\left[\Lambda\frac{\partial}{\partial\Lambda}+\beta\left(g\right)\frac{\partial}{\partial g}-\frac{n}{2}\eta\left(g\right)-l\eta_2\left(g\right)\right]\Gamma^{(l,n)}=\delta_{n0}\delta_{l2}\Lambda^{-\varepsilon}B\left(g\right)\,.
\tag{25.66}
$$

In addition to β and η (equations (25.42)) two new RG functions, $\eta_2(g)$ and $B(g)$, appear:

$$
\eta_2\left(g\right)=-\left.\Lambda\frac{\partial}{\partial\Lambda}\right|_{g_{\mathrm{r}},\mu}\ln\left[Z_2\left(g,\Lambda/\mu\right)/Z\left(g,\Lambda/\mu\right)\right]\,,
\tag{25.67}
$$

$$
B\left(g\right)=\left[\left.\Lambda\frac{\partial}{\partial\Lambda}\right|_{g_{\mathrm{r}},\mu}-2\eta_2\left(g\right)-\varepsilon\right]A\left(g,\Lambda/\mu\right)\,.
\tag{25.68}
$$

Again these two RG functions, because they are calculable in terms of correlation functions which do not depend on μ, cannot depend on the ratio Λ/μ.

Note that for $n = 0$, $l = 2$, the RG equation (25.66) is not homogeneous. This is a consequence of the non-multiplicative character of the renormalization in this case. Multiple insertions of operators of higher dimension like $\phi^4(x)$, lead to even more complicate RG equations. This question have been discussed in Chapter 12 and we want here only to warn the reader against a too naive application of RG ideas.

In the homogeneous case equation (25.66) can be solved by the method of characteristics explained in Section 25.5, exactly in the same way as equation (25.41). With the RG function $\eta_2(g)$ is associated a new scale dependent function $\zeta_2(\lambda)$:

$$\lambda \frac{\mathrm{d}}{\mathrm{d}\lambda} \left[Z^{-n/2}(\lambda) \zeta_2^{-l}(\lambda) \Gamma^{(l,n)} \left(q_i; p_j; g(\lambda), \lambda\Lambda \right) \right] = 0,$$

which therefore satisfies:

$$\lambda \frac{\mathrm{d}}{\mathrm{d}\lambda} \ln \zeta_2(\lambda) = \eta_2[g(\lambda)].$$

It then follows from the arguments of Section 25.5 that the critical behaviour of $\Gamma^{(l,n)}$ is governed by the IR fixed point g^*.

To relate the RG function $\eta_2(g)$ to standard exponents we also introduce the function $\nu(g)$:

$$1/\nu(g) = \eta_2(g) + 2. \tag{25.69}$$

For λ small ζ_2 behaves like:

$$\ln \zeta_2 \sim (1/\nu - 2) \ln \lambda,$$

where the new critical exponent ν is defined by:

$$\nu = \nu(g^*).$$

Combining again the solution of equation (25.66) with simple dimensional analysis we finally obtain the behaviour of $\Gamma^{(l,n)}$:

$$\Gamma^{(l,n)} \left(\lambda q_i; \lambda p_j; g, \Lambda \right) \underset{\lambda \to 0}{\propto} \lambda^{d - n(d-2+\eta)/2 - l/\nu}. \tag{25.70}$$

Using equations (25.67, 25.69) it is straightforward to calculate $\nu(g)$ at one-loop order:

$$2\nu(g) = 1 + \frac{g}{32\pi^2} + O\left(g^2\right), \tag{25.71}$$

and then, setting $g = g^*$ (equation (25.56)), the exponent ν at first order in ε:

$$2\nu = 1 + \frac{\varepsilon}{6} + O\left(\varepsilon^2\right).$$

The $\langle \phi^2 \phi^2 \rangle$ correlation function. The ϕ^2 (energy density) two-point function $\Gamma^{(2,0)}$ satisfies an inhomogeneous RG equation. To solve it one first looks for a particular solution, which can be chosen of the form $\Lambda^{-\varepsilon} C_2(g)$:

$$\beta(g) C_2'(g) - [\varepsilon + 2\eta_2(g)] C_2(g) = B(g). \tag{25.72}$$

The solution is uniquely determined by imposing its *regularity* at $g = g^*$.

The general solution of equation (25.66) then is the sum of this particular solution and the general solution of the homogeneous equation which has a behaviour given by equation (25.70):

$$\Gamma^{(2,0)}(q;g,\Lambda) - \Lambda^{-\varepsilon}C_2(g) \underset{q\to 0}{\sim} K\Lambda^{2/\nu-4}q^{d-2/\nu}. \tag{25.73}$$

Note that the regular contribution depends on g while the constant K is g independent .

It now remains to study the behaviour of correlation functions in the critical domain (25.36) away from T_c and this will be the subject of Chapter 26.

The dimension of the ϕ^2 operator. The scaling behaviour (25.70) determines the dimension of the operator ϕ^2 at the IR fixed point. More generally the connected n-point correlation function with insertions of l scaling fields $\mathcal{O}_i$ of dimensions d_i has, in real space, the scaling behaviour:

$$\left\langle \prod_i \mathcal{O}_i(y_i/\lambda) \prod_j \phi(x_j/\lambda) \right\rangle \underset{\lambda\to 0}{\propto} \lambda^D, \quad \text{with } D = nd_\phi + \sum_i d_i .$$

After Fourier transformation, factorization of the δ-function of momentum conservation and Legendre transformation with respect to the field ϕ one finds (neglecting possible regular terms):

$$\Gamma^{(n)}_{\mathcal{O}_1,\ldots,\mathcal{O}_l}(\lambda q_i,\lambda p_j) \propto \lambda^{-D'},$$

with $D' = D - (n+l-1)d - n(2d_\phi - d) = d - nd_\phi + \sum_i d_i - d$. It follows in particular that $d_{\phi^2} = d - 1/\nu$.

Remark. The physics we intend to describe corresponds to integer values of ε, $\varepsilon = 1,2$. Although we can only prove the validity of all RG results within the framework of the ε-expansion, we shall eventually assume that their validity extends beyond an infinitesimal neighbourhood of dimension 4. The comparison of the results obtained from the summed expansion with experimental or numerical data thus provides a crucial test of the theory (see Chapter 28).

The fixed point hamiltonian. Let us note that the fixed point hamiltonian $\mathcal{H}^*$ has never been explicitly constructed for any non-gaussian theory. Indications from field theory are that such a construction is not easy. Since fixed point correlation functions have been calculated, they define implicitly a class of fixed point hamiltonians. It has also been proposed to define $\mathcal{H}^*$ as a limit of a sequence of hamiltonians $\mathcal{H}_k$ whose correlation functions $W_k^{(n)}$ converge towards fixed point correlation functions $W_*^{(n)}$ in the following way:

$$\left| W_k^{(n)}(\lambda q_i) - W_*^{(n)}(\lambda q_i) \right| = O\left(\lambda^{2k}\right).$$

In Sections 25.3–25.5 we have constructed $\mathcal{H}_1$.

Quantum field theory in particle physics. Let us now comment about the significance of these results for field theory in the particle physics context. From the preceding analysis we understand that the existence of a renormalizable field theory, beyond perturbation theory, relies on two properties:

(i) Mass renormalization must make sense beyond perturbation theory (the existence of a critical temperature).

(ii) An IR fixed point in the RG sense must exist, to ensure that the long distance physics is short-distance insensitive and can be described by a renormalizable quantum field theory.

An inspection of the action (25.30) however shows that, in contrast with conventional QFT, the ϕ^4 coupling constant in the critical phenomena applications has a dependence in Λ given *a priori*. For $d < 4$ the ϕ^4 coupling is very large in terms of the scale relevant for the critical domain. In the usual formulation of QFT instead the *bare* coupling constant is considered as an adjustable parameter: A renormalizable QFT corresponds to a hamiltonian maintained close to the gaussian fixed point by adjusting the coefficients of all relevant and marginal operators. The resulting theory is thus less generic.

These coefficients then introduce a new scale in the theory (called μ^{-1} in the text), much larger than the microscopic scale, in such a way that it is possible to define, in addition to the universal long distance physics, a short distance or large momentum physics with $\mu \ll |p_i| \ll \Lambda$.

Note however that if we demand to particle physics theories to be *natural*, i.e. that coupling constants at the cut-off scale (which can be identified with the scale of some new physics) are numbers of order 1, we are back into the statistical framework.

Note finally that the RG analysis indicates that in four dimensions, in the domain of attraction of the $g = 0$ IR fixed point, the renormalized coupling constant of the ϕ^4 field theory goes to zero logarithmically when the cut-off becomes infinite. Since no other fixed point seems to exist, this leads to the so-called *triviality problem*.

Bibliographical Notes

The modern formulation of the RG ideas is due to

K.G. Wilson, *Phys. Rev.* B4 (1971) 3174, 3184;

and presented in an expanded form in

K.G. Wilson and J. Kogut, *Phys. Rep.* 12C (1974) 75.

The idea of RG transformations was earlier proposed in a simplified form in

L.P. Kadanoff, *Physics* 2 (1966) 263.

The systematic classification of operators can be found in

F.J. Wegner, *Phys. Rev.* B5 (1972) 4529, B6 (1972) 1891, and in Wegner's contribution to *Phase Transitions and Critical Phenomena,* vol. 6, C. Domb and M.S. Green eds. (Academic Press, London 1976).

The idea of the ε-expansion is due to

K.G. Wilson and M.E. Fisher, *Phys. Rev. Lett.* 28 (1972) 240.

After Wilson's original articles, several authors have realized that RG methods, as developed in Quantum Field Theory, could be applied to Critical Phenomena

C. Di Castro, *Lett. Nuovo Cimento.* 5 (1972) 69; G. Mack, *Kaiserslautern 1972,* Lecture Notes in Physics vol. 17, W. Ruhl and A. Vancura eds. (Springer Verlag, Berlin 1972); E. Brézin, J.C. Le Guillou and J. Zinn-Justin, *Phys. Rev.* D8 (1973) 434, 2418; P.K. Mitter, *Phys. Rev.* D7 (1973) 2927; G. Parisi, *Cargèse Lectures 1973,* published in *J. Stat. Phys.* 23 (1980) 49; B. Schroer, *Phys. Rev.* B8 (1973) 4200; C. Di Castro, G. Jona-Lasinio and L. Peliti, *Ann. Phys. (NY)* 87 (1974) 327; F. Jegerlehner and B. Schroer, *Acta Phys. Austr. Suppl.* XI (1973) 389 (Springer-Verlag, Berlin).

The RG equations for the bare theory were first derived in

J. Zinn-Justin, *Cargèse Lectures 1973,* unpublished, incorporated in the contribution of E. Brézin, J.C. Le Guillou and J. Zinn-Justin to *Phase Transitions and Critical Phenomena* vol. 6, quoted in Chapter 24.

Exercises

Exercise 25.1

Study with the formalism of Chapter 25 the tricritical ϕ^6 field theory near three dimensions. We recall that the tricritical point is defined by:

$$\Gamma^{(2)}(p = 0) = 0, \quad \Gamma^{(4)}(p_i = 0) = 0.$$

Calculate in particular the RG functions at leading order. We write the interaction $g\Lambda^{6-2d}\phi^6/6!$ and set $\varepsilon = 6 - 2d$. It is advised to use dimensional regularization.

Extend this analysis to the $u\phi^{2N}/(2N)!$ field theory near the dimension in which it is renormalizable (the upper-critical dimension). It is suggested to set $\varepsilon = 2N - d(N - 1)$.

Exercise 25.2

Long range forces. From a mean field analysis one can show that O(N) symmetric models with two-body long range forces yield, near a continuous phase transition, a $(\phi^2)^2$ field theory with however an unusual, non-local, propagator. For a potential V decaying at large distance r as $V(r) \propto r^{-d-\sigma}$ the effective hamiltonian reads

$$S(\phi) = \int \mathrm{d}^d x \left[\tfrac{1}{2}\phi \cdot (-\Delta)^{\sigma/2}\phi + \tfrac{1}{2}r\phi^2 + \frac{g}{4!}\left(\phi^2\right)^2 \right],$$

with $\sigma < 2$. This means that in momentum space the propagator is

$$\Delta_{ij}(k) = \frac{\delta_{ij}}{k^\sigma}.$$

25.2.1. Find the upper-critical dimension, the equivalent of dimension four for the short range case.

25.2.2. What can be said about the field strength renormalization and the exponent η, and the limit $\sigma \to 2$?

25.2.3. Calculate the β-function at one-loop order and the exponent ν at leading order in the ε-expansion.

26 SCALING BEHAVIOUR IN THE CRITICAL DOMAIN

In Chapter 25 we have established the scaling behaviour of correlation functions at criticality, $T = T_c$. We now study the critical domain which is defined by the property that the correlation length is large with respect to the microscopic scale, but finite.

Using results proven in Chapter 10, we first demonstrate strong scaling above T_c: in the critical domain above T_c, all correlation functions, after rescaling, can be expressed in terms of universal correlation functions, in which the scale of distance is provided by the correlation length.

However, because the correlation length is singular at T_c, this formalism does not allow to cross the critical temperature and thus to describe the whole critical domain. Therefore we then expand correlation functions in formal power series of the deviation $t = (T - T_c)/T_c$ from the critical temperature. The coefficients are critical correlation functions involving $\phi^2(x)$ insertions at zero momentum (as one can infer from the analysis of Chapter 10), whose scaling behaviour has been derived in Section 25.6. Summing the expansion, we obtain RG equations valid for $T \neq T_c$.

To cross the critical temperature, avoiding the critical singularities, it is in addition necessary to explicitly break the symmetry of the hamiltonian. We thus add a small magnetic field to the spin interactions. We then derive RG equations in a field, or at fixed magnetization. In this way we are able to connect continuously correlation functions above and below T_c, and establish scaling laws in the whole critical domain.

In the first part of the chapter we restrict ourselves to Ising-like systems, we then generalize the results to N-component order parameters in Section 26.6. In Section 26.7 we show how to expand the universal 2-point function when T approaches T_c, using the short distance expansion. Finally in the appendix we discuss the energy correlation function when the specific heat exponent α vanishes.

Remark. The temperature is coupled to the total hamiltonian or configuration energy. Therefore a variation of the temperature generates a variation of all terms contributing to the effective hamiltonian. However, as we have shown in 25.2, the most relevant contribution (the most IR singular) corresponds to the $\phi^2(x)$ operator. We can therefore take the difference $t = r - r_c$ between the coefficient of ϕ^2 in (25.34) and its critical value as a linear measure of the deviation from the critical temperature and parametrize the effective hamiltonian as:

$$\mathcal{H}(\phi) = \int \left\{ \frac{1}{2} [\partial_\mu \phi(x)]^2 + \frac{1}{2}(r_c + t)\phi^2(x) + \frac{1}{4!}g\Lambda^\varepsilon \phi^4(x) \right\} \mathrm{d}^d x. \tag{26.1}$$

Dimensional analysis then yields the relation

$$\Gamma^{(n)}(p_i; t, g, \Lambda) = \Lambda^{d-n(d-2)/2}\Gamma^{(n)}(p_i\Lambda^{-1}; t\Lambda^{-2}, g, 1). \tag{26.2}$$

With this parametrization the critical domain corresponds to

$$|t = r - r_c| \ll \Lambda^2.$$

26.1 Strong Scaling above T_c: The Renormalized Theory

We discuss in this section various properties of the critical behaviour above T_c. For $T > T_c$, because the theory is massive, it is possible to introduce a special renormalized theory with renormalization conditions imposed at zero momentum. Generalizing to any dimension $d \leq 4$ the formalism of Chapter 10, we define renormalized correlation functions by:

$$\Gamma_r^{(2)}\left(p; m_r, g_r\right) = m_r^2 + p^2 + O\left(p^4\right),\qquad(26.3)$$

$$\Gamma_r^{(4)}\left(p_i = 0; m_r, g_r\right) = m_r^\varepsilon g_r.\qquad(26.4)$$

This renormalized theory is obtained from the original "bare" theory by taking the large cut-off limit at g_r and m_r fixed:

$$\Gamma_r^{(n)}\left(p_i; m_r, g_r\right) = \lim_{\Lambda \to \infty} Z^{n/2}\left(m_r/\Lambda, g_r\right)\Gamma^{(n)}\left(p_i; t, g, \Lambda\right).\qquad(26.5)$$

The condition (26.4) is written in such a way that g_r is a dimensionless parameter. A simple dimensional analysis then yields for the renormalized correlation functions the relation

$$\Gamma_r^{(n)}\left(p_i; m_r, g_r\right) = m_r^{d-(n/2)(d-2)}\Gamma_r^{(n)}\left(p_i/m_r; 1, g_r\right).\qquad(26.6)$$

Therefore, the correlation length ξ, which characterizes the decay of the connected two-point function, is proportional to m_r^{-1}: it fixes the scale of the zero momentum renormalized theory.

Note that equation (26.5) holds for any dimension $d \leq 4$ and therefore some consequences of this equation are valid beyond the ε-expansion. This is one reason why the consideration of this particular renormalized theory is so useful.

As we have shown in Chapter 10, the renormalized correlation functions satisfy CS equations:

$$\left[m_r\frac{\partial}{\partial m_r} + \beta\left(g_r\right)\frac{\partial}{\partial g_r} - \frac{n}{2}\eta\left(g_r\right)\right]\Gamma_r^{(n)}\left(p_i; m_r, g_r\right) = m_r^2\left[2 - \eta\left(g_r\right)\right]\Gamma_r^{(1,n)}\left(0; p_i; m_r, g_r\right),\qquad(26.7)$$

in which the renormalized ϕ^2 insertions are specified by the condition:

$$\Gamma_r^{(1,2)}\left(0; 0, 0; m_r, g_r\right) = 1.\qquad(26.8)$$

As usual we have given to different RG functions the same name $\{\beta, \eta\}$ because they play the same role in the equations. We examine some consequences of equation (26.7) in Subsection 26.1.3.

26.1.1 Bare and renormalized coupling constants

Let us first discuss the problem of the bare and renormalized coupling constants. Several remarks are here in order:

(i) Since in dimension $d < 4$ the ϕ^4 field theory is super-renormalizable, all correlation functions have a large cut-off limit after a simple mass renormalization, i.e. after one has taken the deviation from the critical temperature as a parameter. The coupling constant and the field amplitude renormalizations are finite and therefore are not required in general. If we only renormalize the mass, we obtain a finite theory which is a function of the bare coupling constant u, the coefficient of $\phi^4(x)$ in the hamiltonian. The coupling constant u, however, has a mass dimension ε. Therefore perturbation theory is really an expansion in powers of u/m_r^ε. It only makes sense if this ratio is kept fixed, which implies that the coupling constant u (which characterizes the deviation from the gaussian fixed point) goes to zero with the inverse correlation length as m_r^ε. This is indeed what is implicitly assumed in the conventional field theory framework. In the critical phenomena situation instead, the coupling constant, which is related to microscopic parameters of the theory, is fixed. This means, as we already stressed in Chapter 25, that after introduction by rescaling of the cut-off Λ, $g = u/\Lambda^\varepsilon$ remains finite (see action (25.34)) when $\Lambda \to \infty$. Therefore, in the critical domain, i.e. in the large cut-off limit, u becomes large for $d < 4$ and the mass renormalized perturbation theory becomes useless. This is the reason why it is necessary to introduce a field amplitude renormalization and a new expansion parameter g_r which, as we verify below, remains finite in this limit.

(ii) We note that by this method (introduction of the renormalized theory) we have taken the large cut-off limit in the following way: $m \ll u^{1/\varepsilon} \ll \Lambda$ instead of $m \ll \Lambda \propto u^{1/\varepsilon}$. We have to assume that the result is the same at leading order.

The renormalized coupling constant. The function $\beta(g_r)$ in (26.7) is given by equation (10.26):

$$\beta(g_r) = m_r \left. \frac{\mathrm{d}}{\mathrm{d}m_r}\right|_{\Lambda,g} g_r(g, m_r/\Lambda). \tag{26.9}$$

Let us introduce the variable $\lambda = m_r/\Lambda$. Considering g_r as a function of λ at g *fixed*, we can rewrite equation (26.9):

$$\lambda \frac{\mathrm{d}}{\mathrm{d}\lambda} g_r(\lambda) = \beta(g_r(\lambda)). \tag{26.10}$$

This equation, which is similar to equation (25.47a), is a flow equation for $g_r(\lambda)$. When the correlation length increases, the ratio m_r/Λ decreases, and thus λ goes to zero. The renormalized coupling constant will be driven towards an IR stable zero of $\beta(g_r)$ if such a zero exists. Assuming the existence of such an IR fixed point:

$$\beta(g_r^*) = 0, \qquad \text{with} \quad \omega = \beta'(g_r^*) > 0, \tag{26.11}$$

we can integrate equation (26.10) in the neighbourhood of g_r^*, and estimate $g_r - g_r^*$:

$$|g_r - g_r^*| \sim (m_r/\Lambda)^\omega. \tag{26.12}$$

Therefore, as a consequence of the special conditions imposed by the statistical origin of the problem, in the critical domain the renormalized coupling constant is close to the IR fixed point value.

26.1.2 Strong scaling

At leading order we can then replace g_r by g_r^*. Let us evaluate the behaviour of the renormalization constant Z (equation (10.27)):

$$m_r \left. \frac{\mathrm{d}}{\mathrm{d}m_r}\right|_{g,\Lambda} \ln Z(g_r, m_r/\Lambda) = \eta(g_r). \tag{26.13}$$

Using the parameter $\lambda = m_r/\Lambda$ and the function $g_r(\lambda)$ of equation (26.10), we can rewrite equation (26.13) as:

$$\lambda \frac{\mathrm{d}}{\mathrm{d}\lambda} \ln Z(\lambda) = \eta(g_r(\lambda)). \tag{26.14}$$

In the case of an IR fixed point we find for $\lambda = m_{\mathrm{r}}/\Lambda$ small:

$$Z \sim (m_{\mathrm{r}}/\Lambda)^{\eta}, \qquad (26.15)$$

where we have called η the value (assumed finite) of the function $\eta(g)$ at the fixed point:

$$\eta = \eta(g_{\mathrm{r}}^{*}). \qquad (26.16)$$

Let us also assume that the renormalized correlation functions $\Gamma_{\mathrm{r}}^{(n)}(p_i; m_{\mathrm{r}}, g_{\mathrm{r}}^{*})$ are finite. Combining then equations (26.5,26.6,26.15) we find in the critical domain the scaling relation:

$$\Gamma^{(n)}(p_i, t, g, \Lambda = 1) \underset{m_{\mathrm{r}} \ll 1, \, |p_i| \ll 1}{\sim} m_{\mathrm{r}}^{d-(n/2)(d-2+\eta)} \Gamma_{\mathrm{r}}^{(n)}(p_i/m_{\mathrm{r}}; 1, g_{\mathrm{r}}^{*}). \qquad (26.17)$$

The parameter Λ has only been introduced to provide a scale for all quantities. Once the form of the critical behaviour has been obtained, we can rescale all dimensional parameters (here momenta, deviation from T_c, and correlation length) to eliminate Λ, as we have done above (see equation (26.2)).

Equation (26.17) provides a proof of *strong scaling* and *universality* in the whole critical domain above T_c. The original correlation functions depend on the detailed form of the interaction. The ϕ^4 theory bare correlation functions depend only on the way the cut-off is introduced and explicitly on p_i, g, and t. The r.h.s. of equation (26.17) involves only the renormalized correlation functions at $g_{\mathrm{r}} = g_{\mathrm{r}}^{*}$ i.e. functions of ratios p_i/m_{r}. Furthermore the result (26.17) holds for any fixed dimension $d \le 4$. Resort to the ε-expansion has been avoided. It has only been necessary to assume the existence of an IR fixed point. Of course within the ε-expansion, using the results of Chapter 11, we can immediately obtain $\beta(g_{\mathrm{r}})$ at leading order and calculate g_{r}^{*} and ω:

$$g_{\mathrm{r}}^{*} = \frac{3}{16\pi^2}\varepsilon + O(\varepsilon^2), \quad \omega = \varepsilon + O(\varepsilon^2). \qquad (26.18)$$

However, as first suggested by Parisi, it is also possible to analyze numerically the perturbative expansion in powers of g_{r} at fixed dimension 3 or 2. Such an analysis convincingly demonstrates the existence of an IR fixed point and allows a precise determination of critical exponents as will be shown in Chapter 28 when we discuss numerical results.

Note finally that the equations (10.28–10.31) can be used to characterize the divergence of m_{r}^{-1}, and thus the correlation length, as a function of the temperature or bare mass. This behaviour will be derived in Section 26.3 by another, simpler, method. It confirms that the relation between m_{r} and t is singular. Therefore the method of this section, based on the introduction of a zero-momentum renormalized theory, does not allow to cross the critical temperature. Indeed all correlation functions are parametrized in terms of the correlation length which is singular at T_c. To avoid this problem we introduce in next section a different formalism which is a natural extension of the formalism of Chapter 25.

Let us conclude this discussion with a few remarks.

(i) We have shown that the renormalized coupling constant g_{r} has a finite limit $g_{\mathrm{r}} = g_{\mathrm{r}}^{*}$ although the bare coupling $u = g\Lambda^{\varepsilon}$ becomes infinite. In addition precisely at g_{r}^{*} the field amplitude renormalization diverges as equation (26.15) shows. The conclusion is that the IR fixed point field theory behaves even below four dimensions as a renormalizable field theory, and a complete renormalization is indeed required.

(ii) A second remark connected to the previous one, is that when u varies from zero to infinity, g_r varies from zero to g_r^*. This property seems to indicate that it is impossible to construct a theory with $g_r > g_r^*$. Since $g_r^* = O(\varepsilon)$ this argument suggests that it is somewhat unlikely that a non-trivial ϕ^4 field theory exists in four dimensions. On the other hand to construct the renormalized theory we have taken the large cut-off limit at u fixed. This procedure is only legitimate if the bare renormalization group has only one IR fixed point. Otherwise other non-trivial theories might be obtained by sending the cut-off and u to infinity at g fixed. This point will be further discussed in Subsection 34.1.1.

26.1.3 Large momentum behaviour

Let us return to equation (26.7) to investigate the behaviour of correlation functions at large momenta, in the critical domain $m_r \ll |p_i| \ll \Lambda$. At leading order we can replace in equation (26.7) g_r by its IR fixed point value g_r^*:

$$\left(m_r \frac{\partial}{\partial m_r} - \frac{n}{2}\eta\right) \Gamma_r^{(n)}(p_i; m_r, g_r^*) = m_r^2 (2 - \eta) \Gamma_r^{(1,n)}(0; p_i; m_r, g_r^*). \qquad (26.19)$$

It follows from scaling relation (26.6) or (26.17) that the large momentum behaviour is directly related to the approach to the critical theory $m_r = 0$. It is thus not too surprising that in order to extract some information from equation (26.19) it is necessary to return to the framework of the ε-expansion. A simple extension of Weinberg's theorem shows that, order by order in g_r and ε, the r.h.s. of equation (26.19) is negligible at large (non-exceptional) momenta, or small masses, as one might naively guess from the factor m_r^2 which has been factorized. By contrast below 4 dimensions, at fixed dimension, there always exists an order at which the r.h.s. ceases to be negligible. Our interpretation is that this is a consequence of expanding perturbation theory around the wrong fixed point, the IR unstable gaussian fixed point. The true asymptotic behaviour of correlation functions is different because it is governed by the non-trivial point $g_r = g_r^*$. Asymptotically the $\Gamma_r^{(n)}$ therefore satisfy:

$$\left(m_r \frac{\partial}{\partial m_r} - \frac{n}{2}\eta\right) \Gamma_r^{(n)}(p_i, m_r) \underset{|p_i| \gg m_r}{\approx} 0. \qquad (26.20)$$

Together with relation (26.6) this leads to:

$$\Gamma_r^{(n)}(\lambda p_i; m_r) \underset{|p_i| \gg m_r}{\sim} \lambda^{d-(n/2)(d-2+\eta)} \Gamma_r^{(n)}(p_i; m_r). \qquad (26.21)$$

Using then the relation (26.5) between bare and renormalized correlation functions, we recover the scaling behaviour of equation (25.61) derived in Section 25.5.

It is also possible in this approach to analyze the large momentum behaviour of the r.h.s. of equation (26.19), and therefore to calculate corrections to the leading behaviour, using the short distance expansion introduced in Chapter 12 (see Section 26.7).

Remark. Note that if g_r is fixed with $g_r < g_r^*$ then the large momentum behaviour of correlation functions is given by the UV fixed point $g_r = 0$, i.e. by perturbation theory. However, as we have already discussed, this corresponds to a situation where the bare coupling constant u goes to zero as m_r^ε, which is rather unnatural.

26.2 Expansion around the Critical Theory

To be able to describe the whole critical domain we now use a different strategy. First we consider the functional integral representation of n-point correlation functions corresponding to the hamiltonian (26.1). If we formally expand it in powers of t, we obtain an expansion in terms of *critical* correlation functions with $\frac{1}{2}\int d^d x\, \phi^2(x)$ (the most IR singular part of the energy operator) insertions (see also Section 10.10). Consequently, to be able to define these correlation functions by their perturbative expansion, we have to now return to the framework developed in Sections 25.4, 25.6 and to the ε-expansion.

However, even so, because the insertion of $\int d^d x\, \phi^2(x)$ is the insertion of the Fourier transform of $\phi^2(x)$ at zero momentum, the corresponding correlation functions are still IR divergent. We thus first replace in the hamiltonian (26.1) the constant t by a field $t(x)$. We can then write the 1PI correlation functions $\Gamma^{(n)}$, as functions of space variables, as:

$$\Gamma^{(n)}\left(x_i; t, g, \Lambda\right) = \sum_{l=0}^{\infty} \frac{1}{l!} \int dy_1 \ldots dy_l\, t\left(y_1\right) \ldots t\left(y_l\right) \Gamma^{(l,n)}\left(y_j; x_i; t = 0, g, \Lambda\right). \quad (26.22)$$

As already noted in Section 10.10, by acting with the functional differential operator $\int dy\, t(y)\delta/\delta t(y)$ on equation (26.22) we generate in the r.h.s. a factor l in front of $\Gamma^{(l,n)}$. It is then easy to verify that equation (25.66) implies (see equation (10.72)):

$$\left[\Lambda\frac{\partial}{\partial \Lambda} + \beta\left(g\right)\frac{\partial}{\partial g} - \frac{n}{2}\eta\left(g\right) - \eta_2\left(g\right)\int t\left(y\right)\frac{\delta}{\delta t\left(y\right)}\right]\Gamma^{(n)}\left(x_i; t, g, \Lambda\right) = 0. \quad (26.23)$$

To calculate $\Gamma^{(n)}$, it is possible to perform a partial summation of perturbation theory in order to introduce the mass renormalized free propagator $\left(p^2 + t\right)^{-1}$, for example by going over to a loopwise expansion. One verifies that in this new perturbation theory the constant t limit leads to a massive theory and no IR divergences are generated anymore. Equation (26.23) then, in momentum representation, becomes:

$$\left[\Lambda\frac{\partial}{\partial \Lambda} + \beta\left(g\right)\frac{\partial}{\partial g} - \frac{n}{2}\eta\left(g\right) - \eta_2\left(g\right)t\frac{\partial}{\partial t}\right]\Gamma^{(n)}\left(p_i; t, g, \Lambda\right) = 0. \quad (26.24)$$

Equation (26.24) is the formal analogue of equation (10.74) and differs from equation (26.7), which also applies to the non-critical (i.e. massive) theory, by the property that it is homogeneous.

26.3 Scaling Laws above T_c

Let us first discuss again the critical behaviour above T_c, which we have already examined in Section 26.1, with this new formalism. As we have done for previous RG equations, we integrate equation (26.24) using the method of characteristics. In addition to the functions $g(\lambda)$ and $Z(\lambda)$ of equations (25.49), we now need a function $t(\lambda)$ which we determine by imposing the consistency between equation (26.24) and:

$$\lambda\frac{d}{d\lambda}\left[Z^{-n/2}(\lambda)\Gamma^{(n)}\left(p_i; t(\lambda), g(\lambda), \lambda\Lambda\right)\right] = 0. \quad (26.25)$$

Consistency implies the set of equations:

$$\lambda \frac{d}{d\lambda} g(\lambda) = \beta\left(g(\lambda)\right), \qquad g(1) = g, \tag{26.26}$$

$$\lambda \frac{d}{d\lambda} \ln t(\lambda) = -\eta_2\left(g(\lambda)\right), \qquad t(1) = t, \tag{26.27}$$

$$\lambda \frac{d}{d\lambda} \ln Z(\lambda) = \eta\left(g(\lambda)\right), \qquad Z(1) = 1. \tag{26.28}$$

Dimensional analysis yields:

$$\Gamma^{(n)}\left(p_i; t\left(\lambda\right), g(\lambda), \lambda\Lambda\right) = \left(\lambda\Lambda\right)^{d-n(d-2)/2} \Gamma^{(n)}\left(\frac{p_i}{\lambda\Lambda}; \frac{t(\lambda)}{\lambda^2\Lambda^2}, g(\lambda), 1\right). \tag{26.29}$$

The critical region is defined in particular by $|t| \ll \Lambda^2$, and this is the source of the IR singular behaviour which appears in perturbation theory. Let us assume that the equation for λ:

$$t(\lambda) = \lambda^2\Lambda^2, \tag{26.30}$$

has a solution. Then the theory at scale λ will no longer be critical. Combining equations (26.25–26.30) we find:

$$\Gamma^{(n)}\left(p_i; t, g, \Lambda\right) = Z^{-n/2}(\lambda) m^{(d-n(d-2)/2)} \Gamma^{(n)}\left(p_i/m; 1, g(\lambda), 1\right), \tag{26.31}$$

where we have introduced the notation:

$$m = \lambda\Lambda. \tag{26.32}$$

The solution of equation (26.27) can be written:

$$t(\lambda) = t \exp\left[-\int_1^\lambda \frac{d\sigma}{\sigma} \eta_2\left(g(\sigma)\right)\right]. \tag{26.33}$$

Substituting this relation into equation (26.30) we then obtain:

$$\ln(t/\Lambda^2) = \int_1^\lambda \frac{d\sigma}{\sigma} \frac{1}{\nu\left(g\left(\sigma\right)\right)}. \tag{26.34}$$

We look for a solution λ in the limit $t/\Lambda^2 \ll 1$. Since $\nu(g)$ is a positive function, at least for g small enough as can be verified in the explicit expression (25.71), equation (26.34) implies that the value of the parameter λ is small, and thus that $g(\Lambda)$ is close to the IR fixed point g^*. In this limit $\nu\left(g\left(\sigma\right)\right)$ can be replaced at leading order by the exponent ν and equation (26.34) can be rewritten:

$$\ln(t/\Lambda^2) \sim \frac{1}{\nu} \ln\lambda. \tag{26.35}$$

Equation (26.28) then yields:

$$Z(\lambda) \sim \lambda^\eta. \tag{26.36}$$

Taking the large Λ, or the small λ limit, and using equations (26.35,26.36) in equation (26.31) we finally obtain:

$$\Gamma^{(n)}\left(p_i; t, g, \Lambda = 1\right) \underset{\substack{t \ll 1 \\ |p_i| \ll 1}}{\sim} m^{(d-n(d-2+\eta)/2)} F_+^{(n)}\left(p_i/m\right), \qquad (26.37)$$

with:

$$m\left(\Lambda = 1\right) = \xi^{-1} \sim t^\nu. \qquad (26.38)$$

Equations (26.37) and (26.17) express the same scaling property. However one new result has been obtained. From equation (26.37) we infer that the quantity m is proportional to the physical mass or inverse correlation length. Equation (26.38) then shows that the divergence of the correlation length $\xi = m^{-1}$ at T_c is characterized by the exponent ν (a result which we could also have derived with the formalism of Section 26.1).

For $t \neq 0$, the correlation functions are finite at zero momentum and behave as:

$$\Gamma^{(n)}\left(0; t, g, \Lambda\right) \propto t^{\nu(d-n(d-2+\eta)/2)}. \qquad (26.39)$$

In particular for $n = 2$ we obtain the inverse magnetic susceptibility:

$$\chi^{-1} = \Gamma^{(2)}\left(p = 0; t, g, \Lambda\right) \propto t^{\nu(2-\eta)}. \qquad (26.40)$$

The exponent which characterizes the divergence of χ is usually called γ. The equation (26.39) establishes the relation between exponents:

$$\gamma = \nu\left(2 - \eta\right). \qquad (26.41)$$

Finally let us note that, since we have assumed that the critical theory exists when t goes to zero, different powers of t have to cancel in equation (26.37). From this observation we immediately recover equation (25.61) in the form:

$$\Gamma^{(n)}\left(\lambda p_i; t, g, \Lambda = 1\right) \underset{\substack{t^\nu \ll \lambda |p_i| \ll 1}}{\propto} \lambda^{d-n(d-2+\eta)/2}. \qquad (26.42)$$

26.4 Correlation Functions with ϕ^2 Insertions

A differentiation of the functional integral with respect to $t(x)$ l times before taking the uniform t limit, generates correlation functions with l $\left[\frac{1}{2}\phi^2(x)\right]$ insertions (in the statistical formulation insertions of the hamiltonian or configuration energy density). By differentiating equation (26.23) with respect to $t\left(y_1\right) \ldots t\left(y_l\right)$ before taking the same limit, we derive RG equations for the corresponding 1PI correlation functions. It is easy to verify that the resulting equation, except for $l = 2$, $n = 0$, is:

$$\left[\Lambda \frac{\partial}{\partial \Lambda} + \beta\left(g\right) \frac{\partial}{\partial g} - \frac{n}{2}\eta\left(g\right) - \eta_2\left(g\right)\left(l + t\frac{\partial}{\partial t}\right)\right] \Gamma^{(l,n)}\left(q_i; p_j; t, g, \Lambda\right) = 0. \qquad (26.43)$$

This equation can be solved exactly in the same way as equation (26.24) and leads, for $t \ll 1$, $|q_i| \ll 1$, $|p_j| \ll 1$, to the critical behaviour:

$$\Gamma^{(l,n)}\left(q_i; p_j; t, g, \Lambda = 1\right) \sim m^{[d-l/\nu-n(d-2+\eta)/2]} F_+^{(l,n)}\left(q_i/m; p_j/m\right), \qquad (26.44)$$

with:

$$m \sim t^{\nu}. \tag{26.45}$$

The discussion then exactly follows the lines of Section 26.3, and we do not repeat it here. Rather we concentrate on the case $n = 0$, $l = 2$, which corresponds to the energy density two-point correlation function. Starting from equation (25.66), and using the method explained above, we obtain:

$$\left[\Lambda \frac{\partial}{\partial \Lambda} + \beta(g) \frac{\partial}{\partial g} - \eta_2(g) \left(2 + t \frac{\partial}{\partial t} \right) \right] \Gamma^{(2,0)}(q; t, g, \Lambda) = \Lambda^{-\varepsilon} B(g). \tag{26.46}$$

The function $\Lambda^{-\varepsilon} C_2(g)$ of equation (25.72) is still a solution of the inhomogeneous equation. The equation (26.44) then is replaced by:

$$\Gamma^{(2,0)}(q; t, g, \Lambda = 1) \underset{\substack{t \ll 1 \\ |q| \ll 1}}{\sim} m^{(d-2/\nu)} F_+^{(2,0)}(q/m) + C_2(g). \tag{26.47}$$

At zero momentum $\Gamma^{(2,0)}(q = 0)$ is also the second derivative with respect to t of the free energy (the connected vacuum amplitude), i.e. the specific heat. Its critical behaviour is thus:

$$\Gamma^{(2,0)}(0; t, g, \Lambda = 1) \sim A^+ t^{-(2-\nu d)} + C_2(g). \tag{26.48}$$

The specific heat exponent is called α. We have therefore derived the scaling law:

$$\alpha = 2 - \nu d. \tag{26.49}$$

Integrating $\Gamma^{(2,0)}$ twice with respect to t, we obtain $\Omega^{-1}\Gamma(t, g, \Lambda)$, the thermodynamic potential per unit volume (calling Ω the volume factor) for a vanishing magnetization (and thus also the free energy in zero magnetic field):

$$\Omega^{-1}\Gamma(M = 0, t, g, \Lambda) = \Lambda^d C_0(g) + \Lambda^{d-2} C_1(g) t + \Lambda^{d-4} C_2(g) \frac{t^2}{2}$$

$$+ \Lambda^d \frac{A^+}{(2-\alpha)(1-\alpha)} \left(\frac{t}{\Lambda^2} \right)^{2-\alpha} + \cdots, \tag{26.50}$$

for $|t| \ll \Lambda^2$.

The three first terms correspond to the beginning of the small t expansion of the regular part of the free energy and depend explicitly on g through three functions C_0, C_1 and C_2 while the fourth term characterizes the leading behaviour of the singular part of the free energy and is universal (it still depends on the normalization of the temperature but this normalization can be cancelled in appropriate ratios).

Since we have completely described the critical behaviour above T_c, we examine in the next section the critical behaviour in the ordered phase $(M \neq 0)$.

26.5 Scaling Laws in a Magnetic Field and Below T_c

In order to pass continuously from the disordered $(T > T_c)$ to the ordered phase $(T < T_c)$, it is necessary to add to the hamiltonian an interaction which explicitly breaks its symmetry. We thus consider correlation functions in the presence of an external magnetic field. Actually it is more convenient to consider correlation functions at fixed magnetization. Therefore we first discuss the relation between field and magnetization, i.e. the equation of state.

As we have already shown in the mean field analysis, in the ordered phase some qualitative differences appear between systems which have a discrete and a continuous symmetry. We shall illustrate these differences with the example of the N-vector model with $O(N)$ symmetry and $(\phi^2)^2$ interaction in Section 26.6.

26.5.1 The equation of state

Let us call M the average value of $\phi(x)$ in a constant field H. The thermodynamic potential per unit volume, as a function of M, is by definition:

$$\Omega^{-1}\Gamma\left(M, t, g, \Lambda\right) = \sum_{n=0}^{\infty} \frac{M^n}{n!}\Gamma^{(n)}\left(p_i = 0; t, g, \Lambda\right). \tag{26.51}$$

The magnetic field H is given by:

$$H = \Omega^{-1}\frac{\partial\Gamma}{\partial M} = \sum_{n=1}^{\infty} \frac{M^n}{n!}\Gamma^{(n+1)}\left(p_i = 0; t, g, \Lambda\right). \tag{26.52}$$

Noting that $n \equiv M\left(\partial/\partial M\right)$, we immediately derive from the RG equation (26.24), the RG equation satisfied by $H\left(M, t, g, \Lambda\right)$:

$$\left[\Lambda\frac{\partial}{\partial\Lambda} + \beta\left(g\right)\frac{\partial}{\partial g} - \frac{1}{2}\eta\left(g\right)\left(1 + M\frac{\partial}{\partial M}\right) - \eta_2\left(g\right)t\frac{\partial}{\partial t}\right] H\left(M, t, g, \Lambda\right) = 0. \tag{26.53}$$

To integrate equation (26.53) by the method of characteristics we have to introduce, in addition to the functions $g(\lambda)$, $t(\lambda)$ and $Z(\lambda)$ of equations (26.26–26.28), a new function $M(\lambda)$ which satisfies:

$$\lambda\frac{\mathrm{d}}{\mathrm{d}\lambda}\ln M(\lambda) = -\tfrac{1}{2}\eta\left[g(\lambda)\right], \qquad M(1) = M. \tag{26.54}$$

Comparing equations (26.54) and (26.28) we see that $M(\lambda)$ is given by:

$$M(\lambda) = MZ^{-1/2}(\lambda).$$

The solution of equation (26.53) can then be written:

$$H\left(M, t, g, \Lambda\right) = Z^{-1/2}(\lambda)H\left[M(\lambda), t(\lambda), g\left(\lambda\right), \lambda\Lambda\right]. \tag{26.55}$$

Dimensional analysis shows that:

$$H\left(M, t, g, \Lambda\right) = \Lambda^{3-\varepsilon/2}H\left(\frac{M}{\Lambda^{1-\varepsilon/2}}, \frac{t}{\Lambda^2}, g, 1\right). \tag{26.56}$$

Applying this relation to the r.h.s. of equation (26.55) we obtain:

$$H\left(M, t, g, \Lambda\right) = (\lambda\Lambda)^{3-\varepsilon/2} Z^{-1/2}(\lambda) H\left(\frac{M(\lambda)}{(\lambda\Lambda)^{1-\varepsilon/2}}, \frac{t(\lambda)}{\lambda^2\Lambda^2}, g(\lambda), 1\right). \qquad (26.57)$$

We again use the arbitrariness of λ to move outside the critical domain, in order to remove the critical singularities in the r.h.s. of equation (26.57). Here a natural choice is:

$$M(\lambda) = (\lambda\Lambda)^{1-\varepsilon/2}, \qquad (26.58)$$

which implies, using the solution of equation (26.54), that:

$$\ln\frac{M}{\Lambda^{1-\varepsilon/2}} = \frac{1}{2}\int_1^\lambda \frac{d\sigma}{\sigma}\left[d - 2 + \eta\left(g\left(\sigma\right)\right)\right]. \qquad (26.59)$$

Remember that the critical domain is defined in particular by:

$$M \ll \Lambda^{1-\varepsilon/2}.$$

For g small, the expression $d - 2 + \eta(g)$ is positive because $d \geq 2$ and $\eta(g) > 0$, which again implies that λ should be close to zero and thus $g(\lambda)$ close to g^*. In this limit equation (26.59) leads to:

$$M\Lambda^{\varepsilon/2-1} \sim \lambda^{(d-2+\eta)/2}. \qquad (26.60)$$

Now equation (26.27) implies:

$$t(\lambda)/\lambda^2 \sim t\lambda^{-1/\nu}, \qquad (26.61)$$

and we have already seen that:

$$Z(\lambda) \sim \lambda^\eta. \qquad (26.62)$$

Replacing $t(\lambda)$ and $Z(\lambda)$ by their asymptotic forms (26.61,26.62) and using equation (26.60) to eliminate λ, we finally obtain:

$$H\left(M, t, g, 1\right) \sim M^\delta f\left(tM^{-1/\beta}\right), \qquad (26.63)$$

with:

$$\beta = \frac{\nu}{2}\left(d - 2 + \eta\right), \qquad (26.64)$$

$$\delta = \frac{d + 2 - \eta}{d - 2 + \eta}. \qquad (26.65)$$

Equation (26.63) establishes the scaling properties of the equation of state. Moreover equations (26.64,26.65) relate the traditional critical exponents which characterize the vanishing of the spontaneous magnetization and the singular relation between magnetic field and magnetization at T_c respectively to the exponents η and ν introduced previously.

Valid for $d < 4$, these latter two relations seem to be inconsistent with the values of the mean field exponents for $d > 4$. To understand this point, it is necessary to remember that for $d > 4$, g^* vanishes, and all terms in H, except the term linear in M, come from corrections to expression (26.63) (see Section 27.1).

Properties of the universal function $f(x)$.

(i) Griffith's analyticity: equation (26.52) shows that H has a regular expansion in odd powers of M for $t > 0$. This implies that when the variable x becomes large and positive, $f(x)$ has the expansion:

$$f(x) = \sum_{p=0}^{\infty} a_p x^{\gamma - 2p\beta}. \tag{26.66}$$

(ii) When M is different from zero, the theory remains massive when t vanishes. In the loop expansion the corresponding massive propagator has the form $\left(p^2 + aM^2\right)^{-1}$ in momentum space. It follows that we can expand $\Gamma(M,t)$ and therefore $H(M,t)$ in powers of t without meeting IR divergences. Consequently $f(x)$ is infinitely differentiable at $x = 0$.

(iii) The appearance of a spontaneous magnetization, below T_c, implies that the function $f(x)$ has a negative zero x_0:

$$f(x_0) = 0, \quad x_0 < 0. \tag{26.67}$$

Then equation (26.63) leads to the relation:

$$M = |x_0|^{-\beta} (-t)^{\beta} \quad \text{for} \quad H = 0, \ t < 0. \tag{26.68}$$

Equation (26.68) gives the behaviour of the spontaneous magnetization when the temperature approaches the critical temperature from below.

26.5.2 Correlation functions in a field

We now examine the behaviour of correlation functions in a field. All expressions will again be written for Ising-like systems. The generalization to the N-vector model with $O(N)$ symmetry is simple and will be briefly discussed in Section 26.6.

The correlation functions at fixed magnetization M are obtained by expanding the generating functional $\Gamma(M(x))$ of 1PI correlation functions, around $M(x) = M$:

$$\Gamma^{(n)}(x_1, \ldots, x_n; t, g, \Lambda) = \frac{\delta^n}{\delta M(x_1) \ldots \delta M(x_n)} \Gamma(M(x), t, g, \Lambda)\bigg|_{M(x)=M}. \tag{26.69}$$

The expansion in powers of M of the r.h.s. of the equation then yields:

$$\Gamma^{(n)}(p_1, \ldots, p_n; t, M, g, \Lambda) = \sum_{s=0}^{\infty} \frac{M^s}{s!} \Gamma^{(n+s)}(p_1, \ldots, p_n, 0, \ldots, 0; t, 0, g, \Lambda). \tag{26.70}$$

From the RG equations satisfied by the correlation functions in zero magnetization (equations (26.24)) it is then easy to derive:

$$\left[\Lambda \frac{\partial}{\partial \Lambda} + \beta(g) \frac{\partial}{\partial g} - \frac{1}{2}\eta(g)\left(n + M \frac{\partial}{\partial M}\right) - \eta_2(g) t \frac{\partial}{\partial t}\right] \Gamma^{(n)}(p_i; t, M, g, \Lambda) = 0. \tag{26.71}$$

This equation can be solved by exactly the same method as equation (26.53):

$$\Gamma^{(n)}(p_i; t, M, g, \Lambda = 1) \sim m^{[d-(d-2+\eta)/2]} F^{(n)}\left(p_i/m, tm^{-1/\nu}\right), \tag{26.72}$$

for $|p_i| \ll 1$, $|t| \ll 1$, $M \ll 1$ and with the definition:

$$m = M^{\nu/\beta}. \tag{26.73}$$

The r.h.s. of equation (26.72) now depends on two different mass scales: $m = M^{\nu/\beta}$ and t^ν.

26.5.3 Correlation functions below T_c

We have argued above that correlation functions are regular functions of t for small t, provided M does not vanish. It is therefore possible to cross the critical point and to then take the zero external magnetic field limit. In the limit M becomes the spontaneous magnetization which is given, as a function of t, by equation (26.68). After elimination of M in favour of t in equation (26.72), one finds the critical behaviour below T_c:

$$\Gamma^{(n)}\left(p_i; t, M\left(t, H = 0\right), g, 1\right) \sim m^{d - n(d-2+\eta)/2} F_-^{(n)}\left(p_i / m\right), \qquad (26.74)$$

with:

$$m = |x_0|^{-\nu}(-t)^{\nu}, \qquad H = 0, \qquad t < 0. \qquad (26.75)$$

We conclude that the correlation functions have exactly the same scaling behaviour above and below T_c. In particular since traditionally one defines below T_c:

$$m^{-1} = \xi \sim (-t)^{-\nu'}, \qquad \left[\Gamma^{(2)}(0)\right]^{-1} = \chi \sim (-t)^{-\gamma'}, \qquad (26.76)$$

we have thus established $\nu' = \nu$ and $\gamma' = \gamma$.

The universal functions $F_+^{(n)}$ and $F_-^{(n)}$, instead, are of course different.

The extension of these considerations to the functions with ϕ^2 insertions, $\Gamma^{(l,n)}$ is straightforward. In particular the same method yields the behaviour of the specific heat below T_c:

$$\Gamma^{(2,0)}\left(q = 0, M\left(H = 0, t\right)\right) - \Lambda^{-\varepsilon} C_2\left(g\right) \underset{\text{for } t < 0}{\sim} A^-\left(-t\right)^{-\alpha}, \qquad (26.77)$$

which similarly proves that the exponent above and below T_c are the same, $\alpha' = \alpha$.

Note that the constant term $\Lambda^{-\varepsilon} C_2(g)$ which depends explicitly on g is the same above and below T_c, in contrast with the coefficient of the singular part.

The derivation of the equality of exponents above and below T_c, relies on the existence of a path which avoids the critical point, along which the correlation functions are regular, and the RG equations everywhere satisfied.

Remark. The universal functions characterizing the behaviour of correlation functions in the critical domain still depend on the normalization of the physical parameters t, H, M, distances or momenta. Physical quantities which are independent of these normalizations are universal pure numbers. Examples are provided by the ratios of the amplitudes of the singularities above and below T_c like A^+/A^- for the specific heat.

26.6 The N-Vector Model

We now generalize previous results to models in which the order parameter is a N-vector and which have symmetries such that the Landau–Ginzburg–Wilson hamiltonian has still the form of a ϕ^4 field theory. We first consider a simple but important example: a model in which the hamiltonian has an $O(N)$ symmetry, and then discuss the general case.

26.6.1 The $O(N)$ symmetric N-vector model

The $O(N)$ symmetric effective hamiltonian has the form:

$$\mathcal{H}(\phi) = \int \left\{ \tfrac{1}{2} \left[\partial_\mu \phi(x) \right]^2 + \tfrac{1}{2} \left(r_c + t \right) \phi^2(x) + \tfrac{1}{4!} g \Lambda^\varepsilon \left(\phi^2(x) \right)^2 \right\} \mathrm{d}^d x \,, \qquad (26.78)$$

in which ϕ is a N-vector.

The RG equations have exactly the same form as the Ising-like case $N = 1$. The coupling constant RG function has the expansion:

$$\beta(g) = -\varepsilon g + \frac{N+8}{48\pi^2} g^2 + O\left(g^3, g^2 \varepsilon\right). \qquad (26.79)$$

At leading order in ε, $\beta(g)$ has a zero g^* which is an IR fixed point:

$$g^* = \frac{48\pi^2}{N+8}\varepsilon + O\left(\varepsilon^2\right), \qquad (26.80)$$

$$\omega \equiv \beta'\left(g^*\right) = \varepsilon + O\left(\varepsilon^2\right), \qquad (26.81)$$

and therefore all the scaling laws derived above and in Chapter 25 can also be proven for the N-vector model with $O(N)$ symmetry. We give the expressions of the other RG functions at leading order in Subsection 26.6.2 by specializing the expressions obtained for the general N-vector model.

Correlation functions in a field or below T_c. The addition of a magnetic field term in a $O(N)$ symmetric hamiltonian has various effects. First, the magnetization and the magnetic field are now vectors. The scaling forms derived previously apply to the modulus of these vectors. Second, the continuous $O(N)$ symmetry of the hamiltonian is broken linearly in the dynamical variables by the addition of a magnetic field (see Chapter 13).

Since the field and the magnetization distinguish a direction in vector space, there now exist 2^n n-point functions, each spin being either along the magnetization or transverse to it. As we have shown in Chapter 13, these different correlation functions are related by a set of identities, called WT identities, which we have discussed there in a general framework. Let us here recall in the case of the $O(N)$ symmetry the simplest one, involving the 2-point function at zero momentum, also directly established in Section 24.4 (equations (24.57,24.59)). In terms of Γ_{T} the inverse 2-point function transverse to $\mathbf{M}$ at zero momentum or the transverse susceptibility, it reads:

$$\Gamma_{\mathrm{T}}(p = 0) = \chi_{\mathrm{T}}^{-1} = \frac{H}{M}. \qquad (26.82)$$

We recognize, in different notations, equation (13.44). As we have already discussed in Section 13.4, it follows from this equation that if H goes to zero below T_c, H/M and therefore Γ_{T} at zero momentum vanish. This last result implies the existence of $N-1$ Goldstone modes corresponding to the spontaneous breaking of the $O(N)$ symmetry into $O(N-1)$.

Note finally that the inverse longitudinal 2-point function $\Gamma_{\mathrm{L}}(p)$ is singular at zero momentum in zero field below T_c as one can infer from its Feynman graph expansion (figure 26.1). This IR singularity is not generated by critical fluctuations but by the Goldstone modes. It is characteristic of continuous symmetries. It implies that the

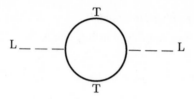

Fig. 26.1   One-loop Goldstone mode contribution to Γ_L.

longitudinal 2-point correlation function does not decrease exponentially in space at large distance. We shall discuss this problem in detail in Chapter 30.

Remark. Previous results also apply to a model in which the hamiltonian has a symmetry smaller than the group $O(N)$ but is still such that the effective hamiltonian has the form (26.78) because the quadratic and quartic group invariants are uniquely determined. Then the $O(N)$ symmetry is dynamically generated in the critical domain. Only a close examination of the leading corrections to the critical behaviour reveals the difference. We have already encountered a similar phenomenon: the hypercubic symmetry of the lattice has led to a $O(d)$ continuum space symmetry in the critical domain.

26.6.2 The general N-vector model

There exist interesting physical systems in which the hamiltonian is not $O(N)$ invariant. A first category consists in systems in which there are several correlation lengths. In such situations generically, when the temperature varies, only one correlation length becomes infinite at a time. Then the components of the dynamic variables which are non-critical do not contribute to the IR singularities. They can be integrated out in much the same way as the auxiliary fields in Pauli–Villars's regularization scheme of Section 9.1. The effect is to renormalize the effective local hamiltonian for the critical components. This remark is related to the decoupling theorem of Particle Physics. We can therefore restrict ourselves to theories with only one correlation length. This second category consists in systems in which the hamiltonian is invariant under a symmetry group, subgroup of $O(N)$, which admits a unique quadratic invariant $\Sigma\phi_i\phi_i$ but several quartic invariants.

The general hamiltonian then has the form:

$$\mathcal{H}(\phi) = \int \mathrm{d}^d x \left\{ \frac{1}{2} \sum_{i=1}^{N} \left[(\partial_\mu \phi_i)^2 + (r_c + t)\,\phi_i^2 \right] + \frac{\Lambda^\varepsilon}{4!} \sum_{i,j,k,l=1}^{N} g_{ijkl}\phi_i\phi_j\phi_k\phi_l \right\}. \tag{26.83}$$

Due to the symmetry, the tensor g_{ijkl} has special properties which imply that the 2-point correlation function $\Gamma_{ij}^{(2)}$ is always proportional to the unit matrix:

$$\Gamma_{ij}^{(2)}(p,t,g) = \delta_{ij}\Gamma^{(2)}(p,t,g). \tag{26.84}$$

This equation implies successive conditions on the tensor g_{ijkl} in perturbation theory.

In Section 11.7 we have discussed the renormalization and RG equations for a general ϕ^4 field theory and we can apply the formalism here.

Renormalization group equations. Let us first sketch the derivation of the RG equations for a multi-component critical theory. Since the field amplitude renormalization constant is independent of the components, the relation between bare and renormalized correlation functions now takes the form:

$$\Gamma_{\mathrm{r};i_1,i_2,\dots,i_n}^{(n)}(p_k,m_\mathrm{r},g_\mathrm{r},\mu) = Z^{n/2}\Gamma_{i_1,i_2,\dots,i_n}^{(n)}(p_k,t,g,\Lambda), \tag{26.85}$$

in which g stands for g_{ijkl} and g_r for $g_{r;ijkl}$.

If we differentiate the equation with respect to Λ at g_r, m_r and μ fixed, we obtain the RG equation:

$$\left(\Lambda\frac{\partial}{\partial\Lambda} + \beta_{i'j'k'l'}\frac{\partial}{\partial g_{i'j'k'l'}} - \frac{n}{2}\eta(g) - \eta_2(g)\,t\frac{\partial}{\partial t}\right)\Gamma^{(n)}_{i_1,i_2,\dots,i_n} = 0\,, \tag{26.86}$$

with the definition:

$$\beta_{i'j'k'l'}\frac{\partial g_{r;ijkl}}{\partial g_{i'j'k'l'}} = -\varepsilon g_{r;ijkl}\,. \tag{26.87}$$

These equations can be integrated by the same method as before. We have to introduce a scale dependent coupling constant $g_{ijkl}(\lambda)$ obeying the flow equation:

$$\lambda\frac{\mathrm{d}}{\mathrm{d}\lambda}g_{ijkl}(\lambda) = \beta_{ijkl}(g(\lambda))\,. \tag{26.88}$$

The long distance properties of such theories are then governed by fixed points solution of the equation:

$$\beta_{ijkl}(g^*) = 0\,. \tag{26.89}$$

The local stability properties of fixed points are governed by the eigenvalues of the matrix $M_{ijkl,i'j'k'l'}$:

$$M_{ijkl,i'j'k'l'} = \frac{\partial\beta_{ijkl}(g^*)}{\partial g_{i'j'k'l'}}\,. \tag{26.90}$$

If the real parts of all eigenvalues are positive the fixed point is locally stable. The global properties depend on the complete solutions of equation (26.88) which determine the basin of attraction in coupling space of each IR stable fixed point. We shall not discuss further this problem here and refer to the literature for details.

The $\phi_i(x)\phi_j(x)$ insertions. It is also useful to consider correlation functions with $\frac{1}{2}\phi_i(x)\phi_j(x)$ insertions. Their renormalization involves a multiplication of each insertion by the matrix $\zeta^{(2)}_{ij,kl}$. This leads to the RG equation:

$$\left(\Lambda\frac{\partial}{\partial\Lambda} + \beta_{i'j'k'l'}\frac{\partial}{\partial g_{i'j'k'l'}} - \frac{n}{2}\eta(g) - \eta_2(g)t\frac{\partial}{\partial t}\right)\Gamma^{(l,n)}_{j_1k_1\dots j_lk_l,i_1\dots i_n}$$

$$- \sum_{m=1}^{l}\eta^{(2)}_{j_mk_m,b_mc_m}\Gamma^{(l,n)}_{j_1k_1\dots b_mc_m\dots j_lk_l,i_1\dots i_n} = 0\,, \tag{26.91}$$

with the definition:

$$\eta^{(2)}_{ij,kl} = -\beta_{i'j'k'l'}\left(\frac{\partial\zeta^{(2)}}{\partial g_{i'j'k'l'}}\left[\zeta^{(2)}\right]^{-1}\right)_{ij,kl}\,. \tag{26.92}$$

Since the insertions of $\Sigma\phi_i\phi_i$, which are generated by a differentiation with respect to t, are multiplicatively renormalized, the matrix $\eta^{(2)}_{ij,kl}(g)$ has δ_{kl} as eigenvector and:

$$\eta^{(2)}_{ij,kk}(g) = \eta_2(g)\,\delta_{ij}\,. \tag{26.93}$$

RG functions. At one-loop order the RG functions are given by:

$$\beta_{ijkl} = -\varepsilon g_{ijkl} + \frac{1}{16\pi^2} \left(g_{ijmn} g_{mnkl} + 2 \text{ terms} \right) + O\left(g^3 \right), \qquad (26.94)$$

$$\eta\left(g \right) \delta_{ij} = \frac{1}{24} \frac{1}{\left(8\pi^2 \right)^2} g_{iklm} g_{jklm} + O\left(g^3 \right), \qquad (26.95)$$

$$\eta^{(2)}_{ij,kl} = -\frac{1}{16\pi^2} g_{ijkl} + O\left(g^2 \right). \qquad (26.96)$$

Condition (26.93) then implies:

$$g_{ijkk} = G \delta_{ij}. \qquad (26.97)$$

Stability of the $O\left(N \right)$ symmetric fixed point. Among the possible fixed points, one always finds, in addition to the trivial gaussian fixed point, the $O\left(N \right)$ symmetric fixed point. We can here study its local stability at leading order in ε. Let us first specialize expressions (26.94–26.96) to the case of the $\left(\phi^2 \right)^2$ field theory with $O\left(N \right)$ symmetry. We then have to substitute:

$$g_{ijkl} = \frac{g}{3} \left(\delta_{ij} \delta_{kl} + \delta_{ik} \delta_{jl} + \delta_{il} \delta_{jk} \right). \qquad (26.98)$$

After a short calculation the expression (26.79) of the β-function is recovered and in addition:

$$\eta\left(\tilde{g} \right) = \frac{\left(N + 2 \right)}{72} \tilde{g}^2 + O\left(\tilde{g}^3 \right), \qquad (26.99)$$

where the notation (11.84) has been used:

$$\tilde{g} = N_d g, \qquad N_d = 2(4\pi)^{-d/2}/\Gamma(d/2).$$

As noted in Section 11.7, introducing the identity matrix $\mathbf{I}$ and the projector $\mathbf{P}$:

$$I_{ij,kl} = \tfrac{1}{2} \left(\delta_{ik} \delta_{jl} + \delta_{il} \delta_{jk} \right), \qquad (26.100)$$

$$P_{ij,kl} = \frac{1}{N} \delta_{ij} \delta_{kl}, \qquad (26.101)$$

we can write the matrix $\eta^{(2)}_{ij,kl}$:

$$\eta^{(2)} = -\left(N\mathbf{P} + 2\mathbf{I} \right) \frac{\tilde{g}}{6} + O\left(\tilde{g}^2 \right). \qquad (26.102)$$

The trace of the matrix $\eta^{(2)}$ yields $\eta_2\left(g \right)$. The second eigenvalue of the matrix $\eta_2'\left(g \right)$, given by its traceless part, corresponds to a symmetry breaking mass term, and as we have discussed at the beginning of this section, describes the crossover from a situation with one correlation length to a situation in which some components of the order parameter decouple. It is traditional to introduce a new function $\varphi\left(g \right)$ and to parametrize it as:

$$\eta_2'\left(g \right) = \frac{\varphi\left(g \right)}{\nu\left(g \right)} - 2. \qquad (26.103)$$

The fixed point value $\varphi = \varphi\left(g^* \right)$ is called the crossover exponent. Finally the stability conditions are given by the eigenvalues of the matrix M. Setting:

$$g_{ijkl} = g^*_{ijkl} + s_{ijkl}, \qquad (26.104)$$

we find at leading order:

$$(Ms)_{ijkl} = -\varepsilon s_{ijkl} + \frac{\varepsilon}{N+8} \left(\delta_{ij} s_{mmkl} + 5 \text{ terms} + 12 s_{ijkl} \right). \qquad (26.105)$$

Taking s_{ijkl} proportional to g^*_{ijkl} we recover the exponent ω. More generally the eigenvectors can be classified according to their trace properties. Let us write s_{ijkl}:

$$s_{ijkl} = u g^*_{ijkl} + \left(v_{ij} \delta_{kl} + 5 \text{ terms} \right) + w_{ijkl}, \qquad (26.106)$$

in which the tensors v_{ij} and w_{ijkl} are traceless:

$$v_{ii} = 0; \qquad w_{ijkk} = 0. \qquad (26.107)$$

The three eigenvalues corresponding to u, w, v are respectively:

$$\omega = \varepsilon + O\left(\varepsilon^2\right), \quad \omega_{\text{anis.}} = \varepsilon \frac{4-N}{N+8} + O\left(\varepsilon^2\right), \quad \omega' = \frac{8\varepsilon}{N+8} + O\left(\varepsilon^2\right). \qquad (26.108)$$

Note that the perturbation proportional to v_{ij} does not satisfy the trace condition (26.97). It therefore lifts the degeneracy between the correlation lengths of the different components of the order parameter. It induces a crossover to a situation in which some components decouple. However one easily verifies that the corresponding eigenvalue ω' leads to effects subleading for ε small with respect to the eigenvalue $\eta'_2(g^*)$. Within the class of interactions satisfying equation (26.97), the relevant eigenvalue is $\omega_{\text{anis.}}$. We find the very interesting result that the $O(N)$ symmetric fixed point is stable against any perturbation for N smaller than some value N_c. This is again an example of dynamically induced symmetry: Correlation functions in the critical domain have a larger symmetry than microscopic correlation functions. The calculation of $\omega_{\text{anis.}}$ at order ε yields:

$$N_c = 4 - 2\varepsilon + O\left(\varepsilon^2\right). \qquad (26.109)$$

Remarks. It can be seen in expression (26.94) that the β-function defines a gradient flow. Actually it has been checked up to three-loop order that this property remains true in the following sense: Calling g_α the set of all coupling constants, one finds:

$$\beta_\alpha = T_{\alpha\beta} \frac{\partial B}{\partial g_\beta},$$

in which $T_{\alpha\beta}$ is a symmetric and, for g_α small, positive matrix. This is the best one can expect, considering the transformation properties of the β-function under reparametrization of the coupling constant space. It follows that the fixed points are extrema of the function B and that the matrix of derivatives of the β-function at a fixed point, being equivalent to a symmetric matrix, has real eigenvalues.

An empirical observation, in many examples, is that when several fixed points are found, then the stablest one corresponds, for g_α small, to the largest value of the exponent of η.

26.7 Asymptotic Expansion of the 2-Point Function in the Critical Domain

In the critical domain, when points are separated by distances much smaller than the correlation length ξ, the correlation functions tend towards the correlation functions of the critical theory $(T = T_c)$

$$\Gamma^{(n)}(p) \underset{\xi^{-1} \ll p \ll \Lambda}{\propto} p^{2-\eta}. \tag{26.110}$$

The r.h.s. is actually the first term of an asymptotic expansion in the variable $p\xi$ for $p\xi$ large. The leading term has been obtained by using the property that at large non-exceptional momenta the derivative $\partial \Gamma^{(n)}(p_1, \ldots, p_n)/\partial t$ is asymptotically negligible with respect to $\Gamma^{(n)}(p_1, \ldots, p_n)$. However since:

$$\frac{\partial}{\partial t}\Gamma^{(n)}(p_1, \ldots, p_n) = \Gamma^{(1,n)}(0; p_1, \ldots, p_n),$$

the derivative $\partial \Gamma^{(n)}/\partial t$ cannot be evaluated with the same method because the momenta are exceptional. As we have explained in Chapter 12, it is necessary to use the short distance expansion (SDE) of operator products. Let us now concentrate on the 2-point function. We have to evaluate $\Gamma^{(1,2)}(0; p, -p)$. However we cannot apply directly the SDE to $\Gamma^{(1,2)}$ because this would involve $\Gamma^{(2,0)}$ which needs additional renormalizations.

We therefore differentiate once more with respect to t:

$$\frac{\partial^2}{(\partial t)^2}\Gamma^{(2)}(p) = \Gamma^{(2,2)}(0,0; p, -p). \tag{26.111}$$

To $\Gamma^{(2,2)}$ we can now apply the SDE:

$$\Gamma^{(2,2)}(0,0; p, -p) \sim B(p)\Gamma^{(3,0)}(0,0,0) \quad \text{for } \xi^{-1} \ll p \ll \Lambda. \tag{26.112}$$

As shown in Section 12.5, $B(p)$ satisfies a RG equation which can be obtained by applying the differential operator D:

$$D \equiv \Lambda\frac{\partial}{\partial\Lambda} + \beta(g) - \left(\frac{1}{\nu(g)} - 2\right)t\frac{\partial}{\partial t}, \tag{26.113}$$

on both sides of equation (26.112) and using the RG equations (26.43).

One finds:

$$\left[D + \nu^{-1}(g) - 2 - \eta(g)\right]B(p) \sim 0. \tag{26.114}$$

Taking the $t = 0$ limit and setting $g = g^*$, the equation becomes:

$$\left(\Lambda\frac{\partial}{\partial\Lambda} + \frac{1}{\nu} - 2 - \eta\right)B(p) \sim 0. \tag{26.115}$$

Equation (26.112) shows also that $B(p)$ has canonical dimension ε. It follows:

$$B(p) \sim \Lambda^\varepsilon (p/\Lambda)^{2-\eta-(1-\alpha)/\nu}. \tag{26.116}$$

Differentiating equation (26.48) with respect to t, we find:

$$\Gamma^{(3,0)}(0,0,0) \sim \Lambda^{-\varepsilon}(t/\Lambda^2)^{-1-\alpha}. \tag{26.117}$$

Integrating equation (26.111) twice with respect to t and using the set of equations (26.112,26.116,26.117) we obtain finally:

$$\Gamma^{(2)}(p)\underset{\xi^{-1}\ll p\ll\Lambda}{=}p^{2-\eta}\left(a+bt\,p^{-1/\nu}+ct^{1-\alpha}p^{-(1-\alpha)/\nu}+\cdots\right). \tag{26.118}$$

A systematic use of the SDE beyond leading order allows us to calculate systematic corrections to expression (26.118).

Note that the effect of the differentiation with respect to t has been simply to generate the regular terms in the temperature which are of order comparable to the singular terms for ε small.

Next to leading terms in a field or below T_c. It is also possible to obtain expressions in a field or below T_c by expanding correlation functions in powers of the magnetization and applying the SDE expansion to each term. The results now are different for Ising-like systems and the N-vector model. For Ising-like systems one finds:

$$\Gamma^{(2)}(p,t,M)=p^{2-\eta}\left[a+b\frac{t}{p^{1/\nu}}+F_1\left(\frac{t}{M^{1/\beta}}\right)\left(\frac{p}{M^{\nu/\beta}}\right)^{-(1-\alpha)/\nu}+\cdots\right], \tag{26.119}$$

in which the function F_1 can be related to the free energy and thus the equation of state by:

$$F_1(x)=\int_1^\infty ds\,s^{\delta-1/\beta}\left[f'(0)-f'\left(\frac{x}{s^{1/\beta}}\right)\right]+\frac{f'(0)}{\delta-(1/\beta)+1}. \tag{26.120}$$

In the $O(N)$ symmetric case the SDE involves a second operator of dimension 2:

$$\mathcal{O}_{ij}\left[\phi(x)\right]=\phi_i(x)\phi_j(x)-\frac{\delta_{ij}}{N}\phi^2(x). \tag{26.121}$$

This operator is multiplicatively renormalizable. Correlation functions with the insertion of $\mathcal{O}_{ij}$ satisfy the RG equations:

$$\left[\Lambda\frac{\partial}{\partial\Lambda}+\beta(g)\frac{\partial}{\partial g}-\frac{n}{2}\eta(g)-\left(\frac{\varphi(g)}{\nu(g)}-2\right)\right]\Gamma^{(n)}_{\mathcal{O}_{ij}}(p_i;g,\Lambda)=0, \tag{26.122}$$

in which the RG function $\varphi(g)$ has been defined in equation (26.103). Consequently a new term is present at the same order in the asymptotic expansion of the 2-point function which becomes:

$$\Gamma^{(2)}(p,t,M)=p^{2-\eta}\left\{\left[a+b\frac{t}{p^{1/\nu}}+F_1\left(\frac{t}{M^{1/\beta}}\right)\left(\frac{p}{M^{\nu/\beta}}\right)^{-(1-\alpha)/\nu}\right]\delta_{ij}\right.$$
$$\left.+F_2\left(\frac{t}{M^{1/\beta}}\right)\left(\frac{p}{M^{\nu/\beta}}\right)^{-d+\varphi/\nu}\left(\frac{\delta_{ij}}{N}-\frac{M_iM_j}{M^2}\right)\right\}+\cdots. \tag{26.123}$$

in which φ is the crossover exponent and F_2 a new universal function which may be calculated in an ε-expansion. At order ε it is given by:

$$F_2(x)=1+\frac{\varepsilon}{2(N+8)}\left[(x+3)\ln(x+3)-(x+1)\ln(x+1)\right]+O\left(\varepsilon^2\right). \tag{26.124}$$

Bibliographical Notes

Most references given at the end of Chapter 25 are also relevant here and will not be repeated. The chapter is inspired by Sections VI, VII, XI and XII of the review article

E. Brézin, J.C. Le Guillou and J. Zinn-Justin, in *Phase Transitions and Critical Phenomena*, vol. 6, already quoted in Chapter 24,

which also contains a more detailed discussion of the multi-component ϕ^4 theory. The scaling form of the equation of state has been proposed in

B. Widom, *J. Chem. Phys.* 43 (1965) 3892, 3898; C. Domb and D.L. Hunter, *Proc. Phys. Soc. (London)* 86 (1965) 1147,

and verified in RG calculations by

E. Brézin, D.J. Wallace and K.G. Wilson, *Phys. Rev. Lett.* 29 (1972) 591, *Phys. Rev.* B7 (1973) 232; G.M. Avdeeva and A.A. Migdal, *JETP Lett.* 16 (1972) 178; see also the contribution of D.J. Wallace in *Phase Transitions and Critical Phenomena* vol. 6 already quoted.

Also relevant is

R.B. Griffiths, *Phys. Rev.* 158 (1967) 176.

The N-vector model with cubic anisotropy has been studied in

D.J. Wallace, *J. Phys.C: Solid State Phys.* 6 (1973) 1390; A. Aharony, *Phys. Rev.* B8 (1973) 3342, 3349, 3358, 3363, 4270, *Phys. Rev. Lett.* 31 (1973) 1494.

See also

I.J. Ketley and D.J. Wallace, *J. Phys. A: Math. Nucl. Gen.* 6 (1973) 1667,

and the contribution of A. Aharony to *Phase Transitions and Critical Phenomena* vol. 6 already quoted.

The general N-vector model is discussed in

E. Brézin, J.C. Le Guillou and J. Zinn-Justin, *Phys. Rev.* B10 (1974) 892.

The decoupling of non-critical or massive modes is discussed in

K. Symanzik, *Acta Phys. Austr. Suppl.* XI (1973) 199; T. Applequist and J. Carazzone, *Phys. Rev.* D11 (1975) 2856.

For the application of the SDE to critical phenomena see

E. Brézin, D. Amit and J. Zinn-Justin, *Phys. Rev. Lett.* 32 (1974) 151; E. Brézin, J.C. Le Guillou and J. Zinn-Justin, *Phys. Rev. Lett.* 32 (1974) 473.

Some results had been anticipated on physical grounds by

M.E. Fisher and J.S. Langer, *Phys. Rev. Lett.* 20 (1968) 665; M.E. Fisher and A. Aharony, *Phys. Rev.* B10 (1974) 2818.

The crossover exponent has been discussed in

E. Riedel and F. Wegner, *Z. Phys.* 225 (1969) 195; M.E. Fisher and P. Pfeuty, *Phys. Rev.* B6 (1972) 1889; K.G. Wilson, *Phys. Rev. Lett.* 28 (1972) 548; Y. Yamazaki and M. Suzuki, *Prog. Theor. Phys.* 50 (1973) 1097; Y. Yamazaki, *Phys. Lett.* 49A (1974) 215.

The effect of long range forces have been considered in

M.E. Fisher, S.K. Ma and B.G. Nickel, *Phys. Rev. Lett.* 29 (1972) 917.

Exercises

Exercise 26.1

Discuss a general situation in which the RG β-function has the form of a gradient flow

$$\beta_i(g) = \frac{\partial V}{\partial g_i}.$$

Exercise 26.2

Renormalization Group for models with cubic symmetry. Consider a model in which the order parameter is a N-component field $\phi(x)$ which has a cubic symmetry:

$$\phi_i \leftrightarrow \phi_j, \quad \phi_i \leftrightarrow -\phi_i.$$

The Landau–Ginzburg action $S(\phi)$ expanded up to fourth order in ϕ then takes the form:

$$S(\phi) = \int \mathrm{d}^d x \left[\tfrac{1}{2} \left(\partial_\mu \phi \right)^2 + \tfrac{1}{2} r \phi^2 + \frac{g}{4!} \left(\phi^2 \right)^2 + \frac{\lambda}{4!} \sum_{i=1,N} \phi_i^4 \right].$$

The term which breaks the $O(N)$ symmetry of the action may be generated in a ferromagnetic model by the interaction between the spin degrees of freedom with the lattice which has only cubic spatial symmetry.

26.2.1. Write the conditions relating g and λ which guaranty the stability of the classical potential.

26.2.2. Calculate the RG functions at leading order.

26.2.3. Find the various IR fixed points and discuss their stability.

26.2.4. Relate the eigenvalues of the stability matrix at the Ising model fixed point to Ising model exponents.

APPENDIX 26
THE SCALING BEHAVIOUR OF THE SPECIFIC HEAT FOR $\alpha = 0$

In the case of the 2D Ising model the exponent α vanishes. In such a situation it is impossible to find a function $C_2(g)\,\Lambda^{-\varepsilon}$ solution of equation (25.72) regular at $g = g^*$. Indeed the equation for $C_2(g)$ is then:

$$\beta(g)\,C_2'(g) + (d - 2/\nu(g))\,C_2(g) = B(g),\qquad (A26.1)$$

with:

$$\alpha = 0 \ \Rightarrow\ \nu(g^*) = 2/d.\qquad (A26.2)$$

Thus at $g = g^*$ the l.h.s. of equation $(A26.1)$ vanishes if $C_2(g)$ is regular, and the r.h.s. does not. The general solution of equation $(A26.1)$ has instead the form:

$$C_2(g) = D(g)\ln|g - g^*| + E(g),\qquad (A26.3)$$

in which $D(g)$ and $E(g)$ are regular at g^* and

$$D(g^*) = B(g^*)/\omega.\qquad (A26.4)$$

As a consequence, the function $\Gamma^{(2,0)}(q;t,g,\Lambda) - \Lambda^{-\varepsilon}C_2(g)$, which satisfies a homogeneous RG equation, is singular at g^*. When $g(\lambda)$ approaches g^*, the theory at scale λ is dominated by this singularity. Since:

$$|g(\lambda) - g^*| \sim \lambda^\omega \sim t^{\omega\nu},\qquad (A26.5)$$

equations (26.47,26.48) are replaced by:

$$\Gamma^{(2,0)}(q;t,g,\Lambda) - \Lambda^{-\varepsilon}C_2(g) \sim \Lambda^{-\varepsilon}\left[F\left(\frac{q}{m}\right) - A_2(g)\ln\frac{t}{\Lambda^2}\right],\qquad (A26.6)$$

$$\Gamma^{(2,0)}(0;t,g,\Lambda) - \Lambda^{-\varepsilon}C_2(g) \sim \Lambda^{-\varepsilon}\left[F(0) - A_2(g)\ln\frac{t}{\Lambda^2}\right],\qquad (A26.7)$$

with:

$$A_2(g) = B(g^*)\,\nu\exp\left[2\int_0^1 d\sigma\frac{\eta_2(g(\sigma)) - \eta_2}{\sigma}\right].$$

Alternative derivation. Alternatively we can solve directly RG equation (26.46) by the method of characteristics. We set:

$$\lambda\frac{\mathrm{d}}{\mathrm{d}\lambda}\zeta^{-2}(\lambda)\left[\Gamma^{(2,0)}(q;t(\lambda),g(\lambda),\lambda\Lambda) - (\lambda\Lambda)^{-\varepsilon}C(\lambda)\right] = 0.\qquad (A26.8)$$

Applying the chain rule to evaluate explicitly the derivative and expressing the compatibility with equation (26.46), we find in addition to equations (26.27–26.28), two equations:

$$\lambda\frac{\mathrm{d}}{\mathrm{d}\lambda}\ln\zeta(\lambda) = \eta_2(g(\lambda)),\qquad\qquad \zeta(1) = 1;\qquad (A26.9)$$

$$\lambda\frac{\mathrm{d}}{\mathrm{d}\lambda}C(\lambda) = [2\eta_2(g(\lambda)) + \varepsilon]\,C(\lambda) + B(g(\lambda)),\quad C(1) = 0.\qquad (A26.10)$$

The solution of the first equation can be written:

$$\zeta(\lambda) = \exp\left[\int_1^\lambda \frac{d\sigma}{\sigma} \eta_2(g(\sigma))\right].$$

(A26.11)

It follows in particular:

$$t(\lambda) = t/\zeta(\lambda).$$

(A26.12)

The solution of equation (A26.10) can then be written:

$$C(\lambda) = \int_1^\lambda \frac{d\sigma}{\sigma} B\left(g(\sigma)\right) \left(\frac{\lambda}{\sigma}\right)^\varepsilon \left(\frac{\zeta(\lambda)}{\zeta(\sigma)}\right)^2.$$

(A26.13)

Using dimensional analysis and imposing the condition (26.30), $t(\lambda) = \lambda^2 \Lambda^2$, we obtain first:

$$\Gamma^{(2,0)}\left(q; t, g, \Lambda\right) = \zeta^{-2}(\lambda)(\lambda\Lambda)^{-\varepsilon}\left[\Gamma^{(2,0)}\left(q/m; 1, g(\lambda), 1\right) - C(\lambda)\right],$$

(A26.14)

with (equation (26.32)):

$$m = \lambda\Lambda.$$

Equation (26.35) gives the asymptotic relation between λ and t. For λ small $\zeta(\lambda)$ has a power behaviour which in terms of the exponent ν, related to η_2 by equation (25.69), has the form:

$$\zeta(\lambda) \sim \lambda^{1/\nu-2} \sim t^{1-2\nu}.$$

(A26.15)

In the absence of the inhomogeneous term we would recover the scaling form (26.44) applied to $l = 2$, $n = 0$. Let us then examine the behaviour of the inhomogeneous term for λ small:

$$\zeta^{-2}(\lambda)\lambda^{-\varepsilon}C(\lambda) = \int_1^\lambda d\sigma\, \sigma^{\alpha/\nu-1}\left[B\left(g(\sigma)\right)\sigma^{2\eta_2(g^*)}\zeta^{-2}(\sigma)\right].$$

(A26.16)

For $\alpha > 0$ the integral in the r.h.s. has a limit when λ goes to zero and becomes a function of g in agreement with equation (26.48). When α vanishes, it becomes necessary to extract one singular term from the integrand before taking the limit. One obtains:

$$\zeta^{-2}(\lambda)\lambda^{-\varepsilon}C(\lambda) = B\left(g^*\right)\exp\left[2\int_0^1 d\sigma\frac{\eta_2(g(\sigma)) - \eta_2}{\sigma}\right]\ln\lambda + \text{ regular terms}$$

$$= A_2(g)\ln\frac{t}{\Lambda^2} + \text{ regular terms}.$$

(A26.17)

Replacing $C(\lambda)$ by this expression in equation (A26.14), we recover equation (A26.6).

27 CORRECTIONS TO SCALING BEHAVIOUR

In previous chapters, while deriving the scaling behaviour of correlation functions, we have always kept only the leading term in the critical region. For instance when we have always kept only the leading term in the critical region. For instance when we have solved the RG equations we have always replaced the effective coupling constant $g(\lambda)$ at scale λ by g^*, neglecting the small difference $g(\lambda) - g^*$, which vanishes only if $g = g^*$. Moreover, to establish RG equations, we had already neglected corrections subleading by powers of Λ, and effects of other couplings of higher canonical dimensions. Subleading terms related to the motion of $g(\lambda)$, which give the leading corrections for ε small, can easily be obtained from the solutions of the RG equations derived previously and will be discussed first. The situations below and at four dimensions (the upper-critical dimension) have to examined separately. The second type of corrections involves additional considerations and will be studied in the second part of this chapter. Finally a physical application is provided by systems with strong dipolar forces which have three as upper-critical dimension.

27.1 Corrections to Scaling: Generic Dimensions

In dimensions $d < 4$, to characterize the corrections to scaling due to an initial value of the ϕ^4 coupling g different from the fixed point value g^*, it is convenient to solve the RG equations in a slightly different manner by introducing a set of coupling constant dependent renormalizations:

$$
\begin{cases}
\ln Z(g) = -\displaystyle\int_{g^*}^{g} \frac{dg'}{\beta(g')} \left[\eta(g') - \eta\right], \\[2mm]
M(g) = M Z^{-1/2}(g), \\[2mm]
t(g) = t \exp\left[\displaystyle\int_{g^*}^{g} \frac{dg'}{\beta(g')} \left(\frac{1}{\nu(g')} - \frac{1}{\nu}\right)\right],
\end{cases}
\tag{27.1}
$$

and a new coupling constant $\tilde{g}$ which characterizes the deviation of g from g^*:

$$
\tilde{g} = (g - g^*) \exp\left[\int_{g^*}^{g} dg' \left(\frac{\omega}{\beta(g')} - \frac{1}{(g' - g^*)}\right)\right].
\tag{27.2}
$$

We then substitute:

$$
\Gamma^{(n)}(p_i; t, M, g, \Lambda) = Z^{-n/2}(g)\Gamma^{(n)}(p_i; t(g), M(g), g^*, \Lambda)\, C^{(n)}(p_i; t(g), M(g), \tilde{g}, \Lambda).
\tag{27.3}
$$

The functions $C^{(n)}$ satisfy the following boundary conditions:

$$
C^{(n)}(p_i; t, M; 0, \Lambda) = 1.
\tag{27.4}
$$

The finite renormalizations (27.1) eliminate trivial deviations from the fixed point theory which correspond simply to a change of normalization of the different scaling variables.

Equation (26.71) then implies:

$$\left[\Lambda\frac{\partial}{\partial\Lambda} + \omega\tilde{g}\frac{\partial}{\partial\tilde{g}} - \frac{\eta}{2}M\frac{\partial}{\partial M} - \left(\frac{1}{\nu} - 2\right)t\frac{\partial}{\partial t}\right]C^{(n)}(p_i, t, M, \tilde{g}, \Lambda) = 0. \tag{27.5}$$

Solving this equation by expanding $C^{(n)}$ in powers of $\tilde{g}$ we obtain:

$$C^{(n)}(p_i, t, M, \tilde{g}, 1) = 1 + \sum_{s=1}^{\infty} \tilde{g}^s t^{s\omega\nu} D_s^{(n)}\left(p_i t^{-\nu}, t M^{-1/\beta}\right), \tag{27.6}$$

in which Λ has been set equal to 1.

The exponent ω, defined as the value of the derivative of $\beta(g)$ at g^*, which characterizes the approach to the fixed point, characterizes also, therefore, the leading corrections to the critical behaviour.

For models with one coupling constant, $\beta(g)$ has the form:

$$\beta(g) = -\varepsilon g + ag^2 + O\left(g^3, g^2\varepsilon\right). \tag{27.7}$$

In such a case one always finds:

$$\omega = \varepsilon + O\left(\varepsilon^2\right), \qquad \omega\nu = \varepsilon/2 + O\left(\varepsilon^2\right). \tag{27.8}$$

Note that to render all terms in the expansion (27.6) dimensionless, in the sense of scaling dimensions, we can assign to $\tilde{g}$ the dimension $-\omega$.

Scaling for $d > 4$. Up to now we have considered corrections to scaling for $d < 4$. In four dimensions or above, the fixed point corresponds to $g^* = 0$, i.e. to the gaussian fixed point, and therefore the leading contributions to all correlation functions, except the 2-point function, come from corrections to scaling since these functions vanish at the fixed point. It is easy to verify that this special feature of the gaussian fixed point explains the apparent contradiction between some RG predictions like relation between exponents involving explicitly the dimension d (called hyperscaling relations) and mean field exponents. For example if we assign to the magnetic field H and the magnetization M the dimensions $\beta\delta/\nu$ and β/ν respectively, the exponents β and δ being given by equations (26.64,26.65), and, according the discussion above, to the coupling constant g the dimension $-\omega$ with:

$$\omega = -\varepsilon \quad \text{for } d > 4, \tag{27.9}$$

all terms in the equation of state have the same dimension. Indeed with these assignments:

$$[H] = [t] + [M] = 3[M] + [g] \iff 2\beta\delta = 2 + 2\beta = 6\beta - \omega,$$

for all dimensions $d > 4$. In exactly four dimensions the situation is more subtle since the $\phi^4(x)$ operator is marginal, and the approach to the fixed point is only logarithmic. This will be the subject of the next section.

27.2 Logarithmic Corrections at the Upper-Critical Dimension

The upper-critical dimension is the dimension at which deviations from mean field theory appear. For our model this dimension is four. In this dimension there generally exists a marginal operator, here $\int \phi^4(x)\mathrm{d}x$, and therefore, as we have indicated in the general discussion of Section 26.1, logarithmic corrections to mean field behaviour are expected. Although the dimension four is not of physical relevance for statistical problems, its study is of special pedagogical value, because exact predictions can be derived from RG arguments. Moreover a ϕ^4 interaction is present in the Higgs sector of the Standard Model of weak electromagnetic interactions. Because the fixed point corresponds to $g^* = 0$, no assumption about the fixed point theory is required. Finally let us note some physical systems have $d = 3$ as upper-critical dimension, for example tricritical systems or ferroelectrics with dipolar uniaxial long range forces. The latter example will be discussed in Section 27.5.

We study below only the equation of state and the specific heat, the generalization to correlation functions being straightforward. Let us rewrite the relation (26.57) for $\varepsilon = 0$:

$$H\left(M,t,g,\Lambda\right) = Z^{-1/2}(\lambda)\left(\lambda\Lambda\right)^3 H\left(\frac{M(\lambda)}{\lambda\Lambda}, \frac{t(\lambda)}{\lambda^2\Lambda^2}, g(\lambda), 1\right). \tag{27.10}$$

We recall the definitions of the various functions:

$$\ln\lambda = \int_g^{g(\lambda)} \frac{\mathrm{d}g'}{\beta\left(g'\right)}, \tag{27.11}$$

$$\ln Z(\lambda) = \int_g^{g(\lambda)} \mathrm{d}g' \frac{\eta\left(g'\right)}{\beta\left(g'\right)}, \tag{27.12}$$

$$M(\lambda)/\lambda\Lambda = \left[\lambda Z^{1/2}(\lambda)\right]^{-1} M/\Lambda, \tag{27.13}$$

$$\ln\frac{t(\lambda)}{\lambda^2\Lambda^2} = \ln\frac{t}{\Lambda^2} - \int_g^{g(\lambda)} \frac{\mathrm{d}g'}{\beta\left(g'\right)\nu\left(g'\right)}. \tag{27.14}$$

It is also necessary to here recall the expansion of the RG functions for g small (equations (26.79, 26.99, 26.102)). In the $\left(\phi^2\right)^2$ field theory, ϕ being a N-component vector field , i.e. in the $O(N)$ symmetric N-vector model:

$$\beta(g) = \frac{(N+8)}{6}\frac{g^2}{8\pi^2} + O\left(g^3\right),$$

$$\eta(g) = \frac{(N+2)}{72}\left(\frac{g}{8\pi^2}\right)^2 + O\left(g^3\right), \tag{27.15}$$

$$\nu^{-1}(g) = 2 - \frac{(N+2)}{6}\frac{g}{8\pi^2} + O\left(g^2\right).$$

As we have done previously we choose λ such that:

$$M(\lambda) = \lambda\Lambda, \tag{27.16}$$

or, in zero magnetization, solution to the equation:

$$t(\lambda) = \lambda^2\Lambda^2. \tag{27.17}$$

In both cases the solution λ of these equations is small in the critical domain. The corresponding effective coupling $g(\lambda)$ is then also small. It is easy to integrate equation (27.11) in this regime. One finds:

$$g(\lambda) = 8\pi^2 \frac{6}{N+8} \frac{1}{|\ln \lambda|} + O\left[\frac{\ln|\ln \lambda|}{(\ln|\lambda|)^2}\right].$$
(27.18)

Due to conditions (27.16) or (27.17), the theory at scale λ is no longer critical; furthermore $g(\lambda)$ is small: therefore $H(\lambda)$ in the r.h.s. of equation (27.10) can be calculated from perturbation theory. Note here the power of the RG method: we started from a theory with a coupling constant of order 1 and perturbative coefficients increasing like powers of $\ln(t/\Lambda^2)$ or equivalent. Direct perturbation theory was obviously meaningless. In contrast the rescaled theory has a small coupling constant $g(\lambda)$ and perturbative coefficients of order 1 due to conditions (27.16) or (27.17).

For example let us use equation (27.16):

$$M(\lambda) = \lambda\Lambda \implies \lambda \sim M/\Lambda,$$

and calculate $t(\lambda)$:

$$\frac{t(\lambda)}{\lambda^2\Lambda^2} \sim K(g)\frac{t}{\lambda^2\Lambda^2} \left(|\ln \lambda|\right)^{-(N+2)/(N+8)} \sim K(g)\frac{t}{M^2}\left|\ln\frac{M}{\Lambda}\right|^{-(N+2)/(N+8)},$$
(27.19)

$K(g)$ being a finite temperature renormalization. Using now the perturbative expansion:

$$H = tM + \frac{g}{6}M^3 + \cdots,$$

and relations (27.10) and (27.18), it is easy to derive:

$$H(M, t, g, \Lambda = 1) = \frac{K(g)\,tM}{|\ln M|^{(N+2)/(N+8)}} + \frac{8\pi^2}{N+8}\frac{M^3}{|\ln M|} + O\left(M^3 \frac{\ln|\ln M|}{|\ln M|^2}\right).$$
(27.20)

At T_c we thus find:

$$H \propto M^3 / |\ln M|.$$
(27.21)

The spontaneous magnetization is given by:

$$M \propto (-t)^{1/2} |\ln(-t)|^{3/(N+8)}.$$
(27.22)

Since expression (27.20) is not uniform for M small, it is necessary to use equation (27.17) to calculate the susceptibility in zero field. We then find:

$$\chi^{-1} \propto |t| \, |\ln|t||^{-[(N+2)/(N+8)]}.$$
(27.23)

Finally the specific heat satisfies the RG equation:

$$\left[\Lambda\frac{\partial}{\partial\Lambda} + \beta(g)\frac{\partial}{\partial g} - \eta_2(g)\left(2 + t\frac{\partial}{\partial t}\right)\right]\Gamma^{(2,0)} = B(g),$$
(27.24)

in which $B(g)$ has the expansion:

$$B(g) = \frac{N}{16\pi^2} + O(g). \tag{27.25}$$

It is easy to verify that a function $C_2(g)$ solution of equation (27.24) and thus of

$$\beta(g)C_2'(g) - 2\eta_2(g)C_2(g) = B(g),$$

is necessarily singular at the origin (see also Appendix A26). For example one can take a solution of the form:

$$C_2(g) = \frac{3N}{(N-4)}\frac{1}{g} + O(1) \quad \text{for } N \neq 4. \tag{27.26}$$

For $N = 4$ an additional logarithmic singularity is present. The combination $\Gamma^{(2,0)} - C_2(g)$ solution of the homogeneous RG equation is for g small dominated by the pole of $C_2(g)$:

$$\Gamma^{(2,0)}(0;t,g,\Lambda) - C_2(g) \sim -\frac{3N}{(N-4)}\frac{1}{g(\lambda)}\zeta^{-2}(\lambda), \tag{27.27}$$

with:

$$\zeta(\lambda) = \exp\left[\int_g^{g(\lambda)} \frac{dg'}{\beta(g')}\eta_2(g')\right] \propto [g(\lambda)]^{-(N+2)/(N+8)}. \tag{27.28}$$

Collecting all factors we obtain the behaviour of the specific heat:

$$\Gamma^{(2,0)}(0;t,g,\Lambda) - C_2(g) \propto |\ln t|^{(4-N)/(N+8)}\left[1 + O\left(\frac{\ln|\ln t|}{\ln t}\right)\right]. \tag{27.29}$$

Remark. It is apparent from these expressions that a parametrization in terms of the variables t or M leads to very complicate expressions. A better way of writing all these results is to introduce a parametric representation in terms of the effective coupling constant $g(\lambda)$. We can calculate λ in terms of $g(\lambda)$ from equation (27.11). Let us parametrize $\beta(g)$ as:

$$\beta(g) = \beta_2 g^2 + \beta_3 g^3 + O\left(g^4\right), \tag{27.30}$$

and set:

$$s = g(\lambda). \tag{27.31}$$

Then λ is given by:

$$\lambda = s^{-\beta_3/\beta_2^2}\, e^{-1/\beta_2 s}\, \tilde{\lambda}(s). \tag{27.32}$$

In equation (27.32) the function $\tilde{\lambda}(s)$ is a regular function of s for s small. The renormalization factor $Z(\lambda)$ is a regular function of s. Finally equation (27.14) yields:

$$\frac{t(\lambda)}{\lambda^2\Lambda^2} = \frac{t}{\Lambda^2}s^{2[2\nu_1 + \beta_3/\beta_2]/\beta_2}\, e^{2/\beta_2 s}\, [\tilde{t}(s)]^{-1}, \tag{27.33}$$

in which $\tilde{t}(s)$ is a regular function of s and $\nu(g)$ has been written:

$$\nu(g) = (2 + \eta_2(g))^{-1} = \tfrac{1}{2} + \nu_1 g + O\left(g^2\right). \tag{27.34}$$

Then if we impose condition (27.17), equation (27.33) parametrizes t as a function of s. At leading order, all critical behaviours will be described by a singular factor of the form occurring in equations (27.32) or (27.33) multiplied by regular series in $s = g(\lambda)$.

27.3 Irrelevant Operators and the Question of Universality

We now examine, in generic dimensions, the contributions coming from irrelevant operators. We again stress that these operators have been found to be irrelevant at the gaussian
fixed point, near four dimensions. We still rely on the assumption that dimensions vary
continuously when the IR fixed point moves away from the gaussian fixed point. Finally
the analysis is local, we consider only the neighbourhood of the fixed point.

Let us first recall power counting arguments for a general theory with an action $S(\phi)$
(for more details see Chapters 8, 12):

$$S(\phi) = \int \mathrm{d}^d x \left[\frac{1}{2} (\partial_\mu \phi)^2_\Lambda + \frac{1}{2} m^2 \phi^2 + \sum_\alpha u_\alpha \mathcal{O}_\alpha (\phi) \right], \tag{27.35}$$

in which $\mathcal{O}_\alpha (\phi)$ is a local monomial of degree n_α in ϕ with k_α derivatives. The dimension
$[u_\alpha]$ of the coupling constant u_α is then:

$$[u_\alpha] = d - k_\alpha - \tfrac{1}{2} n_\alpha (d - 2). \tag{27.36}$$

We treat all interactions in action (27.35) in perturbation theory. Order by order in
the loopwise expansion we evaluate the divergent part of the corresponding Feynman
diagrams and add counterterms to the action to render the theory finite. Since the
action (27.35) contains all possible monomials in the field, any counterterm is a linear
combination of the operators $\mathcal{O}_\alpha (\phi)$.

In order for a product of constants u_β to appear in a counterterm proportional to an
operator $\mathcal{O}_\alpha$, it is necessary and generically sufficient that the condition:

$$\Delta = d - [\mathcal{O}_\alpha] - \sum_l [u_{\beta_l}] \geq 0, \tag{27.37}$$

is satisfied. Then the coefficient $\delta u_\alpha (\Lambda)$ of the counterterm proportional to $\mathcal{O}_\alpha$ diverges
like a positive power of the cut-off Λ (or a power of logarithm if $\Delta = 0$),

$$\delta u_\alpha (\lambda) \sim \Lambda^\Delta. \tag{27.38}$$

Let us now return to the question of irrelevant operators. We restrict ourselves to dimensions $d = 4 - \varepsilon, \varepsilon > 0$ and small, since as we have seen, these are the only dimensions
for which a reliable analysis is possible. The first operators we have for example in mind
are ϕ^6, $\phi^2 (\partial_\mu \phi)^2$... which are operators of dimension 6 in four dimensions.

To introduce the cut-off Λ in our statistical problem we have rescaled the lengths and
the field ϕ (equations (25.31–25.34)). Therefore each coupling constant u_α is the product
of a pure number g_α by a power of the cut-off which gives it its dimension:

$$u_\alpha = g_\alpha \Lambda^{-\delta_\alpha}. \tag{27.39}$$

Equation (27.36) gives for δ_α:

$$\delta_\alpha = -d + k_\alpha + \tfrac{1}{2} n_\alpha (d - 2). \tag{27.40}$$

If δ_α is positive the corresponding operator $\mathcal{O}_\alpha$ leads to a non-renormalizable theory and
we have already stated that it is irrelevant. In the tree approximation the statement

follows from equation (27.39): the operator gives contributions vanishing with a power of the cut-off. However in higher orders the statement is less trivial since divergences at large cut-off coming from the momentum integration can potentially compensate the powers coming from the coupling constants. To discover what happens it is necessary to analyze the counterterms generated by these operators, using equations (27.37) and (27.38).

The total power Δ' of the cut-off which multiplies the operator $\mathcal{O}_\alpha$ in a counterterm is the sum of the power Δ generated by the divergence of perturbation theory (equation (27.37)) and the powers already present in the coefficients u_α (equation (27.39)):

$$\Delta' = \Delta - \sum_l \delta_{\beta_l} = \Delta + \sum_l [u_{\alpha_l}], \qquad (27.41)$$

and therefore:

$$\Delta' = d - [\mathcal{O}_\alpha] = [u_\alpha]. \qquad (27.42)$$

The result is therefore very simple: due to the divergences of perturbation theory, irrelevant operators give indeed non-vanishing contributions, but these contributions can be cancelled by varying the amplitudes of the relevant or marginal terms in the hamiltonian, because $\Delta' \geq 0$ is equivalent to $[u_\alpha] \geq 0$.

Example. The leading new corrections come from operators $\mathcal{O}_i^6$ which have dimension 6 in four dimensions. The corresponding interactions have the form:

$$\Lambda^{2\varepsilon-2} \int \mathrm{d}^d x\, \phi^6(x), \quad \Lambda^{\varepsilon-2} \int \mathrm{d}^d x\, (\phi\partial_\mu\phi)^2, \quad \Lambda^{-2} \int \mathrm{d}^d x\, \phi\partial^4\phi.$$

In terms of the renormalized operators they have the expansion:

$$\int \mathrm{d}^d x\, \mathcal{O}_i^6(x) = \int \mathrm{d}^d x\, \left\{ Z_{ij} \left[\mathcal{O}_j^6(x)\right]_{\mathrm{r}} + A_{ij} \left[\mathcal{O}_j^4(x)\right]_{\mathrm{r}} + B \left[\phi^2(x)\right]_{\mathrm{r}} \right\}.$$

We have denoted by $\mathcal{O}_j^4(x)$ the operators which have dimension 4 in four dimensions: ϕ^4 and $(\partial_\mu\phi)^2$. In the framework of the ε-expansion the coefficients Z_{ij} diverge like powers of $\ln\Lambda$, A_{ij} like Λ^2 and B like Λ^4, up to powers of $\ln\Lambda$. Taking into account the powers of Λ in front of the interaction terms we see that only the contributions proportional to operators of dimensions 4 and 2 are divergent. If we cancel these contributions by subtracting to the operators of dimension 6 a suitable combination of operators of dimensions 4 and 2 we obtain the true new corrections which go to zero like $\Lambda^{-2+O(\varepsilon)}$.

This discussion also clarifies the interpretation of the constants r and g which parametrize the ϕ^4 hamiltonian. These are not the parameters which for instance appear as coefficients of the ϕ^2 or ϕ^4 terms in the mean field analysis, but instead effective parameters taking into account the effect of neglected irrelevant operators. The analysis of previous chapters is however at leading order not modified. Indeed the change in ϕ^2 corresponds only to a modification of the critical temperature which is a non-universal quantity. Moreover below four dimensions we have shown that many physical quantities (universal quantities) do not depend on g either, since g can be replaced by its fixed point value g^*. Finally let us note that a change in the cut-off procedure corresponds generally to a change in the coefficients of the irrelevant part of the propagator $(\phi\Delta^2\phi...)$. The effect of such a change is obtained from the previous analysis also. We can now clarify the concept of universality: below four dimensions all dimensionless quantities in which g

can be replaced by g^* the IR fixed point value, and which do not depend on the normalizations of the field ϕ, the deviation from the critical temperature t, and of the magnetic field are universal. Obvious examples are ratios of amplitude of singularities below and above T_c, ratios of amplitudes of leading corrections to scaling, the rescaled equation of state (relation between H and M), the renormalized correlation functions as defined in Chapters 25,26, etc..

Remark. Another simple consequence of this analysis is the following: Let us add to the hamiltonian an irrelevant operator which breaks a symmetry of the hamiltonian. Then the symmetry of the critical theory will be broken if and only if the irrelevant operator can generate relevant or marginal operators breaking the symmetry. An example is the following: on the lattice operators of the form $\sum_\mu \int \phi(x) \left(\partial_\mu\right)^4 \phi(x) \mathrm{d}x$ which break the $O(d)$ symmetry of the effective $\phi^4(x)$ action are present. However these operators have a hypercubic symmetry and since the only relevant operators they can generate, like $\int \left(\partial_\mu\phi\right)^2 \mathrm{d}^d x$, due to the hypercubic symmetry, are $O(d)$ symmetric, the $O(d)$ symmetry of the critical theory is not broken. Conversely the addition of a naively irrelevant term like $\int \phi^5(x)\mathrm{d}^d x$ to a hamiltonian which is symmetric in $\phi \mapsto -\phi$ generates relevant terms linear in ϕ which are equivalent to the addition of a magnetic field.

27.4 Corrections Coming from Irrelevant Operators. Improved Action

For simplicity we consider the effect, in the critical theory, of an operator $\mathcal{O}_\alpha$ at first order only in the corresponding coupling constant u_α. The following discussion applies in the framework of the ε-expansion and relies on the results of Chapter 12 concerning the renormalization of composite operators.

27.4.1 Corrections to scaling

In Section 27.3 we have shown that an operator $\mathcal{O}_\alpha$ gives contributions equivalent to all operators of lower canonical dimensions. For example $\phi^8(x)$ first generates effects equivalent to $\phi^2(x)$, $\phi^4(x)$ and $(\partial_\mu\phi(x))^2$ and all operators of dimension 6, and then genuine new corrections. To isolate these corrections it is necessary to subtract from the operator a linear combination of all operators which have a lower dimensions at $d = 4$, i.e. to perform an additive renormalization. Note that we can omit all operators which are total derivatives since only the space integrals appear in the action. We define a subtracted operator $\tilde{\mathcal{O}}_\alpha(\phi)$ by:

$$\tilde{\mathcal{O}}_\alpha(\phi) = \mathcal{O}_\alpha(\phi) - \sum_{\beta \text{ such that } [\mathcal{O}_\beta]<[\mathcal{O}_\alpha]} C_{\alpha\beta}(\Lambda, g)\, \mathcal{O}_\beta(\phi). \tag{27.43}$$

For example in the case of operators of dimension 6, like $\phi^6(x)$, one subtracts a linear combination of operators of dimensions 2 and 4:

$$\tilde{\phi}^6(x) = \phi^6(x) - C_1\phi^2(x) - C_2\left[\partial_\mu\phi(x)\right]^2 - C_3\phi^4(x).$$

The coefficients C_1, C_2 and C_3 can be determined by a set of renormalization conditions at zero momentum:

$$\Gamma^{(2)}_{\tilde{\phi}^6}(p) = O\left(p^4 \times \text{ powers of } \ln p\right) \text{ for } p \to 0,$$

$$\Gamma^{(4)}_{\tilde{\phi}^6}(p_i = 0) = 0. \tag{27.44}$$

The first condition implies in particular that the critical temperature is not changed. These conditions are not affected by IR divergences because the correlation functions with an operator insertion have positive dimensions. After such an additive renormalization, the bare operators of dimension 6 are related to the completely renormalized operators $\mathcal{O}_\alpha^{\mathrm{r}}$ by:

$$\mu^{-\delta_\alpha} \int \mathrm{d}^d x\, \mathcal{O}_\alpha^{\mathrm{r}}(x) = \sum_\beta Z_{\alpha\beta}\,(g, \Lambda/\mu)\, \Lambda^{-\delta_\beta} \int \mathrm{d}^d x\, \tilde{\mathcal{O}}_\beta(x), \qquad (27.45)$$

in which μ is the renormalization scale. Additional renormalization conditions at scale μ for the insertion of renormalized operators determine the matrix $Z_{\alpha\beta}$.

The relation between correlation functions $\Gamma_{\tilde{\mathcal{O}}_\alpha}^{(n)}$ with one $\Lambda^{-\delta_\alpha} \int \mathrm{d}^d x\, \tilde{\mathcal{O}}_\alpha$ insertion and the renormalized functions with $\mu^{-\delta_\alpha} \int \mathrm{d}^d x\, \mathcal{O}_\alpha(x)$ insertion is then:

$$\sum_\beta Z_{\alpha\beta} Z^{n/2} \Gamma_{\tilde{\mathcal{O}}_\beta}^{(n)}\,(p_i; g, \Lambda) = \left[\Gamma_{\mathcal{O}_\alpha}^{(n)}\,(p_i, g_{\mathrm{r}}, \mu) \right]_{\mathrm{ren}}. \qquad (27.46)$$

This leads to the RG equations:

$$\sum_\beta \left\{ \left[\Lambda \frac{\partial}{\partial \Lambda} + \beta(g) \frac{\partial}{\partial g} - \frac{n}{2}\eta(g) \right] \delta_{\alpha\beta} - \eta_{\alpha\beta}\,(g) \right\} \Gamma_{\tilde{\mathcal{O}}_\beta}^{(n)} = 0\,, \qquad (27.47)$$

with

$$\eta_{\alpha\beta}(g) = -\sum_\gamma \left(\Lambda \frac{\partial}{\partial \Lambda} Z_{\alpha\gamma} \right) \left(Z^{-1} \right)_{\gamma\beta}. \qquad (27.48)$$

Effects of the insertions of the operators $\tilde{\mathcal{O}}_\alpha$ are then governed by the values of η_α the eigenvalues of the matrix $\eta_{\alpha\beta}\,(g^*)$. From relation (27.45) we see that the renormalization matrix $Z_{\alpha\beta}$ has the form:

$$Z_{\alpha\beta} = \delta_{\alpha\beta}\,(\Lambda/\mu)^{\delta_\alpha}\,\left(1 + O(g) \right). \qquad (27.49)$$

Therefore the eigenvalues η_α are at leading order given by:

$$\eta_\alpha = \delta_\alpha + O\,(\varepsilon)\,.$$

The effects of the irrelevant operators of canonical dimension $d + \delta_\alpha$ in four dimensions, are suppressed by powers $\Lambda^{-\delta_\alpha + O(\varepsilon)}$ of the cut-off Λ. For operators of dimension 6, $\delta = 2$. In an infinitesimal neighbourhood of dimension four, these operators remain irrelevant. Our general analysis, which is based upon the idea that the critical behaviour of ferromagnetic systems can be described by an effective ϕ^4 field theory, remains valid, beyond the ε-expansion, as long as this property remains true.

Remark. Note finally that in some cases the irrelevant effects may be specially important. An example is provided by systems where the initial theory has only a discrete symmetry while the symmetry of the critical theory is continuous. In the low temperature phase the critical theory has Goldstone mode singularities. These singularities are suppressed by irrelevant corrections.

27.4.2 Fixed point in hamiltonian space and improved actions

We have seen that by adding to the $\phi^4(x)$ field theory irrelevant interactions, we could modify correlation functions by terms in the ε-expansion of order $1/\Lambda^2$ (up to logarithms). In the r.h.s. of RG equations (25.43) we have just neglected terms of the same order. Symanzik has shown in perturbation theory that by adding to the hamiltonian the proper linear combination of irrelevant operators, it is possible to cancel exactly these corrections. The coefficients of the linear combination are functions of the ϕ^4 coupling constant g. For example the complete set of operators of dimension 6 can be used to cancel exactly the corrections of order $1/\Lambda^2$ in the r.h.s. of the RG equations (25.43), the operators of dimension 8 to cancel the order $1/\Lambda^4$ and so on. An iteration of this procedure leads to a theory which depends on only one ϕ^4 coupling constant and which satisfies the RG equations exactly. It is actually a "renormalized" theory constructed without using the renormalization procedure, constructed by considering an infinite sequence of hamiltonians.

From the general RG point of view, the hamiltonians which lead to correlation functions satisfying RG equations exactly belong to a one parameter line in hamiltonian space which goes from the gaussian fixed point to the non-trivial IR fixed point.

Conversely by constructing directly the renormalization group for a cut-off field theory, it is possible to prove the existence of the renormalized field theory (see Appendix A10.1).

Finally let us mention that Symanzik has advocated the use of such improved actions (adding for example all terms of dimension 6) for numerical calculations on the lattice. It should mentioned, however, that the applications of this ingenious idea have, up to now, been disappointing. Because the improved action involves more extended interactions on the lattice (like second nearest neighbours) the effective size of the lattice is reduced, increasing finite size effects. The method is thus only applicable to very large lattices. But statistics then becomes in general the dominant problem, and more complicated interactions slow down the numerical calculations.

27.5 Application: Uniaxial Systems with Strong Dipolar Forces

In Section 27.2 we have stressed that the renormalization group leads to exact predictions for critical systems at the upper-critical dimension. Unfortunately in the case of the N-vector model the upper-critical dimension is four and therefore the predictions are not useful for experimental physics. The main application is numerical physics, in the sense that the Higgs sector ($N = 4$) of the Standard Model has been investigated numerically.

Therefore we here present another system on which accurate measurements have been made, and which has dimension three as the upper-critical dimension: a uniaxial ferromagnet or ferroelectric system with strong dipolar forces.

Dipolar forces. We consider a spin system in d dimensions in which the d-component spins S^μ interact both through short range and dipolar forces:

$$-\beta \mathcal{H}(\mathbf{S}) = \sum_{\mathbf{x},\mathbf{x}'} V_{\mu\nu}(\mathbf{x}-\mathbf{x}') S_\mathbf{x}^\mu S_{\mathbf{x}'}^\nu + \gamma (\mathbf{S_x} \cdot \nabla_\mathbf{x})(\mathbf{S_{x'}} \cdot \nabla_{\mathbf{x'}}) \frac{1}{|\mathbf{x}-\mathbf{x}'|^{d-2}}, \qquad (27.50)$$

$V_{\mu\nu}(\mathbf{x})$ being the short range potential. We assume that the long range dipolar forces are strong enough to play a role in the part of the critical domain accessible experimentally. In addition we assume that the lattice is strongly anisotropic in such a way that only one component of the spin $\mathbf{S}$ is critical and the effective hamiltonian can be simplified into:

$$-\beta \mathcal{H}\left(S\right) = \sum_{\mathbf{x},\mathbf{x}'} S_{\mathbf{x}} S_{\mathbf{x}'} \left[V\left(\mathbf{x}-\mathbf{x}'\right) + \gamma \partial_z \partial_{z'} \frac{1}{\left|\mathbf{x}-\mathbf{x}'\right|^{d-2}} \right], \qquad (27.51)$$

in which S now denotes the component of the spin vector $\mathbf{S}$ along the $z \equiv x_d$ axis.

After Fourier transformation the hamiltonian can be written:

$$-\beta \mathcal{H}\left(S\right) = \int \mathrm{d}^d q\, S\left(\mathbf{q}\right) S\left(-\mathbf{q}\right) \left[\tilde{V}\left(\mathbf{q}\right) + \tilde{\gamma} \frac{q_z^2}{q^2} \right], \qquad (27.52)$$

in which we recall that $\tilde{V}\left(\mathbf{q}\right)$ is a regular function of $\mathbf{q}$ which, due to hypercubic symmetry, has the expansion:

$$\tilde{V}\left(\mathbf{q}\right) = a + \frac{b}{2} q^2 + O\left(q^4\right). \qquad (27.53)$$

In the critical domain, in which $|\mathbf{q}|$ is small, the two terms coming from the short range potential and the dipolar forces are of the same order of magnitude:

$$q^2 \sim q_z^2 / q^2. \qquad (27.54)$$

This implies that q_z, the z component of the vector $\mathbf{q}$, is much smaller than the other components $\mathbf{q}_\perp$:

$$|q_z| \sim \left(\mathbf{q}_\perp\right)^2. \qquad (27.55)$$

We can therefore simplify the interaction potential. In the case of an even spin distribution, we finally obtain an effective hamiltonian $\mathcal{H}\left(\phi\right)$ of the form:

$$\mathcal{H}\left(\phi\right) = \tfrac{1}{2} \int \mathrm{d}^d q\, \phi\left(-\mathbf{q}\right) \left(\mathbf{q}_\perp^2 + A_0^2 \frac{q_z^2}{\mathbf{q}_\perp^2} + r_c + t_0 \right) \phi\left(\mathbf{q}\right)$$

$$+ \frac{u_0}{4!} \int \mathrm{d}^d q_1 \dots \mathrm{d}^d q_4\, \delta\left(\textstyle\sum \mathbf{q}_i\right) \phi\left(\mathbf{q}_1\right) \dots \phi\left(\mathbf{q}_4\right). \qquad (27.56)$$

The upper-critical dimension. Usual power counting is now modified because space is no longer isotropic. In units of the transverse components of $\mathbf{q}$, the dimension of q_z is 2 (equation (27.55)):

$$[q_z] = 2 \quad \Rightarrow \quad [z] = -2.$$

The volume element in configuration space $\mathrm{d}\mathbf{x}_\perp \mathrm{d}z$ has thus canonical dimension $-d-1$. This implies that power counting analysis is the same as in the conventional ϕ^4 theory in $(d+1)$ dimensions. In particular the upper-critical dimension is given by:

$$d+1 = 4 \quad \Rightarrow \quad d = 3.$$

The renormalization group method allows to make exact predictions in the physical dimension $d = 3$.

Renormalization group equations. From expression (27.56) we read off the propagator $\Delta\left(\mathbf{q}\right)$:

$$\Delta\left(\mathbf{q}\right) = \frac{q_\perp^2}{\left(\mathbf{q}_\perp^2\right)^2 + A^2 q_z^2 + t\mathbf{q}_\perp^2}. \qquad (27.57)$$

Diagrams calculated with this propagator are regular for $\mathbf{q}$ small, therefore

$$\int d^d q\, \phi\,(-\mathbf{q})\,\frac{q_z^2}{\mathbf{q}_\perp^2}\phi\,(\mathbf{q})\,,$$

will never appear as a counterterm.

The renormalized hamiltonian has thus the form:

$$\mathcal{H}_r = \tfrac{1}{2}\int d^d q\, \phi\,(-\mathbf{q})\left[Z\mathbf{q}_\perp^2 + A^2\,\frac{q_z^2}{\mathbf{q}_\perp^2} + \delta m^2 + Z_2 t\right]\phi\,(\mathbf{q})$$

$$+ \frac{\mu^\varepsilon}{4!}A\,g\,Z_g(g)\int d^d q_1\ldots d^d q_4\,\delta\left(\sum \mathbf{q}_i\right)\phi\,(\mathbf{q}_1)\ldots\phi\,(\mathbf{q}_4)\,, \qquad (27.58)$$

in which μ is the renormalization scale and ε is now defined by:

$$d = 3 - \varepsilon\,. \qquad (27.59)$$

Let us introduce also a bare dimensionless coupling constant:

$$u_0 = \Lambda^\varepsilon A_0 g_0\,. \qquad (27.60)$$

The relation between bare and renormalized correlation functions reads:

$$\Gamma_r^{(n)}\,(p_i; t, g, A, \mu) = Z^{n/2}\Gamma^{(n)}\,(p_i; t_0, g_0, A_0, \Lambda)\,. \qquad (27.61)$$

In addition comparing expressions (27.56) and (27.58) we find:

$$A = Z^{1/2}A_0\,. \qquad (27.62)$$

RG equations follow:

$$\left[\Lambda\frac{\partial}{\partial\Lambda} + \beta\,(g_0)\frac{\partial}{\partial g_0} + \frac{1}{2}\eta\,(g_0)\left(A_0\frac{\partial}{\partial A_0} - n\right) - \eta_2\,(g_0)\,t_0\frac{\partial}{\partial t_0}\right]\Gamma^{(n)}\,(p_i; g_0, A_0, t_0, \Lambda) = 0\,. \qquad (27.63)$$

Two-loop calculation of RG functions. As was explained in Section 25.4, for practical calculations we use the renormalized theory and minimal subtraction (see Chapter 11). In what follows we omit, for convenience, the index r on renormalized functions. The renormalized RG equations are formally identical to equations (27.63):

$$\left[\mu\frac{\partial}{\partial\mu} + \beta(g)\frac{\partial}{\partial g} + \frac{1}{2}\eta\,(g)\left(A\frac{\partial}{\partial A} - n\right) - \eta_2\,(g)\,t\frac{\partial}{\partial t}\right]\Gamma^{(n)}\,(p_i; g, A, t, \Lambda) = 0\,, \quad (27.64)$$

but the RG functions are different at two-loop order.

The calculations are here similar to the dynamical calculations as described in Appendix A17, because if we identify the z direction with time, the propagators have the same denominators. The combinatorial factors of Feynman diagrams are those of the ϕ^4 field theory. Only the expressions of the diagrams differ. We need only their values at vanishing external z components of momenta, and this simplifies the integration over

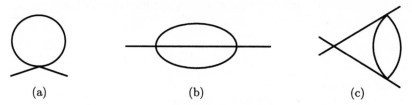

Fig. 27.1   Diagrams needed at two-loop order.

the momenta variables in the z direction (called ω hereafter). Figure 27.1 gives the three diagrams needed at two-loop order. In the massless theory the expressions are

$$(a) \equiv \frac{1}{(2\pi)^d} \int \frac{d\omega\, d^{d-1}k}{\left(k^2 + \frac{\omega^2}{k^2}\right)\left[(p+k)^2 + \frac{\omega^2}{(p+k)^2}\right]}$$

$$= \frac{1}{(2\pi)^{d-1}} \frac{1}{2} \int \frac{d^{d-1}k}{k^2 + (p+k)^2} = \frac{1}{(16\pi)^{(d-1)/2}} \Gamma\left(\frac{\varepsilon}{2}\right) p^{-\varepsilon}.$$

$$(b) \equiv \int \frac{(2\pi)^{-2d}\, d\omega_1 d\omega_2 d^{d-1}k_1 d^{d-1}k_2}{\left(k_1^2 + \frac{\omega_1^2}{k_1^2}\right)\left(k_2^2 + \frac{\omega_2^2}{k_2^2}\right)\left[(k_1+k_2+p)^2 + \frac{(\omega_1+\omega_2)^2}{(k_1+k_2+p)^2}\right]}$$

$$= \frac{1}{(2\pi)^{d-2}} \frac{1}{4} \int \frac{d^{d-1}k_1 d^{d-1}k_2}{k_1^2 + k_2^2 + (k_1+k_2+p)^2} = \frac{1}{(16\pi)^{d-1}} \frac{3}{8} \left(\frac{16}{27}\right)^{(d-1)/2} \Gamma(2-d) p^{2d-4}.$$

After integration over the z components (c) takes the form:

$$(c) \equiv \frac{1}{(2\pi)^{2(d-1)}} \frac{1}{8} \left[C_1(p_1,p_2) + C_1(p_2,p_1) + C_2(p_1,p_2)\right],$$

$$C_1(p,q) = \int \frac{d^{d-1}k_1 d^{d-1}k_2}{\left[k_1^2 + (k_1+p+q)^2\right]\left[k_1^2 + k_2^2 + (k_1+k_2+p)^2\right]},$$

$$C_2(p,q) = \int \frac{d^{d-1}k_1 d^{d-1}k_2}{\left[(k_1+p+q)^2 + k_2^2 + (k_1+k_2+p)^2\right]} \frac{1}{\left[k_1^2 + k_2^2 + (k_1+k_2+p)^2\right]}.$$

The final result is:

$$C_1(\varepsilon) \simeq \frac{1}{(16\pi)^{d-1}} \left(\frac{2}{\sqrt{3}}\right)^\varepsilon \frac{\Gamma(1+\varepsilon)}{\varepsilon^2} (|p_1+p_2|)^{-2\varepsilon} + O(1),$$

$$C_2(\varepsilon) = \frac{2}{3} \frac{1}{(16\pi)^2} \frac{1}{\varepsilon} + O(1).$$

The renormalization constants Z_g, Z and Z_2 are then determined up to order g^2:

$$Z_g = 1 + \frac{3N_d}{\varepsilon} g + \left(\frac{9}{\varepsilon^2} - \frac{3}{\varepsilon}\ln\frac{4}{3} - \frac{2}{\varepsilon}\right) N_d^2 g^2 + O(g^3), \tag{27.65}$$

$$Z = 1 - \frac{2}{27} N_d^2 \frac{g^2}{\varepsilon} + O(g^3), \tag{27.66}$$

$$Z_2^{-1} = 1 - N_d \frac{g}{\varepsilon} + \left(-\frac{1}{\varepsilon^2} + \frac{1}{3\varepsilon} + \frac{1}{2\varepsilon}\ln\frac{4}{3}\right) N_d^2 g^2 + O(g^3), \tag{27.67}$$

with:

$$N_d = (16\pi)^{\varepsilon/2-1}\,\Gamma\left(1+\varepsilon/2\right). \tag{27.68}$$

The RG functions follow:

$$\beta(g) = -\varepsilon\left[\frac{\mathrm{d}}{\mathrm{d}g}\ln\left(gZ_gZ^{-3/2}\right)\right]^{-1}, \tag{27.69}$$

$$\eta(g) = \beta(g)\frac{\mathrm{d}}{\mathrm{d}g}\ln Z(g), \tag{27.70}$$

$$\eta_2(g) = \frac{1}{\nu(g)} - 2 = \frac{\mathrm{d}}{\mathrm{d}g}\ln\left(Z_2Z^{-1}\right), \tag{27.71}$$

and therefore:

$$N_d\beta\left(\tilde{g}\right) = -\varepsilon\tilde{g} + 3\tilde{g}^2 - \left(-6\ln\frac{4}{3} + \frac{34}{9}\right)\tilde{g}^3 + O\left(\tilde{g}^4\right), \tag{27.72}$$

$$\eta\left(\tilde{g}\right) = \frac{4}{27}\tilde{g}^2 + O\left(\tilde{g}^3\right), \tag{27.73}$$

$$\eta_2\left(\tilde{g}\right) = -\tilde{g} + \left(\frac{14}{27} + \ln\frac{4}{3}\right)\tilde{g}^2 + O\left(\tilde{g}^3\right), \tag{27.74}$$

with:

$$\tilde{g} = N_d g. \tag{27.75}$$

The scaling behaviour of correlation functions below three dimensions. Dimensional analysis in the critical theory tells us:

$$\Gamma^{(n)}\left(\lambda\mathbf{p}_\perp, \rho p_z; t, g, A, \mu\right) = \lambda^{n+(n-2)(1-d)/2}\rho^{(2-n)/2}\Gamma^{(n)}\left(\mathbf{p}_\perp, p_z; \frac{t}{\lambda^2}, g, A\frac{\rho}{\lambda^2}, \frac{\mu}{\lambda}\right). \tag{27.76}$$

In $d = 3 - \varepsilon$ dimensions the model has IR fixed point $g^*\left(\varepsilon\right)$. At the fixed point we find:

$$\Gamma^{(n)}\left(\mathbf{p}_\perp, p_z, t, A = \mu = 1\right) = t^{\gamma-(n-2)d_\phi}\Gamma^{(n)}\left(\frac{\mathbf{p}_\perp}{t^\nu}, \frac{p_z}{t^{\nu(2-\eta/2)}}\right), \tag{27.77}$$

with

$$\gamma = \nu(2-\eta), \qquad d_\phi = \tfrac{1}{2}(d-1+\eta).$$

At two-loop order the exponents are:

$$\tilde{g}^*\left(\varepsilon\right) = \frac{\varepsilon}{3} + \left(\frac{2}{9}\ln\frac{4}{3} + \frac{34}{243}\right)\varepsilon^2 + O\left(\varepsilon^3\right), \tag{27.78}$$

$$\eta = \frac{4}{243}\varepsilon^2 + O\left(\varepsilon^3\right), \tag{27.79}$$

$$\nu^{-1} = 2 - \frac{\varepsilon}{3} - \left(\frac{1}{9}\ln\frac{4}{3} + \frac{20}{243}\right)\varepsilon^2 + O\left(\varepsilon^3\right), \tag{27.80}$$

$$\omega = \varepsilon - \left(\frac{2}{3}\ln\frac{4}{3} + \frac{34}{81}\right)\varepsilon^2 + O\left(\varepsilon^3\right). \tag{27.81}$$

Logarithmic corrections to mean field behaviour in three dimensions. In three dimensions the RG equations can be solved as indicated in Section 27.2.

For the effective coupling constant at scale λ one finds:

$$\frac{1}{g(\lambda)} \underset{\lambda \to 0}{\sim} \frac{3}{16\pi} \ln \frac{1}{\lambda} \left[1 - 2 \frac{(17 + 27 \ln 4/3)}{81} \frac{\ln |\ln \lambda|}{|\ln \lambda|} + O\left(\frac{1}{\ln \lambda}\right) \right]. \tag{27.82}$$

A short calculation then yields, for example, the susceptibility in zero field:

$$\chi^{-1} \sim C_\pm |\ln t|^{-1/3} \left[1 + \frac{1}{243} \left(108 \ln \frac{4}{3} + 41 \right) \frac{\ln |\ln t|}{|\ln t|} + O\left(\frac{1}{|\ln t|}\right) \right], \tag{27.83}$$

or the specific heat C:

$$C = A_\pm |\ln |t||^{1/3} \left[1 - \frac{1}{243} \left(108 \ln \frac{4}{3} + 41 \right) \frac{\ln |\ln t|}{|\ln t|} + O\left(\frac{1}{\ln |t|}\right) \right], \tag{27.84}$$

with the universal ratio

$$\frac{A_+}{A_-} = \frac{1}{4}. \tag{27.85}$$

The specific heat has been measured in a high precision experiment on the dipolar Ising ferromagnet $LiTbF_4$ by Ahlers *et al.* Fitting the specific heat by:

$$C_+ = \frac{A_+}{b^z} \left\{ [1 + b \ln (a / |t|)]^z - 1 \right\} + B,$$

$$C_- = \frac{A_-}{b^{z'}} \left\{ [1 + b \ln (a / |t|)]^{z'} - \frac{1}{4} \right\} + B,$$

they find:

$$\frac{A_+}{A^-} = 0.244 \pm 0.009,$$

$$z = z' = 0.336 \pm 0.024,$$

results which agree beautifully with the theoretical predictions.

Bibliographical Notes

Correction terms due to irrelevant operators are discussed in
  F. Wegner, *Phys. Rev.* B5 (1972) 4529; B6 (1972) 1891, and in Wegner's contribution to *Phase Transitions and Critical Phenomena*, vol. 6, quoted in Chapter 24; F.J. Wegner and E.K. Riedel, *Phys. Rev.* B7 (1973) 248; E. Brézin, J.C. Le Guillou and J. Zinn-Justin, *Phys. Rev.* D8 (1973) 434, 2418 and in *Phase Transitions and Critical Phenomena* vol. 6 already quoted, Section VIII.
Recent calculations of dimensions of operators can be found in
  S.K. Kehrein and F. Wegner, *Nucl. Phys.* B424 (1994) 521.
The logarithmic corrections in four dimensions for the ϕ^4 theory and the uniaxial system with dipolar forces have been obtained, with the use of the parquet approximation by
  A.I. Larkin and D.E. Khmel'nitskii, *Zh. Eksp. Teor. Fiz.* 56 (1969) 647 (*JETP 29*, 1123).
The idea of improved action has been discussed in
  K. Symanzik, in *Mathematical Problems in Theoretical Physics (Lecture Notes in Physics 153)*, R. Schrader, R. Seiler and D.A. Uhlenbrock eds. (Springer-Verlag, Berlin 1982); *Nucl. Phys.* B226 (1983) 205.
The application of the renormalization group to the uniaxial system with dipolar forces is found in Wegner and Riedel as quoted above and
  A. Aharony, *Phys. Rev.* B8 (1973) 3363; E. Brézin and J. Zinn-Justin, *Phys. Rev.* B13 (1976) 251.
The experimental results are quoted from
  G. Ahlers, A. Kornblit and H.J. Guggenheim, *Phys. Rev. Lett.* 34 (1975) 1227.

Exercises

The tricritical $(\phi^2)^3$ field theory is also IR free in three dimensions. Verify that the RG equations for the $n\phi$, $l\phi^2$, $k\phi^4$ correlation functions can be written:

$$\left\{ \Lambda \frac{\partial}{\partial \Lambda} + \beta(v)\frac{\partial}{\partial v} - k\eta_4(v)u\frac{\partial}{\partial u} - l\left[\eta_2(v)t + \eta_{24}(v)u^2\right]\frac{\partial}{\partial t} - \frac{n}{2}\eta(v) \right\} \Gamma^{(k,l,n)} = 0 \,,$$

except for $l = n = 0, k \leq 3$. Calculate the RG functions and study the logarithmic deviations from mean field theory. A review of tricritical behaviour can be found in

D. Lawrie and S. Sarbach, in *Phase Transitions and Critical Phenomena* vol. 9, C. Domb and J.L. Lebowitz eds. (Academic Press, London 1984).

Many models with second order phase transitions have been investigated, the effective field theories identified and then various universal quantities calculated by field theoretical methods. We can of course here report only a small number of significant results. Therefore in this chapter we consider the important example of the N-vector model, i.e. the $O(N)$ symmetric $\left(\phi^2\right)^2$ field theory,

$$\mathcal{H}(\phi) = \int \left\{ \frac{1}{2}\left[\partial_\mu \phi(x)\right]^2 + \frac{1}{2}\left(r_c + t\right)\phi^2(x) + \frac{1}{4!}g\Lambda^\varepsilon \left(\phi^2\right)^2(x) \right\} \mathrm{d}^d x, \qquad (28.1)$$

in which ϕ is a N-vector.

We present results for critical exponents, the equation of state and some amplitude ratios. We discuss more thoroughly critical exponents because they allow the most detailed and accurate comparison between Field Theory, lattice models and experiments.

Universal quantities have been calculated in Field Theory by two methods: the ε-expansion, which we have systematically discussed in previous chapters, and perturbation theory at fixed dimension (see Section 26.1). In both cases the expansion is divergent for all values of the expansion parameter. The rate of divergence can be obtained from instanton calculus, as we shall explain in Chapters 41 and 42. There exist methods to deal with divergent series (see Sections 41.4–41.6), all of which rely on some additional knowledge about the analytic properties of the function which has been expanded. In the case of the ϕ^4 field theory, the Borel summation has been extensively used because the information drawn from the large order behaviour analysis can easily be incorporated. In the case of the series at fixed dimension Borel summability has actually been rigorously proven. For the ε-expansion Borel summability has been assumed. Finally to sum the series efficiently, it has been necessary to make some plausible assumptions about the analytic properties of the Borel transform itself. Therefore the reliability of the results has been estimated by checking their stability with respect to reasonable variations of the summation method. Furthermore the comparison between the two families of results has provided an internal consistency check of field theory methods. The agreement with series estimates coming from high temperature expansions on lattices is then a verification of the concept of universality in theoretical models. Finally of course the ultimate test comes from the confrontation with experimental results.

The results of the N-vector model do not apply only to ferromagnetic systems. The superfluid Helium transition corresponds to $N = 2$. In the appendix we explain how the $N = 0$ limit is related to the statistical properties of polymers and how the Ising-like $N = 1$ model describes the physics of the liquid–vapour transition.

28.1 The ε-Expansion

Many quantities have been calculated at three-loop order within the ε-expansion. Moreover a series of very ingenious tricks and the use of symbolic manipulations on computers, have allowed the determination of critical exponents in the N-vector model, up to order ε^5. The higher order calculations have been done using dimensional regularization and the minimal subtraction scheme (see Chapter 11).

28.1.1 Critical exponents

Although the RG functions of the $(\phi^2)^2$ theory and therefore the critical exponents are known up to five-loop order, we here give the expansions only up to three loops (four loops for η), referring to the literature for more details. In terms of the variable $\tilde{g}$:

$$\tilde{g} = \frac{2g}{(4\pi)^{d/2}\Gamma(d/2)}, \tag{28.2}$$

the RG functions $\beta(\tilde{g})$ and $\eta_2(\tilde{g})$ at three-loop order, $\eta(\tilde{g})$ at four-loop order are:

$$\beta(\tilde{g}) = -\varepsilon\tilde{g} + \frac{(N+8)}{6}\tilde{g}^2 - \frac{(3N+14)}{12}\tilde{g}^3$$
$$+ \frac{[33N^2 + 922N + 2960 + 96(5N+22)\zeta(3)]}{12^3}\tilde{g}^4 + O\left(\tilde{g}^5\right), \tag{28.3}$$

$$\eta(\tilde{g}) = \frac{(N+2)}{72}\tilde{g}^2\left[1 - \frac{(N+8)}{24}\tilde{g} + \frac{5(-N^2+18N+100)}{576}\tilde{g}^2\right] + O\left(\tilde{g}^5\right), \tag{28.4}$$

$$\eta_2(\tilde{g}) = -\frac{(N+2)}{6}\tilde{g}\left[1 - \frac{5}{12}\tilde{g} + \frac{(5N+37)}{48}\tilde{g}^2\right] + O\left(\tilde{g}^4\right), \tag{28.5}$$

in which $\zeta(s)$ is the Riemann ζ-function:

$$\zeta(3) = 1.20205690315\ldots.$$

The zero $\tilde{g}^*(\varepsilon)$ of the β-function is then:

$$\tilde{g}^*(\varepsilon) = \frac{6\varepsilon}{(N+8)}\left[1 + \frac{3(3N+14)}{(N+8)^2}\varepsilon + \left(\frac{18(3N+14)^2}{(N+8)^4}\right.\right.$$
$$\left.\left. - \frac{33N^2 + 922N + 2960 + 96(5N+22)\zeta(3)}{8(N+8)^3}\right)\varepsilon^2\right] + O\left(\varepsilon^4\right). \tag{28.6}$$

The values of the critical exponents η, γ and ω,

$$\eta = \eta\left(\tilde{g}^*\right), \quad \gamma = \frac{2-\eta}{2+\eta_2\left(\tilde{g}^*\right)}, \quad \omega = \beta'\left(\tilde{g}^*\right),$$

follow:

$$\eta = \frac{\varepsilon^2(N+2)}{2(N+8)^2}\left\{1 + \frac{(-N^2+56N+272)}{4(N+8)^2}\varepsilon + \frac{1}{16(N+8)^4}\left[-5N^4 - 230N^3\right.\right.$$
$$\left.\left. + 1124N^2 + 17920N + 46144 - 384(5N+22)(N+8)\zeta(3)\right]\varepsilon^2\right\} + O\left(\varepsilon^5\right), \tag{28.7}$$

$$\gamma = 1 + \frac{(N+2)}{2(N+8)}\varepsilon + \frac{(N+2)}{4(N+8)^3}\left(N^2 + 22N + 52\right)\varepsilon^2 + \frac{(N+2)}{8(N+8)^5}$$
$$\times\left[N^4 + 44N^3 + 664N^2 + 2496N + 3104 - 48(5N+22)(N+8)\zeta(3)\right]\varepsilon^3$$
$$+ O\left(\varepsilon^4\right), \tag{28.8}$$

$$\omega = \varepsilon - \frac{3(3N+14)}{(N+8)^2}\varepsilon^2 + \frac{[33N^2 + 922N + 2960 + 96(5N+22)\zeta(3)]}{4(N+8)^3}\varepsilon^3$$
$$- 18\frac{(3N+14)^2}{(N+8)^4}\varepsilon^3 + O\left(\varepsilon^4\right). \tag{28.9}$$

All other exponents can be obtained from the scaling relations derived in Chapter 26. Note that the results presented above involve $\zeta(3)$. At higher orders $\zeta(5)$ and $\zeta(7)$ successively appear. In table 28.1 we give the values of the critical exponents γ and η obtained by simply adding the successive terms of the ε-expansion for $\varepsilon = 1$ and $N = 1$.

Table 28.1

Sum of the successive terms of the ε-expansion of γ and η for $\varepsilon = 1$ and $N = 1$.

k	0	1	2	3	4	5
γ	1.000	1.1667	1.2438	1.1948	1.3384	0.8918
η	0.0...	0.0...	0.0185	0.0372	0.0289	0.0545

We immediately observe a striking phenomenon: the sums first seem to settle near some reasonable value and then begin to diverge with increasing oscillations. We shall argue in Chapter 42 that the ε-expansion is divergent for all values of ε. Divergent series can be used for small values of the argument. However only a finite number of terms of the series can be taken into account. The last term added gives an indication of the size of the irreducible error. For the exponents γ and η we therefore conclude from the series:

$$\gamma = 1.244 \pm 0.050, \qquad \eta = 0.037 \pm 0.008,$$

where the errors are only indicative of the uncertainty about the result.

Remark. The definition of the β-function by minimal subtraction has an intrinsic meaning, unlike other definitions, since then $\beta(g)$ can be recalculated in terms of $\omega(\varepsilon)$. Setting:

$$\beta(g) = g\left(-\varepsilon + s(g)\right),$$

then:

$$g(s) = as \exp\left[\int_0^s ds'\left(\frac{1}{\omega(s')} - \frac{1}{s'}\right)\right].$$

where a is an arbitrary normalization of the coupling constant.

28.1.2 The scaling equation of state

The scaling equation of state provides an interesting example of a universal function. Its ε-expansion has been obtained up to order ε^2 for arbitrary N, and order ε^3 for $N = 1$. We set:

$$H = M^\delta f\left(x = t/M^{1/\beta}\right), \tag{28.10}$$

in which the normalizations of x and the function $f(x)$ are such that:

$$f(0) = 1, \qquad f(-1) = 0. \tag{28.11}$$

It is also convenient to set:

$$y = x + 1, \qquad z = x + 3, \qquad \rho = z/4y. \tag{28.12}$$

The expansion up to order ε^2 is then:

$$f(x) = 1 + x + \varepsilon f_1(x) + \varepsilon^2 f_2(x) + O\left(\varepsilon^3\right), \tag{28.13}$$

with:

$$f_1(x) = \frac{1}{2(N+8)}\left[(N-1)y\ln y + 3z\ln z - 9y\ln 3 + 6x\ln 2\right] \qquad (28.14a)$$

$$f_2(x) = \left[\frac{1}{2(N+8)}\right]^2 \{[N-1+6\ln 2 - 9\ln 3 + (N-1)\ln y]\,[3z\ln z + (N-1)y\ln y$$

$$+6x\ln 2 - 9y\ln 3] + \tfrac{1}{2}(10-N)y\left(\ln^2 z - \ln^2 3\right) + 36\left(\ln^2 z - y\ln^2 3 + x\ln^2 2\right)$$

$$-54\ln 2\left(\ln z + x\ln 2 - y\ln 3\right) + 3\ln\frac{27}{4}(N-1)y\ln y + \frac{-4N^2+17N+212}{N+8}$$

$$\times\left[z\ln z + 2x\ln 2 - 3y\ln 3\right] + (N-1)y\ln y\ln z - \tfrac{1}{2}N(N-1)y\ln^2 y$$

$$+\frac{N-1}{N+8}(19N+92)y\ln y - 2(N-1)\left[(x+6)J_1(x) - 6yJ_1(0)\right]$$

$$-6(N-1)\left[J_2(x) - yJ_2(0)\right] + 4(N-1)\left[J_3(x) - yJ_3(0)\right]\}, \qquad (28.14b)$$

where:

$$J_i(x) = I_i(\rho), \qquad (28.15)$$

and:

$$I_1(\rho) = \int_0^\infty \frac{du\,\ln u}{u(1-u)}\left[(1-u/\rho)^{1/2}\,\theta(\rho-u) - 1\right], \qquad (28.16)$$

$$I_2(\rho) = \rho\frac{d}{d\rho}I_1(\rho), \qquad (28.17)$$

$$I_3(\rho) = I_1(\rho) + 2I_2(\rho). \qquad (28.18)$$

The expressions (28.14) are not uniform, and valid only for x of order 1. For x large, i.e. for small magnetization M, the magnetic field has a regular expansion in odd powers of M, i.e. in the variable $x^{-\beta}$ (Section 26.5). It is therefore convenient to introduce Josephson's parametrization which leads to a representation uniform in both limits.

28.1.3 Parametric representation of the equation of state

We set:

$$x = x_0\left(1-\theta^2\right)\theta^{-1/\beta}, \quad \theta > 0, \qquad (28.19)$$

where x_0 is an arbitrary positive constant. More directly we can parametrize M and t in terms of two variables R and θ, setting:

$$\begin{cases} M = R^\beta\theta, \\ t = x_0 R\left(1-\theta^2\right), \\ H = R^{\beta\delta}h(\theta). \end{cases} \qquad (28.20)$$

Then the function $h(\theta)$:

$$h(\theta) = \theta^\delta f\left(x(\theta)\right), \qquad (28.21)$$

is an odd function of θ regular near $\theta = 1$, which is x small, and near $\theta = 0$ which is x large.

For the special choice

$$x_0 = 3\left(3/2\right)^{1/2\beta-1}, \qquad (28.22)$$

the equation of state at order ε takes a rather simple form:

$$h(\theta) = \theta \left(3 - 2\theta^2\right) \left[1 + \frac{\varepsilon(N-1)}{2(N+8)} \ln \left(3 - 2\theta^2\right)\right] + O\left(\varepsilon^2\right). \tag{28.23}$$

For $N = 1$ the equation is specially simple and corresponds to the so-called linear parametric model in which $h(\theta)$ is a cubic odd function of θ. One verifies that it is still possible to adjust x_0 at order ε^2 to preserve this form. However, at order ε^3, which is also known for $N = 1$, the introduction of a term proportional to θ^5 becomes necessary. One finds:

$$h(\theta) = h_0 \theta \left(b^2 - \theta^2\right) \left(1 + c\theta^2\right) + O\left(\varepsilon^4\right), \tag{28.24}$$

in which h_0 is the field normalization constant, and b, c are given by:

$$b^2 = \frac{3}{2}\left(1 - \frac{\varepsilon^2}{12}\right), \qquad c = -\frac{\varepsilon^3}{18}\left(\zeta(3) + \frac{I-1}{4}\right), \tag{28.25}$$

with:

$$I = \int_0^1 dx \frac{\ln\left[x(1-x)\right]}{1 - x(1-x)} = \tfrac{4}{9}\pi^2 - \tfrac{2}{3}\psi'(1/3) = -2.3439072386\ldots. \tag{28.26}$$

The constant x_0 is given by:

$$x_0 = b^{1/\beta}/\left(b^2 - 1\right). \tag{28.27}$$

Remark. In the case $N > 1$ the function $h(\theta)$ still has a singularity on the coexistence curve, due to the presence of Goldstone modes in the ordered phase. The nature of this singularity can be obtained from the study of the non-linear σ-model presented in Chapter 30. We shall show that the behaviour of correlation functions below T_c in a theory with a spontaneously broken continuous symmetry is governed by the zero temperature IR fixed point. Therefore the coexistence curve singularities can be obtained from a low temperature expansion (for more details see Sections 30.3,30.4).

In all cases, as stated above, the essential feature of the parametric representation is that it automatically satisfies the different requirements about the regularity properties of the equation of state and leads to uniform approximations.

The comparison with the numerical results for the Ising model ($N = 1$) and the Heisenberg model $N = 3$ in three dimensions shows that the successive ε and ε^2 corrections improve the mean field theory prediction.

From the parametric representation of the equation of state it is also possible to derive a representation for the singular part of the free energy per unit volume. Setting:

$$F(M,t) \equiv \Omega^{-1}\Gamma_{\text{sg.}}(M,t) = R^{2-\alpha}g(\theta), \tag{28.28}$$

one finds for $g(\theta)$ a differential equation:

$$h(\theta)\left(1 - \theta^2 + 2\beta\theta^2\right) = 2(2-\alpha)\theta g(\theta) + \left(1 - \theta^2\right)g'(\theta). \tag{28.29}$$

The integration constant is obtained by requiring the regularity of $g(\theta)$ at $\theta = 1$. In the same way the inverse magnetic susceptibility is given by:

$$\chi^{-1} = R^\gamma g_2(\theta), \tag{28.30}$$

with:

$$g_2(\theta)\left(1 - \theta^2 + 2\beta\theta^2\right) = 2\beta\delta\theta h(\theta) + \left(1 - \theta^2\right)h'(\theta). \tag{28.31}$$

These expressions can in particular be used to calculate some universal ratios of amplitudes.

28.1.4 Amplitude ratios

Apart from critical exponents some other simple universal numbers have been calculated: the ratios of amplitudes of singularities near T_c. We first consider two examples which can be derived directly from the equation of state.

The specific heat. If we write the singular part of the specific heat, i.e. the $\phi^2(x)$ 2-point correlation function at zero momentum, under the form:

$$C_H = A^{\pm} |t|^{-\alpha}, \qquad t = T - T_c \rightarrow \pm 0, \tag{28.32}$$

the ratio A^+/A^- is universal. This ratio is directly related to the function $g(\theta)$ defined by equation (28.29):

$$\frac{A^+}{A^-} = (b^2 - 1)^{2-\alpha} \frac{g(0)}{g(b)}. \tag{28.33}$$

At order ε^2 one finds:

$$\frac{A^+}{A^-} = 2^{\alpha-2} N \left\{ 1 + \varepsilon + \left[3N^2 + 26N + 100 + (4 - N)(N - 1)\zeta(2) \right. \right.$$

$$\left. \left. -6(5N + 22)\zeta(3) - 9(4 - N)\lambda \right] \frac{\varepsilon^2}{2(N + 8)^2} \right\} + O\left(\varepsilon^3\right), \tag{28.34}$$

with:

$$\zeta(2) = \pi^2/6 = 1.64493406684\ldots,$$

while λ is defined in terms of the integral I given in (28.26):

$$\lambda = -I/2 = \tfrac{1}{3}\psi'(1/3) - \tfrac{2}{9}\pi^2 = 1.17195361934\ldots.$$

The magnetic susceptibility. The magnetic susceptibility in zero field can also be calculated from the function $g_2(\theta)$ defined by equation (28.31). As we know, below T_c the susceptibility diverges for systems with Goldstone modes. We restrict ourselves therefore to $N = 1$. Defining:

$$\chi = C^{\pm} |t|^{-\gamma}, \qquad t \rightarrow \pm 0, \tag{28.35}$$

one obtains:

$$\frac{C^+}{C^-} = \frac{2\left(1 + cb^2\right)\left(b^2 - 1\right)^{1-\gamma}}{\left[1 - b^2\left(1 - 2\beta\right)\right]} \tag{28.36a}$$

$$= \frac{2^{\gamma+1}}{6\beta - 1} \left[1 + \left(\frac{2\lambda + 1}{4} - \zeta(3) \right) \frac{\varepsilon^3}{12} \right] + O\left(\varepsilon^4\right). \tag{28.36b}$$

The ratio C^+/C^- can be expressed, at order ε^2, entirely in terms of critical exponents. This form follows naturally from the parametric representation of the equation of state. The ε^3 relative correction is of the order of only 3%.

The correlation length. Let us here define the correlation length in terms of the ratio of the two first moments of the 2-point correlation function:

$$\Gamma^{(2)}(p) = \Gamma^{(2)}(0)\left(1 + \xi_1^2 p^2\right) + O\left(p^4\right). \tag{28.37}$$

The function ξ_1^2 has the scaling form:

$$\xi_1^2(M,t) = M^{-2\nu/\beta} f_\xi\left(\frac{t}{M^{1/\beta}}\right).$$

(28.38)

It shares otherwise all the properties of the equation of state. It can be written in parametric form:

$$\xi_1^2(M,t) = R^{-2\nu} g_\xi(\theta).$$

(28.39)

At order ε for $N = 1$ for example one finds:

$$g_\xi(\theta) = g_\xi(0)\left(1 - \frac{5}{18}\varepsilon\theta^2\right) + O\left(\varepsilon^2\right).$$

(28.40)

Setting in zero field:

$$\xi_1 = f_1^\pm \left|t\right|^{-\nu}, \qquad t \to \pm 0,$$

(28.41)

quantity which exists only for $N = 1$, one can use the determination of g_ξ to calculate this ratio:

$$f_1^+/f_1^- = 2^\nu\left[1 + \frac{5}{24}\varepsilon + \frac{1}{432}\left(\frac{295}{24} + 2I\right)\varepsilon^2\right] + O\left(\varepsilon^3\right),$$

(28.42)

in which the constant I is given by equation (28.26).

An additional universal constant. To the relation between exponents:

$$2 - \alpha = d\nu,$$

is associated a universal combination which involves only amplitudes of singularities when T_c is approached from above:

$$R_\xi^+ = f_1^+\left(\alpha A^+\right)^{1/d}.$$

(28.43)

Indeed from the definitions (28.28, 28.41)

$$\left(R_\xi^+\right)^d = (1-\alpha)(2-\alpha)t^{\alpha-2}F(0,t)t^{\nu d}(\xi_1)^d = (1-\alpha)(2-\alpha)F(0,t)(\xi_1)^d,$$

and the last product is normalization independent. The ε-expansion of R_ξ^+ is:

$$\left(R_\xi^+\right)^d = \sigma_d\frac{N}{2}\nu(1-\alpha)\left[1 + \eta\left(\frac{-11}{2} + \frac{14}{3}\lambda\right)\right] + O\left(\varepsilon^3\right),$$

(28.44)

with:

$$\sigma_d = \Gamma\left(1 + \varepsilon/2\right)\Gamma\left(1 - \varepsilon/2\right)2/\left(4\pi\right)^{d/2}\Gamma\left(d/2\right).$$

Other universal ratios. It is of course possible to define an infinite number of other universal ratios. We shall here give a few other examples which have been considered in the literature. Let us first define some additional amplitudes. On the critical isotherm, the correlation length behaves as

$$\xi_1 = f_1^c/H^{2/(d+2-\eta)},$$

(28.45)

the magnetic susceptibility as

$$\chi = C^c / H^{1-1/\delta};\tag{28.46}$$

the spontaneous magnetization vanishes as

$$M = B\left(-t\right)^{\beta},\tag{28.47}$$

and the spin–spin correlation function in momentum space at T_c behaves as

$$\chi(p) = \left[\Gamma^{(2)}(p)\right]^{-1} = D p^{\eta-2}.\tag{28.48}$$

One can then define the following universal ratio:

$$R_c = \alpha A^+ C^+ / B^2,\tag{28.49}$$

which corresponds to the relation between exponents

$$\alpha + 2\beta + \gamma = 2.$$

Indeed using this relation we verify that R_c is proportional to $F(0,t)M^{-2}\chi$ which is normalization independent. The ε-expansion of R_c is:

$$R_c = \frac{N}{N+8}2^{-2\beta-1}\varepsilon\left[1+\left(1-\frac{30}{(N+8)^2}\right)\varepsilon\right]+O\left(\varepsilon^3\right).\tag{28.50}$$

Following Fisher and Tarko, one can construct the three following universal combinations:

$$Q_1 = C^c\delta/\left(B^{\delta-1}C^+\right)^{1/\delta},\tag{28.51}$$

$$Q_2 = \left(f_1^c/f_1^+\right)^{2-\eta}C^+/C^c,\tag{28.52}$$

$$Q_3 = D\left(f_1^+\right)^{2-\eta}/C^+,\tag{28.53}$$

which correspond to the relations $\gamma = \beta(\delta - 1)$, the explicit expression of δ and $\gamma = \nu(2-\eta)$. Moreover Q_1 and Q_3 are normalization independent because $H\chi/M$ and $p\xi$ respectively are. For Q_2 this property follows immediately from the definition. Thus all three quantities are universal.

The quantity Q_1 is related to R_χ defined by Aharony and Hohenberg:

$$R_\chi = Q_1^{-\delta}.$$

Their ε-expansions are:

$$R_\chi = 3^{(\delta-3)/2}2^{\gamma+(1-\delta)/2}\left[1+\left(\frac{2\lambda+1}{4}-\zeta(3)\right)\frac{\varepsilon^3}{18}\right]+O\left(\varepsilon^4\right),\tag{28.54}$$

$$Q_2 = 1+\frac{\varepsilon}{18}+\left(\frac{23}{9}+\frac{4}{3}\lambda\right)\frac{\varepsilon^2}{54}+O\left(\varepsilon^3\right),\tag{28.55}$$

$$Q_3 = 1-\left(\frac{8}{3}\lambda+5\right)\frac{\varepsilon^2}{216}+O\left(\varepsilon^3\right).\tag{28.56}$$

Numerical results are given in table 28.10 and compared with various high temperature (HT) series and experimental determinations.

It is worth mentioning that universal ratios of amplitudes of corrections to the leading critical behaviour have also been calculated. Let us write a physical quantity for $t = T - T_c$ small as:

$$f(t) = A_f \, |t|^{-\lambda_f} \left(1 + a_f \, |t|^{\theta} + \cdots \right), \tag{28.57}$$

where the correction exponent θ (also called Δ_1) is given by:

$$\theta = \omega \nu \,. \tag{28.58}$$

The ratio of correction amplitudes a_{f_1}/a_{f_2} corresponding to two different quantities f_1 and f_2 is also universal. A few such ratios have been calculated. Let us give one example corresponding to the correlation length and the susceptibility above T_c:

$$\frac{a_{\chi}^{+}}{a_{\xi}^{+}} = 2 \left\{ 1 - \frac{\varepsilon}{N+8} - \left[\frac{2\lambda}{3(N+8)} - \frac{N^2 - 15N - 124}{2(N+8)^3} \right] \varepsilon^2 \right\} + O\left(\varepsilon^3\right). \tag{28.59}$$

28.2 The Perturbative Expansion at Fixed Dimension

Critical exponents and various universal quantities have also been calculated within the framework defined in Section 26.1, i.e. in the massive $(\phi^2)^2$ field theory, as perturbative series in fixed dimension. For example the RG β-function in three dimensions, for $N = 1$, has the expansion:

$$\beta(\tilde{g}) = -\tilde{g} + \tilde{g}^2 - \tfrac{308}{729}\tilde{g}^3 + 0.3510695978\tilde{g}^4 - 0.3765268283\tilde{g}^5$$
$$+ 0.49554751\tilde{g}^6 - 0.749689\tilde{g}^7 + O\left(\tilde{g}^8\right), \tag{28.60}$$

with the normalization:

$$\tilde{g} = 3g/(16\pi). \tag{28.61}$$

To calculate exponents or other universal quantities, we have first to find the IR stable zero g^* of the function $\beta(g)$ which is given by a few terms of a divergent expansion. A first problem is the absence of a small parameter in which to expand: g^* is a number of order 1. Already at this stage a summation method is required. A further problem arises from the property that RG functions, unlike the universal quantities in the ε-expansion, depend explicitly on the renormalization scheme. On the other hand, because one-loop diagrams have, in 3 dimensions, a simple analytic expression, it has been possible to calculate the RG functions of the N-vector model up to six and partially seven loop order. Estimates of critical exponents are displayed in table 28.4. Universal ratios of amplitude have also been calculated, as well as the equation of state for $N = 1$. Numerical results will be presented in Section 28.7.

Note however that in this framework the calculation of physical quantities in the ordered phase leads to additional technical problems because the theory is parametrized in terms of the disordered phase correlation length m^{-1} which is singular at T_c. Also the normalization of correlation functions is singular at T_c (equation (26.17)).

Let us discuss the example of Ising-like systems ($N = 1$). The free energy F has the form

$$F(M) - F(0) = \frac{m^d}{g} \varphi \left(g^{1/2} m^{1-d/2} \tilde{M}, g \right),$$

in which g has to be set to its fixed point value g^* and $\tilde{M}$ is related to the magnetization M by the field renormalization (26.15):

$$\tilde{M} = M m^{-\eta/2}.$$

The derivative with respect to M of F yields the equation of state under the form

$$H = g^{-1/2} m^{1+d/2-\eta/2} \mathfrak{h}\left(g^{1/2} m^{1-d/2}\tilde{M}, g\right), \quad \mathfrak{h}(z,g) = \varphi'_z(z,g). \tag{28.62}$$

At one-loop order the function $\mathfrak{h}(z,g)$ is given by:

$$\mathfrak{h}(z,g) = z + \frac{z^3}{6} + \frac{gz}{2}\frac{1}{(2\pi)^d}\int\frac{d^d p}{p^2 + 1 + z^2/2} + O\,(2\text{ loops})$$

$$= z + \frac{z^3}{6} + \frac{\pi N_d}{4\sin\pi d/2}gz\left[(1+z^2/2)^{d/2-1} - 1 - \tfrac{1}{4}(d-2)z^2\right]. \tag{28.63}$$

In terms of the deviation from the critical temperature $t = m^{1/\nu}(g^*)^{1/2\beta} \sim T - T_c$, the equation (28.62) takes the form:

$$H = H_0 t^{\beta\delta}\mathfrak{h}(Mt^{-\beta}), \quad (H_0 \text{ being a constant}) . \tag{28.64}$$

This expression is adequate for the description of the disordered phase when $Mt^{-\beta}$ is small but all terms in the loopwise expansion become singular when t goes to zero.

The ordered phase. This does not completely prevent calculations near the coexistence curve, i.e. for $t < 0$. Since at the fixed point g^* all functions have simple power law singularities at T_c, it is possible to proceed by analytic continuation in the complex t-plane. The scaling variable z,

$$z = Mt^{-\beta}, \tag{28.65}$$

picks up a phase below T_c:

$$\text{for } t = |t|\,\mathrm{e}^{i\pi}, \quad z = |z|\,\mathrm{e}^{-i\pi\beta} . \tag{28.66}$$

The scaling variable $H(-t)^{-\beta\delta}$ is then given by:

$$H(-t)^{-\beta\delta} = H_0\,\mathrm{e}^{i\pi\beta\delta}\,\mathfrak{h}(z) = H_0\,|\mathfrak{h}(z)| . \tag{28.67}$$

It is in particular possible to evaluate ratios of amplitudes of singularities above and below T_c: we can calculate the complex zero of $\mathfrak{h}(z,g)$ as a power series in g and substitute it in other quantities. The result is complex but its modulus taken at $g = g^*$ converges towards the correct result. Let us take the example of the magnetic susceptibility for illustration purpose. From equation (28.62) we derive:

$$C^+/C^- = \mathrm{e}^{-i\pi\gamma}\,\mathfrak{h}'\left(z_0\left(g^*\right),g^*\right)/\mathfrak{h}'\left(0,g^*\right) = |\mathfrak{h}'\left(z_0\left(g^*\right),g^*\right)|, \tag{28.68}$$

in which z_0 is the complex zero of $\mathfrak{h}(z,g)$. We thus get a series expansion for C^+/C^-.

However this method does not allow to extrapolate the equation of state to t small. Following the lines of Subsection 28.1.3, it is natural to introduce the parametric representation (28.20),

$$z = x_0^{-\beta}\theta\left(1 - \theta^2\right)^{-\beta},$$

and consider the function $h(\theta)$:

$$h(\theta) = \left(1 - \theta^2\right)^{\beta\delta} \mathfrak{h}\big(z(\theta)\big).$$

However in an expansion at fixed dimension, if we just replace all quantities by their perturbative expansion, the singularity of $h(\theta)$ at $\theta = 1$ (i.e. $t = 0$) does not cancel anymore. Therefore, inspired by results coming from the ε-expansion, one also expands $h(\theta)$ in powers of θ. The method is the following. One first determines by Borel summation, as explained in Section 28.3, the first terms of the expansion of the function $\mathfrak{h}(z)$ in powers of z. As expected the apparent accuracy decreases with increasing degree. One then determines the corresponding coefficients of the expansion of $h(\theta)$ in powers of θ (note $z \sim \theta$). These coefficients are polynomials in $x_0^{-\beta}$. One then adjusts the arbitrary constant x_0 to minimize the last term, as in the ODM method explained in Section 41.6. This strategy has been applied to the $N = 1$ series which are known up to order g^5. A general representation of the equation of state has been obtained. Corresponding results for the amplitude ratios are reported in table 28.10. It would be interesting to apply this method to $N \neq 1$ series.

28.3 The Series Summation

The principles and the theoretical justification of the summation method based on Borel summation and conformal mapping are explained in Section 41.6. We add here only a few more details about the specific implementation used in the case of the calculation of critical exponents and other universal quantities. A few examples of transformed series are displayed in table 28.2 to illustrate the convergence.

The method. Several different variants based on the Borel–Leroy transformation have been implemented and tested. Let $R(z)$ be the function whose expansion has to be summed (z here represents the coupling constant $\tilde{g}$ or ε):

$$R(z) = \sum_{k=0} R_k z^k. \tag{28.69}$$

One transforms the series into:

$$R(z) = \sum_{k=0}^{\infty} B_k(\rho) \int_0^{\infty} t^\rho \, \mathrm{e}^{-t} \, [u(zt)]^k \, \mathrm{d}t, \tag{28.70}$$

$$u(z) = \frac{\sqrt{1 + az} - 1}{\sqrt{1 + za} + 1}. \tag{28.71}$$

The coefficients B_k are calculated by identifying the expansion of the r.h.s. of equation (28.70) in powers of z with the expansion (28.69). The constant a has been determined by the large order behaviour analysis,

$$a(d = 3) = 0.147774232 \times \big(9/(N + 8)\big), \qquad a(d = 2, N = 1) = 0.238659217, \tag{28.72}$$

and ρ is a free parameter, adjusted empirically to improve the convergence of the transformed series by weakening the singularities of the Borel transform near $z = -a$. Moreover, in many cases, a conformal transformation has been made on the initial function

Table 28.2

Series summed by the method based on Borel transformation and mapping for the zero $\tilde{g}^*$ of the $\beta(g)$ function and the exponents γ and ν in the ϕ_3^4 field theory.

k	2	3	4	5	6	7
$\tilde{g}^*$	1.8774	1.5135	1.4149	1.4107	1.4103	1.4105
ν	0.6338	0.6328	0.62966	0.6302	0.6302	0.6302
γ	1.2257	1.2370	1.2386	1.2398	1.2398	1.2398

$R(z)$ in order to send away its closest singularities, and the procedure described above applied instead to the function $\tilde{R}(z)$:

$$\tilde{R}(z) = R\left[z/(1 - \tau z)\right], \tag{28.73}$$

in which τ is also left as an adjustable parameter because the location of all singularities of $R(z)$ is not known. This transformation is necessary in the case of the ε-expansion because the critical exponents, as functions of ε, have close singularities. It has been verified that it also improves the series in fixed dimensions.

Finally in general a third parameter was introduced, which will not be discussed here.

Needless to say, with three parameters and short initial series it becomes possible to find occasionally some transformed series whose apparent convergence is deceptively good. It is therefore essential to vary the parameters in some range around the optimal values to examine the sensitivity of the results upon their variations. Finally it is useful to sum independently series for exponents related by scaling relations. An underestimation of the apparent errors leads to inconsistent results. It is clear from these remarks that the errors quoted in the final results should be considered as somewhat indicative.

The $\left(\phi^2\right)^2$ field theory at fixed dimensions. The RG β-function has been determined up to six-loop order in 3 dimensions, while the series for the dimensions of the fields ϕ and ϕ^2 have recently been extended to seven loops. In 2 dimensions the series are known only up to four loops. They have been analyzed by two methods. In the first method the series of the RG β-function has been first summed and its zero $\tilde{g}^*$ calculated ($\tilde{g} = g(N + 8)/(48\pi)$ for $d = 3$, $\tilde{g} = 3g/8\pi$ for $d = 2$). The series of the other RG functions have then been summed for $\tilde{g} = \tilde{g}^*$. Examples of convergence are given in table 28.2.

The main drawback of this procedure is that the values of the critical exponents depend strongly on the value of $\tilde{g}^*$. Therefore an error in the estimation of $\tilde{g}^*$ biases all exponents. A variant, which avoids this problem, has thus been used as a check. A pseudo-ε parameter has been introduced by setting:

$$\beta(\tilde{g}, \varepsilon) = \tilde{g}(1 - \varepsilon) + \beta(\tilde{g}). \tag{28.74}$$

The two functions $\beta(\tilde{g}, \varepsilon)$ and $\beta(\tilde{g})$ coincide for $\varepsilon = 1$, and the zero of $\beta(\tilde{g}, \varepsilon)$ is calculated as a power series in ε. Critical exponents are then also calculated as series in ε, and these series are summed. However there are some indications that the mapping $\tilde{g} \mapsto \varepsilon$ introduces singularities because the apparent convergence is poorer. It is however gratifying that all variants give consistent results. A comparison of all the results is helpful for the determination of the apparent errors.

The ε-expansion. The ε-expansion has one important advantage: it allows us to connect the results in 3 and 2 dimensions. In particular in the cases $N = 1$ and $N = 0$, it is possible to compare the ϕ^4 results with exact results coming from lattice models and to test both universality and the reliability of the summation procedure. Moreover it is possible to improve the 3 dimensional results by imposing the exact 2 dimensional values or the behaviour near 2 dimensions for $N > 1$. However since the series in ε are shorter than the series at fixed dimension 3, the apparent errors are larger. Finally, as already emphasized, the comparison between the different results is a check of the consistency of Field Theory methods combined with the summation procedures.

28.4 Numerical Estimates of Critical Exponents

Fixed dimension. We give in table 28.3 the results obtained from summed perturbation theory at fixed dimension 2 for $N = 1$ and compare them with the exact values of the Ising model. The apparent errors are large because the series are short. The agreement with the Ising model is satisfactory. Note that $\tilde{g}^*$ is known only from HT series; only analytic corrections to scaling have been found in the Ising model which makes the identification of ω difficult. However an analysis based on conformal invariance predicts a correction exponent $\omega = 4/m$ for ϕ^{2m-2} field theories and for $m > 3$. One may conjecture that the amplitudes of the singularities involving the correction exponent ω vanish when m approaches 3 for $d = 2$, or for $m = 3$ when d approaches 2.

Table 28.3

Estimates of critical exponents in the ϕ_2^4 field theory.

	$\tilde{g}^*$	ω	γ	ν	η
ϕ_2^4	1.85 ± 0.10	1.3 ± 0.2	1.79 ± 0.04	0.96 ± 0.04	0.18 ± 0.04
Ising	1.751 ± 0.005	$4/3$?	1.75	$1.$	0.25

Table 28.4 displays the results obtained from summed perturbation series at fixed dimension 3. The last exponent $\theta = \omega\nu$ characterizes corrections to scaling in the temperature variable (equation (28.58)).

Table 4

Estimates of critical exponents in the $O(N)$ symmetric $(\phi^2)_3^2$ field theory.

N	0	1	2	3
$\tilde{g}^*$	1.413 ± 0.006	1.411 ± 0.004	1.403 ± 0.003	1.390 ± 0.004
g^*	26.63 ± 0.11	23.64 ± 0.07	21.16 ± 0.05	19.06 ± 0.05
γ	1.1596 ± 0.0020	1.2396 ± 0.0013	1.3169 ± 0.0020	1.3895 ± 0.0050
ν	0.5882 ± 0.0011	0.6304 ± 0.0013	0.6703 ± 0.0015	0.7073 ± 0.0035
η	0.0284 ± 0.0025	0.0335 ± 0.0025	0.0354 ± 0.0025	0.0355 ± 0.0025
β	0.3024 ± 0.0008	0.3258 ± 0.0014	0.3470 ± 0.0016	0.3662 ± 0.0025
α	0.235 ± 0.003	0.109 ± 0.004	-0.011 ± 0.004	-0.122 ± 0.010
ω	0.812 ± 0.016	0.799 ± 0.011	0.789 ± 0.011	0.782 ± 0.0013
$\theta = \omega\nu$	0.478 ± 0.010	0.504 ± 0.008	0.529 ± 0.009	0.553 ± 0.012

The ε-expansion. In table 28.5 we give the results coming from the summed ε-expansion for $\varepsilon = 2$ and compare them with exact results.

Table 28.5

Critical exponents in the ϕ_2^4 field theory from the ε-expansion.

	γ	ν	η	β	ω
$N=0$	1.39 ± 0.04	0.76 ± 0.03	0.21 ± 0.05	0.065 ± 0.015	1.7 ± 0.2
Exact	1.34375	0.75	$0.2083\cdots$	$0.0781\cdots$	?
$N=1$	1.73 ± 0.06	0.99 ± 0.04	0.26 ± 0.05	0.120 ± 0.015	1.6 ± 0.2
Ising	1.75	$1.$	0.25	0.125	$1.33\ldots$?

We see in this table that the agreement for $N=0$ and $N=1$ between field theory and lattice models is satisfactory. We feel justified therefore in using a summation procedure of the ε-expansion which automatically incorporates the $d=2$, $\varepsilon=2$ values. Note, however, that in both cases, the identification of ω remains a problem.

Table 28.6 then displays the results for $\varepsilon=1$, both for the ε series (free) and a modified ε series where the $d=2$ results are imposed (bc).

Table 28.6

Critical exponents in the $(\phi^2)_3^2$ field theory from the ε-expansion.

N	0	1	2	3
γ (free)	1.1575 ± 0.0060	1.2355 ± 0.0050	1.3110 ± 0.0070	1.3820 ± 0.0090
γ (bc)	1.1571 ± 0.0030	1.2380 ± 0.0050	1.317	1.392
ν (free)	0.5875 ± 0.0025	0.6290 ± 0.0025	0.6680 ± 0.0035	0.7045 ± 0.0055
ν (bc)	0.5878 ± 0.0011	0.6305 ± 0.0025	0.671	0.708
η (free)	0.0300 ± 0.0050	0.0360 ± 0.0050	0.0380 ± 0.0050	0.0375 ± 0.0045
η (bc)	0.0315 ± 0.0035	0.0365 ± 0.0050	0.0370	0.0355
β (free)	0.3025 ± 0.0025	0.3257 ± 0.0025	0.3465 ± 0.0035	0.3655 ± 0.0035
β (bc)	0.3032 ± 0.0014	0.3265 ± 0.0015		
ω	0.828 ± 0.023	0.814 ± 0.018	0.802 ± 0.018	0.794 ± 0.018
θ	0.486 ± 0.016	0.512 ± 0.013	0.536 ± 0.015	0.559 ± 0.017

Discussion. We can now compare the two sets of results coming from the perturbation series at fixed dimension, and the ε-expansion. First let us emphasize that the agreement is quite spectacular, although the apparent errors of the ε-expansion are in general larger because the series are shorter. Moreover the agreement has improved with longer series.

The best agreement is found for the exponents ν and β. On the other hand the values of η coming from the ε-expansion are systematically larger by about 3×10^{-3}, though the error bars always overlap. The corresponding effect is observed on γ. We notice in tables 28.3, 28.4 that a similar remark applies at $d=2$: the result at fixed dimension, $N=1$, for η is smaller than the result coming from the ε-expansion, however the latter result is closer to the Ising value, even when one takes into account the relative errors.

28.5 Comparison with Lattice Model Estimates

The N-vector with nearest-neighbour interactions has been studied on various lattices. Most of the results for critical exponents come from the analysis of HT series expansion by different types of ratio methods, Padé approximants or differential approximants (see Sections 41.6 and A41.3). Some results have also been obtained from low temperature expansions, computer calculations using stochastic methods, and in low dimensions, transfer matrix methods. Table 28.7 tries to give an idea of the agreement between lattice and Field Theory results. A historical remark is here in order: the agreement between both types of theoretical results has improved as the HT series became longer which is of course encouraging. The main reason is that it has become possible to take into account in the analysis of longer series, the influence of confluent singularities due to corrections to the leading power law behaviour, as predicted by the renormalization group. The effect of this improvement has been specially spectacular for the exponents γ and ν of the 3D Ising model: the longer series obtained by Nickel on the BCC lattice did almost completely eliminate the disturbing small differences which had remained between HT series and RG results.

Table 28.7

Critical exponents in the N-vector model on the lattice.

N	0	1	2	3
γ	1.1575 ± 0.0006	1.2385 ± 0.0025	1.322 ± 0.005	1.400 ± 0.006
ν	0.5877 ± 0.0006	0.631 ± 0.002	0.674 ± 0.003	0.710 ± 0.006
α	0.237 ± 0.002	0.103 ± 0.005	-0.022 ± 0.009	-0.133 ± 0.018
β	0.3028 ± 0.0012	0.329 ± 0.009	0.350 ± 0.007	0.365 ± 0.012
θ	0.56 ± 0.03	0.53 ± 0.04	0.60 ± 0.08	0.54 ± 0.10

We have not listed all available results but rather only typical numbers in order to give a feeling of the consistency between RG and lattice estimates (for $N \neq 1$ α, β are obtained by scaling). The obvious conclusion is that one observes no systematic differences. In particular the agreement is extremely good in the case of the Ising model where the HT series are the most accurate. To the best of our knowledge the N-vector lattice models and the $\left(\phi^2\right)^2$ field theory belong to the same universality class.

28.6 Critical Exponents from Experiments

We have discussed the N-vector model in the ferromagnetic language, even though most of our experimental knowledge comes from physical systems which are non-magnetic, but belong to the universality class of the N-vector model. The case $N = 0$ describes the statistical properties of long polymers, i.e. long not intersecting chains or self-avoiding walks (see Appendix A28.1). The case $N = 1$ (Ising-like systems) describes liquid–vapour transitions in classical fluids, critical binary fluids and uniaxial antiferromagnets. The helium superfluid transition corresponds to $N = 2$. Finally for $N = 3$ the experimental information comes from ferromagnetic systems.

Table 28.8

Critical exponents in fluids and antiferromagnets.

	γ	ν	β	α	θ
(a)	1.236 ± 0.008	0.625 ± 0.010	0.325 ± 0.005	0.112 ± 0.005	0.50 ± 0.03
(b)	$1.23 - 1.25$	0.625 ± 0.006	$0.316 - 0.327$	0.107 ± 0.006	0.50 ± 0.03
(c)	1.25 ± 0.01	0.64 ± 0.01	0.328 ± 0.009	0.112 ± 0.007	

Table 28.9

Ferromagnetic systems.

γ	ν	β	α	θ
1.40 ± 0.03	$0.700 - 0.725$	0.35 ± 0.03	$-0.09- -0.012$	0.54 ± 0.10

Critical exponents and polymers. In the case of polymers only the exponent ν is easily accessible. The best results are:

$$\nu = 0.586 \pm 0.004 \,,$$

in excellent agreement with the RG result.

Ising-like systems $N = 1$. Table 28.8 gives a survey of the experimental situation for critical binary fluids (*a*), liquid–vapour transition in classical fluids (*b*), and antiferromagnets (*c*). For the binary mixtures we quote a weighted world average. In the case of the liquid–vapour transition, we quote a range of experimental results rather than statistical errors for all exponents but ν, the reason being that the values depend much on the method of analysis of the experimental data. The agreement with RG results is clearly impressive.

Helium superfluid transition, $N = 2$. The helium transition allows extremely precise measurements very close to T_c and this explains the accuracy of the determination of critical exponents. The order parameter, however, is not directly accessible and therefore only ν and α have been determined. Most recent reported values are

$$\nu = 0.6705 \pm 0.0006 \,, \quad \nu = 0.6708 \pm 0.0004 \quad \text{and} \quad \alpha = -0.01285 \pm 0.00038 \,.$$

The agreement with RG values is quite remarkable and the accuracy of ν becomes a challenge to field theory.

Ferromagnetic systems, $N = 3$. Finally table 28.9 presents some results concerning magnetic systems.

Table 28.10

Amplitude ratios: Models and binary critical fluids.

	ε-expansion	Fixed dim. $d=3$	Lattice models	Experiment
A^+/A^-	0.527 ± 0.037	0.537 ± 0.019	$\begin{cases} 0.523 \pm 0.009 \\ 0.560 \pm 0.010 \end{cases}$	0.56 ± 0.02
C^+/C^-	4.73 ± 0.16	4.79 ± 0.10	$\begin{cases} 4.75 \pm 0.03 \\ 4.95 \pm 0.15 \end{cases}$	4.3 ± 0.3
f_1^+/f_1^-	1.91	2.04 ± 0.04	1.96 ± 0.01	1.9 ± 0.2
R_ξ^+	0.28	0.270 ± 0.001	0.266 ± 0.001	$0.25 - 0.32$
R_c	0.0569 ± 0.0035	0.0574 ± 0.0020	0.0581 ± 0.0010	0.050 ± 0.015
$R_\xi^+ R_c^{-1/3}$	0.73	0.700 ± 0.014	0.650	$0.60 - 0.80$
R_χ	1.648 ± 0.036	1.669 ± 0.018	1.75	1.75 ± 0.30
Q_2	1.13		1.21 ± 0.04	1.1 ± 0.3
Q_3	0.96		0.896 ± 0.005	

28.7 Amplitude Ratios

Table 28.10 contains a comparison of amplitude ratios as obtained from renormalization group for $N = 1$, lattice calculations for Ising-like models and experiments on binary mixtures. RG values for amplitudes are less accurate than for exponents because the series are shorter. Note that a RG determination of the equation of state is also available.

Some results are available for uniaxial magnetic systems and liquid–vapour transitions. For the latter systems a few reported values are:

$$C^+/C^- = 5. \pm 0.2 \,, \quad R_c = 0.047 \pm 0.010 \,, \quad R_\chi = 1.69 \pm 0.14 \,.$$

For A^+/A^- results range from 0.48 to 0.53. The set of results shows, with indeed large errors, a satisfactory agreement with RG predictions.

Finally let us give a few results concerning ratios of amplitudes of corrections to the leading scaling behaviour. If in addition to the correlation length and the susceptibility amplitudes a_ξ and a_χ we consider also the specific heat amplitude a_C and the coexistence curve magnetization amplitude a_M, we can form three independent ratios. The results for $N = 1$ are given in table 28.11.

Table 28.11

Correction amplitude ratios for $N = 1$.

	ε – expansion	Fixed dim. $d=3$	HT series	Experiment
a_ξ^+/a_χ^+	0.56 ± 0.15	0.65 ± 0.05	0.70 ± 0.03	
a_C^+/a_ξ^+	2.03	1.45 ± 0.11		
a_C^+/a_χ^+	1.02	0.94 ± 0.10	1.96	0.87 ± 0.13
a_C^+/a_C^-	2.54	1.0 ± 0.1		$0.7 - 1.35$
a_χ^+/a_χ^-	0.3	0.315 ± 0.013		
a_C^+/a_M		1.10 ± 0.25		1.85 ± 0.10
a_M/a_χ^+		0.90 ± 0.21		$0.08 - 1.4$

Table 28.12

Amplitude ratios for $N = 2$ and $N = 3$.

	N	Field theory	HT series	Experiment
A^+/A^-	2	1.056 ± 0.004	1.08	$1.054 \pm .001$
R_ξ^+	2	0.36	0.36	
R_c	2	0.123 ± 0.003		
A^+/A^-	3	1.52 ± 0.02	1.52	1.40–1.52
R_ξ^+	3	0.42	0.42	0.45
R_c	3	0.189 ± 0.009	0.165	

A few amplitude ratios have been calculated and measured for Helium ($N = 2$) and ferromagnets ($N = 3$). We give in table 28.12 the examples of A^+/A^-, R_ξ^+ and R_c.

If one takes into account all data (critical exponents, amplitude ratios,...) one is forced to conclude that the RG predictions are remarkably consistent with the whole experimental information available. Considering the variety of experimental situations, this is a spectacular confirmation of the RG ideas and the concept of universality.

Bibliographical Notes

The ε-expansion calculations for exponents were initiated by
K.G. Wilson and M.E. Fisher, *Phys. Rev. Lett.* 28 (1972) 240; K.G. Wilson, *Phys. Rev. Lett.* 28 (1972) 548.
The results at next order ε^3 were reported in
E. Brézin, J.C. Le Guillou, J. Zinn-Justin and B.G. Nickel, *Phys. Lett.* 44A (1973) 227.
Exponents at order ε^5 have been obtained by
A.A. Vladimirov, D.I. Kazakov and O.V. Tarasov, *Zh. Eksp. Teor. Fiz.* 77 (1979) 1035 (*Sov. Phys. JETP* 50 (1979) 521); K.G. Chetyrkin, A.L. Kataev and F.V. Tkachov, *Phys. Lett.* B99 (1981) 147; B101 (1981) 457(E); K.G. Chetyrkin and F.V. Tkachov, *Nucl. Phys.* B192 (1981) 159; K.G. Chetyrkin, S.G. Gorishny, S.A. Larin and F.V. Tkachov, *Phys. Lett.* 132B (1983) 351; D.I. Kazakov, *Phys. Lett.* 133B (1983) 406; S.G. Gorishny, S.A. Larin and F.V. Tkachov, *Phys. Lett.* 101A (1984) 120; H. Kleinert, J. Neu, V. Schulte-Frohlinde, K.G. Chetyrkin and S.A. Larin, *Phys. Lett.* B272 (1991) 39, Erratum B319 (1993) 545.
The Borel summation of Section 28.3 has been first applied on the ε-expansion in
J.C. Le Guillou and J. Zinn-Justin, *J. Physique Lett. (Paris)* 46 (1985) L137; *J. Physique (Paris)* 48 (1987) 19; *ibidem* 50 (1989) 1365.
The use of perturbation series at fixed dimension has been advocated by
G. Parisi, *Cargèse Lectures 1973*, published in *J. Stat. Phys.* 23 (1980) 49.
The calculation of the series expansion for the RG functions has been initiated by Nickel and reported (together with an estimate of $N = 1$ exponents) in
G.A. Baker, B.G. Nickel, M.S. Green and D.I. Meiron, *Phys. Rev. Lett.* 36 (1976) 1351; B.G. Nickel, D.I. Meiron, G.B. Baker, *Univ. of Guelph Report* 1977.
First precise estimates based on Borel summation and large order behaviour were published in
J.C. Le Guillou and J. Zinn-Justin, *Phys. Rev. Lett.* 39 (1977) 95; *Phys. Rev.* B21 (1980) 3976.

The updated 3D results presented in Section 28.4 are taken from

R. Guida and J. Zinn-Justin, *J. Phys. A* 31 (1998) 8103, cond-mat/9803240,

and obtained from corrected ε series, and seven-loop 3D terms for γ and η reported in

D.B. Murray and B.G. Nickel, unpublished Guelph University report (1991).

Six-loop series have been published for generic values of N

S.A. Antonenko and A.I. Sokolov, *Phys. Rev. E51* (1995) 1894.

The scaling equation of state is known to order ε^2 for all N and order ε^3 for $N = 1$

G.M. Avdeeva and A.A. Migdal, *JETP Lett.* 16 (1972) 178; E. Brézin, D.J. Wallace and K.G. Wilson, *Phys. Rev. Lett.* 29 (1972) 591; *Phys. Rev. B7* (1973) 232, D.J. Wallace and R.P.K. Zia, *J. Phys. C* 7 (1974) 3480.

For an early review see

D.J. Wallace, in *Phase Transitions and Critical Phenomena* vol. 6, (references page x).

The parametric representation has been introduced in

P. Schofield, J.D. Litster and J.T. Ho, *Phys. Rev. Lett.* 23 (1969) 1098; B.D. Josephson, *J. Phys. C* 2 (1969) 1113.

Calculations of ratios of amplitudes have been reported in

E. Brézin, J.C. Le Guillou and J. Zinn-Justin, *Phys. Lett.* 47A (1974) 285; H.B. Tarko and M.E. Fisher, *Phys. Rev. Lett.* 31 (926) 1973; *Phys. Rev. B11* (1975) 1217; M.E. Fisher and H.B. Tarko, *Phys. Rev. B11* (1975) 1131; A. Aharony and P.C. Hohenberg, *Phys. Rev. B13* (1976) 3081; *Physica* 86-88B (1977) 611; Y. Okabe and K. Ideura, *Prog. Theor. Phys.* 66 (1981) 1959; J.F. Nicoll and P.C. Albright, *Phys. Rev. B31* (1985) 4576.

Series and estimates for equation of state and ratios of amplitude are reported in

C. Bagnuls and C. Bervillier, *Phys. Rev. B24* (1981) 1226; *J. Phys. A* 19 (1986) L85; C. Bervillier, *Phys. Rev. B34* (1986) 8141; C. Bagnuls, C. Bervillier, D.I. Meiron and B.G. Nickel, *Phys. Rev. B35* (1987) 3585; R. Schloms and V. Dohm, *Phys. Rev. B42* (1990) 6142; H. J. Krause, R. Schloms and V. Dohm, *Z. Physik B79* (1990) 287; F.J. Halfkann and V. Dohm, *Z. Phys. B89* (1992) 79; G. Münster, J. Heitger, *Nucl. Phys. B424* [FS] (1994) 582; R. Guida and J. Zinn-Justin, *Nucl. Phys. B489* [FS] (1997) 626; S.A. Larin, M. Moennigmann, M. Stroesser and V. Dohm, *Phys. Rev.B58* (1998) 3394, cond-mat/9805028; M. Stroesser, S.A. Larin and V. Dohm, *Nucl.Phys. B540* (1999) 654, cond-mat/9806103.

For a review on high temperature series see for example

A.J. Guttmann, *Phase Transitions and Critical Phenomena* vol. 13, C. Domb and J. Lebowitz eds. (Academic Press, New York 1989).

A number of results extracted from high temperature series can be found in

J. Zinn-Justin, *J. Physique* (Paris) 42, (1981) 783; A.J. Guttmann, *J. Phys. A* 20 (1987) 1855; A.J. Liu and M.E. Fisher, *Physica* A156 (1989) 35; A.J. Guttmann and I.G. Enting, *J. Phys. A* 26 (1993) 807, 27 (1994) 8007; G. Bhanot, M. Creutz, U. Glässner and K. Schilling, *Phys. Rev. B49* (1994) 12909; P. Butera and M. Comi, *Phys. Rev. E55* (1997) 6391, hep-lat/9703017, *Phys. Rev. B56* (1997) 8212, 9703018.

Results from Monte Carlo simulations for Ising model and of the SAW are found in

G.S. Pawley, R.H. Swendsen, D.J. Wallace and K.G. Wilson, *Phys. Rev. B29* (1984) 4030; C.F. Baillie, R. Gupta, K.A. Hawick and G.S. Pawley, *Phys. Rev. B45* (1992) 10438; B. Li, N. Madras and A.D. Sokal, *J. Stat. Phys.* 80 (1995) 661; S. Caracciolo, M.S. Causo and A. Pelissetto, *Phys. Rev. E57* (1998) 1215, cond-mat/9703250.

For a review on amplitude ratios see

V. Privman, P.C. Hohenberg and A. Aharony, *Phase Transitions and Critical Phenomena* vol. 14, C. Domb and J. Lebowitz eds. (Academic Press, New York 1991).

Experiment, high temperature series and field theory RG are compared, using the results then available, in the Proceedings of the Cargèse Summer Institute 1980

    Phase Transitions vol. B72, M. Lévy, J.C. Le Guillou and J. Zinn-Justin eds. (Plenum, New York 1982).

The experimental determination of the exponent ν for polymers is taken from

    J.P. Cotton, *J. Physique Lett. (Paris)* 41 (1980) L231.

Experimental determinations of universal quantities for binary fluids are reported in

    D. Beysens, A. Bourgou and P. Calmettes, *Phys. Rev.* A26 (1982) 3589; A. Kumar, H.R. Krishnamurthy and E.S.R. Gopal, *Phys. Rep.* 98 (1983) 57.

For experimental results on Helium superfluid transition see

    L.S. Goldner and G. Ahlers, *Phys. Rev.* B45 (1992) 13129; J.A. Lipa, D.R. Swanson, J. Nissen, T.C.P. Chui and U.E. Israelson, *Phys. Rev. Lett.* 76 (1996) 944.

The relation between statistical properties of polymers and the $N = 0 \left(\phi^2\right)^2$ field theory has been found by

    P.G. de Gennes, *Phys. Lett.* 38A (1972) 339.

We have followed the derivation of

    V.J. Emery, *Phys. Rev.* B11 (1975) 239; B. Duplantier, *C.R. Acad. Sci. Paris* 290B (1980) 199.

The application of RG methods to polymers is presented in

    J. des Cloizeaux, *J. Physique (Paris)* 36 (1975) 281; L. Schäfer and T.A. Witten, *J. Chem. Phys.* 66 (1977) 2121; *J. Physique (Paris)* 41 (1980) 459.

For a comprehensive discussion of both experimental and theoretical aspects of renormalization group and polymer physics see

    J. des Cloizeaux and G. Jannink, *Les Polymères en Solution* (Les Editions de Physique, Paris 1987), English version *Polymers in Solution: Modelling and Structure*, (Oxford Univ. Press 1990).

The exact two-dimensional results have first been conjectured by

    B. Nienhuis, *Phys. Rev. Lett.* 49 (1982) 1062.

For a discussion of the relation between classical fluids at criticality and the ϕ^4 field theory, see for example

    F.W. Wiegel, *Phys. Rep.* 16C (1975) 57.

The lattice–gas model was introduced in

    C.N. Yang and T.D. Lee, *Phys. Rev.* 87 (1952) 404; T.D. Lee and C.N. Yang, *Phys. Rev.* 87 (1952) 410.

Exercises

Exercise 28.1

To the reader who finds the relation between the polymer model and the ϕ^4 field theory somewhat mysterious the following exercise is proposed: Calculate at one or two-loop order $Z^{(2)}(k,t)$ directly by expanding the measure ($A28.1$) in powers of g. Apart the trick which has already been used in the gaussian case the following remark is useful: Write a Fourier representation for the chain self-interaction term:

$$\delta^d\left(\mathbf{r}\left(u_1\right) - \mathbf{r}\left(u_2\right)\right) = \int \frac{\mathrm{d}^d q}{(2\pi)^d} \exp\left[i\mathbf{q}\cdot\int_0^s \mathrm{d}u\left(\theta(u_1 - u) - \theta(u_2 - u)\right)\dot{\mathbf{r}}(u)\right]$$

where $\theta(u)$ is the Heaviside step function. To recover the Feynman diagrams of the ϕ^4 field theory it is suggested to integrate over $\mathbf{r}(u)$, then over all u variables and finally to Laplace transform, keeping the q-integrations which come from the Fourier representation for the end. Some additional inspiration can be found in Section 31.1.

APPENDIX 28
NON-MAGNETIC STATISTICAL MODELS AND THE $(\phi^2)^2$ FIELD THEORY

A28.1 Statistics of Self-Repelling Chains, Approximations

We here show that, as anticipated in Chapter 28, the statistical properties of long polymer chains can be derived from the critical behaviour of the $O(N)$ symmetric $(\phi^2)^2$ field theory in the $N = 0$ limit. More precisely we study the statistical properties of long chains with repulsive contact self-interaction. A discretized form of the model is provided by the self-avoiding random walk (SAW) on a lattice.

We call u the distance along the chain, $\mathbf{r}(u)$ the position in space of the point on the chain of parameter u. We then characterize the chain by a probability distribution for a chain $\mathbf{r}(u)$ of total length s:

$$[d\rho(\mathbf{r}(u))] = [d\mathbf{r}(u)]\exp\left\{-\left[\frac{1}{4}\int_0^s \dot{\mathbf{r}}^2(u)du + \frac{g}{6}\int_0^s du_1 du_2\, \delta^d(\mathbf{r}(u_1) - \mathbf{r}(u_2))\right]\right\}.$$
$$(A28.1)$$

The special case $g = 0$ corresponds to the brownian chain or gaussian random walk, as has been discussed in Chapter 4. The self-avoiding random walk (SAW) model also provides a lattice regularization: at a microscopic scale Λ^{-1} the chain becomes much stiffer than what is implied by expression $(A28.1)$. Note finally that any short range potential would yield, at leading order for long chains, to the same results as the δ-function interaction, (as can be verified by repeating in the more general case the derivation of an equivalent ϕ^4 like field theory, and using the analysis of corrections to scaling as in the case of critical phenomena).

A generating function. Various geometric characteristic properties of the chain can be obtained from the 2-point function $G^{(2)}(\mathbf{k}, s)$:

$$G^{(2)}(\mathbf{k}, s) = \left\langle e^{i\mathbf{k}\cdot(\mathbf{r}(s)-\mathbf{r}(0))}\right\rangle. \qquad (A28.2)$$

The brackets mean average with respect to the distribution $(A28.1)$. Indeed if we expand $G^{(2)}(\mathbf{k},s)$ in powers of $\mathbf{k}$ and use the rotational invariance of distribution $(A28.1)$ we get:

$$G^{(2)}(\mathbf{k}, s) = 1 - \frac{1}{2!}\frac{\mathbf{k}^2}{d}\left\langle(\mathbf{r}(s)-\mathbf{r}(0))^2\right\rangle + \frac{(\mathbf{k}^2)^2}{4!}\frac{1}{d(d+2)}\left\langle(\mathbf{r}(s)-\mathbf{r}(0))^4\right\rangle + \cdots.$$
$$(A28.3)$$

It is actually convenient to consider the Laplace transform of $G^{(2)}(\mathbf{k}, s)$ with respect to s:

$$Z^{(2)}(\mathbf{k}, t) = \int_0^\infty e^{-st}\, G^{(2)}(\mathbf{k}, s)\, ds. \qquad (A28.4)$$

Indeed if we know the 2-point correlation function $Z^{(2)}(\mathbf{k}, t)$, we obtain $G^{(2)}(\mathbf{k}, s)$ by inverting the Laplace transformation:

$$G^{(2)}(\mathbf{k}, s) = \frac{1}{2i\pi}\oint_C e^{st}\, Z^{(2)}(\mathbf{k}, t)\, dt, \qquad (A28.5)$$

in which the contour C is a parallel to the imaginary t axis at the right of all singularities of $Z^{(2)}(\mathbf{k}, t)$. Remember that, since we are going to expand in powers of $\mathbf{k}$ (see equation

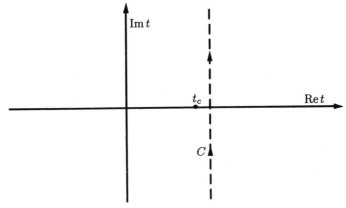

Fig. 28.1

$(A28.3))$, $\mathbf{k}$ has to be considered as infinitesimal. Thus only the location of singularities in the limit $\mathbf{k} \to \mathbf{0}$ is relevant. Let us call t_c the first singularity one meets while trying to displace the contour to the left (figure 28.1). On C the real part $\mathrm{Re}(t)$ of t thus satisfies:

$$\mathrm{Re}\,(t) > \mathrm{Re}\,(t_c).$$

It is easy to verify that the behaviour of $Z^{(2)}(\mathbf{k}, t)$ near the singularity governs the large positive s behaviour of the integral $(A28.5)$ and thus of the moments of $\mathbf{r}(s) - \mathbf{r}(0)$.

A28.1.1 Special examples. Flory's approximation
The gaussian random walk. In expression $(A28.2)$ we can substitute

$$\mathbf{r}(s) - \mathbf{r}(0) = \int_0^s \dot{\mathbf{r}}(u)\mathrm{d}u.$$

We then get

$$G^{(2)}(\mathbf{k}, s) = \int [\mathrm{d}\mathbf{r}(u)] \exp\left\{-\int_0^s \mathrm{d}u \left[\frac{1}{4}\dot{\mathbf{r}}^2(u) - i\mathbf{k}\cdot\dot{\mathbf{r}}(u)\right]\right\}.$$

Shifting $\mathbf{r}$ we find:

$$G^{(2)}(\mathbf{k}, s) = \mathrm{e}^{-s\mathbf{k}^2}, \qquad (A28.6)$$

which implies as it is well-known for the gaussian random walk:

$$\left\langle (\mathbf{r}(s) - \mathbf{r}(0))^{2n} \right\rangle \sim s^n. \qquad (A28.7)$$

The Laplace transformation then yields:

$$Z^{(2)}(\mathbf{k}, t) = \left(t + \mathbf{k}^2\right)^{-1}. \qquad (A28.8)$$

The one dimensional chain. If the chain self-repulsion plays a role, it must increase the average spatial extension of the chain for s large, compared to the gaussian case. This is obvious in the one-dimensional case. The chain is then completely stretched:

$$r(s) = \pm s, \qquad (A28.9)$$

and therefore:

$$\begin{cases} G^{(2)}(k,s) = \cos ks \\ Z^{(2)}(k,t) = \dfrac{t}{t^2 + k^2}. \end{cases} \qquad (A28.10)$$

The one dimensional example shows that the self-avoiding condition can change the statistical properties of the random walk. It also provides an upper bound on the moments. A sensible guess is then in general

$$\left\langle (\mathbf{r}(s) - \mathbf{r}(0))^{2n} \right\rangle \sim s^{2\nu n} \qquad \text{for } s \to \infty,$$

in which the exponent ν is bounded from below by $1/2$, the value for the gaussian chain, and from above by 1, the value for the stretched chain.

The upper-critical dimension. Note that the self-avoiding condition becomes weaker when the dimension d of embedding space increases. Actually a simple argument suggests, as in the case of the critical behaviour, the existence of an "upper-critical" dimension. If we consider a very long chain, we can consider the effect of self-avoiding as the influence of one chain onto another. A brownian chain has Hausdorf dimension two. Therefore above dimension four (two +two) two chains do no more see each other, self-avoiding should no longer play a role and the chain should have the statistical properties of a brownian chain.

Beyond the gaussian model: Flory's approximation. Before solving the problem by more systematic methods let us first describe a simple approximate solution. The approximation is based on an energy balance argument. For s large, the gaussian term should scale as the interaction:

$$\left\langle \int_0^s \dot{\mathbf{r}}^2(u) du \right\rangle \sim s^{2\nu - 1}, \qquad (A28.11)$$

$$\int du_1 du_2 \, \delta^d \left(\mathbf{r}(u_1) - \mathbf{r}(u_2) \right) \sim s^{2 - d\nu}. \qquad (A28.12)$$

Balancing both terms we obtain the equation for ν:

$$2\nu - 1 = 2 - d\nu, \; \Rightarrow \; \nu = \frac{3}{d+2}, \qquad (A28.13)$$

valid for $1 \le d \le 4$. This expression yields for $d = 1$ the exact value $\nu = 1$. It predicts that for $d \ge 4$ the interaction can never balance the gaussian term, since ν is bounded by $1/2$. More surprisingly, for $d = 2$ it yields $\nu = 3/4$ which is also known, from conformal field theory arguments, to be the exact value.

We shall discuss below other dimensions. Flory's approximation relies on intuitive arguments and it is difficult to see how it can be improved. Actually the values it yields for the exponent ν are so good that one may wonder if this result is not exact. We show below that it is indeed only an approximation. For this purpose it is convenient to completely reformulate the theory.

A28.1.2 Equivalence with the ϕ^4 field theory

We first transform the interaction term in equation (A28.1) by writing it as a gaussian integral over an auxiliary field $\sigma(\mathbf{r})$. If $\mathbf{r}(u)$ is a given chain, then

$$\int [\mathrm{d}\sigma(\mathbf{r})] \exp\left[\frac{3}{2g}\int \mathrm{d}^d r\, \sigma^2(r) - \int \mathrm{d}u\, \sigma(\mathbf{r}(u))\right]$$
$$= \exp\left[-\frac{g}{6}\int \mathrm{d}u_1 \mathrm{d}u_2\, \delta^d(\mathbf{r}(u_1) - \mathbf{r}(u_2))\right], \qquad (A28.14)$$

where we have used:

$$\int \mathrm{d}u\, \sigma(\mathbf{r}(u)) = \int \mathrm{d}^d r\, \sigma(r) \int \mathrm{d}u\, \delta^d(\mathbf{r}(u) - \mathbf{r}).$$

Integration over imaginary σ-fields is assumed. After this substitution we identify the $\mathbf{r}(u)$ measure of integration with the integrand in the path integral representation of the evolution operator in imaginary time u of a d-dimensional quantum mechanical system with potential $\sigma(\mathbf{r})$ (see Chapter 2). The 2-point function $Z^{(2)}(\mathbf{k}, t)$ can thus be written:

$$Z^{(2)}(\mathbf{k}, t) = \int [\mathrm{d}\sigma(\mathbf{r})] \exp\left[\frac{3}{2g}\int \mathrm{d}^d r\, \sigma^2(r)\right] \int_0^\infty \mathrm{e}^{-ts}\, \mathrm{d}s$$
$$\times \int \mathrm{d}^d r\, \mathrm{d}^d r'\, \mathrm{e}^{i\mathbf{k}(\mathbf{r}-\mathbf{r}')} \langle \mathbf{r}' | \mathrm{e}^{-sH} | \mathbf{r} \rangle, \qquad (A28.15)$$

in which H is the quantum hamiltonian:

$$H = -\Delta + \sigma(\mathbf{r}). \qquad (A28.16)$$

The Laplace transform can be explicitly evaluated:

$$Z^{(2)}(\mathbf{k}, t) = \int [\mathrm{d}\sigma(\mathbf{r})] \exp\left[\frac{3}{2g}\int \mathrm{d}^d r\, \sigma^2(r)\right] \int \mathrm{d}^d r\, \mathrm{d}^d r'\, \mathrm{e}^{i\mathbf{k}(\mathbf{r}-\mathbf{r}')}$$
$$\times \langle \mathbf{r}' | (-\Delta + t + \sigma)^{-1} | \mathbf{r} \rangle. \qquad (A28.17)$$

We now apply a variant of the so-called *replica trick*:

$$\lim_{N\to 0} \int [\mathrm{d}\phi(\mathbf{r})]\, \phi_1(\mathbf{r})\, \phi_1(\mathbf{r}') \exp\left[-\frac{1}{2}\int \mathrm{d}^d r\left((\partial_\mu \phi)^2 + t\phi^2 + \sigma(r)\phi^2\right)\right]$$
$$= \langle \mathbf{r}' | (-\Delta + t + \sigma)^{-1} | \mathbf{r} \rangle, \qquad (A28.18)$$

in which N is the number of components of the field $\phi(x)$. Indeed the gaussian integral over the field $\phi(r)$ yields the $\phi\phi$ propagator divided by a factor $[\det(-\Delta + t + \sigma)]^{N/2}$ which goes to 1 in the "unphysical" $N = 0$ limit.

Substituting this identity into expression (A28.17) we obtain:

$$Z^{(2)}(\mathbf{k}, t) = \int [\mathrm{d}\phi(r)]\, \phi_1(\mathbf{k})\, \phi_1(-\mathbf{k}) \int [\mathrm{d}\sigma(r)]$$
$$\times \exp\left[\int \mathrm{d}^d r\left(\frac{3}{2g}\sigma^2 - \frac{1}{2}\sigma\phi^2 - \frac{1}{2}(\partial_\mu \phi)^2 - \frac{1}{2}t\phi^2\right)\right]. \qquad (A28.19)$$

We can then integrate over σ:

$$Z^{(2)}(\mathbf{k},t) = \int [\mathrm{d}\phi]\,\phi_1(\mathbf{k})\,\phi_1(-\mathbf{k})\exp\left[-S(\phi)\right],$$

$$S(\phi) = \int \mathrm{d}^d r \left[\frac{1}{2}(\partial_\mu\phi)^2 + \frac{1}{2}t\phi^2 + \frac{g}{4!}(\phi^2)^2\right]. \qquad (A28.20)$$

This is a most remarkable result: the statistical properties of polymers are related to the properties of the $(\phi^2)^2$ field theory in the $N=0$ limit as first noticed by de Gennes.

A28.1.3 RG Approach to SAW and Statistics of Polymers

The large scale statistical properties of the chain are related to the singularities of correlation functions of the $(\phi^2)^2$ field theory, which arise in the massless or critical limit. The short distance stiffness of the chain provides an UV cut-off Λ for the theory. Near t_c, in which t_c is the critical temperature or critical bare mass, the 2-point function $Z^{(2)}(k,t)$ has a scaling behaviour:

$$Z^{(2)}(\mathbf{k},t) \sim (t-t_c)^{-\gamma} f\left[\mathbf{k}(t-t_c)^{-\nu}\right], \qquad (A28.21)$$

in which ν and γ are the $N=0$ limits of the critical exponents of the $(\phi^2)^2$ field theory.

Using expression $(A28.21)$ in integral $(A28.5)$ we then obtain a scaling form for the large s behaviour of $G^{(2)}(s,k)$:

$$G^{(2)}(s,\mathbf{k}) \sim e^{st_c}\,s^{\gamma-1}g\left(|\mathbf{k}|\,s^\nu\right). \qquad (A28.22)$$

Expanding in powers of $\mathbf{k}^2$, we find a first term:

$$\langle 1\rangle \sim e^{st_c}\,s^{\gamma-1}g(0),$$

instead of 1 which shows that the distribution is not correctly normalized. The first term is actually the ratio of the number of configurations of the SAW to the number of configurations of the gaussian random walk. Therefore the properly normalized average $\langle\exp[i\mathbf{k}(r(s)-r(0))]\rangle$ has a scaling behaviour entirely characterized by the exponent ν. For example:

$$\left\langle (\mathbf{r}(s)-\mathbf{r}(0))^2\right\rangle \sim s^{2\nu}. \qquad (A28.23)$$

The exponent can be obtained from the ε-expansion:

$$\nu = \tfrac{1}{2}\left(1 + \tfrac{1}{8}\varepsilon + \tfrac{15}{256}\varepsilon^2 + O\left(\varepsilon^3\right)\right). \qquad (A28.24)$$

Instead Flory's formula $(A28.13)$ predicts at the same order:

$$\nu(\text{Flory}) = \tfrac{1}{2}\left(1 + \tfrac{1}{6}\varepsilon + O\left(\varepsilon^2\right)\right). \qquad (A28.25)$$

Comparing the two expressions, we see immediately that Flory's formula is not exact in general, although, as mentioned before, it correctly predicts the upper-critical dimension $d=4$, and the values for $d=1$ and 2.

For $d=3$ RG calculations yield:

$$\nu = 0.5880 \pm 0.0015,$$

result which can be compared with $\nu(\text{Flory}) = 0.6$. The two values are very close although definitively different. Actually it is easy to estimate the difference approximately. If we simply take into account the orders ε and ε^2 and the property that Flory's approximation is exact for $\varepsilon = 0, 2, 3$ we can write:

$$\nu_{\text{RG}}^{-1} - \nu^{-1}\,(\text{Flory}) = \tfrac{1}{12}\varepsilon \left(1 - \tfrac{1}{2}\varepsilon\right)\left(1 - \tfrac{1}{3}\varepsilon\right)\left(1 - \tfrac{19}{96}\varepsilon\right) + O\left(\varepsilon^3\right).$$

For $\varepsilon = 1$ we find successively $\nu = 0.590$ and order ε and $\nu = 0.592$ at order ε^2, results consistent with the RG value obtained by summing all terms (tables 28.4, 28.6). Experimental results, $\nu = 0.586 \pm 0.004$, HT series and Monte Carlo simulations, $\nu = 0.592 \pm 0.003$, favour the RG prediction compared to Flory's approximation.

Moreover the renormalization group makes many additional predictions and allows the calculation of many universal quantities which cannot be obtained from Flory's argument.

A28.2 Liquid–Vapour Phase Transition and Field Theory

That the liquid–vapour phase transition is the same universality class as the Ising model and can be described by the ϕ^4 field theory is not completely obvious *a priori*. In particular no apparent discrete symmetry is broken. There exist several methods to establish a connection between the liquid–vapour transition and the $N = 1$ ϕ^4 theory: one can either start from a model which is physically not very realistic, the lattice gas model, but rigorously equivalent to the Ising model, or from a true gas model in the continuum, and by a number of approximations for which it is difficult to provide a rigorous justification, obtain directly a ϕ^4 field theory.

The lattice gas model. One assumes that particles are living on a lattice and that at each site i the particle occupation number n_i can only be 0 or 1. The lattice is a schematic way to represent a hard core type interaction. A longer range attractive potential is represented by a nearest neighbour interaction which favours the occupation of neighbour sites. One finally adds a chemical potential term, i.e. a source term coupled to the number of particles:

$$-\beta \mathcal{H}\,(n_i) = J \sum_{\text{n.n.}} n_i n_j - \mu \sum_i n_i \,. \qquad (A28.26)$$

This model is trivially related to the Ising model by the transformation:

$$S_i = 2n_i - 1\,, \qquad (A28.27)$$

where S_i is an Ising spin. One finds in d dimensions:

$$-\beta \mathcal{H}\,(S_i) = \tfrac{1}{4}J \sum_{\text{n.n.}} S_i S_j + \tfrac{1}{2}\,(dJ - \mu) \sum_i S_i + \text{const.}\,. \qquad (A28.28)$$

We recognize the energy of the Ising model with n.n. interaction and magnetic field.

A gas model in the continuum. The introduction of a lattice is somewhat artificial. A more direct method, although it involves a few steps that are hard to justify rigorously, is the following: one considers a real classical gas in the continuum and assumes that the potential between particles is the sum of a short range hard core potential and a longer range attractive two-body potential $V(r)$. The canonical partition function for n particles of mass m in a volume Ω then has the form:

$$Z(n, \beta, \Omega) = \frac{1}{n!} \left(\frac{2\pi m}{\beta h^2} \right)^{nd/2} \int_{\substack{\Omega \\ |\mathbf{r}_i - \mathbf{r}_j| > 2a}} \prod_{i=1}^{n} \mathrm{d}^d r_i \exp \left[-\beta \sum_{i < j} V(\mathbf{r}_i - \mathbf{r}_j) \right], \quad (A28.29)$$

where a is the size of the hard core. In three dimensions a convenient choice for the potential is for example:

$$V(\mathbf{r}) = -J \mathrm{e}^{-\mu r}/4\pi r. \qquad (A28.30)$$

The important hypotheses are that the potential depends only on the distance and is short range in such a way that its Fourier transform is regular at zero momentum. Furthermore its Fourier transform is assumed to have no zero. We can then define the inverse $K(\mathbf{r} - \mathbf{r}')$ of $V(\mathbf{r} - \mathbf{r}')$ in the sense of kernels:

$$\int \mathrm{d}^d r'' K(\mathbf{r} - \mathbf{r}'') V(\mathbf{r}'' - \mathbf{r}') = \delta^d (\mathbf{r} - \mathbf{r}'). \qquad (A28.31)$$

In the case of the Yukawa potential $(A28.30)$, K is the differential operator $J^{-1} \left(\partial^2 - \mu^2 \right)$. Using the simple identity:

$$\exp \left[-\tfrac{1}{2} \beta \sum_{i,j} V(\mathbf{r}_i - \mathbf{r}_j) \right] = \int [\mathrm{d}\phi(r)]$$

$$\times \exp \left[\tfrac{1}{2} \beta^{-1} \int \mathrm{d}^d r \, \mathrm{d}^d r' \phi(\mathbf{r}) K(\mathbf{r} - \mathbf{r}') \phi(\mathbf{r}') + \sum_i \phi(\mathbf{r}_i) \right], \qquad (A28.32)$$

we rewrite the partition function:

$$Z(n, \beta, \Omega) = \mathrm{e}^{n\beta V(0)/2} \int [\mathrm{d}\phi(r)] \exp \left[\tfrac{1}{2} \beta^{-1} \int \mathrm{d}^d r \, \mathrm{d}^d r' \phi(\mathbf{r}) K(\mathbf{r} - \mathbf{r}') \phi(\mathbf{r}') \right]$$

$$\times Z_{\mathrm{h.c.}}(n, \beta, \Omega, \phi), \qquad (A28.33)$$

where $Z_{\mathrm{h.c.}}(n, \beta, \Omega, \phi)$ is the canonical partition function of particles interacting through a hard core two-body potential in a random external potential $\phi(\mathbf{r})$. The fact that in example $(A28.30)$ $V(0)$ diverges is irrelevant since we can modify the potential for $r < a$. We can transform equation $(A28.33)$ into an identity between grand partition functions:

$$Z(z, \beta, \Omega) = \sum_n \frac{z^n}{n!} Z(n, \beta, \Omega), \qquad (A28.34)$$

where we have called z the fugacity. The partition function of the hard core potential in presence of a given external one-body potential cannot be calculated exactly. However a few remarks can be made: since we are interested in distances much larger than the size a of the hard core we can make a local expansion of $\ln Z_{\mathrm{h.c.}}(\phi)$. When ϕ is repulsive the hard core can for example be neglected because the hard core then prevents the collapse of the system. The partition function can thus be represented approximately by the functional integral:

$$Z(\alpha, \beta, \Omega) = \int [\mathrm{d}\phi(r)] \exp \left[\tfrac{1}{2} \beta^{-1} \int \mathrm{d}^d r \, \mathrm{d}^d r' \phi(\mathbf{r}) K(\mathbf{r} - \mathbf{r}') \phi(\mathbf{r}') - U(\phi) \right], \quad (A28.35)$$

where α is a renormalized fugacity and

$$U(\phi) = -\alpha \int d^d r \, e^{\phi(\mathbf{r})} + O\left(\alpha^2\right).$$

This expression has to be understood with an implicit short distance cut-off a. We have therefore been able to write the partition function approximately as a functional integral with an effective local action $S(\phi)$. In the case of the potential ($A28.30$) for example we find:

$$S(\phi) = \int d^d r \left[\tfrac{1}{2} \left(\beta J\right)^{-1} \left(\left(\partial_\mu \phi\right)^2 + \mu^2 \phi^2 \right) + U(\phi) \right]. \qquad (A28.36)$$

Let us discuss this action at tree level following the lines of Section 24.3 where the mean field approximation was considered. First we can shift ϕ in such a way that the minimum of the potential occurs at $\phi = 0$. At high temperature the action has only one homogeneous minimum. By varying the temperature we can reach a situation where a second minimum appears and eventually the two minima become degenerate. However here, since the action has no reflection symmetry, the generic situation corresponds to a first order transition. It is only when, by varying a second thermodynamic variable like the chemical potential, one has been able to cancel the ϕ^3 term in the expansion of the action for ϕ small that one obtains a second order phase transition. But then the terms odd in ϕ are all at least of dimension 5 for $d = 4$ and therefore irrelevant at the phase transition. This explains that the ϕ^4 field theory can describe correctly the liquid–vapour phase transition. The only specific features of this transition with respect to magnetic systems are:

(i) The complicated, although regular, relations of the thermodynamic variables to the more natural, from the symmetry point of view, variables magnetic field and temperature of the magnetic systems. This introduces additional parameters in the description of experimental results.

(ii) The appearance of corrections to scaling due to operators of odd canonical dimensions like ϕ^5.

29 THE $\left(\phi^2\right)^2$ FIELD THEORY IN THE LARGE N LIMIT

In the preceding chapters we have derived the universal properties of critical systems within the framework of the formal $\varepsilon = 4 - d$ expansion. It is therefore reassuring to verify in a particular model that the results obtained in this way remain valid even when ε is no longer infinitesimal. We show in this chapter that, in the case of the $\left(\phi^2\right)^2$ $O(N)$ symmetric field theory, the same universal properties can also be derived at fixed dimension in the large N limit, and more generally order by order in the large N-expansion.

29.1 Introduction

We consider the partition function in an external field or source $\mathbf{J}$:

$$Z(\mathbf{J}) = \int [\mathrm{d}\phi(x)] \exp\left[-S(\phi) + \int \mathrm{d}^d x \, \mathbf{J}(x) \cdot \phi(x)\right], \qquad (29.1)$$

where $S(\phi)$ is a $O(N)$ symmetric action:

$$S(\phi) = \int \left\{\frac{1}{2}[\partial_\mu \phi(x)]^2 + \frac{1}{2} r \phi^2(x) + \frac{u}{4!}[\phi^2(x)]^2\right\} \mathrm{d}^d x. \qquad (29.2)$$

A cut-off Λ, consistent with the symmetry, is implied.

The basic idea of the large N expansion is the same as mean field theory: for N large, the $O(N)$ invariant quantities self-average and therefore have small fluctuations. Thus for example

$$\langle \phi^2(x)\phi^2(y)\rangle \underset{N\to\infty}{\sim} \langle \phi^2(x)\rangle \langle \phi^2(y)\rangle.$$

Therefore one should try to take $\phi^2(x)$ as a dynamical variable. Technically, in the case of the $\left(\phi^2\right)^2$ theory, this can be achieved by using an identity similar to the Hubbard transformation:

$$\exp\left\{-\int \mathrm{d}^d x \left[\frac{1}{2} r \phi^2(x) + \frac{u}{4!}\left(\phi^2\right)^2\right]\right\} \propto \int [\mathrm{d}\lambda] \exp\left[\int \mathrm{d}^d x \left(\frac{3}{2u}\lambda^2 - \frac{3r}{u}\lambda - \frac{1}{2}\lambda\phi^2\right)\right], \qquad (29.3)$$

where the integration contour is parallel to the imaginary axis. This identity allows us to rewrite the interaction term in the functional integral (29.1). The new functional integral is then gaussian in ϕ and the integral over the field ϕ can be performed. The dependence on N of the partition function becomes explicit. Actually it is convenient to separate the components of ϕ into one component σ, and $N - 1$ components π, and integrate only over π (for $T < T_c$ it may even be convenient to integrate over only $N - 2$ components). For N large the difference is negligible. We call $H(x)$ the source for $\sigma(x)$, and set all other components of $J(x)$ to zero:

$$Z(H) = \int [\mathrm{d}\lambda(x)] [\mathrm{d}\sigma(x)] \exp\left[-S_{\mathrm{eff}}(\lambda, \sigma) + \int \mathrm{d}^d x \, H(x)\sigma(x)\right], \qquad (29.4)$$

with:

$$S_{\text{eff}}\left(\lambda,\sigma\right)=\int\left[\frac{1}{2}\left(\partial_\mu\sigma\right)^2-\frac{3}{2u}\lambda^2(x)+\frac{3r}{u}\lambda(x)+\frac{1}{2}\lambda(x)\sigma^2(x)\right]\mathrm{d}^dx$$

$$+\frac{(N-1)}{2}\operatorname{tr}\ln\left[-\Delta+\lambda(x)\right]. \tag{29.5}$$

We now take the large N limit at Nu fixed. With this condition S_{eff} is of order N and the functional integral can be calculated for N large by steepest descent. We expect $\sigma=O(N^{1/2})$, $\lambda\doteq O(1)$.

The λ-field correlation functions. In this formalism it is natural to also calculate λ-field correlation functions. These have a simple interpretation in the initial ϕ-field formalism. Indeed let us add a source j_λ for λ in the action (29.5). Then reintroducing the ϕ-field and integrating over λ we recover instead of action (29.2),

$$S(\phi)-(u/6)\phi^2 j_\lambda+(u/6)j_\lambda^2-r j_\lambda. \tag{29.6}$$

Therefore j_λ generates the ϕ^2 correlation functions, up to a multiplicative factor and a translation of the connected 2-point function.

29.2 The Large N Limit

We look for a uniform saddle point $(\sigma(x),\lambda(x)$ space-independent),

$$\sigma(x)=\sigma,\qquad\lambda(x)=\lambda.$$

Differentiating then action (29.5) with respect to σ and λ we obtain the saddle point equations for N large:

$$\lambda\sigma=0, \tag{29.7a}$$

$$\frac{\sigma^2}{N}-\frac{6}{Nu}\left(\lambda-r\right)+\frac{1}{(2\pi)^d}\int^\Lambda\frac{\mathrm{d}^dp}{p^2+\lambda}=0. \tag{29.7b}$$

Remark. In the large N limit the dominant contributions in perturbation theory are given by chains of "bubble" diagrams of the form displayed in figure 29.1. In this limit they form a geometrical series which is summed by the algebraic techniques explained above.

Fig. 29.1　The dominant diagrams in the large N limit.

29.2.1 The critical domain

The low temperature phase. Below T_c, σ, the average value of the field, does not vanish. Equation (29.7a) then implies $\lambda = 0$ and equation (29.7b) yields:

$$\frac{\sigma^2}{N} = -\frac{6}{Nu}r - \frac{1}{(2\pi)^d}\int^{\Lambda}\frac{\mathrm{d}^d p}{p^2}. \tag{29.8}$$

Note that this equation has a solution only for $d > 2$. This is a manifestation of the Mermin–Wagner–Coleman theorem: in a system with only short range forces a continuous symmetry cannot be broken for $d \leq 2$, in the sense that the average σ of the order parameter necessarily vanishes. This result is also consistent with the analysis of Section 23.5. Physically the would-be Goldstone modes are responsible for this property: being massless, as we know from general arguments and as the propagator in the r.h.s. of (29.8) confirms, they induce an IR instability for $d \leq 2$.

Equation (29.8) has a solution σ provided:

$$\frac{6}{Nu}r + \frac{1}{(2\pi)^d}\int^{\Lambda}\frac{\mathrm{d}^d p}{p^2} < 0. \tag{29.9}$$

The critical value r_c of r is thus:

$$r_c = -\frac{Nu}{6}\frac{1}{(2\pi)^d}\int^{\Lambda}\frac{\mathrm{d}^d p}{p^2}. \tag{29.10}$$

Setting:

$$r = r_c + (u/6)\,\tau\,, \tag{29.11}$$

we can rewrite equation (29.8):

$$\sigma^2 = -\tau \sim (-\tau)^{2\beta} \;\Rightarrow\; \beta = \frac{1}{2}. \tag{29.12}$$

For N large the exponent β remains classical, i.e. mean-field like, in all dimensions.

The high temperature phase. Above T_c, σ vanishes. In expression (29.5) we see that the σ-propagator then becomes

$$\Delta_\sigma = \frac{1}{p^2 + \lambda}. \tag{29.13}$$

Therefore $\lambda^{1/2}$ is at this order the physical mass, i.e. the inverse correlation length ξ^{-1} of the field σ. We thus from now on set

$$\lambda^{1/2} = m. \tag{29.14}$$

Using equations (29.10,29.11) in equation (29.7b) we find:

$$\frac{6}{u} + \frac{N}{(2\pi)^d}\int^{\Lambda}\frac{\mathrm{d}^d p}{p^2\,(p^2 + m^2)} = \frac{\tau}{m^2}. \tag{29.15}$$

(i) For $d > 4$ the integral in (29.15) has a limit for $m = 0$ and therefore at leading order:

$$m^2 = \xi^{-2} \sim \tau \quad\text{and thus}\quad \nu = \tfrac{1}{2}, \tag{29.16}$$

which is the mean field result.

(ii) For $2 < d < 4$ instead, (setting $d = 4 - \varepsilon$) the integral behaves for m small like:

$$\frac{1}{(2\pi)^d} \int^{\Lambda} \frac{d^d p}{p^2 (p^2 + m^2)} = C(d)m^{-\varepsilon} - a(d)\Lambda^{-\varepsilon} + O\left(m^{2-\varepsilon}\Lambda^{-2}\right), \quad (29.17)$$

with

$$C(d) = -\Gamma\left(1 - d/2\right)(4\pi)^{-d/2}. \quad (29.18)$$

The constant $a(d)$ which characterizes the leading correction in equation (29.17), depends on the cut-off procedure.

The l.h.s. of equation (29.15) is thus dominated by the most singular contribution of the integral and at leading order we obtain:

$$m = \xi^{-1} \sim \tau^{1/(2-\varepsilon)}, \quad (29.19)$$

which shows that the exponent ν is not classical:

$$\nu = \frac{1}{2 - \varepsilon} = \frac{1}{d - 2}. \quad (29.20)$$

(iii) For $d = 4$ the l.h.s. is still dominated by the integral:

$$\frac{1}{(2\pi)^4} \int^{\Lambda} \frac{d^4 p}{p^2 (p^2 + m^2)} \underset{m \to 0}{\sim} \frac{1}{8\pi^2} \ln(\Lambda/m).$$

The correlation length has no longer a power law behaviour but instead a mean-field behaviour modified by a logarithm. This is typical of a situation where the gaussian fixed point is stable, in the presence of a marginal operator.

(iv) If we examine equation (29.7b) for $\sigma = 0$ and $d = 2$ we find the correlation length can become large only for $r \to -\infty$. This peculiar situation will be examined when we discuss the non-linear σ-model.

Finally, in the critical limit $\tau = 0$, λ vanishes and thus from the form (29.13) of the σ-propagator we find that the critical exponent η remains classical for all d

$$\eta = 0. \quad (29.21)$$

We verify that the exponents β, ν, η satisfy the scaling relation (26.64).

The scaling equation of state. In a magnetic field H the saddle point equation (29.7a) is modified:

$$m^2 \sigma = H. \quad (29.22)$$

Using equations (29.10,29.11) we can rewrite equation (29.7b):

$$\frac{6}{u}m^2 + \frac{N}{(2\pi)^d}m^2 \int^{\Lambda} \frac{d^d p}{p^2 (p^2 + m^2)} = \sigma^2 + \tau. \quad (29.23)$$

Let us consider only the case $2 < d < 4$. From equation (29.17) we know the behaviour of the l.h.s. for m small. Keeping only the leading term, we can rewrite equation (29.23):

$$NC(d)m^{d-2} = \sigma^2 + \tau. \quad (29.24)$$

Eliminating m between equations (29.22) and (29.24) we finally obtain an equation of state in scaling form (σ is the magnetization):

$$H = h_0 \sigma^\delta f\left(\tau/\sigma^2\right),\tag{29.25}$$

in which h_0 is a normalization constant, δ is given by:

$$\delta = \frac{d+2}{d-2},\tag{29.26}$$

in agreement with the scaling relation (26.65), and the function $f(x)$ by:

$$f(x) = (1+x)^{2/(d-2)}.\tag{29.27}$$

The asymptotic form of $f(x)$ for x large implies $\gamma = 2/(d-2)$ again in agreement with the scaling relation $\gamma = \nu(2-\eta)$. Taking into account the values of the critical exponents γ and β it is then easy to verify that the function f satisfies all required properties like for example Griffith's analyticity (see Section 26.5). In particular the equation of state can be cast into the parametric form of Subsection 28.1.3:

$$\sigma = R^{1/2}\theta,$$
$$\tau = 3R\left(1-\theta^2\right),$$
$$H = h_0 R^{\delta/2}\theta\left(3-2\theta^2\right)^{2/(d-2)}.$$

The specific heat exponent. Differentiating twice action (29.5) with respect to $\lambda(x)$, replacing then the field $\lambda(x)$ by its expectation value m^2, we find the λ-propagator $\Delta_\lambda(p)$ above T_c

$$\Delta_\lambda^{-1}(p) = -\frac{N}{2}\left(\frac{6}{Nu} + \frac{1}{(2\pi)^d}\int^\Lambda \frac{d^d q}{(q^2+m^2)\left[(p-q)^2+m^2\right]}\right).\tag{29.28}$$

The λ-propagator is negative because the λ-field is imaginary. As noted in Section 29.1, it is equal to the ϕ^2 2-point function up to an additive constant. At zero momentum we thus obtain the specific heat. The r.h.s. is dominated for m small by the integral

$$\int^\Lambda \frac{d^d q}{(q^2+m^2)^2} \sim (d/2-1)C(d)m^{-\varepsilon},$$

where $C(d)$ is defined in (29.18). The singular part of the specific heat thus vanishes as m^ε. From the RG analysis we expect $m^{-\alpha/\nu}$. Using equation (29.20) we verify

$$-\alpha/\nu = 4 - d = d - 2/\nu,$$

in agreement with the general relation (26.49).

In the critical theory ($m=0$ at this order) also, for $2 \le d \le 4$ the denominator is dominated at low momentum by the integral, which for $2 < d < 4$ behaves like

$$\frac{1}{(2\pi)^d}\int^\Lambda \frac{d^d q}{q^2(p-q)^2} \underset{2<d<4}{\sim} b(\varepsilon)p^{-\varepsilon},$$

with

$$b\left(\varepsilon\right) = \frac{1}{\left(4\pi\right)^{d/2}}\Gamma\left(2 - d/2\right)\frac{\Gamma^2\left(d/2 - 1\right)}{\Gamma\left(d - 2\right)}, \tag{29.29}$$

and thus:

$$\Delta_\lambda(p) \underset{p\to 0}{\sim} -\frac{2}{Nb(\varepsilon)}p^\varepsilon. \tag{29.30}$$

We again verify consistency with scaling relations. In particular we note that in the large N limit the dimension of the field λ is $[\lambda] = \frac{1}{2}(d - \varepsilon) = 2$, a result important for the $1/N$ perturbation theory.

Remarks.

(i) For $d = 4$ the integral has a logarithmic behaviour and thus still dominates the behaviour of the propagator $\Delta_\lambda \propto 1/\ln(\Lambda/p)$.

(ii) Note therefore that for $d \le 4$ the contributions generated by the term proportional to $\lambda^2(x)$ in (29.5) always are negligible in the critical domain.

29.2.2 Leading corrections to scaling and RG functions

Let us again consider equation (29.15), but now also keep the leading correction to the integral for m small (equation (29.17)). The constant $a(d)$ depends on the specific cut-off procedure, but for ε small satisfies:

$$a(d) \sim \frac{1}{8\pi^2\varepsilon}. \tag{29.31}$$

Equation (29.15) then leads to:

$$m^2 \propto \tau^{1/(2-\varepsilon)}\left[1 + \text{const.}\left(6/Nu - a(d)\Lambda^{-\varepsilon}\right)t^{\varepsilon/(2-\varepsilon)} + \cdots\right]. \tag{29.32}$$

According to our previous analysis of corrections to scaling we identify:

$$\omega\nu = \frac{\varepsilon}{2 - \varepsilon} \qquad \text{or} \qquad \omega = \varepsilon. \tag{29.33}$$

Within the framework of the ε-expansion ω is associated with the leading correction to scaling. In the large N limit ω remains smaller than 2 for $\varepsilon < 2$, extending this property to all dimensions $2 \le d \le 4$. Furthermore we notice that for the special value $u = u^*$:

$$u^* = \frac{6\Lambda^\varepsilon}{Na(d)} = \Lambda^\varepsilon\frac{48\pi^2\varepsilon}{N} + O\left(\varepsilon^2\right), \tag{29.34}$$

the leading correction to the scaling relation (29.19) cancels, together actually with all other corrections decaying like powers proportional to ε.

Remark. For simple regularization schemes or quite generally for d close to four, $a(d)$ is positive and thus u^* is positive as it should. However this is not always the case and it then seems that the leading correction to scaling cannot be cancelled in this way. It is not known whether such a situation can also occur for N finite.

The λ-propagator. In the same way if we keep the leading correction to the λ-propagator in the critical theory we find:

$$\Delta_\lambda(p) = -\frac{2}{N}\left[\frac{6}{Nu} - a(d)\Lambda^{-\varepsilon} + b(\varepsilon)p^{-\varepsilon}\right]^{-1}, \tag{29.35}$$

where we have neglected terms of order Λ^{-2} and $1/N$. The leading corrections to scaling again exactly cancel for $u = u^*$ as expected.

The RG functions. Setting:

$$u = g\Lambda^\varepsilon, \;\Rightarrow\; g^* = u^*\Lambda^{-\varepsilon} = 6/(Na), \qquad \text{and} \quad \tau = ut, \tag{29.36}$$

we calculate the RG functions. The form (29.13) of the σ 2-point function shows that the field is not renormalized at leading order and thus

$$Z_\sigma = 1 \;\Rightarrow\; \eta(g) = 0, \tag{29.37}$$

result consistent with the value (29.21) of the exponent η.

The critical renormalized σ 4-point function at renormalization scale μ is then given by

$$\Gamma_r^{(4)}(p_i \propto \mu) = -3\Delta_\lambda(m = 0, \mu).$$

Differentiating with respect to Λ at fixed renormalized parameters we obtain the β-function (equation (25.42))

$$\beta(g) = -\varepsilon g(1 - g/g^*). \tag{29.38}$$

It is important to note that in (29.35) the relevant term was the first correction to scaling.

We then calculate the $\sigma^2\sigma\sigma$ renormalized correlation function at renormalization scale μ to obtain Z_2 and thus η_2. From the remark (29.6) we find

$$-Z_2\frac{3}{u}\Delta_\lambda(\mu) = 1, \;\Rightarrow\; Z_2\Gamma_r^{(4)}(p_i \propto \mu) = g\Lambda^\varepsilon,$$

and therefore (equation (25.67))

$$\eta_2(g) = -\left.\Lambda\frac{\mathrm{d}}{\mathrm{d}\Lambda}\right|_{\text{ren. fixed}} \ln Z_2 = -\frac{\beta(g)}{g} - \varepsilon = -\varepsilon\frac{g}{g^*}.$$

It follows

$$\nu^{-1}(g) = 2 + \eta_2(g) = 2 - \varepsilon g/g^*, \tag{29.39}$$

Replacing the coupling constant by its IR fixed point value g^* we find results obviously consistent with the equations (29.20, 29.33, 29.34).

It is then easy to verify that m^2 in equation (29.15) is a RG invariant, i.e. satisfies:

$$\left[\Lambda\frac{\partial}{\partial\Lambda} + \beta(g)\frac{\partial}{\partial g} - \eta_2(g)\,t\frac{\partial}{\partial t}\right] m(t, g, \Lambda) = 0. \tag{29.40}$$

Remarks.

(i) Again an IR fixed point g^* exists only if $a(d)$ is positive. Otherwise the RG method fails, at least in this simple form.

(ii) In Chapter 27 we have argued that a perturbation due to irrelevant operators, is equivalent, at leading order in the critical region, to a modification of ϕ^4 coupling. This can be explicitly verified here. The amplitude of the leading correction to scaling has been found to be proportional to $6/Nu - a(d)\Lambda^{-\varepsilon}$ where the value of $a(d)$ depends on the cut-off procedure and thus of contributions of irrelevant operators. Let us call u' the ϕ^4

coupling constant in another scheme where a is replaced by a'. Identifying the leading correction to scaling we find the relation:

$$\frac{6\Lambda^\varepsilon}{Nu} - a(d) = \frac{6\Lambda^\varepsilon}{Nu'} - a'(d),$$

homographic relation which is consistent with the special form (29.38) of the β-function.

The renormalized coupling constants. Let us calculate the renormalized coupling constants both in the massive theory renormalized at zero momentum and the massless theory renormalized at a symmetric point defined by $(p_1 + p_2)^2 = \mu^2$. Keeping the leading correction to scaling we find respectively

$$\frac{1}{g_{\mathrm{r}}(0)} = \frac{N}{12}(d - 2)C(d) + \left(\frac{1}{g} - \frac{1}{g^*}\right)\left(\frac{m}{\Lambda}\right)^\varepsilon$$

$$\frac{1}{g_{\mathrm{r}}(\mu)} = \frac{N}{6}b(\varepsilon) + \left(\frac{1}{g} - \frac{1}{g^*}\right)\left(\frac{\mu}{\Lambda}\right)^\varepsilon.$$

We recover that for generic values of g the renormalized coupling constants are fixed at their IR fixed point value, up to corrections to the leading scaling behaviour. They significantly deviate from the fixed point value, which then provides an upper-bound, only for non generic small values of g.

29.3 The $1/N$ Expansion

Higher order terms in the steepest descent calculation of the functional integral (29.4) generate a systematic $1/N$ expansion. Let us first slightly rewrite action (29.5). We shift the field $\lambda(x)$ by its expectation value m^2 (equation (29.14)), $\lambda(x) \mapsto m^2 + \lambda(x)$:

$$S_{\mathrm{eff}}\left(\sigma, \lambda\right) = \frac{1}{2}\int \mathrm{d}^d x \left[(\partial_\mu \sigma)^2 + m^2 \sigma^2 + \lambda(x)\sigma^2(x) - \frac{3}{u}\lambda^2(x) - \frac{6}{u}\left(m^2 - r\right)\lambda(x)\right]$$

$$+ \frac{(N-1)}{2}\,\mathrm{tr}\ln\left[-\Delta + m^2 + \lambda(x)\right]. \tag{29.41}$$

Power counting. Neglecting all irrelevant terms, we can analyze the action (29.41) from the point of view of power counting. The dimension of the field $\sigma(x)$ is $(d-2)/2$. From the critical behaviour (29.30) of the λ-propagator we have deduced the engineering dimension $[\lambda]$ of the field $\lambda(x)$:

$$2\,[\lambda] - \varepsilon = d \qquad \text{i.e.} \quad [\lambda] = 2\,.$$

The interaction term $\int \lambda(x)\sigma^2(x)\mathrm{d}^d x$ has thus dimension zero. It is easy to verify that the non-local interactions involving the λ-field coming from the expansion of the $\mathrm{tr}\ln$ have all the canonical dimension zero:

$$\left[\mathrm{tr}\left[\lambda(x)\left(-\Delta + m^2\right)^{-1}\right]^k\right] = k\,[\lambda] - 2k = 0\,.$$

This power counting property has the following implication: In contrast with usual perturbation theory, the $1/N$ expansion generates only logarithmic corrections to the leading long distance behaviour for any fixed dimension d, $2 < d \leq 4$. The situation is thus similar to the situation one encounters for the ε-expansion and one expects to be able to calculate universal quantities like critical exponents for example as power series in $1/N$. However, because the interactions are non-local, the results of standard renormalization theory do not immediately apply. We devote Subsection 29.3.2 to a thorough discussion of this problem.

The non-linear σ-model. We have already noticed that the term proportional to $\int \mathrm{d}^d x\, \lambda^2(x)$, which has dimension $4 - d$ for large N in all dimensions, is irrelevant in the critical domain for $d < 4$ and can thus be omitted at leading order (this also applies to $d = 4$ where it is marginal but yields only logarithmic corrections). Actually the constant part in the inverse propagator as written in equation (29.35) plays the role of a large momentum cut-off. Let us thus consider the action (29.41) without the λ^2 term. If we then work backwards, reintroduce the initial field ϕ and integrate over $\lambda(x)$ we find

$$Z = \int [\mathrm{d}\phi(x)]\, \delta\left[\phi^2(x) - \frac{6}{u}\left(m^2 - r\right)\right] \exp\left[-\int \frac{1}{2}\left(\partial_\mu \phi(x)\right)^2 \mathrm{d}^d x\right]. \qquad (29.42)$$

Under this form we recognize the partition function of the $O(N)$ symmetric non-linear σ-model in an unconventional normalization. We have therefore discovered a remarkable correspondence: to all orders in an $1/N$ expansion the renormalized non-linear σ-model is identical to the renormalized $\left(\phi^2\right)^2$ field theory at the IR fixed point. In Section 30.7 we shall solve the σ-model in the large N limit to show more explicitly the relation between the different parametrizations.

Analyticity in the coupling constant and the large momentum expansion. We have seen that, in the framework at the $1/N$ expansion, we can calculate at fixed dimension $d < 4$ in the critical limit ($T = T_c, m^2 = 0$). This implies that the terms of the $1/N$ expansion cannot be expanded in a power series of the coupling constant, at least with integer powers. Note that since the gaussian fixed point is an UV fixed point, the small coupling expansion is also a large momentum expansion. To understand the phenomenon we consider the σ 2-point function at order $1/N$. At this order only one diagram has to be evaluated containing two $\lambda^2 \sigma$ vertices. In the large cut-off limit and after mass renormalization we find:

$$\Gamma_{\sigma\sigma}^{(2)}(p) = p^2 + \frac{2}{N(2\pi)^d} \int \frac{\mathrm{d}^d q}{(6/Nu) + b(\varepsilon)q^{-\varepsilon}} \left(\frac{1}{(p+q)^2} - \frac{1}{q^2}\right) + O\left(\frac{1}{N^2}\right). \qquad (29.43)$$

If we study the integral we find that it has an expansion of the form

$$\sum_k \alpha_k u^k p^{2-k\varepsilon} + \beta_k u^{(2+2k)/\varepsilon} p^{-k}. \qquad (29.44)$$

The terms with integer powers of u can be obtained by expanding the integral and calculating each term by dimension regularization. α_k has poles at $\varepsilon = (2l + 2)/k$ for which the corresponding power of p^2 is $-l$, i.e. an integer. One verifies that β_l has a pole at the same value of ε and that the singular contributions cancel in the sum.

29.3.1 The exponent η at order $1/N$

For illustration purpose let us calculate the exponent η at order $1/N$. We can obtain it from the σ 2-point function at $1/N$ order, at T_c and at the IR fixed point. Expression (29.43) becomes:

$$\Gamma^{(2)}_{\sigma\sigma}(p) = p^2 + \frac{2}{Nb(\varepsilon)} \int \frac{d^d q}{(2\pi)^d} q^\varepsilon \left[\frac{1}{(p+q)^2} - \frac{1}{q^2} \right] + O\left(\frac{1}{N^2}\right). \tag{29.45}$$

As explained above, after mass renormalization the q-integral is logarithmically divergent in any dimension. We assume a cut-off Λ whose precise form is irrelevant at this order. The integral then behaves for large cut-off or small momentum like:

$$\int \frac{d^d q}{(2\pi)^d} q^\varepsilon \left[(p+q)^{-2} - q^{-2} \right] \sim A p^2 \ln(\Lambda/p). \tag{29.46}$$

Comparing with the expansion of the 2-point function at the IR fixed point:

$$\Gamma^2_\sigma(p) \sim p^{2-\eta} \sim p^2 \left(1 - \eta \ln p + O\left(1/N^2\right) \right), \tag{29.47}$$

we conclude:

$$\eta = 2A/(Nb(\varepsilon)).$$

Expression (29.46) shows that the coefficient of $\ln p$ can be obtained by evaluating the divergent part of the integral. The latter contribution corresponds to large values of the integration momentum q and is thus regular in p. We can thus expand the integral for p small. We find:

$$\int \frac{d^d q}{(2\pi)^d} q^\varepsilon \left[(p+q)^{-2} - q^{-2} \right] \sim p^2 \int^\Lambda \frac{d^d q}{q^d (2\pi)^d} \left(\frac{4(p\cdot q)^2}{p^2 q^2} - 1 \right). \tag{29.48}$$

Using rotational invariance we can substitute:

$$\frac{4(p\cdot q)^2}{p^2 q^2} \mapsto \frac{4}{d}.$$

The divergent part of the integral is then:

$$\int \frac{d^d q}{(2\pi)^d} q^\varepsilon \left[(p+q)^{-2} - q^{-2} \right] \sim \frac{2\varepsilon}{4-\varepsilon} \frac{1}{(4\pi)^{d/2}\Gamma(d/2)} p^2 \ln \Lambda. \tag{29.49}$$

The value of the exponent η follows. It is convenient to set:

$$X_1 = \frac{2\Gamma(d-2)}{\Gamma(d/2)\Gamma(2-d/2)\Gamma^2(d/2-1)} = \frac{4\sin(\pi\varepsilon/2)\Gamma(2-\varepsilon)}{\pi\Gamma(1-\varepsilon/2)\Gamma(2-\varepsilon/2)}. \tag{29.50}$$

Then:

$$\eta = \frac{\varepsilon}{N(4-\varepsilon)} X_1 + O\left(\frac{1}{N^2}\right). \tag{29.51}$$

29.3.2 An alternative critical theory

To be able to use the standard results of renormalization theory we reformulate the critical theory to deal with the non-local interactions. Omitting for $d < 4$ the irrelevant term proportional to $\int \lambda^2(x) d^d x$ in the effective action (29.41), and reintroducing the field $\phi(x)$, we can write the functional integral:

$$Z = \int [d\lambda(x)]\,[d\phi(x)] \exp\left[-S\left(\phi, \lambda\right)\right],\tag{29.52}$$

$$S\left(\phi, \lambda\right) = \int d^d x \left[\frac{1}{2}\left(\partial_\mu \phi\right)^2 + \frac{1}{2}m^2\phi^2 + \frac{1}{2}\lambda\phi^2 - \frac{3}{u}\left(m^2 - r\right)\lambda\right].\tag{29.53}$$

We now replace the action by a modified action S_v obtained by adding a term quadratic in λ which at tree level yields a λ- propagator of the form (29.30):

$$S_v\left(\phi, \lambda\right) = \int d^d x \left[\frac{1}{2}\left(\partial_\mu \phi\right)^2 + \frac{1}{2}m^2\phi^2 + \frac{1}{2}\lambda\phi^2 + \frac{1}{2v^2}\lambda(-\partial^2)^{-\varepsilon/2}\lambda - c\lambda\right],\tag{29.54}$$

where we have set $c = 3(m^2 - r)/u$. In the limit $v \to \infty$ the initial action is recovered. However, now standard field theory methods can be applied. It is convenient to rescale λ into $v\lambda$. The renormalized critical action then reads:

$$[S_v]_{\text{ren}} = \frac{1}{2}\int d^d x \left[Z_\phi \left(\partial_\mu \phi\right)^2 + \delta m^2 \phi^2 + v_r Z_v \lambda \phi^2 + \lambda(-\partial^2)^{-\varepsilon/2}\lambda - 2c_r\lambda\right].\tag{29.55}$$

Note that in generic dimensions the λ-field is not renormalized because counterterms are local while the λ inverse propagator is non-local. This situation is also encountered in systems with long range forces. It follows that the RG equations for the critical theory take the form:

$$\left[\Lambda\frac{\partial}{\partial\Lambda} + \beta_{v^2}(v)\frac{\partial}{\partial v^2} - \frac{n}{2}\eta(v)\right]\Gamma^{(l,n)} = 0.\tag{29.56}$$

We can then calculate the RG functions as power series in $1/N$. It is easy to verify that v^2 has to be taken of order $1/N$. Therefore to generate a $1/N$ expansion one has first to sum the multiple insertions of one-loop λ 2-point function contributions which form a geometrical series. From the calculation of the $\phi\phi$ and $\phi\phi\lambda$ correlation functions at order $1/N$, we derive

$$\eta(v) = -\frac{\varepsilon v^2}{4 - \varepsilon}X_1 D^{-1}(v),\tag{29.57}$$

$$\beta_{v^2}(v) = -\frac{8v^4}{4 - \varepsilon}X_1 D^{-1}(v) + 4N(1 - \varepsilon)v^6 X_1 D^{-2}(v),\tag{29.58}$$

with

$$D(v) = 2/b(\varepsilon) - Nv^2.$$

The solution to the RG equations (29.56) can be written:

$$\Gamma^{(l,n)}(\tau p, v, \Lambda) = Z^{-n/2}(\tau)\tau^{d-2l-n(d-2)/2}\Gamma^{(l,n)}(p, v(\tau), \Lambda),\tag{29.59}$$

with the usual definitions

$$\tau\frac{dv^2}{d\tau} = \beta(v(\tau)), \qquad \tau\frac{d\ln Z}{d\tau} = \eta(v(\tau)).$$

We are interested in the neighbourhood of the fixed point $v^2 = \infty$. Then the RG function $\eta(v)$ approaches the exponent η obtained by direct calculation. The flow equation for the coupling constant becomes:

$$\tau \frac{\mathrm{d}v^2}{\mathrm{d}\tau} = \rho v^2,$$

with

$$\rho = \frac{4(3 - \varepsilon)(2 - \varepsilon)}{4 - \varepsilon} X_1 .$$

Thus

$$v^2(\tau) \sim \tau^\rho,$$

We then note that the $\lambda\lambda$ 2-point function is proportional to v^2 for v large

$$\Gamma_{\lambda\lambda}(p) = p^{-\varepsilon} D(v) + O\left(N^{-1}\right) \sim -Nv^2 p^{-\varepsilon}.$$

It follows that, although the field λ is not renormalized, still the power law is modified. We find

$$\Gamma_{\lambda\lambda}(\tau p) \sim (\tau p)^{-\varepsilon} v^2(\tau) \sim \tau^{\alpha/\nu},$$

with

$$\frac{\alpha}{\nu} = -\varepsilon + \frac{4(2 - \varepsilon)(3 - \varepsilon)}{(4 - \varepsilon)N} X_1 + O\left(\frac{1}{N^2}\right).$$

29.3.3 Other results

The calculations beyond the order $1/N$ are rather technical. The reason is easy to understand: Because the effective field theory is renormalizable in all dimensions $2 \leq d \leq 4$, the dimensional regularization which was so useful for practical calculations no longer works. Therefore either one keeps a true cut-off or one introduces new more sophisticated regularization schemes. For details the reader is referred to the literature.

Generic dimensions. The exponents γ and η are known up to order $1/N^2$ and $1/N^3$ respectively in arbitrary dimensions but the expressions are too complicated to be reproduced here. We give instead the expansions of γ and ν only up to order $1/N$, which can be directly deduced from the results of the preceding subsections:

$$\gamma = \frac{1}{1 - \varepsilon/2}\left(1 - \frac{3}{2N}X_1\right) + O\left(\frac{1}{N^2}\right), \tag{29.60}$$

$$\nu = \frac{1}{2 - \varepsilon}\left(1 - \frac{2(3 - \varepsilon)}{(4 - \varepsilon)N}X_1\right) + O\left(\frac{1}{N^2}\right), \tag{29.61}$$

The exponents ω and $\theta = \omega\nu$, governing the leading corrections to scaling, have also been calculated:

$$\omega = \varepsilon\left(1 - \frac{2(3 - \varepsilon)^2}{(4 - \varepsilon)N}X_1\right) + O\left(\frac{1}{N^2}\right), \tag{29.62}$$

$$\theta = \omega\nu = \frac{\varepsilon}{2 - \varepsilon}\left(1 - \frac{2(3 - \varepsilon)}{N}X_1\right) + O\left(\frac{1}{N^2}\right). \tag{29.63}$$

Note that the exponents are regular functions of ε up to $\varepsilon = 2$ and free of renormalon singularities at $\varepsilon = 0$ (see Section 42.6).

The equation of state and the spin–spin correlation function in zero field are also known at order $1/N$, but since the expressions are complicated we refer the reader to the literature for details.

Three dimensional results. Let us give the expansion of η in three dimensions at the order presently available:

$$\eta = \frac{\eta_1}{N} + \frac{\eta_2}{N^2} + \frac{\eta_3}{N^3} + O\left(\frac{1}{N^4}\right),$$

with

$$\eta_1 = \tfrac{8}{3\pi^2}, \quad \eta_2 = -\tfrac{8}{3}\eta_1^2, \quad \eta_3 = \eta_1^3\left[-\tfrac{797}{18} - \tfrac{61}{24}\pi^2 + \tfrac{27}{8}\psi''(1/2) + \tfrac{9}{2}\pi^2\ln 2\right],$$

$\psi(x)$ being the logarithmic derivative of the Γ function.

The exponent γ is known only up to order $1/N^2$:

$$\gamma = 2 - \tfrac{24}{N\pi^2} + \tfrac{64}{N^2\pi^4}\left(\tfrac{44}{9} - \pi^2\right) + O\left(\tfrac{1}{N^3}\right).$$

Note that the $1/N$ expansion seems to be rapidly divergent and certainly a direct summation of these terms does not provide very good estimates of critical exponents in 3 dimensions for useful values of N.

29.4 Generalizations

We now briefly explain how the algebraic method presented in Section 29.1 can be generalized to actions which have a more complicated dependence in one or several vector fields. In Chapter 24 we have used an identity similar to the identity (29.3) to construct the mean field expansion in the special case of two-body interactions. We have shown in Appendix A24 how this method can be generalized. An analogous strategy also works here. In the case of a general $O(N)$ symmetric field theory, we introduce pairs of fields and Lagrange multipliers for all independent $O(N)$ invariant scalar products constructed from the many-component fields.

Let us first take the example of one field ϕ and assume that the interaction is an arbitrary function of the only invariant $\phi^2(x)$

$$S(\phi) = \int d^d x \left\{ \tfrac{1}{2}[\partial_\mu \phi(x)]^2 + V(\phi^2) \right\}. \tag{29.64}$$

We then introduce two fields $\rho(x)$ and $\lambda(x)$ and use the identity:

$$\exp\left[-\int d^d x\, V(\phi^2)\right] \propto \int [d\rho(x)\, d\lambda(x)] \exp\left\{-\int d^d x\left[\tfrac{1}{2}\lambda(\phi^2 - \rho) + V(\rho)\right]\right\}. \tag{29.65}$$

In the special case in which $V(\rho)$ is a quadratic function, the integral over ρ can be performed. In all cases, however, the identity (29.65) transforms the action into a quadratic form in ϕ and therefore the integration over ϕ can be performed and the dependence in N becomes explicit. The effective action is then:

$$S_{\text{eff}} = \int d^d x \left[\tfrac{1}{2}(\partial_\mu \sigma)^2 + V(\rho) + \tfrac{1}{2}\lambda(\sigma^2 - \rho)\right] + \tfrac{1}{2}(N-1)\,\text{tr}\ln(-\Delta + \lambda), \tag{29.66}$$

and the saddle point equations become:

$$\lambda\sigma = 0\,, \quad V'(\rho) = \tfrac{1}{2}\lambda\,, \quad \sigma^2 = \rho - \frac{N}{(2\pi)^d}\int\frac{\mathrm{d}^d p}{p^2+\lambda}\,. \qquad (29.67)$$

The solutions of these equations can be studied in the same spirit as equation $(29.7b)$.

If the action is an $O(N)$ invariant function of two fields ϕ_1 and ϕ_2 the potential depends on the three scalar products $\phi_1\cdot\phi_2$, ϕ_1^2 and ϕ_2^2. Then three pairs of fields are required.

Bibliographical Notes

As shown by Stanley the large N-limit of the classical N-vector model coincides with the spherical model solved by Berlin and Kac

T.H. Berlin and M. Kac, *Phys. Rev.* 86 (1952) 821; H.E. Stanley, *Phys. Rev.* 176 (1968) 718.

Early work on calculating critical properties includes

R. Abe, *Prog. Theor. Phys.* 48 (1972) 1414; 49 (1973) 113, 1074, 1877; S.K. Ma, *Phys. Rev. Lett.* 29 (1972) 1311; *Phys. Rev.* A7 (1973) 2172; M. Suzuki, *Phys. Lett.* 42A (1972) 5; *Prog. Theor. Phys.* 49 (1973) 424, 1106, 1440; R.A. Ferrel and D.J. Scalapino, *Phys. Rev. Lett.* 29 (1972) 413; K.G. Wilson, *Phys. Rev.* D7 (1973) 2911.

The equation of state to order $1/N$ is derived in

E. Brézin and D.J. Wallace, *Phys. Rev.* B7 (1973) 1967.

The spin–spin correlation in zero field is obtained in

M.E. Fisher and A. Aharony, *Phys. Rev. Lett.* 31 (1973) 1238; A. Aharony, *Phys. Rev.* B10 (1974) 2834; R. Abe and S. Hikami, *Prog. Theor. Phys.* 51 (1974) 1041.

The exponent ω has been calculated to order $1/N$ in

S.K. Ma, *Phys. Rev.* A10 (1974) 1818.

See also the contributions of S.K. Ma and E. Brézin, J.C. Le Guillou and J. Zinn-Justin to

Phase Transitions and Critical Phenomena vol. 6, C. Domb and M.S. Green eds. (Academic Press, London 1976).

The consistency of the $1/N$ expansion to all orders has been proven in

I. Ya Aref'eva, E.R. Nissimov and S.J. Pacheva, *Commun. Math. Phys.* 71 (1980) 213; A.N. Vasil'ev and M.Yu. Nalimov, *Teor. Mat. Fiz.* 55 (1983) 163.

For a recent review see

J. Zinn-Justin, *Vector models in the large N limit: a few applications*, hep-th/9810198.

At present the longest $1/N$ series for exponents and amplitudes are found in

I. Kondor and T. Temesvari, *J. Physique Lett. (Paris)* 39 (1978) L99; Y. Okabe and M. Oku, *Prog. Theor. Phys.* 60 (1978) 1277, 1287; 61 (1979) 443; A.N. Vasil'ev, Yu.M. Pis'mak and Yu.R. Honkonen, *Teor. Mat. Fiz.* 46 (1981) 157; 50 (1982) 195.

See also

I. Kondor, T. Temesvari and L. Herenyi, *Phys. Rev.* B22 (1980) 1451.

Renormalization of operators is discussed in

K. Lang and W. Rühl, *Nucl. Phys.* B400 (1993) 597; *Z. Phys.* C61 (1994) 459.

The case of long range forces has been discussed in

S.K. Ma, *Phys. Rev.* A7 (1973) 2172.

Exercises

Exercise 29.1

Discuss the logarithmic corrections to mean field theory at the upper-critical dimension $d = 4$ in the large N expressions of Chapter 29. More specifically, consider the correlation length, the equation of state and the specific heat.

Exercise 29.2

Use the expressions of Section 29.4 to generalize the previous study to a potential $V(\phi^2)$ and look for a tricritical point.

30 FERROMAGNETIC ORDER AT LOW TEMPERATURE: THE NON-LINEAR σ-MODEL

In Chapters 24–26 we have studied the critical properties of lattice models described in terms of N-vector spin variables $\mathbf{S}_i$ of unit length on each lattice site i, interacting through a short range ferromagnetic $O(N)$ symmetric two-body interaction V_{ij}. In zero field the partition function of such models can be written:

$$Z = \int \prod_i d\mathbf{S}_i \, \delta \left(\mathbf{S}_i^2 - 1 \right) \exp \left[-\mathcal{H} \left(\mathbf{S} \right) / T \right], \qquad (30.1)$$

in which the configuration energy $\mathcal{H}$ is:

$$\mathcal{H} \left(\mathbf{S} \right) / T = - \sum_{ij} V_{ij} \mathbf{S}_i \cdot \mathbf{S}_j / T \,. \qquad (30.2)$$

To derive the critical properties of the N-vector model we have proceeded in the following way: we have first found an approximation for the partition function, the mean field approximation, valid in high dimensions. We have characterized the critical properties of all physical quantities within this approximation scheme. We have then shown that the mean field approximation was the first term in a systematic expansion. Examining the first correction to mean field we have discovered the role of the dimension four, dimension at which the validity of mean field theory ceases. We have finally explained how a summation of the leading IR divergences to all orders in the expansion leads to an effective ϕ^4 field theory whose critical properties can be analyzed by RG methods in $4 - \varepsilon$ dimension, i.e. near the so-called upper-critical dimension. The results one derives in this way have been presented in Chapters 25, 26.

Here, instead, we discuss these models from the point of view of a low temperature expansion. The methods we employ, however, apply only to continuous symmetries, here to $N \geq 2$. They rely on the property that models with continuous symmetries, in contrast to models with discrete symmetries, have a non-trivial long distance physics at any temperature below T_c, due to the massless Goldstone modes.

We first prove universal properties of the low temperature, ordered, phase at fixed temperature. Then, in the non-abelian case, $N > 2$, we show that additional information about critical properties can be obtained, by analyzing the instability of the ordered phase at low temperature and near two dimensions, due to Goldstone mode interactions. The special role of two dimensions, the lower-critical dimension, in models with continuous symmetries has already been stressed in Chapter 23.

The analysis is based on the following observation: The N-vector model (30.1,30.2) can be considered as a lattice regularization of the non-linear σ-model introduced in Chapter 14. The low temperature expansion of the lattice model is the perturbative expansion of the regularized field theory (Subsection 14.3.2). The formal properties of the field theory have been discussed in Chapter 14. We have explained that it is renormalizable in dimension two. RG equations, valid in two and more generally $2 + \varepsilon$ dimension follow. Their solutions will help us to understand the long distance behaviour of correlation functions.

It is somewhat surprising that two different continuum field theories, the $(\phi^2)^2$ and the non-linear σ-model describe the long distance physics of the same lattice model. Actually we have already noted in Section 29.3, that within the framework of the $1/N$-expansion both field theories are directly related. This property, totally mysterious at the classical level, emphasizes the essential nature of quantum fluctuations.

In the appendix we briefly discuss a model of self-interacting fermions, the Gross–Neveu model, which shares some of these properties.

Before we begin our general discussion, let us here confirm by a different approach that at fixed temperature $T < T_c$, the non-linear σ-model also emerges directly in the analysis of the large distance behaviour in the ordered phase of the $\left[(\phi^2)\right]^2$ field theory.

30.1 The $\left[\phi^2(x)\right]^2$ Field Theory at Low Temperature

At low temperature, i.e. T fixed, $T < T_c$, in a system in which a discrete symmetry is spontaneously broken the connected correlation functions decrease exponentially. The situation is quite different, as we have already noted, when the symmetry is continuous because Goldstone modes are present. Let us consider the $(\phi^2)^2$ field theory:

$$S\left(\phi\right) = \int \mathrm{d}^d x \left[\tfrac{1}{2} \left(\partial_\mu \phi\right)^2 + \tfrac{1}{2} r \phi^2 + \frac{1}{4!} u \left(\phi^2\right)^2 \right], \qquad (30.3)$$

in the low temperature phase ($r < r_c$). We change variables in the functional integral:

$$Z = \int [\mathrm{d}\phi] \exp[-S(\phi)],$$

setting

$$\phi(x) = \rho(x)\hat{\phi}(x) \quad \text{with} \quad \hat{\phi}^2(x) = 1. \qquad (30.4)$$

The functional integral becomes:

$$Z = \int \left[\rho^{N-1}(x)\mathrm{d}\rho\left(x\right)\right] \left[\mathrm{d}\hat{\phi}(x)\right] \exp\left[-S\left(\rho, \hat{\phi}\right)\right], \qquad (30.5)$$

with:

$$S\left(\rho, \hat{\phi}\right) = \int \mathrm{d}^d x \left\{ \tfrac{1}{2}\rho^2(x) \left[\partial_\mu \hat{\phi}\left(x\right)\right]^2 + \tfrac{1}{2}\left[\partial_\mu \rho(x)\right]^2 + \tfrac{1}{2}r\rho^2 + \frac{1}{4!}u\rho^4 \right\}. \qquad (30.6)$$

In the ordered phase, below T_c, the field $\rho(x)$ has a non-zero expectation value and is massive; its dynamics is therefore not critical. The integration over the $\rho(x)$ field generates an effective local action $S_{\text{eff}}(\hat{\phi})$ for the field $\hat{\phi}$:

$$\exp\left[-S_{\text{eff}}\left(\hat{\phi}\right)\right] = \int \left[\rho^{N-1}(x)\mathrm{d}\rho(x)\right] \exp\left[-S\left(\rho, \hat{\phi}\right)\right]. \qquad (30.7)$$

Moreover the functional integral (30.7) can be calculated perturbatively. Let us call M the expectation value of $\rho(x)$ at the tree level:

$$\rho(x) = M + \rho'(x). \qquad (30.8)$$

In terms of ρ' the action (30.6) reads:

$$S\left(\rho', \hat{\phi}\right) = \int \mathrm{d}^d x \left\{ \tfrac{1}{2} \left(M^2 + 2M\rho' + \rho'^2 \right) \left[\partial_\mu \hat{\phi}(x) \right]^2 + \tfrac{1}{2} \left[\partial_\mu \rho'(x) \right]^2 \right.$$
$$\left. + \tfrac{1}{2} r \left(M + \rho' \right)^2 + \frac{1}{4!} u \left(M + \rho' \right)^4 \right\}.$$

Neglecting all fluctuations of the field ρ', we obtain S_{eff} at leading order:

$$S_{\mathrm{eff}}^{(0)} \left(\hat{\phi} \right) = \tfrac{1}{2} M^2 \int \mathrm{d}^d x \left[\partial_\mu \hat{\phi}(x) \right]^2. \tag{30.9}$$

We recognize the classical action of the non-linear σ-model studied in Chapter 14.

Loop corrections coming from the integration over ρ' renormalize the coefficient M^2 in (30.9). They also generate additional ϕ-interactions. However, as long as we explore momenta much smaller than the ρ-mass or distances much larger than the corresponding correlation length, the effective action resulting from the integration over the ρ-field can be expanded in local terms. The term dominant at large distance is the term with only two derivatives. Due to the $O(N)$ symmetry it is proportional to the leading order effective action $S_{\mathrm{eff}}^{(0)}(\hat{\phi})$ and renormalizes only the coefficient M^2 which appears in equation (30.9). The other interactions have four derivatives or more and correspond to irrelevant operators. Note that for a temperature T close to T_c, the domain in momentum space in which these arguments apply is

$$|p_i| \ll (T_c - T)^\nu.$$

In such a limit the non-linear σ-model (30.9) completely describes the long distance properties of the $\left(\phi^2\right)^2$ field theory at T fixed, $T < T_c$. In addition the coefficient in front of the effective action becomes large at low temperature like in the lattice model.

30.2 IR Divergences, Spontaneous Symmetry Breaking and the Role of Dimension Two

In Chapter 14 we have shown that the perturbative phase of the non-linear σ model is automatically a phase in which the $O(N)$ symmetry is spontaneously broken, and $(N-1)$ components of $\mathbf{S}_i$, called hereafter π_i, are massless Goldstone modes. This leads to IR divergences for $d \leq 2$ which we reexamine in the context of phase transitions.

(i) We have argued in Section 23.5 that in the N-vector model, for $d > 2$, the $O(N)$ symmetry is spontaneously broken at low temperature. This argument is consistent with the property that for $d > 2$ perturbation theory which also predicts spontaneous symmetry breaking (SSB), is not IR divergent. At $T < T_c$ fixed, the large distance behaviour of the theory is dominated by the massless or spin wave excitations. On the other hand nothing can be said, in perturbation theory, of a possible critical region $T \sim T_c$.

(ii) For $d \leq 2$ we know from the Mermin–Wagner theorem (see also the analysis of Section 23.5) that SSB with ordering ($\langle \mathbf{S} \rangle \neq 0$) is impossible in a model with a continuous symmetry and short range forces, and this is again consistent with the appearance of IR divergences in perturbation theory. For $d \leq 2$ the critical temperature T_c vanishes and perturbation theory makes sense only in presence of an IR cut-off which breaks explicitly the symmetry and orders the spins. Therefore nothing can be said about the long distance properties of the unbroken theory directly from perturbation theory.

To go somewhat beyond perturbation theory we shall use field theory RG methods. It is therefore necessary to first define the model in the dimension in which it is renormalizable. Because the non-linear σ-model is renormalizable in two dimensions IR divergences have to be dealt with. We therefore introduce an IR cut-off in the form of a magnetic field. We then proceed in formal analogy with the case of the $(\phi^2)^2$ field theory, i.e. study the theory in $2 + \varepsilon$ dimension as a double series expansion in the temperature T and ε. In this way the perturbative expansion is renormalizable and RG equations follow.

30.3 The Non-Linear σ-Model: RG Equations

By three different methods we have shown that the large distance physics of the N-vector model can be described below T_c by the non-linear σ-model. We now study this model from the point of view of renormalization and renormalization group. To generate perturbation theory we parametrize the field $\hat{\phi}$ of previous section by (see Section 14.1):

$$\hat{\phi} = \begin{cases} \sigma(x) \\ \boldsymbol{\pi}(x) \end{cases}$$

and eliminate locally the field $\sigma(x)$ by:

$$\sigma(x) = \left(1 - \boldsymbol{\pi}^2(x)\right)^{1/2}.$$

As we have done for the $(\phi^2)^2$ model, we scale all distances in order to measure momenta in units of the inverse lattice spacing Λ. We then consider the partition function in an external magnetic field h:

$$Z(h) = \int \left[\left(1 - \boldsymbol{\pi}^2(x)\right)^{-1/2} \mathrm{d}\boldsymbol{\pi}(x)\right] \exp\left[-S(\boldsymbol{\pi}, h)\right], \tag{30.10}$$

with

$$S(\boldsymbol{\pi}, h) = \frac{\Lambda^{d-2}}{t} \int \mathrm{d}^d x \left\{ \frac{1}{2} \left[(\partial_\mu \boldsymbol{\pi}(x))^2 + \frac{(\boldsymbol{\pi} \cdot \partial_\mu \boldsymbol{\pi}(x))^2}{1 - \boldsymbol{\pi}^2(x)} \right] - h\sqrt{1 - \boldsymbol{\pi}^2(x)} \right\}. \tag{30.11}$$

The properties of this action have been studied, in slightly different notations, in Chapter 14. We here directly borrow the results. We have shown that it is renormalizable in two dimensions and characterized the form of the renormalized action. We recall that the presence of a non-vanishing magnetic field h is required, for $d \leq 2$, both to select the classical minimum of the action around which to expand perturbation theory and to provide the theory with an IR cut-off. The renormalized action has the form:

$$S_{\mathrm{r}}(\boldsymbol{\pi}_{\mathrm{r}}) = \frac{\mu^{d-2} Z}{2 t_{\mathrm{r}} Z_t} \int \mathrm{d}^d x \left[(\partial_\mu \boldsymbol{\pi}_{\mathrm{r}})^2 + (\partial_\mu \sigma_{\mathrm{r}})^2 \right] - \frac{\mu^{d-2}}{t_{\mathrm{r}}} h_{\mathrm{r}} \int \sigma_{\mathrm{r}}(x) \mathrm{d}^d x, \tag{30.12}$$

in which μ is the renormalization scale and:

$$\sigma_{\mathrm{r}}(x) = \left[Z^{-1} - \boldsymbol{\pi}_{\mathrm{r}}^2 \right]^{1/2}. \tag{30.13}$$

Note that the renormalization constants can be chosen h independent. This is automatically realized in the minimal subtraction scheme.

The relation:

$$\pi_{\mathrm{r}}(x) = Z^{-1/2}\pi(x), \tag{30.14}$$

implies

$$\mu^{d-2}\frac{h_{\mathrm{r}}}{t_{\mathrm{r}}} = \Lambda^{d-2}Z^{1/2}\frac{h}{t}. \tag{30.15}$$

With our conventions the coupling constant, which is proportional to the temperature, is dimensionless. The relation between the cut-off dependent and the renormalized correlation functions is:

$$Z^{n/2}\left(\Lambda/\mu, t\right)\Gamma^{(n)}\left(p_i; t, h, \Lambda\right) = \Gamma^{(n)}_{\mathrm{r}}\left(p_i; t_{\mathrm{r}}, h_{\mathrm{r}}, \mu\right). \tag{30.16}$$

Differentiating with respect to Λ at renormalized parameters fixed, we obtain the RG equations:

$$\left[\Lambda\frac{\partial}{\partial\Lambda} + \beta(t)\frac{\partial}{\partial t} - \frac{n}{2}\zeta(t) + \rho(t)h\frac{\partial}{\partial h}\right]\Gamma^{(n)}\left(p_i; t, h, \Lambda\right) = 0. \tag{30.17}$$

We have assumed that the renormalization constants, and thus the RG functions defined by:

$$\Lambda\frac{\partial}{\partial\Lambda}\bigg|_{\text{ren. fixed}} t = \beta(t),$$

$$\Lambda\frac{\partial}{\partial\Lambda}\bigg|_{\text{ren.fixed}} (-\ln Z) = \zeta(t), \tag{30.18}$$

$$\Lambda\frac{\partial}{\partial\Lambda}\bigg|_{\text{ren. fixed}} \ln h = \rho(t),$$

have been chosen h independent. The coefficient of $\partial/\partial h$ can be derived from equation (30.15) which implies (taking the logarithm of both members):

$$0 = h^{-1}\Lambda\frac{\partial}{\partial\Lambda}h + d - 2 - \frac{1}{2}\zeta(t) - \frac{\beta(t)}{t}, \tag{30.19}$$

and therefore:

$$\rho(t) = 2 - d + \frac{1}{2}\zeta(t) + \frac{\beta(t)}{t}. \tag{30.20}$$

To be able to discuss correlation functions involving the σ-field, we also need the RG equations satisfied by the connected correlation functions $W^{(n)}$:

$$\left[\Lambda\frac{\partial}{\partial\Lambda} + \beta(t)\frac{\partial}{\partial t} + \frac{n}{2}\zeta(t) + \left(\frac{1}{2}\zeta(t) + \frac{\beta(t)}{t} - \varepsilon\right)h\frac{\partial}{\partial h}\right]W^{(n)} = 0, \tag{30.21}$$

in which we now have set:

$$d = 2 + \varepsilon. \tag{30.22}$$

In Section 15.6 we have calculated the β-function at leading order for all symmetric spaces among which the σ-model is one of the simplest examples. In this model the two RG functions can be obtained at one-loop order from a calculation of the 2-point function $\Gamma^{(2)}$:

$$\Gamma^{(2)}\left(p\right) = \frac{\Lambda^{\varepsilon}}{t}\left(p^2 + h\right) + \left[p^2 + \frac{1}{2}\left(N-1\right)h\right]\frac{1}{(2\pi)^d}\int^{\Lambda}\frac{d^dq}{q^2 + h} + O(t). \tag{30.23}$$

Applying the RG equation (30.17) to $\Gamma^{(2)}$ and identifying the coefficients of p^2 and h, we derive two equations which determine $\beta(t)$ and $\zeta(t)$ at one-loop order

$$\beta(t) = \varepsilon t - \frac{(N-2)}{2\pi} t^2 + O\left(t^3, t^2\varepsilon\right), \tag{30.24a}$$

$$\zeta(t) = \frac{(N-1)}{2\pi} t + O\left(t^2, t\varepsilon\right). \tag{30.24b}$$

30.4 Discussion of the RG Equations

From the expression of $\beta(t)$ of equation (30.24a) we immediately conclude:

For $d \leq 2$ $(\varepsilon \leq 0)$, $t = 0$ is an unstable IR fixed point, this IR instability being induced by the vanishing mass of the would-be Goldstone bosons. The spectrum of the theory thus is not given by perturbation theory and the perturbative assumption of spontaneous symmetry breaking at low temperature is inconsistent. As mentioned before, this result agrees with rigorous arguments. Note that since the model depends only on one coupling constant, $t = 0$ is also a UV stable fixed point (the property of large momentum asymptotic freedom). Section 30.6 contains a short discussion of the physics in two dimensions for $N > 2$. The abelian case $N = 2$ is special and will be examined in Chapter 32.

For $d > 2$, i.e. $\varepsilon > 0$, $t = 0$ is a stable IR fixed point, the $O(N)$ symmetry is spontaneously broken at low temperature in zero field. The effective coupling constant, which determines the large distance behaviour, approaches the origin for all temperatures $t < t_c$, t_c being the first non-trivial zero of $\beta(t)$. Therefore the large distance properties of the model can be obtained from the low temperature expansion and renormalization group, replacing the perturbative parameters by effective parameters obtained by solving the RG equations.

The critical temperature. Finally we observe that, at least for ε positive and small, and $N > 2$, the RG function $\beta(t)$ has a non-trivial zero t_c:

$$t_c = \frac{2\pi\varepsilon}{N-2} + O\left(\varepsilon^2\right) \Rightarrow \beta\left(t_c\right) = 0, \text{ and } \beta'\left(t_c\right) = -\varepsilon + O\left(\varepsilon^2\right). \tag{30.25}$$

Since t_c is an unstable IR fixed point, it is by definition a critical temperature. Consequences of this property are studied below. Let us only immediately note that t_c is also a UV fixed point, i.e. it governs the large momentum behaviour of the renormalized theory. The large momentum behaviour of correlation functions is not given by perturbation theory but by the fixed point. As a consequence the perturbative result that the theory can not be rendered finite for $d > 2$ with a finite number of renormalization constants, cannot be trusted.

We now discuss more precisely the solutions of the RG equations.

30.4.1 Integration of the RG equations: $d > 2$, $t < t_c$

We first examine the implications of the RG equations for the large distance behaviour of correlation functions for $d > 2$ where $t = 0$ is an IR fixed point. As usual we solve equation (30.17) by introducing a scaling parameter λ and looking for a solution of the form:

$$\Gamma^{(n)}\left(p_i, t, h, \Lambda\right) = Z^{-n/2}(\lambda)\Gamma^{(n)}\left(p_i, t(\lambda), h(\lambda), \lambda\Lambda\right). \tag{30.26}$$

Compatibility with equation (30.17) implies:

$$\ln \lambda = \int_t^{t(\lambda)} \frac{dt'}{\beta(t')}, \tag{30.27a}$$

$$\ln Z(\lambda) = \int_t^{t(\lambda)} dt' \frac{\zeta(t')}{\beta(t')}, \tag{30.27b}$$

$$h(\lambda) = \lambda^{2-d} Z^{1/2}(\lambda) \frac{t(\lambda)}{t} h. \tag{30.27c}$$

With our conventions $\Gamma^{(n)}$ has the dimension d and h dimension 2. Taking into account dimensional analysis we then rewrite relation (30.26):

$$\Gamma^{(n)}(p_i, t, h, \Lambda) = Z^{-n/2}(\lambda)(\lambda\Lambda)^d \Gamma^{(n)}\left(p_i/\Lambda\lambda, t(\lambda), h(\lambda)/(\lambda\Lambda)^2, 1\right). \tag{30.28}$$

For $h \ll \Lambda^2$ perturbation theory has IR singularities. By choosing λ solution of the equation

$$h(\lambda) = (\lambda\Lambda)^2, \tag{30.29}$$

we ensure that the perturbation expansion in the effective theory at scale λ is no longer IR singular.

It is easy to verify that, at least for t small, $h \ll \Lambda^2$ implies $\lambda \to 0$. Let us then introduce three functions of the temperature $M_0(t)$, $\xi(t)$ and $K(t)$:

$$M_0(t) = \exp\left[-\frac{1}{2}\int_0^t \frac{\zeta(t')}{\beta(t')}dt'\right], \tag{30.30}$$

$$\xi(t) = \Lambda^{-1} t^{1/\varepsilon} \exp\left[\int_0^t \left(\frac{1}{\beta(t')} - \frac{1}{\varepsilon t'}\right)dt'\right]. \tag{30.31}$$

$$K(t) = M_0(t)\left[\Lambda\xi(t)\right]^{d-2}/t = 1 + O(t). \tag{30.32}$$

Solving then (30.29) we find

$$\lambda \sim K^{1/2}(t)h^{1/2}\Lambda^{-1}. \tag{30.33}$$

Because $t = 0$ is an IR fixed point, the scale-dependent temperature $t(\lambda) \to 0$ and thus the leading terms in the small h and small momenta limit can be calculated perturbatively. Using equations (30.27a, b) we obtain the behaviours of $t(\lambda)$ and $Z(\lambda)$:

$$t(\lambda) \sim \lambda^{d-2} t K(t) M_0^{-1} \sim t\left[K(t)\right]^{d/2} M_0^{-1} h^{(d-2)/2}\Lambda^{2-d}, \tag{30.34}$$

$$Z(\lambda) \sim M_0^2(t). \tag{30.35}$$

It follows

$$\Gamma^{(n)}(p_i, t, h, \Lambda) \sim M_0^{-n}(t)[K(t)h]^{d/2}\Gamma^{(n)}\left(\frac{p_i}{[K(t)h]^{1/2}}, \frac{t[K(t)]^{d/2}}{M_0(t)}\left(\frac{h}{\Lambda^2}\right)^{(d-2)/2}, 1, 1\right). \tag{30.36}$$

Actually it is easy to verify directly, using dimensional analysis in the form

$$\left(\Lambda\frac{\partial}{\partial\Lambda} + 2h\frac{\partial}{\partial h} + p_i\frac{\partial}{\partial p_i}\right)\Gamma^{(n)} = d\Gamma^{(n)},$$

that equation (30.36) gives the general solution of equation (30.17).

Let us apply this result to the determination of the singularities near the coexistence curve, i.e. at t fixed below the critical temperature when the magnetic field h goes to zero.

The coexistence curve. The magnetization satisfies:

$$M(t, h, \Lambda) \equiv \langle \sigma(x) \rangle = \Lambda^{-\varepsilon} t \frac{\partial \Gamma^{(0)}}{\partial h} = Z^{1/2}(\lambda) M\big(t(\lambda), 1, 1\big), \qquad (30.37)$$

($\Gamma^{(0)}$ is the magnetic field dependent free energy). At one-loop order in a field it is given by:

$$M = 1 - \frac{N-1}{2} \Lambda^{-\varepsilon} t \frac{1}{(2\pi)^d} \int^{\Lambda} \frac{d^d q}{q^2 + h} + O\left(t^2\right).$$

Thus using the relation (30.37) we find:

$$M(t, h, \Lambda = 1) = M_0(t) - \frac{N-1}{2} t \left[K(t)\right]^{d/2} h^{(d-2)/2} \frac{\Gamma(1 - d/2)}{(4\pi)^{d/2}} + O\left(h, h^{d-2}\right).$$

This result shows that $M_0(t)$ is the spontaneous magnetization and gives an interpretation to the logarithm found in equation (28.23) and to the singularity for $x = -1$ found in equation (29.27).

30.4.2 Scaling form of correlation functions, the critical domain

It is convenient to rewrite equation (30.36) as

$$\Gamma^{(n)}\left(p_i; t, h, \Lambda\right) = \xi^{-d}(t) M_0^{-n}(t) F^{(n)}\left(p_i \xi(t), h/h_0(t)\right). \qquad (30.38)$$

with

$$h_0(t) = t M_0^{-1}(t) \xi^{-d}(t) \Lambda^{2-d}. \qquad (30.39)$$

Equation (30.38) shows that $\xi(t)$ has in zero field the nature of a correlation length.

For the connected correlation functions the same analysis leads to:

$$W^{(n)}\left(p_i; t, H, \Lambda\right) = \xi^{d(n-1)}(t) M_0^n(t) G^{(n)}\left(p_i \xi(t), h/h_0(t)\right). \qquad (30.40)$$

Let us calculate the induced magnetization by:

$$M = \Lambda^{2-d} t \frac{\partial \Gamma^{(0)}}{\partial h} = M_0(t) F^{(0)}\left(h/h_0(t)\right). \qquad (30.41)$$

Inversion of this relation yields the scaling form of the equation of state:

$$h = h_0(t) f\left(\frac{M}{M_0(t)}\right), \qquad (30.42)$$

and the 1PI correlation functions can thus be written in terms of the magnetization as:

$$\Gamma^{(n)}\left(p_i, t, M, \Lambda\right) = \xi^{-d}(t) M_0^{-n}(t) F^{(n)}\left(p_i \xi(t), \frac{M}{M_0(t)}\right). \qquad (30.43)$$

The equations (30.42,30.43) are consistent with the equations (26.63,26.72): the appearance of two different functions $\xi(t)$ and $M_0(t)$ corresponds to the existence of two independent critical exponents ν, β in the $(\phi^2)^2$ field theory. They extend, in the large distance limit, the scaling form of correlation functions, valid in the critical region, to all temperatures below t_c. There is however one important difference between the RG equations of the $(\phi^2)^2$ theory and of the σ-model: the $(\phi^2)^2$ theory depends on two coupling constants, the coefficient of ϕ^2 which plays the role of the temperature, and the coefficient of $(\phi^2)^2$ which has no equivalent here. The correlation functions of the continuum $(\phi^2)^2$ theory have the exact scaling form (30.43) only at the IR fixed point. In contrast, in the case of the σ-model, it has been possible to eliminate all corrections to scaling corresponding to irrelevant operators order by order in perturbation theory. We are therefore led to a remarkable conclusion: the correlation functions of the $O(N)$ non-linear model are identical to the correlation functions of the $(\phi^2)^2$ field theory at the IR fixed point. This conclusion is supported by the analysis of the scaling behaviour performed within the $1/N$ expansion (see equation (29.42)).

The critical domain: critical exponents. Let us now study more precisely what happens when t approaches t_c (for $N > 2$). The function $\xi(t)$ diverges as:

$$\xi(t) \sim \Lambda^{-1}\left(t_c - t\right)^{1/\beta'(t_c)}. \tag{30.44}$$

Comparing with the scaling form (26.35) we conclude that the correlation length exponent ν is given by

$$\nu = -\frac{1}{\beta'(t_c)}. \tag{30.45}$$

For d close to 2 the exponent ν thus behaves like:

$$\nu \sim 1/\varepsilon. \tag{30.46}$$

The function $M_0(t)$ vanishes at t_c:

$$\ln M_0(t) = -\frac{1}{2}\frac{\zeta(t_c)}{\beta'(t_c)} \ln(t_c - t) + \text{const.}. \tag{30.47}$$

This yields the exponent β and thus also η through the scaling relation $\beta = \frac{1}{2}\nu(d-2+\eta)$:

$$\eta = \zeta(t_c) - \varepsilon. \tag{30.48}$$

A leading order we find:

$$\eta = \frac{\varepsilon}{N-2} + O\left(\varepsilon^2\right). \tag{30.49}$$

We finally note that the singularity of $\Gamma^{(n)}$ coming from the prefactor $\xi^{-d}M_0^{-n}$ indeed agrees near t_c with the result of equation (26.37).

The nature of the correlation length $\xi(t)$. The length scale $\xi(t)$ is a cross-over scale between two different behaviours of correlation functions. For distances large compared to $\xi(t)$, the behaviour of correlation functions is governed by the Goldstone modes (spin wave excitations) and can thus be deduced from the perturbative low temperature expansion. However when t approaches t_c, $\xi(t)$ becomes large. There then exist distances large with respect to the microscopic scale but small with respect to $\xi(t)$ in which correlation functions have a critical behaviour. In this situation we can construct continuum correlation functions consistent on all scales, the critical behaviour being also the large momentum behaviour of the renormalized field theory.

General comment. From the consideration of the low temperature expansion we have been able to describe, for theories with a continuous symmetry, not only the complete structure of the low temperature phase, and this was expected, but also the critical behaviour near two dimensions in the non-abelian case.

What is somewhat surprising in this result is that perturbation series is only sensitive to the local structure of the sphere $\mathbf{S}^2 = 1$ while the restoration of symmetry involves the sphere globally. This explains the peculiarity of the abelian case $N = 2$ because locally a circle cannot be distinguished from a non-compact straight line. For $N > 2$ the sphere has instead a local characteristic curvature. Still different regular compact manifolds may have the same local metric, and therefore the same perturbation theory. They all have the same low temperature physics. However the previous results concerning the critical behaviour are physically relevant only if they are still valid when ε is not infinitesimal and t approaches t_c, a condition which cannot be checked directly. In particular the low temperature expansion misses in general terms decreasing like $\exp{(\mathrm{const.}/t)}$ which may in some cases be essential for the physics. Finally let us note that, at least, we have found a direct connection between the $(\phi^2)^2$ and the σ-model through the large N expansion (see Section 29.3). This gives us some confidence that the previous considerations are valid for the N-vector model at least for N sufficiently large. On the other hand the physics of $N = 2$ is not well reproduced (see Chapter 32). Cardy and Hamber have speculated about the RG flow for N close to 2 and dimension d close to 2, incorporating phenomenologically the Kosterlitz–Thouless in their analysis.

30.5 Results beyond One-Loop

For explicit calculations of the RG functions beyond one-loop, it is convenient to use the renormalized action (30.12), dimensional regularization and minimal subtraction. For example the one-loop renormalized 2-point function reads:

$$\Gamma^{(2)}(p) = \frac{\mu^\varepsilon}{t} \left(\frac{Zp^2}{Z_t} + hZ^{1/2} \right) + \left[p^2 + \tfrac{1}{2}(N-1)h \right] \frac{1}{(2\pi)^d} \int \frac{\mathrm{d}^d q}{q^2 + h} + O(t). \quad (30.50)$$

This expression of $\Gamma^{(2)}$ yields the renormalization constants at one-loop order. In the minimal subtraction scheme, we find:

$$Z = 1 + (N-1)\frac{N_d t}{\varepsilon} + O\left(t^2\right),$$

$$Z_t = 1 + (N-2)\frac{N_d t}{\varepsilon} + O\left(t^2\right),$$

$$(30.51)$$

in which N_d is the usual loop factor:

$$N_d = \frac{2}{(4\pi)^{d/2}\Gamma(d/2)} = \frac{1}{2\pi} + O(\varepsilon).$$

The RG functions in the renormalized theory follow:

$$\beta(t) = \varepsilon t \left(1 + t\frac{d\ln Z_t}{dt}\right)^{-1} = \varepsilon t - \frac{(N-2)}{2\pi}t^2 + O\left(t^3\right),$$

$$\zeta(t) = \beta(t)\frac{d\ln Z}{dt} = \frac{(N-1)}{2\pi}t + O\left(t^2\right).$$

(30.52)

At the one-loop order we recover, as expected, the expressions (30.24). We now give directly the two RG functions and the critical exponents at the order presently available, i.e. four loops, without presenting the details of the calculation. For convenience we rescale the coupling constant t

$$\tilde{t} = tN_d.$$

(30.53)

At this order $\tilde{\beta}$, the corresponding β-function, reads:

$$\tilde{\beta}(\tilde{t}) = \varepsilon\tilde{t} - (N-2)\tilde{t}^2 \left[1 + \tilde{t} + \frac{(N+2)}{4}\tilde{t}^2 + b\tilde{t}^3\right] + O\left(\tilde{t}^6\right),$$

(30.54)

in which the quantity b is a constant:

$$b = -\frac{1}{12}\left(N^2 - 22N + 34\right) + \frac{3}{2}\zeta(3)(N-3).$$

(30.55)

We recall that the value of the numerical constant $\zeta(3)$ (which should not be confused with the function $\zeta(t)$ of equation (30.24b)) is:

$$\zeta(3) = 1.2020569\ldots.$$

The anomalous dimensions of the composite operator of spin l (see Section 14.8) is given by:

$$\zeta_l(\tilde{t}) = a\tilde{t}\left\{1 + (N-2)\tilde{t}^2 \left[\frac{3}{4} + \left(\frac{5-N}{3} + \frac{1+a}{2}\zeta(3)\right)\tilde{t}\right]\right\} + O\left(\tilde{t}^5\right),$$

(30.56)

with:

$$a = -(N+l-2)l/2.$$

(30.57)

The case $l = 1$ corresponds to the field itself. The function $\zeta(t)$ given at one-loop order by equation (30.24b) is related to ζ_1 by:

$$\zeta(t) = -2\zeta_1(t).$$

From these expressions the values of the critical exponents η and ν follow. Defining:

$$\tilde{\varepsilon} = \varepsilon/(N-2),$$

(30.58)

we obtain:

$$\eta = \tilde{\varepsilon} + (N-1)\tilde{\varepsilon}^2 \left\{-1 + \frac{N}{2}\tilde{\varepsilon} + \left[-b + (N-2)\left(\frac{2-N}{3} + \frac{3-N}{4}\zeta(3)\right)\right]\tilde{\varepsilon}^2\right\} + O\left(\tilde{\varepsilon}^5\right),$$

$$\nu(d-2) = 1 - \tilde{\varepsilon} + \frac{(4-N)}{2}\tilde{\varepsilon}^2 + \frac{1}{4}\left[N^2 - 10N + 18 + 18(3-N)\zeta(3)\right]\tilde{\varepsilon}^3 + O\left(\tilde{\varepsilon}^4\right).$$

30.6 The Dimension Two

Dimension two is of special interest from the particle physics point of view. The RG function $\beta(t)$ is then:

$$\beta(t) = -\frac{(N-2)}{2\pi} t^2 + O\left(t^3\right). \tag{30.59}$$

The non-linear σ-model for $N > 2$ is the simplest example of a so-called asymptotically free field theory (UV free) since the first coefficient of the β-function is negative, in contrast with the ϕ^4 field theory. Therefore the large momentum behaviour of correlation functions is entirely calculable from perturbation theory and RG arguments. There is, however a counterpart, the theory is IR unstable and thus, in zero field h, the spectrum of the theory is not perturbative. Contrary to perturbative indications, it consists of N massive degenerate states since the $O(N)$ symmetry is not broken. Asymptotic freedom and the non-perturbative character of the spectrum are also properties of QCD in four dimensions, the theory of strong interactions (see Chapters 33,34).

If we now define a function $\xi(t)$ by:

$$\xi(t) = \mu^{-1} \exp\left[\int^t \frac{dt'}{\beta(t')}\right], \tag{30.60}$$

we can integrate the RG equations in the same way and we find that $\xi(t)$ is the correlation length in zero field. In addition we can use the explicit expression of the β-function (equation (30.54)) to calculate the correlation length or the physical mass for small t:

$$\xi^{-1}(t) = m(t) = K\mu t^{-1/(N-2)} e^{-2\pi/[(N-2)t]} (1 + O(t)). \tag{30.61}$$

However the exact value of the integration constant K, which gives the physical mass in the RG scale, can only be calculated by non-perturbative techniques.

Finally the scaling forms (30.38,30.40) imply that the perturbative expansion at fixed magnetic field is valid, at low momenta or large distances, and for $h/h_0(t)$ large.

Elitzur's conjecture. The $O(N)$ symmetric action (30.9) ($h = 0$) has a sphere of degenerate classical minima. To define perturbation theory we have been forced to add a linear symmetry breaking term to the action, which selects one particular classical minimum. We have claimed, and it is easy to verify, that for $d \le 2$ the correlation functions have IR divergences when the parameter h goes to zero, a property that is consistent with the absence of SSB for $d \le 2$. However to calculate in perturbation theory another option is available, which will be used systematically in the case of instanton calculations (see Chapters 37–43): one does not introduce a symmetry breaking term but instead a set of *collective coordinates* which parametrizes the set of classical minima. One then expands in perturbation theory around one fixed minimum but treats perturbatively only the modes of the field that do not correspond to a global rotation. One finally sums over all classical minima. Clearly, after this last summation, only $O(N)$ invariant correlation functions survive. As already mentioned in Section 14.4, it has been conjectured by Elitzur and proven by David that in two dimensions the $O(N)$ invariant correlation functions obtained by this procedure have a regular low temperature expansion: this means that if we calculate $O(N)$ invariant correlation functions by perturbation theory with a non-vanishing field and take the limit $h = 0$, this limit is IR finite. The subtlety of this problem when compared to the instanton case is that in the infinite volume the zero momentum singularity due to the choice of one classical minimum is not an isolated pole (see also Section 3.4 and Chapter 37).

30.7 The Large N Limit

In Chapter 29 we have shown that within the framework of the large N expansion the non-linear σ-model was obtained from the ϕ^4 field theory by neglecting terms irrelevant in the large distance limit. In order to more explicitly show the correspondence between the set of parameters used in the two models, let us solve the σ-model in the large N limit. Using the linear formalism of Section 14.9, we rewrite the partition function:

$$Z = \int [\mathrm{d}\phi(x)\mathrm{d}\lambda(x)] \exp\left[-S(\phi,\lambda)\right], \tag{30.62}$$

with:

$$S(\phi,\lambda) = \frac{1}{2t} \int \mathrm{d}^d x \left[(\partial_\mu \phi)^2 + \lambda\left(\phi^2 - 1\right)\right]. \tag{30.63}$$

Integrating, as was done in Chapter 29, over $N-1$ components of ϕ and calling σ the remaining component, we obtain:

$$Z = \int [\mathrm{d}\sigma(x)\mathrm{d}\lambda(x)] \exp\left[-S_{\mathrm{eff}}(\sigma,\lambda)\right], \tag{30.64}$$

with:

$$S_{\mathrm{eff}}(\sigma,\lambda) = \frac{1}{2t} \int \left[(\partial_\mu \sigma)^2 + \left(\sigma^2(x) - 1\right)\lambda(x)\right]\mathrm{d}^d x + \frac{1}{2}(N-1)\,\mathrm{tr}\ln\left[-\Delta + \lambda(x)\right]. \tag{30.65}$$

The large N limit is now taken at tN fixed. The saddle point equations, analogous to equations (29.7), are:

$$m^2 \sigma = 0, \tag{30.66}$$

$$\sigma^2 = 1 - \frac{(N-1)t}{(2\pi)^d} \int^\Lambda \frac{\mathrm{d}^d p}{p^2 + m^2}, \tag{30.67}$$

where we have set $\langle\lambda(x)\rangle = m^2$. At low temperature σ is different from zero and thus m, which is the mass of the π-field, vanishes. Equation (30.67) gives the spontaneous magnetization:

$$\sigma^2 = 1 - \frac{(N-1)t}{(2\pi)^d} \int^\Lambda \frac{\mathrm{d}^d p}{p^2}. \tag{30.68}$$

At t_c, σ vanishes:

$$\frac{1}{t_c} = \frac{(N-1)}{(2\pi)^d} \int^\Lambda \frac{\mathrm{d}^d p}{p^2}. \tag{30.69}$$

Therefore equation (30.68) can be rewritten:

$$\sigma^2 = 1 - t/t_c. \tag{30.70}$$

Above t_c, σ instead vanishes and m, which is now the common mass of the π- and σ-field, is for $2 < d < 4$ given by:

$$\frac{1}{t_c} - \frac{1}{t} = m^{d-2} \frac{(N-1)}{(2\pi)^d} \int \frac{\mathrm{d}^d p}{p^2\left(p^2+1\right)} + O\left(m^2\Lambda^{d-4}\right). \tag{30.71}$$

We recover the scaling form of the correlation length $\xi = 1/m$. From the equations (30.70,30.71), we can also derive the RG functions at leading order for N large:

$$\beta(t) = \varepsilon t - \frac{N}{2\pi}t^2, \qquad \zeta(t) = \frac{N}{2\pi}t. \tag{30.72}$$

The calculation of other physical quantities and the expansion in $1/N$ follow from the considerations of Chapter 29.

Two dimensions and the question of Borel summability. For $d = 2$ the critical temperature vanishes and the parameter m has the form:

$$m \sim \Lambda\, e^{-2\pi/(Nt)}, \tag{30.73}$$

in agreement with the RG considerations. Note that the field 2-point function takes in the large N-limit the form:

$$\Gamma^{(2)}(p) = p^2 + m^2. \tag{30.74}$$

The mass term vanishes to all orders in the expansion in powers of the coupling constant t, preventing any perturbative calculation of the mass of the field. The perturbation series is trivially not Borel summable. Most likely this property is also true for the model at finite N. On the other hand if we break the $O(N)$ symmetry by a magnetic field, adding a term $h\sigma$ to the action, the physical mass becomes calculable in perturbation theory.

Corrections to scaling and the dimension four. In equation (30.71) we have neglected corrections to scaling. If we take into account the leading correction we get instead:

$$m^2\left(m^{d-4} - K(d)\Lambda^{d-4}\right) \propto t - t_c,$$

where $K(d)$ is a constant which explicitly depends on the cut-off procedure and can thus be varied by changing contributions of irrelevant operators. By comparing with the results of Subsection 29.2.2, we discover that, although the non-linear σ-model superficially depends on one parameter less than the corresponding ϕ^4 field theory, actually this parameter is hidden in the cut-off function. This remark becomes important in the four dimensional limit where most leading contributions come from the leading corrections to scaling. For example for $d = 4$ equation (30.71) takes a different form, the dominant term in the r.h.s. is proportional to $m^2 \ln m$. We recognize in the factor $\ln m$ the effective ϕ^4 coupling at mass scale m. Beyond the $1/N$ expansion, to describe with perturbation theory and renormalization group the physics of the non-linear σ model it is necessary to introduce the operator $\int d^d x\, \lambda^2(x)$, which irrelevant for $d < 4$, becomes marginal, and to return to the ϕ^4 field theory.

30.8 Generalizations

We have shown in Chapter 15 that from the point of view of the renormalization group, the properties of the non-linear σ-model generalize to all models defined on symmetric spaces. They are all UV free in two dimensions (see Section 15.6), and therefore have a phase transition at a critical temperature of order ε in $2 + \varepsilon$ dimension. However the identification of the correlation functions of these models with the correlation functions of a ϕ^4 type theory at an IR fixed point is in general not easy. In particular the connection through a large N expansion does not exist in general. It is likely that for some of these models the transition found from the $2 + \varepsilon$ expansion is actually a first order transition for ε non-infinitesimal. Let us mention that the β-function has been calculated up to four loops for a large class of symmetric spaces.

Symmetric spaces corresponding to non-compact groups. Previous considerations can be formally extended to symmetric spaces G/H in which the group G is non-compact, but the subgroup H remains compact. Symmetry groups obtained by complexification of compact group generate such spaces. A simple example is $O(M,N)/O(M)/O(N)$, in which $O(M,N)$ is a pseudo-orthogonal group. From the point of view of perturbation theory the only difference is that the sign of the coupling has changed. However this means that these models are no longer UV free but IR free in two dimensions. They have a non-trivial IR fixed point in $d = 2 - \varepsilon$ dimension. The physics is thus completely different: the situation bears some analogies with the behaviour of the ϕ^4 theory in four dimensions. The existence of massless modes below 2 dimensions is not in contradiction with rigorous theorems because the symmetry group is not compact. Such models play a role in the theory of localization.

Bibliographical Notes

The renormalization group properties of the non-linear σ-model have been discussed in
  A.M. Polyakov, *Phys. Lett.* 59B (1975) 79; E. Brézin and J. Zinn-Justin, *Phys. Rev. Lett.* 36 (1976) 691; *Phys. Rev.* B14 (1976) 3110; W.A. Bardeen, B.W. Lee and R.E. Shrock, *Phys. Rev.* D14 (1976) 985.
Higher order calculations of critical exponents are due to
  S. Hikami, *Nucl. Phys.* B215[FS7] (1983) 555; W. Bernreuther and F.J. Wegner, *Phys. Rev. Lett.* 57 (1986) 1383; F. Wegner, *Nucl. Phys.* B316 (1989) 663.
Speculations about a RG flow near dimension 2 consistent with the $N = 2$ Kosterlitz–Thouless transition and the non-linear σ-model results for $N > 2$ can be found in
  J.L. Cardy and H.W. Hamber, *Phys. Rev. Lett.* 45 (1980) 499.
Long range forces have been discussed in
  E. Brézin, J.C. Le Guillou and J. Zinn-Justin, *J. Phys. A: Math. Gen.* 9 (1976) L119.
For an introduction to the problem of localization in electron systems see for example
  E. Fradkin, G. Parisi in *Recent Advances in Field Theory and Statistical Mechanics*, Les Houches 1982, R. Stora and J.-B. Zuber eds. (North Holland, Amsterdam 1984); F. Wegner, *Lecture Notes in Physics* 216 (1985) 141.
Renormalization of composite operators is discussed in
  F.J. Wegner, *Z. Phys.* B78 (1990) 33; G.E. Castilla and S. Chakravarty, *Phys. Rev. Lett.* 71 (1993) 384.
For the Gross–Neveu model discussed in the appendix see
  K.G. Wilson, *Phys. Rev.* D7 (1973) 2911; D.J. Gross and A. Neveu, *Phys. Rev.* D10 (1974) 3235; A.B. Zamolodchikov and Al.B. Zamolodchikov, *Phys. Lett.* 72B (1978) 481; W. Wentzel, *Phys. Lett.* 153B (1985) 297; J.A. Gracey, *Nucl. Phys.* B367 (1991) 657.
The relation with its linearized extension is discussed in
  A. Hasenfratz, P. Hasenfratz, K. Jansen, J. Kuti and Y. Shen, *Nuclear Physics* B365 (1991) 79; J. Zinn-Justin, *Nuclear Physics* B367 (1991) 105.
Finally all these techniques have also been applied to the supersymmetric extension of the non-linear σ-model. A recent reference is
  J.A. Gracey, *Nucl. Phys.* B352 (1991) 183.

APPENDIX 30
THE LINEARIZED GROSS–NEVEU MODEL

To illustrate the techniques developed in Chapters 29,30, we discuss in the appendix the Gross–Neveu (GN) model which is described in terms of a $U(N)$ symmetric action for a set of N massless Dirac fermions $\{\psi^i, \bar{\psi}^i\}$:

$$S\left(\bar{\psi}, \psi\right) = -\int \mathrm{d}^d x \left[\bar{\psi} \cdot \slashed{\partial}\psi + \tfrac{1}{2}G\left(\bar{\psi} \cdot \psi\right)^2\right].$$

The GN model has in even dimensions a discrete chiral symmetry:

$$\psi \mapsto \gamma_S \psi, \quad \bar{\psi} \mapsto -\bar{\psi}\gamma_S, \tag{A30.1}$$

which prevents the addition of a fermion mass term while in odd dimensions a mass term breaks space parity (see Appendix A5). Actually the two symmetry operations can be written in a form

$$\mathbf{x} = \{x_1, x_2, \ldots, x_d\} \mapsto \tilde{\mathbf{x}} = \{-x_1, x_2, \ldots, x_d\}, \quad \begin{cases} \psi(x) \mapsto \gamma_1 \psi(\tilde{x}), \\ \bar{\psi}(x) \mapsto \bar{\psi}(\tilde{x})\gamma_1 \end{cases},$$

valid in all dimensions.

This model illustrates the physics of spontaneous fermion mass generation and, in even dimensions, chiral symmetry breaking. However, as in the case of the non-linear σ model, the perturbative GN model describes only one phase and the symmetry breaking mechanism is non-perturbative. It is therefore useful to first consider a more general model, the linearized GN model, which has the same chiral and $U(N)$ symmetries, and whose action is:

$$S\left(\bar{\psi}, \psi, \sigma\right) = \int \mathrm{d}^d x \left[-\bar{\psi} \cdot \left(\slashed{\partial} + g\sigma\right)\psi + \tfrac{1}{2}\left(\partial_\mu \sigma\right)^2 + \tfrac{1}{2}m^2\sigma^2 + \frac{\lambda}{4!}\sigma^4\right], \tag{A30.2}$$

where σ is an additional scalar field. The action still has a reflection symmetry, σ transforming into $-\sigma$ when the fermions transform by $(A30.1)$. We recognize the model that we have introduced in Section 11.8 to illustrate the calculation of RG functions at one-loop order in a model with bosons and fermions. In contrast with the GN model, the symmetry breaking can then be discussed by perturbative methods. An analogous situation has already been encountered when comparing the $(\phi^2)^2$ field theory with the non-linear σ model. Even more, the GN model is renormalizable in two dimensions and the linearized model in dimension four. The main difference is that the GN perturbative model describes the symmetric phase while the perturbative non-linear σ model describes the broken phase (but in both cases the phases are massless). We now show that the physics of the GN model can indeed be discussed by the same methods as ferromagnetic systems, that is RG equations near two and four dimensions, and large N expansion.

The phase transition. Examining the action $(A30.2)$ we see that in the tree approximation when m^2 is negative the chiral symmetry is spontaneously broken. The σ expectation value gives a mass to the fermions, a mechanism reminiscent of the standard model of weak-electromagnetic interactions:

$$m_\psi = g\langle\sigma\rangle, \tag{A30.3}$$

while the σ mass then is:

$$m_\sigma^2 = \frac{\lambda}{3g^2}m_\psi^2. \tag{A30.4}$$

To study the model beyond the tree approximation we now discuss RG equations near four dimensions.

A30.1 RG Equations near Four Dimensions

The model $(A30.2)$ is trivial above four dimensions, renormalizable in four dimensions and can thus be studied near dimension 4 by RG techniques. Calling μ the renormalization scale, setting $d = 4 - \varepsilon$, we can write the renormalized action:

$$S_{\mathrm{r}}\left(\bar{\psi}, \psi, \sigma\right) = \int \mathrm{d}^d x \left[-Z_\psi \bar{\psi} \cdot \partial\!\!\!/\psi - \mu^{\varepsilon/2} g Z_g \sigma \bar{\psi} \cdot \psi \right.$$
$$\left. + \tfrac{1}{2} Z_\sigma \left(\partial_\mu \sigma\right)^2 + \tfrac{1}{2}(m_c^2 + Z_m t)\sigma^2 + \mu^\varepsilon Z_\lambda \frac{\lambda}{4!}\sigma^4 \right], \qquad (A30.5)$$

where m_c^2 is the critical bare mass squared, the critical temperature in statistical language, and t characterizes the deviation from the critical temperature. The RG equations for the renormalized functions are:

$$\left[\mu\frac{\partial}{\partial\mu} + \beta_{g^2}\frac{\partial}{\partial g^2} + \beta_\lambda\frac{\partial}{\partial\lambda} - \tfrac{1}{2}l\eta_\psi - \tfrac{1}{2}n\eta_\sigma - \eta_m t\frac{\partial}{\partial t} \right] \Gamma^{(l,n)} = 0. \qquad (A30.6)$$

Remark. Note that in this section for convenience we set for the γ matrices $\mathrm{tr}\,\mathbf{1} = 4$ as in four dimensions. To extrapolate the results to other dimensions one has to replace everywhere N by $N' = \mathrm{tr}\,\mathbf{1}N/4$.

The RG functions. The RG functions have been calculated at one-loop order in Section 11.8:

$$\beta_\lambda = -\varepsilon\lambda + \frac{1}{8\pi^2}\left(\frac{3}{2}\lambda^2 + 4N\lambda g^2 - 24Ng^4\right), \qquad (A30.7)$$

$$\beta_{g^2} = -\varepsilon g^2 + \frac{2N+3}{8\pi^2}g^4. \qquad (A30.8)$$

In four dimensions the origin $\lambda = g^2 = 0$ is IR stable. For $d = 4 - \varepsilon$ one finds a non-trivial fixed point:

$$g_*^2 = \frac{8\pi^2\varepsilon}{2N+3}, \quad \lambda_* = 8\pi^2\varepsilon\tilde{\lambda}_*, \qquad (A30.9)$$

with

$$\tilde{\lambda}_* = \frac{1}{3(2N+3)}\left[-(2N-3) + \sqrt{4N^2 + 132N + 9}\right]. \qquad (A30.10)$$

The matrix of derivatives of the β-functions has two eigenvalues ω, ω',

$$\omega = \varepsilon, \quad \omega' = \varepsilon\sqrt{4N^2 + 132N + 9}/(2N + 3), \qquad (A30.11)$$

and thus the fixed point is IR stable. The first eigenvalue is always the smallest.

The field renormalization RG functions are at the same order:

$$\eta_\sigma = \frac{2N}{8\pi^2}g^2, \quad \eta_\psi = \frac{1}{16\pi^2}g^2. \qquad (A30.12)$$

At the fixed point one finds

$$\eta_\sigma = \frac{2N\varepsilon}{2N+3}, \quad \eta_\psi = \frac{\varepsilon}{2(2N+3)}. \qquad (A30.13)$$

The RG function η_m (called η_2 in Section 11.8) corresponding to the mass operator is at one-loop order:

$$\eta_m = -\frac{\lambda}{16\pi^2} - \eta_\sigma \,,$$

and thus the exponent ν:

$$\nu = (2 + \eta_m)^{-1} = \left(2 - \frac{\varepsilon}{2}\tilde{\lambda}_* - \frac{2N\varepsilon}{2N+3}\right)^{-1} = \frac{1}{2} + \frac{\varepsilon}{8}\tilde{\lambda}_* + \frac{N\varepsilon}{2(2N+3)}.$$

Finally we can evaluate the ratio of masses ($A30.4$) at the fixed point:

$$\frac{m_\sigma^2}{m_\psi^2} = \frac{\lambda_*}{3g_*^2} = \frac{-(2N-3) + \sqrt{4N^2 + 132N + 9}}{9}.$$

The existence of an IR fixed point has a remarkable consequence: If we assume, as we do in the theory of phase transitions, that the coupling constants are generic at the cut-off scale, then *the ratio of fermion and scalar masses is fixed*, provided the σ expectation value is much smaller than the cut-off. Note, however, that in four dimensions the coupling dependent corrections vanish only logarithmically with the cut-off.

A30.2 Large N Limit

We can examine the large N behaviour of previous expressions. For example we find $\lambda_* \sim 48\pi^2/N$ and $d - 2 + \eta_\sigma = 2$. This reminds us the $(\phi^2)^2$ field theory and suggests a study of the large N limit. For this purpose we integrate over $N - 1$ fermion fields, rescale for convenience $g\sigma$ into σ, and get the effective action:

$$S\left(\bar{\psi}, \psi, \sigma\right) = \int \mathrm{d}^d x \left[-\bar{\psi}\left(\slashed{\partial} + \sigma\right)\psi + \frac{1}{2g^2}\left(\partial_\mu\sigma\right)^2 + \frac{m^2}{2g^2}\sigma^2 + \frac{\lambda}{4!g^4}\sigma^4\right] \tag{$A30.14$}$$
$$- (N-1)\operatorname{tr}\ln\left(\slashed{\partial} + \sigma\right).$$

To take the large N limit we assume σ finite and $g^2, \lambda = O(1/N)$.

The expectation value of σ is now given by a *gap* equation:

$$\frac{m^2}{g^2}\sigma + \frac{\lambda}{6g^4}\sigma^3 - N'\frac{\sigma}{(2\pi)^d}\int^\Lambda \frac{\mathrm{d}^d q}{q^2 + \sigma^2} = 0\,, \tag{$A30.15$}$$

where N' is the number of fermions, i.e. $N' = NK_d$ with K_d the trace of the identity matrix (see Appendix A5). The critical temperature or bare mass is thus given by:

$$\frac{m_c^2}{g^2} = N'\frac{1}{(2\pi)^d}\int^\Lambda \frac{\mathrm{d}^d q}{q^2}\,, \tag{$A30.16$}$$

which shows that the fermions favour the chiral transition. In particular when d approaches 2 we observe that $m_c^2 \to +\infty$ which implies that the chiral symmetry is always broken in 2 dimensions. Using equation ($A30.16$) inside ($A30.15$) we obtain an equation for the expectation value of σ, in the low temperature phase, as a function of temperature. Setting $t = (m^2 - m_c^2)/g^2$ we find:

$$t + \frac{\lambda}{6g^4}\sigma^2 + N'\frac{\sigma^2}{(2\pi)^d}\int^\Lambda \frac{\mathrm{d}^d q}{q^2\left(q^2 + \sigma^2\right)} = 0\,. \tag{$A30.17$}$$

For $d \leq 4$ the last term is more singular than the second one for σ small. In particular for $d < 4$:

$$\sigma \propto (-t)^{1/(2-\varepsilon)}, \tag{A30.18}$$

as in the non-linear σ-model.

Since, at leading order, the fermion mass $m_\psi = \sigma$, it immediately follows that the exponent ν is also given by:

$$\nu \sim \beta \sim 1/(d-2) \Rightarrow \eta_\sigma = \varepsilon. \tag{A30.19}$$

We read from the effective action that at this order $\eta_\psi = 0$.

Finally from the effective action we can calculate the σ-propagator at leading order. We observe that the term coming from the fermion determinant is again dominant at small momentum near the critical temperature. Thus at T_c

$$\langle \sigma \sigma \rangle \sim \frac{N' b(\varepsilon)}{2p^{d-2}}, \tag{A30.20}$$

with

$$b(\varepsilon) = \frac{1}{(4\pi)^{d/2}} \Gamma(2 - d/2) \frac{\Gamma^2(d/2 - 1)}{\Gamma(d-2)}. \tag{A30.21}$$

The result is consistent with the value of η_σ found above.

More generally, using the saddle point equation, one finds for the inverse σ-propagator in the massive phase:

$$\Delta_\sigma^{-1}(p) = \frac{p^2}{g^2} + \frac{\lambda}{3g^4}\sigma^2 + \frac{N'}{2(2\pi)^d}\left(p^2 + 4\sigma^2\right) \int^\Lambda \frac{d^d q}{(q^2 + \sigma^2)\left[(p+q)^2 + \sigma^2\right]}. \tag{A30.22}$$

In the scaling limit the propagator vanishes for $p^2 = -4\sigma^2$ which is just the $\bar\psi\psi$ threshold. Thus, in this limit, $m_\sigma = 2m_\psi$. Taking into account corrections to scaling, the σ mass lies above or below the two fermion threshold depending on the value of λ.

Let us finally note that from the behaviour of the propagator at the critical point, $\Delta_\sigma(p) \propto p^{2-d}$, we derive the canonical dimension $[\sigma]$ of the field σ in the large N expansion: $[\sigma] = 1$.

Corrections to scaling and the IR fixed point. The IR fixed point is determined by demanding the cancellation of the leading corrections to scaling. The integral in $(A30.17)$ is given for large values of the cut-off Λ by equation (29.17):

$$\frac{1}{(2\pi)^d} \int^\Lambda \frac{d^d q}{q^2(q^2 + \sigma^2)} = C(d)\sigma^{-\varepsilon} - a(d)\Lambda^{-\varepsilon} + O\left(\frac{\sigma^{2-\varepsilon}}{\Lambda^2}\right), \tag{A30.23}$$

$(a(\varepsilon) \sim 1/8\pi^2\varepsilon)$. Demanding the cancellation of the coefficient of σ^2, we obtain a relation between λ and g^2

$$g_*^4 = \frac{\lambda_*}{6N'} \frac{\Lambda^\varepsilon}{a(d)} = \frac{\lambda_* \varepsilon \pi^2}{3N} + O\left(\varepsilon^2\right),$$

a result consistent with the results of the ε-expansion.

In the same way it is possible to calculate the leading correction to the σ-propagator $(A30.22)$. Demanding the cancellation of the leading correction we obtain

$$\frac{p^2}{g_*^2} + \frac{\lambda_*}{3g_*^4}\sigma^2 - \tfrac{1}{2}N'\left(p^2 + 4\sigma^2\right) a(d)\Lambda^{-\varepsilon} = 0.$$

The coefficient of σ^2 cancels from the previous relation and the cancellation the coefficient of p^2 yields

$$g_*^2 = \frac{2\Lambda^\varepsilon}{N'a(d)} = \frac{\Lambda^\varepsilon 4\pi^2 \varepsilon}{N} + O\left(\varepsilon^2\right),$$

in agreement with the ε-expansion for N large.

The fermion anomalous dimension at order $1/N$. Let us now calculate the dimension η_ψ at first non-trivial order. The fermion 2-point function at the critical point is then:

$$\Gamma_{\bar\psi\psi}(p) = i\slashed{p} + \frac{2i}{N'b(\varepsilon)(2\pi)^d} \int^\Lambda \frac{d^d q (\slashed{p} + \slashed{q})}{q^{d-2}(p+q)^2}.$$

The explicit form of the cut-off Λ is irrelevant at this order, and we just have to calculate the coefficient of the term $\slashed{p} \ln \Lambda$. This can be achieved by expanding the integrand for q large:

$$\frac{(\slashed{p} + \slashed{q})}{q^{d-2}(p+q)^2} = \frac{\slashed{p}}{q^d} - \frac{2p \cdot q\slashed{q}}{q^{d+2}} + \cdots .$$

Then using

$$\int d^d q \, f(q^2) q_\mu q_\nu = \frac{\delta_{\mu\nu}}{d} \int d^d q \, f(q^2) q^2,$$

we find

$$\frac{1}{(2\pi)^d} \int^\Lambda \frac{d^d q \, (\slashed{p} + \slashed{q})}{q^{d-2}(p+q)^2} \sim \frac{2}{\Gamma(d/2)(4\pi)^{d/2}} \left(\frac{d-2}{d}\right) \slashed{p} \ln \Lambda.$$

The exponent η_ψ at order $1/N$ follows

$$\eta_\psi = \frac{(d-2)^2}{N'd} \frac{\Gamma(d-1)}{\Gamma^3(d/2)\Gamma(2-d/2)}.$$

For $d = 4 - \varepsilon$ we find $\eta_\psi \sim \varepsilon/4N$, result consistent with $(A30.13)$ for N large. For $d = 2 + \varepsilon$ instead one finds $\eta_\psi \sim \varepsilon^2/4N$.

The relation to the GN model for dimensions $2 \leq d \leq 4$. We have seen that the terms $(\partial_\mu \sigma)^2$ and σ^4 of the effective action which have a canonical dimension 4, are irrelevant in the IR critical region for $d \leq 4$. We recognize a situation already encountered in the $(\phi^2)^2$ field theory in the large N limit. In the scaling region it is possible to omit them and one then finds the action:

$$S\left(\bar\psi, \psi, \sigma\right) = \int d^d x \left[-\bar\psi \cdot \left(\slashed{\partial} + \sigma\right)\psi + \frac{m^2}{2g^2}\sigma^2\right]. \qquad (A30.24)$$

The integral over the σ field can explicitly be performed and yields the action of the GN model:

$$S\left(\bar\psi, \psi\right) = -\int d^d x \left[\bar\psi \cdot \slashed{\partial}\psi + \frac{g^2}{2m^2}\left(\bar\psi \cdot \psi\right)^2\right].$$

The GN model and the linearized model are thus equivalent for the large distance physics. In the GN model, in the large N limit, the σ particle appears as a $\bar\psi\psi$ boundstate at threshold.

Conversely, it would seem that the GN model depends on a smaller number of parameters than its renormalizable extension. Again this problem is only interesting in four dimensions where corrections to scaling, i.e. to free field theory, are important. However, if we examine the divergences of the term $\text{tr} \ln (\partial\!\!\!/ + \sigma)$ in the effective action $(A30.14)$ relevant for the large N limit, we find a local polynomial in σ of the form:

$$\int \mathrm{d}^4x \left[A\sigma^2(x) + B\left(\partial_\mu \sigma\right)^2 + C\sigma^4(x) \right].$$

Therefore the value of the determinant can be modified by a local polynomial of this form by changing the way the cut-off is implemented: additional parameters, as in the case of the non-linear σ-model, are hidden in the cut-off procedure. Near two dimensions these operators can be identified with $(\bar\psi\psi)^2, [\partial_\mu(\bar\psi\psi)]^2, (\bar\psi\psi)^4$. It is clear that by changing the cut-off procedure we change the amplitude of higher dimension operators. These bare operators in the IR limit have a component on all lower dimensional renormalized operators.

Note finally that we could have added to the linear model an explicit breaking term linear in the σ field, which becomes a fermion mass term in the GN model, and which would have played the role of the magnetic field of the ferromagnets.

A30.3 The Gross–Neveu Model near Two Dimensions

The Gross–Neveu model plays with respect to model $(A30.2)$ the role the non-linear σ-model plays with respect to the ϕ^4 field theory. In perturbation theory it is renormalizable in two dimensions, and describes only one phase. The main difference is that the role of the spontaneously broken and the explicitly symmetric phase are interchanged. This is due to the property that it is always the massless phase which is unstable in low dimensions.

In perturbation theory the GN model is renormalizable in two dimensions. In this dimension the symmetry group is really $O(2N)$, as one verifies after some relabelling of the fields. Therefore the $(\bar\psi\psi)^2$ interaction is multiplicatively renormalized. As a function of the cut-off Λ the bare correlation functions thus satisfy the RG equations:

$$\left[\Lambda\frac{\partial}{\partial\Lambda} + \beta(G)\frac{\partial}{\partial G} - \frac{n}{2}\eta_\psi(G) \right] \Gamma^{(n)}\left(p_i; G, \Lambda\right) = 0. \tag{A30.25}$$

A direct calculation of the β-function in $d = 2 + \varepsilon$ dimension yields:

$$\beta(G) = \varepsilon G - (N' - 2)\frac{G^2}{2\pi} + (N' - 2)\frac{G^3}{4\pi^2} + O\left(G^4\right), \tag{A30.26}$$

where in generic dimensions $N' = NK_d$ (K_d is here the identity in the space of γ matrices): N' counts the total number of fermionic variables. For $d = 2$ $N' = 2N$.

Note that $N' = 2$ corresponds, as we shall see in Chapter 32, to the special case of the Thirring model (because for $N' = 2$ $(\bar\psi\gamma_\mu\psi)^2 = -2(\bar\psi\psi)^2$) for which the β-function vanishes identically in two dimensions. The latter model is the equivalent of the $O(2)$ σ-model to which it is actually related (see Chapter 32).

Finally the field RG function $\eta_\psi(G)$ is

$$\eta_\psi(G) = \frac{N' - 1}{8\pi^2}G^2 - \frac{(N' - 1)(N' - 2)}{32\pi^3}G^3 + O\left(G^4\right). \tag{A30.27}$$

A30.3.1 Two Dimensions

For $d = 2$ the model is asymptotically free. The spectrum, however, is non-perturbative, and many arguments lead to the conclusion that the chiral symmetry is always broken and a fermion mass generated. S-matrix considerations have then led to the conjecture that, for N finite, the spectrum is:

$$m_n = M \frac{2(N-1)}{\pi} \sin \left(\frac{n\pi}{2(N-1)} \right), \quad n = 1, 2 \ldots < N, \ N > 2,$$

where the parameter M is a mass-scale whose dependence on the coupling constant can be derived from RG arguments. This result is consistent with the spectrum for N large evaluated by semiclassical methods. In particular the ratio of the masses of the fundamental fermion and the lowest lying boson is:

$$\frac{m_\sigma}{m_\psi} = 2 \cos \left(\frac{\pi}{2(N-1)} \right). \tag{A30.28}$$

Note that the two first values of N are special, the model $N = 2$ is conjectured to be equivalent to two decoupled sine-Gordon models.

A30.3.2 Dimension $d = 2 + \varepsilon$

As in the case of the σ-model, asymptotic freedom implies the existence of a non-trivial UV fixed point G_c, in $2 + \varepsilon$ dimension

$$G_c = \frac{2\pi}{N'-2} \varepsilon \left(1 - \frac{\varepsilon}{N'-2} \right) + O\left(\varepsilon^3\right).$$

G_c is also the critical coupling constant for the transition between a phase in which the chiral symmetry is spontaneously broken and a massless small G phase.

At the fixed point one finds the correlation length exponent ν:

$$\nu^{-1} = -\beta'(G_c) = \varepsilon - \frac{\varepsilon^2}{N'-2} + O\left(\varepsilon^3\right), \tag{A30.29}$$

and the fermion field dimension $[\psi]$:

$$2[\psi] = d - 1 + \eta_\psi(G_c) = 1 + \varepsilon + \frac{N'-1}{2(N'-2)^2} \varepsilon^2 + O\left(\varepsilon^3\right). \tag{A30.30}$$

As for the σ-model the existence of a non-trivial UV fixed point shows that the large momentum behaviour is not given by perturbation theory above two dimensions. However, to study this problem in higher dimensions another method is required, like the $1/N$ expansion which we have considered in the previous section.

31 TWO-DIMENSIONAL MODELS AND BOSONIZATION METHOD

In Chapter 30 we have discussed the generic $O(N)$ non-linear σ-model. We have noticed that in two dimensions the case $N = 2$ is special because the RG β-function vanishes. This particular case is specially important and will be examined in Chapter 32: The $O(2)$ invariant ferromagnet in two dimensions provides an example of the celebrated Kosterlitz–Thouless phase transition. However, a consistent and simple discussion of this model requires a new technique: We have to establish relations, special to two dimensions, between fermion and boson models, a method called bosonization. The derivation involves several steps which we illustrate in this chapter with the help of various two-dimensional models which are physically interesting in their own right.

We first study the free massless boson and fermion fields. We evaluate the determinant of the covariant fermion derivative in presence of an external gauge field. This will allow to prove the basic relations of the bosonization technique. We then discuss the sine-Gordon (SG) model. We solve the Schwinger model, QED with massless fermions. Finally we demonstrate the equivalence between the SG model and two fermion models with current–current interaction: the Thirring model and another model with two species of fermions.

Massless (unitary) field theories are conformal invariant. In 2 dimensions this is a very strong property because the conformal group is infinite dimensional. A powerful technology has been developed to study them. A general discussion of conformal field theory (CFT) goes much beyond the scope of this work and the interested reader is referred to the abundant literature. Here we limit ourselves to simpler considerations. The functional integral technique will be used everywhere although in some cases the operator formulation may lead to more elegant derivations.

The appendix contains some additional remarks concerning the $SU(N)$ Thirring model and solitons in the SG model.

Finally note that unlike in most other chapters, we here work with correlation functions expressed in terms of space variables rather their Fourier transform. Renormalization then takes the form of defining singular distributions (in the mathematical sense).

31.1 The Free Massless Scalar Field

Before discussing various models it is necessary to examine the peculiar properties of massless free fields in two dimensions. Let us consider the action S for a free massless scalar field $\varphi(x)$:

$$S(\varphi) = \frac{1}{2} \int \mathrm{d}^2 x \, [\partial_\mu \varphi(x)]^2 \,. \tag{31.1}$$

The action is invariant under constant translations of the field

$$\varphi(x) = \varphi'(x) + \theta \,. \tag{31.2}$$

To this symmetry corresponds a conserved current J_μ^V

$$J_\mu^V(x) = \partial_\mu \varphi(x). \tag{31.3}$$

Note a peculiarity of dimension two: there exists another trivially conserved current

$$J_\mu^A(x) = \epsilon_{\mu\nu}\partial_\mu\varphi(x) = \epsilon_{\mu\nu}J_\nu^V(x), \qquad (31.4)$$

where $\epsilon_{\mu\nu}$ is the antisymmetric tensor with $\epsilon_{12} = 1$.

The propagator. The field propagator $\Delta(x)$ is:

$$\Delta(x) = \frac{1}{(2\pi)^2} \int \frac{\mathrm{d}^2 p}{p^2} e^{ipx}. \qquad (31.5)$$

We see that in dimension two the propagator of the massless boson field is IR divergent. The fluctuations which translate the field $\varphi(x)$ by an almost constant function are not damped by the action (31.1) and are responsible for this divergence.

To discuss the nature of the IR divergence we give a small mass m to the field φ. For m small the massive propagator behaves like:

$$\begin{aligned}
\Delta(x,m) &= \frac{1}{(2\pi)^2} \int \frac{\mathrm{d}^2 p}{p^2 + m^2} e^{ipx} \\
&= -\frac{1}{4\pi} \left(\ln(m^2 x^2/4) + 2\gamma \right) + O(m),
\end{aligned} \qquad (31.6)$$

where $\gamma = -\psi(1)$ is Euler's constant. At $x = 0$ the massive propagator is UV divergent. Introducing a UV cut-off Λ we find:

$$\Delta(0,m) \underset{\Lambda\to\infty}{=} \frac{1}{2\pi} \ln(\Lambda/m) + K + o\left(\Lambda^{-1}\right),$$

where K is a constant which depends on the cut-off procedure. For convenience we absorb it hereafter in the definition of Λ, setting:

$$\Delta(0,m) = \frac{1}{2\pi} \left[\ln(2\Lambda/m) - \gamma \right] + o\left(\Lambda^{-1}\right). \qquad (31.7)$$

IR finite correlation functions. Let us now show that, although φ itself has IR divergent correlation functions, some local functions of φ have IR finite correlation functions. For instance the correlation functions of exponentials of the field φ are given by

$$\left\langle \prod_{i=1}^n e^{i\kappa_i\varphi(x_i)} \right\rangle = \int [\mathrm{d}\varphi] \exp\left\{ -\frac{1}{2} \int \mathrm{d}^2 x \left[(\partial_\mu\varphi)^2 + m^2\varphi^2 \right] + i\sum_i \kappa_i\varphi(x_i) \right\}, \qquad (31.8)$$

where the κ_i's are arbitrary coefficients. We set:

$$J(x) = i\sum_i \kappa_i \, \delta(x - x_i). \qquad (31.9)$$

The integral in equation (31.8) takes the form:

$$\int [\mathrm{d}\varphi] \exp\left\{ -\int \mathrm{d}^2 x \left[\tfrac{1}{2}(\partial_\mu\varphi)^2 + \tfrac{1}{2}m^2\varphi^2(x) - J(x)\varphi(x) \right] \right\}.$$

We recognize the first functional integral we have calculated in field theory:

$$\left\langle \prod_{i=1}^{n} e^{i\kappa_i \varphi(x_i)} \right\rangle = \exp \cdot \left[\tfrac{1}{2} \int \mathrm{d}^2 x \, \mathrm{d}^2 y \, J(x) \Delta(x - y, m) J(y) \right],$$

$$= \exp \left[-\tfrac{1}{2} \sum_{i,j} \kappa_i \kappa_j \Delta \left(x_i - x_j, m \right) \right]. \tag{31.10}$$

Replacing the propagator by its small m expansion (equation (31.6)) and collecting all IR divergent terms, we find:

$$\sum_{i,j} \kappa_i \kappa_j \Delta \left(x_i - x_j, m \right) = -\frac{1}{2\pi} \left[\left(\sum_i \kappa_i \right)^2 (\ln m - \ln 2 + \gamma) - \sum_i \kappa_i^2 \ln \Lambda \right.$$

$$\left. + \sum_{i \neq j} \kappa_i \kappa_j \ln |x_i - x_j| \right] + O(m). \tag{31.11}$$

Therefore when m goes to zero only the products such that $\sum_i \kappa_i = 0$ survive. This result is related to the simple property that in two dimensions the Coulomb potential created by point-like charges decreases at large distance only if the total system is neutral.

Finally in equation (31.11) we see that the UV divergences can be removed by a multiplicative renormalization ζ_i of the composite fields $e^{i\kappa_i \varphi(x)}$:

$$e^{i\kappa_i \varphi(x)} = \zeta_i \left[e^{i\kappa_i \varphi(x)} \right]_{\text{ren.}}, \quad \zeta_i = (\Lambda/\mu)^{-\kappa_i^2/(4\pi)}, \tag{31.12}$$

where μ is a renormalization scale. Thus finally:

$$\left\langle \prod_{i=1}^{n} e^{i\kappa_i \varphi(x_i)} \right\rangle \Bigg|_{\text{ren.}} = \prod_{i<j} (\mu |x_i - x_j|)^{\kappa_i \kappa_j / 2\pi}. \tag{31.13}$$

Note that the correlation functions of exponentials of the field decay algebraically at large distance.

Remark. In a translation (31.2) of the φ-field the field $e^{i\kappa_i \varphi}$ becomes

$$e^{i\kappa_i \varphi(x)} \mapsto e^{i\kappa_i \theta} e^{i\kappa_i \varphi(x)}.$$

All non-vanishing correlation functions are invariant under this transformation, a result in agreement with the analysis in previous chapters of spontaneous breaking of continuous symmetries in two dimensions.

The two charge example. A particular case of equation (31.13) is of special interest, when the κ_i's take only two values $\pm \kappa$. Then only products with an equal number of $\pm$ signs do not vanish

$$\left\langle \prod_{i=1}^{n} e^{i\kappa[\varphi(x_i) - \varphi(y_i)]} \right\rangle \Bigg|_{\text{ren.}} = \frac{\prod_{i<j} (\mu |x_i - x_j|)^{\kappa^2/2\pi} (\mu |y_i - y_j|)^{\kappa^2/2\pi}}{\prod_{i,j} (\mu |x_i - y_j|)^{\kappa^2/2\pi}}. \tag{31.14}$$

The 2-point function is given by:

$$\left\langle e^{i\kappa \varphi(x)} e^{-i\kappa \varphi(0)} \right\rangle \propto x^{-\kappa^2/2\pi}. \tag{31.15}$$

The dimension of the operator $e^{i\kappa\varphi(x)}$ follows

$$[e^{i\kappa\varphi(x)}] = \kappa^2/4\pi \, . \tag{31.16}$$

Note that for $\kappa^2/2\pi \geq 2$ the 2-point function is singular at short distance in the sense of distributions. It has to be renormalized as the UV divergences of its Fourier transform show

$$\int d^2x \frac{e^{ipx}}{x^{\kappa^2/2\pi}} = \pi \frac{\Gamma(1 - \kappa^2/4\pi)}{\Gamma(\kappa^2/4\pi)} \left(\frac{p^2}{4}\right)^{\kappa^2/4\pi - 1} \, .$$

This expression is valid for $\kappa^2 < 4\pi$. At $\kappa^2 = 4\pi$ a divergent constant has to be subtracted. For $4\pi \leq \kappa^2 < 8\pi$ it is thus defined only up an arbitrary additive renormalization. For $\kappa^2 = 8\pi$ a second additive renormalization proportional to p^2 is required. We shall again meet these two special values when we discuss the SG model.

The currents. The currents (31.3,31.4) provide another example of fields with IR finite correlation functions. For instance in Fourier space the 2-point functions are:

$$\left\langle \tilde{J}^V_\mu(k) \tilde{J}^V_\nu(-k) \right\rangle = k_\mu k_\nu / k^2 \, , \tag{31.17a}$$

$$\left\langle \tilde{J}^A_\mu(k) \tilde{J}^A_\nu(-k) \right\rangle = \delta_{\mu\nu} - k_\mu k_\nu / k^2 \, . \tag{31.17b}$$

After Fourier transformation the 2-point functions become

$$\left\langle J^V_\mu(x) J^V_\nu(0) \right\rangle = \frac{1}{2\pi x^2} \left(\delta_{\mu\nu} - 2\frac{x_\mu x_\nu}{x^2}\right) \, , \tag{31.18a}$$

$$\left\langle J^A_\mu(x) J^A_\nu(0) \right\rangle = -\frac{1}{2\pi x^2} \left(\delta_{\mu\nu} - 2\frac{x_\mu x_\nu}{x^2}\right) \, . \tag{31.18b}$$

We find proportional results (as imposed by current conservation), although in Fourier space the functions are different: The r.h.s. involves a singular distribution ambiguous at $x = 0$, defined up to the addition of a $\delta^{(2)}(x)$ function.

It is also easy to calculate the effect of the insertion of the current in a correlation function of exponentials, since this involves calculating a one-point function in an external source. We find

$$\left\langle J^V_\mu(x) \prod_{i=1}^n e^{i\kappa_i\varphi(x_i)} \right\rangle = i \sum_i \kappa_i \partial^x_\mu \Delta(x - x_i) \left\langle \prod_{i=1}^n e^{i\kappa_i\varphi(x_i)} \right\rangle$$

$$= -\frac{i}{2\pi} \sum_i \kappa_i \frac{(x - x_i)_\mu}{(x - x_i)^2} \left\langle \prod_{i=1}^n e^{i\kappa_i\varphi(x_i)} \right\rangle \, .$$

Complex coordinates. It is a peculiarity of the dimension two that in real time massless fields can be decomposed into left and right moving components. In euclidean space this corresponds to a description of the $x = \{\xi_1, \xi_2\}$ plane in terms of complex coordinates

$$z = \xi_1 + i\xi_2 \, , \qquad \bar{z} = \xi_1 - i\xi_2 \, , \tag{31.19}$$

which leads to the action

$$S(\varphi) = \int dz d\bar{z} \, \partial_z \varphi \partial_{\bar{z}} \varphi \, , \tag{31.20}$$

and to a decomposition of the field into analytic and anti-analytic components

$$\varphi\left(\xi_1, \xi_2\right) = \varphi_+(z) + \varphi_-(\bar{z})$$

with the propagators

$$\Delta(\xi_1, \xi_2) = \langle\varphi_+(z)\varphi_+(0)\rangle + \langle\varphi_-(\bar{z})\varphi_-(0)\rangle,$$

$$\langle\varphi_+(z)\varphi_+(0)\rangle = -\frac{1}{4\pi}\ln z + \text{const.}, \quad \langle\varphi_-(\bar{z})\varphi_-(0)\rangle = -\frac{1}{4\pi}\ln\bar{z} + \text{const.}.$$

These complex variables are particularly well suited for exploring the consequences of conformal symmetry: the action (31.20) is obviously invariant in the change $z = f(z'), \bar{z} = \bar{f}(z')$. They will appear naturally in the discussion of massless fermions.

31.2 The Free Massless Dirac Fermion

Let us consider the massless Dirac fermion action:

$$S\left(\bar{\psi}, \psi\right) = -\int \mathrm{d}^2 x\, \bar{\psi}(x)\slashed{\partial}\psi(x). \tag{31.21}$$

The massless classical action has two $U(1)$ symmetries corresponding to phase and chiral phase transformations

$$\begin{cases} \psi(x) = \mathrm{e}^{i(\theta_S\gamma_S + i\theta)}\,\psi'(x), \\ \bar{\psi}(x) = \bar{\psi}'(x)\,\mathrm{e}^{i(\gamma_S\theta_S - i\theta)}. \end{cases} \tag{31.22}$$

A peculiarity of dimension two is that the corresponding vector and axial currents

$$J_\mu(x) = \bar{\psi}(x)\gamma_\mu\psi(x), \qquad J_\mu^S(x) = i\bar{\psi}(x)\gamma_S\gamma_\mu\psi(x), \tag{31.23}$$

are related since equation $(A5.2)$ implies:

$$i\gamma_S\gamma_\mu = -\epsilon_{\mu\nu}\gamma_\nu, \tag{31.24}$$

$$\Rightarrow \quad J_\mu^S = -\epsilon_{\mu\nu}J_\nu. \tag{31.25}$$

The free massless fermion propagator is (equation (5.58)):

$$[\Delta_\psi]_{\alpha\beta}(x) \equiv \langle\bar{\psi}'_\alpha(x)\psi'_\beta(0)\rangle = \frac{1}{4\pi^2}\int \mathrm{d}^2 p\, \mathrm{e}^{-ipx}[i\slashed{p}]_{\beta\alpha}^{-1} = -\frac{1}{2\pi}[\slashed{x}]_{\beta\alpha}^{-1}. \tag{31.26}$$

Let us now decompose the massless fermion into left and right moving components $\frac{1}{2}(1\pm\gamma_S)\psi$, $\frac{1}{2}\bar{\psi}(1\mp\gamma_S)$ (see Appendix A5.4). We denote by $\psi_\pm, \bar{\psi}_\pm$ the corresponding components: In the representation in which γ_S is diagonal we write

$$\psi = \begin{pmatrix} \psi_+ \\ \psi_- \end{pmatrix}, \quad \bar{\psi} = \begin{pmatrix} \bar{\psi}_- \\ \bar{\psi}_+ \end{pmatrix}. \tag{31.27}$$

With this parametrization the vector current becomes

$$\bar{\psi}(x)\gamma_1\psi(x) = \bar{\psi}_+\psi_+ + \bar{\psi}_-\psi_-, \quad \bar{\psi}(x)\gamma_2\psi(x) = i\left(\bar{\psi}_+\psi_+ - \bar{\psi}_-\psi_-\right). \tag{31.28}$$

In the complex notation (31.19) the action (31.21) becomes

$$S\left(\bar{\psi}, \psi\right) = -\int \mathrm{d}z\mathrm{d}\bar{z}\,\left(\bar{\psi}_+\partial_{\bar{z}}\psi_+ + \bar{\psi}_-\partial_z\psi_-\right).$$

(31.29)

In the euclidean formalism the left and right movers become analytic and anti-analytic fields. The corresponding propagators Δ_ψ^+ and Δ_ψ^- then read:

$$\Delta_\psi^+ \equiv \left\langle \bar{\psi}_+\left(\bar{z}, z\right)\psi_+(0,0)\right\rangle = -\frac{1}{2\pi z},$$

(31.30)

$$\Delta_\psi^- \equiv \left\langle \bar{\psi}_-\left(\bar{z}, z\right)\psi_-(0,0)\right\rangle = -\frac{1}{2\pi\bar{z}}.$$

(31.31)

The $\psi_\pm$ correlation functions. Let us calculate the $2n$-point correlation functions of $\psi_\pm, \bar{\psi}_\pm$. From Wick's theorem for fermions (equation (1.96)) we obtain

$$\left\langle \prod_{i=1}^n \bar{\psi}_+(x_i)\psi_+(x_i')\right\rangle = \det \Delta_\psi^+(z_i - z_j') = \left(\frac{-1}{2\pi}\right)^n \det \frac{1}{z_i - z_j'},$$

(31.32a)

$$\left\langle \prod_{i=1}^n \bar{\psi}_-(x_i)\psi_-(x_i')\right\rangle = \det \Delta_\psi^-(\bar{z}_i - \bar{z}_j') = \left(\frac{-1}{2\pi}\right)^n \det \frac{1}{\bar{z}_i - \bar{z}_j'},$$

(31.32b)

where here x_i stands for the pair $\{z_i, \bar{z}_i\}$.

In he r.h.s. we find Cauchy determinants which satisfy the simple identity:

$$(-1)^{n+1} \det \frac{1}{z_i - z_j'} = \frac{\prod_{i<j}\left(z_i - z_j\right)\left(z_i' - z_j'\right)}{\prod_{i,j}(z_i - z_j')},$$

(31.33)

and its complex conjugated. This identity allows to relate boson and fermion quantum field theories in two dimensions.

31.2.1 Bilinear operator correlations and boson–fermion correspondence

Let us now consider the two composite operators

$$\sigma_+(x) = \bar{\psi}_-(x)\psi_+(x)\,,\qquad \sigma_-(x) = \bar{\psi}_+(x)\psi_-(x),$$

(31.34)

linear combinations of the scalar and pseudoscalar operators:

$$\bar{\psi}(x)\psi(x) = \sigma_+(x) + \sigma_-(x)\,,\qquad \bar{\psi}(x)\gamma_5\psi(x) = \sigma_+(x) - \sigma_-(x).$$

(31.35)

From the form of the action it is clear that only correlation functions of a product of an equal number of σ_+ and σ_- fields do not vanish (conservation of chirality). Furthermore they factorize into the form:

$$\left\langle \prod_{i=1}^n \sigma_+\left(x_i\right)\sigma_-\left(x_i'\right)\right\rangle = (-1)^n \left\langle \prod_{i=1}^n \bar{\psi}_+\left(x_i'\right)\psi_+\left(x_i\right)\right\rangle \left\langle \prod_{i=1}^n \bar{\psi}_-\left(x_i\right)\psi_-\left(x_i'\right)\right\rangle$$

(31.36)

where again x_i stands for the pair $\{z_i, \bar{z}_i\}$.

We can now use the identities (31.32,31.33) and find

$$\left\langle \prod_{i=1}^{n} \sigma_+(x_i)\,\sigma_-(x_i') \right\rangle = \left(\frac{1}{2\pi}\right)^{2n} \frac{\prod_{i<j}|x_i-x_j|^2\,|x_i'-x_j'|^2}{\prod_{i,j}|x_i-x_j'|^2}. \tag{31.37}$$

The r.h.s. has short distance divergences associated with the required additive renormalization of the $\sigma_+\sigma_-$ 2-point function.

Comparing with equation (31.14) we discover an identity between averages with the free massless fermion and boson actions:

$$\left\langle \prod_{i=1}^{n} \mu^2\, e^{i\kappa(\varphi(x_i)-\varphi(x_i'))} \right\rangle\Bigg|_{\varphi\,\text{ren.}} = (2\pi)^{2n} \left\langle \prod_{i=1}^{n} \sigma_+(x_i)\,\sigma_-(x_i') \right\rangle_{\psi}$$

for $\kappa^2 = 4\pi$. This equation can be translated into a correspondence between operators

$$2\pi\sigma_\pm(x) \mapsto \mu\left[e^{\pm i\sqrt{4\pi}\varphi(x)}\right]_{\text{ren.}} = \Lambda\, e^{\pm i\sqrt{4\pi}\varphi(x)}, \tag{31.38}$$

with the definition (31.7). This remarkable relation between local theories of bosons and fermions is special to two dimensions: in higher dimensions spin degrees of freedom distinguish between bosons and fermions.

31.2.2 The massive free fermion and the sine-Gordon model

Let us consider the partition function $Z_\psi(M_+, M_-)$

$$Z_\psi(M_\pm) = \int [\mathrm{d}\psi\mathrm{d}\bar\psi]\exp[-S(\bar\psi,\psi)], \tag{31.39}$$

with the action

$$S(\bar\psi,\psi) = -\int \mathrm{d}^2x\left[\bar\psi(x)\,\slashed\partial\psi(x) + M_+(x)\sigma_+(x) + M_-(x)\sigma_-(x)\right] \tag{31.40}$$

If we expand Z_ψ in powers of the sources $M_\pm$, the term of degree n in $M_\pm$ in the expansion is an average with the action (31.21) of the form

$$\frac{1}{n!}\left\langle \left[\int \mathrm{d}^2x\left(M_+(x)\sigma_+(x) + M_-(x)\sigma_-(x)\right)\right]^n \right\rangle. \tag{31.41}$$

Evaluating the averages we obtain an expansion of the partition function $Z_\psi(M)$ of the form

$$Z_\psi(M) = \sum_n \frac{1}{(n!)^2}\int \prod_i \mathrm{d}^2x_i\,\mathrm{d}^2y_i\, \frac{M_+(x_i)M_-(y_i)\prod_{i<j}|x_i-x_j|^2\,|y_i-y_j|^2}{(2\pi)^2\,\prod_{i,j}|x_i-y_j|^2}. \tag{31.42}$$

We can also directly use the correspondence (31.38). The term (31.41) is replaced by

$$\frac{1}{n!}\left(\frac{\mu}{2\pi\zeta}\right)^n \left\langle \left[\int \mathrm{d}^2x\left(M_+(x)\,e^{i\sqrt{4\pi}\varphi(x)} + M_-(x)\,e^{-i\sqrt{4\pi}\varphi(x)}\right)\right]^n \right\rangle_{\varphi} \tag{31.43}$$

where ζ is the renormalization constant (31.12) for $\kappa^2 = 4\pi$:

$$\zeta = \mu/\Lambda \ \Rightarrow \ \mu/\zeta = \Lambda\,, \tag{31.44}$$

with the definition (31.7).

We can now sum this expansion in the boson theory. We find that the partition function $Z_\psi(M)$ is identical to the partition function of a boson field φ with the action

$$S(\varphi) = \int \mathrm{d}^2x \left\{ \tfrac{1}{2}\left[\partial_\mu\varphi(x)\right]^2 - \frac{\Lambda}{2\pi}\left[M_+(x)\,\mathrm{e}^{i\sqrt{4\pi}\varphi(x)} + M_-(x)\,\mathrm{e}^{-i\sqrt{4\pi}\varphi(x)}\right]\right\}. \tag{31.45}$$

If we take the constant $M_+ = M_- = M$ limit we find another remarkable correspondence: the free massive fermion theory is equivalent to a particular case of the sine-Gordon model. This relation will be discussed in Section 31.3 and generalized in Section 31.5.

31.2.3 The gauge invariant fermion determinant

Let us now consider the fermion action in an external gauge field B_μ:

$$S(\bar\psi, \psi, B) = -\int \mathrm{d}^2x\, \bar\psi(x)\left[\slashed{\partial} + i\slashed{B}(x)\right]\psi(x). \tag{31.46}$$

Because the vector and axial current are related, the classical action (31.46) is invariant not only under phase gauge transformations but also under chiral gauge transformations:

$$\psi(x) = \mathrm{e}^{i\gamma_S\varphi(x)}\,\psi'(x), \qquad \bar\psi(x) = \bar\psi'(x)\,\mathrm{e}^{i\gamma_S\varphi(x)}\,. \tag{31.47}$$

Indeed this transformation generates the term:

$$\bar\psi'(x)i\slashed{\partial}\varphi(x)\gamma_S\psi'(x) = \bar\psi'(x)\epsilon_{\mu\nu}\gamma_\nu\partial_\mu\varphi(x)\gamma_S\psi'(x), \tag{31.48}$$

which is cancelled by the transformation:

$$B_\mu(x) = B'_\mu(x) - i\epsilon_{\mu\nu}\partial_\nu\varphi(x)\,. \tag{31.49}$$

The field B_μ is a gauge field for two sets of gauge transformations. Since it has only two components it can be completely eliminated by gauge transformations from the classical action, which is thus equivalent to a free field action. Indeed let us parametrize B_μ as

$$B_\mu(x) = -\left[\partial_\mu\chi(x) + i\epsilon_{\mu\nu}\partial_\nu\varphi(x)\right], \tag{31.50}$$

which implies

$$i\epsilon_{\mu\nu}\partial^2\varphi(x) = \partial_\mu B_\nu - \partial_\nu B_\mu\,, \quad \text{or} \quad \partial^2\varphi(x) = -i\epsilon_{\mu\nu}\partial_\mu B_\nu\,, \tag{31.51}$$

and set

$$\begin{cases} \psi(x) = \mathrm{e}^{i(\chi(x)+\gamma_S\varphi(x))}\,\psi'(x), \\ \bar\psi(x) = \bar\psi'(x)\,\mathrm{e}^{i(-\chi(x)+\gamma_S\varphi(x))}\,. \end{cases} \tag{31.52}$$

The action (31.46) then becomes

$$S(\bar\psi, \psi, B) = -\int \mathrm{d}^2x\, \bar\psi'(x)\slashed{\partial}\psi'(x). \tag{31.53}$$

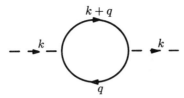

Fig. 31.1   The one-loop contribution to the current 2-point function.

Quantum theory: the anomaly. As we have more generally discussed in Section 20.3 the chiral gauge symmetry is broken at the quantum level by the chiral anomaly. Here we will recover the anomaly by a direct calculation. In Appendix A31.1.1 we show the consistency of the result obtained below with the general form derived in Section 20.3 (equation (20.106)).

The integral over the fermions fields generates a determinant which has an expansion in terms of one-loop diagrams. In two dimensions only the term quadratic in B_μ shown in figure 31.1 is divergent.

To regularize the determinant we use Pauli–Villars's type regularization, introducing a massive spinorial boson field ϕ:

$$S\left(\bar\psi, \psi, B_\mu, \phi\right) = -\int \mathrm{d}^2 x \left[\bar\psi(x)\left(\slashed\partial + i\slashed B\right)\psi(x) - \bar\phi(x)\left(\slashed\partial + i\slashed B + M\right)\phi(x)\right]. \quad (31.54)$$

With this addition the theory is finite (and thus no longer anomalous) but the ϕ mass breaks chiral symmetry explicitly. Let us parametrize the 2-component gauge field in terms of 2 scalar fields as in equation (31.50) and perform the corresponding gauge transformations (31.52) both on $\bar\psi, \psi$ and $\bar\phi, \phi$. The action becomes

$$S\left(\bar\psi', \psi', B_\mu, \phi'\right) = -\int \mathrm{d}^2 x \left[\bar\psi'(x)\slashed\partial\psi'(x) - \bar\phi'(x)\left(\slashed\partial + M\,\mathrm{e}^{2i\gamma_S\varphi(x)}\right)\phi'(x)\right]. \quad (31.55)$$

We now integrate over the fields $\bar\phi', \phi'$. The result is a functional $\mathcal{D}(\varphi)$,

$$\mathcal{D}^{-1}(\varphi) = \det\left(\slashed\partial + M\,\mathrm{e}^{2i\gamma_S\varphi(x)}\right), \quad (31.56)$$

which has to be evaluated for M large. The *bosonization* technique is useful in this respect. Indeed $\mathcal{D}^{-1}(\varphi)$ can be written as a fermion integral

$$\frac{1}{\mathcal{D}(\varphi)} = \int [\mathrm{d}\psi\mathrm{d}\bar\psi] \exp \int \mathrm{d}^2 x \left[\bar\psi(x)\slashed\partial\psi(x) + \mathrm{e}^{2i\varphi(x)}\,\sigma_+(x) + \mathrm{e}^{-2i\varphi(x)}\,\sigma_-(x)\right]. \quad (31.57)$$

In Subsection 31.2.2 we have derived a fermion–boson equivalence. We can thus replace the fermion average with the action (31.40) by a boson average with the action (31.45) and $M_\pm = \mathrm{e}^{\pm 2i\varphi}$. Calling $\vartheta(x)$ the SG field we find a modified SG model (31.45) with ϑ replaced by $\vartheta + \varphi/\sqrt{\pi}$ in the interaction:

$$S(\vartheta, \varphi) = \int \mathrm{d}^2 x \left\{\frac{1}{2}\left[\partial_\mu\vartheta(x)\right]^2 - \frac{M\Lambda}{\pi}\cos\left(\sqrt{4\pi}\vartheta + 2\varphi\right)\right\}. \quad (31.58)$$

We now translate ϑ setting $\vartheta + \varphi/\sqrt{\pi} \mapsto \vartheta$. The action becomes

$$S'(\vartheta) = \int \mathrm{d}^2 x \left\{\frac{1}{2}\left[\partial_\mu\vartheta(x) - \partial_\mu\varphi(x)/\sqrt{\pi}\right]^2 - \frac{M\Lambda}{\pi}\cos\sqrt{4\pi}\vartheta(x)\right\}. \quad (31.59)$$

In the large M limit the ϑ-field becomes very massive and has thus vanishing fluctuations around $\vartheta(x) = 0$. At leading order we can set $\vartheta = 0$ and thus find the finite result

$$\mathcal{D}^{-1}(\varphi) \propto \exp\left[-\frac{1}{2\pi}\int d^2x\big(\partial_\mu\varphi(x)\big)^2\right]. \tag{31.60}$$

Since $\mathcal{D}(\varphi)$ does not go to a constant for M large, the chiral gauge symmetry is broken by quantum fluctuations: this is the simplest example of a chiral anomaly.

The action (31.46) is thus equivalent to

$$S(\bar\psi',\psi',B_\mu) = -\int d^2x\left[\bar\psi'(x)\slashed{\partial}\psi'(x) + \frac{1}{2\pi}\int d^2x\big(\partial_\mu\varphi(x)\big)^2\right]. \tag{31.61}$$

31.2.4 Current correlation functions

Note first that the expressions (31.32) are singular when $x_i \to x_i'$, therefore they cannot be used to define the current without a short distance regularization. We can avoid this problem because we note that in the action (31.46) the gauge field acts as a source for the current: Differentiating with respect to B_μ yields iJ_μ.

Current 2-point function. From the result (31.60) expanded at second order in φ and the relation (31.51) we obtain the current 2-point function

$$\left\langle \tilde J_\mu(k)\tilde J_\nu(-k)\right\rangle = \frac{1}{\pi}\left(\delta_{\mu\nu} - \frac{k_\mu k_\nu}{k^2}\right),$$

$$\langle J_\mu(x)J_\nu(0)\rangle = -\frac{1}{2\pi^2 x^2}\left(\delta_{\mu\nu} - 2\frac{x_\mu x_\nu}{x^2}\right),$$

expressions proportional to the 2-point functions (31.17*b*,31.18*b*) of the bosonic current. Note that the second expression is a distribution singular at $x=0$ where it is ambiguous.

Using then equation (31.25) we obtain the axial current 2-point function

$$\left\langle \tilde J_\mu^S(k)\tilde J_\nu^S(-k)\right\rangle = \frac{1}{\pi}\frac{k_\mu k_\nu}{k^2},$$

a result identical to (31.17*a*). In the space representation both currents are proportional but their Fourier representations are different: we see that only the vector current is conserved.

Current insertion. Let us now calculate $\sigma_\pm$ correlations with one current $J_\mu(x)$ insertion. In the transformation (31.52) the operators $\sigma_\pm(x)$ become $\sigma_\pm(x)\,\mathrm{e}^{\pm 2i\varphi(x)}$. To obtain the current insertion we expand at first order in φ. This yields a factor

$$2\epsilon_{\mu\nu}\int d^2x\,\partial_\mu B_\nu(x)\sum_i [\Delta(x-x_i)-\Delta(x-x_i')].$$

Differentiating with respect to $B_\mu(x)$ we finally obtain

$$\left\langle J_\mu(x)\prod_{i=1}^n \sigma_+(x_i)\sigma_-(x_i')\right\rangle = \left\langle \prod_{i=1}^n \sigma_+(x_i)\sigma_-(x_i')\right\rangle$$

$$\times \sum_i \frac{i}{\pi}\epsilon_{\mu\nu}\left(\frac{(x-x_i)_\nu}{(x-x_i)^2} - \frac{(x-x_i')_\nu}{(x-x_i')^2}\right). \tag{31.62}$$

This expression is proportional to the boson current insertion $J_\mu^A(x)$ for $\kappa_i = \pm\sqrt{4\pi}$. In the same way the axial current J_μ^A is proportional to J_μ^S:

$$J_\mu^A(x) \;\mapsto\; \sqrt{\pi}\, J_\mu(x), \quad J_\mu^V(x) \;\mapsto\; \sqrt{\pi}\, J_\mu^S(x). \tag{31.63}$$

The second result is not surprising: a translation of φ $\varphi \mapsto \varphi + \theta$ multiplies $\mathrm{e}^{\pm i\sqrt{4\pi}\varphi}$ by $\mathrm{e}^{\pm i\sqrt{4\pi}\theta}$. To J_μ^S is associated the transformation $\sigma_\pm \mapsto \sigma_\pm \,\mathrm{e}^{\pm 2i\theta}$.

31.3 The Sine-Gordon Model

The sine-Gordon (SG) model is defined by the action $S(\vartheta)$

$$S(\vartheta) = \int \mathrm{d}^2x \left\{ \tfrac{1}{2} \left[\partial_\mu \vartheta(x) \right]^2 - \frac{\alpha_0}{\kappa^2} \cos \kappa\vartheta(x) \right\}, \tag{31.64}$$

where $\vartheta(x)$ is a scalar boson field. Note one can choose $\alpha_0 > 0$ without loss of generality. Depending on the question we want to investigate we will sometimes normalize the SG field differently, setting $\theta = \kappa\vartheta$ and thus

$$S(\theta) = \frac{1}{t} \int \mathrm{d}^2x \left\{ \tfrac{1}{2} \left[\partial_\mu \theta(x) \right]^2 - \alpha_0 \cos \theta(x) \right\} \qquad \text{with } t = \kappa^2. \tag{31.65}$$

The model has been extensively studied. The classical equations of motion are completely integrable. This allows to obtain finite energy solutions of the real-time equation of motions, *solitons*, and to infer the semiclassical spectrum. This integrability survives quantization and thus, for example, the spectrum and the S- matrix can be exactly obtained, confirming the semiclassical analysis. The relevant techniques, however, are outside the scope of this work, and will not be presented here. We merely want to study the model for some of its algebraic and RG properties.

31.3.1 Perturbative expansion

For κ small the field θ fluctuates around one of the minima $\theta = 2n\pi$ of $-\cos\theta$. We can choose one of them, $\theta = 0$, to expand perturbation theory because they are all equivalent and cannot be connected (see Section 40.3). Note, however, that the degeneracy of the classical minimum is responsible for the existence of solitons. Expanding $\cos\theta$ in powers of θ we see that the θ-field is massive with a mass $\alpha^{1/2} + O(\kappa^2)$.

The model is super-renormalizable. From the discussion of Appendix A10.2 we know that divergences, in the perturbative expansion in powers of κ, arise only from the self-contractions of the interaction term $\cos\theta$. These divergences have actually been calculated in the Section 31.1. Using equation (31.12), we can set:

$$\mathrm{e}^{\pm i\theta(x)} = \mathrm{e}^{\pm i\kappa\vartheta(x)} = Z_\theta^{1/2} \left[\mathrm{e}^{\pm i\theta(x)} \right]_{\text{ren.}}, \tag{31.66a}$$

$$\alpha_0 = Z_\theta^{-1/2} \alpha, \tag{31.66b}$$

$$\text{with} \quad Z_\theta = (\Lambda/\mu)^{-\kappa^2/(2\pi)}, \tag{31.66c}$$

in which μ is the renormalization scale.

31.3.2 RG equations

The theory is super-renormalizable and thus the β-function vanishes. The field $e^{\pm i\theta(x)}$ and the coupling constant α RG functions are given in terms of Z_θ. We find

$$\eta(t = \kappa^2) = \mu \frac{\partial}{\partial\mu} \ln Z_\theta(t, \Lambda/\mu) = \frac{t}{2\pi} . \tag{31.67}$$

The renormalized n-point correlation function $W^{(n)}$ of $e^{\pm i\theta(x)}$ thus satisfies the RG equation:

$$\left[\mu \frac{\partial}{\partial\mu} + \frac{n}{2}\eta(t) + \frac{1}{2}\eta(t)\alpha\frac{\partial}{\partial\alpha} \right] W^{(n)}(p_i, t, \alpha) = 0 . \tag{31.68}$$

Using the dimensional relation

$$\mu \frac{\partial}{\partial\mu} + 2\alpha\frac{\partial}{\partial\alpha} + p_i\frac{\partial}{\partial p_i} = 2(1 - n) ,$$

we obtain the scaling relation:

$$W^{(n)}(p_i) = \alpha^{t/(8\pi - t)} \alpha^{1-n} F^{(n)}\left(p_i \alpha^{-4\pi/(8\pi - t)}, t \right) . \tag{31.69}$$

In particular, for the 1-point function and the mass scale or θ-mass we find:

$$\left\langle e^{\pm i\theta(x)} \right\rangle = W^{(1)} \sim \alpha^{t/(8\pi - t)} , \qquad m_\theta \sim \alpha^{4\pi/(8\pi - t)} . \tag{31.70}$$

In these relations we see that two values of $t = \kappa^2$ play a special role:

(i) For $\kappa^2 = 4\pi$ the propagator (31.15) becomes singular at short distance, in the sense of distributions (or UV divergent in momentum space). Therefore the term of order α^2 in the expansion of the SG partition function requires a new additive renormalization. This yields an additive renormalization of the free energy proportional to α^2.

Also the mass scale m_θ and $\langle e^{i\theta} \rangle$ become linear in α, results which have simple interpretations in terms of the correspondence (31.45) with a free massive fermion model. Indeed $m_\theta \propto m_\psi = \alpha$ and

$$\left\langle e^{i\theta} \right\rangle \propto \left\langle \bar\psi(x)\psi(x) \right\rangle = \frac{1}{2\pi} \operatorname{tr} \int d^2p \frac{1}{i\not{p} + \alpha} \sim \alpha \ln(\Lambda^2/\alpha^2).$$

The UV divergence is the same that we have discussed above since by differentiating with respect to α we obtain the 2-point function at zero-momentum. It leads to a logarithmic correction to the linear behaviour. We shall further discuss this relation in Section 31.5. For $\kappa^2 > 4\pi$ generically the leading contribution to $W^{(1)}$ comes from short distance effects and remains linear in α.

(ii) For $\kappa^2 = 8\pi$ the quantities $\langle e^{i\theta} \rangle$ and m_θ in equations (31.70) vanish identically for α small. Let us find an interpretation of this result by calculating the dimension of the operator $\cos\theta(x)$ and more generally of $\cos l\theta(x)$:

$$[\cos l\theta] = l^2 \kappa^2/4\pi . \tag{31.71}$$

Therefore these operators, which are relevant for κ small and give a mass to the θ-field, become irrelevant beyond a finite value of κ:

$$l^2 \kappa^2/4\pi = 2 \implies \kappa^2 = 8\pi/l^2 . \tag{31.72}$$

For $\kappa^2 > 8\pi$ the last operator $\cos l\theta$ also becomes irrelevant, which explains why no mass is generated. At $\kappa^2 = 8\pi$ the interaction $\int \mathrm{d}^2x \cos\theta(x)$ is marginal. Translated into field theory language this means that the theory is just renormalizable; for $\kappa^2 > 8\pi$ the theory is no longer renormalizable. We shall discuss in Chapter 32 in more detail this peculiar transition. Note, however, that this analysis is only valid perturbatively in α since at $\kappa^2 = 8\pi$ new divergences are generated at each order in α and thus the dimension of operators can be modified.

31.4 The Schwinger Model

We now consider two-dimensional QED, with one Dirac fermion coupled to an abelian gauge field. The bosonization technique will allow to solve the massless model exactly and will provide us with some interesting information about the massive model.

31.4.1 The massless model

The corresponding action reads

$$S\left(\bar{\psi}, \psi, A_\mu\right) = \int \mathrm{d}^2x \left[\tfrac{1}{4}F_{\mu\nu}^2(x) - \bar{\psi}(x)\left(\slashed{\partial} + ie\slashed{A}\right)\psi(x)\right]. \tag{31.73}$$

This model, first discussed by Schwinger, exhibits the simplest example of a *chiral anomaly*, illustrates both *confinement* (see Chapter 33) and *spontaneous chiral symmetry breaking* in 2 dimensions.

The field theory is super-renormalizable by power counting in a covariant gauge. The only divergent diagram corresponds to the one-loop contribution to the gauge field 2-point function (figure 31.1) and comes from the expansion of the fermion determinant which we have already discussed in Subsection 31.2.3. In Subsection 31.2.3 we have shown that the fermion part of the action with $B_\mu = eA_\mu$, after the gauge transformations (31.52) and with the parametrization (31.50), is equivalent to the free field action (31.61). We have here simply to add the contribution coming from $F_{\mu\nu}^2$

$$\frac{1}{4}F_{\mu\nu}^2 = \frac{1}{2e^2}\partial_\mu\varphi\partial^2\partial_\mu\varphi.$$

In terms of the fields $\bar{\psi}', \psi', \varphi$ the Schwinger model action then takes the free field form (in the gauge $\partial_\mu A_\mu = \partial^2\chi/e = 0$)

$$S(\bar{\psi}', \psi', \varphi) = \int \mathrm{d}^2x \left[-\bar{\psi}'\slashed{\partial}\psi' - \frac{1}{2\pi}\left(\partial_\mu\varphi\right)^2 + \frac{1}{2e^2}\partial_\mu\varphi\partial^2\partial_\mu\varphi\right]. \tag{31.74}$$

Let us calculate the φ-field propagator in momentum space

$$\Delta_\varphi(p) = \pi\left(\frac{1}{p^2 + e^2/\pi} - \frac{1}{p^2}\right). \tag{31.75}$$

This expression shows that the scalar φ-field propagates two modes corresponding to a positive metric neutral massive field with mass

$$m = e/\sqrt{\pi}\,, \tag{31.76}$$

and a massless mode with negative metric. The appearance of a non-vanishing mass is a direct consequence of the chiral anomaly.

Finally, using the representation (31.50) we can calculate the gauge field transverse 2-point function. We find

$$\Delta^{(2)}_{\mu\nu}(p) = \left(\delta_{\mu\nu} - p_\mu p_\nu/p^2\right) \frac{1}{p^2 + m^2}, \tag{31.77}$$

a result we verify in Appendix A31.1.2 by a direct one-loop calculation.

We can now also bosonize the free fermion $\bar{\psi}', \psi'$, introducing a free massless boson ϑ. The action becomes

$$S(\vartheta, \varphi) = \int \mathrm{d}^2 x \left[\frac{1}{2}\left(\partial_\mu \vartheta\right)^2 - \frac{1}{2\pi}\left(\partial_\mu \varphi\right)^2 + \frac{1}{2\pi m^2}\partial_\mu \varphi \partial^2 \partial_\mu \varphi\right]. \tag{31.78}$$

We then translate ϑ, $\vartheta + \varphi/\sqrt{\pi} \mapsto \vartheta$. The action becomes

$$S(\vartheta, \varphi) = \int \mathrm{d}^2 x \left[\frac{1}{2}\left(\partial_\mu \vartheta\right)^2 - \frac{1}{\sqrt{\pi}}\partial_\mu \varphi \partial_\mu \vartheta + \frac{1}{2\pi m^2}\partial_\mu \varphi \partial^2 \partial_\mu \varphi\right].$$

We then integrate over φ and finally obtain

$$S(\vartheta) = \tfrac{1}{2} \int \mathrm{d}^2 x \left[\left(\partial_\mu \vartheta(x)\right)^2 + m^2 \vartheta^2(x)\right]. \tag{31.79}$$

It is remarkable that the fermion contribution just cancels the massless boson field, leaving only the massive boson. Starting from the representation (31.52) it is easy to verify that, in these transformations the chiral components $\sigma_\pm$ of the neutral composite field $\bar{\psi}\psi$ are mapped to

$$\sigma_\pm(x) = \bar{\psi}_\mp(x)\psi_\pm(x) \longmapsto \frac{\Lambda}{2\pi} \mathrm{e}^{\pm i\sqrt{4\pi}\vartheta(x)}. \tag{31.80}$$

Remark. Before integration over φ the action is formally invariant when we translate ϑ by a constant. Therefore the form (31.79) corresponds to a choice of boundary conditions on the inverse of the Laplace operator. A different choice leads to the replacement $\vartheta(x) \mapsto \vartheta(x) - \vartheta_\infty$. From the correspondence (31.80) we see that in general space reflection symmetry is then broken, except for $\vartheta_\infty = 0 \pmod{\sqrt{\pi}/2}$. Finally this modification formally corresponds to adding a topological term proportional to $\partial^2 \varphi \propto \epsilon_{\mu\nu} F_{\mu\nu}$ to the action density, i.e. a constant electric field. For a discussion of the physical effects of such a modification we refer to the literature.

31.4.2 Neutral correlation functions: confinement and chiral symmetry breaking

We note that the gauge field (31.77) 2-point function has no cut corresponding to fermion intermediate states. This is a sign of *confinement* because it means that no charged particles are emitted by the neutral particle: The electromagnetic forces between particles of opposite charge are strong enough to prevent the separation of charged particles: these cannot be observed as free particles (for details see Chapter 33).

This observation is confirmed by the form of $\sigma_\pm(x)$ correlation functions. Indeed they can be calculated with the massive free action (31.79) where no massless field appears.

Let us now calculate the expectation values of $\sigma_\pm$. From the correspondence (31.80) and the equation (31.10) we obtain

$$\langle\sigma_\pm\rangle = \frac{\Lambda}{2\pi}\left\langle e^{\pm i\sqrt{4\pi}\vartheta(x)}\right\rangle = \frac{\Lambda}{2\pi}\,e^{-2\pi\Delta(0,m)}\,.$$

Taking into account the definition (31.7) we find a finite result

$$\langle\sigma_\pm\rangle = \frac{e^\gamma}{4\pi}m \quad\Rightarrow\quad \langle\bar\psi(x)\psi(x)\rangle = \frac{e^\gamma}{2\pi}m\,. \qquad (31.81)$$

Since $\bar\psi\psi$ is a composite field which is not chiral invariant, the non-vanishing result shows that global chiral symmetry is spontaneously broken. Note that spontaneous symmetry breaking with ordering is here possible because the electromagnetic interaction generates long range forces.

Let us now calculate from equation (31.10) the 2-point functions of $\sigma_\pm$

$$\langle\sigma_{\epsilon_1}(x_1)\sigma_{\epsilon_2}(x_2)\rangle = \langle\sigma\rangle^2\,e^{-4\pi\epsilon_1\epsilon_2\Delta(x_1-x_2,m)}\,. \qquad (31.82)$$

It follows

$$\langle\bar\psi(x)\psi(x)\bar\psi(0)\psi(0)\rangle = \langle\bar\psi\psi\rangle^2\cosh\big(4\pi\Delta(x,m)\big)\,, \qquad (31.83a)$$

$$\langle\bar\psi(x)\gamma_S\psi(x)\bar\psi(0)\gamma_S\psi(0)\rangle = \langle\bar\psi\psi\rangle^2\sinh\big(4\pi\Delta(x,m)\big)\,. \qquad (31.83b)$$

These expressions have several remarkable properties: The 2-point functions have only singularities associated with the massive field. If we expand the exponentials in powers of the propagator and Fourier transform we find that the expression (31.83b) has a pole at $k^2 = -m^2$ and cuts at $k^2 = -(2n-1)^2m^2$, $n > 1$, in momentum space, while the expression (31.83a) has cuts at $k^2 = -(2n)^2m^2$. Only the massive neutral boson appears in the intermediate states but no charged fermions (the confinement property), and moreover the boson is a pseudoscalar since it appears as a simple pole only in the $\bar\psi\gamma_S\psi$ 2-point function.

Finally the propagator has a short distance behaviour given by (31.6), since $\Delta(m,x)$ is a function only of mx. Therefore the 2-point functions $\langle\sigma_+(x)\sigma_+(0)\rangle$ and $\langle\sigma_-(x)\sigma_-(0)\rangle$ go to zero as x^2 while

$$\langle\sigma_+(x)\sigma_-(0)\rangle \sim \frac{1}{4\pi^2x^2}\,,$$

i.e. like in a free massless fermion theory. This property reflects the *asymptotic freedom* (at large momentum) of super-renormalizable theories.

These are all properties we also expect in the true physical world with quarks and gluons.

31.4.3 The massive Schwinger model

Let us now briefly discuss the effect of the addition of a fermion mass term. Of course chiral symmetry is no longer an issue since the mass term explicitly breaks chiral symmetry. However there is still the problem of confinement: Will charged particles appear in the spectrum?

We thus consider the action:

$$S\left(\bar\psi,\psi,A_\mu\right) = \int \mathrm{d}^2x\left[\tfrac{1}{4}F_{\mu\nu}^2(x) - \bar\psi(x)\left(\slashed\partial + ie\slashed A + M\right)\psi(x)\right]\,. \qquad (31.84)$$

We now perform the transformations (31.47) and (31.50) ($B_\mu = eA_\mu$) and find:

$$S(\bar\psi',\psi',\varphi) = -\int \mathrm{d}^2x \left[\bar\psi'\left(\slashed\partial + M\,\mathrm{e}^{2i\gamma_S\varphi}\right)\psi' + \frac{1}{2e^2}\partial_\mu\varphi\left(m^2-\partial^2\right)\partial_\mu\varphi\right]. \quad (31.85)$$

The model is no longer free and cannot be solved exactly. However the action can be further transformed, as shown in Subsection 31.2.2. We call ϑ the boson field associated by bosonization with $\bar\psi',\psi'$. We then use the equivalence between the actions (31.40) and (31.45). The fermion action is replaced by the action (31.58)

$$S(\vartheta,\varphi) = \int \mathrm{d}^2x \left\{\tfrac12[\partial_\mu\vartheta(x)]^2 - \frac{M\Lambda}{\pi}\cos\left(\sqrt{4\pi}\vartheta + 2\varphi\right) - \frac{1}{2e^2}\partial_\mu\varphi\left(m^2-\partial^2\right)\partial_\mu\varphi\right\}.$$

We now translate ϑ setting $\vartheta+\varphi/\sqrt\pi \mapsto \vartheta$. After translation the action becomes quadratic in φ. We can thus integrate over φ and find a "massive" SG action for ϑ:

$$S(\vartheta) = \int \mathrm{d}^2x \left\{\tfrac12[\partial_\mu\vartheta(x)]^2 + \tfrac12 m^2\vartheta^2(x) - \alpha_0\cos\sqrt{4\pi}\vartheta(x)\right\}, \quad (31.86)$$

with the correspondence (31.80) and

$$\alpha_0 = M\Lambda/\pi.$$

Physical consequences. We see in action (31.86) that at least for M small the result obtained for $M=0$ survives: The spectrum consists in a massive neutral boson of mass squared $m^2 + O(mM)$. No charged particles appear in the spectrum. Note on the other hand that for M large we expect a non-relativistic analysis to be valid. Then we have a set of fermions interacting through a one-dimensional Coulomb potential. This potential raises, at large distance, linearly with the distance between charged particles and thus charged particles can never be separated.

31.5 The Massive Thirring Model

In Section 31.2, comparing equations (31.13,31.37), we have discovered identities between free massless fermion averages of powers of $\bar\psi\psi$ and averages of exponentials of free massless boson fields. This has led to the correspondence derived in Subsection 31.2.2 between the free massive fermion and the SG model for $\kappa^2 = 4\pi$. The same identities will allow us to establish a relation between the general quantum SG and the massive Thirring model. The Thirring model is described in terms of Dirac fermions with an action:

$$S(\bar\psi,\psi) = -\int \mathrm{d}^2x\left[\bar\psi\left(\slashed\partial + m_0\right)\psi - \tfrac12 gJ_\mu J_\mu\right], \quad (31.87)$$

where

$$J_\mu = \bar\psi\gamma_\mu\psi. \quad (31.88)$$

For $m_0 = 0$ the Thirring model retains the $U(1)$ symmetries (31.22) of the free massless action. Note that the interaction is the only local interaction possible since we have only 4 fermion variables. Finally power counting shows that the model is renormalizable in two dimensions and the coupling constant g thus dimensionless.

The massive Thirring model can be mapped onto the general SG model. The correspondence between the two models can be summarized by the relations:

$$1 + g/\pi = 4\pi/\kappa^2 \,, \tag{31.89a}$$

$$J_\mu \longmapsto \frac{1}{2\pi} \epsilon_{\mu\nu} \partial_\nu \theta \,, \tag{31.89b}$$

$$\bar\psi\psi \longmapsto \frac{\Lambda}{\pi} \cos\theta \,. \tag{31.89c}$$

The chiral invariant model, $m_0 = 0$, is mapped onto a free boson theory, as we have already discussed in Section 31.3, and thus exactly soluble.

We derive below this correspondence, using the bosonization identities established in previous sections.

31.5.1 Bosonization: the sine-Gordon model

The first step of the bosonization is to introduce a vector field A_μ and rewrite the interaction term as a gaussian integral over A_μ:

$$\tfrac{1}{2} g J_\mu J_\mu \longmapsto A_\mu^2/2g + iA_\mu J_\mu \,. \tag{31.90}$$

The fermion action becomes a quadratic action of charged fermions coupled to an abelian gauge field:

$$S(A_\mu, \bar\psi, \psi) = -\int \mathrm{d}^2x \left[\bar\psi \left(\slashed\partial + i\slashed A + m_0 \right) \psi - A_\mu^2/2g \right] \,. \tag{31.91}$$

Bosonization. The action then differs from the massive Schwinger model action (31.84) only by the kinetic gauge field term. As we discussed in Subsection 31.2.2, the action can thus be replaced by a purely bosonic action. Parametrizing the 2-component field A_μ in terms of 2 scalar fields χ and φ as in (31.50) we find

$$\int \mathrm{d}^2x \, A_\mu^2(x) = \int \mathrm{d}^2x \left[\left(\partial_\mu \chi(x) \right)^2 - \left(\partial_\mu \varphi(x) \right)^2 \right] \,.$$

After the gauge transformations (31.52) the action (31.91) then becomes:

$$S(\chi, \varphi, \bar\psi', \psi') = -\int \mathrm{d}^2x \, \bar\psi' \left(\slashed\partial + m_0 \, \mathrm{e}^{2i\gamma_S \varphi} \right) \psi'$$
$$+ \frac{1}{2} \int \mathrm{d}^2x \left[\frac{1}{g} \left(\partial_\mu \chi \right)^2 - \left(\frac{1}{g} + \frac{1}{\pi} \right) \left(\partial_\mu \varphi \right)^2 \right] \,, \tag{31.92}$$

where we have taken into account the anomaly.

The field χ decouples from neutral correlation functions. The kinetic term of the field φ has the wrong sign. Averages of products of $\mathrm{e}^{\pm i\varphi}$ when the numbers of $+$ and $-$ signs are different are thus IR divergent. Fortunately, they always appear multiplied by fermion averages which, as we have indicated in Subsection 31.2.1, due to chiral symmetry vanish.

The correspondence between the massive Thirring and SG models now follows directly from the analysis of Subsection 31.2.2. We have shown that the fermion action can be replaced by a SG action for $\kappa^2 = t = 4\pi$ of the form (31.58)

$$\int \mathrm{d}^2x \, \bar\psi' \left(\slashed\partial + m_0 \, \mathrm{e}^{2i\gamma_S \varphi} \right) \psi' \longmapsto \int \mathrm{d}^2x \left[\tfrac{1}{2} \left(\partial_\mu \vartheta \right)^2 - \frac{m_0 \Lambda}{\pi} \cos\left(\sqrt{4\pi}\vartheta + 2\varphi \right) \right] \,. \tag{31.93}$$

In the special case of the massless Thirring model $m_0 = 0$ we obtain an equivalent free boson theory

$$S(\vartheta, \varphi, \chi) = \tfrac{1}{2} \int d^2x \left[(\partial_\mu \vartheta)^2 + \frac{1}{g} (\partial_\mu \chi)^2 - \left(\frac{1}{g} + \frac{1}{\pi} \right) (\partial_\mu \varphi)^2 \right]. \tag{31.94}$$

The massless Thirring model can thus be solved exactly and we calculate a few correlation functions in the next subsection.

For $m_0 \neq 0$, we change variables $\varphi \mapsto \theta$, setting

$$\sqrt{4\pi}\vartheta + 2\varphi = \theta \,.$$

The integral over ϑ is then gaussian,

$$S(\vartheta, \theta) = \tfrac{1}{2} \int d^2x \left[(\partial_\mu \vartheta)^2 - \frac{2m_0 \Lambda}{\pi} \cos \theta - \frac{\pi + g}{4\pi g} (\partial_\mu \theta - \sqrt{4\pi} \partial_\mu \vartheta)^2 \right],$$

and can be explicitly performed. We finally obtain

$$S(\theta) = \int d^2x \left[\frac{1}{8\pi} \left(1 + \frac{g}{\pi} \right) (\partial_\mu \theta)^2 - \frac{m_0 \Lambda}{\pi} \cos \theta \right], \tag{31.95}$$

which establishes the correspondence between the massive Thirring model and the sine-Gordon model. Comparing with the action in the normalization (31.65) we see that the relation between parameters is $(t = \kappa^2)$

$$\frac{4\pi}{t} = 1 + \frac{g}{\pi} \,,$$

proving relations (31.89a, c). Since the coupling constant t is not renormalized this correspondence establishes that the coupling constant g is not renormalized either and thus the RG $\beta(g)$-function vanishes.

The field $e^{\pm 2i\varphi}$ instead has to be renormalized, the fermion mass thus needs a renormalization. With the regularization and renormalization at mass scale μ of Section 31.1, we find:

$$m_0 = m(\mu/\Lambda)^{g/(g+\pi)}.$$

The relation between renormalized parameters is then:

$$\alpha/\kappa^2 = m\mu/\pi \,.$$

Let us now briefly discuss the special massless case.

31.5.2 The massless Thirring model

The massless Thirring model has a $U(1)$ chiral symmetry. Adding a source term B_μ for the current J_μ we find the effective action $S(A_\mu)$ for the remaining A_μ field:

$$S(A_\mu) = \tfrac{1}{2} \int d^2x \left[\frac{1}{g} A_\mu^2 + \frac{1}{\pi} (A_\mu - iB_\mu) \left(\delta_{\mu\nu} - \frac{\partial_\mu \partial_\nu}{\partial^2} \right) (A_\nu - iB_\nu) \right]. \tag{31.96}$$

We can then integrate over the A_μ field to obtain the current 2-point correlation function. In momentum space

$$\langle J_\mu(k) J_\nu(-k) \rangle = \frac{1}{\pi + g} \left(\delta_{\mu\nu} - \frac{k_\mu k_\nu}{k^2} \right). \tag{31.97}$$

All other connected correlation functions vanish. If we instead calculate the 2-point correlation function of $\epsilon_{\mu\nu} \partial_\nu \theta$ with the free action (31.95) for $m_0 = 0$ we find:

$$\langle \epsilon_{\mu\rho} \partial_\rho \theta(k) \epsilon_{\nu\sigma} \partial_\sigma \theta(-k) \rangle = \frac{4\pi^2}{\pi + g} \left(\delta_{\mu\nu} - \frac{k_\mu k_\mu}{k^2} \right), \tag{31.98}$$

result consistent with the two relations (31.89a, b).

All ψ-field correlation functions can be calculated explicitly. In particular the bosonized form of the $\bar{\psi}(x)\psi(x)$ correlation functions depends only on $\theta(x)$ and thus can be calculated with the action (31.95) for $m_0 = 0$. The one-point function $\bar{\psi}(x)\psi(x)$ vanishes. The $\bar{\psi}\psi$ 2-point function is given by

$$\langle \bar{\psi}(x)\psi(x)\bar{\psi}(0)\psi(0) \rangle \propto \langle \cos\theta(x) \cos\theta(0) \rangle \propto x^{-\lambda},$$

with: $\lambda = t/2\pi = 2\pi/(g + \pi)$.

31.5.3 Discussion

RG properties. The massless Thirring model exhibits a scaling behaviour for all values of the coupling constant g. This is a consequence of the vanishing of the $\beta(g)$-function: This model provides an example of a line of IR fixed points. Since the $\bar{\psi}\psi$ 2-point function decreases at large distance the chiral symmetry is not spontaneously broken, although the correlation function decreases only algebraically.

As we have discussed in the case of the SG model, at the value $\kappa^2 = 8\pi$, i.e. $g = \pi/2$, something new happens, the mass term becomes irrelevant. This suggests the possibility of a phase transition, a question we examine in the next chapter. The transition point, however, is neither in the perturbative domain of the Thirring model nor of the SG model. The properties of the bosonization method suggest that it should be possible to map the SG model with $\kappa^2 = 8\pi$ on a free model by doubling the number of fermions. We indeed exhibit in next section a two fermion model which is also equivalent to the SG model.

Mass spectrum. The correspondence between the Thirring and SG models allows, in particular, to easily calculate physical quantities in the SG model for g small, i.e. for t close to 4π, or in the Thirring model for g large. Moreover we see that for $g > 0$, i.e. $t < 4\pi$, where the potential between fermions is attractive, the spectrum of the theory consists at least in one Dirac fermion and a boson bound state, corresponding to the field of the SG model (the fermion appears semiclassically in the SG model as a soliton). Actually the exact boundstate mass spectrum is given by:

$$m_n = 2m(\alpha, t) \sin(n\gamma/16), \quad n = 1, 2 \ldots < 8\pi/\gamma,$$

$$\text{with} \quad \gamma = \frac{t}{1 - t/8\pi} = \frac{8\pi}{1 + 2g/\pi},$$

where $m(\alpha, t)$ is the fermion mass (see also equation (32.21)):

$$m(\alpha, t) = f(t) \frac{8}{\gamma} \alpha^{1/2 + \gamma/16\pi}, \quad f(0) = 1. \tag{31.99}$$

The lowest mass $n = 1$ corresponds to the SG field; the remaining boundstates can also be considered as collective excitations of the SG particle.

31.6 A Two-Fermion Model

The derivation of the correspondence between the massive Thirring and the SG models indicates that it is possible to construct a fermion model equivalent to the SG model, such that $\kappa^2 = 8\pi$ corresponds to a free field theory, by introducing two Dirac fermions and by choosing the 4-point interaction appropriately. We now describe such a model.

31.6.1 The model

We consider a generalized massless Thirring model with two fermions coupled through a current-current interaction. The action is:

$$S\left(\bar{\psi}^a, \psi^a\right) = -\int \mathrm{d}^2 x \left[\bar{\psi}^a \slashed{\partial} \psi^a - \tfrac{1}{2} g_{ab} j^a_\mu j^b_\mu - f \bar{\psi}^1 \gamma_\mu \psi^2 \bar{\psi}^2 \gamma_\mu \psi^1\right], \qquad (31.100)$$

where the matrix g_{ab} has the form

$$\mathbf{g} = \frac{1}{2}\begin{pmatrix} g'+g & g'-g \\ g'-g & g'+g \end{pmatrix}, \qquad (31.101)$$

and the current j^a_μ is

$$j^1_\mu = \bar{\psi}^1 \gamma_\mu \psi^1, \quad j^2_\mu = \bar{\psi}^2 \gamma_\mu \psi^2. \qquad (31.102)$$

This action is the most general renormalizable action with the following symmetries:

(i) A $U(1)$ chiral symmetry which prevents a fermion mass term:

$$\begin{cases} \psi^a(x) \mapsto \mathrm{e}^{i\gamma_5\theta}\,\psi^a(x), \\ \bar{\psi}^a(x) \mapsto \bar{\psi}^a\,\mathrm{e}^{i\gamma_5\theta}. \end{cases} \qquad (31.103)$$

(ii) Separate fermion number conservation for each type of fermions and finally symmetry between fermions 1 and 2.

Note that the model has an additional $SU(2)$ invariance when $g = f$ since the interaction can also be written

$$-\tfrac{1}{4}\left[g'J^0_\mu J^0_\mu + gJ^3_\mu J^3_\mu + f\left(J^1_\mu J^1_\mu + J^2_\mu J^2_\mu\right)\right],$$

where J^0_μ and $\mathbf{J}_\mu$ are the $U(1)$ and the $SU(2)$ currents respectively

$$J^0_\mu = \bar{\psi}^a \gamma_\mu \psi^a, \qquad \mathbf{J}_\mu = \bar{\psi}^a \boldsymbol{\sigma}_{ab} \gamma_\mu \psi^b.$$

We have denoted by $\boldsymbol{\sigma}$ the three Pauli matrices acting on the $SU(2)$ indices.

Due to the $U(1)$ chiral invariance of action (31.100), the coupling g' is not renormalized, as will be derived in next subsection. The RG β-functions at one-loop order are (see Appendix A31.2.2):

$$\begin{cases} \beta_g = -2f^2/\pi, \\ \beta_f = -2fg/\pi. \end{cases} \qquad (31.104)$$

The equivalence with the SG model is derived below. The correspondence is summarized by the relations:

$$8\pi/\kappa^2 = 1 + g/\pi, \qquad \alpha_0 = \kappa^2 \frac{f\Lambda^2}{\pi^2}, \qquad (31.105)$$

$$\cos\theta \mapsto \frac{\pi^2}{\Lambda^2} \bar{\psi}^1 \gamma_\mu \psi^2 \bar{\psi}^2 \gamma_\mu \psi^1. \qquad (31.106)$$

From these expressions we see that for t close to 8π and α small, f and g are both small. The study of the phase transition is reduced to standard perturbation theory with fermion 4-point renormalizable interactions.

Since the transition taking place at $g = f = 0$ will be discussed extensively in next chapter, let us here note only that in the fermion language the transition occurs between two phases, a massless phase as seen in perturbation theory, and a massive phase which exhibits the property of asymptotic freedom at large momentum.

31.6.2 Derivation

The general idea is to separate the action (31.100),

$$S\left(\bar{\psi}^a, \psi^a\right) = -\int \mathrm{d}^2x \left(\bar{\psi}^a \slashed{\partial} \psi^a - \tfrac{1}{2} g_{ab} j_\mu^a j_\mu^b - f\bar{\psi}^1 \gamma_\mu \psi^2 \bar{\psi}^2 \gamma_\mu \psi^1\right),$$

into the sum of two terms $S = S_0 + S_1$:

$$S_0\left(\bar{\psi}^a, \psi^a\right) = -\int \mathrm{d}^2x \left(\bar{\psi}^a \slashed{\partial} \psi^a - \tfrac{1}{2} g_{ab} j_\mu^a j_\mu^b\right),$$

$$S_1\left(\bar{\psi}^a, \psi^a\right) = f\int \mathrm{d}^2x \, \bar{\psi}^1 \gamma_\mu \psi^2 \bar{\psi}^2 \gamma_\mu \psi^1.$$

The first term S_0 contains all interaction terms which have a separate chiral invariance for each species of fermions (interaction of Thirring model type) and is treated in much the same way as the massless Thirring action: It can be transformed into a free field action.

The remainder S_1 is expanded in perturbation theory like the mass term in the Thirring model, and yields the interaction.

We thus write the interaction terms of S_0 as generated by gaussian integrals over two vector fields $A_\mu^\pm$. The action then becomes:

$$S_0\left(A_\mu^\pm, \bar{\psi}, \psi\right) = -\int \mathrm{d}^2x \left[\bar{\psi}\left(\slashed{\partial} + i\slashed{A}^+ + i\sigma_3 \slashed{A}^-\right)\psi - \left(A_\mu^+\right)^2/g' - \left(A_\mu^-\right)^2/g\right],$$

where $[\sigma_3]_{ab}$ is a Pauli matrix and acts on the vector ψ^b. Parametrizing the vector fields as in equation (31.50):

$$A_\mu^\pm = -\partial_\mu \chi^\pm - i\epsilon_{\mu\nu}\partial_\nu \varphi^\pm. \tag{31.107}$$

and performing the corresponding gauge transformations (31.52) on both fermion fields we obtain (see equation (31.92)):

$$S_0 = -\int \mathrm{d}^2x \left[\bar{\psi}^a \slashed{\partial} \psi^a + \frac{\pi+g}{\pi g}\left(\partial_\mu \varphi^-\right)^2 + \frac{\pi+g'}{\pi g'}\left(\partial_\mu \varphi^+\right)^2 - \frac{1}{g}\left(\partial_\mu \chi^-\right)^2\right.$$
$$\left. - \frac{1}{g'}\left(\partial_\mu \chi^+\right)^2\right], \tag{31.108}$$

(we have omitted the primes on the fermions) which shows that g' indeed is not renormalized.

For S_1, to be able to use directly the identities derived for the Thirring model, it is convenient to rewrite the f-term using a Fierz transformation (see Appendix A5.9):

$$\bar{\psi}^1 \gamma_\mu \psi^2 \bar{\psi}^2 \gamma_\mu \psi^1 = -\bar{\psi}^1 \psi^1 \bar{\psi}^2 \psi^2 + \bar{\psi}^1 \gamma_S \psi^1 \bar{\psi}^2 \gamma_S \psi^2. \tag{31.109}$$

Introducing the chiral components of the mass operators

$$\varsigma^1_\pm(x) = \bar{\psi}^1_\mp(x)\psi^1_\pm(x), \quad \varsigma^2_\pm(x) = \bar{\psi}^2_\mp(x)\psi^2_\pm(x),$$

we can rewrite the r.h.s.:

$$-\bar{\psi}^1\psi^1\bar{\psi}^2\psi^2 + \bar{\psi}^1\gamma_S\psi^1\bar{\psi}^2\gamma_S\psi^2 = -2\left(\varsigma^1_+\varsigma^2_- + \varsigma^2_+\varsigma^1_-\right). \tag{31.110}$$

Therefore, in a perturbative expansion in powers of f, the integrals over the two fermions factorize and each term is just the square of the corresponding term in the expansion in powers of m in the Thirring model.

In the transformations (31.52) the chiral components $\varsigma^a_\pm$ become

$$\varsigma^1_\pm \mapsto \varsigma^1_\pm \, e^{\pm 2i(\varphi_+ + \varphi_-)}, \quad \varsigma^2_\pm \mapsto \varsigma^2_\pm \, e^{\pm 2i(\varphi_+ - \varphi_-)}.$$

Using then identity (31.110) we can rewrite the f term

$$f\bar{\psi}^1\, e^{2i\gamma_S\varphi^-}\,\gamma_\mu\psi^2\bar{\psi}^2\, e^{-2i\gamma_S\varphi^-}\,\gamma_\mu\psi^1 = -2f\left(e^{4i\varphi-}\,\varsigma^1_+\varsigma^2_- + e^{-4i\varphi-}\,\varsigma^2_+\varsigma^1_-\right). \tag{31.111}$$

We now associate two bosons ϑ^a with the fermions $\bar{\psi}^a, \psi^a$. In the expansion in powers of f we see that we can simply replace the quantities $\varsigma^a_\pm$ by their bosonic counterparts. We then obtain a boson action

$$S_{\text{bos}} = \int d^2x \left\{ \tfrac{1}{2}\partial_\mu\vartheta^a\partial_\mu\vartheta^a - \frac{f\Lambda^2}{\pi^2}\cos\left[\sqrt{4\pi}\left(\vartheta^1 - \vartheta^2\right) + 4\varphi_-\right] \right.$$
$$\left. -\frac{\pi+g}{\pi g}\left(\partial_\mu\varphi^-\right)^2 - \frac{\pi+g'}{\pi g'}\left(\partial_\mu\varphi^+\right)^2 + \frac{1}{g}\left(\partial_\mu\chi^-\right)^2 + \frac{1}{g'}\left(\partial_\mu\chi^+\right)^2 \right\}$$

Changing variables, for example $\varphi_- \mapsto \theta$:

$$\theta = \sqrt{4\pi}\left(\vartheta^1 - \vartheta^2\right) + 4\varphi_-,$$

and integrating over the fields ϑ^i (and $\varphi^+, \chi_\pm$) we find a simple equivalent SG model

$$S(\theta) == \int d^2x \left[\frac{1}{16\pi}\left(1 + \frac{g}{\pi}\right)(\partial_\mu\theta)^2 - \frac{f\Lambda^2}{\pi^2}\cos\theta\right]. \tag{31.112}$$

Comparing with the form (31.65), we see that the SG model has a parameter κ

$$8\pi/\kappa^2 = 1 + g/\pi. \tag{31.113}$$

Therefore the value $\kappa^2 = 8\pi$ corresponds to a free massless fermion theory.

Bibliographical Notes

Some references on the sine-Gordon model and its exact integrability are

L.D. Faddeev and L.A. Takhtajan, *Teor. Mat. Fiz.* 21 (1974) 160; R.F. Dashen, B. Hasslacher and A. Neveu, *Phys. Rev.* D11 (1975) 3424; L.D. Faddeev and V. Korepin, *Phys. Rep.* 42C (1978) 1; A.B. Zamolodchikov and Al. B. Zamolodchikov, *Ann. Phys.* 120 (1979) 253.

Some references on the Schwinger model and its relation to confinement are

J. Schwinger, *Phys. Rev.* 128 (1962) 2425; J.H. Lowenstein and J.A. Swieca, *Ann. Phys. (NY)* 68 (1971) 172; A. Casher, J. Kogut and L. Susskind, *Phys. Rev.* D10 (1974) 732; S. Coleman, R. Jackiw and L. Susskind, *Ann. Phys. (NY)* 93 (1975) 267; S. Coleman, *Ann. Phys. (NY)* 101 (1976) 239.

The correlation functions of the massless Thirring model are given in

R. Klaiber, *Lectures in Theoretical Physics* 10A (1967) 141, A. Barut and W. Brittin eds. (Gordon and Breach, London 1968).

The correspondence between the massive Thirring and the sine-Gordon models has been established in

S. Coleman, *Phys. Rev.* D11 (1975) 208 8.

For the bosonization of the $SU(N)$ Thirring model see

T. Banks, D. Horn and H. Neuberger, *Nucl. Phys.* B108 (1976) 119.

For further references on bosonization techniques see

E. Witten, *Commun. Math. Phys.* 92 (1984) 455; I. Affleck, *String, Fields and Critical Phenomena*, Les Houches 1988, E. Brézin and J. Zinn-Justin eds. (Elsevier, Amsterdam 1989).

For a discussion of the solitons in the sine-Gordon model see for example

R. Rajaraman, *Solitons and Instantons* (North-Holland, Amsterdam 1982),

and references therein. The spectrum and the S-matrix of the $SU(N)$ chiral Gross–Neveu model are discussed in

J.H. Lowenstein, *Recent Advances in Field Theory and Statistical Mechanics*, Les Houches 1982, J.-B. Zuber and R. Stora eds. (Elsevier 1984).

Exercises

Exercise 31.1

Study in the 2D free massless scalar field theory the short distance expansion of the product of operators $e^{i\kappa_1 \varphi(x_1)} e^{i\kappa_2 \varphi(x_2)}$.

Exercise 31.2

Consider 2D QED in the case of N identical charged massless Dirac fermions. Denoting the fermion fields by $\psi, \bar{\psi}$ we write the action

$$S\left(\bar{\psi}, \psi, A_\mu\right) = \int \mathrm{d}^2 x \left[\tfrac{1}{4} F_{\mu\nu}^2(x) - \bar{\psi}(x) \cdot (\slashed{\partial} + ie\slashed{A})\,\psi(x)\right]. \qquad (31.114)$$

Bosonize the model and identify the symmetry breaking operator with non-zero expectation value.

Exercise 31.3

To explicitly verify the çonformal invariance of the free massless boson 2D theory calculate the two-point function $\left\langle e^{ik_2\phi(x,y)} e^{ik_1\phi(0,0)} \right\rangle$ in the cylindrical geometry, i.e. with a finite size L and periodic boundary conditions in one direction (see also Section 36.5 for the problem of the zero mode).

Exercise 31.4

Find the most general analytic function $f(z)$ which satisfies an identity of the form (31.33)

$$(-1)^{n+1} \det \frac{1}{f(z_i - z_j')} = \frac{\prod_{i<j} f\left(z_i - z_j\right) f(z_i' - z_j')}{\prod_{i,j} f(z_i - z_j')}. \tag{31.115}$$

One assumes $f'(0) \neq 0$.

APPENDIX 31
A FEW ADDITIONAL RESULTS

A31.1 The Schwinger Model

In this section we examine the role of the anomaly in the Schwinger model from two other point of views.

A31.1.1 The general anomaly

The anomaly of the axial current $J_\mu^S = i\bar{\psi}\gamma_S\gamma_\mu\psi$ is (equation (20.106)):

$$\partial_\mu J_\mu^S = -\frac{ie}{2\pi}\epsilon_{\mu\nu}F_{\mu\nu}\,.$$

On the other hand the vector current $J_\mu = \bar{\psi}\gamma_\mu\psi$ is exactly conserved:

$$\partial_\mu J_\mu = 0\,.$$

We have observed that in dimension two the two currents are related (equation (31.25)),

$$J_\mu^S = -\epsilon_{\mu\nu}J_\nu\,,$$

and thus

$$\partial_\mu J_\nu - \partial_\nu J_\mu = \frac{ie}{\pi}F_{\mu\nu}\,.$$

The gauge field equation of motion yields:

$$ieJ_\mu + \partial_\nu F_{\nu\mu} = 0\,.$$

Combining these equations we get:

$$\left(-\partial^2 + (e^2/\pi)\right)J_\mu = 0\,.$$

This equation shows that the current J_μ and thus the curvature $F_{\mu\nu}$ are free fields of mass $m = e/\sqrt{\pi}$, in agreement with the result derived in Subsection 31.4.1.

A31.1.2 The Schwinger model: one-loop calculation

We know from the general analysis that we can regularize the fermion determinant while preserving the QED gauge invariance. However then we cannot preserve in general the chiral gauge invariance. For example in dimensional regularization the anticommutation properties of γ_S together with the relation (31.24) cannot be maintained (see Subsection 9.3.3). Let us here explicitly work out the one-loop contribution to the A_μ field 2-point function (figure 31.1) in the Feynman gauge. Using dimensional regularization which preserves non-chiral gauge invariance and giving temporarily a mass M to the fermion field to avoid IR problems, we find:

$$\Gamma_{\mu\nu}^{(2)}(k) = k^2\delta_{\mu\nu} - e^2\int\frac{\mathrm{d}^d q}{(2\pi)^d}\,\mathrm{tr}\left[\gamma_\mu\frac{M - i\slashed{q}}{M^2 + q^2}\gamma_\nu\frac{M - i(\slashed{q} + \slashed{k})}{M^2 + (q+k)^2}\right]\,. \tag{A31.1}$$

It is easy to verify that the dimensionally regularized one-loop contribution is transverse as expected. Setting:

$$\int \frac{d^d q}{(2\pi)^d} \, \text{tr} \left[\gamma_\mu \frac{M - i\slashed{q}}{M^2 + q^2} \gamma_\nu \frac{M - i(\slashed{q} + \slashed{k})}{M^2 + (q+k)^2} \right] = D \left(\delta_{\mu\nu} - \frac{k_\mu k_\mu}{k^2} \right), \qquad (A31.2)$$

and taking the trace of both members we obtain:

$$D(k)(d-1) = \text{tr}\, \mathbf{1} \int \frac{d^d q}{(2\pi)^d} \frac{dM^2 + (d-2)\left(q^2 + q \cdot k\right)}{(q^2 + M^2)\left[(q+k)^2 + M^2\right]}. \qquad (A31.3)$$

The r.h.s. contains a term proportional to M^2 which is UV convergent and which goes to zero with M for $\mathbf{k} \neq 0$. The second term is proportional to $d-2$. In the $d=2$ limit only the divergent part of the integral survives:

$$\int \frac{d^d q}{(2\pi)^d} \frac{q^2}{(q^2 + M^2)^2} \underset{d \to 2}{\sim} \frac{1}{2\pi(2-d)} \quad \Rightarrow \quad D = -1/\pi. \qquad (A31.4)$$

The gauge field inverse 2-point function at one-loop order follows:

$$\Gamma^{(2)}_{\mu\nu}(k) = \left(k^2 \delta_{\mu\nu} - k_\mu k_\nu \right) \left(1 + \frac{e^2}{\pi k^2} \right) + k_\mu k_\nu, \qquad (A31.5)$$

a result consistent with the exact expression (31.77).

A31.2 The $SU(N)$ Thirring Model

In Section 31.6 we have shown that the $SU(2)$ invariant Thirring model is equivalent to a bosonic model, the SG model. It is clear from the derivation of this property that more general models can be bosonized. Let us here consider in particular the $SU(N)$ Thirring model, a model described in terms of N Dirac fermions, with the action:

$$S\left(\bar{\psi}, \psi\right) = -\int d^2 x \left[\bar{\psi} \cdot \slashed{\partial} \psi - \tfrac{1}{2} f \left(\bar{\psi} \cdot \gamma_\mu \psi \right)^2 - \tfrac{1}{2} g \bar{\psi}^i \gamma_\mu \psi^j \, \bar{\psi}^j \gamma_\mu \psi^i \right] \qquad (A31.6)$$

$$= -\int d^2 x \left[\bar{\psi} \cdot \slashed{\partial} \psi - \frac{1}{2N} \left((g + Nf) J^0_\mu J^0_\mu + g J^\alpha_\mu J^\alpha_\mu \right) \right], \qquad (A31.7)$$

where J^0_μ and J^α_μ are the $U(1)$ and $SU(N)$ currents:

$$J^0_\mu = \bar{\psi} \cdot \gamma_\mu \psi, \quad J^\alpha_\mu = \bar{\psi}^i \gamma_\mu t^\alpha_{ij} \psi^j.$$

This model has a $SU(N) \times U(1)$ vector together with a chiral $U(1)$ symmetry. After a Fierz transformation the second interaction term becomes:

$$\bar{\psi}^i \gamma_\mu \psi^j \, \bar{\psi}^j \gamma_\mu \psi^i = -\bar{\psi}^i \psi^i \, \bar{\psi}^j \psi^j + \bar{\psi}^i \gamma_S \psi^i \, \bar{\psi}^j \gamma_S \psi^j.$$

We then recognize the action of chiral invariant Gross–Neveu model. Strictly speaking the Gross–Neveu model corresponds to the special case $f = 0$ but such a model is not multiplicatively renormalizable. Another current interaction has to be added for renormalization purpose. Due to the global chiral invariance of the model the coupling constant $g + Nf$ associated with the current J_μ^0 is not renormalized. The remaining RG β_g-function in $d = 2 + \varepsilon$ at one-loop order is then:

$$\beta_g = \varepsilon g - \frac{N}{\pi} g^2, \tag{A31.8}$$

showing that the model is asymptotically free and has a non-perturbative spectrum for g positive and small. For $g < 0$ instead the model is IR free: the fermions remain massless.

The model can be bosonized like the Thirring model to yield a generalized SG model of the form:

$$S(\boldsymbol{\theta}) = \int \mathrm{d}^2 x \left\{ \frac{1}{4\pi^2} \left[(\pi + g) \sum_i \left(\partial_\mu \theta^i \right)^2 + f \left(\sum_i \partial_\mu \theta^i \right)^2 \right] - G \sum_{i,j} \cos(\theta^i - \theta^j) \right\}, \tag{A31.9}$$

with

$$G = g \frac{\Lambda^2}{\pi^2}. \tag{A31.10}$$

The physics of the $SU(N)$ model is again the physics of spontaneous mass generation. The fermions are massless in perturbation theory but become massive for $g > 0$ (where the forces between fermions are attractive) with $\langle \bar{\boldsymbol{\psi}} \cdot \boldsymbol{\psi} \rangle = 0$. Conflict with the Mermin–Wagner theorem is avoided because the would-be massless Goldstone boson which is associated the $U(1)$ chiral transformation, and thus with the translation $\theta^i \mapsto \theta^i + \theta$ decouples from the other fields. This is another example of the Kosterlitz–Thouless mechanism (see Chapter 32).

Note that the spectrum has been exactly obtained:

$$m_n \propto \frac{N}{\pi} \sin \left(\frac{n\pi}{N} \right), \quad n = 1, 2 \ldots < N,$$

n odd corresponding to fermions and n even to bosons. This result confirms that fermions become massive.

A31.2.1 Derivation

The method is a simple extension of the method explained in Section 31.6 in the $SU(2)$ case. Introducing the currents j_μ^i:

$$j_\mu^i = \bar{\psi}^i \gamma_\mu \psi^i, \tag{A31.11}$$

(no sum over i) we can rewrite action (A31.6) as:

$$S\left(\bar{\boldsymbol{\psi}}, \boldsymbol{\psi}\right) = -\int \mathrm{d}^2 x \left[\bar{\boldsymbol{\psi}} \cdot \partial\!\!\!/ \boldsymbol{\psi} - \tfrac{1}{2} g_{ij} j_\mu^i j_\mu^j - \tfrac{1}{2} g \sum_{i \neq j} \bar{\psi}^i \gamma_\mu \psi^j \bar{\psi}^j \gamma_\mu \psi^i \right], \tag{A31.12}$$

with:

$$g_{ij} = g \delta_{ij} + f. \tag{A31.13}$$

We then introduce vector fields A_μ^i to rewrite the action

$$S\left(\bar{\boldsymbol{\psi}}, \boldsymbol{\psi}, \mathbf{A}_\mu\right) = -\int \mathrm{d}^2 x \left[\sum_i \bar{\psi}^i \left(\partial\!\!\!/ + i A\!\!\!/^i \right) \psi^i - \tfrac{1}{2} \tilde{g}_{ij} A_\mu^i A_\mu^j - \tfrac{1}{2} g \sum_{i \neq j} \bar{\psi}^i \gamma_\mu \psi^j \bar{\psi}^j \gamma_\mu \psi^i \right], \tag{A31.14}$$

with

$$\tilde{g}_{ij} \equiv [g^{-1}]_{ij} = \frac{1}{g}\left(\delta_{ij} - \frac{f}{g+Nf}\right). \qquad (A31.15)$$

Combining the usual $U(1)$ gauge transformations on the fermions with a parametrization of the vector fields:

$$\psi^i(x) \mapsto \mathrm{e}^{i\chi^i(x)+i\gamma_S\varphi^i(x)}\,\psi^i(x),$$
$$\bar{\psi}^i(x) \mapsto \bar{\psi}^i\,\mathrm{e}^{-i\chi^i(x)+i\gamma_S\varphi^i(x)},$$
$$A^i_\mu = -\partial_\mu\chi^i - i\epsilon_{\mu\nu}\partial_\nu\varphi^i.$$

we find:

$$S\left(\bar{\psi},\psi,\varphi,\chi\right) = \int \mathrm{d}^2x\left[-\bar{\psi}^i\slashed{\partial}\psi^i + \tfrac{1}{2}\tilde{g}_{ij}\partial_\mu\chi^i\partial_\mu\chi^j - \tfrac{1}{2}\left(\tilde{g}_{ij} + \frac{\delta_{ij}}{\pi}\right)\partial_\mu\varphi^i\partial_\mu\varphi^j\right.$$
$$\left. - \tfrac{1}{2}g\sum_{i\neq j}\bar{\psi}^i\,\mathrm{e}^{i\gamma_S(\varphi^i-\varphi^j)}\,\gamma_\mu\psi^j\,\bar{\psi}^j\,\mathrm{e}^{-i\gamma_S(\varphi^i-\varphi^j)}\,\gamma_\mu\psi^i\right]. \qquad (A31.16)$$

Introducing the chiral components of the mass operator $\bar{\psi}^i\psi^i$

$$\varsigma^i_\pm = \bar{\psi}^i_\pm\psi^i_\pm,$$

and using the identity (31.111) we can write

$$\bar{\psi}^i\,\mathrm{e}^{i\gamma_S(\varphi^i-\varphi^j)}\,\gamma_\mu\psi^j\,\bar{\psi}^j\,\mathrm{e}^{-i\gamma_S(\varphi^i-\varphi^j)}\,\gamma_\mu\psi^i\mathop{=}_{i\neq j} -2\left(\mathrm{e}^{2i(\varphi^i-\varphi^j)}\,\sigma^i_+\sigma^j_- + \mathrm{e}^{2i(\varphi^j-\varphi^i)}\,\sigma^j_+\sigma^i_-\right).$$

We then expand the four-fermion term in perturbation theory. As in the $SU(2)$ model we see that the average over the fermions $\bar{\psi}^i,\psi^i$ can be replaced by an average over bosons ϑ^i. After summation this leads to a SG interaction of the form

$$-g\frac{\Lambda^2}{\pi^2}\sum_{i\neq j}\cos\left[\sqrt{4\pi}\left(\vartheta^i-\vartheta^j\right)+2\left(\varphi^i-\varphi^j\right)\right].$$

Changing then variables, $\varphi^i \mapsto \theta^i$, setting

$$2\varphi^i + \sqrt{4\pi}\vartheta^i = \theta^i,$$

and integrating on all fields but the θ^i's we find the action (A31.9).

A31.2.2 The RG β-Function with a 4-Fermion Current Interaction

We leave as an exercise to calculate the RG β-function in the case of a 4-fermion current-current interaction:

$$S\left(\bar{\psi}^a,\psi^a\right) = -\int\mathrm{d}^2x\left(\bar{\psi}^a\slashed{\partial}\psi^a - \tfrac{1}{2}g_{abcd}\bar{\psi}^a\gamma_\mu\psi^b\bar{\psi}^c\gamma_\mu\psi^d\right), \qquad (A31.17)$$

where g_{abcd} is chosen to be symmetric in the exchange $(ab)\leftrightarrow(cd)$. The result is:

$$\pi\beta_{abcd} = g_{aicj}g_{ibjd} - g_{aijd}g_{ibcj} + O\left(g^3\right). \qquad (A31.18)$$

A31.3 Solitons in the Sine-Gordon model

Solitons correspond to finite energy solutions of the real-time classical equations of motion. Soliton calculus should be thought as the field theory generalization of the WKB method. Solitons have a particle interpretation, the energy of the soliton being its mass in the semiclassical limit. The sine-Gordon model being classically integrable, the whole soliton spectrum can be derived since also time-dependent solutions can be obtained. The semiclassical results have been confirmed by the exact quantum analysis. Let us as a simple example calculate the mass associated with the static solution. The equation of motion associated with action (31.64) is a differential equation in the space coordinate x:

$$\theta'' = \alpha \sin \theta \, .$$

Finite energy solutions necessarily connect minima of the potential. Thus

$$\theta = 4 \arctan e^{\sqrt{\alpha} x} \, .$$

The corresponding energy is then the space integral of the lagrangian density, (after subtraction of the vacuum energy):

$$M_{\text{sol.}} = \frac{1}{t} \int \mathrm{d}x \left[\tfrac{1}{2} \left(\partial_x \theta \right)^2 + \alpha (1 - \cos \theta) \right] = 8\sqrt{\alpha}/t + O(1) \, ,$$

which coincides with the Thirring fermion mass (31.99) for t small.

32 THE $O(2)$ 2D CLASSICAL SPIN MODEL

Having established in Chapter 31 a few properties of two-dimensional models we now have the necessary technical background to discuss the abelian $O(2)$ spin model near and in two dimensions. As we have seen in Chapter 30, at low temperature its long distance properties can be described in terms of the $O(2)$ non-linear σ-model. We recall that the $O(2)$ case is special because the RG β-function reduces in the low temperature expansion to the dimensional term $(d-2)t$ and therefore the properties, from the RG point of view, are quite different. In particular in 2 dimensions the $O(2)$ model is not asymptotically free. The origin of this difference can be found in the local structure of the manifold: For $N = 2$ the $O(N)$ sphere reduces to a circle which is locally a flat manifold, i.e. which cannot be distinguished from a straight line. Therefore if we parametrize the spin $\mathbf{S}(x)$ as:

$$\mathbf{S}(x) = \begin{cases} \cos\theta(x) \\ \sin\theta(x) \end{cases} \tag{32.1}$$

the euclidean action (30.11) in zero field becomes:

$$S(\theta) = \frac{\Lambda^{d-2}}{2t} \int \mathrm{d}^d x \left[\partial_\mu \theta(x)\right]^2, \tag{32.2}$$

i.e. a free massless field action. This explains the form of the RG β-function. Although the theory is free, we are not interested in the correlation functions of the field θ but instead of $\sin\theta$ or $\cos\theta$ or equivalently $\mathrm{e}^{\pm i\theta}$. Even in a free field theory these fields have to be renormalized and thus have a non-trivial anomalous dimension for all temperatures (equations (30.24b,31.12)).

The simplicity of action (32.2) however leads to a mystery: since the field θ is massless the correlation length remains infinite for all t. By contrast a simple high temperature analysis of the corresponding spin model on the lattice shows that the correlation length is finite for t large enough. A phase transition at finite temperature is necessary to explain this phenomenon. The action (32.2) therefore cannot represent the long distances properties of the lattice model for all temperatures.

We shall find a clue in the property that θ is a cyclic variable, condition which is not incorporated in action (32.2) and is indeed irrelevant at low temperature. At higher temperatures, however, classical configurations with singularities at isolated points around which θ varies by a multiple of 2π become important. The action of these configurations (vortices) can be identified with the energy of a Coulomb gas. The neutral Coulomb gas exhibits a transition between a low temperature of bound neutral molecules and a high temperature phase of a plasma of free charges.

Remarkably enough the Coulomb gas can be mapped onto the sine-Gordon (SG) model, mapping in which the low and high temperature regions of the two models are exchanged. This will help us to understand some properties of the transition, the famous Kosterlitz–Thouless phase transition (KT), which separates an infinite correlation length phase *without ordering* (the low temperature phase of the $O(2)$ model) from a finite correlation length phase. Additional information will be obtained from the equivalence between the SG model and several fermion models, established in Sections 31.5, 31.6.

In the appendix we recall the relation between the statistical mechanics of 1D non-relativistic fermions and the quantum field theory of massless relativistic Dirac fermions in 1+1 dimensions.

32.1 The Spin Correlation Functions

Let us first calculate the spin correlation functions in d dimensions with the free action (32.2):

$$\left\langle \prod_{i=1}^{n} e^{i\epsilon_i \theta(x_i)} \right\rangle = \int [\mathrm{d}\theta] \exp\left[-\frac{\Lambda^{d-2}}{2t} \int \mathrm{d}^d x \, (\partial_\mu \theta)^2 + i \sum_i \epsilon_i \theta(x_i) \right], \qquad (32.3)$$

with $\epsilon_i = \pm 1$. The method has been explained in Section 31.1. The result is:

$$\left\langle \prod_{i=1}^{n} e^{i\epsilon_i \theta(x_i)} \right\rangle = \exp\left[-\tfrac{1}{2} t \Lambda^{2-d} \sum_{i,j} \epsilon_i \epsilon_j \Delta(x_i - x_j) \right]. \qquad (32.4)$$

in which $\Delta(x)$ is the massless propagator:

$$\Delta(x) = \frac{1}{(2\pi)^d} \int \frac{\mathrm{d}^d p}{p^2} e^{ipx} = \Gamma(d/2 - 1) \frac{x^{2-d}}{4\pi^{d/2}} \qquad \text{for } d > 2. \qquad (32.5)$$

$\Delta(0)$ diverges with the cut-off Λ:

$$\int \frac{\mathrm{d}^d p}{p^2} \sim \Lambda^{d-2}.$$

The limit $d = 2$ is singular, as we have discussed in Section 31.1, because the field is massless (IR divergence).

Remark. To define renormalized correlation functions, we have to introduce a renormalization scale μ and the corresponding coupling constant t_r:

$$t\Lambda^{2-d} = t_r \mu^{2-d},$$

and cancel the divergent terms in the exponential in the r.h.s. of equation (32.4):

$$\sum_{i,j} \epsilon_i \epsilon_j \Delta(x_i - x_j) = n\Delta(0) + 2 \sum_{i<j} \epsilon_i \epsilon_j \Delta(x_i - x_j). \qquad (32.6)$$

As noted in Section 31.1 the fields $e^{\pm i\theta(x)}$, which are composite fields in terms of $\theta(x)$, require a field renormalization Z:

$$Z = e^{-t_r \mu^{2-d} \Delta(0)}. \qquad (32.7)$$

Dimension 2. Let us now examine more closely what happens when the dimension approaches 2 (at fixed cut-off). The propagator Δ has an IR divergence. Setting $d = 2 + \varepsilon$ we find:

$$\Delta(x) = \frac{1}{2\pi\varepsilon} - \frac{1}{4\pi} \left(\ln x^2 + \ln \pi + \gamma \right) + O(\varepsilon), \qquad (32.8)$$

$$\Delta(0) = \frac{1}{2\pi\varepsilon} + \frac{1}{2\pi} \ln \Lambda + \text{ terms finite when } \varepsilon \to 0 \text{ or } \Lambda \to \infty \qquad (32.9)$$

where $\gamma = -\psi(1)$ is Euler's constant. It follows that the sum of the divergent contributions in the exponential of equation (32.4) takes the form:

$$-t\frac{\Lambda^{2-d}}{4\pi\varepsilon}\sum_{i,j}\epsilon_i\epsilon_j = -t\frac{\Lambda^{2-d}}{4\pi\varepsilon}\left(\sum_i\epsilon_i\right)^2.\tag{32.10}$$

Therefore, only the correlation functions such the sum $\sum\epsilon_i$ vanishes have a finite non-vanishing limit when d goes to 2. The other correlation functions vanish at $d = 2$. This result has a simple interpretation in the $O(2)$ non-linear σ-model: all non $O(2)$ invariant correlation functions vanish. In particular

$$\langle\mathbf{S}(x)\rangle = 0.$$

This result is in agreement with the absence of spontaneous symmetry breaking with ordering in 2 dimensions. Note however that this property does not depend on the compact nature of the $O(2)$ group. All correlation functions invariant under the translation $\theta(x)$ by a constant have a finite limit. In particular, as shown in Section 31.1, we can replace the signs ϵ_i by any set of numbers with a vanishing sum.

Let us write more explicitly the limit in the non-trivial case:

$$\left\langle\prod_{i=1}^{n}e^{i\epsilon_i\theta(x_i)}\right\rangle \propto \prod_{i<j}(\Lambda|x_i - x_j|)^{\epsilon_i\epsilon_j t/2\pi}.\tag{32.11}$$

This result is rather surprising: although the $O(2)$ symmetry is not broken the correlation functions decay algebraically at large distance and the correlation length is thus infinite. In addition the power behaviour depends continuously on the temperature: in the RG sense we have a line of fixed points and this is consistent with the property that the RG β-function vanishes identically. Applying equation (32.11) to the 2-point function

$$\left\langle e^{i\theta(x)}e^{-i\theta(0)}\right\rangle \sim x^{-t/2\pi},\tag{32.12}$$

we obtain the value of the exponent η:

$$\eta = t/2\pi.\tag{32.13}$$

It remains however to understand the absence of phase transition in this model. In particular no invariant relevant operator can be constructed which would modify action (32.2). Since we do not have yet all the necessary ingredients to examine this question we postpone it until Section 32.4. Let us finally note that $t > 4\pi$ implies $\eta > 2$. This is an unphysical result from the point of view of the $O(2)$ spin model since it implies that, in momentum space, the 2-point correlation function vanishes at low momentum. In the corresponding lattice model the correlation function is then dominated by a regular constant (non-critical) contribution. $t = 4\pi$ is thus the *a priori* largest possible value of t for which the action (32.2) can represent the $O(2)$ model below T_c.

32.2 Correlation Functions in a Field

Let us now add a magnetic field term in action (32.2):

$$S(\theta) = \frac{\Lambda^{d-2}}{t} \int d^d x \left\{ \tfrac{1}{2} [\partial_\mu \theta(x)]^2 - h \cos \theta(x) \right\}. \tag{32.14}$$

We recognize the SG action studied in Section 31.3. Since we know the 2 RG functions:

$$\beta(t) = (d-2)t, \qquad \zeta(t) = t/2\pi, \tag{32.15}$$

we can use the expressions of Chapter 30 to find the scaling form of correlation functions. For $d > 2$ the two functions $M_0(t)$ and $\xi(t)$ defined by equations (30.30, 30.31) are:

$$M_0(t) = \exp\left[-\frac{1}{2} \int_0^t \frac{\zeta(t')}{\beta(t')} dt' \right] = \exp\left(-\frac{t}{4\pi\varepsilon} \right), \tag{32.16}$$

$$\xi(t) = \Lambda^{-1} t^{1/\varepsilon} \exp\left[\int_0^t \left(\frac{1}{\beta(t')} - \frac{1}{\varepsilon t'} \right) dt' \right] = \Lambda^{-1} t^{1/\varepsilon}, \tag{32.17}$$

and therefore (equation (30.40)):

$$W^{(n)}(p_i, t, h) = t^{(n-1)d/(d-2)} e^{-nt/4\pi\varepsilon} F^{(n)}\left(p_i t^{1/\varepsilon}, h e^{-t/4\pi\varepsilon} t^{-2/\varepsilon} \right). \tag{32.18}$$

Dimension 2. In dimension 2 the situation again is different since the β-function vanishes. In perturbation theory the action (32.14) is super-renormalizable, which corresponds to the relevance of the magnetic field for long distance properties. The RG equations have been written in Chapter 30 and Section 31.3. For any temperature t the correlation functions have a scaling behaviour. Let us rewrite RG equation (30.21) in this particular case:

$$\left[\Lambda \frac{\partial}{\partial \Lambda} + \frac{n}{2} \zeta(t) + \frac{1}{2} \zeta(t) h \frac{\partial}{\partial h} \right] W^{(n)} = 0. \tag{32.19}$$

Using the dimensional relation:

$$\Lambda \frac{\partial}{\partial \Lambda} + 2h \frac{\partial}{\partial h} + p_i \frac{\partial}{\partial p_i} = 2(1-n), \tag{32.20}$$

we obtain the scaling form of correlation functions:

$$W^{(n)}(p_i) = h^{t/(8\pi - t)} h^{1-n} F^{(n)}\left(p_i h^{-4\pi/(8\pi - t)} \right). \tag{32.21}$$

This scaling form is consistent with the general scaling form of the ϕ^4 field theory at T_c in a magnetic field and the value of the exponent η (equation (32.13)).

In particular, calling M the magnetization, we have:

$$M = W^{(1)} \sim h^{t/(8\pi - t)}. \tag{32.22}$$

We have already noticed in Section 31.3 the special values $t = 4\pi$ and $t = 8\pi$. At $t = 4\pi$ the relation between M and h is linear up to a logarithmic correction, $M \propto -h \ln h$. This relation has been explained by the equivalence with the free massive Thirring model. As argued above when examining the behaviour of the spin 2-point function, for $t > 4\pi$ the model no longer represents the long distance physics of the lattice $O(2)$ ferromagnet.

32.3 The Coulomb Gas in 2 Dimensions

The partition function $\mathcal{Z}(T, z)$ of a globally neutral gas of particles of charges ± 1 interacting via a Coulomb potential $V(x)$ in d dimensions can be written:

$$\mathcal{Z}(T, z) = \sum_{n=0}^{\infty} \frac{z^{2n}}{(n!)^2} \int \left(\prod_i d^d x_i \, d^d y_i \right) \exp\left[-E_n(x_i, y_i)/T \right],$$

$$E_n(x_i, y_i) = \sum_{i<j\leq n} V\left(|x_i - x_j|\right) + V\left(|y_i - y_j|\right) - \sum_{i,j\leq n} V\left(|x_i - y_j|\right),$$

where T is the temperature and z the gas fugacity.

We note that the Coulomb potential $V(x)$ in d dimensions is also the propagator (32.5). Therefore the r.h.s. of equation (32.4) is, up to a multiplicative factor, the Boltzmann weight for this gas. Comparing then with expression (31.42), we see that we can represent the partition function as a functional integral over a field $\phi(x)$ with the action of the SG model:

$$\mathcal{Z}(T, z) = \int [d\phi] \exp\left\{ -\int d^d x \left[\tfrac{1}{2} T (\partial_\mu \phi(x))^2 - 2z \cos \phi(x) \right] \right\}. \qquad (32.23)$$

We also here recognize the transformation $(A28.32)$ applied to the Coulomb potential. The identity is true in any dimension, however we will in this chapter discuss only two dimensions. The properties of the SG model will then allow us to understand the physics of the Coulomb gas. The correspondence with the notation of Section 31.3 is:

$$T = 1/t, \quad z = \alpha/2t.$$

Remark. We have seen that for $t = 4\pi$ the free energy becomes infinite in the absence of a short distance cut-off. This implies that for $t \geq 4\pi$ i.e. $T \leq 1/4\pi$ the Coulomb gas is only stable if the charged particles have a hard core.

The phase transition. We have shown in Section 31.3 that the quantum SG model must undergo a phase transition at $t = 8\pi$ (for α small) between a phase with a finite correlation length at low t, i.e. at high temperature T in the Coulomb gas, and a phase with infinite correlation length at high t, i.e. at low T in the Coulomb gas. In the Coulomb gas language the nature of these phases is obvious (but not the existence of a transition at finite strictly positive temperature). At high T the gas is composed of free charges. At low T the system approaches the classical ground state: pairs of positive and negative charges are tightly bound (in pairs with finite polarizability). In the Coulomb gas language the $\sin \phi$ 2-point correlation function is the charge density correlation function. Therefore the correlation length characterizes the decay of the correlation between the charges. It characterizes also the decay of the effective potential between 2 infinitesimal external charges. In the free charge phase the correlation length is finite which means that the electrostatic potential is screened, the correlation length being the screening length. In the phase of molecular bound states instead no screening occurs, the effective potential is proportional to the bare potential, and the correlation function has a power law decay.

We shall now try to study in more detail this phase transition, which seems at first sight to be unrelated to the physics of the $O(2)$ model and we shall come back to the

$O(2)$ model afterwards. Since the transition point corresponds to a finite value of the coupling constant T^{-1} the renormalization and RG properties of the model do not follow from simple perturbative considerations. The derivation of the RG equations within the SG framework thus involves a series of intuitive arguments which are not easily made rigorous. Fortunately we have derived in Chapter 31 a remarkable relation between the SG model and a two fermion model which is a free field theory just at the transition point. This allows to use the fermion model perturbation theory to gain some further insight into the problem and in particular to justify the RG equations obtained more directly.

32.3.1 Renormalization and renormalization group

At $T = 1/8\pi$ (and $z \to 0$) the SG model is just renormalizable. To study its renormalization properties for T close to $1/8\pi$ we note that the deviation $T^{-1} - 8\pi$ plays a role analogous to the difference $d - 4$ in the ϕ^4 field theory. Therefore we can try to calculate RG quantities in a double z and $8\pi - T^{-1}$ expansion. It is convenient to introduce the two dimensionless coupling constants:

$$\begin{cases} u = 1 - 1/8\pi T\,, \\ v = 2z/T\Lambda^2\,. \end{cases} \tag{32.24}$$

We assume that the set $\{u, v\}$ is multiplicatively renormalizable. To these coupling constants correspond two RG β-functions. The property that the free average of the product of an odd number of operators $e^{\pm i\phi}$ vanishes (chiral symmetry in the equivalent fermion formulation) implies a parity symmetry in v. The RG function β_v is given at leading order in v in terms of the function $\zeta(t)$ of equations (32.15). Thus:

$$\beta_v = v\left[\tfrac{1}{2}\zeta\left(T^{-1}\right) - 2\right] + O(v^3) = -2uv + O(v^3)\,. \tag{32.25}$$

We know that the function β_u vanishes at $v = 0$. Therefore β_u starts at order v^2. The sign of the coefficient of v^2 is fixed by the requirement that $u > 0$ and $u < 0$ cannot be connected, for v infinitesimal, by RG transformations. The exact value is normalization dependent. We choose:

$$\begin{aligned} \beta_u &= -2v^2\,, \\ \beta_v &= -2uv\,. \end{aligned} \tag{32.26}$$

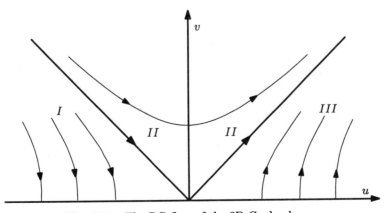

Fig. 32.1   The RG flow of the 2D Coulomb gas.

The RG flow. At this order it is easy to find curves which are RG invariant:

$$\lambda \frac{\mathrm{d}}{\mathrm{d}\lambda}\left(u^2(\lambda) - v^2(\lambda)\right) = 2u\beta_u - 2v\beta_v = 0\,. \tag{32.27}$$

For $|u| > |v|$ we parametrize these hyperbolas by:

$$u = a\frac{1 + s^2}{1 - s^2} \quad |s| < 1\,, \tag{32.28}$$

$$v = a\frac{2s}{1 - s^2}\,. \tag{32.29}$$

The RG equation for the parameter s is then:

$$\lambda \frac{\mathrm{d}}{\mathrm{d}\lambda}s(\lambda) = -2as(\lambda)\,,$$

The solution is:

$$s(\lambda) = s_0\lambda^{-2a}\,. \tag{32.30}$$

In the long distance limit λ goes to zero. Therefore if a is positive, i.e. if initially $u > 0$ or $T > 1/8\pi$, then $s(\lambda)$ increases as well as $u(\lambda)$ and $v(\lambda)$ until one leaves the perturbative regime. This is consistent with the relevance of v for large T. If instead we start with $T < 1/8\pi$, which implies $u < 0$ and $a > 0$, then $s(\lambda)$ goes to zero for small λ. $v(\lambda)$ then goes to zero and $u(\lambda)$ to a finite limit. This is consistent with the irrelevance of v for $T < 1/8\pi$. Thus the half-line $T < 1/8\pi$, $z = 0$, is a half-line of IR fixed points.

For $|u| < |v|$ we parametrize the hyperbolas instead by:

$$u = a\frac{2s}{1 - s^2}, \quad |s| < 1\,, \tag{32.31}$$

$$v = a\frac{1 + s^2}{1 - s^2}\,. \tag{32.32}$$

We then find:

$$\lambda \frac{\mathrm{d}}{\mathrm{d}\lambda}s(\lambda) = -a(1 + s^2)\,.$$

The solution can be written:

$$\arctan s(\lambda) = -a\ln\lambda + \arctan s_0\,. \tag{32.33}$$

Therefore, irrespective of the sign of a, i.e. the initial sign of u, $u(\lambda)$ and $v(\lambda)$ always increase until one leaves the perturbative regime. This again is a region in which v is relevant. The conclusions are that the half-plane $v > 0$ is divided in three regions (see figure 32.1) separated by the lines $u = \pm v$. Region I is an infinite correlation length phase, the low temperature phase of the Coulomb problem. Regions II and III are both finite correlation phases with free charges. The line $u + v = 0$ is thus the line of phase transition. Regions II and III differ by the property that in region III the field theory is asymptotically free at short distance while in region II the field theory is non-trivial at both long and short distances.

32.3.2 The correlation length near the phase transition

Let us now characterize the behaviour of the correlation length in the high temperature phase when one approaches the phase transition. We set:

$$v + u = \tau \,, \tag{32.34}$$

$$v - u = 2w \,. \tag{32.35}$$

With this parametrization τ plays the role of the deviation $T - T_c$ from the critical temperature.

The correlation length is a RG invariant of mass dimension -1. It therefore satisfies:

$$\xi\left(\tau(\lambda), w(\lambda)\right) = \lambda \xi\left(\tau(1), w(1)\right). \tag{32.36}$$

To find the relation between τ and ξ we look for a value of λ such that $\xi(\lambda)$ is of order 1 when $\xi(1) = \xi$ is large:

$$\lambda \sim 1/\xi \,.$$

We have shown that the quantity τw is a RG invariant:

$$\tau(\lambda) w(\lambda) = \tau w \,.$$

The RG equation for $\tau(\lambda)$ then reads:

$$\lambda \frac{\mathrm{d}\tau}{\mathrm{d}\lambda} = -2\tau w - \tau^2(\lambda) \,. \tag{32.37}$$

A short calculation gives the function $\tau(\lambda)$

$$\arctan\left(\tau(\lambda)/\sqrt{2w\tau}\right) - \arctan\left(\tau/\sqrt{2w\tau}\right) = -\sqrt{2w\tau}\,\ln \lambda \,. \tag{32.38}$$

We evaluate the l.h.s. when τ is small and $\tau(\lambda)$ is of order 1. The equation then becomes:

$$\pi/2 = -\sqrt{2w\tau}\,\ln \lambda \,.$$

We thus obtain:

$$\xi \sim \exp\left(\frac{\pi}{2\sqrt{2w\tau}}\right), \tag{32.39}$$

a result that is characteristic of the 2D Coulomb gas phase transition.

Remark. The RG β-functions have also been calculated to third order. The result is:

$$\beta_u = -2v^2 - 4v^2 u \,, \tag{32.40}$$

$$\beta_v = -2uv + 5v^3 \,. \tag{32.41}$$

None of the conclusions drawn from the analysis at leading order is qualitatively affected.

32.4 $O(2)$ Non-Linear σ-Model and Coulomb Gas

We have seen that the action (32.2) cannot represent the $O(2)$ spin model for all temperatures and we feel that somehow we have to introduce the condition that the $\theta(x)$ is a cyclic variable. We also know from the analysis of Section 30.1 that at higher temperature the non-linear σ-model approximation fails because there may be points where the field $\phi(x)$ in the $(\phi^2)^2$ field theory vanishes. The cost in energy is minimized when these points are isolated because this corresponds to point defects rather than line defects. Then in a turn around these points the direction of the field changes by an angle multiple of 2π. Let us therefore consider a configuration of the field $\theta(x)$ which is the sum of a smooth background $\theta_1(x)$ and a configuration $\theta_V(x, x_i)$ solution of the classical field equation and regular everywhere except at isolated points x_i where $\theta_V(x, x_i)$ changes by a multiple of 2π:

$$\theta(x) = \theta_1(x) + \theta_V(x, x_i), \qquad (32.42)$$

with:

$$\theta_V(x, x_i) = \sum_i n_i \arctan \frac{(x - x_i)_2}{(x - x_i)_1}, \quad n_i \in \mathbb{Z}. \qquad (32.43)$$

The terminology is that $\theta_V(x, x_i)$ is a sum of vortex excitations located at points x_i and of vorticity n_i. Vortices are topological excitations in the sense that they cannot be removed by a continuous deformation of the field $\theta(x)$. The energy of the configuration is (setting $\Lambda = 1$):

$$S(\theta) = \frac{1}{2t} \int d^2x \, [\partial_\mu \theta_1(x) + \partial_\mu \theta_V(x)]^2. \qquad (32.44)$$

We now use the identities:

$$\partial_\mu \theta_V(x) = \sum_i n_i \epsilon_{\mu\nu} \frac{(x - x_i)_\nu}{(x - x_i)^2} \qquad (32.45)$$

$$= \sum_i n_i \epsilon_{\mu\nu} \partial_\nu \ln |x - x_i|. \qquad (32.46)$$

We observe from expression (32.44) that the energy can be finite only if:

$$\sum_i n_i = 0.$$

An integration by parts shows that the cross term in the r.h.s. of equation (32.44) vanishes and yields:

$$\int d^2x \, [\partial_\mu \theta_V(x)]^2 = \int d^2x \sum_{ij} n_i n_j \partial_\mu \ln |x - x_i| \, \partial_\mu \ln |x - x_j|$$

$$= -2\pi \sum_{ij} n_i n_j \ln |x_i - x_j|. \qquad (32.47)$$

We recognize the energy of a neutral Coulomb gas of charges n_i and temperature $T = t/4\pi^2$. We know from the analysis of Subsection 31.3.2 (equation (31.71)) that the most relevant terms correspond to $n_i = \pm 1$ which we assume from now on. Of course the fugacity A of the equivalent Coulomb gas has to be calculated from a microscopic model.

Only results which are independent of its explicit value can be obtained by this method. The relation between the Coulomb gas temperature $t/4\pi^2$ and the temperature t' of the equivalent SG model is:

$$t't = 4\pi^2. \tag{32.48}$$

In particular by introducing an auxiliary field $\theta_2(x)$ and using identity (32.4) we can write an effective action:

$$S(\theta_1, \theta_2) = \int d^2x \left[\frac{1}{2t} (\partial_\mu \theta_1)^2 + \frac{t}{8\pi^2} (\partial_\mu \theta_2)^2 - \frac{At}{4\pi^2} \cos \theta_2 \right]. \tag{32.49}$$

The analysis of the Coulomb transition then shows that $t = \pi/2$ is the transition temperature. For $t > \pi/2$, the $\cos \theta$ interaction is relevant and the correlation length finite. For $t < \pi/2$, $\cos \theta$ is irrelevant and no mass is generated.

32.5 The Critical 2-point Function in the $O(2)$ Model

We now want to compute the 2-point correlation function of the $O(2)$ model near $t = \pi/2$. We therefore have to calculate the average:

$$\left\langle e^{i(\theta_1(y) + \theta_V(y))} e^{-i(\theta_1(x) + \theta_V(x))} \right\rangle = |x - y|^{-t/2\pi} \left\langle e^{i(\theta_V(y) - \theta_V(x))} \right\rangle_{\text{CG}}. \tag{32.50}$$

To find an interpretation to this average in terms of a modification of the energy of the Coulomb gas we make a few transformations:

$$\theta_V(y) - \theta_V(x) = \int_x^y ds_\mu \partial_\mu \theta_V(s)$$

$$= \frac{2\pi}{t} \sum_i n_i \frac{t}{2\pi} \int_x^y ds_\mu \epsilon_{\mu\nu} \partial_\nu \ln |s - x_i|. \tag{32.51}$$

To perform the Coulomb gas average in the r.h.s. of equation (32.50) we want to write the sum of the term (32.51) and the Coulomb gas energy (32.50) as arising from the free field average in a source. The source $J_2(s)$ for the free field $\theta_2(s)$:

$$J_2(s) = J(s) + K(s), \qquad \begin{cases} J(s) = i \sum_i n_i \delta(s - x_i), \\[2mm] K(s) = \dfrac{t}{2\pi} \displaystyle\int_x^y ds'_\mu \epsilon_{\mu\nu} \partial_\nu \delta(s - s'), \end{cases} \tag{32.52}$$

generates the two terms we need: After integration over θ_2 the JJ term gives the Coulomb gas energy, the JK cross term gives the term (32.51), but in addition a KK contribution $D(x, y)$ arises of the form:

$$D(x, y) \equiv \frac{2\pi^2}{t} \int ds ds' K(s) K(s') \Delta(s - s')$$

$$= -\frac{t}{4\pi} \int_x^y ds_\mu ds'_\nu \epsilon_{\mu\rho} \epsilon_{\nu\sigma} \partial_\rho^s \partial_\sigma^{s'} \ln |s - s'|. \tag{32.53}$$

We can thus write the Coulomb average (32.50) as a free field average in the source (32.52) provided we multiply the expression by e^{-D}. To evaluate $D(x, y)$ we use the identity:

$$\epsilon_{\mu\rho} \epsilon_{\nu\sigma} = \delta_{\mu\nu} \delta_{\rho\sigma} - \delta_{\mu\sigma} \delta_{\nu\rho},$$

and find:

$$D(x,y) = -\frac{t}{2} \int_x^y ds \cdot ds' \delta(s - s') + \frac{t}{4\pi} \int_x^y ds_\mu ds'_\nu \partial_\mu^s \partial_\nu^{s'} \ln |s - s'|. \qquad (32.54)$$

The integration over s and s' can be performed. Only one term coming from the second integral, gives a contribution which depends on $x - y$, the others are cut-off dependent. The final result is:

$$D(x,y) = -\frac{t}{2\pi} \ln |x - y| + \text{ const. }. \qquad (32.55)$$

If we replace the Coulomb gas average in equation (32.50) by the free field average with the source (32.52), multiplied by e^{-D}, something remarkable happens: The last factor just cancels the factor coming from the integration over θ_1. Therefore the spin correlation functions can be entirely calculated from the SG model, the spin field being represented by the non-local field appearing in expression (32.52) (see also equation (31.89b)):

$$\exp\left[i\theta(x)\right] \longmapsto \exp\left[-\frac{t}{2\pi} \int^x ds_\mu \epsilon_{\mu\nu} \partial_\nu \theta_2(s)\right]. \qquad (32.56)$$

The critical 2-point function. To calculate the 2-point function at the critical temperature we can now use the RG considerations of Section 32.3. Criticality corresponds to the special line $u = -v$ in the notation of Section 32.3. Note that the relation between t and u on the critical line is:

$$t = \pi/2(1 - u). \qquad (32.57)$$

The origin $u = v = 0$ is an IR fixed point. The behaviour of the correlation function can thus be obtained from perturbation theory. At leading order in the parameter A of equation (32.49) the 2-point function is still given by equation (32.12).

RG equations. The critical spin 2-point correlation function $W^{(2)}$ satisfies the RG equation:

$$\left[\Lambda \frac{\partial}{\partial \Lambda} + \beta(t) \frac{\partial}{\partial t} + \zeta(t)\right] W^{(2)}(x, t, \Lambda) = 0. \qquad (32.58)$$

The functions $\zeta(t)$ and $\beta(t)$ are (equations (32.15,32.26,32.57)):

$$\zeta(t) = t/2\pi + O\left(t - \frac{\pi}{2}\right)^2,$$
$$\beta(t) = -\frac{4}{\pi}\left(t - \frac{\pi}{2}\right)^2 + O\left(t - \frac{\pi}{2}\right)^3. \qquad (32.59)$$

For $t < \pi/2$ the theory is IR free. The effective coupling constant at scale λ behaves like:

$$t(\lambda) = \frac{\pi}{2}\left(1 + \frac{1}{2\ln\lambda}\right) + o\left(\frac{1}{\ln\lambda}\right) \text{ for } \lambda \to 0. \qquad (32.60)$$

$W^{(2)}(x, t)$ being dimensionless, the solution of the RG equation can be written:

$$W^{(2)}(x/\lambda, t) = Z^2(\lambda) W^{(2)}(x, t(\lambda)), \qquad (32.61)$$

with:

$$Z(\lambda) = \exp\left[\frac{1}{2} \int_1^\lambda \frac{d\sigma}{\sigma} \zeta(t(\sigma))\right] \sim \lambda^{1/8} |\ln \lambda|^{1/16}. \qquad (32.62)$$

Therefore, at large distance, the critical 2-point correlation function $W^{(2)}$ behaves like:

$$W^{(2)}(x, t) \underset{x \to \infty}{\sim} x^{-1/4}(\ln x)^{1/8}. \qquad (32.63)$$

This is the celebrated KT result for the phase transition of the classical 2D XY model.

32.6 The Generalized Thirring Model

Let us now use the results of Section 31.6 to justify the RG functions (32.26) and thus the RG flow near the phase transition. The correspondence with the SG model, in the notation of the Coulomb gas (32.23) is:

$$z \sim f, \qquad 8\pi T = 1 + g/\pi. \qquad (32.64)$$

From these expressions we see that for T close to $1/8\pi$ and z small, f and g are both small. The study of the phase transition is reduced to standard perturbation theory with fermion 4-point renormalizable interactions. The β-functions at one-loop order are (equation $(A31.18)$):

$$\begin{cases} \beta_g = -2f^2/\pi, \\ \beta_f = -2fg/\pi. \end{cases} \qquad (32.65)$$

We recognize the RG functions (32.26) in a different parametrization.

In the fermion language, the KT phase transition corresponds to the spontaneous breaking of the abelian $U(1)$ chiral symmetry. The phase diagram is simplest in the case of the $SU(2)$ symmetric model. Then for g positive the force between fermions is attractive. The model is asymptotically free and the spectrum, which is non-perturbative, contains massive particles, a massive fermion and a boson bound state: The chiral symmetry is spontaneously broken. However, the average of the order parameter $\langle \bar\psi\psi \rangle = 0$, vanishes in agreement with the Mermin–Wagner theorem. For $g < 0$ instead, the force between fermion is repulsive and the chiral symmetry is unbroken.

The same picture is valid for the $SU(N)$ Thirring model (see Appendix A31.2).

For the two parameter model we have the three regions of figure 32.1, a region I in which fields are massless and the chiral is unbroken, and regions II and III where the chiral symmetry is spontaneously broken.

Note finally that in the language of the simple massive Thirring model of Section 31.5 the situation is reversed. We have a transition at $g = -\pi/2$ (for m small) between a perturbative massive phase and a massless phase where the mass operator $\bar\psi\psi$ becomes irrelevant.

Bibliographical Notes

The low temperature phase of the model was considered by
   V.L. Berezinskii, *Zh. Eksp. Teor. Fiz.* 59 (1970) 907 (*Sov. Phys. JETP* 32 (1970) 493).
Relevant is also the article
   J. Villain, *J. Phys. C: Solid State Phys.* 6 (1973) 581.
The properties of the Kosterlitz–Thouless transition have been studied in
   J.M. Kosterlitz and D.J. Thouless, *J. Phys. C: Solid State Phys.* 6 (1973) L97; J.M. Kosterlitz, *J. Phys. C: Solid State Phys.* 7 (1974) 1046; J.V. José, L.P. Kadanoff, S. Kirkpatrick and D.R. Nelson, *Phys. Rev.* B16 (1977) 1217.
For the relation between the Coulomb gas and the $O(2)$ XY model see
   B. Nienhuis in *Phase Transitions and Critical Phenomena* vol. 11, C. Domb and J.L. Lebowitz eds. (Academic Press, New York 1987).
Our discussion of the Kosterlitz–Thouless transition follows the line of
   P.B. Wiegmann, *J. Phys. C: Solid State Phys.* 11 (1978) 1583; D.J. Amit, Y.Y. Goldschmidt and G. Grinstein, *J. Phys. A: Math. Gen.* 13 (1980) 585.
For the application of these ideas to one-dimensional conductors (Appendix A32) see the review article
   J. Solyom, *Advances in Physics* 28 (1979) 201.

APPENDIX 32
NON-RELATIVISTIC FERMIONS IN ONE DIMENSION

A non-relativistic propagator in Fourier space for a particle of momentum k, energy E and mass m reads

$$\Delta(E, k) = \frac{1}{E - k^2/(2m)}.$$

In many physical processes only those non-relativistic fermions with energies remaining close to the Fermi energy E_F play a role. In momentum space the Fermi sea is in general a sphere:

$$k_F^2 = 2mE_F,$$

or some deformation of a sphere depending on the space structure (lattice...). However in one space dimension the Fermi momenta take only two values $\pm\sqrt{2mE_F}$. Near the Fermi surface we can expand energies and momenta:

$$E = E_F + i\omega, \qquad k = \pm\sqrt{2mE_F} + q,$$

and approximate locally the dispersion curve by straight lines. The propagator can be simplified

$$\Delta_\varepsilon(\omega, q) = \frac{1}{i\omega - \varepsilon q\sqrt{2E_F/m}}, \qquad \varepsilon = \pm 1.$$

We recognize the propagator of a massless relativistic fermion, the two values $\varepsilon = \pm 1$ corresponding to right and left movers. The quantity $\sqrt{2E_F/m}$ plays the role of the speed of light. If we combine the two components $\varepsilon = \pm 1$ we obtain a massless Dirac fermion. Finally the non-relativistic spin appears as an external quantum number. Therefore a spin $1/2$ fermion yields a doublet of massless Dirac fermions. The two fermion model considered in Sections 31.6, 32.6 thus describes self-interacting one-dimensional non-relativistic fermions. All results derived from the point of view of the KT phase transition have an interpretation in systems like the Luttinger liquid (the equivalent of the Fermi liquid of higher dimensions), or one-dimensional conductors. For a systematic discussion of such models the reader is referred to the literature.

33 CRITICAL PROPERTIES OF GAUGE THEORIES

In this chapter we consider lattice models with gauge symmetry and study their properties from the point of view of phase transitions and spectrum structure. We concentrate on pure lattice gauge theories (without fermions) and study them mainly with lattice methods. Physically this means that we cannot investigate many properties of a realistic theory like QCD where fermions are coupled through a gauged $SU(3)$ colour group, but we can still try to answer one important question:

Does the theory generate confinement, i.e. a force between charged particles increasing at large distances, so that heavy quarks in the fundamental representation cannot be separated?

More generally can one find charged (from the gauge group point of view) asymptotic states like massless vector particles in the theory?

Other problems which we do not consider here, can also be discussed in this framework: for example the appearance of massive group singlet bound states in the spectrum (gluonium), the question of a deconfinement transition at finite physical temperature in QCD.

We first construct lattice models with gauge symmetry. We show that, as anticipated, they provide a lattice regularization of the continuum gauge theories studied in Chapters 18–21: the low temperature or small coupling expansion of the lattice model is a regularized continuum perturbation theory. We then discuss pure gauge theories (without matter fields) on the lattice. We discover that gauge theories have properties quite different from the ferromagnetic systems we have studied so far. In particular the absence of a local order parameter will force us to examine the behaviour of a non-local quantity, a functional of loops called hereafter Wilson's loop to distinguish between the confined and deconfined phases. Results will be obtained in the high temperature or strong coupling limit and in the mean field approximation.

33.1 Gauge Invariance on the Lattice

Let us follow the arguments developed in Chapter 19 to construct a lattice gauge theory. We assume that we start from a model possessing a global (rigid) symmetry group G, and we want to make it gauge invariant.

Let φ_i be the dynamical variables on the site i on which acts an orthogonal representation $\mathcal{D}(G)$ of the group G:

$$\varphi_{\mathbf{g}} = \mathbf{g}\varphi, \qquad \mathbf{g} \in \mathcal{D}(G). \tag{33.1}$$

The model is gauge invariant if it is possible to make an independent group transformation on each lattice site i without changing the partition function. For the measure of integration on φ_i as well as for all the terms in the action which depend only on one site, global invariance implies local invariance as in the continuum. Problems arise only when terms connect different sites.

Let us imagine for the sake of argument that the action has only nearest neighbour interaction terms of the form $\varphi_i \cdot \varphi_j$, i and j being nearest neighbour sites on the lattice. Such terms are obviously not invariant since the transformations of φ_i and φ_j are different:

$$\varphi_i \varphi_j \mapsto \varphi_i{}^T\mathbf{g}_i\, \mathbf{g}_j \varphi_j\,, \tag{33.2}$$

(where $^T\mathbf{g}$ means $\mathbf{g}$ transposed). To make it locally invariant it is necessary to introduce a parallel transporter depending on both sites i, j, $\mathbf{U}_{ij}$ which belongs to the representation $D\,(G)$ and transforms like:

$$\mathbf{U}_{ij} \mapsto \mathbf{g}_i \mathbf{U}_{ij}{}^T\mathbf{g}_j\,. \tag{33.3}$$

Then the quantity

$$\varphi_i \mathbf{U}_{ij} \varphi_j\,, \tag{33.4}$$

is gauge invariant. There is an important difference between the lattice model and the continuum theory: in the continuum the dynamical variables are the gauge fields which are associated with the Lie algebra, while on the lattice they are the parallel transporters associated with the Lie group.

The abelian case. In continuum field theory, in the abelian $U(1)$ case, we have already explicitly constructed the parallel transporter (equation (18.30)). Let us consider the product $\varphi^*\,(x)\,\varphi\,(y)$ of two complex fields at different points. We call A_μ the gauge field. Then the following expression is gauge invariant:

$$\varphi^*\,(x)\,\varphi\,(y) \exp\left[ie \int_x^y A_\mu\,(s)\,\mathrm{d}s_\mu\right]\,, \tag{33.5}$$

in which e is the gauge coupling constant, and the gauge field is integrated over some differentiable curve going from x to y. Indeed in a gauge transformation:

$$\left\{ \begin{array}{l} \varphi\,(x) \mapsto \mathrm{e}^{i\Lambda(x)}\,\varphi\,(x)\,, \\[2mm] A_\mu\,(x) \mapsto A_\mu\,(x) - \dfrac{1}{e}\partial_\mu\Lambda\,(x)\,, \end{array} \right. \tag{33.6}$$

the variation of the phase factor compensates the variation of $\varphi^*\,(x)\,\varphi\,(y)$ since

$$e \int_x^y A_\mu\,(s)\,\mathrm{d}s_\mu \mapsto e \int_x^y A_\mu\,(s)\,\mathrm{d}s_\mu - [\Lambda(y) - \Lambda(x)]\,. \tag{33.7}$$

The gauge field dependent phase factor which we have introduced is of course an element of the representation of the $U(1)$ group acting on $\varphi(x)$, which depends on the two points x and y.

The non-abelian case. In the non-abelian case the explicit relation is more complicated because the gauge field $\mathbf{A}_\mu^\alpha\,(x)\,t_\alpha$ is an element of the Lie algebra of G and the matrices representing the field at different points do not commute. It can be formally written as:

$$\mathbf{U}\,(x,y) = \mathrm{P}\left\{\exp\left[\int_x^y \mathbf{A}_\mu^\alpha\,(s)\,t^\alpha \mathrm{d}s_\mu\right]\right\}\,, \tag{33.8}$$

in which the symbol P means path ordered integral.

Let us now return to our lattice theory. It is clear that if we associate to each link $\ell(\equiv ij)$ of the lattice a gauge element $\mathbf{U}_\ell$ with the convention

$$\mathbf{U}_{ij} = \left(^T\mathbf{U}\right)_{ji} = \left(\mathbf{U}^{-1}\right)_{ji}\,, \tag{33.9}$$

consistent with the transformation law (33.3), then we can always render all products gauge invariant by introducing products on curves of these group elements.

33.2 The Pure Gauge Theory

We now discuss the pure gauge theory and its formal continuum limit as obtained from a low temperature, strong coupling expansion.

33.2.1 Action and partition function

We now have to construct a gauge invariant interaction for the gauge elements **U**. It follows from the transformation law (33.3) that only the traces of the products of **U**'s on closed loops are gauge invariant. On a hypercubic lattice the shortest loop is a square, called hereafter a *plaquette*. In what follows we consider a pure gauge action of the form:

$$S\left(\mathbf{U}\right) = - \sum_{\text{plaquettes}} \operatorname{tr} \mathbf{U}_{ij}\mathbf{U}_{jk}\mathbf{U}_{kl}\mathbf{U}_{li}, \qquad (33.10)$$

in which β_p is the plaquette coupling. The appearance of products of parallel transporters along closed loops is not surprising since we know quite generally that the curvature tensor $\mathbf{F}_{\mu\nu}$ which appears in the pure gauge action of the continuum theory is associated with infinitesimal transport along a closed loop. Note that each plaquette appears with both orientations in such a way that the sum is real when the group is unitary.

The partition function. We can then write a partition function corresponding to the action (33.10):

$$Z = \int \prod_{\text{links}\{ij\}} d\mathbf{U}_{ij} \ e^{-\beta_p S(\mathbf{U})}. \qquad (33.11)$$

We integrate over $\mathbf{U}_{ij}$ with the group invariant (de Haar) measure associated with the group G. In contrast to continuum gauge theories, expression (33.11) is well-defined on the lattice (at least as long as the volume is finite) because, since the group is compact, the volume of the group is finite. Therefore it is not necessary to fix the gauge and a *completely gauge invariant formulation* of the theory is possible.

33.2.2 Low temperature analysis

We want to understand the precise connection between the lattice theory (33.11) and the continuum field theory. For this purpose let us investigate the lattice theory at low temperature, i.e. at large positive β_p. In this limit the partition function is dominated by minimal energy configurations.

Let us show that the minimum of the energy corresponds to matrices **U** gauge transform of the identity. We start from a first plaquette 1234. Without loss of generality we can set:

$$\mathbf{U}_{12} = \mathbf{g}_1^{-1}\mathbf{g}_2 \qquad \mathbf{g}_1, \mathbf{g}_2 \in \mathcal{D}\left(G\right). \qquad (33.12)$$

The matrix $\mathbf{g}_1$ is arbitrary and $\mathbf{g}_2$ is calculated from $\mathbf{U}_{12}$ and $\mathbf{g}_1$. Then we can also set

$$\mathbf{U}_{23} = \mathbf{g}_2^{-1}\mathbf{g}_3, \qquad \mathbf{U}_{34} = \mathbf{g}_3^{-1}\mathbf{g}_4. \qquad (33.13)$$

These relations define first $\mathbf{g}_3$, then $\mathbf{g}_4$. The minimum of the action is obtained when the real part of all traces is maximum, i.e. when the products of the group elements on a plaquette are 1. (The trace of an orthogonal matrix **U** is maximum when all its eigenvalues are 1).

In particular:

$$\mathbf{U}_{12}\mathbf{U}_{23}\mathbf{U}_{34}\mathbf{U}_{41} = \mathbf{1}, \qquad (33.14)$$

which yields:

$$\mathbf{U}_{41} = \mathbf{g}_4^{-1}\mathbf{g}_1 \,. \tag{33.15}$$

If we now take an adjacent plaquette the argument can be repeated for all links but one which has already been fixed. In this way we can show that the minimum of the action is a pure gauge. Thus when the coupling constant β_p becomes very large, all group elements are constrained to stay, up to a gauge transformation, close to the identity. From this analysis we learn that the minimum of the potential is highly degenerate at low temperature, since it is parametrized by a gauge transformation, which corresponds to a finite number of degrees of freedom per site. This unusual property of lattice gauge theories corresponds to the property that the gauge action in classical mechanics determines the motion only up to a gauge transformation. To perform a low temperature expansion it is necessary to "fix" the gauge in order to sum over all minima.

Low temperature expansion. Let us choose a gauge such that the minimum of the energy corresponds to all matrices $\mathbf{U} = \mathbf{1}$. At low temperature the matrices $\mathbf{U}$ are then close to the identity:

$$\mathbf{U}\left(x, x + an_\mu\right) = 1 - a\mathbf{A}_\mu\left(x\right) + O\left(a^2\right), \tag{33.16}$$

in which a is the lattice spacing, x the point on the lattice, and n_μ the unit vector in the direction μ. We know from the discussion of Section 19.1 that the matrix $\mathbf{A}_\mu(x)$ is the connection or gauge field. We have already shown that the transformation law (33.3) of the parallel transporter implies for $\mathbf{A}_\mu\left(x\right)$ at leading order in the lattice spacing:

$$\mathbf{A}_\mu\left(x\right) \mapsto \mathbf{g}\left(x\right)\partial_\mu\mathbf{g}^{-1}\left(x\right) + \mathbf{g}\left(x\right)\mathbf{A}_\mu\left(x\right)\mathbf{g}^{-1}\left(x\right),$$

which is the usual gauge transformation.

Let us now expand the lattice action for small fields. To simplify the calculation we parametrize the orthogonal matrix $\mathbf{U}$ associated with the link $(x, x + an_\mu)$ in terms of the antisymmetric matrix $\mathbf{A}_\mu(x)$ as:

$$\ln\mathbf{U}\left(x + an_\mu, x\right) = -a\mathbf{A}_\mu\left(x + \tfrac{1}{2}an_\mu\right) + O\left(a^3\right). \tag{33.17}$$

We shall see in what follows that we need $\mathbf{U}$ up to order a^2. With the parametrization given above equation (33.9) implies that the term of order a^2 vanishes. Let us define the antisymmetric matrix $\mathbf{F}_{\mu\nu}\left(x\right)$ by:

$$\begin{aligned}
e^{-a^2\mathbf{F}_{\mu\nu}(x)} &= \mathbf{U}\left(x, x + an_\nu\right)\mathbf{U}\left(x + an_\nu, x + a\left(n_\mu + n_\nu\right)\right) \\
&\quad \times \mathbf{U}\left(x + a\left(n_\mu + n_\nu\right), x + an_\mu\right)\mathbf{U}\left(x + an_\mu, x\right). \tag{33.18}
\end{aligned}$$

To calculate $\mathbf{F}_{\mu\nu}\left(x\right)$ we introduce the definition (33.17) and use repeatedly the Baker–Hausdorf formula:

$$\ln\left(e^A e^B\right) = A + B + \frac{1}{2}[A, B] + \cdots. \tag{33.19}$$

Applied to the product of several factors it takes the form:

$$\ln\left(e^{A_1} e^{A_2}\ldots e^{A_n}\right) = \sum_i A_i + \frac{1}{2}\sum_{i<j}[A_i, A_j] + \cdots, \tag{33.20}$$

and therefore:

$$a^2 \mathbf{F}_{\mu\nu}(x) = a \left[\mathbf{A}_\mu \left(x + \tfrac{1}{2} a n_\mu \right) + \mathbf{A}_\nu \left(x + a n_\mu + \tfrac{1}{2} a n_\nu \right) - \mathbf{A}_\mu \left(x + a n_\nu + \tfrac{1}{2} a n_\mu \right) \right.$$
$$\left. - \mathbf{A}_\nu \left(x + \tfrac{1}{2} a n_\nu \right) \right] + a^2 \left[\mathbf{A}_\mu(x), \mathbf{A}_\nu(x) \right] + O\left(a^3 \right). \tag{33.21}$$

At leading order we recover the curvature tensor:

$$\mathbf{F}_{\mu\nu}(x) = \partial_\mu \mathbf{A}_\nu - \partial_\nu \mathbf{A}_\mu + [\mathbf{A}_\mu, \mathbf{A}_\nu] + O(a). \tag{33.22}$$

We obtain one term in the plaquette action by taking the trace of expression (33.18). Since $\mathbf{F}_{\mu\nu}$ is an antisymmetric matrix $\operatorname{tr} \mathbf{F}_{\mu\nu}$ vanishes. Thus

$$\operatorname{tr} e^{-a^2 \mathbf{F}_{\mu\nu}(x)} = \operatorname{tr} \mathbf{1} + a^4 \operatorname{tr} \mathbf{F}_{\mu\nu}^2(x) + O\left(a^6 \right). \tag{33.23}$$

This result shows that the leading term of the small field expansion of the plaquette action (33.10) is the standard gauge action studied in Chapter 19. The relation between β_p and the bare coupling constant e_0 of continuum gauge theories is thus:

$$a^4 \beta_p \sim e_0^{-2}. \tag{33.24}$$

As anticipated in Chapters 18,19, we conclude that the low temperature expansion, in a fixed gauge, of lattice gauge theories provides indeed a lattice regularization of continuum gauge theories. We have here discussed only the pure gauge action, but the result generalizes immediately to matter fields. Higher order terms in the small field expansion yield additional interactions needed to maintain gauge invariance on the lattice. This is not surprising: we have already shown that the gauge invariant extension of Pauli–Villars's regularization also introduces additional interactions.

33.3 Wilson's Loop and Confinement

In Subsection 20.2.2 we have calculated the RG β-functions for non-abelian gauge theories and shown that pure gauge theories are asymptotically free in 4 dimensions, which means that the origin in the coupling constant space is an UV fixed point and also implies that the effective interaction increases at large distance. Therefore, as in the case of the 2D non-linear σ-model, the spectrum of a non-abelian gauge theory cannot be determined from perturbation theory. To explain the non-observation of free quarks, it has been conjectured that the spectrum of the symmetric phase consists only in neutral states, i.e. states which are singlets for the group transformations.

Clearly it would be convenient to identify a local order parameter, i.e. a local observable whose average would distinguish between the QED phase of abelian gauge theories, in which charge states can be produced, from the so-called *confined* phase. However, in gauge theories such a local order parameter does not exist (see Elitzur's theorem). This property follows from the simple remark that physical observables correspond to gauge invariant operators which are neutral by construction. Furthermore we have seen in the study of continuum gauge theories (Chapters 18,19) that the only gauge independent quantities corresponding to non-gauge invariant operators are the S-matrix elements. Since it is very difficult to determine S-matrix elements beyond perturbation theory, it has been suggested by Wilson to study, in pure gauge theories, a gauge invariant non-local quantity, the energy of the vacuum in presence of largely separated static charges. Let us thus first study this quantity in pure abelian gauge theories, in which, in the continuum, all calculations can be done explicitly.

33.3.1 Wilson's loop in continuum abelian gauge theories

In continuum field theory, in order to calculate the average energy, it is necessary to introduce the gauge hamiltonian, and therefore convenient to work in the temporal gauge. We have constructed a wave function for two static point-like charges, in the temporal gauge, in Section 18.4 (equation (18.48)):

$$\psi\left(A\right) = \exp\left[-ie \oint_{C_0} A_i\left(s\right) \mathrm{d}s_i\right],$$ (33.25)

in which the charges are located at both ends of the curve C_0.

By evaluating the behaviour for large time T of the average:

$$W\left(C_0\right) = \left\langle \psi \left| \mathrm{e}^{-HT} \right| \psi \right\rangle,$$ (33.26)

in which H is the gauge hamiltonian in the temporal gauge, we obtain the energy $E\left(C_0\right)$ of the vacuum in presence of static charges:

$$W\left(C_0\right) \underset{T \to \infty}{\sim} \mathrm{e}^{-TE(C_0)}.$$ (33.27)

If the charges are separated by a distance R, we expect E to depend only on R and not on C_0.

The loop functional $W\left(C_0\right)$ has a representation in terms of a functional integral:

$$W\left(C_0\right) = \left\langle \exp\left[-ie \oint_{C_0'} A_\mu\left(s\right) \mathrm{d}s_\mu\right]\right\rangle,$$

C_0', which is now defined in space and time, is the union of two curves, which coincide with C_0 at time 0, and with $-C_0$ at time T respectively. The average here means average over gauge field configurations.

Since in the temporal gauge the time component of A_μ vanishes, we can add to C_0' two straight lines in the time direction which join the ends of the curves $C_0(t = 0)$ and $C_0(t = T)$. $W(C_0)$ then becomes a functional of a closed loop C (see figure 33.1):

$$W\left(C_0\right) \equiv W(C) = \left\langle \exp\left[-ie \oint_C A_\mu\left(s\right) \mathrm{d}s_\mu\right]\right\rangle.$$ (33.28)

The advantage of the representation (33.28) is that it is explicitly gauge invariant since it represents the average of the parallel transporter along a closed loop in space and time.

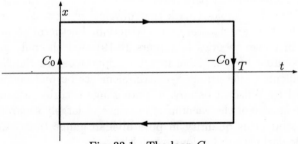

Fig. 33.1   The loop C.

The question of confinement is related to the behaviour of the energy E when the separation R between charges becomes large. In a pure abelian gauge theory in the continuum, which is a free field theory, the expression (33.28) can be evaluated explicitly. To simplify the calculation we take for C_0 also a straight line and use the Feynman gauge. The quantity $W(C)$ is then given by:

$$W(C) = \int [\mathrm{d}A_\mu] \exp\left[-S(A_\mu) + \int \mathrm{d}^d x\, J_\mu(x)\, A_\mu(x)\right], \tag{33.29}$$

with:

$$S(A_\mu) = \tfrac{1}{2} \int \mathrm{d}^d x\, [\partial_\mu A_\nu(x)]^2, \tag{33.30}$$

and:

$$J_\mu(x) = -ie \oint_C \delta(x - s)\, \mathrm{d}s_\mu. \tag{33.31}$$

The result is:

$$\ln W(C) = -\frac{1}{8\pi^{d/2}} \Gamma\left(\frac{d}{2} - 1\right) e^2 \oint_{C \times C} \mathrm{d}s_1 \cdot \mathrm{d}s_2\, |s_1 - s_2|^{2-d}. \tag{33.32}$$

The integral in the r.h.s. exhibits a short distance singularity, and a short distance cut-off has to be provided. Furthermore to normalize the r.h.s. of equation (33.32), we divide it by $W(C)$ taken for $R = a$, a being a fixed distance. Let us now write more explicitly the integrals:

$$\oint_{C \times C} \frac{\mathrm{d}s_1 \cdot \mathrm{d}s_2}{2\,|s_1 - s_2|^{d-2}} = \int_0^T |u - t|^{2-d}\, \dot{\mathrm{d}}u\, \mathrm{d}t + \int_0^R |x - y|^{2-y}\, \mathrm{d}x\, \mathrm{d}y$$
$$- \int_0^R \left[(x - y)^2 + T^2\right]^{1-d/2} \mathrm{d}x\, \mathrm{d}y - \int_0^T \left[(t - u)^2 + R^2\right]^{1-d/2} \mathrm{d}t\, \mathrm{d}u. \tag{33.33}$$

The first term in the r.h.s. is cancelled by the normalization. The second term is independent of T and therefore negligible for large T. It is actually related to the scalar product of the wave function $\psi(A)$ and the ground state eigenfunction. The third term decreases with T for $d > 2$ which we now assume. Only the last term increases with T:

$$\int_0^T \left\{ \left[(t - u)^2 + R^2\right]^{1-d/2} - \left[(t - u)^2 + a^2\right]^{1-d/2} \right\} \mathrm{d}t\, \mathrm{d}u$$
$$\sim \sqrt{\pi}\, \frac{\Gamma((d-3)/2)}{\Gamma(d/2 - 1)} \left(R^{3-d} - a^{3-d}\right) T. \tag{33.34}$$

Therefore the vacuum energy $E(R)$ in presence of the static charges has the form:

$$E(R) - E(a) = \frac{e^2}{4\pi^{(d-1)/2}} \Gamma((d-3)/2) \left(a^{3-d} - R^{3-d}\right). \tag{33.35}$$

We recognize the Coulomb potential between two charges.

For $d \leq 3$ the energy of the vacuum increases without bound when the charges are separated, and free charges cannot exist.

For $d = 3$ the potential increases logarithmically.

For $d = 2$ the Coulomb potential increases linearly with distance.

In more general situations the method that we have used above to determine the energy is complicated because we have to take the large T limit first and then evaluate the large R behaviour. It is more convenient to take a square loop, $T = R$, and evaluate the large R behaviour of $W(C)$. We here obtain:

$$\ln W\left[C\left(R\right)\right] - \ln W\left[C\left(a\right)\right] = \frac{1}{2\pi^{d/2}}\Gamma\left(\frac{d}{2} - 1\right)e^2\left\{\int_0^R \left[\left(u - t\right)^2 + R^2\right]^{1-d/2} du\, dt\right.$$

$$\left. - \int_0^a \left[\left(u - t\right)^2 + a^2\right]^{1-d/2} du\, dt - \int_a^R \left|u - t\right|^{2-d} du\, dt\right\}. \quad (33.36)$$

For $d > 3$, dimensions in which the Coulomb potential decreases, the r.h.s. is dominated by terms which correspond to the region $|s_1 - s_2| \ll R$ in equation (33.32):

$$\ln W\left[C\left(R\right)\right] - \ln W\left[C\left(a\right)\right] \sim \text{const.} \times R. \quad (33.37)$$

This is called the perimeter law since $\ln W(C)$ is proportional to the perimeter of C.

Instead for $d \leq 3$, $\ln W(C)$ increases as R^{4-d}. The reason is that two charges separated on C by a distance of order R, feel a potential of order R^{d-3}.

In particular for $d = 2$, $\ln W(C)$ increases like R^2, i.e. like the area of the surface enclosed by C: this is the area law.

33.3.2 Non-abelian gauge theories

It follows from the discussion of Section 19.3 that in the temporal gauge the wave function corresponding to two opposite point-like static charges is also related to a parallel transporter along a curve joining the charges. The same arguments as in the abelian case, show that the average of the operator e^{-TH} in the corresponding state is given by the average, in the sense of the functional integral, of the parallel transporter along a closed loop:

$$W\left(C\right) = \left\langle \mathrm{P}\exp\left[-i\oint_C \mathbf{A}_\mu\left(s\right)\mathrm{d}s_\mu\right]\right\rangle, \quad (33.38)$$

in which we recall that the symbol P means path ordering since the matrices $\mathbf{A}_\mu\left(s\right)$ at different points do not commute.

If we calculate $W(C)$ in perturbation theory we find of course at leading order the same results as in the abelian case. However we know from renormalization group, that we cannot trust perturbation theory at large distances. Therefore to get a qualitative idea about the phase structure we first use the lattice model to calculate $W(C)$ in the large coupling or high temperature limit $\beta_p \to 0$.

Strong coupling expansion for Wilson's loop. We here assume that the group we consider has a *non-trivial centre*. We shall take the explicit example of gauge elements on the lattice belonging to the fundamental representation of $SU(N)$ (whose centre is $\mathbb{Z}_N$, with elements the identity multiplied by roots z of unity, $z^N = 1$).

We calculate $W(C)$ by expanding the integrand in expression (33.11) in powers of β_p. We choose for simplicity for the loop C a rectangle although the generalization to other contours is easy.

Any non-vanishing contribution must be invariant by the change of variables $U_\ell \mapsto z_\ell U_\ell$ where z_ℓ belongs to the centre. Let us consider one link belonging to the loop and multiply the corresponding link variable $\mathbf{U}(x, x + an_\mu)$ by z_0. We now consider the set of all link variables $\mathbf{U}(x + y, x + y + an_\mu)$ which are obtained by a translation y in the hyperplane perpendicular to n_μ and also multiply them by z_y. Another link belonging to the loop belongs to the set but with opposite orientation. Plaquettes involving such variables involve them in pairs. For a result to be invariant and thus non-vanishing we need that the number of times each link variable appears in the direction n_μ minus the number of times it appears in the direction $-n_\mu$ vanishes (mod N). Thus we start adding plaquettes to satisfy this condition at point x. However the addition of one plaquette does not change the total difference between the numbers of links in the $+n_\mu$ and $-n_\mu$ directions. Therefore always at least one condition remains unsatisfied until the plaquettes reach the other link of the loop. We can then repeat the arguments for the remaining links of the loop and the new non integrated remaining links of the plaquettes. The number of required plaquette variables to get a non-vanishing result, is at least equal to the area of the rectangle, the minimal area surface having the loop as boundary. We can then perform the integrations which are just factorized group integrations. In this way we get a contribution to $W(C)$ proportional to $(\beta_p)^A$, in which A is the number of plaquettes. The largest contribution corresponds to plaquettes covering the minimal area surfaces bounded by the loop. It is indeed obtained by covering the rectangle with plaquettes in such a way that each link variable appears only twice in either orientation. For a rectangular loop $R \times T$ we just get

$$W(C) \sim e^{RT \ln \beta_p} . \tag{33.39}$$

This results indicates that the potential between the static charges is linearly rising at large distance. Static charges creating the loop cannot simply be screened by the gauge field, in which case we would again get a perimeter law.

Remarks.

(i) If the centre is trivial, it is possible to form a tube along the loop and this implies a perimeter law. If for example the group is $SO(3)$, in the decomposition of a product of two spin 1 representations, we again find a spin 1 which can be coupled to a third spin 1 to form a scalar. Thus two plaquettes can be glued to the same link of the loop without constraint on the orientation of the plaquette.

(ii) The asymptotic form (33.39) is also valid for the abelian $U(1)$ lattice gauge theory. Therefore, in four dimensions, Wilson's loop has a perimeter law at any order in the weak coupling expansion and an area law at large coupling. We expect a phase transition between a low coupling Coulomb phase, described by a free field theory, and a strong coupling confined phase. This phase transition has been observed in numerical simulations. It seems to be first order, but this question has not been definitively settled. The existence of the transition is related to the compact nature of the $U(1)$ group which is only relevant on the lattice (lattice QED based on group elements is also called compact QED). Defects in which the group element varies by a multiple of 2π around a plaquette govern the dynamics of the transition. They correspond in the continuum to magnetic monopoles. In four dimensions monopole loops yield, for dimensional reasons, logarithmic contributions to the action, a situation reminiscent of the two-dimensional Coulomb gas discussed in Chapter 32. The separation of vortices in the Kosterlitz–Thouless (KT) phase transition is here replaced by the separation of magnetic monopoles.

The string tension. The coefficient in front of the area is called the string tension σ,

$$\sigma\left(\beta_p\right) \underset{\beta_p \to 0}{\sim} -\ln \beta_p. \tag{33.40}$$

If no phase transition occurs when β_p varies from zero to infinity, the gauge theory leads to confinement. In this case the behaviour of the string tension for β_p small is predicted by the renormalization group. Since σ has the dimension of a mass squared one finds:

$$\sigma\left(e_0\right) \sim \left(e_0^2\right)^{-\beta_2/\beta_3^2} \exp\left(-1/\beta_2 e_0^2\right). \tag{33.41}$$

in which e_0^2 is related to β_p by the equation (33.24) and β_2, β_3 are two first coefficients of the RG β-function which are given in equation (34.51). A physical quantity relevant to the continuum limit can then be obtained by dividing $\sqrt{\sigma}$ by its asymptotic behaviour. Let us define Λ_L as:

$$\Lambda_L = a^{-1}\left(\beta_2 e_0^2\right)^{-\beta_3/2\beta_2^2} \exp\left(-1/2\beta_2 e_0^2\right), \tag{33.42}$$

then $\Lambda_L/\sqrt{\sigma}$ has a continuum limit. When one calculates σ by non-perturbative lattice methods, the verification of the scaling behaviour (33.41) indicates that the result is relevant to the continuum field theory and not only a lattice artifact.

It is possible to systematically expand σ in powers of β_p. The possibility of verifying that confinement is realized in the continuum limit, depends on the possibility of analytically continuing the strong coupling expansion up to the origin. Unfortunately, theoretical arguments lead to believe that, independently of the group, the string tension is affected by a singularity associated with the roughening transition, transition which however is not related to bulk properties. At strong coupling the contributions to the string tension come from smooth surfaces. When e_0^2 decreases (β_p increases), one passes through a critical point e_{0R}^2, after which the relevant surfaces become rough. At the singular coupling e_{0R}^2, the string tension does not vanish but has a weak singularity. Still at this point the strong coupling expansion diverges. Therefore it is impossible to extrapolate to arbitrarily small coupling. The usefulness of the strong coupling expansion then depends on the position of the roughening transition with respect of the onset of weak coupling behaviour. Notice that numerically in the neighbourhood of the roughening transition, rotational symmetry is approximately restored (at least at large enough distance).

One can also calculate other quantities which are associated to bulk properties, and are therefore not affected by roughening singularities, such as the free energy (the connected vacuum amplitude) or the plaquette–plaquette correlation function. However even for these quantities the extrapolation is not easy because the transition between strong and weak coupling behaviours is in general very sharp. This is confirmed by results coming from Monte Carlo simulations and is interpreted as indicating the presence of singularities in the complex β_P plane close to the real axis. From the numerical point of view it seems that the plaquette–plaquette correlation function is the most promising case for strong coupling expansion.

Remark. We note that the potential between static charges in the confined phase is linearly increasing in the same way as the Coulomb potential in one space dimension. This leads to the following physical picture: in QED the gauge field responsible of the potential has no charge and propagates essentially like a free field isotropically in all space directions. Conservation of flux on a sphere then yields the R^{2-d} force between the charges. However in the non-abelian case the attractive force between the gauge particles generates instead a flux tube between static charges in such a way that the force remains the same as in one space dimension.

Gauge symmetry breaking: Elitzur's theorem. Let us add a simple comment about the absence of a local order parameter in gauge theories. We have seen that in the temporal gauge the ground state is invariant under space dependent gauge transformations. This property is incompatible with the existence of a local order parameter which is necessarily non-gauge invariant. Therefore the question is: can a phase transition on the lattice lead to a spontaneous breaking of gauge invariance? The answer to this question can obtained by generalizing the arguments developed for ordinary symmetries in Sections 23.2 and 23.5. We consider the transition probability at low temperature between two states, concentrated one around the minimal energy configuration $\mathbf{A}_\mu = 0$ and the other one around a pure non-trivial gauge. If the gauge function is different from zero only in a finite space volume, the cost in energy is the same as in a one dimensional system and therefore the transition probability always remains finite independently of the number of space dimensions. Therefore the quantum mechanical ground state is gauge invariant. Note that this argument does not apply to gauge transformations which do not vanish at large distances. Therefore it does not forbid a spontaneous breaking of the global symmetry associated with the gauge group.

33.4 Mean Field Theory

We have shown that the pure gauge lattice model yields at low temperature or coupling the continuum gauge theory. The continuum model allows, in perturbation theory, the separation of charges at large distances. On the contrary, at high temperature, charges are confined in the lattice model.

It is therefore necessary to investigate the possibility of phase transitions in lattice gauge theories. In the case of spin models, mean field theory has given us a semi-quantitative understanding of the phase structure at least for $d > 2$. It is therefore natural to study gauge theories in the mean field approximation.

The general mean field formalism has been described in Appendix A24. We introduce two sets of real matrices ϕ_ℓ and H_ℓ, in which the index ℓ stands for link. We then rewrite the partition function Z:

$$Z = \int \prod_\ell \mathrm{d}U_\ell \exp\left[-\beta_p S\left(U\right)\right], \qquad (33.43)$$

in which $S\left(U\right)$ is the lattice action (33.10), as:

$$Z = \int \prod_{\text{links } \ell} \mathrm{d}\phi_\ell \, \mathrm{d}H_\ell \, \mathrm{d}U_\ell \exp\left[-\beta_p S\left(\phi\right) + \sum_{\text{links}} \operatorname{tr} H_\ell\left(\phi_\ell - U_\ell\right)\right]. \qquad (33.44)$$

The introduction of the variables ϕ allows to express the action in terms of an average link variable. Since the average of an orthogonal (unitary) matrix is not orthogonal (resp. not unitary), we have defined ϕ as an arbitrary real (resp. complex) matrix. The variables H_ℓ represent directly at leading order the mean field which approximates the effect of the plaquette interaction.

The integral over the matrices U now factorizes into a product of integrals over each link variable:

$$\int dU \, e^{-\operatorname{tr} HU} = e^{-\rho(H)},$$ (33.45)

in which $\rho(H)$ is thus a $G \times G$ group invariant function of H (H transforming under independent right and left multiplication).

The partition function becomes:

$$Z = \int \prod_\ell dH_\ell \, d\phi_\ell \exp\left\{ - \left[\beta_p S(\phi) + \sum_\ell (\rho(H_\ell) - \operatorname{tr} H_\ell \phi_\ell) \right] \right\}.$$ (33.46)

We then look for saddle points in the variables H and ϕ. Since H and ϕ are general real or complex matrices, we expect to find many saddle points. However both for simplicity and symmetry reasons we look for solutions in which H_ℓ and ϕ_ℓ are constant on the lattice and multiple of the identity (up to a gauge transformation):

$$\phi_\ell = \varphi I, \qquad H_\ell = h I,$$ (33.47)

in which I is the identity matrix. Calling $S(\varphi, h)$ the lattice action per link we then find:

$$S(h, \varphi) = \operatorname{tr} I \left[-\tfrac{1}{2}(d-1)\beta_p \varphi^4 + V(h) - h\varphi \right],$$ (33.48)

in which we have defined $V(h)$ by:

$$V(h) = \frac{\rho(hI)}{\operatorname{tr} I},$$ (33.49)

with $V(h) = -\frac{1}{4}h^2 + O(h^4)$ for $SU(2)$. The saddle point equations are:

$$\varphi = V'(h), \qquad h = -2(d-1)\beta_p \varphi^3.$$ (33.50)

We can eliminate φ and obtain:

$$h = -2(d-1)\beta_p \left[V'(h) \right]^3.$$ (33.51)

For h small, $V'(h)$ is at least linear in h (as in $SU(2)$). We realize immediately the essential difference with the spin models we had considered so far. The r.h.s. of equation (33.51) is at least cubic in h instead of being linear. Thus the equation has never a non-trivial solution arbitrarily close to zero. For β_p small there exists only the trivial solution $h = 0$, which, according to the strong coupling or high temperature analysis, corresponds to the confined phase in which Wilson's loop follows an area law. For a critical value β_c, h jumps from zero to a finite value, indicating a *first order* phase transition. We recall that at a first order transition the correlation length, at least above the transition, remains finite. Therefore the neighbourhood of the transition temperature does not define a continuum field theory, in contrast with the non-linear σ-model. Above β_c the average of Wilson's loop is given by:

$$W(C) = \left\langle \operatorname{tr} \prod_{\text{all links } \ell \in C} \phi_\ell \right\rangle \sim \phi^{P(C)},$$ (33.52)

in which $P(C)$ is the perimeter of the loop. Therefore Wilson's loop follows a perimeter law and the phase is deconfined. Above β_c we are in the low temperature phase which can be described by a continuum field theory and perturbation theory.

Discussion. As we have shown in Appendix A24, mean field theory is valid in high dimensions. Continuum field theory tells us that the zero temperature ($\beta_p = \infty$) is an IR stable fixed point for $d > 4$. Thus the mean field result can only apply for $d > 4$. However we would naively expect a second order phase transition in $4 + \varepsilon$ dimensions with a critical temperature of order ε, or $\beta_p \sim 1/\varepsilon$, in analogy with the non-linear σ-model. The open question is whether in any integer dimension $d > 4$ the transition is really second order.

For $d \leq 4$ the zero temperature is a UV fixed point. The simplest consistent scheme is one in which the critical temperature vanishes and the model always remains in the confined high temperature phase. The dimension $d = 4$ for gauge theories plays the role of the dimension $d = 2$ for the non-linear σ-model. The large momentum behaviour of correlation functions can be determined from perturbation theory, but no analytical method yields directly their low momentum behaviour and therefore for example the spectrum of the theory. The only other analytical piece of information available is the small coupling expansion in a finite volume of the eigenstates of the quantum hamiltonian, which one can try to extrapolate by numerical methods towards the infinite volume limit (see Chapter 36). However, again there is numerical evidence of a sharp transition between the finite volume and infinite volume results, making the extrapolation difficult. The most promising quantities seem to be ratios of masses. This lack of reliable analytical methods explains the popularity of numerical simulations based on stochastic methods of Monte Carlo type in this problem.

Monte Carlo methods. We shall not describe in any detail the numerical methods which have been used in lattice gauge theories. They are generalizations of the methods which we have briefly described in the case of simpler systems in Appendix A4.4. In pure gauge theories the existence of phase transitions has been investigated for many lattice actions. For the gauge group $SU(3)$, relevant to the physics of Strong Interactions, the string tension has been carefully measured, the plaquette–plaquette correlation function has been studied to determine the mass of low lying gluonium states. Finally calculations have been performed at finite physical temperature, i.e. on a 3+1 dimensional lattice in the limit in which the size of the lattice remains finite in one dimension, this size being related to the temperature. In this way the temperature of a deconfinement transition has been determined.

Fermions in numerical simulations. One important qualitative feature of Strong Interaction physics is the approximate spontaneous breaking of chiral symmetry (see Section 13.6). However we have already shown in Section 9.2 that non-trivial problems arise when one tries to construct a chiral invariant lattice action. One has the choice only between writing an action which is not explicitly chiral symmetric and in which one tries to restore chiral symmetry by adjusting the fermion mass term (Wilson's fermions), or writing a chiral symmetric action with too many fermions (staggered or Kogut–Susskind fermions). This is a theoretical problem. An important practical difficulty also arises with fermions: because it is impossible to simulate numerically fermions, it is necessary to integrate over fermions explicitly. This generates an effective gauge field action which contains a contribution proportional to the fermion determinant and is therefore no longer local. The speed of numerical methods crucially depends on the locality of the action. This explains that most numerical simulations with fermions have been up to now performed in the so-called quenched approximation in which the determinant is neglected. This approximation corresponds to the neglect of all fermion loops and bears some similarity with the eikonal approximation. In this approximation the approximate spontaneous breaking of chiral symmetry has been verified by measuring the decrease of the pion mass for decreasing quark masses. Owing to the difficulty of the problem the numerical study of dynamical fermions is still in a preliminary stage.

Bibliographical Notes

Lattice gauge theories have been introduced in
   K.G. Wilson, *Phys. Rev.* D10 (1974) 2445; F.J. Wegner, *J. Math. Phys.* 12 (1971) 2259.
Compact QED is studied in
   A.M. Polyakov, *Phys. Lett.* 59B (1975) 82.
High temperature expansion is discussed in
   R. Balian, J.M. Drouffe and C. Itzykson, *Phys. Rev.* D11 (1975) 2098; J.M. Drouffe and J.-B. Zuber, *Phys. Rep.* 102 (1983) 1.
Elitzur's theorem is derived in
   S. Elitzur, *Phys. Rev.* D12 (1975) 3978.
For a review of the early development of lattice gauge theories see
   J.B. Kogut in *Recent Advances in Field Theory and Statistical Mechanics*, Les Houches 1982, J.-B. Zuber and R. Stora eds. (North-Holland, Amsterdam 1984); *Rev. Mod. Phys.* 55 (1983) 775.
Calculations of the spectrum in non-abelian gauge theories can be found in
   M. Lüscher, *Nucl. Phys.* B219 (1983) 233; M. Lüscher and G. Münster, *Nucl. Phys.* B232 (1984) 445; J. Koller and P. van Baal, *Nucl. Phys.* 302 (1988) 1; P. van Baal, *Nucl. Phys.* B307 (1988) 274.
Monte Carlo simulations were initiated by
   M. Creutz, *Phys. Rev. Lett.* 45 (1980) 313.
Early articles on lattice gauge theories and Monte Carlo simulations are reprinted in
   Lattice Gauge Theories and Monte Carlo Simulations, C. Rebbi ed. (World Scientific, Singapore 1983).
See also
   M. Creutz, L. Jacobs and C. Rebbi, *Phys. Rep.* 95 (1983) 201; I. Montvay, *Rev. Mod. Phys.* 59 (1987) 263.
For Monte Carlo methods in statistical physics see for example
   Monte Carlo Methods in Statistical Physics, K. Binder ed. (Springer-Verlag, New York 1979).
Relevant are also the lectures
   M. Lüscher in *Critical Phenomena, Random Systems, Gauge Theories*, Les Houches 1984, K. Osterwalder and R. Stora eds. (Elsevier, Amsterdam 1986) and *Fields, Strings and Critical Phenomena*, Les Houches 1988, E. Brézin and J. Zinn-Justin eds. (Elsevier, Amsterdam 1989).
Finite temperature transitions are discussed in
   B. Svetitsky, *Phys. Rep.* 132 (1986) 1; E. Marinari, *Phys. Rep.* 184 (1989) 131.
For pure gauge theories in two dimensions see for example
   D.J. Gross and E. Witten, *Phys. Rev.* D21 (1980) 446.

Exercises

Exercise 33.1

The Landau–Ginzburg model of superconductivity. The Landau–Ginzburg model of a superconductor is, in field theory language, the abelian Higgs model, i.e. a field theory for a charged scalar field coupled to an electromagnetic gauge field (for more details see the classical books on superconductivity). We propose to study by RG methods the nature of the phase transition:

$$S(A_\mu, \phi) = \int \mathrm{d}^d x \left(\frac{1}{4} F_{\mu\nu}^2 + |D_\mu \phi|^2 + M^2 |\phi|^2 + \frac{1}{6} g |\phi|^4 \right),$$

with:

$$D_\mu = \partial_\mu + ie A_\mu, \qquad F_{\mu\nu} = \partial_\mu A_\nu - \partial_\nu A_\mu.$$

Calculate the RG β-functions at one-loop order, solve the flow equations for the coupling constants in 4 and in $4 - \varepsilon$ dimensions. Discuss the RG flow, the locations and stability of the fixed points.

APPENDIX 33
GAUGE THEORY AND CONFINEMENT IN TWO DIMENSIONS

Two dimensions is from the point of view of gauge theories peculiar in the sense that a gauge field has no real dynamical degrees of freedom. Still the gauge field generates a force between particles which, as we have seen in Section 33.3, leads to confinement, even in the abelian case. An example of such a situation has been encountered in the Schwinger model in Section 31.4. This is a question that we here examine in more detail on the lattice for pure gauge theories.

Lattice gauge theories in two dimensions. Let us consider a pure gauge lattice action (33.10) in two dimensions. The partition function is given by:

$$Z = \int \prod_{\text{links}\{ij\}} d\mathbf{U}_{ij} \exp \left(\beta \sum_{\text{plaquettes}} \operatorname{tr} \mathbf{U}_{ij} \mathbf{U}_{jk} \mathbf{U}_{kl} \mathbf{U}_{li} \right). \tag{A33.1}$$

We assume free boundary conditions in the time direction (to avoid closed loop variables surviving due to the topologic properties of the two dimensional space manifold). We can then use the gauge invariance to set to 1 all link variables in the time direction. This is the equivalent of the temporal gauge of the continuum theory. However in two dimensions this has the remarkable effect of decoupling all links in the orthogonal direction. The partition function factorizes. If we call L the size in this direction, we find:

$$Z = Z_t^L, \tag{A33.2}$$

with:

$$Z_t = \int \prod_i dU_i \exp \left(\beta \sum_i \operatorname{tr} U_{i+1}{}^T U_i \right), \tag{A33.3}$$

in which i is now the coordinate in the time direction. We recognize the partition function of a simple one dimensional model with two-body nearest neighbour interactions. The partition function can be calculated by the transfer matrix method. In the case of free space boundary conditions we also can set:

$$V_{i+1} = U_{i+1}{}^T U_i. \tag{A33.4}$$

The integrals over the matrices V_i factorize and we find:

$$Z(\beta) = z^\Omega(\beta), \tag{A33.5}$$

in which Ω is the area of the lattice, and $z(\beta)$ is given by:

$$z(\beta) = \int dU \exp \left(\beta \operatorname{tr} U \right). \tag{A33.6}$$

We hereafter assume that the volume of the group has been normalized to 1. The function $z(\beta)$ is, as expected, a regular positive function: no phase transition occurs in two dimensions. It is also easy to calculate the average of Wilson's loop. Calling R and T the sizes of the loop in space and time, one finds:

$$W(C) = e^{-RT\sigma(\beta)}, \tag{A33.7}$$

with:

$$\sigma(\beta) = -\ln\left(\frac{z'(\beta)}{Nz(\beta)}\right). \qquad (A33.8)$$

We have again assumed that the centre of the group is non-trivial and called N the trace of the unit matrix. As expected, Wilson's loop has an area law for all groups and all couplings in two dimensions.

For β small, $z(\beta)$ has the expansion:

$$z(\beta) = 1 + z_2\beta^2 + O\left(\beta^4\right). \qquad (A33.9)$$

This yields for the string tension:

$$\sigma(\beta) \sim -\ln\beta, \qquad (A33.10)$$

in agreement with expression (33.39). For β large, $z(\beta)$ can be calculated by steepest descent which is also perturbation theory, and one finds:

$$\ln z(\beta) = N\left(\beta - K\ln\beta\right) + O\left(\beta^{-1}\right), \qquad (A33.11)$$

in which K is a constant. Therefore:

$$\sigma(\beta) \sim K/\beta, \qquad (A33.12)$$

in agreement with perturbation theory in the continuum. Finally let us note that since in two dimensions the gauge field has no physical degrees of freedom, no particle propagates, and no gluonium state can be found.

Up to now we have mainly discussed the IR behaviour of field theories. In this chapter we use the RG equations to characterize instead the large momentum behaviour of renormalized field theories.

It has been observed experimentally that the quarks, fundamental particles of the theory of Strong Interactions, behave like free particles at the shortest distances presently accessible. Therefore the discussion of the large momentum behaviour and the identification of field theories which behave approximately as free field theories at short distance is directly relevant for Particle Physics. Examining the large momentum behaviour of all field theories renormalizable in four dimensions, we show that only theories having a non-abelian gauge symmetry can be asymptotically free, i.e. behave as a free field theory at large euclidean momenta (Coleman–Gross theorem). We begin our investigation with scalar ϕ^4-like field theories and then add fermions and gauge fields. As an application we calculate the total cross section of electron–positron annihilation into hadrons at large momentum.

Another more theoretical reason for discussing the large momentum behaviour is the apparent connection between the existence of non-trivial renormalized quantum field theories and the presence of UV fixed points. The absence of identified UV fixed points in theories like the ϕ^4 field theory or QED leads to the so-called triviality problem which we examine in Subsection 34.1.1. If we then consider the ϕ^4 field theory as the main responsible of the Higgs mass, we obtain a semi-quantitative bound on the Higgs mass (see Subsection 34.1.3).

34.1 The $\left(\phi^2\right)^2$ Field Theory: Large Momentum Behaviour and Triviality

Before discussing the problem of large momentum behaviour of correlation functions in renormalized field theories, let us first emphasize a difference between the ϕ^4 field theory as it arises in Statistical Mechanics, and was discussed in Chapters 25 and 26, and the ϕ^4 theory as considered from the traditional point of view of Particle Physics: in the case of phase transitions, the "bare" coupling constant g_0 is fixed and related to the initial microscopic physics, and the renormalized coupling constant g is a function of g_0, the mass m and the cut-off Λ. Instead in quantum field theory one traditionally assumes that the renormalized coupling constant g is fixed, and g_0 varies when the cut-off Λ increases.

Below four dimensions the ϕ^4 field theory is super-renormalizable and, at g fixed, g_0 has a finite limit when Λ becomes infinite:

$$g_0 = m^\varepsilon f(g). \tag{34.1}$$

The coupling constant $g = 0$ is a UV stable fixed point and the large momentum behaviour of correlation functions is simply given by the leading order of perturbation theory. There is only one special point, the IR fixed point $g = g^*$, where a non-trivial scaling behaviour is obtained.

In four dimensions instead the situation is different, the ϕ^4 field theory is renormalizable and the determination of the large momentum behaviour of correlation functions relies on the study of RG equations, like for example the CS equations. At large momentum the r.h.s. of the CS equations can be neglected, reducing them to the homogeneous equations (10.50) for the asymptotic massless theory (see Section 10.9):

$$\left[m\frac{\partial}{\partial m} + \beta\left(g\right)\frac{\partial}{\partial g} - \frac{N}{2}\eta\left(g\right) \right] \Gamma_{\text{as.}}^{(n)}\left(p_i; m, g\right) = 0 \,. \tag{34.2}$$

Equation (34.2) can be solved by the method of characteristics which we have introduced in Section 25.5. Calling λ the scaling factor, we look for a solution of equation (34.2) of the form:

$$\Gamma_{\text{as.}}^{(n)}\left(p_i; m, g\right) = Z^{N/2}\left(\lambda\right)\Gamma_{\text{as.}}^{(n)}\left(p_i; \lambda m, g\left(\lambda\right)\right), \tag{34.3}$$

with:

$$g(1) = g\,; \qquad Z(1) = 1\,. \tag{34.4}$$

Differentiating the r.h.s. with respect to λ and using equation (34.2), we find:

$$\lambda\frac{\mathrm{d}}{\mathrm{d}\lambda}g\left(\lambda\right) = \beta\left(g\left(\lambda\right)\right), \tag{34.5a}$$

$$\lambda\frac{\mathrm{d}}{\mathrm{d}\lambda}\ln Z\left(\lambda\right) = -\eta\left(g\left(\lambda\right)\right). \tag{34.5b}$$

The solutions of equations (34.5) are:

$$\ln\lambda = \int_g^{g(\lambda)}\frac{\mathrm{d}g'}{\beta\left(g'\right)} \tag{34.6}$$

$$Z\left(\lambda\right) = \exp\left[-\int_1^\lambda\frac{\mathrm{d}\sigma}{\sigma}\eta\left(g\left(\sigma\right)\right) \right]. \tag{34.7}$$

Rescaling in equation (34.3) λm into m, and using the dimensional equation (26.6), we obtain for $d = 4$:

$$\Gamma_{\text{as.}}^{(n)}\left(\lambda p_i; m, g\right) = \lambda^{4-n}Z^{n/2}\left(\lambda\right)\Gamma_{\text{as.}}^{(n)}\left(p_i; m, g\left(\lambda\right)\right). \tag{34.8}$$

This equation shows that $g\left(\lambda\right)$ is the effective coupling at scale λ. If we start with a coupling constant g small enough such that perturbation theory is applicable, then $\beta(g)$ is positive and when λ increases, $g\left(\lambda\right)$ increases. If $\beta\left(g\right)$ has an UV zero, i.e. a zero g^* at which the derivative $\beta'\left(g^*\right)$ is negative, $g\left(\lambda\right)$ will have g^* as limit. If $\beta\left(g\right)$ has no zero, the behaviour of $g\left(\lambda\right)$ depends on the behaviour of $\beta\left(g\right)$ for g large. If the integral $u(g)$:

$$u(g) = \int^{+\infty}\frac{\mathrm{d}g}{\beta(g)}, \tag{34.9}$$

diverges, $g\left(\lambda\right)$ tends towards $+\infty$. If instead this integral converges, the equation (34.6) has no solution for λ large.

In the case of generic phase transitions, the special form of the bare coupling constant, as a function of the cut-off, has imposed to g to sit on an IR fixed point.

34.1.1 Existence of the renormalized ϕ_4^4 field theory

We now show that the existence of a non-trivial renormalized ϕ^4 field theory in four dimensions relies on the same properties. The problem is to be able to find, for arbitrarily large values of the cut-off Λ, a bare coupling constant g_0 which yields a given renormalized coupling constant g. This problem of course has always a formal perturbative solution but we want to discuss this question beyond perturbation theory. Using the RG equations, we can determine the behaviour of renormalization constants at fixed renormalized parameters when the cut-off varies.

Setting:

$$g_0 = g_0(s, g), \quad s = \frac{m}{\Lambda}, \tag{34.10}$$

and using the definition of the β-function (for example equation (10.26)), we find:

$$0 = \left(s\frac{\partial}{\partial s} + \beta(g)\frac{\partial}{\partial g} \right) g_0(s, g). \tag{34.11}$$

The solution of the equation can be expressed in terms of the effective coupling constant $g(\lambda)$, defined by equation (34.5a), as:

$$g_0(s/\lambda, g) = g_0(s, g(\lambda)), \tag{34.12}$$

When Λ increases, the parameter λ increases. For g small, $\beta(g)$ is positive and thus $g(\lambda)$ increases until perturbation theory is no longer valid. The large cut-off behaviour of renormalization constants is related to the UV behaviour of the renormalized theory (not surprisingly) and the possible existence of UV fixed points. If a UV fixed point exists, $g(\lambda)$ flows into the fixed point. Numerical studies of the ϕ^4 field theory on the lattice strongly suggest that no such UV fixed point exists. In the absence of a fixed point we have to again examine the convergence of integral (34.9). If the integral diverges, we can say that there is a UV fixed point at infinity. If the integral converges, the equation (34.12) has no solution for large enough values of the cut-off and the renormalized theory does not exist.

The problem can also be examined from the point of view of the bare theory. We have shown that for the values of the bare coupling constant g_0 which are in the domain of application of perturbation theory, the corresponding renormalized coupling constant vanishes at large cut-off. Again the existence of a non-trivial renormalized theory relies on the existence of a non-trivial IR unstable fixed point. As we discussed in Chapter 30 in the case of the non-linear σ-model, when a theory is IR free, a non-trivial theory at smaller distance implies the existence of an intermediate scale large compared with the microscopic scale but small compared with the correlation length, at which a crossover between a trivial IR behaviour and a non-trivial UV behaviour takes place. Renormalization group tells us that this scale must have the form (see equation (30.31)):

$$\xi(g_0) = \Lambda^{-1} \exp\left[\int^{g_0} \frac{dg'}{\beta(g')} \right], \tag{34.13}$$

in which $\beta(g_0)$ is now the β-function of the bare theory. Therefore the length scale ξ diverges only at a UV fixed point.

An equivalent condition is obtained by calculating the effective coupling constant at scale $\mu \ll \Lambda$ (which can also be chosen as the renormalized coupling constant):

$$\ln(\Lambda/\mu) = \int_g^{g_0} \frac{dg'}{\beta(g')}. \tag{34.14}$$

If we increase Λ at g fixed, we find a solution g_0 only if the there exists a value at which the integral diverges, i.e. an UV fixed point. If the integral converges for g_0 large, $\Lambda/\mu \to \infty$ implies that g goes to the IR fixed point $g = 0$.

In the case of a one component system (Ising-like) we have some definite information about this question. In the limit of infinite bare coupling constant, the lattice regularized ϕ^4 theory becomes the Ising model. We have presented conclusive numerical evidence in Chapter 28 that in two and three dimensions the Ising model and the ϕ^4 belong to the same universality class. This means that no IR unstable fixed point exists in these dimensions. In four dimensions the evidence is somewhat weaker. This is expected within the RG framework since the approach to scaling is only logarithmic. In higher dimensions again the Ising model and the ϕ^4 fall in the same universality class. It would thus be highly surprising if only in four dimensions these models would behave differently, with moreover rather close exponents. Finally Monte Carlo numerical simulations are completely consistent with triviality. Some of these remarks can also be extended to the $\left(\phi^2\right)^2$ theory for a number of components $N \leq 4$. Furthermore we have calculated the β-function in the large N limit in Chapter 29 and found that it is proportional to g_0^2. Therefore no fixed point exists in the domain of the $1/N$ expansion, i.e. for g_0 of order $1/N$. On the other hand for large g_0 the $\left(\phi^2\right)^2$ theory becomes the non-linear σ-model which we have examined in the large N limit in Section 30.7. Again we have found that it becomes a free field theory in four dimensions. Therefore we have strong evidence that no continuum renormalized $\left(\phi^2\right)^2$ theory exists in four dimensions. Rigorous results would also imply triviality if it were possible to prove the divergence of the field renormalization.

One may then wonder about the meaning of the correlation functions as defined by renormalized perturbation theory. We shall argue in Chapter 42 that there are intrinsic difficulties in the reconstruction of correlation functions from the knowledge of their perturbative expansion. However, irrespective of this problem, it follows from the RG arguments that to a finite renormalized coupling constant can only correspond a complex bare coupling constant, the imaginary part vanishing at all orders in perturbation theory. As a consequence, the correlation functions, depending on the summation procedure, will either be complex, or will not satisfy the field equations beyond perturbation theory.

34.1.2 The ϕ^4 field theory for negative coupling

In the previous analysis we have assumed that the renormalized coupling constant is always positive. It has been argued, although not very convincingly, that the renormalized coupling constant could also be negative. If we start from $g < 0$ and small, the RG equations still tell us that g increases, but it now converges towards $g = 0$ which is thus a UV fixed point. This procedure then defines a non-trivial theory. However a question arises immediately: does this theory correspond to a hamiltonian bounded from below? We have shown in Section 6.6 that the 1PI functional is related to the average of the hamiltonian in states with fixed field average. We have seen in Section 26.5 that RG equations relate the thermodynamical potential for small magnetization to IR fixed points. Not surprisingly the large constant field behaviour of the 1PI functional is governed by UV fixed points. Following the arguments of Section 26.5, we can derive the RG equation satisfied by $\Gamma(\varphi)$, the thermodynamical potential at constant field and per unit volume. We consider for simplicity a massless theory renormalized at scale μ, and subtract $\Gamma(\varphi)$ at $\varphi = 0$. We obtain the equation:

$$\left[\mu \frac{\partial}{\partial \mu} + \beta\left(g\right)\frac{\partial}{\partial g} - \frac{1}{2}\eta\left(g\right)\varphi\frac{\partial}{\partial \varphi}\right]\Gamma\left(\varphi, g, \mu\right) = 0\,. \tag{34.15}$$

Solving the equation by the method of characteristics and using dimensional analysis we find:

$$\Gamma\left(\lambda\varphi, g, \mu\right) = \lambda^4 \Gamma\left(Z^{1/2}(\lambda)\varphi, g(\lambda), \mu\right), \qquad (34.16)$$

in which $g(\lambda)$ and $Z(\lambda)$ are defined by equations (34.4,34.5) in terms of the corresponding RG functions. We see that to study the large field behaviour of $\Gamma(\varphi)$ we have to increase λ and therefore study the UV limit of the theory. If we start from $g < 0$ and small, $g(\lambda)$ approaches the origin for λ large. The field renormalization $Z(\lambda)$ then tends towards 1, and $\Gamma(\varphi)$ can be taken from perturbation theory. Therefore, in the case of the $O(N)$ invariant $\left(\phi^2\right)^2$ field theory, φ being then the length of the vector φ, we obtain:

$$\Gamma\left(\lambda\varphi, g, \mu\right) \sim \frac{1}{4!}\lambda^4 g(\lambda)\varphi^4, \qquad (34.17)$$

with:

$$g(\lambda) \sim -\frac{48\pi^2}{(N+8)\ln\lambda}. \qquad (34.18)$$

The consequences of this result are obvious: by increasing λ, Γ can be made arbitrarily large and negative. It follows that the corresponding hamiltonian is not bounded from below.

34.1.3 Non-asymptotic freedom. A bound on the Higgs mass

A first conclusion emerges, important for Particle Physics: the $\left(\phi^2\right)^2$ field theory cannot be asymptotically free at large momenta and is therefore, alone, not a good candidate to represent the physics of Strong Interactions at experimentally accessible short distances.

A second conclusion seems unavoidable, considering the arguments we have given above and the numerical evidence available: the $\left(\phi^2\right)^2$ field theory is a free field theory in the infinite cut-off limit. This a problem for the Higgs sector of the Standard Model of weak electromagnetic interactions which contains a $\left(\phi^2\right)^2$ interaction. Of course then other coupling constants contribute to the ϕ^4 coupling β-function, and the RG flow may be different, as we discuss in the coming sections. However the most likely conclusion is that in the Standard Model this problem cannot be avoided. Considering the phenomenological success of the model this conclusion is somewhat surprising. However we realize that the model has only been tested in some limited range of energies. Therefore, demanding that the field theory should be consistent on all scales has no real physical justification. If we keep the cut-off large but fixed, at the cut-off scale the theory will break down but at lower scales it will give reasonable answers. Correspondingly the renormalized coupling will be allowed to vary in a limited range, which goes to zero (logarithmically) for large cut-off. In this sense the Standard Model is an effective low energy field theory, in which the cut-off, reflection of a large mass scale of new physics, cannot be completely eliminated, even though the theory is perturbatively renormalizable: The theory is not consistent on all scales.

This conclusion is specially unavoidable if the ϕ^4 Higgs coupling is not very small. Indeed since the gauge and Yukawa couplings are quite small (even in the case of the top quark), the Higgs coupling then probably dominates the Higgs sector and a purely ϕ^4 analysis is relevant.

An upper bound on the Higgs mass. The Higgs field through its various couplings gives masses to all other fields. The observed masses determine the corresponding couplings.

Only the Higgs mass and thus the Higgs self-coupling are unknown parameters. Note, however, that it is likely that the renormalized ϕ^4 coupling g is such that perturbation theory remains at least semi-quantitatively applicable. Otherwise the successes of the Standard Model are difficult to understand. In the perturbative regime the Higgs mass increases with g. To obtain an upper-bound on the Higgs mass one has to examine what happens when g increases. As we have argued above for g large enough probably the Higgs mass is mostly determined by the Higgs self-coupling. We therefore examine below the pure ϕ^4 field theory. Since we remain in the perturbative regime RG arguments are still applicable. A bound on m_H, the Higgs mass, can then be derived.

We solve equation (34.14), using the expansion of the β-function,

$$\beta\left(g_0\right) = \beta_2 g_0^2 + \beta_3 g_0^3 + O\left(g_0^4\right),$$

for g small. We find:

$$\ln(\Lambda/\mu) = \frac{1}{\beta_2 g} + \frac{\beta_3}{\beta_2^2}\ln g + K\left(g_0\right) + O(g), \qquad (34.19)$$

where the function $K(g_0)$, according to the previous discussion, is bounded but can only be determined by non-perturbative methods. For g small we can use perturbation theory to relate the ϕ (=Higgs) expectation value, which is known from the Z mass ($\langle\phi\rangle \sim 250\text{GeV}$), and the Higgs mass (see Section 27.2). At leading order we find:

$$m_H^2 = \tfrac{1}{3}g\left\langle\phi\right\rangle^2 + O\left(g^2\right). \qquad (34.20)$$

To minimize higher order corrections we choose for g the renormalized coupling constant at scale $\langle\phi\rangle$. We can then eliminate g between equations (34.20) and (34.19), and find

$$\ln\left(\frac{\Lambda}{\langle\phi\rangle}\right) = \frac{1}{3\beta_2}\frac{\langle\phi\rangle^2}{m_H^2} + \frac{2\beta_3}{\beta_2^2}\ln\left(\frac{m_H}{\langle\phi\rangle}\right) + \tilde{K}\left(g_0\right) + O(g).$$

If we assume that we can neglect in the r.h.s. all terms but the two first ones, we obtain a relation between the two ratios $\Lambda/\langle\phi\rangle$ and $m_H/\langle\phi\rangle$. Moreover if the Higgs is really associated to a physical particle its mass must be smaller than the cut-off (which at this point only represents the onset of new physics beyond the Standard Model). Taking for the two coefficients of the β-function the values from equation(28.3) for $N = 4$, $8\pi^2\beta_2 = 2$, $\beta_3/\beta_2^2 = -13/24$ we obtain an upper bound for m_H

$$m_H < 2.6\left\langle\phi\right\rangle \ \Rightarrow\ m_H < 640\text{ GeV},$$

value which can be compared with computer simulation values which vary in the range 670–700GeV. Note that the corresponding value of g is such that perturbation theory at leading order should still be semi-quantitatively correct.

Conversely if the know the physical coupling constant at scale μ we can infer from equation (34.19) an upper bound on the cut-off or scale of new physics. Of course this bound is very sensitive to small corrections since the equation determines $\ln(\Lambda/\mu)$.

34.2 The General ϕ^4 Field Theory

We now study the large momentum behaviour in a general ϕ^4-like field theory in four dimensions. We consider an action for a massless field $\phi_i(x)$ which at the tree level has the form:

$$S(\phi) = \int \mathrm{d}^d x \left[\frac{1}{2} \sum_i (\partial_\mu \phi_i)^2 + \frac{\mu^{4-d}}{4!} \sum_{ijkl} g_{ijkl} \phi_i \phi_j \phi_k \phi_l \right]. \tag{34.21}$$

The corresponding RG equations have been established in Section 11.7. The n-point correlation functions satisfy equation (11.98):

$$\left(\mu \frac{\partial}{\partial \mu} + \beta_{i'j'k'l'} \frac{\partial}{\partial g_{i'j'k'l'}} \right) \Gamma^{(n)}_{i_1 i_2 \ldots i_n} - \frac{1}{2} \sum_{m=1}^{n} \eta_{i_m j_m} \Gamma^{(n)}_{i_1 i_2 \ldots j_m \ldots i_n} = 0. \tag{34.22}$$

We also need the explicit leading order of β_{ijkl} (equation (11.103)) in 4 dimensions:

$$\beta_{ijkl} = \frac{1}{16\pi^2} \left(g_{ijmn} g_{mnkl} + g_{ikmn} g_{mnjl} + g_{ilmn} g_{mnjk} \right) + O\left(|g|^3 \right). \tag{34.23}$$

Furthermore we know that η_{ij} is of order g^2. Following the arguments given in Chapter 26, it is easy to verify that the thermodynamical potential per unit volume in a constant field φ, and subtracted at $\varphi = 0$, satisfies:

$$\left(\mu \frac{\partial}{\partial \mu} + \beta_{i'j'k'l'} \frac{\partial}{\partial g_{i'j'k'l'}} - \frac{1}{2} \eta_{ij} \varphi_i \frac{\partial}{\partial \varphi_j} \right) \Gamma(\varphi) = 0. \tag{34.24}$$

We again solve this equation by the method of characteristics (see Chapters 25 and 26). Introducing a scale parameter λ, scale dependent coupling constants $g_{ijkl}(\lambda)$ and field renormalization matrix $Z_{ij}(\lambda)$ defined by:

$$\lambda \frac{\mathrm{d}}{\mathrm{d}\lambda} g_{ijkl}(\lambda) = \beta_{ijkl}(g(\lambda)), \qquad g_{ijkl}(1) = g_{ijkl}, \tag{34.25}$$

$$\left(\lambda \frac{\mathrm{d}Z^{1/2}}{\mathrm{d}\lambda} Z^{-1/2} \right)_{ij} = -\frac{1}{2} \eta_{ij}(g(\lambda)), \qquad Z_{ij}(1) = \delta_{ij}, \tag{34.26}$$

we obtain:

$$\Gamma(\lambda \varphi_i, g_{ijkl}, \mu) = \lambda^4 \Gamma\left(Z_{ij}^{1/2}(\lambda) \varphi_j, g_{ijkl}(\lambda), \mu \right). \tag{34.27}$$

We know from the analysis of the previous section that the large field behaviour of $\Gamma(\varphi)$ is governed by the UV fixed points of the theory. About the zeros of the RG β-function, little is known in the general case. However there is a problem of interest for Particle Physics we can investigate: the existence of asymptotically free field theories. If for some initial value g_{ijkl} of the renormalized coupling constant the effective coupling constant $g_{ijkl}(\lambda)$ flows into the origin for large scale, then $\Gamma(\lambda \varphi)$ can be calculated from perturbation theory for λ large:

$$\Gamma(\lambda \varphi) \sim \lambda^4 \frac{g_{ijkl}(\lambda)}{4!} \varphi_i \varphi_j \varphi_k \varphi_l. \tag{34.28}$$

We have immediately taken into account that for g_{ijkl} small the renormalization matrix Z_{ij} goes to 1. The boundness of the hamiltonian implies therefore that $g_{ijkl}(\lambda)\varphi_i\varphi_j\varphi_k\varphi_l$ must be a non-negative quartic form. Let us examine the motion of the quantity $v(\lambda)$,

$$v(\lambda) = g_{ijkl}(\lambda)\varphi_i\varphi_j\varphi_k\varphi_l, \tag{34.29}$$

using the explicit form of the β function given by equation (34.23):

$$\lambda\frac{\mathrm{d}}{\mathrm{d}\lambda}v(\lambda) = \frac{3}{16\pi^2}\varphi_i\varphi_j g_{ijmn}(\lambda)\,\varphi_k\varphi_l g_{klmn}(\lambda). \tag{34.30}$$

The r.h.s. of the equation is a sum of squares. Therefore $v(\lambda)$ is a positive increasing function. This is clearly incompatible with the assumed property that, at least for λ large enough, all functions $g_{ijkl}(\lambda)$ go to zero. The only other possibility is that all terms in the r.h.s. vanish:

$$\varphi_i\varphi_j g_{ijmn}(\lambda) = 0, \qquad \forall m, n. \tag{34.31}$$

However $v(\lambda)$ then vanishes identically. The argument is valid for all vectors φ. This implies that we started from a free field theory.

We can conclude therefore that no stable ϕ^4-like theory can be asymptotically free.

34.3 Theories with Scalar Bosons and Fermions

Let us now consider theories renormalizable in four dimensions and involving only scalar bosons and fermions. We first discuss a theory with one scalar and one Dirac fermion field for which we have already calculated the RG β-functions in Section 11.8.

A simple example. In the case of one scalar boson field coupled to one spin $1/2$ Dirac field through an interaction term of the form:

$$S_{\text{Int}} = \int \mathrm{d}^4x\left[-u\bar{\psi}(x)\psi(x)\phi(x) + \frac{g}{4!}\phi^4(x)\right], \tag{34.32}$$

the RG β-functions at one-loop order in four dimensions are given by equations (11.139):

$$\begin{cases} \beta_{u^2} = \dfrac{5}{8\pi^2}u^4 + O\left(u^6, u^2 g^2\right), \\[2mm] \beta_g = \dfrac{1}{8\pi^2}\left(\dfrac{3}{2}g^2 + 4gu^2 - 24u^4\right) + O\left(g^3, g^2 u^2, gu^4, u^6\right). \end{cases} \tag{34.33}$$

We see that the fermions also generate negative contributions to the ϕ^4 coupling RG function, which is no longer obviously positive. Instead β_{u^2} now is strictly positive. Therefore for u^2 small the running coupling constant $u^2(\lambda)$ increases for large λ. Since u^2 is positive, it grows in absolute value and the theory cannot be asymptotically free. There is however one case which must be examined separately: if u is of order g then the two-loop contribution of order $u^2 g^2$ is comparable to the one-loop term. This two-loop term which comes entirely from Z_ϕ, the ϕ-field renormalization in the purely ϕ^4 theory, is given by equation (11.78) and has to be added in equation (11.133). The function β_{u^2} then becomes:

$$\beta_{u^2} = \frac{5}{8\pi^2}u^4 + \frac{1}{48\left(8\pi^2\right)^2}u^2 g^2 + O\left(u^6, u^2 g^3\right). \tag{34.34}$$

Therefore this additional term is also positive and the conclusion is the same.

The general case. The most general interaction, renormalizable in four dimensions, has the form:

$$S_{\text{Int}}(\phi, \bar{\psi}, \psi) = \int \mathrm{d}^4 x \left[\frac{1}{4!} g_{ijkl} \phi_i \phi_j \phi_k \phi_l - \bar{\psi}_a \left(u^i_{ab} + i\gamma_5 v^i_{ab} \right) \phi_i \psi_b \right], \qquad (34.35)$$

in which $\left(u^i \right)_{ab}$ and $\left(v^i \right)_{ab}$ are hermitian matrices (see Appendix A5). Since the diagrams contributing to the RG functions have been calculated in the one component case in Section 11.8, we just have to take into account the additional geometrical factors. It is convenient to set:

$$w^i_{ab} = u^i_{ab} + iv^i_{ab}. \qquad (34.36)$$

A remark simplifies the calculation. If one calculates the Feynman diagrams in the massless theory, then each time a $\bar{\psi}\psi\phi$ vertex commutes with a fermion propagator, the matrix $\mathbf{w}^i$ is changed into its hermitian conjugate $\mathbf{w}^{i\dagger}$. The renormalization constants are then:

$$\mathbf{Z}_\psi = \mathbf{I} - \frac{1}{16\pi^2\varepsilon} \mathbf{w}^{i\dagger}\mathbf{w}^i + O(2 \text{ loops}), \qquad (34.37)$$

$$(\mathbf{Z}_\phi)_{ij} = \delta_{ij} - \frac{1}{8\pi^2\varepsilon} \operatorname{tr}\left(\mathbf{w}^{i\dagger}\mathbf{w}^j + \mathbf{w}^{j\dagger}\mathbf{w}^i \right) + O(2 \text{ loops}), \qquad (34.38)$$

($\mathbf{I}$ is the identity matrix) while the divergent part of the $\bar{\psi}\psi\phi$ 3-point function is proportional to $\mathbf{w}^j\mathbf{w}^{i\dagger}\mathbf{w}^j$. A short calculation then leads to the expression of the RG β_w-function:

$$16\pi^2\beta^i_w = \tfrac{1}{2} \left(\mathbf{w}^j\mathbf{w}^{j\dagger}\mathbf{w}^i + \mathbf{w}^i\mathbf{w}^{j\dagger}\mathbf{w}^j \right) + \operatorname{tr}\left(\mathbf{w}^{i\dagger}\mathbf{w}^j \right) \mathbf{w}^j + \operatorname{tr}\left(\mathbf{w}^i\mathbf{w}^{j\dagger} \right) \mathbf{w}^j + 2\mathbf{w}^j\mathbf{w}^{i\dagger}\mathbf{w}^j. \qquad (34.39)$$

Let us write the flow equation for the quantity $\operatorname{tr}\mathbf{w}^{i\dagger}\mathbf{w}^i$:

$$8\pi^2\lambda \frac{\mathrm{d}}{\mathrm{d}\lambda} \operatorname{tr}\mathbf{w}^{i\dagger}\mathbf{w}^i = \tfrac{1}{2} \left(\operatorname{tr}\mathbf{w}^j\mathbf{w}^{j\dagger}\mathbf{w}^i\mathbf{w}^{i\dagger} + \operatorname{tr}\mathbf{w}^{j\dagger}\mathbf{w}^j\mathbf{w}^{i\dagger}\mathbf{w}^i \right) + \operatorname{tr}\mathbf{w}^{i\dagger}\mathbf{w}^j \operatorname{tr}\mathbf{w}^{i\dagger}\mathbf{w}^j$$
$$+ \operatorname{tr}\mathbf{w}^i\mathbf{w}^{j\dagger} \operatorname{tr}\mathbf{w}^{i\dagger}\mathbf{w}^j + 2\operatorname{tr}\mathbf{w}^j\mathbf{w}^{i\dagger}\mathbf{w}^j\mathbf{w}^{i\dagger}. \qquad (34.40)$$

The matrices $\mathbf{w}^i\mathbf{w}^{i\dagger}$ and $\mathbf{w}^{i\dagger}\mathbf{w}^i$ are positive therefore the two first terms in the r.h.s. of equation (34.40), being of the form of the trace of the square of a positive matrix, are positive. The fourth term is larger than the third one. Indeed we have:

$$\operatorname{tr}\mathbf{w}^{i\dagger}\mathbf{w}^j \operatorname{tr}\mathbf{w}^{i\dagger}\mathbf{w}^j = \operatorname{Re}\left(\operatorname{tr}\mathbf{w}^{i\dagger}\mathbf{w}^j \operatorname{tr}\mathbf{w}^{i\dagger}\mathbf{w}^j \right)$$
$$\leq \left(\operatorname{tr}\mathbf{w}^{i\dagger}\mathbf{w}^j \right)^* \left(\operatorname{tr}\mathbf{w}^{i\dagger}\mathbf{w}^j \right) = \operatorname{tr}\mathbf{w}^i\mathbf{w}^{j\dagger} \operatorname{tr}\mathbf{w}^{i\dagger}\mathbf{w}^j. \qquad (34.41)$$

It follows that:

$$8\pi^2\lambda \frac{\mathrm{d}}{\mathrm{d}\lambda} \operatorname{tr}\mathbf{w}^{i\dagger}\mathbf{w}^i \geq 2 \left[\left(\operatorname{tr}\mathbf{w}^{i\dagger}\mathbf{w}^j \right) \left(\operatorname{tr}\mathbf{w}^{i\dagger}\mathbf{w}^j \right) + \operatorname{tr}\mathbf{w}^j\mathbf{w}^{i\dagger}\mathbf{w}^j\mathbf{w}^{i\dagger} \right]. \qquad (34.42)$$

Let us introduce the 4-point vertex M_{abcd}:

$$M_{abcd} = w^i_{ab} w^i_{cd}. \qquad (34.43)$$

It is easy to verify that equation (34.42) can then be rewritten:

$$8\pi^2\lambda\frac{\mathrm{d}}{\mathrm{d}\lambda}\,\mathrm{tr}\,\mathbf{w}^{i\dagger}\mathbf{w}^i \geq (M_{abcd}+M_{adbc})\left(M^*_{abcd}+M^*_{adbc}\right),\qquad(34.44)$$

which proves that the r.h.s. is positive. We conclude that $\mathrm{tr}\,\mathbf{w}^{i\dagger}\mathbf{w}^i(\lambda)$ is a positive increasing function for λ large, except as before if the two-loop contribution coming from the ϕ-field renormalization can cancel the one-loop term we have just considered. This two-loop term is proportional to $g_{iklm}g_{klmj}w^j_{ab}$, with a positive coefficient as we have seen before. It contributes in equation (34.40) by a term proportional to $g_{iklm}g_{klmj}w^j_{ab}w^{i\dagger}_{ba}$ which is positive:

$$g_{iklm}g_{klmj}w^j_{ab}w^{i\dagger}_{ab} = \left(g_{iklm}w^{i*}_{ab}\right)\left(g_{jklm}w^j_{ab}\right).\qquad(34.45)$$

Therefore the inequality remains valid and the conclusion is unchanged: a field theory containing only scalar bosons and fermions cannot be asymptotically free in four dimensions. If such a theory has no other UV fixed point, it is probably "trivial" in the sense of the triviality of the ϕ^4 field theory.

34.4 Gauge Theories

We have calculated in Section 18.9 the RG β-function for QED. In four dimensions, in a theory with n_F fermions and n_B bosons of charge e the β function reads:

$$\beta\left(e^2\right) = (4n_F + n_B)\frac{1}{3}\frac{e^4}{8\pi^2} + O\left(e^6\right),\qquad(34.46)$$

therefore QED is IR free in four dimensions and like the ϕ^4 theory it is doubtful that it exists as a theory consistent at all scales. Of course, since the physical coupling constant is very small, the physical predictions of QED are not affected by this possible inconsistency whose effects are much too small. In other words the size of the cut-off required to define QED can be enormous (the Planck mass?).

In Subsection 20.2.2 we have calculated the β-function for purely non-abelian gauge theories and found:

$$\beta(g^2) = -\frac{g^4}{8\pi^2}\frac{11}{6}C(G) + O\left(g^6\right),\qquad(34.47)$$

in which $C(G)$ is the Casimir of the group G. As we have already emphasized, non-abelian gauge theories corresponding to semi-simple groups are asymptotically free in four dimensions. What we have learned in addition in this chapter is quite remarkable: only theories possessing a non-abelian gauge symmetry may share this property. In the language of critical phenomena these theories are the only ones for which dimension four is the lower critical dimension, in the same sense as dimension two is the lower critical dimension for theories which have a global continuous symmetry.

Gauge theories and fermions. In Subsection 20.2.2 we have also calculated the contribution of fermions to the β-function. If the fermions belong to the representation R and $T(R)$ is the trace of the square of the generators of the Lie algebra in the representation:

$$\mathrm{tr}\,t^a t^b = -\delta_{ab}T(R),\qquad(34.48)$$

the β-function reads (equation (20.73)):

$$\beta\left(g^2\right) = -\left[\frac{11}{3}C(G) - \frac{4}{3}T(R)\right]\frac{g^4}{8\pi^2} + O\left(g^6\right).\qquad(34.49)$$

This result is sometimes expressed in terms of the equivalent of the fine structure constant $\alpha_s = g^2/4\pi$.

Before any calculation we knew that the contribution of fermions would be positive as in the abelian case. Therefore a gauge theory with enough fermions is no longer asymptotically free. Actually with the conventional normalization used for this problem:

$$C(G) = N \quad \text{for } SU(N), \qquad T(R) = \tfrac{1}{2}N_F,$$

in which N_F is the number of *flavours*, i.e. the number of fermion multiplets belonging to the fundamental representation of $SU(N)$. For the physical *colour group* $SU(3)$, asymptotic freedom imposes:

$$N_F < \frac{33}{2}. \tag{34.50}$$

For completeness let us here also give the β-function at two-loop order:

$$\beta\left(g^2\right) = \beta_2 g^4 + \beta_3 g^6 + O\left(g^8\right), \tag{34.51}$$

with:

$$\beta_2 = -\left[\frac{11}{3}C(G) - \frac{4}{3}T(R)\right]\frac{1}{8\pi^2}, \tag{34.52}$$

$$\beta_3 = -\left[\frac{17}{6}C^2(G) - C(R)T(R) - \frac{5}{3}C(G)T(R)\right]\left(\frac{1}{8\pi^2}\right)^2. \tag{34.53}$$

We recall that, as shown in Section 10.11, these two first coefficients are independent of the renormalization scheme. For the $SU(N)$ group, with N_F flavours in the fundamental representation one finds:

$$\beta_3 = -\left(\frac{17N^2}{6} - \frac{5NN_F}{6} - \frac{N_F\left(N^2-1\right)}{4N}\right)\left(\frac{1}{8\pi^2}\right)^2. \tag{34.54}$$

Gauge fields and scalar bosons. The situation is more complicated in the case of a theory also containing scalar bosons, like in the Higgs model. In general the scalar fields have a tendency to destroy asymptotic freedom. First they yield a positive contribution to the gauge coupling β-function which, as can be seen in equation (18.128), is $1/4$ of the fermion contribution corresponding to the same representation. However, more important, they introduce a ϕ^4 coupling which, as we know from the analysis of Section 34.2, by itself does not allow asymptotic freedom. Let us assume for simplicity that we need only one ϕ^4 coupling constant u. The corresponding β-function has at one-loop order the form:

$$\beta_u = au^2 + 2bug^2 + cg^4. \tag{34.55}$$

We know that a is positive. It is easy to verify that the contributions to c of the two diagrams (a) of figure 34.1 are positive. If the gauge group is $SU(N)$ and the scalar field belongs to the fundamental representation, a short calculation shows that b is negative (diagrams (b) of figure 34.1).

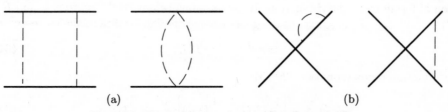

(a)   (b)

Fig. 34.1   Dashed lines correspond to gauge fields.

From the form (34.54) of the β-function we see that when g and u are small, if g^2 is much smaller than u or the converse, then u increases. Therefore the only possibility for asymptotic freedom is that u and g^2 remain of the same order. It is then natural to introduce the ratio:

$$v = u/g^2. \tag{34.56}$$

Setting:

$$\beta_{g^2} = dg^4 + \cdots, \tag{34.57}$$

and using equation (10.80) which relates the β-functions in different parametrizations:

$$\beta_j(g) \frac{\partial \tilde{g}_i}{\partial g_j} = \tilde{\beta}_i(\tilde{g}), \tag{34.58}$$

we can calculate β_v:

$$\beta_v = \beta_{g^2} \frac{\partial v}{\partial g^2} + \beta_u \frac{\partial v}{\partial u}. \tag{34.59}$$

We obtain:

$$\beta_v = g^2 \left[av^2 + (2b - d)v + c \right]. \tag{34.60}$$

The theory can only be asymptotically free if the second degree polynomial in v has positive zeros. This requires:

$$(b - d/2)^2 > ac, \qquad d - 2b > 0. \tag{34.61}$$

The two zeros v_1 and v_2 are then positive. Assuming:

$$0 < v_1 < v_2,$$

we verify that v_1 is UV stable while v_2 is IR stable. Therefore, depending on the starting point in the (g, u)-plane, the effective coupling constants may be driven towards the origin. The conditions (34.61) are rather stringent. In particular one can verify by explicit calculation of the coefficients (a, b, c, d) that it is impossible to add enough scalars belonging to the fundamental representation of $SU(N)$ to give masses to all vector bosons.

A general analysis of a system involving gauge fields, fermions and scalar bosons is rather complex. However a few results have been obtained:

(i) It is necessary to render the coefficient d of equation (34.57) small by adding enough fermions.

(ii) Generically it is impossible to give a mass to all gauge fields through the Higgs mechanism without losing asymptotic freedom. There exist however theories in which one can find a manifold of measure zero in the space of coupling constants which leads to asymptotic freedom. This situation requires a fine tuning of the Yukawa-type interactions between scalars and fermions. Only in some supersymmetric theories is this fine tuning automatically realized and therefore natural.

The conclusion is that most probably the weak electromagnetic sector in Particle Physics is not asymptotically free. This is completely consistent with the observation that the gauge couplings are small, as one would expect in an IR free theory.

34.5 Applications: The Theory of Strong Interactions

At the shortest distances presently experimentally accessible, Strong Interactions are well described by a set of fermions, quarks, transforming under the fundamental representation of the group $SU(3)$, and interacting via $SU(3)$ gauge fields. Presently six quarks have been observed indirectly, corresponding to six *flavours*. The sixth quark (the top) has been confirmed only recently at Fermilab, but its existence had been generally postulated because flavours are paired in successive generations. LEP and Fermilab recent experiments yield a top quark with a mass in the 150–200 GeV range, and exclude an additional generation of quarks associated with light leptons (neutrinos of mass below 45 GeV). With this six quarks the theory is still asymptotically free and asymptotic freedom alone would leave much room for additional quarks.

Asymptotic freedom has first emerged to provide an explanation for the experimental observations of point-like structure in deep inelastic scattering. One measures the inclusive cross section for the scattering of leptons (electrons, muons or neutrinos) off nucleons to give leptons plus any number of unobserved hadrons at large momentum transfer. In this way one probes the matrix elements between nucleon states of the product of two electromagnetic or weak currents near the light cone. We have shown in Chapter 12 that information about the behaviour near the light cone can be obtained from RG arguments. It is clear from the discussion of Section 27.2, which can be immediately transposed to UV stable fixed points, that this behaviour is characterized by logarithmic deviations from a free field behaviour. Asymptotic freedom thus provides a simple and elegant explanation to the results obtained in deep inelastic scattering experiments. We here do not give a detailed discussion of the theoretical predictions and refer the interested reader to the abundant literature. We rather examine as an illustration a somewhat simpler example: electron–positron annihilation.

Electron–positron annihilation. The total cross section of annihilation of electron–positron pairs into hadrons is related, at leading order in the electromagnetic charge, to the expectation value of the product of two hadronic electromagnetic currents J_μ. Since the electromagnetic current is conserved, we can write in momentum space:

$$\langle J_\mu(q) J_\nu(-q) \rangle = \left(\delta_{\mu\nu} q^2 - q_\mu q_\nu \right) F\left(q^2\right). \qquad (34.62)$$

The cross section is proportional to $\text{Im}\, F\left(q^2\right)$ for $q^2 < 0$. At large values of q^2, in an asymptotically free theory, the behaviour of $F\left(q^2\right)$ can be estimated from renormalization group and perturbation theory. Strictly speaking we can only obtain the behaviour of $F\left(q^2\right)$ at short distance, i.e. in the euclidean region q^2 large and positive. This behaviour is valid in any direction in the complex q^2-plane at the possible exception of the physical region $q^2 < 0$ because $F\left(q^2\right)$ has a cut on the negative real axis corresponding to the intermediate hadron states we are looking for. It is easy to construct analytic functions which decrease faster than any power in all directions of the complex plane and oscillate on the cut; an example is provided by the function $\exp\left(-\sqrt{q^2}\right)$. We hereafter ignore this difficulty but remember that the behaviour we obtain might only apply to some local average of $F\left(q^2\right)$ which smoothes such oscillations.

The electromagnetic current $J_\mu(q)$ requires no renormalization because it is exactly conserved, and has thus no anomalous dimension (see Appendix A13.2). However the product of two currents which have dimension 3 has dimension 2. Since, due to current conservation, we have been able to extract a factor of dimension 2, $F\left(q^2\right)$ is logarithmically divergent. This situation is similar to the ϕ^2 2-point function in the ϕ^4 field theory.

The 2-point function needs an additive renormalization which induces an inhomogeneous term in the RG equations (see Section 10.7). Calling μ the renormalization scale, and associating a parameter m to the masses of the quarks, we can write the RG equation:

$$\left(\mu\frac{\partial}{\partial\mu} + \beta\frac{\partial}{\partial g} + \eta_m m\frac{\partial}{\partial m}\right)F\left(q^2, g, m, \mu\right) = B(g).\tag{34.63}$$

The solution of a similar equation has already been discussed in Section 27.2 (see equations (27.24–A26.10)). Let us call $C(g)$ a particular solution of the inhomogeneous equation:

$$\beta(g)\frac{\partial}{\partial g}C(g) = B(g).\tag{34.64}$$

Then $F\left(q^2\right) - C$ satisfies a homogeneous equation which can be solved in the usual way:

$$F\left((\lambda q)^2, g, m, \mu\right) - C(g) = F\left(q^2, g(\lambda), m(\lambda)/\lambda, \mu\right) - C\left(g(\lambda)\right).\tag{34.65}$$

Since the theory is asymptotically free, the effective coupling constant $g(\lambda)$ goes to zero for large λ. The l.h.s. term $B(g)$ does not vanish for $g = 0$ (even in a free theory $F\left(q^2\right)$ is divergent),

$$B(g) = B_0 + B_1 g^2 + O\left(g^4\right),\tag{34.66}$$

and thus $C(g)$ behaves for g small like:

$$C(g) = -\frac{B_0}{2\beta_2 g^2} + \left(\frac{B_1}{\beta_2} - \frac{B_0\beta_3}{\beta_2^2}\right)\ln g + O(1).\tag{34.67}$$

Therefore at large scale λ the r.h.s. of equation (34.65) is dominated by the singular terms of $C\left(g(\lambda)\right)$:

$$F\left(\lambda^2 q^2, g, m, \mu\right) = -C(g(\lambda)) + O(1).\tag{34.68}$$

From the definition of $g(\lambda)$ and equation (34.64) it follows that:

$$\lambda\frac{\mathrm{d}}{\mathrm{d}\lambda}C\left(g(\lambda)\right) = B\left(g(\lambda)\right),\tag{34.69}$$

and therefore, using the expansion of $g(\lambda)$ for λ large:

$$C\left(g(\lambda)\right) = B_0\ln\lambda - \frac{B_1}{2\beta_2}\ln\ln\lambda + O(1).\tag{34.70}$$

The final result is:

$$F\left(q^2\right) = -\frac{B_0}{2}\ln\left(\frac{q^2}{\mu^2}\right) + \frac{B_1}{2\beta_2}\ln\ln\left(\frac{q^2}{\mu^2}\right) + O(1), \quad\text{for } q^2 \to \infty.\tag{34.71}$$

It follows that:

$$\mathrm{Im}\left(F\left(q^2\right)\right) = -\frac{B_0}{2}\pi + \frac{B_1}{2\beta_2}\frac{\pi}{\ln\left(q^2/\mu^2\right)} + o\left(\frac{1}{\ln\left(q^2/\mu^2\right)}\right), \quad\text{for } q^2 \to \infty.\tag{34.72}$$

One usually expresses this result in terms of the ratio $R\left(q^2\right)$ of the cross section for $e_+\,e_-$ into hadrons to $e_+\,e_-$ into $\mu_+\mu_-$. The latter cross section is given by the imaginary part of the one-loop correction to the photon inverse propagator due to muons. The expression of the corresponding diagram has been given in Section 18.9 (equation (18.109)). It behaves for q^2 large as $\ln q^2$, its imaginary part is just a constant. In a free quark theory, the hadronic cross section for q^2 is given in terms of the same diagram, the only difference being the coefficient which involves the charges Q_i of the quarks. Therefore for q^2 large and negative, this ratio is just the sum of the squares of the quark charges, the charge of the electron being taken as unit. In an asymptotically free theory the result is the same at leading order. The gauge interaction between quarks leads to logarithmic corrections to the leading term:

$$R\left(q^2\right) = \sum_i Q_i^2 \left[1 - \frac{B_1}{B_0 \beta_2} \frac{1}{\ln\left(q^2/\mu^2\right)} + o\left(\frac{1}{\ln\left(q^2/\mu^2\right)}\right)\right]. \tag{34.73}$$

A two-loop calculation yields the coefficient B_1. The final result is usually expressed in terms of the effective coupling constant $g(q/\mu)$ at scale q/μ:

$$R\left(q^2\right) = \sum_i Q_i^2 \left[1 + \frac{1}{4\pi^2} g^2\left(q/\mu\right) + O\left(g^4\left(q/\mu\right)\right)\right]. \tag{34.74}$$

For the $SU(3)$ colour group and for the six flavours already observed the coefficient of the leading term is:

$$\sum_i Q_i^2 = 3 \times \left(3 \times \left(\frac{2}{3}\right)^2 + 3 \times \left(\frac{1}{3}\right)^2\right) = 5. \tag{34.75}$$

This result is valid when q is large compared to all quark masses. In fact in experiments one measures R for momenta large compared to some quark masses and comparable or smaller than others. If the masses are well separated one expects, and indeed observes, R to be slowly varying in intermediate regions and close to the value obtained by taking into account only the quarks of smaller masses. For instance all measurements have been made below the top threshold and therefore the largest relevant value is $11/3$.

Bibliographical Notes

For a discussion of the applications of renormalization group to Strong Interaction Physics see
  D.J. Gross in *Methods in Field Theory*, Les Houches 1975, already quoted in Chapter 11.
The RG β-function for gauge theories has been calculated at one-loop order by
  H.D. Politzer, *Phys. Rev. Lett.* 30 (1973) 1346; D.J. Gross and F. Wilczek, *Phys. Rev. Lett.* 30 (1973) 1343;
The general analysis of RG properties of four dimensional field theories is due to
  S. Coleman and D.J. Gross, *Phys. Rev. Lett.* 31 (1973) 851.
See also
  S. Coleman in *Properties of the Fundamental Interactions*, Erice 1973, A. Zichichi ed., reprinted in *Aspects of Symmetries* (Cambridge University Press, Cambridge 1985).
Applications of asymptotically free theories to the moments of the deep inelastic structure functions are considered by
  H. Georgi and H.D. Politzer, *Phys. Rev.* D9 (1974) 416; D.J. Gross and F. Wilczek, *Phys. Rev.* D9 (1974) 920.

An elementary textbook is

M. Le Bellac, *Des Phénomènes Critiques aux Champs de Jauge* (Editions du CNRS, 1988), english version: *Quantum and Statistical Field Theory* (Oxford Univ. Press, Oxford 1992).

Calculation of the RG β-function to higher orders is due to

W.E. Caswell, *Phys. Rev. Lett.* 33 (1974) 244; D.R.T. Jones, *Nucl. Phys.* B75 (1974) 531; A.A. Belavin and A.A. Migdal, *JETP Lett.* 19 (1974) 181; E.Sh. Egoryan and O.V. Tarasov, *Teor. Math. Phys.* 41 (1979) 863; O.V. Tarasov, A.A. Vladimirov and A.Y. Zharkov, *Phys. Lett.* 93B (1980) 429.

For a discussion of the triviality of ϕ^4 see also

D.J.E. Callaway, *Phys. Rep.* 167 (1988) 241.

The upper bound on the Higgs mass is discussed for example in

M. Lüscher and P. Weisz, *Nucl. Phys.* B318 (1989) 705.

35 CRITICAL DYNAMICS

Up to now we have been concerned only with statistical properties of critical systems at equilibrium. We shall now study their time-dependent properties (note that from the point of view of Particle Physics this time is completely unphysical and can be thought of as, for example, the computer time in Monte Carlo simulations).

Typical quantities of interest are relaxation rates towards equilibrium, time-dependent correlation functions and transport coefficients.

The main motivation for such a study is that, in systems in which the dynamics is *local* (on short time scales a modification of a dynamic variable has an influence only locally in space), when the correlation length of a system becomes large, a large time scale emerges which characterizes the rate of time evolution. This phenomenon called *critical slowing down* leads to universal behaviour and scaling laws for time-dependent quantities.

Note however that many techniques we have developed in Chapter 17 also apply to general local dynamical equations without reference to a possible equilibrium state.

In contrast to the situation in static critical phenomena, there is however no clean and systematic derivation of the dynamical equations governing the time evolution in the critical domain. One reason is that it is often necessary to dynamically couple the order parameter to other variables which have automatically a slow dynamics and which correspond to conserved densities, i.e. to densities whose integral over whole space is a constant of motion: typical such quantities are energy, momentum, angular momentum.... Examples of such couplings will be given later.

One can argue however that the dynamics of all these quantities can be described by coupled Langevin equations of the type considered in Chapters 4 and 17. We have already shown in Section 4.3 that the equilibrium distribution does not determine the driving force in the Langevin equation. Only the dissipative couplings which are generated by the derivative of the effective hamiltonian are related to the static properties. Indeed the Langevin equation:

$$\dot{\varphi}_i(t) = -\frac{1}{2}\beta\Omega_{ij}\frac{\delta\mathcal{H}}{\delta\varphi_j} + F_i(\boldsymbol{\varphi}(t)) + \nu_i(t), \qquad (35.1)$$

in which $\nu_i(t)$ is the usual gaussian noise:

$$\langle\nu_i(t)\rangle = 0, \qquad \langle\nu_i(t)\nu_j(t')\rangle = \Omega_{ij}\delta(t-t'), \qquad (35.2)$$

leads to the equilibrium distribution $e^{-\beta\mathcal{H}}$ if the "streaming" term $F_i(\boldsymbol{\varphi})$ satisfies the conservation equation:

$$\frac{\partial}{\partial\varphi_i}\left[F_i(\boldsymbol{\varphi})e^{-\beta\mathcal{H}(\boldsymbol{\varphi})}\right] = 0, \qquad (35.3)$$

as can be verified immediately on the corresponding Fokker–Planck equation (see for example equations (4.27)).

This equation does not determine the streaming term. As a direct consequence, with each static universality class are associated an infinite number of dynamical universality classes and Critical Dynamics has no longer the beautiful simplicity of the statics.

Correlation and response functions. As mentioned above we can be interested in the critical behaviour of relaxation towards equilibrium, time dependent correlation functions:

$$W_{i_1\ldots i_n}^{(n)}(t_1,\ldots,t_n) = \langle \varphi_{i_1}(t_1),\ldots,\varphi_{i_n}(t_n)\rangle_\nu, \tag{35.4}$$

and also response functions which characterize the response of the system to infinitesimal time-dependent perturbations. They can be generated by adding to the hamiltonian $\mathcal{H}$ $(\varphi(t))$ a source:

$$\mathcal{H}(\varphi(t)) \mapsto \mathcal{H}(\varphi(t)) - \int dt\, h(t)\mathcal{O}(\varphi(t)). \tag{35.5}$$

The variation $R^{(n)}$ of the correlation function $W^{(n)}$ under an infinitesimal perturbation proportional to the function $\mathcal{O}(\varphi(t))$ is then given by:

$$R_{i_1\ldots i_n}^{(n)}(t_0;t_1,\ldots,t_n) = \frac{\delta}{\delta h(t_0)}\langle \varphi_{i_1}(t_1;h),\ldots,\varphi_{i_n}(t_n;h)\rangle\bigg|_{h=0}. \tag{35.6}$$

The causality of the Langevin equation implies:

$$R^{(n)} = 0 \quad \text{for } t_i < t_0 \quad \forall\, 1 \leq i \leq n. \tag{35.7}$$

If we take h constant, then the modification (35.5) is just a modification of the hamiltonian allowing at equilibrium to generate $\mathcal{O}(\varphi)$ correlation functions. Therefore:

$$\lim_{T\to+\infty} \int dt\, R_{i_1,\ldots,i_n}^{(n)}(t;T,T,\ldots,T) = \langle \mathcal{O}(\varphi)\varphi_{i_1},\ldots,\varphi_{i_n}\rangle, \tag{35.8}$$

in which the r.h.s. is now an average taken with the equilibrium distribution $e^{-\beta\mathcal{H}(\varphi)}$. We can of course calculate higher order derivatives which respect to $h(t)$ and generalize properties (35.7,35.8).

35.1 RG Equations near 4 Dimensions: The Purely Dissipative Case

We first discuss a model with a purely dissipative dynamics and without conservation laws. In the classification of the review article of Halperin and Hohenberg, which we follow in this chapter, we consider model A. We have developed in Chapter 17 most of the technology we need to derive RG equations for the dynamics. Moreover we have already calculated in Appendix A17 the dynamic RG functions both for the ϕ^4 field theory and the non-linear σ-model.

The N-vector model near 4 dimensions. We first consider a dissipative dynamics for the N-vector model:

$$\dot{\varphi}(x,t) = -\frac{\Omega_0}{2}\left[(-\Delta + r_0)\varphi(x,t) + g_0\frac{\Lambda^\varepsilon}{3!}\varphi\varphi^2\right] + \nu(x,t). \tag{35.9}$$

This model has been briefly discussed in Appendix A17.1. It leads to a supersymmetric effective action $S(\phi)$ which in terms of the superfield

$$\phi = \varphi + \bar{\theta}\mathbf{c} + \bar{\mathbf{c}}\theta + \bar{\theta}\theta\boldsymbol{\lambda}, \tag{35.10}$$

can be written:

$$S(\phi) = \int \mathrm{d}t \mathrm{d}\theta \mathrm{d}\bar{\theta} \left[\int \mathrm{d}^d x \frac{2}{\Omega_0} \frac{\partial \phi}{\partial \theta} \left(\frac{\partial \phi}{\partial \bar{\theta}} - \theta \frac{\partial \phi}{\partial t} \right) + \mathcal{H}(\phi) \right], \qquad (35.11)$$

in which $\mathcal{H}$ characterizes the equilibrium distribution:

$$\mathcal{H}(\varphi) = \int \mathrm{d}^d x \left[\tfrac{1}{2} (\partial_\mu \varphi)^2 + \tfrac{1}{2} r_0 \varphi^2 + g_0 \frac{\Lambda^\varepsilon}{4!} (\varphi^2)^2 \right]. \qquad (35.12)$$

Supersymmetry and the fluctuation–dissipation theorem. Note that the addition to the static action of a perturbation proportional to the order parameter itself:

$$\mathcal{H}(\varphi(x,t)) \mapsto \mathcal{H}(\varphi(x,t)) - \int \mathrm{d}t \, \mathrm{d}^d x \, \mathbf{h}(x,t) \cdot \boldsymbol{\varphi}(x,t), \qquad (35.13)$$

is equivalent to the addition of a term to the noise in equation (35.9):

$$\boldsymbol{\nu}(x,t) \mapsto \boldsymbol{\nu}(x,t) + \tfrac{1}{2}\Omega \mathbf{h}(x,t). \qquad (35.14)$$

Therefore differentiations with respect to $\mathbf{h}(x,t)$ generate $\boldsymbol{\lambda}(x,t)$ field correlation functions.

In Subsection 17.5.2 we have derived the supersymmetric WT identities for the connected time-dependent correlation functions (equation (17.47)):

$$\sum_{j=1}^n \left(\frac{\partial}{\partial \theta_j} + \bar{\theta}_j \frac{\partial}{\partial t_j} \right) W^{(n)}(k_i, t_i, \bar{\theta}_i, \theta_i) = 0. \qquad (35.15)$$

We have then obtained the general form of a 2-point function consistent with supersymmetry and causality (equation (17.52)):

$$W^{(2)}(k,t) = \left\{ 1 + \tfrac{1}{2}(\bar{\theta} - \bar{\theta}') [\theta + \theta' - (\theta - \theta') \epsilon(t-t')] \frac{\partial}{\partial t} \right\} A(k, t-t').$$

In our example $A(t)$ is in addition an even real function of t. In terms of the Fourier transform over time:

$$B(k,\omega) = \int_0^{+\infty} e^{i\omega t} A(k,t) \mathrm{d}t, \qquad (35.16)$$

the Fourier transform of $W^{(2)}$ takes the form:

$$W^{(2)}(k,\omega,\theta,\theta') = B(k,\omega) + B^*(k,\omega) - i\omega(\bar{\theta} - \bar{\theta}')[\theta' B(k,\omega) + \theta B^*(k,\omega)]$$
$$+ A(k,0)\delta^2(\boldsymbol{\theta} - \boldsymbol{\theta}'), \qquad (35.17)$$

in which we denote by B^* the complex conjugate of B and by $\boldsymbol{\theta}$ the set $\{\theta, \bar{\theta}\}$.

The sum $B + B^*$ which is the $\langle \varphi\varphi \rangle$ connected 2-point function, is a real function. The response function $\langle \lambda\varphi \rangle$ is the coefficient of $\bar{\theta}'\theta'$. Taking its imaginary part we find the relation:

$$\mathrm{Im}\left[\frac{\delta}{\delta h(k,\omega)} \langle \varphi(-k,-\omega) \rangle \right]\bigg|_{h=0} \equiv \langle \lambda(k,\omega)\varphi(-k,-\omega) \rangle$$
$$= \frac{\omega}{2} \langle \varphi(k,\omega)\varphi(-k,-\omega) \rangle, \qquad (35.18)$$

known under the name of fluctuation–dissipation theorem.

RG equations at T_c. We have shown in Subsection 17.5.3 that the static and super-symmetric dynamic theories have the same upper critical dimension. Fluctuations are therefore only relevant for dimensions $d \leq 4$. The renormalized action $S_r(\phi)$ then has the form:

$$S_r(\phi) = \int d\theta \, d\bar\theta \, dt \left[\int d^d x \frac{2}{\Omega} Z_\omega \frac{\partial\phi}{\partial\theta} \left(\frac{\partial\phi}{\partial\bar\theta} - \theta \frac{\partial\phi}{\partial t} \right) + \mathcal{H}_r(\phi) \right], \tag{35.19}$$

in which ϕ is now the renormalized field and $\mathcal{H}_r(\phi)$ is the static renormalized hamiltonian. It is characterized by the appearance of a new independent renormalization constant Z_ω and therefore a new RG function $\eta_\omega(g)$ which in the minimal subtraction scheme and in the renormalized theory is given by:

$$\eta_\omega(g) = \beta(g) \frac{d}{dg} \ln(Z_\omega/Z). \tag{35.20}$$

The RG equations for the critical theory then read:

$$\left[\mu \frac{\partial}{\partial\mu} + \beta(g) \frac{\partial}{\partial g} + \eta_\omega(g) \Omega \frac{\partial}{\partial\Omega} - \frac{n}{2} \eta(g) \right] \Gamma^{(n)}(p_i, \omega_i, \boldsymbol{\theta}_i, \mu, \Omega, g) = 0, \tag{35.21}$$

in which μ is the renormalization scale.

At the IR fixed point g^*, they reduce to:

$$\left(\mu \frac{\partial}{\partial\mu} + \eta_\omega \Omega \frac{\partial}{\partial\Omega} - \frac{n}{2} \eta \right) \Gamma^{(n)}(p_i, \omega_i, \boldsymbol{\theta}_i, \mu, \Omega) = 0, \tag{35.22}$$

with:

$$\eta_\omega = \eta_\omega(g^*).$$

Dimensional analysis yields the relation:

$$\Gamma^{(n)}(\lambda p_i, \omega_i, \boldsymbol{\theta}_i, \mu, \Omega) = \lambda^{d - n(d-2)/2} \rho^{1-n} \Gamma^{(n)} \left(p_i, \frac{\omega_i}{\rho}, \boldsymbol{\theta}_i \sqrt{\rho}, \frac{\mu}{\lambda}, \frac{\Omega\lambda^2}{\rho} \right). \tag{35.23}$$

Combining both equations, setting:

$$z = 2 + \eta_\omega, \tag{35.24}$$

and choosing:

$$\rho = \Omega\lambda^z \mu^{-\eta_\omega}, \tag{35.25}$$

we find the dynamic scaling relations:

$$\Gamma^{(n)}(\lambda p_i, \omega_i, \boldsymbol{\theta}_i, \mu = 1, \Omega) = \lambda^{d - n(d-2+\eta)/2 - z(n-1)} \Omega^{1-n} F^{(n)} \left(p_i, \frac{\omega_i}{\Omega} \lambda^{-z}, \boldsymbol{\theta}_i \Omega^{1/2} \lambda^{z/2} \right) \tag{35.26}$$

A few algebraic manipulations yield the equivalent relation for connected correlation functions:

$$W^{(n)}(\lambda p_i, \omega_i, \boldsymbol{\theta}_i, \mu = 1, \Omega = 1) = \lambda^{(d+z)(1-n) + n(d-2+\eta)/2} G^{(n)} \left(p_i, \omega_i \lambda^{-z}, \boldsymbol{\theta}_i \lambda^{z/2} \right). \tag{35.27}$$

For the φ-field 2-point correlation function, which corresponds to $\theta = 0$, we therefore get:

$$W^{(2)}\left(p,\omega,\boldsymbol{\theta}=0\right)\sim\frac{1}{p^{2-\eta+z}}G^{(2)}\left(\frac{\omega}{p^z}\right).$$ (35.28)

The equal-time correlation function is obtained by integrating over ω. We thus verify the consistency with the static scaling.

The dynamic critical 2-point function thus depends on a frequency scale which vanishes at small momentum like p^z or a time scale which diverges like p^{-z}. The RG function η_ω for this model has been calculated in Section A17.2 (see equation $(A17.21)$):

$$\eta_\omega\left(\tilde{g}\right)=\frac{(N+2)}{72}\left(6\ln\frac{4}{3}-1\right)\tilde{g}^2+O\left(\tilde{g}^3\right),$$ (35.29)

with as usual:

$$g=\frac{2}{\Gamma\left(d/2\right)\left(4\pi\right)^{d/2}}\tilde{g}.$$ (35.30)

The dynamic critical exponent z follows:

$$z=2+\frac{N+2}{2\left(N+8\right)^2}\left(6\ln\frac{4}{3}-1\right)\varepsilon^2+0\left(\varepsilon^3\right).$$ (35.31)

At this order it can also be written:

$$z=2+c\eta+O\left(\varepsilon^3\right),$$ (35.32)

with:

$$c=6\ln(4/3)-1.$$ (35.33)

The scaling behaviour of the response function can be obtained by considering for example the coefficient of $\bar{\theta}\theta$ in $W^{(2)}$.

Correlation functions above T_c in the critical domain. Let us just write the RG equations at the IR fixed point:

$$\left(\mu\frac{\partial}{\partial\mu}+\eta_\omega\Omega\frac{\partial}{\partial\Omega}-\eta_2\sigma\frac{\partial}{\partial\sigma}-\frac{n}{2}\eta\right)\Gamma^{(n)}\left(p_i,\omega_i,\boldsymbol{\theta}_i,\sigma,\mu,\Omega\right)=0,$$ (35.34)

in which σ is a measure of the deviation from the critical temperature:

$$\sigma\sim T-T_c.$$ (35.35)

The dimensional relation (35.23) becomes:

$$\Gamma^{(n)}\left(p_i,\omega_i,\boldsymbol{\theta}_i,\sigma,\mu,\Omega\right)=\lambda^{d-n(d-2)/2}\rho^{1-n}\Gamma^{(n)}\left(\frac{p_i}{\lambda},\frac{\omega_i}{\rho},\boldsymbol{\theta}_i\sqrt{\rho},\frac{\sigma}{\lambda^2},\frac{\mu}{\lambda},\frac{\Omega\lambda^2}{\rho}\right).$$ (35.36)

Combining this equation with the RG equation (35.34) and choosing:

$$\lambda=\sigma^\nu\mu^{\nu\eta_2}\sim\xi^{-1},\qquad\rho=\Omega\mu^{-\eta_\omega}\lambda^z\sim\xi^{-z},$$ (35.37)

in which ξ is the correlation length, we finally obtain:

$$\Gamma^{(n)}\left(p_i,\omega_i,\boldsymbol{\theta}_i,\sigma,\mu,\Omega\right)\sim\xi^{-d+n(d-2+\eta)/2+z(n-1)}F^{(n)}\left(p_i\xi,\omega_i\xi^z,\boldsymbol{\theta}_i\xi^{-z/2}\right).$$ (35.38)

We now see that all times are measured in terms of a *correlation time* τ which diverges at the critical temperature as ξ^z:

$$\tau\sim\xi^z.$$ (35.39)

35.2 RG Equations near 2 Dimensions: The Purely Dissipative Case

Near two dimensions the N-vector model is described by a Langevin equation of the form (see Section 17.7, Appendix A17.3):

$$\dot{\varphi} = -\tfrac{1}{2}\Omega\beta\left[-\Delta\varphi + \varphi\left(\varphi\cdot\Delta\varphi\right)\right] + \nu - \varphi\left(\varphi\cdot\nu\right),\tag{35.40}$$

$\nu(x,t)$ being the noise of Section 35.1 and φ satisfying:

$$\varphi^2 = 1.\tag{35.41}$$

Setting:

$$\beta = g_0/\Lambda^{d-2},\tag{35.42}$$

we can derive from equation (35.40) an effective action $S(\phi)$:

$$S(\phi) = \frac{\Lambda^{d-2}}{g_0}\int \mathrm{d}\theta\mathrm{d}\bar\theta\mathrm{d}t\left[\int \mathrm{d}^d x \frac{2}{\Omega_0}\frac{\partial\phi}{\partial t}\cdot\left(\frac{\partial\phi}{\partial\bar\theta} - \theta\frac{\partial\phi}{\partial t}\right) + \mathcal{H}(\phi)\right],\tag{35.43}$$

with:

$$\mathcal{H}(\phi) = \tfrac{1}{2}\int \mathrm{d}^d x\left[\partial_\mu\phi\left(x,t\right)\right]^2.\tag{35.44}$$

The superfield ϕ satisfies the constraint:

$$\phi^2 = 1.\tag{35.45}$$

In the superfield form (35.43) we see that the dynamical actions for $\left(\phi^2\right)^2$ field theory and the non-linear σ-model are related in the same way as the corresponding static hamiltonians.

The renormalized action $S_{\mathrm{r}}(\phi)$ then reads:

$$S_{\mathrm{r}}(\phi) = \frac{\mu^\varepsilon}{g}\int \mathrm{d}\theta\mathrm{d}\bar\theta\mathrm{d}t\left[\int \mathrm{d}^d x \frac{2}{\Omega}Z_\omega\frac{\partial\phi}{\partial\bar\theta}\cdot\left(\frac{\partial\phi}{\partial\bar\theta} - \theta\frac{\partial\phi}{\partial t}\right) + \mathcal{H}_{\mathrm{r}}(\phi)\right],\tag{35.46}$$

in which μ is the renormalization scale and $d = 2 + \varepsilon$.

In the minimal subtraction scheme, the new RG function $\eta_\omega(g)$ is given by:

$$\eta_\omega(g) = \beta(g)\frac{\mathrm{d}}{\mathrm{d}g}\ln\left(Z_\omega Z_g/Z\right).\tag{35.47}$$

The RG equations in zero magnetic field then read (see equation (30.17)):

$$\left[\mu\frac{\partial}{\partial\mu} + \beta(g)\frac{\partial}{\partial g} + \eta_\omega(g)\Omega\frac{\partial}{\partial\Omega} - \frac{n}{2}\zeta(g)\right]\Gamma^{(n)}\left(p_i,\omega_i,\boldsymbol{\theta}_i,g,\mu,\Omega\right) = 0.\tag{35.48}$$

Dimensional analysis yields (equation (35.23)):

$$\Gamma^{(n)}\left(p_i,\omega_i,\boldsymbol{\theta}_i;g,\mu,\Omega\right) = \lambda^d\rho^{1-n}\Gamma^{(n)}\left(\frac{p_i}{\lambda},\frac{\omega_i}{\rho},\boldsymbol{\theta}_i\sqrt{\rho};\frac{\Omega\lambda^2}{\rho},\frac{\mu}{\lambda}\right).\tag{35.49}$$

We introduce in addition to the functions $M_0(g)$ and $\xi(g)$:

$$M_0(g) = \exp\left[-\frac{1}{2}\int_0^g \frac{\zeta(g')}{\beta(g')}\mathrm{d}g'\right],\tag{35.50}$$

$$\xi(g) = \mu^{-1}g^{1/\varepsilon}\exp\left[\int_0^g\left(\frac{1}{\beta(g')}-\frac{1}{\varepsilon g'}\right)\mathrm{d}g'\right],\tag{35.51}$$

(see equations (30.30,30.31)) the new RG function:

$$\tau(g) = \Omega^{-1}\xi^2(g)\exp\left[\int_0^g \frac{\eta_\omega(g')}{\beta(g')}\mathrm{d}g'\right].\tag{35.52}$$

Combining equations (35.48–35.52) and choosing:

$$\lambda = \xi^{-1}(g),\qquad \rho = \tau^{-1}(g),\tag{35.53}$$

we obtain:

$$\Gamma^{(n)}(p_i,\omega_i,\boldsymbol{\theta}_i;g,\mu,\Omega) = \tau^{n-1}\xi^{-d}M_0^{-n}F^{(n)}\left(p_i\xi,\omega\tau,\boldsymbol{\theta}\tau^{-1/2}\right).\tag{35.54}$$

Near the critical temperature g^*, this expression agrees with the scaling form (35.38). The exponent z is then given by:

$$z = 2 + \eta_\omega(g^*).\tag{35.55}$$

Equation $(A17.41)$ gives $\eta_\omega(g)$ at two-loop order. The exponent z is then:

$$z = 2 + \left(1 - \ln(4/3)\right)\frac{\varepsilon^2}{N-2} + O\left(\varepsilon^3\right),\tag{35.56}$$

or again in terms of η:

$$z = 2 + \left(1 - \ln(4/3)\right)\varepsilon\eta + O\left(\varepsilon^3\right).\tag{35.57}$$

35.3 Conserved Order Parameter (Model B)

A simple modification of equation (35.9) ensures that the order parameter is conserved, i.e.:

$$\frac{\mathrm{d}}{\mathrm{d}t}\int \mathrm{d}^d x\, \varphi(x,t) = 0.\tag{35.58}$$

We write the equation:

$$\dot\varphi(x,t) = \frac{\Omega_0}{2}\Delta_x\frac{\delta\mathcal{H}}{\delta\varphi(x,t)} + \nu(x,t).\tag{35.59}$$

For the Langevin equation to generate the same equilibrium distribution, the noise 2-point function now has to be modified accordingly:

$$\langle \nu_i(x,t)\nu_j(x',t')\rangle = -\delta_{ij}\Omega_0\,\delta(t-t')\,\Delta_x\delta(x-x').\tag{35.60}$$

The effective action $S(\phi)$ is still supersymmetric:

$$S(\phi) = \int \mathrm{d}\theta \mathrm{d}\bar{\theta} \mathrm{d}t \left[-\int \mathrm{d}^d x \frac{2}{\Omega_0} \frac{\partial \phi}{\partial \theta} \cdot \Delta^{-1} \left(\frac{\partial \phi}{\partial \bar{\theta}} - \theta \frac{\partial \phi}{\partial t} \right) + \mathcal{H}(\phi) \right]. \qquad (35.61)$$

The appearance of a non-local term in the action reminds us of the effective field theory for uniaxial systems with dipolar forces.

Power counting is now different. Since the propagator reads:

$$\Delta(k, \omega, \theta, \theta') = \frac{\Omega k^2 \left[1 - \frac{1}{2} i\omega \left(\bar{\theta} - \bar{\theta}' \right) (\theta + \theta') + \frac{1}{4} \Omega k^2 \left(k^2 + r \right) \delta^2 (\theta - \theta') \right]}{\omega^2 + \frac{1}{4} \Omega^2 \left(k^2 \right)^2 \left(k^2 + r \right)^2}, \qquad (35.62)$$

ω has the dimension of k^4. Above 4 dimensions, the characteristic frequency diverges like k^4 which means that the dynamical exponent z is 4 instead of 2 as in model A.

Above two dimensions Feynman diagrams calculated with the propagator (35.62) are not singular at zero momentum for $\omega \neq 0$. Therefore no counterterm singular in $\mathbf{k}$ can be generated and thus Ω_0 in the effective action remains unrenormalized. The renormalized action $S_\mathrm{r}(\phi)$ is:

$$S_\mathrm{r}(\phi) = \int \mathrm{d}\theta \, \mathrm{d}\bar{\theta} \, \mathrm{d}t \left[-\int \mathrm{d}^d x \frac{2}{\Omega} \frac{\partial \phi}{\partial \theta} \cdot \Delta^{-1} \left(\frac{\partial \phi}{\partial \bar{\theta}} - \theta \frac{\partial \phi}{\partial t} \right) + \mathcal{H}_\mathrm{r}(\phi) \right]. \qquad (35.63)$$

The field amplitude renormalization implies the relation:

$$\Omega_0 = Z\Omega, \qquad (35.64)$$

and therefore the RG function $\eta_\omega(g)$ is:

$$\eta_\omega(g) = -\beta(g) \frac{\mathrm{d}}{\mathrm{d}g} \ln Z = -\eta(g). \qquad (35.65)$$

The analysis then becomes quite similar to the previous case. However relation (35.65) implies that the exponent z is no longer an independent exponent but given instead by:

$$z = 4 - \eta. \qquad (35.66)$$

35.4 Relaxational Model with Energy Conservation (Model C)

Still in the framework of the N-vector model we now assume that the total energy is conserved. We know that in the critical domain the most singular part of the energy is φ^2. We shall therefore couple a field $e(x)$ to $\varphi^2(x)$ and write an equilibrium hamiltonian $\mathcal{H}(\varphi(x), e(x))$:

$$\mathcal{H}(\varphi, e) = \int \mathrm{d}^d x \left[\frac{1}{2} (\partial_\mu \varphi)^2 + \frac{1}{2} r\varphi^2 + \frac{u}{4!} (\varphi^2)^2 + \frac{1}{2} \Lambda^{3/2} v e(x) \varphi^2(x) + \frac{1}{2} e^2(x) \right]. \qquad (35.67)$$

The static properties are not affected by this modification. Let us call $K(x)$ the source for the correlation functions of $e(x)$. If we integrate over $e(x)$ we obtain a new effective hamiltonian $\mathcal{H}(\varphi, K)$:

$$\mathcal{H}(\varphi, K) = \int \mathrm{d}^d x \left[\frac{1}{2} (\partial_\mu \varphi)^2 + \frac{1}{2} r\varphi^2 + \frac{1}{4!} \left(u - 3\Lambda^\epsilon v^2 \right) (\varphi^2)^2 + \Lambda^{\epsilon/2} \frac{v}{2} K\varphi^2 - \frac{K^2}{2} \right]. \qquad (35.68)$$

The *e*-field correlations are given in terms of the usual static φ^2 correlation functions and are thus the standard energy correlation functions. The *e*-field amplitude and the v coupling constant renormalizations can be expressed in terms of φ^2 insertion and $\langle \varphi^2 \varphi^2 \rangle$ additive renormalization constants. More precisely we have the relation:

$$\langle e\,(k)\,e\,(-k)\rangle_{\mathrm{c}} = 1 + \Lambda^\varepsilon v^2 \left\langle \left(\tfrac{1}{2}\varphi^2\right)(k)\left(\tfrac{1}{2}\varphi^2\right)(-k)\right\rangle_{\mathrm{c}}. \tag{35.69}$$

All other correlation functions of the field $e\,(x)$ are identical to the correlation functions of $\Lambda^{\varepsilon/2} v \varphi^2\,(x)\,/2$. From the RG equations satisfied by the correlation functions, we can derive the RG equations for $e\,(x)$ insertions. At T_c they take the form:

$$\left[\Lambda\frac{\partial}{\partial\Lambda} + \beta\,(g)\,\frac{\partial}{\partial g} + \beta_v\,(g,v)\,\frac{\partial}{\partial v} - \frac{l}{2}\eta_e\,(g,v) - \frac{n}{2}\eta\,(g)\right]\Gamma_e^{(l,n)}\,(q,p;g,v,\Lambda) = 0\,, \tag{35.70}$$

in which $\Gamma_e^{(l,n)}$ is the correlation function with $l\ e(x)$ insertions.

Applying this equation to the case $l + n > 2$ and using equation (25.66) we obtain the relation:

$$\eta_e\,(g,v) = \varepsilon + 2\frac{\beta_v}{v} + 2\eta_2\,(g)\,. \tag{35.71}$$

Then applying equation (35.70) to relation (35.69) and again using equation (25.66) we find:

$$\varepsilon + \frac{2\beta_v}{v} + 2\eta_2\,(g) = v^2 B\,(g)\,, \tag{35.72}$$

and therefore:

$$\eta_e = v^2 B\,(g)\,, \tag{35.73a}$$

$$\beta_v = -\frac{v}{2}\left[\varepsilon + 2\eta_2 - v^2 B\,(g)\right]\,. \tag{35.73b}$$

At the IR fixed point g^* the function β_v becomes:

$$\beta_v = -\frac{v}{2}\left[\frac{\alpha}{\nu} - v^2 B\,(g^*)\right]\,. \tag{35.74}$$

At leading order in the ε-expansion $B\,(g^*)$ is positive:

$$B\,(g^*) = \frac{N}{16\pi^2} + O\,(\varepsilon)\,.$$

Therefore two cases have to be envisaged:

(i) $\alpha < 0$:
The origin $v = 0$ is the unique IR fixed point and the coupling to $e\,(x)$ is irrelevant, we are back to the model A of Section 35.1.

(ii) $\alpha > 0$:
The IR fixed point is non-trivial, it corresponds to:

$$v = v^* \equiv \pm \left[\frac{\alpha}{\nu B\,(g^*)}\right]^{1/2}\,. \tag{35.75}$$

At leading order α vanishes for $N = 4$:

$$\alpha = \frac{4 - N}{2\,(N + 8)}\varepsilon + O\,(\varepsilon^2)\,. \tag{35.76}$$

In three dimensions numerical calculations (see Chapter 28) show that α is already slightly negative for $N = 2$. For $\alpha > 0$ the dynamics differs from the simple dynamics of Section 35.1 as we shall see below.

Of course in both cases the values of η_e lead to a behaviour consistent with previous results concerning the $\langle \varphi^2 \varphi^2 \rangle$ correlation functions.

The Langevin equation. Let us therefore examine the dynamics of the model in the case $\alpha > 0$. The Langevin equation has the form:

$$\dot{\varphi} = -\frac{\Omega}{2} \frac{\delta \mathcal{H}}{\delta \varphi (x,t)} + \nu (x,t) , \tag{35.77}$$

$$\dot{e} = \frac{\Omega'}{2} \Delta \frac{\delta \mathcal{H}}{\delta e (x,t)} + \nu' (x,t) . \tag{35.78}$$

The new gaussian noise $\nu' (x,t)$ is defined by:

$$\langle \nu' \rangle = 0 , \qquad \langle \nu'(x,t)\nu'(x',t') \rangle = -\Omega'\delta (t - t') \, \partial_x^2 \delta (x - x') . \tag{35.79}$$

The effective action now written in terms of 2 superfields $\phi (x,t)$ and $E (x,t)$ is still supersymmetric and reads:

$$S (\phi, E) = \int d\theta d\bar{\theta} dt \left\{ \int d^d x \left[-\frac{2}{\Omega'} \frac{\partial E}{\partial \theta} \Delta^{-1} \left(\frac{\partial E}{\partial \bar{\theta}} - \theta \frac{\partial E}{\partial t} \right) + \frac{2}{\Omega} \frac{\partial \phi}{\partial \theta} \cdot \left(\frac{\partial \phi}{\partial \bar{\theta}} - \theta \frac{\partial \phi}{\partial t} \right) \right] \right.$$
$$\left. + \mathcal{H} (\phi, E) \right\}. \tag{35.80}$$

For the reasons we have already given, the $\langle EE \rangle$ correlation function is not singular at zero momentum, the parameter Ω' is unrenormalized. However the renormalization of Ω is now modified by the presence of loops with E fields. Setting again:

$$\Omega = \Omega_{\rm r} / Z_\omega , \tag{35.81}$$

at one-loop order we get:

$$Z_\omega = 1 - \frac{\Omega}{\Omega + \Omega'} N_d v^2 \ln \Lambda , \tag{35.82}$$

while for Ω' we have:

$$\Omega' = Z_e \Omega'_{\rm r} , \tag{35.83}$$

in which Z_e is the e-field renormalization. The corresponding RG function η_e is given by equation (35.73a).

The RG equations then read:

$$\left[\Lambda \frac{\partial}{\partial \Lambda} + \beta (g) \frac{\partial}{\partial g} + \beta_v \frac{\partial}{\partial v} + \eta_\omega \Omega \frac{\partial}{\partial \Omega} + \eta'_\omega \Omega' \frac{\partial}{\partial \Omega'} - \frac{l}{2} \eta_e - \frac{n}{2} \eta \right]$$
$$\Gamma_e^{(l,n)} (q, p; g, v, \Lambda, \Omega, \Omega') = 0 . \tag{35.84}$$

Equation (35.83) leads to the identity:

$$\eta'_\omega = \eta_e , \tag{35.85}$$

while at leading order:

$$\eta_\omega = \frac{1}{8\pi^2} \frac{\Omega}{\Omega + \Omega'} v^2.$$

(35.86)

To separate variables we set:

$$s = \Omega'/\Omega,$$

(35.87)

and take for example s and Ω' as independent variables. We then have:

$$\eta_\omega \Omega \frac{\partial}{\partial \Omega} + \eta_e \Omega' \frac{\partial}{\partial \Omega'} = \eta_e \Omega' \frac{\partial}{\partial \Omega'} + \beta_s \frac{\partial}{\partial s},$$

(35.88)

with:

$$\beta_s = \eta_e - \eta_\omega.$$

(35.89)

At the IR fixed in the (g, v) plane, the function β_s takes the form:

$$\beta_s = \frac{\alpha}{\nu} \left[1 - \frac{2}{N} \frac{1}{1+s} \right] s + O\left(\varepsilon^2\right).$$

(35.90)

The fixed points in the variables s are:

$$s^* = 0, \qquad s^* = 2/N - 1, \qquad s = \infty.$$

(35.91)

Since α is positive, the fixed point $s = 0$ is stable for $N > 2 + O(\varepsilon)$. However the $s = 0$ limit is a peculiar limit and it is not clear whether this result is consistent with the ε-expansion.

The fixed point $s = \infty$ is never stable for $\alpha > 0$. It also corresponds to decoupling of the E sector.

Finally for $N < 2$, $s^* = 2/N - 1$ is the stable fixed point. The ratio Ω'/Ω is finite and therefore the dynamics of E and ϕ are coupled. The function η_e corresponds to η_ω in model A and therefore, since η_e at the fixed point has the value α/ν, the exponent z is:

$$s^* = \frac{2}{N} - 1 + O(\varepsilon) \Rightarrow z = 2 + \frac{\alpha}{\nu}.$$

(35.92)

35.5 A Non-Relaxational Model (Model E)

General remarks. Let us consider the Langevin equation (35.1) (we have set $\beta = 1$):

$$\dot{\varphi}(t) = -\tfrac{1}{2} \Omega_{ij} \frac{\delta \mathcal{H}}{\delta \varphi_j(t)} + F_i(t) + \nu_i(t),$$

(35.93)

with a noise (35.2):

$$\langle \nu_i(t) \nu_j(t') \rangle = \Omega_{ij} \, \delta(t - t').$$

If the function F_i satisfies equation (35.3):

$$\frac{\partial}{\partial \varphi_i} \left[F_i(\varphi) e^{-\mathcal{H}(\varphi)} \right] = 0,$$

the equilibrium distribution is $e^{-\mathcal{H}}$. A particular solution of this equation is provided by:

$$F_i(\varphi) = \frac{\partial}{\partial \varphi_j} R_{ij}(\varphi) - R_{ij}(\varphi) \frac{\partial \mathcal{H}}{\partial \varphi_j} , \tag{35.94}$$

where R_{ij} is an antisymmetric matrix:

$$R_{ij} = -R_{ji} .$$

In particular the mode-coupling of Kawasaki and Kadanoff–Swift has this form. In concrete examples the matrix R_{ij} is linear in the field φ_i and associated with transformations corresponding to symmetries of the hamiltonian.

Example. We now give an example of a model with non-dissipative couplings. Since the effective action in such cases is no longer supersymmetric, we expect the number of independent renormalizations to increase substantially and therefore the analysis to become more complex. Moreover, by losing the supersymmetry we lose a powerful and elegant technique to solve the renormalization problem.

We consider the so-called model E: the order parameter is a complex field $\varphi(x)$, and as in previous section there is a conserved density $e(x)$. The Langevin equation however reads:

$$\frac{\partial \varphi}{\partial t} = -\frac{\Omega}{2} \frac{\delta \mathcal{H}}{\delta \varphi^*(x,t)} - is\varphi \frac{\delta \mathcal{H}}{\delta e(x,t)} + \nu(x,t) , \tag{35.95}$$

$$\frac{\partial e}{\partial t} = \frac{\Omega'}{2} \Delta \frac{\delta \mathcal{H}}{\delta e(x,t)} + is \left[\varphi^* \frac{\delta \mathcal{H}}{\delta \varphi^*(x,t)} - \varphi \frac{\delta \mathcal{H}}{\delta \varphi(x,t)} \right] + \nu'(x,t) . \tag{35.96}$$

The hamiltonian is:

$$\mathcal{H}(\varphi,e) = \int d^d x \left[|\partial_\mu \varphi|^2 + r |\varphi|^2 + \frac{1}{3!} u |\varphi|^4 + \frac{1}{2} e^2 \right] . \tag{35.97}$$

The noise 2-point functions are the same as in Section 35.4 (for the $N = 2$ case):

$$\begin{aligned}
\langle \nu(x,t) \nu^*(x',t') \rangle &= \Omega \delta(x-x') \delta(t-t') , \\
\langle \nu\nu \rangle = \langle \nu^*\nu^* \rangle &= 0 , \\
\langle \nu'(x,t) \nu'(x',t') \rangle &= -\Omega' \Delta_x \delta(x-x') \delta(t-t') .
\end{aligned} \tag{35.98}$$

It is easy to verify that this model provides one example of Langevin equation (35.93) with a streaming term of the form (35.94) which ensures that $e^{-\mathcal{H}(\varphi)}$ remains the equilibrium distribution.

The model has a $U(1)$ symmetry corresponding to the multiplication of φ by a phase. From the invariance of the hamiltonian under an infinitesimal $U(1)$ transformation follows:

$$\int d^d x \left(\varphi^* \frac{\delta \mathcal{H}}{\delta \varphi^*(x,t)} - \varphi \frac{\delta \mathcal{H}}{\delta \varphi(x,t)} \right) = 0 ,$$

and therefore $e(x,t)$ is a conserved quantity.

To simplify the discussion, we have omitted the generically expected $e |\varphi^2|$ coupling and this generates an additional reflection symmetry $e(x) \mapsto -e(x)$. According to the analysis of Section 35.4, this simplification is really justified only if the exponent of the specific heat α is negative. The analysis can be generalized to the case where the $e\left|\varphi^2\right|$ coupling is included. The interested reader is referred to the literature.

Renormalization. Power counting tells us that the theory is renormalizable in four dimensions and that the canonical dimensions of the fields are:

$$[\varphi] = 1, \quad [e] = 2.$$

From the general results of Section 17.4, based on the BRS symmetry of the effective action, we know that the Langevin equations renormalize as predicted by power counting. The form of the renormalized equations is further restricted by the $U(1)$ symmetry, the parity symmetry and the conservation of $e(x, t)$. However these conditions are not sufficiently restrictive. They do not forbid a term proportional to $\partial^2\left(\varphi^*\varphi\right)$ in equation (35.96) and do not imply the equality of the coupling constants s in equations (35.95) and (35.96). We here need a rather indirect argument: since the regularized dynamic theory has the regularized static theory as equilibrium distribution the same must be true for the renormalized theories. We then have three renormalization constants given by the statics, Z_φ, $Z_e = 1$ and Z_u. In addition we have to renormalize Ω, Ω' and s. However a WT identity follows from the remark that $e(x, t)$ is coupled to phase transformations on the field φ. Indeed if we perform the transformation:

$$\varphi(x, t) = e^{i\alpha(t)}\,\varphi'(x, t), \quad e(x, t) = e'(x, t) - \dot{\alpha}(t)/s,$$

equation (35.95) is unchanged while an additional field independent term is added to the r.h.s. of equation (35.96): $\ddot{\alpha}/s$. Therefore the renormalization of s is connected to the renormalization of the Lagrange multiplier λ_e associated with e in the effective action. It is then easy to verify that s is not renormalized.

The RG β-functions. The model depends on three independent dimensionless coupling constants which we can choose to be u and:

$$v = s^2/\Omega\Omega', \quad w = \Omega/\Omega'. \tag{35.99}$$

The function β_u is given by the statics and determines the IR fixed point value $u^* = 3\varepsilon/40\pi^2 + O\left(\varepsilon^2\right)$. From the previous discussion, and dimensional considerations, it follows that the two other β-functions can be written in $4 - \varepsilon$ dimensions:

$$\beta_v = -v\left(\varepsilon + \eta_\omega + \eta_{\omega'}\right), \tag{35.100}$$

$$\beta_w = w\left(\eta_\omega - \eta_{\omega'}\right). \tag{35.101}$$

The coupling constant v has one obvious fixed point value $v = 0$ which decouples $\varphi(x, t)$ and $e(x, t)$. Then $\eta_{\omega'} = 0$ and η_ω assumes the value of model A (equation (35.29)) and is thus positive. The stability matrix constructed with the derivatives of the β-functions has an eigenvalue ω given by:

$$\omega = (\partial\beta_v/\partial v) = -\left(\varepsilon + \eta_\omega\right) < 0,$$

showing that model A is unstable with respect to the introduction of the coupling s.

If v does not vanish, equation (35.100) implies:

$$\varepsilon + \eta_\omega + \eta_{\omega'} = 0. \tag{35.102}$$

Equation (35.101) has three type of solutions, $w = \infty$, $w = 0$ or $\eta_\omega = \eta_{\omega'}$. To find the stable fixed points we then need the RG functions at leading order:

$$\eta_\omega = -\frac{v}{8\pi^2(1+w)}, \qquad \eta_{\omega'} = -\frac{v}{16\pi^2}. \tag{35.103}$$

It follows:

(i) The fixed point $w = \infty$ is unstable because the stability matrix has one negative eigenvalue $-\varepsilon$.

(ii) The fixed point $w = 0$, at leading order, also appears to be unstable because one eigenvalue is negative $\omega = -\varepsilon/3$. However the next term in the ε-expansion has been calculated and is positive. Therefore one cannot exclude that this fixed point becomes stable for $\varepsilon = 1$. This fixed point exhibits an interesting violation of dynamic scaling since the renormalized ratio of the time scales Ω^{-1} and Ω'^{-1} vanishes.

(iii) The last fixed point $w = \varepsilon/2 + O\left(\varepsilon^2\right)$ corresponds to normal dynamic scaling since the two time scales are relate in a finite way. All eigenvalues of the stability matrix are positive at leading order in ε. In this case $\eta_\omega = \eta_{\omega'}$ and equation (35.102) holds. Therefore, to all orders:

$$\eta_\omega = \eta_{\omega'} = -\varepsilon/2. \tag{35.104}$$

The exponent z is then exactly calculable:

$$z = d/2. \tag{35.105}$$

For a more detailed discussion we refer the interested reader to the literature.

Bibliographical Notes

The classification of the various dynamic models is taken from the review article
   P.C. Hohenberg and B.I. Halperin, *Rev. Mod. Phys.* 49 (1977) 435.
The effective action associated with the Langevin equation has been introduced in
   P.C. Martin, E.D. Siggia and H.A. Rose, *Phys. Rev.* A8 (1978) 423.
For an early discussion of the renormalization of dynamic theories in a field theory language see for example
   C. De Dominicis and L. Peliti, *Phys. Rev.* B18 (1978) 353,
and references therein.
The non-linear σ-model has been considered in
   R. Bausch, H.K. Janssen and Y. Yamazaki, *Z. Phys.* B37 (1980) 163.
For a review on early work on the critical dynamics of He see
   P.C. Hohenberg, *Physica* 109 & 110B (1982) 1436.
A recent review with many references can be found in
   V. Dohm, *J. Low Temp. Phys.* 69 (1987) 51.
The exponent z for model A is given at order ε^3 in
   N.V. Antonov and A.N. Vasil'ev, *Theor. Math. Phys.* 60 (1984) 671.
The dynamics of various other models has also been discussed, like the interface model of Wallace and Zia:
   R. Bausch, V. Dohm, H.K. Janssen and R.K.P. Zia, *Phys. Rev. Lett.* 47 (1981) 1837,
   J.C. Niel, *Phys. Rev.* B37 (1988) 2156,
and the Yang–Lee edge singularity
   N. Breuer and H.K. Janssen, *Z. Phys.* B41 (1981) 55.
Finally the critical behaviour in driven diffusive systems has been studied in
   H.K. Janssen and B. Schmittmann, *Z. Phys.* B64 (1986) 503, K. Leung and J.L. Cardy, *J. Stat. Phys.* 44 (1986) 567.

36 FIELD THEORY IN A FINITE GEOMETRY: FINITE SIZE SCALING

Many numerical calculations, like Monte Carlo or transfer matrix calculations, are performed with systems in which the size in several or all dimensions is finite. To extrapolate the results to the infinite system, it is thus necessary to have some idea about how the infinite size limit is reached. In particular in a system in which the forces are short range no phase transition can occur in a finite volume, or in a geometry in which the size is infinite only in one dimension. This indicates that the infinite size extrapolation is somewhat non-trivial. We present in this chapter an analysis of the problem in the case of second order phase transitions. We first discuss it from the RG point of view (in the case therefore of second order phase transitions) and derive the existence of a finite size scaling. We then distinguish between the finite volume geometry (in explicit calculations we take the example of the hypercube) and the cylindrical geometry in which the size is finite in all dimensions except one. We explain how to modify the methods used in the case of infinite systems to calculate the new universal quantities appearing in finite size effects, for example in $d = 4 - \varepsilon$ or $d = 2 + \varepsilon$ dimensions. Special properties of the commonly used periodic boundary conditions are emphasized. Finally both static and dynamical finite size effects are discussed.

Note finally that two-dimensional models with one finite size and periodic boundary conditions also describe finite temperature 2D quantum field theory.

The appendix contains a few remarks about finite size effects in the several phase region when the correlation length is finite, and about the calculation of one-loop finite size Feynman diagrams.

36.1 Renormalization Group in Finite Geometries

We always assume in this chapter that the size of our system is characterized by one length L which is large in the microscopic scale, for example much larger than the lattice spacing in lattice models. When the correlation length is also large, the universal properties of the system can be described by a continuum field theory. We use boundary conditions which do not break translation symmetry to avoid surface effects which are of a different nature. Periodic boundary conditions certainly satisfy such a criterion. Depending of the specific symmetries of a model, other boundary conditions are also available (like anti-periodic boundary conditions for Ising-like systems).

The crucial observation which explains finite size scaling is that the renormalization theory which leads to RG equations is completely *insensitive to finite size effects* since renormalizations are entirely due to *short distance singularities*. As a consequence RG equations are not modified. However their solution is different because correlation functions now depend on one additional dimensional parameter L. We discuss below the solution of RG equations both in the examples of the ϕ^4 field theory and of the non-linear σ-model.

36.1.1 The $(\phi^2)^2$ field theory for $d < 4$

Let us first consider the example of the field theory action

$$S(\phi) = \int \left\{ \frac{1}{2} \left[\partial_\mu \phi(x) \right]^2 + \frac{1}{2} \left(r_c + t \right) \phi^2(x) + \frac{1}{4!} u \left(\phi^2(x) \right)^2 \right\} \mathrm{d}^d x \,, \qquad (36.1)$$

where t characterizes the deviation from the critical temperature. In terms of the dimensionless coupling constant $g = u\Lambda^{-\varepsilon}$ the corresponding correlation functions satisfy for $d < 4$ the RG equations (Section 26.5):

$$\left[\Lambda \frac{\partial}{\partial \Lambda} + \beta(g) \frac{\partial}{\partial g} - \frac{1}{2} \eta(g) \left(n + M \frac{\partial}{\partial M} \right) - \eta_2(g) t \frac{\partial}{\partial t} \right] \Gamma^{(n)}(p_i; t, M, g, L, \Lambda) = 0 \,. \qquad (36.2)$$

They can be solved in the usual way by setting:

$$\Gamma^{(n)}(p_i; t, M, g, L, \Lambda) = Z^{-n/2}(\lambda) \Gamma^{(n)}(p_i; t(\lambda), M(\lambda), g(\lambda), L, \lambda\Lambda) \,, \qquad (36.3)$$

the various functions of λ being defined by (equations (26.26–26.28,26.54))

$$\lambda \frac{\mathrm{d}}{\mathrm{d}\lambda} g(\lambda) = \beta\big(g(\lambda)\big), \qquad\qquad g(1) = g \,,$$

$$\lambda \frac{\mathrm{d}}{\mathrm{d}\lambda} t(\lambda) = 2 - \nu^{-1}\big(g(\lambda)\big), \qquad t(1) = t \,,$$

$$\lambda \frac{\mathrm{d}}{\mathrm{d}\lambda} Z(\lambda) = \eta\big(g(\lambda)\big), \qquad\qquad Z(1) = 1 \,,$$

$$M(\lambda) = Z^{-1/2}(\lambda) M \,.$$

However the presence of the new length scale L now modifies the dimensional relations:

$$\Gamma^{(n)}(p_i; t, M, g, L, \Lambda) = \Lambda^{d - n(d-2)/2} \Gamma^{(n)}\left(\frac{p_i}{\Lambda}; \frac{t}{\Lambda^2}, \frac{M}{\Lambda^{1-\varepsilon/2}}, g, L\Lambda, 1 \right). \qquad (36.4)$$

We use this relation in the r.h.s. of equation (36.3) and then choose λ such that:

$$\lambda L \Lambda = 1 \quad \text{or} \quad \lambda = 1/L\Lambda \,. \qquad (36.5)$$

When the product ΛL becomes large, λ goes to zero and therefore $g(\lambda)$ approaches the IR fixed point g^*. This implies for the various functions of λ the behaviour:

$$Z(\lambda) \sim (L\Lambda)^{-\eta} \,,$$

$$\frac{t(\lambda)}{\lambda^2 \Lambda^2} \sim \frac{t}{\Lambda^2} (L\Lambda)^{1/\nu} \,, \qquad\qquad (36.6)$$

$$\frac{M(\lambda)}{(\lambda\Lambda)^{1-\varepsilon/2}} \sim \frac{M}{\Lambda^{1-\varepsilon/2}} (L\Lambda)^{(d-2+\eta)/2} \,.$$

We have therefore derived the scaling of finite size correlation functions:

$$\Gamma^{(n)}(p_i; t, M, g, L, 1) \sim L^{-d + n(d-2+\eta)/2} \Gamma^{(n)}\left(L p_i; t L^{1/\nu}, M L^{\beta/\nu}, g^*, 1, 1 \right). \qquad (36.7)$$

It is characterized by the appearance of a new scaling variable $tL^{1/\nu}$ which is a power of ξ/L.

From equation (36.7) the usual infinite size scaling form is recovered by expressing that $\Gamma^{(n)}(L)$ has a limit for $L \gg \xi \sim t^{-\nu}$. On the other hand, in the limit $\xi \gg L$, the correlation functions have a regular expansion in t even in zero magnetization in the finite volume or cylindrical geometry, since no phase transitions can occur in both cases (for short range interactions).

Note that all combinations which are independent of the normalization of the field ϕ, of the temperature t, and of the magnetic field are universal for the reasons explained in the infinite volume case, once the geometry and boundary conditions are fixed.

Adapting the usual analysis of corrections to scaling to this new situation, one immediately finds that the leading corrections to the scaling form (36.7) have near $d = 4$ the form of a scaling function multiplied by a factor $L^{-\omega}$.

Examples. Let us now give two examples of quantities which have been considered in practical calculations. In a finite volume V the ratio $\mathcal{R}_4$:

$$\mathcal{R}_4 = \frac{m_4}{m_2^2} \quad \text{with} \quad m_{2p} = \left\langle \left(\frac{1}{V} \left| \int \phi(x)\mathrm{d}^d x \right| \right)^{2p} \right\rangle, \tag{36.8}$$

has been used to determine the critical temperature. It has in zero magnetic field the scaling form:

$$\mathcal{R}_4(t, L) \sim g\left(tL^{1/\nu} \right). \tag{36.9}$$

By calculating $\mathcal{R}_4(t, L)$ for different values of L and looking for a temperature at which it is independent of L, one finds the critical temperature $t = 0$, provided the corrections in $L^{-\omega}$ are negligible. The quantity $f(0)$ is a universal number which in principle can be calculated from the continuum field theory. We examine this problem later.

In the cylindrical geometry, the correlation length ξ_L in the infinite direction is another quantity of interest. From equation (36.7) one concludes:

$$\xi_L \sim LX\left(tL^{1/\nu} \right). \tag{36.10}$$

In particular at $t = 0$, ξ_L grows like L and the ratio ξ_L/L is universal. Since ξ_L is related to the ratio of the two largest eigenvalues of the transfer matrix (equation (23.12)) λ_0 and λ_1 we learn also:

$$\lambda_0/\lambda_1 = 1 + L/g(0).$$

With this knowledge, it is interesting to return to the analysis of the existence of phase transitions in Chapter 23.

Since for $t > 0$, ξ_L goes to a constant for large L, and since for $t < 0$ it grows faster than L, as can be easily verified, the ratio ξ_L/L can be used to determine the critical temperature in transfer matrix calculations.

36.1.2 Low temperature expansion and finite size effects

We have shown in Chapter 30 that in models with continuous symmetries, at low temperature the long distance behaviour is described by the effective interactions between Goldstone modes. In the case of the N-vector model the universal physical observables can be derived from the low temperature or low coupling expansion of the $O(N)$

non-linear σ-model. Let us thus shortly examine the problem of finite size effects in this framework. Previous considerations concerning RG equations also apply to the RG equations derived for the σ-model: the equations are not modified, only the solution is changed by the finite size: The general solution (30.38,30.40) of the RG equations (30.21) now depends on an additional scaling variable $L/\xi(t)$ where for $d > 2, t < t_c$ the length $\xi(t)$ is defined by equation (30.31):

$$\xi(t) = \Lambda^{-1} t^{1/(d-2)} \exp\left[\int_0^t \left(\frac{1}{\beta(t')} - \frac{1}{(d-2)t'}\right) dt'\right], \tag{36.11}$$

and for $d = 2$ by

$$\xi(t) \propto \Lambda^{-1} \exp\left[\int^t \frac{dt'}{\beta(t')}\right]. \tag{36.12}$$

Another convenient way to parametrize the solution is to introduce a size-dependent temperature t_L, obtained by solving the equation:

$$\lambda \frac{d}{d\lambda} t(\lambda) = \beta[t(\lambda)] \qquad t(1) = t, \tag{36.13}$$

at scale $\lambda = 1/\Lambda L$:

$$t_L \equiv t(1/\Lambda L). \tag{36.14}$$

Then

$$\ln(\Lambda L) = \int_{t_L}^t \frac{dt'}{\beta(t')}, \tag{36.15}$$

which shows in particular that t_L is a function of t and L only through the expected combination $L/\xi(t)$.

When ΛL increases, $\lambda = 1/L\Lambda$ goes to zero. For $d > 2$ and $t < t_c$ fixed, t_L approaches the IR fixed point $t = 0$:

$$t_L \sim \left(\xi(t)/L\right)^{d-2}. \tag{36.16}$$

Therefore finite size effects can be calculated from the low temperature expansion and renormalization group.

At t_c, and more generally in the critical domain, physical quantities can be calculated in an $\varepsilon = d - 2$ expansion, as has been shown in Chapter 30. Since t_c is a RG fixed point $t_L(t_c) = t_c$.

Finally calculations can be performed in *two dimensions* even in zero magnetic field h, L providing an IR cut-off. However, because $t = 0$ is then a UV fixed point, t_L goes to zero *for $L/\xi(t)$ small*,

$$t_L \sim \frac{2\pi}{(N-2)\ln(\xi(t)/L)}, \tag{36.17}$$

and this is the limit in which physical quantities can be calculated.

Notice here that a field theory in two dimensions with a finite size in one of the dimensions, and periodic boundary conditions is also a finite temperature field theory. Therefore the considerations of the Subsection 36.5.2 are also relevant for finite temperature.

36.2 Perturbation Theory in Finite Geometries

The scaling properties (36.7) are valid for all boundary conditions, but the explicit universal finite size expressions change. Even the technical details of the calculation when the temperature approaches T_c vary. In this chapter we mainly consider periodic boundary conditions, but we briefly indicate in the next subsection how the method can be adapted to other examples. In Section 3.4 we have actually given one example of twisted boundary conditions.

The characteristic feature of finite geometries is that, in Fourier space, the momenta which correspond to directions in which the size of the system is finite are quantized. As mentioned above, to avoid extra complications due to surface effects, we here consider only boundary conditions which do not break translation symmetry. For generic systems only periodic boundary conditions satisfy this condition. However for systems with symmetries other boundary conditions are possible.

36.2.1 Periodic boundary conditions and the zero mode

In the case of a d dimensional hypercube of linear size L with periodic boundary conditions, the quantized momenta p_μ have the form:

$$p_\mu = \frac{2\pi}{L} k_\mu, \qquad k_\mu \in \mathbb{Z}^d. \tag{36.18}$$

When the products $tL^{1/\nu}$ or L/ξ are positive and not small, the usual methods of calculation of the infinite volume are applicable and finite size effects due to momentum quantization are only quantitative, decreasing like $\exp[-\text{const.}\ L/\xi]$. When the product $tL^{1/\nu}$ is negative and not small (the ordered phase) the physics of the infinite and finite systems are very different, and this problem will be examined later. At T_c in a finite volume the propagator has an isolated pole at $p_\mu = 0$ and therefore IR divergences will appear in perturbation theory for $\xi \gg L$.

In the cylindrical geometry one component of the momentum ω varies continuously but still when all other components vanish, massless Feynman diagrams receive a divergent contribution of the form $\int d\omega/\omega^2$. Finally a geometry in which the sizes in two dimensions or three dimensions (this is the case of finite temperature quantum field theory) among d are infinite still leads to IR divergences. We shall not discuss these latter cases further in this chapter because the IR problem cannot be solved exactly.

As a consequence even in high dimensions, for which in the infinite geometry mean field theory is exact, IR divergences are generated. To overcome this difficulty it is necessary to separate the zero momentum Fourier components of the field. The components $k_\mu \neq 0$ can be treated by the methods developed in the infinite geometry (perturbation theory and RG), the component $k_\mu = 0$ whose fluctuations are damped at T_c only by interaction terms, has to be treated exactly. In the case of the finite volume we therefore construct an effective integral over the component $\tilde\phi\,(p_\mu = 0)$ by integrating out all other components. In the cylindrical geometry we construct an effective quantum mechanical problem corresponding to the component $\tilde\phi\,(\mathbf{p}_\mathrm{T} = \mathbf{0}, \tau)$, denoting by τ the coordinate in the infinite direction. Finally the geometry with two or three infinite dimensions leads to effective 2D or 3D theories. Note that similar considerations will apply to the zero modes of instanton calculations (see Chapters 37–43).

We examine in the next section the two first geometries separately, beginning with the simplest case of the periodic hypercube.

36.2.2 Twisted boundary conditions

For systems with symmetries additional boundary conditions do not break translation invariance: conditions such that the values of the order parameter at both boundaries (for each direction in space) differ by a constant group transformation (often called twisted boundary conditions). For instance for Ising-like systems one can use antiperiodic boundary conditions, for the N-vector model with $O(N)$ symmetry one can impose a rotation of a given angle around some axis.

In such cases the quantized momenta p_μ are shifted by some additional constants.

Let us verify this explicitly for a field theory invariant under the transformations of a Lie group G.

We impose to the field ϕ the boundary conditions

$$\phi(x_1, x_2, \ldots, x_\mu + L, \ldots, x_d) = e^{A_\mu}\, \phi(x_1, x_2, \ldots, x_\mu, \ldots, x_d),$$

where the A_μ's are constant (i.e. space independent) commuting matrices, $[A_\mu, A_\nu] = 0$, elements of the Lie algebra of G (A_μ is a curvature-free gauge field).

To return to the situation of periodic boundary conditions, we perform a gauge transformation on the field ϕ. We set

$$\phi(\mathbf{x}) = e^{A_\mu x_\mu/L}\, \phi'(\mathbf{x}).$$

The new field ϕ' then satisfies periodic boundary conditions. However the derivatives have now been replaced by covariant derivatives:

$$\partial_\mu \mapsto \partial_\mu + A_\mu/L\,.$$

After Fourier transformation the derivatives yield quantized momenta of the form $\partial_\mu \mapsto (2i\pi k_\mu + A_\mu)/L$. For a unitary or orthogonal group the matrices A_μ are antihermitian and thus have imaginary eigenvalues $i\theta_\mu^\alpha$. Writing the field ϕ' in a base where all A_μ (which commute) are diagonal, we finally see that the effect of twisted boundary conditions has been to generate a set of quantized momenta of the form

$$p_\mu = (2\pi k_\mu + \theta_\mu^\alpha)/L\,, \qquad k_\mu \in \mathbb{Z}^d,$$

for each α component of the field. We can clearly choose all angles θ_μ^α to belong to the interval $[-\pi, \pi]$.

36.3 The ϕ^4 Field Theory: The Periodic Hypercube

We first study the cases of dimensions $d > 4$ and $d = 4 - \varepsilon$ in the framework of the effective $(\phi^2)^2$ field theory. As explained above, we expand $\phi(x)$ in Fourier components, separating the zero momentum component:

$$\phi(x) = \left(\frac{2\pi}{L}\right)^d \left(\tilde{\phi}(0) + \sum_{\mathbf{p}=2\pi\mathbf{k}/L}' e^{i\mathbf{p}\cdot\mathbf{x}}\, \tilde{\phi}(\mathbf{p})\right), \qquad \mathbf{k} \in \mathbb{Z}^d. \qquad (36.19)$$

The symbol $\sum'$ means that the value $\mathbf{k} = \mathbf{0}$ has been omitted in the summation. The integration over the $\tilde{\phi}_{\mathbf{k}}$, $\mathbf{k} \neq \mathbf{0}$, modes is performed as in the infinite geometry limit: this generates a perturbative expansion which has RG properties. An integral remains on the last $\mathbf{k} = \mathbf{0}$ modes which must be calculated exactly. Note that the first part of the procedure is formally equivalent to the shift of the expectation value of the field $\phi(x)$ in the infinite geometry. The main difference, apart from the replacement of integrals by discrete sums in Feynman diagrams, is that the average $L^{-d} \int \phi(x) \mathrm{d}^d x$ here remains a dynamical variable (see also the discussion of Section 6.6).

As an illustration let us calculate averages of the form (36.8), moments of the average spin per unit volume distribution, in a spin system. We set

$$\varphi = L^{-d} \int \phi(x) \mathrm{d}^d x = (2\pi/L)^d \tilde{\phi}(0) \,,$$

and

$$\exp[-\Sigma(\varphi)] = \mathcal{N}^{-1} \int \prod_{\mathbf{k} \neq 0} \mathrm{d}\tilde{\phi} \,(\mathbf{p} = 2\pi \mathbf{k}/L) \exp[-S(\phi)],$$

where $S(\phi)$ is the action (36.1) and the normalization $\mathcal{N}$ is chosen such that $\Sigma(0) = 0$ for $t = 0$. The moments are then given by

$$m_\sigma = Z^{-1} \int \mathrm{d}\varphi |\varphi|^\sigma \exp[-\Sigma(\varphi)], \tag{36.20}$$

where Z is the partition function

$$Z = \int \mathrm{d}\varphi \exp[-\Sigma(\varphi)].$$

For $\sigma = 2p$ they are related to zero momentum correlation functions:

$$m_{2p} = \left\langle \left[L^{-2d} \int \mathrm{d}^d x \, \mathrm{d}^d y \, \phi(x) \cdot \phi(y) \right]^p \right\rangle = (2\pi/L)^{2pd} \left\langle \left(\tilde{\phi}^2(0) \right)^p \right\rangle. \tag{36.21}$$

It follows from the discussion of Section 6.6 that in the infinite volume limit $\Sigma(\varphi) = \Gamma(\varphi) - \Gamma(0)$ where $\Gamma(\varphi)$ is the thermodynamic potential as obtained in perturbation theory. It satisfies the RG equation

$$\left[\Lambda \frac{\partial}{\partial \Lambda} + \beta(g) \frac{\partial}{\partial g} - \frac{1}{2}\eta(g)\varphi \frac{\partial}{\partial \varphi} - \eta_2(g)t \frac{\partial}{\partial t} \right] \Sigma(t, \varphi, g, L, \Lambda) = R(t, g, \Lambda) \,, \tag{36.22}$$

where we recall that R is a second degree polynomial in t. Though for $\sigma \geq 0$ generic, m_σ is not related to correlation functions, it satisfies a usual RG equation which can be easily derived from the equation (36.22):

$$\left[\Lambda \frac{\partial}{\partial \Lambda} + \beta(g) \frac{\partial}{\partial g} + \frac{1}{2}\sigma\eta(g) - \eta_2(g)t \frac{\partial}{\partial t} \right] m_\sigma(t, g, L, \Lambda) = 0 \,. \tag{36.23}$$

Another quantity of interest is the specific heat $C(L, t)$

$$C(L, t) = L^{-d} \frac{\partial^2}{(\partial t)^2} \ln Z \,. \tag{36.24}$$

36.3.1 Dimensions d > 4

We first discuss the case $d > 4$, which in the infinite geometry is very simple. To obtain the zero mode integral at leading order, we simply neglect the $\tilde{\phi}_k$, $\mathbf{k} \neq \mathbf{0}$, modes altogether. The action (36.1) then reduces to

$$\Sigma\left(\varphi\right) = L^d \left[\frac{1}{2}t\varphi^2 + \frac{1}{4!}u\left(\varphi^2\right)^2\right]. \tag{36.25}$$

After the change of variables

$$\varphi \mapsto \left(uL^d\right)^{-1/4}\varphi, \tag{36.26}$$

we find that the moments m_σ can be written:

$$m_\sigma(L,t) = \left(uL^d\right)^{-\sigma/4} f_\sigma\left(tL^{d/2}u^{-1/2}\right), \tag{36.27}$$

in which $f_\sigma(z)$ is given by

$$f_\sigma(z) = \frac{g_{\sigma+N}(z)}{g_N(z)}, \tag{36.28}$$

with

$$g_\sigma(z) = \int_0^\infty d\varphi\, \varphi^{\sigma-1} \exp\left[-\left(\frac{1}{2}z\varphi^2 + \frac{1}{4!}\varphi^4\right)\right]. \tag{36.29}$$

Equation (36.27) shows that above four dimensions the finite size scaling relations, proven for a non-trivial fixed point, and which predict instead for the moment m_σ a behaviour

$$m_\sigma = L^{-\sigma(d-2)/2}\tilde{f}_\sigma(tL^2),$$

do not hold. In particular, instead of the argument $tL^2 \propto (L/\xi)^2$ one finds $tL^2L^{(d-4)/2}$. The extra factor $L^{(d-4)/2}$ arises because the leading order result depends explicitly on u, which has a dimension $4 - d$, and characterizes the violation of the naive scaling (see the problem of hyperscaling in Section 27.1).

The result (36.27) leads to the prediction of some universal quantities. For instance let us calculate the moments at $T = T_c$ ($t = 0$). Evaluating integral (36.29) explicitly, we find:

$$g_\sigma(0) = \tfrac{1}{4}(24)^{\sigma/4}\Gamma\left(\sigma/4\right).$$

Dimensionless ratios are universal, for example

$$\mathcal{R}_\sigma(T = T_c) = \frac{m_\sigma}{(m_2)^{\sigma/2}} = \frac{\Gamma\left(\frac{\sigma+N}{4}\right)\left[\Gamma\left(\frac{N}{4}\right)\right]^{\sigma/2-1}}{\left[\Gamma\left(\frac{N+2}{4}\right)\right]^{\sigma/2}}. \tag{36.30}$$

In particular for Ising-like systems the quantity $\mathcal{R}_4$ is:

$$\mathcal{R}_4 = \Gamma^4\left(1/4\right)/(8\pi^2) = 2.1884\cdots. \tag{36.31}$$

Another quantity of interest is m_2/m_1^2:

$$\mathcal{R}_1^{-2} = \frac{m_2}{m_1^2} = \sqrt{2}. \tag{36.32}$$

Note finally that at leading order the specific heat C (equation (36.24)) is simply given by

$$C(L,t) = \frac{1}{4}L^d \left(m_4 - m_2^2\right) = \frac{1}{4u}\left(f_4 - f_2^2\right) + \text{const.} \,,$$

where the constant comes from the regular part of the free energy.

Remark. The expression (36.27) is *a priori* only valid for $tL^{d/2}$ finite. It is however easy to verify that it has both for $t < 0$ and $t > 0$ fixed the correct $L \to \infty$ behaviour. Indeed for $t < 0$ one finds

$$m_\sigma(L,t) \to (-6t/u)^{\sigma/2} \equiv [M_0(t)]^\sigma \,,$$

where $M_0(t)$ is the infinite size spontaneous magnetization at this order. For $t > 0$ one obtains instead

$$m_\sigma(L,t) \sim \frac{\Gamma[(\sigma+N)/2]}{\Gamma(N/2)}\left(2\chi(t)L^{-d}\right)^{\sigma/2} \,,$$

where $\chi(t) = 1/t$ is the infinite size magnetic susceptibility.

The correction terms to this leading behaviour however are incorrect. To calculate them it is necessary to take into account higher order contributions.

Higher order corrections. It now remains to show that higher order contributions do not invalidate the leading order results. The contribution to $\Sigma(\varphi)$ coming from loop diagrams contains a part divergent (UV) for large cut-off, $\delta\Sigma_{\text{div}}(\varphi,t)$, which is the same as in the infinite volume limit and which, according to the analysis of Chapter 27, is regular in φ^2 and t. It is actually a polynomial at any finite order in perturbation theory:

$$\delta\Sigma_{\text{div}} = L^d \sum_{k,m} \Sigma_{km} t^m \varphi^{2k}.$$

After the change of variables (36.26), and taking into account that t is of order $L^{-d/2}$ we find that a term proportional to $\varphi^{2k}t^m$ gives a contribution of order $L^{d(2-m-k)/2}$. Therefore only the terms $m + k \leq 2$ have to be discussed. The terms $k > 0$ simply renormalize r_c, t and u. The two remaining terms are proportional to t and t^2. They cancel in correlation functions. The latter yields a constant non-universal contribution to the specific heat.

Let us now examine the UV finite contributions, generated by the low momentum region. For instance the one-loop corrections are:

$$\Sigma_{1\,\text{loop}}(\varphi,L,t,u) = \frac{1}{2}\sum_{\mathbf{k}\neq 0} \text{tr}\ln\left[\delta_{ij} + \frac{1}{(2\pi\mathbf{k}/L)^2}\left(\left(t + \frac{u}{6}\varphi^2\right)\delta_{ij} + \frac{u}{3}\varphi_i\varphi_j\right)\right]$$

$$= \frac{1}{2}\sum_{\mathbf{k}\neq 0}\left[\ln\left(1 + \frac{t + \frac{1}{2}u\varphi^2}{(2\pi\mathbf{k}/L)^2}\right) + (N-1)\ln\left(1 + \frac{t + \frac{1}{6}u\varphi^2}{(2\pi\mathbf{k}/L)^2}\right)\right]. \quad (36.33)$$

After subtraction of the UV divergent terms, $\Sigma_{1\,\text{loop}}$ can be expanded in powers of φ and t. Moreover $uL^2\varphi^2$, after the change of variables (36.26), is of order $u^{1/2}L^{(4-d)/2}$. In the same way equation (36.27) shows that the combination tL^2 is also of order $u^{1/2}L^{(4-d)/2}$ in the scaling region. It follows that a term proportional to $\varphi^{2k}t^m$ is of order $(uL^{4-d})^\kappa$ with $\kappa = \frac{1}{2}(k+m)$ and thus goes to zero for L large.

The argument can easily be generalized to higher orders in the loop expansion because at L fixed perturbation theory is IR convergent and all terms are regular functions for φ and t small. Moreover we note that each loop in the perturbative expansion yields a factor u which is the loop expansion parameter, and thus a factor uL^{4-d} for dimensional reasons. It follows that a term of l-loop order proportional to $\varphi^{2k}t^m$ is of order $(uL^{4-d})^\kappa$ with $\kappa = l - 1 + \frac{1}{2}(k + m)$.

The conclusion is that for $d > 4$ the effective action (36.25) can be simply derived from mean field theory, as in the infinite volume limit, the only modification coming from the last integration over the average field (36.20). In particular the expression (36.30) is indeed universal.

Finally the leading term at two-loop order has $\kappa = 3/2$ and thus the only significant corrections at one-loop order correspond to $m + k \le 2$. A short calculation then shows that for the dimensionless ratios considered above the leading corrections can be reproduced by replacing t by a quantity t_L which has the form

$$t_L = \left(1 + A_1 L^{4-d}\right) t + A_2 L^{2-d},$$

A_1, A_2 being two constants. The specific heat receives an additional contribution of order L^{4-d}.

36.3.2 Dimensions $d = 4 - \varepsilon$

We now use the RG arguments presented in Section 36.1: instead of calculating physical quantities as function of $\{t, M, u, L\}$, we can, in the critical domain, set $L = 1$, $u = u^*$ the IR fixed point value, then replace t by $tL^{1/\nu}$, M (if we introduce a magnetic field) by $ML^{\beta/\nu}$ and thus φ by $\varphi L^{\beta/\nu}$ in $\Sigma(\varphi)$.

At leading order the function $\Sigma(\varphi)$ is

$$\Sigma(\varphi, L = 1, t, u^*) = \tfrac{1}{2}t\varphi^2 + \frac{u^*}{4!}\varphi^4.$$

At the same order u^* can be replaced by its value at order ε:

$$u^* = \frac{48\pi^2\varepsilon}{N + 8} + O\left(\varepsilon^2\right). \qquad (36.34)$$

Then replacing t by $tL^{1/\nu}$ and φ by $\varphi L^{\beta/\nu}$, and integrating over φ we find the moments m_σ at leading order

$$m_\sigma(L, t) = \frac{L^{-\sigma(d-2+\eta)/2}}{(u^*)^{\sigma/4}} f_\sigma\left(tL^{1/\nu}(u^*)^{-1/2}\right), \qquad (36.35)$$

where f_σ has been defined in (36.28). The equation shows that the ε-expansion is not uniform. The method used here, in which the zero mode is treated separately, gives the correct leading order only if t is assumed to be of order $\varepsilon^{1/2}$ (this condition is realized in particular at $t \propto T - T_c = 0$).

Note the appearance of powers of $(u^*)^{1/2}$ which, for ε small, is equivalent to $\varepsilon^{1/2}$. This suggests that physical quantities will have an expansion in powers of $\varepsilon^{1/2}$ rather than ε. The analysis of higher order corrections confirms this observation. Let us exhibit this phenomenon in the one-loop approximation.

One-loop calculation. The integration over the $\mathbf{k} \neq 0$ components of the field in the gaussian approximation yields the one-loop contribution $\Sigma_{1\,\mathrm{loop}}$ to Σ given in equation (36.33). The divergent part of the one-loop term has first to be subtracted. We then set $L = 1$, $u = u^*$. As we have already discussed at the end of the preceding subsection, at L fixed all terms in perturbation theory can then be expanded in powers of φ^2 and t. After the change of variables (36.26) φ^2 has a coefficient proportional to $u^{*1/2} \sim \varepsilon^{1/2}$. In the same way, t is of order $\varepsilon^{1/2}$. A term contributing to the l-loop order and proportional to $\varphi^{2k} t^m$ is of order $\varepsilon^{l-1+(k+m)/2}$. The leading two-loop correction will come from the term proportional to φ^2 and thus be of order $\varepsilon^{3/2}$. Therefore at one-loop order only the terms proportional to φ^2, $\varphi^2 t$, $\varphi^4\, t$ and t^2 have to be considered. The form of $\Sigma(\varphi)$ at one-loop order will thus be

$$\Sigma\left(\varphi, L = 1, t, u^*\right) - \Sigma(0, 1, t, u^*) = \tfrac{1}{2}\big(t(1 + a_1 u^*) + a_2 u^*\big)\varphi^2 + u^*(1 + a_3 u^*)\frac{\varphi^4}{4!}.$$

Note that the correction to the coefficient of φ^4 can be eliminated by a rescaling of φ. The only relevant effect is to change a_1 into $a_1 - a_3/2$. But the complete coefficient of t can be absorbed into a finite change of normalization of t. The conclusion is the only correction relevant at one-loop order is related to the coefficient a_2 which we now calculate.

The φ^2 contribution. The coefficient $\tilde{a}_2$ of $u^* \varphi^2/2$ in the expansion of expression (36.33) is:

$$\tilde{a}_2 = \frac{N+2}{6} \sideset{}{'}\sum_{k_\mu} \frac{1}{(2\pi\mathbf{k})^2}. \tag{36.36}$$

The coefficient a_2 is obtained from $\tilde{a}_2$ by subtracting the infinite size contribution, which is a critical temperature shift.

Let us introduce a useful function (related to elliptic functions, see Appendix A36.2):

$$A(s) = \sum_{n=-\infty}^{+\infty} \mathrm{e}^{-sn^2}. \tag{36.37}$$

As shown in Appendix A36.2, for $s \to 0$

$$A(s) - (\pi/s)^{1/2} \sim 2(\pi/s)^{1/2}\, \mathrm{e}^{-\pi^2/s}, \tag{36.38}$$

where $(\pi/s)^{1/2}$ is the infinite size limit.

In terms of this function we can formally write:

$$\sideset{}{'}\sum_{k_\mu} \frac{1}{\mathbf{k}^2} = \int_0^{+\infty} \mathrm{d}s\, \left(A^d(s) - 1\right). \tag{36.39}$$

The integral in the r.h.s. converges exponentially for s large, but the behaviour (36.38) shows that it diverges for s small. However if we subtract the infinite size limit we obtain a finite result:

$$\sideset{}{'}\sum_{k_\mu} \frac{1}{\mathbf{k}^2} - \int \frac{\mathrm{d}^d p}{\mathbf{p}^2} = \int_0^{+\infty} \mathrm{d}s\, \left[A^d(s) - 1 - (\pi/s)^{d/2}\right]. \tag{36.40}$$

The coefficient a_2 follows:

$$a_2 = \frac{N+2}{24\pi^2} \int_0^{+\infty} ds \left[A^d(s) - 1 - (\pi/s)^{d/2} \right]. \tag{36.41}$$

Introducing rescaled temperature $t' \propto t$ and field ϕ we can rewrite the moments at one-loop order:

$$m_\sigma(L, t') = L^{-\sigma(d-2+\eta)/2} f_\sigma \left(t' L^{1/\nu} + b \right), \tag{36.42}$$

where the constant b is given by (36.34,36.41),

$$b = a_2 u^{*1/2} = \frac{N+2}{\sqrt{N+8}} \frac{(3\varepsilon)^{1/2}}{6\pi} \int_0^{+\infty} ds \left[A^4(s) - 1 - \pi^2/s^2 \right] + O\left(\varepsilon^{3/2} \right). \tag{36.43}$$

The ratio $\mathcal{R}_4 (T_c)$. From expression (36.42), setting $t = 0$ we immediately derive the universal dimensionless ratio $\mathcal{R}_4(T = T_c)$ at order $\varepsilon^{1/2}$:

$$\mathcal{R}_4 (T_c) = \frac{g_{4+N}(b) g_N(b)}{[g_{2+N}(b)]^2}, \tag{36.44}$$

g_σ being defined by equation (36.29). Using the value of the integral:

$$\int_0^{+\infty} ds \left[A^4(s) - 1 - \pi^2/s^2 \right] = -1.7650848012\ldots,$$

we obtain in three dimensions for $N = 1$:

$$\mathcal{R}_4 (T_c) = 1.800\ldots . \tag{36.45}$$

This result should be compared to the mean field value 2.188 and a Monte Carlo numerical estimate 1.6. The agreement is comparable to other results at order ε.

The specific heat. The t^2 terms, both divergent and convergent in Σ have to be taken into account. A short calculation yields

$$C(L, t) = L^{\alpha/\nu} \left[\frac{1}{4} \left(f_4(t' L^{1/\nu} + b) - f_2^2(t' L^{1/\nu} + b) \right) + \frac{3N}{4-N} \right] + \text{const.} ,$$

where again the constant comes from the non-universal regular contribution.

36.4 The ϕ^4 Field Theory: The Cylindrical Geometry

In what follows we consider a system infinite in one dimension hereafter called time and of finite size L with periodic boundary conditions in the remaining $d-1$ space dimensions. To isolate the zero modes we now expand the fields in Fourier components in the $d-1$ space dimensions:

$$\phi(x, \tau) = \sum_{k_\mu \in \mathbb{Z}^{d-1}} (2\pi/L)^{d-1} e^{i2\pi k \cdot x/L} \phi_k(\tau). \tag{36.46}$$

Let us again consider only the simple example of correlation functions of space integrals

$$\varphi(\tau) = L^{1-d} \int d^{d-1}x \, \phi(x, \tau).$$

These can be calculated using only the effective action $\Sigma(\varphi)$ obtained by integrating over all components $\phi_k(\tau)$, $k \neq 0$:

$$\exp[-\Sigma[\varphi(\tau)] = \mathcal{N}^{-1} \int \prod_{\mathbf{k}\neq 0} [d\phi(2\pi k/L) \exp[-S(\phi)],$$

where the normalization $\mathcal{N}$ is now chosen in such a way that $\Sigma(0) = 0$. The effective action Σ has, at leading order, the form of a quantum mechanical action, function of the time-dependent path $\varphi(\tau)$.

We illustrate the method with the calculation of the finite size correlation length ξ_L. We first examine the case $d > 4$ which is described by mean field theory in the infinite volume limit.

36.4.1 Dimensions $d > 4$

The leading order approximation is obtained by neglecting all corrections due the integration over the $\mathbf{k} \neq \mathbf{0}$ components of the field:

$$\Sigma[\varphi(\tau)] = L^{d-1} \int d\tau \left[\frac{1}{2}(\dot\varphi)^2 + \frac{t}{2}\varphi^2 + \frac{u}{4!}(\varphi^2)^2 \right]. \tag{36.47}$$

To this action is associated the quantum mechanical hamiltonian H (see Chapter 2):

$$H = \frac{\mathbf{p}^2}{2L^{d-1}} + L^{d-1} \left[\frac{t}{2}\mathbf{q}^2 + \frac{u}{4!}(\mathbf{q}^2)^2 \right]. \tag{36.48}$$

The finite size correlation length ξ_L is related to the two lowest eigenvalues E_0 and E_1 of H by:

$$\xi_L = (E_1 - E_0)^{-1}. \tag{36.49}$$

After the canonical transformation:

$$q \mapsto L^{-(d-1)/3} u^{-1/6} q, \qquad p \mapsto L^{(d-1)/3} u^{1/6} p, \tag{36.50}$$

the eigenvalues E_i can be expressed in terms of the eigenvalues $\widehat{E}_i(z)$ of the hamiltonian $\widehat{H}$:

$$\widehat{H} = \tfrac{1}{2}\mathbf{p}^2 + \tfrac{1}{2}z\mathbf{q}^2 + \frac{1}{4!}(\mathbf{q}^2)^2, \tag{36.51}$$

by:

$$E_i(t, u, L) = u^{1/3} L^{-(d-1)/3} \widehat{E}_i \left(tL^{2(d-1)/3} u^{-2/3} \right). \tag{36.52}$$

The finite size correlation length ξ_L has thus the form:

$$\xi_L = u^{-1/3} L^{(d-1)/3} X \left(tL^{2(d-1)/3} u^{-2/3} \right), \tag{36.53}$$

where the function $X(z)$ is defined by:

$$X(z) = \left(\widehat{E}_1(z) - \widehat{E}_0(z)\right)^{-1}. \tag{36.54}$$

In terms of the infinite size correlation length ξ_∞, this expression can be rewritten:

$$\xi_L(t) = \xi_\infty(t) Y\left(\frac{L}{\xi_\infty(t)} L^{(d-4)/3}\right), \qquad Y(z) = \sqrt{z} X(z). \tag{36.55}$$

This expression exhibits a violation of the naive extension of the scaling from (36.10) proven for $d < 4$. The reason is here, as before, that in case of the gaussian fixed point, the amplitude of expected leading terms vanishes and the first non-trivial contributions correspond to what are for $d < 4$ only corrections to scaling (see Section 27.1).

Note that the canonical transformation (36.50) is equivalent in the language of the effective action (36.47) to:

$$\varphi \mapsto L^{-(d-1)/3} u^{-1/6} \varphi, \qquad \tau \mapsto L^{(d-1)/3} u^{-1/3} \tau, \tag{36.56}$$

as can be verified directly. The second transformation can also be written in terms of the Fourier variable ω associated with τ:

$$\omega \mapsto L^{-(d-1)/3} u^{1/3} \omega. \tag{36.57}$$

Finally we observe that, in the limit of interest, t has to be taken of order $u^{2/3} L^{2(1-d)/3}$.

Loop corrections. Let us now show that loop corrections due to the integration over the non-zero modes do not modify the scaling form (36.55) at leading order. These corrections are of two types:

(i) Corrections corresponding due to UV divergent terms already present in the infinite volume limit. At any finite order divergences generate a polynomial in φ and its derivatives. After the changes of variables (36.56) a monomial with $2k$ powers of φ and $2n$ derivatives is of order L^κ with $\kappa = (d-1)(4 - 2k - 2n)/3$. Therefore terms which are absent from the leading order action have vanishing coefficients. The simplest example is ϕ^6 which induces a correction proportional to $L^{d-1} q^6$ in the hamiltonian H and thus proportional to $L^{-2(d-1)/3}$ in $\widehat{H}$. The remaining terms with $2k + n \le 4$ simply renormalize the coefficients of the leading order action.

(ii) Corrections which are finite in the large cut-off limit. The one-loop correction $\Sigma_{1\,\text{loop}}(\varphi)$ to the effective action, analogous to expression (36.33), is given by:

$$\Sigma_{1\,\text{loop}}(\varphi) = \frac{1}{2} \sum_{k_\mu}' \left\{ \mathrm{tr} \ln\left[1 + \frac{u}{2}\left(-\partial_\tau^2 + (2\pi\mathbf{k}/L)^2 + t\right)^{-1}\varphi^2(\tau)\right] \right.$$
$$\left. + (N-1)\,\mathrm{tr} \ln\left[1 + \frac{u}{6}\left(-\partial_\tau^2 + (2\pi\mathbf{k}/L)^2 + t\right)^{-1}\varphi^2(\tau)\right] \right\}. \tag{36.58}$$

The action now becomes a non-local functional of the field $\varphi(\tau)$ and it would seem at first sight that we can no longer use a hamiltonian formalism to evaluate corrections to the mean field approximation.

However we first notice that φ appears, after the change (36.56), only in the combination $uL^2\varphi^2$ which after the change (36.56) is of order $u^{2/3} L^{2(4-d)/3}$, which goes to zero for L large. Therefore because L provides an IR cut-off we can expand in powers of φ. In the same way t appears only in the combination tL^2 which, according to the form (36.53), is of order $u^{2/3}L^{2(4-d)/3}$ and thus also goes to zero for L large. For the same reason we can expand in powers of t. Finally we face the problem of the non-local operator $[-L^2\partial_\tau^2 + (2\pi\mathbf{k})^2]^{-1}$. After the change of variables (36.56), the differential operator $L^2\partial_\tau^2$ becomes also of order $u^{2/3}L^{2(4-d)/3}$ and thus is small compared to $\mathbf{k}^2$. We then know from general arguments that the action (36.58) has a local expansion. Actually this problem is related to the classical problem of expanding the resolvent of the Schrödinger operator at high energy:

$$\mathrm{tr}\left[-\mathrm{d}_q^2 + U(q) - E\right]^{-1} = \sum_{n=0} E^{-1/2-n}\Pi_n(U),$$

where the $\Pi_n(U)$ are local polynomials in the potential U.

We have thus shown that for L large the effective action Σ has, at one-loop order, a local expansion. Similar arguments apply at higher orders in the loop expansion, and therefore $\Sigma(\varphi)$ has a local expansion to any finite order. Moreover we have a kind of power counting property

$$\varphi^2(\tau) \sim t \sim \omega^2 \sim -\partial_\tau^2 \sim \left(uL^{4-d}\right)^{2/3}. \tag{36.59}$$

In addition each new loop brings a factor uL^{4-d}. Therefore at loop-order l a term with $2k$ powers of φ, m powers of t and $2n$ derivatives is of order $(uL^{4-d})^\kappa$ with $\kappa = l + 2(k + m + n - 2)/3$. For $d > 4$ all loop corrections have a negative power of L. The dominant correction for large L comes from the term proportional to $\int \mathrm{d}\tau\,\varphi^2(\tau)$ which vanishes like $L^{-2(d-4)/3}$.

Remark. Here again, although the scaling relation (36.53) has been proven only for $tL^{2(d-1)/3}$ finite, it has the correct large argument behaviour. For $t > 0$

$$X(z) \underset{z\to+\infty}{\sim} z^{-1/2} \;\Rightarrow\; \xi_L \sim t^{-1/2}.$$

For $t < 0$ the behaviour changes drastically depending whether the symmetry is continuous or discrete as we discuss in detail in Subsections 36.5.2 and A36.1.2. For $N > 2$ we observe that the lowest eigenvalues of the hamiltonian (36.51) can, for $z \to -\infty$, be obtained by approximating $\hat{H}$ by the angular moment part, fixing the radial coordinate at $|\mathbf{q}| = \sqrt{-6z}$. The corresponding eigenvalues are then $\ell(\ell + N - 2)/(-12z)$. It follows

$$X(z) \underset{z\to-\infty}{\sim} -\frac{12z}{N-1} \;\Rightarrow\; \frac{\xi_L}{L} \sim -\frac{12}{u}\frac{t}{N-1}L^{d-2} \propto \frac{L^{d-2}}{\xi_\infty^2(t)},$$

a result which can be compared with (36.90).

For $N = 1$ instead, instantons are responsible for the splitting between the two lowest-lying states (Chapter 40). A WKB analysis yields

$$\ln X(z) \sim 2(-2z)^{3/2} \;\Rightarrow\; \ln\xi_L \sim 2(-2t)^{3/2}\frac{L^{d-1}}{u} = \frac{2}{u}\frac{L^{d-1}}{\xi_\infty^3(t)}, \tag{36.60}$$

a result which can be compared with (A36.18).

36.4.2 Dimensions $d = 4 - \varepsilon$

For $d = 4 - \varepsilon$, at leading order, we can replace u by its IR fixed point value u^* which is of order ε. We then use the preceding considerations to analyze the leading corrections for ε small. We have seen that the expansion parameter is actually uL^{4-d}, which for $d > 4$ is small because L is large, while here it is small because u is of order ε. Therefore the power counting argument given above transforms into an argument about the powers of ε. An interaction term generated at loop order l with m powers of t, $2k$ fields ϕ and $2n$ time derivatives is multiplied by ε^κ:

$$\kappa = l + 2(k + m + n - 2)/3, .$$

The leading one-loop term is thus proportional to $\int d\tau\, \varphi^2(\tau)$ and of order $\varepsilon^{1/3}$. Note that the leading two-loop correction, which we neglect in this calculation, is of order $\varepsilon^{4/3}$. Since the one-loop contribution to the two-point function is a constant, no term proportional to $\dot\varphi^2$ is generated. The one-loop terms proportional to $\int d\tau\, \left(\varphi^2(\tau)\right)^2$ and $t \int d\tau\, \varphi^2(\tau)$ are of order ε . The $\int d\tau\, \left(\varphi^2(\tau)\right)^3$ term and the term with two derivatives coming from the 4-point function are of order $\varepsilon^{5/3}$ and can be neglected.

From now on the discussion closely follows the lines of Section 36.3. We set $L = 1$ and $u = u^*$. Calling $a_1 u^*$ the coefficients of $t \int d\tau\, \varphi^2(\tau)$, $a_2 u^*$ of $\int d\tau\, \varphi^2(\tau)$ and $a_3 u^{*2}$ of $\int d\tau\, \left(\varphi^2(\tau)\right)^2$, we obtain a one-loop corrected hamiltonian (36.48):

$$H = \frac{\mathbf{p}^2}{2L^{d-1}} + L^{d-1}\left[\frac{1}{2}\left(t(1 + a_1 u^*) + a_2 u^*\right)\mathbf{q}^2 + \frac{1}{4!}\left(u^* + a_3 u^{*2}\right)\left(\mathbf{q}^2\right)^2\right].$$

Using equation (36.53) and the RG result (36.10), we find the finite size correlation length ξ_L

$$\xi_L(t') = L\left(u^* + a_3 u^{*2}\right)^{-1/3} X\left(t'L^{1/\nu} + b\right), \tag{36.61}$$

$$b = a_2 u^*\left(u^* + a_3 u^{*2}\right)^{-2/3} = a_2\left(\frac{48\pi^2\varepsilon}{N+8}\right)^{1/3} + O\left(\varepsilon^{4/3}\right),$$

where t'

$$t' = u^{*-2/3}\left[1 + t(a_1 - \tfrac{2}{3}a_3)\right],$$

is a renormalized temperature.

As in the case of the hypercubic geometry we observe that the contribution coming from a_3, which is of the same order as the two-loop contribution ($\varepsilon^{4/3}$), is negligible at this order. We thus only need the coefficient $\tilde a_2$ of $u^* \int d\tau\, \varphi^2/2$.

The coefficient of φ^2. The coefficient $\tilde a_2$ of $u^* \int d\tau\, \varphi^2/2$ in the expansion of the expression (36.58) is:

$$a_2 = \frac{N+2}{6}\int \frac{d\omega}{2\pi} {\sum_{k_\mu}}' \frac{1}{\omega^2 + (2\pi\mathbf{k})^2}. \tag{36.62}$$

We can integrate over ω. We then use the formal identity:

$${\sum_{k_\mu}}' \frac{1}{|\mathbf{k}|} = \frac{1}{\sqrt\pi}\int_0^{+\infty} \frac{ds}{\sqrt s}\left(A^{d-1}(s) - 1\right), \tag{36.63}$$

in which the function $A(s)$ has been defined by equation (36.37). The integral has a divergence for s small which, as we have seen in Subsection 36.3.2, is cancelled by the critical temperature shift. We then find:

$$a_2 = \frac{N+2}{24\pi^{3/2}} \int_0^{+\infty} \frac{ds}{\sqrt{s}} \left[A^3(s) - 1 - (\pi/s)^{3/2} \right] + O\left(\varepsilon\right).$$

(36.64)

The finite size correlation length at T_c. Substituting the value of a_2 in (36.61) we obtain in particular the finite size correlation length at T_c at one-loop order:

$$\frac{\xi_L}{L} = \left(\frac{48\pi^2 \varepsilon}{N+8} \right)^{-1/3} X \left[K \pi^{-5/6} \frac{N+2}{12} \left(\frac{6\varepsilon}{N+8} \right)^{1/3} \right] (1 + O(\varepsilon)),$$

(36.65)

with:

$$K = \int_0^{+\infty} \frac{ds}{\sqrt{s}} \left[A^3(s) - 1 - (\pi/s)^{3/2} \right] = -5.0289788.$$

(36.66)

36.5 Finite Size Effects in the Non-Linear σ-Model

In models with continuous symmetries, below T_c, the propagator corresponding to the Goldstone modes has a pole at zero momentum. In a finite volume or in a cylindrical geometry with periodic boundary conditions, the zero mode leads to IR divergences for all temperature below T_c. We thus have to separate the zero mode and treat it non-perturbatively. Let us consider the action of the non-linear σ-model (see Chapter 30):

$$S(\phi) = \frac{\Lambda^{d-2}}{2t} \int d^d x \, [\partial_\mu \phi(x)]^2,$$

(36.67)

with:

$$\phi^2(x) = 1.$$

(36.68)

The non-linear σ-model can be used to describe the IR behaviour at fixed temperature below T_c and fixed dimension $2 < d < 4$ in which case a cut-off Λ is required, or the whole physical range up to T_c but only within the framework of the $\varepsilon = d - 2$ expansion. In the latter case we shall trade the cut-off scale for a renormalized scale μ and use dimensional regularization.

We shall examine separately the two cases of the hypercubic and cylindrical geometries.

36.5.1 The hypercubic geometry

The zero mode can here be associated with the set of collective coordinates which parametrize the degenerate classical minima of the potential. The integration over the zero mode is then equivalent to an average over all directions, which restores the $O(N)$ symmetry broken by the choice of a classical minimum of the action around which perturbation theory is expanded. In the case of $O(N)$ invariant observables, the averaging is trivial and the result is simply that the zero mode has to be omitted in the perturbative expansion. We recall that for $d \geq 2$ these observables then have a finite limit when the size L becomes infinite. Let us, for illustration purpose, first consider the second moment m_2 of the spin distribution (equation (36.20)):

$$m_2 = L^{-d} \left\langle \int d^d x \, \phi(x) \cdot \phi(0) \right\rangle.$$

(36.69)

In terms of the functions (30.30,30.31) m_2 has the scaling form:

$$m_2 = M_0^2(t)\mu_2\big(L/\xi(t)\big). \tag{36.70}$$

For $d > 2$ m_2 is finite at the critical temperature t_c and thus

$$\mu_2(z) \underset{z\to 0}{\propto} z^{-(d-2+\eta)}.$$

To calculate m_2 we parametrize the field $\phi(x)$ in terms of $\pi(x)$:

$$\phi(x) = \left\{ \sqrt{1-\pi^2(x)}\,, \quad \pi(x) \right\}. \tag{36.71}$$

It is convenient to set

$$L^{-2d} \int d^d x\, d^d y\, \phi(x)\cdot\phi(y) = 1 - 2\mathcal{O}. \tag{36.72}$$

At leading order

$$\mathcal{O} = \tfrac{1}{4} L^{-2d} \int d^d x\, d^d y\, [\pi(x)-\pi(y)]^2, \quad \text{with} \tag{36.73}$$

$$\langle \pi_i(x)\pi_j(0)\rangle = \delta_{ij}\Delta(x), \qquad \Delta(x) \equiv t\Lambda^{2-d}L^{-d} \sum_{\mathbf{p}=2\pi\mathbf{k}/L}' \frac{e^{i\mathbf{p}\cdot\mathbf{x}}}{\mathbf{p}^2}. \tag{36.74}$$

We therefore find at one-loop order

$$m_2 = 1 - 2\langle\mathcal{O}\rangle = 1 + t(N-1)(\Lambda L)^{2-d} \int d^d x \sum_{\mathbf{k}}' \frac{(e^{i2\pi\mathbf{k}\cdot\mathbf{x}/L}-1)}{(2\pi k)^2} + O\left(t^2\right). \tag{36.75}$$

After integration over x the expression becomes:

$$m_2 = 1 - \frac{t(N-1)}{4\pi^2}(\Lambda L)^{2-d} \sum_{\mathbf{k}}' \frac{1}{k^2} + O\left(t^2\right), \tag{36.76}$$

a regularization being implied. At this order for $d > 2$ the divergence is removed by dividing m_2 by the square of the spontaneous magnetization M_0

$$M_0 = 1 - \frac{t\Lambda^{2-d}}{2(2\pi)^d} \int \frac{d^d p}{p^2} + O\left(t^2\right) = 1 - \frac{t\Lambda^{2-d}}{2(2\pi)^d} \int_0^\infty ds \left(\frac{\pi}{s}\right)^{d/2}. \tag{36.77}$$

As usual we transform the sum over $\mathbf{k}$ into an integral involving the function (36.37). Taking into account equation (36.77) we find for the scaling function (36.70):

$$\mu_2 = 1 - \frac{t(N-1)}{4\pi^2}(\Lambda L)^{2-d} \int_0^\infty ds \left(A^d(s) - 1 - (\pi/s)^{d/2}\right) + O\left(t^2\right). \tag{36.78}$$

RG arguments tell us that we can set $L=1$ and replace t by the size-dependent coupling t_L (equation (36.14)), which at leading order at $d > 2$ is given by

$$t_L = t(\Lambda L)^{2-d} + O\left(t^2\right),$$

We then obtain

$$\mu_2 = 1 - \frac{t_L(N-1)}{4\pi^2} \int_0^\infty ds \left(A^d(s) - 1 - (\pi/s)^{d/2} \right) + \cdots ,$$

expression valid for $2 < d < 4$ and $t < t_c$ fixed. Since then t_L goes to zero for L large, this expression shows how μ_2 approaches 1. Using the estimate (36.16) we find

$$\mu_2 - 1 \sim -\frac{N-1}{4\pi^2} \int_0^\infty ds \left(A^d(s) - 1 - (\pi/s)^{d/2} \right) \left(\frac{\xi(t)}{L} \right)^{d-2} . \tag{36.79}$$

To generate a universal amplitude we again consider the ratio (36.30) $\mathcal{R}_\sigma = m_\sigma/(m_2)^{\sigma/2}$. With the definition (36.72)

$$\left(L^{-d} \left| \int d^d x\, \phi(x) \right| \right)^\sigma = 1 - \sigma \mathcal{O} + \frac{\sigma(\sigma-2)}{2} \mathcal{O}^2 + \cdots .$$

Therefore

$$\mathcal{R}_\sigma - 1 \sim \tfrac{1}{2}\sigma(\sigma-2) \left[\langle \mathcal{O}^2 \rangle - (\langle \mathcal{O} \rangle)^2 \right] .$$

Using equations (36.73,36.74) we obtain

$$\mathcal{R}_\sigma = 1 + \tfrac{1}{8}\sigma(\sigma-2)(N-1)L^{-2d} \int d^d x\, d^d y \left[\Delta(x) - \Delta(y) \right]^2 ,$$

$$= 1 + \tfrac{1}{4}\sigma(\sigma-2)(N-1)\frac{t^2(\Lambda L)^{4-2d}}{(2\pi)^4} \int_0^\infty ds\, s \left[A^d(s) - 1 \right] + O\left(t^3 \right) .$$

We then use RG arguments, set $L = 1$ and replace t by t_L:

$$\mathcal{R}_\sigma = 1 + \dot\sigma(\sigma-2)\frac{N-1}{4} \frac{t_L^2}{(2\pi)^4} \int_0^\infty ds\, s \left[A^d(s) - 1 \right] . \tag{36.80}$$

For $d > 2$ and $t < t_c$ fixed, t_L goes to zero for L large. Using the estimate (36.16) we thus find how $\mathcal{R}_\sigma$ goes to 1:

$$\mathcal{R}_\sigma = 1 + \sigma(\sigma-2)\frac{N-1}{4} \frac{1}{(2\pi)^4} \int_0^\infty ds\, s \left[A^d(s) - 1 \right] \left(\frac{\xi(t)}{L} \right)^{2d-4} \tag{36.81}$$

The neighbourhood of the critical temperature can be studied only within the $\varepsilon = d - 2$ expansion. Setting $t_L = t_c$ we obtain the universal ratio $\mathcal{R}_\sigma$ at order ε^2:

$$\mathcal{R}_\sigma = 1 + \varepsilon^2 \sigma(\sigma-2)\frac{N-1}{4(N-2)^2} \frac{1}{(2\pi)^2} \int_0^\infty ds\, s \left[A^2(s) - 1 \right] . \tag{36.82}$$

Induced magnetization in a small field. In a small magnetic field h, i.e. such that hL^d is small, we can easily calculate the magnetization M at leading order. Since $\langle \phi(x) \rangle$ is not an $O(N)$ invariant observable, the average over the zero mode is not trivial. RG equations predict:

$$M(h, t, L) = M_0(t)m \left[\frac{hM_0(t)}{t}\xi^d(t), \frac{L}{\xi(t)} \right] . \tag{36.83}$$

At leading order, the partition function in a field is given by:

$$Z(h,t,L) = \int d^N \phi_0 \, \delta \left(\phi_0^2 - 1 \right) \exp \left(L^d \, \mathbf{h} \cdot \boldsymbol{\phi}_0 / t \right)$$

$$= \frac{1}{\pi} \int_0^\pi d\theta \, \sin^{N-2} \theta \, \exp \left(L^d h \cos \theta / t \right). \tag{36.84}$$

The integral is a modified Bessel function. The magnetization is the logarithmic derivative of Z. At this order M depends only on the scaling variable v:

$$v = \frac{h M_0(t) L^d}{t}, \tag{36.85}$$

and is given by:

$$M(h,t,L) = M_0(t) \frac{d}{dv} \ln \int_0^\pi d\theta \, \sin^{N-2} \theta \, e^{v \cos \theta}. \tag{36.86}$$

36.5.2 The cylindrical geometry

Let us now consider, as an illustration, the calculation of the finite size correlation length in a cylindrical geometry. At leading order the action of the zero mode is the action of the $O(N)$ rigid rotator which we have already discussed in Section 3.4:

$$S(\varphi) = \frac{\Lambda^{d-2} L^{d-1}}{2t} \int d\tau \, \dot{\varphi}^2(\tau). \tag{36.87}$$

From equation (3.61) we infer the correlation length at low temperature, at leading order:

$$\xi_L(t) = \frac{2}{N-1} \frac{\Lambda^{d-2} L^{d-1}}{t}. \tag{36.88}$$

Combining this result with the RG arguments of Subsection 36.1.2, we can rewrite this equation

$$\frac{\xi_L(t)}{L} \sim \frac{2}{(N-1)t_L}. \tag{36.89}$$

For $d > 2$ in the ordered phase t_L goes to zero and we find the scaling form:

$$\frac{\xi_L(t)}{L} \underset{\text{for } 2<d<4}{\sim} \frac{2}{N-1} \left(\frac{L}{\xi_\infty(t)} \right)^{d-2}. \tag{36.90}$$

We immediately learn that in a system in which a continuous symmetry is broken, at any temperature below t_c, the finite size correlation length grows like L^{d-1}. This behaviour has to be contrasted with the behaviour of the correlation length in the case of discrete symmetry (see Subsection A36.1.2).

The critical temperature is a RG fixed point and thus $t_L = t_c$. We then obtain the universal ratio ξ_L/L at leading order in $\varepsilon = d - 2$

$$\frac{\xi_L(t_c)}{L} \underset{d \to 2}{\sim} \frac{N-2}{(N-1)\pi\varepsilon}.$$

Finally in two dimensions for $L/\xi_\infty(t)$ small we find

$$\frac{\xi_L(t)}{L} \sim \frac{N-2}{(N-1)\pi} \ln(\xi_\infty(t)/L).$$

One-loop corrections. We now calculate the one-loop corrections to this result for $d > 2$ below t_c, at t_c in an ε-expansion or for $d = 2$ in a low temperature expansion. Since we need the size-dependent temperature t_L defined by (36.14) at one-loop order, let us relate it directly to the divergence of one-loop diagrams. In what follows it will be convenient to define the renormalized coupling constant in terms of the coefficient of θ^2 in the angle-dependent free energy. At one-loop order the relation between bare t and renormalized coupling t_r at scale μ takes naturally the form

$$\frac{\mu^{d-2}}{t_r} = \Lambda^{d-2}\left(\frac{1}{t} - \frac{\beta_2(d)}{d-2} + O(t)\right), \quad \beta_2(d) = \frac{N-2}{2\pi} + O(d-2).$$

Differentiating with respect to Λ, at renormalized parameters fixed, we obtain an equation for the RG $\beta(t)$ function (Section 30.3) which yields

$$\beta(t) = (d-2)t - \beta_2(d)t^2 + O\left(t^3\right).$$

Solving then equation (36.14) perturbatively we find

$$\frac{1}{t_L} = \frac{(\Lambda L)^{d-2}}{t} - \frac{\beta_2(d)}{d-2}\left[(\Lambda L)^{d-2} - 1\right] + O(t). \tag{36.91}$$

We note also that for $d-2$ small $t_c \sim (d-2)/\beta_2(d)$.

To calculate the one-loop corrections to (36.89) we extend to higher dimensions the method explained in Section 3.4. Instead of explicitly constructing the effective hamiltonian H, we evaluate the eigenvalues of the hamiltonian directly from the functional integral. We consider a system hypercubic in $d-1$ dimensions of linear size L, and of size β in the "time" direction, with $\beta \gg L$. In the time direction we impose fixed twisted boundary conditions (which of course break translation invariance):

$$\phi(\tau = 0, \mathbf{x}) = \phi_1, \quad \phi(\tau = \beta, \mathbf{x}) = \phi_2, \quad \mathbf{x} \in \mathbb{R}^{d-1}, \tag{36.92}$$

in which ϕ_1 and ϕ_2 are two constant vectors such that:

$$\phi_1 \cdot \phi_2 = \cos\theta, \quad 0 \le \theta < \pi. \tag{36.93}$$

The corresponding partition function is given by:

$$Z(L, \beta, \theta) = \left\langle \phi_2 \left| e^{-\beta H} \right| \phi_1 \right\rangle. \tag{36.94}$$

By projecting $Z(L, \beta, \theta)$ over the hyperspherical polynomials $P_l^N(\cos\theta)$ (see equation (3.58)), we obtain the eigenvalues of H. We choose a frame such that:

$$\begin{aligned} \phi_1 &= [1, 0; \mathbf{0}], \\ \phi_2 &= [\cos\theta, \sin\theta; \mathbf{0}]. \end{aligned} \tag{36.95}$$

We then parametrize the field $\phi(\tau, x)$ by:

$$\phi(\tau, x) = \begin{cases} \cos(\theta\tau/\beta)\,\sigma_1(\tau, x) - \sin(\theta\tau/\beta)\,\sigma_2(\tau, x), \\ \sin(\theta\tau/\beta)\,\sigma_1(\tau, x) + \cos(\theta\tau/\beta)\,\sigma_2(\tau, x), \\ \pi(\tau, x), \end{cases} \tag{36.96}$$

in which the field $\pi(\tau, x)$ here has only $N - 2$ components. With this parametrization the boundary conditions (36.92) take the simple form:

$$\sigma_1 = 1, \quad \sigma_2 = 0, \quad \pi = 0 \quad \text{for } \tau = 0 \text{ and } \tau = \beta, \tag{36.97}$$

and the solution to the classical field equation satisfying (36.97) is then $\sigma_1 = 1, \sigma_2 = 0,$ $\pi = 0$. Finally the transformation (36.96) is a rotation. Therefore the three fields σ_1, σ_2, π still satisfy the constraint:

$$\sigma_1^2 + \sigma_2^2 + \pi^2 = 1, \tag{36.98}$$

and the integration measure in the functional integral is left invariant.

The action $S(\phi)$ in the new fields reads:

$$S(\sigma_1, \sigma_2, \pi) = \frac{\Lambda^\varepsilon}{2t} \int d\tau\, d^{d-1}x \left[\frac{\theta^2}{\beta^2}(\sigma_1^2 + \sigma_2^2) + \dot{\sigma_1}^2 + \dot{\sigma_2}^2 + \dot{\pi}^2 \right.$$
$$\left. + 2\frac{\theta}{\beta}(\dot{\sigma}_2\sigma_1 - \dot{\sigma}_1\sigma_2) + (\partial_i\sigma_1)^2 + (\partial_i\sigma_2)^2 + (\partial_i\pi)^2 \right]. \tag{36.99}$$

To calculate the effects of small fluctuations around the classical minimum $\sigma_1 = 1, \sigma_2 = 0, \pi = 0$, we eliminate the field σ_1 using the constraint (36.98):

$$\sigma_1 = \left(1 - \sigma_2^2 - \pi^2\right)^{1/2}, \tag{36.100}$$

and expand the action in powers of σ_2 and π. The quadratic part of the action $S_2(\sigma_2, \pi)$ needed for the one-loop calculation is then:

$$S_2(\sigma_2, \pi) = \frac{\theta^2\Lambda^\varepsilon}{2t\beta}L^{d-1} + \frac{\Lambda^\varepsilon}{2t} \int d\tau\, d^{d-1}x \left[-\frac{\theta^2}{\beta^2}\pi^2 + \dot{\sigma}_2^2 + \dot{\pi}^2 + (\partial_i\sigma_2)^2 + (\partial_i\pi)^2 \right]. \tag{36.101}$$

At this order the integration over σ_2 gives a factor independent of θ which can be absorbed into the normalization of the functional integral. The integral over π yields a determinant to the power $(N - 2)/2$. Hence the result at one-loop order is:

$$-\ln\frac{Z(L, \beta, \theta)}{Z(L, \beta, 0)} = \frac{\theta^2}{2t\beta}\Lambda^\varepsilon L^{d-1} + \tfrac{1}{2}(N - 2)\,\text{tr}\ln\left[\left(-\frac{\theta^2}{\beta^2} - \partial_\tau^2 - \partial_i^2 \right)\left(-\partial_\tau^2 - \partial_i^2 \right)^{-1} \right]. \tag{36.102}$$

The leading term of the free energy shows that the relevant values of θ in the θ integral are of order $t^{1/2}$. At one-loop order we thus need $\ln Z$ only at order θ^2. We set

$$-\ln\frac{Z(L, \beta, \theta)}{Z(L, \beta, 0)} = \theta^2 F_2(t, L, \beta) + O\left(\theta^4\right). \tag{36.103}$$

The eigenmodes of the operator $-\partial_\tau^2 - \partial_i^2$, given the boundary conditions (36.97) in the time direction, and periodic boundary conditions in the remaining $d-1$ dimensions, are

$$\frac{m^2\pi^2}{\beta^2} + \left(\frac{2\pi\mathbf{k}}{L}\right)^2, \quad \mathbf{k} \in \mathbb{Z}^{d-1},$$

and m is a positive integer. It follows:

$$F_2(t, L, \beta) = \frac{L^{d-1}}{2t\beta}\Lambda^\varepsilon - \frac{N-2}{2\beta^2}\sum_{\mathbf{k}}\sum_{m=1}^{\infty}\left[\frac{m^2\pi^2}{\beta^2} + \left(\frac{2\pi\mathbf{k}}{L}\right)^2\right]^{-1}. \tag{36.104}$$

The sums have to be understood with a large momentum cut-off. We now evaluate expression (36.104) for β large. It is convenient to treat the $\mathbf{k} = \mathbf{0}$ mode separately. We eventually find:

$$\sum_{\mathbf{k}}\sum_{m=1}^{\infty}\left[\frac{m^2\pi^2}{\beta^2} + \left(\frac{2\pi\mathbf{k}}{L}\right)^2\right]^{-1} \underset{\beta\to\infty}{=} \frac{\beta^2}{6} + \sum_{\mathbf{k}}'\frac{\beta L}{4\pi|\mathbf{k}|} + O(1). \tag{36.105}$$

It follows

$$F_2(t, L, \beta) = -\frac{N-2}{12} + \frac{L}{2\beta}\left(\frac{(\Lambda L)^{d-2}}{t} - \frac{N-2}{4\pi}\sum_{\mathbf{k}}'\frac{1}{|\mathbf{k}|}\right).$$

Comparing this expression with expression (3.57), expanded in powers of θ, we find that the only effect of the one-loop corrections is to rescale the parameter β. Using equations (3.60,3.61) we conclude:

$$\frac{\xi_L(t)}{L} = \frac{2}{N-1}\left(\frac{(\Lambda L)^{d-2}}{t} - \frac{N-2}{4\pi}\sum_{\mathbf{k}}'\frac{1}{|\mathbf{k}|}\right).$$

The cut-off dependence can be is eliminated by subtracting the infinite size limit. We use the identity:

$$\sum_{\mathbf{k}}'\frac{1}{|\mathbf{k}|} - \int\frac{d^{d-1}\mathbf{p}}{|\mathbf{p}|} = \frac{1}{\sqrt{\pi}}\int_0^\infty\frac{ds}{\sqrt{s}}\left[A^{d-1}(s) - 1 - (\pi/s)^{(d-1)/2}\right]. \tag{36.106}$$

Therefore:

$$\frac{\xi_L}{L} = \frac{2}{N-1}\left[(L\Lambda)^{d-2}\left(\frac{1}{t} - \frac{\beta_2(d)}{d-2}\right) - \frac{N-2}{4\pi^{3/2}}\int_0^\infty\frac{ds}{\sqrt{s}}\left(A^{d-1}(s) - 1 - \left(\frac{\pi}{s}\right)^{(d-1)/2}\right)\right]$$
$$+ O(t).$$

We finally rewrite this expression in terms of the size-dependent coupling constant t_L given by (36.91)

$$\frac{\xi_L(t)}{L} = \frac{2}{N-1}\left[\frac{1}{t_L} - \frac{\beta_2(d)}{d-2} - \frac{N-2}{4\pi^{3/2}}\int_0^\infty\frac{ds}{\sqrt{s}}\left(A^{d-1}(s) - 1 - \left(\frac{\pi}{s}\right)^{(d-1)/2}\right)\right] + O(t_L). \tag{36.107}$$

We now specialize this expression to various situations. For $d > 2$ and $t < t_c$ fixed or $d = 2$, $t \to 0$, we obtain the correction to the leading behaviour when t_L goes to zero. To obtain the result at t_c we have to expand in $\varepsilon = d - 2$. The subtracted integral diverges for s large for $d \to 2$

$$\int_0^\infty \frac{ds}{\sqrt{s}} \left(A^{d-1}(s) - 1 - (\pi/s)^{(d-1)/2} \right) = \int_0^\infty \frac{ds}{\sqrt{s}} \left(A(s) - 1 - \theta(\pi - s)\sqrt{\pi/s} \right) - \frac{2\sqrt{\pi}}{\varepsilon}.$$

Note that the expression in the r.h.s. of (36.107) has thus, as expected, a limit for $d \to 2$. The remaining integral can be explicitly calculated

$$\frac{1}{\sqrt{\pi}} \int_0^\infty \frac{ds}{\sqrt{s}} \left(A(s) - 1 - \theta(\pi - s)\sqrt{\pi/s} \right) = \gamma - \ln 4\pi,$$

where γ is Euler's constant. At the fixed point t_c we have $t_L = t_c$. We need t_c at order ε^2, and it is easy to verify that the combination $1/t_c - \beta_2/\varepsilon$ has a limit independent of the regularization scheme. Using the expression (30.54), coming from the minimal subtraction scheme, we then find the universal ratio:

$$\frac{\xi_L(t_c)}{L} = \frac{N-2}{N-1} \frac{1}{\pi\varepsilon} \left[1 + \frac{\varepsilon}{N-2} - \frac{\varepsilon}{2}(\gamma - \ln 4\pi) + O\left(\varepsilon^2\right) \right]. \tag{36.108}$$

36.6 Finite Size Effects and Dynamics

We have discussed dynamics in Chapters 17 and 35 from the renormalization group point of view. Previous considerations immediately apply to the dynamics in a finite volume. Like in the case of static properties, finite size effects are characterized by the dependence in the scaling variable $L/\xi(T)$. For example if the IR fixed point is not gaussian, in the critical domain, the correlation time $\tau(T, L)$ in a finite volume of linear size L has a scaling form which generalizes equation (35.39):

$$\tau(T, L) = L^z f \left[(T - T_c) L^{1/\nu} \right]. \tag{36.109}$$

Physical quantities, for example in a periodic hypercube, can be calculated by methods analogous to those used in the case of the cylindrical geometry in previous sections, the time being now the physical time instead of one of the spatial directions. We here give a few examples of calculations of correlation times in the simple case of a purely dissipative model without conservation laws, based on the static N-vector model.

Notation. Because in this section t denotes time, we denote by $\Delta T \propto T - T_c$ the deviation from the critical temperature.

36.6.1 The $\left(\varphi^2\right)^2$ theory

We consider the Langevin equation (35.9) with gaussian white noise $\boldsymbol{\nu}$:

$$\dot{\varphi}(x, t) = -\frac{\Omega}{2} \left[(-\Delta + r_c + \Delta T) \varphi(x, t) + \frac{u}{3!} \varphi \varphi^2 \right] + \nu(x, t), \tag{36.110}$$

and:

$$\langle \nu_i(t) \rangle = 0, \qquad \langle \nu_i(t) \nu_j(t') \rangle = \Omega \delta_{ij} \delta(t - t'). \tag{36.111}$$

We recall that to the Langevin equation (36.110) we can associate a supersymmetric dynamic action $S(\phi)$:

$$S(\phi) = \int dt\, d\theta\, d\bar{\theta} \left[\int d^d x \frac{2}{\Omega} \frac{\partial \phi}{\partial \theta} \cdot \left(\frac{\partial \phi}{\partial \bar{\theta}} - \theta \frac{\partial \phi}{\partial t} \right) + \mathcal{H}(\phi) \right], \tag{36.112}$$

in which $\mathcal{H}(\varphi)$ is the static euclidean action:

$$\mathcal{H}(\varphi) = \int d^d x \left[\frac{1}{2} (\partial_\mu \varphi)^2 + \frac{1}{2} (r_c + \Delta T)\varphi^2 + \frac{u}{4!} (\varphi^2)^2 \right], \tag{36.113}$$

and ϕ the superfield:

$$\phi = \varphi + \bar{\theta} \mathbf{c} + \bar{\mathbf{c}} \theta + \bar{\theta}\theta \lambda. \tag{36.114}$$

In a periodic hypercube, at the critical temperature we have to separate the zero mode. At leading order the effective action for the zero mode is just obtained calculating the action (36.112) for a space-independent field. We then recognize the action associated with a stochastic differential equation of a type studied in Chapter 4. The corresponding Fokker–Planck hamiltonian H_{FP}, in hermitian form, is (equation (4.40)):

$$H_{\mathrm{FP}} = \frac{\Omega}{2L^d} \left[\mathbf{p}^2 + \tfrac{1}{4} (\nabla E(\mathbf{q}))^2 - \tfrac{1}{2} \Delta E(\mathbf{q}) \right], \tag{36.115}$$

with, at leading order:

$$E(\mathbf{q}) = L^d \left[\tfrac{1}{2} \Delta T \mathbf{q}^2 + \frac{u}{4!} (\mathbf{q}^2)^2 \right]. \tag{36.116}$$

Rescaling $\mathbf{q}$:

$$\mathbf{q} \mapsto \mathbf{q} \left(L^d u \right)^{-1/4}, \tag{36.117}$$

we find that the eigenvalues E_i of H_{FP}, as functions of L and the deviation of the critical temperature ΔT, take the form:

$$E_i = \Omega u^{1/2} L^{-d/2} e_i \left[L^{d/2} \Delta T u^{-1/2} \right]. \tag{36.118}$$

The scaling for $d > 4$ of the relaxation time τ, which is the inverse of the difference of the two first eigenvalues, follows:

$$\tau(T, L) = \Omega^{-1} u^{-1/2} L^{d/2} f \left[L^{d/2} \Delta T u^{-1/2} \right]. \tag{36.119}$$

As for the finite size correlation length (see equation (36.53)), the naive extrapolation of the scaling form valid for $d < 4$ is incorrect. An analysis similar to the one performed in the static case shows that loop corrections do not modify the scaling form for $d > 4$.

As for the correlation length, let us examine the behaviours of τ when $\Delta T L^{d/2}$ is large. For $\Delta T > 0$ we find the expected limit $\tau^{-1} \to \Omega \Delta T$. For $\Delta T < 0$ and $N > 1$ we obtain

$$\tau \sim -\frac{12}{(N-1)u} \Omega^{-1} L^d \Delta T.$$

Note, however, that for $N = 1$ the behaviour found in this way is incorrect, as we discuss in Subsection A36.1.3.

The relaxation time in $d = 4 - \varepsilon$ dimensions. Let us now calculate the relaxation time in $4 - \varepsilon$ dimensions, at the IR fixed point. The calculations are performed in the minimal subtraction scheme, using the supersymmetric formulation of Appendix A17.1. Following the lines of Section 36.4, we perform a rescaling in the action, equivalent to (36.117):

$$\phi \mapsto \phi \left(L^d u\right)^{-1/4} , \quad t \mapsto t \left(L^d / u\right)^{1/2} , \quad \boldsymbol{\theta} \mapsto \boldsymbol{\theta} \left(L^d / u\right)^{1/4} , \tag{36.120}$$

and count the powers of ε. We again verify that only the coefficients of ϕ^2, $\Delta T \phi^2$ and $\left(\phi^2\right)^2$ are relevant at one-loop order. In addition the scaling of the time t and Grassmann variables $\boldsymbol{\theta}$ shows that we need only the contributions proportional to $(1, \Delta T) \int dt d\theta d\bar{\theta} \, \phi^2$ and $\int dt \, d\theta \, d\bar{\theta} \left(\phi^2\right)^2$. It is easy to verify that the calculation then becomes identical to the static calculation in a finite volume. The main result is given by equation (36.43). The relaxation time, at the IR fixed point, in the one-loop approximation, is then:

$$\tau(\Delta T, L) = \Omega'^{-1} L^z f \left(L^{1/\nu} \Delta T' + b\right) , \tag{36.121}$$

in which $f(z)$ is the function implicitly defined by equation (36.119), and Ω'^{-1} and $\Delta T'$ are a renormalized time scale and a renormalized deviation from the critical temperature respectively.

36.6.2 The non-linear σ-model

With the non-linear σ-model, we can calculate for example the relaxation time at fixed temperature below T_c, close to T_c in a $d - 2$ expansion or in two dimensions. The form of the dynamic action has been given in Appendix A17.3:

$$S(\phi) = \frac{\Lambda^\varepsilon}{g} \int d\theta \, d\bar{\theta} dt \int d^d x \left[\frac{2}{\Omega} \frac{\partial \phi}{\partial \theta} \cdot \left(\frac{\partial \phi}{\partial \bar{\theta}} - \theta \frac{\partial \phi}{\partial t}\right) + \frac{1}{2} (\partial_\mu \phi)^2\right] , \tag{36.122}$$

with:

$$\phi^2 = 1 . \tag{36.123}$$

To avoid confusions we here denote the coupling constant or temperature by g.

The RG equations of this model have been discussed in Section 35.2. The relaxation time, defined in the infinite volume by equation (35.52), satisfies the RG equation:

$$\left(\Lambda \frac{\partial}{\partial \Lambda} + \beta(g) \frac{\partial}{\partial g} + \eta_\omega(g) \Omega \frac{\partial}{\partial \Omega}\right) \tau(\Lambda, L, g, \Omega) = 0 . \tag{36.124}$$

The finite size relaxation time τ satisfies the dimensional relation:

$$\tau(\Lambda, L, g, \Omega) = \Omega^{-1} \Lambda^{-2} \tau(1, \Lambda L, g, 1) . \tag{36.125}$$

The RG equation (36.124) can thus be rewritten:

$$\left(L \frac{\partial}{\partial L} + \beta(g) \frac{\partial}{\partial g} - 2 - \eta_\omega(g)\right) \tau(\Lambda, L, g, \Omega) = 0 . \tag{36.126}$$

We have now set $\Lambda = 1$ and simplify the notation $\tau(\Lambda, L, g, \Omega) \mapsto \tau(L, g)$ because the dependence in Ω is trivial. Solving this equation by the method of characteristics, we find:

$$\tau(L, g) = L^2 \zeta(L, g) \tau\left(1, g_L\right), \tag{36.127}$$

with the notation $g_L \equiv g(1/L)$ (see Subsection 36.1.2) and

$$\ln(L) = \int_{g_L}^{g} \frac{dg'}{\beta(g')}, \tag{36.128}$$

$$\zeta(L,g) = \exp\left[\int_{g_L}^{g} \frac{\eta_\omega(g')}{\beta(g')} dg'\right]. \tag{36.129}$$

The equation (36.127) can also be rewritten:

$$\tau(L,g) = L^2 \zeta(L,g) \mathcal{T}(L/\xi(g)), \tag{36.130}$$

in which $\xi(g)$ is the infinite volume correlation length. As we have already shown in Section 36.1, at fixed coupling $g < g_c$, the effective temperature g_L goes to zero. Therefore τ can be derived from perturbation theory.

Note finally that the function $\eta_\omega(g)$ begins at order g^2 (see Section A17.3). Since we calculate only at one-loop order, the function $\zeta(L,g)$ can be replaced in what follows by a constant renormalization factor $\zeta(g)$:

$$\zeta(g) = \exp\left[\int_0^{g} \frac{\eta_\omega(g')}{\beta(g')} dg'\right]. \tag{36.131}$$

To calculate τ we can use the method explained in Subsection 36.5.2. The only difference appears in the form of the solution of the field equation which satisfies the boundary conditions. If we impose:

$$\begin{aligned}
\phi(\tau = 0, \boldsymbol{\theta} = \mathbf{0}, x) &= [1, 0; \mathbf{0}], \\
\phi(\tau = \beta, \boldsymbol{\theta} = \mathbf{0}, x) &= [\cos\alpha, \sin\alpha; \mathbf{0}],
\end{aligned} \tag{36.132}$$

the rotation which transforms the solution of the field equation into a constant is:

$$\phi(t, \boldsymbol{\theta}, x) = \begin{cases} \cos\left[\frac{\alpha}{\beta}\left(t + \frac{\bar{\theta}\theta}{2}\right)\right] \sigma_1(t, \boldsymbol{\theta}, x) - \sin\left[\frac{\alpha}{\beta}\left(t + \frac{\bar{\theta}\theta}{2}\right)\right] \sigma_2(t, \boldsymbol{\theta}, x), \\ \sin\left[\frac{\alpha}{\beta}\left(t + \frac{\bar{\theta}\theta}{2}\right)\right] \sigma_1(t, \boldsymbol{\theta}, x) + \cos\left[\frac{\alpha}{\beta}\left(t + \frac{\bar{\theta}\theta}{2}\right)\right] \sigma_2(t, \boldsymbol{\theta}, x), \\ \pi(t, \boldsymbol{\theta}, x). \end{cases} \tag{36.133}$$

We then note that, as a consequence of equation (36.123), the terms of the action which explicitly depend on α are

$$\frac{\alpha^2}{2\beta^2 \Omega g} \left(L^d \beta - \int dt\, d^d x\, \pi^2(t, \boldsymbol{\theta} = \mathbf{0}, x)\right).$$

We immediately obtain the form of the relaxation time, at leading order for large L, from the classical contribution and renormalization group:

$$\tau(L,g) \sim \Omega'^{-1}(g) \frac{L^2}{g_L}, \quad \text{with } \Omega'(g) = \Omega \zeta^{-1}(g) \frac{N-1}{2}, \tag{36.134}$$

where $\zeta(g)$ is defined by equation (36.131).

Furthermore, in the one-loop order calculation, we can omit the field σ_2 and replace all components of $\pi(t, \theta, x)$ other than $\pi(t, 0, x)$ by their classical value. In terms of this last field the action reduces to:

$$S(\pi) = \frac{\alpha^2}{2\beta\Omega g} L^d + \frac{1}{2\Omega g} \int dt\, d^d x \left[\dot{\pi}^2 + \frac{\Omega^2}{4} (\triangle\pi)^2 - \frac{\alpha^2}{\beta^2} \pi^2 \right]. \tag{36.135}$$

The integration over the π-field yields the partition at one-loop order. Expanding for α small we find

$$F_2(L, \beta) \sim -\frac{1}{\alpha^2} \ln \frac{Z(L, \beta, \alpha)}{Z(L, \beta, 0)}$$

$$= \frac{L^d}{2\beta\Omega g} - \frac{N-2}{2\beta^2} \sum_{\mathbf{k}} \sum_{m=1}^{\infty} \left[\frac{m^2\pi^2}{\beta^2} + \left(\frac{2\pi^2\Omega\mathbf{k}^2}{L^2} \right)^2 \right]^{-1}. \tag{36.136}$$

From here on we follow very closely the method of Subsection 36.5.2. For large β the r.h.s. has the expansion:

$$F_2(L, \beta) = -\frac{N-2}{12} + \frac{L^2}{2\beta\Omega} \left(\frac{L^{d-2}}{g} - \frac{N-2}{4\pi^2} {\sum_{\mathbf{k}}}' \frac{1}{\mathbf{k}^2} \right). \tag{36.137}$$

The relaxation time follows

$$\frac{\Omega\tau(g, L)}{L^2} = \frac{2}{N-1} \left(\frac{L^{d-2}}{g} - \frac{N-2}{4\pi^2} {\sum_{\mathbf{k}}}' \frac{1}{\mathbf{k}^2} \right). \tag{36.138}$$

The sum has to be understood with a cut-off. As we have explained in Subsection 36.5.2, if we subtract to the sum its infinite size limit we obtain a finite result

$$\sum_{\mathbf{k}}' \frac{1}{\mathbf{k}^2} - \int \frac{d^d\mathbf{p}}{\mathbf{p}^2} = \int_0^\infty ds \left[A^d(s) - 1 - (\pi/s)^{d/2} \right].$$

We then introduce the size-dependent temperature g_L (equation (36.91) with $t \mapsto g$ and $\Lambda = 1$) and find

$$\frac{\tau(g, L)}{L^2} = \Omega'^{-1}(g) \left(\frac{1}{g_L} - \frac{\beta_2(d)}{d-2} - \frac{N-2}{4\pi^2} \int_0^\infty ds \left[A^d(s) - 1 - (\pi/s)^{d/2} \right] \right), \tag{36.139}$$

with $\Omega'(g)$ defined in (36.134). We have thus obtained the first correction to the leading term for $g < g_c$. Note again that this expression has a finite limit when $d \to 2$. Indeed

$$\int_0^\infty ds \left[A^d(s) - 1 - (\pi/s)^{d/2} \right] = -\frac{2\pi}{d-2} + \int_0^\infty ds \left[A^2(s) - 1 - \theta(\pi - s)\frac{\pi}{s} \right]. \tag{36.140}$$

Hence we also obtain the form of the leading correction for $d = 2$ and $\xi(t)/L$ large. Finally we can calculate the value at g_c in an ε-expansion but the result is proportional to a time scale.

Bibliographical Notes

Early articles and reviews about finite size effects are

M.E. Fisher in *Critical Phenomena, Proceedings of the International School of Physics Enrico Fermi*, Varenna 1971, M.S. Green ed. (Academic Press, New York 1972); M.E. Fisher and M.N. Barber, *Phys. Rev. Lett.* 28 (1972) 1516; Y. Imry and D. Bergman, *Phys. Rev.* A3 (1971) 1416.

See also

M.N. Barber in *Phase Transitions and Critical Phenomena* vol. 8, C. Domb and J. Lebowitz eds. (Academic Press, New York 1983),

and the papers collected in

Finite Size Scaling, J.L. Cardy ed. (North-Holland, Amsterdam 1988).

The field theoretical analysis has been performed in

E. Brézin, *J. Physique (Paris)* 43 (1982) 15; E. Brézin and J. Zinn-Justin, *Nucl. Phys.* B257 [FS14] (1985) 867; J. Rudnick, H. Guo and D. Jasnow, *J. Stat. Phys.* 41 (1985) 353; M. Lüscher, *Phys. Lett.* 118B (1982) 391; *Nucl. Phys.* B219 (1983) 233.

The correlation length is calculated with twisted boundary conditions in

E.G. Floratos and D. Petscher, *Nucl. Phys.* B252 (1985) 689.

A RG analysis is also given in

V. Privman and M.E. Fisher, *J. Stat. Phys.* 33 (1983) 385.

The almost planar interface model is studied in

D.J. Wallace and R.K.P. Zia, *Phys. Rev. Lett.* 43 (1979) 808; D.J. Wallace in *Recent Advances in Field Theory and Statistical Mechanics*, Les Houches 1982, R. Stora and J.-B. Zuber eds. (North-Holland, Amsterdam 1984).

Finite size scaling laws in Critical Dynamics have been postulated in

M. Suzuki, *Prog. Theor. Phys.* 58 (1977) 1142.

A systematic RG analysis is given in

J.C. Niel and J. Zinn-Justin, *Nucl. Phys.* B280 [FS18] (1987) 335; Y.Y. Goldschmidt, *Nucl. Phys.* B280[FS18] (1987) 340.

APPENDIX 36

A36.1 Discrete Symmetries and Finite Size Effects

We have already characterized finite size effects in the case of second order phase transitions, in the critical domain. We now consider, for completeness, a few examples of finite size effects in the absence of critical fluctuations: first order phase transitions, general phase transitions below the critical temperature. We recall that in the case of discrete symmetries, the infinite size correlation length is finite in the ordered phase.

A36.1.1 Finite volume

We here consider first order phase transitions in which the order parameter jumps from one constant value to another. We restrict ourselves to the calculation of homogeneous quantities like the magnetization. Then the basic function we need is the free energy $-S(\varphi)$ at fixed field average

$$\exp\left[-\mathfrak{S}(\varphi)\right] = \int [d\phi(x)] \, \delta\left(\varphi - L^{-d}\int d^dx \, \phi(x)\right) \exp\left[-S(\phi)\right], \qquad (A36.1)$$

a quantity whose physical meaning has already been discussed in Section 6.6. In the case of first order transitions, fluctuations are not critical because the correlation length remains finite at the transition. Therefore the integration over the zero-momentum component does not solve a non-existent zero-mode problem, but instead the symmetry restoration problem.

The free energy, in presence of a constant magnetic field h, is given by:

$$e^{W(h)} = \int d\varphi \, \exp\left[-\mathfrak{S}(\varphi, L) + L^d \beta h \varphi\right]. \qquad (A36.2)$$

In a translation invariant finite system, $\mathfrak{S}$ at large size L behaves like:

$$\mathfrak{S}(\varphi) \sim L^d \Sigma(\varphi), \qquad (A36.3)$$

in which $\Sigma(\varphi)$ is independent of L.

For L large the integral can thus be calculated by steepest descent. In the case of a unique saddle point one finds

$$W(h)/L^d = -\Sigma(\varphi) + \beta h \varphi, \qquad (A36.4)$$

with

$$\Sigma'(\varphi) = \beta h.$$

Note that $\mathfrak{S}(\varphi, L)$ is such that the corrections for L large to $(A36.4)$ are exponentially small in L because the infinite size correlation length remains finite.

Degenerate minima. When instead several saddle points are found, $W(h)$ is the sum of saddle point contributions. As an example let us assume that we are discussing an Ising-like system $(\mathfrak{S}(\varphi) = \mathfrak{S}(-\varphi))$, in the ordered phase, in an infinitesimal magnetic field h. The minimum is then almost degenerate. We call $\pm M_0$ the two minima of $\Sigma(\varphi)$ (the generalization to any discrete set of minima is straightforward):

$$\Sigma(\varphi) = \Sigma(M_0) + \tfrac{1}{2}\Sigma''(M_0)(\varphi - M_0)^2 + O(\varphi - M_0)^3. \qquad (A36.5)$$

The free energy is now the sum of the two saddle point contributions:

$$W(h) - W(0) = \ln \cosh\left(\beta h L^d M_0\right) + \tfrac{1}{2}\beta^2 h^2 L^d / \Sigma''(M_0). \qquad (A36.6)$$

The magnetization M and the zero-field susceptibility χ are then given by:

$$M = \langle\varphi\rangle = (\beta L^d)^{-1} W'(h) = M_0 \tanh\left(\beta h L^d M_0\right) + \beta h / \Sigma''(M_0), \qquad (A36.7)$$
$$\chi = \beta / \Sigma''(M_0), \qquad (A36.8)$$

where χ is here defined as $\chi = \langle\varphi^2\rangle - (\langle|\varphi|\rangle)^2$.

In equation $(A36.7)$, for hL^d finite, the second term is negligible, and the finite volume magnetization takes the universal form:

$$M = M_0 \tanh\left(\beta h L^d M_0\right). \qquad (A36.9)$$

Let us note that in zero-field the situation is more subtle, as we discuss in next section. For $|\varphi| < M_0$ the dominant configurations are no longer almost uniform configurations but clusters of one phase inside the other. Let us call ρ the fraction of the total volume occupied by the $-M_0$ phase, $0 < \rho < 1$. The cost in energy is then proportional to the surface tension $\sigma(T)$ multiplied by the minimal area which is of the order $L^{d-1}\rho^{(d-1)/d}$. The average magnetization is $M_0(1 - 2\rho)$. For ρ small the region of transition between uniform and non-uniform configurations corresponds to the balance

$$\Sigma''(M_0)M_0^2\rho^2 L^d \propto \sigma(T)\rho^{(d-1)/d}L^{d-1} \quad \Rightarrow \quad \rho \propto \left(\frac{\xi(T)}{L}\right)^{d/(d+1)},$$

where the factor $\xi(T)$ is implied by dimensional considerations. When $L > \xi(T)$ $S(\varphi)$ is at leading order of the form

$$\mathfrak{S}(\varphi, L) = L^d \Sigma(M_0) + \text{const. } \sigma(T) L^{d-1}(M - M_0)^{(d-1)/d}.$$

Finally for ρ large enough in a finite volume with periodic boundary conditions the dominant configuration consists in two phases separated by two flat interfaces and then the cost in energy becomes constant and equal for an hypercube to $2\sigma(T)L^{d-1}$.

A36.1.2 Finite size correlation length in Ising-like systems below T_c

In Chapter 40 we shall relate the restoration of discrete symmetries in one dimension to the existence of instantons. Assuming the results of Chapter 40 known, we here indicate how these arguments extend to scalar field theories in higher dimensions in order to demonstrate the complete parallel between the low temperature analysis of Section 23.2 for the Ising model and the instanton analysis. The form of the finite size correlation length below T_c will follow.

We consider therefore the example of an effective euclidean action in d dimensions of the form:

$$S(\phi) = \int \mathrm{d}^d x \left[\tfrac{1}{2}(\partial_\mu \phi(x))^2 + V(\phi(x))\right], \qquad (A36.10)$$

where $V(\phi)$ is a potential which is invariant in the reflection:

$$\phi(x) \mapsto -\phi(x), \qquad (A36.11)$$

and has degenerate minima. It could for instance have the form of the double-well potential

$$V(\phi) \sim (\phi^2 - M_0^2)^2. \qquad (A36.12)$$

We want to evaluate the correlation ξ_L in the cylindrical geometry. We can deduce it from the ratio r (see equations (23.23) and (40.6)):

$$r = \operatorname{tr} P \, e^{-\beta H} / \operatorname{tr} e^{-\beta H}, \qquad (A36.13)$$

in which we recall that P is the operator which implements the reflection $(A36.11)$. The path integral representation of $\operatorname{tr} P e^{-\beta H_L}$ is

$$\operatorname{tr} P \, e^{-\beta H_L} = \int [d\phi(x,t)] \exp[-S(\phi)], \qquad (A36.14)$$

$$\text{with} \quad \phi(x, -\beta/2) = -\phi(x, \beta/2), \qquad (A36.15)$$

where in $\mathbb{R}^d$ we have chosen a time direction and called t the corresponding variable. With the boundary conditions $(A36.15)$ the leading configurations correspond to instanton configuration in the time direction

$$\phi_c(x, t) = f_c(t), \qquad (A36.16)$$

in which in the large β limit the function $f_c(t)$ interpolates between the two minima of the potential. The corresponding ratio r then has the form:

$$\ln r \sim -\sigma(T) L^{d-1}. \qquad (A36.17)$$

We have identified the classical action $S(f_c)$ with a surface tension which is traditionally denoted by $\sigma(T)$. This result is analogous to results obtained in Sections 23.2 for Ising-like systems. Instantons replace, in continuous systems, the walls of the lattice models. From r we derive the correlation length:

$$\ln \xi_L(T) \sim \sigma(T) L^{d-1}. \qquad (A36.18)$$

This behaviour of the finite size correlation length, characteristic of the breaking of discrete symmetries, has to be contrasted with the power law behaviour (36.90) found in the case of continuous symmetries. The equation $(A36.18)$ is valid for temperatures $T < T_c$, and the surface tension vanishes at T_c. The behaviour near T_c is given by equation (36.60) for $d > 4$ and by RG arguments for $2 < d < 4$:

$$\sigma(T) \propto (T_c - T)^{3/2} \propto [\xi_\infty(T)]^3 \text{ for } d > 4,$$
$$\propto (T_c - T)^{\nu(d-1)} \propto [\xi_\infty(T)]^{d-1} \text{ for } d < 4.$$

Remark. In the case of one dimensional instantons, to integrate out the fluctuations around the saddle point, one has to introduce collective coordinates corresponding to the position of the saddle point. In d dimensions, the position of the wall is defined by a function $\theta(x)$, which has also to be considered as a set of collective coordinates. Translation invariance implies that the action can only depend on the derivatives of $\theta(x)$. It has thus the form:

$$S_{\text{eff}}(\theta) \sim \int d^{d-1}x \left[(\partial_i \theta(x))^2 + O\left(|\partial_i \theta(x)|^4 \right) \right].$$ (A36.19)

The term with only two derivatives gives the leading contribution as long as the surface tension is large. When one approaches the critical point, terms with more derivatives become important. General euclidean invariance tells us that the effective action begins with a term proportional to the area of the wall and has thus the form:

$$S_{\text{eff}}(\theta) \sim \int d^{d-1}x \left[1 + (\partial_i \theta(x))^2 \right]^{1/2}.$$ (A36.20)

It has been conjectured by Wallace and Zia that such this model, the almost planar interface model, could describe the critical properties of the Ising model in $d = 1 + \varepsilon$ dimensions. It is easy to verify that the model has an UV fixed point of order ε and that the correlation exponent ν can be calculated as a series in ε. These properties remind us the non-linear σ-model.

A36.1.3 Dynamics in the ordered phase

Let us consider a purely dissipative dynamics associated with a static action of the form (A36.10), for example:

$$\mathcal{A}(\varphi) = \int d^d x \left[\tfrac{1}{2} (\partial_\mu \varphi(x))^2 + g \left(\varphi^2 - M_0^2 \right)^2 \right].$$ (A36.21)

We assume that our system evolves inside a hypercube of linear size L with periodic boundary conditions.

It is shown in Section 40.6 that the difference between the two first eigenvalues of the corresponding Fokker–Planck hamiltonian is of the order of

$$e^{-\Delta\mathcal{A}}, \quad \text{with} \quad \Delta\mathcal{A} = (\mathcal{A}_{\text{max}} - \mathcal{A}_{\text{min}}),$$

in which $\mathcal{A}_{\text{min}}$ is the value of $\mathcal{A}(\varphi)$ at the degenerate minima and $\mathcal{A}_{\text{max}}$ the value at the saddle point which separates them. We have already briefly analyzed this problem in Section 23.4. Let us start from a configuration in which φ is closed to M_0, and create a bubble of the phase $\varphi = -M_0$. Since the cost in energy is proportional to the area of the bubble, the saddle point will correspond to the situation in which the hypercube is evenly divided between the two phases. Due to the periodic boundary conditions, the minimal area surface which evenly divides the hypercube consists in two parallel sections perpendicular to the axes. Such a configuration corresponds to an instanton anti-instanton pair of the static action. Calling σ the instanton action in one dimension we thus find that the relaxation time τ_L, which is the inverse of the second eigenvalue, behaves like:

$$\tau_L \propto e^{2\sigma L^{d-1}}.$$ (A36.22)

Comparing with equation (A36.18) we find a simple relation between the relaxation time in a cubic geometry and the finite size correlation length ξ_L in a cylindrical geometry:

$$\tau_L \propto \xi_L^2.$$ (A36.23)

A36.2 Perturbation Theory in a Finite Volume

Let us add a few remarks concerning the calculation of Feynman diagrams in a finite volume. It is convenient to introduce Schwinger's parameters and write the momentum space propagator $\Delta(p)$,

$$\Delta(p) = \frac{1}{p^2 + \mu^2} = \int_0^\infty ds \, e^{-s(p^2+\mu^2)},$$

a method already used in the infinite volume limit. After this transformation gaussian integrals over momenta are replaced by infinite sum over integers which can no longer be calculated exactly. However, dimensional continuation can be defined, and the infinite size limit studied. In the chapter we have considered only simple one-loop diagrams D_γ which can be written

$$D_\gamma \equiv L^{-d} \sum_{p=2\pi k/L} (p^2 + \mu^2)^{-\sigma/2} = \frac{L^{-d}}{\Gamma(\sigma/2)} \int_0^\infty ds \, s^{\sigma/2-1} \sum_{p=2\pi k/L} e^{-s(p^2+\mu^2)}.$$

In terms of the function $A(s)$ defined by (36.37),

$$A(s) = \sum_{n=-\infty}^{+\infty} e^{-sn^2}, \tag{A36.24}$$

the sums can be written

$$D_\gamma = \frac{L^{-d}}{\Gamma(\sigma/2)} \int_0^\infty ds \, s^{\sigma/2-1} e^{-s\mu^2} A^d(4s\pi^2/L^2).$$

Poisson's formula is useful in this context. Let $f(x)$ be a function which has a Fourier transform

$$\tilde{f}(k) = \int dx \, f(x) \, e^{i2\pi kx}.$$

Then from

$$\sum_{k=-\infty}^{+\infty} e^{i2\pi kx} = \sum_{l=-\infty}^{+\infty} \delta(x-l),$$

follows Poisson's formula

$$\sum_{k=-\infty}^{+\infty} \tilde{f}(k) = \sum_{l=-\infty}^{+\infty} f(l). \tag{A36.25}$$

Applying this relation to the function e^{-sx^2} one finds the identity:

$$A(s) = (\pi/s)^{1/2} A\left(\pi^2/s\right). \tag{A36.26}$$

This identity shows, in particular, that the infinite size limit is approached exponentially when the mass μ is finite

$$A(s) - \int dp \, e^{-sp^2} \sim 2(\pi/s)^{1/2} e^{-\pi^2/s} \Rightarrow D_\gamma(L) - D_\gamma(L=\infty) \underset{L\to\infty}{\propto} (\mu L)^{(\sigma-d-1)/2} e^{-\mu L}.$$

More generally the function

$$A(s; \nu, \lambda) = e^{-s\nu^2} \theta_3(\lambda + i\nu s/\pi, e^{-s}) = \sum_n e^{-s(n+\nu)^2 + 2i\pi n\lambda}, \tag{A36.27}$$

where θ_3 is an elliptic Jacobi's function, satisfies

$$A(s; \nu, \lambda) = A(s; -\nu, -\lambda) = (\pi/s)^{1/2} A(\pi^2/s; \lambda, -\nu).$$

The exponential approach to the infinite size limit in the case of twisted boundary conditions follows.

37 INSTANTONS IN QUANTUM MECHANICS: THE ANHARMONIC OSCILLATOR

Up to now, to expand euclidean functional integrals in perturbation theory, in the absence of sources we have always looked for a saddle point (a solution to the classical equations of motion) corresponding to a constant field and then expanded the action around such a solution. However classical field equations may have non-constant solutions. In stable theories the value of the classical action for these solutions is larger than for the constant solutions because the gradient term always gives a positive contribution. If the difference in action is infinite the solutions give no contribution to the functional integral. However we shall find examples in which the difference is finite. These solutions are called, in the case of euclidean field theories, *instanton solutions*. We shall show that they are related to the physics of barrier penetration. The expansion around these non-trivial solutions leads to a semiclassical calculation of tunnelling effects, alternative to the WKB method of Quantum Mechanics, which can be easily generalized to Field Theory.

Dyson had conjectured that the behaviour of perturbation theory at large orders in QED was related to the vacuum instability for negative values of e^2 (e being the electric charge). Considering the quantum mechanical example of the quartic anharmonic oscillator, Bender and Wu showed that the behaviour of perturbation theory at large orders could indeed be obtained from a semiclassical calculation of barrier penetration effects occurring when the sign of the coupling constant is changed. We shall show in Chapter 41 how these results can be recovered from instantons, and how they can be generalized to a large class of potentials (provided one considers also complex finite action solutions). As first shown by Lipatov, instanton calculus allows also to extend this analysis to field theory. We shall therefore derive in Chapters 41, 42 the large order behaviour of the perturbative expansion of several field theories. We shall show how this knowledge can be used to efficiently sum perturbation theory. Remarkably enough the main application of these summation methods is the accurate calculation of critical exponents. The results have been given in Chapter 28.

Finally let us mention that, although we shall only deal with euclidean field theory, many aspects of the techniques we shall develop apply also to the calculation of effects coming from finite energy solutions of the real time field equations, called *soliton* solutions in the literature.

In this chapter we explain the role of instantons in unstable theories on the simple example of the quartic anharmonic oscillator in quantum mechanics and calculate their contribution at leading order.

The appendix contains some additional remarks about collective coordinates.

37.1 The Anharmonic Oscillator for Negative Coupling: Preliminary Considerations

Let us consider the hamiltonian H of the quartic anharmonic oscillator:

$$H = -\tfrac{1}{2} \left(\mathrm{d}/\mathrm{d}q \right)^2 + \tfrac{1}{2}q^2 + \tfrac{1}{4}gq^4. \tag{37.1}$$

The ground state energy $E_0(g)$ can be obtained from the large β limit of the partition function $\operatorname{tr} e^{-\beta H}$:

$$E_0\left(g\right) = \lim_{\beta \to +\infty} -\frac{1}{\beta} \ln \operatorname{tr} e^{-\beta H}.$$

Moreover a systematic expansion of the partition function for β large also yields the energies of the excited states. We can thus derive the eigenvalues of H from the path integral representation of the partition function:

$$\operatorname{tr} e^{-\beta H} = \int_{q(-\beta/2)=q(\beta/2)} [dq(t)] \exp\left[-S\left(q(t)\right)\right], \tag{37.2}$$

where $S\left(q\right)$ is the euclidean action:

$$S\left(q\right) = \int_{-\beta/2}^{\beta/2} \left[\tfrac{1}{2}\dot{q}^2(t) + \tfrac{1}{2}q^2(t) + \tfrac{1}{4}gq^4(t)\right] dt. \tag{37.3}$$

Generalization of arguments applicable to finite dimensional integrals indicates that the path integral (37.2) defines a function of g analytic in the half plane $\operatorname{Re}\left(g\right) > 0$. In this domain the integral is dominated for g small by the saddle point $q(t) \equiv 0$. Therefore it can be calculated by expanding the integrand in powers of g and integrating term by term. This leads to the perturbative expansion of the partition function and thus of the ground state $E_0\left(g\right)$ by taking the large β limit.

Remarks.

(i) Note that we always expand in g before taking the large β limit. Since $E_N(g)$, the Nth eigenvalue of H, satisfies:

$$E_N\left(g\right) = N + \tfrac{1}{2} + O\left(g\right),$$

the perturbative expansion can be written:

$$\begin{aligned} \operatorname{tr} e^{-\beta H} &= \sum_{N=0} e^{-\beta E_N(g)} \\ &= \sum_{N=0} e^{-(N+1/2)\beta} \sum_{k=0} \frac{1}{k!}(-\beta)^k \left(E_N - \tfrac{1}{2} - N\right)^k. \end{aligned} \tag{37.4}$$

We observe: that $E_N(g)$ can be deduced from the coefficient of $e^{-(N+1/2)\beta}$, that the coefficient of g^k is a polynomial of degree k in β.

(ii) As we have already mentioned when discussing the ϕ^4 field theory, by rescaling $q(t)$:

$$q(t) \mapsto q(t)g^{-1/2},$$

we factorize the whole dependence in g in front of the action:

$$S\left(q\right) = \frac{1}{g}S\left(q\sqrt{g}\right). \tag{37.5}$$

The coupling constant g plays the same formal role as $\hbar$ in the semiclassical expansion.

Negative coupling. For $g < 0$ the hamiltonian is unbounded from below for all values of g. Therefore the energy levels, considered as analytic functions of g, must have a singularity at $g = 0$ and the perturbation series is always divergent.

A wave function $\psi(t)$, localized at initial time $t = 0$ (t is here the *real physical time* of the Schrödinger equation) in the well of the potential near $q = 0$, decays due to barrier penetration. To calculate the decay rate of the wave function $\psi(t)$ we can use the following method: For g positive the time-dependent solution $\psi_0(t)$ of the Schrödinger equation associated with the ground state energy E_0 behaves like:

$$\psi_0(t) \sim e^{-iE_0 t}.$$

We then proceed by analytic continuation in the complex g plane to go to $g < 0$. After analytic continuation E_0 becomes complex and thus $\psi_0(t)$ decreases exponentially with time at a rate

$$|\psi_0(t)| \sim e^{-|\operatorname{Im} E_0|t}.$$

$|\operatorname{Im} E_0|$ is the inverse life-time of the wave function $\psi(t)$. Actually the decay of $\psi(t)$ also involves the imaginary parts of the continuations of all excited states. However we expect on intuitive grounds that, when the real part of the energy increases, the corresponding lifetime decreases (this can easily be verified by examples). Thus at large times only the component corresponding to the ground state survives. We shall therefore hereafter calculate $\operatorname{Im} E_0$ for g small and negative.

To now understand how we can define and evaluate $E_0(g)$ for g negative, we first study this problem on the example of a simple integral: the "zero-dimensional ϕ^4 field theory".

37.2 A Simple Integral

We consider the integral which counts the number of Feynman diagrams contributing to the partition function in the ϕ^4 field theory:

$$I = \frac{1}{\sqrt{2\pi}} \int_{-\infty}^{+\infty} e^{-\left(x^2/2 + gx^4/4\right)} \, dx. \tag{37.6}$$

For g positive and small, the integral is dominated by the saddle point at the origin:

$$I = 1 + O(g). \tag{37.7}$$

The function $I(g)$ is analytic in a cut plane. To be able to perform the analytic continuation of the integral to $g < 0$, it is necessary to rotate the contour of integration C as one changes the argument of g:

$$C: \ \operatorname{Arg} x = -\tfrac{1}{4} \operatorname{Arg} g \qquad (\mathrm{mod} \ \pi),$$

so that $\operatorname{Re}\left(gx^4\right)$ remains positive. As a consequence, one obtains two different expressions for $I(g)$ depending on the direction in which one has rotated in the g-plane:

$$\text{for } g = -|g| + i0: \quad I(g) = \int_{C_+} e^{-\left(x^2/2 + gx^4/4\right)} \, dx$$

$$\text{with } C_+: \quad \operatorname{Arg} x = -\frac{\pi}{4} \quad (\mathrm{mod} \ \pi), \tag{37.8}$$

$$\text{for } g = -|g| - i0: \quad I(g) = \int_{C_-} e^{-\left(x^2/2 + gx^4/4\right)} \, dx$$

$$\text{with } C_-: \quad \operatorname{Arg} x = \frac{\pi}{4} \quad (\mathrm{mod} \ \pi). \tag{37.9}$$

For $g \to 0_-$, the two integrals are still dominated by the saddle point at the origin since the contribution of the other saddle points:

$$x + gx^3 = 0 \Rightarrow x^2 = -1/g\,, \tag{37.10}$$

are of the order

$$\mathrm{e}^{-\left(x^2/2 + gx^4/4\right)} \sim \mathrm{e}^{1/4g} \ll 1\,. \tag{37.11}$$

Let us however calculate the discontinuity of $I(g)$ on the cut, which is given by the difference between the two integrals:

$$I\left(g + i0\right) - I\left(g - i0\right) = 2i\,\mathrm{Im}\,I\left(g\right) = \frac{1}{\sqrt{\pi}} \int_{C_+ - C_-} \mathrm{e}^{-\left(x^2/2 + gx^4/4\right)}\,\mathrm{d}x\,. \tag{37.12}$$

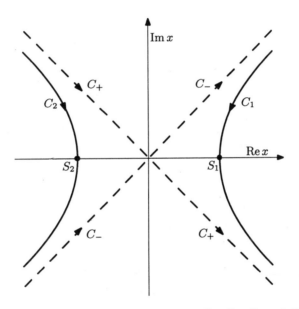

Fig. 37.1   The contours of integration C_+, C_-, C_1 and C_2.

It corresponds to the contour $C_+ - C_-$ which, as figure 37.1 shows, can be deformed into the sum of the contours C_1 and C_2 passing through the non-trivial saddle points S_1 and S_2: $x = \pm 1/\sqrt{-g}$. This means that the contribution of the saddle point at the origin cancels, and that the integral is now dominated by these saddle points:

$$\mathrm{Im}\,I(g) \sim \mathrm{e}^{1/4g}\,. \tag{37.13}$$

Thus for g negative and small, the real part of the integral is given by perturbation theory, while the exponentially small imaginary part is given by the contribution of non-trivial saddle points.

Let us now try to use the same strategy for the path integral (37.2).

37.3 Quantum Mechanics: The Saddle Point

Inspired by the previous example we rotate the contour in the functional $q(t)$ space, while we change the argument g from g positive to g negative:

$$q(t) \mapsto q(t)\,e^{-i\theta},$$

in which θ is time independent. If we return to the definition of the path integral as a limit of integrals in a discretized time (see Chapter 2), we understand that this is a quite sensible procedure.

There is however one difference with the case of the simple integral: we have to stay in a domain in which $\mathrm{Re}\left[\dot{q}^2(t)\right] > 0$, since the kinetic part $\int \dot{q}^2(t)dt$, as we have discussed in Chapter 42, favours the smooth paths and ensures therefore the existence of the continuum limit of the discretized path integral.

For g negative we thus integrate along the path:

$$\mathrm{Arg}\,q(t) = -\theta, \quad \tfrac{1}{8}\pi < \theta < \tfrac{1}{4}\pi, \tag{37.14}$$

which satisfies the two conditions:

$$\mathrm{Re}\left[gq^4(t)\right] > 0, \qquad \mathrm{Re}\left[\dot{q}^2(t)\right] > 0. \tag{37.15}$$

For g small, the two path integrals corresponding to the two analytic continuations are here also dominated by the saddle point at the origin

$$q(t) = 0,$$

but if we calculate the difference between the two integrals, this contribution cancels. We have to look for non-trivial saddle points, which are solutions of the euclidean equation of motion:

$$-\ddot{q}(t) + q(t) + gq^3(t) = 0 \quad (g < 0), \tag{37.16}$$
$$q(-\beta/2) = q(\beta/2). \tag{37.17}$$

The contribution of the constant saddle point:

$$q^2(t) = -1/g, \tag{37.18}$$

is of the order of $e^{\beta/4g}$ and therefore negligible in the large β limit. We have to look for solutions which have an action which remains finite for $\beta \to +\infty$. These are called *instantons*.

The solutions of equations (37.16,37.17) represent a periodic motion in *real time* in the potential $-V(q)$

$$V(q) = \tfrac{1}{2}q^2 + \tfrac{1}{4}gq^4. \tag{37.19}$$

It is clear that solutions to the equation of motion exist which correspond to oscillations around the minima of $-V$, $q = \pm\sqrt{-1/g}$. Integrating once equation (37.16) we find:

$$\tfrac{1}{2}\dot{q}^2 - \tfrac{1}{2}q^2 - \tfrac{1}{4}gq^4 = \epsilon,$$

with $\epsilon < 0$. Calling q_0 and q_1 the points with $q > 0$ where the velocity $\dot{q}$ vanishes, we find for the period of such a solution:

$$\beta = 2 \int_{q_0}^{q_1} \frac{dq}{\sqrt{q^2 + \frac{1}{2}gq^4 + 2\epsilon}}.$$

β can become large only if the constant ϵ and thus q_0 go to zero. The classical trajectory then comes closer and closer to the origin. In the infinite β limit the classical solution becomes:

$$q_c(t) = \pm \left(-\frac{2}{g}\right)^{1/2} \frac{1}{\cosh(t - t_0)}. \tag{37.20}$$

The corresponding classical action is:

$$S[q_c(t)] = -\frac{4}{3g} + O\left(e^{-\beta}/g\right). \tag{37.21}$$

Since the euclidean action is invariant under time translations, the classical solution depends on a free parameter t_0 which for β finite varies between 0 and β,

$$0 \le t_0 < \beta.$$

Therefore, in contrast to the simple integral, we do not find two degenerate saddle points but two one-parameter families.

Notice also that we could have considered trajectories oscillating n times around $q^2 = -1/g$ in the time interval β. It is easy to verify that the corresponding action in the large β limit becomes:

$$S(q_c) = -n4/3g, \tag{37.22}$$

and yields therefore a contribution proportional to $e^{n4/3g}$. For g small the path integral is dominated by the term $n = 1$.

37.4 Instanton Contribution at Leading Order

The gaussian approximation. To evaluate the contribution, at leading order, of the saddle points let us expand the action around a saddle point and perform the gaussian integration (one-loop order). Setting:

$$q(t) = q_c(t) + r(t),$$

we find:

$$\operatorname{Im} \operatorname{tr} e^{-\beta H} \propto \frac{1}{i} e^{4/3g} \int [dr(t)] \exp\left(-\frac{1}{2}\int dt_1 dt_2\, r(t_1)\, M(t_1, t_2)\, r(t_2)\right),$$

$$\propto \frac{1}{i} e^{4/3g} (\det M)^{-1/2},$$

for $g \to 0_-$, $\beta \to \infty$. The operator M is given by:

$$M(t_1, t_2) = \frac{\delta^2 S}{\delta q_c(t_1)\, \delta q_c(t_2)} = \left[-\left(\frac{d}{dt_1}\right)^2 + 1 + 3gq_c^2(t_1)\right] \delta(t_1 - t_2). \tag{37.23}$$

The zero mode. Let us differentiate equation (37.16) with respect to t:

$$-\left(\frac{\mathrm{d}}{\mathrm{d}t}\right)^2 \dot{q}_c(t) + \dot{q}_c(t) + 3g q_c^2(t)\dot{q}_c(t) = 0. \tag{37.24}$$

Since the function $\dot{q}_c(t)$ is square integrable, this equation implies that $\dot{q}_c(t)$ is an eigenvector of M with eigenvalue zero:

$$M\dot{q}_c = 0. \tag{37.25}$$

Hence the naive gaussian approximation yields a result, proportional to $(\det M)^{-1/2}$, which is infinite!

We should not be too surprised by this result: as we have noted above, due to translation invariance in time, we have two one-parameter families of continuously connected degenerate saddle points. An infinitesimal variation of $q(t)$ which corresponds to a variation of the parameter t_0, i.e. proportional to $\dot{q}_c$, leaves the action unchanged. The problem we here face is by no means special to path integrals as the following example shows.

37.4.1 Zero modes in simple integrals

Let us again consider a simple integral:

$$I = \int \mathrm{d}^\nu \mathbf{x} \, e^{\mathbf{x}^2 - g(\mathbf{x}^2)^2}. \tag{37.26}$$

in which $\mathbf{x}$ is an ν-component vector, and the integrand is $O(\nu)$ invariant. For g small, this integral can be calculated by steepest descent. The saddle point is given by

$$\mathbf{x}_c\left(1 - 2g\mathbf{x}_c^2\right) = 0, \tag{37.27}$$

that is

$$|\mathbf{x}_c| = (2g)^{-1/2}. \tag{37.28}$$

We here also find a $\nu - 1$ parameter family of degenerate saddle points, since only the length of the vector $\mathbf{x}_c$ is determined by the saddle point equation. If we single out one saddle point and evaluate its contribution in the gaussian approximation, we are led to calculate the determinant of the matrix $M_{\alpha\beta}$:

$$M_{\alpha\beta} = 8g x_\alpha x_\beta, \tag{37.29}$$

which is the projector on $\mathbf{x}$ and has therefore $(\nu - 1)$ zero eigenvalues.

The solution to this problem is here obvious: it is necessary to calculate the integral over angular variables exactly; only the integral over the radial variable can be evaluated by the steepest descent method.

In the case of the path integral it is also necessary to integrate exactly over the parameters which describe the saddle points, in the example above the time translation parameter. A suitable set of integration variables has to be found in which the time parameter explicitly appears. This is the method of so-called *collective coordinates*.

Remark. We have already studied theories possessing a symmetry group in which the classical minimum is not invariant under the group, for example the $\left(\phi^2\right)^2$ field theory in the ordered phase:

$$S[\phi] = \frac{1}{2}\int \mathrm{d}^d x \left[\frac{1}{2}\left(\partial_\mu \phi(x)\right)^2 + \frac{r}{2}\phi^2(x) + \frac{1}{4!}g\left(\phi^2(x)\right)^2\right]. \tag{37.30}$$

In this case we have generally chosen one classical minimum and made a systematic expansion around it. However this is a sensible procedure only when the symmetry is spontaneously broken. We have actually seen in Chapter 30 that the absence of symmetry breaking manifests itself in perturbation theory by the appearance of IR singularities. In the case of instanton solutions we have shown above that the propagator in an instanton field has an isolated pole at the origin. Therefore all terms in perturbation theory are infinite. We conclude that the time translation symmetry is not spontaneously broken and that it is necessary to sum over all degenerate saddle points (see also Chapter 23).

37.4.2 Collective coordinates and gaussian integration

In order to take a variable associated with time translation as an integration variable, we set:

$$
\begin{cases}
q(t) = q_c\,(t - t_0) + r\,(t - t_0)\,, \\
r(t) = \displaystyle\sum_{n=1}^{\infty} q_n f_n(t)\,,
\end{cases}
\tag{37.31}
$$

in which t_0 is no longer a parameter, but forms, together with the set $\{q_n\}$ the new set of integration variables. We impose to the functions $\{f_n\}$ to form an orthonormal basis up to a point. An infinitesimal variation of t_0 adds to $q(t)$ a term proportional to $\dot{q}_c$. If we want our new set $\{t_0, \{q_n\}\}$ to consist only of independent variables, we have to impose a constraint. We choose:

$$
\int \dot{q}_c\,(t - t_0)\,f_n\,(t - t_0)\,\mathrm{d}t = 0 \quad \forall n\,.
\tag{37.32}
$$

We see that the normalized eigenvectors of the hermitian operator $M\,(t_1, t_2)$ corresponding to the non-vanishing eigenvalues satisfy all the conditions imposed to the functions $f_n(t)$ since they are orthogonal to the last eigenvector $\dot{q}_c$ and form with $\dot{q}_c$ a complete basis. With this choice the argument of the exponential in the r.h.s. takes a very simple form:

$$
\tfrac{1}{2} \int \mathrm{d}t_1 \mathrm{d}t_2\, r\,(t_1)\, M\,(t_1, t_2)\, r\,(t_2) = \tfrac{1}{2} \sum_{n>0} m_n q_n^2\,,
\tag{37.33}
$$

in which $\{m_n\}$ forms the set of all non-vanishing eigenvalues of M.

It is useful at this point to remember that the functional measure $[\mathrm{d}q(t)]$ can be defined in the continuum as the flat integration measure on the coefficients c_m of the expansion of $q(t)$ on an orthonormal basis of square integrable functions (see equation (2.41)):

$$
q(t) = \sum_{m=0}^{\infty} c_m g_m(t), \qquad g_m(t) \in \mathcal{L}^2,
$$

$$
[\mathrm{d}q(t)] = \mathcal{N} \prod_{m=0}^{\infty} \mathrm{d}c_m\,.
$$

The jacobian of the transformation which goes from the set $\{c_m\}$ to the set $\{t_0, \{q_n\}\}$ is given by (see Appendix A37):

$$
J = \left| \int f_0\,(t - t_0)\, \frac{\mathrm{d}}{\mathrm{d}t_0} \left[q_c\,(t - t_0) + \sum_{n=1}^{\infty} f_n\,(t - t_0)\, q_n \right] \mathrm{d}t \right|\,.
\tag{37.34}
$$

At leading order in g we can neglect the dependence in q_n:

$$J \sim \|\dot{q}_c\| = \left[\int \dot{q}_c^2(t)\mathrm{d}t \right]^{1/2}. \tag{37.35}$$

Since the integrand does not depend on t_0, the integration over the collective coordinate t_0 simply yields a factor β. The integration over the variables q_n yields $(\det' M)^{-1/2}$, in which $\det' M$ is the product of all non-vanishing eigenvalues of M, i.e. the determinant of M in the subspace orthogonal to $\dot{q}_c$.

The normalization. To normalize the path integral, we compare it to its value at $g = 0$ (the harmonic oscillator), which, in the large β limit, reduces to $\mathrm{e}^{-\beta/2}$. At $g = 0$ the operator M becomes the operator M_0:

$$M_0(t_1, t_2) = \left[-\left(\frac{\mathrm{d}}{\mathrm{d}t_1} \right)^2 + 1 \right] \delta(t_1 - t_2). \tag{37.36}$$

When comparing the instanton contribution and the reference path integral corresponding to the harmonic oscillator, we have to remember that we have excluded one mode from the gaussian integration in the case of the instanton. We have therefore a difference of one gaussian integration between the two path integrals. We have to divide by a factor:

$$\int_{-\infty}^{+\infty} \mathrm{e}^{-\lambda^2/2} \, \mathrm{d}\lambda = (2\pi)^{1/2},$$

the instanton contribution.

Taking into account the two families of saddle points and collecting all factors we obtain:

$$\operatorname{Im} \operatorname{tr} \mathrm{e}^{-\beta H} \sim \frac{2}{2i} \left[\det' M \, (\det M_0)^{-1} \right]^{-1/2} J \frac{\beta}{\sqrt{2\pi}} \, \mathrm{e}^{-\beta/2} \, \mathrm{e}^{4/3g}. \tag{37.37}$$

37.4.3 The result at leading order

In Appendix A38.2, we shall compare this calculation with the corresponding WKB calculation. We shall indirectly show that for all systems for which we can solve explicitly the classical equations of motion with arbitrary boundary conditions, we can also explicitly calculate the determinant of the operator governing the small fluctuations around the classical trajectory. In the special case considered here, M happens to be the hamiltonian corresponding to an exactly soluble Bargmann potential, and one finds:

$$\det(M + \varepsilon)(M_0 + \varepsilon)^{-1} = \frac{\sqrt{1 + \varepsilon} - 1}{\sqrt{1 + \varepsilon} + 1} \frac{\sqrt{1 + \varepsilon} - 2}{\sqrt{1 + \varepsilon} + 2}. \tag{37.38}$$

Simple arguments tell us that the ground state wave function has no node, the wave function of the first excited state one node, etc. The wave function $\dot{q}_c(t)$ vanishes once at the turning point, therefore it does not correspond to the ground state. There exists one other state which is the ground state, and M has therefore one negative eigenvalue, as is apparent on expression (37.38).

It can be verified, by expressing the determinant of M as a product of eigenvalues that:

$$\det' M \, (\det M_0)^{-1} = \lim_{\varepsilon \to 0} \frac{1}{\varepsilon} \det \left(M + \varepsilon \right) \left(M_0 + \varepsilon \right)^{-1} . \qquad (37.39)$$

Then from (37.38) we conclude

$$\det' M \, (\det M_0)^{-1} == -\frac{1}{12} . \qquad (37.40)$$

We still have an ambiguity of sign since we have to take the square root of expression (37.38). This ambiguity can only be resolved by following the analytic continuation from g positive to g negative. However in both cases the square root of the determinant is imaginary, so that the final result is real, as expected.

The jacobian J is easily calculable, but the important point is that J^2 is proportional, as is the classical solution itself, to $1/g$. As we shall see later, this is the first example of a general situation: each time the instanton solution breaks some continuous symmetry of the classical action, the solution depends on parameters generated by the action of the symmetry group on the solution. Each parameter has to be taken as an integration variable and the corresponding jacobian generates as a factor the loop expansion parameter to the power $-1/2$. Here we find:

$$J = \frac{2}{\sqrt{3}} \frac{1}{\sqrt{-g}} . \qquad (37.41)$$

The expression then becomes:

$$\operatorname{Im} \operatorname{tr} e^{-\beta H} = -\beta e^{-\beta/2} \frac{4}{\sqrt{2\pi}} \frac{1}{\sqrt{-g}} e^{4/3g} \left[1 + O\left(g, e^{-\beta} \right) \right] , \quad \text{for } g \to 0_- , \ \beta \to \infty . \qquad (37.42)$$

For β large, the l.h.s. has the form:

$$\operatorname{Im} \operatorname{tr} e^{-\beta H} \sim \operatorname{Im} e^{-\beta E_0(g)} \equiv \operatorname{Im} e^{-\beta (\operatorname{Re} E_0(g) + i \operatorname{Im} E_0(g))} , \quad \text{for } g \to 0_- , \ \beta \to \infty . \qquad (37.43)$$

We have learned that for g small the imaginary part of E_0 is exponentially small. Since the small g limit has always to be taken before the large β limit, we can write:

$$\operatorname{Im} \operatorname{tr} e^{-\beta H} \sim -\beta \operatorname{Im} \left(E_0(g) \right) e^{-\beta \operatorname{Re} E_0(g)} \sim -\beta e^{-\beta/2} \operatorname{Im} E_0 . \qquad (37.44)$$

Equation (37.42) then leads to:

$$\operatorname{Im} E_0(g) = \frac{4}{\sqrt{2\pi}} \frac{e^{4/3g}}{\sqrt{-g}} \left[1 + O(g) \right] , \quad g \to 0_- . \qquad (37.45)$$

A systematic expansion around the saddle point then generates an expansion in powers of g.

Remark. We have obtained the behaviour, for g small and negative, of the imaginary part of the ground state energy and therefore the decay rate of a state localized in the unbounded potential corresponding to the anharmonic oscillator with negative coupling. In Chapter 41 we shall show that this result also leads to an evaluation of the large order behaviour of perturbation series for the anharmonic oscillator.

Bibliographical Notes

Common references to Chapters 37–43 are

J. Zinn-Justin, *Recent Advances in Field Theory and Statistical Mechanics*, Les Houches 1982, R. Stora and J.-B. Zuber eds. (Elsevier, Amsterdam 1984),

which we follow closely and which also contains additional references and

S. Coleman, *The Whys of Subnuclear Physics*, Erice 1977, A. Zichichi ed. (Plenum, New York 1979).

For an early investigation of the analytic properties of the anharmonic oscillator see

C.M. Bender and T.T. Wu, *Phys. Rev.* 184 (1969) 1231.

The content of Chapter 37 is based on

E. Brézin, J.C. Le Guillou and J. Zinn-Justin, *Phys. Rev.* D15 (1977) 1544.

The method of collective coordinates was first described in this context by

J. Zittartz and J.S. Langer, *Phys. Rev.* 148 (1966) 741; J.S. Langer, *Ann. Phys. (NY)* 41 (1967) 108.

The analytic continuation in the g-plane is discussed in

G. Parisi, *Phys. Lett.* 66B (1977) 167; E.B. Bogolmony, *Phys. Lett.* 67B (1977) 193.

For the calculation of the determinant arising in the gaussian integration around the instanton see for example

R.G. Newton, *Scattering Theory of Waves and Particles* (Mc Graw-Hill, New York 1966, 2nd edn. Springer-Verlag, New York 1982) p. 433.

Rigorous results concerning the spectrum of the quartic anharmonic oscillator can be found in

J.J. Loeffel, A. Martin, B. Simon and A.S. Wightman, *Phys. Lett.* 30B (1968) 1656; S. Graffi, V. Grecchi and B. Simon, *Phys. Lett.* 32B (1970) 631; B. Simon, *Functional Integration and Quantum Physics* (Academic Press, New York 1979).

Exercises

Exercise 37.1

Calculate the imaginary part of the ground state energy for the hamiltonian

$$H = -\tfrac{1}{2}\left(\mathrm{d}/\mathrm{d}q\right)^2 + \tfrac{1}{2}q^2 + \tfrac{1}{2}gq^{2N},$$

for g negative.

A result useful for all exercises: if $M(t_1, t_2)$ is of the form:

$$M(t_1, t_2) = \left[-\left(\frac{\mathrm{d}}{\mathrm{d}t_1}\right)^2 + 1 - \frac{\lambda(\lambda+1)\omega^2}{\cosh^2 \omega t_1}\right]\delta(t_1 - t_2),$$

in the notation of equation (37.38) we have:

$$\det(M+\varepsilon)(M_0+\varepsilon)^{-1} = \frac{\Gamma(1+z)\,\Gamma(z)}{\Gamma(1+\lambda+z)\,\Gamma(z-\lambda)}, \quad \text{with } z = \sqrt{1+\varepsilon}/\omega.$$

Exercise 37.2

Calculate the imaginary part of the ground state energy for the hamiltonian with $O(\nu)$ internal symmetry

$$H = -\tfrac{1}{2}\left(\mathrm{d}/\mathrm{d}\mathbf{q}\right)^2 + \tfrac{1}{2}\mathbf{q}^2 + \tfrac{1}{2}g\left(\mathbf{q}^2\right)^N.$$

APPENDIX 37
THE JACOBIAN, CONSTRAINTS

In the appendix we first calculate the jacobian of the transformation from $q(t)$ to the set t_0, q_n and then indicate how collective coordinates can also be introduced by a constraint equation.

A37.1 The Jacobian

Let us expand $q(t)$ on a complete set of real orthonormal functions (in the $\mathcal{L}^2$ sense) satisfying periodic boundary conditions on $[0, \beta]$,

$$q(t) = \sum_{m=0}^{\infty} c_m g_m(t), \qquad (A37.1)$$

$$\delta_{nm} = \int_0^\beta dt\, g_m(t) g_n(t). \qquad (A37.2)$$

In the explicit evaluation of Section 2.3 (see equation (2.41)), we have defined the functional measure as

$$[dq(t)] = \mathcal{N} \prod_{m=0}^{\infty} dc_m, \qquad (A37.3)$$

in which $\mathcal{N}$ is the usual constant normalization factor. We now change variables

$$q(t) = q_c (t - t_0) + \sum_{n=1}^{\infty} f_n (t - t_0)\, q_n, \qquad (A37.4)$$

in such a way that the set $\{\dot{q}_c (t - t_0), f_n (t - t_0)\}$ forms an orthogonal basis, and the f_n's are normed. The new variables are t_0 and the set $\{q_n\}$.

Let us calculate the variables c_m in terms of the new variables:

$$c_m = \int dt\, g_m(t) q_c (t - t_0) + \sum_{n=1}^{\infty} q_n \int dt\, g_m(t) f_n (t - t_0). \qquad (A37.5)$$

The jacobian of the transformation is the determinant of the matrix

$$\left[\frac{\partial c_m}{\partial t_0}, \frac{\partial c_m}{\partial q_n} \right], \qquad (A37.6)$$

with:

$$\frac{\partial c_m}{\partial t_0} = - \int dt\, g_m(t) \dot{q}_c (t - t_0) - \sum_{n=1}^{\infty} q_n \int dt\, g_m(t) \dot{f}_n (t - t_0),$$

$$\frac{\partial c_m}{\partial q_n} = \int dt\, g_m(t) f_n (t - t_0). \qquad (A37.7)$$

Because we calculate at leading order only we can neglect the dependence of the jacobian on the $\{q_n\}$. Since the set

$$\left\{ \frac{\dot{q}_c (t - t_0)}{|\dot{q}_c|}, f_n \right\},$$

forms an orthonormal basis, the determinant of the matrix

$$\left[\int dt \, g_m(t) \frac{\dot{q}_c \, (t-t_0)}{|\dot{q}_c|} \, , \quad \int dt \, g_m(t) f_n \, (t-t_0) \right], \tag{A37.8}$$

is the determinant of an orthogonal matrix which is one. Therefore the jacobian of our transformation is just $\|\dot{q}_c \, (t-t_0)\|$:

$$\|\dot{q}_c\| = \left\{ \int dt \, [\dot{q}_c \, (t-t_0)]^2 \right\}^{1/2}. \tag{A37.9}$$

A37.2 Constraint Equations

Another method can be used to introduce collective coordinates, which yields identical results. Inspired by the Faddeev–Popov quantization method of gauge theories, we start from the identity:

$$1 = \int dt_0 \left[\int dt \, \dot{q}_c(t) \dot{q}(t+t_0) \right] \delta \left[\int dt \, \dot{q}_c(t) \big(q(t+t_0) - q_c(t) \big) \right]. \tag{A37.10}$$

We insert $(A37.10)$ into the path integral and replaces the δ-function by its Fourier representation. Calling λ the variable which imposes $(A37.10)$, we can write the total action $S[q(t), \lambda]$:

$$S[q(t), \lambda] = S[q(t)] + i\lambda \int dt \, \dot{q}_c(t) \big(q(t+t_0) - q_c(t) \big). \tag{A37.11}$$

The variation of the total action $S[q(t), \lambda]$ then vanishes for $q(t) = q_c(t-t_0)$, $\lambda = 0$. The determinant generated by the gaussian integration around the saddle point is now a 2×2 matrix with non-diagonal elements

$$\frac{\delta \partial}{\delta q(t) \partial \lambda} S[q, \lambda] = i\dot{q}_c(t-t_0).$$

This modifies the determinant

$$\det \frac{\delta^2 S}{\delta q(t_1) \delta q(t_2)} \mapsto \det \left\{ \frac{\delta^2 S}{\delta q(t_1) \delta q(t_2)} + \dot{q}_c(t_1 - t_0) \dot{q}_c(t_2 - t_0) \right\}.$$

The modified operator has all the same eigenvalues as the initial one but one: The eigenvalue which corresponds to the eigenvector $\dot{q}_c$ is now $\|\dot{q}_c\|^2$ instead of 0. Taking into account this modification of the determinant and the first factor in $(A37.10)$ at leading order one exactly recovers the jacobian of previous section. Note finally that the factor $(2\pi)^{-1/2}$ in (37.37) arises as the ratio of a factor $(2\pi)^{1/2}$ generated by the additional gaussian integration due to λ and the factor $1/(2\pi)$ of the Fourier representation of the δ-function.

We now want to calculate for a general class of potentials, still in Quantum Mechanics, the decay rate of a wave packet located at initial time at a relative minimum of a potential, generalizing the method of Chapter 37. To guide our intuition, we again imagine that we start from a situation in which a given minimum of a potential is an absolute minimum, and after some analytic continuation becomes a relative minimum of the potential. As we have discussed in the previous chapter, the corresponding ground state energy then becomes complex in the analytic continuation, and its imaginary part yields the inverse lifetime of a wave packet concentrated around the minimum of the potential and decaying through barrier penetration. In the semiclassical limit the decay rate is again related to finite action, i.e. instanton solutions of the euclidean classical equations. We shall calculate, at leading order, the contribution of instantons to the imaginary part of the energy.

The appendix contains, to compare, analogous calculations using WKB methods, and some remark about the average action in path integrals.

38.1 The Instanton Contribution

The instanton solution. We consider hamiltonians of the form:

$$H = -\tfrac{1}{2} \left(\mathrm{d}/\mathrm{d}q \right)^2 + g^{-1}V\left(q\sqrt{g}\right). \qquad (38.1)$$

The function $V(q)$ is an analytic function of q which, for q small, behaves like:

$$V(q) = \tfrac{1}{2}q^2 + O\left(q^3\right). \qquad (38.2)$$

Again in the hamiltonian (38.1) the potential has been parametrized in such a way that g plays the formal role of $\hbar$.

The path integral representation of the partition function $\mathrm{tr}\, e^{-\beta H}$ is:

$$\mathrm{tr}\, e^{-\beta H} = \int_{q(-\beta/2)=q(\beta/2)} [\mathrm{d}q(t)] \exp\left[-S\big(q(t)\big)\right], \qquad (38.3)$$

$$S\left[q(t)\right] = \int_{-\beta/2}^{\beta/2} \left[\tfrac{1}{2}\dot{q}^2(t) + g^{-1}V\left(q(t)\sqrt{g}\right)\right]\mathrm{d}t. \qquad (38.4)$$

In the situation that we are considering we know that instanton solutions exist: because $q = 0$ is only a relative minimum of the potential, the function $V(q)$ which we have assumed regular and thus continuous, has at least another zero. For β infinite, one instanton solution $q_c(t)$ starts from the origin at time $-\infty$, is reflected on the zero of the potential, and comes back to the origin at time $+\infty$.

Let us write the euclidean equation of motion:

$$\ddot{q}_c(t) = \frac{1}{\sqrt{g}} \frac{\partial V}{\partial q}\left(q_c\sqrt{g}\right). \qquad (38.5)$$

Integrating once, we obtain for a finite action solution (for $\beta = \infty$):

$$\tfrac{1}{2}\dot{q}_c^2(t) - g^{-1}V\left(q_c(t)\sqrt{g}\right) = 0\,. \tag{38.6}$$

Calling x_0 the relevant zero of $V(x)$, we can write the corresponding action:

$$S(q_c) = \int_{-\infty}^{+\infty} \dot{q}_c^2(t)\mathrm{d}t = \frac{A}{g}\,, \tag{38.7}$$

$$A = 2\int_0^{x_0} \sqrt{2V(x)}\mathrm{d}x\,. \tag{38.8}$$

We notice that the classical action is positive and proportional to g^{-1}.

The gaussian integration. To calculate the instanton contribution at leading order, we have to integrate around the saddle point $q_c(t)$ in the gaussian approximation. However, as we have shown in previous section, we must before separate a collective coordinate corresponding to time translation. This yields a factor β and a jacobian J:

$$J = \left[\int_{-\infty}^{+\infty} \dot{q}_c^2(t)\,\mathrm{d}t\right]^{1/2} = (A/g)^{1/2}\,. \tag{38.9}$$

Since the function $\dot{q}_c(t)$, which is an eigenfunction of $\delta^2 S/(\delta q_c)^2$ with eigenvalue zero, has a node at the turning point $x_0/\sqrt{g}$, there exists an eigenfunction associated with a negative eigenvalue. Consequently, the determinant which appears in the gaussian integration around the saddle point is negative.

Collecting all factors we obtain:

$$\mathrm{Im\,tr}\,\mathrm{e}^{-\beta H} \sim \frac{1}{2}\frac{\beta}{\sqrt{2\pi}}\,\mathrm{e}^{-\beta/2}\sqrt{\frac{A}{g}}\left[\det M_0\left(-\det' M\right)^{-1}\right]^{1/2}\mathrm{e}^{-A/g}$$
$$\text{for } g \to 0\,, \quad \beta \to \infty\,. \tag{38.10}$$

This yields for the imaginary part of the "ground state" energy E_0:

$$\mathrm{Im}\,E_0 = \frac{1}{2}\sqrt{\frac{A}{2\pi g}}\left[\det M_0\left(-\det' M\right)^{-1}\right]^{1/2}\mathrm{e}^{-A/g}\,. \tag{38.11}$$

The definitions of M_0, M, $\det'$ are the same as in the preceding chapter:

$$M(t_1, t_2) = \delta^2 S/\delta q_c(t_1)\,\delta q_c(t_2)\,, \qquad M_0(t_1, t_2) = \left[-\left(\mathrm{d}/\mathrm{d}t_1\right)^2 + 1\right]\delta(t_1 - t_2)\,, \tag{38.12}$$

and $\det'$ means determinant in the subspace orthogonal to $\dot{q}_c$.

To complete the calculation we shall explicitly evaluate $\det' M$ in quantum mechanics. In Appendix A38.2, to compare, we shall recall how the calculation is done with the WKB method.

38.2 Calculation of the Determinant: The Shifting Method

For reasons which will become apparent later, we first calculate the general matrix element:

$$\langle x' | e^{-\beta H} | x \rangle = \int_{q(-\beta/2)=x'}^{q(\beta/2)=x} [dq(t)] \exp\left[-S(q(t))\right]. \tag{38.13}$$

Let us call $S_c(x', x; \beta)$ the classical action corresponding to the classical solution with the prescribed boundary conditions:

$$\begin{cases} S_c(x', x; \beta) = \int_{-\beta/2}^{\beta/2} \left[\frac{1}{2}\dot{q}_c^2(t) + g^{-1}V\left(q_c(t)\sqrt{g}\right)\right] dt, \\ q_c(-\beta/2) = x', \quad q_c(\beta/2) = x. \end{cases} \tag{38.14}$$

Setting:

$$q(t) = q_c(t) + r(t), \tag{38.15}$$

we obtain, at leading order, the path integral:

$$\langle x' | e^{-\beta H} | x \rangle = e^{-S_c} \int_{r(-\beta/2)=0}^{r(\beta/2)=0} [dr(t)] \exp\left[-\Sigma(r(t))\right],$$

$$\Sigma(r(t)) = \int_{-\beta/2}^{\beta/2} \frac{1}{2} \left[\dot{r}^2(t) + V''\left(q_c\sqrt{g}\right)r^2(t)\right] dt. \tag{38.16}$$

We now calculate the gaussian integral over $r(t)$ using the so-called *shifting method*. The main drawback of this method is that it involves a dangerous change of variables and the final result is, as we shall see, slightly ill-defined. On the other hand it allows a rather straightforward evaluation of the determinant. The idea behind the calculation is that if we know the solutions of the classical equation of motion for arbitrary boundary conditions, we can construct a canonical transformation which maps any hamiltonian system onto a standard one (here we choose a free hamiltonian). For details see Appendix A38.1.

38.2.1 The shifting method
We first set:

$$V''\left(\sqrt{g}q_c(t)\right) = \ddot{\kappa}(t)/\kappa(t). \tag{38.17}$$

We know at least one possible choice for $\kappa(t)$. If we differentiate the equation of motion (38.5), we find:

$$\left(\frac{d}{dt}\right)^2 \dot{q}_c(t) = V''\left(q_c\sqrt{g}\right)\dot{q}_c(t). \tag{38.18}$$

If $\dot{q}_c(t)$ does not vanish on the classical trajectory, we choose

$$\kappa(t) = \dot{q}_c(t). \tag{38.19}$$

Otherwise we look for a linear combination of the two independent solutions of equation (38.17), $\dot{q}_c(t)$ and

$$\dot{q}_c(t)\int^t \frac{d\tau}{[\dot{q}_c(\tau)]^2},$$

---

which does not vanish on the classical trajectory.

The action in expression (38.16) then becomes:

$$\int_{-\beta/2}^{\beta/2} \frac{1}{2}\left[\dot{r}^2(t) + \frac{\ddot{\kappa}(t)}{\kappa(t)}r^2(t)\right] dt = \int_{-\beta/2}^{\beta/2} \frac{1}{2}\left[\dot{r}^2(t) - \dot{\kappa}(t)\frac{d}{dt}\left(\frac{r^2(t)}{\kappa(t)}\right)\right] dt. \tag{38.20}$$

In the integration by parts, the integrated terms vanish due to the boundary conditions. The evaluation of the r.h.s. of equation (38.20) leads to the remarkable identity:

$$\int_{-\beta/2}^{\beta/2} \frac{1}{2}\left[\dot{r}^2(t) + \frac{\ddot{\kappa}(t)}{\kappa(t)}r^2(t)\right] dt = \int_{-\beta/2}^{\beta/2} \frac{1}{2}\left[\dot{r}(t) - \frac{\dot{\kappa}}{\kappa}r(t)\right]^2 dt. \tag{38.21}$$

This suggests an obvious linear change of variable, $r(t) \mapsto \sigma(t)$:

$$\dot{r}(t) - \frac{\dot{\kappa}(t)}{\kappa(t)}r(t) = \dot{\sigma}(t), \qquad \sigma(-\beta/2) = 0, \tag{38.22}$$

which transforms the hamiltonian of the time-dependent harmonic oscillator, into the free hamiltonian:

$$\int_{-\beta/2}^{\beta/2} \frac{1}{2}\left[\dot{r}^2(t) + \frac{\ddot{\kappa}(t)}{\kappa(t)}r^2(t)\right] dt = \int_{-\beta/2}^{\beta/2} \frac{1}{2}\left[\dot{\sigma}^2(t)\right] dt. \tag{38.23}$$

The change of variables (38.22) reminds us of the Langevin equation we discussed in Chapter 4, $r(t)$ playing the role of the dynamical variable and $\dot{\sigma}$ the role of the noise. We are therefore not too surprised to encounter the same difficulty as in the derivation of the Fokker–Planck equation. Integrating equation (38.22), we obtain

$$r(t) = \kappa(t)\int_{-\beta/2}^{t} d\tau\, \frac{\dot{\sigma}(\tau)}{\kappa(\tau)} = \sigma(t) + \kappa(t)\int_{-\beta/2}^{t} d\tau\, \sigma(\tau)\frac{\dot{\kappa}(\tau)}{\kappa^2(\tau)}. \tag{38.24}$$

The jacobian J of this transformation is formally the determinant of the kernel (see Section 4.5)

$$J = \det\frac{\delta r(t_2)}{\delta\sigma(t_1)} = \det\left[\delta(t_1 - t_2) + \theta(t_2 - t_1)\,\kappa(t_2)\frac{\dot{\kappa}(t_1)}{\kappa^2(t_1)}\right]. \tag{38.25}$$

Using:

$$\ln\det(1 + M) = \operatorname{tr}\ln(1 + M) = \operatorname{tr} M - \tfrac{1}{2}\operatorname{tr} M^2 + \cdots, \tag{38.26}$$

we get:

$$\ln J = \theta(0)\int_{-\beta/2}^{\beta/2} dt\,\frac{\dot{\kappa}(t)}{\kappa(t)}. \tag{38.27}$$

As we have shown in Section 4.2, and already used in Section 4.6, symmetry considerations lead to the prescription:

$$\theta(0) = \tfrac{1}{2}. \tag{38.28}$$

Integrating over t we then obtain the jacobian J:

$$J = \sqrt{\frac{\kappa(\beta/2)}{\kappa(-\beta/2)}}. \tag{38.29}$$

In the path integral we still have to impose the boundary condition:

$$0 = r\left(\beta/2\right) = \kappa\left(\beta/2\right) \int_{-\beta/2}^{\beta/2} dt\, \frac{\dot{\sigma}(t)}{\kappa(t)}. \tag{38.30}$$

This can be achieved by introducing a δ-function for which, as usual, we use a Fourier representation

$$\delta\bigl(r(\beta/2)\bigr) = \frac{1}{\kappa(\beta/2)} \int \frac{d\lambda}{2\pi} \exp\left(i\lambda \int_{-\beta/2}^{\beta/2} dt\, \frac{\dot{\sigma}(t)}{\kappa(t)} \right).$$

The complete expression then reads:

$$\langle x' | e^{-\beta H} | x \rangle \sim e^{-S_c} \int_{\sigma(-\beta/2)=0} [d\sigma(t)] \frac{d\lambda}{2\pi} \frac{1}{\sqrt{\kappa(\beta/2)\kappa(-\beta/2)}} e^{-S(\sigma,\lambda)}, \tag{38.31}$$

with

$$S(\sigma, \lambda) = \int_{-\beta/2}^{\beta/2} dt\, \left(\tfrac{1}{2}\dot{\sigma}^2(t) - i\lambda \frac{\dot{\sigma}(t)}{\kappa(t)} \right). \tag{38.32}$$

To eliminate the term linear in $\dot{\sigma}(t)$ in equation (38.32), we shift $\dot{\sigma}(t)$,

$$\dot{\sigma}(t) = i\frac{\lambda}{\kappa(t)} + \dot{\sigma}'(t). \tag{38.33}$$

After this shift the path integral becomes

$$\langle x' | e^{-\beta H} | x \rangle \sim e^{-S_c} \int_{\sigma(-\beta/2)=0} [d\sigma(t)] \frac{d\lambda}{2\pi} \frac{1}{\sqrt{\kappa(\beta/2)\kappa(-\beta/2)}}$$
$$\times \exp\left[-\frac{1}{2}\lambda^2 \int_{-\beta/2}^{\beta/2} \frac{dt}{\kappa^2(t)} - \frac{1}{2} \int_{-\beta/2}^{\beta/2} \dot{\sigma}^2(t) dt \right]. \tag{38.34}$$

We integrate over λ:

$$\langle x' | e^{-\beta H} | x \rangle \sim e^{-S_c(x',x;\beta)} \left[\kappa\left(\beta/2\right) \kappa\left(-\beta/2\right) \int_{-\beta/2}^{\beta/2} \frac{dt}{\kappa^2(t)} \right]^{-1/2} \mathcal{N}(\beta). \tag{38.35}$$

The constant $\mathcal{N}(\beta)$ does not depend on x and x', and is proportional to the matrix element $\langle 0 | e^{-\beta H_0} | 0 \rangle$ in which H_0 is the free hamiltonian:

$$\langle x' | e^{-\beta H_0} | x \rangle = (2\pi\beta)^{-1/2} e^{-(x'-x)^2/2\beta}. \tag{38.36}$$

To determine $\mathcal{N}(\beta)$ we set $H = H_0$ in equation (38.35) and note that in this case $\kappa(t)$ is a constant.

The final result is:

$$\langle x' | e^{-\beta H} | x \rangle \sim e^{-S_c(x',x;\beta)} \left[2\pi\kappa\left(\beta/2\right) \kappa\left(-\beta/2\right) \int_{-\beta/2}^{\beta/2} \frac{dt}{\kappa^2(t)} \right]^{-1/2}. \tag{38.37}$$

We leave as an exercise to show that the result (38.37) is formally independent of the particular linear combination of the two solutions of equation (38.17) one has chosen. To obtain a more explicit expression we then substitute for example $\kappa(t) = \dot{q}_c(t)$. We integrate the equation of motion (38.5), taking into account the boundary conditions:

$$\tfrac{1}{2}\dot{q}_c^2 = g^{-1}\left[V\left(q_c(t)\sqrt{g}\right) + E\right],\tag{38.38}$$

and therefore

$$\beta = \int_{x'\sqrt{g}}^{x\sqrt{g}} \frac{dq}{\left[2\left(E + V(q)\right)\right]^{1/2}}.\tag{38.39}$$

Differentiating equation (38.39) with respect to β we obtain:

$$1 = -\int_{x'\sqrt{g}}^{x\sqrt{g}} \frac{dq}{\left[2\left(E + V(q)\right)\right]^{3/2}}\frac{\partial E}{\partial \beta},\tag{38.40}$$

which can be rewritten:

$$\frac{\partial E}{\partial \beta} = -\left[\int_{-\beta/2}^{\beta/2} \frac{dt}{\kappa^2(t)}\right]^{-1}.\tag{38.41}$$

The result (38.37) can then also be written

$$\left\langle x' \left| e^{-\beta H} \right| x \right\rangle \sim e^{-S_c(x',x;\beta)} \frac{1}{\sqrt{2\pi \dot{q}_c(\beta/2)\dot{q}_c(-\beta/2)}} \left(\frac{-\partial E}{\partial \beta}\right)^{1/2}.\tag{38.42}$$

We leave as an exercise to verify that one also has the identity

$$\kappa\left(\beta/2\right)\kappa\left(-\beta/2\right)\int_{-\beta/2}^{\beta/2} \frac{dt}{\kappa^2(t)} = \left(-\frac{\partial^2 S_c}{\partial x \partial x'}\right)^{-1}.\tag{38.43}$$

Substituting equation (38.43) into equation (38.37) one then obtains Van Vleck's formula:

$$\left\langle x' \left| e^{-\beta H} \right| x \right\rangle \sim \left(-\frac{1}{2\pi}\frac{\partial^2 S_c}{\partial x \partial x'}\right)^{1/2} \exp\left[-S_c(x',x;\beta)\right].\tag{38.44}$$

Several degrees of freedom. The calculation of the instanton contribution by the shifting method can be generalized to $d > 1$ degrees of freedom provided one can find a non-singular matrix K solution of the equation:

$$\ddot{K}_{ij} = \frac{\partial V\left(\mathbf{q}_c(t)\right)}{\partial q_i \partial q_k} K_{kj}.\tag{38.45}$$

The change of variables (38.22) then takes the form:

$$\dot{\mathbf{r}} - \dot{K}K^{-1}\mathbf{r} = \dot{\boldsymbol{\sigma}}.\tag{38.46}$$

The matrix K can be chosen in such a way that $\dot{K}K^{-1}$ is symmetric. It is then easy to verify that all arguments can be repeated and one finally obtains an expression similar to (38.42):

$$\left\langle \mathbf{x}' \left| e^{-\beta H} \right| \mathbf{x} \right\rangle \sim \left\{(2\pi)^d \det\left[K\left(\frac{\beta}{2}\right)K\left(-\frac{\beta}{2}\right)\int_{-\beta/2}^{\beta/2} dt\, \left(^T K\right)^{-1}K^{-1}\right]\right\}^{-1/2}$$
$$\times \exp\left[-S_c\left(\mathbf{x}',\mathbf{x};\beta\right)\right].\tag{38.47}$$

This expression is again equivalent to Van Vleck's formula (see Appendix A38.2) and can be derived in the same conditions, i.e. if the classical equations of motion can be solved for arbitrary initial and final conditions. For more than one degree of freedom this is no longer the generic situation and corresponds instead to the special class of integrable hamiltonians. A simple example is provided by $O(d)$ invariant systems.

38.2.2 The partition function

In order to calculate $\operatorname{tr} e^{-\beta H}$ we can now impose periodic boundary conditions. Then

$$[\dot{q}_c(\beta/2)\dot{q}_c(-\beta/2)]^{-1/2} = \left\{\frac{2}{g}\left[V\left(x\sqrt{g}\right) + E\right]\right\}^{-1/2}. \tag{38.48}$$

Integrating over x we obtain the trace. Using equation (38.39) we find

$$\int \mathrm{d}x \left[\frac{2}{g}\left(V\left(x\sqrt{g}\right) + E\right)\right]^{-1/2} = \beta. \tag{38.49}$$

We now collect all factors and obtain a more explicit expression:

$$\operatorname{Im} \operatorname{tr} e^{-\beta H} \sim \frac{\beta}{2i}\left(-\frac{\partial E}{\partial \beta}\frac{1}{2\pi g}\right)^{1/2} e^{-A(\beta)/g}, \tag{38.50}$$

where $E(\beta)$ and $A(\beta)$ are defined by:

$$\beta = 2\int_{x_-}^{x_+} \frac{\mathrm{d}x}{\left[2\left(E\left(\beta\right) + V\left(x\right)\right)\right]^{1/2}}, \tag{38.51}$$

$$A\left(\beta\right) = 2\int_{x_-}^{x_+} \mathrm{d}x\left[2\left(E\left(\beta\right) + V\left(x\right)\right)\right]^{1/2} - \beta E\left(\beta\right). \tag{38.52}$$

The quantities x_+ and x_- are the zeros of $[E + V\left(x\right)]$. Notice the useful relation

$$\partial A/\partial \beta = -E\left(\beta\right). \tag{38.53}$$

It is clear, at least for β large enough, that $E\left(\beta\right)$ is a negative increasing function of β. Therefore $-\partial E/\partial \beta$ is negative and the result is real as expected:

$$\operatorname{Im} \operatorname{tr} e^{-\beta H} \sim -\frac{\beta}{2}\left(\frac{\partial E}{\partial \beta}\frac{1}{2\pi g}\right)^{1/2} e^{-A(\beta)/g} \quad \text{for } g \to 0. \tag{38.54}$$

This completes the calculation at finite β.

Remark. At β finite the calculation is valid only above some critical value β_c. Indeed when β decreases, x_+ and x_- approach a common value x_0 which corresponds to a maximum of $V(x)$:

$$\begin{cases} V\left(x\right) \sim V_0 - \frac{1}{2}\omega^2\left(x - x_0\right)^2 + O\left[\left(x - x_0\right)^3\right], \\ V_0 > 0. \end{cases} \tag{38.55}$$

Let us parametrize E, x_+ and x_-,

$$E = -V_0 + \tfrac{1}{2}\omega^2\varepsilon^2, \qquad x_\pm = x_0 \pm \varepsilon, \tag{38.56}$$

then

$$\beta = \int_{x_0-\varepsilon}^{x_0+\varepsilon} \frac{\mathrm{d}x}{\left[\omega^2\varepsilon^2 - \omega^2\left(x - x_0\right)^2\right]^{1/2}}, \quad \Rightarrow \lim_{\varepsilon\to 0} \beta = \beta_c = 2\pi/\omega. \tag{38.57}$$

For $\beta \leq \beta_c$, no instanton solution can be found and it is on the contrary the perturbative expansion around the classical extremum $x = x_0$ of the potential which is relevant.

38.3 The Large β Limit

Writing equation (38.51) as

$$\beta = 2 \int_{x_-}^{x_+} \left\{ [2\left(V\left(x\right)+E\right)]^{-1/2} - \left(x^2+2E\right)^{-1/2} + \left(x^2+2E\right)^{-1/2} \right\} dx , \qquad (38.58)$$

we can explicitly evaluate the last term and neglect E in the difference between the first two terms. This leads to

$$E(\beta) \sim -2C\,\mathrm{e}^{-\beta}, \qquad C = x_+^2 \exp\left[2 \int_0^{x_+} \left(\frac{1}{\sqrt{2V(x)}} - \frac{1}{x} \right) dx \right], \qquad (38.59)$$

where x_+ is now the zero of the potential. In the same way equation (38.52) becomes:

$$A(\beta) = a - 2C\,\mathrm{e}^{-\beta} + O\left(\mathrm{e}^{-2\beta}\right), \qquad a = 2 \int_0^{x_+} \sqrt{2V(x)} . \qquad (38.60)$$

Substituting into equation (38.54) we obtain at leading order

$$\mathrm{Im}\,\mathrm{e}^{-\beta E_0(g)} \underset{g\to 0}{\sim} \frac{\beta}{2}\,\mathrm{e}^{-\beta/2} \left(\frac{C}{\pi g} \right)^{1/2} \mathrm{e}^{-a/g}, \qquad (38.61)$$

and thus

$$\mathrm{Im}\,E_0(g) \underset{g\to 0}{\sim} -\frac{1}{2} \left(\frac{C}{\pi g} \right)^{1/2} \mathrm{e}^{-a/g} . \qquad (38.62)$$

We have here calculated only the imaginary part of the would-be ground state energy. To obtain the imaginary part of the excited levels we have to keep the correction of order $\mathrm{e}^{-\beta}$ in $A(\beta)$ for β large. We then expand $\exp[-g^{-1}A(\beta)]$ in powers of $\mathrm{e}^{-\beta}$. The coefficient of $\mathrm{e}^{-N\beta}$ in the expansion yields the imaginary part of the Nth level at leading order.

Two Remarks.

(i) We have assumed that we have only one instanton solution corresponding to a given zero of the potential. If we find several instanton solutions corresponding to different zeroes of the potential, we have to look for the solution of minimal action, which gives the largest contribution in the small coupling limit.

(ii) At the end of Chapter 37 we have argued that the imaginary part of the energy levels which we evaluate, is the inverse lifetime of a state whose wave function is originally concentrated near the bottom of the unstable minimum of the potential. This interpretation is not problematic for potentials which are either unbounded or have a continuous spectrum in which case the complex energy level corresponds to a resonance in the potential. For potentials which have a pure discrete spectrum (and all eigenvalues are real) the situation is more puzzling. The intuitive solution to this puzzle is that in the semiclassical limit, the lifetime of the metastable state is very long. For times which are not too long, the decay process is exponential and ignores effects coming from the shape of the potential outside of the barrier. Eventually inverse tunnelling will occur and the decay law will be modified.

Bibliographical Notes

The calculation of the instanton contribution for an arbitrary potential can be found in

    E. Brézin, J.C. Le Guillou and J. Zinn-Justin, *Phys. Rev.* D15 (1977) 1558.

The shifting method is exposed in

    I.M. Gel'fand and A.M. Yaglom, *J. Math. Phys.* 1 (1960) 48; R.F. Dashen, B. Hass-lacher and A. Neveu, *Phys. Rev.* D10 (1974) 4114.

The appendix is inspired by the classical paper

    J.H. Van Vleck, *Proc. Natl. Acad. Sci. USA* 14 (1928) 178.

Exercises

Exercise 38.1

Use the shifting method, rather than the method of Section 2.3, to calculate the path integral giving the matrix elements of the euclidean evolution operator $U(x', x; \beta)$ for the simple harmonic oscillator

$$U(x', x; \beta) = \int_{r(0)=x}^{r(\beta)=x'} [dr(t)] \exp\left[-S\big(r(t)\big)\right],$$

$$S(r) = \tfrac{1}{2} \int_0^\beta dt \left(\dot{r}^2(t) + \omega^2 r^2(t)\right).$$

Exercise 38.2

38.2.1. Calculate the imaginary part of the Nth level above the ground state at leading order.

38.2.2. Calculate the imaginary part of the ground state energy for a system with $O(\nu)$ internal symmetry, with the action

$$S(\mathbf{q}) = \int dt \left[\tfrac{1}{2}\dot{\mathbf{q}}^2 + \mathcal{V}\left(g\mathbf{q}^2\right)/g\right].$$

Exercise 38.3

Establish, using the results of Appendix A38.2, the semiclassical expression of the scattering amplitude for an $O(d)$ invariant hamiltonian with a radial potential $V(|\mathbf{x}|)$:

$$H = \tfrac{1}{2}\mathbf{p}^2 + V(r), \quad r = |\mathbf{x}|.$$

APPENDIX 38
THE WKB METHOD: MISCELLANEOUS REMARKS

A38.1 The Classical Equations of Motion

We consider general hamiltonian systems for which the classical equations of motion can be solved with arbitrary boundary conditions. This means in particular that we can explicitly calculate the classical action on the trajectory as a function of initial and final positions $\mathbf{x}'$ and $\mathbf{x}$, and times T' and T. The classical action $\mathcal{A}$ corresponding to a hamiltonian $H(p,q,t)$ is:

$$\mathcal{A}(\mathbf{p},\mathbf{q}) = \int_{T'}^{T} dt\, [\mathbf{p}(t)\cdot\dot{\mathbf{q}}(t) - H(\mathbf{p}(t),\mathbf{q}(t);t)], \qquad (A38.1)$$

with $q_i(T') = x_i'$ and $q_i(T) = x_i$.

The action A_c corresponding to a trajectory $\{\mathbf{p}_c(t),\mathbf{q}_c(t)\}$ in phase space solution of the classical of motions is a function of the initial and final positions, and time:

$$A_c(\mathbf{x}',\mathbf{x};T) = \mathcal{A}(\mathbf{p}_c,\mathbf{q}_c). \qquad (A38.2)$$

In this appendix, in order to obtain the usual expressions of classical mechanics, we shall work in *real time*. The analytic continuation of all expressions to imaginary time is straightforward.

As is well known and will be verified below, the classical action A_c satisfies the Hamilton–Jacobi equations.

Preliminary Remarks. Let us first recall a few classical results. Let us call $S(\mathbf{Q},\mathbf{q};t)$ a function satisfying the Hamilton–Jacobi equations:

$$\frac{\partial S}{\partial t} = -H\left(\frac{\partial S}{\partial \mathbf{q}},\mathbf{q};t\right), \qquad (A38.3)$$

with $S(\mathbf{Q},\mathbf{Q};T') = 0$ and with additional implicit boundary conditions at $t = T'$ which will be explained below. We then use S to generate a time-dependent canonical transformation in phase space, transforming the set $(\mathbf{p},\mathbf{q})$ into $(\mathbf{P},\mathbf{Q})$:

$$p_i = \partial S/\partial q_i \qquad P_i = -\partial S/\partial Q_i. \qquad (A38.4)$$

The implicit boundary conditions come from the conditions:

$$t = T' \to P_i = p_i \quad \text{and} \quad Q_i = q_i. \qquad (A38.5)$$

We make the transformation $(A38.4)$ in the action $(A38.1)$

$$\mathcal{A}(\mathbf{p},\mathbf{q}) = \int_{T'}^{T}\left(\frac{\partial S}{\partial q_i}\dot{q}_i + \frac{\partial S}{\partial t}\right)dt. \qquad (A38.6)$$

The quantities $\mathbf{q}$ and $\mathbf{Q}$ are now considered as time-dependent. Expression $(A38.6)$ can be rewritten as

$$\mathcal{A}(\mathbf{p},\mathbf{q}) = \int_{T'}^{T}\frac{d}{dt}S(\mathbf{q}(t),\mathbf{Q}(t);t)\,dt - \int_{T'}^{T}\frac{\partial S}{\partial Q_i}\dot{Q}_i dt. \qquad (A38.7)$$

Using again relations ($A38.4$), we finally obtain

$$\mathcal{A}\left(\mathbf{p},\mathbf{q}\right) = S\left(\mathbf{x},\mathbf{Q}(T);T\right) + \int_{T'}^{T} P_i \dot{Q}_i \mathrm{d}t.$$ (A38.8)

The equations of motion are now trivial,

$$\dot{Q}_i = 0 \Longrightarrow Q_i(t) = Q_i\left(T'\right).$$ (A38.9)

The conditions ($A38.5$) determine the solution

$$Q_i(t) = x_i'.$$ (A38.10)

We have therefore shown that

$$A_c\left(\mathbf{x}',\mathbf{x};T\right) \equiv S\left(\mathbf{x},\mathbf{x}';T\right),$$ (A38.11)

and found a canonical transformation which maps the initial hamiltonian system onto a trivial one with a vanishing hamiltonian. By performing an additional inverse transformation based on a standard hamiltonian like a free hamiltonian of the form

$$H = \tfrac{1}{2}\sum p_i^2,$$ (A38.12)

or a harmonic oscillator

$$H = \tfrac{1}{2}\sum_i \left(p_i^2 + q_i^2\right),$$ (A38.13)

we can map the original hamiltonian system onto any convenient hamiltonian.

Let us also verify that transformation ($A38.4$) leaves the measure in phase space invariant. We perform the transformation in two steps. First we go from p_i to Q_i:

$$\prod_i \mathrm{d}p_i \mathrm{d}q_i = \prod_i \mathrm{d}q_i \mathrm{d}Q_i \det \frac{\partial^2 S}{\partial q_i \partial Q_j}.$$ (A38.14)

We now eliminate q_i in favour of P_i:

$$\frac{\partial q_i}{\partial P_j} = \left[\frac{\partial P}{\partial q}\right]_{ij}^{(-1)} = -\left[\frac{\partial^2 S}{\partial q \partial Q}\right]_{ij}^{(-1)}.$$ (A38.15)

Therefore the second jacobian cancels the first one (note that a similar argument directly shows the invariance already proven above of the symplectic form $\mathrm{d}p_i \wedge \mathrm{d}q_i$). This analysis suggests that we can perform two transformations ($A38.4$) on the path integral representation of the evolution operator, to reduce it to a path integral corresponding to a standard system for which the evolution operator is exactly known:

$$\langle \mathbf{x}' | U\left(T',T\right) | \mathbf{x}\rangle = \int \prod_i [\mathrm{d}q_i(t)\mathrm{d}p_i(t)] \exp\left[i\mathcal{A}\left(\mathbf{p},\mathbf{q}\right)\right],$$ (A38.16)

with $q_i\left(T'\right) = x_i',\quad q_i\left(T\right) = x_i.$

In this way, it would seem that we are able to calculate exactly the evolution operator of any system for which we know how to solve the classical equations of motion with arbitrary boundary conditions. This would in particular apply to systems with one degree of freedom, with a hamiltonian H of the form

$$H = \tfrac{1}{2}p^2 + V(q) \,. \tag{A38.17}$$

Unfortunately it is easy to verify that the result is wrong at least in Quantum Mechanics. Actually the whole procedure is somewhat ill-defined. This comes from the fact that changes of variables on the path integral on phase space are even more ambiguous than transformations on ordinary path integrals in configuration space. We have given some indications about this problem in Section 3.1. Let us just mention that if we discretize time, we discover that the transformation is not really canonical because to a variable $q(t_k)$ corresponds a momentum variable $p(t_k + \Delta t)$ of a slightly displaced time. This effect, invisible in the naive continuum limit, completely changes the result. It is thus necessary to work on the discretized form of the phase space path integral. It should be possible in this way to derive the semiclassical result which we shall now establish, but, to our knowledge, no one has yet done so. On the other hand, as often, the situation seems to be more favourable in field theory. The problem we have described above comes mainly from commutation of quantum operators. We have emphasized in Chapter 5 that the commutators are infinite in field theory and disappear in the renormalization. Therefore one expects, as this has been verified on examples, that the semiclassical approximations of classically integrable field theories reproduce features of the exact solution.

A38.2 The WKB Method

We now explicitly write the Schrödinger equation for the evolution operator

$$H\left(\frac{\hbar}{i}\frac{\partial}{\partial \mathbf{x}}, \mathbf{x}; t\right) U(\mathbf{x}', \mathbf{x}; t) = i\hbar \frac{\partial U}{\partial t}(\mathbf{x}', \mathbf{x}; t) \,, \tag{A38.18}$$

with the definitions

$$U(\mathbf{x}', \mathbf{x}; t) = \langle \mathbf{x}' | \mathbf{U}(t) | \mathbf{x} \rangle \,, \tag{A38.19}$$

and the boundary condition

$$\mathbf{U}(t = T') = \mathbf{1} \,. \tag{A38.20}$$

In order to write equation $(A38.18)$, it has been necessary to associate with the classical function $H(\mathbf{p}, \mathbf{q}, ; t)$ a quantum operator.

We shall see below how much the semiclassical result depends upon the quantization. We have reintroduced the quantity $\hbar$ which we usually set equal to one, to make the expansion parameter explicit. The ansatz will be to set

$$U(\mathbf{x}', \mathbf{x}; t) = G(\mathbf{x}', \mathbf{x}; t)\, e^{iA(\mathbf{x}', \mathbf{x}; t)/\hbar} \left[1 + O(\hbar)\right] \,. \tag{A38.21}$$

Introducing the ansatz into equation $(A38.18)$ and keeping the two first terms in $\hbar$ we obtain two equations. The first equation is

$$H\left(\frac{\partial A}{\partial \mathbf{x}}, \mathbf{x}; t\right) = -\frac{\partial A}{\partial t} \,. \tag{A38.22}$$

This equation involves only the classical hamiltonian and is the Hamilton–Jacobi equation for the classical action on the classical trajectory. Together with the boundary conditions implied by the condition (A38.20), it determines A completely. The derivation of the second equation involves some more work. Let us first note that

$$H\left(\frac{\hbar}{i}\frac{\partial}{\partial \mathbf{x}}, \mathbf{x}; t\right) G = GH - i\hbar\frac{\partial H}{\partial p_i}\frac{\partial G}{\partial x_i} + O\left(\hbar^2\right). \qquad (A38.23)$$

Here again only the classical hamiltonian is needed. The term containing $\partial H/\partial p_i$ is already multiplied by a factor $\hbar$, thus we can replace the operator p_i by $\partial A/\partial x_i$. For the first term we now use the identity:

$$\begin{aligned}
\mathrm{e}^{-iA/\hbar}\, H\, \mathrm{e}^{iA/\hbar} = {}&H\left(\frac{\partial A}{\partial \mathbf{x}}, \mathbf{x}; t\right) - \frac{i\hbar}{2}\frac{\partial^2 H}{\partial p_j \partial q_j}\left(\frac{\partial A}{\partial \mathbf{x}}, \mathbf{x}; t\right) \\
&- \frac{i\hbar}{2}\frac{\partial^2 H}{\partial p_j \partial p_k}\left(\frac{\partial A}{\partial \mathbf{x}}, \mathbf{x}; t\right)\frac{\partial^2 A}{\partial x_j \partial x_k} + O\left(\hbar^2\right).
\end{aligned} \qquad (A38.24)$$

The second term in the r.h.s. comes from commuting all the derivatives completely on the right. It relies on the assumption that the classical hamiltonian H is real and can be quantized to generate a hermitian operator. Indeed let us first assume that we have symmetrized all monomials:

$$p^n q^m \to \tfrac{1}{2}\left(p^n q^m + q^m p^n\right). \qquad (A38.25)$$

Then a contribution to this term arises each time an operator p of $p^n q^m$ acts on q^m and the factor $\frac{1}{2}$ comes from the symmetrization. If we choose another hermitian quantization procedure, we can start commuting all operators p and q until the hamiltonian is again a sum of terms (A38.25). Each commutation introduces a factor $i\hbar$. Since the difference between the two expressions is hermitian, it can only involve $(i\hbar)^2$, which can be neglected at this order.

The third term in expression (A38.24) arises from two derivatives acting on the action. The factor $1/2$ is a counting factor. We then obtain an equation for G:

$$\frac{\partial H}{\partial p_i}\frac{\partial G}{\partial x_i} + \frac{1}{2}\left(\frac{\partial^2 H}{\partial p_j \partial q_j} + \frac{\partial^2 H}{\partial p_j \partial p_k}\frac{\partial^2 A}{\partial x_j \partial x_k}\right)G = -\frac{\partial G}{\partial t}. \qquad (A38.26)$$

Let us now differentiate equation (A38.22) with respect to x_i' and x_j successively:

$$\frac{\partial^2 A}{\partial x_i' \partial x_k}\frac{\partial H}{\partial p_k} = -\frac{\partial^2 A}{\partial t \partial x_i'}, \qquad (A38.27)$$

$$\frac{\partial^3 A}{\partial x_i' \partial x_j \partial x_k}\frac{\partial H}{\partial p_k} + \frac{\partial^2 A}{\partial x_i' \partial x_k}\frac{\partial^2 A}{\partial x_j \partial x_l}\frac{\partial^2 H}{\partial p_k \partial p_l} + \frac{\partial^2 A}{\partial x_i' \partial x_k}\frac{\partial^2 H}{\partial p_k \partial q_j} = -\frac{\partial^3 A}{\partial t \partial x_i' \partial x_j}. \qquad (A38.28)$$

Let us introduce a matrix notation

$$M_{ij} = \frac{\partial^2 A}{\partial x_i' \partial x_j}, \quad H_{ij} = \frac{\partial^2 H}{\partial p_i \partial p_j}, \quad \tilde{H}_{ij} = \frac{\partial^2 H}{\partial p_i \partial q_j}. \qquad (A38.29)$$

The equation (A38.28) can then be rewritten:

$$\frac{\partial \mathbf{M}}{\partial x_k}\frac{\partial H}{\partial p_k} + \mathbf{MHM} + \mathbf{M}\tilde{\mathbf{H}} = -\frac{\partial \mathbf{M}}{\partial t}. \qquad (A38.30)$$

All multiplications are meant in a matrix sense. We now multiply equation ($A38.30$) by $\mathbf{M}^{-1}$ on the left and take the trace:

$$\frac{\partial H}{\partial p_k}\operatorname{tr}\mathbf{M}^{-1}\frac{\partial \mathbf{M}}{\partial x_k} + \operatorname{tr}\left(\mathbf{H}\mathbf{M}+\widetilde{\mathbf{H}}\right) = -\operatorname{tr}\mathbf{M}^{-1}\frac{\partial \mathbf{M}}{\partial t}. \qquad (A38.31)$$

Let us note that:

$$\frac{\partial}{\partial z}\ln\det\mathbf{M}(z) = \frac{\partial}{\partial z}\operatorname{tr}\ln\mathbf{M}(z) = \operatorname{tr}\frac{\partial\mathbf{M}}{\partial z}\mathbf{M}^{-1}. \qquad (A38.32)$$

The equation ($A38.31$) can then be rewritten as

$$\frac{\partial H}{\partial p_i}\frac{\partial}{\partial x_i}\ln\det\mathbf{M} + \frac{\partial^2 H}{\partial p_j\partial q_j} + \frac{\partial^2 H}{\partial p_j\partial p_k}\frac{\partial^2 A}{\partial x_j\partial x_k} = -\frac{\partial}{\partial t}\ln\det\mathbf{M}, \qquad (A38.33)$$

while equation ($A38.26$) can be written as

$$\frac{\partial H}{\partial p_i}\frac{\partial}{\partial x_i}\ln G + \frac{1}{2}\frac{\partial^2 H}{\partial p_j\partial q_j} + \frac{1}{2}\frac{\partial^2 H}{\partial p_j\partial p_k}\frac{\partial^2 A}{\partial x_j\partial x_k} = -\frac{\partial}{\partial t}\ln G. \qquad (A38.34)$$

A comparison between these two equations shows that a solution to equation ($A38.34$) is

$$\ln G = \tfrac{1}{2}\ln\det\mathbf{M} + \text{const.}. \qquad (A38.35)$$

Taking into account the boundary conditions we finally obtain Van Vleck's formula:

$$\langle \mathbf{x}'|U(T',T)|\mathbf{x}\rangle \sim \frac{1}{(2\pi i\hbar)^{n/2}}\left(-\det\frac{\partial^2 A}{\partial\mathbf{x}\partial\mathbf{x}'}\right)^{1/2}e^{(i/\hbar)A}. \qquad (A38.36)$$

It is straightforward to derive from this equation the corresponding expression for imaginary time.

A38.3 The Average Action in Path Integrals

Since we have emphasized finite action solutions, one could have the impression that the result of the path integral is dominated by finite action paths. As the two examples below show, this is never the case. Instead the path integral is dominated by paths which are in the neighbourhood of finite action solutions.

(i) *To a functional integral contribute only infinite action configurations.* Let us give a very simple example to illustrate this point:

$$I = \lim_{N\to\infty} I_N, \quad I_N = (2\pi)^{-N/2}\int dx_1\cdots dx_N \exp\left(-\frac{1}{2}\sum_{i=1}^N x_i^2\right). \qquad (A38.37)$$

Obviously $I = 1$. Setting

$$\sum_{i=1}^N x_i^2 = R^2, \qquad (A38.38)$$

we obtain

$$I_N = (2\pi)^{-N/2}\frac{2\pi^{N/2}}{\Gamma(N/2)}\int_0^\infty R^{N-1}dR\, e^{-R^2/2}. \qquad (A38.39)$$

Finite action means that R is bounded:

$$I_N(R_0) = \frac{2^{1-N/2}}{\Gamma(N/2)} \int_0^{R_0} R^{N-1} dR\, e^{-R^2/2} < \left(\frac{R_0^2}{2}\right)^{N/2} \frac{1}{\Gamma(1+N/2)}. \qquad (A38.40)$$

Therefore:

$$\lim_{N\to\infty} I_N(R_0) = 0. \qquad (A38.41)$$

More precisely the integral $(A38.39)$ can be calculated by steepest descent. The saddle point R_s is

$$R_s = N^{1/2}, \qquad (A38.42)$$

and the classical actions A which are relevant to the path integral satisfy:

$$A = \frac{N}{2} + O\left(N^{1/2}\right). \qquad (A38.43)$$

(ii) *Calculation of the average action in Quantum Mechanics.*

$$\operatorname{tr} e^{-\beta H} = \int_{q(-\beta/2)=q(\beta/2)} [dq(t)] \exp\left[-\mathcal{A}(q(t))\right], \qquad (A38.44)$$

$$\mathcal{A}[q(t)] = \int_{-\beta/2}^{\beta/2} dt \left[\frac{1}{2}\dot{q}^2(t) + g^{-1}V\left(q(t)\sqrt{g}\right)\right]. \qquad (A38.45)$$

To calculate the average action we use the expression

$$\langle \mathcal{A}\rangle = -\frac{\partial}{\partial\lambda}\bigg|_{\lambda=1} \ln \int [dq(t)] \exp\left[-\lambda\mathcal{A}(q(t))\right]. \qquad (A38.46)$$

We now rescale $q(t)$ in this expression

$$\lambda^{1/2} q(t) \mapsto q(t).$$

We get a factor coming from the jacobian. If we assume that we have discretized the time in the path integral with a spacing $\Delta\beta$ we find

$$\langle \mathcal{A}\rangle = -\frac{\partial}{\partial\lambda}\bigg|_{\lambda=1} \ln \left\{\lambda^{-\frac{1}{2}\beta/\Delta\beta} \operatorname{tr} \exp\left[-\beta H(g/\lambda)\right]\right\}. \qquad (A38.47)$$

In the large β limit this yields

$$\langle \mathcal{A}\rangle = \frac{1}{2}\frac{\beta}{\Delta\beta} - \beta g \frac{\partial E_0}{\partial g}. \qquad (A38.48)$$

The first term $\frac{1}{2}\beta/\Delta\beta$ is just the term $N/2$ we found in the previous example. The second term is still proportional to the "volume" β but no longer depends on the discretization.

39 UNSTABLE VACUA IN FIELD THEORY

We begin with this chapter a semiclassical study of barrier penetration effects in quantum field theory. We first discuss a scalar field theory with a ϕ^4 interaction, generalization of the quartic anharmonic oscillator considered in Chapter 37, in the dimensions in which it is super-renormalizable, i.e. two and three dimensions. New problems arise in four dimensions, the dimension in which the theory is only renormalizable, which will be examined later.

We have shown that in quantum mechanics barrier penetration is associated with classical motion in imaginary time; we here consider therefore quantum field theory in its euclidean formulation.

We have argued that the lifetime of metastable states is related to the imaginary part of the "ground state" energy. However, for later purpose, it is useful to calculate the imaginary part not only of the vacuum amplitude but also of correlation functions. We therefore start from the functional integral representation of euclidean correlation functions. As in quantum mechanics, barrier penetration in the semiclassical limit is related the existence of instanton solutions which we try to characterize. We then explain how to evaluate the instanton contributions at leading order, the main new problem arising from UV divergences.

In Section 39.2 we consider more general scalar field theories, of a form analogous to the quantum mechanical models discussed in Chapter 38.

In the case of the vacuum amplitude we find that the instanton contribution is proportional to the space-time volume. Dividing by the volume we therefore obtain the probability per unit time and unit volume of the metastable pseudo-vacuum to decay. The last section is devoted to a brief discussion of a speculative cosmological application of this result.

39.1 The ϕ^4 Field Theory for Negative Coupling

We consider the d-dimensional field theory for a scalar field ϕ, corresponding to the action $S(\phi)$

$$S(\phi) = \int \mathrm{d}^d x \left(\frac{1}{2} \left(\partial_\mu \phi(x) \right)^2 + \frac{1}{2} m^2 \phi^2(x) + \frac{1}{4!} g m^{4-d} \phi^4(x) \right), \qquad (39.1)$$

m being the mass and g the dimensionless coupling constant (the power of m which appears in front of the interaction term ϕ^4 takes care of the dimension).

The complete n-point correlation function has the functional representation:

$$Z^{(n)}(x_1, \ldots, x_n) = \int [\mathrm{d}\phi(x)] \, \phi(x_1) \, \phi(x_2) \ldots \phi(x_n) \exp[-S(\phi)]. \qquad (39.2)$$

We normalize the functional integral with respect to the vacuum amplitude (partition function) at $g = 0$, to avoid introducing a non-trivial g dependence through the normalization. Following the method used in Chapter 37, we assume that we start from positive values of g and proceed by analytic continuation to define the functional integral for g negative. The imaginary part of correlation functions is given by the difference between the continuations above and below the negative g-axis. For g small, only non-trivial saddle points contribute to the imaginary part. Therefore we look for non-trivial finite action solutions of the euclidean field equations, i.e. instanton configurations, and then calculate the corresponding contributions.

39.1.1 Instantons: Classical solutions and classical action

The instanton solutions. The field equation corresponding to the action (39.1) is:

$$\left(-\Delta + m^2\right)\phi(x) + \tfrac{1}{6}gm^{4-d}\phi^3(x) = 0. \tag{39.3}$$

Let us set (g is negative):

$$\phi(x) = (-6/g)^{1/2}m^{d/2-1}f(mx). \tag{39.4}$$

In terms of f the classical action (39.1) reads

$$S(f) = -\frac{6}{g}\int \mathrm{d}^d x \left[\tfrac{1}{2}\left(\partial_\mu f\right)^2 + \tfrac{1}{2}f^2 - \tfrac{1}{4}f^4\right]. \tag{39.5}$$

The function $f(x)$ then satisfies a parameter-free equation,

$$\left(-\Delta + 1\right)f(x) - f^3(x) = 0. \tag{39.6}$$

It can be shown (for details see Appendix A39) that the solution with the smallest action is spherically symmetric. We therefore choose an arbitrary origin x_0 and set:

$$r = |x - x_0|. \tag{39.7}$$

A function $f(x)$ which depends only on the radial variable r satisfies the differential equation

$$\left[-\left(\frac{\mathrm{d}}{\mathrm{d}r}\right)^2 - \frac{d-1}{r}\frac{\mathrm{d}}{\mathrm{d}r} + 1\right]f(r) - f^3(r) = 0. \tag{39.8}$$

Interpreting r as a time, we note that this equation describes the motion of a particle in a potential $-V(f)$

$$V(f) = \tfrac{1}{2}f^2 - \tfrac{1}{4}f^4, \tag{39.9}$$

submitted in addition to a viscous damping force.

Since we look for finite action solutions we have to impose the boundary condition

$$f(r) \to 0 \quad \text{for} \quad r \to \infty. \tag{39.10}$$

Equation (39.8) shows that if $f(r)$ goes to zero at infinity it goes exponentially. The equation has solutions even in r, which are thus determined by the value of f at the origin. For a generic value of $f(0)$, the corresponding solution tends at infinity towards a minimum of the potential $f = \pm 1$. The condition (39.10) is satisfied only for a discrete set of initial values of $f(0)$. It can moreover be shown that the minimal action solution corresponds to the function for which $|f(0)|$ is minimal in the set, and which vanishes only at infinity.

The classical action. Let us denote by f_c a solution. Since g is dimensionless the corresponding classical action has the form

$$S(\phi_c) \equiv S(f_c) = -A/g, \tag{39.11}$$

in which A is a pure number, whose sign can be determined by the following argument:

If $S[f_c(x)]$ is finite so is $S[f_c(\lambda x)]$. After the change of variables $\lambda x \mapsto x$, $S[f_c(\lambda x)]$ takes the form

$$-\tfrac{g}{6} S[f_c(\lambda x)] = \lambda^{2-d} \int \mathrm{d}^d x\, \tfrac{1}{2} (\partial_\mu f_c)^2 + \lambda^{-d} \int \mathrm{d}^d x\, \left(\tfrac{1}{2} f_c^2 - \tfrac{1}{4} f_c^4\right).$$

This expression is stationary at $\lambda = 1$ because $f_c(x)$ is a solution of the field equation. This yields the identity:

$$\int \mathrm{d}^d x\, \left(f_c^2 - \tfrac{1}{2} f_c^4\right) = \frac{2-d}{d} \int \mathrm{d}^d x\, (\partial_\mu f_c)^2 . \tag{39.12}$$

We can use this relation to eliminate the potential term. We then get an expression for A

$$A = \frac{6}{d} \int [\partial_\mu f(x)]^2\, \mathrm{d}^d x > 0, \tag{39.13}$$

which shows that A is always positive.

Remark. It is clear that the relation (39.12) has an analogue for arbitrary potentials. In the case of interaction terms which are proportional to a simple power of ϕ, an additional relation can be derived.

If $S(f_c)$ is finite, so is $S(\lambda f_c)$. Moreover $S(\lambda f_c)$ is stationary at $\lambda = 1$ because f_c is a solution of the field equation. Let us write the corresponding relation,

$$\int \mathrm{d}^d x\, \left[(\partial_\mu f_c)^2 + f_c^2 - f_c^4\right] = 0. \tag{39.14}$$

We then find the relations

$$A = \frac{6}{d} \int [\partial_\mu f(x)]^2\, \mathrm{d}^d x = \frac{3}{2} \int f^4(x)\mathrm{d}^d x = \frac{6}{4-d} \int f^2(x)\mathrm{d}^d x .$$

We immediately note that these relations can only be true for $d < 4$ and thus the dimension four is singular (see Appendix).

Let us here give the numerical results for $d = 2$ and $d = 3$:

$$d = 2: \quad f_c(0) = 2.20620086465 \quad A = 35.102689573 \tag{39.15}$$
$$d = 3: \quad f_c(0) = 4.33738767997 \quad A = 113.383507815 \tag{39.16}$$

39.1.2 The gaussian integration

We want to perform the gaussian integration in the neighbourhood of the saddle point. This involves studying the eigenvalues of the operator $\mathbf{M}$:

$$\langle x\,|\mathbf{M}|\,x'\rangle \equiv M\,(x,x') = \frac{\delta^2 S}{\delta\phi_c(x)\delta\phi_c\,(x')}, \tag{39.17}$$

$$= \left[(-\Delta_x + m^2) + \tfrac{1}{2}gm^{4-d}\phi_c^2\,(x)\right]\delta\,(x-x')\,, \tag{39.18}$$

or, introducing the solution $f(x)$,

$$M\,(x,x') = \left[(-\Delta_x + m^2) - 3m^2 f^2\,(mx)\right]\delta\,(x-x')\,. \tag{39.19}$$

Differentiating the equation of motion (39.3) with respect to x_μ, we learn that the functions $\partial_\mu\phi_c$ are d eigenvectors of $\mathbf{M}$ with vanishing eigenvalue:

$$(-\Delta + m^2)\,\partial_\mu\phi_c(x) + \tfrac{1}{2}gm^{4-d}\phi_c^2(x)\partial_\mu\phi_c(x) = 0 \quad\Longleftrightarrow\quad \mathbf{M}\,\partial_\mu\phi_c = 0\,. \tag{39.20}$$

This property had to be expected. Due to translation symmetry we have found a family of degenerate saddle points $\phi_c(x)$ depending on d parameters $x_{0\mu}$ (equation (39.7)). As in quantum mechanics, we have to sum over all saddle points. We thus change variables, taking the collective coordinates $x_{0\mu}$ as d of our new variables. For instance we can set:

$$\phi(x) = \phi_c\,(x - x_0) + \sum_{n \geq 0}\varphi_n f_n\,(x - x_0)\,, \tag{39.21}$$

with the constraints:

$$\int \mathrm{d}^d x\,\partial_\mu\phi_c(x)f_n(x) = 0 \qquad \forall\, n\,,\mu\,, \tag{39.22}$$

in which $\{x_{0\mu}, \varphi_n\}$ are the new variables. We then integrate over $x_{0\mu}$ exactly and evaluate the integral over the other variables φ_n by the steepest descent method.

The gaussian integration involves the determinant of $\mathbf{M}$ in the subspace orthogonal to the zero eigenvalue sector. The change of variables generates a jacobian J, which, at leading order, is:

$$J = \prod_{\mu=1}^{d} \|\partial_\mu\phi_c\,(x)\|\,. \tag{39.23}$$

Since the solution is spherically symmetric, we can rewrite J as

$$J = \left[\frac{1}{d}\int \mathrm{d}^d x \sum_\mu (\partial_\mu\phi_c)^2\right]^{d/2}\,. \tag{39.24}$$

Using the relation (39.13) we can express the jacobian in terms of the classical action

$$J = (-A/g)^{d/2}\,. \tag{39.25}$$

Note one important feature of this expression: each component of x_0 has generated a factor $(-g)^{-1/2}$. Finally in expression (39.2) we can replace at leading order the field $\phi(x)$ by $\phi_c(x)$ in the product $\prod_{i=1}^{n}\phi\,(x_i)$. Since we normalize the functional integral

(39.2) by comparing it with the functional integral of the free theory, we still have to divide by a factor $\sqrt{2\pi}$ for each missing gaussian integration. Collecting all factors we find

$$\operatorname{Im} Z^{(n)}\left(x_1, \ldots, x_n\right) = \frac{1}{2i}\left(\frac{A}{2\pi}\right)^{d/2} \Omega \frac{e^{A/g}}{(-g)^{(d+n)/2}} F_n\left(x_1, \ldots, x_n\right), \qquad (39.26)$$

with

$$F_n\left(x_1, \ldots, x_n\right) = m^{d+n(d-2)/2} 6^{n/2} \int \mathrm{d}^d x_0 \prod_{i=1}^n f\left(m\left(x_i - x_0\right)\right), \qquad (39.27)$$

and

$$\langle x \left|\mathbf{M}_0\right| x'\rangle = \left(-\Delta_x + m^2\right)\delta\left(x - x'\right), \qquad (39.28a)$$

$$\Omega = \left(\det' \mathbf{M}\mathbf{M}_0^{-1}\right)^{-1/2}\bigg|_{m=1} = \lim_{\epsilon \to 0}\epsilon^{-d}\det\left[(\mathbf{M}+\epsilon)\mathbf{M}_0^{-1}\right]\big|_{m=1}. \qquad (39.28b)$$

While for the vacuum amplitude the integration over x_0 generates a factor proportional to the volume, for non-trivial correlation functions the integration restores translation invariance.

Wave function arguments of the kind used for the Schrödinger equation show that $\partial_\mu \phi_c$ is not the ground state of $\mathbf{M}$. There is one state with a negative eigenvalue so that the final result is real as expected. In Appendix A39 we give a proof of this property using Sobolev inequalities.

Discussion. A few comments concerning expression (39.26) are here in order. We have obtained a result for the complete correlation functions, improperly normalized, for convenience, with respect to the free field theory. We notice, however, that, because $\phi_c(x)$ is proportional to $1/\sqrt{-g}$, the imaginary part of the n-point function increases with n for g small. This shows that at leading order the correlation functions normalized with respect to the partition function corresponding to the complete action (39.1) have the same behaviour as those renormalized with respect to the free field theory.

Moreover, for the same reason, when we consider a complete n-point function, the imaginary part coming from disconnected parts is subleading by at least a power of g. If we call $W^{(n)}\left(x_1, \ldots, x_n\right)$ the connected n-point function, we thus find at leading order:

$$\operatorname{Im} W^{(n)} \sim \operatorname{Im} Z^{(n)},$$

a result that is consistent with the observation that the explicit expression (39.26) is indeed connected. To pass from connected correlation functions to 1PI functions, we have first to subtract the reducible contributions which involve functions with a smaller number of arguments and which are therefore negligible at leading order, and then to amputate the remaining part. Again for the same reason only the real part of the propagator matters; to amputate expression (39.26) we therefore have simply to multiply it by the product of the inverse free propagators corresponding to each external line. Introducing $\tilde{f}(p/m)$, the Fourier transform of $f(mx)$, and writing the n-point 1PI function $\tilde{\Gamma}^{(n)}$ in momentum space representation we obtain:

$$\operatorname{Im} \tilde{\Gamma}^{(n)}\left(p_1, \ldots, p_n\right) \sim -\frac{1}{2i}\left(\frac{A}{2\pi}\right)^{d/2}\Omega\frac{e^{A/g}}{(-g)^{(d+n)/2}} m^{d-n(d/2+1)}\prod_{i=1}^n \sqrt{6}\tilde{f}\left(\frac{p_i}{m}\right)\left(p_i^2 + m^2\right).$$

$$(39.29)$$

The structure, at leading order, of the imaginary part of the n-point function is particularly simple in momentum space, in particular it depends only on the square of the momenta and not of their scalar products.

Up to this point the discussions of the ϕ^4 field theory and of the anharmonic oscillator have been remarkably similar. Now comes one important difference: as we shall see the determinant of the operator $\mathbf{M}$ is actually UV divergent and we shall have to deal with this new problem.

39.1.3 Renormalization

To define properly the ϕ^4 in 2 and 3 dimensions, we have first to introduce a UV cut-off and then to add to the classical action a mass counterterm before taking the infinite cut-off limit: the cut-off dependent action S_Λ has the form:

$$S_\Lambda(\phi) = \int \mathrm{d}^d x \left[\tfrac{1}{2}\phi(x)\left(-\Delta + \Delta^2/\Lambda^2 + m^2\right)\phi(x) + \tfrac{1}{4!}g\phi^4(x) + \tfrac{1}{2}\delta m^2(\Lambda)\phi^2(x) \right].$$
(39.30)

Let us examine, at cut-off Λ large but fixed, the effect of these modifications. The additional term

$$\frac{1}{\Lambda^2}\int \phi(x)\Delta^2\phi(x)\mathrm{d}^d x,$$

modifies the equation of motion but when Λ becomes large this modification becomes small like $1/\Lambda^2$.

The counterterm on the other hand increases with the cut-off but is proportional to at least one power of g. Hence, because we take the small g limit before taking the large cut-off limit, the counterterm does not contribute to the classical equation of motion.

On the other hand, if we calculate the contribution of the counterterm to the classical action we find that the one-loop counterterm, which is proportional to g, gives a contribution of order 1 in g because $\phi_c(x)$ is proportional to $1/\sqrt{-g}$. It therefore generates an additional multiplicative factor. Let us now consider the determinant of $\mathbf{M}$ in the regularized theory:

$$M(x, x') = \left[\left(-\Delta + \Delta^2/\Lambda^2 + m^2\right) + \tfrac{1}{2}gm^{4-d}\phi_c^2(x)\right]\delta(x - x'),$$
(39.31)

We can expand it in powers of $\phi_c^2(x)$ using the identity $\operatorname{tr}\ln = \ln\det$:

$$\ln\det\mathbf{M} = \ln\det\left(-\Delta + \Delta^2/\Lambda^2 + m^2\right)$$
$$- \sum_{k=1}^{\infty} \frac{1}{k} \operatorname{tr}\left[\tfrac{1}{2}gm^{4-d}\phi_c^2(x)\left(\Delta - \Delta^2/\Lambda^2 - m^2\right)^{-1}\right]^k.$$
(39.32)

The first term is cancelled by the free determinant $\det\mathbf{M_0}$. All terms for $k \geq 2$, are UV finite in 2 and 3 dimensions. Let us rewrite more explicitly the $k = 1$ term:

$$\tfrac{1}{2}\operatorname{tr} gm^{4-d}\phi_c^2(x)\left(-\Delta + \Delta^2/\Lambda^2 + m\right)^{-1} = \tfrac{1}{2}gm^{4-d}D\int \mathrm{d}^d x\,\phi_c^2(x),$$
(39.33)

in which D is the free regularized propagator at coinciding arguments,

$$D = \frac{1}{(2\pi)^d}\int \frac{\mathrm{d}^d p}{p^2 + p^4/\Lambda^2 + m^2}.$$
(39.34)

The determinant of the operator $\mathbf{M}$ therefore contains a factor which diverges at large cut-off:

$$\exp\left[-\tfrac{1}{4}gm^{4-d}D\int \mathrm{d}^d x\,\phi_c^2(x)\right].\tag{39.35}$$

This factor exactly cancels the infinite factor coming from the counterterm, so that the final expression for the imaginary part is finite. The fact that really we have to calculate $\det'\mathbf{M}$ does not change this argument, because UV divergences are insensitive to the omission of a finite number of eigenvalues of $\mathbf{M}$, as the second form (39.28b) explicitly shows.

The decay of the false vacuum. The special case $n = 0$ corresponds to the imaginary part of the vacuum amplitude. We have expanded perturbation theory around the minimum $\phi = 0$ of the potential. The perturbative ground state corresponds to a wave function concentrated around small fields. However, because we have expanded around a relative minimum of the potential, this state actually is unstable. We have calculated its decay rate due to barrier penetration. We note that the integral over $x_{0\mu}$ in equation (39.27) yields a space-time volume factor. To obtain a finite decay amplitude we have to divide the result by this volume factor. We thus obtain the probability per unit time and *unit volume* for the would-be ground state ("false vacuum") of the theory to decay. Some implications of such a result will be discussed in a slightly more general context in Section 39.3.

39.2 General Potentials: Instanton Contributions

We now extend the analysis of Chapter 38 to more general scalar field theories, using the techniques developed in Section 39.1. We consider an euclidean action of the form:

$$S\left(\phi\right)=\int \mathrm{d}^d x\left(\tfrac{1}{2}\left(\partial_\mu\phi\right)^2+g^{-1}V\left(\phi\sqrt{g}\right)\right),\tag{39.36}$$

in which the potential $V(\phi)$ has a stable and an unstable minimum, and is of the type discussed in Chapter 38. Assuming that at some initial time the quantum mechanical state corresponds to a field concentrated around the unstable minimum of the potential, the "false" vacuum, we want to evaluate semiclassically the probability for the false vacuum to decay into the true vacuum of the theory. The calculation, at leading order, again involves the determination of an instanton solution and a gaussian integration around the instanton.

39.2.1 Calculation of the instanton contribution

We define the field in such a way that the unstable minimum corresponds to $\phi = 0$. The discussion of the existence of an instanton solution is similar to the one given in Section 39.1. A theorem establishes, under mild assumptions, that spherically symmetric solutions give the minimal action. We therefore look for such a solution and set:

$$r=|x-x_0|,\quad f(r)=\sqrt{g}\phi_c(x).$$

The classical equation of motion reduces to:

$$\frac{\mathrm{d}^2 f}{\mathrm{d}r^2}+\frac{d-1}{r}\frac{\mathrm{d}f}{\mathrm{d}r}=V'\left(f\right).\tag{39.37}$$

This again is the equation governing the motion of a particle in a potential $V(f)$ and submitted to a viscous damping force. The solution depends on its value at the origin $f(0)$.

Let us call f_+ the absolute minimum of the potential. If we choose $f(0)$ too close to f_+, $f(r) - f(0)$ will remain very small until r becomes very large. When r is large, the damping force is small so that energy is almost conserved and the particle will overshoot. If $f(0)$ is too close to zero, the particle will lose too much energy and therefore undershoot, the asymptotic value $f(r)$ then corresponding to the maximum of $V(f)$. Thus somewhere in between we expect to find a value $f(0)$, which corresponds to a solution which goes to zero at infinity and therefore has a finite action. The virial theorem, as used in Section 39.1, implies that the corresponding action is positive: Indeed let $\phi_c(x)$ be a solution. We calculate $S[\phi_c(\lambda x)]$,

$$S[\phi_c(\lambda x)] = \lambda^{2-d} \int \frac{1}{2}(\partial_\mu \phi_c)^2 \, \mathrm{d}^d x + \lambda^{-d} \frac{1}{g} \int V(\phi_c \sqrt{g}) \, \mathrm{d}^d x . \tag{39.38}$$

Expressing that the action is stationary for $\lambda = 1$ yields the relation

$$(2-d) \int \frac{1}{2}(\partial_\mu \phi_c)^2 \, \mathrm{d}^d x - \frac{d}{g} \int \mathrm{d}^d x \, V(\phi_c(x)\sqrt{g}) = 0 . \tag{39.39}$$

Therefore

$$S[\phi_c(x)] = \frac{A}{g} , \tag{39.40}$$

with again:

$$A = \frac{1}{d} \int (\partial_\mu f)^2 \, \mathrm{d}^d x > 0 . \tag{39.41}$$

The second derivative of $S[\phi_c(\lambda x)]$ is

$$\frac{\mathrm{d}^2}{(\mathrm{d}\lambda)^2} S[\phi_c(\lambda x)]|_{\lambda=1} = (2-d) \int [\partial_\mu \phi_c(x)]^2 \, \mathrm{d}^d x . \tag{39.42}$$

This equation yields another proof at least for $d \geq 3$ that the second derivative of the action at the saddle point has at least one negative eigenvalue. To get a more precise result, one has to resort to a general theorem established by Coleman *et al* that states that $\phi_c(x)$ corresponds to an absolute minimum of $S(\phi)$ at fixed integral of the potential $\int \mathrm{d}^d x \, V(\sqrt{g}\phi_c(x))$. Indeed let us assume that the operator

$$\frac{\delta^2 S}{\delta\phi_c(x)\delta\phi_c(x')}$$

has two negative eigenvalues. One can then add to $\phi_c(x)$ a linear combination of the corresponding two eigenvectors leaving at first order the integral of the potential unchanged, and decreasing $S(\phi)$. This contradicts the theorem. Thus $\delta^2 S/\delta\phi_c(x)\,\delta\phi_c(x')$ has at most one negative eigenvalue.

Since equation (39.42) or more generally the argument about the eigenvalue, shows that it has at least one, $\delta^2 S/(\delta\phi_c)^2$ has one and only one negative eigenvalue.

We now have to take x_0 as an integration variable. This leads to a jacobian J which, as we have seen in Section 39.1, can be written as

$$J = \prod_{\mu=1}^{d} \|\partial_\mu \phi_c\| = \left[\frac{1}{d} \int \mathrm{d}^d x \sum_\mu (\partial_\mu \phi_c)^2 \right]^{d/2} = \left(\frac{A}{g} \right)^{d/2} , \tag{39.43}$$

where equations (39.40,39.41) have been used.

All other details of the calculation can be borrowed from the ϕ^4 example, and we finally obtain an explicit expression for the imaginary part of the n-point correlation function.

Let us just make a few remarks concerning the renormalization. (If we suppose the theory super-renormalizable or renormalizable).

Renormalization. To construct the renormalized theory we can proceed in an inductive way: we regularize the theory, we then add the counterterms which make the theory finite order by order in a loopwise expansion, i.e. here an expansion in powers of g. The renormalized action $S_{\rm r}(\phi)$ has the form:

$$S_{\rm r}(\phi) = \frac{1}{g} S_0(\phi\sqrt{g}) + S_1(\phi\sqrt{g}) + \cdots + g^{L-1} S_L(\phi\sqrt{g}) + \cdots .$$

At leading order only the one-loop counterterms matter in the instanton calculation. To evaluate them we expand the effective potential $\Gamma(\phi)$ (the generating functional of 1PI correlation functions) at one-loop order using the regularized action, and calculate its divergent part. In Chapter 5 we have derived:

$$\Gamma_{1\,\text{loop}}(\phi) = S(\phi) + \tfrac{1}{2}\,{\rm tr}\ln\frac{\delta^2 S}{\delta\phi(x)\delta\phi(x')}. \tag{39.44}$$

To render correlation functions finite we have to subtract to the regularized action the divergent part of the one loop term of $\Gamma(\phi)$:

$$S_1(\phi\sqrt{g}) = -\tfrac{1}{2}\left({\rm tr}\ln\frac{\delta^2 S}{\delta\phi\delta\phi}\right)_{\rm div.}.$$

When evaluated for $\phi = \phi_c$, this contribution exactly cancels the divergence in the determinant coming from the gaussian integration around the saddle point. This argument can be generalized to arbitrary orders.

39.3 The Decay of the False Vacuum: Cosmological Interpretation

In preceding sections we have in particular determined the probability for a "false vacuum" of a Field Theory to decay through barrier penetration effects. It has been speculated that such a phenomenon could be linked to the dynamics of the early universe. When the universe started to cool down, some symmetries started to be spontaneously broken. Some region might have been trapped in the wrong phase. The false vacuum must eventually decay in the true vacuum, but if the process is slow enough, it might have occurred at a much later time when the universe was already cool. This is the kind of physical application that we here have in mind.

According to the previous discussion, if the universe is in the wrong vacuum, there is some probability at each point in space for some bubble of true vacuum to be created, and if the bubble is large enough, it becomes favourable for it to expand, absorbing eventually the whole space. To discuss what happens once a bubble has been created, it is useful to consider first the analogous problem in ordinary quantum mechanics.

Quantum Mechanics. In the language of particle physics, a semiclassical description of the decay process would be the following: a particle is sitting in the well of the potential corresponding to the unstable minimum. At a given time, it makes a quantum jump and reappears outside of the barrier, at the point where the potential has the same value as in the bottom of the well, with zero velocity (by energy conservation). Then its further trajectory can be entirely described by Classical Mechanics.

Field Theory. Let us apply the same ideas to the field theoretical model we discuss here. At time zero the system makes a quantum jump. According to the previous discussion, the value of the field at time zero is then (with the choice $x_{0\mu} = 0$)

$$\phi\left(t = 0, \mathbf{x}\right) = \phi_c\left(x_d = 0, \mathbf{x}\right), \qquad \left(\mathbf{x} = x_1, \ldots, x_{d-1}\right), \tag{39.45}$$

and its time derivative vanishes,

$$\partial_t \phi\left(t = 0, \mathbf{x}\right) = 0. \tag{39.46}$$

At a later time $\phi\left(t, \mathbf{x}\right)$ then obeys the *minkowskian* equation of motion

$$\left[\nabla_i^2 - \partial_t^2\right] \phi\left(t, \mathbf{x}\right) = \frac{1}{\sqrt{g}} V'\left(\sqrt{g}\phi\right). \tag{39.47}$$

The first equation (39.45) tells us that the same function describes the form of the instanton in euclidean space, and its shape in ordinary $(d-1)$ space when it materializes. Let us now consider the continuation in real time of the solution of the euclidean equation of motion $\phi_c\left[\left(\mathbf{x}^2 - t^2\right)^{1/2}\right]$ (since $\phi_c(r)$ is an even function, the sign in front of the square root is irrelevant). It satisfies the conditions (39.45,39.46) and obviously obeys the equation of motion (39.47). It is therefore the solution of our problem for positive times.

Since the size of the bubble is given by microphysics, the interior of the bubble corresponds to small values of r on a macroscopic scale,

$$0 \leq \mathbf{x}^2 - t^2 = r^2 \ll 1.$$

Therefore after a short time the bubble starts expanding at almost the speed of light.

Bibliographical Notes

The first calculations of instanton contributions in Field Theory are due to
> L.N. Lipatov, *JETP Lett.* 24 (1976) 157, *Sov. Phys. JETP* 44 (1976) 1055, *JETP Lett.* 25 (1977) 104, *Sov. Phys. JETP* 45 (1977) 216.

We here have followed the method presented in
> E. Brézin, J.C. Le Guillou and J. Zinn-Justin, *Phys. Rev.* D15 (1977) 1558; J. Zinn-Justin, *Phys. Rep.* 70 (1981) 109.

To simplify expressions we have used scaling arguments
> G.H. Derrick, *J. Math. Phys.* 5 (1964) 1252.

The field equations for the ϕ^4 field theory have been solved numerically by Zinn-Justin unpublished, and the determinant calculated in
> E. Brézin and G. Parisi, *J. Stat. Phys.* 19 (1978) 269.

The relativistic generalization of the method of collective coordinates is described in
> J.-L. Gervais and B. Sakita, *Phys. Rev.* D11 (1975) 2943; V.E. Korepin, P.P. Kulish and L.D. Faddeev, *JETP Lett.* 21 (1975) 138; N.H. Christ and T.D. Lee, *Phys. Rev.* D12 (1975) 1606.

The standard early reference on vacuum decay is

M.B. Voloshin, I. Yu. Kobzarev and L.B. Okun, *Yad. Fiz.* 20 (1974) 1229 (*Sov. J. Nucl. Phys.* 20 (1975) 644).

The instanton approach to this problem has been developed in

S. Coleman, *Phys. Rev.* D15 (1977) 2929; C.G. Callan and S. Coleman, *Phys. Rev.* D16 (1977) 1762; M. Stone, *Phys. Rev.* D14 (1976) 3568, *Phys. Lett.* 67B (1977) 186; S. Coleman and F. De Luccia, *Phys. Rev.* D21 (1980) 3305; P.J. Steinhardt, *Phys. Rev.* D24 (1981) 842.

Properties of the minimal action are proven in

S. Coleman, V. Glaser and A. Martin, *Commun. Math. Phys.* 58 (1978) 211.

For a discussion of Sobolev inequalities see for example

E.M. Stein, *Singular Integrals and Differentiability Properties of Functions* (Princeton University Press, Princeton 1970); J.R. Klauder, *Acta Phys. Austr. Suppl.* XI (1973) 341.

The relevance of Sobolev inequalities for the instanton problem in the ϕ^4 field theory has been emphasized by

G. Parisi, *Phys. Lett.* 66B (1977) 167; 68B (1977) 361.

Exercises

Exercise 39.1

Find an estimate for the instanton action in the ϕ^4 field theory in $d = 2, 3, 4$ dimensions by using a first gaussian trial function, then a function proportional to the one-dimensional solution, $f(r) \propto 1/\cosh(r/r_0)$, and compare with the exact results given in Section 28.3, for $d = 2, 3$ and in Section 42.1 for $d = 4$.

Exercise 39.2

Consider the unphysical field theory in d dimensions, whose action is

$$S(\phi) = \int \mathrm{d}^d x \left[\frac{1}{2} \phi(x) \, \mathrm{e}^{-\Delta} \, \phi(x) + \frac{1}{4} g \phi^4(x) \right],$$

in which Δ is just the Laplace operator in d dimensions. In this theory the free propagator is in momentum space e^{-p^2}, thus the theory is UV finite in any dimension, and the Feynman diagrams can be calculated algebraically. It is also possible to calculate analytically the contribution of the presumably dominant instanton solution.

Ansatz: Look for a solution of the form $\mathrm{e}^{-\alpha x^2}$. Solution in Bervillier *et al.*, *Phys. Rev.* D17 (1978) 2144.

APPENDIX 39
SOBOLEV INEQUALITIES

Let us consider the following functional $R(\varphi)$,

$$R(\varphi) = \frac{\left\{ \int \mathrm{d}^d x \left[(\partial_\mu \varphi(x))^2 + \varphi^2(x) \right] \right\}^2}{\int \varphi^4(x) \mathrm{d}^d x}. \tag{A39.1}$$

If the dimension d is not larger than four:

$$d \leq 4, \tag{A39.2}$$

Sobolev inequalities tell us that

$$R(\varphi) \geq R > 0. \tag{A39.3}$$

In addition, for $d < 4$, there exists a spherically symmetric, zero-free function $\varphi_c(x)$ which saturates the bound

$$R(\varphi_c) = R, \tag{A39.4}$$

and is a solution of the variational equation

$$\delta R / \delta \varphi_c(x) = 0. \tag{A39.5}$$

Dimensional smaller than four. Let us write this last equation explicitly,

$$(-\Delta + 1)\,\varphi(x) - \varphi^3(x) K = 0, \tag{A39.6}$$

in which we have defined

$$K = \int \mathrm{d}^d x \left[(\partial_\mu \varphi_c)^2 + \varphi_c^2 \right] \bigg/ \int \varphi_c^4(x) \mathrm{d}^d x. \tag{A39.7}$$

This equation is, up to a rescaling of $\varphi_c(x)$, the equation of motion (39.6). Both equations become identical if we choose the scale of $\varphi_c(x)$ which is otherwise arbitrary such that

$$K = 1. \tag{A39.8}$$

For each instanton solution we have derived the identity (39.14), which for $f(x)$ can be written as

$$\int \mathrm{d}^d x \left[(\partial_\mu f(x))^2 + f^2(x) \right] = \int \mathrm{d}^d x\, f^4(x). \tag{A39.9}$$

Therefore we can also write the classical action as

$$A = \frac{3}{2} \int \mathrm{d}^d x\, f^4(x) = \frac{3}{2} \frac{\left\{ \int \mathrm{d}^d x \left[(\partial_\mu f(x))^2 + f^2(x) \right] \right\}^2}{\int f^4(x) \mathrm{d}^d x}. \tag{A39.10}$$

The smallest action solution thus corresponds to the minimum of $R(\varphi)$:

$$A = 3R/2, \qquad (A39.11)$$

and the solution $f(x)$ we are looking for is given by

$$f(x) = \varphi_c(x) \quad \text{for} \quad K = 1. \qquad (A39.12)$$

The introduction of the functional $R(\varphi)$ has the following advantage: the functional $S(\phi)$ is obviously not bounded from below. But if we restrict ourselves to fields $\phi(x)$ solutions of the equation of motion with finite action, then $S(\phi)$ can be related to the functional $R(\varphi)$ which is bounded from below for all fields.

Let us then derive the property that the operator $\mathbf{M}$ has one and only one negative eigenvalue from the form of $R(\varphi)$ and the assumption that φ_c corresponds to an absolute minimum of R. In the operator sense:

$$\frac{\delta^2 R}{\delta\varphi_c(x)\delta\varphi_c(x')} \geq 0. \qquad (A39.13)$$

Calculating the second derivative of R explicitly, we obtain:

$$\frac{\delta^2 R}{\delta\varphi_c(x)\delta\varphi_c(x')} = 4\left\{\left[-\Delta + 1 - 3\varphi_c^2(x)\right]\delta(x-x') + \frac{2\varphi_c^3(x)\varphi_c^3(x')}{\int\varphi_c^4(y)\mathrm{d}^d y}\right\}. \qquad (A39.14)$$

We have again set $K = 1$. Let us write below $M(x,x')$ in terms of $f(x)$ or $\varphi_c(x)$ for $m = 1$,

$$M(x,x') = \left[-\Delta + 1 - 3\varphi_c^2(x)\right]\delta(x-x'). \qquad (A39.15)$$

We therefore derive the relation:

$$M(x',x) = \frac{1}{4}\frac{\delta^2 R}{\delta\varphi_c(x)\delta\varphi_c(x')} - 2\left(\frac{\varphi_c^3(x')\,\varphi_c^3(x)}{\int\varphi_c^4(y)\mathrm{d}^d y}\right). \qquad (A39.16)$$

(i) Since $R(\varphi)$ is invariant in the change $\varphi_c(x)$ in $\lambda\varphi_c(x)$, φ_c is an eigenvector of $\delta^2 R \big/ (\delta\varphi_c)^2$ with eigenvalue zero, thus

$$\int\varphi_c(x')\,\varphi_c(x)M(x',x)\,\mathrm{d}^d x\,\mathrm{d}^d x' = -2\int\varphi_c^4(x) < 0. \qquad (A39.17)$$

The operator $\mathbf{M}$ has at least one negative eigenvalue.

(ii) Since $\mathbf{M}$ is the sum of a positive operator and a projector of rank 1, it can have at most one negative eigenvalue. Indeed if it had two negative eigenvalues, we could find a linear combination of the corresponding two eigenvectors which would decrease $\mathbf{M}$ at an average of the projector fixed. This would imply that $\delta^2 R/(\delta\varphi_c)^2$ is not a positive operator.

We conclude that $\mathbf{M}$ has *one and only one* negative eigenvalue.

Dimension four. Let us calculate $R[\varphi(\lambda x)]$ for $d \leq 4$. Then, changing λx in x' in the various integrals, we obtain:

$$R[\varphi(\lambda x)] = \frac{\left\{\int\mathrm{d}^d x\left[\lambda^{2-d}(\partial_\mu\varphi)^2 + \lambda^{-d}\varphi^2\right]\right\}^2}{\lambda^{-d}\int\varphi^4\mathrm{d}^d x}. \qquad (A39.18)$$

We can now write

$$R = \min_{\{\varphi(x)\}} \min_{\lambda} R\left[\varphi(\lambda x)\right]. \tag{A39.19}$$

The minimum in λ of expression $(A39.18)$ is obtained for

$$\lambda = \left[\frac{d}{(4-d)} \frac{\int \varphi^2(x) \mathrm{d}^d x}{\int (\partial_\mu \varphi)^2 \mathrm{d}^d x}\right]^{1/2}, \tag{A39.20}$$

and equation $(A39.19)$ becomes

$$R = \min_{\{\varphi(x)\}} \frac{16}{d^{d/2} (4-d)^{(4-d)/2}} \frac{\left(\int (\partial_\mu \varphi)^2 \mathrm{d}^d x\right)^{d/2} \left(\int \varphi^2(x) \mathrm{d}^d x\right)^{2-d/2}}{\int \varphi^4 \mathrm{d}^d x}. \tag{A39.21}$$

For $d = 4$, we see that the solution is $\lambda = \infty$ and expression $(A39.21)$ is just the equivalent of expression $(A39.1)$ for the massless ϕ^4 field theory. Since for $d = 4$ the massless ϕ^4 is scale invariant the contribution of the mass term can be arbitrarily decreased by a rescaling of the variable x.

We can draw two interesting conclusions from this analysis: the minimal value of $R(\varphi)$ is the same in four dimensions for the massive and the massless theory. The same will apply to $S(\phi)$.

The minimum of $R(\varphi)$ will be obtained from a solution of the massless field equation. The massive field equation has no solution.

These remarks will become useful when we discuss the ϕ^4 field theory in Chapter 42.

40 DEGENERATE CLASSICAL MINIMA AND INSTANTONS

We consider in this chapter a somewhat different situation in which instantons play a role: classical actions with degenerate isolated minima. The simplest examples are provided by quantum mechanics in the form of hamiltonian systems corresponding to potentials with degenerate minima. Classically the state of minimum energy corresponds to a particle sitting at any of the minima of the potential. The position of the particle breaks (spontaneously) the symmetry of the system. In quantum mechanics on the contrary, we expect the modulus of the ground state wave function to be large near all minima of the potential, as a consequence of barrier penetration effects. We discuss in this chapter this phenomenon on two typical examples: the double-well potential and the periodic cosine potential.

In field theory the problem is more subtle as the study of phase transitions has shown. However the presence of instantons again indicates that the classical minima are connected and that the symmetry between them is not spontaneously broken. Examples of instantons of this type are provided in two dimensions by the $CP(N-1)$ models and in four dimensions by $SU(2)$ gauge theories.

Finally in Section 40.6 we discuss the role of instantons in the context of stochastic dynamics.

40.1 The Double-Well Potential

We first discuss the simple example of the double-well potential:

$$H = -\tfrac{1}{2}\left(\mathrm{d}/\mathrm{d}x\right)^2 + \tfrac{1}{2}x^2\left(1 - x\sqrt{g}\right)^2. \tag{40.1}$$

The potential has two degenerate minima located at the origin and at $1/\sqrt{g}$. In addition the hamiltonian commutes with the reflection operator P which exchanges x and $g^{-1/2} - x$. This last property is not essential for the existence of instanton solutions. It is just a simplifying feature, which moreover is present in most examples of physical interest.

The structure of the ground state. Due to the symmetry of the potential, we can expand around each of the minima of the potential, and we find the same perturbative expansion to all orders. It would therefore seem that the quantum hamiltonian also has a doubly degenerate ground state, corresponding to two eigenfunctions respectively concentrated around each of the classical minima of the potential. Actually we know that, due to barrier penetration effects, this is not the case and the true eigenstates are also eigenstates of the reflection operator P, the ground state being an even state. The reflection symmetry is not spontaneously broken in quantum mechanics in the case of regular potentials: as we have extensively discussed, correlation functions constructed with a hamiltonian of this type have, from the point of view of phase transitions, the properties of correlation functions of the 1D Ising model (see Section 23.1). Nevertheless if we use the quantity $\mathrm{tr}\,\mathrm{e}^{-\beta H}$ to calculate the ground state energy, since we expand in g small first, we find it difficult to separate the two first eigenstates in the large β limit:

$$\mathrm{tr}\,\mathrm{e}^{-\beta H} \sim \mathrm{e}^{-\beta H_+} + \mathrm{e}^{-\beta H_-} \sim 2\,\mathrm{e}^{-\beta(E_+ + E_-)/2}\cosh\beta\left(E_+ - E_-\right)/2$$

$$\text{for } g \to 0, \quad \beta \to \infty. \tag{40.2}$$

The partition function is dominated by the perturbative expansion of the half sum $\frac{1}{2}(E_+ + E_-)$, and is only sensitive to the non-perturbative difference between the eigenvalues E_+ and E_- at order $(E_+ - E_-)^2$:

$$-\frac{1}{\beta}\ln \operatorname{tr} e^{-\beta H} = \frac{1}{2}(E_+ + E_-) - \frac{1}{\beta}\ln 2 + O\left[e^{-\beta}, \beta\left(E_+ - E_-\right)^2\right]$$

$$\text{for } g \to 0, \quad \beta \to \infty. \tag{40.3}$$

To calculate the difference $(E_+ - E_-)$, it is therefore much easier to evaluate the quantity $\operatorname{tr} P e^{-\beta H}$ (see Section 23.2). In the same limits we now find:

$$\operatorname{tr} P e^{-\beta H} \sim e^{-\beta E_+} - e^{-\beta E_-} \sim -2\sinh \tfrac{1}{2}\beta\left(E_+ - E_-\right) e^{-\beta(E_+ + E_-)/2} \tag{40.4}$$

$$g \to 0, \quad \beta \to \infty.$$

Since $E_+ - E_-$ vanishes in perturbation theory, the r.h.s. is dominated by

$$\operatorname{tr} P e^{-\beta H} \sim -\beta e^{-\beta/2}\left(E_+ - E_-\right)\left[1 + O\left(g, e^{-\beta}\right)\right]. \tag{40.5}$$

It is actually more convenient to consider the ratio between the quantities (40.2) and (40.4):

$$\operatorname{tr} P e^{-\beta H} / \operatorname{tr} e^{-\beta H} \sim -\tfrac{1}{2}\beta\left(E_+ - E_-\right)\left[1 + O\left(e^{-\beta}, (E_+ - E_-)^2\right)\right]. \tag{40.6}$$

By evaluating this ratio, we can distinguish between a situation in which the ground state is degenerate and the symmetry spontaneously broken, and a situation in which quantum fluctuations restore the symmetry and lift the degeneracy between the two lowest lying states. Since the ratio vanishes in perturbation theory, we have to look for non-perturbative effects: they are here due to instantons.

Instanton contributions. The path integral representation of $\operatorname{tr} P e^{-\beta H}$ differs from the representation of the partition function only by boundary conditions:

$$\operatorname{tr} P e^{-\beta H} = \int_{q(-\beta/2)+q(\beta/2)=g^{-1/2}} [dq(t)]\exp\left[-S(q)\right], \tag{40.7}$$

with

$$S(q) = \int_{-\beta/2}^{\beta/2}\left[\tfrac{1}{2}\dot{q}^2(t) + \tfrac{1}{2}q^2(t)\left(1 - \sqrt{g}q(t)\right)^2\right]dt. \tag{40.8}$$

The path integral representation of $\operatorname{tr} e^{-\beta H}$ is dominated by the trivial saddle points:

$$q(t) = 0 \quad \text{or} \quad q(t) = g^{-1/2},$$

and this leads to the usual perturbative expansion. However these paths do not contribute to the path integral (40.7) because they do not satisfy the boundary conditions. This is not too surprising since we have already seen that the difference $E_+ - E_-$ vanishes to all orders in an expansion in powers of g. We therefore have to look for non-trivial solutions of the equation of motion which have a finite action in the infinite β limit. The boundary conditions then impose

$$q(\mp\infty) = 0 \quad \text{and} \quad q(\pm\infty) = g^{-1/2}.$$

The non-degeneracy of the ground state thus depends on the existence of an instanton solution connecting the two minima of the potential.

The euclidean equation of motion is

$$-\ddot{q}(t) + q(t)\left(1 - q(t)\sqrt{g}\right)\left(1 - 2q(t)\sqrt{g}\right) = 0. \tag{40.9}$$

In the infinite β limit, this equation indeed has two solutions with finite classical action which we call generically instantons or instanton and anti-instanton when it is necessary to distinguish between them:

$$q_c^{\pm}(t) = \frac{1}{\sqrt{g}} \frac{1}{\left(1 + e^{\mp(t-t_0)}\right)}. \tag{40.10}$$

Let us calculate the corresponding classical action. It is easy to adapt the method of Section 38.2 to the present problem. From equations analogous to (38.51,38.54), we obtain the expansions of the classical energy and action for β large:

$$E\left(\beta\right) = -2\,\mathrm{e}^{-\beta} + O\left(\mathrm{e}^{-2\beta}\right), \tag{40.11}$$

$$S\left(q_c\right) = \frac{1}{g}\left[\frac{1}{6} - 2\,\mathrm{e}^{-\beta} + O\left(\mathrm{e}^{-2\beta}\right)\right]. \tag{40.12}$$

The determinant resulting from the integration around the saddle point can also be evaluated by the method explained in Chapter 38. The only noticeable modification comes from the fact that $\dot{q}_c(t)$ has no zero. It is thus corresponds to the ground state of the hamiltonian $\partial^2 S/\left(\partial q_c\right)^2$, which is therefore a positive operator. The final result is real, as expected. We can use expression (38.50) to obtain it, except that no $1/2i$ factor is needed here, but instead one has to multiply by a factor 2 since the two solutions q_c^+ and q_c^- give identical contributions:

$$\mathrm{tr}\,P\mathrm{e}^{-\beta H} \sim \frac{2}{\sqrt{\pi g}}\beta\,\mathrm{e}^{-\beta/2}\,\mathrm{e}^{-1/6g}\left(1 + O\left(g\right)\right), \quad g \to 0, \quad \beta \to \infty. \tag{40.13}$$

Using then equation (40.6) we find the asymptotic behaviour of $E_+ - E_-$ for g small,

$$E_+ - E_- \underset{g \to 0}{=} -\frac{2}{\sqrt{\pi g}}\,\mathrm{e}^{-1/6g}\left(1 + O\left(g\right)\right). \tag{40.14}$$

The difference is exponentially small in $1/g$ and this is consistent with the property that it vanishes to all orders in g.

40.2 The Periodic Cosine Potential

We now consider a slightly more complicated problem:

$$H = -\tfrac{1}{2}\left(\mathrm{d}/\mathrm{d}x\right)^2 + g^{-1}\left(1 - \cos x\sqrt{g}\right). \tag{40.15}$$

The potential is periodic and has therefore an infinite number of degenerate minima. We can expand perturbation series around each of these minima and therefore, to all orders in perturbation theory, the quantum mechanical hamiltonian also has an infinite number of degenerate ground states. Actually we know that the spectrum of the hamiltonian H is continuous and has, at least for g small enough, a band structure: this property, for g small, is again due to barrier penetration effects.

40.2.1 The structure of the ground state

To make our remarks more precise let us introduce the translation operator T, corresponding to an elementary translation of one period $2\pi/\sqrt{g}$

$$[T, H] = 0. \tag{40.16}$$

The operators T and H can be diagonalized simultaneously. Since we want the eigenfunctions of H to be bounded at infinity, the eigenvalues of T must be pure phases. Each eigenfunction of H is thus characterized by an angle φ (pseudo-momentum) eigenvalue of T:

$$T \,|\varphi\rangle = \mathrm{e}^{i\varphi} \,|\varphi\rangle. \tag{40.17}$$

In the limit $g \to 0$, H has the spectrum of the harmonic oscillator, each state being infinitely degenerate. In the basis in which T is diagonal, all the eigenstates corresponding to the same eigenvalue are indexed by the angle φ. When $g \neq 0$, each point of the spectrum becomes a band. In a band the energy eigenvalue is a periodic function of φ which can be expanded in a Fourier series:

$$E\,(\varphi) = \sum_{-\infty}^{+\infty} E_l \,\mathrm{e}^{il\varphi}, \quad E_l = E_{-l}. \tag{40.18}$$

All coefficients E_l except E_0 vanish to all orders in a perturbative expansion in g.

Let us now consider the partition function which is here $\operatorname{tr}' T \mathrm{e}^{-\beta H}$. The notation tr' has the following meaning: since the diagonal matrix elements of $\mathrm{e}^{-\beta H}$ in configuration space are periodic functions, we integrate only over one period.

The large β limit selects the lowest band and we obtain (see Appendix A40.1):

$$\operatorname{tr}' \mathrm{e}^{-\beta H} \underset{\beta \to \infty}{\sim} \frac{1}{2\pi} \int_0^{2\pi} \mathrm{d}\varphi \; \mathrm{e}^{-\beta E(\varphi)}. \tag{40.19}$$

Like in the case of the double-well potential, we observe that it is difficult to determine the dependence on φ of the energy levels from the partition function. Let us instead consider:

$$\operatorname{tr}' T \mathrm{e}^{-\beta H} \underset{\beta \to \infty}{\sim} \frac{1}{2\pi} \int_0^{2\pi} \mathrm{d}\varphi \; \mathrm{e}^{i\varphi} \, \mathrm{e}^{-\beta E(\varphi)}. \tag{40.20}$$

For g small $E(g) - E_0(g)$ vanishes faster than any power of g. We can therefore expand equation (40.20):

$$\operatorname{tr}' T \mathrm{e}^{-\beta H} \sim \mathrm{e}^{-\beta E_0} \int \frac{\mathrm{d}\varphi}{2\pi} \, \mathrm{e}^{i\varphi} \left[1 - \beta \,(E - E_0) + \cdots\right] \quad \text{for } g \to 0, \quad \beta \to \infty. \tag{40.21}$$

The integration over φ selects E_1:

$$\operatorname{tr}' T \mathrm{e}^{-\beta H} \sim -\beta \mathrm{e}^{-\beta E_0} \, E_1\,(g), \quad g \to 0, \quad \beta \to \infty. \tag{40.22}$$

This equation can be more conveniently rewritten:

$$\operatorname{tr}' T \mathrm{e}^{-\beta H} \,/\, \operatorname{tr}' \mathrm{e}^{-\beta H} \sim -\beta E_1\,(g). \tag{40.23}$$

As indicated above, if E_1 does not vanish this implies that the translation symmetry is not spontaneously broken.

Remark. To evaluate the other Fourier series coefficients $E_2, E_3, \ldots$, for g small, the most convenient method is to consider $\operatorname{tr}' T^n \mathrm{e}^{-\beta H}$ for $n = 2, 3, \ldots$. This evaluation leads to additional problems which we shall solve in Chapter 43.

40.2.2 The instanton contributions

The path integral representations of the partition function $\mathrm{tr}'\, \mathrm{e}^{-\beta H}$ and of $\mathrm{tr}'\, T\, \mathrm{e}^{-\beta H}$ again differ only by the boundary conditions. The operator T has the effect of translating the argument q in the matrix element $< q'|\,\mathrm{tr}'\, \mathrm{e}^{-\beta H}\,|q>$ before taking the trace:

$$\mathrm{tr}'\, T\, \mathrm{e}^{-\beta H} = \int_{q(\beta/2)=q(-\beta/2)+2\pi/\sqrt{g}} [dq(t)] \exp\left[-S(q)\right], \qquad (40.24)$$

$$S(q) = \int_{-\beta/2}^{\beta/2} \left[\tfrac{1}{2}\dot{q}^2(t) + g^{-1}\left(1 - \cos\sqrt{g}q(t)\right)\right] dt. \qquad (40.25)$$

We recall that $q(-\beta/2)$ varies over only one period of the potential. For β large and g small, due to the boundary conditions, the path integral is dominated by instanton configurations which connect two consecutive minima of the potential. Solving the equation of motion explicitly we find:

$$q_c(t) = \frac{4}{\sqrt{g}} \tan^{-1} \mathrm{e}^{(t-t_0)}, \qquad (40.26)$$

and the corresponding classical action, in the infinite β limit, is:

$$S(q_c) = 8/g. \qquad (40.27)$$

For all potentials for which the minima can be exchanged by a reflection, the analogue of expression (38.59) is

$$E(\beta) \sim -\mathrm{e}^{-\beta}\frac{x_0^2}{2}\exp\left[2\int_0^{x_0/2}\left(\frac{1}{\sqrt{2V(x)}} - \frac{1}{x}\right)dx\right], \qquad (40.28)$$

in which x_0 is the location of the other minimum. Applying equation (40.28) to the analogue of equation (38.50) we obtain:

$$-\beta\,\mathrm{e}^{-\beta/2}\, E_1(g) \underset{g\to 0}{\sim} \frac{4\beta\,\mathrm{e}^{-\beta/2}}{\sqrt{\pi g}}\, \mathrm{e}^{-8/g}, \qquad (40.29)$$

or:

$$E_1(g) \underset{g\to 0}{\sim} -\frac{4}{\sqrt{\pi g}}\, \mathrm{e}^{-8/g}. \qquad (40.30)$$

Although we shall not evaluate E_n for $n \geq 2$ here, it is easy to see that the corresponding boundary conditions for $\mathrm{tr}\, T^n\, \mathrm{e}^{-\beta H}$ which are:

$$q(\beta/2) = q(-\beta/2) + n\frac{2\pi}{\sqrt{g}}, \qquad$$

select an instanton solution which for β large has an action $8n/g$. Therefore E_1 gives the dominant non-perturbative contribution for g small, and

$$E_\varphi(g) = E_0(g) - \frac{8}{\sqrt{\pi g}}\, \mathrm{e}^{-8/g}\left[1 + O(g)\right]\cos\varphi + O\left(\mathrm{e}^{-16/g}\right). \qquad (40.31)$$

Conclusions. We have shown on two examples that, as anticipated, in a theory in which, at the classical level, a discrete symmetry is spontaneously broken because the classical potential has degenerate minima, the existence of instanton solutions implies that quantum fluctuations restore the symmetry. An analogous analysis has been made on the lattice in Sections 23.1,23.2 where we have also shown that, in contrast, spontaneous symmetry breaking of discrete symmetries is possible in higher dimensions.

Note that, in contrast to discrete symmetries, where quantum fluctuations lead to exponentially small effects in $1/\hbar$ or the equivalent coupling constant, in the case of continuous symmetries the effects of quantum fluctuations show up already at first order in perturbation theory as a consequence of the Goldstone phenomenon (see Chapter 30). Finally, as we briefly discuss below, while in theories in which the dynamical variables live in flat euclidean space, instantons are always associated with a degeneracy of the classical minimum of the potential, this is no longer necessarily the case when the space has curvature or is topologically non-trivial.

40.2.3 Instantons in stable theories with unique classical vacuum

Let us again consider the example of the cosine potential but with a different interpretation. We now assume that the coordinate x represents a point on a circle of radius $2\pi/\sqrt{g}$. The hamiltonian then represents a $O(2)$ rotator in a potential (Section 3.3) or a one-dimensional classical spin chain in a magnetic field. The classical minimum is no longer degenerate because all minima are identified to one point on the circle. The quantum ground state is equally unique since the Hilbert space consists in strictly periodic eigenfunctions ($\varphi = 0$). Still instanton solutions exist but they start from and return to the same classical minimum, winding around the circle. They are stable because the circle is topologically non-trivial. They generate the same exponentially small corrections to the perturbative expansion that we have described before.

Finally to gain some further insight in the problem we can generalize the problem to the $O(N)$ rotator of Section 3.4. The classical action becomes

$$S(\mathbf{x}) = \frac{1}{g} \int dt \left[\tfrac{1}{2}\dot{\mathbf{x}}^2(t) + 1 - x_1(t) \right]. \tag{40.32}$$

The classical solutions are the same, however the degeneracy and the stability properties are different. For $N > 2$ the solutions which wind around the sphere have $N-2$ directions of instability. Their contributions have to discussed in the framework of the large order behaviour of perturbation theory (see Chapter 41).

40.3 Instantons in Stable Boson Field Theories: General Remarks

We now briefly discuss instantons in stable theories, connecting for example degenerate classical minima. The most interesting examples correspond unfortunately to scale invariant classical theories and the evaluation of the instanton contributions at leading order, which formally follows the lines presented in Chapter 39, leads to difficulties due both to UV and IR divergences. Some of them will be examined in Chapter 42. Since for the two examples we consider in Sections 40.4,40.5, they have not been satisfactorily solved yet, we here restrict ourselves to classical considerations.

We first make a few elementary and general remarks about the possible existence of instantons.

Scalar field theories. Let us assume that the action has the form

$$S(\phi) = \int \left[\tfrac{1}{2} g_{ij}(\phi) \partial_\mu \phi^i \partial_\mu \phi^j + V(\phi) \right] \mathrm{d}^d x, \tag{40.33}$$

in which ϕ^i is a multicomponent scalar boson field, $g_{ij}(\phi)$ a positive matrix (positive definite almost everywhere) and

$$\min_{\{\phi\}} V(\phi) = 0. \tag{40.34}$$

Let us write the analogue of equation (39.39):

$$(2 - d) \int \tfrac{1}{2} \left(g_{ij}(\phi) \partial_\mu \phi_c^i \partial_\mu \phi_c^j \right) \mathrm{d}^d x = d \int V(\phi_c) \, \mathrm{d}^d x.$$

We see that this equation has no solution for $d > 2$. For $d = 2$, it has only solutions for which

$$V(\phi_c(x)) = 0. \tag{40.35}$$

The condition (40.34) then implies that $\phi_c(x)$ is for all x a minimum of the potential:

$$\frac{\partial V(\phi_c)}{\partial \phi} = 0,$$

and therefore $\phi_c(x)$ is a solution of the field equations:

$$\frac{\delta}{\partial \phi^k(y)} \int \tfrac{1}{2} g_{ij}(\phi) \, \partial_\mu \phi^i \partial_\mu \phi^j \mathrm{d}^2 x = 0.$$

These two equations are in general incompatible, except if $V(\phi)$ vanishes identically. In the latter case action (40.33) corresponds to a two-dimensional model on a Riemannian manifold. A particular class of such models, models based on homogeneous spaces, has been discussed in Chapters 14, 15. Among them, the $CP(N-1)$ models are known to admit instanton solutions and we shall describe them in Section 40.4.

Gauge theories. If, in addition to scalar fields, the theory contains gauge fields A_μ^a, the gauge invariant action has the form

$$S[\phi, \mathbf{A}_\mu] = \int \mathrm{d}^d x \left[\tfrac{1}{4} F_{\mu\nu}^a F_{\mu\nu}^a + \tfrac{1}{2} (D_\mu \phi_i)^2 + V(\phi) \right]. \tag{40.36}$$

Let us assume that there exists a finite action solution $\{\phi^c, \mathbf{A}_\mu^c\}$ (in which $\mathbf{A}_\mu^c$ is not a pure gauge), and let us calculate the action for $\lambda \mathbf{A}_\mu^c(\lambda x)$ and $\phi^c(\lambda x)$. We obtain:

$$S\left[\phi^c(\lambda x), \lambda \mathbf{A}_\mu^c(\lambda x)\right] = \lambda^{4-d} \int \tfrac{1}{4} (\mathbf{F}_{\mu\nu})^2 \, \mathrm{d}^d x + \lambda^{2-d} \tfrac{1}{2} \int (D_\mu \phi)^2 \, \mathrm{d}^d x$$

$$+ \lambda^{-d} \int V(\phi) \, \mathrm{d}^d x. \tag{40.37}$$

Stationarity at $\lambda = 1$ implies:

$$(4 - d) \tfrac{1}{4} \int \mathrm{d}^d x \, (\mathbf{F}_{\mu\nu})^2 + (2 - d) \tfrac{1}{2} \int (D_\mu \phi)^2 \, \mathrm{d}^d x - d \int V(\phi) \, \mathrm{d}^d x = 0. \tag{40.38}$$

We see that no solution can exist for $d > 4$, since a sum of negative terms cannot vanish. For $d = 4$ we get two conditions:

$$V(\phi) = 0, \tag{40.39a}$$

$$D_\mu \phi = 0. \tag{40.39b}$$

Writing the field equations, we conclude that $\mathbf{A}_\mu^c$ is the solution of the pure gauge field equations. As we show in Section 40.5, instantons can indeed be found in pure gauge theories. Equation (40.39b), which is now an equation for ϕ^c, then leads to the integrability conditions:

$$[D_\mu, D_\nu] = F_{\mu\nu} \implies \left(F_{\mu\nu}^a\right)^c t_{ij}^a \phi_j^c = 0, \tag{40.40}$$

in which the matrices t^a are the generators of the Lie algebra. The conditions (40.40) together with the equation (40.39a) show that in general the system has only the trivial solution $\phi^c = 0$.

40.4 Instantons in $CP(N-1)$ Models

To illustrate previous considerations let us first give one example of instantons in a two-dimensional model, the $CP(N-1)$ model which is based on one of the symmetric spaces presented in Appendix A15.4.3. We just describe the nature of the solutions and refer the reader to the literature for a deeper and more detailed analysis.

We consider a set of N complex fields φ_α, subject to the condition

$$\bar{\varphi}_\alpha \varphi_\alpha = 1. \tag{40.41}$$

In addition two vectors φ_α and φ_α' are equivalent if

$$\varphi_\alpha'(x) = e^{i\Lambda(x)} \varphi_\alpha(x). \tag{40.42}$$

These conditions characterize the manifolds $CP(N-1)$ which are isomorphic to the symmetric spaces $U(N)/U(1)/U(N-1)$ and thus a special case of complex Grassmannian manifolds. One form of the unique classical action is:

$$S(\varphi) = \frac{1}{g} \int \mathrm{d}^d x \, \overline{D_\mu \varphi} \cdot D_\mu \varphi, \tag{40.43}$$

in which the covariant derivative D_μ is:

$$D_\mu = \partial_\mu - \bar{\varphi} \cdot \partial_\mu \varphi. \tag{40.44}$$

The field $\bar{\varphi} \cdot \partial_\mu \varphi$ is a composite gauge field for the $U(1)$ gauge invariance defined by the transformation (40.42). Note that the $CP(1)$ model is locally isomorphic to the $O(3)$ non-linear σ-model, with the identification

$$\phi = \bar{\varphi}_\alpha \sigma_{\alpha\beta} \varphi_\beta. \tag{40.45}$$

The existence of locally stable non-trivial minima of the action, in two dimensions, follows from simple considerations. The following inequality holds

$$\int \mathrm{d}^2 x \, |D_\mu \varphi \mp i\varepsilon_{\mu\nu} D_\nu \varphi|^2 \geq 0, \tag{40.46}$$

($\varepsilon_{\mu\nu}$ being the antisymmetric tensor). Expanding this expression we obtain

$$S(\varphi) \geq g^{-1}Q(\varphi),\tag{40.47}$$

with

$$Q(\varphi) = \left| i\varepsilon_{\mu\nu} \int d^2x \, D_\mu\varphi \cdot \overline{D_\nu\varphi} \right|.\tag{40.48}$$

The integrand in the r.h.s. is a pure divergence. Indeed using equation (40.41), we can transform equation (40.48) into

$$Q(\varphi) = \left| i\varepsilon_{\mu\nu} \int d^2x \, \partial_\mu \left(\bar\varphi \cdot \partial_\nu\varphi \right) \right|.\tag{40.49}$$

The quantity $Q(\varphi)$ is a topological charge: it depends only on the behaviour of the classical solution for $|x|$ large. Finiteness of the action implies that at large distances φ_α becomes a gauge transform of a constant vacuum solution,

$$\varphi_\alpha(x) \underset{|x|\to\infty}{\sim} e^{i\Lambda(x)} v_\alpha, \quad v_\alpha\bar v_\alpha = 1.\tag{40.50}$$

The topological charge measures the variation of the angle $\Lambda(x)$ on a large circle, which is a multiple of 2π,

$$Q(\varphi) = \left| \int dx_\mu \partial_\mu\Lambda(x) \right| = 2\pi n, \quad n \geq 0,\tag{40.51}$$

$$\Longrightarrow S(\varphi) \geq 2\pi n/g.\tag{40.52}$$

We have been thus led to the consideration of the homotopy classes of mappings from $U(1)$, i.e. S_1 to S_1.

The local minimum corresponds to the equality

$$S(\varphi) = 2\pi n/g,\tag{40.53}$$

which implies that the classical solutions must satisfy first order partial differential equations (self-duality equations):

$$D_\mu\varphi = \pm i\varepsilon_{\mu\nu}D_\nu\varphi.\tag{40.54}$$

It is easy to show that the vectors solution of the equations (40.54) are proportional to holomorphic or anti-holomorphic (depending on the sign) vectors in the variable $z = x_1 + ix_2$. Using then equation (40.41) and the equivalence (40.42) the holomorphic solution can be cast into the form (up to a gauge transformation)

$$\varphi_\alpha = P_\alpha(z)\Big/\sqrt{P\cdot\bar P},\tag{40.55}$$

where the property (40.50) implies that the quantities $P_\alpha(z)$ are polynomials in z without common roots. This form is consistent with the conformal invariance of the classical field theory. The anti-holomorphic solution corresponds to interchange φ and $\bar\varphi$. Translating the $CP(1)$ minimal solution (polynomials of degree 1) in the $O(3)$ σ model language one finds

$$\phi_1 = \frac{z+\bar z}{1+\bar z z}, \quad \phi_2 = i\frac{z-\bar z}{1+\bar z z}, \quad \phi_3 = \frac{1-\bar z z}{1+\bar z z}.$$

We recognize the stereographic mapping of the sphere S_2 onto the plane.

40.5 Instantons in the $SU(2)$ Gauge Theory

Let us now give an example of instantons in four dimensions. According to the analysis of Section 40.3, we can restrict ourselves to gauge theories. Actually it is sufficient to consider the gauge group $SU(2)$ since a general theorem states that for a Lie group containing $SU(2)$ as a subgroup the instantons are those of the $SU(2)$ subgroup.

In $O(3)$ notation the gauge action is:

$$S\left(A_\mu\right) = \frac{1}{4g} \int \left[\mathbf{F}_{\mu\nu}(x)\right]^2 \mathrm{d}^4 x \,, \tag{40.56}$$

with

$$\mathbf{F}_{\mu\nu} = \partial_\mu \mathbf{A}_\nu - \partial_\nu \mathbf{A}_\mu + \mathbf{A}_\mu \times \mathbf{A}_\nu \,. \tag{40.57}$$

The existence and some properties of instantons in this theory follow from simple considerations very similar to those presented for the $CP(N-1)$ model.

Let us define the dual of the tensor $\mathbf{F}_{\mu\nu}$ by

$$\tilde{\mathbf{F}}_{\mu\nu} = \tfrac{1}{2}\varepsilon_{\mu\nu\rho\sigma}\mathbf{F}_{\rho\sigma} \,. \tag{40.58}$$

Then the inequality

$$\int \mathrm{d}^4 x \left[\mathbf{F}_{\mu\nu}(x) \pm \tilde{\mathbf{F}}_{\mu\nu}(x)\right]^2 \geq 0 \,, \tag{40.59}$$

yields

$$\int \mathrm{d}^4 x \left(\mathbf{F}_{\mu\nu}\right)^2 \geq \left| \int \mathrm{d}^4 x \, \mathbf{F}_{\mu\nu} \cdot \tilde{\mathbf{F}}_{\mu\nu} \right| \,. \tag{40.60}$$

The quantity $\mathbf{F}_{\mu\nu} \cdot \tilde{\mathbf{F}}_{\mu\nu}$ again is a pure divergence:

$$\mathbf{F}_{\mu\nu} \cdot \tilde{\mathbf{F}}_{\mu\nu} = \partial_\mu V_\mu \,, \tag{40.61}$$

with

$$V_\mu = 2\varepsilon_{\mu\nu\rho\sigma}\left[\mathbf{A}_\nu \cdot \partial_\rho \mathbf{A}_\sigma + \tfrac{1}{3}\mathbf{A}_\nu \cdot (\mathbf{A}_\rho \times \mathbf{A}_\sigma)\right] \,. \tag{40.62}$$

As in the case of the $CP(N-1)$ model, the r.h.s. of equation (40.60) is a topological quantity, which depends only on the behaviour of the gauge fields at large distances. We have met it already in Subsection 20.3.3 (equation (20.110)) in the calculation of the axial anomaly. We have in particular shown that its values are quantized (equation (20.115)).

The finiteness of the action, implies that the classical solution should asymptotically become a pure gauge, i.e. with our conventions,

$$\tfrac{1}{2}\mathbf{A}_\mu \cdot \boldsymbol{\sigma} = g\left(x\right)\partial_\mu g^{-1}\left(x\right) + O\left(|x|^{-2}\right) \quad |x| \to \infty \,, \tag{40.63}$$

in which $\boldsymbol{\sigma}$ are Pauli matrices and $g\left(x\right)$ is an element of $SU(2)$. Since $SU(2)$ is topologically equivalent to S_3, we are now led to the study of the homotopy classes of mappings from S_3 to S_3, which are also classified by an integer called the winding number.

The simplest one to one mapping corresponds to an element $g(x)$ of the form

$$\mathbf{g}\left(x\right) = \frac{x_4 + i\mathbf{x} \cdot \boldsymbol{\sigma}}{r}, \quad r = \left(x_4^2 + \mathbf{x}^2\right)^{1/2}, \tag{40.64}$$

and thus

$$A_m^i = 2\left(x_4\delta_{im} + \varepsilon_{imk}x_k\right)r^{-2}, \quad A_4^i = -2x_i r^{-2}. \tag{40.65}$$

It follows that:

$$\int d^4x\, \mathbf{F}_{\mu\nu} \cdot \tilde{\mathbf{F}}_{\mu\nu} = \int d\Omega\, \hat{n}_\mu \cdot V_\mu = 32\pi^2, \tag{40.66}$$

in which $d\Omega$ is the measure on the sphere and $\hat{n}_\mu$ the unit vector normal to the sphere.

If we compare this result with equation (20.115) we see that we have indeed found the minimal action solution. In general we then expect

$$\int d^4x\, \mathbf{F}_{\mu\nu} \cdot \tilde{\mathbf{F}}_{\mu\nu} = 32\pi^2 n, \tag{40.67}$$

and therefore

$$S\left[A_\mu\right] \geq 8\pi^2 n/g. \tag{40.68}$$

The equality, which corresponds to a local minimum of the action, is obtained for fields satisfying the self-duality equations

$$\mathbf{F}_{\mu\nu} = \pm\tilde{\mathbf{F}}_{\mu\nu}, \tag{40.69}$$

which are first order partial differential equations. The one-instanton solution, which depends on a scale parameter λ, is

$$\begin{cases} A_m^i = \dfrac{2}{r^2 + \lambda^2}\left(x_4\delta_{im} + \varepsilon_{imk}x_k\right), \quad m = 1,2,3, \\[2ex] A_4^i = \dfrac{-2x_i}{r^2 + \lambda^2}. \end{cases} \tag{40.70}$$

The structure of the semiclassical vacuum. In contrast to our analysis of the quantum mechanical problems, we have here discussed the existence of instantons without reference to the structure of the classical vacuum. To find an interpretation of the instantons in gauge theories, it is convenient to express the results in the temporal gauge $A_4 = 0$. In this gauge the classical minima of the potential correspond to space components gauge fields which are pure gauge functions of the three space variables:

$$\tfrac{1}{2}\mathbf{A}_m \cdot \boldsymbol{\sigma} = \mathbf{g}(x_i)\partial_m\mathbf{g}^{-1}(x_i). \tag{40.71}$$

However we know that the wave functions have to satisfy Gauss's law. This implies that in the absence of external static charges the wave function is invariant under infinitesimal gauge transformations (Section 19.3). Therefore two gauge functions $\mathbf{g}(x_i)$ which are infinitesimally close do not correspond to different vacua. This implies that the structure of the classical vacuum is connected to the homotopy classes of mappings of the group elements $\mathbf{g}$ into compactified $\mathbb{R}^3$, i.e. again of S_3 into S_3 and that the classical potential has a periodic structure reminiscent of the periodic cosine potential. One can verify that the instanton solution (40.70), transported into the temporal gauge by a gauge transformation, connects vacua with different winding numbers. Therefore, as in the case of the periodic potential, if we want to project onto a proper quantum mechanical eigenstate, the "θ-vacuum", we have to add, in analogy with expressions (43.65, 43.66), a term to the classical action of gauge theories:

$$S_\theta\left(\mathbf{A}_\mu\right) = S\left(\mathbf{A}_\mu\right) + \frac{i\theta}{32\pi^2}\int d^4x\, \mathbf{F}_{\mu\nu} \cdot \tilde{\mathbf{F}}_{\mu\nu}, \tag{40.72}$$

and we can then integrate over all fields $\mathbf{A}_\mu$ without restriction. A similar property holds for the $CP(N-1)$ models. At least at the semiclassical level, the field theory depends on an additional parameter, the angle θ. For non-vanishing values of θ the additional term violates CP conservation, and is at the origin of the strong CP violation problem.

Introduction of fermions. If we now consider the case of QCD, gauge fields coupled to fermions $\mathbf{Q}$, $\bar{\mathbf{Q}}$ through an action

$$ S\left(\mathbf{A}_\mu, \bar{\mathbf{Q}}, \mathbf{Q}\right) = -\int \mathrm{d}^4x \left[\frac{1}{4g^2} \operatorname{tr} \mathbf{F}_{\mu\nu}^2 + \sum_{f=1}^{N_f} \bar{\mathbf{Q}}_f \left(\not{D} + m_f\right) \mathbf{Q}_f\right], $$

several comments can be made. First, if one fermion field is massless, then according to the analysis of Subsection 20.3.4, the determinant resulting from the fermion integration vanishes because the Dirac operator has a vanishing eigenvalue. Then the instantons do not contribute to the functional integral and the strong CP violation problem is solved. However it seems that it is inconsistent with experimental data to assume that one quark is massless.

Second, in the limit in which the quarks are massless, the symmetry of the action is the chiral $U(N_f) \times U(N_f)$ group which has an axial $U(1)$ subgroup. As we already discussed in Section 20.5, if the instantons contribute, they solve the $U(1)$ problem, i.e. the absence of a Goldstone boson associated with the almost spontaneous breaking of the axial $U(1)$ current.

The quantum mechanical vacuum. In both examples, $CP(N-1)$ models, and non-abelian gauge theories, the classical theory is scale invariant. Therefore the solutions depend on a scale parameter λ which has to be taken as a collective coordinate in the quantization. This leads to very serious problems as the analysis of the massless ϕ_4^4 field theory will reveal (see Chapter 42). In an asymptotically free theory, like the two theories considered above, the difficulties come from the infrared region, i.e. from instantons of large size for which the semiclassical approximation is no longer legitimate because the interaction increases at large distance (see Chapters 33,34). Only with an IR cut-off is the calculation possible. A solution is to consider a system in a finite volume.

The role of instantons in the infinite volume is thus not clearly understood. The only piece of information presently available concerns the $O(3)$ non-linear σ-model, whose instantons are derived from those of the $CP(1)$-model. It has been recently rather indirectly argued, by mapping the σ-model onto a one dimensional quantum spin chain, that instantons are only relevant for $\theta = \pi$ but then lead to a quite different physics.

40.6 Instantons and Stochastic Differential Equations

In this section we discuss the role of instantons in the context of stochastic dynamics. We consider the problem of evaluating the decay probability of a metastable state by thermal fluctuations. At first sight one could imagine that this topic should have considered already in Chapter 39, simultaneously with the problem of decay by quantum fluctuations. We show here that technically the problem has a more direct relation with quantum ground state degeneracy. At the end of the section we also briefly examine the role of instantons when the equilibrium distribution has degenerate minima.

We recall (see Chapter 4) that with the Langevin equation (4.37)

$$ \dot{q}_i = -\tfrac{1}{2}\partial_i E\left(\mathbf{q}\right) + \nu_i(t), \tag{40.73} $$

with gaussian noise (4.2):

$$[\mathrm{d}\rho\,(\nu)] = [\mathrm{d}\nu]\exp\left[-\tfrac{1}{2}\Omega^{-1}\int\nu_i^2(t)\,\mathrm{d}t\right],$$

is associated the *hermitian* hamiltonian H (4.40):

$$H = \tfrac{1}{2}\left[-\Omega\Delta + \tfrac{1}{4}\Omega^{-1}\left(\nabla E\,(\hat{\mathbf{q}})\right)^2 - \tfrac{1}{2}\Delta E\,(\hat{\mathbf{q}})\right],\qquad(40.74)$$

and therefore the effective action:

$$\mathcal{A}(\mathbf{q}) = \tfrac{1}{2}\Omega^{-1}\int\left[\dot{\mathbf{q}}^2 + \tfrac{1}{4}\left(\nabla E\right)^2 - \tfrac{1}{2}\Omega\Delta E\right]\mathrm{d}t\,.\qquad(40.75)$$

The classical limit here corresponds to the small Ω (also low temperature) limit. At leading order in a semiclassical analysis the term ΔE can thus be omitted. It follows that to each extremum of the function $E(\mathbf{q})$ corresponds a minimum of the classical action (40.75). In particular, if $E(\mathbf{q})$ has more than one extremum, the classical action has instanton solutions. Let us discuss using a simple example the role of instanton solutions in this context.

An example. Let us consider the example of a function $E(q)$ which is such that the corresponding distribution $\mathrm{e}^{-E(q)/\Omega}$ is not normalizable:

$$E(q) = q^2 - 2q^3/3\,.\qquad(40.76)$$

Physically we then know that if we put a particle at time 0 at the relative minimum $q = 0$ of the distribution, then after some time the particle will escape from the well, as a result of the thermal fluctuations described by the Langevin equation. Our problem is to evaluate in the small Ω limit the escape probability per unit time. We know that it is given by the inverse of the smallest eigenvalue of the hamiltonian (40.74)(see Section 4.3). When the distribution is not normalizable this eigenvalue is strictly positive.

However to all orders in a perturbative expansion in powers of Ω, $\mathrm{e}^{-E/(2\Omega)}$ is the ground state eigenvector associated with the eigenvalue 0. It follows that the eigenvalue must be related to non-perturbative effect. We indeed note that the function $E(q)$ has two extrema located at $q = 0$ and $q = 1$. Hence $E'^2(q)$ has two minima. We can suspect that the instantons connecting these minima provide a solution to our problem.

The classical equation of motion. Since we only look for solutions which connect minima of the potential in infinite time, we can immediately integrate a first time the equation of motion and find:

$$\dot{q} = \pm E'(q)/2\,.\qquad(40.77)$$

Therefore the instanton action is:

$$\mathcal{A}(q) = \Omega^{-1}\int\mathrm{d}t\,\dot{q}^2 = \tfrac{1}{2}\Omega^{-1}\int\mathrm{d}t\,\dot{q}\,E'(q) = \tfrac{1}{2}\Omega^{-1}\left(E(1) - E(0)\right).\qquad(40.78)$$

However this is not the end of the story. Indeed, in Section 4.5 we have shown that the degeneracy between the minima and maxima of the function $E(q)$ is lifted by the first quantum correction. Therefore the two minima of the action are not really degenerate and no instanton can connect them. What really happens is that we have to consider only closed trajectories passing through the origin. If we consider a finite time interval β we can find such trajectories. In the infinite β limit they decompose into a succession of instantons and anti-instantons. The limit of the classical action is an even multiple of the instanton action. The leading contribution is thus:

$$\mathcal{A}(q) = (E(1) - E(0)) / \Omega. \tag{40.79}$$

These arguments obviously are independent of the precise form of the function $E(q)$. Therefore we conclude quite generally that if the function $E(q)$ has a relative minimum where $E = E_{\min}$ separated from a lower minimum (possibly $E = -\infty$) by a local maximum $E = E_{\max}$, then the eigenvalue corresponding to a eigenfunction concentrated around the first minimum is of the order $\mathrm{e}^{-\Delta E/\Omega}$ in which ΔE is the variation of the function E:

$$\Delta E = E_{\max} - E_{\min}.$$

The time $\mathrm{e}^{\Delta E/\Omega}$ has the following interpretation: it characterizes the exponential decay of the probability of finding $q(t)$ near the origin when the initial conditions at $t = 0$ are $q(t = 0) = 0$.

Note finally that in order to complete the calculation of the eigenvalue it is necessary to use multi-instanton techniques which will be developed in Chapter 43.

Generalization. Previous analysis can be generalized to several degrees of freedom. The classical equation of motion then takes the form:

$$\ddot{q}_i = \partial_i \partial_j E \partial_j E. \tag{40.80}$$

If we consider only finite action solutions and infinite time, the equation can be integrated:

$$\dot{q}_i = \pm \partial_i E. \tag{40.81}$$

The classical action is then:

$$\mathcal{A}(q) = \Omega^{-1} \int \mathrm{d}t \, \dot{\mathbf{q}}^2 = \tfrac{1}{2}\Omega^{-1} \int \mathrm{d}t \, \dot{\mathbf{q}} \cdot \nabla E = \tfrac{1}{2}\Omega^{-1} \left(E_{\max} - E_{\min} \right), \tag{40.82}$$

in which $E_{\min}$ and $E_{\max}$ are respectively the values of the function $E(\mathbf{q})$ at the relative minimum and the saddle point one has to pass to be able to go lower.

Quantum Field Theory. We now consider a dynamics governed by a purely dissipative Langevin equation, which formally converges towards an equilibrium distribution corresponding to the path or the functional integral of a d dimensional euclidean quantum field theory (see also Subsection A36.1.3). The effective action then is $d+1$ dimensional. Let us again examine the problem of an unstable field theory with euclidean action

$$S(\phi) = \Omega^{-1} \int \mathrm{d}^d x \left[\tfrac{1}{2}(\partial_\mu \phi)^2 + \tfrac{1}{2}m^2\phi^2 + \tfrac{1}{6}g\phi^3 \right]. \tag{40.83}$$

We know that a quantum state concentrated around the minimum $\phi(x) = 0$ will decay due to quantum fluctuations, and we have calculated the rate by instanton methods.

We now want to evaluate the decay probability due to thermal fluctuations. The relevant effective action at leading order reads:

$$\mathcal{A}(\phi) = \tfrac{1}{2}\Omega^{-1} \int \mathrm{d}^d x \, \mathrm{d}t \left[\dot{\phi}^2 + \tfrac{1}{4}\left(-\Delta\phi + m^2\phi + \tfrac{1}{2}g\phi^2 \right)^2 \right]. \tag{40.84}$$

Formally the discussion follows the same line as in the case of a finite number of degrees of freedom. The problem is to identify the minimum and the maximum of the action. The minimum is easy to find: $\phi \equiv 0$. The maximum requires some more thought. It does not correspond to a constant field configuration: $\phi(x) = -2m^2/g$. Indeed it is sufficient that some part of the field starts passing the barrier. This means that the relevant maximum of the action instead corresponds to a static instanton configuration. Using the arguments of the preceding subsection, the result follows:

$$\tau \sim \exp[S_{\text{inst.}}/\Omega].$$

Degenerate minimum. Another problem arises when the function $E(q)$ has a degenerate minimum. Let us assume that the corresponding distribution is normalizable. Then the ground state eigenvalue vanishes. The interesting question is how to calculate the difference between the two first eigenvalues, difference which vanishes to all orders in perturbation theory. This is the problem we have solved in Section 40.1 for the double-well potential. However here the set-up is slightly different because, if $E(q)$ is regular, as we always assume, the two minima are necessarily separated by a maximum and therefore $E'^2(q)$ are three minima. An example is:

$$E(q) = q^2(1-q)^2, \tag{40.85}$$

and therefore:

$$E'^2(q) = 4(1-2q)^2 q^2 (1-q)^2. \tag{40.86}$$

This time we look for instanton solutions which connect $q = 0$ to $q = 1$. However in the infinite time limit only instantons which go from 0 to 1/2 or 1/2 to 1 survive. From the analysis of the previous problem we guess that the relevant configurations will correspond to glue together two instantons. Therefore the difference between the two leading eigenvalues, which is also the second eigenvalue ϵ_1, is again of the form:

$$\epsilon_1 \sim \exp[-\left(E_{\max} - E_{\min}\right)/\Omega]. \tag{40.87}$$

in which $E_{\min}$ and $E_{\max}$ are respectively the values of the function $E(q)$ at the degenerate minima and at the maximum which connects them.

Bibliographical Notes

Polyakov has emphasized the double-well as the typical instanton problem in
  A.M. Polyakov, *Nucl. Phys.* B120 (1977) 429.
The role of instantons in the question of spontaneous symmetry breaking of discrete symmetries has been discussed in
  J. Zinn-Justin, *Basko Polje Summer School 1976*, Saclay preprint DPh-T/76/99, unpublished.
We have used scaling arguments to discuss the existence of instantons
  G.H. Derrick, *J. Math. Phys.* 5 (1964) 1252.
References about instantons in the $CP(N-1)$ model include
  A. Jevicki *Nucl. Phys.* B127 (1977) 125; D. Förster, *Nucl. Phys.* B130 (1977) 38; M. Lüscher, *Phys. Lett.* 78B (1978) 465; A. D'Adda, P. Di Vecchia and M. Lüscher, *Nucl. Phys.* B146 (1978) 63; B152 (1979) 125; H. Eichenherr, *Nucl. Phys.* B146 (1978) 215; V.L. Golo and A. Perelemov, *Phys. Lett.* 79B (1978) 112; A.M. Perelemov, *Phys. Rep.* 146 (1987) 135.

For instantons in gauge theories see

A.A. Belavin, A.M. Polyakov, A.S. Schwartz and Yu S. Tyupkin, *Phys. Lett.* 59B (1975) 85; G. 't Hooft, *Phys. Rev. Lett.* 37 (1976) 8; *Phys. Rev.* D14 (1976) 3432 (Erratum *Phys. Rev.* D18 (1978) 2199); R. Jackiw and C. Rebbi, *Phys. Rev. Lett.* 37 (1976) 172; C.G. Callan, R.F. Dashen and D.J. Gross, *Phys. Lett.* 63B (1976) 334; A.A. Belavin and A.M. Polyakov, *Nucl. Phys.* B123 (1977) 429; F.R. Ore, *Phys. Rev.* D16 (1977) 2577; S. Chadha, P. Di Vecchia, A. D'Adda and F. Nicodemi, *Phys. Lett.* 72B (1977) 103; T. Yoneya, *Phys. Lett.* 71B (1977) 407; I.V. Frolov and A.S. Schwarz, *Phys. Lett.* 80B (1979) 406; E. Corrigan, P. Goddard and S. Templeton, *Nucl. Phys.* B151 (1979) 93.

For a recent discussion of the $U(1)$ problem see

G. 't Hooft, *Phys. Rep.* 142 (1986) 357.

A possible solution to the strong CP violation problem, which has been shown since then to lead to difficulties with experiment, was proposed in

R.D. Peccei and H.R. Quinn, *Phys. Rev.* D16 (1977) 1791,

and discussed in

S. Weinberg, *Phys. Rev. Lett.* 40 (1978) 223; F. Wilczek, *Phys. Rev. Lett.* 40 (1978) 279.

Another field theory with interesting properties is the two-dimensional abelian Higgs model

C.G. Callan, R.F. Dashen and D.J. Gross, *Phys. Lett.* 66B (1977) 375.

APPENDIX 40

A40.1 Trace Formula for Periodic Potentials

We consider a hamiltonian H corresponding to a real periodic potential $V(x)$ with period τ:

$$V(x+\tau) = V(x). \tag{A40.1}$$

Eigenfunctions $\psi_\varphi(x)$ are then also eigenfunctions of the translation operator T:

$$T\psi_\varphi(x) \equiv \psi_\varphi(x+\tau) = e^{i\varphi}\,\psi_\varphi(x). \tag{A40.2}$$

Let us put the system in a box of size $N\tau$ with periodic boundary conditions. This implies a quantization of φ

$$e^{iN\varphi} = 1 \Rightarrow \varphi = \varphi_p \equiv \frac{2\pi p}{N}, \quad 0 \le p < N. \tag{A40.3}$$

Calling $\psi_{p,n}$ the eigenfunction of H corresponding to the band n and the pseudo-momentum φ_p, we can write the matrix elements of $T\,e^{-\beta H}$:

$$\left\langle x'\left|T\,e^{-\beta H}\right|x\right\rangle = \sum_{p,n} \psi_{p,n}^*(x')\,e^{-\beta E_n(\varphi_p)}\,\psi_{p,n}(x+\tau), \tag{A40.4}$$

and therefore:

$$\left\langle x'\left|T\,e^{-\beta H}\right|x\right\rangle = \sum_{p,n} \psi_{p,n}^*(x')\,e^{-\beta E_n(\varphi_p)+i\varphi_p}\,\psi_{p,n}(x). \tag{A40.5}$$

This implies for the diagonal elements:

$$\left\langle x\left|T\,e^{-\beta H}\right|x\right\rangle = \sum_{p,n} |\psi_{p,n}(x)|^2\,e^{-\beta E_n(\varphi_p)+i\varphi_p}. \tag{A40.6}$$

Because in a translation of a period (equation $(A40.2)$), the eigenfunctions are multiplied by a phase, equation $(A40.6)$ shows that the l.h.s. is a periodic function of x with periodicity τ. Therefore:

$$\int_0^\tau \left\langle x\left|T\,e^{-\beta H}\right|x\right\rangle \mathrm{d}x = \sum_{n,p} e^{i\varphi_p - \beta E_n(\varphi_p)}\frac{1}{N}\int_0^{N\tau} |\psi_{p,n}(x)|^2\,\mathrm{d}x. \tag{A40.7}$$

The functions $\psi_{p,n}(x)$ are orthonormal over $N\tau$:

$$\int_0^\tau \left\langle x\left|T\,e^{-\beta H}\right|x\right\rangle \mathrm{d}x = \frac{1}{N}\sum_{n,p} e^{i\varphi_p - \beta E_n(\varphi_p)}. \tag{A40.8}$$

We now take the large N limit and obtain expression (40.51):

$$\int_0^\tau \left\langle x\left|T\,e^{-\beta H}\right|x\right\rangle \mathrm{d}x = \frac{1}{2\pi}\sum_n \int_0^{2\pi} e^{i\varphi - \beta E_n(\varphi)}\,\mathrm{d}\varphi, \tag{A40.9}$$

since reality implies:

$$E_n(\varphi) = E_n(-\varphi). \tag{A40.10}$$

41 PERTURBATION THEORY AT LARGE ORDERS AND INSTANTONS. THE SUMMATION PROBLEM

In Chapter 37 we have discussed the analytic structure of the ground state energy $E(g)$ of the anharmonic oscillator. We have argued that $E(g)$ is analytic in a cut-plane, and calculated by instanton methods its imaginary part on the cut for g small and negative. On the other hand, perturbation theory yields $E(g)$ for g small as a power series in g:

$$E(g) = \sum_{k=0}^{\infty} E_k g^k. \tag{41.1}$$

We explain in this chapter how the small g behaviour of $\operatorname{Im} E(g)$ is related to the behaviour of the coefficients E_k when the order k becomes large. We then generalize the method to the class of potentials for which we have calculated instanton contributions. The same method can be readily applied to boson field theories, while the extension to field theories involving fermions requires, as we show, some additional considerations.

We already know that the expansion (41.1) is divergent for all values of g. This implies that, even for g small, the series does not determine the function $E(g)$ uniquely. We thus examine the implications of the large order behaviour for the problem of the summation of the series. Finally we describe a few practical methods commonly used to sum divergent series of the type met in quantum mechanics and quantum field theory. Some of these methods have been successfully applied to the $\left(\phi^2\right)^2$ field theory in 2 and 3 dimensions and have led to the accurate predictions of critical exponents displayed in Chapter 28.

41.1 Quantum Mechanics

41.1.1 Real instanton solutions

The anharmonic oscillator. We first consider the ground state energy of the quartic anharmonic oscillator. Since $E(g)$ is analytic in the cut-plane and behaves like $g^{1/3}$ for g large, it has a Cauchy representation for $E(g)$ of the form

$$E(g) = \frac{1}{2} + \frac{g}{\pi} \int_{-\infty}^{0} \frac{\operatorname{Im} E\left(g'\right) \mathrm{d}g'}{g'\left(g' - g\right)}. \tag{41.2}$$

This equation leads to an integral representation for the coefficients E_k obtained by expanding the integrand in powers of g:

$$E_k = \frac{1}{\pi} \int_{-\infty}^{0} \frac{\operatorname{Im} E(g) \, \mathrm{d}g}{g^{k+1}} \qquad \text{for } k > 0. \tag{41.3}$$

Let us now examine what happens when k, the order in the expansion, becomes large. Due to the factor g^{-k}, the dispersion integral (41.3) is dominated by the small negative g values. In Chapter 37 we have calculated $\operatorname{Im} E(g)$ for g small and negative. We can here use this result to estimate the large k behaviour of E_k:

$$E_k \underset{k \to \infty}{\sim} \frac{1}{\pi} \int^{0^-} \left(\frac{8}{\pi}\right)^{1/2} \frac{1}{\sqrt{-g}} \frac{e^{4/3g}}{g^{k+1}} \mathrm{d}g. \tag{41.4}$$

The integration can be done explicitly and we find:

$$E_k = (-1)^{k+1} \left(\frac{6}{\pi^3}\right)^{1/2} \left(\frac{3}{4}\right)^k \Gamma\left(k + \frac{1}{2}\right) \left(1 + O\left(\frac{1}{k}\right)\right). \qquad (41.5)$$

Successive corrections to the semiclassical result yield a series in powers of g which, integrated, generates a systematic expansion in powers of $1/k$.

General potentials. The same argument is applicable to the situation described in Chapter 38. We can calculate the energy of the unstable state in power series of the coupling constant g by making a systematic expansion around the relative minimum of the potential. On the other hand we can, as above, derive from the knowledge of the imaginary part of the energy level for small coupling, an estimate of the behaviour of the perturbative coefficients at large order. Let us consider the action:

$$S(q) = \int dt \left[\tfrac{1}{2}\dot{q}^2(t) + g^{-1}V\left(q\sqrt{g}\right)\right]. \qquad (41.6)$$

The analogue of the dispersion relation (41.3) is:

$$E_k \sim \frac{1}{\pi} \int_0^\infty \frac{\operatorname{Im} E(g)}{g^{k+1}} dg.$$

The behaviour of $\operatorname{Im} E(g)$ for g small is given by expression (38.62). Integrating near $g = 0$ we obtain

$$E_k \sim -\frac{1}{2\pi^{3/2}} x_+ \exp\left[\int_0^{x_+} \left(\frac{1}{\sqrt{2V(x)}} - \frac{1}{x}\right) dx\right] A^{-(k+1/2)} \Gamma\left(k + \frac{1}{2}\right), \qquad (41.7)$$

where A is the classical action

$$A = 2 \int_0^{x_+} \sqrt{2V(x)} \, dx. \qquad (41.8)$$

We now see generic features emerge: at large orders, the perturbative coefficients E_k behave like

$$E_k \underset{k \to \infty}{\sim} C k^{b-1} k! \, A^{-k}. \qquad (41.9)$$

The factor $k!$ is universal and characteristic of the semiclassical or loop expansion. It shows that the perturbation series is a divergent series. The factor A^{-k} depends only on the action, since it is the action of the classical solution; in particular it also characterizes the behaviour at large orders of the excited energy levels or of correlation functions. The power k^b comes from the power of g in front of the result. It in particular depends on the number of continuous symmetries broken by the classical solution, but it would also change if we considered an excited state rather than the ground state. This can be verified by explicitly calculating the imaginary parts of the energy of the excited levels as explained in Chapter 37, and using equation (41.3). The parameter b is in general a half integer. Finally there is a constant multiplicative factor c which depends in a more complicated way on all the specific features of the expanded quantity.

Discussion. In both examples we have been able to calculate the large order behaviour of perturbation series from the decay rate due to barrier penetration of an unstable minimum of the potential. For the potentials considered in Chapter 38 the action A is positive and therefore all terms in the perturbative expansion have the same sign. The same property holds for the anharmonic oscillator in the unstable case, i.e. when g is negative. However for g positive, in which case perturbation series has been expanded around the stable minimum of the potential, we observe that the perturbative coefficients oscillate in sign. Also we note that for $g > 0$ the instanton solution becomes purely imaginary. This will help to understand how to obtain the large behaviour in the generic stable case.

41.1.2 Complex instantons

We have up to now characterized the large order behaviour of perturbation theory in two cases, in the generic case in which we expand around a relative minimum of the potential, and in one special case in which we were expanding around an absolute minimum of the potential, but which by analytic continuation in the coupling constant could be transformed into the unstable one. We now consider actions of the form (41.6), in which the potential $V(q)$ is an entire function of q and satisfies the condition:

$$V(q) = \tfrac{1}{2}q^2 + O\left(q^3\right),$$

and assume that perturbation theory is expanded around $q = 0$, the absolute minimum of the potential. Then clearly no real instanton solutions can be found. Following the example of the anharmonic oscillator, we are going to assume that we can introduce parameters in the potential which allow us to go to an unstable situation by analytic continuation in these parameters. We can then obtain the large order behaviour from expression (41.7). We then use the inverse analytic continuation and return to the original situation. If nothing dramatic happens, the large behaviour of the original expansion will be given by the analytic continuation of expression (41.7).

We can now formulate the principles of the large order behaviour calculation in the original theory: to the complex zeros (at finite or infinite distance) of the potential $V(q)$ are associated complex instanton solutions, with, in general, complex (or exceptionally negative) action. These instanton solutions are candidates to give contributions to the large order behaviour. From expression (41.7) we see that the action(s) with the smallest modulus (when the action is complex, there will be at least two complex conjugate actions) gives the dominant contribution to the large order behaviour. Note that the difference we have found between the anharmonic oscillator and the unstable case is generic. In the stable case, the classical action is non-real positive, and the phase of the perturbative coefficients at large order varies with the order.

When we discuss the question of summability of divergent series we shall see that this difference in the behaviour of the series plays an essential role.

A special case. The previous discussion does not immediately apply to the case of potentials with degenerate minima. Let us indeed consider such a potential as the limit of a potential which has two minima at which the values of the potential are very close. From the explicit form of the action, we see that the classical action has a limit which is twice the action of the instanton which connects the two minima of the potential

$$A = 2\int_0^{x_+} \sqrt{2V\left(x\right)}\mathrm{d}x\,.$$

However the amplitude in front of the expression (41.7) diverges when x_+ is an extremum of the potential. This result can be easily understood. When the values at the two minima approach each other, the time spent close to the second minimum of the potential by the classical trajectory corresponding to the instanton solution diverges. Therefore fluctuations which tend to change this time leave the action almost stationary. Correspondingly one eigenvalue of the operator second derivative of the action at the saddle point goes to zero, and this explains the divergence of expression (41.7) in this case. Only by allowing this time to fluctuate, and introducing an additional time collective coordinate can we obtain the correct answer. Let us also note that here, as in the case of unstable minima, the classical action is positive. This is the source of serious difficulties when one tries to sum the perturbation series. We shall discuss these problems in Chapter 43.

41.2 Scalar Field Theory

In Chapter 39 we have evaluated the contributions of instantons to the decay rate of metastable states. We can here apply these results to large order behaviour estimates. In the case of the ϕ^4 field theory the discontinuity across the cut of the n-point function reads (equation (39.26)):

$$\operatorname{disc} Z^{(n)}(x_1, \ldots, x_n) \sim \left(\frac{A}{2\pi}\right)^{d/2} \frac{\mathrm{e}^{-A/g}}{g^{(d+n)/2}} \left(\det M' M_0^{-1}\right)_{\mathrm{ren}}^{-1/2} F_n(x_1, \ldots, x_n), \quad (41.10)$$

with

$$F_n(x_1, \ldots, x_n) = m^{d+n(d-2)/2} 6^{n/2} \int \mathrm{d}^d x_0 \prod_{i=1}^{n} f(m(x_i - x_0)). \quad (41.11)$$

Using previous arguments, we can immediately translate this result into a large order behaviour estimate for correlation functions

$$\left\{Z^{(n)}(x_1, \ldots, x_n)\right\}_k = \frac{1}{2i\pi} \int \frac{\mathrm{d}g}{g^{k+1}} \operatorname{disc} Z^{(n)}(x_1, \ldots, x_n),$$

and therefore

$$\left\{Z^{(n)}(x_1, \ldots, x_n)\right\}_k \underset{k \to \infty}{\sim} \frac{1}{2i\pi} \frac{1}{(2\pi)^{d/2}} F_n(x_1, \ldots, x_n) A^{-n/2-k} \Gamma(k + d/2 + n/2).$$

$$(41.12)$$

In a general scalar boson field theory, if we can find instanton solutions, the same arguments will lead to:

$$\left\{Z^{(n)}(x_1, \ldots, x_n)\right\}_k \underset{k \to \infty}{\sim} \sum_{\substack{\text{dominant} \\ \text{saddle points}}} C_n(x_1, \ldots, x_n) k^{b-1} A^{-k} k!, \quad (41.13)$$

in which:

(i) A is the instanton action, which is in general complex;
(ii) $b = \frac{1}{2}(n + \delta)$ and δ is the number of symmetries broken by the classical solution;
(iii) $C_n(x_1, \ldots, x_n)$, which does not depend on k, contains the whole dependence in the external arguments.

Example: the RG β-function in the $(\phi^2)^2$ in 3 dimensions. The large order behaviour has been determined by solving numerically the field equations to find the classical action A and calculating numerically the determinant. The predictions of the asymptotic

Table 41.1

The coefficients β_k of the coupling constant renormalization group function $\beta(g)$ divided by the large order estimate for the $O(N)$ symmetric ϕ_3^4 field theory.

k	2	3	4	5	6	7
$N=0$	3.53	1.55	1.185	1.022	0.967	0.951
$N=1$	3.98	1.75	1.32	1.120	1.050	1.023
$N=2$	4.82	2.09	1.53	1.29	1.20	1.15
$N=3$	6.14	2.58	1.86	1.55	1.41	1.35

formulae have been compared with the terms of the series which have been calculated (see Section 28.2). The agreement is quite reasonable and gives us some confidence that the large order behaviour estimates are indeed correct (see table 41.1).

41.3 Field Theories with Fermions

In the case of boson field theories we have related the large order behaviour of perturbation theory to the decay of the false vacuum for, in general, unphysical values of the coupling constant. We expect therefore some modifications if we consider a system of self-interacting fermions, or of fermions interacting with bosons which themselves have no self-interaction. (Actually the first case can be reduced to the second one by introducing an auxiliary boson field.) Indeed the Pauli principle will make the decay of the false vacuum more difficult because it is not possible to have many fermions in the same state to produce a classical field, and this effect should be especially strong in low dimensions. Of course if the bosons have a self-interaction, this interaction will drive the decay of the vacuum, and the fermions will no longer play any role.

Seen from the point of view of integrals, the difference between fermions and bosons is also immediately apparent. We have shown that the simple integral counting the number of Feynman diagrams, which is also the ϕ^4 field theory in $d=0$ dimensions, already has the characteristic $k!$ behaviour at large orders. Let us instead consider a zero-dimensional fermion theory, i.e. an integral over a finite number of fermion degrees of freedom:

$$I(\lambda) = \int \prod_{i=1}^{N} \mathrm{d}\bar{\xi}_i \mathrm{d}\xi_i \exp\left[\bar{\xi}_i D_{ij}\xi_j + \lambda C_{ijkl}\bar{\xi}_i\bar{\xi}_j\xi_k\xi_l\right]. \tag{41.14}$$

The quantities ξ_i and $\bar{\xi}_i$ are anticommuting variables and D_{ij} and C_{ijkl} are a set of numbers. Because we assume that we have a finite number of anticommuting variables, if we expand the exponential, we get only a finite number of non-vanishing terms in the expansion. Therefore $I(\lambda)$ is a polynomial in λ.

41.3.1 Example of a Yukawa-like field theory

We now consider the vacuum amplitude of the following theory:

$$Z = \int [\mathrm{d}\phi(x)] \, [\mathrm{d}\bar{\psi}(x)] \, [\mathrm{d}\psi(x)] \exp\left[-S\left(\phi, \bar{\psi}, \psi\right)\right], \tag{41.15}$$

in which the action is:

$$S\left(\phi, \bar{\psi}, \psi\right) = \int \mathrm{d}^d x \left[-\bar{\psi} \left(\slashed{\partial} + M + \sqrt{g}\phi \right) \psi + \tfrac{1}{2} \left(\partial_\mu \phi \right)^2 + \tfrac{1}{2} m^2 \phi^2 \right]. \tag{41.16}$$

In the functional integral we have denoted by $\phi(x)$ the scalar boson field, by $\bar{\psi}(x)$ and $\psi(x)$ the fermion fields. The parameter g is a loopwise expansion parameter. Since there is no classical limit for a fermion field, the expression (41.15) is not very well adapted to the study of the vacuum decay. In fact we expect the fermion fields to generate an effective interaction for the boson field $\phi(x)$, and this effective interaction will lead to the decay of the vacuum. This suggests that we should integrate over the ψ and $\bar{\psi}$ variables, and study the instantons of the effective theory for $\phi(x)$. In addition the zero-dimensional example has shown that the fermion integration gives some hints about the analytic structure of the theory. After the integration over ψ and $\bar{\psi}$ we obtain:

$$Z = \int [\mathrm{d}\phi(x)] \det \left[\slashed{\partial} + M + \sqrt{g}\phi(x) \right] \exp \left[-\frac{1}{2} \int \mathrm{d}^d x \left((\partial_\mu \phi)^2 + m^2 \phi^2 \right) \right]. \tag{41.17}$$

We are faced with a new difficulty arising from the integration, the effective action is now non-local in $\phi(x)$, and leads to non-local field equations. However, because we are concerned only with the determination of the large behaviour, we can simplify the effective action. The determinant generated by the fermion integration is, at least for the class of relevant $\phi(x)$ fields, an entire function of the coupling constant $\sqrt{g}$. As a consequence, essential singularities can only be generated by the range of the integration. It is sufficient to calculate the contribution to the functional integral of large fields $\phi(x)$. This situation has to be contrasted with what would have happened if $\psi(x)$ and $\bar{\psi}(x)$ would have been commuting variables. The integration then would have generated the inverse of the determinant function which has singularities for all zeros in g of the determinant. These singularities would have yielded essential singularities in the coupling constant after integration. Finally let us note that this difference, determinant versus inverse determinant, is responsible for the minus sign for each fermion loop in perturbation theory, which allows for cancellations.

41.3.2 Evaluation of the fermion determinant for large fields

As an exercise we first solve a similar problem in which, however, some slight additional complications due to the spin structure are absent.

The Fredholm determinant of a Schrödinger operator for large potentials. Let us evaluate, in the limit of large (smooth) potentials $V(x)$, the Fredholm determinant of the Schrödinger operator:

$$D(V) = \det \left[-\Delta + \mu^2 + V(x) \right] \left[-\Delta + \mu^2 \right]^{-1}. \tag{41.18}$$

On intuitive grounds we expect the determinant to converge towards a local functional. The limit can be obtained for instance by the following method: We rewrite the determinant using the identity

$$\operatorname{tr} \ln H H_0^{-1} = -\operatorname{tr} \int_0^\infty \frac{\mathrm{d}t}{t} \left[\mathrm{e}^{-Ht} - \mathrm{e}^{-H_0 t} \right], \tag{41.19}$$

applied to the case:

$$H = -\Delta + \mu^2 + V(x)\,, \qquad H_0 = -\Delta + \mu^2. \tag{41.20}$$

When $|V(x)|$ becomes large, the integral over t is dominated by the small t region. The evaluation for t small of the evolution operator e^{-tA}, corresponding to the Schrödinger operator A, is a problem we have faced in Section 2.2, in the construction of the path integral representation. We have seen that for t small we can replace in $V(x)$ the operator x by a constant. Using directly equation (2.14) we obtain

$$\operatorname{tr} e^{-Ht} - \operatorname{tr} e^{-H_0 t} \underset{t\to 0}{\sim} \frac{1}{(4\pi t)^{d/2}} \int d^d x \left[e^{-t[V(x)+\mu^2]} - e^{-t\mu^2} \right]. \tag{41.21}$$

As we have indicated there, this evaluation is valid only if $V(x)$ is at least continuous. It then follows

$$\operatorname{tr} \ln H H_0^{-1} \sim -\frac{1}{(4\pi)^{d/2}} \int_0^\infty \frac{dt}{t^{1+d/2}} \int d^d x \left[e^{-t[V(x)+\mu^2]} - e^{-t\mu^2} \right]. \tag{41.22}$$

For $d < 2$ we can also integrate over t and finally obtain:

$$\operatorname{tr} \ln H H_0^{-1} \sim -\frac{\Gamma(-d/2)}{(4\pi)^{d/2}} \int d^d x \left[\left(V(x) + \mu^2 \right)^{d/2} - \mu^d \right]. \tag{41.23}$$

For $d \geq 2$ we know that the quantity (41.18), which has the form of a one-loop diagram in a scalar field theory, has to be renormalized. For $d = 2$ we have to add a mass counterterm. Since $V(x)$ is large we can neglect μ. We then obtain the evaluation

$$\ln D(V) \sim \lim_{d\to 2} \left\{ -\frac{1}{(4\pi)^{d/2}} \int d^d x \left[\Gamma(-d/2) V^{d/2}(x) + \Gamma(1 - d/2) V(x) \right] \right\},$$

$$\sim -\frac{1}{4\pi} \int d^2 x\, V(x) \ln V(x). \tag{41.24}$$

In the same limit we obtain for $d = 3$

$$\ln D(V) \sim -\frac{1}{6\pi} \int d^3 x\, V^{3/2}(x). \tag{41.25}$$

For $d = 4$ a counterterm quadratic in $V(x)$ is required. We then find:

$$\ln D(V) \sim \frac{1}{32\pi^2} \int d^4 x\, V^2(x) \ln V(x). \tag{41.26}$$

The fermion determinant. The same method can be used to evaluate the fermion determinant $D(g)$:

$$D(g) \equiv \det \left(\slashed{\partial} + M + \sqrt{g}\phi(x) \right) \left(\slashed{\partial} + M \right)^{-1}. \tag{41.27}$$

Using reflection symmetry we first rewrite $D(g)$:

$$\ln D(g) = \tfrac{1}{2} \operatorname{tr} \ln \left[\slashed{\partial} + M + \sqrt{g}\phi(x) \right] \left[-\slashed{\partial} + M + \sqrt{g}\phi(x) \right] \left(-\partial^2 + M^2 \right)^{-1},$$

(the order of factors does not matter in a determinant). Evaluating the product we find:

$$\ln D(g) = \tfrac{1}{2} \operatorname{tr} \ln \left[-\partial^2 + (M + \sqrt{g}\phi)^2 + \sqrt{g}\partial\!\!\!/\,\phi \right] (-\partial^2 + M^2)^{-1} .$$

We then again use identity (41.19) with now:

$$H = -\partial^2 + (M + \sqrt{g}\phi)^2 + \sqrt{g}\partial\!\!\!/\,\phi , \tag{41.28}$$
$$H_0 = -\partial^2 + M^2 . \tag{41.29}$$

For g large and $\phi(x)$ smooth enough, we can neglect the term $\sqrt{g}\partial\!\!\!/\,\phi$ in (41.28). Then the trace over γ matrices yields a factor $N = \operatorname{tr} \mathbf{1}$ and the remaining part of the calculation is identical to the case of the Schrödinger equation with the potential:

$$V(x) = g\phi^2(x) + 2\sqrt{g}M\phi(x) . \tag{41.30}$$

Substituting this expression into equation (41.23) we obtain the large field behaviour:

$$\ln D(g) \sim -\frac{N}{2} \frac{\Gamma(-d/2)}{(4\pi)^{d/2}} \int \mathrm{d}^d x \left[(M + \sqrt{g}\phi(x))^d - M^d \right] . \tag{41.31}$$

41.3.3 The large order behaviour

We can now study the essential singularity of the theory at $g = 0$ small from the properties of the effective local action:

$$S_{\text{eff}}(\phi) = \int \mathrm{d}^d x \left[\frac{1}{2} (\partial_\mu \phi)^2 + \frac{1}{2} m^2 \phi^2 + \frac{N}{2} \frac{\Gamma(-d/2)}{(4\pi)^{d/2}} g^{d/2} \phi^d(x) \right] . \tag{41.32}$$

It should be understood that for d even, the necessary counterterms are provided to make the action finite. We have to look for instanton solutions of the corresponding field equations. Since the particular model is not interesting in itself, we shall not do it here. Let us just assume that we have found a solution and determine the dependence in the coupling constant g. We rescale the field ϕ to factorize the dependence on g in front of the classical action:

$$\phi(x) \mapsto \frac{1}{g^{d/2(d-2)}} \phi(x) . \tag{41.33}$$

The classical action calculated for a solution takes thus the form:

$$S[\phi_c] = (A/g)^{d/d-2} . \tag{41.34}$$

Introducing this form into the Cauchy formula we find:

$$Z_k \underset{k\to\infty}{\sim} \int_0^\infty \frac{e^{-(A/g)^{d/d-2}}}{g^{k-1}} \mathrm{d}g . \tag{41.35}$$

The integration yields the large order estimate:

$$Z_k \sim A^{-k} \Gamma\left[k(d-2)/d \right] . \tag{41.36}$$

We observe that, as expected, this theory is less divergent than a purely bosonic field theory. However for d large the Pauli principle becomes decreasingly effective and we recover the boson result (in a cut-off field theory). For $d = 2$ expression (41.36) becomes:

$$Z_k \sim A^{-k} \left(\ln k\right)^k ,\tag{41.37}$$

in agreement with rigorous bounds which imply:

$$|Z_k| < (k!)^\varepsilon \quad \text{for all } \varepsilon > 0 .\tag{41.38}$$

Remark. To compare the contribution of boson interactions and fermion interactions, we have implicitly assumed that we were dealing only with loopwise expansions. In such circumstances, the boson contributions always dominate the large order behaviour. If we group the diagrams in another way, the answer can be different. Let us again consider the theory defined by the action (41.16) in four dimensions. In four dimensions, this theory cannot be renormalized without the addition of a $\lambda\phi^4$ counterterm. Thus renormalization forces us to introduce a boson self-interaction. But it is consistent with renormalization to consider λ as being of order g^2. Then both interaction terms $\bar{\psi}\psi\phi$ and ϕ^4 give contributions of the same order to the large order behaviour.

41.3.4 The case of QED

Of course the potentially most interesting application of the previous analysis is QED, which satisfies all previous requirements. One additional complication then arises. The fermion integration leads to the determinant

$$D(e) = \det\left(\slashed{\partial} + m + ie\slashed{A}\right) .\tag{41.39}$$

To estimate $D(e)$ for large charge e we can use the equations (18.66,18.67) (we assume d even):

$$D^2(e) = \det\left(m^2 - D_\mu^2 - \tfrac{1}{2}e\sigma_{\mu\nu}F_{\mu\nu}\right) .\tag{41.40}$$

In the large e limit the last term, which is of order e, is negligible with respect to D_μ^2 which is of order e^2:

$$\ln D(e) \sim \tfrac{1}{2}N_d \operatorname{tr} \ln\left(m^2 - D_\mu^2\right) .\tag{41.41}$$

We then use the representation (41.19) to evaluate the determinant of the electromagnetic Schrödinger operator. The path integral representation of e^{-tA} has been given in Section 3.2. The determination of the large coupling constant behaviour is however more subtle than before. The electromagnetic term in the path integral depends only on the geometry of the loop one integrates along and not on the time spent on the loop trajectory (see equation (3.19)). Therefore the large coupling constant limit does not select the short time contributions in the representation (41.19). As a consequence the determinant no longer generates a local term. A direct calculation of the determinant has not been performed. The difficulty seems to be related to the fact that, due to gauge invariance, the gauge degree of freedom of the gauge field cannot be considered as slowly varying. It has therefore been conjectured, on the basis of studying the determinant for special gauge fields, that the determinant is equivalent for large e to

$$\ln D(e) \sim C(d) \int \mathrm{d}^d x\, |e[A_{\mathrm{T}}]_\mu|^d ; \quad C^{-1}(d) = d(4\pi)^{(d-1)/2}\Gamma\big((d+1)/2\big),$$

where $[A_\text{T}]_\mu$ is the transverse part of A_μ:

$$[A_\text{T}]_\mu(x) = A_\mu(x) - \partial^{-2}\partial_\mu\partial_\nu A_\nu(x).$$

This result is gauge invariant, as it should, but non-local except in the gauge $\partial_\mu A_\mu = 0$. It agrees for $d = 2$ with the exact result (31.60) obtained from the abelian anomaly ($C(2) = 1/2\pi$). For $d = 4$, the case of physical interest, $C(4) = 1/12\pi^2$. The effective classical field theory then is scale invariant. Arguments related to conformal invariance can be used to construct some ansatz for the instanton solutions. Two kind of solutions have been explored by Balian *et al.*, or Bogolmony and Fateev. Taking the minimal action solution one obtains an evaluation of the form:

$$Z_k \sim (-1)^k A^{-k}\Gamma(k/2), \quad A = 4.886, \tag{41.42}$$

the expansion parameter being: $\alpha = e^2/4\pi$. It is worthwhile mentioning that this evaluation is probably not very useful as a practical mean to predict new orders in QED for several reasons. First the theory is not asymptotically free and thus has a potential renormalon problem, which can be understood by inserting in a Feynman diagram the one-loop corrected photon propagator. Second the cancellation coming from the sign of fermion loops does not seem to be very effective at low orders. Therefore an alternative calculation, which leads to a large order behaviour at a fixed number of fermion loops, seems to be more useful. Predictions of this kind made for diagrams with one fermion loop, seem to agree well with numerical estimates.

41.4 Divergent Series, Borel Summability

Asymptotic series. Let us consider a function $f(z)$, analytic in a sector S:

$$|\operatorname{Arg} z| \leq \alpha/2, \quad |z| \leq |z_0|. \tag{41.43}$$

We assume that $f(z)$ has in S the following asymptotic expansion:

$$f(z) = \sum_0^\infty f_k z^k. \tag{41.44}$$

This means that the series (41.44) diverges for all non-trivial values of z and that in S the following bound is satisfied

$$\left| f(z) - \sum_{k=0}^N f_k z^k \right| \leq C_{N+1}|z|^{N+1} \quad \text{for all } N, \tag{41.45}$$

in which:

$$C_N|z|^N \underset{\text{for } N\to+\infty}{\longrightarrow} \infty \quad \forall z \neq 0.$$

Though the series (41.44) diverges, it is possible to use it to estimate the function $f(z)$ for $|z|$ small. At $|z|$ fixed we can look for a minimum in the bound (41.45) when N varies. If $|z|$ is small enough, the bound first decreases with N and then, since the series is divergent, finally increases. If we truncate the series at the minimum, we get the best possible estimate of $f(z)$, with a finite error $\varepsilon(z)$. Let us assume for definiteness that the coefficients C_N have the form:

$$C_N = M\,A^{-N}\,(N!)^{\beta}\,. \tag{41.46}$$

We can then estimate $\varepsilon(z)$ explicitly and find:

$$\varepsilon(z) = \min_{\{N\}} C_N |z|^N \sim \exp\left[-\beta\,(A/|z|)^{1/\beta}\right]. \tag{41.47}$$

We see that an asymptotic series does not in general define a unique function. Indeed if we have found one function, we can add to it any function analytic in the sector (41.43) and smaller than $\varepsilon\,(z)$ in the whole sector. The new function still satisfies the condition (41.45). However there is one situation in which the asymptotic series defines a unique function. If the angle α satisfies: $\alpha \geq \pi\beta$, then a classical theorem about analytic functions tells us that a function analytic in the sector and bounded by $\varepsilon(z)$ in the whole sector vanishes identically. In particular in the special case $\beta = 1$, which is typical for perturbative expansions, we find

$$\alpha \geq \pi\,. \tag{41.48}$$

In the marginal case in which the series is asymptotic only in the open interval $|\operatorname{Arg} z| \in (-\pi\beta/2, \pi\beta/2)$, additional conditions have to be imposed to prove uniqueness.

Borel transformation. We specialize from now on to $\beta = 1$, since it is the most useful example, but the generalization to arbitrary β is straightforward. Under the condition (41.48), the function $f(z)$ is uniquely defined by the series. In addition there then exist methods to "sum" the series, which means that one can reconstruct the function from the knowledge of the terms of the series. One set of methods is based upon the Borel transformation.

Let us define the Borel transform $B_f(z)$ of $f(z)$ by

$$B_f(z) = \sum_0^{\infty} B_k z^k \equiv \sum_0^{\infty} \frac{f_k}{k!} z^k. \tag{41.49}$$

The bound (41.45) and the estimate (41.46) give us a bound on the coefficients f_k:

$$|f_k\,/k!| < M\,A^{-k}. \tag{41.50}$$

Thus $B_f(z)$ is analytic at least in a circle of radius A and uniquely defined by the series. Furthermore, in the sense of power series:

$$f(z) = \int_0^{\infty} \mathrm{e}^{-t}\,B_f\,(zt)\,\mathrm{d}t\,. \tag{41.51}$$

As a consequence of the inequality (41.48), it can be shown that $B_f(z)$ is also analytic in a sector

$$|\operatorname{Arg} z| \in [0, \tfrac{1}{2}(\alpha - \pi)[\,, \tag{41.52}$$

and does not increase faster than an exponential in the sector, so that integral (41.51) converges for $|z|$ small enough and inside the sector

$$|\operatorname{Arg} z| < \alpha/2\,.$$

In addition it can be shown that the r.h.s. of equation (41.51) satisfies a bound of type (41.45). Hence this integral representation yields the unique function which has the asymptotic expansion (41.44) in the domain (41.43).

41.5 Large Order Behaviour and Borel Summability

We have learned that, for a large class of potentials in quantum mechanics and for a number of field theories, the perturbative coefficients f_k in the loopwise expansion of any physical quantity $f(g)$,

$$f(g) = \sum_{k=0}^{\infty} f_k g^k, \tag{41.53}$$

behave at large orders as

$$f_k \sim ck^{b-1}a^k k! . \tag{41.54}$$

Therefore the coefficients B_k of the Borel transform B_f (equation (41.49)) of $f(g)$, behave as

$$B_k \sim ck^{b-1}a^k . \tag{41.55}$$

This asymptotic estimate tells us that the singularity of $B_f(g)$ closest to the origin is located at the point $g = 1/a$. Consequently, the Borel transformation does not exist if the classical action $A = 1/a$ is positive. The perturbation series in such theories is not Borel summable. Let us, in the light of this result, discuss the various situations we have encountered:

(i) The field equations have no real instanton solutions. This is in particular the case if we have expanded around the unique absolute minimum of the potential. If complex instanton solutions exist, the corresponding classical action is non-positive, and the perturbative expansion is presumably Borel summable. It is only a presumption because various features of the perturbative expansion, invisible at large orders, could prevent Borel summability. The perturbative expansion could contain for instance contributions all of the same sign, growing faster than any exponential of the order k, but much smaller than $k!$ (for example $\sqrt{k!}$). Then $B_f(z)$ would grow too rapidly for large argument z ($\ln B_f(z) \sim z^2$ in the example) and the Borel integral would not converge at infinity.

(ii) We have found real instantons in the theory because we expanded around a relative minimum of the potential: the perturbative expansion is not Borel summable.

However in this case we can provide one additional piece of information useful for determining the solution: the unstable situation can be considered as coming from a stable situation by analytic continuation. Therefore a possible solution could be to integrate in the Borel transform just above the cut which is on the real positive axis. As a consequence, from a real perturbative expansion we would obtain a complex result, but this is exactly what we expect. It is easy to verify that the imaginary part is for g small exactly what we have calculated directly. Actually this is only the solution of the problem in the simplest case, when no other instanton singularities cross the contour of integration in the analytic continuation.

(iii) There are real instantons connecting degenerate classical minima.

The theory is not Borel summable. Integration above or below the axis yields a complex result for a real quantity. This cannot be the correct prescription. The half sum of the integral above and below is real, but even in the simple example of the double well-potential, one can verify numerically, and argue analytically, that it is not the correct solution. We shall show, in quantum mechanics, that the additional information needed to determine the sum of the perturbative expansion is provided by the consideration of many instanton contributions (see Chapter 43). The corresponding problem has not been solved in field theory examples yet.

Remarks. We have given field theory examples of such a situation in Sections 40.4, 40.5: the two-dimensional $CP(N-1)$ models and four-dimensional $SU(2)$ gauge theory. In these models real instantons connect degenerate minima of the classical action and the corresponding classical action is positive. Therefore the coefficients of the perturbative expansion contain a non-Borel summable contribution. This contribution does not necessarily dominate the large order behaviour, because, as the example of of the ϕ_4^4 massless field theory in Chapter 42 will illustrate, when a field theory is classically scale invariant, the perturbative expansion might be dominated by contributions unobtainable by semiclassical methods, and related to the UV or IR singularities.

41.6 Practical Summation Methods

Various practical summation methods rely upon a Borel transformation.

The Borel transformation reduces the problem of determining the function to the analytic continuation of the Borel transform. The Borel transform is given by a Taylor series in a circle and an analytic continuation of the series in a neighbourhood of the real positive axis is required. This analytic continuation can be performed by many methods and the optimal choice depends somewhat on the additional information one possesses about the function. Let us give two examples:

Padé approximants. In the absence of accurate knowledge of the positions of all singularities of the Borel transform in the complex plane, one can use the Padé approximation. From the series one derives Padé approximants which are rational functions P_M/Q_N satisfying:

$$B_f(z) = \frac{P_M(z)}{Q_N(z)} + O\left(z^{N+M+1}\right), \qquad (41.56)$$

where P_M and Q_N are polynomials of degrees M and N respectively. If one knows K terms of the series, one can construct all $[M, N]$ Padé approximants with $N + M \leq K$. This method is well adapted to meromorphic functions. The main disadvantage of the method is that for a rather general class of functions Padé approximants are known to converge only in measure and thus spurious poles may occasionally appear close to or on the real positive axis.

Even if the Padé approximants converge, this property may lead to instabilities in the results.

Conformal mapping. If we know the domain of analyticity of the Borel transform, we can find a mapping which preserves the origin, and maps the domain of analyticity onto a circle. In the transformed variable the new series converges in the whole domain of analyticity. Let us explain the method on an example.

If the Borel transform is analytic in a cut-plane, the cut running along the real negative axis from $-\infty$ to $-1/a$, we can map the cut-plane onto a circle of radius 1:

$$z \mapsto u, \quad u(z) = \frac{\sqrt{1+az}-1}{\sqrt{1+az}+1}. \qquad (41.57)$$

From the original series for the Borel transform we derive a series in powers of the new variable u:

$$B_f(z) = \sum \frac{f_k}{k!} z^k,$$

$$B_f[z(u)] = \sum_0^\infty B_k u^k. \qquad (41.58)$$

Introducing this expansion in the Borel transformation we obtain a new expansion for $f(z)$,

$$f(z) = \sum_0^\infty B_k I_k(z), \tag{41.59}$$

in which the functions $I_k(z)$ have the integral representation:

$$I_k(z) = \int_0^\infty e^{-t} [u(zt)]^k dt. \tag{41.60}$$

It is possible to study the natural domain of convergence of this new expansion. Using for $u(z)$ the explicit expression (41.57), we can evaluate $I_k(z)$ for k large by steepest. The saddle point equation is:

$$-1 + \frac{k}{t} \frac{1}{\sqrt{1 + azt}} = 0, \tag{41.61}$$

which for k large yields:

$$t \sim k^{2/3}/(az)^{1/3}. \tag{41.62}$$

It follows that $I_k(z)$ behaves for k large as:

$$I_k(z) \sim \exp\left[-3k^{2/3}/(az)^{1/3}\right]. \tag{41.63}$$

Three situations can now arise:

(i) The coefficients B_k either decrease or at least do not grow too rapidly,

$$|B_k| < M\, e^{\varepsilon k^{2/3}} \quad \text{for all } \varepsilon > 0.$$

Then the expansion (41.59) converges at least in the region

$$\operatorname{Re} z^{-1/3} > 0 \;\Rightarrow\; |\mathrm{Arg}| < 3\pi/2. \tag{41.64}$$

This in particular implies that the function $f(z)$ must be analytic in the corresponding region which contains a part of the second sheet.

(ii) The coefficients behave like

$$B_k \sim \exp\left(ck^{2/3}\right) \quad \text{for } k \text{ large}. \tag{41.65}$$

The domain of convergence is then:

$$\operatorname{Re} z^{-1/3} > \tfrac{1}{3} ca^{1/3}. \tag{41.66}$$

This condition implies analyticity in a finite domain containing a part of the second sheet since for $|z|$ small the r.h.s. is negligible.

(iii) The coefficients B_k grow faster than $\exp(ck^{2/3})$. This is quite possible since the only constraint on the coefficients B_k is that the series (41.58) has a radius of convergence 1. For instance the coefficients B_k could grow like $\exp(ck^{4/5})$. In such a situation the new series is also divergent. Such a situation arises when the singularities on the boundary of the domain of analyticity are too strong. One should map a smaller part of the domain of analyticity onto a circle.

Application to the calculation of critical exponents. In the summation method based on Borel transformation and mapping it is easy to incorporate the information coming from the large order behaviour analysis. This is one reason why it has been used quite systematically in the framework of the ϕ^4 field theory to calculate critical exponents and other universal quantities. In Chapter 28 we have given values for various critical exponents, obtained by applying variants of the Borel summation method to the known terms of the perturbative expansion, i.e. six successive terms in fixed dimension 3 and up to order ε^5 for the ε-expansion.

Let us now summarize the information available for in the ϕ^4 field theory that justifies the use of this summation method.

(i) The Borel summability of perturbation theory in ϕ_2^4 and ϕ_3^4 has been rigorously established.

(ii) The large order behaviour has been determined in all cases and compares favourably with the first terms of the series available (see Section 41.2).

(iii) Since all known instanton solutions in the ϕ^4 theory give negative actions, it is plausible that the Borel transform is analytic in a cut plane, the location and nature of the singularity closest to the origin being given by the large order estimates.

Consequently the methods based upon a Borel transformation and a conformal mapping of the cut-plane onto a circle, have appeared as ideal candidates to sum the perturbation series and the ε-expansion.

For completeness let us finally give at least one example of a summation method not based on a Borel transformation.

Order dependent mappings (ODM). The ODM method requires, to be applicable, some knowledge of the analyticity properties of the function itself. As we have discussed, the series diverge because the function has singularities accumulating at the origin. However the strengths of the singularities have to decrease fast enough for the function to have a series expansion. In the examples we have met, the discontinuity of the function decreases exponentially near the origin. The idea is then to pretend that the function is analytic, in addition to its true domain of analyticity, in a small circle centred at the origin of adjustable radius ρ and to map this extended domain onto a circle centred at the origin, keeping the origin fixed. If the function would really be analytic in such a domain, the expansion in the transformed variable would converge in the whole domain of analyticity and our problem would be solved. Since the original series is in fact only asymptotic, the series in the transformed variable is also asymptotic. However, as a result of this transformation, the coefficients of the new series now depend on an adjustable parameter ρ.

Let us assume for instance that $f(z)$ is analytic in a cut plane. We then use the mapping

$$z = 4\rho u / (1 - u)^2. \tag{41.67}$$

The transformed series has the form

$$f(z(u)) = \sum_0^\infty P_k(\rho) u^k, \tag{41.68}$$

in which the coefficients $P_k(\rho)$ are polynomials of degree k in the parameter ρ. In more general situations, one can often use a mapping of the form

$$z = \rho h(u), \qquad h(u) = O(u). \tag{41.69}$$

The kth order approximation is obtained by truncating the series at order k, and choosing ρ as one of the zeros of the polynomial $P_k(\rho)$. The zero cannot actually be chosen arbitrarily, but roughly speaking must be the zero of largest modulus for which the derivative $P'_k(\rho)$ is small. The idea behind the method is the following: with the original series the best approximation is obtained by truncating the series at z fixed, at an order dependent on z such the modulus of the last term taken into account is minimal. By introducing an additional parameter we have modified the situation. We fix first the order of truncation and then try to adjust the parameter ρ in such a way that, at z again fixed, the last term taken into account is minimal.

The kth order approximant has the form

$$\{f(z)\}_k = \sum_{l=0}^{k} P_l(\rho_k) [u(z)]^l, \qquad P_k(\rho_k) = 0. \tag{41.70}$$

It can be shown under certain conditions that if the terms of the original series grow like

$$f_k = O\left[(k!)^\beta\right], \tag{41.71}$$

then the sequence ρ_k decreases like

$$\rho_k = O\left(1/k^\beta\right). \tag{41.72}$$

Such a method has been successfully applied to test problems like the quartic anharmonic oscillator with a mapping:

$$z = \rho u/(1-u)^{3/2},$$

and to one physical example, the hydrogen atom in a strong magnetic field.

Bibliographical Notes

The connection between barrier penetration effects in the semiclassical limit and large behaviour in perturbation theory was first realized by

F.J. Dyson, *Phys. Rev.* 85 (1952) 631; C.M. Bender and T.T. Wu, *Phys. Rev.* 184 (1969) 1231; D7 (1973) 1620; *Phys. Rev. Lett.* 37 (1976) 117; T. Banks, C.M. Bender and T.T. Wu, *Phys. Rev.* D8 (1973) 3346; T. Banks and C.M. Bender, *Phys. Rev.* D8 (1973) 3366;

These authors calculate the barrier penetration coefficients in Quantum Mechanics using standard WKB methods. The idea of using instead path integrals can be found in an early paper

C.S. Lam, *Nuovo Cimento*, 55A (1968) 258;

however the field theory calculation (the ϕ_4^4 theory) is due to Lipatov already quoted

L.N. Lipatov, *JETP Lett.* 24 (1976) 157, *Sov. Phys. JETP* 44 (1976) 1055, *JETP Lett.* 25 (1977) 104, *Sov. Phys. JETP* 45 (1977) 216.

The method has been generalized in

E. Brézin, J.C. Le Guillou and J. Zinn-Justin, *Phys. Rev.* D15 (1977) 1558,

and then applied to many theories like the abelian gauge field coupled to scalars

C. Itzykson, G. Parisi and J.-B. Zuber, *Phys. Rev. Lett.* 38 (1977) 306; A.P. Bukhvostov and L.N. Lipatov, *Zh. Eksp. Teor. Fiz.* 73 (1977) 1658;

to the ϕ^3 field theory with Potts symmetry

A. Houghton, J.S. Reeve and D.J. Wallace, *Phys. Rev.* B17 (1978) 2956;

theories with degenerate minima (see Chapter 43)

E. Brézin, G. Parisi and J. Zinn-Justin, *Phys. Rev.* D16 (1977) 408;

non-abelian gauge theories

E.B. Bogomolny and V.A. Fateyev, *Phys. Lett.* 71B (1977) 93; L.N. Lipatov, A.P. Bukhvostov and E.I. Malkov, *Phys. Rev.* D19 (1979) 2974.

The idea of evaluating the large order behaviour by first calculating the discontinuity of functions is due to

G. Parisi, *Phys. Lett.* 66B (1977) 167; E.B. Bogomolny, *Phys. Lett.* 67B (1977) 193.

Let us also quote the review article

E.B. Bogomolny, V.A. Fateyev and L.N. Lipatov, *Sov. Sci. Rev.*, I.M. Khalatnikov ed., section A: Phys. Rev. 2 (1980) 247,

and the reprint volume

Large Order Behaviour of Perturbation Theory, Current Physics vol. 7, J.C. Le Guillou and J. Zinn-Justin eds., (North-Holland, Amsterdam 1990).

Large order in field theories with fermions has first studied by

G. Parisi, *Phys. Lett.* 66B (1977) 382.

QED has been considered in

C. Itzykson, G. Parisi and J.-B. Zuber, *Phys. Rev.* D16 (1977) 996; R. Balian, C. Itzykson, J.-B. Zuber and G. Parisi, *Phys. Rev.* D17 (1978) 1041; E.B. Bogomolny and V.A. Fateyev, *Phys. Lett.* 76B (1978) 210; E.B. Bogomolny and Yu A. Kubyshin, *Yad. Fiz.* 34 (1981) 1535; 35 (1982) 202 (*Sov. J. Nucl. Phys.* 34 (1981) 853; 35 (1982) 114).

The Yukawa interaction in two dimensions has been discussed in

M.P. Fry, *Phys. Lett.* 80B (1978) 65.

The mathematical discussion of divergent series can be found in

E. Borel, *Leçon sur les Séries Divergentes*, 2nd edn (Gauthier-Villars, Paris 1928); G. Hardy, *Divergent Series* (Oxford University Press, Oxford 1949).

For a textbook concerning Padé approximants see

G.A. Baker and P. Graves-Morris, *Encyclopedia of Mathematics and its Applications* vols. 13 and 14, Gian-Carlo Rota ed. (Addison-Wesley, New York 1981).

The method of summation of series by Borel transformation and mapping was proposed by

J.J. Loeffel, *Saclay Report*, DPh-T/76/20 unpublished.

The summation method based on order-dependent mappings has been studied in

R. Seznec and J. Zinn-Justin, *J. Math. Phys.* 20 (1979) 1398; J.C. Le Guillou and J. Zinn-Justin, *Ann. Phys. (NY)* 147 (1983) 57,

and proofs of convergence in special cases given in

R. Guida, K. Konishi and H. Suzuki, *Ann. Phys. (NY)* 241 (1995) 152.

The Borel summability of the ϕ^4 in two and three dimensions has been established in

J.P. Eckmann, J. Magnen and R. Sénéor, *Commun. Math. Phys.* 39 (1975) 251; J.S. Feldman and K. Osterwalder, *Ann. Phys. (NY)* 97 (1976) 80; J. Magnen and R. Sénéor, *Commun. Math. Phys.* 56 (1977) 237; J.-P. Eckmann and H. Epstein, *Commun. Math. Phys.* 68 (1979) 245.

For differential approximants see

A.J. Guttmann and G.S. Joyce, *J. Phys. A: Math. Gen.* 5 (1972) L81; M.E. Fisher and J.H. Chen, *Phase Transitions*, Cargèse 1980, vol. B72, M. Lévy, J.C. Le Guillou and J. Zinn-Justin eds. (Plenum, New York 1982),

and references therein.

APPENDIX 41

A41.1 Large Order Behaviour for Simple Integrals

Consider the integral of Chapter 37:

$$I(g) = \frac{1}{\sqrt{2\pi}} \int_{-\infty}^{+\infty} \exp\left[-\left(\frac{x^2}{2} + \frac{gx^4}{4}\right)\right] dx. \qquad (A41.1)$$

We can expand $I(g)$ in power series,

$$I(g) = \sum_{0}^{\infty} I_k g^k. \qquad (A41.2)$$

The coefficients I_k count the number of vacuum Feynman diagrams with the proper weights in a ϕ^4 field theory. We have argued in Chapter 37 that the imaginary part of $I(g)$ for g negative and small was dominated by the non-trivial saddle points

$$x^2 = -1/g. \qquad (A41.3)$$

The contribution of the saddle points yields

$$\operatorname{Im} I(g) \underset{g \to 0_-}{\sim} 2^{-1/2} e^{1/4g}. \qquad (A41.4)$$

Therefore I_k behaves for k large as:

$$I_k \underset{k \to \infty}{\sim} \frac{1}{\pi\sqrt{2}} (-4)^k (k-1)!. \qquad (A41.5)$$

This result leads to the following interpretation of the large order behaviour formulae obtained in the Chapter 41: in the case of the anharmonic oscillator and the ϕ^4 field theory, the number of Feynman diagrams is of the order of $4^k k!$ for k large and a typical diagram behaves at large orders as $(4A)^{-k}$ where k is the order but also, up to an additive constant, the number of loops and A the classical action.

A41.2 Non-Loopwise Expansions

Although we have discussed large order behaviour estimates only for loopwise expansions, it is straightforward to generalize the analysis for perturbation expansions in different parameters. For example let us consider the action $S(q)$,

$$S[q] = \int \left[\tfrac{1}{2}\dot{q}^2 + \tfrac{1}{2}q^2 + \lambda V(q)\right] dt, \qquad (A41.6)$$

in which $V(q)$ is a polynomial interaction. If we study the large order behaviour of the expansion in powers of λ, it is easy to verify that we need only to consider the instanton solutions of an action in which $V(q)$ has been replaced by its term of highest degree. Let us assume that

$$V(q) = \sum_{2}^{2N} V_n q^n. \qquad (A41.7)$$

Then simply by rescaling, we see that the classical solution $q_c(t)$ has the form

$$q_c(t) = \lambda^{-1/(2N-2)} f(t), \qquad (A41.8)$$

and the term of degree n in $V(q)$ will give a contribution to the classical action proportional to $\lambda^{1-n/(2N-2)}$. We verify that the term of highest degree gives indeed the largest contribution to the action for λ small. The saddle point in λ in the dispersion relation for large order k is of the order

$$\lambda \sim k^{-(N-1)}.$$

Thus the term of degree n in the potential generates a factor of the form

$$\exp\left[c_n k^{2/(n+2-2N)}\right],$$

which is relevant, at leading order, only for $n \geq 2N - 2$.

A41.3 Linear Differential Approximants

Padé approximants provide the simplest example of a general class of approximants, which are obtained as solutions to equations (algebraic or differential) with polynomial coefficients. These polynomials are chosen to be the polynomials of lowest degree for which the solution of the equation has the same power series expansion up to a given order as the function one wants to approximate. To be more concrete let us give the example of the linear differential approximants.

Let $f(z)$ be a function for which we know a power series expansion. We can construct approximants $\bar{f}(z)$ to this function by looking for solutions of the differential equation:

$$\sum_{n=0}^{N} P_n(z) \left(\frac{\partial}{\partial z}\right)^n \bar{f}(z) = R(z), \qquad (A41.9)$$

in which the polynomials $P_n(z)$ and $R(z)$ form a set of polynomials of lowest possible total degree chosen such that

$$f(z) - \bar{f}(z) = O\left(z^k\right). \qquad (A41.10)$$

In the generic situation the degrees $[P_n]$ and $[R]$ of the polynomials P_n and R satisfy:

$$\sum_{n=0}^{N} [P_n] + [R] = k. \qquad (A41.11)$$

The advantage of these kinds of approximants is that they are extremely flexible. It is possible to use a lot of additional information one possesses about the function by imposing additional constraints on the polynomials P_n and R.

Furthermore while Padé approximants generate only approximants with poles, the more general approximants can have a large class of new singularities. There is of course a price to pay: this approximation is much more unstable. It is necessary to select among the large number of approximants one can construct, those for which one has some reasons to believe that they are especially well adapted to the original function one wants to approximate.

Due to the generality of the problem, a systematic study of this class of approximants is lacking. Notice that the method can be generalized to power series in more than one variable. One then writes partial differential equations with polynomial coefficients in several variables.

42 INSTANTONS: THE ϕ^4 FIELD THEORY IN DIMENSION FOUR

We have shown in Appendix A39, using Sobolev inequalities, that in dimension four the massive field equation has no instanton solution and the relevant instanton is a solution of the massless field equation instead. We therefore first study the massless ϕ^4 theory and comment at the end about the massive theory. The price to pay for such a simplification is the appearance of some subtle infrared (IR) problems.

The ϕ^4 field theory in dimension four is just renormalizable. We have thus to deal at leading order with the one-loop coupling constant renormalization in addition to the mass renormalization. This fact, together with the scale invariance of the classical theory, leads to the appearance of an effective coupling constant at the scale of the instanton, and therefore the calculation of the contribution of the instanton depends on global renormalization group properties of the theory. Finally we shall discover that, as a consequence of their large momenta properties, individual diagrams at order k grow themselves like $k!$, introducing some new complications in the large order behaviour analysis. Similarly IR singularities in the massless theory also yield contributions of order $k!$, but with a different sign.

Note that we here explain the calculation only for one-component ϕ^4 field theory, but the extension to the $O(N)$ symmetric model is simple.

42.1 The Euclidean Equation of Motion. The Instanton Action

The euclidean action of the massless theory ϕ^4 theory can be written:

$$S(\phi) = \int d^4x \left[\tfrac{1}{2} \left(\partial_\mu \phi \right)^2 + \tfrac{1}{4} g \phi^4 \right], \tag{42.1}$$

and the corresponding equation of motion reads:

$$-\Delta \phi(x) + g \phi^3(x) = 0. \tag{42.2}$$

Note the unconventional normalization of the coupling constant. To return to the usual convention one has to set $g \mapsto g/6$.

We know that the solution of minimal action is spherically symmetric, thus we set:

$$\phi(x) = \frac{1}{\sqrt{-g}} f(r), \tag{42.3}$$

with

$$r = |x - x_0|. \tag{42.4}$$

We then obtain a differential equation

$$-\left[\left(\frac{d}{dr} \right)^2 + \frac{3}{r} \frac{d}{dr} \right] f(r) = f^3(r). \tag{42.5}$$

We now use the scale invariance of the classical theory (the theory is actually conformal invariant, see Appendix A42.2). If $\phi(x)$ is the solution to the equation, then $\psi(x)$ is also a solution with

$$\phi(x) = \lambda\psi(\lambda x).\tag{42.6}$$

This suggests the following change:

$$f(r) = e^{-t}\,h(t),\qquad r = e^t,\tag{42.7}$$

which transforms equation (42.5) into:

$$\ddot{h}(t) = h(t) - h^3(t).\tag{42.8}$$

We recognize the equation of motion of the anharmonic oscillator that we have solved in Chapter 37:

$$h_c(t) = \pm\frac{\sqrt{2}}{\cosh\,(t - t_0)}.\tag{42.9}$$

The solution $\phi_c(x)$ of equation (42.2) then is

$$f(r) = \pm\frac{2\sqrt{2}\lambda}{1 + \lambda^2 r^2},\tag{42.10a}$$

$$\Rightarrow \phi_c(x) = \pm\frac{1}{\sqrt{-g}}\frac{2\sqrt{2}\lambda}{1 + \lambda^2\,(x - x_0)^2},\tag{42.10b}$$

where we have defined $\lambda = e^{-t_0}$. The corresponding classical action $S(\phi_c)$ is

$$S(\phi_c) = -A/g,\qquad A = 8\pi^2/3.\tag{42.11}$$

With the standard normalization of g one finds $A = 16\pi^2$.

Because the classical theory is scale invariant, the instanton solution now depends on a scale parameter λ, in addition to the four translation parameters $x_{0\mu}$. We have therefore to introduce five collective coordinates to calculate the instanton contribution.

42.2 The Instanton Contribution at Leading Order

The general strategy. The second derivative $M(x, x')$ of the action at the saddle point is

$$M(x, x') = \frac{\delta^2 S}{\delta\phi_c(x)\delta\phi_c\,(x')} = \left[-\Delta - \frac{24\lambda^2}{(1 + \lambda^2 x^2)^2}\right]\delta^{(4)}(x - x').\tag{42.12}$$

To find the eigenvalues the operator of $\mathbf{M}$, one has to solve a 4-dimensional Schrödinger equation with a spherically symmetric potential. We immediately note at this stage two serious problems. The operator $\mathbf{M}$ has of course five eigenvectors, $\partial_\mu\phi_c(x)$ and $(d/d\lambda)\,\phi_c(x)$, with eigenvalue zero. The last of these eigenvectors is not normalizable with the natural measure of this problem,

$$\int\left[\frac{d}{d\lambda}\phi_c(x)\right]^2 d^4x = \infty.\tag{42.13}$$

This is an IR problem which arises because the theory is massless.

Another difficulty comes from the mass counterterm which has to be added to the action. It has the form:

$$\tfrac{1}{2}\delta m_0^2 \int d^4x \, \phi_c^2(x) = \infty.$$ (42.14)

The integral of $\phi_c^2(x)$ is also IR divergent, and this IR divergence is expected to cancel with an IR divergence of det $\mathbf{M}$. Thus we need in general some kind of IR regularization. In the particular case of the dimensional regularization, this problem is postponed to two-loop order.

These problems will be solved in several steps. First we realize that we do not need the eigenvalues of $\mathbf{M}$ but only the determinant $\det' \mathbf{M} \mathbf{M}_0^{-1}$ (equations (39.28)). We can multiply $\mathbf{M}$ and $\mathbf{M}_0$ by the same operator. A specific choice which makes full use of the scale invariance of the classical theory, then transforms $\mathbf{M}$ into an operator whose eigenvalues can be calculated analytically. Because the calculations are somewhat tedious, we here indicate only the various steps, without giving all details.

The transformation. We extend the transformation (42.7) to arbitrary fields, setting:

$$\phi(x) = e^{-t} h(t, \hat{n}), \qquad \text{with} \quad t = \ln |x|, \qquad \hat{n}_n = \frac{x^\mu}{|x|}.$$ (42.15)

The classical action then becomes

$$S(\phi) = \tilde{S}(h) = \int dt \, d\Omega \left[\frac{1}{2} \left(\dot{h} - h \right)^2 + h \mathbf{L}^2 h + \frac{1}{4} g h^4 \right].$$ (42.16)

The symbol $\int d\Omega$ means integration over the angular variables $\hat{n}$, and $\mathbf{L}^2$ is the square of the angular momentum operator with eigenvalues $l(l+2)$ and degeneracy $(l+1)^2$. Expression (42.16) can be rewritten

$$\tilde{S}(h) = \int dt \, d\Omega \left\{ \frac{1}{2} \left[\dot{h}^2 + h \left(\mathbf{L}^2 + 1 \right) h \right] + \frac{1}{4} g h^4 \right\}.$$ (42.17)

The integral of $\dot{h} h$ vanishes due to boundary conditions.

With the parametrization

$$\lambda = e^{-t_0}, \qquad \mathbf{x}_0 = e^{t_0} \mathbf{v},$$

the classical solution (42.10b) transforms into $h_c(t)$:

$$\sqrt{-g} h_c(t) = \frac{\pm 2\sqrt{2}}{e^{(t-t_0)} - 2\mathbf{v} \cdot \mathbf{n} + e^{-(t-t_0)} (\mathbf{v}^2 + 1)}.$$ (42.18)

We note that in these new variables translations take a complicated form, unlike dilatation which simply corresponds to a translation of the variable t.

The second derivative of the classical action at the saddle point now takes the form (for $t_0 = x_{0\mu} = 0$)

$$\mathbf{M} = \frac{\delta^2 S}{\delta h_c \delta h_c} = -\left(\frac{d}{dt} \right)^2 + \mathbf{L}^2 + 1 - \frac{6}{\cosh^2 t}.$$ (42.19)

The natural measure associated to this hamiltonian problem is

$$\int dt \, d\Omega,$$

which in the original language means

$$\int \frac{d^4 x}{x^2}.$$

This measure is not translation invariant, and thus the jacobian resulting from the introduction of collective coordinates, and the determinant depend individually on $x_{0\mu}$. However the product of the corresponding contributions to the final result should not, thus we perform the calculation for $x_{0\mu} = 0$.

42.2.1 The jacobian

With the new measure $d\phi_c / d\lambda$ is normalizable:

$$J_1 = \left[\int \frac{d^4 x}{x^2} \left(\frac{d}{d\lambda} \phi_c(x) \right)^2 \right]^{1/2}, \tag{42.20}$$

$$= \left[\frac{16\pi^2}{(-g)} \int_0^\infty r dr \frac{(1 - \lambda^2 r^2)^2}{(1 + \lambda^2 r^2)^4} \right]^{1/2}. \tag{42.21}$$

This leads to a first factor:

$$J_1 = \frac{1}{\lambda} \sqrt{\frac{8}{3}} \frac{\pi}{\sqrt{-g}}. \tag{42.22}$$

The second jacobian J_2 comes from the collective coordinates $x_{0\mu}$:

$$J_2 = \left[\frac{1}{4} \int \frac{d^4 x}{x^2} \sum_{\mu=1}^4 (\partial_\mu \phi_c)^2 \right]^2, \tag{42.23}$$

$$= \frac{1}{g^2} \left[16\pi^2 \int_0^\infty \frac{r^3 dr \lambda^6}{(1 + \lambda^2 r^2)^4} \right]^2 = \frac{\lambda^4}{g^2} \times \frac{16}{9} \pi^4. \tag{42.24}$$

The complete jacobian J is thus

$$J = J_1 J_2 = \frac{\lambda^3}{(-g)^{5/2}} \pi^5 \times \frac{32\sqrt{2}}{9\sqrt{3}}. \tag{42.25}$$

42.2.2 The determinant

Using the result of equation (37.38) it is possible to calculate the determinant of **M** for each value l of the angular momentum:

$$M_l = - \left(\frac{d}{dt} \right)^2 + (1 + l)^2 - \frac{6}{\cosh^2 t}. \tag{42.26}$$

The determinant is thus:

$$\det\left(M_l + \varepsilon\right)\left(M_{0l} + \varepsilon\right)^{-1} = \frac{\left[\sqrt{\varepsilon + (l+1)^2} - 1\right]\left[\sqrt{\varepsilon + (l+1)^2} - 2\right]}{\left[\sqrt{\varepsilon + (l+1)^2} + 2\right]\left[\sqrt{\varepsilon + (l+1)^2} + 1\right]}, \qquad (42.27)$$

in which M_{0l} is the operator of the corresponding free theory. As we know, this determinant is UV divergent and we have to renormalize it. However let us first calculate formally the unrenormalized determinant:

$$l \geq 2: \qquad\qquad \det M_l M_{0l}^{-1} = \frac{l\,(l-1)}{(l+2)\,(l+3)}, \qquad (42.28)$$

$$l = 1: \qquad \lim_{\varepsilon \to 0} \frac{1}{\varepsilon} \det\left(M_1 + \varepsilon\right)\left(M_{01} + \varepsilon\right)^{-1} = \frac{1}{48}, \qquad (42.29)$$

$$l = 0: \qquad \lim_{\varepsilon \to 0} \frac{1}{\varepsilon} \det\left(M_{l=0} + \varepsilon\right)\left(M_{0l=0} + \varepsilon\right)^{-1} = -\frac{1}{12}. \qquad (42.30)$$

As expected the determinant is negative and we obtain the formal expression

$$\det{}' \mathbf{M} \mathbf{M}_0^{-1} = -\frac{1}{12} \times \left(\frac{1}{48}\right)^4 \times \prod_{l=2}^{\infty} \left[\frac{l\,(l-1)}{(l+2)\,(l+3)}\right]^{(l+1)^2}. \qquad (42.31)$$

Renormalization. In these variables, the UV divergences appear as divergences of the infinite product on l. Let us thus use in an intermediate step a maximum value L of l as a cut-off. From the general analysis we know the UV divergent part of $\ln \det \mathbf{M}$ is completely contained in the two first terms of the expansion in powers of ϕ_c^2. We therefore proceed in the following way: the determinant of the operator $\mathbf{M}(s)$,

$$\mathbf{M}(s) = -\left(\frac{\mathrm{d}}{\mathrm{d}t}\right)^2 - \frac{s(s+1)}{\cosh^2 t}, \qquad (42.32)$$

is exactly known:

$$\det\left[\mathbf{M}(s) + z\right]\left[\mathbf{M}_0 + z\right]^{-1} = \frac{\Gamma\left(1 + \sqrt{z}\right)\Gamma\left(\sqrt{z}\right)}{\Gamma\left(1 + s + \sqrt{z}\right)\Gamma\left(\sqrt{z} - s\right)}. \qquad (42.33)$$

Setting:

$$s(s+1) = 6\gamma, \qquad (42.34)$$

it is easy to expand $\ln \det \mathbf{M}(s)$ in powers of γ. We deduce immediately from this expansion, the expansion up to second order of $\ln \det \mathbf{M}$ in powers of the potential $-6/\cosh^2 t$ in the representation (42.27). We then subtract these two terms from $\ln \det \mathbf{M}$ as obtained from the representation (42.31). It is easy to verify that indeed the large L limit of the subtracted quantity:

$$\left\{\det{}' \mathbf{M} \mathbf{M}_0^{-1}\right\}_{\mathrm{ren}}^{-1/2} = \lim_{L \to +\infty} i2\sqrt{3} \times (48)^2 \prod_{l=2}^{L} \left[\frac{(l+2)\,(l+3)}{(l-1)}\right]^{(l+1)^2/2} \prod_{l=0}^{L} \mathrm{e}^{-3(l+1)}$$

$$\times \prod_{l=0}^{L} \mathrm{e}^{-18(l+1)^2} \left[\sum_{k=l+1}^{\infty} \frac{1}{k^2} - \frac{1}{l+1} - \frac{1}{2\,(l+1)^2}\right], \qquad (42.35)$$

is finite. We set:

$$\left\{\det' \mathbf{M}\mathbf{M}_0^{-1}\right\}_{\text{ren}}^{-1/2} = iC_1 \,. \tag{42.36}$$

Taking into account the jacobians, the factor $(2\pi)^{-1/2}$ for each collective mode, the factor $(2i)^{-1}$ and a factor two for the two saddle points, we get a first factor C_2 of the form

$$C_2 = \frac{\lambda^3}{(-g)^{5/2}} \times \pi^5 \times \frac{32\sqrt{2}}{9\sqrt{3}} \times \frac{C_1}{(2\pi)^{5/2}} \,, \tag{42.37}$$

which we write as

$$C_2 = \frac{\lambda^3}{(-g)^{5/2}} C_3 \,. \tag{42.38}$$

We then have to add to the classical action the two terms we have subtracted above to $\ln \det \mathbf{M}$. However we can now write them in the normal space representation, regularized as we have regularized the perturbative correlation functions, and take into account the one-loop counterterms. The first term in the expansion in powers of ϕ_c^2 is exactly cancelled by the mass counterterm, as we have already discussed. The second term in the expansion, which is the one-loop contribution to the four point function, is logarithmically divergent. In next section we calculate explicitly the finite difference between this term and the coupling constant counterterm which cancels the divergence.

42.3 The Coupling Constant Renormalization

The terms we want to calculate involve the renormalized 4-point function. We have to choose a renormalization scheme: we assume therefore that we have renormalized the field theory by minimal subtraction after dimensional regularization. The renormalization constants have been calculated in Section 11.6. Notice the different normalization of the coupling constant. Let us write the contribution δS_2 which we have to add to the action, coming from the subtraction of $\ln \det \mathbf{M}$ and the one-loop coupling renormalization constant:

$$\delta S_2 = \frac{9}{4}\frac{N_d}{\varepsilon}g^2 \int \phi_c^4(x)\mathrm{d}^4x - \frac{9}{4}g^2 \operatorname{tr}\left[\phi_c^2\left(-\Delta\right)^{-1}\phi_c^2\left(-\Delta\right)^{-1}\right], \tag{42.39}$$

in which N_d is the usual geometrical factor:

$$N_d = 2(4\pi)^{-d/2}/\Gamma(d/2) \,. \tag{42.40}$$

and $d = 4 - \varepsilon$. The expression can be rewritten:

$$\delta S_2 = -\frac{9}{4}g^2 \int \mathrm{d}^4x\,\mathrm{d}^4y\,\frac{\mathrm{d}^4p}{(2\pi)^4}\,\mathrm{e}^{ip(x-y)}\,\phi_c^2(x)\phi_c^2(y)\lim_{d\to4}\left(\int\frac{\mathrm{d}^dq}{(2\pi)^d}\frac{\mu^\varepsilon}{q^2(p-q)^2} - \frac{N_d}{\varepsilon}\right), \tag{42.41}$$

in which μ is the renormalization scale. The integral over $\mathbf{q}$ has been performed in Section 9.3 (see equation (9.29)):

$$\int\frac{\mathrm{d}^dq}{(2\pi)^d}\frac{1}{q^2(p-q)^2} - \frac{N_d}{\varepsilon} = \frac{1}{8\pi^2}\left(\frac{1}{2} - \ln p\right) + O(\varepsilon) \,. \tag{42.42}$$

Let us also introduce the Fourier transform of the function $f^2(r)$ ($f(r)$ is given by equation (42.10a)):

$$v(p) = \frac{1}{(2\pi)^4} \int d^4x \frac{8\,e^{ipx}}{(1+x^2)^2}.$$
(42.43)

The solution $\phi_c(x)$ depends on the scale λ. Rescaling the variables x, y, and p, we can then write the total expression more explicitly:

$$\delta S_2 = -\frac{9\pi^2}{2} \int d^4p\, v^2(p) \left[\frac{1}{2} - \ln\left(\frac{\lambda p}{\mu} \right) \right].$$
(42.44)

From the definition of $v(p)$ we deduce after a short calculation:

$$\int d^4p\, v^2(p) = \frac{2}{(3\pi^2)},$$
(42.45)

$$\int d^4p \, \ln p\, v^2(p) = \frac{2}{3\pi^2} \left(\ln 2 + \gamma + \frac{1}{6} \right),$$
(42.46)

in which γ is Euler's constant: $\gamma = -\psi(1) = 0.577215\ldots$. We then obtain:

$$\delta S_2 = 3 \ln \lambda/\mu - \ln C_4,$$
(42.47)

with:

$$\ln C_4 = 1 - 3 \ln 2 - 3\gamma.$$
(42.48)

We note that the r.h.s. of equation (42.47) now depends on the scale parameter λ. The interpretation of this result is the following: the coupling constant renormalization breaks the scale invariance of the classical theory, and therefore the scale parameter λ remains in the expression. Moreover the term proportional to $\ln \lambda$ together with the contribution from the classical action can be rewritten as:

$$\frac{8\pi^2}{3g} - 3 \ln \lambda/\mu = \frac{8\pi^2}{3g(\lambda)} + O(g),$$
(42.49)

in which $g(\lambda)$ is the effective coupling at the scale λ, solution of the renormalization group equation,

$$\frac{dg(\lambda)}{d\ln\lambda} = \beta[g(\lambda)],$$
(42.50)

with

$$\beta(g) = \frac{9}{8\pi^2}g^2 + O(g^3).$$
(42.51)

This property is expected. The renormalization of the perturbative expansion renders the instanton contribution, before integration over dilatation, finite. Consequently this contribution should satisfy a renormalization group equation, and the coupling constant g can be present only in the combination $g(\lambda)$, since λ fixes the scale in the calculation (for more details see Appendix A42.1).

42.4 The Imaginary Part of the n-Point Function

We can now write the complete contribution to the imaginary part of the n-point function,

$$\operatorname{Im} Z^{(n)} (x_1, \ldots, x_n) \underset{g \to 0_-}{\sim} C_5 \int \mathrm{d}^4 x_0 \int_0^\infty \frac{\mathrm{d}\lambda}{\lambda} \lambda^4 \prod_{i=1}^n \frac{2\sqrt{2}\lambda}{1 + \lambda^2 (x_i - x_0)^2} \frac{\mathrm{e}^{8\pi^2/3g(\lambda)}}{(-g)^{(n+5)/2}},$$
(42.52)

where we have set:

$$C_5 = C_3 C_4 .$$

It is convenient to Fourier transform expression (42.52), introducing

$$u(p) = 2\sqrt{2} \int \mathrm{e}^{ipx} \frac{\mathrm{d}^4 x}{1 + x^2} .$$
(42.53)

Then, after factorizing the δ-function of momentum conservation,

$$. \operatorname{Im} \tilde{Z}^{(n)} (p_1, \ldots, p_n) \sim \frac{C_5}{(-g)^{(n+5)/2}} \int_0^\infty \mathrm{d}\lambda \, \lambda^{3-3n} \prod_{i=1}^n u \left(\frac{p_i}{\lambda} \right) \mathrm{e}^{8\pi^2/3g(\lambda)} .$$
(42.54)

We can express this result on 1PI correlation functions $\tilde{\Gamma}^{(n)} (p_1, \ldots, p_n)$:

$$\operatorname{Im} \tilde{\Gamma}^{(n)} (p_1, \ldots, p_n) \sim \frac{C_5}{(-g)^{(n+5)/2}} \int_0^\infty \frac{\mathrm{d}\lambda}{\lambda} \lambda^{4-n} \prod_{i=1}^n \frac{p_i^2}{\lambda^2} u \left(\frac{p_i}{\lambda} \right) \mathrm{e}^{8\pi^2/3g(\lambda)} .$$
(42.55)

It is straightforward to verify that $p^2 u(p)$ goes to a constant for $|p|$ small.

In contrast to the super-renormalizable case, because the theory is only renormalizable the final result is not totally explicit, but involves instead a final integration over dilatations whose convergence is not obvious. Let us now discuss this point.

The small instanton contribution. Small instantons correspond to λ large. For λ large, the integral behaves like:

$$\int^\infty \mathrm{d}\lambda \, \lambda^{3-n} \mathrm{e}^{8\pi^2/3g(\lambda)},$$
(42.56)

and therefore we have to examine the behaviour of $g(\lambda)$ for λ large. From equation (42.51) we see that the theory is asymptotically because for g negative, i.e. $g(\lambda)$ goes to zero for λ large. Thus perturbation theory is applicable and we can use the approximation (42.49). The argument remains true even if we take g slightly complex. Thus the integral has the form

$$\int^\infty \mathrm{d}\lambda \, \lambda^{-n}.$$
(42.57)

We see that the power behaviour in λ depends explicitly on the coefficient of the g^2 term of the $\beta(g)$-function. Without the contribution coming from $g(\lambda)$, the integral (42.57) would have a UV divergence similar to the one found in the corresponding perturbative expansion. Due to the additional power of λ coming from $g(\lambda)$, only the vacuum amplitude is divergent.

The convergence of the dilatation integral is thus better than expected: indeed the renormalization constants are now themselves given by divergent series and are complex for g negative. Their imaginary part contributes directly to the imaginary part of

$\tilde{\Gamma}^{(n)}\left(p_1,\ldots,p_n\right)$ for $n \leq 4$. In the ϕ^6 field theory in dimension 3 for example, these contributions cancel the divergences coming from the integral over λ. Here instead the integrals over λ are finite at this order. This implies in particular that in the minimal subtraction scheme the imaginary parts of the renormalization constants vanish at leading order. In another renormalization scheme (fixed momentum subtraction for example) these imaginary parts are finite at leading order.

The large instanton contribution. Let us now examine the convergence of the λ integral for λ small. The behaviour of $g\left(\lambda\right)$ is totally unknown. On the other hand, it is easy to verify that the factors $u\left(p_i/\lambda\right)$ decrease exponentially for λ small. Thus, if the behaviour of $g\left(\lambda\right)$ is not too dramatic, the integrals will converge and it will be justified to replace $g\left(\lambda\right)$ by the expansion (42.49). For the vacuum amplitude, this argument does not apply, and so the result is unknown.

This analysis shows that, although this calculation seems to be a simple formal extension of the calculation for lower dimensions, coupling constant renormalization introduces a set of new problems which are not all completely under control. The fact that the theory is massless only makes matters worse. Consideration of the massive theory improves the situation in this respect, but the instanton calculation becomes more complicated.

42.5 Instanton Contribution to Large Order Behaviour

As a byproduct of the calculation, we get the instanton contribution to the large order behaviour:

$$\left\{\Gamma^{(n)}\left(p_1,\ldots,p_n\right)\right\}_k = \frac{1}{\pi}\int_{-\infty}^{0}\frac{\operatorname{Im}\Gamma^{(n)}\left(p_1,\ldots,p_n\right)}{g^{k+1}}dg. \qquad (42.58)$$

This yields a result of the form

$$\left\{\Gamma^{(n)}\left(p_1,\ldots,p_n\right)\right\}_k \underset{k\to\infty}{\sim} C_n\left(p_1,\ldots,p_n\right)\int^{0-}\frac{\mathrm{e}^{8\pi^2/3g}}{(-g)^{n+5/2}}\frac{dg}{g^{k+1}}. \qquad (42.59)$$

After integration we obtain

$$\left\{\Gamma^{(n)}\left(p_1,\ldots p_n\right)\right\}_k \sim C_n\left(p_1,\ldots,p_n\right)(-1)^k\left(\frac{3}{8\pi^2}\right)^{n+3+k}\Gamma\left(k+\frac{n}{2}+\frac{5}{2}\right). \qquad (42.60)$$

From this expression, it is straightforward to derive the large order behaviour of various renormalization group functions in for example the fixed momentum subtraction scheme. A comparison between large order behaviour and explicit calculations can be found in table 42.1, in the case of the RG β-function.

Table 42.1

The coefficients β_k of the RG β-function divided by the asymptotic estimate, in the case of the $O(N)$ symmetric ϕ_4^4 field theory.

k	2	3	4	5
$N=1$	0.10	0.66	1.08	1.57
$N=2$	0.06	0.49	0.87	1.32
$N=3$	0.04	0.33	0.66	1.09

The large order behaviour of Wilson–Fisher ε-expansion, which is important for the theory of Critical Phenomena, can instead only be guessed at because, as discussed above, the RG functions in the minimal subtraction scheme vanish at leading order. A calculation of the next order would be necessary and this has yet not been done. Since at leading order the fixed point constant $g^*(\varepsilon)$ is:

$$g^*(\varepsilon) \sim \left(8\pi^2\right)/(N+8),$$

except if for some unknown reason the accident of leading order persists, the ε-expansion is likely to involve a factor $(-3/(N+8))^k k!$ multiplied by an unknown power of k.

42.6 UV and IR Contributions to Large Order Behaviour

Implicit in the large order behaviour calculation is the assumption that the singularities of correlation functions come entirely, in the neighbourhood of the origin, from barrier penetration effects. If this assumption is certainly correct in quantum mechanics, if there are very strong indications that it is valid for super-renormalizable theories, it is much more questionable for renormalizable theories, not to mention massless renormalizable theories. Let us first discuss the large momentum problem and then the IR problem of massless theories.

UV singularities: renormalons. If the arguments go through, without modifications, for the regularized field theory, they become extremely formal for the renormalized theory in the infinite cut-off limit. We have already seen that even in the naive calculation, non-trivial questions arise about the global renormalization group properties of the theory. Direct investigation of the perturbative expansion raises new questions and suggests that UV singularities yield additional contributions to the large order behaviour.

Let us consider the $\left(\phi^2\right)^2$ field theory in dimension four, in which ϕ is an N component vector, and the model has an $O(N)$ symmetry,

$$S[\phi] = \int \mathrm{d}^4x \left[\frac{1}{2}\left(\partial_\mu\phi\right)^2 + \frac{m^2}{2}\phi^2 + \frac{g}{4}\left(\phi^2\right)^2\right]. \tag{42.61}$$

A systematic $1/N$ expansion allows us to rearrange the perturbative expansion by performing partial summations. At leading order, the 4-point function is replaced by the sum of the bubble diagrams,

$$\frac{2g}{1+gNI(p)}, \tag{42.62}$$

where the bubble diagram is:

$$I(p) = \frac{1}{(2\pi)^4}\int \frac{\mathrm{d}^4q}{\left[(p+q)^2+m^2\right][q^2+m^2]} - \text{subtraction}. \tag{42.63}$$

The important point is that for large momenta $I(p)$ behaves like

$$I(p) \sim \frac{1}{8\pi^2}\ln\frac{m}{|p|}, \quad |p|\to\infty. \tag{42.64}$$

Therefore the sum of the bubble diagrams has a singularity for g small and positive at momentum

$$|p| \simeq m\,\mathrm{e}^{8\pi^2/Ng}. \tag{42.65}$$

This is one example of the famous Landau ghost. Since the theory is IR free, and not asymptotically free, this singularity occurs for positive values of the coupling constant. Once this sum of bubbles is inserted at next order, for instance in the 2-point function, it produces a cut for g small and positive,

$$F_2(p) = \int \frac{\mathrm{d}^4 q}{\left[(p+q)^2 + m^2\right]\left[1 + NgI(q)\right]} - \text{subtractions}. \tag{42.66}$$

After subtraction, and for $|q|$ large, the integrand behaves like

$$F_2\,(p) = \int_{|q| \gg 1} \frac{\mathrm{d}q}{q^3} \frac{1}{1 + \dfrac{Ng}{8\pi^2} \ln \dfrac{m}{q}} + \cdots . \tag{42.67}$$

Setting

$$t = \ln(q/m), \tag{42.68}$$

we transform equation (42.67) into

$$\int^\infty \mathrm{d}t\ \mathrm{e}^{-2t} \frac{1}{1 - Ngt/(8\pi^2)}. \tag{42.69}$$

This yields an imaginary contribution to the correlation functions for g small and positive of the form $\exp\left(-16\pi^2/Ng\right)$. Alternatively, we could have considered individual diagrams containing bubble insertions. By expanding for instance equation (42.66) in powers of g, we would have concluded that these diagrams behave like $\left(N/16\pi^2\right)^k k!$ at large order k. Therefore, in contrast to super-renormalizable theories in which an individual diagram behaves like a power in k and the $k!$ comes from the number of diagrams, here individual diagrams have a $k!$ behaviour, without the sign oscillations characteristic of the semiclassical result.

Further investigations show that if a non-perturbative contribution exists, it should satisfy the homogeneous RG equations. Let us for simplicity consider the case of a dimensionless ratio of correlation functions $R(p/m, g)$ without anomalous dimensions,

$$\left[m\frac{\partial}{\partial m} + \beta\,(g)\,\frac{\partial}{\partial g}\right] R\left(\frac{p}{m}, g\right) = 0. \tag{42.70}$$

The RG equation tells us that the function $R(p/m, g)$ is actually a function of only one variable $s(g)p/m$, in which $s(g)$ then satisfies:

$$\beta\,(g)\,s'\,(g) = s\,(g), \tag{42.71}$$

which after integration yields:

$$s\,(g) \sim \exp\left[\int^g \frac{\mathrm{d}g'}{\beta\,(g')}\right]. \tag{42.72}$$

For g small, $s\,(g)$ behaves like:

$$\beta\,(g) = \beta_2 g^2 + O\left(g^3\right), \quad \text{with } \beta_2 = \frac{N+8}{8\pi^2}, \tag{42.73}$$

$$s\,(g) \simeq \exp\left[-\frac{1}{\beta_2 g}\right]. \tag{42.74}$$

This is exactly the singularity that we obtained from the set of bubbles in the large N limit.

Since the correlation function depends only on the mass squared, only $s^2(g)$ can enter the calculation, and the contribution to the large order behaviour is of the form

$$\int_0^{} \frac{e^{-2/\beta_2 g}}{g^{k+1}} dg \sim \left(\frac{\beta_2}{2}\right)^k k! . \tag{42.75}$$

This potential contribution has to be compared with the semiclassical result (42.60).

All these problems are of course related to the question of the existence of the renormalized ϕ^4 field theory in four dimensions. If the theory does not exist, then probably the sum of perturbation theory is complex for g positive, and these singular terms, sometimes called *renormalon* effects, are the small coupling evidence of this situation. More generally, the existence of renormalons shows that the perturbation series is not Borel summable and does not define unique correlation functions.

Finally let us note that, at leading order in the $1/N$ expansion, for the Wilson–Fisher ε-expansion, and thus also for suitably defined renormalization group functions, the renormalon singularities cancel. We conjecture on this basis and on the basis of the numerical evidence presented in Chapter 28 that the ε-expansion is free of renormalon singularities.

Massless renormalizable theories. Let us again illustrate the problem with the $(\phi^2)^2$ field theory in the large N limit. We now work in a massless theory with fixed cut-off Λ. We evaluate the contribution of the small momentum region to the mass renormalization constant. The bubble diagram (42.63) behaves like

$$I(p) \sim \frac{1}{8\pi^2} \ln(\Lambda/p).$$

The sum of bubbles yields a contribution to the mass renormalization proportional to

$$\int^\Lambda \frac{d^4 q}{q^2 (1 + N g I(q))} = \int \frac{d^4 q}{q^2 \left(1 + \frac{N}{8\pi^2} g \ln(\Lambda/q)\right)} .$$

Expanded in powers of g this yields a contribution of order $(-1)^k (N/16\pi^2)^k k!$ for large order k. This contribution has the sign oscillations of the semiclassical term. More generally for finite N one finds $(-\beta_2/2)^k k!$. IR singularities yield an additional Borel summable contribution to the large order behaviour.

For massless, but asymptotically free theories the role of the IR and UV regions are interchanged. UV renormalons are expected yielding additional singularities to the Borel transform on the real negative axis, while IR contributions destroy Borel summability. When these theories have real instantons like QCD or the $CP(N-1)$ models, the Borel transform has also semiclassical singularities on the real positive axis.

42.7 The Massive Theory

We have shown in Appendix A39 that the massive field equations can have no solution, and that the minimum of the action can be obtained from the massless theory. To study the massive theory, we thus start from the instanton solution of the massless theory, with its scale parameter λ. However we notice a difficulty: as explained in Section 42.2 the integral of ϕ_c^2 is IR divergent. We have thus to modify the field configuration at large distances, by connecting it smoothly to the solution of the massive free equation with mass m. Qualitatively speaking we consider a configuration which up to a distance R, $\lambda R \gg 1$, $mR \ll 1$, is $\lambda \phi_c(\lambda x)$ and for $|x| > R$, is proportional to the free massive solution. An analogous problem will be discussed in Chapter 43 in the case of multi-instanton configuration. Although the theory is no longer scale invariant, λ has to be kept as a collective coordinate. The mass term then acts as an IR cut-off, and restrict the domain of integration in λ to values large with respect to m. The classical action has the form

$$S_m(\phi_c) = -\frac{1}{g}\left(\frac{8\pi^2}{3} + 8\pi^2 \frac{m^2}{\lambda^2}\ln\frac{\lambda}{m}\right) \quad \text{for} \quad \lambda \gg m, \qquad (42.76)$$

where the $\ln m$ term is directly related to the initial IR divergence of the ϕ^2 integral.

The remaining part of the calculation closely follows the calculation for the massless case and the reader is referred to the literature for details.

In the massless theory the instanton contribution to the vacuum energy could not be evaluated without some knowledge of the non-perturbative IR behaviour of the β-function. In the massive theory the problem is solved because the λ integral is cut at a scale $m/\sqrt{-g}$. For correlation functions the integral will be cut by the largest between momenta and $m/\sqrt{-g}$. This implies that the limits $m \to 0$ and $g \to 0$ do not commute.

In the large order estimate the λ integral will be cut at a scale of order $m\sqrt{k}$.

Bibliographical Notes

References to the work of Lipatov and its generalization by Brézin, Le Guillou and Zinn-Justin has already been given in previous chapters. See also

A.J. McKane and D.J. Wallace, *J. Phys. A: Math. Gen.* 11 (1978) 2285,

and for a discussion of the massive theory

Y. Frishman and S. Yankielowicz, *Phys. Rev.* D19 (1979) 540; I. Affleck, *Nucl. Phys.* B191 (1981) 429.

Renormalons have been discussed in

G. Parisi in *Cargèse Lectures 1977*, vol. B39 (Plenum, New York 1979) and *Phys. Lett.* 76B (1978) 65; B. Lautrup, *Phys. Lett.* 69B (1977) 109;

G. 't Hooft, *Erice Lectures 1977*, A. Zichichi ed. (Plenum, New York 1979); P. Olesen, *Phys. Lett.* 73B (1978) 327.

Renormalons have IR analogues in massless field theories

G. Parisi, *Nucl. Phys.* B150 (1979) 163; F. David, *Nucl. Phys.* B209 (1982) 433; B234 (1984) 237; B263 (1986) 637.

Exercises

Generalize the calculations of Chapter 42 to the massless ϕ^{2n} field theory in the dimension in which it is renormalizable:

$$S(\phi) = \int d^d x \left[\tfrac{1}{2} (\partial_\mu \phi)^2 + \frac{g}{2(n-1)^2} \phi^{2n} \right].$$

The eigenvalues of $\mathbf{L}^2$ in d dimensions are $l(l + d - 2)$ with a degeneracy $\delta_l(d)$:

$$\delta_l(d) = \frac{(d + 2l - 2)\Gamma(l + d - 2)}{\Gamma(l + 1)\Gamma(d - 1)}.$$

We also recall that if $M(t_1, t_2)$ is of the form:

$$M(t_1, t_2) = \left[-\left(\frac{d}{dt_1} \right)^2 + 1 - \frac{s(s+1)\omega^2}{\cosh^2 \omega t_1} \right] \delta(t_1 - t_2),$$

in the notation of equation (37.38) we have:

$$\det(M + \varepsilon)(M_0 + \varepsilon)^{-1} = \frac{\Gamma(1 + z)\Gamma(z)}{\Gamma(1 + s + z)\Gamma(z - s)}, \tag{42.77}$$

with

$$z = \sqrt{1 + \varepsilon}/\omega. \tag{42.78}$$

APPENDIX 42

A42.1 Instantons and RG Equations

The instanton contribution to the 1PI n-point function can be written as:

$$\operatorname{Im} \Gamma^{(n)}\left(p_i; \mu, g\right)=\int_0^{\infty} \frac{\mathrm{d}\lambda}{\lambda} F^{(n)}\left(p_i; \mu, g, \lambda\right), \qquad (A42.1)$$

in which μ represents the subtraction scale, and λ the dilatation parameter. The counterterms which renormalize the perturbative expansion, also render $F^{(n)}$ finite for reasons we have already explained. Therefore $F^{(n)}$ satisfies a RG equation:

$$\left[\mu\frac{\partial}{\partial\mu}+\beta\left(g\right)\frac{\partial}{\partial g}-\frac{n}{2}\eta\left(g\right)\right]F^{(n)}\left(p_i; \mu, g, \lambda\right)=0, \qquad (A42.2)$$

which, integrated by the method of characteristics, yields:

$$F^{(n)}\left(p_i; \mu; g; \lambda\right)=Z^{-n/2}\left(\tau\right)F^{(n)}\left(p_i; \mu\tau, g\left(\tau\right), \lambda\right), \qquad (A42.3)$$

with the definitions:

$$\begin{aligned}
\ln \tau &=\int_g^{g(\tau)} \frac{\mathrm{d}g'}{\beta\left(g'\right)}, \\
\ln Z &=\int_g^{g(\tau)} \frac{\eta\left(g'\right)}{\beta\left(g'\right)}.
\end{aligned} \qquad (A42.4)$$

The coupling constant $g\left(\tau\right)$ is the effective coupling constant at the scale τ.
   Ordinary dimensional considerations now tell us that:

$$F^{(n)}\left(p_i; \mu; g; \lambda\right)=\lambda^{4-n} F^{(n)}\left(\frac{p_i}{\lambda}, \frac{\mu}{\lambda}, g, 1\right). \qquad (A42.5)$$

Applied to the r.h.s. of equation $(A42.3)$, this identity yields:

$$F^{(n)}\left(p_i; \mu; g; \lambda\right)=Z^{-n/2}\left(\tau\right)\lambda^{4-n} F^{(n)}\left(\frac{p_i}{\lambda}, \frac{\mu\tau}{\lambda}, g\left(\tau\right)\right). \qquad (A42.6)$$

The choice:

$$\tau=\lambda/\mu,$$

finally leads to the relation:

$$F^{(n)}\left(p_i; \mu; g; \lambda\right)=\left[Z\left(\frac{\lambda}{\mu}\right)\right]^{-n/2}\lambda^{4-n} F^{(n)}\left[\frac{p_i}{\lambda}; g\left(\frac{\lambda}{\mu}\right)\right]. \qquad (A42.7)$$

A42.2 Conformal Invariance

The scale invariance of the classical ϕ_4^4 field theory has allowed us to obtain an analytic instanton solution. Moreover by introducing the special coordinates (t, n_μ) we have been able to use the results obtained for the anharmonic oscillator in Chapter 37, and calculate explicitly the instanton contribution at leading order. Actually the scale invariant classical ϕ_4^4 theory is also conformal invariant (see Appendix A13.5). This property, which also holds for other scale invariant field theories like gauge theories, can be used more directly to calculate the instanton contribution. The conformal group is isomorphic to $SO(5,1)$. It is expected that the minimal action solution will be invariant under a maximal compact subgroup of $SO(5,1)$, i.e. $SO(5)$. It is then convenient to perform a stereographic mapping of $\mathbb{R}^4$ onto the sphere S_4 to simplify the $SO(5)$ transformations. One sets:

$$\xi^\mu = \frac{2x^\mu}{1+\mathbf{x}^2}, \qquad \xi^5 = \frac{1-\mathbf{x}^2}{1+\mathbf{x}^2}, \qquad (A42.8)$$

so that

$$\sum_{a=1}^{5} \xi^a \xi^a = 1. \qquad (A42.9)$$

It is also useful to introduce a field which has simple transformation properties under $SO(5)$. In the ϕ_4^4 theory, the conformal transformation properties of the ϕ-field lead to set:

$$\phi = \frac{1}{1+\mathbf{x}^2}\psi. \qquad (A42.10)$$

Let us rewrite the classical action (42.1) in terms of these new variables. We perform the transformations in two steps: first we keep the variables x^μ, but now considered as coordinates on S_4, and perform only the substitution $(A42.10)$. The metric $g_{\mu\nu}$ on S_4 in the coordinates x^μ is:

$$g_{\mu\nu} = 4\frac{\delta_{\mu\nu}}{(1+\mathbf{x}^2)^2}. \qquad (A42.11)$$

The invariant measure on the sphere involves the square root of the determinant of the metric $\mathbf{g}$ (see Section 22.5):

$$\sqrt{\det \mathbf{g}} = \frac{16}{(1+\mathbf{x}^2)^4}. \qquad (A42.12)$$

Finally, after an integration by parts the kinetic term can be rewritten:

$$\int \mathrm{d}^4x \, (\partial_\mu\phi)^2 = \int \mathrm{d}^4x \left[\frac{(\partial_\mu\psi)^2}{(1+\mathbf{x}^2)^2} + \frac{8\psi^2}{(1+\mathbf{x}^2)^4}\right]. \qquad (A42.13)$$

The classical action then reads:

$$S(\psi) = \int \mathrm{d}^4x \sqrt{\det \mathbf{g}} \left(\frac{1}{8}g^{\mu\nu}\partial_\mu\psi\partial_\nu\psi + \frac{1}{4}\psi^2 + \frac{g}{64}\psi^4\right). \qquad (A42.14)$$

In this covariant form the change of coordinates $(A42.8)$ is straightforward and hardly necessary. One solution of minimal action is a constant:

$$\psi^2 = -1/8g. \qquad (A42.15)$$

The classical action is proportional to the surface of S_4 which is $8\pi^2/3$. The operator $\mathbf{M}$ second derivative of the action is given by (see Section 3.4):

$$\mathbf{M} = \tfrac{1}{4}\mathbf{L}^2 - 1. \qquad (A42.16)$$

in which $\mathbf{L}$ is the angular momentum in 5 space dimensions. The eigenvalues of $\mathbf{L}^2$ are $l(l+3)$ with a degeneracy δ_l:

$$\delta_l = \frac{1}{6}\frac{(2l+3)\Gamma(l+3)}{\Gamma(l+1)}. \qquad (A42.17)$$

The form of $\mathbf{M}$ shows that it has 0 as eigenvalue, corresponding to $l = 1$, with degeneracy 5, in agreement with the considerations of Chapter 42. We leave up to the reader, as an exercise, to verify the other results of Chapter 42.

43 MULTI-INSTANTONS IN QUANTUM MECHANICS

A linear combination of instanton solutions is not a solution of the equations of motion since these equations are non-linear. However, in the limit in which all instantons are largely separated, such a configuration renders the action almost stationary because each instanton solution differs from a constant solution only by exponentially small corrections at large distances (in a field theory this is only true if the theory is massive). We here examine in the context of Quantum Mechanics the significance of such multi-instanton *quasi-solutions*. The generalization to Field Theory, however, is non-trivial and has still to be worked out.

We have found in preceding chapters several situations in which we can expect multi-instantons to play a role. In the case in which instantons are found, the ground state energy E_0 takes the form:

$$E_0 = E_0^{(0)} + E_0^{(1)} + \cdots,$$

in which $E_0^{(0)}$ and $E_0^{(1)}$ are the perturbative and one-instanton contributions respectively, and the dots represent possible multi-instanton contributions. The ground state energy has been derived from a semiclassical calculation at β large of the partition function which then has the form

$$\operatorname{tr} e^{-\beta H} \sim e^{-\beta E_0} \sim e^{-\beta E_0^{(0)}} \sum_{n=0}^{\infty} \frac{(-\beta)^n}{n!} \left(E_0^{(1)} \right)^n. \tag{43.1}$$

Thus the existence of an one-instanton contribution to the energy implies the existence of n-instanton contributions to the partition function proportional to β^n.

When we have evaluated $\operatorname{tr} e^{-\beta H}$ at finite β we always have kept only the solution which describes the classical trajectory once. We have argued that the other solutions, in which the trajectory is described n-times, have in the large β limit an action n-times larger, and therefore give subleading contributions to the path integral. In the infinite β limit these configurations have the properties we expect from a n-instanton. However, there is a subtlety: naively we expect these configurations to give a contribution of order β because a classical solution depends only on one time parameter, in contrast with the nth term in (43.1) which is of order β^n.

Another example is provided by the estimation of the large order behaviour of perturbation theory for potentials with degenerate minima. We have seen that when we start from a situation in which the minima are almost degenerate, we obtain in the degenerate limit a typical two-instanton contribution but with an infinite multiplicative coefficient. This divergence has the following interpretation: In the degenerate limit the classical solution decomposes into the superposition of an instanton and an anti-instanton infinitely separated and thus fluctuations which tend to change the distance between the instanton and the anti-instanton induce a vanishingly small variation of the action. It follows that, to properly study the limit, one has to introduce a second collective coordinate which describes these fluctuations, although there is no corresponding symmetry of the action. It can then also be understood where, in the first example, the factor β^n comes from. Although a given classical trajectory can only generate a factor β, these new configurations depend on n independent collective coordinates over which one has to integrate.

To summarize: we know that n-instanton contributions do exist. However these contributions do not in general correspond to solutions of the classical equation of motion. They correspond to configurations of largely separated instantons connected in a way which we shall discuss, which become solutions of the equation of motion only asymptotically, in the limit of infinite separation. These configurations depend on n times more collective coordinates than the one-instanton configuration.

In Sections 43.1, 43.2 we first study explicitly two examples which we have already considered in Chapter 40: the *double-well* potential and the *periodic cosine* potential. We then discuss general potentials with degenerate minima. We also calculate the large order behaviour in the case of the $O(\nu)$ symmetric anharmonic oscillator. Finally, from the results obtained for the many instanton contributions, we are led to conjecture the exact form of the semiclassical expansion for potentials with degenerate minima.

The appendix contains in particular some remarks about constrained instanton solutions and a simple example of a non-Borel summable expansion.

43.1 The Double-Well Potential

We first consider the hamiltonian of the double-well potential, already discussed in Section 40.1:

$$H = -\frac{1}{2}\left(\frac{\mathrm{d}}{\mathrm{d}q}\right)^2 + \frac{1}{g}V(q\sqrt{g}), \quad V(q) = \frac{1}{2}q^2(1-q)^2. \tag{43.2}$$

We have seen that, in the infinite β limit, the instanton solutions are (equation (40.10)):

$$q_\pm(t) = \frac{1}{\sqrt{g}}f\bigl(\mp(t-t_0)\bigr), \tag{43.3}$$

$$f(t) = 1/\bigl(1+\mathrm{e}^t\bigr) = 1 - f(-t), \tag{43.4}$$

where the constant t_0 characterizes the instanton position.

43.1.1 The two-instanton configuration

We first construct the two-instanton configuration (really an instanton–anti-instanton). We look for a configuration depending on an additional time parameter, the separation between instantons, which in the limit of infinite separation decomposes into two instantons, and which for large separations minimizes the variation of the action (figure 43.1). For this purpose we could introduce a constraint in the path integral fixing the separation between instantons, and solve the equation of motion with a Lagrange multiplier for the constraint (for details see Appendices A37.2 and A43.3). We use instead a method which, at least at leading order, is simpler and shows that the result is universal.

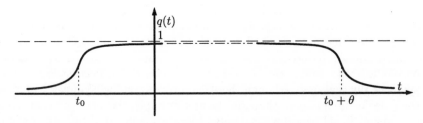

Fig. 43.1   The two-instanton configuration.

We consider a configuration $q_c(t)$ which is the sum of instantons separated by a distance θ, up to an additive constant adjusted in such a way as to satisfy the boundary conditions. It is convenient to introduce some notation:

$$u(t) = f(t - \theta/2), \tag{43.5}$$
$$\underline{u}(t) = f(-t - \theta/2), \tag{43.6}$$
$$v(t) = 1 - \underline{u}(t) = u(t + \theta). \tag{43.7}$$

where $f(t)$ is the function (43.4). We then take

$$q_c(t) = \frac{1}{\sqrt{g}}(u + \underline{u} - 1) = \frac{1}{\sqrt{g}}(u - v). \tag{43.8}$$

(Again time translation by a constant t_0 generates a set of degenerate configurations). This path has the following properties: It is continuous and differentiable and when θ is large it differs, near each instanton, from the instanton solution only by exponentially small terms of order $e^{-\theta}$. Although the calculation of the corresponding action is straightforward, let us perform it step wise to show the generality of the previous ansatz. The action of the path (43.8) is

$$S(q_c) = \int dt \left[\tfrac{1}{2}\dot{q}_c^2 + \frac{1}{g}V(q_c\sqrt{g}) \right]$$
$$= 2 \times \frac{1}{6g} + \frac{1}{g}\int dt \left[\underline{u}\dot{u} + V(u + \underline{u} - 1) - V(\underline{u}) - V(u) \right]. \tag{43.9}$$

The parity of q_c allows to restrict the integration to the region $t > 0$, where v is small. After an integration by parts of the term $\dot{u}\underline{u} = -\dot{v}\dot{u}$ we find

$$S(q_c) = \frac{1}{3g} + \frac{2}{g}\left\{ v(0)\dot{u}(0) + \int_0^{+\infty} dt \left[v\ddot{u} + V(u - v) - V(u) - V(v) \right] \right\}. \tag{43.10}$$

We now expand in powers of v, and take into account the equation of motion for u. The leading terms will be of order $e^{-\theta}$ and we thus stop at order v^2. We obtain:

$$S(q_c) = \frac{1}{3g} + \frac{2}{g}v(0)\dot{u}(0) + \frac{2}{g}\left\{ \int_0^{+\infty} dt \left[\tfrac{1}{2}v^2 V''(u) - V(v) \right] \right\}. \tag{43.11}$$

The function v decreases exponentially away from the origin so the main contributions to the integral come from the neighbourhood of $t = 0$, where $V''(u) \sim 1$. $V(v)$ can be replaced at leading order by $\tfrac{1}{2}v^2$. Therefore the integrals cancel and we are left with the integrated term:

$$v(0)\dot{u}(0) \sim -e^{-\theta},$$

and thus

$$S(q_c) = g^{-1}\left[\frac{1}{3} - 2e^{-\theta} + O(e^{-2\theta}) \right]. \tag{43.12}$$

It will become clearer later why we need the classical action only up to order $e^{-\theta}$. It is convenient in the calculation to keep β large but finite. Symmetry between θ and $\beta - \theta$ implies

$$S(q_c) = g^{-1}\left[\frac{1}{3} - 2e^{-\theta} - 2e^{-(\beta-\theta)} \right]. \tag{43.13}$$

As a check, we can calculate the extremum of $S(q)$ at β fixed and we obtain:

$$\theta_c = \beta/2 \,, \quad \Rightarrow \quad S(q_c) = g^{-1}\left[\frac{1}{3} - 4\,\mathrm{e}^{-\beta/2} + O\left(\mathrm{e}^{-\beta}\right)\right].$$

In Chapter 40, for the same hamiltonian, we have found for β large (equation (40.12)):

$$S(q_c) = g^{-1}\left[\frac{1}{6} - 2\,\mathrm{e}^{-\beta} + O\left(\mathrm{e}^{-2\beta}\right)\right]. \tag{43.14}$$

Both results are consistent. Indeed to compare them we have to replace β by $\beta/2$ in equation (43.14) and multiply the action by a factor 2, since the action corresponds to a trajectory described twice in the total time β.

The variation of the action. We now show that if we infinitesimally (for θ large) modify the configuration to further decrease the variation of the action, the change $r(t)$ of the path will be of order $\mathrm{e}^{-\theta}$ and the variation of the action of order $\mathrm{e}^{-2\theta}$ at least. Setting

$$q(t) = q_c(t) + r(t), \tag{43.15}$$

we find, expanding the action up to second order in $r(t)$,

$$S(q_c + r) = S(q_c) + \int\left[\dot{q}_c(t)\dot{r}(t) + \frac{1}{\sqrt{g}}V'(q_c(t)\sqrt{g})\,r(t)\right]dt$$

$$+ \frac{1}{2}\int dt\left[\dot{r}^2(t) + V''(q_c\sqrt{g})\,r^2(t)\right] + O\left([r(t)]^3\right). \tag{43.16}$$

In the term linear in $r(t)$, we integrate by parts $\dot{r}(t)$, in order to use the property that $q_c(t)$ approximately satisfies the equation of motion. In the term quadratic in $r(t)$, we replace V'' by one, since we expect $r(t)$ to be large only far from the instantons. We then verify that the term linear in r is of order $\mathrm{e}^{-\theta}$ while the quadratic term is of order 1. Shifting r to eliminate the linear term would then give a contribution of order $\mathrm{e}^{-2\theta}$ and thus negligible at the order we consider.

43.1.2 The n-instanton configuration

We now consider a succession of n instantons, separated by times θ_i,

$$\sum_{i=1}^{n}\theta_i = \beta. \tag{43.17}$$

At leading order we need consider "interactions" only between nearest neighbours. This is an essential simplifying feature of quantum mechanics. The classical action $S_c(\theta_i)$ can then be directly inferred from expression (43.13):

$$S_c(\theta_i) = \frac{1}{g}\left[\frac{n}{6} - 2\sum_{i=1}^{n}\mathrm{e}^{-\theta_i} + O\left(\mathrm{e}^{-(\theta_i+\theta_j)}\right)\right]. \tag{43.18}$$

Other interactions are negligible because they are of higher order in $\mathrm{e}^{-\theta}$.

Note that for n even the n-instanton configuration contributes to $\operatorname{tr}\mathrm{e}^{-\beta H}$, while for n odd it contributes to $\operatorname{tr}\left(P\mathrm{e}^{-\beta H}\right)$ (P is the parity operator). Then calculating

$$Z_\varepsilon = \tfrac{1}{2}\operatorname{tr}\left[(1+\varepsilon P)\,\mathrm{e}^{-\beta H}\right], \tag{43.19}$$

we obtain for $\varepsilon = +1$ and $\varepsilon = -1$ contributions to the even and odd eigenstate energies respectively.

Remark. Since we keep in the action all terms of order $\mathrm{e}^{-\beta}$ we expect to find the contributions not only to the two lowest energies but also to all energies which remain finite when g goes to zero (see the remark after equation (38.62) in Section 38.3).

43.1.3 The n-instanton contribution

We have calculated the n-instanton action. Let us now evaluate, at leading order, the contribution to the path integral of the neighbourhood of the classical path. Although the configuration is not a solution of the equation of motion, we have constructed it in such a way that we can neglect the linear terms in the gaussian integration. The second derivative of the action at the classical path $M(t', t)$,

$$M(t', t) = \left[-\left(\frac{d}{dt} \right)^2 + V'' \left(\sqrt{g} q_c(t) \right) \right] \delta(t - t'), \tag{43.20}$$

is a hamiltonian with a potential which consists of n wells almost identical to the well arising in the one-instanton problem, and which are largely separated. At leading order the corresponding spectrum is therefore the spectrum arising in the one-instanton problem n-times degenerate. Corrections are exponentially small in the separation (for details see Appendix A43.1). At the same time, by introducing n collective time variables, we have suppressed n times the zero eigenvalue, and generated the jacobian of the one-instanton case to the power n. We can therefore write the n-instanton contribution to Z_ε (equation (43.19)):

$$Z_\varepsilon^{(n)} = \mathrm{e}^{-\beta/2} \frac{\beta}{n} \left(\frac{\mathrm{e}^{-1/6g}}{\sqrt{\pi g}} \right)^n \int_{\theta_i \geq 0} \delta \left(\sum \theta_i - \beta \right) \prod_i d\theta_i \exp \left[\frac{2}{g} \sum_{i=1}^n \mathrm{e}^{-\theta_i} \right]. \tag{43.21}$$

All the factors have already been explained except the factor β which comes from the integration over a global time translation, and the factor $1/n$ which arises because the configuration is invariant under a cyclic permutation of the θ_i. The factor $\mathrm{e}^{-\beta/2}$ is the usual normalization factor.

Let us define the quantity λ, the "fugacity" of the instanton gas:

$$\lambda = \frac{\varepsilon}{\sqrt{\pi g}} \, \mathrm{e}^{-1/6g}, \tag{43.22}$$

which is half of the one-instanton contribution at leading order. It is convenient to introduce the sum $\Sigma(\beta, g)$ of the leading order n-instanton contributions:

$$\Sigma(\beta, g) = \mathrm{e}^{-\beta/2} + \sum_{n=1}^{\infty} Z_\varepsilon^{(n)}(\beta, g). \tag{43.23}$$

If we neglect the instanton interaction (the dilute gas approximation), we can integrate over the θ_i's and calculate the sum

$$\Sigma(\beta, g) = \mathrm{e}^{-\beta/2} \left[1 + \beta \sum_{n=1}^{\infty} \frac{\lambda^n}{n} \frac{\beta^{n-1}}{(n-1)!} \right] = \mathrm{e}^{-\beta(1/2 - \lambda)}. \tag{43.24}$$

We recognize the perturbative and one-instanton contribution, at leading order, to $E_\varepsilon(g)$, the ground state and the first excited state energies:

$$E_\varepsilon(g) = \tfrac{1}{2} + O(g) - \frac{\varepsilon}{\sqrt{\pi g}} \, \mathrm{e}^{-1/6g} \left(1 + O(g) \right). \tag{43.25}$$

Discussion. To go beyond the one-instanton approximation we have to take into account the interaction between instantons. We then face a problem: examining expression (43.21), we discover that the interaction between instantons is *attractive*. For g small the dominant contributions to the integral therefore come from configurations in which the instantons are close. For such configurations the concept of instanton is no longer valid, since such configurations cannot be distinguished from the fluctuations around the constant or the one-instanton solution.

We should not be too surprised by this phenomenon. Indeed the large order behaviour analysis has indicated that the perturbative expansion in the case of potentials with degenerate minima is not Borel summable. An ambiguity is expected at the two-instanton order. But if the perturbative expansion is ambiguous at the two-instanton order, we should not expect to be able to calculate a contribution of the same order or smaller. To proceed any further we first have to make a choice about how we define the sum of perturbation theory. It is possible to show that the perturbation series is Borel summable for g negative by relating it to the perturbative expansion of the $O(2)$ anharmonic oscillator. We therefore define the sum of the perturbation series as the analytic continuation of this Borel sum from g negative to $g = |g| \pm i0$. This corresponds in the Borel transformation to integrate above or below the real positive axis. We then note that for g *negative* the interaction between instantons is *repulsive*, and expression (43.21) becomes meaningful. We therefore calculate, for g small and negative, both the sum of the perturbation series and the instanton contributions, and perform an analytic continuation to g positive of all quantities consistently. In the same way the perturbative expansion around each multi-instanton configuration is also non-Borel summable and has to be summed by the same procedure.

43.1.4 The calculation

Let us again write the n-instanton contribution:

$$Z_{\varepsilon}^{(n)} \sim \frac{\beta}{n} e^{-\beta/2} \lambda^n \int_{\theta_i \geq 0} \delta \left(\sum \theta_i - \beta \right) \prod_{i=1}^{n} d\theta_i \exp \left[\frac{2}{g} \sum_{i=1}^{n} e^{-\theta_i} \right]. \qquad (43.26)$$

To factorize the integral over the variables θ_i, we replace the δ-function by an integral representation:

$$\delta \left(\sum_{i=1}^{n} \theta_i - \beta \right) = \frac{1}{2i\pi} \int_{-i\infty-\eta}^{i\infty-\eta} ds \, \exp \left[-s \left(\beta - \sum_{i=1}^{n} \theta_i \right) \right], \quad \eta > 0. \qquad (43.27)$$

In terms of the function

$$I(s) = \int_{0}^{+\infty} e^{s\theta - \mu e^{-\theta}} \, d\theta, \qquad (43.28)$$

integral (43.26) can be rewritten:

$$Z_{\varepsilon}^{(n)} \sim \frac{\beta e^{-\beta/2}}{2i\pi} \frac{\lambda^n}{n} \int_{-i\infty-\eta}^{i\infty-\eta} ds \, e^{-\beta s} \left[I(s) \right]^n, \quad \text{with } \mu = -2/g. \qquad (43.29)$$

By giving to s a small negative real part, we have ensured the convergence of integral (43.28). To evaluate the integral (43.28) we set:

$$\mu e^{-\theta} = t, \qquad (43.30)$$

and the integral becomes:

$$I(s) = \int_0^\mu \frac{dt}{t} \left(\frac{\mu}{t}\right)^s e^{-t} = \int_0^{+\infty} \frac{dt}{t} \left(\frac{\mu}{t}\right)^s e^{-t} + O\left(e^{-\mu}/\mu\right), \qquad (43.31)$$

for μ positive and large, i.e $g \to 0_-$. Up to an exponentially small correction we thus obtain:

$$I(s) \sim \mu^s \Gamma(-s). \qquad (43.32)$$

The generating function $\Sigma(\beta, g)$ of the leading order multi-instanton contributions (43.23) then becomes:

$$\Sigma(\beta, g) = -\frac{\beta e^{-\beta/2}}{2i\pi} \int_{-i\infty-\eta}^{i\infty-\eta} ds \, e^{-\beta s} \sum_n \frac{\lambda^n}{n} \mu^{ns} \left[\Gamma(-s)\right]^n. \qquad (43.33)$$

The sum can easily be performed:

$$\Sigma(\beta, g) = -\frac{\beta e^{-\beta/2}}{2i\pi} \int_{-i\infty-\eta}^{i\infty-\eta} ds \, e^{-\beta s} \ln\left[1 - \lambda \mu^s \Gamma(-s)\right]. \qquad (43.34)$$

We set:

$$E = s + 1/2, \qquad \phi(E) = 1 - \lambda \mu^{E-1/2} \Gamma(1/2 - E), \qquad (43.35)$$

We then integrate $\beta e^{-\beta s}$ by parts and obtain

$$\Sigma(\beta, g) = -\frac{1}{2i\pi} \int_{-i\infty}^{+i\infty} dE \, e^{-\beta E} \frac{\phi'(E)}{\phi(E)}. \qquad (43.36)$$

The asymptotic behaviour of the Γ-function (given by the Stirling formula) ensures the convergence of the integral, and moreover the contour can be deformed to enclose the poles of the integrand in the half-plane $\mathrm{Re}(E) > 0$. Integrating we obtain a sum of residues:

$$\Sigma(\beta, g) = \sum_{N \geq 0} e^{-\beta E_N}, \qquad (43.37)$$

where the energies E_N are solutions of the equation:

$$\phi(E) = 1 - \lambda \mu^{E-1/2} \Gamma(1/2 - E) = 0. \qquad (43.38)$$

Since λ is small, a zero E of this equation is close to a pole of $\Gamma(1/2 - E)$:

$$E_N = N + \tfrac{1}{2} + O(\lambda), \quad N \geq 0. \qquad (43.39)$$

We can then expand the solutions of equation (43.38) in a systematic power series in λ:

$$E_N(g) = \sum E_N^{(n)}(g) \lambda^n. \qquad (43.40)$$

We obtain at once the many instanton contributions to all energy levels $E_N(g)$ of the double-well potential at leading order. It is convenient to rewrite equation (43.38) as

$$\frac{e^{-1/6g}}{\sqrt{2\pi}} \left(-\frac{2}{g}\right)^E \Gamma(\tfrac{1}{2} - E) = -\varepsilon i \Leftrightarrow \frac{\cos \pi E}{\pi} = \varepsilon i \frac{e^{-1/6g}}{\sqrt{2\pi}} \left(-\frac{2}{g}\right)^E \frac{1}{\Gamma(\tfrac{1}{2} + E)}. \qquad (43.41)$$

For example the one-instanton contribution is

$$E_N^{(1)}(g) = -\frac{\varepsilon}{N!} \left(\frac{2}{g}\right)^{N+1/2} \frac{e^{-1/6g}}{\sqrt{2\pi}} (1 + O(g)). \qquad (43.42)$$

The two-instanton contribution then is:

$$E_N^{(2)}(g) = \frac{1}{(N!)^2} \left(\frac{2}{g}\right)^{2N+1} \frac{e^{-1/3g}}{2\pi} \left[\ln(-2/g) - \psi(N+1) + O(g \ln g)\right], \qquad (43.43)$$

where ψ is the logarithmic derivative of the Γ-function. The appearance of a factor $\ln g$ can be simply understood by noting that the interaction terms are only relevant for $g^{-1} e^{-\theta}$ of order 1, i.e. θ of order $-\ln g$.

More generally it can be verified that the n-instanton contribution has at leading order the form:

$$E_N^{(n)}(g) = -\left(\frac{2}{g}\right)^{n(N+1/2)} \left(\frac{e^{-1/6g}}{\sqrt{2\pi}}\right)^n \left[P_n^N(\ln(-g/2)) + O\left(g(\ln g)^{n-1}\right)\right], \qquad (43.44)$$

in which $P_n^N(\sigma)$ is a polynomial of degree $n - 1$. For example for $N = 0$ we find:

$$P_2(\sigma) = \sigma + \gamma, \qquad P_3(\sigma) = \frac{3}{2}(\sigma + \gamma)^2 + \frac{\pi^2}{12}, \qquad (43.45)$$

in which γ is Euler's constant $\gamma = -\psi(1) = 0.577215....$

Discussion. When we now perform our analytic continuation from g negative to g positive, two things happen: the Borel sums become complex and get an imaginary part exponentially smaller by about a factor $e^{-1/3g}$ than the real part. Simultaneously the function $\ln(-2/g)$ also becomes complex and gets an imaginary part $\pm i\pi$. Since the sum of all the contributions is real, the imaginary parts should cancel. This argument leads to an evaluation of the imaginary part of the Borel sum of the perturbation series, or of the expansion around one instanton for example.

From the imaginary part of P_2 we derive

$$\operatorname{Im} E^{(0)}(g) \sim \frac{1}{\pi g} e^{-1/3g} \operatorname{Im}\left[P_2\left(\ln\left(-g/2\right)\right)\right], \qquad (43.46)$$

and therefore:

$$\operatorname{Im} E^{(0)}(g) \sim -\frac{1}{g} e^{-1/3g}. \qquad (43.47)$$

Expanding now $E^{(0)}(g)$ in perturbation series

$$E^{(0)}(g) = \sum_k E_k^{(0)} g^k, \qquad (43.48)$$

and again using a dispersion relation to calculate the coefficients $E_k^{(0)}$, we obtain the large order behaviour of the perturbative expansion:

$$E_k^{(0)} \underset{k \to \infty}{\sim} \frac{1}{\pi} \int_0^\infty \operatorname{Im}\left[E^{(0)}(g)\right] \frac{dg}{g^{k+1}}, \quad \Rightarrow \quad E_k^{(0)} \sim -\frac{1}{\pi} 3^{k+1} k!. \qquad (43.49)$$

From the imaginary part of P_3 we can derive the large order behaviour of the expansion of $E^{(1)}(g)$:

$$E^{(1)}(g) = -\frac{1}{\sqrt{\pi g}} e^{-1/6g} \left(1 + \sum_{k}^{\infty} E_k^{(1)} g^k\right). \qquad (43.50)$$

We express that the imaginary part of $E^{(1)}(g)$ and $E^{(3)}(g)$ have to cancel at leading order:

$$\text{Im}\, E^{(1)}(g) \sim -\left(\frac{e^{-1/6g}}{\sqrt{\pi g}}\right)^3 \text{Im}\,[P_3(\ln(-g/2))]. \qquad (43.51)$$

For the coefficients $E_k^{(1)}$ we write a dispersion integral:

$$E_k^{(1)} = \frac{1}{\pi} \int_0^\infty \left\{\text{Im}\left[E^{(1)}(g)\right] \sqrt{\pi g}\, e^{1/6g}\right\} \frac{dg}{g^{k+1}}. \qquad (43.52)$$

Using then equations (43.45) and (43.51) we find

$$E_k^{(1)} \sim -\frac{1}{\pi} \int_0^\infty 3\left(\ln\frac{2}{g} + \gamma\right) e^{-1/3g} \frac{dg}{g^{k+2}}. \qquad (43.53)$$

At leading order for k large, we can replace g by its saddle point value $1/3k$ in $\ln g$ and finally obtain:

$$E_k^{(1)} = -\frac{3^{k+2}}{\pi} k! \left[\ln 6k + \gamma + O\left(\frac{\ln k}{k}\right)\right]. \qquad (43.54)$$

Both results (43.49) and (43.54) have been checked against the numerical behaviour of the corresponding series for which about 100 terms are known.

To check the real part of P_2, the following quantity has been evaluated numerically:

$$\Delta(g) = 4\frac{\left\{\frac{1}{2}(E_+ + E_-) - \text{Re}\left[\text{Borel sum } E^{(0)}(g)\right]\right\}}{(E_+ - E_-)^2 (\ln 2g^{-1} + \gamma)}. \qquad (43.55)$$

In this expression E_+ and E_- are respectively the ground state and the first excited state energy. In the sum $(E_+ + E_-)$ the contributions corresponding to an odd number of instantons cancel. Therefore the numerator is dominated for g small by the real part of the two instanton contribution. The difference $(E_+ - E_-)$, as we know, is dominated by the one-instanton contribution. Using the various expressions given above, it is easy to verify that $\Delta(g)$ should go to 1 when g goes to zero. Table 43.1 shows what happens in the range of values of g for which the numerical calculation was reasonably accurate.

Table 43.1

The ratio $\Delta(g)$ as a function of g: For g small it should behave like $1 + 3g$. (The value for $g = 0.05$ is 0.938807.)

g	0.024	0.026	0.028	0.030	0.032	0.034	0.036	0.038	0.040
$\Delta(g)$	1.01..	1.0134	1.0098	1.0064	1.0018	0.9967	0.9904	0.9835	0.9760

43.2 The Periodic Cosine Potential

To avoid the proliferation of big integer factors it is convenient to use a somewhat non-standard normalization of the coupling constant . We write the hamiltonian

$$H = -\frac{1}{2}\left(\frac{\mathrm{d}}{\mathrm{d}q}\right)^2 + \frac{1}{16g}\left(1 - \cos 4q\sqrt{g}\right).$$

(43.56)

As we have already discussed in Chapter 40, to each state of the harmonic oscillator is associated a band for g small. A state in the band is characterized by an angle φ,

$$T\psi_\varphi(q) \equiv \psi_\varphi\left(q + \frac{\pi}{2\sqrt{g}}\right) = e^{i\varphi}\,\psi_\varphi(q).$$

(43.57)

The operator T is the operator which translates by one period of the potential.

If we expand in a Fourier series the corresponding energy $\mathcal{E}(g,\varphi)$

$$\mathcal{E}(g,\varphi) = \sum_{l=-\infty}^{+\infty} \mathcal{E}_l(g)\,e^{il\varphi}, \qquad \mathcal{E}_l = \mathcal{E}_{-l},$$

(43.58)

then, as we have seen, for g small $\mathcal{E}_l(g)$ is dominated by l-instanton configurations. In particular for the ground state energy in the φ sector, $E_0(g,\varphi)$, the $l = 1$ term behaves like

$$E_{0,l=1}(g) \sim \frac{1}{\sqrt{\pi g}}\,e^{-1/2g}.$$

(43.59)

The partition function in the φ sector. Let us recall a useful formula. We first define

$$Z(\beta,g,\varphi) = \sum_N e^{-\beta E_N(g,\varphi)},$$

(43.60)

the partition function in the sector corresponding to angle φ. In particular for β large

$$Z(\beta,g,\varphi) \underset{\beta\to\infty}{\sim} e^{-\beta E_0(g,\varphi)}.$$

(43.61)

Then in the notation of Chapter 40 we have

$$Z(\beta,g,\varphi) = \sum_{l=-\infty}^{+\infty} e^{-il\varphi}\,\mathrm{tr}'\left(T^l\,e^{-\beta H}\right).$$

(43.62)

This is the expression we shall use to calculate $Z(\beta,g,\varphi)$. The path integral representation of $\mathrm{tr}'\left(T^l\,e^{-\beta H}\right)$ is

$$\mathrm{tr}'\left(T^l\,e^{-\beta H}\right) = \int_{q(\beta/2)=q(-\beta/2)+l2\pi/\sqrt{g}} [dq(t)]\exp\left[-S(q)\right],$$

(43.63)

with

$$S(q) = \int_{-\beta/2}^{\beta/2} dt\left[\frac{1}{2}\dot{q}(t)^2 + \frac{1}{16g}\left(1 - \cos 4q\sqrt{g}\right)\right].$$

(43.64)

Note that for β large the sum (43.62) can also be written:

$$e^{-\beta E_0(g,\varphi)} \underset{\beta \to \infty}{\sim} \int [dq(t)] \exp\left\{ -S(q) + i\frac{\sqrt{g}}{2\pi}\varphi \left[q\left(\beta/2\right) - q\left(-\beta/2\right) \right] \right\}. \tag{43.65}$$

Indeed the configurations for which $q\left(\beta/2\right) - q\left(-\beta/2\right)$ is not a multiple of the period are suppressed in the large β limit, since their classical action is necessarily infinite. The additional term can also be trivially rewritten as the integral of a local density:

$$q\left(+\beta/2\right) - q\left(-\beta/2\right) = \int_{-\beta/2}^{+\beta/2} dt\, \dot{q}(t). \tag{43.66}$$

These expressions have a straightforward generalization in the case of the θ-vacuum of the $CP(N-1)$ model and gauge theories (see Sections 40.4, 40.5).

The n-instanton configurations. There is one important difference between this example, and the double well potential. In the previous case each configuration was a succession of instantons and anti-instantons. Here in contrast we can at each step decide to go to the next minimum of the potential or to go one step back. We assign therefore a sign $\varepsilon = +1$ to an instanton and a sign $\varepsilon = -1$ to an anti-instanton. A straightforward calculation, similar to the calculation presented above (for details see Appendix A43.2) yields the following interaction term between two consecutive instantons of types ε_1 and ε_2 separated by a distance θ_{12},

$$\frac{2\varepsilon_1\varepsilon_2}{g}\, e^{-\theta_{12}}. \tag{43.67}$$

The force between instantons of the same kind is repulsive, while it is attractive for different kinds. Let us again call λ the one-instanton contribution at leading order,

$$\lambda = \frac{1}{\sqrt{\pi g}}\, e^{-1/2g}. \tag{43.68}$$

With this notation the n-instanton contribution reads

$$Z^{(n)}(\beta, g, \varphi) = \beta\, e^{-\beta/2} \frac{\lambda^n}{n} \int_{\theta_i \geq 0} \delta\left(\sum_{i=1}^{n} \theta_i - \beta \right) J_n(\theta_i), \tag{43.69}$$

with

$$J_n(\theta_i) = \sum_{\varepsilon_i = \pm 1} \exp\left(\sum_{i=1}^{n} -\frac{2}{g}\varepsilon_i\varepsilon_{i+1}\, e^{-\theta_i} - i\varepsilon_i\varphi \right). \tag{43.70}$$

The additional term $-i\varepsilon_i\varphi$ comes from the formula (43.62). We have identified ε_{n+1} and ε_1.

In contrast with the case of the double-well potential, the interaction between instantons contains both attractive and repulsive terms. Thus we have to begin with g complex to perform the analytic continuation of both the Borel sums and the instanton contributions.

Following the same steps as in the case of the double-well potential we then obtain:

$$Z^{(n)}(\beta, g, \varphi) = \frac{\beta}{2i\pi}\, e^{-\beta/2} \frac{\lambda^n}{n} \oint ds\, e^{-\beta s}\, \Gamma^n(-s)$$

$$\times \sum_{\{\varepsilon_i = \pm 1\}} \exp\left[\sum_{i=1}^{n} -i\varepsilon_i\varphi - s\ln\left(\frac{g}{2}\varepsilon_i\varepsilon_{i+1} \right) \right]. \tag{43.71}$$

Let us introduce the notation:

$$\sigma = \ln(g/2) , \tag{43.72}$$

and decide to make the analytic continuation from above so that:

$$\ln\left(\frac{g}{2}\varepsilon_i\varepsilon_{i+1}\right) = \sigma - \frac{1}{2}i\pi\left(1 - \varepsilon_i\varepsilon_{i+1}\right). \tag{43.73}$$

Expression (43.71) can then be rewritten:

$$Z^{(n)}(\beta, g, \varphi) \sim \frac{\beta}{2i\pi}\, e^{-\beta/2}\, \frac{\lambda^n}{n} \oint ds\; e^{-\beta s}\left[\Gamma\left(-s\right)e^{-\sigma s}\right]^n$$

$$\times \sum_{\{\varepsilon_i = \pm 1\}} \exp\left[\sum_{i=1}^{n} -i\varepsilon_i\varphi + \frac{i\pi}{2}s\left(1 - \varepsilon_i\varepsilon_{i+1}\right)\right]. \tag{43.74}$$

The summation over the set $\{\varepsilon_i\}$ corresponds to the calculation of the partition function of a one-dimensional Ising model whose transfer matrix $\mathbf{M}$ is:

$$\mathbf{M} = \begin{bmatrix} e^{-i\varphi} & e^{i\pi s} \\ e^{i\pi s} & e^{i\varphi} \end{bmatrix}. \tag{43.75}$$

The sum then is just $\operatorname{tr} \mathbf{M}^n$. The eigenvalues $m_\pm$ are:

$$m_\pm = \cos\varphi \pm \left(e^{2i\pi s} - \sin^2\varphi\right)^{1/2}. \tag{43.76}$$

Expression (43.74) then becomes:

$$Z^{(n)}(\beta, g, \varphi) \sim \frac{\beta\, e^{-\beta/2}}{2i\pi}\, \frac{\lambda^n}{n} \oint ds\; e^{-\beta s}\left[\Gamma\left(-s\right)e^{-\sigma s}\right]^n\left(m_+^n + m_-^n\right). \tag{43.77}$$

The sum $\Sigma\left(\beta, g\right)$ of all leading order multi-instanton contributions can now be calculated:

$$\Sigma\left(\beta, g\right) = e^{-\beta/2} - \frac{\beta\, e^{-\beta/2}}{2i\pi} \oint ds\; e^{-\beta s} \ln\left\{\left[1 - \lambda\Gamma\left(-s\right)e^{-\sigma s}\, m_+\left(s\right)\right]\right.$$

$$\left. \times \left[1 - \lambda\Gamma\left(-s\right)e^{-\sigma s}\, m_-\left(s\right)\right]\right\}. \tag{43.78}$$

The argument of the logarithm can also be written as:

$$\left[1 - \lambda\Gamma\left(-s\right)e^{-\sigma s}\, m_+\left(s\right)\right]\left[1 - \lambda\Gamma\left(-s\right)e^{-\sigma s}\, m_-\left(s\right)\right]$$

$$= 1 - \lambda\Gamma\left(-s\right)e^{-\sigma s}\left[2\cos\varphi + \lambda\frac{2i\pi\, e^{(i\pi - \sigma)s}}{\Gamma\left(1 + s\right)}\right]. \tag{43.79}$$

An integration by parts of $\beta\, e^{-\beta s}$ in integral (43.78) yields our final result:

$$\Sigma\left(\beta, g\right) = \sum_{N=0}^{\infty} e^{-\beta E_N(g)}, \tag{43.80}$$

in which $E_N(g) = \frac{1}{2} + s_N(\lambda, \sigma)$ is a solution expandable in powers of λ of the equation:

$$\left(\frac{2}{g}\right)^{-E}\frac{e^{1/2g}}{\Gamma(\frac{1}{2} - E)} + \left(-\frac{2}{g}\right)^{E}\frac{e^{-1/2g}}{\Gamma(\frac{1}{2} + E)} = \frac{2\cos\varphi}{\sqrt{2\pi}}. \tag{43.81}$$

Note the symmetry in the change $g, E \mapsto -g, -E$. This symmetry is, however, slightly fictitious because the equation is actually quadratic in $\Gamma(\frac{1}{2} - E)$ and only one root, corresponding to m_+, is relevant for $g > 0$.

43.3 General Potentials with Degenerate Minima

We now consider a general analytic potential having two degenerate minima located at the origin and another point x_0:

$$\begin{cases} V(x) = \tfrac{1}{2}x^2 + O\left(x^3\right), \\ V(x) = \tfrac{1}{2}\omega^2 \left(x - x_0\right)^2 + O\left(\left(x - x_0\right)^3\right). \end{cases} \tag{43.82}$$

For definiteness we assume $\omega > 1$.

In such a situation the classical equation of motion have instanton solutions connecting the two minima of the potential. However there is no ground state degeneracy beyond the classical level. Therefore the one-instanton solution does not contribute anymore to the path integral. Only periodic classical paths are relevant: the leading contribution now comes from the two-instanton configuration.

To calculate the potential between instantons and the normalization of the path integral it is convenient to first calculate the contribution at β finite of a trajectory described n times and take the large β limit of this expression. Using the expressions derived in Chapter 38 we find:

$$\left\{\mathrm{tr}\, e^{-\beta H}\right\}_{(n)} = i\left(-1\right)^n \frac{\beta}{n\sqrt{\pi g}} \sqrt{\frac{\omega C}{n\left(1+\omega\right)}}\, e^{-\omega\beta/[2n(1+\omega)]}\, e^{-nA(\beta)/g}, \tag{43.83}$$

with the definitions:

$$C = x_0^2 \omega^{2/(1+\omega)} \exp\left\{\frac{2\omega}{1+\omega}\left[\int_0^{x_0} dx \left(\frac{1}{\sqrt{2V(x)}} - \frac{1}{x} - \frac{1}{\omega\left(x_0 - x\right)}\right)\right]\right\}, \tag{43.84}$$

and

$$A\left(\beta\right) = 2\int_0^{x_0} \sqrt{2V(x)}dx - 2C\frac{\left(1+\omega\right)}{\omega}\, e^{-(\beta/n)\omega/(1+\omega)} + \cdots. \tag{43.85}$$

Note that n has not the same meaning here as in Section 43.1. Since ω is different from 1, the one-instanton configuration does not contribute and n instead counts the number of instanton anti-instanton pairs in the language of Section 43.1. Therefore n here actually corresponds to $2n$ in the $\omega = 1$ limit.

43.3.1 The n-instanton action

We now call θ_i the successive amounts of time the classical trajectory spends near x_0, and φ_i near the origin. The n-instanton action takes the form

$$A\left(\theta_i, \varphi_j\right) = na - \sum_{i=1}^{n} \left(C_1\, e^{-\omega\theta_i} + C_2\, e^{-\varphi_i}\right), \tag{43.86}$$

with $\sum_{i=1}^{n}\left(\theta_i + \varphi_i\right) = \beta$ and

$$a = 2\int_0^{x_0} \sqrt{2V(x)}dx. \tag{43.87}$$

By comparing the value of the action at the saddle point

$$\theta_i = \frac{\beta}{n(1+\omega)}, \qquad \varphi_i = \frac{\omega\beta}{n(1+\omega)}, \tag{43.88}$$

with the expression (43.85), we see that we can choose

$$C_1 = 2C/\omega, \qquad C_2 = 2C, \tag{43.89}$$

by adjusting the definitions of θ and φ.

43.3.2 The n-instanton contribution

The n-instanton contribution then has the form:

$$\{\operatorname{tr} e^{-\beta H}\}_{(n)} = \beta e^{-\beta/2} \frac{e^{-na/g}}{(\pi g)^n} N_n \int_{\theta_i,\varphi_i \geq 0} \delta \left(\sum_i \theta_i + \varphi_i - \beta \right)$$

$$\times \exp \left[\sum_{i=1}^{n} \frac{1}{2} (1-w) \theta_i - \frac{1}{g} A(\theta,\varphi) \right]. \tag{43.90}$$

The additional term $\sum_i \frac{1}{2} (1-w) \theta_i$ in the integrand comes from the determinant generated by the gaussian integration around the classical path. The normalization can be obtained by performing a steepest descent integration over the variables θ_i and φ_i and compare the result with expression (43.83). The result is:

$$N_n = \frac{(C\sqrt{w})^n}{n}. \tag{43.91}$$

The factor $1/n$ comes from the symmetry of the action under cyclic permutations of the θ_i and φ_i.

We now set:

$$\lambda = \frac{e^{-a/g}}{\pi g} C \sqrt{w}, \qquad \mu = -\frac{2C}{g}. \tag{43.92}$$

As in Section 43.1, we introduce an integral representation for the δ-function:

$$\delta \left(\sum_{i=1}^{n} (\theta_i + \varphi_i) - \beta \right) = \frac{1}{2i\pi} \int_{-i\infty-\eta}^{+i\infty+\eta} ds \, \exp \left[-s\beta + s \sum_{i=1}^{n} (\theta_i + \varphi_i) \right], \tag{43.93}$$

with $n > 0$.

Expression (43.90) can be rewritten as:

$$\{\operatorname{tr} e^{-\beta H}\}_{(n)} = \beta e^{-\beta/2} \frac{\lambda^n}{n} \frac{1}{2i\pi} \int_{-i\infty-\eta}^{+i\infty-\eta} ds \, e^{-s\beta} \left[I(s) J(s) \right]^n, \tag{43.94}$$

where we have now defined

$$I(s) = \int_0^{+\infty} e^{\theta s - \mu e^{-\theta}} d\theta, \tag{43.95}$$

$$J(s) = \int_0^{+\infty} \exp \left\{ \left[\frac{1}{2} (1-w) + s \right] \theta - \frac{\mu}{w} e^{-w\theta} \right\} d\theta. \tag{43.96}$$

In the small g limit these integrals can be evaluated:

$$I(s) = \mu^s \Gamma(-s), \tag{43.97}$$

$$J(s) = \frac{1}{\sqrt{\mu w}} \left(\frac{\mu}{w} \right)^{(s+1/2)/w} \Gamma \left(\tfrac{1}{2} - (s + \tfrac{1}{2})/w \right). \tag{43.98}$$

We call as before $\Sigma(\beta, g)$ the generating functional of the many-instanton contributions.

Summing over n and integrating by parts, we obtain $\Sigma(\beta, g)$ as a sum of residues:

$$\Sigma(\beta, g) = \sum_\alpha e^{-\beta E_\alpha}, \tag{43.99}$$

in which the values $E_\alpha = \frac{1}{2} + s_\alpha$ are the solutions expandable for g small of the equation:

$$\left(\frac{\mu}{\omega}\right)^{E/\omega} \Gamma\left(\tfrac{1}{2} - E/\omega\right) \mu^E \Gamma\left(\tfrac{1}{2} - E\right) \frac{e^{-a/g}}{2\pi} = -1. \tag{43.100}$$

We now notice that we find two series of energy levels corresponding to the poles of the two Γ-functions:

$$E_N = N + \tfrac{1}{2} + O(\lambda), \tag{43.101}$$
$$E_N = \left(N + \tfrac{1}{2}\right)\omega + O(\lambda). \tag{43.102}$$

The same expression contains the instanton contributions to the two different sets of levels.

One can verify that the many-instanton contributions are singular for $\omega = 1$. But if one directly sets $\omega = 1$ in equation (43.100) one obtains:

$$\mu^{2E} \Gamma^2\left(\tfrac{1}{2} - E\right) \frac{e^{-a/g}}{2\pi} = -1,$$

equation which can be rewritten as

$$\mu^E \Gamma\left(\tfrac{1}{2} - E\right) \frac{e^{-a/(2g)}}{\sqrt{2\pi}} = \pm i. \tag{43.103}$$

This is exactly the set of two equations obtained in Section 43.1.

43.3.3 Large order estimates of perturbation theory

Expression (43.100) can be used to determine the large order behaviour of perturbation theory by calculating the imaginary part of the leading instanton contribution and writing a dispersion integral as we have done in Section 41.1.1. For the energy $E_N(g) = N + \tfrac{1}{2} + O(g)$ one finds

$$\operatorname{Im} E_N(g) \sim K_N g^{-(N+1/2)(1+1/\omega)} e^{-a/g}, \tag{43.104}$$

with

$$K_N = \frac{(-1)^{N+1}}{2\pi N!} \omega^{-(N+1/2)/\omega} (2C)^{(N+1/2)(1+1/\omega)} \sin\left[\pi\left(N + \tfrac{1}{2}\right)(1 + 1/\omega)\right]$$
$$\times \Gamma\left[\tfrac{1}{2} - \left(N + \tfrac{1}{2}\right)/\omega\right]. \tag{43.105}$$

From the imaginary part of $E_N(g)$ one deduces at large order k:

$$E_{Nk} \sim K_N \frac{\Gamma\left(k + (N+1/2)(1+1/\omega)\right)}{a^{k+(N+1/2)(1+1/\omega)}} \left(1 + O\left(k^{-1}\right)\right). \tag{43.106}$$

Note that this expression, in contrast with the instanton contribution to the real part, is uniform in the limit $\omega = 1$ in which the result (43.49) is recovered.

43.4 The $O(\nu)$ Symmetric Anharmonic Oscillator

It is interesting to consider a last example, the analytic continuation of the energy levels of the $O(\nu)$ symmetric anharmonic oscillator for $g < 0$:

$$H = -\tfrac{1}{2}\nabla^2 + \tfrac{1}{2}\mathbf{q}^2 + g\left(\mathbf{q}^2\right)^2 . \tag{43.107}$$

We first discuss the $\nu = 2$ example, and the generalization will be straightforward.

The $O(2)$ anharmonic oscillator. The instanton solution is of the form

$$\mathbf{q}(t) = \mathbf{u}f(t) , \tag{43.108}$$

in which $\mathbf{u}$ is a fixed unit vector. The one-instanton contribution to the ground state energy is:

$$E^{(1)}(g) = \frac{4i}{g}\, e^{1/3g}\left(1 + O(g)\right) , \quad \text{for } g \to 0_- . \tag{43.109}$$

It is easy to calculate the instanton interaction. The n-instanton action then is:

$$A\left(\theta_i\right) = -\frac{1}{3}n - 4\sum_i e^{-\theta_i}\cos\varphi_i , \tag{43.110}$$

in which θ_i is the distance between two successive instantons and φ_i the angle between them:

$$\cos\varphi_i = \mathbf{u}_i \cdot \mathbf{u}_{i+1} . \tag{43.111}$$

It is convenient to consider the quantity (Sections 3.3, 23.5):

$$\operatorname{tr}\left[R(\alpha)\,e^{-\beta H}\right] = \int [d\mathbf{q}(t)]\exp\left(-S(q)\right), \text{ with } \hat{\mathbf{q}}\left(-\tfrac{1}{2}\beta\right)\cdot\hat{\mathbf{q}}\left(\tfrac{1}{2}\beta\right) = \cos\alpha . \tag{43.112}$$

The matrix $R(\alpha)$ is a rotation matrix which rotates vectors by an angle α. It leads to the boundary condition that $\mathbf{q}(t)$ at initial and final times should differ by an angle α.

The r.h.s. of equation (43.112) can be rewritten:

$$\operatorname{tr}\left[R(\alpha)\,e^{-\beta H}\right] = \sum_{l,N} e^{-il\alpha - \beta E_{l,N}} . \tag{43.113}$$

In this expression l is the angular momentum. The boundary condition imposed on the path integral (43.112) implies a constraint on the many-instanton configuration:

$$\sum_{i=1}^{n}\varphi_i = \alpha , \tag{43.114}$$

constraint which can be implemented through the identity:

$$\delta\left(\sum_{i=1}^{n}\varphi_i - \alpha\right) = \frac{1}{2\pi}\sum_{l=-\infty}^{+\infty}\exp\left[il\left(\sum_{i=1}^{n}\varphi_i\right) - i\alpha l\right] . \tag{43.115}$$

The n-instanton contribution to expression (43.113) then takes the form:

$$\left\{\operatorname{tr}\left[R\left(\alpha\right)e^{-\beta H}\right]\right\}_{(n)} \sim \frac{\lambda^n}{2i\pi n}\beta e^{-\beta}\int ds\, e^{-s\beta}\sum_{l=-\infty}^{+\infty} e^{-il\alpha}\left[I_l\left(s\right)\right]^n, \tag{43.116}$$

where we have set:

$$\lambda = \frac{4i}{g}e^{1/3g}, \qquad \mu = -\frac{4}{g}, \tag{43.117}$$

$$I_l\left(s\right) = \frac{1}{2\pi}\int_0^{2\pi}d\varphi\int_{-\infty}^{+\infty}d\theta\,\exp\left(s\theta + il\varphi - \mu e^{-\theta}\cos\varphi\right). \tag{43.118}$$

We call $\Sigma_l\left(\beta,g\right)$ the generating function of n-instanton contributions at fixed angular momentum l:

$$\Sigma_l\left(\beta,g\right) = \sum_n \frac{\lambda^n}{2i\pi n}\beta e^{-\beta}\int ds\, e^{-s\beta}\left[I_l\left(s\right)\right]^n. \tag{43.119}$$

Let us evaluate $I_l\left(s\right)$. The integration over θ yields:

$$I_l\left(s\right) = \mu^s\Gamma\left(-s\right)\int_0^{2\pi}\frac{d\varphi}{2\pi}e^{il\varphi}\left(\cos\varphi\right)^s. \tag{43.120}$$

Performing the last integral and using various relations among Γ functions, we finally obtain:

$$I_l\left(s\right) = \mu^s\, e^{\frac{1}{2}i\pi(s+1)}\frac{\Gamma\left(\frac{1}{2}\left(l-s\right)\right)2^{-s-1}}{\Gamma\left(1+\frac{1}{2}\left(l+s\right)\right)}. \tag{43.121}$$

It is easy to verify that:

$$I_l(s) = I_{-l}(s). \tag{43.122}$$

From equation (43.119) we now derive the result:

$$\operatorname{tr} e^{-\beta H_l} = \sum_{N=0}^{\infty} e^{-\beta E_{N,l}}, \tag{43.123}$$

with $E_{N,l} = s_{N,l} + 1$ the solution of the equation:

$$e^{1/(3g)}\left(-\frac{2}{g}\right)^E e^{i\pi(E+l)/2}\frac{\Gamma\left(\frac{1}{2}\left(l+1-E\right)\right)}{\Gamma\left(\frac{1}{2}\left(l+1+E\right)\right)} = 1, \tag{43.124}$$

which satisfies

$$E_{N,l} = l + 2N + 1 + O(g), \qquad N \geq 0. \tag{43.125}$$

Note that checks about these expressions are provided by the surprising perturbative relation between the $O(2)$ anharmonic oscillator with negative coupling and the double-well potential.

The $O(\nu)$ symmetric hamiltonian. One can extend this result to the general $O(\nu)$ case since, at fixed angular momentum l, the hamiltonian depends only on the combination $l + \frac{1}{2}\nu$. Hence, making in equation (43.124) the corresponding substitution, we obtain:

$$i\, e^{1/(3g)}\left(-\frac{2}{g}\right)^E e^{i\pi(E+l+\nu/2)/2}\frac{\Gamma\left(\frac{1}{2}\left(l+\nu/2-E\right)\right)}{\Gamma\left(\frac{1}{2}\left(l+\nu/2+E\right)\right)} = 1, \tag{43.126}$$

At leading order in λ we recover the imaginary part of the energy levels for g small and negative:

$$\operatorname{Im} E_{N,l} \underset{g \to 0_-}{=} -\frac{1}{N!}\frac{1}{\Gamma\left(\frac{1}{2}\nu + l + N\right)}\left(\frac{2}{g}\right)^{(\nu/2)+l+2N} e^{1/3g}\left(1 + O\left(g\right)\right). \qquad (43.127)$$

Using the Cauchy formula we can derive from this expression large order estimates for perturbation theory. At next order in λ we obtain the two-instanton contribution which is related by the same dispersion relation to the large order behaviour of the perturbative expansion around one instanton.

43.5 A General Conjecture

Up to now we have considered instanton contributions only at leading order. However the form of the result is extremely suggestive and has led us to conjecture the general form of the semiclassical expansion for potentials with degenerate minima.

To be specific let us explain the conjecture for the double-well potential, although it can be easily generalized to the other problems discussed in this chapter (see Appendix A43.5).

We introduce the function $D(E,g)$

$$D(E,g) = E + \sum_{k=1}^{\infty} g^k D_{k+1}(E), \qquad (43.128)$$

in terms of which the perturbation expansion for an energy level $E_N^{(0)}$ can be obtained by inverting:

$$N + \tfrac{1}{2} = D\left(E^{(0)}, g\right). \qquad (43.129)$$

To take into account instanton contributions we write instead another equation:

$$\frac{1}{\sqrt{2\pi}}\Gamma\left(\tfrac{1}{2} - D\right)\left(-\frac{2}{g}\right)^{D(E,g)} e^{-A(g,E)/2} = \varepsilon i, \qquad (43.130)$$

with

$$A(g,E) = \frac{1}{3g} + \sum_{k=1}^{\infty} g^k A_{k+1}(E). \qquad (43.131)$$

The functions $A(E,g)$, $D(E,g)$ can be calculated by expanding the corresponding WKB series, which is an expansion at gE fixed, for E small (for details see Appendix A43.5). The coefficients D_k and A_k are polynomials of degree k in E of defined parity.

If we solve (43.130) in the one-instanton approximation and substitute into equation (43.128) we find

$$E = E^{(0)}(g) - \varepsilon\left(\frac{2}{g}\right)^N \frac{1}{N!}\frac{e^{-A(g,E^{(0)})/2}}{\sqrt{\pi g}}\frac{\partial D}{\partial E}\left(E^{(0)}\right)^{-1}. \qquad (43.132)$$

We see that the knowledge of the two functions D and A is equivalent to the knowledge and the perturbative and one-instanton expansions for all levels and to all orders.

If we now systematically expand equation (43.130) we find for an energy level the following expansion,

$$E\left(g\right) = \sum_{0}^{\infty} E_l^{(0)} g^l + \sum_{n=1}^{\infty} \left(\frac{1}{\sqrt{\pi g}}\, e^{-1/6g}\right)^n \sum_{k=0}^{n-1} (\ln(-2/g))^k \sum_{l=0}^{\infty} \varepsilon_{nkl} g^l. \qquad (43.133)$$

All the series in powers of g appearing in this expansion are determined by the perturbative expansion of A and D. This phenomenon has recently found an explanation in the framework of the theory of *resurgent* functions.

Moreover we conjecture that all these series have to be summed for g negative first, and the value of each instanton contribution for g positive is then obtained by analytic continuation. The property that the infinite number of perturbation series around all instantons are related may, at least in quantum mechanics, simplify the problem of the summation of the many-instanton contributions.

Bibliographical Notes

The two-instanton contribution and its application to large order behaviour for potential with degenerate minima has been discussed in

E. Brézin, G. Parisi and J. Zinn-Justin, *Phys. Rev.* D16 (1977) 408; E.B. Bogomolny, *Phys. Lett.* 91B (1980) 431.

The multi-instanton contributions have been calculated in

J. Zinn-Justin, *Nucl. Phys.* B192 (1981) 125; B218 (1983) 333; *J. Math. Phys.* 22 (1981) 511; 25 (1984) 549.

In these articles a relation between the $O(2)$ anharmonic oscillator and the double-well potential was conjectured, which was later proven in

A.A. Andrianov, *Ann. Phys. (NY)* 140 (1982) 82.

Multi-instantons in the $O(3)$ non-linear σ-model have been considered in

V.A. Fateev, I.V. Frolov and A.S. Schwarz, *Nucl. Phys.* B154 (1979) 1, *Yad. Fiz.* 30 (1979) 1134; B. Berg and M. Lüscher, *Commun. Math. Phys.* 69 (1979) 57; A.P. Bukhvostov and L.N. Lipatov, *Nucl. Phys.* B180[FS2] (1981) 116.

The peculiar equations (43.128,43.130) which imply relations between the perturbative expansions around multi-instantons have been recently proven using the theory of *resurgent* functions in

F. Pham, *Resurgence, Quantized Canonical Transformation and Multi-Instanton*, Algebraic Analysis, vol. II (1988), *C.R. Acad. Sci. Paris* 309 (1989) 999; E. Delabaere et H. Dillinger, thesis Université de Nice (Nice 1991).

For the WKB method as discussed in the appendix see also

A. Voros, *Annales IHP* 29 (1983) 3.

Exercises

Exercise 43.1

Complete the calculation of Appendix 40.6 by evaluating explicitly the ground state energy at leading order for the hamiltonian H:

$$H = \tfrac{1}{2} \left[-\Omega (\mathrm{d}/\mathrm{d}q)^2 + \tfrac{1}{4} \Omega^{-1} \left(E'(q)\right)^2 - \tfrac{1}{2} E''(q) \right],$$

in which Ω is the expansion parameter and $E(q) = q^2 - 2q^3/3$.

APPENDIX 43

A43.1 Multi-instantons: The Determinant

We can write the operator M defined by equation (43.20) as:

$$M = -\left(\frac{\mathrm{d}}{\mathrm{d}t}\right)^2 + 1 + \sum_{i=1}^{n} v\left(t - t_i\right), \qquad (A43.1)$$

in which $v(t)$ is a potential localized around $t = 0$, and t_i are the positions of the instantons:

$$v(t) = O\left(\mathrm{e}^{-|t|}\right), \quad |t| \to \infty. \qquad (A43.2)$$

We want to calculate

$$\det M\, M_0^{-1} = \det\left\{1 + \left[-\left(\mathrm{d}/\mathrm{d}t\right)^2 + 1\right]^{-1} \sum_{i=1}^{n} v\left(t - t_i\right)\right\}. \qquad (A43.3)$$

Using the identity $\ln\det = \mathrm{tr}\ln$, we expand the r.h.s. in powers of $v(t)$:

$$\ln\det M M_0^{-1} = \sum_{k=1}^{\infty} \frac{(-1)^{k+1}}{k} \int \left[\Delta\left(u_1 - u_2\right) \sum_{i_1=1}^{n} v\left(u_2 - t_{i_1}\right) \Delta\left(u_2 - u_3\right) \cdots \Delta\left(u_k - u_1\right)\right.$$
$$\left. \times \sum_{i_k=1}^{n} v\left(u_1 - t_{i_k}\right)\right] \prod_{j=1}^{k} \mathrm{d}u_j, \qquad (A43.4)$$

with the definition:

$$\Delta(t) = \left\langle 0 \left| \left[-\left(\mathrm{d}/\mathrm{d}t\right)^2 + 1\right]^{-1} \right| t \right\rangle \sim \frac{1}{2}\,\mathrm{e}^{-|t|} \quad \text{for } 1 \ll t \ll \beta. \qquad (A43.5)$$

It is clear from the behaviour of $v(t)$ and $\Delta(t)$, that when the instantons are largely separated, only the terms in which one retains from each potential the same instanton contribution survive. Therefore

$$\ln\det M M_0^{-1} = n \sum_{k=1}^{\infty} \frac{(-1)^{k+1}}{k} \int \Delta\left(u_1 - u_2\right) v\left(u_2\right) \cdots \Delta\left(u_k - u_1\right) v\left(u_1\right) \prod_{j=1}^{k} \mathrm{d}u_j,$$
$$\text{for } |t_i - t_j| \gg 1. \qquad (A43.6)$$

We recognize n times the logarithm of the one-instanton determinant.

A43.2 The Instanton Interaction

We assume, as in Section 43.3, that the potential has two degenerate minima at the points $x = 0$ and $x = x_0$ with:

$$\begin{cases} V(x) = \frac{1}{2}x^2 + O(x^3) \\ V(x) = \frac{1}{2}\omega^2 (x - x_0)^2 + O\left((x - x_0)^3\right). \end{cases} \qquad (A43.7)$$

Let us write the one-instanton solution $q_c(t)$ which goes from 0 to $q_0 = x_0/\sqrt{g}$ as

$$q_c(t) = f(t)/\sqrt{g}. \qquad (A43.8)$$

We choose the function $f(t)$ in such a way that it satisfies:

$$\begin{cases} x_0 - f(t) \sim \sqrt{C}\, e^{-\omega t}/\omega & \text{for } t \to +\infty \\ \quad f(t) \sim \sqrt{C}\, e^{t} & \text{for } t \to -\infty. \end{cases} \qquad (A43.9)$$

By solving the equation of motion, it is easy to calculate the constant C:

$$C = x_0^2 \omega^{2/(1+\omega)} \exp\left\{ \frac{2\omega}{1+\omega} \left[\int_0^{x_0} dx \left(\frac{1}{\sqrt{2V(x)}} - \frac{1}{x} - \frac{1}{\omega(x_0 - x)} \right) \right] \right\}. \qquad (A43.10)$$

We recognize the constant (43.84).

We now construct instanton–anti-instanton pair configurations $q(t)$ which correspond to trajectories starting from, and returning to, $q = q_0$ or $q = 0$. Since we want also to consider the case of two successive instantons, we assume, but only in this last case, that $V(x)$ is an even function and has therefore a third minimum at $x = -x_0$.

According to the discussion of Chapter 43, we can take as a two-instanton configuration:

$$q_1(t) = \frac{1}{\sqrt{g}} \left(f_+(t) + \varepsilon f_-(t) \right), \qquad \varepsilon = \pm 1, \qquad (A43.11)$$

with:

$$f_+(t) = f(t - \theta/2), \qquad f_-(t) = f(-t - \theta/2). \qquad (A43.12)$$

θ is a measure of the instanton separation. The case $\varepsilon = 1$ corresponds to an instanton–anti-instanton pair starting from $q = q_0$ at time $-\infty$, approaching $q = 0$ at intermediate times, and returning to q_0. The case $\varepsilon = -1$ corresponds to a sequence of two instantons going from $-q_0$ to q_0. Finally for the classical trajectory which goes instead from the origin to q_0 and back we can take:

$$q_2(t) = \left[f(t + \theta/2) + f(\theta/2 - t) - x_0 \right]/\sqrt{g}. \qquad (A43.13)$$

Let us now calculate the classical action corresponding to $q_1(t)$. We separate the action into two parts, corresponding at leading order to the two instanton contributions:

$$S(q_1) = S_+(q_1) + S_-(q_1), \qquad (A43.14)$$

with:

$$S_+ (q_1) = \int_0^{+\infty} \left[\frac{1}{2}\dot{q}_1^2 + \frac{1}{g}V\left(\sqrt{g}q_1(t)\right) \right] dt,$$

$$S_- (q_1) = \int_{-\infty}^0 \left[\frac{1}{2}\dot{q}_1^2 + \frac{1}{g}V\left(\sqrt{g}q_1(t)\right) \right] dt. \tag{A43.15}$$

The value $t = 0$ of the separation point is somewhat arbitrary and can be replaced by any value which remains finite when θ becomes infinite. We then use the properties that for θ large $f_+(t)$ is small for $t < 0$, and $f_-(t)$ is small for $t > 0$, to expand both terms. For example for S_+ we find:

$$S_+ (q_1) = \frac{1}{g}\int_0^{+\infty} dt \left\{ \left[\tfrac{1}{2}\dot{f}_+^2(t) + V\left(f_+(t)\right) \right] + \varepsilon \left[\dot{f}_-(t)\dot{f}_+(t) + V'\left(f_+(t)\right)f_-(t) \right] \right.$$

$$\left. + \tfrac{1}{2}\left[\dot{f}_-^2(t) + V''\left(f_+(t)\right)f_-^2(t) \right] \right\}. \tag{A43.16}$$

Since $f_-(t)$ decreases exponentially, only values of t small compared to $\theta/2$ contribute to the last term of equation (A43.16) which is proportional to V''. For such values of t, we have:

$$\tfrac{1}{2}V''\left(f_+(t)\right)f_-^2 \sim V\left(f_-(t)\right). \tag{A43.17}$$

For the terms linear in $f_-(t)$, we integrate by parts the kinetic term and use the equation of motion

$$\ddot{f}(t) = V'\left[f(t)\right]. \tag{A43.18}$$

Only the integrated term survives and yields:

$$\int_0^{+\infty} dt \left[\dot{f}_-(t)\dot{f}_+(t) + V'\left(f_+(t)\right)f_-(t) \right] = -\dot{f}\left(-\theta/2\right)f\left(-\theta/2\right). \tag{A43.19}$$

The contribution S_- can be evaluated by exactly the same method. We note that the sum of the two contributions reconstructs twice the classical action a. We then find:

$$S (q_1) = \frac{1}{g}\left[2a - 2\varepsilon f\left(-\theta/2\right)\dot{f}\left(-\theta/2\right) + \cdots \right], \tag{A43.20}$$

with

$$a = \int_0^{x_0} \sqrt{2V(x)}\,dx. \tag{A43.21}$$

Replacing, for θ large, f by its asymptotic form (A43.9), we finally obtain the classical action:

$$S (q_1) = g^{-1}\left[2a - 2C\varepsilon\,e^{-\theta} + O\left(e^{-2\theta}\right) \right], \tag{A43.22}$$

and therefore the instanton interaction.

Following the same steps, we can calculate the classical action corresponding to $q_2(t)$. The result is:

$$S (q_2) = \frac{1}{g}\left\{ 2a - 2\left[f\left(\theta/2\right) - x_0\right]\dot{f}\left(\theta/2\right) + \cdots \right\}, \tag{A43.23}$$

which for θ large is equivalent to:

$$S (q_2) = \frac{1}{g}\left[2a - 2(C/\omega)\,e^{-\omega\theta} \right]. \tag{A43.24}$$

Finally, if we consider the case of a finite time interval β with periodic boundary conditions, we can combine both results to find the action of a periodic trajectory passing close to $q = 0$ and $q = q_0$:

$$S (q) = g^{-1}\left[2a - 2C\left(e^{-\beta+\theta} + e^{-\omega\theta}/\omega\right) \right], \tag{A43.25}$$

in agreement with equations (43.86,43.89).

A43.3 Multi-instantons from Constraints

Although multi-instanton configurations do not correspond to solutions of the equation of motion, it is nevertheless possible to modify the classical action by introducing constraints and integrating over all possible constraints, generalizing the method of Appendix A37.2. The main problem with such a method is to find a system of constraints which are both theoretically reasonable, and convenient for practical calculations.

One can for instance fix the positions of the instantons by introducing in the path integral (in the example of the double-well):

$$1 = \int \prod_{i=1}^{n} \left[\int \mathrm{dt}\, \dot{q}_{\varepsilon_i}^2 (t - t_i) \right] \delta \left[\int \mathrm{dt}\, \dot{q}_{\varepsilon_i} (t - t_i) \big(q(t) - q_{\varepsilon_i}(t - t_i) \big) \right] \mathrm{dt}_i , \qquad (A43.26)$$

where t_i are the instanton positions and ε_i a successions of $\pm$ indicating instantons and anti-instantons. One then uses an integral representation of the δ-functions, so that the path integral becomes

$$\left(\frac{\|\dot{q}_+\|^2}{2i\pi} \right)^n \int \prod_{i=1}^{n} \mathrm{dt}_i \mathrm{d}\lambda_i \int [\mathrm{d}q(t)] \prod_{i=1}^{n} \exp\left[-S\left(q, \lambda_i \right) \right] \quad \text{with}$$

$$S\left(q, \lambda_i \right) = S(q) + \sum_{i=1}^{n} \lambda_i \int \mathrm{dt}\, \dot{q}_{\varepsilon_i} (t - t_i) \big(q(t) - q_{\varepsilon_i}(t - t_i) \big). \qquad (A43.27)$$

The arguments of Appendix A37.2 can then be generalized to recover the results of Section 43.1.

A43.4 A Simple Example of Non-Borel Summability

Let us try to illustrate the problem of non-Borel summability with the example of a simple integral, which shares some of the features of the problem in quantum mechanics which we have studied in Chapter 43. We consider the function:

$$I(g) = \frac{1}{\sqrt{2\pi}} \int_{-\infty}^{+\infty} \mathrm{d}q \, \exp\left[-\frac{1}{g} V\left(q\sqrt{g} \right) \right], \qquad (A43.28)$$

where $V(x)$ is an entire function with an absolute minimum at $x = 0$, $V(0) = 0$. For g small $I(g)$ can be calculated by steepest descent, expanding V around $q = 0$:

$$I(g) = \sum_{k \geq 0} I_k g^k. \qquad (A43.29)$$

It is easy to write a finite dimensional integral of the form $(A43.28)$ as a generalized Borel or Laplace transform,

$$I(g) = \frac{1}{\sqrt{2\pi}} \int \mathrm{d}q \, \mathrm{dt} \, \delta \left[V\left(q\sqrt{g} \right) - t \right] e^{-t/g} . \qquad (A43.30)$$

We integrate over q,

$$I(g) = \frac{1}{\sqrt{2\pi g}} \int_0^{\infty} \mathrm{dt} \, e^{-t/g} \sum_i \frac{1}{|V'\left[x_i(t) \right]|}, \qquad (A43.31)$$

in which $\{x_i(t)\}$ are the solutions of the equation

$$V[x_i(t)] = t. \qquad (A43.32)$$

When the function $V(x)$ is monotonous both for x positive and negative, the equation (A43.32) has two solutions for all values of t and the equation (A43.31) is directly the Borel representation of the function $I(g)$, which has a Borel summable power series expansion.

We now assume instead that $V(x)$ has a second local minimum which gives a negligible contribution to $I(g)$ for g small. A simple example is

$$V(x) = \tfrac{1}{2}x^2 - \tfrac{1}{3a}x^3(1+a) + \tfrac{1}{4a}x^4, \quad \tfrac{1}{2} < a < 1, \qquad (A43.33)$$

which has a minimum at $x = 1$. Between its two minima the potential $V(x)$ has a maximum, located at $x = a$, whose contribution dominates the large order behaviour of the expansion in powers of g

$$I_k \underset{k\to\infty}{\propto} \Gamma(k)\,[V(a)]^{-k}, \qquad V(a) > 0, \qquad (A43.34)$$

(in the example (A43.33) $V(a) = a^2(1-a/2)/6$) and the series is not Borel summable.

The *naive* Borel transform of $I(g)$ is obtained by retaining in equation (A43.31) only the roots of equation (A43.32) which exists for t small. The singularities of the Borel transform then correspond to the zeros of $V''(x)$.

For the potential (A43.33) the expression (A43.31) has the form

$$I(g) = \frac{1}{\sqrt{2\pi g}} \int_0^{+\infty} dt\; e^{-t/g} \left[\frac{1}{|V'(x_1(t))|} + \frac{\theta(V(a)-t)}{|V'(x_2(t))|} + \frac{\theta(V(a)-t)\theta(t-V(1))}{|V'(x_3(t))|} \right.$$
$$\left. + \frac{\theta(t-V(1))}{|V'(x_4(t))|} \right], \qquad (A43.35)$$

with the definitions (see figure 43.2): $x_1(t) \le 0 \le x_2(t) \le a \le x_3(t) \le 1 \le x_4(t)$.

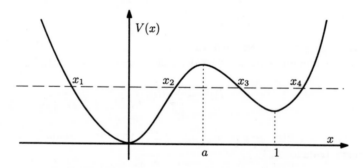

Fig. 43.2   The four roots of equation (A43.32).

The idea of the analytic continuation is to integrate each contribution up to $t = +\infty$ following a contour which passes below or above the cut along the positive real axis. This means that we consider $x_2(t)$ to be solution of the equation:

$$V[x_2(t)] = t \pm i\varepsilon. \qquad (A43.36)$$

The sign is arbitrary. Let us for instance choose the positive sign. We then have to subtract this additional contribution. We proceed in the same way for $x_3(t)$ for $t > V(a)$. Since $x_2(t)$ and $x_3(t)$ meet at $t = V(a)$, the analytic continuation will correspond to take for $x_3(t)$ the other solution:

$$V[x_3(t)] = t \mp i\varepsilon. \qquad (A43.37)$$

We therefore have to subtract from the total expression the contributions of two roots of the equation. But it is easy to verify that this is just the contribution of the saddle point located at $x = a$, which corresponds to a maximum of the potential.

We have therefore succeeded in writing expression $(A43.35)$ as the sum of three saddle point contributions (see figure 43.3). There is some arbitrariness in this decomposition which here corresponds to the choice $\varepsilon = \pm 1$.

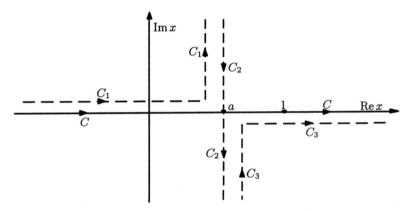

Fig. 43.3   The different contours in the x plane.

In the complex x plane, we have replaced the initial contour C on the real positive axis, by a sum of three contours C_1, C_2 and C_3 corresponding to the three saddle points located at $0, a, 1$.

A43.5 Multi-Instantons and WKB Approximation

Let us give a more general form of the conjecture presented in Section 43.5 and indicate how a few properties of the functions $A(E, g)$ and $D(E, g)$ can be obtained from the WKB expansion.

A43.5.1 The conjecture

The double-well potential. For the double-well potential we have conjectured the equation

$$\frac{1}{\sqrt{2\pi}} \Gamma\left(\tfrac{1}{2} - D\right) \left(-\frac{2}{g}\right)^{D(E,g)} e^{-A(g,E)/2} = \varepsilon i, \qquad (A43.38)$$

with

$$D(E, g) = -D(-E, -g) = E + \sum_{k=1}^{\infty} g^k D_{k+1}(E), \qquad (A43.39)$$

$$A(g, E) = -A(-E, -g) = \frac{1}{3g} + \sum_{k=1}^{\infty} g^k A_{k+1}(E). \qquad (A43.40)$$

The coefficients $D_k(E)$ and $A_k(E)$ are polynomials of E of degree k. This conjecture has been generalized to a potential with two asymmetric wells:

$$\frac{1}{2\pi}\Gamma(\tfrac{1}{2} - D_1(E,g))\left(-\frac{C}{g}\right)^{D_1(E,g)}\Gamma(\tfrac{1}{2} - D_2(E,g))\left(-\frac{C}{g}\right)^{D_2(E,g)} e^{-A(g,E)} = -1\,,$$

$$(A43.41)$$

where $D_1(E,g)$ and $D_2(E,g)$ are determined by the perturbative expansions around each of the two minima of the potential and C is a numerical constant.

The cosine potential. For the cosine potential $\frac{1}{16}(1 - \cos 4q)$ the conjecture is

$$\left(\frac{2}{g}\right)^{-D}\frac{e^{A(E,g)/2}}{\Gamma(\tfrac{1}{2} - D)} + \left(\frac{-2}{g}\right)^{D}\frac{e^{-A(g,E)/2}}{\Gamma(\tfrac{1}{2} + D)} = \frac{2\cos\varphi}{\sqrt{2\pi}}\,. \qquad (A43.42)$$

The $O(\nu)$-symmetric anharmonic oscillator. For the $O(\nu)$ anharmonic oscillator the conjecture corresponds to the expansion of the energy levels for $g < 0$ and, in particular, yields the instanton contributions to the large order behaviour.

$$i\,e^{-A(E,g)}\left(-\frac{2}{g}\right)^{D} e^{i\pi(D+l+\nu/2)/2}\frac{\Gamma\left(\tfrac{1}{2}\left(l + \nu/2 - D\right)\right)}{\Gamma\left(\tfrac{1}{2}\left(l + \nu/2 + D\right)\right)} = 1\,. \qquad (A43.43)$$

A43.5.2 The WKB approximation

Consider the Schrödinger equation:

$$-\tfrac{1}{2}\psi'' + \left(V(\sqrt{g}q)/g\right)\psi = E\psi\,. \qquad (A43.44)$$

We first assume that $q = 0$ is the absolute minimum of $V(q)$ and moreover $V(q) \sim \tfrac{1}{2}q^2$ for q small. Setting $x = \sqrt{g}q$ the equation can be rewritten

$$-g^2\psi'' + 2V(x)\psi = 2gE\psi\,. \qquad (A43.45)$$

The WKB expansion is an expansion for $g \to 0$ at Eg fixed, in contrast with the perturbative expansion where E is fixed. It can be constructed by introducing the corresponding Riccati equation, setting:

$$S(x) = -g\psi'/\psi\,, \qquad (A43.46)$$

where S satisfies:

$$gS'(x) - S^2(x) + S_0^2(x) = 0\,, \qquad S_0^2(x) = 2V(x) - 2gE\,. \qquad (A43.47)$$

One then expands systematically in powers of g, at Eg fixed, starting from $S(x) = S_0(x)$. It is convenient to decompose $S(x)$ into an odd and even part setting:

$$S(x,g,E) = S_+(x,g,E) + S_-(x,g,E)\,, \qquad S_\pm(x,-g,-E) = \pm S_\pm(x,g,E)\,. \qquad (A43.48)$$

It follows:

$$gS'_- - S_+^2 - S_-^2 + S_0^2 = 0\,, \qquad (A43.49)$$

$$gS'_+ - 2S_+S_- = 0\,, \qquad (A43.50)$$

which allows to write the wave function in terms of S_+ only:

$$\psi = (S_+)^{-1/2} \exp\left[-\frac{1}{g}\int_{x_0}^{x} \mathrm{d}x'\, S_+(x')\right]. \qquad (A43.51)$$

The spectrum can then be determined by the condition

$$\frac{1}{2i\pi}\oint_C \mathrm{d}z\frac{\psi'(z)}{\psi(z)} = N, \qquad (A43.52)$$

where N is the number of nodes of the eigenfunction and C a contour which encloses them. In the semiclassical limit C encloses the cut of $S_0(x)$ which joins the two turning points solutions of $S_0(x) = 0$ (x_1, x_2 in figure 43.2). In terms of S_+ equation ($A43.52$) becomes:

$$-\frac{1}{2i\pi g}\oint_C \mathrm{d}z\, S_+(z) = N + \tfrac{1}{2}. \qquad (A43.53)$$

If we replace S_+ by its WKB expansion and expand each term in a power series of Eg we get the function $D(E,g)$ (the perturbative expansion):

$$-\frac{1}{2i\pi g}\oint_C \mathrm{d}z\, S_+(z) = D(E,g) = -D(-E,-g). \qquad (A43.54)$$

Potentials with degenerate minima. In the case of potentials with degenerate minima two functions D_1 and D_2 (in the notation ($A43.41$)) appear, corresponding to the expansions around each minimum. An additional contour integral arises corresponding to barrier penetration effects. The expansion for Eg small of its WKB expansion yields the function $A(g,E)$:

$$\frac{1}{g}\oint_{C'} \mathrm{d}z\, 2S_+(z) = A(E,g) - \sum_{i=1}^{2}\ln\Gamma\left(\tfrac{1}{2} - D_i(E,g)\right) + D_i(E,g)\ln(-g/C), \qquad (A43.55)$$

where C' encloses $[x_2, x_2]$ in figure 43.2. In the WKB expansion the functions $\Gamma(\tfrac{1}{2} - D_i)$ appear in the form of an asymptotic expansion for D_i large. Still a calculation of $A(E,g)$ at a finite order in g requires the WKB expansion and the asymptotic expansion of the Γ function only at a finite order.

$O(\nu)$ symmetric potentials. These expressions can be generalized to the case of $O(\nu)$ symmetric potentials. The perturbative expansion can be obtained by inverting a relation of the form

$$\mu + 2N + 1 = D(E,g,\mu), \qquad \mu = l + \nu/2 - 1, \qquad (A43.56)$$

where the function $D(E,g,\mu)$ is given by a contour integral surrounding all zeros of the wave function on the real axis (including the negative real axis) of the even part (in the sense of equation ($A43.48$)) of $-g(\psi_l'/\psi_l + (\nu - 1)/2|q|)$. The following properties can then verified

$$D(E,g,\mu) = -D(-E,-g,\mu), \qquad D(E,g,\mu) = D(E,g,-\mu),$$

and the coefficient of order g^k in the expansion of D is a polynomial of degree $[(k+1)/2]$ in μ. In the WKB expansion the functions D and A again correspond to different contour integrals around turning points.

Chapter 1: Algebraic Preliminaries

Exercise 1.1

One considers the fermion integral

$$\int \left(\prod_i d\theta_i d\bar\theta_i \right) \exp \left(\sum_{i,j=1}^{n} a_{ij}\bar\theta_i\theta_j + \sum_i s_i\bar\theta_i\theta_i \right).$$

Integration directly yields the l.h.s. of equation (1.99). Expanding in perturbation theory the term proportional to the s_i's and using fermion's Wick theorem (1.96) yields the r.h.s..

Exercise 1.2

First

$$D(ABC) = DA\,BC + P(A)D(BC)$$
$$= DA\,BC + P(A)DB\,C + P(A)P(B)DC.$$

Also

$$D(ABC) = D(AB)C + P(AB)DC$$
$$= DA\,BC + P(A)DB\,C + P(AB)DC.$$

Comparing both results one finds

$$P(A)P(B) = P(AB).$$

Exercise 1.3

It is of course sufficient to characterize the action of $\mathcal{D}$ on x^p. We can set $Dx = 1$. We have to use the basic relation

$$D(AB) = DAB + H(A)DB,$$

where H being a homomorphism of algebra, consistent with the grading is characterized by a complex constant h

$$H(x) = hx,$$

while

$$D(x^p) = c_p x^{p-1}.$$

It is now easy to verify that the solution is

$$D(x^p) = \frac{1 - h^p}{1 - h} x^{p-1}.$$

Then the consistency condition $D(x^n) = 0$ implies $h^n = 1$ with $h \neq 1$.

Note that $D^n = 0$ and also in the sense of operators on $\mathfrak{A}$

$$\frac{1 - h^{p+1}}{1 - h^p} xD + Dx = 0.$$

Finally an linear integration operator on $\mathfrak{A}$ satisfying

$$DI = ID = 0,$$

is

$$I = D^{n-1}.$$

Exercise 1.4

Let us express the condition for any pair of generators. It is then sufficient to consider the sum of any two of them like x_1 and x_2. The result relies on the proof by induction of the identity:

$$(x_1 + x_2)^p = \sum_{m=0}^{p} x_1^m x_2^{p-m} \prod_{l=0}^{m-1} \frac{1 - c^{p-l}}{1 - c^{l+1}}.$$

Specializing this result for $p = n$ we obtain the necessary condition. We then note that any linear combination of x_1 and x_2 also satisfies

$$x_3(a_1 x_1 + a_2 x_2) = c(a_1 x_1 + a_2 x_2)x_3,$$

with the condition (1.100). We can then prove the result for any triplet of generators. The final result follows from this second induction.

Chapter 2: Euclidean Path Integrals in Quantum Mechanics

Exercise 2.1

2.1.1. One finds:

$$Z = (2\sinh{(\beta/2)})^{-1}\left(1 - 3\lambda \int_{-\beta/2}^{\beta/2} dt\, K^2\,(t, t) + O\left(\lambda^2\right)\right),$$

in which:

$$K(t, t) = \tfrac{1}{2}\text{cotanh}\,(\beta/2)\,.$$

Therefore:

$$Z = [2\sinh{(\beta/2)}]^{-1}\left[1 - \frac{3}{4}\lambda\beta\left(\text{cotanh}^2\,(\beta/2)\right) + O\left(\lambda^2\right)\right].$$

Using two algebraic identities:

$$\sum_{n=0}^{\infty} \exp\left[-\left(n + \tfrac{1}{2}\right)\beta\right] = 1/\left[2\sinh{(\beta/2)}\right],$$

and, differentiating twice with respect to β,

$$\sum_{n=0}^{\infty} \left(n + \tfrac{1}{2}\right)^2 \exp\left[-\left(n + \tfrac{1}{2}\right)\beta\right] = \frac{2\,\text{cotanh}^2\,(\beta/2) - 1}{8\sinh{(\beta/2)}},$$

one can verify that Z is the expansion to first order in λ of $\Sigma\exp\left(-\beta E_n\right)$ with:

$$E_n = n + \tfrac{1}{2} + \tfrac{3}{2}\lambda\left(n^2 + n + \tfrac{1}{2}\right) + O\left(\lambda^2\right).$$

2.1.2. Setting $\nu = n + \tfrac{1}{2}$, the result can be written

$$E = \nu + \tfrac{3}{2}\left(\nu^2 + \tfrac{1}{4}\right)\lambda - \tfrac{1}{4}\nu\left(17\nu^2 + \tfrac{67}{4}\right)\lambda^2 + O\left(\lambda^3\right).$$

Note that more generally for the potential

$$V(q) = \tfrac{1}{2}q^2 + v_1 q^4 + v_2 q^6 + O\left(q^8\right),$$

one finds

$$n + \tfrac{1}{2} = E - v_1\left(\tfrac{3}{2}E^2 + \tfrac{3}{8}\right) + E^3\left(\tfrac{35}{4}v_1^2 - \tfrac{5}{2}v_2\right) + E\left(\tfrac{85}{16}v_1^2 - \tfrac{25}{8}v_2\right).$$

2.1.3. For $N = 6$ and $N = 8$ the results are respectively

$$E_n = n + \tfrac{1}{2} + \tfrac{5}{2}\lambda(n + \tfrac{1}{2})\left(n^2 + n + \tfrac{3}{2}\right)$$

$$E_n = n + \tfrac{1}{2} + \tfrac{35}{8}\lambda\left(n^4 + 2n^3 + 5n^2 + 4n + \tfrac{3}{2}\right).$$

Exercise 2.2

Let $K(q, q')$ be the kernel

$$K(q, q') = \exp\left[-\tfrac{1}{2}a\left(q^2 + q'^2\right) + bqq'\right].$$ (44.1)

It is convenient to parametrize the coefficients a, b as:

$$a = \rho\frac{\cosh\theta}{\sinh\theta}, \quad b = \frac{\rho}{\sinh\theta} \quad \Rightarrow \rho = \sqrt{a^2 - b^2}, \quad \theta = \ln\left(\frac{a}{b} + \sqrt{\frac{a^2}{b^2} - 1}\right).$$

One then easily verifies that $K^p(q, q')$ has the form (44.1) with θ replaced by $p\theta$ (up to a global normalization).

Starting with the approximation,

$$a = \frac{1}{\varepsilon} + \lambda\varepsilon, \quad b = \frac{1}{\varepsilon} + \varepsilon(\lambda - \tfrac{1}{2}),$$

where λ is a free parameter and $\varepsilon = \beta/n$ one finds

$$\rho = 1 + \tfrac{1}{2}(\lambda - \tfrac{1}{4})\varepsilon^2 + O\left(\varepsilon^4\right)$$
$$\theta = \varepsilon\left[1 - \tfrac{1}{2}(\lambda - \tfrac{5}{12})\varepsilon^2\right] + O\left(\varepsilon^5\right).$$

Comparing $K^n(q, q')$ with the exact result (2.38) one sees that the corrections come from the deviations of ρ from 1 and $n\theta$ from β, and are of order $1/n^2$. No value of λ cancels both corrections. A good compromise is $\lambda = 1/3$ which minimizes both corrections. We note that this value corresponds to the result (2.13) obtained from the small ε expansion.

Exercise 2.3

We first briefly explain how the S-matrix can be calculated exactly. We consider the corresponding Schrödinger equation:

$$-\tfrac{1}{2}\psi'' + \frac{\lambda}{\cosh^2 x}\psi = \tfrac{1}{2}k^2\psi.$$ (44.2)

We first set

$$\psi(x) = \frac{1}{(2\cosh x)^\alpha}\varphi, \quad \alpha = -ik, \quad k > 0.$$

We find:

$$-\varphi'' + 2\alpha\tanh x\varphi' + \frac{2\lambda + \alpha(\alpha + 1)}{\cosh^2 x}\varphi = 0.$$

For later purpose it is useful to note

$$\frac{1}{(e^x + e^{-x})^\alpha} \sim \begin{cases} x \to -\infty & : e^{\alpha x} \\ x \to +\infty & : e^{-\alpha x} \end{cases}$$

We then set $t = \tanh x$ and find

$$(1 - t^2)\varphi'' - 2(1 + \alpha)\varphi' - \left(2\lambda + \alpha(\alpha + 1)\right)\varphi = 0.$$

The solutions of this equation are hypergeometric functions, which in special cases degenerate into Jacobi's polynomials. The solution regular at $t = -1$ for example is

$$\varphi(t) = F\left(a, b; \tfrac{1}{2}(a + b + 1); \tfrac{1}{2}(1 + t)\right),$$

with a, b given by:

$$a = \alpha + \tfrac{1}{2} - \tfrac{1}{2}\sqrt{1 - 8\lambda}, \qquad b = \alpha + \tfrac{1}{2} + \tfrac{1}{2}\sqrt{1 - 8\lambda}.$$

For $\operatorname{Re} \tfrac{1}{2}(1 + a + b) > \operatorname{Re} b > 0$, which implies $\lambda > 0$, we have Euler's formula

$$\varphi(t) = 2^b \int_1^\infty \left(s^2 - 1\right)^{(a-b-1)/2} (s - t)^{-a} ds.$$

We find

$$\varphi(-1) = \frac{\Gamma(\alpha + 1 - d)\Gamma(d)}{\Gamma(1 + \alpha)}, \qquad \text{with } d = \tfrac{1}{2}(a - b + 1) = \tfrac{1}{2}\left(1 - \sqrt{1 - 8\lambda}\right).$$

Using $a = d + \alpha$, we can write the behaviour of $\varphi(t)$ near $t = 1$ as

$$\varphi(t) \underset{t \to 1}{\sim} \frac{\Gamma(\alpha - d + 1)\Gamma(-\alpha)}{\Gamma(1 - d)} + 2^\alpha \frac{\Gamma(d)\Gamma(\alpha)}{\Gamma(d + \alpha)}(1 - t)^{-\alpha}.$$

The singular contribution corresponds to the incoming wave. Note that

$$(1 - t)^{-\alpha} \sim 2^{-\alpha} e^{2\alpha x}.$$

We conclude that the forward and backward S-matrix elements, S_+ and S_- are:

$$S_+ = \frac{\Gamma(d + \alpha)\Gamma(\alpha + 1 - d)}{\Gamma(1 + \alpha)\Gamma(\alpha)}, \qquad S_- = \frac{\sin \pi d}{\sin \pi \alpha} \frac{\Gamma(d + \alpha)\Gamma(\alpha + 1 - d)}{\Gamma(1 + \alpha)\Gamma(\alpha)}.$$

These expressions, established for $\lambda > 0$ have an analytic continuation for $\lambda < 0$ and give the complete answer.

We can check the unitarity relation for d real and α imaginary

$$|S_+|^2 + |S_-|^2 = -\frac{\sin^2 \pi \alpha}{\sin \pi(d - \alpha)\sin \pi(d + \alpha)}\left(1 - \frac{\sin^2 \pi d}{\sin^2 \pi \alpha}\right) = 1.$$

For $\lambda > 1/8$ d is complex but the relation $d^* + d = 1$ leads to the same identity.

Note that the potential is reflectionless for integer values of d, i.e. for $\lambda = -\tfrac{1}{2}n(n - 1)$, situation in which the potential is attractive and the S-matrix a rational fraction of $\alpha = ik$

$$S(\alpha) = (-1)^{n-1} \prod_{m=1}^{n-1} \frac{m + \alpha}{m - \alpha}.$$

The S-matrix has $n - 1$ poles corresponding to $n - 1$ bound states.

Note finally that in the normalization of exercise 2.1 we have the relation

$$n + \tfrac{1}{2} = E(1 + g^2/4)^{-1/2} + \tfrac{1}{2}g\left(n^2 + n + \tfrac{1}{2}\right), \qquad \text{with } g^2 = -1/2\lambda,$$

for the potential $\frac{1}{2}\tanh^2(q)$. This provides a check of the corresponding perturbative results.

Semiclassical limit. The semiclassical limit is obtained by letting k and λ becoming simultaneously large $k \propto \sqrt{\lambda}$. Then $d \sim -\sqrt{-2\lambda}$ and the Γ functions are replaced by their Stirling's approximation. Let us consider the different cases.

(i) $\lambda > 0$, $\sigma = \sqrt{2\lambda}/k < 1$ with the choice $d = -i\sqrt{2\lambda}$. We find:

$$\ln S_+ \sim -ik\left[(1+\sigma)\ln(1+\sigma) + (1-\sigma)\ln(1-\sigma)\right],$$
$$\ln S_- \sim -ik\left[(1+\sigma)\ln(1+\sigma) + (1-\sigma)\ln(1-\sigma)\right] - \pi k(1-\sigma).$$

We note that, as expected, the classically forbidden backward scattering is exponentially suppressed.

(ii) $\lambda > 0$, $\sigma = \sqrt{2\lambda}/k > 1$ with the choice $d = -i\sqrt{2\lambda}$. We find:

$$\ln S_+ \sim -ik\left[(\sigma+1)\ln(\sigma+1) - (\sigma-1)\ln(\sigma-1)\right] - \pi k(\sigma-1),$$
$$\ln S_- \sim -ik\left[(\sigma+1)\ln(\sigma+1) - (\sigma-1)\ln(\sigma-1)\right].$$

We now note that the classically forbidden forward scattering is exponentially suppressed.

(iii) $\lambda < 0$, $\sigma = \sqrt{-2\lambda}/k$. We find:

$$\ln S_+ \sim -ik\left[(1+i\sigma)\ln(1+i\sigma) + (1-i\sigma)\ln(1-i\sigma)\right],$$
$$\ln S_- \sim -ik\left[(1+i\sigma)\ln(1+i\sigma) + (1-i\sigma)\ln(1-i\sigma)\right] - \pi k.$$

The asymptotic form of S_- is only valid for d non integer. We note again that the classically forbidden backward scattering is exponentially suppressed.

Semiclassical calculation. We now perform the explicit calculation of the corresponding path integral for the evolution operator by steepest descent. We first assume a generic analytic potential, decreasing fast enough for large arguments. We consider the hamiltonian

$$H = \tfrac{1}{2}p^2 + V(x).$$

Forward scattering. We assume that the energy is larger than the maximal value of the potential so that classical forward scattering is possible. We then solve the equation of motion. Integrating once we obtain:

$$\tfrac{1}{2}\dot{x}^2(\tau) + V(x) = \tfrac{1}{2}\kappa^2,$$

with the boundary conditions

$$x(\tau') = x', \quad x(\tau'') = x''.$$

We set $X = x'' - x'$, $T = \tau'' - \tau'$. We know from the general analysis of Section 2.6 that we are interested in trajectories which correspond to classical scattering, with $X/T = k$ finite when $\tau' \to -\infty$, $\tau'' \to \infty$. In these limits the boundary condition

$$T = \int_{x'}^{x''} \frac{dx}{\sqrt{\kappa^2 - 2V(x)}},$$

leads to

$$\kappa = k + \frac{1}{T} \int_{-\infty}^{+\infty} \mathrm{d}x \left(\frac{k}{\sqrt{k^2 - 2V(x)}} - 1 \right) + O\left(T^{-2}\right).$$

The action of the trajectory is then

$$\mathcal{A} = \tfrac{1}{2}\kappa^2 T - 2 \int V(x) \mathrm{d}\tau = \tfrac{1}{2}\kappa^2 T - 2 \int_{x'}^{x''} \frac{V(x)\mathrm{d}x}{\sqrt{\kappa^2 - 2V(x)}}$$

$$= \tfrac{1}{2}k^2 T + \int_{-\infty}^{+\infty} \mathrm{d}x \left(\sqrt{k^2 - 2V(x)} - k \right) + O\left(T^{-1}\right).$$

We then calculate the Fourier transform. The result only depends on $x'' - x'$ and thus a factor of momentum conservation $\delta(k'' - k')$ follows. The remaining integral over X is calculated by steepest descent. At leading order for T large we need only consider the terms of order T for the saddle point. We find

$$X = k'T \;\Rightarrow\; k' = k.$$

As shown in Section 2.6 the terms proportional to T cancel with the free motion factors, and we finally obtain the semiclassical result.

$$\ln S_+(k) = i \int_{-\infty}^{+\infty} \mathrm{d}x \left(\sqrt{k^2 - 2V(x)} - k \right).$$

In the special case of the hamiltonian (44.2) a short calculation yields for $\ln S_+(k)$ a result in agreement with the expansion of the exact result.

Backward scattering. We now assume that the energy is lower than the maximum value of the potential in such a way that a classical reflection is allowed. If we imagine starting from $-\infty$ we obtain after a similar calculation the classical action:

$$\mathcal{A} = \frac{1}{2T} \left(x' + x'' - 2x_0 \right)^2 + 2 \int_{-\infty}^{x_0} \mathrm{d}x \left(\sqrt{k^2 - 2V(x)} - k \right) + O\left(T^{-1}\right),$$

where x_0 is the reflection point $k^2 = 2V(x_0)$, and $k = (2x_0 - x' - x'')/T$.

Since the result only depends on the combination $x' + x''$ we find after Fourier transformation as expected a factor $\delta(k'' + k')$. The remaining integral over $X = x' + x''$, calculated by steepest descent, yields:

$$2x_0 - X = k'T, \quad k' = k,$$

and thus

$$\ln S_-(k) = 2i \int_{-\infty}^{x_0} \mathrm{d}x \left(\sqrt{k^2 - 2V(x)} - k \right) + 2ikx_0.$$

The application to the hamiltonian (44.2) again shows agreement with the expansion of the exact expression. Note finally that the results in the forbidden region can be obtained by a proper analytic continuation in the potential.

Chapter 3: Path Integrals in Quantum Mechanics: Generalization

Exercise 3.1

For level n and angular momentum l, setting $\mu = l + \nu/2 - 1$, one finds

$$E_{nl} = 2n + 1 + \mu + \tfrac{1}{2}\lambda \left[3(2n+1)^2 + 6\mu(2n+1) + 2\mu^2 + 1\right] + O\left(\lambda^2\right).$$

Exercise 3.2

We shall explain the idea in the case of the lagrangian

$$\mathcal{L}(q,\dot{q}) = \tfrac{1}{2}g_{ij}(q)\dot{q}^i\dot{q}^j.$$

The short time evolution operator can then be written

$$U(q'',q') = \mathcal{N}(q'')\exp\left[\frac{i}{2\varepsilon}g_{ij}(q'')\left(q''^i - q'^i\right)\left(q''^j - q'^j\right)\right].$$

We then use the unitarity relation

$$\int dq'\, U(q'',q')U^*(q,q') = \prod_i \delta(q^i - q''^i).$$

We use the property that for ε small q' can be replaced by q or q'' in g_{ij} and $\mathcal{N}$. The remaining integral over q' is then gaussian. One finds

$$U(q'',q')U^*(q,q') = \mathcal{N}(q)\mathcal{N}(q'')\exp\left[\frac{i}{2\varepsilon}g_{ij}(q)\left(2q'^i - q^i - q''^i\right)\left(q''^j - q^j\right)\right].$$

The quadratic term in q' has cancelled, and the resulting integration yields a δ-function:

$$\int dq'\, U(q'',q')U^*(q,q') = \mathcal{N}^2(q)\det^{-1}\left(g_{ij}(q)\right)\prod_i \delta(q^i - q''^i).$$

Unitarity thus implies

$$\mathcal{N}^2(q) = \det\left(g_{ij}(q)\right),$$

as we have found by direct calculation.

Exercise 3.3

If we start from the discretized free evolution operator we obtain of course the action

$$S_1 = \sum_k \frac{1}{2\varepsilon}\left[f(q_{k+1}) - f(q_k)\right]^2,$$

with a measure $\prod_k f'(q_k)dq_k$.

If we change variables in the continuum path integral we find the action

$$S = \tfrac{1}{2}\int dt\, \dot{q}^2(t)f'^2(q(t)),$$

which we could imagine discretizing in a hermitian way as

$$S_2 = \sum_k \frac{1}{2\varepsilon}(q_{k+1} - q_k)^2 f'^2\big((q_{k+1} + q_k)/2\big),$$

with the same measure as previously.

Let us thus calculate the difference between the terms in S_2 and S_1. We find

$$S_1 - S_2 = \sum_k \frac{1}{12\varepsilon}(q_{k+1} - q_k)^4 f'(q_k)f'''(q_k).$$

We that since $q_{k+1} - q_k = O(\varepsilon^{1/2})$ this term is of order ε, i.e. comparable to a potential term. Returning to the initial variables and treating this difference at first order in perturbation we find that the two actions differ by the potential

$$S_2 - \int dt\, \dot{q}^2(t) \sim -\int dt\, f'''(q(t))[f'(q(t))]^{-3},$$

which exhibits the difference in the order of operators in the two quantization methods.

Exercise 3.4

3.4.1. Let us first consider the effect of a translation on the evolution operator. In the path integral we change variables, setting

$$\mathbf{q}(t) \mapsto \mathbf{q}(t) + \mathbf{a},$$

where $\mathbf{a}$ is a constant vector. The boundary conditions then become

$$\mathbf{q}(-\beta/2) = \mathbf{q}' - \mathbf{a}, \quad \mathbf{q}(\beta/2) = \mathbf{q}'' - \mathbf{a}, \qquad \beta = t'' - t'.$$

The action receives an additive contribution $\frac{1}{2}ie\mathbf{B} \times \mathbf{a}(\mathbf{q}'' - \mathbf{q}')$ due to the magnetic field, which leads to the relation

$$\langle \mathbf{q}''\,|U(\beta)|\,\mathbf{q}'\rangle = \exp[\tfrac{1}{2}ie(\mathbf{B} \times \mathbf{a})(\mathbf{q}'' - \mathbf{q}')]\,\langle \mathbf{q}'' - \mathbf{a}\,|U(\beta)|\,\mathbf{q}' - \mathbf{a}\rangle.$$

We can use this relation for $\mathbf{a} = \mathbf{q}'$ to simplify calculations. The additional phase then becomes $\frac{1}{2}ie\mathbf{B} \cdot (\mathbf{q}' \times \mathbf{q}'')$. We have to calculate $\langle \mathbf{q}'' - \mathbf{q}'\,|U(\beta)|\,\mathbf{0}\rangle$.

We thus solve the classical equation of motion. The motion along $\mathbf{B}$ being free, we restrict ourselves to the plane perpendicular to $\mathbf{B}$. We find

$$S_c = \frac{eB}{4\tanh(eB\beta/2m)}\,(\mathbf{q}'' - \mathbf{q}')^2.$$

We then evaluate the determinant resulting from the gaussian integration. We have to calculate the eigenvalues of the differential operator

$$-m\frac{d^2}{(dt)^2}\delta_{ij} + ieB_k\epsilon_{kij}\frac{d}{dt},$$

with the boundary conditions that the eigenfunctions vanish at $\pm\beta/2$. After Fourier transformation this operator takes the form

$$\begin{pmatrix} m\omega^2 & -\omega eB \\ \omega eB & m\omega^2 \end{pmatrix} ,$$

where ω is the corresponding frequency. The eigenvalues λ_n then follow:

$$\lambda_n = n^2\pi^2/\beta^2 + e^2B^2/4m^2 ,$$

each eigenvalue being twice degenerate. The product of eigenvalues divided by the value in zero field then follows from the identity (2.35). Normalizing with respect to the free hamiltonian one finally obtains:

$$\langle \mathbf{q}'' | U(\beta) | \mathbf{q}' \rangle = \frac{eB}{4\pi\sinh(\beta eB/2m)} \exp\left[\tfrac{1}{2}ie\mathbf{B}\cdot(\mathbf{q}'\times\mathbf{q}'')\right] \exp[-S_c].$$

3.4.2. The classical action then becomes

$$S_c = \frac{m\omega'}{2\sinh\beta\omega'} \left[(\mathbf{q}'^2 + \mathbf{q}''^2)\cosh\beta\omega' - 2\mathbf{q}'\mathbf{q}''\cosh\left(\frac{eB\beta}{2m}\right) \right.$$
$$\left. -2i\sinh\left(\frac{eB\beta}{2m}\right)\hat{\mathbf{B}}\cdot(\mathbf{q}'\times\mathbf{q}'') \right],$$

where

$$\hat{\mathbf{B}} = \mathbf{B}/B , \qquad \omega' = \sqrt{\omega^2 + e^2B^2/4m^2} .$$

The gaussian integration then yields the factor $\mathcal{N}$

$$\mathcal{N} = \frac{eB}{4\pi\sinh(\beta\omega')} .$$

Exercise 3.5

Let us first calculate the classical action of the various trajectories which satisfy the proper boundary conditions. It is possible to go from q' to q'' directly or after being reflected an arbitrary number of times by the two walls. One find two series

$$S_c = (q'' - q' + 2n\pi)^2/2T , \qquad S_c = (q'' - q' + 2n\pi)^2/2T .$$

After a shift of the classical trajectory the remaining gaussian integral is independent of the path and yields a constant normalization. If we would just add the two series we would find

$$U(q'', q'; T) = (2\pi T)^{-1/2} \left(\sum_n e^{i(q''-q'+2n\pi)^2/2T} + e^{i(q''+q'+2n\pi)^2/2T} \right).$$

The sum can be evaluated, as in Section 3.3 using Poisson's formula and the result is

$$U(q'', q'; \beta) = \sum_k e^{iTk^2/2} \left(e^{ik(q''-q')} + e^{ik(q''+q')} \right). \tag{44.3}$$

This sum does not coincide with the exact result which can be written

$$U(q'', q'; \beta) = \sum_k e^{iTk^2/2} \left(e^{ik(q''+q')} - e^{ik(q''-q')} \right).$$

The eigenfunctions $\psi(q)$ corresponding to expression (44.3) are $\cos kq$ and thus satisfy the boundary conditions $\psi'|_{q=0,\pi} = 0$ while the correct eigenfunctions satisfy $\psi|_{q=0,\pi} = 0$. The correct result is obtained by multiplying the amplitude by a sign for each reflection. This phase factor which can only be obtained by a careful study of the limit of a smooth potential converging towards the square well using the semiclassical expressions of Section A38.2.

Exercise 3.6

3.6.1. We first write the equation of motion

$$\dot{z}(t) + \omega z(t) = j(t),$$
$$-\dot{\bar{z}}(t) + \omega \bar{z}(t) = \bar{j}(t).$$

The solutions are

$$z(t) = z\,e^{-\omega(t-t')} + \int_{t'}^{t} d\tau\, e^{-\omega(t-\tau)}\, j(\tau),$$

$$\bar{z}(t) = \bar{z}''\,e^{-\omega(t''-t)} + \int_{t}^{t''} d\tau\, e^{\omega(\tau-t)}\, \bar{j}(\tau).$$

Taking into account the equation of motion the classical action simplifies

$$S_c = -\bar{z}'' z(t'') - \int_{t'}^{t''} \bar{j}(t) z(t) dt\,.$$

We then substitute the explicit solutions. After shifting $\bar{z}(t)$, $z(t)$ by the classical solution we have to perform a gaussian integration which depends only on t', t''. It yields a normalization which can be directly inferred from the ground state energy.

3.6.2. It is convenient to introduce the notation

$$\sigma = \int dt\, e^{i\omega t}\, j(t), \quad \bar{\sigma} = \int dt\, e^{-i\omega t}\, \bar{j}(t).$$

The unitarity relation reads

$$\int \frac{dz'd\bar{z}'}{2i\pi} \bar{S}(\bar{z}, z')\, e^{-\bar{z}'z'}\, S(\bar{z}', z) = e^{\bar{z}z}\,.$$

We note

$$2\,\mathrm{Re}\left[\int_{-\infty}^{+\infty} dt d\tau\, \bar{j}(t)\theta(t-\tau)\, e^{-i\omega(t-\tau)}\, j(\tau)\right] = \bar{\sigma}\sigma\,.$$

We then have to evaluate

$$K(\bar{z}, z) = \int \frac{dz'd\bar{z}'}{2i\pi} \exp\left[\bar{z}z' - i\left(\bar{\sigma}z' + \sigma\bar{z}\right) - \bar{z}'z' + i\left(\bar{\sigma}z + \sigma\bar{z}'\right) + \bar{z}'z - \bar{\sigma}\sigma\right].$$

This a simple gaussian integral which leads to the result.

3.6.3. In the first case the interaction term yields a contribution to the action

$$\tfrac{1}{4}\lambda \int dt \left(\bar{z}(t) + z(t)\right)^4.$$

At this order only the term proportional to $(\bar{z}z)^2$ contributes. Therefore, using Wick's theorem, we find

$$\mathrm{tr}\,e^{-\beta H} = \frac{1}{2\sinh(\omega\beta/2)}\left[1 - 3\lambda \int dt\, \langle \bar{z}(t)z(t)\rangle^2 + O\left(\lambda^2\right)\right],$$

where $\langle \bar{z}(t)z(t)\rangle$ is given by the propagator (3.77) at time 0,

$$\langle \bar{z}(t)z(t)\rangle \equiv \Delta(0) = \frac{1+e^{-\beta}}{1-e^{-\beta}}\theta(0).$$

With the convention $\theta(0) = \frac{1}{2}$ we find

$$\operatorname{tr} e^{-\beta H} = \frac{1}{2\sinh(\omega\beta/2)}\left[1 - \tfrac{3}{4}\lambda\beta\operatorname{cotanh}^2(\beta/2) + O\left(\lambda^2\right)\right],$$

which coincides with the result of exercise 2.1.1.

In the second case we note quite generally (as a consequence of Baker–Hausdorf's formula)

$$e^{x(a+a^\dagger)} = e^{xa^\dagger}\,e^{xa}\,e^{x^2/2}.$$

It follows

$$\left(a+a^\dagger\right)^4 = a^{\dagger 4} + a^4 + 4a^\dagger a^3 + 4a^{\dagger 3}a + 6a + 6a^\dagger + 3 + 12a^\dagger a + 6a^{\dagger 2}a^2.$$

We now replace the quantum operators $a, a^\dagger$ by the classical variables $z, \bar{z}$. At this order only the terms with an equal number of a and $a^\dagger$ factors contribute. We then obtain

$$\operatorname{tr} e^{-\beta H} = \frac{1}{2\sinh(\omega\beta/2)}\left[1 - \tfrac{3}{4}\lambda\beta\lambda\left(1 + 4\Delta(0) + 4\Delta^2(0)\right) + O\left(\lambda^2\right)\right].$$

We thus find that to recover the correct result we have to assign to $\theta(0)$ the value $\theta(0_+) = 0$.

Exercise 3.7

We first calculate the norm of $\psi(z)$ and find $e^{\alpha\bar{\alpha}}$. We can then represent the position and momentum operators by

$$\hat{q} \mapsto (2\omega)^{-1/2}\left(\bar{z} + \frac{\partial}{\partial\bar{z}}\right), \qquad \hat{p} \mapsto i(\omega/2)^{1/2}\left(\bar{z} - \frac{\partial}{\partial\bar{z}}\right).$$

A short calculation then yields

$$\langle \hat{q}\rangle = (2/\omega)^{1/2}\operatorname{Re}\alpha, \quad \langle \hat{q}\rangle = (2\omega)^{1/2}\operatorname{Im}\alpha, \quad (\Delta q)^2 = \tfrac{1}{2\omega}, \quad (\Delta p)^2 = \tfrac{\omega}{2}.$$

Chapter 4: Stochastic Differential Equations

Exercise 4.1

To derive coupled differential equations for moments we differentiate the moments with respect to time and use the equation.

$$\dot{x} = -\tfrac{1}{2}ax^n(t) + \nu(t). \tag{44.4}$$

Setting $\langle x^k\rangle = m_k$ we get

$$\dot{m}_k = -\tfrac{1}{2}akm_{k+n-1} + k\left\langle x^{k-1}(t)\nu(t)\right\rangle.$$

To evaluate the last term we use the same strategy as in the derivation of the Fokker–Planck equation and find

$$\langle x^{k-1}(t)\nu(t)\rangle = \tfrac{1}{2}\Omega(k-1)m_{k-2}.$$

Therefore

$$\dot{m}_k = -\tfrac{1}{2}akm_{k+n-1} + \tfrac{1}{2}k(k-1)\Omega m_{k-2}.$$

In particular if we assume the existence of an equilibrium we find the relations

$$m_{k+n-1} = (k-1)\Omega m_{k-2}/a\,,$$

which determines all moments once m_k, $k = 1, ..., n$ are known.   4.1.1. The Fokker–Planck reads:

$$\dot{P}(x,t) = \frac{1}{2}\frac{\partial}{\partial x}\left[\Omega\frac{\partial P}{\partial x} + ax^n P\right].\qquad(44.5)$$

The possible equilibrium distribution is the static solution

$$P(x) \propto \exp\left(-\frac{a}{\Omega(n+1)}x^{n+1}\right).$$

For $P(x)$ to be normalizable n must be odd and a positive. Transforming the FP hamiltonian into a hermitian hamiltonian one verifies it has a complete discrete spectrum and the convergence towards equilibrium is exponential.

4.1.2. What can be said if the parameter a and $x(t)$ are complex and $\nu(t)$ is still the real noise (4.2). Begin with $n = 1$; for $n > 2$ this is a research problem.

For $n = 1$ we can solve the Langevin equation

$$x(t) = x(0)\,e^{-at/2} + \int_0^t dt'\,\nu(t')\,e^{a(t'-t)/2}\,.$$

It is clear that the condition for the existence of an equilibrium distribution is $\operatorname{Re} a > 0$. Then all averages are determined by the two-point function.

Exercise 4.2

The solution of the stochastic differential equation can be written

$$x(t) = A_0\,e^{-r_1 t} + B_0\,e^{-r_2 t} + \frac{1}{a\,(r_1 - r_2)}\int_0^t d\tau\left(e^{-r_1(\tau-t)} - e^{-r_2(\tau-t)}\right)\nu(\tau),$$

where r_1, r_2 are the solutions of the equation

$$ar^2 - br + c = 0\,.$$

First it is clear that an equilibrium can be reached only if both r_1, r_2 have a positive real part, which implies $b/a > 0$, $c/a > 0$. Then at large times we can forget about the boundary conditions at $t = 0$, or equivalently we can integrate the equation from $t = -\infty$.

   1. *The two-point function.* Assuming that the gaussian noise distribution is normalized as

$$\langle \nu(t)\nu(t')\rangle = 2\delta(t - t'),$$

we can calculate the two-point correlation function:

$$\langle x(t)x(t')\rangle = \frac{1}{a^2\left(r_2^2 - r_1^2\right)}\left(\frac{1}{r_1}\,\mathrm{e}^{-r_1|t-t'|} - \frac{1}{r_2}\,\mathrm{e}^{-r_2|t-t'|}\right). \qquad (44.6)$$

Note that here, in contrast with the ordinary Langevin equation, also the $\langle \dot{x}(t)\dot{x}(t')\rangle$ correlation function is continuous at $t = t'$ and thus the typical paths are smoother.

2. The equilibrium distribution. The equilibrium distribution is of course gaussian in the two variables $x(t)$ and $\dot{x}(t)$. It is thus characterized by

$$\langle x^2(t)\rangle = 1/\left[a^2 r_1 r_2\left(r_1 + r_2\right)\right] = 1/(bc),$$
$$\langle \dot{x}^2(t)\rangle = 1/\left[a^2\left(r_1 + r_2\right)\right] = 1/(ab).$$
$$\langle \dot{x}(t)x(t)\rangle = 0\,.$$

3. The autocorrelation time. We observe that there are two possible situations:

(i) Both r_1 and r_2 are real and then at large times the largest value gives negligible contributions. The second order differential equation can be well approximated by an ordinary Langevin equation. The autocorrelation time τ is given by

$$\tau = \frac{b + \sqrt{b^2 - 4ac}}{2c}.$$

(ii) r_1 and r_2 are complex conjugated. Both terms then contribute and the behaviour differs from the ordinary Langevin equation. The autocorrelation time τ is then

$$\tau = 2a/b\,.$$

(iii) The border case $r_1 = r_2$ is specially interesting and corresponds to critical damping. At b, c fixed (and thus also $\langle x^2(t)\rangle$) it leads to the fastest decay of correlation or relaxation towards equilibrium:

$$\tau_{\mathrm{min}} = \tfrac{1}{2}(b/c) = \tfrac{1}{2}\tau(a = 0).$$

Chapter 5: Functional Integrals in Field Theory

Exercise 5.1

5.1.1. The field equation of motion reads:

$$\left[\left(-\partial_x^2 + m^2\right)\frac{\delta}{\delta J(x)} + \frac{g}{2}\left(\frac{\delta}{\delta J(x)}\right)^2\right]Z(J) = J(x)Z(J).$$

5.1.2. Calling $Z^{(n)}$ the n-point correlation function we successively obtain

$$(-\partial_x^2 + m^2)Z^{(1)}(x) + \tfrac{1}{2}gZ^{(2)}(x, x) = 0\,,$$
$$(-\partial_x^2 + m^2)Z^{(2)}(x, y) + \tfrac{1}{2}gZ^{(3)}(x, x, y) = \delta(x - y),$$
$$(-\partial_x^2 + m^2)Z^{(3)}(x, y_1, y_2) + \tfrac{1}{2}gZ^{(4)}(x, x, y_1, y_2) = \delta(x - y_1)Z^{(1)}(y_2)$$
$$+ \delta(x - y_2)Z^{(1)}(y_1),$$
$$(-\partial_x^2 + m^2)Z^{(4)}(x, y_1, y_2, y_3) + \tfrac{1}{2}gZ^{(5)}(x, x, y_1, y_2, y_3) = \delta(x - y_1)Z^{(2)}(y_2, y_3)$$
$$+ \delta(x - y_2)Z^{(2)}(y_3, y_1) + \delta(x - y_2)Z^{(2)}(y_3, y_1).$$

5.1.3. The first equation yields $Z^{(1)}(x)$ which by translation invariance is constant from the 2-point function at leading order. Denoting by $\Delta(x-y)$ the propagator we obtain

$$Z^{(1)}(x) = -\frac{g}{2m^2}\Delta(0).$$

From the last equation we obtain the 4-point function at leading order, which is, as expected a disconnected sum of products of 2-point functions (property of the gaussian measure).

$$Z^{(4)}(y_1, y_2, y_3, y_4) = \Delta(y_1 - y_2)\Delta(y_3 - y_4) + \text{ 2 terms}.$$

We introduce these expressions in the equation for $Z^{(3)}$, and find

$$Z^{(3)}(y_1, y_2, y_3) = -g\int \mathrm{d}x\, \Delta(x-y_1)\Delta(x-y_2)\Delta(x-y_2)$$
$$-\frac{g}{2m^2}\Delta(0)\left(\Delta(y_1-y_2) + \text{ 2 terms}\right).$$

We can now calculate the two-point function

$$Z^{(2)}(y_1 - y_2) = \Delta(y_1 - y_2) - \tfrac{1}{2}g\int \mathrm{d}x\Delta(y_1-x)Z^{(3)}(x,x,y_2)$$

$$= \Delta(y_1 - y_2) + \tfrac{1}{2}g^2\int \mathrm{d}x\mathrm{d}z\,\Delta(y_1-x)\Delta^2(x-z)\Delta(z-y_2)$$

$$+ \frac{g^2}{2m^2}\int \mathrm{d}x\,\Delta(y_1-x)\Delta(x-y_2) + \frac{g^2}{4m^4}\Delta^2(0)$$

Exercise 5.2

5.2.1. The equations of motion are

$$\left[\not\partial + M + g\frac{\delta}{\delta J(x)}\right]\frac{\delta}{\delta\bar\eta(x)}Z(\bar\eta, \eta, J) = \eta(x)Z(\bar\eta, \eta, J) \qquad (44.7a)$$

$$\left[^T(-\not\partial + M) + g\frac{\delta}{\delta J(x)}\right]\frac{\delta}{\delta\eta(x)}Z(\bar\eta, \eta, J) = \bar\eta(x)Z(\bar\eta, \eta, J) \qquad (44.7b)$$

$$\left[(-\partial_x^2 + m^2)\frac{\delta}{\delta J(x)} + g\frac{\delta}{\delta\eta(x)}\frac{\delta}{\delta\bar\eta(x)}\right]Z(\bar\eta, \eta, J) = J(x)Z(\bar\eta, \eta, J), \qquad (44.7c)$$

where ^{T} means transposed.

5.2.2. Setting $J = \eta = \bar\eta = 0$ in (44.7c) we immediately obtain the ϕ-field average:

$$\left(-\partial_x^2 + m^2\right)\langle\phi(x)\rangle - g\langle\bar\psi(x)\psi(x)\rangle = 0,$$

and thus, because $\langle\phi(x)\rangle$ is x independent (translation invariance)

$$\langle\phi(x)\rangle = \frac{g}{m^2}\operatorname{tr}\Delta_{\mathrm{F}}(0).$$

If instead we first differentiate with respect to J we find

$$\left(-\partial_x^2 + m^2\right)\langle\phi(x)\phi(y)\rangle - g\langle\phi(y)\bar\psi(x)\psi(x)\rangle = \delta(x - y).$$

Differentiating (44.7a) with respect to $\eta(y)$ yields

$$(\slashed{\partial}_x + M)_{\alpha\gamma} \langle \bar\psi_\beta(y)\psi_\gamma(x)\rangle + g\langle \phi(x)\bar\psi_\beta(y)\psi_\alpha(x)\rangle = \delta(x-y)\delta_{\alpha\beta}.$$

We also need equations (44.7c) differentiated with respect to η and $\bar\eta$, and (44.7a) with respect to $\bar\eta$ and twice η

$$(-\partial_x^2 + m^2)\langle \phi(x)\bar\psi(y_1)\psi(y_2)\rangle - g\langle \bar\psi(y_1)\psi(y_2)\bar\psi(x)\psi(x)\rangle = 0\,,$$

$$(\slashed{\partial}_x + M)\langle \bar\psi(y_3)\psi(x)\bar\psi(y_1)\psi(y_2)\rangle + g\langle \phi(x)\bar\psi(y_3)\psi(x)\bar\psi(y_1)\psi(y_2)\rangle$$
$$= \delta(x-y_3)\langle \bar\psi(y_1)\psi(y_2)\rangle - \delta(x-y_1)\langle \bar\psi(y_3)\psi(y_2)\rangle\,.$$

In the last equation we set $g = 0$, replace the fermion 2-point function by the propagator, and obtain the fermion 4-point function at leading order (in the gaussian approximation). From the previous equation we then find the $\phi\bar\psi\psi$ vertex

$$\langle \phi(x)\bar\psi_\beta(z)\psi_\alpha(y)\rangle = \langle\phi\rangle\,(\Delta_{\mathrm F})_{\alpha\beta}\,(x-y) - g\int du\,\Delta_{\mathrm B}(x-u)\Delta_{\mathrm F}\,[\Delta_{\mathrm F}(y-u)\Delta_{\mathrm F}(u-z)]_{\alpha\beta}\,.$$

We then substitute in the equations for the 2-point functions. The final answers are:

$$\langle \phi(x)\phi(y)\rangle = \Delta_{\mathrm B}(x-y) + \langle\phi\rangle^2 - g^2\int dz_1\,dz_2\,\Delta_{\mathrm B}(x-z_1)$$
$$\times \mathrm{tr}\,[\Delta_{\mathrm F}(z_1-z_2)\Delta_{\mathrm F}(z_2-z_1)]\,\Delta_{\mathrm B}(z_2-y),$$

$$\langle \bar\psi(y)\psi(x)\rangle = \Delta_{\mathrm F}(x-y)(1 - g\langle\phi\rangle) + g^2\int dz_1\,dz_2\,\Delta_{\mathrm F}(x-z_1)\Delta_{\mathrm F}(z_1-z_2)$$
$$\times \Delta_{\mathrm F}(z_2-y)\Delta_{\mathrm B}(z_1-z_2),$$

where $\Delta_{\mathrm B}$ and $\Delta_{\mathrm F}$ are the boson and fermion propagators.

Chapter 6: Generating Functions of Correlation Functions

Exercise 6.1

The field equation of motion reads:

$$\left[(-\partial^2 + m^2)\frac{\delta}{\delta J(x)} + \frac{g}{2}\left(\frac{\delta}{\delta J(x)}\right)^2\right]Z(J) = J(x)Z(J).$$

For the generating functional of connected correlation functions it becomes

$$\left[(-\partial^2 + m^2)\frac{\delta}{\delta J(x)} + \frac{g}{2}\left(\frac{\delta}{\delta J(x)}\right)^2\right]W(J) + \frac{g}{2}\left(\frac{\delta W}{\delta J(x)}\right)^2 = J(x).$$

Finally for the 1PI functional we find

$$\frac{\delta\Gamma}{\delta\varphi(x)} = (-\partial^2 + m^2)\,\varphi(x) + \tfrac{1}{2}g\varphi^2(x) + \frac{g}{2}\left(\frac{\delta^2\Gamma}{\delta\varphi(x_1)\delta\varphi(x_2)}\right)^{-1}\Bigg|_{x_1=x_2=x}. \qquad (44.8)$$

If we omit the inverse two-point function we can integrate and obtain Γ in the tree approximation:

$$\Gamma(\varphi) = S(\varphi).$$

Then iterating the equation we obtain the derivative of the one-loop functional:

$$\frac{\delta\Gamma_{1\,\text{loop}}}{\delta\varphi(x)} = \frac{g}{2}\left[\left(-\partial^2 + m^2 + +g\varphi(x_1)\right)\delta(x_1 - x_2)\right]^{-1}\Big|_{x_1=x2=x}.$$

Expanding this expression in powers of g we find

$$\frac{\delta\Gamma_{1\,\text{loop}}}{\delta\varphi(x)} = \tfrac{1}{2}g\Delta(0) - \tfrac{1}{2}g^2 \int dy\,\Delta^2(x-y)\varphi(y)$$

$$+ \tfrac{1}{2}g^3 \int dy dz\,\Delta(x-y)\varphi(y)\Delta(y-z)\varphi(z)\Delta(z-x) + O\left(g^4\right),$$

where we have called Δ the propagator. Comparing then with the expansion of the 1PI functional we conclude

$$\Gamma_1^{(1)} = \tfrac{1}{2}g\Delta(0)$$
$$\Gamma_1^{(2)} = -\tfrac{1}{2}g^2\Delta^2(x-y)$$
$$\Gamma_1^{(3)} = g^3\Delta(x-y)\Delta(y-z)\Delta(z-x).$$

Note finally that in expression (44.8) we recognize the functional derivative of the expected tr ln:

$$g\left[\left(-\partial^2 + m^2 + +g\varphi(x_1)\right)\delta(x_1 - x_2)\right]^{-1}\Big|_{x_1=x2=x} = \frac{\delta}{\delta\varphi(x)}\,\text{tr}\ln\left[-\partial^2 + m^2 + g\varphi(.)\right].$$

Exercise 6.2

The equation for the complete correlation functions has been given in Chapter 5. For the connected functions we obtain

$$\left[\left(-\partial^2 + m^2\right)\frac{\delta}{\delta J(x)} + \frac{g}{6}\left(\frac{\delta}{\delta J(x)}\right)^3\right]W(J) + \frac{g}{6}\left(\frac{\delta W}{\delta J(x)}\right)^3 + \frac{g}{2}\frac{\delta W}{\delta J(x)}\frac{\delta^2 W}{(\delta J(x))^2} = J(x).$$

For the 1PI functional we finally obtain

$$\frac{\delta\Gamma}{\delta\varphi(x)} = \left(-\partial^2 + m^2\right)\varphi(x) + \tfrac{1}{6}g\varphi^3(x) + \frac{g}{2}\varphi(x)\left(\frac{\delta^2\Gamma}{\delta\varphi(x_1)\delta\varphi(x_2)}\right)^{-1}\Big|_{x_1=x_2=x}$$

$$+ \tfrac{1}{6}gW^{(3)}(J),$$

where the three-point connected function $W^{(3)}(J)$ has the symbolic form

$$W^{(3)}(J) \equiv \frac{\delta^3 W(J)}{(\delta J)^3} = \Gamma^{(3)}(\varphi)\left(\Gamma^{(2)}(\varphi)\right)^{-1}\left(\Gamma^{(2)}(\varphi)\right)^{-1}\left(\Gamma^{(2)}(\varphi)\right)^{-1}.$$

One easily verifies that the last term does not contribute at one-loop order. Then, after integration, one finds at tree order

$$\Gamma(\varphi) = S(\varphi),$$

and thus at one-loop order

$$\frac{\delta\Gamma_{1\,\mathrm{loop}}}{\delta\varphi(x)} = \frac{g}{2}\varphi(x)\left(\frac{\delta^2 S}{\delta\varphi(x_1)\delta\varphi(x_2)}\right)^{-1}\Bigg|_{x_1=x_2=x},$$

which, after integration, yields the general tr ln term

$$\Gamma_{1\,\mathrm{loop}} = \tfrac{1}{2}\operatorname{tr}\ln\frac{\delta^2 S}{\delta\varphi\delta\varphi}.$$

Exercise 6.3

The equations of motion for the partition function are given in the exercise 5.2.1. From them one derives

$$\frac{\delta\Gamma}{\delta\bar\psi(x)} = \left[\slashed\partial + M + g\phi(x)\right]\psi(x) + gW^{(2)}_{J\bar\eta}(x,x)$$

$$\frac{\delta\Gamma}{\delta\phi(x)} = \left(-\partial_x^2 + m^2\right)\phi(x) - g\bar\psi(x)\psi(x) + gW^{(2)}_{\eta\bar\eta}(x,x),$$

where we have defined

$$W^{(2)}_{J\bar\eta}(x,y) = \frac{\delta^2 W}{\delta J(x)\delta\bar\eta(y)}, \qquad W^{(2)}_{\eta\bar\eta}(x,y) = \frac{\delta^2 W}{\delta\eta(x)\delta\bar\eta(y)}.$$

The second functional derivatives of W form an operator and a 3×3 matrix in the $\{J,\eta,\bar\eta\}$ space, inverse of the analogous matrix for Γ. The explicit expressions in terms of Γ are too complicated to be reproduced here.

Chapter 9: Regularization Methods

Exercise 9.1

The equation reads

$$\left[\left(-\partial_x^2 + m^2\right)\frac{\delta}{\delta J_i(x)} + \frac{g}{6N}\frac{\delta}{\delta J_i(x)}\left(\frac{\delta}{\delta J_j(x)}\right)^2\right]Z(J) = J_i(x)Z(J).$$

9.1.1. Using the $O(N)$ invariance (assumed not spontaneously broken) we obtain

$$\left(-\partial_x^2 + m^2\right)Z^{(2)}(x-y) + \frac{g}{6N^2}Z^{(4)}_{iijj}(y,x,x,x) = \delta(x-y).$$

9.1.2. If the action is quadratic, the four-point function reduces to the sum of products of two-point functions:

$$Z^{(4)}_{ijkl}(x_1,x_2,x_3,x_4) = \delta_{ij}\delta_{kl}Z^{(2)}(x_1-x_2)Z^{(2)}(x_3-x_4) + \text{2 terms}.$$

Introducing this relation in the equation we find:

$$\left[(-\partial_x^2 + m^2) + \frac{g(N+2)}{6N} Z^{(2)}(0) \right] Z^{(2)}(x - y) = \delta(x - y).$$

We set

$$\mu^2 = m^2 + \frac{g(N+2)}{6N} Z^{(2)}(0),$$

and thus after Fourier transformation

$$\tilde{Z}^{(2)}(p) = \frac{1}{p^2 + \mu^2} \;\Rightarrow\; \mu^2 = m^2 + \frac{g(N+2)}{6N} \frac{1}{(2\pi)^d} \int^\Lambda \frac{d^dp}{p^2 + \mu^2} \, .$$

We note that this equation is meaningful only in the presence of a cut-off Λ for dimensions $d \geq 2$, which we indicate symbolically. It yields a relation between the "bare" mass m, the renormalized or physical mass μ and the cut-off. This relation becomes exact in the large N limit as the analysis of Chapter 29 shows.

Exercise 9.2

Calling η and $\bar{\eta}$ the sources for $\bar{\psi}$ and ψ respectively we find:

$$\left[\partial_x \frac{\delta}{\delta \bar{\eta}_i(x)} + g \frac{\delta}{\delta \bar{\eta}_i(x)} \frac{\delta}{\delta \eta_k(x)} \frac{\delta}{\delta \bar{\eta}_k(x)} \right] Z(\bar{\eta}, \eta) = \eta_i(x) Z(\bar{\eta}, \eta).$$

A relation between the two and four-point functions follows by differentiating with respect to $\eta_j(y)$ and then setting the sources to zero:

$$\partial_x \left\langle \psi_i(x) \bar{\psi}_j(y) \right\rangle + g \left\langle \psi_i(x) \bar{\psi}_j(y) \psi_k(x) \bar{\psi}_k(x) \right\rangle = -\delta(x - y) \delta_{ij} \, .$$

Again assuming that the functions are calculated with a gaussian measure we can factorize the four-point function

$$\left\langle \psi(x) \bar{\psi}_j(y) \psi_k(x) \bar{\psi}_k(x) \right\rangle = \left\langle \psi_i(x) \bar{\psi}_j(y) \right\rangle \left\langle \psi_k(x) \bar{\psi}_k(x) \right\rangle - \left\langle \psi_k(x) \bar{\psi}_j(y) \right\rangle \left\langle \psi_i(x) \bar{\psi}_k(x) \right\rangle .$$

The function $< \psi(x) \bar{\psi}(x) >$ being a constant, we can solve the equation. After Fourier transformation, and taking into account $U(N)$ invariance, we find

$$\left\langle \psi_i(p) \bar{\psi}_j(-p) \right\rangle = \delta_{ij} \frac{1}{i\not{p} + \mu} \, ,$$

where we have set

$$\mu = \frac{g(N-1)}{N} \left\langle \psi_k(x) \bar{\psi}_k(x) \right\rangle = g(N-1) \frac{\operatorname{tr} \mathbf{1}}{(2\pi)^d} \mu \int^\Lambda \frac{d^dp}{p^2 + \mu^2} \, .$$

We have here assumed that the regularization was chiral invariant. We see that this equation, called often the gap equation, has always the solution $\mu = 0$ which preserves the chiral symmetry. However, in $d > 2$ dimensions, when the coupling satisfies

$$g > g_c, \qquad g_c^{-1} = (N-1) \frac{\operatorname{tr} \mathbf{1}}{(2\pi)^d} \int^\Lambda \frac{d^dp}{p^2} \, ,$$

i.e. is sufficiently attractive, there exists another solution which breaks the chiral symmetry

$$\frac{1}{g} = \frac{(N-1) \operatorname{tr} \mathbf{1}}{(2\pi)^d} \int^\Lambda \frac{d^dp}{p^2 + \mu^2} \, ,$$

and which, in addition, has a lower action. Within this approximation scheme, this model thus provides an example of non-perturbative spontaneous symmetry breaking and fermion mass generation, the Nambu–Jona-Lasinio mechanism, due to fermion pair condensation. Here again the approximation can be justified only in the large N limit (see Appendix A30).

Chapter 10: Introduction to Renormalization Theory

Exercise 10.1

One verifies that the result is

$$\Gamma^{(2)}_{\text{r,as.}} = p^2 + C_1 p^2 \frac{g_{\text{r}}}{1 - \beta_2 g_{\text{r}} \ln(p/m_{\text{r}})}.$$

This expression has a pole when the momentum reaches the value

$$p^2 = m_{\text{r}}^2\, e^{2/\beta_2 g_{\text{r}}},$$

(for $\beta_2 > 0$) often called the Landau pole. This singularity shows that even if g_{r} is small, perturbation theory will break at sufficiently high momentum. To go beyond this simple remark, more sophisticated RG arguments are required (see Chapter 25). Conversely if in some other theory $\beta_2 < 0$ then perturbation theory breaks down at sufficiently low momentum, indicating that the perturbative prediction of a massless spectrum is not reliable.

Chapter 11: Dimensional Regularization and Minimal Subtraction

Exercise 11.1

Let us take the example of the coupling constant renormalization. Then the bare coupling constant g is related to the renormalized coupling constant by

$$g = g_{\text{r}} + \ln \Lambda/m_{\text{r}} G_2(g_{\text{r}}) + (\ln \Lambda/m_{\text{r}})^2 G_3(g_{\text{r}}) + \cdots,$$

where $G_n(g_{\text{r}})$ is of order g_{r}^n. We can then calculate the β-function by differentiating with respect to m_{r} at g, Λ fixed

$$0 = \beta(g_{\text{r}}) \left(1 + \sum_{n=1}(\ln \Lambda/m_{\text{r}})^n G'_{n+1}(g_{\text{r}})\right) - \sum_{n=0}(n + 1)(\ln \Lambda/m_{\text{r}})^n G_{n+2}(g_{\text{r}}).$$

Since the β function is finite we conclude

$$\beta(g_{\text{r}}) = G_2(g_{\text{r}}), \qquad G_{n+2}(g_{\text{r}}) = \frac{1}{n+1} G'_{n+1}(g_{\text{r}})\beta(g_{\text{r}}).$$

In analogy with dimensional regularization all quantities are determined by the coefficient of $\ln(\Lambda/m_{\text{r}})$.

Exercise 11.2

Pseudoscalar Yukawa interaction. Consider the action

$$S\left(\phi, \psi, \bar\psi\right) = \int \mathrm{d}^d x \left\{ -\bar\psi(x) \left[\slashed\partial + M + ig\gamma_S \phi(x)\right] \psi(x) + \tfrac{1}{2}\left(\partial_\mu \phi(x)\right)^2 \right.$$
$$\left. + \tfrac{1}{2} m^2 \phi^2(x) + \frac{1}{4!}\lambda \phi^4(x) \right\}, \tag{44.9}$$

in which ϕ is a pseudoscalar boson field and $\psi, \bar\psi$ Dirac fermions.

The factor i in front of γ_S is imposed by invariance under reflection hermiticity, i.e. euclidean hermitian conjugation followed by space reflection (see Appendix A5.6) which implies:

$$\gamma_1 \gamma_S^\dagger \left(ig\phi(x)\right)^* \gamma_1 = \gamma_S i g \phi(x).$$

In addition the action is reflection symmetric (see Appendix A5.5):

$$\psi(x) \mapsto \gamma_1 \psi\left(\tilde{x}\right),$$
$$\bar{\psi}(x) \mapsto \bar{\psi}\left(\tilde{x}\right)\gamma_1,$$
$$\phi(x) \mapsto -\phi\left(\tilde{x}\right),$$

with $\tilde{x} = (-x_1, x_2, \dots, x_d)$.

The transformation of $\phi(x)$ is imposed by the factor γ_S in front of the $\phi\bar{\psi}\psi$ interaction and justifies the denomination pseudoscalar for the field $\phi(x)$.

Calculations can be done using dimensional regularization and minimal subtraction because it is never necessary to evaluate any trace of products of γ_S by an even number of γ_μ matrices.

11.2.1. One finds in the notation of Section 11.8,

$$
I_1(p) = \left[\frac{N_d}{\varepsilon} - \frac{1}{16\pi^2}\ln\left(\frac{Mm}{\mu^2}\right)\right]\left(M + \frac{i\not{p}}{2}\right) - \frac{1}{32\pi^2}\frac{i\not{p}}{p^2}\left(M^2 + m^2\right)\ln\left(\frac{M}{m}\right)
$$
$$
- \frac{1}{16\pi^2}\left[M + \frac{i\not{p}}{2p^2}\left(p^2 + m^2 - M^2\right)\right]\left[\frac{1}{p^2}\sigma_+\sigma_- \ln\left(\frac{\sigma_+ + \sigma_-}{\sigma_+ - \sigma_-}\right)\right.
$$
$$
\left. - 1 - \ln\left(\frac{M}{m}\right)\left(\frac{M^2 - m^2}{p^2}\right)\right] + O\left(\varepsilon\right), \tag{44.10}
$$

where we have set:

$$\sigma_\pm\left(p^2\right) = \sqrt{p^2 + \left(M \pm m\right)^2}. \tag{44.11}$$

11.2.2. The divergent part of the ϕ 2-point function is:

$$I_2(p) = 4\left(M^2 + \frac{p^2}{2}\right)\frac{N_d}{\varepsilon} + O\left(1\right). \tag{44.12}$$

11.2.3. We call M_0, m_0, g_0, λ_0 the bare parameters, M, m, g, λ the renormalized parameters, μ the renormalization scale.

Substituting in the Feynman diagrams and identifying the divergent parts we obtain equations for the renormalization constants:

$$Z_\psi\left(M_0 + i\not{p}\right) + N_d g^2 \left(M + \frac{i\not{p}}{2}\right)\frac{1}{\varepsilon} = \text{finite } O\left(2 \text{ loops}\right),$$

$$\mu^{-\varepsilon/2}g_0 Z_\psi Z_\phi^{1/2} - N_d\frac{g^3}{\varepsilon} = \text{finite } O\left(2 \text{ loops}\right),$$

$$Z_\phi\left(p^2 + m_0^2\right) + N_d\left[-\frac{\lambda}{2}\frac{m^2}{\varepsilon} + 4\left(M^2 + \frac{p^2}{2}\right)\frac{g^2}{\varepsilon}\right] = \text{finite } O\left(2 \text{ loops}\right),$$

$$\mu^{-\varepsilon}\lambda_0 Z_\phi^2 + N_d\left(-\frac{3}{2}\frac{\lambda^2}{\varepsilon} + \frac{24g^4}{\varepsilon}\right) = \text{finite } O\left(2 \text{ loops}\right).$$

It follows that:

$$Z_\psi = 1 - N_d \frac{g^2}{2\varepsilon} + O\,(2\text{ loops}), \tag{44.13}$$

$$Z_\phi = 1 - N_d \frac{2Ng^2}{\varepsilon} + O(2\text{ loops}), \tag{44.14}$$

$$g_0 = m^{\varepsilon/2} g \left(1 + N_d \frac{(2N+3)}{2} \frac{g^2}{\varepsilon}\right) + O(2\text{ loops}), \tag{44.15}$$

$$\lambda_0 = m^\varepsilon \left[\lambda + \frac{N_d}{\varepsilon}\left(\frac{3}{2}\lambda^2 + 4N\lambda g^2 - 24Ng^4\right)\right] + O(2\text{ loops}), \tag{44.16}$$

$$m_0^2 = m^2 \left[1 + \frac{N_d}{\varepsilon}\left(\frac{\lambda^2}{2} + 2g^2\right)\right] + O(2\text{ loops}). \tag{44.17}$$

The β-function. When bare and renormalized coupling constants are related by an equation of the form (11.115), the coupling constant RG functions $\beta_i\,(\lambda)$ are given by:

$$\beta_i\,(\lambda) = \mu\frac{\partial}{\partial\mu}\bigg|_{\lambda_0} \lambda_i, \tag{44.18}$$

and thus using equation (11.115):

$$0 = \mu\frac{\partial}{\partial\mu}\lambda_{0i}\bigg|_{\lambda_0} = \mu^\varepsilon\left(\varepsilon f_i\,(\lambda) + \beta_j\,(\lambda)\frac{\partial f_i}{\partial\lambda_j}\right). \tag{44.19}$$

A short calculation then yields:

$$\beta_{g^2} = -\varepsilon g^2 + N_d 5g^4 + \cdots, \tag{44.20}$$

$$\beta_\lambda = -\varepsilon\lambda + N_d \left(\frac{3}{2}\lambda^2 + 4\lambda g^2 - 24g^4\right) + \cdots. \tag{44.21}$$

11.2.4. For $M = 0$ let us set

$$\psi = \mathrm{e}^{-i\pi\gamma_S/4}\,\psi', \qquad \bar\psi = \bar\psi'\,\mathrm{e}^{-i\pi\gamma_S/4}.$$

Then the action becomes identical to the action of the model (11.113). This explains why the renormalization constants are the same.

A few additional results are:

$$M_0 = M\left(1 - \frac{1}{2}g^2\frac{N_d}{\varepsilon}\right) + O(2\text{ loops}), \tag{44.22}$$

$$m_0^2 = m^2\left[1 + \left(\frac{1}{2}\lambda + 2g^2\right)\frac{N_d}{\varepsilon}\right] - 4M^2g^2\frac{N_d}{\varepsilon} + O(2\text{ loops}). \tag{44.23}$$

Exercise 11.3

The general problem is to prove that renormalized correlation functions have a finite limit when $\varepsilon = 4 - d$ goes to zero.

Dimensional analysis is modified and the dimensions of correlation functions is now $[\Gamma^{l,n}] = d - n(d-2)/2 - 2l$.

To define a dimensionless coupling constant we have to modify the third renormalization condition (10.9)

$$\tilde{\Gamma}^{(4)}(0,0,0,0) = m_{\mathrm{r}}^{\varepsilon} g_{\mathrm{r}} \,.$$

With this definition the coefficient functions of the CS equations are only functions of g_{r} and ε.

Equation (10.33) is modified and becomes

$$\varepsilon g_{\mathrm{r}} + \beta(g_{\mathrm{r}}) - 2g_{\mathrm{r}}\eta(g_{\mathrm{r}}) = m_{\mathrm{r}}^{2-\varepsilon}(2-\eta)\Gamma_{\mathrm{r}}^{(1,4)}(0;0,0,0,0).$$

To take care of the additional term in the β-function it is useful to define

$$\tilde{\beta}(g_{\mathrm{r}}) = \beta(g_{\mathrm{r}}) + \varepsilon g_{\mathrm{r}} = O\left(g_{\mathrm{r}}^2\right).$$

In the previous equation only $\tilde{\beta}$ appears.

Then we keep in the r.h.s. of our induction equation (10.35) $\tilde{\beta}$ and add the remaining part in the l.h.s.:

$$\left[m_{\mathrm{r}}\frac{\partial}{\partial m_{\mathrm{r}}} - \varepsilon g_{\mathrm{r}}\frac{\partial}{\partial g_{\mathrm{r}}} \right] \Gamma_{\mathrm{r}}^{(l,n)}.$$

With this modification many arguments remain valid. Let us solve for example the equation

$$\left[m_{\mathrm{r}}\frac{\partial}{\partial m_{\mathrm{r}}} - \varepsilon g_{\mathrm{r}}\frac{\partial}{\partial g_{\mathrm{r}}} \right] \Gamma_{\mathrm{r}}^{(4)} = m_{\mathrm{r}}^{\varepsilon} g_{\mathrm{r}} f^{(4)}\left(p_i/m_{\mathrm{r}}, g_{\mathrm{r}}, \varepsilon\right),$$

where the l.h.s. is assumed to be finite when $\varepsilon \to 0$ and satisfies

$$f^{(4)}\left(0, g_{\mathrm{r}}, \varepsilon\right) = m_{\mathrm{r}}^{-\varepsilon}\frac{1}{g_{\mathrm{r}}}\frac{\partial}{\partial m_{\mathrm{r}}}(g_{\mathrm{r}}m_{\mathrm{r}}^{\varepsilon}) = 0 \,.$$

The integration of the equation then yields

$$\Gamma_{\mathrm{r}}^{(4)} = m_{\mathrm{r}}^{\varepsilon} g_{\mathrm{r}} - m_{\mathrm{r}}^{\varepsilon} g_{\mathrm{r}} \int_0^1 \frac{\mathrm{d}\rho}{\rho} f^{(4)}\left(\rho p_i/m_{\mathrm{r}}, \rho^{\varepsilon} g_{\mathrm{r}}, \varepsilon\right).$$

The integral converges at $\rho = 0$ because $f^{(4)}$ is expandable in powers of g_{r} and is of order p^2 for p small. Hence no singularity in ε is generated.

Chapter 12: Renormalization of Composite Operators

Exercise 12.1

12.1.1. Let us first note that, unlike in the ϕ^4 theory, the action is no longer reflection symmetric in ϕ and the ϕ has an non-vanishing expectation value. It is thus convenient to shift ϕ by a constant in such a way that $\langle\phi\rangle = 0$. The action then takes the form

$$S(\phi) = \int d^6x \left[\tfrac{1}{2}\left(\partial_\mu\phi\right)^2 + \tfrac{1}{2}m^2\phi^2 + \frac{1}{3!}g\phi^3 - c\phi \right].$$

In six dimensions the field ϕ has dimension two and the 1PI functions $\Gamma^{(l,n)}$ have dimension $6 - 2n - 2l$. Power counting then tells us that the renormalized action has the form

$$S_r(\phi_r) = \int d^6x \left[\tfrac{1}{2}Z_\phi\left(\partial_\mu\phi_r\right)^2 + \tfrac{1}{2}(m_r^2 + \delta m^2)\phi_r^2 + \frac{1}{3!}g_r Z_g \phi_r^3 - C\phi_r \right].$$

The relations between bare and renormalized quantities are then

$$Z_\phi^{1/2}\phi_r = \phi, \quad g = g_r Z_g Z_\phi^{-3/2}, \quad m^2 = (m_r^2 + \delta m^2)Z_\phi^{-1}, \quad C = cZ_\phi^{1/2}.$$

Possible renormalization conditions are $\langle\phi_r(x)\rangle = 0$ and

$$\tilde{\Gamma}_r^{(2)}(p = 0) = m_r^2,$$

$$\frac{\partial}{\partial p^2}\tilde{\Gamma}_r^{(2)}(p)\big|_{p=0} = 1,$$

$$\tilde{\Gamma}_r^{(3)}(0,0,0) = g_r.$$

The other primitively divergent functions are $\Gamma^{(1,1)}$, $\Gamma^{(1,2)}$ and $\Gamma^{(2,1)}$.

We now write the equation of motion of the bare theory. We define the insertion of the ϕ^2 operator as the correlation function of $-\tfrac{1}{2}\phi^2$. Then we find

$$(-\partial^2 + m^2)\frac{\delta Z}{\delta J(x)} - gZ_{\phi^2(x)} = \left(J(x) + c\right)Z(J).$$

The equation for $W(J)$ is then

$$(-\partial^2 + m^2)\frac{\delta W}{\delta J(x)} - gW_{\phi^2(x)} = J(x) + c.$$

After Legendre transformation we obtain

$$\frac{\delta\Gamma}{\delta\phi(x)} + c = (-\partial^2 + m^2)\phi(x) + g\Gamma_{\phi^2(x)}(\phi).$$

The equation for the renormalized 1PI functional is then

$$\frac{\delta\Gamma_r}{\delta\phi(x)} + C = Z_\phi^{1/2}\left[(-\partial^2 + m^2)Z_\phi^{1/2}\phi(x) + g_r Z_g Z_\phi^{-3/2}\Gamma_{\phi^2(x)}(\phi Z_\phi^{1/2})\right].$$

This equation relates the renormalization constants due to the insertion of the $\phi^2(x)$ operator to the renormalization constants already present in the action: $\phi^2(x)$ is not really a new operator but is related to $\phi(x)$. Introducing the notation (in Fourier space)

$$\Gamma^{(n)}_{\phi^2}(q; p_1, \ldots, p_n) \equiv \Gamma^{(1,n)}(q; p_1, \ldots, p_n),$$

we can define renormalized correlation functions with one insertion as

$$\Gamma^{(1,n)}_{\mathrm{r}}(q; p_1, \ldots, p_n) = Z_g Z_\phi^{n/2-1} \Gamma^{(1,n)}(q; p_1, \ldots, p_n) \text{ for } n > 1,$$

which implies

$$\Gamma^{(n+1)}_{\mathrm{r}}(q, p_1, \ldots, p_n) = g_{\mathrm{r}} \Gamma^{(1,n)}_{\mathrm{r}}(q; p_1, \ldots, p_n).$$

In particular for $n = 2$ we find a renormalization condition for the function $\Gamma^{(1,2)}$

$$\Gamma^{(1,2)}_{\mathrm{r}}(0, 0, 0) = 1.$$

For $n = 1$ instead we set

$$\Gamma^{(1,1)}_{\mathrm{r}}(p) = Z_g Z_\phi^{-1/2} \Gamma^{(1,1)}(p) + \left((Z_\phi - 1)p^2 + Z_\phi m^2 - m_{\mathrm{r}}^2\right)/g_{\mathrm{r}},$$

which corresponds to

$$\Gamma^{(2)}_{\mathrm{r}}(p) = p^2 + m_{\mathrm{r}}^2 + g_{\mathrm{r}} \Gamma^{(1,1)}_{\mathrm{r}}(p),$$

and the renormalization conditions

$$\Gamma^{(1,1)}_{\mathrm{r}}(p) = O\left(p^4\right).$$

12.1.2. The relations between renormalized and bare functions lead to CS equations. Defining the operator

$$D_{\mathrm{CS}} = m_{\mathrm{r}} \frac{\partial}{\partial m_{\mathrm{r}}} + \beta(g_{\mathrm{r}}) \frac{\partial}{\partial g_{\mathrm{r}}} - \frac{n}{2}\eta(g_{\mathrm{r}}) - l\eta_2(g_{\mathrm{r}}),$$

we find for $n > 1$:

$$D_{\mathrm{CS}} \Gamma^{(l,n)}_{\mathrm{r}}(q_1, \ldots, q_l; p_1, \ldots, p_n) = m_{\mathrm{r}}^2 \sigma(g_{\mathrm{r}}) \Gamma^{(l+1,n)}_{\mathrm{r}}(0, q_1, \ldots, q_l; p_1, \ldots, p_n),$$

valid except for $n = 1, l = 1, 2$. Using renormalization conditions one immediately finds $\sigma(g_{\mathrm{r}}) = 2 - \eta(g_{\mathrm{r}})$. Moreover, since we have related the renormalization of ϕ^2 to the renormalization of ϕ and g, the corresponding RG function $\eta_2(g_{\mathrm{r}})$ is related to $\eta(g_{\mathrm{r}})$ and $\beta(g_{\mathrm{r}})$

$$\eta_2(g_{\mathrm{r}}) = \tfrac{1}{2}\eta(g_{\mathrm{r}}) - \beta(g_{\mathrm{r}})/g_{\mathrm{r}}.$$

A simple algebra yields the CS equations for the remaining functions.

Exercise 12.2

We start from the action (11.3) in the minimal subtraction scheme to establish the relation between dimension four operators. Following the method of Section 12.2 we find

$$-Z_\phi \Gamma_{\phi(x)\partial^2\phi(x)} + Z_2(g)m^2\Gamma_{\phi^2(x)} + \frac{1}{3!}gm^\varepsilon Z_g\Gamma_{\phi^4(x)} = \phi(x)\frac{\delta\Gamma}{\delta\phi(x)}. \qquad (44.24)$$

By definition of $Z_2(g)$ the insertion of $Z_2\phi^2$, where we recall that ϕ is the renormalized field, is finite

$$[\phi^2(x)]_r = Z_2(g)\phi^2(x).$$

Since the r.h.s. of equation (44.24) is finite we immediately conclude

$$[\phi(x)\partial^2\phi(x)]_r = Z_\phi\phi(x)\partial^2\phi(x) - \frac{1}{3!}gm^\varepsilon Z_g\phi^4(x).$$

Differentiating with respect to g we obtain

$$\frac{\partial\Gamma}{\partial g} = \tfrac{1}{2}Z'_\phi(g)\Gamma_{(\partial_\mu\phi)^2} + \tfrac{1}{2}m^2 Z'_2(g)\Gamma_{\phi^2} + \frac{1}{4!}m^\varepsilon\big(Z_g(g) + gZ'_g(g)\big)\Gamma_{\phi^4}. \qquad (44.25)$$

The l.h.s. is finite. We conclude that the insertion at zero momentum of the sum of the operators in the r.h.s. is also finite. We can thus define

$$[\phi^4(x)]_r = \big(Z_g(g) + gZ'_g(g)\big)\phi^4(x) - 12m^{-\varepsilon}Z'_\phi(g)\phi(x)\partial^2\phi(x)$$
$$+ 12m^{2-\varepsilon}Z'_2(g)\phi^2(x) + m^{-\varepsilon}A(g)\partial^2\phi^2(x).$$

Note that we can also differentiate with respect to the mass:

$$m\frac{\partial\Gamma}{\partial m} = m^2 Z_2(g)\Gamma_{\phi^2} + \frac{1}{4!}\varepsilon m^\varepsilon g Z_g(g)\Gamma_{\phi^4}. \qquad (44.26)$$

Equation (44.25) can be rewritten

$$\beta(g)\frac{\partial\Gamma}{\partial g} = \tfrac{1}{2}\eta(g)Z_\phi\Gamma_{(\partial_\mu\phi)^2} + \tfrac{1}{2}m^2(\eta(g) + \eta_2(g))Z_2(g)\Gamma_{\phi^2} + \frac{1}{4!}m^\varepsilon\big(2\eta(g) - \varepsilon\big)gZ_g\Gamma_{\phi^4}.$$

By integrating equation (44.24) over x and eliminating the dimension four operator insertions between (44.24, 44.26, 44.25) we recover the CS equations for correlation functions.

Moreover we can then directly calculate the elements of the 2×2 matrix which appears in the CS equations with the insertion of the two operators of dimension four. Indeed the renormalized $(\partial_\mu\phi)^2$ operator at zero momentum from (44.24) just corresponds $\int dx\,\phi(x)\delta/\delta\phi(x)$, i.e. to the multiplication of $\Gamma^{(n)}$ by n, an operation which commutes with the CS differential operator. Thus the corresponding matrix elements are zero. For what concerns ϕ^4 at zero momentum, from (44.25) we see that it is equivalent to $m^{-\varepsilon}\partial/\partial g$. The commutation of this operator with the CS operator D_{CS}

$$D_{CS} = m\frac{\partial}{\partial m} + \beta(g)\frac{\partial}{\partial g} - \tfrac{1}{2}n\eta(g),$$

yields

$$[D_{CS}, m^{-\varepsilon}\partial/\partial g] = -\varepsilon - \beta'(g) + \tfrac{1}{2}n\eta'(g).$$

The diagonal element is thus $-\varepsilon - \beta'(g)$ and the component on the operator $\phi\delta/\delta\phi = n$ is $\tfrac{1}{2}\eta'(g)$. æ æ

Chapter 13: Symmetries and Renormalization

Exercise 13.1

The action can be written

$$S(\mathbf{M}) = \int d^4x \left[\mathrm{tr}\left(\partial_\mu \mathbf{M}^\dagger \partial_\mu \mathbf{M} + m^2 \mathbf{M}^\dagger \mathbf{M} + \tfrac{1}{2}h\left(\mathbf{M}^\dagger \mathbf{M}\right)^2 \right) + \tfrac{1}{2}g\left(\mathrm{tr}\,\mathbf{M}^\dagger \mathbf{M}\right)^2 \right.$$
$$\left. - u\left(\det \mathbf{M}^\dagger + \det \mathbf{M}\right) \right]. \tag{44.27}$$

An interaction proportional to $\det \mathbf{M}$ is allowed because for a 3×3 matrix it is equivalent to a cubic interaction, owing to the identity:

$$6 \det \mathbf{M} = (\mathrm{tr}\,\mathbf{M})^3 + 2\,\mathrm{tr}\,\mathbf{M}^3 - 3\,\mathrm{tr}\,\mathbf{M}\,\mathrm{tr}\,\mathbf{M}^2.$$

Note that for the last two terms in (44.27) we have made a conventional choice of global phase for $\mathbf{M}$. When u (which we can choose positive) vanishes the symmetry group becomes instead $U(3) \times U(3)$ (cf. the $U(1)$ anomaly). Finally since

$$\mathrm{tr}\left(\mathbf{M}^\dagger \mathbf{M}\right)^2 \le \left(\mathrm{tr}\,\mathbf{M}^\dagger \mathbf{M}\right)^2 \le 3\,\mathrm{tr}\left(\mathbf{M}^\dagger \mathbf{M}\right)^2,$$

we have the restrictions or g, h positive or if one is negative

$$h - 3|g| \ge 0, \quad \text{or} \quad g - |h| \ge 0.$$

Spontaneous symmetry breaking. A general complex 3×3 matrix $\mathbf{M}$ can be written

$$\mathbf{M} = e^{i\theta/3}\,\mathbf{U}_1^\dagger \Lambda \mathbf{U}_2,$$

where U_1 and U_2 are two $SU(3)$ matrices, and Λ is diagonal with positive matrix elements λ_i. Only the determinants depend on the angle θ:

$$\det \mathbf{M}^\dagger + \det \mathbf{M} = 2\cos\theta \det \Lambda.$$

For $u > 0$ the minimum of the corresponding term is at $\theta = 0$. Minimizing the potential with respect to λ_i we then get

$$\left(m^2 + g\,\mathrm{tr}\,\Lambda^2\right)\lambda_i^2 + h\lambda_i^4 - u\det\Lambda = 0 \text{ or } \left(m^2 + g\,\mathrm{tr}\,\Lambda^2\right)\Lambda^2 + h\Lambda^4 - u\det\Lambda = 0. \tag{44.28}$$

We now assume that the parameters are such that the equation has non-trivial roots. We then find two cases

(i) $h > 0$: At $\det \Lambda$, $\mathrm{tr}\,\Lambda^2$ fixed, the equation has only one positive root in λ_i^2, and since $\lambda_i > 0$ all three λ_i are equal, $\lambda_i = \lambda$, the matrix Λ is proportional to the identity and the residual symmetry is $SU(3)$:

$$\mathbf{U}_1^\dagger \lambda \mathbf{1} \mathbf{U}_2 = \lambda \mathbf{1} \ \Rightarrow \ \mathbf{U}_1 = \mathbf{U}_2.$$

Equation (44.28) becomes:

$$(3g + h)\lambda^2 - u\lambda + m^2 = 0.$$

We know $3g + h \geq 0$ and thus the equation has positive solutions either if $m^2 < 0$ or if $u^2 - 4m^2(3g + h) \geq 0$.

(ii) $h < 0$: Again at $\det \Lambda, \operatorname{tr} \Lambda^2$ fixed, if the equation has one positive root in λ_i^2, it has two, and the λ_i can take two different values. Let us call λ the value which appears twice λ' the other one. Thus a new pattern of SSB is possible where $SU(3) \times SU(3)$ is broken down generically to $U(2)$:

$$\mathbf{U}_1^\dagger \Lambda \mathbf{U}_2 = \Lambda \text{ with } \Lambda = \begin{pmatrix} \lambda & 0 & 0 \\ 0 & \lambda & 0 \\ 0 & 0 & \lambda' \end{pmatrix}$$

$$\Rightarrow \ \mathbf{U}_1 = \mathbf{U}_2 = \mathbf{U} \text{ with } U_{i3} = U_{3i} = 0 \text{ for } i = 1, 2 \,.$$

Expressing that $\lambda \neq \lambda'$ we derive from equation (44.28) a simple relation

$$h\lambda' + u = 0 \,. \tag{44.29}$$

(iii) There is an exceptional case when the double value λ vanishes which leads to $SU(2) \times SU(2) \times U(1)$. Equation (44.28) then reduces to

$$(g + h)\lambda'^2 + m^2 = 0 \,.$$

Exercise 13.2

(i) $SU(3)$ has 8 generators. In case (i) the number of Goldstone particles is thus $8+8-8 = 8$. To identify them we consider infinitesimal $SU(3)$ transformations which do not belong to the $SU(3)$ subgroup. We can parametrize them in terms of a traceless hermitian matrix $\mathbf{H}$:

$$\mathbf{U}_1 = 1 - i\tfrac{1}{2}\mathbf{H} \,, \qquad \mathbf{U}_2 = 1 + i\tfrac{1}{2}\mathbf{H} \,.$$

Acting on the vacuum expectation value Λ we get $i\lambda\mathbf{H}$, i.e. a traceless antihermitian matrix. The conclusion are that the massless states correspond to an octet of pseudoscalar particles. There are three types of massive states, one scalar and one pseudoscalar singlet, one scalar octet.

Remark. Note that in the special case $u = 0$ the symmetry group becomes $SU(3) \times SU(3) \times U(1)$ and the same vacuum expectation value leads to a nonet of Goldstone pseudoparticles. Since such a symmetry group is generated in QCD by assuming the quarks $\mathbf{u}, \mathbf{d}, \mathbf{s}$ massless, this leads to the famous $U(1)$ problem.

Let us now consider the mass matrix corresponding to action (44.27). We set $\mathbf{M} = \lambda + (\Sigma + i\Pi)/\sqrt{2}$ and expand the different terms up quadratic order in the fields:

$$\operatorname{tr} \mathbf{M}^\dagger \mathbf{M} - \operatorname{tr} \Lambda^2 = \operatorname{tr} \left(\sqrt{2}\lambda\Sigma + \tfrac{1}{2}\Pi^2 + \tfrac{1}{2}\Sigma^2 \right)$$

$$\operatorname{tr} \left(\mathbf{M}^\dagger \mathbf{M} \right)^2 - \operatorname{tr} \Lambda^4 = \lambda^2 \operatorname{tr} \left(2\sqrt{2}\lambda\Sigma + \Pi^2 + 3\Sigma^2 \right) + \cdots$$

$$\left(\operatorname{tr} \mathbf{M}^\dagger \mathbf{M} \right)^2 - \left(\operatorname{tr} \Lambda^2 \right)^2 = \lambda^2 \operatorname{tr} \left(6\sqrt{2}\lambda\Sigma + 3\Pi^2 + 3\Sigma^2 \right) + 2\lambda^2 \left(\operatorname{tr} \Sigma \right)^2 + \cdots$$

$$\det \mathbf{M}^\dagger + \det \mathbf{M} - 2 \det \Lambda = \tfrac{1}{2}\lambda \left[\operatorname{tr} \left(2\sqrt{2}\lambda\Sigma + \Pi^2 - \Sigma^2 \right) - \left(\operatorname{tr} \Pi \right)^2 + \left(\operatorname{tr} \Sigma \right)^2 \right] + \cdots \,,$$

where we have used for the last equation $\ln \det = \operatorname{tr} \ln$. By definition of λ the coefficient of $\operatorname{tr} \Sigma$ must vanish. We then observe that $\operatorname{tr} \Pi^2$ only appears in the combination $2\sqrt{2}\lambda\Sigma +$

Π^2. Thus the coefficient of $\operatorname{tr}\Pi^2$ also vanishes. The only remaining dependence in Π is through $(\operatorname{tr}\Pi)^2$. Thus, as anticipated, the determinants give a mass to the $SU(3)$ pseudoscalar singlet. Finally we note that the scalar fields are all massive but with a different mass for the singlet and octet. Eliminating the parameter m^2 with the help of equation (44.28) we find

$$m_{\Pi_1}^2 = 3u\lambda \quad m_{\Sigma_1}^2 = (6g + 2h)\lambda^2 - u\lambda, \quad m_{\Sigma_8}^2 = 2h\lambda^2 + 2u\lambda.$$

(ii) We now expect $8 + 8 - 4 = 12$ massless Goldstone particles. To identify them we now set

$$\mathbf{U}_1 = 1 - i\tfrac{1}{2}\mathbf{H}_1, \quad \mathbf{U}_2 = 1 + i\tfrac{1}{2}\mathbf{H}_2,$$

where $\mathbf{H}_1, \mathbf{H}_2$ are traceless hermitian matrices, and impose

$$(\mathbf{H}_1)_{ij} = (\mathbf{H}_2)_{ij} \quad \text{for } i, j \in \{1, 2\} \text{ and } i = j = 3.$$

Acting on Λ we find that the Goldstone particles form a singlet, two doublets and a triplet of $SU(2)$ for what concerns the pseudoscalars, and two doublets of scalars. Again we verify these properties on the mass matrix. We set $\mathbf{M} = \lambda + (\Sigma + i\Pi)/\sqrt{2}$ and expand the different terms up to quadratic order in the fields:

$$\operatorname{tr}\mathbf{M}^\dagger\mathbf{M} - \operatorname{tr}\Lambda^2 = \operatorname{tr}\left(\sqrt{2}\Lambda\Sigma + \tfrac{1}{2}\Pi^2 + \tfrac{1}{2}\Sigma^2\right)$$

$$\operatorname{tr}\left(\mathbf{M}^\dagger\mathbf{M}\right)^2 - \operatorname{tr}\Lambda^4 = \operatorname{tr}\left(2\sqrt{2}\Lambda^3\Sigma + 2\Lambda^2\Pi^2 - (\Lambda\Pi)^2 + 2\Lambda^2\Sigma^2 + (\Lambda\Sigma)^2\right) + \cdots$$

$$\left(\operatorname{tr}\mathbf{M}^\dagger\mathbf{M}\right)^2 - \left(\operatorname{tr}\Lambda^2\right)^2 = \operatorname{tr}\Lambda^2 \operatorname{tr}\left(2\sqrt{2}\Lambda\Sigma + \Pi^2 + \Sigma^2\right) + 2\left(\operatorname{tr}\Lambda\Sigma\right)^2 + \cdots$$

$$\det\mathbf{M}^\dagger + \det\mathbf{M} - 2\det\Lambda = \tfrac{1}{2}\det\Lambda\left[\operatorname{tr}\left(2\sqrt{2}\Lambda^{-1}\Sigma + \left(\Lambda^{-1}\Pi\right)^2 - \left(\Lambda^{-1}\Sigma\right)^2\right)\right.$$

$$\left. - \left(\operatorname{tr}\Lambda^{-1}\Pi\right)^2 + \left(\operatorname{tr}\Lambda^{-1}\Sigma\right)^2\right] + \cdots.$$

Using eqs. (44.28,44.29), one verifies that all terms depending on the Π field cancel except the term proportional to $u(\operatorname{tr}\Lambda^{-1}\Pi)^2$ which corresponds to a $SU(2)$ singlet. In the same way one checks that the coefficients of the terms $|\Sigma_{i3}|^2$, $i = 1, 2$, vanish.

(iii) The last case can be treated in a completely similar way. The 9 Goldstone bosons are 5 pseudoscalars, a singlet and a $(\tfrac{1}{2}, 0) \oplus (0, \tfrac{1}{2})$ representation of $SU(2) \times SU(2)$ and 4 scalars in the same representation.

Exercise 13.3

With the linear symmetry breaking term, the residual symmetry is the diagonal $SU(2)$ subgroup of case (ii) of exercise 13.2. The matrix $\mathbf{C}$ is a diagonal matrix similar to Λ. Its addition allows to use λ, λ' as free parameters, related to $\mathbf{C}$ by:

$$\mathbf{C} = \left(m^2 + g\operatorname{tr}\Lambda^2\right)\Lambda + h\Lambda^3 - u\det\Lambda\Lambda^{-1}.$$

One finds for the pseudoscalar particles the following analytic expressions:

$$m_{\Pi_3}^2 \equiv m_\pi^2 = m^2 + g\left(2\lambda^2 + \lambda'^2\right) + h\lambda^2 - u\lambda',$$

$$m_{\Pi_2}^2 \equiv m_K^2 = m_\pi^2 + (u + h\lambda')(\lambda' - \lambda),$$

$$m_\eta^2 = m_\pi^2 + \tfrac{1}{2}h\left(\lambda'^2 - \lambda^2\right) + \tfrac{3}{2}u\lambda' - \tfrac{1}{2}\left[\left(h\left(\lambda^2 - \lambda'^2\right) + u\lambda'\right)^2 + 8u^2\lambda^2\right]^{1/2},$$

$$m_{\eta'}^2 = m_\pi^2 + \tfrac{1}{2}h\left(\lambda'^2 - \lambda^2\right) + \tfrac{3}{2}u\lambda' + \tfrac{1}{2}\left[\left(h\left(\lambda^2 - \lambda'^2\right) + u\lambda'\right)^2 + 8u^2\lambda^2\right]^{1/2}.$$

Again in the limit $u \to 0$ one finds results inconsistent with experimental data since then $m_\eta = m_\pi$ and $m_K \sim m_{\eta'}$.

For the scalars the masses are

$$m^2_{\Sigma_3} = m^2_\pi + 2h\lambda^2 + 2u\lambda',$$

$$m^2_{\Sigma_2} \equiv m^2_\kappa = m^2_\pi + (u + h\lambda')(\lambda + \lambda'),$$

$$m^2_{\Sigma_1\pm} = m^2_\pi + \tfrac{1}{2}h(\lambda^2 + 3\lambda'^2) + g(2\lambda^2 + \lambda'^2) + \tfrac{1}{2}u\lambda'g \pm \tfrac{1}{2}\sqrt{(A-C)^2 + 4B^2},$$

with we have defined

$$A = (3h + 4g)\lambda^2 - u\lambda'$$

$$B = \sqrt{2}\lambda(2g\lambda' - u)$$

$$C = (3h + 2g)\lambda'^2.$$

Consistency with experimental masses requires

$$\lambda' > \lambda, \qquad -\frac{1}{\lambda'} < \frac{h}{u} < \frac{1}{\lambda' - \lambda}.$$

If we determine the parameters by imposing the physical values of the pseudoscalar masses we find the solution

$$\lambda'/\lambda = 1.80..., \quad u\lambda/m_\pi = 10.4..., \quad h\lambda^2/m^2_\pi = 2.42... .$$

The last coupling constant only appears in the scalar masses and is thus poorly determined $g\lambda^2/m^2_\pi \sim 10$. Finally the matrix $\mathbf{C}$ has on its diagonal $\lambda m^2_\pi, \lambda m^2_\pi, (\lambda + \lambda')m^2_K - \lambda m^2_\pi$.

Chapter 14: The Non-Linear σ-Model: An Example of Non-Linearly Realized Symmetries

Exercise 14.1

One first verifies that the space dependence plays no role and can thus be omitted. Then

$$[\Delta_i, \Delta_j] = \left[\frac{\partial S}{\partial \pi_i}\frac{\partial^2 S}{\partial \pi_j \partial H} + \frac{\partial S}{\partial H}\frac{\partial^2 S}{\partial \pi_i \partial \pi_j}\right]\frac{\partial}{\partial H} + \left[\frac{\partial S}{\partial \pi_i}\frac{\partial^2 S}{\partial H \partial H} + \frac{\partial S}{\partial H}\frac{\partial^2 S}{\partial \pi_i \partial H}\right]\frac{\partial}{\partial \pi_j}$$
$$- (i \leftrightarrow j).$$

Differentiating equation (14.35) with respect to π_j and H respectively we obtain

$$\frac{\partial S}{\partial \pi_i}\frac{\partial^2 S}{\partial \pi_j \partial H} + \frac{\partial S}{\partial H}\frac{\partial^2 S}{\partial \pi_i \partial \pi_j} + H\delta_{ij} = 0,$$

$$\frac{\partial S}{\partial \pi_i}\frac{\partial^2 S}{\partial H \partial H} + \frac{\partial S}{\partial H}\frac{\partial^2 S}{\partial \pi_i \partial H} + \pi_i = 0.$$

It follows

$$[\Delta_i, \Delta_j] = -\Delta_{ij}.$$

To calculate the commutator $[\Delta_{ij}, \Delta_k]$ we note that

$$\Delta_{ij}S = 0, \qquad \Delta_{ij}\frac{\partial S}{\partial H} = 0,$$

as a consequence of the $O(N-1)$ invariance of the action. Also

$$\Delta_{ij}\frac{\partial S}{\partial \pi_k} - \frac{\partial}{\partial \pi_k}\Delta_{ij}S = -\delta_{ik}\frac{\partial S}{\partial \pi_j} + \delta_{jk}\frac{\partial S}{\partial \pi_i}.$$

Therefore

$$[\Delta_{ij}, \Delta_k] = \left[\delta_{jk}\frac{\partial S}{\partial \pi_i} - \delta_{ik}\frac{\partial S}{\partial \pi_j}\right]\frac{\partial}{\partial H} + \frac{\partial S}{\partial H}\left(\delta_{jk}\frac{\partial}{\partial \pi_i} - \delta_{ik}\frac{\partial}{\partial \pi_j}\right) = \delta_{jk}\Delta_i - \delta_{ik}\Delta_j.$$

Exercise 14.2

We consider the partition function of the $O(4) \sim SU(2) \times SU(2)$ non-linear σ-model

$$Z = \int \left[(f^2 - \pi^2(x))^{-1/2} \mathrm{d}\pi(x) \right] \exp\left[-S(\pi, c) \right], \qquad (44.30)$$

with the action

$$S(\pi, c) = \int \mathrm{d}^d x \left\{ \frac{1}{2} \left[(\partial_\mu \pi(x))^2 + \frac{(\pi \cdot \partial_\mu \pi(x))^2}{f^2 - \pi^2(x)} \right] - c\sqrt{f^2 - \pi^2(x)} \right\}. \qquad (44.31)$$

It what follows it will be convenient to set $c = \mu^2 f$.

We propose to study this model at one-loop order, to verify in particular that it can be renormalized with only two renormalization constants Z, Z_f as proven in Chapter 14.

It will be convenient to set $d = 2 + \varepsilon$ and to define a dimensionless coupling constant g:

$$f^{-2}\Gamma(1 - \varepsilon/2)\mu^\varepsilon (4\pi)^{-\varepsilon/2} = g.$$

Dimensional regularization will everywhere be used.

    14.2.1.

$$\Delta_2(p) = \frac{1}{p^2 + \mu^2},$$

$$V^{(4)} = \frac{1}{8f^2} \delta_{i_1 i_2} \delta_{i_3 i_4} \left((p_1 + p_2)^2 + \mu^2 \right),$$

$$V^{(6)} = \frac{1}{16f^4} \delta_{i_1 i_2} \delta_{i_3 i_4} \delta_{i_5 i_6} \left((p_1 + p_2)^2 + \mu^2 \right).$$

14.2.2. The renormalized action reads

$$S_\mathrm{r}(\pi) = \frac{Z Z_f}{2} \int \mathrm{d}^d x \left[(\partial_\mu \pi)^2 + (\partial_\mu \sigma)^2 \right] - \mu^2 f \int \sigma(x) \mathrm{d}^d x, \qquad (44.32)$$

in which σ is now given by:

$$\sigma(x) = \left[f^2 Z^{-1} - \pi^2 \right]^{1/2}.$$

Thus in terms of the renormalized fields Δ_2^{-1} gets an extra factor $Z_f Z$, $V^{(4)}$ a factor $Z_f Z^2$ and $V^{(6)}$ a factor $Z_f Z^3$. Moreover μ^2 is replaced by $\mu^2 Z^{-1/2} Z_f^{-1}$.

    14.2.3. *The vacuum expectation.* $\langle \sigma \rangle$ is given by

$$= f Z^{-1/2} - \tfrac{1}{2} f^{-1} \langle \pi^2 \rangle = f \left(Z^{-1/2} - \frac{3}{2f^2} I \right),$$

where I is an integral which appears everywhere at one-loop

$$I = \frac{1}{(2\pi)^d} \int \frac{\mathrm{d}^d p}{p^2 + \mu^2} = \mu^{d-2} \frac{1}{(4\pi)^{d/2}} \Gamma(1 - d/2).$$

If we then choose

$$Z = 1 + \frac{3g}{2\pi\varepsilon} + O\left(g^2\right),$$

we get $\langle \sigma \rangle = f$.

The 2-point function. The 2-point function is proportional to δ_{ij} which we omit in what follows. Then

$$\Gamma^{(2)}(p) = Z Z_f p^2 + Z^{1/2}\mu^2 + \frac{1}{f^2}\left(p^2 + \tfrac{3}{2}\mu^2\right)\frac{1}{(2\pi)^d}\int \frac{d^d q}{q^2 + \mu^2} + O\left(f^{-4}\right).$$

We verify that the value of Z obtained above renders the coefficient of μ^2 finite. From the coefficient of p^2 we find that if we choose

$$Z_f = 1 - \frac{2g}{2\pi\varepsilon},$$

the 2-point is finite and reduces to its tree approximation

$$\Gamma^{(2)}(p) = p^2 + \mu^2 + O\left(f^{-4}\right).$$

14.2.4. For symmetry reasons the 4-point function $\tilde{W}^{(4)}_{\text{amp.}}$ has the form

$$\left(\tilde{W}^{(4)}_{\text{amp.}}\right)_{ijkl} = C\left(p_1, p_2, p_3, p_4\right)\delta_{ij}\delta_{kl} + C\left(p_1, p_3, p_2, p_4\right)\delta_{ik}\delta_{jl} + C\left(p_1, p_4, p_3, p_2\right)\delta_{il}\delta_{kj},$$

(44.33)

where $C\left(p_1, p_2, p_3, p_4\right)$ is symmetric in the exchange $p_1 \leftrightarrow p_2$ with $p_3 \leftrightarrow p_4$ and $\{p_1, p_2\} \leftrightarrow \{p_3, p_4\}$. In what follows we only give C.

It is convenient to introduce the notation

$$p_{ij} = p_i + p_j.$$

At tree level $\Gamma^{(4)}$ is immediately obtained from the 4-point vertex:

$$C = -\frac{1}{f^2}\left(Z_f Z^2 p_{12}^2 + Z^{3/2}\mu^2\right) = -\frac{1}{f^2}\left[\left(1 + \frac{4g}{2\pi\varepsilon}\right)p_{12}^2 + \left(1 + \frac{9}{2}\frac{g}{2\pi\varepsilon}\right)\mu^2\right].$$

At one-loop both the 4-point and the 6-point interactions contribute. For what concerns the 6-point vertex we have just to contract 2 external legs. One finds

$$\delta_1 C = -\frac{1}{f^4}\left(5p_{12}^2 + p_1^2 + p_2^2 + p_3^2 + p_4^2 + \tfrac{13}{2}\mu^2\right)I.$$

Then we have the contributions of two 4-point vertices. The following identity is useful

$$\frac{1}{(2\pi)^d}\int \frac{d^d q\, q_\mu q_\nu}{\left[(q + p/2)^2 + \mu^2\right]\left[(q - p/2)^2 + \mu^2\right]} = S\delta_{\mu\nu} + Tp_\mu p_\nu,$$

with

$$(d-1)S = \tfrac{1}{2}I - \tfrac{1}{4}\left(p^2 + 4\mu^2\right)B(p)$$

$$T = \frac{1}{(d-1)p^2}\left[\tfrac{1}{2}(d-2)I + \tfrac{1}{4}\left(p^2 + 4\mu^2\right)B(p)\right]$$

where we defined the bubble diagram B

$$B(p) = \frac{1}{(2\pi)^d} \int \frac{d^d q}{(q^2 + \mu^2)\left[(p+q)^2 + \mu^2\right]}.$$

Note the useful relation

$$\tfrac{1}{2}(d-2)I = -\mu^2 B(0).$$

We then find that C gets four contributions, which in the picture of faithful diagrams correspond to a bubble, a triangle and a square box direct and crossed. The sum of the two first contributions is

$$f^4\delta_2 C = \tfrac{3}{2}\left(p_{12}^2 + \mu^2\right)^2 B\left(p_{12}\right) + \left(p_{12}^2 + \mu^2\right)\left[2I - \left(p_{12}^2 - \tfrac{1}{2}\left(p_1^2 + p_2^2 + p_3^2 + p_4^2\right)\right)B\left(p_{12}\right)\right].$$

The direct box yields

$$f^4\delta_3 C = \frac{1}{d-1}\left[\left(p_{14}^2 - p_1^2 - p_4^2\right) - \tfrac{1}{2}(d-2)\left(p_{13}^2 + p_1^2 + p_4^2 - p_2^2 - p_3^2\right.\right.$$
$$\left.\left. - \frac{(p_1^2 - p_3^2)(p_2^2 - p_4^2)}{p_{13}^2}\right)\right]I + \left(p_1^2 + p_4^2\right)I - \frac{1}{4(d-1)}\left(p_{13}^2 + 4\mu^2\right)$$
$$\times\left[2\left(p_{14}^2 - p_1^2 - p_4^2\right) + \left(p_{13}^2 + p_1^2 + p_4^2 - p_2^2 - p_3^2 - \frac{(p_1^2 - p_3^2)(p_2^2 - p_4^2)}{p_{13}^2}\right)\right]$$
$$\times B\left(p_{13}\right) + \tfrac{1}{4}\left(p_{13}^2 - p_1^2 - p_3^2\right)\left(p_{13}^2 - p_2^2 - p_4^2\right)B\left(p_{13}\right).$$

The terms have been written the way they appear in the calculation. They can be slightly rearranged

$$f^4\delta_3 C = \frac{1}{d-1}\left[p_{14}^2 + \tfrac{1}{2}(d-2)\left(p_1^2 + p_2^2 + p_3^2 + p_4^2 - p_{13}^2 + \frac{(p_1^2 - p_3^2)(p_2^2 - p_4^2)}{p_{13}^2}\right)\right]I$$
$$- \frac{(p_{13}^2 + 4\mu^2)}{4(d-1)}\left(2p_{14}^2 + p_{13}^2 - p_1^2 - p_2^2 - p_3^2 - p_4^2 - \frac{(p_1^2 - p_3^2)(p_2^2 - p_4^2)}{p_{13}^2}\right)$$
$$\times B\left(p_{13}\right) + \tfrac{1}{4}\left(p_{13}^2 - p_1^2 - p_3^2\right)\left(p_{13}^2 - p_2^2 - p_4^2\right)B\left(p_{13}\right).$$

The last contribution is simply $\delta_3 C(p_1, p_2, p_4, p_3)$.

After renormalization the total one-loop contribution can be written

$$f^4\delta C = \tfrac{1}{2}\left(p_{12}^2 + \mu^2\right)\left(p_{12}^2 + p_1^2 + p_2^2 + p_3^2 + p_4^2 + 3\mu^2\right)B\left(p_{12}\right)$$
$$+ \frac{1}{4(d-1)}\left(2p_{14}^2 + p_{13}^2 - p_1^2 - p_2^2 - p_3^2 - p_4^2 - \frac{(p_1^2 - p_3^2)(p_2^2 - p_4^2)}{p_{13}^2}\right)$$
$$\times\left[4\mu^2 B(0) - \left(p_{13}^2 + 4\mu^2\right)B\left(p_{13}\right)\right] + \tfrac{1}{4}\left(p_{13}^2 - p_1^2 - p_3^2\right)\left(p_{13}^2 - p_2^2 - p_4^2\right)B\left(p_{13}\right)$$
$$+ \frac{1}{4(d-1)}\left(2p_{13}^2 + p_{14}^2 - p_1^2 - p_2^2 - p_3^2 - p_4^2 - \frac{(p_1^2 - p_4^2)(p_2^2 - p_3^2)}{p_{14}^2}\right)$$
$$\times\left[4\mu^2 B(0) - \left(p_{14}^2 + 4\mu^2\right)B\left(p_{14}\right)\right] + \tfrac{1}{4}\left(p_{14}^2 - p_1^2 - p_4^2\right)\left(p_{14}^2 - p_2^2 - p_3^2\right)B\left(p_{14}\right),$$

expression finite for all $d < 4$.

14.2.5. It is convenient to introduce some additional notation to take into account the $O(3)$ symmetry. Let us set

$$\Gamma_{ij}^{(2)}(p) = \delta_{ij} G(p),$$

$$\tilde{\Gamma}_{ij}^{(3)}(p_1, p_2; p_3) = \delta_{ij} V(p_1, p_2; p_3),$$

in which our conventions are that indices correspond of course to π-fields, and in mixed π–σ correlation functions the arguments of the π-fields are placed first.

In the non-linear σ-model we have the relation

$$\tilde{\Gamma}_{ijkl}^{(4)}(p_1, p_2, p_3, p_4) = \left(\tilde{W}_{\mathrm{amp.}}^{(4)}\right)_{ijkl}(p_1, p_2, p_3, p_4),$$

for which we use the parametrization (44.33).

They are two ways to derive WT identities for correlation functions. One can either start from the WT identities for connected correlation functions which are the same as for the linear σ-model, or from the WT identities for the 1PI functional. In the latter case differentiating respectively with respect first to π (and setting $H(x) = c$), and then π and H one finds, after Fourier transformation,

$$G(0)\langle\sigma\rangle = c, \tag{44.34}$$

$$\tilde{\Gamma}^{(2)}(p)\tilde{\Gamma}_{ij}^{(2)}(p) + \delta_{ij} = \langle\sigma\rangle\,\tilde{\Gamma}_{ij}^{(3)}(0, -p; p). \tag{44.35}$$

The second equation can be rewritten

$$\tilde{\Gamma}^{(2)}(p)G(p) + 1 = \langle\sigma\rangle V(0, -p; p). \tag{44.36}$$

Finally differentiating three times with respect to π we find a relation between 3 and 4-point correlation functions:

$$-\langle\sigma\rangle\,\tilde{\Gamma}_{ijkl}^{(4)}(0, p_2, p_3, p_4) + \tilde{\Gamma}_{ij}(p_2)\tilde{\Gamma}_{kl}^{(3)}(p_3, p_4; p_2) + \; 2 \text{ terms} = 0, \tag{44.37}$$

which implies

$$\langle\sigma\rangle C(0, p_2, p_3, p_4) = G(p_2) V(p_3, p_4; p_2).$$

Setting $p_2^2 = -m_\pi^2$ one recovers from the last equation Adler's consistency conditions

$$C(0, p_2(p_2^2 = -m_\pi^2), p_3, p_4) = 0.$$

Eliminating V between eqs. (44.36) and (44.37) one derives Weinberg's relation

$$\frac{\partial}{\partial p^2}\left((\langle\sigma\rangle)^2 C(0, p, 0, -p) + G(p)\right)\Big|_{p^2=-m_\pi^2} = 0.$$

The verification of these identities on the explicit expressions is the straightforward.

A few additional details can be found in

D. Bessis and J. Zinn-Justin, *Phys. Rev.* D5 (1972) 1313; J. Zinn-Justin, in the proceedings of Les Houches summer school 1971, C. Itzykson ed., Gordon and Breach, New-York 1973,

where the model is studied as a limit of the linear σ-model.

Chapter 15: Models on Homogeneous Spaces in Two Dimensions

Exercise 15.1

Let us first verify that equation (15.42) is indeed equivalent to equation (15.43):

$$[\Delta_\alpha, \Delta_\beta] = D_a^\alpha \frac{\partial}{\partial \pi_a} D_b^\beta \frac{\partial}{\partial \pi_b} - D_a^\beta \frac{\partial}{\partial \pi_a} D_b^\alpha \frac{\partial}{\partial \pi_b}.$$

The only surviving terms correspond to the most left derivatives acting on D and thus

$$[\Delta_\alpha, \Delta_\beta] = D_a^\alpha \frac{\partial D_b^\beta}{\partial \pi_a} \frac{\partial}{\partial \pi_b} - D_a^\beta \frac{\partial D_b^\alpha}{\partial \pi_a} \frac{\partial}{\partial \pi_b}.$$

With the explicit form (15.128) we find

$$D_a^\alpha \frac{\partial D_b^\beta}{\partial \pi_a} = [t_{ac}^\alpha \pi_c + t_{aj}^\alpha \sigma_j] \left(t_{ba}^\beta + t_{bk}^\beta \frac{\partial \sigma_k}{\partial \pi_a} \right).$$

Antisymmetrizing in α, β we find commutators of generators t^α except for the terms proportional to $\partial \sigma_k / \partial \pi_a$. However we then use equation (15.29). This reduce again the calculation to commutation relations and we find the anticipated result.

Exercise 15.2

Let χ^i be another set of coordinates on the manifold. Then the infinitesimal group transformations take the form

$$\delta_\omega \chi^i = \frac{\partial \chi^i}{\partial \phi^j} D_\alpha^j \omega_\alpha \equiv \partial_j \chi^i D_\alpha^j \omega_\alpha.$$

We want thus to prove that the new functions Δ_α^i:

$$\Delta_\alpha^i = \partial_j \chi^i D_\alpha^j,$$

satisfy the commutation relations (15.43).

$$\Delta_\alpha^j \frac{\partial \Delta_\beta^i}{\partial \chi^j} = \partial_k \chi^j D_\alpha^k \left[\partial_l \chi^j \right]^{-1} \left(\partial_l \partial_m \chi^i D_\beta^m + \partial_m \chi^i \partial_l D_\beta^m \right)$$

$$= D_\alpha^l \left(\partial_l \partial_m \chi^i D_\beta^m + \partial_m \chi^i \partial_l D^m \beta \right).$$

We now see that the first term is symmetric in the exchange $\alpha \leftrightarrow \beta$ and thus cancels in the commutator, while the second term is proportional to the term in (15.43). The result follows. Actually using the results of Chapter 22, one can verify that, as in the case of the curl of a vector, the relation (15.43) can be rewritten by replacing the derivative by a covariant derivative when the Christoffel connection is symmetric.

Exercise 15.3

Let us verify the identity (A15.7) for the tensor

$$g^{ij} = D^i_\beta(\varphi) D^j_\beta(\varphi).$$

In terms of g^{ij} it reads

$$0 = D^i_\alpha \frac{\partial g^{jk}}{\partial \varphi^i} - g^{jl} \frac{\partial D^k_\alpha}{\partial \varphi^l} - g^{kl} \frac{\partial D^j_\alpha}{\partial \varphi^l}$$

$$= D^k_\beta \left(D^i_\alpha \partial_i D^j_\beta - D^i_\beta \partial_i D^j_\alpha \right) + D^j_\beta \left(D^i_\alpha \partial_i D^k_\beta - D^i_\beta \partial_i D^k_\alpha \right).$$

We now use the commutation relations (15.43) we find

$$0 = D^k_\beta f_{\alpha\beta\gamma} D^j_\gamma + D^j_\beta f_{\alpha\beta\gamma} D^k_\gamma,$$

which is satisfied because the structure constants are antisymmetric in $\beta \leftrightarrow \gamma$.

Exercise 15.4

The Lie algebra of $O(4)$ is the same as the Lie algebra of $O(3) \times O(3)$. The infinitesimal transformations of parameters ω, ω' can be written

$$\delta \pi_i = \epsilon_{ijk} \omega_j \pi_k + \omega'_i \sigma, \quad \Rightarrow \quad \delta \sigma = -\omega'_i \pi_i,$$

where we have chosen π_i and $\sigma = \sqrt{1 - \pi^2}$ as the coordinates on the sphere. We thus find

$$\gamma^{ij} = \sum_k \left(\epsilon_{ikl} \pi_l \epsilon_{jkm} \pi_m + \delta_{ik} \sigma \delta_{jk} \sigma \right) = \delta_{ij} - \pi_i \pi_j,$$

or calculating the inverse

$$g_{ij} = \delta_{ij} + \pi_i \pi_j / (1 - \pi^2),$$

in which we recognize the metric on S_3.

Chapter 18: Abelian Gauge Theories

Exercise 18.1

Let us differentiate Gauss's law

$$\frac{1}{i} \partial_i \frac{\delta}{\delta A_i(x)} \Psi(\mathbf{A}) = J_d(x,t) \Psi(\mathbf{A}),$$

with respect to time, and use Schrödinger's equation

$$\frac{1}{i} \partial_i \frac{\delta}{\delta A_i(x)} \mathcal{H} \Psi(\mathbf{A}) = i \dot{J}_d(x,t) \Psi(\mathbf{A}) + J_d(x,t) \mathcal{H} \Psi(\mathbf{A}). \tag{44.38}$$

We then use the property that in the absence of source the generator of time-independent gauge transformations commutes with the hamiltonian $\mathcal{H}$

$$\left[\frac{1}{i} \partial_i \frac{\delta}{\delta A_i(x)}, \mathcal{H} \right] = i \partial_i J_i.$$

After commutation in the l.h.s. of (44.38) and again application of Gauss's law we find:

$$-i \partial_i J_i \Psi(\mathbf{A}) = i \dot{J}_d(x,t) \Psi(\mathbf{A}),$$

which follows from current conservation.

Exercise 18.2

The photon propagator reads

$$\Delta_{\mu\nu}(k) = \frac{\delta_{\mu\nu}}{k^2 + m^2}. \tag{44.39}$$

The fermion propagator reads:

$$\Delta_{\mathrm{F}}(p) = \frac{(M - i\not{p})}{p^2 + M^2}, \tag{44.40}$$

and finally there is the $\bar{\psi}\psi A$ vertex V_μ:

$$V_\mu(q; p_1, p_2) = -ie\gamma_\mu. \tag{44.41}$$

At one-loop order, in the Feynman gauge, and for a massless vector field, the fermion 2-point function is then given by:

$$\Gamma^{(2)}(p) = M + i\not{p} + \frac{e^2}{(2\pi)^d} \int \frac{d^d k}{k^2} \gamma_\mu \left[M + i(\not{p} + \not{k})\right]^{-1} \gamma_\mu.$$

We then use the identities:

$$\gamma_\mu \gamma_\mu = d, \quad \gamma_\mu \not{v} \gamma_\mu = (2 - d)\not{v}.$$

Projecting then $\Gamma^{(2)}$ onto its components

$$\Gamma^{(2)}(p) = A\left(p^2\right) + i\not{p}B\left(p^2\right), \tag{44.42}$$

we can replace $\not{k}$ by $\not{p}\,k \cdot p/p^2$. We obtain:

$$A = M\left[1 + \frac{e^2 d}{(2\pi)^d} \int \frac{d^d k}{k^2\left(M^2 + (k + p)^2\right)}\right],$$

$$B = 1 + \frac{e^2(d - 2)}{(2\pi)^d p^2} \int \frac{d^d k\left(p^2 + k \cdot p\right)}{k^2\left(M^2 + (k + p)^2\right)}.$$

Replacing $M + i\not{p}$ by $Z_\psi(Z_M M + i\not{p})$ we derive the renormalization constants in the minimal subtraction scheme:

$$Z_M = 1 - 3e^2\frac{N_d}{\varepsilon}, \quad Z_\psi = 1 - e^2\frac{N_d}{\varepsilon}.$$

In the notation of Section 18.9 it follows ($\alpha = e^2/4\pi$):

$$\eta_M = \frac{3}{2\pi}\alpha, \quad \eta_\psi = \frac{1}{2\pi}\alpha.$$

Exercise 18.3

We now consider the QED action (18.26), with a mass term for the photon and quantized in a general covariant gauge

$$S\left(A_\mu, \bar\psi, \psi\right) = \int \mathrm{d}^d x \left[\tfrac{1}{4} F_{\mu\nu}^2 + \tfrac{1}{2} m^2 A_\mu^2 + \frac{1}{2\xi}\left(\partial_\mu A_\mu\right)^2 - \bar\psi\left(\slashed{D} + M\right)\psi\right]. \qquad (44.43)$$

18.3.1. From identity (18.81) we derive

$$k_\mu \Gamma_{\mu\nu}^{(2)}(k) = k_\nu\left(k^2/\xi + m^2\right),$$

$$k_\mu \Gamma_\mu^{(3)}(k; p; q) = e\Gamma^{(2)}(p + k; q) - e\Gamma^{(2)}(p; q + k),$$

$$k_\mu W_{\mu\nu}^{(2)}(k) = \xi \frac{k_\nu}{k^2 + \xi m^2},$$

$$\left(k^2/\xi + m^2\right) k_\mu W_\mu^{(3)}(k; p; q) = eW^{(2)}(p + k; q) - eW^{(2)}(p; q + k).$$

The first equation implies

$$\Gamma_{\mu\nu}^{(2)}(k) = \Gamma_{\mathrm{T}}(k)\left(k^2 \delta_{\mu\nu} - k_\mu k_\nu\right) + m^2 \delta_{\mu\nu} + k_\mu k_\nu/\xi. \qquad (44.44)$$

This identity for $\Gamma_\mu^{(3)}$ can made even more explicit by parametrizing $\Gamma^{(2)}$ as:

$$\Gamma^{(2)}(p) = A\left(p^2\right) + i\slashed{p} B\left(p^2\right), \qquad (44.45)$$

where:

$$A = M + O(\alpha), \quad B = 1 + O(\alpha).$$

18.3.2. The inverse fermion 2-point function $\Gamma^{(2)}(p)$ at one-loop order is

$$\Gamma^{(2)}(p) = M + i\slashed{p} + e^2 \int \frac{\mathrm{d}^d k}{(2\pi)^d} \frac{\gamma_\mu\left(M + i\slashed{k} - i\slashed{p}\right)\gamma_\nu}{(k - p)^2 + M^2}$$

$$\left(\frac{\delta_{\mu\nu}}{k^2 + m^2} + \frac{(\xi - 1)k_\mu k_\nu}{(k^2 + m^2)(k^2 + \xi m^2)}\right). \qquad (44.46)$$

We set

$$\Gamma^{(2)}(p) = MA_1(p) + i\slashed{p} A_2(p). \qquad (44.47)$$

We then use a few standard tricks:

$$\gamma_\mu \gamma_\mu = d, \quad \gamma_\mu \slashed{p} \gamma_\mu = (2 - d)\slashed{p}, \quad \slashed{k}\slashed{p}\slashed{k} = 2pk\slashed{k} - k^2\slashed{p},$$

$$\int \mathrm{d}^d k \,\slashed{k} f\left(k^2, p^2, kp\right) = \slashed{p} \int \mathrm{d}^d k \frac{kp}{p^2} f\left(k^2, p^2, kp\right),$$

Repeated use of the trivial identity

$$2pk = k^2 + m_1^2 - \left[(k - p)^2 + m_2^2\right] + p^2 + m_2^2 - m_1^2,$$

leads to

$$\int \frac{\mathrm{d}^d k}{(2\pi)^d} \frac{2kp}{(k^2 + m_1^2)[(k - p)^2 + m_2^2]} = I(m_2) - I(m_1) + \left(p^2 + m_2^2 - m_1^2\right) B(p; m_1, m_2),$$

$$\int \frac{\mathrm{d}^d k}{(2\pi)^d} \frac{(2kp)^2}{(k^2 + m_1^2)[(k - p)^2 + m_2^2]} = \left(3p^2 + m_2^2 - m_1^2\right) I(m_2)$$

$$\overset{*}{-} \left(p^2 + m_2^2 - m_1^2\right) I(m_1) + \left(p^2 + m_2^2 - m_1^2\right)^2 B(p; m_1, m_2).$$

It is easy to obtain the coefficient of M in eq. (44.47). Setting $m' = m\sqrt{\xi}$ one finds

$$A_1 = 1 + e^2 \left[(d-1)B(p; M, m) + \xi B(p; M, m') \right].$$

Let us then write A_2 as the sum of terms

$$A_2 = 1 + \frac{e^2}{2p^2} \left(A_{21} + A_{22} \right),$$

where A_{21} corresponds to $\xi = 1$, i.e. is the result in the Feynman gauge, and A_{22} the remaining part. Using the previous remarks A_{21} can easily be evaluated

$$A_{21} = (2-d) \left[I(M) - I(m) - \left(p^2 - M^2 + m^2 \right) B(p; M, m) \right].$$

The remaining contribution can be written

$$A_{22} = (\xi - 1) \int \frac{d^d k}{(2\pi)^d} \frac{2k^2 p^2 + 2(kp)k^2 - 4(kp)^2}{[(k-p)^2 + M^2](k^2 + m^2)(k^2 + \xi m^2)}.$$

We then use the identities

$$\frac{(\xi - 1)k^2}{(k^2 + m^2)(k^2 + \xi m^2)} = \frac{\xi}{k^2 + \xi m^2} - \frac{1}{k^2 + m^2}$$

$$\frac{(\xi - 1)}{(k^2 + m^2)(k^2 + \xi m^2)} = \frac{1}{m^2} \left(\frac{1}{k^2 + m^2} - \frac{1}{k^2 + \xi m^2} \right).$$

We obtain

$$
\begin{aligned}
A_{22} = {} & \left\{ \xi \left[I(M) - I(m') + \left(3p^2 + M^2 - m'^2 \right) B(p; M, m') \right] \right. \\
& \left. - \left[I(M) - I(m) + \left(3p^2 + M^2 - m^2 \right) B(p; M, m) \right] \right\} + (1 - \xi)I(M) \\
& + \frac{1}{m^2} \left\{ \left[-\left(p^2 + M^2 - m'^2 \right) I(m') + \left(p^2 + M^2 - m'^2 \right)^2 B(p; M, m') \right] \right. \\
& \left. + \left[\left(p^2 + M^2 - m^2 \right) I(m) - \left(p^2 + M^2 - m^2 \right)^2 B(p; M, m) \right] \right\}.
\end{aligned}
$$

Our final expression for A_2 is then:

$$
\begin{aligned}
A_2 = {} & 1 + e^2 \frac{2-d}{2p^2} \left[I(M) - I(m) - \left(p^2 - M^2 + m^2 \right) B(p; M, m) \right] \\
& + \frac{e^2}{2m^2 p^2} \left\{ \left(p^2 + M^2 \right) \left[I(m) - I(m') \right] + \left[\left(p^2 + M^2 \right)^2 + \left(p^2 - M^2 \right) m'^2 \right] \right. \\
& \left. \times B(p; M, m') - \left[\left(p^2 + M^2 \right)^2 + \left(p^2 - M^2 \right) m^2 \right] B(p; M, m) \right\}.
\end{aligned}
$$

18.3.3. At one-loop order the fermion mass M_F is given by

$$M_F = M \frac{A_1(iM)}{A_2(iM)}.$$

A short calculation leads to

$$\frac{M_F}{M} = 1 + e^2 \left\{ \frac{2-d}{2M^2} \left[I(M) - I(m) \right] + \left[2 + (d-2) \frac{m^2}{2M^2} \right] B(iM; M, m) \right\}.$$

One verifies that the gauge dependence has cancelled. Indeed the fermion mass is a physical observable and it has been shown that physical observables are gauge independent.

18.3.4. At one-loop order $\Gamma_\mu^{(3)}$ reads

$$\Gamma_\mu^{(3)}(q;p_1;p_2) = -ie\gamma_\mu$$
$$+ (-ie)^3 \int \frac{d^d k}{(2\pi)^d} \gamma_\nu \left[M + i(\not{k}+\not{p}_1)\right]^{-1} \gamma_\mu \left[M + i(\not{k}-\not{p}_2)\right]^{-1} \gamma_\rho [\Delta_\xi]_{\nu\rho}(k).$$

Let us now calculate the one-loop contribution to $q_\mu \Gamma_\mu^{(3)}(q;p_1;p_2)$. We note

$$q + p_1 + p_2 = 0 \;\Rightarrow\; i\not{q} = \left[M + i(\not{k}-\not{p}_2)\right] - \left[M + i(\not{k}+\not{p}_1)\right].$$

It follows:

$$q_\mu \Gamma_\mu^{(3)}(q;p_1;p_2) = e\left[(M + i\not{p}_1) - (M - i\not{p}_2)\right]$$
$$+ e^3 \int \frac{d^d k}{(2\pi)^d} \gamma_\nu \left\{\left[M + i(\not{k}+\not{p}_1)\right]^{-1} - \left[M + i(\not{k}-\not{p}_2)\right]^{-1}\right\} \gamma_\rho [\Delta_\xi]_{\nu\rho}(k).$$

We recognize in the r.h.s. the difference between two inverse propagators.
Calculation of the vertex. Let us set

$$\Gamma_\mu^{(3)}(q;p_1;p_2) = -ie\gamma_\mu + (-ie)^3 C_\mu(q;p_1;p_2).$$

If we specialize to the Feynman gauge ($\xi = 1$) C_μ becomes

$$C_\mu = \int \frac{d^d k}{(2\pi)^d} \frac{\gamma_\nu (M - i(\not{k}+\not{p}_1))\gamma_\mu(M - i(\not{k}-\not{p}_2))\gamma_\nu}{\left[(k+p_1)^2 + M^2\right](k^2+m^2)\left[(k-p_2)^2 + M^2\right]}.$$

We then write C_μ as the sum of three terms:

$$C_\mu = M^2 C_{1\mu} + iM C_{2\mu} + C_{3\mu},$$

where the denominators of the integrands of C_1, C_2, C_3 successively are

$$(2-d)\gamma_\mu,$$
$$(2-d)\left[(\not{k}+\not{p}_1)\gamma_\mu + \gamma_\mu(\not{k}-\not{p}_2)\right] - 2\gamma_\mu(\not{k}+\not{p}_1) - 2(\not{k}-\not{p}_2)\gamma_\mu,$$
$$- 2\gamma_\mu(\not{k}-\not{p}_2)(\not{k}+\not{p}_1) + 2(\not{k}+\not{p}_1)(\not{k}-\not{p}_2)\gamma_\mu - (2-d)(\not{k}+\not{p}_1)\gamma_\mu(\not{k}-\not{p}_2).$$

Chapter 21: Renormalization of Gauge Theories. General Formalism

Exercise 21.1

21.1.1. Because the gauge variation of the gauge condition (21.91) is field-dependent

$$\delta A_\mu = \partial_\mu \Lambda \;\Rightarrow\; \delta G(A) = \partial^2 \Lambda + 2ie A_\mu \partial_\mu \Lambda\,,$$

the Faddeev–Popov determinant is no longer a constant and it is necessary to introduce ghost fields. To express the constraint it is also necessary to introduce a Lagrange multiplier field $\lambda(x)$. The quantized action can then be written:

$$S(A,\lambda,\bar{C},C) = \int d^d x \left[\tfrac{1}{4} F_{\mu\nu}^2 + \lambda \left(\partial_\mu A_\mu + ie A_\mu^2 \right) - \bar{C} \left(\partial^2 + 2ie A_\mu \partial_\mu \right) C \right]. \quad (44.48)$$

Note the relation

$$\partial^2 + 2ie A_\mu \partial_\mu - \left(\partial_\mu + ie A_\mu \right)^2 = -ie \left(\partial_\mu A_\mu + ie A_\mu^2 \right) = 0\,.$$

21.1.2. One immediately notes a drastic simplification with respect to the case of non-abelian gauge symmetries: The BRS transformations are linear in the fields and therefore no sources for composite fields are required. If we add to the action (44.48) the following source terms

$$S_{\text{sources}} = - \int d^d x \left(J_\mu A_\mu + K\lambda + \bar{\eta} C + \bar{C}\eta \right),$$

we find that the generating functional W for connected correlation functions satisfies

$$\int d^d x \left(J_\mu(x) \partial_\mu \frac{\delta}{\delta\eta(x)} + \eta(x) \frac{\delta}{\delta K(x)} \right) W(J, K, \bar{\eta}, \eta) = 0\,.$$

Because the BRS transformations are linear the Legendre transformation is straightforward. The 1PI functional Γ satisfies

$$\mathcal{D}\Gamma(A_\mu, \lambda, \bar{C}, C) = 0 \quad \text{with} \quad \mathcal{D} \equiv \int d^d x \left(\partial_\mu C \frac{\delta}{\delta A_\mu(x)} + \lambda(x) \frac{\delta}{\delta \bar{C}(x)} \right). \quad (44.49)$$

21.1.3.
The additional terms in (44.48) generated by the quantization procedure are the BRS variation of Φ:

$$\Phi(A, \lambda, \bar{C}, C) = \int d^d x \, \bar{C}(x) G(A)$$

$$\Rightarrow \mathcal{D}\Phi = \int d^d x \left[\lambda \left(\partial_\mu A_\mu + ie A_\mu^2 \right) - \bar{C} \left(\partial^2 + 2ie A_\mu \partial_\mu \right) C \right]. \quad (44.50)$$

Exercise 21.2

21.2.1. The $\bar{C}C$ propagator is $1/k^2$ in momentum space. The fields A_μ and λ are coupled at the gaussian level and the propagators thus form a 2×2 matrix:

$$\begin{pmatrix} \left(\delta_{\mu\nu} - \dfrac{k_\mu k_\nu}{k^2}\right)\dfrac{1}{k^2} & i\dfrac{k_\mu}{k^2} \\ -i\dfrac{k_\mu}{k^2} & 0 \end{pmatrix} \tag{44.51}$$

21.2.2. First let us note that the large momentum behaviour of the propagators shows that the quantized action (44.48) is renormalizable by power counting in four dimensions and that the canonical dimensions of fields are

$$d = 4 \;\Rightarrow\; [A] = 1, \quad [\lambda] = 2, \quad [\bar{C}] + [C] = 2\,.$$

Let us then write the most general form of the renormalized action just taking into account power counting and ghost number conservation. A term quartic in the ghost fields is allowed but by power counting it is necessarily proportional to $[\bar{C}(x)C(x)]^2$ and thus vanishes because ghosts are fermions. Then

$$S_{\rm r} = \Sigma(A) + \int \mathrm{d}^d x \left[-\tfrac{1}{2}a\lambda^2 + \lambda G_{\rm r}(A,\bar{C},C) - \bar{C}L(A)C \right], \tag{44.52}$$

where $G_{\rm r}$ has dimension two and is thus at most linear in $\bar{C}C$, $L(A)$ has also dimension two and a is a constant.

We have in full generality shown that if the 1PI functional Γ satisfies linear WT identities the renormalized action satisfies the same. Thus

$$\int \mathrm{d}^d x \left(\partial_\mu C \frac{\delta}{\delta A_\mu(x)} + \lambda(x)\frac{\delta}{\delta \bar{C}(x)} \right) S_{\rm r}(A_\mu, \lambda, \bar{C}, C) = 0\,.$$

We now introduce the decomposition (44.52) and identify the coefficients of powers of λ:

$$\lambda^2 : \frac{\delta G_{\rm r}}{\delta \bar{C}} = 0, \qquad \lambda : \frac{\delta G_{\rm r}}{\delta A_\mu}\partial_\mu C - L(A)C = 0\,,$$

$$\int \mathrm{d}^d x\, \bar{C}\, \partial_\mu C \frac{\delta L(A)}{\delta A_\mu(x)}C = 0\,, \qquad \int \mathrm{d}^d x\, \partial_\mu C \frac{\delta \Sigma(A)}{\delta A_\mu(x)} = 0\,.$$

From the first equation it follows that $G_{\rm r}$ is only a function of A_μ and from the second that $L(A)$ is determined as a function of $G_{\rm r}$. The most general form of $G_{\rm r}$ compatible with power counting is

$$G_{\rm r} = Z_\lambda \left(\partial_\mu A_\mu + ieZ_e A_\mu^2 + Z_G \right)\,.$$

Thus the renormalized ghost action is

$$S_{\rm ghosts} = -\int \mathrm{d}^d x\, Z_\lambda \bar{C}(x) \left(\partial^2 + 2ieZ_e A_\mu \partial_\mu \right) C(x)\,.$$

The third equation is now automatically satisfied and the last equation tells us that $\Sigma(A)$ is gauge invariant. The renormalized action thus reads (with the change in notation $a \mapsto \xi_0 Z_\lambda^2$)

$$
\mathcal{S}_{\mathrm{r}} = \int \mathrm{d}^d x \left[Z_A \tfrac{1}{4} F_{\mu\nu}^2 - \tfrac{1}{2} \xi_0 Z_\lambda^2 \lambda^2 + Z_\lambda \lambda \left(\partial_\mu A_\mu + ie Z_e A_\mu^2 + Z_G \right) \right.
$$
$$
\left. - Z_\lambda \bar{C} \left(\partial^2 + 2ie Z_e A_\mu \partial_\mu \right) C \right].
\tag{44.53}
$$

21.2.3. We see in action (44.53) that a term proportional to λ^2 has been generated by renormalization. The condition $G(A)$ is thus not stable under renormalization but is replaced by a gaussian distribution centred around $G(A) = 0$. Since the action is quadratic in λ the gaussian integral over λ results in an action where λ is replaced by the solution of the corresponding field equation:

$$
Z_\lambda \lambda = \left(\partial_\mu A_\mu + ie Z_e A_\mu^2 + Z_G \right) / \xi_0,
$$

and thus

$$
\mathcal{S}_{\mathrm{r}}(A, \bar{C}, C) = \int \mathrm{d}^d x \left[Z_A \tfrac{1}{4} F_{\mu\nu}^2 + \tfrac{1}{2} \left(\partial_\mu A_\mu + ie Z_e A_\mu^2 \right)^2 / \xi_0 - Z_\lambda \bar{C} \left(\partial^2 + 2ie Z_e A_\mu \partial_\mu \right) C \right].
\tag{44.54}
$$

21.2.4. The induced variation of Φ is

$$
\delta \Phi = i \delta e \int \mathrm{d}^d x \, \bar{C}(x) A_\mu^2(x).
$$

The variation of the generating functional $Z(J)$ of gauge field correlation functions follows:

$$
\delta Z(J) = -i \delta e \int [\mathrm{d}A \mathrm{d}\lambda \mathrm{d}\bar{C} \mathrm{d}C] \exp \left[-\mathcal{S}(A, \lambda, \bar{C}, C) + \int \mathrm{d}^d y \, J_\mu A_\mu \right] \mathcal{D} \int \mathrm{d}^d x \, A_\mu^2(x) \bar{C}(x).
$$

We then integrate by parts to let $\mathcal{D}$ act on the action and source:

$$
\delta Z(J) = i \delta e \int [\mathrm{d}A \mathrm{d}\lambda \mathrm{d}\bar{C} \mathrm{d}C] \int \mathrm{d}^d x \, A_\mu^2(x) \bar{C}(x) \mathcal{D} \exp \left[-\mathcal{S}(A, \lambda, \bar{C}, C) + \int \mathrm{d}^d y \, J_\mu A_\mu \right]
$$
$$
= i \delta e \int [\mathrm{d}A \mathrm{d}\lambda \mathrm{d}\bar{C} \mathrm{d}C] \int \mathrm{d}^d x \, A_\mu^2(x) \bar{C}(x) \int \mathrm{d}^d y \, \partial_\mu C(y) J_\mu(y)
$$
$$
\times \exp \left[-\mathcal{S}(A, \lambda, \bar{C}, C) + \int J_\mu A_\mu \right].
$$

Differentiating then twice with respect to J and setting $J = 0$ we obtain the variation of the two-point function:

$$
\delta W_{\mu\nu}^{(2)}(x, y) = i \delta e \left\langle \int \mathrm{d}^d z \, A_\rho^2(z) \bar{C}(z) A_\mu(x) \partial_\nu^y C(y) \right\rangle,
$$

where the average symbol means functional average with the weight $e^{-\mathcal{S}}$. Covariance and translation invariance tell us that the two-point function $\left\langle \int \mathrm{d}^d z \, A_\rho^2(z) \bar{C}(z) A_\mu(x) C(y) \right\rangle$ must be of the form

$$
\left\langle \int \mathrm{d}^d z \, A_\rho^2(z) \bar{C}(z) A_\mu(x) C(y) \right\rangle = \partial_\mu f(x - y).
$$

It follows after Fourier transformation that the variation of $W_{\mu\nu}^{(2)}$ has only a longitudinal component and thus the transverse part is independent of e. Therefore

$$
\tilde{W}_{\mu\nu}^{(2)}(k) = \left(\delta_{\mu\nu} - \frac{k_\mu k_\nu}{k^2} \right) \frac{1}{k^2} + k_\mu k_\nu w(k, e).
$$

In particular $Z_A = 1$.

Chapter 22: Classical and Quantum Gravity

Exercise 22.1

By a change in the parametrization we can set N to 1. It is then possible to construct a hamiltonian

$$p_\mu = \frac{\partial \mathcal{L}}{\partial \dot{x}^\mu} = -\dot{x}_\mu - eA_\mu(x),$$

and thus

$$\mathcal{H}(p,x) = p_\mu \dot{x}^\mu - \mathcal{L} = \tfrac{1}{2}\left[m^2 - (p + eA)^2\right].$$

In addition we must impose the N equation of motion as a constraint:

$$\dot{x}^\mu \dot{x}_\mu = m^2 \ \Rightarrow\ \mathcal{H} = 0.$$

After quantization we thus look for wave functions $\psi(x)$ which satisfy

$$\left[(p + eA)^2 - m^2\right]\psi = 0.$$

We recognize the Klein–Gordon equation which is, in this framework, the analogue of the Wheeler–DeWitt equation.

In this example we recognize that τ is not really a time but simply an auxiliary parameter, and the absence of evolution in this parameter is not too surprising: the Klein–Gordon equation is the real dynamical equation.

If we consider instead the lagrangian $\mathcal{L}'$ we could have the illusion that conjugated momenta p_μ can be defined because $\mathcal{L}'$ depends on all time derivatives:

$$p_\mu = \frac{\partial \mathcal{L}'}{\partial \dot{x}^\mu} = \mp m \frac{\dot{x}_\mu}{\sqrt{\dot{x}^2}} - eA_\mu.$$

However we note that this equation implies

$$(p_\mu + eA_\mu)^2 = m^2,$$

and therefore the conjugated momenta are not independent. We thus instead define a temporal gauge by the condition

$$x^0(\tau) \equiv \tau.$$

We then restrict ourselves again to space components and introduce the notation $\hat{p}, \hat{A}$ for the vectors of components $p_\mu, A_\mu, \mu > 0$, and $\hat{v}$ for $\dot{x}^\mu, \mu > 0$. We find

$$\hat{p} = \pm \frac{m\hat{v}}{\sqrt{1 - \hat{v}^2}} - e\hat{A} \ \Rightarrow\ \pm m\hat{v} = \frac{\hat{p} + e\hat{A}}{\sqrt{(\hat{p} + e\hat{A})^2 + m^2}}.$$

The hamiltonian follows

$$\mathcal{H} = \pm\sqrt{(\hat{p} + e\hat{A})^2 + m^2} + eA_0.$$

We see that if we take into account the two signs we recover the Klein–Gordon equation. We note also that only one sign corresponds to a hamiltonian bounded from below.

Exercise 22.2

22.2.1. *Random walks.* Let us call $P_n(z)$ the probability to be at point z at time n. It satisfies a recursion relation (in d dimensions)

$$P_n(z) = \frac{a}{2d} \sum_{\mu=1}^{2d} P_{n-1}(z - e_\mu),$$

where the points $z - e_\mu$ are all the neighbours of z on the lattice. The Fourier transform $\tilde{P}_n(k)$ of $P_n(z)$ then satisfies

$$\tilde{P}_n(k) = \frac{a}{d} \left(\sum_{\mu=1}^{d} \cos k_\mu \right) \tilde{P}_{n-1}(k) \qquad P_0(z) = \delta(z - x).$$

After a short calculation we find

$$U(y, x) = (1 - a) \sum_{n=0}^{\infty} P_n(y) = \frac{1}{(2\pi)^d}(1 - a) \int d^d k \frac{e^{ik(y-x)}}{1 - (a/d) \sum_\mu \cos k_\mu}.$$

In the long distance scaling limit the integral is dominated by the small values of k. The probability is non-vanishing only if a is close to 1. We thus set $a = 1 - m^2/2$ and expand the cosines

$$1 - (a/d) \sum_\mu \cos k_\mu \sim \tfrac{1}{2}(m^2 + k^2).$$

It follows that $U(y, x)$ has the form of a propagator of a scalar particle of mass m:

$$U(y, x) = \frac{1}{(2\pi)^d} m^2 \int d^d k \frac{e^{ik(y-x)}}{k^2 + m^2}. \tag{44.55}$$

This interpretation of the euclidean propagator associates random walks to Feynman diagrams.

22.2.2. The expression can be rewritten

$$U(y, x) = \sum_n \langle y|T^n|x \rangle = \langle y|[1 - T]^{-1}|x \rangle,$$

with

$$\langle y|T|x \rangle = e^{-\mu|y-x|}.$$

After a Fourier transformation the operator acts by multiplication:

$$\left\langle p'|\tilde{T}|p \right\rangle = \delta^{(d)}(p' - p)\tilde{T}(p),$$

with

$$\tilde{T}(p) = \int d^d x\, e^{ipx}\, e^{-\mu|x|}.$$

The method of calculation is quite similar to the preceding example. The continuum limit is again obtained when the parameter μ is adjusted in such a way that the lowest eigenvalue $\tilde{T}(0)$ approaches 1. The long distance asymptotic scaling form depends only on the expansion of $\tilde{T}(p)$ for p small:

$$\tilde{T}(p) = 1 - K(m^2 + p^2) + O\left(p^4\right).$$

We recover the Feynman propagator (44.55) and note that the result is quite robust (using the language of critical phenomena we can call it "universal") since it does not depend on the explicit form of $\langle y|T|x\rangle$ except that it is a function of $|x-y|$ and decreases fast enough at large separation.

22.2.3. We face the problem of parametrization and moreover at first sight we know of no direct recipe to calculate a path integral with such an action.

To solve the problem of the reparametrization invariance we can choose the length along the path s:

$$(ds)^2 = d\mathbf{x} \cdot d\mathbf{x}.$$

The path integral then can be formally written

$$U(y,x) = \int_0^\infty ds\, e^{-ms} \int_{x(0)=x}^{x(s)=y} [dx(\sigma)]\delta\left(\dot{\mathbf{x}}(\sigma)\cdot\dot{\mathbf{x}}(\sigma)-1\right).$$

As usual we introduce a Fourier representation for the δ-function

$$U(y,x) = \int_0^\infty ds\, e^{-ms} \int [dx(\sigma)][d\lambda(\sigma)]\exp\left[-\tfrac{1}{2}\int_0^s d\sigma\,(\dot{\mathbf{x}}\cdot\dot{\mathbf{x}}-1)\,\lambda(\sigma)\right], \qquad (44.56)$$

where the integration over λ runs parallel to the imaginary axis. We now change variables $\sigma \mapsto t$, setting

$$\frac{d\sigma}{dt} = \lambda(\sigma).$$

The action in (44.56) becomes

$$\int_0^s d\sigma\,(\dot{\mathbf{x}}(\sigma)\cdot\dot{\mathbf{x}}(\sigma)-1)\,\lambda(\sigma) = \int_0^T dt\left[\left(\frac{d\mathbf{x}}{dt}\right)^2 - \lambda^2\right],$$

with $0 = t(\sigma=0)$, $T = t(s)$. Using then

$$s = \int_0^s d\sigma = \int_0^T \lambda(\sigma(t))dt,$$

we find

$$U(y,x) = \int_0^\infty dT \int_{x(0)=x}^{x(T)=y} [dx(t)][d\lambda(t)]\lambda(T)$$

$$\times \exp\left[-\int_0^T dt\left(\tfrac{1}{2}\dot{\mathbf{x}}(t)\cdot\dot{\mathbf{x}}(t) - \tfrac{1}{2}\lambda^2(t) + m\lambda(t)\right)\right].$$

The integral over λ is gaussian and can be performed. We finally obtain

$$U(y,x) = m\int_0^\infty dT \int_{x(0)=x}^{x(T)=y} [dx(t)]\exp\left[-\tfrac{1}{2}\int_0^T dt\left(\dot{\mathbf{x}}(t)\cdot\dot{\mathbf{x}}(t) + m^2\right)\right].$$

Under this form we recognize a natural representation of the propagator (44.55) in the continuum, when we consider the operator $-d^2 + m^2$ as a quantum hamiltonian:

$$U(y, x) = m \int_0^\infty dT \, \langle y| \, e^{-TH} \, |x\rangle = m \, \langle y| \, H^{-1} \, |x\rangle \,, \quad H = -d^2 + m^2.$$

However the geometrical meaning in terms of a length of a path and the invariance under parametrization are no longer obvious. Returning to the exercise 22.1 we understand that this action results from the invariant action integral of the lagrangian (22.114) (for $A = 0$) by gauge fixing. æ

Chapter 24: Mean Field Theory

Exercise 24.1

We use everywhere the notation of Chapter 24. We find

$$A(\phi) = -\ln(\lambda + \cosh\varphi)\,, \quad \text{with } \lambda = 1/2p - 1 > 0\,.$$

The magnetization is then

$$M = -A'(\varphi) = \frac{\sinh\varphi}{\lambda + \cosh\varphi}\,.$$

Hence $B(M)$, the Legendre transformed of $-A(\varphi)$ satisfies

$$B'(M) = \varphi(M) = \ln\left(\frac{\lambda M + \sqrt{1 - M^2 + \lambda^2 M^2}}{1 - M}\right).$$

From this expression we derive

$$B''(M) = \frac{1}{1 - M^2}\left(1 + \frac{\lambda}{\sqrt{1 - M^2 + \lambda^2 M^2}}\right),$$
$$B(M) = \tfrac{1}{2}(1 + M)\ln D(M) + \tfrac{1}{2}(1 - M)\ln D(-M)\,,$$

with

$$D(M) = \frac{(\lambda^2 - 1)\, M - 1 + \lambda\sqrt{1 - M^2 + \lambda^2 M^2}}{\lambda^2 - 1}\,.$$

The thermodynamical potential $\Gamma(M)$ is given by

$$\Gamma(M) = -VM^2 + B(M).$$

Note that the cases $p = \tfrac{1}{2}$ ($\lambda = 0$) and $p = \tfrac{1}{4}$ ($\lambda = 1$) are identical in the mean field approximation, they correspond to $l = 1$ and $l = 2$ respectively in the notation of Section 24.2.

It is useful to expand $B(M)$ up to order M^6:

$$B(M) = (1 + \lambda)\left[\tfrac{1}{2}M^2 + \tfrac{1}{24}(1 + \lambda)(2 - \lambda)M^4 + \tfrac{1}{240}(1 + \lambda)^2\left(8 - 9\lambda + 3\lambda^2\right)M^6\right]$$
$$+ O\left(M^8\right).$$

Finally we note

$$B''(M) \geq B''(0) \quad \text{for } \lambda \leq 2.$$

We immediately conclude that for $\lambda < 2$ or $p > 1/6$ $\Gamma(M)$ remains convex until V reaches the critical value V_c

$$V_c = \tfrac{1}{2}(1+\lambda) = 1/(4p),$$

which defines the critical temperature of a second order phase transition. The line of second order phase transition terminates at the special point $\lambda = 2$ which corresponds to a different physics and is called a tricritical point. For $V - V_c$ and M small we can expand $\Gamma(M)$

$$\Gamma(M) = (V_c - V)M^2 + \tfrac{9}{40}M^6 + O\left(M^8\right).$$

The tricritical point is characterized by new exponents like $\beta = 1/4$, $\delta = 5$, new universal ratios like

$$C_+/C_- = 4.$$

Finally the whole scaling region is not limited to $\lambda = \lambda_c = 2$ but contains also $|\lambda - \lambda_c|$ small. After rescaling of temperature, field, magnetization and the additional thermodynamical variable $\lambda - \lambda_c$ the scaling equation of state can be cast into the universal form

$$H = \frac{\partial \Gamma}{\partial M} = M^5 f(x,y), \quad f(x,y) = 1 + x + y$$
$$x = (V - V_c)/M^4, \qquad y = (\lambda_c - \lambda)/M^2.$$

Exercise 24.2

The antiferromagnetic Ising model. In the absence of a magnetic field, the antiferromagnetic Ising model on a bipartite lattice (a lattice which can be decomposed into two sublattices such that in the two-spin interaction the two spins belong to different sublattices) can be simply related to a ferromagnetic Ising model by setting

$$\sigma_i = \epsilon_i S_i. \tag{44.57}$$

In the new variables the interaction reads

$$-\beta \mathcal{H} = \sum_{ij} V_{ij} \sigma_i \sigma_j,$$

and is thus ferromagnetic. The alternating spins σ_i define an order parameter for the antiferromagnetic system.

In mean field theory the thermodynamic potential $\Gamma(M_i)$ reads

$$\Gamma(M_i) = -\sum_{ij} V_{ij} M_i M_j + \sum_i B(M_i).$$

A constant magnetic field after the transformation (44.57) becomes an alternating field $\epsilon_i H$. In such a field the minimum of the potential corresponds to the sum a constant and alternating magnetization

$$M_i = M + \epsilon_i \widetilde{M}.$$

The equation

$$H\epsilon_i = \frac{\partial \Gamma}{\partial M_i}$$

for small field H reduces to two equations

$$H = 2V\widetilde{M} + a\widetilde{M} + O\left(\widetilde{M}^3, \widetilde{M}M^2\right)$$

$$0 = -2VM + aM + \tfrac{1}{6}bM^3 + \tfrac{1}{2}b\widetilde{M}^2M + O\left(M^5, \widetilde{M}M^3, \widetilde{M}^4M\right),$$

We see in the first equation that $\widetilde{M}$ always remains linear in H

$$\widetilde{M} \sim H/(2V + a).$$

Substituting this solution in the second equation we observe that the only effect, at leading order, is to modify the critical temperature

$$V_c - \tfrac{1}{2}a \sim \tfrac{1}{16}bH^2/a^2.$$

The analysis confirms that since a constant magnetic field is not coupled to the order parameter, it does not destroy the phase transition, but simply modifies non-universal features, like the critical temperature.

Exercise 24.3

Mean field theory and effective Landau–Ginzburg action for models with Z_N symmetry. An order parameter for the system is the expectation value $M = \langle e^{i\theta} \rangle$. The variable M is thus a complex number, on which the symmetry group acts by multiplication. Symmetry arguments, tell us that the thermodynamical potential, is a real function of the invariant polynomials M^*M, which is $U(1) \equiv O(2)$ invariant, M^N and M^{*N}. If we assume the existence of a second order phase transition, we can expand the potential for M small. Several cases then have to be considered:

(i) If $N \geq 5$ then the terms which are not functions only of M^*M can be neglected and the scaling theory is the same as for the $O(2)$ symmetric model (a model with a continuous symmetry). Only corrections to the leading terms reveal the difference. Note however that the correction terms play here a more important role than usual, since they suppress below T_c the would-be Goldstone modes of the $O(2)$ symmetry.

(ii) If $N = 4$ instead we find the most general model depending on two order parameters M_1, M_2 and symmetric in $M_1 \leftrightarrow M_2$, which is equivalent to the $O(2)$ model with cubic anisotropy.

(iii) For $N = 3$ a cubic term is allowed. Then we expect generically a first order phase transition.

Let us now calculate the expansion of the function $A(\varphi, \varphi^*)$ up the fourth order:

$$e^{-A(\varphi, \varphi^*)} = \frac{1}{N} \sum_n \exp\left(e^{2in\pi/N}\varphi + e^{-2in\pi/N}\varphi^*\right),$$

and thus

$$-A(\varphi, \varphi^*) = \varphi^*\varphi + \tfrac{1}{6}\left(\varphi^3 + \varphi^{*3}\right)\delta_{N3} + \tfrac{1}{24}\left(\varphi^4 + \varphi^{*4}\right)\delta_{N4} - \tfrac{1}{4}\left(\varphi^*\varphi\right)^2.$$

After Legendre transformation

$$-A(\varphi, \varphi^*) + B(M, M^*) = M^*\varphi + M\varphi^*,$$

we obtain

$$B(M, M^*) = M^*M + \tfrac{1}{4}\left(M^*M\right)^2 - \tfrac{1}{6}\left(M^3 + M^{*3}\right)\delta_{N3} - \tfrac{1}{24}\left(M^4 + M^{*4}\right)\delta_{N4}.$$

The thermodynamical potential at fourth order then is

$$\Gamma(MM^*) = VM^*M - B(M, M^*),$$

confirming the general analysis.

Other results. The $N = 3$ model is known to have a second order phase transition in 2 dimensions (with exponents derived from CFT). In 3 dimensions the nature of the transition (first or second order) has been an object of debate and of heavy numerical investigations in recent time. The motivation was the following: The Z_N group is the centre of the $SU(N)$ group and is involved in the question of confinement (Wilson's loop arguments). The confinement phase transition at high physical temperature in lattice QCD in four dimensions is thus related to the phase transition of the Z_3 model in 3 dimensions (the physical gauge group is $SU(3)$). Most recent data suggests that the transition is first order.

Chapter 25: General Renormalization Group Analysis

Exercise 25.1

The weight factor for the $2N$-point function at leading ($N - 1$-loop) order is $(2N - 1)!/(N!)^2/(N - 1)!$. For the 2-point function one finds $1/(2N - 1)!$. These factors correspond to the symmetry factors of the corresponding diagrams. To then calculate the diagrams in the massless theory it is convenient to use real space variables because the diagrams are just powers of the propagator. One starts from

$$\frac{1}{p^2} = \frac{\Gamma(d/2 - 1)}{4\pi^{d/2}}\int d^d x\, x^{2-d}\,e^{-ipx}.$$

Calling $K(d)$ the combination $\Gamma(d/2 - 1)\pi^{-d/2}/4$ one finds for the diagram with $m + 1$ propagator attached to two points in which flows a momentum p:

$$D_{m+1}(p) = K^{m+1}(d)\pi^{d/2}\frac{\Gamma\left(1 - m(d-2)/2\right)}{\Gamma\left((m+1)(d-2)/2\right)}\left(\frac{2}{p}\right)^{2-m(d-2)}.$$

The $2N$-point function corresponds to $m = N - 1$ and the two-point function to $m = 2N - 2$.

For $N = 3$ $K(3) = 1/(4\pi)$. The weight of the contribution to the 6-point function is $5/3$ and the divergent part then $5C/3\varepsilon$ with $C = 1/(16\pi^2)$. For the 2-point function the results are respectively $1/120$ and $C^2/(12\varepsilon)$. After a short calculation one obtains the RG functions at leading order

$$\beta(g) = -\varepsilon g + C\frac{5}{3}g^2, \qquad \eta(g) = C^2\frac{1}{720}g^2.$$

Exercise 25.2

25.2.1. *Long range forces.* We set $u = \Lambda^{2\sigma-d}g$. The function $\beta(g)$ comes from the divergent part of a unique bubble diagram. One finds

$$\beta(g) = -\varepsilon g + \tfrac{N+8}{6}N_{2\sigma}g^2 + O\left(g^3, \varepsilon g^2\right), \qquad N_{2\sigma} = 2(4\pi)^{-\sigma}/\Gamma(\sigma).$$

At this order the calculation of Z_2 and thus η_2 involves the same diagram. One thus finds

$$\eta_2(g^*) = -\varepsilon\tfrac{N+2}{N+8}$$
$$\nu^{-1} = \sigma + \eta_2 = \sigma - \varepsilon\tfrac{N+2}{N+8}$$
$$\gamma = \nu\sigma = 1 + \frac{\varepsilon}{\sigma}\frac{N+2}{N+8}.$$

Chapter 26: Scaling Laws in the Critical Domain

Exercise 26.1

First note that the gradient flow property is parametrization dependent. The properties we derive here will be valid more generally for all equations which are obtained from a gradient flow by reparametrization.

The corresponding flow of the coupling constant takes the form

$$\frac{dg_i}{d\ln\lambda} = \beta_i(g) = \frac{\partial V}{\partial g_i}. \qquad (44.58)$$

Let us multiply the equation by $dg_i/d\ln\lambda$. Then

$$\left(\frac{dg_i}{d\ln\lambda}\right)^2 = \frac{\partial V}{\partial g_i}\frac{dg_i}{d\ln\lambda} = \frac{dV}{d\ln\lambda} > 0.$$

We recover the essential property of a gradient flow: fixed points are minima or maxima of the potential V; pathologies like limiting cycles are excluded. Moreover if we calculate the matrix B of derivatives at a fixed point we find

$$B_{ij} = \frac{\partial\beta_i}{\partial g_j} = \frac{\partial^2 V}{\partial g_i\partial g_j},$$

which is a symmetric matrix. Therefore all its eigenvalues are real.

Exercise 26.2

26.2.1. *Renormalization group for models with cubic symmetry.* The stability conditions are: $g + \lambda \geq 0$, $Ng + \lambda \geq 0$, the second relation relying on Schwartz's inequality.

26.2.2. The two β functions are

$$\beta(g) = -\varepsilon g + \frac{1}{8\pi^2}\left(\frac{N+8}{6}g^2 + g\lambda\right)$$
$$\beta(\lambda) = -\varepsilon\lambda + \frac{1}{8\pi^2}\left(2g\lambda + \frac{3}{2}\lambda^2\right)$$

26.2.3. One finds four fixed points, the gaussian fixed point $g = \lambda = 0$, the Ising fixed point $g = 0, \lambda = 16\varepsilon\pi^2/3$, the isotropic fixed point $\lambda = 0, g = 48\varepsilon\pi^2/(N+8)$ and a new cubic fixed point

$$g = \frac{16\pi^2\varepsilon}{N}, \qquad \lambda = \frac{16\pi^2(N-4)\varepsilon}{3N}.$$

Note that for $N \geq 1$ the cubic fixed point satisfies the stability condition $g + \lambda \geq 0$. Also all fixed points have $g \geq 0$. The lines $\lambda = 0$ and $g = 0$ cannot be crossed.

To find the local stability of the four fixed points one has to calculate the eigenvalues of the matrix M of derivatives of the β-functions

$$M = -\varepsilon + \frac{1}{8\pi^2}\begin{pmatrix} \frac{N+8}{3}g + \lambda & g \\ 2\lambda & 2g + 3\lambda \end{pmatrix}.$$

The set of eigenvalues are

$$\begin{array}{ll} \text{gaussian} & -\varepsilon, -\varepsilon, \\ \text{Ising} & -\frac{1}{3}\varepsilon, \varepsilon, \\ \text{isotropic} & \frac{4-N}{N+8}\varepsilon, \varepsilon, \\ \text{cubic} & \varepsilon, \frac{N-4}{3N}\varepsilon. \end{array}$$

We recover a particular example of a more general result, the isotropic fixed point is stable for $N < N_c$ with $N_c = 4 + O(\varepsilon)$. The cubic fixed point is stable only for $N > N_c$. At $N = N_c$ the two fixed points merge and exchange role. The Ising fixed point has always one direction of instability. Note that for some initial values of g, λ (like $g < 0$) the effective coupling constants are not attracted to a fixed point but run away. It is generally believed that this signals a first order phase transition.

26.2.4. The perturbation of the Ising model by the isotropic $O(N)$ symmetric interaction involves the operators $(\phi^2)^2$. At the fixed point all ϕ components are decoupled. Therefore the $(\phi^2)^2$ is equivalent to a linear combination of ϕ_1^4 and $\phi_1^2\phi_2^2$ which factorizes. The corresponding exponents are ω_{Ising} and $(\alpha/\nu)_{\text{Ising}}$.

Chapter 28: Calculation of Universal Quantities

Polymers. Let us calculate the 2-point function $Z^{(2)}(k,s)$ at one-loop order. At leading order we have found

$$G^{(2)}(k,s) = \langle e^{i\mathbf{k}\cdot[\mathbf{r}(s)-\mathbf{r}(0)]}\rangle = e^{-k^2|s|}$$

$$\Rightarrow Z^{(2)}(k,s) = \int_0^\infty ds\, e^{-st}\, G^{(2)}(k,s) = \frac{1}{k^2+t}.$$

The one-loop contribution to $Z^{(2)}(k,t)$ reads

$$\left[Z^{(2)}(k,t)\right]_{1\,\text{loop}} = -\frac{g}{6}\int \frac{d^d q}{(2\pi)^d}\int_0^\infty ds\, e^{-st}\, \zeta(k,q,s),$$

$$\zeta(k,q,s) = \int_0^s du_1 du_2\, \langle e^{i\mathbf{k}\cdot[\mathbf{r}(s)-\mathbf{r}(0)]}\, e^{i\mathbf{q}\cdot[\mathbf{r}(u_2)-\mathbf{r}(u_1)]}\rangle$$

$$= 2\int_0^s du_2\int_0^{u_2} du_1\, e^{-k^2(s-u_2+u_1)-(k+q)^2(u_2-u_1)}.$$

We have used the property that the average of a product of exponentials is given by a simple gaussian integral, as we discuss in 31.1.

It is then convenient to change variables, $s, u_2, u_1 \mapsto s', u', u_1$:

$$s - u_2 + u_1 = s', \qquad u_2 - u_1 = u'.$$

The integrals over u_1, s', u' can be explicitly performed. Shifting q in the remaining integral, $q + k \mapsto q$, we then obtain

$$Z^{(2)}(k, t) = \frac{1}{k^2 + t} - \frac{g}{3} \frac{1}{(k^2 + t)^2} \frac{1}{(2\pi)^d} \int \frac{d^d q}{q^2 + t},$$

which is indeed the one-loop contribution to the 2-point function in the $(\phi^2)^2$ field theory in the $N = 0$ limit.

Chapter 29: The $\left(\phi^2\right)^2$ Field Theory in the Large N Limit

Exercise 29.1

The correlation length above T_c is given by the limit of equation (29.15):

$$\tau \sim \frac{N}{4\pi^2} m^2 \ln(\Lambda/m) \;\Rightarrow\; \xi^2 = m^{-2} \sim -\frac{N}{16\pi^2} \frac{\ln \tau}{\tau}.$$

One verifies that this result is in agreement with the analysis of Section 27.2. Similarly the equation of state (equation (29.23)) becomes

$$\sigma(\sigma^2 + \tau) = \frac{N}{16\pi^2} H \ln\left(\Lambda^2 \sigma/H\right),$$

or:

$$H \sim \frac{32\pi^2}{N} \frac{\sigma(\sigma^2 + \tau)}{\ln(\Lambda/\sigma)},$$

again in agreement with the large N limit of the general result. Finally the specific heat above T_c is proportional to $1/\ln m$, i.e. at leading order to $1/\ln \tau$ as expected.

We stress again that even in four dimensions the bare coupling constant has played no role and the same results would have obtained from the $O(N)$ non-linear σ-model.

Exercise 29.2

Let us explore further the saddle point equations (29.67), for a general potential V

$$\lambda \sigma = 0, \tag{44.59a}$$

$$V'(\mu) = \tfrac{1}{2}\lambda, \tag{44.59b}$$

$$\sigma^2 = \mu - \frac{N}{(2\pi)^d} \int \frac{d^d p}{p^2 + \lambda}. \tag{44.59c}$$

At low temperature $\sigma \neq 0$ and thus λ vanishes. Equation (44.59c) has a solution only for $\mu > \mu_c$,

$$\mu_c = \frac{N}{(2\pi)^d} \int \frac{d^d p}{p^2}, \;\Rightarrow\; \sigma = \sqrt{\mu - \mu_c}.$$

Equation (44.59b) then yields the critical temperature. Setting $V(\mu) = W(\mu) + \frac{1}{2}r\mu$, we find

$$r_c = -2W'(\mu_c).$$

To find the exponent β we need the relation between the r and μ near the critical point. In a generic situation we find

$$t = r - r_c \sim -2W''(\mu_c)(\mu - \mu_c) \;\Rightarrow\; \beta = \tfrac{1}{2}.$$

This is the case of a simple critical point which we have examined in the section. A new situation arises if we can adjust a parameter of the potential in such a way that $W''(\mu_c) = 0$. This provides an example of a tricritical point. It can be achieved only if the potential V is at least cubic. Higher critical points can be obtained when more derivatives vanish. However this is no longer interesting for continuous symmetries since mean-field behaviour is then obtained for $d \geq 3$ (and symmetry breaking is possible only for $d > 2$).

At high temperature $\sigma = 0$ and λ, which is the σ-field mass squared, is given by

$$2W'(\mu) + r = \lambda, \qquad \mu - \mu_c = -\frac{N}{(2\pi)^d}\lambda \int \frac{d^d p}{p^2(p^2 + \lambda)}.$$

For $d < 4$ we find $\mu - \mu_c \sim \lambda^{(d-2)/2}$. In the tricritical case then

$$\lambda = t + \text{const.} \times W'''(\mu_c)\lambda^{d-2}.$$

For $d = 3$ we find a mean field behaviour without logarithmic corrections. We recall that $d = 3$ is the upper-critical dimension.

Discussion. In the mean field approximation the function $W(\mu) = \mu^3$ is not bounded from below, however $\mu = 0$ is the minimum because by definition $\mu \geq 0$. Here instead we are in the situation where $W(\mu) \sim (\mu - \mu_c)^3$ but μ_c is positive. Thus this extremum of the potential is unstable. The tricritical point and the $(\phi^2)^2$ cannot be studied by the large N limit method.

Chapter 31: Two-dimensional Models and Bosonization Method

Exercise 31.1

Distinguishing two operators in equation (31.13) we find

$$\left\langle \prod_{i=1}^{n} e^{i\kappa_i\varphi(x_i)} \right\rangle\bigg|_{\text{ren.}} = (\mu|x_1 - x_2|)^{\kappa_1\kappa_2/2\pi} \prod_{i>2}(\mu|x_1 - x_i|)^{\kappa_i\kappa_1/2\pi}$$

$$\times (\mu|x_2 - x_i|)^{\kappa_i\kappa_2/2\pi} \prod_{2<i<j}(\mu|x_i - x_j|)^{\kappa_i\kappa_j/2\pi}.$$

Taking the limit $x_2 \to x_1$ we see that the first factor is the singular coefficient function and thus

$$e^{i\kappa_1\varphi(x_1)} e^{i\kappa_2\varphi(x_2)} \underset{x_2 \to x_1}{\propto} C_{\kappa_1,\kappa_2}(x_1 - x_2) e^{i(\kappa_1+\kappa_2)\varphi(x_1)},$$

with

$$C_{\kappa_1,\kappa_2}(x) \sim x^{\kappa_1\kappa_2/2\pi}.$$

Moreover the behaviour of the coefficient function C is related in a way expected from the discussion of Section 12.7 to the dimensions of the three operators (see equation (31.16)) since

$$\kappa_1\kappa_2/2\pi = (\kappa_1 + \kappa_2)^2/4\pi - \kappa_1^2/4\pi - \kappa_2^2/4\pi = [e^{i(\kappa_1+\kappa_2)\varphi}] - [e^{i\kappa_1\varphi}] - [e^{i\kappa_2\varphi}].$$

Exercise 31.2

The model (31.114) possesses, in addition to gauge invariance, a $U(N) \times U(N)$ chiral symmetry because the fermions are massless. The model is a simple extension of the Schwinger model and can be solved exactly. As we shall see, however, for higher values of N its structure is somewhat more complicated.

The first step is the evaluation of anomaly. This is simple because the fermion determinant is replaced by the determinant to the power N. The generalization of the free action (31.61) is thus

$$S(\bar\psi', \psi', B_\mu) = -\int d^2x \left[\bar\psi'(x) \cdot \partial\!\!\!/\psi'(x) + \frac{N}{2\pi} \int d^2x \big(\partial_\mu \varphi(x)\big)^2 \right]. \tag{44.60}$$

Adding the gauge field free action which is unchanged we obtain

$$S(\bar\psi', \psi', \varphi) = -\int d^2x \left[\bar\psi' \cdot \partial\!\!\!/\psi' + \frac{1}{2e^2}\partial_\mu\varphi \left(m^2 - \partial^2\right)\partial_\mu\varphi \right], \tag{44.61}$$

with

$$m^2 = Ne^2/\pi. \tag{44.62}$$

The φ propagator has the form:

$$\langle \varphi(x)\varphi(0)\rangle = \frac{1}{4\pi^2}\int d^2p \, e^{ipx} \frac{\pi}{N}\left(\frac{1}{p^2+m^2} - \frac{1}{p^2}\right), \tag{44.63}$$

and therefore propagates a neutral massive state together with a negative metric massless state, as in the Schwinger model. Since this boson is free it cannot produce charged fermions and this is an indication of confinement.

Let us now bosonize the fermion fields $\bar\psi', \psi'$, introducing N scalar bosons ϑ_i. The equivalent boson action is

$$S(\vartheta_i, \varphi) = \tfrac{1}{2}\int d^2x \left[\partial_\mu\vartheta_i\partial_\mu\vartheta_i - \frac{1}{e^2}\partial_\mu\varphi\left(m^2 - \partial^2\right)\partial_\mu\varphi \right]. \tag{44.64}$$

Then we proceed as for the Schwinger model, translating $\vartheta_i + \varphi/\sqrt{\pi} \mapsto \vartheta_i$ and integrate over φ. The action takes the form

$$S(\vartheta_i) = \tfrac{1}{2}\int d^2x \left[\partial_\mu\vartheta_i\partial_\mu\vartheta_i + \frac{m^2}{N}\left(\textstyle\sum_i\vartheta_i\right)^2 \right]. \tag{44.65}$$

Through all these transformations the chiral components $\sigma_\pm^i$ of $\bar\psi_i\psi_i$ become simply

$$\sigma_\pm^i(x) = \bar\psi_\mp^i(x)\psi_\pm^i(x) \longmapsto \frac{\Lambda}{2\pi} e^{\pm i\sqrt{4\pi}\vartheta_i(x)}. \tag{44.66}$$

The first remark is that in contrast to the Schwinger model, besides one massive field $\sum_i\vartheta_i$, we have in addition $N-1$ massless boson fields.

The $\sigma_\pm^i$ correlation functions. Due to the presence for $N > 1$ of massless bosons averages of the form $\langle\sigma_\pm\rangle$ all vanish. However let us calculate the average of the operator

$$\left\langle \prod_{i=1}^N \sigma_+^i(x)\right\rangle = \frac{\Lambda^N}{(2\pi)^N}\left\langle e^{i\sqrt{4\pi}\sum_i\vartheta_i(x)}\right\rangle = \frac{\Lambda^N}{(2\pi)^N}e^{-2N\pi\Delta(0,m)} = \big(\langle\sigma\rangle_{N=1}\big)^N.$$

Let us write the operator in the fermion representation we see

$$\prod_{i=1}^{N}\sigma_+^i(x) = (-1)^{N(N-1)/2}\prod_{i=1}^{N}\bar{\psi}_-^i(x)\prod_{i=1}^{N}\psi_+^i(x).$$

Due to the antisymmetry of the fermions, if we perform a $U(N)\times U(N)$ transformation of matrices U_+, U_-, the operator transforms by multiplication by $\det U_+/\det U_-$. The operator is thus invariant under the $SU(N)\times SU(N)\times U(1)$ group. Its non-vanishing expectation value breaks only the remaining $U(1)$ chiral component. .

Finally the confirmation of confinement is here a more complicated problem since spectrum considerations are not sufficient.

Exercise 31.3

We first expand the field $\phi(x,y)$ in Fourier modes:

$$\phi(x,y) = \sum_n e^{2i\pi ny/L}\,\phi_n(x).$$

The corresponding action takes the form

$$S(\phi) = \tfrac{1}{2}L\int dx \sum_n \left(\phi_n'(x)\phi_{-n}'(x) + (4\pi^2 n^2/L^2)\phi_n(x)\phi_{-n}(x)\right).$$

As usual in the finite size problem we have to consider separately the zero momentum mode which would naively lead to IR divergences. The zero mode corresponds to a free particle in quantum mechanics. A simple calculation shows that its contribution to the two-point function is a factor

$$(2\pi)\delta(k_1+k_2)\,e^{-\frac{1}{2}k_1^2|x|/L}.$$

The δ-function is expected from the analysis of Section 31.1. The propagator for the other modes then takes the form

$$\Delta_L(x,y) = \frac{1}{2\pi L}\sum_{n\neq0}e^{2i\pi ny/L}\int d\omega\,\frac{e^{i\omega x}}{\omega^2+(2\pi n/L)^2},$$

$$= \frac{1}{4\pi}\sum_{n>0}\frac{1}{n}e^{-2\pi n|x|/L}\left(e^{2i\pi ny/L}+e^{-2i\pi ny/L}\right)$$

$$= -\frac{1}{4\pi}\ln\left[\left(1-e^{-2\pi|x|/L}e^{2i\pi y/L}\right)\left(1-e^{-2\pi|x|/L}e^{-2i\pi y/L}\right)\right].$$

The propagator at coinciding points is UV divergent:

$$\Delta_L(0,0) = \frac{1}{2\pi}\ln(L\Lambda) + O(1).$$

Performing the gaussian integration we obtain the two-point function

$$\left\langle e^{ik_2\phi(x,y)}e^{ik_1\phi(0,0)}\right\rangle \propto \delta(k_1+k_2)\,e^{-\frac{1}{2}k_1^2|x|/L}\,e^{-k_1^2\ln(L\Lambda)/2\pi}\,e^{k_1^2\Delta_L(x,y)}$$

$$\propto \delta(k_1+k_2)\left[\Lambda^2L^2\sinh\left(\frac{\pi w}{L}\right)\sinh\left(\frac{\pi\bar{w}}{L}\right)\right]^{-k_1^2/(4\pi)},$$

where we have set $w = x + iy$, and $\bar{w}$ is the complex conjugate of w.

We now observe that the conformal transformation $z = e^{2\pi w/L}$, maps the complex plane onto a periodic strip of size L. The propagator in the initial z variable is

$$\Delta(z, \bar{z}) = -\frac{1}{4\pi} \ln z\bar{z} .$$

Then the propagator in the new variables between the points $w = x + iy$ and $w = 0$ has the form

$$\Delta = -\frac{1}{4\pi} \ln \left(1 - e^{2\pi w/L}\right) \left(1 - e^{2\pi \bar{w}/L}\right),$$

which coincides with the result obtained by direct calculation. One can actually verify that in a conformal mapping $z \mapsto w$ the two-point function of $e^{ik\phi(z)}$ transforms by the substitution $z \mapsto w$ and is multiplied by a factor coming the dimension $k_1^2/(8\pi)$ of the operator

$$\left(\frac{\partial z}{\partial w}\right)^{-k^2/8\pi} \left(\frac{\partial \bar{z}}{\partial \bar{w}}\right)^{-k^2/8\pi} .$$

Exercise 31.4

Let us first determine the function $f(z)$ by exploiting the identity for $n = 2$. Multiplying by the denominators one finds

$$f(z_1 - z_1')f(z_2 - z_2') - f(z_1 - z_2')f(z_2 - z_1') = f(z_1 - z_2)f(z_1' - z_2').$$

Setting $z_1 = z_2 = 0$ one finds $f(0) = 0$. Setting then $z_2 = z_2'$ one concludes that $f(z)$ is an odd function. One finally differentiates twice with respect to z_1' and once with respect to z_2'. One sets $z_2 = z_1' = z_2' = 0$. A differential equation for f follows

$$f''(z)f'(0) - f(z)f'''(0) = 0 .$$

If $f'''(0)$ vanishes we recover the linear function we have considered so far. However we have another solution

$$f(z) = \sinh(\alpha z).$$

It is then easy to verify that the function satisfies the identity for all n by noting

$$\frac{1}{\sinh(\alpha(z - z'))} = \frac{2 e^{\alpha z} e^{\alpha z'}}{e^{2\alpha z} - e^{2\alpha z'}} .$$

We can factorize the numerators in front of the determinant, and in the denominators we recognize the Cauchy determinant with variables $Z = e^{2\alpha z}$. Therefore we can also bosonize a system in a strip or at finite temperature.

Chapter 33: Critical Properties of Gauge Theories

The Landau–Ginzburg model of superconductivity. Introducing the parameters $\tilde{g}, \tilde{\alpha}$

$$\tilde{g} = g/8\pi^2, \qquad \tilde{\alpha} = e^2/8\pi^2,$$

we obtain for the corresponding β-functions

$$\beta_{\tilde{g}} = -\varepsilon\tilde{g} + \tfrac{5}{3}\tilde{g}^2 - 6\tilde{g}\tilde{\alpha} + 18\tilde{\alpha}^2, \tag{44.67a}$$

$$\beta_{\tilde{\alpha}} = -\varepsilon\tilde{\alpha} + \tfrac{1}{3}\tilde{\alpha}^2. \tag{44.67b}$$

Setting $t = \ln\lambda$, where λ is the scale parameter, and $\tilde{g} = \tilde{\alpha}s$, we can write the flow equations at $d = 4$:

$$\frac{d\tilde{\alpha}}{dt} = \tfrac{1}{3}\tilde{\alpha}^2,$$

$$\frac{ds}{dt} = \tilde{\alpha}\left(\tfrac{5}{3}s^2 - \tfrac{19}{3}s + 18\right).$$

The first gives the flow of $\tilde{\alpha}$ which goes to zero at large distance ($t \to -\infty$), $\tilde{\alpha} \sim -3/t$. One verifies that the factor in the r.h.s. of (44.67a) never vanishes. Therefore $s(t)$ decreases at large distance and passes through the origin for a finite t. Since stability implies that $s(t)$ should remain positive, we face a case of runaway coupling constant, a strong indication of a first order phase transition.

One then verifies that for the same reason in $d = 4 - \varepsilon$ there is no solution to the IR fixed point equations and again one predicts a first order phase transition. This first order character is not observed experimentally, the belief being that the critical domain is too small.

Finally by increasing enough the number N of charged scalar fields eventually one reaches a situation where the transition becomes second order, a result which can also be checked in the large N limit.

Chapter 37: Instantons in Quantum Mechanics

Exercise 37.1

A few elements of the solution are:
(i) The classical solution

$$|q_c|^{N-1}(t) = \frac{1}{\sqrt{-g}}\frac{1}{\cosh[(N-1)(t-t_0)]}.$$

(ii) The classical action S_c

$$S_c = A(N)/(-g)^{1/(N-1)},$$

$$A(N) = \frac{\sqrt{\pi}\Gamma(N/(N-1))}{2\Gamma((3N-1)/2(N-1))} = 4^{1/(N-1)}\frac{\Gamma^2(N/(N-1))}{\Gamma(2N/(N-1))}.$$

(iii) The operator M, second functional of the action at the saddle point, is

$$M = -d_t^2 + 1 - \frac{N(2N-1)}{\cosh^2[(N-1)t]}.$$

Then, in the notation of equation (37.38), we have:

$$\det\left(M + \varepsilon\right)\left(M_0 + \varepsilon\right)^{-1} = \frac{\Gamma\left(1 + z\right)\Gamma\left(z\right)}{\Gamma\left(1 + \lambda + z\right)\Gamma\left(z - \lambda\right)},$$

with

$$\lambda = N/(N - 1), \qquad z = \sqrt{1 + \epsilon}/(N - 1).$$

Then

$$\det\left(M + \varepsilon\right)\left(M_0 + \varepsilon\right)^{-1} \sim -2^{-(N+1)/(N-1)}A(N)\varepsilon.$$

(iv) It again has one zero eigenvalue associated with time translation. The jacobian J induced by taking time as a collective coordinate is

$$J^2 = \int dt \, \dot{q}^2(t) = A/(-g)^{1/(N-1)}.$$

Using equation (37.37) and collecting all factors, we obtain the imaginary part of the ground state energy for negative g:

$$\operatorname{Im} E(g) = C(-g)^{-\beta} \exp\left[-A(N)/(-g)^{1/(N-1)}\right], \qquad (44.68)$$

with

$$\beta = 1/[2(N - 1)], \qquad C = \frac{2^{1/(N-1)}}{\sqrt{\pi}}.$$

Exercise 37.2

We only explain the main new features of the solution, compared to the $\nu = 1$ case. The classical solutions are

$$\mathbf{q}_c(t) = \mathbf{n}\, q_c(\nu = 1, t),$$

where $\mathbf{n}$ is an arbitrary constant unit vector. Therefore the classical action is the same as in the $\nu = 1$ case. The main difference comes from the degeneracy of the saddle point: We have to include the unit vector $\mathbf{n}$ in our set of collective coordinates. This leads to a new jacobian and to a factor which is the surface of the ν dimensional sphere.

The second derivative of the action is

$$M_{ij} = \left(-d_t^2 + 1\right)\delta_{ij} - N\left(\delta_{ij} + 2(N - 1)n_i n_j\right)\frac{1}{\cosh^2[(N - 1)t]}.$$

The determinant can be expressed in terms of the determinants of a transverse and a longitudinal operator (with respect to the vector $\mathbf{n}$)

$$M_{\mathrm{L}} = -d_t^2 + 1 - \frac{N(2N - 1)}{\cosh^2[(N - 1)t]},$$

$$M_{\mathrm{T}} = -d_t^2 + 1 - \frac{N}{\cosh^2[(N - 1)t]}.$$

The longitudinal determinant is the same as before. Its zero mode corresponds to time translation. The transverse determinant which appears $\nu - 1$ times corresponds to $\lambda = 1/(N - 1)$ and its zero mode is related to rotations of the vector $\mathbf{n}$:

$$\det\left(M_{\mathrm{T}} + \varepsilon\right)\left(M_0 + \varepsilon\right)^{-1} \sim \frac{N + 1}{N - 1}2^{-(N+1)/(N-1)}A(N)\varepsilon.$$

Note that all eigenvalues of $M_{\rm T}$ are positive, in such a way that the final result remains real as expected.

The eigenvectors corresponding to a rotation are all unit vectors orthogonal to **n** multiplied by $q_c(\nu = 1, t)$. The corresponding jacobian is thus

$$J_{\rm T} = \left(\int dt \, q_c^2(t) \right)^{(\nu-1)/2} = \left(\frac{N+1}{N-1} A \right)^{(\nu-1)/2} \frac{1}{(-g)^{(\nu-1)/2(N-1)}} \, .$$

We have of course to divide by a factor $\sqrt{2\pi}$ for each zero-mode. Finally the integration over **n** yields the surface of the sphere. Taking also into account the factor $1/2i$ to obtain the imaginary part we find a additional factor:

$$\frac{1}{2i} \frac{2\pi^{\nu/2}}{\Gamma(\nu/2)} \frac{1}{(2\pi)^{\nu/2}} \, .$$

Collecting all factors we finally obtain, in the notation (44.68):

$$\beta = \nu/[2(N-1)], \qquad C = \frac{1}{\Gamma(\nu/2)} 2^{\nu/(N-1)} \, .$$

Chapter 38: Instantons in Quantum Mechanics: Generalization

Exercise 38.1

The action $S(r)$ can be rewritten

$$S(r) = \tfrac{1}{2} \int_0^\beta dt \big(\dot{r}(t) + \omega r(t)\big)^2 + \tfrac{1}{2}\omega \left(x^2 - x'^2\right).$$

We then change variables ($\kappa(t) = {\rm e}^{-\omega t}$)

$$r(t) = \sigma(t) - \omega \, {\rm e}^{-\omega t} \int_0^t d\tau \, \sigma(\tau) \, {\rm e}^{\omega \tau} = {\rm e}^{-\omega t} \left(x + \int_0^t d\tau \, \dot{\sigma}(\tau) \, {\rm e}^{\omega \tau} \right),$$

which corresponds to the boundary condition $\sigma(0) = x$.

The jacobian J of the transformation is (equation (38.29))

$$J = {\rm e}^{-\omega\beta/2} \, .$$

We have still to impose the condition $r(\beta) = x'$. We introduce a variable λ to write the corresponding δ-function in exponential form. We obtain the action

$$S(\sigma, \lambda) = \tfrac{1}{2} \int_0^\beta dt \, \dot{\sigma}^2(t) + i\lambda \left[x' - x \, {\rm e}^{-\omega\beta} - {\rm e}^{-\omega\beta} \int_0^\beta dt \, {\rm e}^{\omega t} \, \dot{\sigma}(t) \right] + \tfrac{1}{2}\omega \left(x^2 - x'^2\right).$$

Shifting then $\sigma(t)$

$$\dot{\sigma}(t) = \dot{\sigma}'(t) + i\lambda \, {\rm e}^{\omega(t-\beta)}, \qquad \sigma'(0) = x \, ,$$

we find

$$S(\sigma, \lambda) = \tfrac{1}{2} \int_0^\beta dt\, \dot\sigma^2(t) + \frac{\lambda^2}{4\omega}\left(1 - e^{-2\omega\beta}\right) + i\lambda\left(x' - x\,e^{-\omega\beta}\right) + \tfrac{1}{2}\omega\left(x^2 - x'^2\right).$$

The integral over $\sigma(t)$ is now

$$\int_{\sigma(0)=x} [d\sigma(t)] \exp\left[-\tfrac{1}{2}\int_0^\beta dt\, \dot\sigma^2(t)\right] = \frac{1}{\sqrt{2\pi\beta}} \int ds\, \exp[-\tfrac{1}{2}(x-s)^2/\beta] = 1.$$

The integral over λ then leads to

$$U(x', x; \beta) = \frac{1}{\sqrt{2\pi}} \frac{\sqrt{\omega}}{\sqrt{\sinh\omega\beta}} \exp[-\tfrac{1}{2}\omega A(x, x')],$$

with

$$A(x, x') = x^2 - x'^2 + 2\frac{\left(x' - x\,e^{-\omega\beta}\right)^2}{1 - e^{-2\omega\beta}} = \frac{(x^2 + x'^2)\cosh\omega\beta - 2xx'}{\sinh\omega\beta},$$

which is the correct result. The factor $(2\pi)^{-1/2}$ is the product of a factor $1/(2\pi)$ from the representation of the δ-function by the factor $(2\pi)^{1/2}$ of the gaussian integration.

Exercise 38.2

All the relevant equations can already be found in Section 38.3. To calculate the imaginary part of the $N$the level we have to identify the coefficient of $e^{-(N+1/2)\beta}$ in the large β expansion of the partition function. The leading contribution then comes from the term with the largest power of $1/g$. If we expand $E(\beta)$ and $A(\beta)$ in (38.50) we observe that the leading contribution comes from the correction of order $e^{-\beta}$ in $A(\beta)$. From equation (38.60)

$$A(\beta) = a - 2C\,e^{-\beta} + O\left(e^{-2\beta}\right),$$

we then find

$$\operatorname{Im} E_N(g) \sim \frac{1}{N!}\left(\frac{2C}{g}\right)^N \operatorname{Im} E_0(g).$$

We notice that, as expected, the imaginary part increases with N and thus the lifetime decreases. Of course eventually for N large enough the expression is no longer be valid and has to be replaced by the corresponding WKB valid for large quantum numbers.

   38.2.1. Barrier penetration effects are related to euclidean trajectories with go from the origin to a turning point and bounce back. If the action is $O(\nu)$ invariant such trajectories will take the form of the $O(1)$ trajectories multiplied by a constant unit vector $\mathbf{n}$:

$$\mathbf{q}_c(t) = \mathbf{n}\, q_c(t).$$

Therefore the classical action is independent of ν and the analysis of the $\nu = 1$ can be repeated. The main difference comes from the gaussian integration. Note first that it is simpler to calculate the contribution to $\operatorname{tr} e^{-\beta H}$ than to the matrix elements, because the classical solutions are simpler. The classical solution, being periodic, is determined by one boundary condition for example the value $\mathbf{x} = \mathbf{q}(-\beta/2)$. The variable $\mathbf{x}$ plays the role of collective coordinates corresponding to time translation and direction in $\mathbf{q}$ space.

We have then to find a non-singular matrix K satisfying equation (38.45). The equation for K takes the form

$$\ddot{K}_{ij} = 2\mathcal{V}'\left(g\mathbf{q}^2\right)K_{ij} + 4gn_iq_c^2(t)\mathcal{V}\left(gq_c^2(t)\right)n_kK_{kj}.$$

We have to find enough independent solutions. Here the problem is rather simple. We decompose the matrix $\mathbf{K}$ as

$$K_{ij} = K_{\mathrm{L}}n_in_j + (\delta_{ij} - n_in_j)\,K_{\mathrm{T}}.$$

For what concerns K_{L} one solution is $\dot{q}_c(t)$ and the problem is the same as in the $\nu = 1$ case. For K_{T} we recognize the equation of motion and therefore a solution is $q_c(t)$ itself. The calculation is then an extension of the $\nu = 1$ example and the result is, with the notation (38.59):

$$\mathrm{Im}\,E_0(g) \sim \frac{1}{\Gamma(\nu/2)}\left(-\frac{2C}{g}\right)^{\nu/2}\mathrm{e}^{a/g}.$$

Exercise 38.3

As we have seen in Section 2.6 we have to evaluate the evolution operator with classical scattering boundary conditions. We recall the expression of the evolution operator at leading WKB order (Van Vleck's formula for $\hbar = 1$), as given in A38.2:

$$\langle\mathbf{x}''\,|U\,(T'',T')|\,\mathbf{x}'\rangle \sim \frac{1}{(2\pi i)^{n/2}}\left(-\det\frac{\partial^2\mathcal{A}}{\partial\mathbf{x}'\partial\mathbf{x}''}\right)^{1/2}\exp\left[i\mathcal{A}\left(\mathbf{x}'',\mathbf{x}';T'',T'\right)\right],$$

where $\mathcal{A}$ is the classical action evaluated for the solutions of the equation of motion.

The classical equation of motions, after integration, yield

$$\tfrac{1}{2}\dot{\mathbf{x}}^2 + V(r) = \tfrac{1}{2}\kappa^2, \qquad r^2\dot{\theta} = l,$$

where κ, l are two constants of the motion and θ characterizes the angular position of the vector $\mathbf{x}$ in the plane of the motion. Taking as boundary conditions $\mathbf{x}(t') = \mathbf{x}'$, $\mathbf{x}(t'') = \mathbf{x}''$, setting $T = t'' - t'$ and

$$\kappa(r) = \left[\kappa^2 - 2V(r) - \frac{l^2}{r^2}\right]^{1/2},$$

we can write the two equations determining κ, l

$$\int_L \frac{\mathrm{d}r}{\kappa(r)} = T, \qquad l\int_L \frac{\mathrm{d}r}{r^2\kappa(r)} = \alpha,$$

where α is the angle between $\mathbf{x}'$ and $\mathbf{x}''$ and L means the domain of integration

$$\int_L \equiv \int_{r_0}^{r'} + \int_{r_0}^{r''}, \qquad \kappa(r_0) = 0.$$

Finally the classical action $\mathcal{A}$ is given by

$$\mathcal{A} = \int \mathrm{d}t\left[\tfrac{1}{2}\dot{\mathbf{x}}^2 - V(r)\right] = \tfrac{1}{2}T\kappa^2 - 2\int_L \frac{V(r)\mathrm{d}r}{\kappa(r)} = \tfrac{1}{2}T\kappa^2 + l\alpha + \int_L \mathrm{d}r\left(\kappa(r) - \frac{\kappa^2}{\kappa(r)}\right).$$

It is useful to calculate the variation of the action when the end-points vary

$$\delta \mathcal{A} = l\delta\alpha + \kappa(r'')\delta r'' - \kappa(r')\delta r'.$$

As a consequence:

$$\frac{\partial \mathcal{A}}{\partial x_i'} = \frac{l}{r'\sin\alpha}\left(\cos\alpha\frac{x_i'}{r'} - \frac{x_i''}{r''}\right) - \kappa(r')\frac{x_i'}{r'},$$

$$\frac{\partial \mathcal{A}}{\partial x_i''} = \frac{l}{r''\sin\alpha}\left(\cos\alpha\frac{x_i''}{r''} - \frac{x_i'}{r'}\right) + \kappa(r'')\frac{x_i''}{r''}.$$

A few other formula are useful:

$$\frac{\partial^2 \alpha}{\partial x_i'\partial x_j''} = -\frac{1}{r'r''\sin\alpha}\left[\delta_{ij} - \frac{x_i'x_j'}{r'^2\sin^2\alpha} - \frac{x_i''x_j''}{r''^2\sin^2\alpha} + \frac{(x_i'x_j'' + x_i''x_j')\cos\alpha}{r'r''\sin^2\alpha}\right].$$

We note that the result is a projector orthogonal to $\mathbf{x}'$ and $\mathbf{x}''$. Finally we can derive

$$\det\frac{\partial^2 \mathcal{A}}{\partial x_i'\partial x_j''} = \frac{1}{r'r''}\left(\frac{-\partial \mathcal{A}}{\partial\alpha}\frac{1}{r'r''\sin\alpha}\right)^{d-2}\left(\frac{\partial^2 \mathcal{A}}{\partial\alpha\partial r'}\frac{\partial^2 \mathcal{A}}{\partial\alpha\partial r''} - \frac{\partial^2 \mathcal{A}}{\partial\alpha\partial\alpha}\frac{\partial^2 \mathcal{A}}{\partial r'\partial r''}\right).$$

A short calculation then yields

$$\det\frac{\partial^2 \mathcal{A}}{\partial x_i'\partial x_j''} = \left(-\frac{l}{r'r''\sin\alpha}\right)^{d-2}\frac{1}{r'r''\kappa(r')\kappa(r'')I}\frac{\partial l}{\partial\alpha},$$

with

$$I = \frac{-1}{\kappa}\frac{\partial}{\partial\kappa}\bigg|_l \int_L \frac{dr}{\kappa(r)}.$$

Large time expansion. Knowing from the general analysis that the boundary conditions should correspond to classical scattering, we can expand for T large (we assume that $V(r)$ decreases fast enough). Setting

$$k = (r' + r'' - 2r_0)/T, \qquad k(r) = \left[k^2 - V(r) - \frac{l^2}{r^2}\right]^{1/2},$$

where r_0 is replaced by the limiting value solution of $k(r_0) = 0$ we find

$$\frac{kT}{\kappa} + 2\int_{r_0}^{\infty} dr\left(\frac{1}{k(r)} - \frac{1}{k}\right) + O\left(T^{-1}\right) = T,$$

and thus

$$\kappa = k + \frac{2}{T}\int_{r_0}^{\infty} dr\left(\frac{k}{k(r)} - 1\right) + O\left(T^{-2}\right).$$

It follows after a short calculation

$$\alpha = 2l\int_{r_0}^{\infty}\frac{dr}{r^2 k(r)},$$

$$\mathcal{A} = \tfrac{1}{2}Tk^2 + 2\int_{r_0}^{\infty} dr\,[k(r) - k] + l\alpha.$$

In this limit l is related to the impact parameter b by $b = l/k$. We then calculate the Fourier transform of the evolution operator. At leading order for T large the action depends only on $r'+r''$. We can thus integrate over the angles. We find Bessel's functions with large arguments $p'r'$ and $p''r''$ which can be replaced by their asymptotic forms. These correspond to the angle between $\mathbf{x}$ and $\mathbf{p}$ being 0 or π. The integral over $r' - r''$ then yields the energy conservation delta-function, $p'^2 = p''^2$. The integral over $r' + r''$ then implies $k = p'$. Thus finally we can substitute

$$p'^2 = p''^2 = k^2, \qquad k^2 \cos\alpha = -\mathbf{p}' \cdot \mathbf{p}''.$$

All these terms give only a phase ϕ obtained by adding to the result the contribution $\frac{1}{2}p'^2$ coming from the free terms

$$\phi = 2 \int_{r_0}^{\infty} dr\,[k(r) - k] + l\alpha - 2ikr_0,$$

while the amplitude completely given by the determinant. Thus the final result for the cross-section is proportional to

$$\frac{\partial\sigma(k,\alpha)}{\partial\alpha} \propto \frac{\partial b}{\partial\alpha} b^{d-2},$$

which is the result obtained directly by a classical calculation.

Chapter 39: Unstable Vacua in Field Theory

Exercise 39.1

The expression (A39.21) is independent of the normalization of φ and the scale of distances. Therefore we just substitute

$$\varphi(r) = e^{-r^2/2},$$

and find $A(d) = 24\pi^{d/2}(4-d)^{(d-4)/2}$ and thus

$$A(2) = 37.699\ldots, \qquad A(3) = 133.639\ldots, \qquad A(4) = 24\pi^2.$$

For the second trial function one finds instead

$$d = 2, \qquad A(2) = 12\pi \ln 2 \frac{\ln 2 + \frac{1}{2}}{2\ln 2 - \frac{1}{2}} = 35.178\ldots$$

$$d = 3, \qquad A(3) = \frac{8\pi}{3} \frac{(\zeta(2)+2)^{3/2}\,\zeta^{1/2}(2)}{\zeta(2)-1} = 115.934\ldots$$

$$d = 4, \qquad A(4) = \frac{\pi^2}{16} \frac{(9\zeta(3)+24\ln 2)^2}{9\zeta(3)-12\ln 2} = \pi^2 \times 18.840\ldots,$$

where $\zeta(s)$ is Riemann's ζ function.

This can be compared with the exact results $A(2) = 35.10268\ldots$, $A(3) = 113.38350\ldots$ and $A(4) = 16\pi^2$. We see that both variational estimates are better for smaller dimensions, but also that the 1D trial function is much better.

Chapter 42: Instantons: The ϕ^4 Field Theory in Dimension Four

Exercise 42.1

We first note that the dimension in which the theory is renormalizable, and thus the classical theory scale invariant, is $d = 2n/(n-1)$. The calculations then are a simple generalization of the ϕ^4 field theory.

The dominant instanton still corresponds to a field depending only on a radial coordinate r. We set

$$\phi(x) = (-g)^{-1/(2n-2)} f(r).$$

The classical action becomes

$$S(\phi) = \frac{S_d}{(-g)^{1/(n-1)}} \Sigma(f)$$

$$\Sigma(f) = \int_0^\infty r^{d-1} dr \left(\tfrac{1}{2} f'^2(r) - \tfrac{1}{2(n-1)^2} f^{2n}(r) \right),$$

with $S_d = 2\pi^{d/2}/\Gamma(d/2)$. The action is invariant under the scale transformation $f(r) \mapsto \lambda^{(d-2)/2} f(\lambda r)$. Therefore we set $r = e^t$ and

$$f(r) = b^{1/2}(d)\, e^{-(d-2)t/2}\, h(t), \qquad b(d) = \left[(d-2)^2 (2n!)/8 \right]^{(d-2)/2}. \qquad (44.69)$$

The action then becomes

$$\Sigma(h) = \tfrac{1}{2} b(d) \int_{-\infty}^{+\infty} dt \left[\dot{h}^2(t) + \tfrac{1}{(n-1)^2} \left(h^2(t) - h^{2n}(t) \right) \right].$$

We recognize a quantum mechanical problem, which we have solved in the exercises of Chapter 37. The instanton solution of the equation of motion is

$$|h|^{n-1}(t) = \frac{1}{\cosh[(t - t_0)]}.$$

The classical action S_c is

$$S_c = A(d)/(-g)^{1/(n-1)},$$

$$A(d) = 2^{d-3} S_d\, b(d)(d-2) \frac{\Gamma^2(d/2)}{\Gamma(d)} = 2(2n!)^{(d-2)/2} \left(\frac{\pi}{2} \right)^{d/2} (d-2)^{d-1} \frac{\Gamma(d/2)}{\Gamma(d)}.$$

If now more generally we change variables in the action

$$\mathbf{x} \mapsto t = \ln r, \qquad \mathbf{n} = \mathbf{x}/r,$$

and allow the field (44.69) to depend on $\mathbf{n}$ we find a scaled action

$$\Sigma(h) = \tfrac{1}{2} b(d) \int d\Omega dt \left[\dot{h}^2(t) + h \mathbf{L}^2 h + \tfrac{1}{(n-1)^2} h^2(t) - \tfrac{1}{(n-1)^2} h^{2n}(t) \right],$$

where $\mathbf{L}^2$ is the angular momentum operator with eigenvalues $l(l+2/(n-1))$. The second derivative $\mathbf{M}$ of the action at the saddle point at angular momentum l is proportional to:

$$\mathbf{M}_l = -d_t^2 + \left(l + 1/(n-1) \right)^2 - \frac{n(2n-1)}{\cosh^2(t)}.$$

In the notation (42.77,42.78) we find

$$\lambda = n/(n-1) \qquad z = \left[(l + 1/(n-1))^2 + \varepsilon\right]^{1/2}.$$

We observe that $\mathbf{M}$ has vanishing eigenvalues for $l = 0$ and $l = 1$, as in the simple ϕ^4 example, corresponding to the dilatation and translation modes respectively. One has to introduce collective coordinates, and calculate the jacobian. One important difference between the $n = 2$ and $n > 2$ case is the absence of one-loop renormalization constant of the coupling constant. The remaining part of the calculation can be found in

E. Brézin, J.C. Le Guillou and J. Zinn-Justin, *Phys. Rev.* D15 (1977) 1544, where results are more generally given for the $(\phi^2)^n$ $O(N)$ symmetric theory.

Chapter 43: Multi-instantons in Quantum Mechanics

Exercise 43.1

The corresponding euclidean action is

$$S(q) = \frac{1}{2\Omega} \int dt \left[\dot{q}^2(t) + \tfrac{1}{4}\left(E'(q)\right)^2 - \tfrac{1}{2}E''(q)\right].$$

At leading order for Ω small $E''(q)$ can be neglected in the classical equation of motion. However, evaluated for the classical solution, it will yield a multiplicative factor in the result. The potential $\frac{1}{4}\left(E'(q)\right)^2$ is exactly the double-well potential. Thus the classical solution is the same. The subtle point is that, due to the $E''(q)$ contribution the wells are not really degenerate and the leading term in the ground state energy is a instanton-antiinstanton pair. A similar situation has been examined in Section 43.3 and the calculation is quite similar. Here we have $\omega = 1$ but we have to add to the instanton action terms corresponding to the values of E'' near the well multiplied by the time spent in the well. In the notation of Section 43.3 we find

$$\left\{\operatorname{tr} e^{-\beta H}\right\}_{(n)} = \frac{\beta}{n} e^{-\beta/2} \frac{e^{-n/3g}}{(\pi g)^n} \int_{\theta_i, \varphi_i \geq 0} \delta\left(\sum_i \theta_i + \varphi_i - \beta\right)$$

$$\times \exp\left\{\sum_{i=1}^n \left[\frac{1}{2}(\theta_i - \varphi_i) + \frac{2}{g}\left(e^{-\theta_i} + e^{-\varphi_i}\right)\right]\right\}. \qquad (44.70)$$

The effect of these additional terms is simply to shift by $\pm\frac{1}{2}$ the argument of the function $I(s)$. The spectral equation then reads

$$\left(-\frac{2}{g}\right)^{2E} \Gamma(-E)\Gamma(1-E)\frac{e^{-1/3g}}{2\pi} = -1. \qquad (44.71)$$

Let us extract from this equation the ground state energy

$$E \sim \frac{e^{-1/3g}}{2\pi}\left(1 + O(g)\right).$$

By expanding to next order in E we also obtain the instanton contribution to the imaginary part of E at leading order

$$\operatorname{Im} E(g) \sim \frac{e^{-2/3g}}{2\pi},$$

a result consistent with the property that the leading contribution to E is already of order $\mathrm{e}^{-1/3g}$. This result is peculiar to the ground state. Excited states have a non-trivial perturbative expansion. Note that at this order the equation seems to indicate that starting from the first excited states the energies are pairwise degenerate in perturbation theory and that the instanton contribution corresponds to the splitting between states in the two wells. Actually this is a feature of the leading order. There is no symmetry between the wells and the degeneracy will lifted in perturbation theory. In the Γ-functions E will be replaced by two different functions. The calculation of the instanton contribution, which is proportional to $\mathrm{e}^{-1/3g}/g^{2N+1}$, requires the knowledge of these functions at order g. æ

INDEX